World Islands in Prehistory

International Insular Investigations

V Deia International Conference of Prehistory

Edited by

W. H. Waldren

and

J. A. Ensenyat

BAR International Series 1095
2002

This title published by

Archaeopress
Publishers of British Archaeological Reports
Gordon House
276 Banbury Road
Oxford OX2 7ED
England
bar@archaeopress.com
www.archaeopress.com

BAR S1095

World Islands in Prehistory: International Insular Investigations
V Deia Conference of Prehistory

ISBN 1 84171 473 9

Printed in England by The Basingstoke Press

V DEIA INTERNATIONAL CONFERENCE OF PREHISTORY

WORLD ISLANDS IN PREHISTORY

International Insular Investigations

***

V DEIA CONGRESO INTERNACIONAL DE PREHISTORIA

ISLAS DEL MUNDO EN LA PREHISTORIA

Investigaciones Internacionales sobre Islas

BAJO LA PRESIDENCIA DE HONOR DE

SU MAJESTAD LA REINA DOÑA SOFIA

***

HONORARY PRESIDENT

HER ROYAL MAJESTY QUEEN SOFIA of SPAIN

Deia Archaeological Museum and Research Centre
September 13 - 18, 2001

Museo Arqueológico y Centro de Investigaciones de Deia
del 13 al 18 de Septiembre de 2001

AMB LA COLLABORACION DE

CONSELL DE MALLORCA

AJUNTAMENT DE DEIA

AJUNTAMENT DE SÓLLER

AJUNTAMENT DE VALLDEMOSSA

CASAL DE CULTURA MUSEU DE SÓLLER

FOREWORD

****

These papers form the fifth in the series of thematic Deia International Conferences of Prehistory dedicated to bringing Balearic Prehistoric investigation and research out of insular and regional contexts and into the mainstream of Continental European prehistory. Much of the success of this objective is owed to the British Archaeological Reports (BAR) and its editor-publishers, Dr. Anthony Hands and Dr. David Davison, both of whom have been responsible over the years for the resulting publication and large collection of papers dedicated to the subject of prehistory in this area of the Western Mediterranean.

The thematic nature of these volumes has ranged from early settlement in this sector of the western Mediterranean (Ist Deia Conference) to questions regarding the definition and understanding of the Bell Beaker Culture in the western Mediterranean (2nd Deia-Oxford Conference), to the question and problems of archaeological techniques, technology and theory in prehistory (3rd Deia Conference) and the subject of ritual, rites and religion in prehistory (4th Deia Conference), along with the present status and state of world islands in prehistory (5th Deia Conference).

It is not at all a question of accident that the organization and locale of these conferences has taken place on an island and in a small village in the mountains of the Baleareic Island of Mallorca, for it has been the idea of the organizers (the Deia Archaeological Museum and Research Centre (DAMARC Foundation), over the years, that conferences of this nature are perhaps enjoyed more and give more room for reflection and contemplation in such quiet surroundings than in those held in crowded and distracting cities. There is also the underlying belief that there is a great need for active local museums and centers of research dedicated to the spread of data and information gathered from local investigations that are capable of not only adding to the general local knowledge of what went on in prehistory, but also capable of altering as well as adding to some of what is already known and what may not be known farther afield. Hopefully too, perhaps what is also demonstrated may help to change local and other governmental opinions as to the importance and real wealth of prehistory by way of a closer study of regional heritage, as well as gained through the support of small regional museums and centers of prehistoric investigation.

A small, privately owned and funded entity, the Deia Archaeological Museum and Research Centre (DAMARC Foundation) was founded in 1962 to study Balearic prehistoric ecology and culture. It is supported over the years by foreign grants, such as the Centre for Field Research, Earthwatch Institute and other research foundations, scientific and academic institutions and facilities, such as the Smithsonian Institution and National Geographic Society, as well as public donations.

SPECIAL ACKNOWLEDGEMENTS AND GRATITUDE

****

These must be extended to the families of Dr. Don Juan Cirera and Sr. Don Pedro Coll, owners of the properties of San Mas and Son Oleza, whose long interest, patience, generosity and friendship have made possible the excavations of the Valldemossa sites over the last twenty years. To each of these wonderful men and their families, we give our special thanks and extend our life-long esteem. Our deep appreciation to Mr. Michael Douglas for his contribution and the use of the Costa Nord Cultural Center.

W. Waldren, for the Organizational Committee

This publication is dedicated
to

Dr ANTONIO ARRIBAS PALAU

friend, colleague and mentor

******

OPENING ADDRESS

Very soon after the creation of the Universe in the "Big Bang" some 13,000 million years ago, all the newly formed, expanding matter became segregated into cosmic islands, and has remained so ever since. When we look around the Universe today, we see billions of island universes (galaxies), nebulae, supernovae, stars and, nearer home, we can observe solar systems, planets, comets, moons. Within our own modest but ambitious solar system, no two planets are identical or even similar, and that also applies to the moons of the planets. Materially, every object in space is an island with its own characteristic properties, all loosely held together by the cosmic glue of time, space and gravitation, and governed by the universal laws of Physics and Chemistry and, dare I say it, perhaps even of Biology. Even the Earth's surface itself is studded with islands of all shapes and sizes, and all manner of mechanisms of formation. As a geologist myself, I can tell you that this has been the case for at least 3800 million years. Some islands consist of continental crust, some of oceanic crust, some are entirely volcanic, whilst others are purely biological Many islands – of any type -- result from rise and fall of sea-level. We are sitting on such an island right now (Mallorca), and long may it remain so! In their overall combination of geographic, geological, biological, climatic, scenic and environmental settings, one can almost say that no two islands, or at least groups of islands, are exactly the same. That is the reason why they are so incredibly important in influencing the course of events on the surface of the Earth in the past, present and future. Every island has a different story to tell – the permutations and combinations are almost infinite. Furthermore, their prehistoric past has had an incalculable influence on their historic past and on the present. That is what we shall learn about at this Conference from international, multidisciplinary experts, and I look forward with keenest anticipation to the forthcoming scientific sessions.

STEPHEN MOORBATH
Department of Earth Sciences,
University of Oxford,
Oxford, England

LAS HUELLAS HUMANAS SEGÚN UNA TEORÍA DE RAMON LLULL

Ramon Llull fue un filósofo y teólogo que nació en Palma de Mallorca, en 1232, y escribió más de 250 libros en árabe, latín y catalán. Cruzó el Mediterráneo varias veces y tuvo múltiples contactos con gente de muchas culturas y creencias. Fundó el Monasterio de Miramar, donde se celebra el presente Congreso titulado "WORLD ISLANDS IN PREHISTORY 2001". Desarrolló una teoría general para todas las ciencias que denominó "Arte", que es un sistema lógico para el tratamiento del conocimiento y, dentro de esta teoría general, estableció una idea original del hombre que, para la tarea de los investigadores de la prehistoria, puede ser interesante hoy.

Ramon Llull fue un pensador neoplatónico y cristiano, por lo que creyó que el Cosmos es una proyección del poder, la sabiduría y el amor de Dios. La contribución original, en un aspecto concreto del conocimiento ontológico para la Ciencia, es su "teoría de los correlativos". Esta descubre que en cada ente, es decir en cada ser material o inmaterial, o en cada ser animado o inanimado, hay tres componentes que constituyen un ternario: el agente, el paciente y la correlación entre ambos; el ternario es consecuencia de la Trinidad de Dios. En el caso del hombre su expresión ternaria podría ser: "forma hominificativa", "materia hominificable" y "hominificar" (acción). Como puede apreciarse aquí mezcla la teoría hilemórfica (materia y forma) procedente de Aristóteles, con la suya propia. El alma sería la forma y el cuerpo la materia y la correlación entre ambos constituiría el ser humano viviente. Esta teoría explicaría el ser del hombre. Pero no se trata en esta comunicación de desarrollar una teoría antropológica ni metafísica, sino una aplicación a los estudios arqueológicos.

Extrapolando desde el conocimiento del ser del hombre de Llull a la investigación sobre la manifestación del hombre, sería la investigación de la "antropofanía" : las huellas o la impronta que deja el hombre en el medio que le rodea.

La pregunta que nos podemos hacer ahora, es : ¿resulta interesante la "teoría de los correlativos" para los arqueólogos?.

La respuesta podría ser: los arqueólogos buscan los restos humanos, lo que queda de sus cuerpos, lo que es equivalente al primer elemento del ternario. El segundo elemento aparece en los objetos que hallan en los yacimientos arqueológicos (vasijas, armas, pinturas, restos de animales, etc.). Pero la principal tarea de los arqueólogos consiste en el tercer elemento del ternario: la investigación del vivir de los hombres prehistóricos, es decir, cómo los hombres prehistóricos vivieron, cómo se establecieron las relaciones entre los objetos de los asentamientos, los restos de animales, etc. y los hombres a los que pertenecieron lo huesos humanos hallados.

La importancia de esta obviedad, que parece que no añade nada nuevo a la Ciencia, estriba, desde nuestro punto de vista, en la categoría de la relación del tiempo continuo que estuvieron relacionados, porque la teoría luliana pide una radical unidad de relación. De manera que si seguimos a Ramon Llull en esta aplicación, su teoría exige al arqueólogo el encontrar la solución de los enigmas de la relación.

Ramon Llull predijo, con dos siglos de anticipación, el descubrimiento de América, manejando la teoría de los correlativos (cuestión 154). La citada teoría se aplica así: uno de los tres elementos son las costas de Europa, otro elemento es el agua del mar que fluye y refluye, golpeando esas costas europeas, y el tercer elemento es exigido para completar el ternario: las costas del otro lado del mar, las americanas. Este fue el argumento luliano que los Franciscanos Juan Pérez, Fray Diego de Deza y Fray Antonio Marchena utilizaron en La Rápida (Huelva) para convencer a la Reina Isabel la Católica, para que facilitara los medios del primer viaje de Colón, como así ocurrió.

La teoría de los correlativos, hoy día, podría ser un instrumento de investigación científica, cuando aparecen dos elementos relacionados. Siempre hay un tercer elemento que cierra

JOSÉ M. SEVILLA,
Magister de la "Maioricensis Schola Lulistica". Miramar – Mallorca (Islas Baleares – España)

Islands are idiosyncratic by nature, showing great variability in material resources, in the life forms and human societies they nurtured and developed throughout time. Their inherent and immense variability in size, geographic location and mainland proximity, geology and natural resources guarantees their individuality.

Islands are also places of paradox in that, despite their variability, they share common, unwritten, almost irrevocable natural laws which control and shape their destinies. In turn these are often quite different from mainland situations. They are inhabited by mainly predator-less, limited animal populations that are often highly specialized and oddly modified in function. Dwarfism of the macrofauna, giantism of the microfauna, and such phenomena as flightless birds are some of the adaptations found quite commonly throughout the islands of the world. In turn, these evolutionary processes and mechanisms of change and adaptation are controlled by forces, for need of a better definition, largely flotsam and jetsam in nature, where island population and outside influences arrive sporadically from many different directions either by accident or design.

In the study of islands, it is the normal view to observe the island from the shore, where little detail of internal or reciprocal interaction present on them can be seen. It is only those who have worked on islands who are aware of the complex worlds they are, and the unique problems involved in their study and the forces that originally shaped them. For islands in their inherent isolation are highly individual worlds where their differences and similarities, both obvious and subtle, make them excellent places from which to observe and reflect the outside world, often in advance of world events.

At a time when global attention is focused on ecological conditions as they transpire continentally, island environmental studies should be paramount, for here lie many of the signposts of the future. As the world's islands will undoubtedly feel the first serious effects of global warming and rising sea levels, they form a potential barometer for changing global conditions. With this in mind, the current conference on world islands in prehistory is timely and important. From their fossil records and evidence of the past climactic conditions, there are many lessons to be learned and information to be gained, regarding both the present and the future, evidence most of which will assuredly change or even disappear with the environmental alteration of the islands themselves.

ORGANIZATIONAL COMMITTEE

******

William H. Waldren, Jacqueline D. Waldren,
Josef Ensenyat Alcover, Jaume Ensenyat i Julia, Jelle Reumer, Mark Van Strydonck
Steve Davis, Stephen Moorbath,
Clare Durst, Bill and Alison Gardner Shelby,
Heidi and Phil Weaver, Mary Nemecek Peterson

CONTENTS

1 Conceptions of the Mediterranean: Islands of the Mind 1
Jacqueline Waldren

2 Rocks and Itineraries: Sea and Land Perspectives on an Aegean Island 7
Christina Marangou

3 Constructing Hypothetical Systems of Exchange: Doing Kula in the Danish Mesolithic 19
George Nash

4 Early Human Impacts in Madagascar 31
Michael Parker-Pearson, Ramilisonina, Retsihisatse, J.L Schwenninger and Helen Smith

5 Imagining Islands 39
Tamara Kohn

6 Innovation, Continuity and Insular Development in the Isle of Man 44
P.J. Davey, J.J. Innes, and Sinéad McCartan

7 Island Archaeology and Issues of Political Centralization: Micronesian Evidence 57
William S. Ayers

8 New Coals on Old Fires: the Question of Early Balearic Island Settlement 68
W. Waldren, J. Ensenyat Alcover, and J. Orvay

9 Pacific Prehistory: an Example of Cultural Evolution from the Island of Cikobia (Macuata, Fiji) 91
Christophe Sand, Frédérique Valentin

10 Akanthou-Arkosyko (Tatlısu-Çiftlikdüzü): the Anatolian Connections in the 9th Millennium BC 98
Müge Sevketoglu

11 "The Island Factor": Insularity as a Variable in the Archaeological Study of the Social Dynamics of Hunter-Gatherers 107
Jordi Estévez, Assumpció Vila y Xavier Terradas

12 Social and Ideological Changes in the Balearic Islands during the Later Prehistory 117
Vincente Lull, Rafael Micó, Cristina Rihuete Herrada, Roberto Risch

13 Islands of Indifference 127
Rosemary McKechnie

14 The Insularity of Malta 135
David Trump and Bridget Trump

15 The Role of Return Migrants and Urban Migrants Associations in the Revival and Reinvention of Tradition: a Greek Example 139
Margaret E. Kenna

16 Links in the Chain: Evidence of Sustained Prehistoric Contact and Cultural Interaction Between the Balearic Islands and Continental Europe 152
William H. Waldren

17 Negotiating Island Identities: Ceramic Production in the Cyclades during the Middle and Late Bronze Age 186
Ina Berg

CONTENTS PAGE, cont.

18 Stable Isotopes (^{13}C and ^{15}N) and Diet: Animal and Human Bone Collagen from Prehistoric Sites on Mallorca, Menorca and Formentera (Balearic Islands, Spain) 189
Mark Van Strydonck, Mathieu Boudini and Anton Ervynck

19 Putting Meat on the Bone: An Investigation into Palaeodiet in the Balearic Islands Using Carbon and Nitrogen Stable Isotope Analysis 198
Michael Davis

20 New Contributions to the Study of the Neolithic Sea/Landscape and Human Interaction on the Aeolian Islands (Sicily, Italy) 217
Elena Falvia Castagnino Berlinghieri

21 Island Neolithics: Animal Exploitation in the Aceramic Neolithic of Cyprus 233
Sheelagh Frame

22 Effects of Environmental Alteration on Easter Island (Rapa Nui) from the Moai Era to the Modern Era, AD 1000-2000 239
Joan A. Wozniak

23 Social Interactions on Cyprus in the Prehistoric Production of Metals: Kissonerga and a Possible Starting Point for a New Religion 249
Maria Rosaria Belgiorno

24 Interpreting Distance and Difference: Inter-Cultural Contacts, Symbolic Meanings, and the Character of the Cycladic "Expansion" in the Aegean During the Early Bronze Age 255
Fanis Mavridis

25 Saffron Crocus Domestication in Bronze Age Crete 267
Moshe Negbi and Ora Negbi

26 Economic Strategies and Limited Resources in the Balearic Insular Ecosystem: The Myth of an Indigenous Animal Farming Society in the First Millennium BC 275
Jordi Hernandez-Gasch, J. Nadal, A. Malgosa, and A. Ales

27 Landscape Management of Forest Resources in the Balearic Islands during the II-I Millennium BCE 292
Raquel Piqué and M. Noguera

28 A Case History: Evidence of Ancient Animal, Water and Land Management, Exploitation and Depletion, Son Oleza Chalcolithic Old Settlement, Valledemossa, Mallorca, Baleares, Spain 301
William H. Waldren

29 Sources of Water Ancient and Modem: Linking Archaeology and Hydrology in South-West Viti Levu, Fiji 312
Jeffrey Davies, Janette Davies

30 L'archipel de Molène (Finistère, France) Mise au Point d'un Inventaire des Sites Préhistoriques 234
Yvan Pailler, Y. Sparfel, E. Yven, S. Cassen, P. Gouletquer, M. Le Goffic, A. Leroy, G. Marchand, A. Tresset et E Yven.

31 La Colonización del Archipiélago Canario: ¿un Proceso Mediterráneo? 337
Pablo Atoche Peña

CONTENTS PAGE, cont.

32 El Control del Territorio Durante la Edad del Bronce en el Área de Dorgali (Nuoro, Cerdeña) 355
Liliana Spanedda, Trinidad Nájera Colino, Juan Antonio Cámara Serrano

33 Decoración, Representaciones Figuradas y Áreas Rituales en la Prehistoria Reciente Sarda: Acumulación, Control del Territorio y Jerarquización 373
Juan Antonio Cámara Serrano y Liliana Spanedda

34 Major Patterns and Processes in Biodiversity: Taxonomic Diversity on Islands Explained in Terms of Sympatric Speciation 395
Jean de Vos, Alexandra Van der Geer

35 The Role of Islands in Pushing the Pleistocene Extinction Envelope: The Strange Case of the Cypriot Pygmy Hippos 406
Alan H. Simmons

36 The Wrangel Dwarf Mammoths Were No Island Endemics 415
J.W.F. Reumer, D. Mol and J. de Vos

37 A Life History and Climate Change Solution to the Evolution and Extinction of Insular Dwarfs: A Cypriot Experience 420
Timothy Bromage, Wendy Dirks, Hediye Erdjument-Bromage, Mathias Huck, Ottmar Kulmer, Rasime Öner, Oliver Sandrock, Freidemann Schrenk

38 Pleistocene Faunas of Sicily: a Review 428
Laura Bonfiglio, Antonella C. Marra, Federico Masini, M. Pavia, Daria Petruso

39 Vertebrate Assemblages of Central-Western Mediterranean Islands during the Pliocene and Quaternary: Reflecting on Extinction Events 437
Federico Masini, Laura Bonfiglio, Laura Abazzi, Massimo Delfino, Flaviano Fanfani, Marco Ferretti, Tassos Kotsakis, Daria Petruso, Cinzia Marra, Danilo Torre

40 The Stratigraphic Sequence of Gonnesa (SW Sardinia): Palaeoenvironmental, Palaeontological and Archaeological Evidence 445
Rita Melis, Maria Rita Palombo, Marghetita Mussi

41 S. Maria Is Acquas, A New Pre-Neolithic Site: South-Western Sardinia 454
Rita Melis, Margherita Mussi

42 Lower Palaeolithic in Sardinia 462
David Trump

43 The Animals of Prehistoric Cyprus 465
Lilian Karali Giannakopouluo

44 Los Grupos Domesticos en la Prehistoria de Mallorca. El Edificio Alfa del Puig Morter (Sineu, Mallorca) y las Practicas Sociales del Horizonte de Son Ferragut 472
Pedro Castro Martínez, T. Escoriza Mateu y M. Encamación Sanahuja

45 Balance y Nuevas Perspectivas en la Investigación Prehistórica de las Islas Pitiusas 484
Benjamí Costa Ribas, Víctor M. Guerrero Ayusa

CONTENTS PAGE, cont.

46 Biniparratx Petit (Sant Lluís): a Research and Re-evaluation Project in the Southeast of the Island of Minorca 502
Víctor Guerrero Ayuso, Joan Sanmartí Grego, Jordi Hernández Gasch, J. Simón Gornés Hachero, Joana Mª. Gual Cerdó, Antoni López Pons, Joan De Nicolás Mascaró

47 Análisis Comparativo del Patrón de Asentamiento de la Cultura Baleárica 517
Javier Aramburu-Zabala

48 Aegean Island Caves: Diachronic Cult Places 529
Anna Papamanoli-Quest

49 Insularity and Identity in Prehistoric Cyprus 537
J. Clarke

50 Indications of a possible Astronomical Orientation of an Architectural Feature in the Prehistoric Sanctuary of Son Mas, Mallorca 545
M. Van Strydonck, W. Waldren, M. Hoskin, J. Ensenyat Alcover, J. Orvay

1

CONCEPTIONS OF THE MEDITERRANEAN: ISLANDS OF THE MIND

JACQUELINE WALDREN

Department of Social Anthropology,
University of Oxford, Oxford England
EMAIL: jacqueline.waldren@anthro.ox.ac.uk

INTRODUCTION

When Bill Waldren began to organise this conference, he asked if I might like to get together a panel of social anthropologists to add diverse perspectives on islands to those of the archaeologists, paleontologists and geologists that would be contributing their expertise to the theme of Islands in Prehistory. I sent out the following message to a small group that had recently been formed to study common issues in island tourism asking " if this conference would not be a good moment to present some of our research on islands in a manner which might be useful and informative for archaeologists (and for one another). I suggested we might consider some of the following issues: eg. concepts of identity, space and place; demographic changes, impact of tourism and/or emigration/immigration, geographical, geological, ecological and/ architectural environments; heritage and authenticity, museum exhibitions, 'cultural tourism'; use, control and politicization of natural resources (water, air, sea, sky...); material culture, tradition, history, connectedness, etc." Any suggestions were welcome.

The response was enthusiastic and within a short time, I had offers of papers from quite a few island specialists. Margaret Kenna, Sean Damer, Tamara Kohn and Rosemary McKechnie, all of whom had many years of fieldwork and research on various islands, were selected. Unfortunately Sean was unable to attend due to health issues. I imagined it would be best to present a panel of anthropology as a 'kick-off' to the conference but on reading the various abstracts it seemed that the anthropology papers actually pursued issues presented by the other scientists and it made sense to split them among the different panels. Margaret Kenna's paper on the revival and re-invention of traditions by Greek return migrants and urban migrants offered a wonderful contemporary example of the colonizations, survivals, social and ideological changes, negotiating island identites, innovation, continuity and insular developments in prehistory described by Atoche for Canary islands, Smith and Mulville for the Western Isles, Llull for the Balearic Islands,, Davey for the Isle of Man and Berg for the Cyclades, respectively. Kohn, Waldren and McKechnie's papers questioned the concepts of insularity, isolation or remoteness suggesting that inner views of island worlds sometimes differ markedly from those of observers.

They all agreed that an island is not just an area surrounded by the sea, but a social, political, economic, cultural unit with its own character and development, an integral part of a greater unit which may include other islands and mainlands. Island systems, are unique laboratories to gain knowledge and understanding of people, culture, and environment al relationships in local, international and global terms. A rise in heritage, identity and environmental consciousness has made conservation of the ancient, the natural and the unique of major importance to discerning people. However what is 'natural' or 'authentic' and worthy of conserving is a disputed question. The inter-disciplinary approach to islands offered all of us a chance to query the history of our subjects, the splitting apart of anthropology and archaeology and the re-union of the two subjects under the rubric of cultural anthropology.

Anthropology started off as a museum science in the nineteenth century but it changed its focus and relied more on impirical sources after the fieldwork revolution of the early twentieth century when it became a social and cultural discipline taught in universities and practised in diverse areas of the under-developed world. Today we have anthropology of museums (Bouquet 2001) and modern cities as well as less familiar places.

Archaeology too has evolved from an early consentration on artifacts, assemblages and cultural definitions based on material remains to 'concerns with the nature of the dynamic relationships between human communities and their environment.'. The active role of human agency is recognised in the manipulation and transformation of the environment and aspects of the environment are used in 'the articulation and transformation of social relations.' (Patton 1996:26).

Insularity is one of these environmental variables. However, we cannot assume that it acts in similar ways around the world. If my understanding is correct, the 'raison d'etre' of this conference was to share the many interpretations of the effect of insularity on human and animal populations and the impact these have on islands. We hoped to define similarities and reveal differences which would enrich the knowledge we all have developed in our particular researches. The participants documented the changing conceptions of islands, places and spaces described from prehistory into the present and re-assessed the importance of understanding the social, political, economic and religious pressures that have affected local and regional identities of island people over the centuries. It became clear that the historian can travel the world from his/her armchair having journeys of the imagination but social anthropologists and archaeologists need to go to their islands to develop the experience and insights needed to describe 'island life' in its many forms.

Today,with tourism a major industry on most Mediterranean islands, we need to remember that islands are not just vacation paradises, but important places in which many varieties of human, botanical and animal social, political and biological evolution, prehistory and history have taken place. Ethnicity and nationalism are important issues in the world today and anthropologists try to reveal the historical processes through which identites are contructed and transformed by competing groups and the way the distant past is used by many to support an ideal of continuity over time.

One of the issues that is saliant to our pursuits is the value placed on archaeologiacal remains and research by living populations and governments of the various islands. The ideological naturalization of modern political communities have often appropriated archaeology to authenticate the 'invention of traditions' (Hobsbawm & Ranger 1983) in reconstructing or re-animating national identities. In the case of appeals to ancient Celtic Identity in Modern Europe, Deitler noted, 'an understanding of this complex process requires exploration of the ways language, objects, places and persons have been differentially emphasized to evoke antiquity and authenticity in the process of constructing and manipulating emotionally and symbolically charged traditions of Celtic identity.' (1994:584). It is notable that in ancient Mediterranean states, Celtic was used to describe 'barbarians' in contrast to the Greek and Roman self-definition as 'civilized' (Ibid:586, Chapman 1982). The modern Celticism obviously has re-interpreted these stereotypical images of the ancient classical world and given them more positive connotations. An additional factor this suggests is the political ramifications of anthropological and archaeological research. Anderson's Imagined Communities' draw strongly on the archaeological past to support their claims to uniqueness and longevity and as Crumley (1991:3) noted,'the most effective expression of ethnicity requires an anchor to a particular geography' and archaeology provides that anchor by tying sites to ancient events and people.11 However, as David Sutton discovered on the Greek Island of Kalymnos, 'there are different ways of constructing the relationship between past, present and future ...there are culturally diverse ways in which the past is seen to inform action and consciousness in the present and different rules for use of history'(1998).

THE MEDITERRANEAN: HISTORY

So when we consider the work of Ferdinand Braudel in his detailed history of The Mediterranean and the Mediterranean World in the Age of Philip II we recognise the importance he placed on the contrasting perspectives from islands linked in a chain and yet socially enclosed. He tells us that the Mediterranean islands are more numerous and above all more important than is generally supposed...'Large or small their significance lies in providing indispensable landfalls on the the sea routes and affording stretches of comparatively calm water to which shipping is attracted, either between islands or between islands and mainland coasts.' 'There is hardly a stretch of the Mediterranean shore which is not broken up into islands, islets and rocks...All islands have areas affected by the general life of the sea, and at the same time looking inward to that world which the historian, preoccupied with texts concerning history, does not notice at first glance: the withdrawn and insecure way of life, the biology isolated under a bell jar of which naturalists have long been aware.' (1981:151).

'A great problem for most islands is how to live off their own resources, off the soil, the orchards, the flocks, and if that was not possible, how to look outwards and assure their future needs. Balearics could not support their military or merchant towns and imported grain from Oran and cereals from north Africa and Sicily. As well as the threat of famine there was also the risk from the sea which in the mid-sixteenth century was more warlike than ever.' According to Braudel, the Balearics, Corsica, Sicily and Sardinia were besieged territories. They had constantly to be defended, watch towers had to be built, fortifications erected, extended and equipped with artillery, either pieces sent in from outside or cast on the spot by the primitive methods of bell founders.(153)

The Balearics, Majorca, Minorca, Ibiza and the scarcely accessible rock of Formentera have always been of considerable importance: an entire shipping sector revolved around them. (149) Some accidental change of ruler or of fortune may bring to the island's shores an entirely different civilization and way of life, with its dress, customs, and language, which the island may receive and preserve intact over several centuries, bearing living witness to forgotten revolutions. (Examples from Majorca: Romans came in 123 BC built roads, aqueducts, names, religion, etc. ; Moors 900-1200. surnames and village names, canalization of water, norias, olives, etc. and Catalan/Aragonese Conquests 1229 language, customs, linages, property rights, fueros, etc.

Braudel claims 'the Mediterranean lands were a series of regions isolated from one another, yet trying to make contact with one another .' In spite of the days of travel by boat that separated them, there was a perpetual coming and going between them. 'Like an enlarged photograph the history of the islands affords one of the most rewarding ways of approaching an explanation of Mediterranean life. How has each Mediterranean province been able to preserve its own irreducible character, its own violently regional flavour in the midst of such an extraordinary mixture of races, religions, customs and civilizations?(161)

ISOLATION?

'Isolation' is a relative phenomenon. That the sea surrounds the islands and cuts them off from the rest of the world more effectively than any other environment is certainly true whenever they are really situated outside the normal sea routes. But when they are integrated into shipping routes, and ...become links in a chain, they are on the contrary actively involved in the dealings of the outside world, less cut off from them than some mountain areas. (150)

Archduke Luis Salvadore of Austria2 , sailed the Mediterranean and around the world in search of romantic adventures and scientific observations during the nineteenth century. His prolific writing and studies of most of the islands within the Mediterranean basin warrant a separate paper. He often fixed his gaze on places ignored, or unknown by most European readers of the period. One of these was the Columbret isles which he described in a small book entitled Columbretes published in Leipzig, Austria in 1895. Interviewing one of the three lighthouse keepers and their spouses, the only inhabitants of the island, he asks them if they feel isolated and faraway from others? One woman responded that, 'when an island is situated near (39 miles away) a populated coastal area where one can go if they chose once every two weeks on the boat that brings them supplies, they feel close to others'. He notes that, 'Isolation did not seem to depress the lighthouse keepers in this beautiful place...furthermore they had the church to offer them comfort and some of their prayers surely went straight to heaven...'(1895:335)

The age of discovery showed us that 'the remote' was actually compounded of imaginary as well as real places, yet they were all of equal conceptual reality or unreality before the differences were revealed. As Edwin Ardener noted, 'A place is remote to those who have not been there, but to those within it is just another place' (in Chapman (ed) 1989)

Books about traveling/voyaging on the Mediterranean were quite the vogue in the nineteenth century. Although references to Majorca had been made earlier, the first complete book dedicated to the islands by a foreign author was Souvenir d'un voyage d'art a Majorque by Lauren's a writer from Montpellier. Published in 1840, it presented the wonders of Mallorca to a European audience. Framed as a story of a voyage - a mixture of scientific and poetic information like many of the publications of the period dedicated to the study and description of 'nature' - it presented philosophical reflections and personal confessions inspired by the beauty and calm he found on the islands. It painfully reminded George Sand of her previous winter there with Frederic Chopin and is said to have prompted her to retrieve her notes from that trip and write Un hivern a Majorque.

ISLANDS OF THE MIND

Many writers carry as much from personal experience, reading and illusions into their travels as they put from their travels into their books so that what we get is a place and people who exist only as the author perceived them, an exotic landscape invented to test, discover or calm their souls. Nevertheless each book offers us historic narrative, impressions, and experiences of particular places and times. The references we have are notes, memories, reconstructions of selected details chosen by each author. Most of these early travelers were attracted to the Mediterranean islands in the mid-nineteenth century in search of unspoiled islands, ' paradise' in contrast to the violent and inhuman progress occurring on the continent. The landscape, the climate, the clear light created the image they sought and the calm and purity of everyday existence led them to feel they had returned to a 'state of nature' (Salvadore, Lauren and others).

The possibility of observing and participating in a social system where face-to-face personal relations still persisted seemingly uncontaminated by the flow of history gave an illusion of mystery to conceptions of 'the Mediterranean'. 'The Mediterranean' was perceived to be one of the privileged places where one might travel to find lost horizons or hidden paradises where 'continuity and the persistence of Mediterranean modes of thought' could be identified. (Peristiany 1966, Bestard 1986). What were these mediterranean modes of thought so idealised by early travelers?

Paul Sant Cassia suggests that Mediterranean identity 'may have been 'manufactured' through the circulation of ancient texts, (Strabo, Pliny, Braudel, etal) and until the nineteenth century scholars from different Mediterranean lands freely circulated from one city to the other of whom Ibn Khaldoun, Ibn Sina (Avicenna) and Gemistos Plethanoas are only some of the most notable examples. In their own times, right up to Mosque and Pareto (in some ways a revitalization of Platonic theory of the ideal City) their texts certainly influenced policy makers and politicians (including modern leaders such as Gaddafi, cf Davis, 1987) and contributed to 'distinctive forms of political organisation.' (1991:5)

Anthropologists in the 1950s shaped the anthropology of the Mediterranean as a 'culture area' based on their research in rural communities in the northern mediterranean countries. They identified 'honour and shame' as the codes relevant to these social systems which had been largely superseded in North-Western Europe; a value-system associated mainly with unchanging traditions, social behaviour and social status. They suggested that prestige, reputation, and honour constituted shared discourse and aided social order in the countries which bordered the Mediterranean.' However as Goddard, Llobera and Shore note, 'what should have been a construct identified with a cluster of cultural features which had a limited but meaningful geographical application became an unmanageable contraption which not only led many anthropologists astray, but generated countless sterile controversies.' (1994:15 fn14)

Sant Cassia supports this and states that 'Certainly the belief in the Mediterranean as possessing some degree of cultural unity due to common intertwined histories as well as heirs of (often opposed) civilizations has changed across time not just among anthropologists and other scholars, but also among Mediterranean peoples themselves. It could hardly be otherwise, as we are dealing with societies with their own histories and ideologies in complex interrelationships among themselves and the West. Many Mediterranean languages have similar roots, historical and political identities were once closely tied. Mediterranean colonialism, whether Catalan, Mallorcan, Spanish, Venetian, Ottoman or Italian affected cultural life, literature, agriculture, land tenure patterns and the production of specific crops, and encouraged intermarriage.'

What seems to have happened since the Mediterranean as a category entered anthropological discourse, is that as the category 'Europe' has extended to encompass Mediterranean societies, it has become synonymous with 'modernity' and 'the Mediterranean' has become increasingly viewed in negative terms by Mediterranean people themselves. Boundaries are being redrawn between the northern and southern shores of the Mediterranean. The north is European, the south is Arab/Muslim.

The 20th C science of language from Saussure, Wittgenstein and others has questioned the verbal structures that determine all representations of reality. Anthropology has taken this on and stresses that history can have discontinuities and continuity, knowledge may be temporary or in motion so the answers we find today may be different in the future or in the past

Anthrony Shelton notes that: 'In the past, anthropologists claimed to describe immediate social facts and thereby give direct access to a cultures operative discourses' however, early descriptions could never really correspond to a foreign culture's discourse because they were inevitably mediated through methodological protocols and disciplinary procedures devised by a discourse external to the culture they claimed to represent.' (2001:142). Today we often identify multivocality whereas in the past this was restrained in traditional ethnographies by giving one voice a pervasive authorial function and to others the role of 'sources of information', informants to be quoted or paraphrased. In the case of the Balearics, Muntaner (1990:69)describes the complexity of island identities:

Of course the islanders as a whole share many general characteristics with other inhabitants of the Mediterranean basin. The traits common to all of these seaside populations (particularly reserved hospitality and a marked respect for tradition) are the consecuences of a history of the symbiosis of a primarily agricultural economy with a seaborne trade. Like the islands of Corsica and Sardinia the Balearics as a whole have avoided becoming too dependent on this sea commerce, largely because a large part of the population

FIGURE 1. The Prehistoric Taula Sanctuary of Torralba den Salort, Alayor, Menorca: an example of a tourism resource and attraction.

has always lived completely ignoring the sea -surprisingly, at times,never leaving the island's interior even as far as to venture into the capital (invariably the major port). But beneath the general characteristics of Mediterraneans are other more specific traits at family level...A Mallorquin is accommodating and self-contented, with an inbuilt prudency that avoids any risk. In an island where everything is at hand the islanders seldom experiment without fear of the unexpected or of rapid change. As a permanent presence the sea slows one down...permits one to go slowly.

Many of the Majorcans with whom I worked explained their reticence to mix with outsiders to the fact that their history was one of drawing inward to avoid the many threats that continuously arrived from 'outside'. Insecurity was a keynote of island life in the past, both in terms of agriculture and subsistance as well as invasions, conquests and sackings by Moors, Arago-Catalan expansion, pirates or bandits. Today this insecurity concerns loss of identity and island values with the onslaught of foreign investment in extensive areas of the island. Many have profited from these incursions however few have become less weary of "outsiders" (JWaldren 1996)

New forms of ethnography are beginning to alter the power relations of the past by recognising informants as co-authors and the ethnographer as much scribe, archivist as well as interpreting observer. Today we attempt to theorize on gender debates, the historical political construction of identities, and self/other relations, and we probe the gendered position that make all accounts of, or by, other people inescapably partial.

Culture in this new idiom is not an object to be described, neither is it a unified corpus of symbols and meanings that can be definitively interpreted. Culture is contested, temporal and emergent. Representation both by insiders and outsiders is implicated in this emergence. Culture may be captured in its traditional, authentic guise in a picture of a Majorcan or Corsican woman, her beads, her peasant costume worn to perform folk dances. Despite the fact that this is a performance rather than an everyday atire and she may change, wear western clothes, be forced to live in a town or whatever, her picture will remain intact. Similariy, monuments are cleaned and made accessible to tourists and the tourism industry relies on such images to attract people to adventure into the past, they market culture. If part of the tourist quest is a quest for the authentic and tourism is a kin to a sacred experience a form of ritual journey from ordinary space to a spatially separated non-ordinary one for a finite period (MacCannell 1989), then visiting archaeological monuments on Menorca (or at Stonehenge, Sardinia or Malta) can be a means of experiencing another past, walking among the ruins, imagining the life lived by those who inhabited those spaces so long ago. However, these symbols of past cultures maynot be relevant to the local people who hardly notice them on their daily routes to work or leisure. Pasts can be invented, re-constructed or erased by politicians, tourist companies or local people depending on the required effect. This adaptation of archaeological pasts to present various contemporary aspirations needs to be recognised by those who study the past. The 'image' various groups want to project to one another and tourists entails some sort of compromise between the various political parties, 'local'village and town dwellers, and international agencies and visitors, on what each means by 'environment, culture, heritage or identity' . In the Balearics, each of the islands has its own priorities (J. Waldren 1997:124).

The island of Menorca has developed its archaeological monuments so that they can be visited by tourists with parking places for buses and kiosks selling drinks and postcards. This 'development' was stimulated by the government in the hopes of capturing more of the income from tourism that often went to the foreign tour companies and commercial establishments. The creation and recreation of a site as an 'attraction' is established by a system of markers which stage certain aspects of the setting as 'out of real time and place' and isolates it from the 'ordinary' through guide books, postcards, and tourist brochures and leaflets (Lindkund 1997).

The much larger island of Majorca has witnessed the destruction of major archaeological sites for highways, airstrips and other forms of development and despite continued research by competant scientists, has only just begun to recognise the possibilities of preserving archaeological precincts and developing some as tourist attractions. Again, this is a partial project supported by various inland villages that have no other 'tourist attractions'. Golf courses, marinas, airstrips, hotels and apartments continue to take precident over archaeological remains. Artefact reproductions are sold side by side with t-shirts, liquor, sweets and assundry tourist items that all say 'made in Majorca'. Often a postcard is as close as most visitors ever get to the islands' archaeological remains.

Elisabeth Edwards work on old photographs and postcards consider the impact of this 'vicarious' visiting. She found that 'the postcard becomes a sacred relic a souvenir from another time and place which reminds us of the things we were looking for in others and photography as a medium is not about photography as such but about its potential to question, arouse curiosity, tell in different voices or see through different eyes. The photograph is an object image that one can hold in one's hand, inscribe, cut with kitchen scissors, paste into albums, send to friends bringing living images across the world. Postcards too provide images of people and places. The power of these images resides in their creation of fragments which come to stand as wholes, reifying culture in the endless repetition of images. While photographs and postcards may not be anthropological in terms of a fully informed and integrated theoretical position, they nonetheless constitute documents of culture or cultural documents whose legitimacy is drawn from the fact their creators are attempting to communicate values and negotiate realities which are integral to human experience and consciousness' (1992). Archaeological artifacts bought over the counter in a souvenir shop have similar impacts.

This paper has tried to suggest through historic, literary and anthropological sources the complexity of identifying places and people, of recognising the varied perceptions of time, space, physical and social boundaries among different peoples at various periods in prehistory, history and the present and the fluidity of these identifications. Historical records often romanticise or harmonise events and local conceptions of history may be composed of fantasy and imagined 'realities'. Prehistoric sites are often associated with recent historical events: eg.the Moors or the Civil War or the plague in Majorca...Although often subjective and selective, narrative histories are useful in re-constructing various perceptions of time past and present.. Oral history accounts or narratives often include silences and pauses and these can be interpreted in different ways.Categories of the Mediterranean, Europe, insularity, isolation, home, near, far or remote are not uniformly defined. Instead they have different meanings to different people and in various circumstances can differ among the same people.

REFERENCES

ARDENER, E.
1987. The Voice of Prophecy and Other Essays, ed.M.Chapman. Oxford: Blackwell
BALLESTER,
BESTARD, J.
1986. Casa y Familia.Palma, Mallorca:Institut d'Estudis Balears
BOUQUET, M, ed.
2001. Academic Anthropology and the Museum: Back to the Future. Oxford: Berghahn Books.
BRAUDEL, F.
1981. The Mediterranean and the Mediterranean World in the Age of Phillip II. Vol. I. London: William Collins & Sons & Co. Ltd.
CHAPMAN, M.
1982. Semantics and the Celt. In Semantic Anthropology. D.Parkin,ed. pp123-144. New York: Academic Press
CRUMLEY, C.
1991. Region, Nation, History. Excursus 4:3-8
DAVIS, J.
1989. The Social Relations of the Production of History, in E. Tonkin, M. McDonald and M. Chapman eds, History and Ethnicity. London: Routledge.
DEITLER, M.
1994. Our Ancestors the Gauls:Archaeology, Ethnic nationalism, and the Manipulation of Celtic Identity in Modern Europe, in American Anthropologist. Vol 96. No3 September 1994, pp584-605.
EDWARDS, E, ed.
1992. Anthropology and Photography 1860-1920. New Haven and London: Yale University Press
GODDARD, V., J. LLOBERA & C. SHORE eds.
1994. The Anthropology of Europe:Identities and Boundaries in Conflict, Oxford:Berg
HOBSBAWM, E. & T. RANGER eds.
1983. The Invention of Tradition, Cambridge: Cambridge University Press.
LAURENS, J.
1840. Souvenir d'un voyage d'art a l'ile de Majorque, Paris:A. Bertrand.
MACCANNELL, D.
1989. The Tourist: A New Theory of the Leisure Class. New York:Schocken Books.
MUNTANER, P.
1990. The islander in Insight guides to Mallorca & Ibiza, Menorca & Formentera.
Singapore:Apa
PATTON,M. 1996. Islands in Time: Island Geography and Mediterranean Prehistory. London and New York: Routledge
PERISTIANY, JG. ed.
1965. Honour & Shame: Values of the Mediterranean, London: Weidenfeld & Nicolson
SAND, G.
1977. A Winter in Majorca 1838-39. L.Ripoll ed. (1840. Un Hiver a Majorque)
SANT CASSIA, P. ed.
1991. Authors in Search of a Character: Personhood, Agency and Identity in the Mediterranean in Journal of mediterranean Studies Vol 1 No.1. Malta: Mediterranean Institute.
SHELTON, S.
2001. Unsettling the meaning: critical museology, art and anthropological discourse in Academic Anthropology and the Museum,Bouquet, M. ed. Oxford:Berghahn books
SUTTON, D.
1998. Memories Cast in Stone. Oxford: Berg Publishers.
WALDREN, J.
1996. Insiders and Outsiders: Paradise and Reality in Mallorca. Oxford: Berghahn Books.
1997. 'The Road to Ruin: The Politics of Development in the Balearic Islands' in S. Abram. & J. Waldren, eds. Anthropological Perspectives on Local Development. London and New York: Routledge.

2

ROCKS AND ITINERARIES: SEA AND LAND PERSPECTIVES ON AN AEGEAN ISLAND

CHRISTINA MARANGOU

EMAIL: christina.manangou@eudoramail.com

INTRODUCTION

The island of Lemnos, 477 km2 (approximately 30 km east-west and 24 km north-south), is situated in the north-east of the Aegean sea, at a distance of approximately 50 miles to the west of the entrance of the Dardanelles straits and 38 miles to the East of the Athos peninsula (Chalkidiki, Central Macedonia) (FIGURE 1). On the western coast, Myrina, the capital and the main port of the island, has been a harbour in various periods of the island's history and could possibly be Homer's "well-built town of Lemnos". The northern side of the port consists of a rocky peninsula called Kastro, projecting towards the west, with a maximum height of 116m (Admiralty Chart 1998). Its volcanic rocks are Lower Miocene dacites, while the low area around and both sides of the ridge that joins it to the main island contains Holocene coastal deposits; a fault cuts across this ridge (Geological map 1993). Continuity of occupation in the area, attested by surface finds belonging to several periods shows that the site had some importance, even though its meaning may have changed considerably during time.

The paper presents the on-going study of the rock-cut features and engravings of the Kastro site . Initially, rock-cut features may have been present on the whole peninsula, but a lot was destroyed by later structures. The preserved features are dispersed. They are more dense mostly on parts of the peninsula which were not built over in the Middle Ages and subsequent periods. They have been identified, till now, in various sectors, on different altitudes starting from a few meters above present sea-level. There are several types of features, isolated or in combinations:

(1) - larger and smaller cavities, "wells" or "cisterns" and conduits, the latter sometimes isolated and sometimes along flights of steps, or close to "rooms",

(2) - flights of steps: steps of different dimensions, stepped passageways, leading uphill, but also restricted groups of small steps of obscure use,

(3) - "room" angles or "platforms" protected by 2 or 3 walls, in isolation or combined to other features, such as platforms and post-holes.

(4) - niches, and

(5) – engravings (carvings), located either on vertical surfaces of niches or on walls or floors of "rooms"/ "platforms". Carvings are mostly linear, but include exceptionally representational themes, at least once a ship.

SEA ITINERARIES AND SEA MARKS

Because of its geographical situation, the island of Lemnos was related to faraway places across the sea at least since the Final Neolithic, during which it was inhabited according to archaeological evidence (Archondidou 1994; Patton 1996, 50). It was involved in mythical travels, such as that of Jason and the Argonauts, who stopped here on their way to the Black sea and the Golden Fleece, as well as in the Trojan war. Navigation of sailing ships was regulated by the north-eastern Aegean current (FIGURE 1), originating from the straits of the Dardanelles (Papageorgiou 1998, 429), passing either side of Lemnos (Zodiatis 1993; Morton 2001, 43), and continuing towards the west and the south. The predominant northern winds help ships follow this current. Besides, a Mediterranean current coming from the west and south moves along by the western coast of Turkey, towards the same area of the north-east (Papageorgiou 1998, 431). Lemnos was therefore in the middle of sea-routes linking the North and the South Aegean. With the exception of an area of "deadly" shoals near its eastern coast, on which the Hellespont current can throw ships (cf. Choiseul Gouffier 1828, 203), the island offered havens and landing places –many more than the nearby Imvros- protecting the boats against winds, tempest or enemies (Choiseul Gouffier 1828, 212). It offered, thus, convenient stops for mariners, good ports and good forts, as it was noted by Bordone (1528, LIX verso).

While approaching by sea, the Lemnos landscape is bare to use sounding leads, but also to memorise landmarks and coastal profiles as they appeared from the sea (Taylor 1956, 37). Landmarks visible at a safe distance from coastal dangerous waters (McGrail 1996, 314) were extremely important. Such were prominent headlands, in particular when they were high, recognisable from their distinctive shapes and silhouettes (Morton 2001, 185-189). Sea-marks would be specially appreciated when coming close to a low and indistinguishable land, and all the more so, when navigating in the open sea, as early navigation was necessarily imprecise (Taylor 1956, 63).

Headlands, the points of land protruding furthest into the sea, were the places from which ships left the coastline to sail across the open sea and at which they landed after the

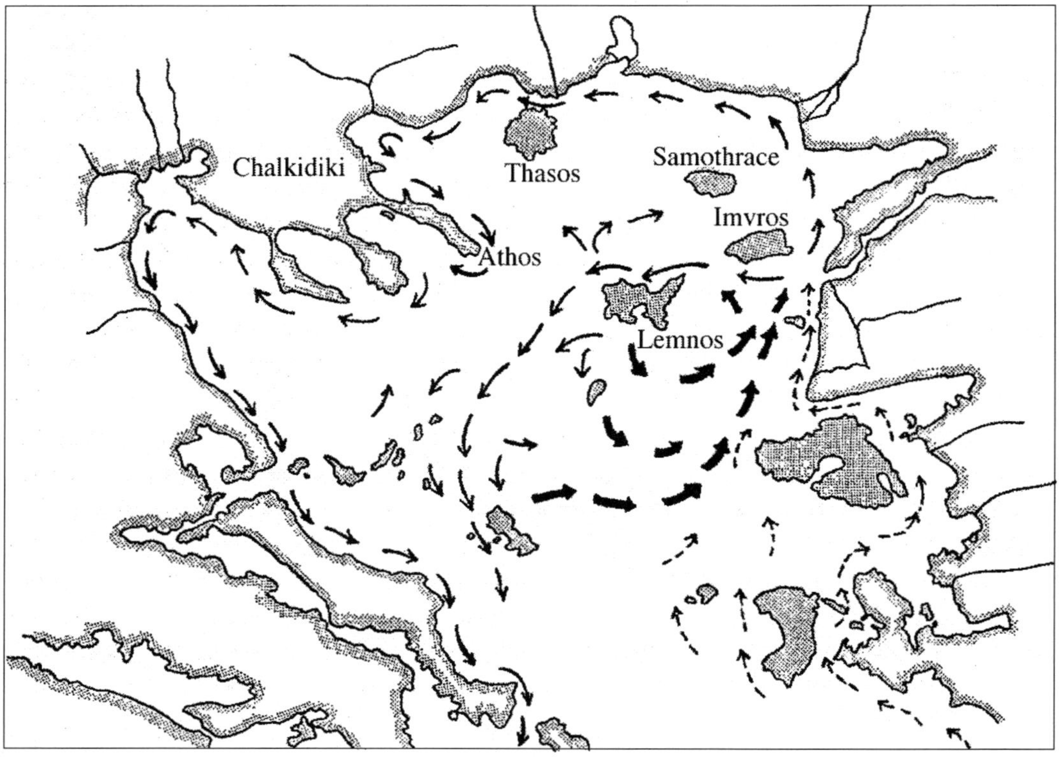

FIGURE 1. Northern Aegean Sea currents in winter (adapted from Papageorgiou 1997, 432, fig. 6b).

and low compared to smaller islands of the area, such as Thasos and in particular Samothrace, which is a wooded mountain (height: approximately 1600 m) emerging from the sea (Conze 1860, 77; cf. Fredrich 1906, 242). Seen from a distance, Lemnos is a long and narrow stripe of land in the horizon, justifying the adjective tenuis given to it by the poet Valerius Flaccus (Argon. II, 431). In fact, the western half of the island is hilly, the eastern one being lower and sandy. When approaching, one can see two mounts, the highest one being Skopia (428 m) in the north-west, the remainder looking low and uniform (Dappert 1703, 243; cf. Fredrich 1906, 242; Tozer 1890, 235).

In early times pilots had to know the winds and the stars, crossing (Morton 2001, 163-167). Sailing conditions around headlands are complicated, as dangers result from their topography, submerged rocks and reefs, and meteorological and oceanographic conditions (Morton 2001, 69-81). Promontories and headlands are mentioned very often in ancient Greek literature; headlands were the points at which ancient mariners came closest to the coast during travel, but they also might have important religious or military functions, which might be interrelated (Morton 2001, 68). Such natural sea-marks not only helped coastal navigation, but also, for the same reasons, were sacred places. They were often dedicated to divinities in historical times, and temples or altars were built on them, containing gods images (Wachsmuth 1967, 394-403). In such places, either the sanctuaries themselves, or, in other cases, tombs or towers could have functioned as sea-marks in coastal waters, or/and as distance markers or milestones (Morton 2001, 193-200, 205). In some cases, a sanctuary on sacred promontories might be unroofed and without images (Romero Recio 2000, 31-32). Offerings of steering oars simply on the rocks of a coast are also attested (Romero Recio 2000, 31-32, 25 and note 231). The promontories' navigational aid function in dangerous places was therefore combined to appeal for supernatural protection (Morton 2001, 202-203, note 84; 310-312).

Promontories of shining limestone, or of dark, black or red rocks (Morton 2001, 189-192) would be even more visible . The Kastro headland is easily distinguished by its form and colour, in contrast to the surrounding low areas (FIGURE 2). In fact, its rocks as well as the rocks of the two promontories northwards (Petassos; FIGURE 3) and Southwards (a church of Saint Nicholas, the patron of mariners, is built on its summit), are dark brown or grey-brown dacites (Davis 1959, 27, 66; Geological map 1993), while the two bays both sides of the Kastro consist of light coloured coastal deposits, sands, containing clay, as well as, in the northern bay, humus (Davis 1959, 6, 49; Geological map 1993). These dark elevations would constitute useful landmarks for mariners in order to find the entrance to the bays. The possibility to choose between two sheltered areas to land the ships on the flat sand offered protection from opposing winds (Conze 1860, ; cf. Morton 2001, 124) (FIGURE 3). The southern side with the present port protects from the predominant northern winds, although not from western winds.

The Athos headland, 2033 m high, of which 400 m are hidden because of the curvature of the earth (Armao 1951, 42, 66, 67), is an impressive seamark on the continent, facing Lemnos (FIGURE 4), already described by Apollonius Rhodius (Argonautica I, 598-608) during the Argonauts's approach to Lemnos. Athos, as well as Samothrace, could be visible from a distance of up to 100 miles from ancient ships (Morton 2001, 144, note 4). When seen from the western coast of Lemnos, Athos is an an omnipresent feature in the skyline, in particular from the Kastro, from where also the tips of the other two Chalcidiki promontories may be seen on clear days; it seems even closer and clearer in winter. Athos has certainly been an early appeal for crossing. According to an obscure fragment (Sophocles), its shadow reached as far as Myrina, overshadowing the back of a cow in its market: travellers have often commented on this (see further).

FIGURE 2. The Kastro headland from the west, as seen from an approaching ship.

OPTICAL SIGNS AND PERIL FROM THE SEA

The dominant position of the Kastro makes it the natural capital of the island (Tozer 1890). It was used, not only against peril from the sea, but also in order to control the islanders (Belon 1578, 57). The Medieval castle, on the northern part of the Kastro, remained close to the sea, in spite of the pirates, because it was situated on a naturally protected place (Conze 1860, 107). The efficiency of its position is confirmed by Myrina's long resistance to the Athenians in the 6th century BC. The view from the summit permits visibility across the sea over a long distance, including the whole western coast from north to south, thus towards all directions from which ships could approach.

As a matter of fact, ships could belong to enemies or to pirates. According to Belon (1588, 189-190), there was no summit of mount in the islands, including Lemnos, or on coasts without guards, who kept watch in order to catch sight of approaching pirates. The guards used to light fires and warn the islanders by means of their smoke during the day and their light at night. The number of fires lit corresponded to the number of suspicious boats, one fire usually meaning security. Guards from other mountain summits, without visibility of the sea, would light the same number of fires, transmitting the information quickly over long distances. Similar practices were attested on various coasts of the Mediterranean (Ormerod 1978, 43-45). The names given to mounts or hills near the coast is eloquent: in Lemnos, besides the highest mount, Skopia, other summits are called Skopos or Vigla, all these meaning watcher or look-out place .

Sea raiding was not a novelty of the Middle Ages. Homer mentions it, and piratical expeditions are probable in the Bronze Age. Thucidides and Herodotus mention Tyrrhenian pirates, in particular Pelasgians from Lemnos raiding Attica (Ormerod 1978, 20, 153). Strabo describes small and quick piratical boats with 25-30 oars like the triakontors, based in the Black Sea (Ormerod 1978, 26-27). Hidden structures on high coastal promontories or sheltered on land ward flanks, as it happens on the Kastro, could have belonged to populations who tried to protect themselves from the pirates, or to the pirates themselves (Rauh et al. 2000, 175). Pirates knew the routes followed usually by ships and expected them at promontories (and at the mouths of gulfs), since these were the coastal areas where ships approached more and would be expected to appear regularly (Morton 2001, 176-177). Pirates not only attacked ships, but also plundered shores (Ormerod 1978, 31). This has been a constant fear for islanders, in particular the Lemnians till the end of the 19th century, if not the beginning of the 20th. Human memory still recalls herder-guards watching from the summits for corsairs .

FIGURE 3. The harbour of Myrina and the Kastro from the south-east. To the right: the promontory of Petassos

Optical signs, including smoke and fire, permitting transmission of information or warning over long distances (Aschoff 1977, 24), were not invented in the Middle Ages either (Riepl 1972, 34 and note 1, 43-90). There is ample evidence about lighthouses and beacons from the Hellenistic and Roman periods (Buchwald 1912; Hague 1973; Blackman 1982). It has been argued that fires on temple roofs may have also functioned as lighthouses (Romero Recio 2000, 33). It is not known when exactly started the use of beacons and other onshore lights as navigational aids and leading marks to indicate the presence of land, mark important points, changes of directions, dangerous reefs etc (Morton 2001, 210-214). Nevertheless, there is evidence about use of optical signs very early.

According to Leiner (1982, 73-79), the Mari archives

(1750BC) mention simple or double fire relays used from one town to another. Homer (Iliad 18. 210-213; Leiner 1982, 88-89) compares the glittering of Achilles' helmet to fire signs from besieged islands calling for help from neighbouring islands. This would mean that alarm signals were known at least since the 8th c BC, when the Iliad was written (Riepl 1972, 50-52; Aschoff 1977, 24, 25). Fire-beacons or torches on top of seven intermediate mountains transmitted the news of the capture of Troy from mount Ida near Troy to the palace of Agamemnon in the Argolid, in the homonymous tragedy by Aeschylus (Agam. 283-311). In these were included the Ermaion Lepas in Lemnos (on its north-eastern promontory), and Athos (Tozer 1971, 328-330; Leiner 1982, 59-68); this means that fire signal relays were known at least since the 6th century (Aschoff 1977, 24, 25, fig. 2). In Apollonius' Argonautica, when the heroes lose their way at sea, the god Apollo, standing on a nearby mountain peak, raises his bow, which gives out a dazzling gleam of light, and enables them to make their way to a nearby island (Clare 2000, 10). Ulysses sees fires (beacons?) while at last approaching Ithaca (Ormerod 1978, 44). Besides, illusionary signs that deceive mariners, present in myths as well as in folk tales, were used till the 19th century in various parts of Europe for looting, when it was the ships that were in danger from the islanders instead than the other way round.

FIGURE 4. View from the Kastro towards the west and mount Athos.

LANDING ON DANGEROUS COASTS

Although the open sea is visible from some rock-cut features of the Kastro, and they might be connected to keeping watch for the arrival of ships, most features could not be navigational sea-marks, since they face the harbour or the land. Nevertheless, some could be landing signposts. Ancient mariners feared hidden reefs, shoals, shallows, sandbanks (Taylor 1956, 4), currents or winds in unknown waters, specially when approaching coasts, turning around promontories or entering harbours; divinities protecting landing or coming onshore were more numerous than those protecting embarking (Wachsmuth 1967, 459-461). One of the epic adjectives of Lemnos, ·ÌÈ¯ı·ÏfiÂÛÛ· (amichthaloessa) (Homer, Iliad 24.753), if not meaning smoke shroud, foggy or dark, because of the god of metallurgy, Hephaistos, working on his favourite island, it has been explained as difficult to come ashore, with harbours of difficult access (Fredrich 1906, 252, note 1 with references). Both interpretations would, in any case, mean low visibility and difficulties for approaching mariners .

This seems in contradiction to the fact that the havens of Lemnos protected boats from winds, unless one thinks of the difficulties to enter them, and in general, to approach rocky Aegean shores, particularly in the dark. Difficulties to find the entrance into the harbours of Lemnos are mentioned by Fredrich (1906, 252, note 1). Coronelli (1696, 273-275; Tourptsoglou-Stefanidou 1986, 209-210) warns about a dangerous shoal in the middle of the entrance to the Myrina harbour, 6 m under the surface, which was demolished in the 50s'. Two shoals are mentioned in this area in the Mediterranean Pilot (1955, 483). Approach to land was difficult after dark, according to a description of the beginning of the 19th century: the boats were guided by the lights of a café, in order to find the entrance to the narrow and shallow harbour, sailing by the coasts which seemed high, rocky, and abrupt (von Richter 1822).

Pilots guiding in unknown waters are mentioned by Homer (Morton 2001, 250-252), local pilots in the Periplus of the Erythraean Sea (1st century aD) (Taylor 1956, 60-61). Landmarks of the principal harbours are given in the Le Compasso da navigare (1250 a D; Taylor 1956, 107). Pillars or temples or towers on shores were used as such landmarks (Morton 2001, 197-200; cfr. Ormerod 1978, 41) (see also above). It is not known if they had flares lit on them at night (Taylor 1956, 62-63). The modern lighthouse is located on the outer (western) side of the Kastro, at about 70 m (Admiralty Chart 1998; Mediterranean Pilot 1955, 484).

FIGURE 5. A ship engraving in the present port area (photograph by T. Marangos, 1950s').

Oared ships, in particular, with their limited load capacity in contrast to cargoes, and with their large human power, needed to go ashore if possible every night, for the oarsmen to rest and in order to get supplies (Morton 2001, 243, 277-278), especially of water. An oared ship carving (2,20 m long, with 12 oars on the visible side) on a rock-cut structure (FIGURE 5), close to the old coastline, in the outer part of the port of Myrina, at 12 m amsl., may be significant in this respect. It is located near the area where sailing boats were repaired on the sand till the beginning of the 20th century (Marangou 1995, 312, figs. 8-11). The carved niche could have been used as a pilotage or landing mark. Besides practical usefulness, symbolic value could be attributed to such a conspicuous feature.

Availability of water, its supply being always a primary need for boats, was an important factor for the choice of anchorages or harbours, all the more so in an island where water is scarce till present times. Water management could then explain some cavities, the conduits and possible wells or cisterns on the Kastro, near the harbour. In small-scale old practices for water management, rain water may be collected along slopes, using underground drainage tunnels; water may be divided into open-air ditches, which follow paths flowing beneath walls or along them (Laureano 1998).

ROCK-CUT ITINERARIES AND MAZES

However, the above hypotheses could not explain all the rock-cut features of the Kastro, several of which are mixed within the scenery and invisible, even concealed or secret, accessible only through narrow paths. Even flights of steps, which should make communication and interconnections between the various structures easier, may be hidden. Was this protection from external dangers (from the sea?) or concealment from unauthorised persons?

In fact, in one of the sectors of the site, an ensemble of flights of steps, on and around a large rock complex (Rock A) with an east-south-east – west-north-west orientation are difficult to understand. On Rock A we can distinguish

from south-east to north-west the following features:
(a) a "room"/“platform” with niches on both sides of a large wall, and a series of three peaks,
(b) a first large, stepped passage cut through perpendicularly to the Rock's direction, that is, from South to North,
(c) two flights of steps, one on each side (north and south) of the hill, in slightly diverging orientation towards the west, stopping in the rock (dead ends) (FIGURE 6),
(d) a second passage leading from the South to a flattened area in the middle of the Rock, and
(e) a final abrupt and inaccessible elevation (60 to 70 m high). An ascending passage leads from the northern side to a flattened space, and again from here to a rounded space, closed around by rocks, except for the restricted passageway. Behind this last passage, further towards the West, another flight of steps leads again upwards, a few meters from the summit. However, the last and highest accessible point has not yet been identified, for security reasons.

Rock A bears therefore a number of rock-cut flights of steps or even stepped carvings big enough for one foot, ascending mostly from south-east to north-west. These ascending passages of limited access lead up on the Rock, with some intermediate stops. Narrow paths across and around this enhanced Rock show that the latter was in fact the focus of attention. It is a conspicuous landmark, dominant in the landscape, offering an open view to a considera-

FIGURE 6. Flight of steps carved in the rock, from the east.

ble distance, including the surrounding hills. If the Venetian (and Genovese) castle was built (1207 aD) over an earlier Byzantine fort on the highest, northern part of the Kastro, this southern outcrop could control the fort, because of its strategic position, on a way from where one could bring help from the sea (Coronelli 1696, pl. XV, drawing 1, cited by Tourptsoglou-Stefanidou 1986, 209, 210, fig. 1). The fact that its orientation is roughly WNW-ESE (setting and rising sun), as it is directed towards the Athos and the summer sunset, may have some significance: the sun sets exactly behind of the Athos summit twice a year, in May and July. Early travellers, such as Belon (1578, 58-59), observed this, in connection to the above-mentioned verse of Sophocles. Many monuments command a view of natural features that might have framed the movements of the sun and moon; but they were often linked to an equally extensive view in other directions (Bradley 1998, 121-122). The view from the highest attained, till now, structure is not directed towards the open sea, and the lower platforms are rather oriented towards the inner port and the land. The northern side of rock A, at least its lower (eastern) part, is in fact hidden from the sea .

There is obviously restricted access to the higher parts of this natural observation post. One person at a time could be able to climb upstairs, and should have to come back down the same way. Paths are sometimes so steep and narrow that it would have been easy to prevent access of unauthorised people. Besides, one would need to previously acquire information about the itineraries, presumably from persons already knowing that there is a way leading uphill (which is not obvious). The use of restricted passages has been interpreted (concerning doorways and portals of megalithic tombs) as an attempt to emphasise the transitional nature of such locations (Carleton 1998, 14). What we see here suggests directed movement, possibly cyclic, whose patterns could be daily, monthly, yearly, or longer cycles (cf. Carleton 1998, 13). Access to the summit of Rock A involves risks, as it is physically difficult to climb along the higher precipitous cliffs, in particular because of the frequent, even in summer, strong winds, which can almost throw you down the cliffs. This island has been called AÓÂÌfiÂÛÛ· (anemoessa), the windy, which explains for a large part the heavy erosion.

Several interpretations of this scheme are possible. The idea of a complicated itinerary –a maze- in the literal sense (adventurous travel) and the symbolic sense (initiation) exists in Homer's epic poetry, possibly even since the end of the Bronze Age. The adventures of the heroes may have been conceived as initiation ordeals (or rites of passage: Clare 2000, 4 about the Odyssey). The initiation of the Argonauts in Samothrace, mentioned by Apollodorus (Argonautica) in the Hellenistic period, as well as the homecoming of Ulysses might go back to epics preceding Homer and conveying elements of a long tradition possibly of Phenician origin concerning the dangers and the ethics of life on sea (Daumas 1998, 157). The Argonautica have been considered as a collection of geographical and aetiological data on Mediterranean and Black Sea locations and even as a treatise on navigational routes (Clare 2000, 6).

A maze is incised on the verso of a Mycenean tablet from Pylos (a 1200 BC) and the words mistress or goddess of the labyrinth are written in Linear B texts at Knossos (da-pu2-ti-to-jo po-ti-ni-ja) (Calame 1996, 240). A labyrinth is a place in form of a winkle, an architectural space with turns, may be unroofed or without upper floor (according to the lexicographers), and should not necessarily be horizontal; it has one entrance and one has to turn back on one's steps in order to come out (Detienne 1989, 25-26). The episode of Theseus in the Knossos labyrinth has often been interpreted as an initiation process (Calame 1996, 432). The panhellenic diffusion of the episode of Theseus and the Minotaure in iconography as well as in literature in the 7th century BC shows that we have to do with a very old narrative (Calame 1996, 424). There was also a mention of a Lemnian labyrinth (Plinius Hist. Natur. XXXVI, XIX 84-94) . Stone mazes in the Scandinavian 16th century aD have been interpreted as related to pilotage and to anchorage sites by Westerdahl (1995). These labyrinthine patterns are not dissimilar from the Kastro sinuous ascending itineraries carved in the rocks .

In any case, connection of Lemnos with the sea is not restricted to practical and technical aspects, but also to cult. On an extremity of the northern coast, in front of Samothrace, the Lemnos sanctuary of the Kaveiri (Kaveirion), possibly the earliest known in Greece (end of the 8th century B.C.) (Beschi 1994b, 36-37), is overlooking the sea and the route of the ships coming from the north. The Kaveiri, besides their other qualities, were sea gods. Their mystic cult is attested in the whole north-eastern Aegean, besides Lemnos, in Imvros, Samothrace and the Troad (Conze 1860), its beginnings possibly related to the Tyrrhenians' arrival in the northern islands, possibly after the end of the 2d millennium (Beschi 1994b, 37). It may be of significance that in the oldest phase (before the 7th century BC) of the Samothracian Kaveirion the cult, of "primitive character", was performed near rocks elevated like altars, according to Lehmann, Williams and Spittle (1982, 295; cf. Beschi 1994a, 67). One could also wonder whether the torches used during the nightly Kaveirian ceremonies in Samothrace or Lemnos ("the rocks are thronged with torchbearing mystics": Nonnus Dionysiaca 13.393-407, cited by Lewis 1959, 32, frg. 69; cf. Cicero de Natura deorum 1,42,119, cited by Lewis 1959, 91, frg. 200) had any connection with fire-beacons or some kind of signals, as a verse might suggest: "from the firepeak rock of Lemnos the two Cabiri in arms answered the stormy call beside the mystic torch of Samos" (Nonnus Dionysiaca 14.17-22, cited by Lewis 1959, 77, frg. 166). During the Kaveirian mysteries, a ship was sent to Delos to bring sacred light while all lights went off on Lemnos (Philostratus Heroic 740, cited by Moschidis 1907, 115 and Dumézil 1924, 15, 25). Navigation, lights and rocks are interrelated in the island's mythical past.

POSSIBLE PARALLELS AND DATING PROBLEMS

Some typological parallels of rock-cut features have been investigated, in particular in Thrace, in Northern Greece and Bulgaria, although the mounts on which they were found are located inland. Several hypotheses have been proposed for their interpretation, including those of dwelling places, ritual spaces, inaccessible observation posts and rupestrian sanctuaries possibly related to a cult of the sun; relationship to funerary megalithic monuments has also been suggested (Triandaphyllos 1973, 1984, Delev 1984). In some cases, as at Malko Gradishte (eastern Rhodopes), some structures have been interpreted as a ritual complex: a "processional" stairway starts at the side of a tomb or "sanctuary" and climbs to the top of the highest cliff (Delev 1984, 31; Fol 1982, 536, phot. 149-156; Anon. 2000, 174). These vestiges, which have had a long tradition, have been tentatively dated by various authors from the 12th or 13th to the 6th or 7th century BC (Delev 1984; Fol 1982, 540), sometimes from the "2d millennium" (Anon. 2000, 174). They are more often dated from the end of the Late Bronze Age and the Early Iron age (approximately1100-800 BC) (Triandaphyllos 1973; 1984; 1994, 42-47), which include the so-called "Dark Ages" of Greece, but the dating problem has not been resolved till now. Few Mycenean (Late Bronze Age, last part of the 2d millennium) sherds have been found on Lemnos, and the absence of finds from the intermediate (before the Geometric) period (10th - 8th century BC) should be due to the limited extent of excavations and not to desertification (Archontidou 1994; Beschi 1994a).

Rock-cut structures are attested in early harbour works in the Eastern Mediterranean, but there are obvious difficulties in dating them without stratigraphic evidence (Blackman 1982, 92; Frost 1991, 325), although in some cases they have been considered very early, including of Bronze Age (cf. Frost 1963, 66). On the other hand, ship engravings on rocks, plaques or walls of temples in the Eastern Mediterranean are often dated from the Late Bronze Age, mostly 13-12th century BC (Basch and Artzy 1986; Basch 1987, 144-145; Artzy 1994 and 1999). On Aegean islands engravings on limestone plaques (Doumas 1990), or on rocks and settlement enclosures (Anon. 2001), have been dated respectively from the Early Bronze Age and Final Neolithic . A Bronze Age dating would not be impossible for the Kastro ship engraving (FIGURE 5) for typological reasons (Marangou 1999). Nevertheless, we will have to wait more evidence for the dating of the Kastro features.

DISCUSSION AND CONCLUSIONS

Independently of dating problems, Kastro is a complex site. As we saw, some of its features could be related to the sea and the harbour, others to the hill flanks or the summit. Some were conspicuous, meant to be perceived from some distance, or to be used for observation, others were hidden and access to them was restricted. A number of features, facing either the land, or the sea, command the lower flanks without offering much space for inhabitants, spectators or participants (Marangou 1998). Some others were functional, possibly used as dwellings for the living or the dead, or as landmarks, the use of others is unknown, and could be secular or symbolic. We do not yet know, though, the specific historical and cultural context from which these "binary" concepts should derive (cf. Knapp and Ashmore 1999, 6).

These could be varied responses to different prerequisites, depending on whether the features are oriented seawards or land ward; in the outer part of the harbour, or even the open sea, or towards the inner port and the main island; on lower or higher altitudes, hidden, of restricted access or conspicuous; with a large view or "protected" behind natural rocks. There are probably sectors, boundaries or territories with different functions, all the more so, since linear features, paths and flights of steps, are being used for intercommunication. As a matter of fact, the rock-cut setting has been created here within the natural environment. The man-made features are linked with natural features in the landscape and they use the landscape, although, at the present stage of the study, it is too early to understand their global organisation in relationship to each other and to physical and/or symbolic space (cf. Sognnes 1996, 22).

Nevertheless, the evidence presents common traits through the centuries, including navigation, pilotage, harbours, conspicuous or commanding landmarks and out of sight or inaccessible structures. The functional and the non-obviously- or non-exclusively-functional are combined in this modified, partly artificial, partly natural landscape. Environmental limitations and resources on the one hand, and human concerns, ecological knowledge and attempts at adaptation to local natural conditions on the other are interlinked.

It has been argued (Taçon 1999, 36-37) that certain landscape features invoke common responses in human beings, feelings of awe, power, beauty, respect, enrichment etc. This would have happened in particular in places of natural transformation, junctions or points of change between geology, hydrology, and vegetation, as sudden changes in elevation, near unusual landscape features, such as a prominent peak, and with large views of interesting and varied landscape features. Rock art sites may be situated near such naturally defined "sacred" locations. More often they overlook, indicate the approach to or mark the limits of the more sacred and restricted landscape zones (Taçon 1999, 40).

Inherent particularities of the Kastro landscape and of natural features in its environment have marked man, as attested in classical, medieval and post-medieval texts and in present life. Man has manipulated natural features, defining spaces, tracing trajectories, constructing structures, which are interrelated in involved patterns. This landscape could have been at the same time constructed, conceptualized and possibly ideational (Knapp and Ashmore 1999, 10-13). Whatever the real meanings, enhancement of this place transmitted some information and influenced man's behaviour (cf. Hartley and Wolley Vawser 1998, 189) with the help of "special attention" markers that single out significant features of the landscape (Bradley 1998), such as prominent landmarks, outstanding and unusual features of the natural landscape .

Human actions are framed by architectural features that guide movements and control the sequence in which different spaces can be experienced. On the Kastro rocks such architectural features influenced and even directed repeatedly man's behaviour in space – and in some cases they were addressed to a restricted group of people – possibly not to the unhauthorised ones. We are back then to the "labyrinthine patterns".

If we should conclude with a few words, we could end with the main question for on-going and future research: what could such contradictory situations and labyrinthine patterns mean on an island located in the middle of mythical, as well as real, sinuous itineraries of sea-travel, currents and winds, and which is at the same time tenuis and difficult to distinguish and of difficult approach?

Perhaps one day we will know the answer.

ACKNOWLEDGEMENTS

I wish to thank warmly the Greek Archaeological Service and the Council for the Monuments of the Islands for granting the permit for the study and publication of the rock-cut features and engravings of the Myrina Kastro, the Institute for Aegean Prehistory for financial support of a part of this study and Miss Honor Frost for her precious help and continuous encouragement.

REFERENCES

ADMIRITY MAP ,
1998. Plans in the Northern Aegean Sea. Admiralty Chart: no. 1636. Taunton: Hydrographic Office.
ANON.,
2000. Ancient Thrace. International Foundation Europa Antiqua. Sofia: Bulgarian Academy of Sciences, Institute of Thracology .
ANON., 2001. Kathimerini, Internet edition [Online], October 2001. Available from: http//www.ekathimerini.com/news [Accessed: 20 October 2001].
ARMAO, E.,
1951. In giro per il Mar Egeo con Vincenzo Coronelli. Firenze.
ARTZY, M.,
1994. On boats, on rocks and on "nomads of the sea". CMS News (Report no. 21). Haifa: University of Haifa.
ARTZY, M., 1999. Carved ship graffiti – an ancient ritual? In: H. Tzalas, ed., Tropis V, Proceedings of the 5th International Symposium on Ship Construction in Antiquity (1993). Athens: Hellenic Institute for the Preservation of Nautical Tradition, 21-29.
ASCHOFF, V.,
1977. Optische Nachrichtenübertragung im klassischen Altertum. NTZ 30 (1), 23-28.
BARTOLOMEO VENETIANO,
1515. Cinquanta carte del mare Egeo e Adratico. Venice.
BASCH, L., 1981. Ex-votos marins dans l'antiquité. In: Anon., Ex-votos marins dans le monde de l'antiquité à nos jours. Paris: Musée de la Marine, 1981, 37-43.
BASCH, L. 1987. Le Musée imaginaire de la Marine Antique. Athens: Hellenic Institute for the Preservation of Nautical Tradition.
BASCH, L. AND ARTZY, M.,
1986. Ship graffiti at Kition. In: Kition V, The Prephenician levels, Appendix II. Nicosia, 1986, 322-336.
Belon, P., du Mans, 1578. Observations de plusieurs singularités et choses mémorables, trouvées en Grèce, Asie, Judée, Egypte, Arabie et autres pays estrangers. Paris.
BELON, P., DU MANS,
1588. Observations de plvsievrs singularitez et choses mémorables, trouvées en Grèce, Asie, Judée, Egypte, Arabie & autres pays estranges, rédigées en trois liures. Reueuz de nouveau et augmentez de Figures. Paris: Hierosme de Marnef, & la veufue Guillaume Cauellar.
BLACKMAN, D.J.,
1982. Ancient harbours in the Mediterranean. Part 1. The International Journal of Nautical Archaeology and Underwater Exploration 11(2), 79-104.
BORDONE, B.,
1528. Libro nel qual si ragiona de tutte l'isole del mondo. Venice.
BRADLEY, R.,
1998. The significance of monuments on the shaping of human experience in Neolithic and Bronze Age Europe. London and New York: Routledge.
BUCHWALD, M., 1912. Leuchtfeuer im Altertum. Weltverkehr und Weltwirtschaft 1, 78-84.
CALAME, C.,
1996. Thésée et l'imaginaire athénien. Légende et culte en Grèce antique. Lausanne: Payot.
CARLETON, JONES,
1998. The archaeology of perception and the senses. Archaeological Review from Cambridge 15(1), 7-22.
CHOISEUL-GOUFFIER, M.G.F.A.,
1823-1828. Voyage pittoresque de la Grèce. (3 vols.). Brussels.
CLARE, R.J.,
2000. Epic itineraries: the sea and seafaring in the Odyssey of Homer and the Argonautica of Apollonius Rhodius. In G.J. Oliver, R. Brock, T.J. Cornell, S. Hodkinson, eds., The Sea in Antiquity. Oxford: BAR Int. Ser. no. 899, 1-12.
CONZE, A.,
1986 [1860]. Reise auf den Inseln des Thrakischen Meeres. 2d ed. Amssterdam: A. Hakkert.
Coronelli, 1696, Isolario. Venice.
DAPPERT, O.,
1703. Description exacte des îles de l'archipel et de quelques autres adjacentes, dont les principales sont Chypre, Rhodes, Candie, Samos, Chio, Negrepont, Lemnos, Paros, Delos, Patmos avec un grand nombre d'autres. Amsterdam.
DAVIS, E.N.,
1959. Die Vulkangesteine der Insel Lemnos. Annales Géologiques des Pays Helléniques 11, 1-82, tables I-IV, map.
DE LAUNAY, L.,
1897. Chez les Grecs de Turquie. Paris.
DELEV, P.,
1984. Megalithic Thracian tombs in South-Eastern Bulgaria. Anatolica XI, 1-37.
DETIENNE, M., 1989. L'écriture d'Orphée. Paris: Gallimard.
DAUMAS, M., 1998. Recherches sur l'iconographie du culte des Cabires. Paris: De Boccard.
DUMÉZIL, G., 1924. Le crime des Lemniennes. Rites et légendes du Monde Egéen. Paris: Librairie Orientaliste Paul Geuthner.
FOL, A., ED.,
1982. Megalithi Thraciae. Pars II. Thracia pontica. Monumenta Thraciae Antiquae III. Sofia: Nauka i Izkustvo.
FREDRICH, C.,
1906. Lemnos II. Topographisches und archäologisches. Mittei-

lungen des Deutschen Archäologisches Instituts Athen (31), 241-255, pl. XIX.

FROST, H., 1963. Under the Mediterranean. London.

FROST, H., 1991. Proto-Harbours of the East Mediterranean. In: Anon., Quatrième Symposium International Thracia Pontica IV (Sozopol 1988). Sofia: Institut de Thracologie et al., 323-338.

GEOLOGICAL MAP,
1993. Geological map of Greece. Limnos island (1:50 000). Athens: Institute of Geology and Mineral Exploration.

HAGUE, D.B., 1973. Lighthouses. In: D.J.Blackman, ed., Marine Archaeology. The Colston Papers. London: Butterworths, 293-314.

HARTLEY, R., AND WOLLEY VAWSER, A., 1998. Spatial behaviour and learning in the prehistoric environment of the Colorado River drainage (South-eastern Utah), western North America. In: Chr. Chippindale and P.S.C. Taçon, eds., The Archaeology of rock-art. Cambridge: Cambridge University Press, 185-211.

KNAPP, A. B. AND W. ASHMORE,
1999. Archaeological Landscapes: Constructed, Conceptualized, Ideational. In: W. Ashmore and A.B. Knapp, eds., Archaeologies of Landscape. Contemporary Perspectives. Massachusetts and Oxford: Blackwell, 1999, 1-30.

LAUREANO, P.,
1998. The Oasis Model: Symbiosis, Intensification, Autopoiesis. In: M. Pearce and M. Tosi, eds., Papers from the EAA Third Annual Meeting at Ravenna 1997: Vol. I. Pre- and Protohistory. Oxford: BAR Int. Series no. 717, 71-74.

LEHMANN, P.WILLIAMS AND D. SPITTLE,
1982. Samothrace: Vol. 5. The Temenos. Bollingen Series LX.5. Princeton: Princeton University Press.

LEINER, W.,
1982. Die Signaltechnik der AntikeStuttgart: W. Leiner.

LEWIS, N., 1959. Samothrace. The Ancient Literary sources. In: K. Lehmann, ed., Samothrace. Excavations conducted by the Institute of Fine Arts of New York University: Vol. I. London: Routledge and Kegan Paul, 1959.

MARANGOS, TR.,
1954-1982. Unpublished archive (documents, drawings, photographs, maps, notes) and oral communications.

MARANGOU, CHR.,
1995. A rock engraving in Lemnos (preliminary study). In: H. Tzalas, ed., Tropis III, Proceedings of the 3d International Symposium on Ship Construction in Antiquity (1989). Athens: Hellenic Institute for the Preservation of Nautical Tradition, 309-319.

MARANGOU, CHR.,
1998. Rock art in the port of Myrina, Lemnos island (North-East Aegean, Greece). In: M. Simoes de Abreu , dir., Proceedings of the International Rock Art Conference (Vila Real, September 1998) (in press).

MARANGOU, CHR.,
1999a. Gravure rupestre dans le port de Myrina, île de Lemnos. In: IFRAO, ed., Proceedings,of the International Rock Art Congress (Torino 1995), Symposium 12d, Rock Art and the Mediterranean sea. Survey Bolletino CesMAP. On compact disc. [CD-ROM]. Available from: Pinerolo: Centro Studi e Museo d'Arte Preistorica di Pinerolo.

MARANGOU, CHR.,
1999b. The Myrina ship re-examined. In: H. Tzalas, ed., Tropis VII, Proceedings of the 7th international Symposium on Ship Construction in Antiquity (Pylos 1999). Athens: Hellenic Institute for the Preservation of nautical Tradition (in press).

MARANGOU, CHR.,
1999c. Rock-art and landscape in Myrina, island of Lemnos, Greece. In: L. Loendorf, ed., Proceedings of the International Rock Art Congress (Ripon, Wisconsin 1999), Session Landscape, Place, and Rock-art (in press).

MARANGOU, CHR.,
2000. Maze of rock steps: an intriguing puzzle. Paper presented at the 6th Annual Meeting of the European Association of Archaeologists (Lisbon 2000), Session: The Anthropology of Rock-art, Abstracts vol.: 49 (in preparation).

MCGRAIL, S., 1996: Navigational techniques in Homer's Odyssey. In: H. Tzalas, ed., Tropis IV, Proceedings of the 4th international Symposium on Ship Construction in Antiquity (1991). Athens: Hellenic Institute for the Preservation of nautical Tradition, 311- 320.

MEDITERRANEAN PILOT
1955. Mediterranean Pilot , vol. IX. GB: Hydrographic Office.

MIKOV, V.,
1928-29. Gravures rupestres en Bulgarie. Bulletin de l'Institut Archéologique Bulgare V, 291-308.

MORTON, J.,
2001, The role of the physical environment in ancient Greek seafaring. Mnemosyne Supplementum 213. Leiden-Boston-Köln: Brill.

ORMEROD, H.
1978. Piracy in the Ancient World. An essay in Mediterranean History. Liverpool: Liverpool University Press.

PAPAGEORGIOU, D.,
1998. In: Chr.Doumas and V. La Rosa, eds., Proceedings of the International Symposium "Poliochni and the Early Bronze Age in the North Aegean" (Athens 1996). Athens: Scuola Archeologica Itlaliana di Atene and University of Athens, 424-442.

PATTON, M.,
1996, Islands in time. London and New York: Routledge.

PLINIUS,
1981, Histoire Naturelle. Paris: Les Belles Lettres.

ROMERO RECIO, M.,
2000. Cultos Marítimos y Religiosidad de Navegantes en el mundo Griego Antiguo. Oxford: BAR Int.Ser. no. 897.

Sealy, F.L.W., 1918-1919. Lemnos. Annual of the British School of Archaeology at Athens XXIII, 148-174, pl. XV.

RAUH, N.K., R.F. TOWNSEND, M. HOFF, AND L. WANDSNIDEr,
2000. Pirates in the Bay of Pamphylia: an Archaeological Inquiry. In: G.J. Oliver, R. Brock, T.J. Cornell and S. Hodkinson, eds., The Sea in Antiquity. Oxford: BAR Int. Ser. no. 899, 151-180.

RICHTER, OTTO FRIEDRICH, VON,
1822. Introduction. In: G. Ewers, Wallfahrten auf dem Morgenland. Berlin, V-XII.

RIEPL, W.,
1972 [1913]. Das Nachrichtenwesen des Altertums mit besonderer Rücksicht auf die Römer. 2d edit. Hildesheim and New York: Georg Olms Verlag.

SOGNNES, K., 1996. Recent rock art research in Northern Europe. In: P.Bahn and A.Fossati, eds., Rock Art Studies: News of the World I. Recent Developments in Rock Art Research. Oxford: Oxbow Monograph no. 72, 15-28.

TAÇON, P.,
1999. Identifying Ancient Sacred Landscapes in Australia: From Physical to Social. In: W. Ashmore and A.B. Knapp, eds., Archaeologies of Landscape. Contemporary Perspectives. Massachusetts and Oxford: Blackwell , 33-57.
TAYLOR, E.G.R.,
1956. The Haven-finding Art. A History of Navigation from Odysseus to Captain Cook. London: Hollis and Carter.
TOZER, H.F.,
1890. The islands of the Aegean. Oxford.
TOZER, H.F.,
1971. A History of Ancient Geography. 2d ed. New York: Biblo and Tannen.
TRIANDAPHYLLOS, D.,
1994. La Thrace Antique. In: Anon., Thrace. Athens. Secretariat Général de la région de Macédoine Orientale et de Thrace, 35-97.
WACHSMUTH, D., 1967. Untersuchung zu den antiken Sakralhandlungen bei Seereisen. Dissertation. Berlin: Freie Universität Berlin.
WESTERDAHL, C.,
1995. Stone maze symbols and navigation. A hypothesis on the origin of coastal stone mazes in the north. The International Journal of Nautical Archaeology 24(4), 267-277.
WOOLNER, D.,
1957. Graffiti of ships at Tarxien, Malta. Antiquity XXXI, 60-67.
ZODIATIS, G.,
1993. Water circulation in the Aegean Sea. In: Proceedings of the 4th National Symposium on Oceanography and Fisheries (Rhodes island (Greece), April 1993), 43-46.

3

CONSTRUCTING HYPOTHETICAL SYSTEMS OF EXCHANGE: DOING KULA IN THE DANISH MESOLITHIC

GEORGE NASH

Centre for the Historic Environment, University of Bristol, England
EMAIL: georgejayne@lineone.net

INTRODUCTION

For this paper, I explore possible similarities between the economic and symbolic exchange of the «kula ring» of Melanesia and the possible symbolism linked to the motif decoration and distribution of portable artefacts from the Danish Mesolithic.

The possible exchange network present within the south Scandinavian Mesolithic is set within a back-drop of dramatic environmental and geomorphological change. The sudden rise in sea level, up to 30m during the early Mesolithic in Scandinavia, caused by ice-sheet deterioration, resulted in rising temperatures and precipitation. Many new islands in the Baltic were created, especially in the lowland areas east of the present day shore-line around eastern Jutland. Isostatic rebound (which is still ongoing) dramatically changed the land surfaces of Zealand and Jutland. Surface rise along with rising sea levels created the islands of Funen and Langeland (Jarman, Bailey & Jarman 1982). Archaeological evidence suggests that many islands were inhabited by small hunter/fisher/gatherer societies who fully utilised the rich esturine and coastal resources.

At the same time, decorated portable artefacts made mainly from bone and antler begin possibly to circulate around the landscape (Nash 1998, 2001b). These artefacts, although variably structured in design, are highly decorative and altogether unique. It would appear that artefacts were being carried and used as the population moved around seasonally. During the late Kongemose and Ertebølle periods populations appear to become more sedentary and territorial. At the same time, mobile art becomes less complex and more standardised (Andersen 1980). Here again, artefacts may circulate as part of an exchange and feasting system between less mobile populations (Nash 1998).

Obviously, one cannot say for sure whether such a system existed. Certainly, the mechanisms, a by-product of the ongoing archaeological investigations within the region, suggests that society was heavily influenced by the sea. In order to answer some of the questions concerning possible contact and exchange between groups of hunter,/fisher/gatherers, one needs to turn towards historical and present day societies that use exchange by a way of establishing social and political interaction. One particular group of contemporary fisher/gatherers using a complex exchange network, yet not fully uncorrupted by western ideology are the people of the Massin Group of Islands, located in Melanesia.

This once non-western pre-industrial exchange system, referred to as the «kula ring», is situated north of Papua New Guinea, and consists of a collection of islands, the main group being the Trobriand Islands. Bronislaw Malinowski, the initiator of interest in the «kula ring», did much to record the daily life of the Trobriand islanders (1922). Participant observation methods, used by Malinowski, gave a complex insight into the social and political behaviour of various social ranks, especially on the more significant islands within the Trobriand Island group (Kiriwina, Kitava and Vakuta), where an exchange network was in operation (Leach & Leach 1983; Munn 1976, 1986; Weiner 1985). Objects of exchange included shell valuables in the form of necklaces and armlets.

By observing the way in which the material culture of the «kula ring» circulates, one can postulate similar exchanges occurring during the Mesolithic.

MOVING AROUND THE MESOLITHIC

Within Mesolithic southern Scandinavia (Skåne) there is a small, but significant mobile assemblage that include decorated and shaped amber pieces. The literature concerning this small assemblage is limited (Andersen 1980; Brondsted 1957; Clark 1967; Clark 1936, 1975; Müller 1896, 1918; Nash 1998; Torbrügge 1968; Vang Petersen 1998; Vebæk 1939). Similar to much of the art from this period, decorated and polished amber has been assigned as archaeological miscellanea. The majority of amber has elaborate geometric decoration and are sometimes shaped into representational zoomorphs. Nearly all has been found in Denmark and probably date from the early Mesolithic or what is termed the Maglemose and Kongemose traditions 2. Of the known assemblage, a significant percentage has drilled holes which suggests these artefacts were probably used as pendants (Clark 1975:147, Müller 1918, Nash 2001b; Rying 1981, Vang Petersen 1998).

There is an interesting dichotomy between changes in social and economic behaviour throughout the north-west European Mesolithic, especially within southern Scandinavia (Clark 1936, 1975; Fischer 1995; Price & Brown 1985; Nash 1998; Tilley 1996). Firstly, during the Maglemose (7,500-5,500 BC) and early Kongemose (5,500-5,600 BC), a highly mobile population possibly exists, using a multitude of seasonal resources throughout the landscape. Together with economic strategies, political and symbolic mechanisms appear to be major factors within the social infrastructure throughout the Mesolithic. Evidence for this can be clearly seen in the decoration of amber, antler, bone, flint and wood (Andersen 1980; Clark 1936; 1975; Fischer 1974, 1995; Müller 1896, 1918; Nash 1998; Tilley 1996). These artefacts, although variably structured in design, are nevertheless highly decorative, with each design being unique (Nash 1998). It would appear that these artefacts are being carried and used as populations move

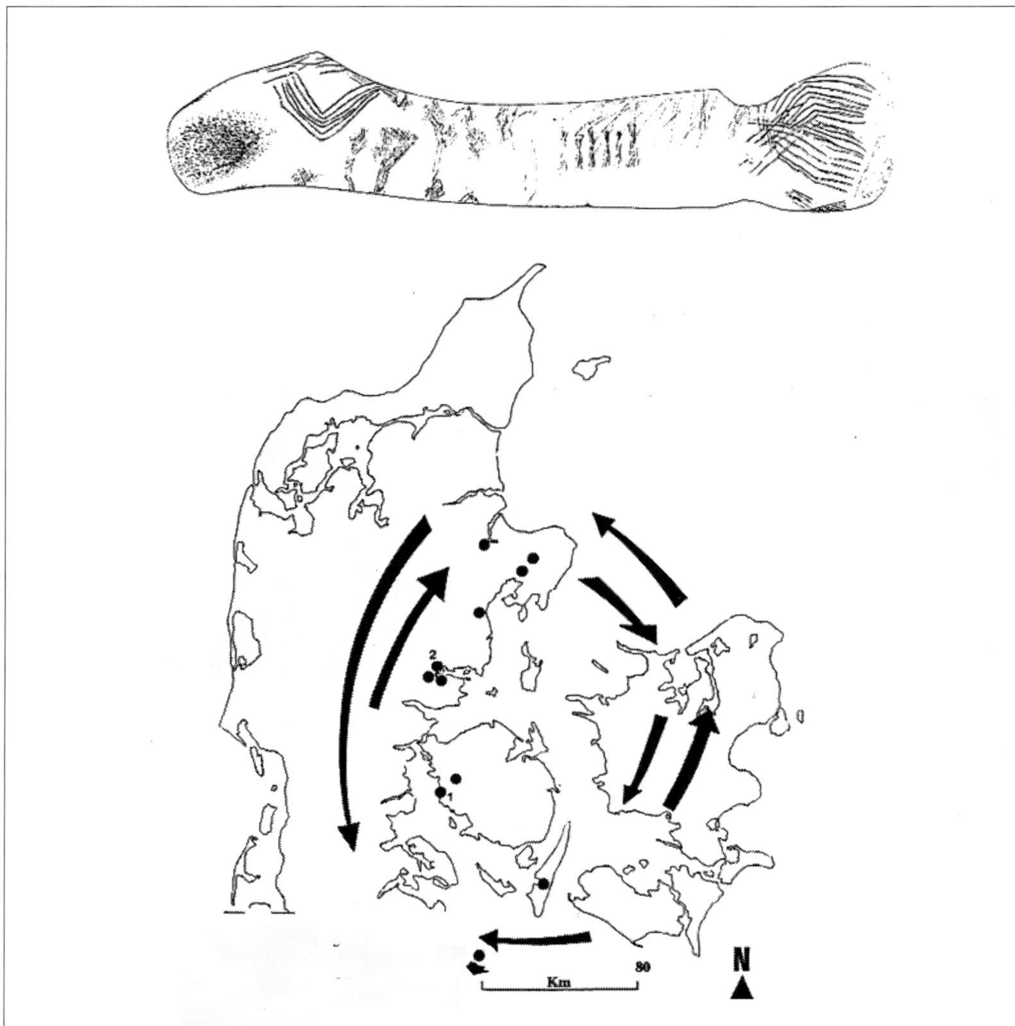

FIGURE 1. THE ERTEBØLLE RING
Represented in the distribution and possible circulation of the 'wheatsheaf' motif
(data adapted from Andersen 1980 and 1986)
1. The submerged settlement of Tybrind vig.Funten; 2. Multiple settlement region of Horsensfjord, East Jutland

around the landscape on a seasonal basis. In complete contrast, populations become sedentary and territorial during the Late Kongemose (5,000-4,500 BC) and Ertebølle (4,500-3,200 BC) phases (Hodder 1990:178-215; Zvelebil 1986:5-15). Portable artefacts of these periods seem to have been circulated as part of an exchange and feasting system between less mobile populations.

The emergence of enormous shell middens during the Early Ertebølle may indicate feasting locations that are linked in with the exchange of artefacts. I say this as a number of decorated artefacts are located within or near middens (Andersen 1980; Jørgensen 1956; Rying 1981). One could visualise shell middens as being major focal points within the landscape. During the latter phase of the Mesolithic, design coding on decorated bone, antler and amber, although complex in structure, is nevertheless rather standardised and more simple in terms of motifs than decorated artefacts from the Early Mesolithic. Furthermore, certain motifs, in particular the «wheatsheaf» motif or «sheaf of grain» motif (Andersen 1980:6-60), are being used and repeated on a number of artefacts that are distributed within a 'ring' situated within the coastal region of East Jutland (FIGURE 1 and Appendix 1). The distribution of this motif suggests that an exchange/contact network similar to that of the «kula ring» exchange system may be in operation. This artefacts assemblage possibly forms part of a cultural exchange package that includes bone combs and bird bone points, Limhamn greenstone axes, straight and curved antler harpoons and T-shaped red deer antler axes (Vang Petersen 1984). The cultural package of kula involves raw materials, the converted artefact, transportation and, most importantly, people. Similarly, one could argue that these components exist during the Early and Late Scandinavian Mesolithic. This being the case, the analogies which I shall draw on do allow us to rethink the Danish Mesolithic within a fresh and alternative framework to the traditional environmentally-based concepts.

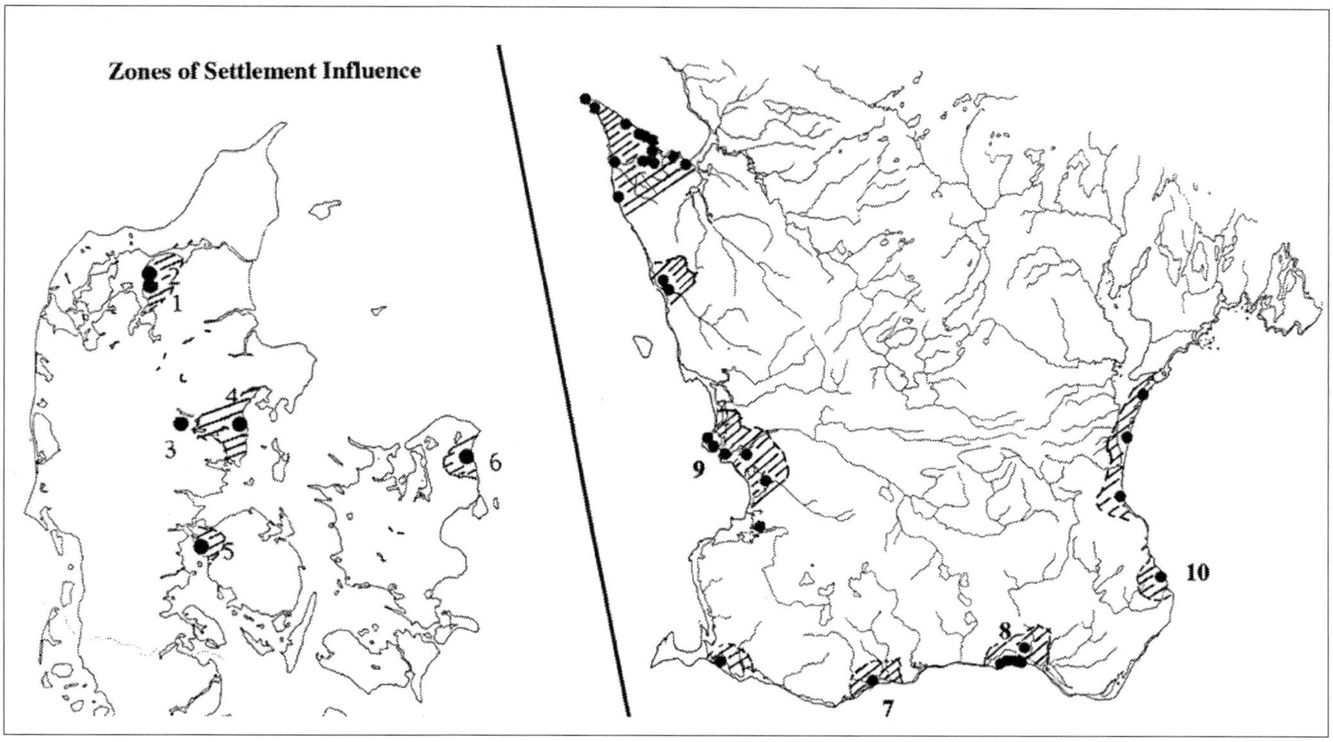

FIGURE 2. DISTRIBUTION OF ERTEBØLLESETTLEMENT SITES IN DENMARK AND SKÅNE, SOUTHERN SWEDEN (modified after Jennbert 1984 and Tilley 1996) 1 Ertebølle, 2 Bjørnsholm, 3 Ringsløster, 4.Norseminde, 5 Tybring vig, 6 Bøgebakken (Vedback), 7 Skateholm, 8 Bredasten, 9loddesborg, 10 Vik.

DOING KULA

Research into the «kula ring» over the past 90 years or so has received much valued interest. Malinowski was the first to recognise the symbolic and ritualistic importance of exchange in the lifestyle of the Trobriand islanders. In his monumental work 'Argonauts of the Western Pacific' (1922), he adopted a functionalist approach using participant observation methods. More recently, Malinowski's research has been expanded upon by a number of researchers, notably Campbell (1983), Gregory (1983), Leach & Leach (1983), Munn (1976; 1983), Strathern (1983) and the feminist approach of Weiner (1983; 1985; 1989). The literature covers a multitude of data, and goes far beyond exchange and contact within kula. However, it is the intricacies of the exchange system with which I am primarily concerned in this paper, in particular the various idiosyncratic components that construct each prestige item.

IDEOLOGY AND EXCHANGE

The basic concept of gift exchange involves three components: to give, to receive and to repay (Mauss 1954). Receiving and repaying (reciprocity) rely on obligation: the more that is given, the more one is obliged to repay. By giving and receiving, one creates a moral, as well as an economic obligation. Exchange, above all, establishes communication between groups and individuals. By exchanging gifts, one creates alliances. These are usually linked to kinship and marriage, which strengthen and bond those concerned. Foremost within Trobriand society is the need to give, rather than to receive, because to give is to acquire inowski 1922; Campbell 1983).

The concept of temporary ownership and circulation of kula valuables may also be extended to motif distribution on bone and antler during the Scandinavian Mesolithic, especially during the Ertebølle. The distribution of certain decorated bone, antler and amber artefacts, suggests a possible circulation network. The clustering of particular motifs may well be associated with areas of high economic ac-

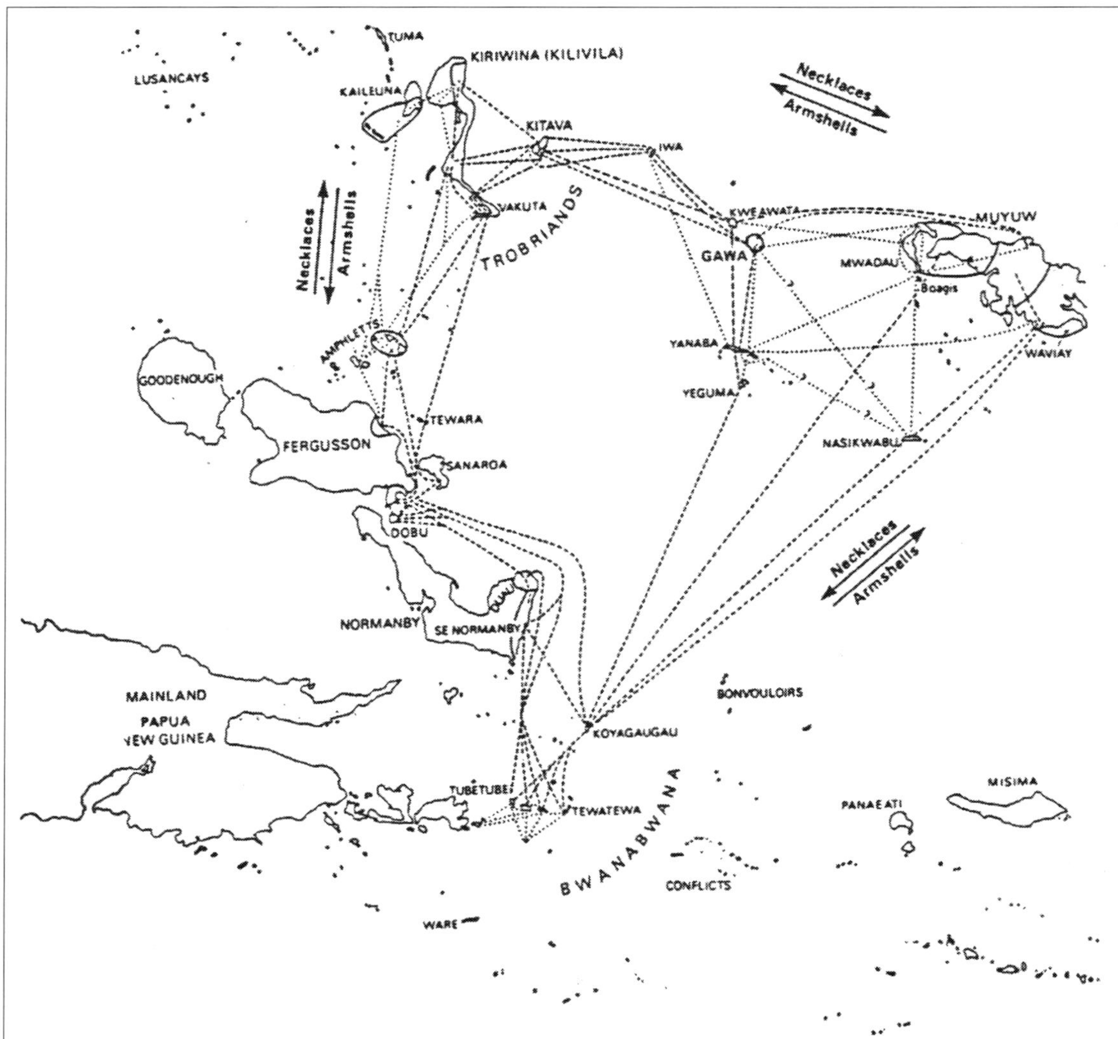

FIGURE 4. THE MASSIM REGION, NORTH PAPUA NEW GUINES
Trading exchange network of the Kula Ring. Circulation of kula valuables ensures success for exchange as well as gaining history, prestige and power (Source:Leach and Leach 1983).

power and prestige (Weiner 1989). The exchange of shell valuables is confined to men only, although once an exchange has taken place, both the recipient's and the provider's immediate families benefit (Weiner 1986). However, shell valuables are never privately owned, nor are they kept in the recipient's possession for very long (Campbell 1983:230-1). Moreover, Leach (1983:1-5) state that valuables act as 'objects of kula', rather than objects of ownership. Moreover, the exchange of shell valuables between individuals stimulates competition. By participating in kula, one immediately gains rank, power and prestige (Mal-

vity, in particular settlement (FIGURE 2). The complex system of exchanging symbolic items within the kula ignores the many boundaries of linguistic and cultural affiliation (Weiner 1989). However, problems exist with inter-island contact. Both politics and distance allow only limited island interaction, and this is mainly achieved through kula (Leach 1983:1-2). The same could be inferred for Denmark during the Mesolithic. It would appear that prestige goods acted as political mechanisms for exchange (Nash 1998). This can be seen in the restricted nature of particular items, for example, the distribution of Lilhann greenstone axes of Zealand and Southern Skåne, or circular cut scapulae on Jutland (Vang Petersen 1984). The ceremonial exchange of kula prestige items, in this case cowrie shell necklaces and shell armlets, takes place within a 'ring' that encompasses an area of 800km2, most of which is open, hazardous ocean. Expeditions are undertaken in large, decorated, wooden outrigger canoes, powered by 1983:230-3). However, one must look further than just the rank and status of shell valuables, since every aspect of kula is bound up with magic and symbolism.

EXCHANGING GOODS AND KULA VALUABLES.

Necklaces (vaiguwa), one of two prestige exchange items, are exchanged only with armlets (mwari). The items move

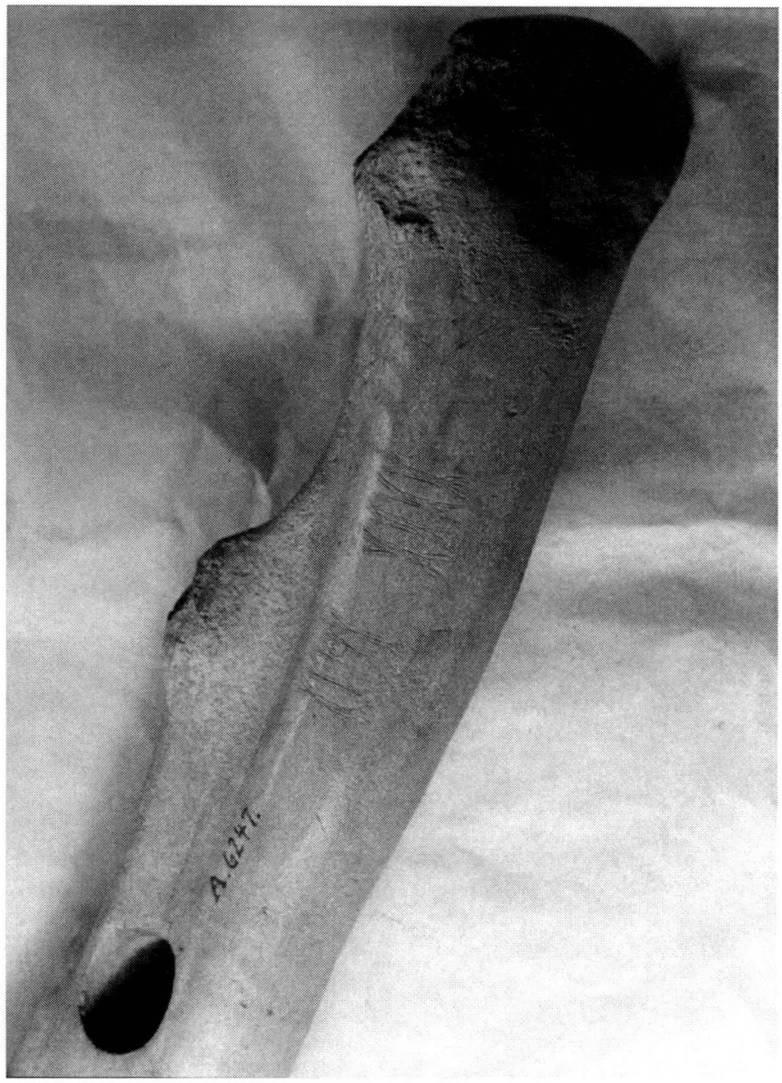

ERETØBOLLE BONE ARTEFACT WITH THE WHEATSHEAF DESIGN
(photo G.H. Nash)

lanteen-rigged sails, and can extend vast distances, lasting many weeks, or even months (Munn 1976:39-42). The whole exchange system appears to act as a major incentive to local and regional groups, who organise themselves around kula (Leach 1983; Strathern 1983; Weiner 1989), although exchange is usually conducted on a male, one-to-one personal basis (Campbell 1983:229-230). This exchange relationship will last many years, ensuring regular and successful contact between exchange partners (Weiner 1985). Shell valuables are ranked according to their condition and the status of the exchange partners (Campbell in opposite directions ensuring a constant flow of exchange. Vaiguwa (also referred to in the literature as soulava) circulate in a clockwise direction, whereas mwari revolve anti-clockwise. There are approximately 16 island groups involved in the exchange network, including a small peninsula of the Papua New Guinea mainland (FIGURE 3). Primarily, this complex pattern of exchange acts as a stimulus for contact between different island groups. It also cements relationships, as well as separating individuals within that relationship, establishing social as well as personal identity (Weiner 1985). Personal identity is recognised mainly through the rank of both the kula valuable and its owner. Those with the highest rank, especially the boat captains, do much to control and manipulate the participants of kula. This control above all enhances political status, as well as prestige. Apart from armlets and necklaces, other goods, less symbolic, are exchanged, including coconuts, sago, vegetables, fish, baskets and mats. These, and the ceremonial canoes (waga), are part of a complex exchange package (Munn 1976:39-40). Other exchanges also exist, usually between inland and coastal peoples on some of the larger islands, especially within the Trobriand Island group. These informal transactions (known as gimwali) are considered as important as the inter-island exchanges, but are symbolically placed at a lower level of intensity than kula. Local exchanges involve such vital and symbolic items as yams and pigs, both of which are important commodities of personal prestige. Internal transactions occur more frequently than the kula exchanges and commit both men and women to a wider exchange system (Munn 1986:105). Obviously the exchange items used within the kula Ring are purely for prestige purposes, although they could be considered as being linked to other reasons for inter-island contact, such as marriage, politics and social interaction. Each prestige item is exchanged and linked to status. Both amulets and necklaces are graded accordingly (Campbell 1983:229-31).

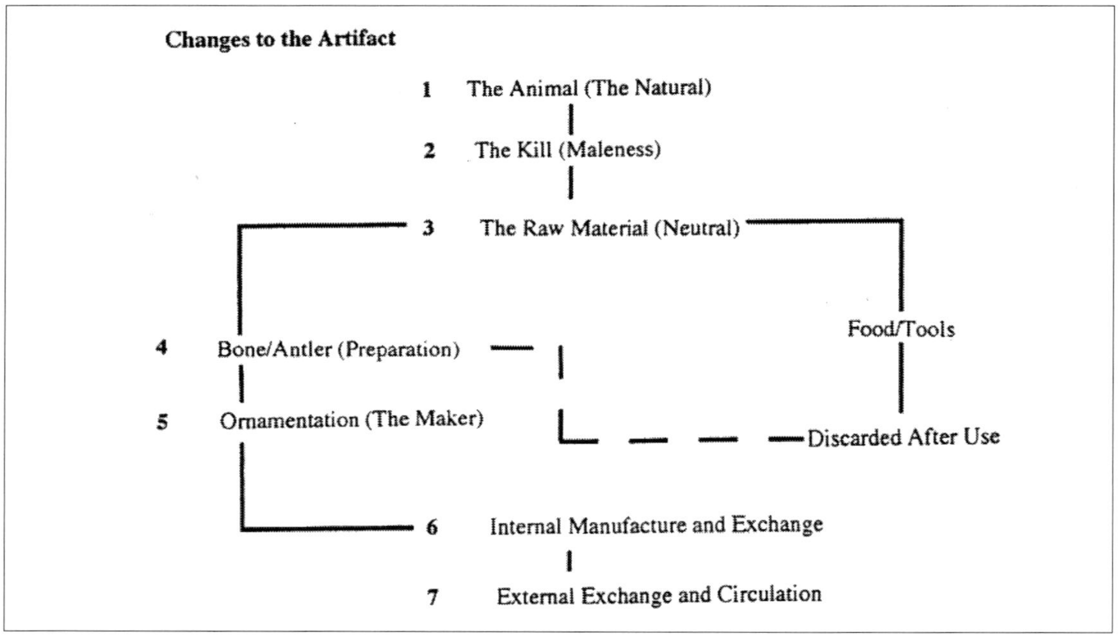

FIGURE 4 CONVERSION PLANE
(based on Munn 1977)

PRESTIGE AND WEALTH

As shell valuables circulate they gain more prestige and value. In order to grade these items according to rank, the age, weight, texture, colour and size of each component are considered. Both mwari and vaiguwa are graded into approximately 10 classes (ibid. 230). The string (or utuna) that threads the shells to form either necklaces or amulets is also ranked. The length of the string signifies status: the longer the string, the more ranked the shell valuable is. Moreover, the string must show signs of wear. Wear and age initially signify antiquity, and, in turn, varying degrees of prestige. Evidence of wear and antiquity can also be seen as important components in assessing the possible prestige incidental to Mesolithic portable art in Denmark, in particular scraped and polished bone and antler. Amber, too, shows clear signs of extensive wear. A large number of amber pendants, especially from the Maglemose, possess two or even three thread holes. It is obvious that, through time, each hole has been worn and fragmented by continuous use. As one hole was worn away the pendant has re-drilled. The drilling and re-drilling of amber suggests long periods of use and, perhaps, circulation (Nash 2001b). Coupled with the circulation of amber is decorated bone and antler. The usage, age and addition of motifs to individual bone and antler artefacts in many ways create an identity, similar to that of shell valuables, whereby, through time and space, each valuable is recognised and develops a signature. According to Mauss (1923), the identification of individual shell valuables represents an inalienable wealth whereby an artefact, although undergoing exchange, will, in fact, be inextricably linked to the original owner (termed by Mauss as immeuble). The item may be exchanged many times, but the signature indicates original ownership and, above all, creates a personal history. Although each shell valuable is passed from generation to generation, the original owner or owner's family may claim legitimate and eventual ownership (Weiner 1985). This restricted distribution establishes control and limits the extent of exchange; the artefact, therefore, can never fall into undesirable hands. By establishing a signature, one has paramount control over distribution and subsequent exchange.

One can link the idea of age and ownership to the scraping and polishing of bone and antler throughout the Danish Mesolithic. The texture of both the cowrie shells (buna) and the decorated bone appears to be an important component in portraying great age and usage. By scraping and polishing bone, the artefact immediately becomes aged, thus instantly producing prestige and status. Equally, buna are aged according to the texture of the shell. Campbell (1983) states that a recipient of mwari will observe the kala ureri (the red striation on the forehead of a buna) prior to exchange: the presence of striations signifies antiquity. By decorating the artefact or displaying a shell valuable in this way, the artefact becomes encoded and recognised. A possible analogy exists with the «wheatsheaf» motif in the Danish Mesolithic (Plate 1). As well as representing kinship identity, this motif may also signify varying degrees of antiquity and, through inter-group contact, the artefact becomes a tool, thus binding different social groups together.

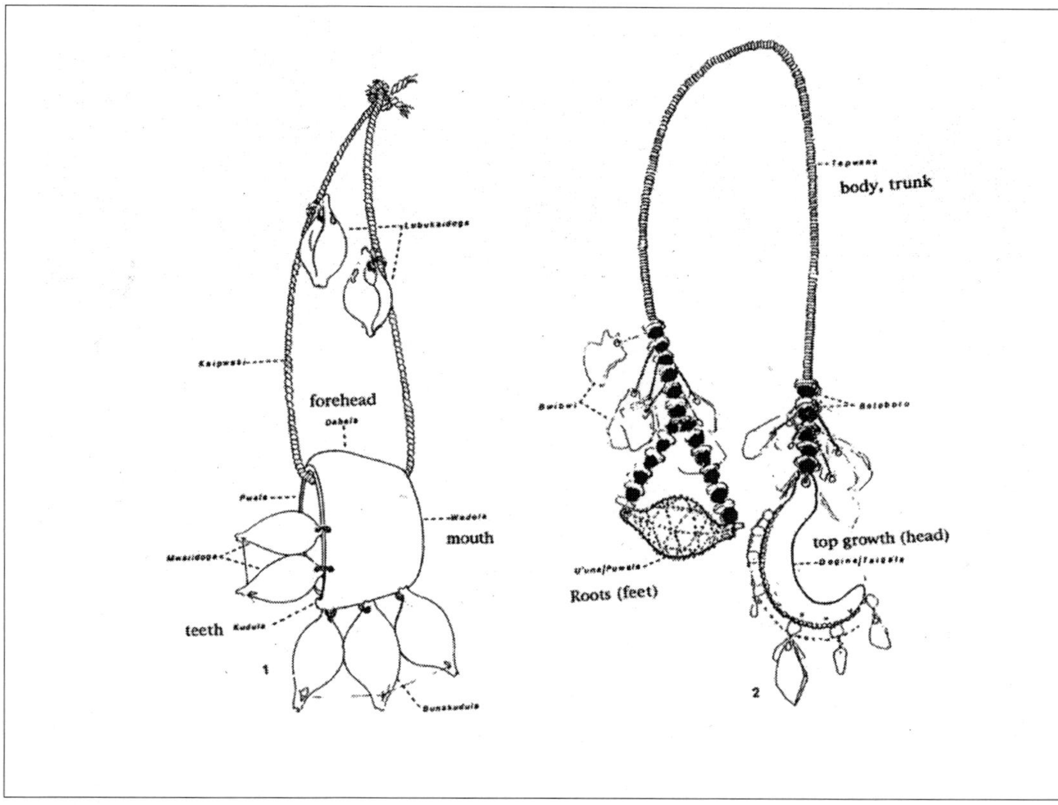

FIGURE 5 THE IDIOSYNCRATIC COMPONENTS THAT FORM KULA VALUABLES (source: Campbell 1983).

SYMBOLISM OF THE BODY AND Buna.

At all levels of exchange within the «kula ring», symbolism appears to be paramount. For example, Munn (1976:40) argues that there are various levels of symbolic change related to boat (waga) construction. The canoe eventually becomes part of the gift package while gift/exchange is in progress. The gawa canoe goes through a number of physical and social changes (what Munn refers to as conversion planes), from the tree (the natural) through to production (the raw material) to initial exchange (maker to user) and to eventual kula exchange. The initial exchange between maker and user is usually internal. That is, the maker and the user are of the same kinship group or from the same island. I, in turn, have used a similar model to explain the various levels of symbolism in the production and ornamentation of bone and antler, from the hunted animal to the decorated artefact (FIGURE 4). Here, the maker (inscriber) and the user of decorated artefact could either be one of the same or a relative or member of the same social group. The artefact is transformed at least six times (to include the stages of production as well as stages of exchange). At each stage, the artefact may change its symbolic identity and meaning. However, in the case of the kula exchange system the historical ownership of the artefact belongs to the original distributor (Weiner 1983).

The model illustrates the transformation of bone and antler from the animal to the artefact and eventual exchange. In order to attain the symbolism at all levels, the artefact must undergo changes in meaning. This change, a metaphorical metamorphosis, transforms a mundane item into something special and revered. Similar to kula shell valuables, each component would possess an intimate and powerful meaning. Taking this argument further, I now discuss briefly the body symbolism of kula shell valuables, and possible analogies with decorated bone, antler and amber.

There are a number of criteria for shell valuables in a successful exchange; for example, the colour, texture, weight, number of components per valuable, size of component and age. Malinowski (1922) and, recently, Campbell (1983) an Munn (1983) have suggested that both vaiguwa and mwari are gender encoded (men wear armlets, women wear necklaces), at a more basic level, the components that make up both mwari and vaiguwa are symbolically encoded; mwari to male body parts and vaiguwa to the tree (FIGURE 5). Campbell (1983:233-5) suggests that tree terminology for vaiguwa appears to be metaphorical. Particular components are referred to as 'roots', the 'trunk' and 'top growth', each often referring also to feet, torso and head. The use of naming particular components using tree terminology corresponds with the fact that many components on vaiguwa derive from plants. For example, black banana seeds (botoboto) are threaded and attached to snail-shell discs. The terminology used for mwari, although pertaining to the body, does in fact focus mainly on male parts (e.g. the frontal edge of the mwari shell is referred to as the pwala, or testicles). Each component on mwari and vaiguwa is named, ranked and possesses a personal history. The number of bunakudula (small cowrie shells) suspended around the dabala (forehead) determines the status and rank of both the owner and the shell valuable: the more shells present, the greater the rank and history (up to six bunakadula can be suspended, or added over time). Likewise, one could argue that the number of motifs, their position and the type of motif used on bone or antler within the Danish Mesolithic, may pertain to the rank and status of particular individuals. Many of the more highly decorated artefacts from Denmark, especially from the Maglemose, are unique, suggesting personal identity and ownership. Similarly, many artefacts also possess small drilled holes. These artefacts, although functional in form, may nevertheless be used as pendants, or form part of a necklace (Müller 1918). Equally, the same argument may be applied to artefacts with larger drilled (hafted) holes. What is difficult to substantiate is whether these artefacts are gender encoded, although I would argue that most of the earlier Mesolithic decorated bone and antler may symbolise hunting and, above all, 'maleness' (Nash 1998; Hodder 1990). Much evidence suggests that, despite a few examples portraying clear female figures, the overwhelming majority of representative figures are male. However, can body symbolism apply to artefacts that possess abstract or animal designs? In most cases, artefacts with human figures are inscribed vertically along the shaft, with the head positioned at the butt or hafted end. By positioning a human figure in this way, all bone and antler components become parts of the body. For example, the point or tip would represent the feet, the central part of the shaft the torso, and the butt the head (similar to the metaphorical terminology used for vaiguwa). By positioning the artefact in this way (point or tip towards the body), all decorated bone and antler may be read in relation to the body (Nash 1998, 2001a).

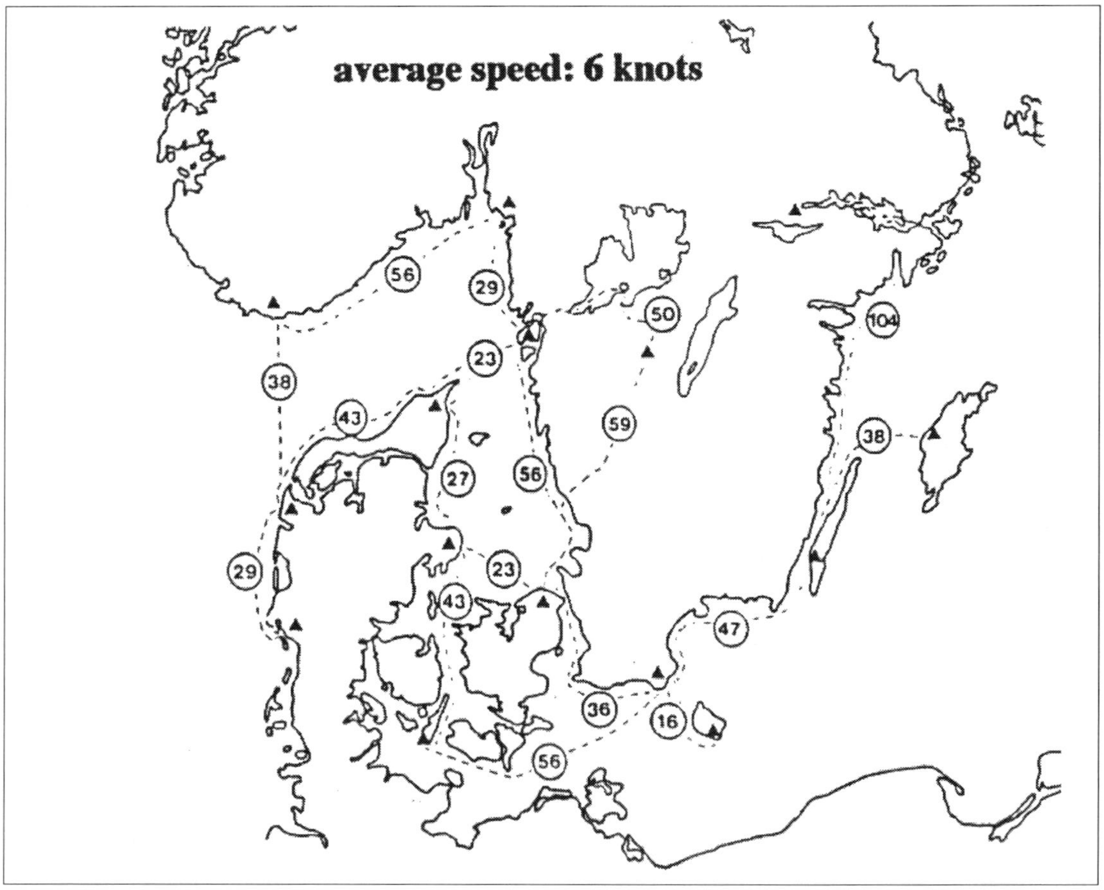

FIGURE 6 ESTIMATED TIME TAKEN BETWEEN MAJOR SETTLEMENT AREAS BY CANOE (IN HOURS)
(sources: Blomqvist 1989)

Associated with the engendered components of kula shell valuables is colour. The natural colouring of individual shells is symbolically segregated into three categories: white, black and red. White shells are regarded as pure, young and uncontaminated, whereas black shells are impure, undesirable, ugly, dangerous and associated with death. Red shells are the most high-ranking and most sought-after. They represent sexual desire and virility and are closely associated with marriage and fertility. The colour symbolism of shell valuables is extended into the daily social and cultural lives of their users. For example, black spondylus shells are used on vaiguwa, and are deemed unattractive; the same belief is used for people with black skins: they are considered 'ugly and undesirable as partners' (Campbell 1983). Alternatively, red is symbolically used by young, unmarried women to attract a partner and indicates a man's willingness for sexual intercourse. In many hunter-gatherer societies, similar colour principles operate. The symbolic use of colour during the Danish Mesolithic, especially on bone or antler, is almost impossible to determine. There is, however, a possible colour symbolism linked with the burial evidence at Skateholm (southern Skåne) and Vedbæk (north-west Zealand). Here, red ochre has been scattered over graves at both cemeteries (Albrethsen & Petersen 1976; Larsson 1989). If we extend contemporary beliefs that red symbolises fertility, blood and sex, then rebirth is created through death (an extension of the life-cycle). Taking this argument further, bone and antler is initially always white. I would suggest that its whiteness represents the metaphysical change from young and wild (of adolescent stags) to the newness of the scraped and decorated artefact. The artefact, if given to a juvenile, creates purity and is initially without history. By creating ownership, both the owner and the artefact acquire history and subsequent status. Of course, as the artefact ages, so the colour changes eventually from white to black. It is possible that the change in colour would determine the rank and prestige of each artefact, as well as the degree of polishing and the complexity of design (Nash 1998).

INSTRUMENTS OF EXCHANGE.

The major components of exchange are commodities. Linked with commodities are ideas, inter-personal relations and socio-political ideology. I have suggested previously (Nash 1998; 2001b) that one of the main mechanisms of exchange, especially during the latter part of the Danish Mesolithic, was the use of decorated bone and antler. Also included within this exchange package are canoes and decorated paddles (Andersen 1985; 1986), decorated flint (Fischer 1974), stone and amber (Andersen 1980; Nash 2001b). Coupled with these prestige items, we can also envisage ceremonial, political and economic interaction. However, as changes occur in social behaviour, so the rules governing the significance of motif ornamentation change. For example, design coding and individual motifs of the Maglemose (from Zealand) are never present during the Ertebølle. Designs and separate motifs through space and time possibly represent different ideas of symbolism. This may also reflect changes in settlement (from inland to coastal habitation) and, subsequently, the availability of subsistence resources.

From the Early Mesolithic to the emergence of agriculture, the design structuring of bone, antler and amber appears to change. Designs become more abstract through time to the point when, during the Late Ertebølle, bone and antler are scraped and polished only (Andersen 1980:53). The eventual demise of ornamentation on bone and antler occurs around 3,500 BC, at about the same time as Trichterrandbecher (TRB) pottery is introduced (from the Danubian regions to the south). Andersen (1980) argues that pottery, a new prestige item, replaced the prestige of decorated bone and antler. As economic and social organisation changed, with the transition from fishing and foraging to agriculture, so did the perception and importance of the ideology of the past. Prestige items associated with hunting and gathering were replaced by items affiliated to agriculture. However, it could be argued that many simple motifs, such as triangles, chevrons and zig-zags, used on ornamented bone and antler, are also incorporated within TRB pottery designs (Hodder 1990; Nash 1998).

Earlier, I suggested the possibility of motifs being added to over time, possibly in the context of a circulatory prestige network of long duration, similar in structure to that of the complex circulation of kula valuables. Another function worth considering is the addition of single motifs through exchange. Motifs would be applied to the artefact over time by individual exchange partners. Although in temporary ownership, the artefact would gain recognition and prestige the more it passed to other individuals or groups. The idea of temporary ownership is not uncommon in contemporary hunter/gatherer and non-western tribal groups. In the case of the kula exchange system, valuables gain prestige and power the longer they remain in circulation. For example, power and prestige is gained through the addition of buna to the shell valuable (Campbell 1983). Similarly, with decorated bone and antler, various additions to total surface decoration, plus the even distribution of particular motifs, especially the «wheatsheaf» motif and the intricate dot/drill examples, possibly indicate a process of seasonal exchange between permanently based groups. Each allied group within the 'ring' would place a 'signature' of ownership on the artefact, this signature being a personalised motif. The distribution of the «wheatsheaf» motif within the Eastern Jutland ring may signify a possible exchange network between social groups within this area. Moreover, the number of motifs on a single artefact may indicate the number of visits from exchange between interacting groups, or even the number of seasons between visits, hence the multiples of three, four and five motifs present on at least six artefacts from this area. This would, above all, keep the artefact in circulation for long periods of time.

I have suggested that there are a number of possibilities for the wide circulation of this design, all relating to contact and exchange. Similarly, the circulation and exchange of shell necklaces and armlets within the «kula ring» is also confined to a definite and prescribed circuit. Possibly similar in structure to the 16 or so kula areas, the East Jutland ring was possibly organised into various tribal chiefdoms, specifically during the Middle and Late Ertebølle. In order to establish status and, above all, stability, each tribal group would have to become sedentary. Coupled with social stratification would be the need for external contact and exchange. During the Ertebølle, contact would be maintained and conducted via sea and river expeditions.

GAWA CANOES v THE TYBRIND VIG CANOES.

Throughout the Danish Mesolithic, water transportation appears to be paramount, especially with the dramatic changes to sea level. During the Ertebølle, the receding shorelines left eastern Denmark with a matrix of new islands (approximately 450). Archaeological evidence show a society partly dependent on river and sea transportation. Boats were used probably for fishing and the hunting of marine mammals, as well as communication (Andersen 1985; Malm 1995). Excavations at the settlement site of Tybrind Vig, located on the west coast of Funen has uncovered two boats. Associated with the Tybrind Vig canoes are two highly decorated paddles, as well as decorated bone and antler pieces and fish traps (Andersen 1985, 1986). The wooden paddles are decorated with an embossed abstract design which has been embossed into the wooden paddle using a 'template' which has then been finally lacquered (Andersen 1985:65) . Both Tybrind Vig canoes possibly reveal evidence of long-haul sea or river expeditions. The shape, size and hull length of both Tybrind Vig canoes suggest that they were used for sea-borne expeditions and both boats were located in reed swamps that were once open sea. Although fragmentary , each canoe is of a thin-hull construction which is streamlined to a pointed bow. The hull design would allow manoeuvrability across calm waters. Andersen argues that the size and design of each canoe suggests that they could carry a considerable cargo, but could only be used on calm waters. A large hearth (0.60m x 0.35m) is located at the stern of each boat. Environmental evidence reveals that food was possibly cooked on a 'makeshift' stone and sandy-clay surface (ibid. 1985:65-7). Shell fragments, charcoal and small splinters of flint were embedded into the inner hull surface. The preparation and cooking of food would be important on long voyages, especially if expeditions between settlements were to run for days rather than hours.

Blomqvist (1989:168) has highlighted the approximate distances and times taken to travel between the mainland and the islands of Furen, Langeland, Jutland and Zealand (FIGURE 6). The distance between Langeland and Skåne, for example, would have taken a canoe approximately 56 hours at a speed of six knots. Similarly, a journey between eastern Jutland and north-east Zealand would have taken 23 hours. One assumes, judging by the deposition of certain prestige items such as T-shaped red deer antler axes and Limhamn axes, that these journeys were more than probable (Vang Petersen 1984:7-18; c.f. Tilley 1996:53).

The preparation and cooking of food would be important on long voyages, especially if expeditions between settlements were to run for days rather than hours. A similar type of hearth exists on canoes in the marshland areas of the central highland region of Papua New Guinea. Here, women (with accompanying children) hunt for fish, and will use these hearths on expeditions that last for more than a few hours (Brown 1978). The shape, size and length, and especially the hull construction, of both Tybrind Vig canoes suggest that they were primarily used for sea expeditions. Both were located in reed swamps that were once coastline. Although fragmented, the canoes are of a thin-plank construction, with a streamlined hull and pointed bow. The design would allow the canoe to manoeuvre quickly across calm waters, Moreover, due to their size and design, Andersen (1986) argues that large cargoes could have been carried. Again, the canoe size and construction link closely with that of the kula canoes, in that secondary commodities would account for much of the cargo transported.

Munn states that canoes used in kula from part of the symbolic package; not only is it a means of transport, but a waga also becomes in itself a prestige exchange item (1976:41). The final stage in its production is devoted to beautifying the canoe through elaborate ceremonies. There is no evidence for the decoration of canoes from the Danish Mesolithic, but possibly associated with the two Tybrind Vig canoes are three highly decorative paddles (Andersen 1985, 1986; Jensen 1982). These paddles are thin and spade-like in shape. The frailty of both the canoes and the paddles suggests that any rigorous use would cause considerable damage. Fishing, therefore, may not have been their function. Indeed, both canoes show evidence for damage and successive repair, possibly due to splitting resulting from age and repeated use (Andersen 1985).

The intricacy of the decorated paddles suggests limited or special usage. One might associate both paddles and canoes with ceremonial or socio-political/economic gatherings. The decorated lacquered surface of the paddle blades would deteriorate if used daily. I would suggest, therefore, that the paddles retained two important functions. Firstly, they were utilitarian tools which were used only when entering neighbouring territorial waters. Secondly, the paddles were decorated in order that they may be seen and recognised (assuming, that is, that the canoes were also decorated). The inter-relationships between designs on paddles, bone and antler suggests further the evidence of symbolic contact/gift exchange and, more importantly, a mechanism for the interaction of ideas. Each motif, each design structure, acts as a symbol for social discourse, similar to the gift exchange used in kula. Associated with communication and exchange is the mystique and symbolism that is bound up with the designs. Within kula, armlets, necklaces and boats speak a material language known only to communicating participants (Campbell 1983:229). I would suggest that a similar language exists with decorated portable artefacts throughout the Danish Mesolithic, especially during the Ertebølle. This secret knowledge would be linked to power and prestige. The shift from inland to coastal habitation, together with the increase in population through the Mesolithic, would have placed heavy constraints on localised sedentary populations (Price & Brown 1985; Rowley-Conwy 1981). By manipulating power and subsequent control, stability and history are created. This can clearly be seen through the analogies with contemporary, non-western exchange systems.

SUMMARY

It would appear that during the Maglemose, portable decorated artefacts were being worn and carried around the landscape. The unique designs inscribed on bone, antler and amber suggest intimacy between the artefact and its owner. Both artefacts and inscribed designs probably represent 'maleness', even though some possess female figures. These designs are possibly linked to fertility, virility and male prowess. The raw material, too (bone or antler), can be perceived as gender encoded. Young stags represent male and the wild. One can visualise both the carrier and the decorated artefact imitating the quarry (red deer or elk), thus linking the symbolism of 'what is male' to the mundane task of hunting for food. Quite the reverse happens during the Ertebølle. Here, artefacts circulate from individual to individual. Moreover, the circulation of artefacts coincides with changes in social and economic behaviour. Connected to these changes are the dramatic changes in sea level. All of these factors would have had a profound effect on the way in which people communicated and exchanged resources. Although the rules applying to artefact decoration have changed through time and space, the artefact nevertheless remained an important device for power and prestige throughout the Mesolithic. The change to single motifs, as well as complete designs, is intimately linked to changes in ideology and peoples' perception and view of the world. For example, during the Ertebølle, the application of simple standardised motifs may indicate social ties between local and regional groups involved with feasting cycles in and around shall middens. The motifs became simple on many artefacts, especially from East Jutland. Although the same occurred during the Maglemose, the rules governing design structure might suggest something completely different. The rising sea levels would instigate the need for sea-borne social contact. The medium for this would be the artefact, a visual and recognised symbol, transportation as well as utilitarian item. These factors, above all, would activate and stimulate a social and symbolic package and link together participants in a broad network of social interaction.

BIBLIOGRAPHY

ALBRETTHSEN, S. & PETERSEN, E. B.,
1976. 'Excavation of a Mesolithic Cemetery at Vebæk, Denmark', Acta Archaeologica No 47, 5-22.
ANDERSEN, S. H.,
1980. 'Ertebølle Art: New Finds of Patterned Ertebølle Artefacts from East Jutland', Kuml, No 4, 6-60.
ANDERSEN, S. H..,
1985. 'Tybrind Vig: A Preliminary Report on a Submerged Ertebølle Settlement on the West Coast of Fyn', Journal of Danish Archaeology, No 4, 52-69.
ANDERSEN, S. H.,
1986. 'Mesolithic Dug-outs and Paddles from Tybrind Vig', Denmark, Acta Archaeologica, No 57, 87-105.
BLOMQVIST, L,
1989. Megalitgravarna i Sverige. Typ, Tid, Rum och Social Miljö, Theses and Papers in Archaeology 1. Stockholm.
BRONDSTED, J.,
1957. Danmarks Oldtid. Vol 1, 2nd ed.Gyldenal Copenhagen.
CAMPBELL, S. F.,
1983. 'Attaining Rank: A classification of Kula Exchange Valuables' in J. W. Leach & E. R. Leach (eds.), The Kula: New Perspectives on Massin Exchange. Cambridge. Cambridge University Press. 229-47.
CLARK, J. G. D.,
1936. The Mesolithic Settlement of Northern Europe. Cambridge. Cambridge University Press.
CLARK, J. G. D.,
1975. The Earlier Stone Age Settlement of Scandinavia. Cambridge. Cambridge University Press. 99-239.
CONKEY, M.W.,
1985. 'Ritual communication, social elaboration and the variable trajectories of Palaeolithic material culture' in T.D. Price & J.A. Brown (eds.), Prehistoric Hunter-Gatherers: The emergence of social and cultural complexity. Academic Press. 299-323.
FISCHER, A.,
1974. 'An ornamented Flint-Core from Holmegåard v, Zealand, Denmark', Acta Archaeologica, No 45, 155-159.
FISCHER, A. (ed.) 1995. Man & Sea in the Mesolithic: Coastal settlement above and below present sea level. Oxbow Monograph Series No 53.
GREGORY, C.,
1983. 'Kula gift Exchange and Capitalist Commodity Exchange: A comparison' in J.W. Leach & E. R. Leach (eds.), The Kula: New Perspectives on Massin Exchange. Cambridge. Cambridge University Press. 103-117.
HODDER, I.,
1990. The domestication of Europe, London: Blackwell Press. 178-215.
JARMAN, M. R., BAILEY, G.N. & JARMAN, H.N.,
1982. Early European Agriculture. Cambridge. Cambridge University Press. 72-94.
JENSEN, J.,
1982. The Prehistory of Denmark. London, Methuen Press.
JØRGENSEN, S. 1956. 'Kongemosen: Endnu en Åamose-Boplads fra Ældre Stenalder'. Kuml, 23-40.
LARSSON, L.,
1989. 'Big Bog and Poor Man: Mortuary Practices in Mesolithic societies in Southern Sweden' in T B Larsson & Lundmark (eds.) Approaches to Swedish Prehistory, BAR International Series 500. 211-223.
LEACH, E. R & LEACH, J. W. (eds.),
1983. The Kula: New Perspectives on Massin Exchange. Cambridge. Cambridge University Press.
MALINOWSKI, B.,
1922. Argonauts of the Western Pacific, London, Routledge and Kegan Paul.
MALM, T.,
1995. 'Excavating submerged Stone Age sites in Denmark - the Tybrind Vig example' in A. Fischer (ed.) Man & Sea in the Mesolithic: Coastal settlement above and below present sea level. Oxbow Monograph No 53. 385-396.
MÜLLER, S.,
1896. 'Nye Stenalders Former'. Aarboger, 303-419.
MÜLLER, S.,
1918. Stenalders Kunst i Denmark, Copenhagen. 1-17.
MUNN, N.,
1976. The spatiotemporal transformations of Gawa canoes, Jour-

nal de la societe des Oceanistes 32, 39-53.
MUNN, N.,
1983. 'Gawan Kula: Spatiotemporal control and the symbolism of Influence' in J. W. Leach & E. R Leach (eds.).The Kula: New Perspectives on Massin Exchange, Cambridge. Cambridge University Press. 277-307.
MUNN, N.,
1986. The Fame of Gawa: A symbolic study of value transformation in a Massin society. Cambridge. Cambridge University Press.
NASH, G.H. ,
998. Exchange, Status and Mobility: Mesolithic Portable Art of Southern Scandinavia. Oxford. BAR International Series 710.
NASH, G.H.,
2001a Things moving around a landscape: the symbolism of south Scandinavian Mesolithic portable art, in D. Gheorghiu (ed.) Material, Virtual and Temporal Compositions: On the relationships between objects. Oxford, BAR International Series 953. 9-16.
NASH, G.H.,
2001b. 'Expressing sexuality, fertility and childbirth in art: engendering south Scandinavian Mesolithic portable artefacts, in L. Bevan (ed.), Indecent Exposure: Sexuality, society and the Archaeological Record. Edinburgh: Cruithne Press . 17-49.
PRICE, T. D. & BROWN, J. A.,
1985. 'Aspects of Hunter-Gatherer Complexity' in T. D. Price and J. A Brown (eds.). Prehistoric Hunter-Gatherers: The Emergence of Complexity, New York. Academic Press. 3-19.
ROWLEY-CONWY, P.,
1981. 'Mesolithic Danish Bacon: Permanent and temporary sites in the Danish Mesolithic' in A. Sheridan & G. Bailey (eds.), Economic Archaeology: Towards an Integration of Ecological and Social Approaches, BAR International Series No 96. 51-65.
RYING, B. 1981. Denmark, A Prehistory, DK Books.
STRATHERN, A.,
1983. 'The Kula in comparative perspective' in J. W. Leach & E. R. Leach (eds.), The Kula: New Perspectives of Massim Exchange. Cambridge University Press. 73-102.
TILLEY, C.,
1996. The Ethnography of the Neolithic: Early Prehistoric Societies in Southern Scandinavian. Cambridge. Cambridge University Press. 43-52.
TORBRÜGGE, W.,
1968. Prehistoric European Art. Abrams.
VANG PETERSEN, P.,
1984. 'Chronological and Regional variation in the late Mesolithic of Eastern Denmark'. Journal of Danish Archaeology No 3, 7-18.
VANG PETERSEN, P.,
1998. Rav, hjortetak og mesolitisk magi: Danefæ fra jægerstenalderen. Nationalmuseets Arbejdsmark. 87-100.
VEBÆK, C. L.,
1939. 'New finds of Mesolithic Ornamented Bone and Antler Artefacts in Denmark'. Acta Archaeologica 9, 205-223.
WEINER, A. B.,
1983. 'A World of Made is not of Born: Doing Kula in Kiriwina' in J. W. Leach & E. R. Leach (eds.), The Kula: New Perspectives of Massim Exchange. Cambridge. Cambridge University Press. 147-170.
WEINER, A. B.,
1985. 'Inalienable Wealth'. American Ethnologist, The American Anthropological Association. 220-227.
WEINER, A. B.,
1989. 'The Trobrianders of Papua New Guinea'. Case Studies in Cultural Anthropology, Holt, Rinehart & Winston, Inc.
ZVELEBIL, M.,
1986. 'Mesolithic Societies and the Transition to farming: Problems of time, scale and organization' in M. Zvelebil (ed.), Hunters in Transition, Cambridge. Cambridge University Press. 167-188.

4

EARLY HUMAN IMPACTS IN MADAGASCAR

MIKE PARKER PEARSON, RAMILISONINA, RETSIHISATSE, JEAN-LUC SCHWENNINGER AND HELEN SMITH

University of Sheffield,
Sheffield, England
EMAIL: m.parker-pearson@sheffield.ac.uk

INTRODUCTION

The human story of Madagascar begins with a paradox. How is it that this fourth largest island in the world was settled only in the last 2000-1400 years ago, when most of the large islands of the world were colonised many thousands, if not tens of thousands of years before? Madagascar lies only a thousand miles from Africa's cradle of humankind, and just 186 miles off the coast of Mozambique (figure 1). Over 65 million years ago it split off from India and Africa and has evolved a wide range of unique habitats, many of whose remarkable species of fauna and flora are found nowhere else in the world. Surprisingly, most of its fauna and flora - including the lemurs - are thought to have arrived since the split, by floating and island-hopping (Preston-Mafham 1991).

Yet why did human beings take so long to reach Madagascar when their adeptness at employing technology to traverse such distances over water already had a considerable antiquity? The traditional answer is that the trade winds blow westwards from the ocean to the East African mainland and, as a result, it was only with the adoption of the outrigger sail that boats could tack against the wind. It is clear that people from Southeast Asia were among the earliest settlers in Madagascar, voyaging on a 3000 mile sea migration across - or around the edges of - the Indian Ocean (Vérin 1990: 31-50). Around the same time, African communities from the Swahili coast made their way to Madagascar (Vérin 1976). It is out of these two groups, together with Arabs, Indians and Chinese that Madagascar's populations have formed (Vérin 1981).

Whatever the reason for this latecoming, the archaeology and history of Madagascar provide a remarkable window into a highly compressed sequence of social, ecological and economic transformations. These have included initial landings and early settlement by people from Southeast Asia and East Africa in the first millennium AD, urban and proto-urban development (900-1300), trading and integration with Swahili and Arabs (900-1500), state formation and the development of the Merina empire (1500-1800), contact with Europeans (1500 onwards), the emergence of pirate bases (1680-1720), and a brief period of colonial French rule from 1896 until independence in 1960.

Secondly, it provides an opportunity to study the long-term human impact on a unique island ecology - before it has been entirely transformed by deforestation and human introductions. The ecological diversity and unique species of Madagascar have enthralled biologists, ecologists, zoologists and other environmental scientists; the island has a surprising number of significant habitats from lowland evergreen rainforest, to montane dry forest, rainforest and bushland, rupicolous shrubland, dry deciduous forest and semi-arid spiny forest. Lemurs, fosa (a cat-like predator), tenrecs and Darwin's predictor moth are among the unusual animal species. Although Madagascar was reputedly the home of a man-eating tree (Mackal 1983), its flora are none the less extraordinary, with the tall spine-covered fronds of Didiereaceae trees, the poisonous sap of the famata (Euphorbia) tree, and the irritant leaves of the stinging-nettle tree. Species like the rosy periwinkle have economic value for their medicinal properties whilst wildlife agencies have campaigned for decades to establish an effective conservation infrastructure and to protect the unique species and habitats which are in danger of destruction by human encroachment. Madagascar has a fast-growing population whose needs for timber and forest products are rapidly outstripping the ability to conserve forest habitats. The extent of soil erosion, caused by agriculture and forest clearance, can be seen in satellite remote-sensing images, in which the island is seen to 'bleed' as its characteristically red soil discolours the waters of Madagascar's major rivers as they debouch around its coasts (Green and Sussman 1990).

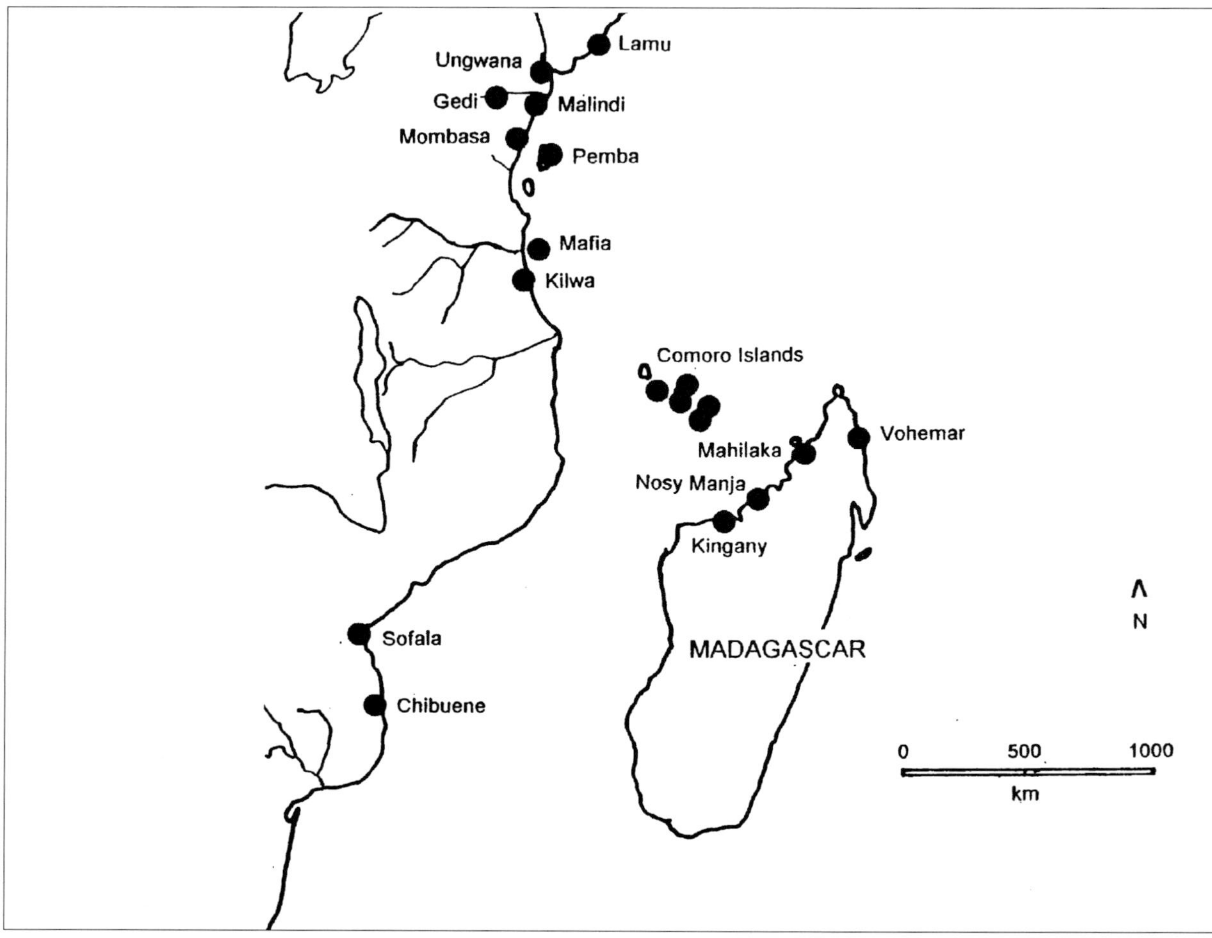

FIGURE 1. THE ISLAND OF MADAGASCAR, off the East African coast, showing some of the main Swahili ports of the early second millennium AD

Thirdly, the archaeology and history of Madagascar have much to teach us about the superficially contradictory characteristics of island communities - isolated and yet linked by the sea, possessing strong islander identities and yet riven with cultural differences, and developing idiosyncratic, insular monument styles whilst participating in wider cultural worlds. Islands are often environments in which strong island-wide identities can be constructed in opposition to the foreign, and in which certain cultural practices may become exaggerated and foregrounded. Both of these aspects are prominent features of Malagasy history, jointly embodied within shared indigenous concepts of Madagascar as 'the island of the ancestors' (Mack 1986). There are, of course, strong dialect differences and variations on island-wide themes such as monumental tomb building, ancestor veneration, funerary extravagence and the symbolic significance of cattle. This is particularly the case for the communities in the south and southeast which have been described as 'islands within an island' (Deschamps 1960; Rakotoarisoa 1998) and yet Madagascar is united by a single language and common culture in a way that contrasts dramatically with the hundreds of languages and ethnicities of that slightly larger island of New Guinea. In historical perspective, Madagascar provides an insight not only into the changing degrees of 'islandness' of its communities but also into the extent to which communities have looked inwards or outwards over the centuries.

In the last 40 years the archaeology of ancient settlements has been explored in 20 study areas within most of the different regions of Madagascar (Dewar and Wright 1993; Wright et al. 1996: fig. 1), notably in the north by Robert Dewar (1984; 1996; 1997), Pierre Vérin (1986), Henry Wright (Wright et al. 1996) and Chantal Radimilahy (1998), in the central highlands by Victor Raharijaona (1988), Henry Wright and Susan Kus (Wright and Kus

1976) and Zoe Crossland (2001), in the southeast by Jean-Aimé Rakotoarisoa (1998), and in the south by Georges Heurtebize (1986), Réné Battistini, Pierre Vérin (1963; 1965; 1967), Chantal Radimilahy (1980, 1988) and ourselves (Clark et al. 1998; Parker Pearson 1992; 1997a; Parker Pearson et al. 1999a and b). Despite this wealth of survey and excavation data, however, there are very few sites which can be dated much before 1000 BP. This paper examines the long-term history of Madagascar particularly concentrating on the extreme south of the island, a region whose relationships with the outside world (both within and beyond Madagascar) became increasingly remote until the 20th century.

EARLIEST HUMAN ARRIVALS

The date range for the earliest human arrivals is vague and problematic. Palynological sequences of the third millennium BP from western Madagascar have produced horizons of forest clearance in association with charcoal (Burney 1997) but, in the absence of other corroborative evidence for a human presence, these are explained as the results of bush fires produced by lightning strikes (Dewar and Wright 1993). The earliest radiocarbon dates associated with human activity come from pygmy hippopotamus bones bearing metal cutmarks at Ambolisatra and Lamboharana, two sub-fossil coastal sites in southwest Madagascar (MacPhee and Burney 1991). These date to around 2000 BP (1970*50 bp [AA-2895] and 1740*50 bp [TO-1437]) but there is not universal agreement that they provide a reliable 'first contact' date. The bones were recovered in the 19th century from poorly recorded contexts and the support for these dates hinges on the acceptance that the cut marks were made when the bones were fresh. Experimental cutting of green bones and sub-fossil bones indicated that the observed cut marks were consonant with the former. Yet the matter may be more complex than it first appears. During a visit to another sub-fossil site which may be of even greater antiquity, we observed cutmarks on some of the bones which had been dug out of travertine by villagers digging a well. Our excitement was tempered by the fact that the newly exposed bones were in an extraordinarily fresh condition as a result of their anaerobic and calcareous environment. They beg the question as to whether the Ambolisatra and Lamboharana assemblages were 'cut' by iron tools on immediate exposure in the 19th century prior to their deterioration as they dried out after removal from their matrix.

Two of the key problems of the Ambolisatra and Lamboharana assemblages is that they not only derive from a context which is non-cultural (without unequivocally associated human artefacts or other indicators of a human presence) but also that they remain consistently earlier by at least 500 years than any new dates for human activity. Even if they are accepted as ancient by-products of human agency, the Ambolisatra and Lamboharana events may have resulted only from brief visitations from the African mainland rather than from predation by an already settled population (Dewar and Wright 1993). In northwestern Madagascar, radiocarbon dates place the origins of human settlement largely within the second half of the first millennium AD at Irodo (Dewar 1996), Andavakoera (1680*65 bp [Beta-29946] and 1300*80 bp [Beta-18424]; Dewar 1996), Nosy Mangabe (1250*60 bp [SMU-2501]) and Sandrakatsy (1140*60 and 1240*50 bp [SMU 2076 and 2359]; Wright and Fanony 1992). Intriguingly, oral histories refer to an original landing on this northwest coast in the vicinity of Maroantsetra (Vérin 1981: 702).

Until recently, archaeological research in southern Madagascar had failed to locate any traces of settlement prior to the 9th century AD. During a programme of coastal survey covering over 30km of dunes and river mouths between 1991 and 2000, we finally found a small group of early settlements at the mouth of the Menarandra, one of the four rivers of the south. One site, Enijo, on the west side of the rivermouth, produced pottery of Swahili TIW (Triangle Incised Ware) type, dating to c. AD 600-900 (Chami 1994; 1998). Another three settlements on the east bank - Ankitsake I and II and Ambatofanañira III - contained Swahili Plain Ware of the period c. AD 900-1200, contemporary with but different from local Malagasy wares, and may provide evidence for the presence of a small Swahili enclave or entrepot at the rivermouth (figure 2). What is of interest with the TIW pottery is that it indicates direct contact and potentially even colonisation from the Swahili coast. This may have implications for the initial settlement of the south of Madagascar by African groups as opposed to the Southeast Asian roots of people in northern and central Madagascar.

When the Enijo site is considered in the context of the early radiocarbon dates for Lamboharena and Ambolisatra as well as for a lost and problematic site at Sarodrano (1460*90 bp [Gak-928]; Battistini and Vérin 1965; 1967), the possibility emerges for early colonisation along the southwest coast at the same time as, or before, colonisation of the northeast of Madagascar. If the sites of the southwest coast are indeed earlier, then they support aspects of Ferrand's hypothesis of an African or Bantu period preceding an Indonesian period (1908), albeit in two different regions of the island. The only extensive coastal zone not to have been surveyed in any detail is the central west coast and, until this is done, there is the possibility that Madagascar was settled in the first millennium AD from two different directions, to the north from the Comoro islands and to the south from the Mozambique coast.

MEGAFAUNAL EXTINCTIONS AND HUMAN ARRIVAL: THE CASE OF THE ELEPHANT BIRD

Prior to human arrival, Madagascar had an exotic range of megafauna which included Elephant Birds (Aepyornis and Mullerornis), archaeolemur, pygmy hippopotamus, and giant tortoise. Climatic change towards increasing aridity in the south around 2000 BP may have affected numbers of some of these species (Burney 1993; 1997) but the arrival of people in the first millennium AD seems most likely to have been instrumental in their extinction. Unlike New Zealand and other parts of the world, there is no evidence of kill sites or other unequivocal indicators of human predation on these megafauna.

Our own research in the extreme south of Madagascar has, amongst other things, focused on the extinction of some of these species - especially the Elephant Bird - and the earliest arrivals of people. The Elephant Birds were flightless ratites, of which Aepyornis maximus was the largest, standing 3m tall and weighing half a ton. Although not quite as tall as the New Zealand moa, it was considerably bigger and was the largest bird in the Holocene world. It probably gave rise to the legend of the roc or rukh, the giant bird which was known to Arab and Chinese merchants in the Indian Ocean and which Marco Polo reported as living in Madagascar and capable of flying with an elephant in its talons (Polo 1298/99 [1908: 393]; Ramiandrasoa 1967). The Elephant Birds' date of extinction is uncertain but there is a mid-17th century secondhand account of a bird larger than an ostrich - the vorompatres - living in the uninhabited parts of southern Madagascar (Flacourt 1661).

Traces of human exploitation of Elephant Birds are hard to find even though the coastal dunes of parts of Madagascar are littered with their broken eggshells. Frustratingly, finds of Elephant Bird bones are restricted to a handful of subfossil sites which appear to date to well before human arrival. On the south coast there is a group of middens at Talaky in which the eggshells are incorporated into deposits containing marine shell and pottery (Battistini et al. 1963). The chronological range of theses middens is from the 9th to 15th centuries AD. Talaky has been thought to be a site where people consumed the contents of the eggs. On an island-wide scale, such predation on the eggs rather than on the birds might explain their extinction. However, our excavation results and radiocarbon dates indicate that the eggshells were most probably laid centuries before they were incorporated in the midden. The range of radiocarbon dates for Elephant Bird eggshells at Talaky is from 1835*30 BP (OxA-8274) to 1240*35 BP (OxA-8270), consistently earlier than dates on charcoal from the same contexts (1080*35 BP (OxA-8118) to 530*35 BP (OxA-8136)). Historical sources indicate that the eggs were valued in the last few centuries as containers for liquids, especially Malagasy rum, and thus might be curated if found whole and intact.

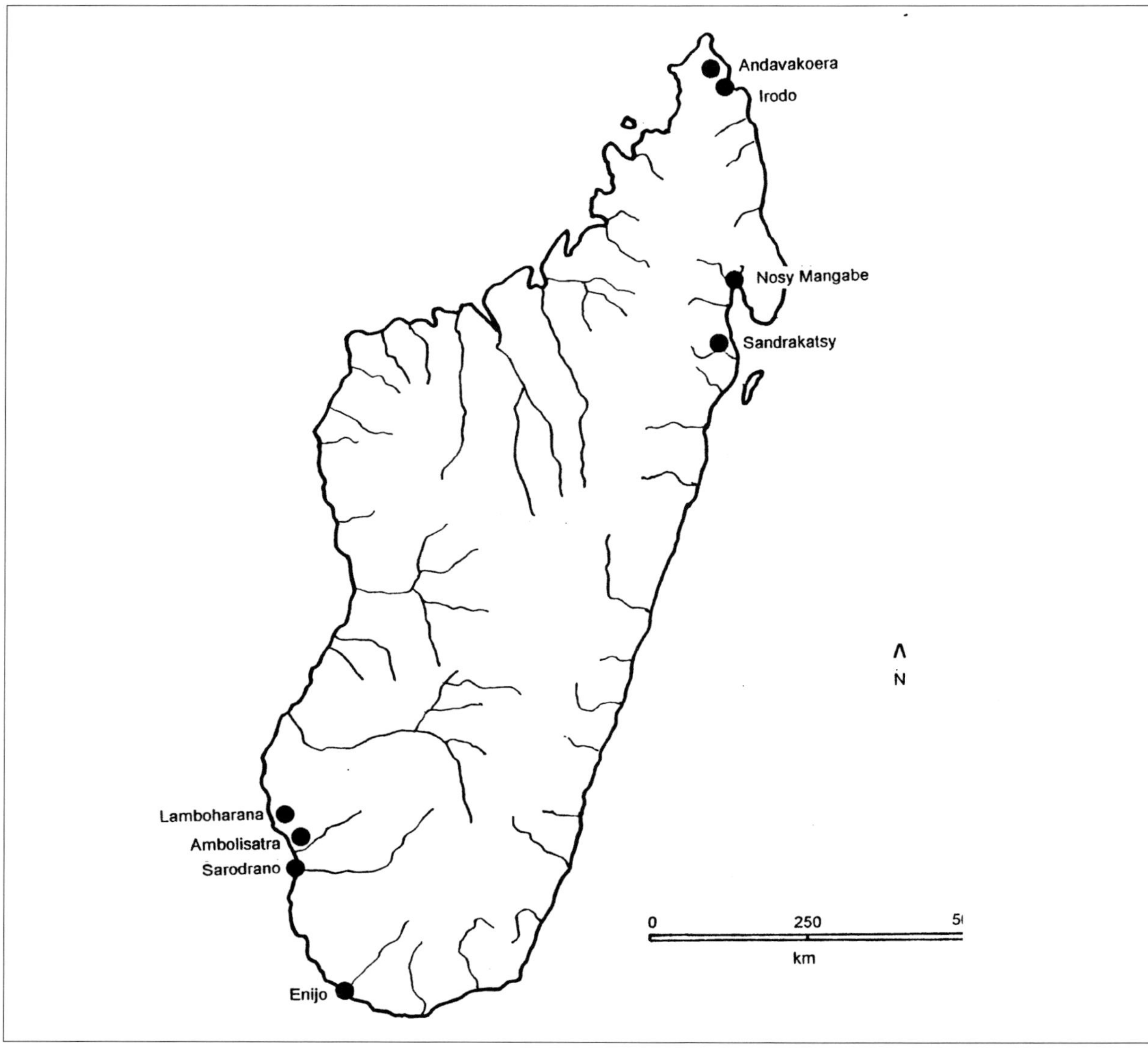

FIGURE 2. EARLIEST SITES OF THE FIRST MILLENNIUM AD IN MADAGASCAR

Elsewhere in southern Madagascar the traces of human-megafauna interaction remain elusive. Other sites of the late 1st and early 2nd millennium AD have yielded egg-shells in association with pottery and other indicators of human settlement. Yet there is still no trace of any Elephant Bird bones amongst the faunal assemblages on these sites. It is possible that the birds and their eggs were not predated as a food resourse but were exterminated either by design or as a result of ecological competition with cattle, sheep and goats.

THE MANDA CIVILISATION OF SOUTHERN MADAGASCAR: ISLAMIC AND CHINESE CONNECTIONS

Madagascar's involvement in the trading network of Swahili city states during their most prosperous period between 1000 and 1500 is well known from research into north-coast trading centres (Vérin 1986) and excavations on the major urban site of Mahilaka (Radimilahy 1998). There were similar large sites along the rivers of the south of Madagascar (Emphoux 1979; 1981; Radimilahy 1988: 145-57) but their development has generally been considered as secondary and peripheral to those of the north coast, even to the extent of deriving their imported ceramics from these northern coastal civilisations (Radimilahy 1997: fig. 8.1). However, recent research suggests that these southern communities were probably involved more directly in trade with the remainder of the western Indian Ocean.

The extreme south of Madagascar is a semi-arid environment in which the flow of water in its rivers is controlled by the rhythm of the seasons; riverbeds which are largely dry for most of the year are filled with water during the short rainy season between December and March. In such an inauspicious landscape, prone to drought and famine, a civilisation of proto-urban enclosed settlements, known as manda (which translates as 'enclosure' or 'fortification') sprung up between the 9th and 13th centuries. Each of the four great rivers of the south has one of these manda on its banks, normally in the upper reaches at least 50km from the sea. From west to east, Andasy Meriñe sits aside the Linta, Mandan d'Remananga is in the upper reaches of the Menarandra, Andranosoa is in the middle Manambovo, and Manda Meriñe at Mafelefo is in a tributary of the Mandrare. These are large sites of 30-15ha, within the top category of Class One Swahili towns (Kusimba 1999: 121-23) though nothing survives above ground of their (assumed) wooden buildings. The ramparts of Andasy Meriñe are impressive earthworks which may have supported a stone wall along

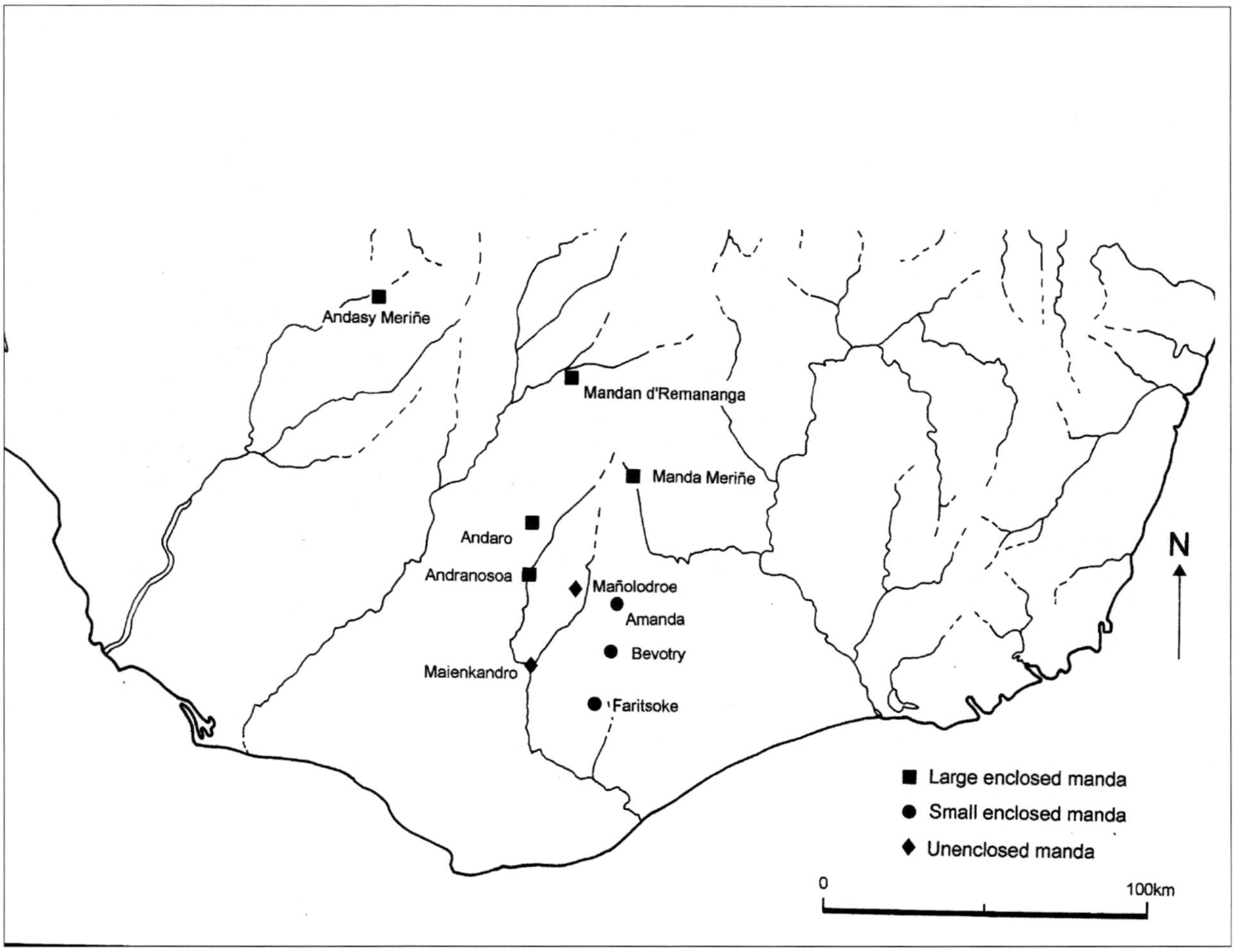

FIGURE 3. THE MANDA ENCLOSURES AND OTHER MAJOR 10TH-13TH CENTURY AD SETTLEMENTS:

their tops but the stone ramparts of Mandan d'Remananga, Andranosoa and Manda Meriñe appear to have been more symbolic than defensive. These sites have produced small but significant quantities of sgraffito and other ceramics from the Persian Gulf and China.

Within the Manambovo catchment, the one river system in the extreme south where sites of this period have been surveyed in detail, an interesting pattern of a hierarchical site structure emerges. The Manambovo basin has produced a distribution of six manda and related large open sites which are spaced at approximately 14km intervals (figure 3). The largest sites are Andranosoa and Andaro. Like their counterparts on the other river systems, they have produced sgraffiato and occasional Chinese ceramics. Yet such numbers of imports are small, less than 1% of the total surface assemblage, and thus less than the proportions found on the large urban sites of the Swahili coast and northern Madagascar. Andaro is a 27ha site comprising two stone enclosures which site aside a tributary of the Manambovo upstream of Andranosoa. Its local ceramics are of a style which can be dated to the 9th-10th centuries and it may be earlier than Andranosoa which is dated by radiocarbon to the 11th-13th centuries. At the second level is Maienkandro (9 ha), an open site which sits at the junction of the Manambovo and its largest tributary and has produced imported ceramics but in smaller quantities than the large sites. At the next level are the small manda (0.5-3 ha) which lack imports but whose quantities of local fine wares - haematite-burnished pottery - hint at activities beyond those of ordinary village life. At the bottom of the settlement hierarchy are small open sites, normally less than 0.5 ha in size, on which local fine wares are either rare or lacking. These sites are located close to streambeds and river banks throughout the Manambovo's dendritic system. They appear to have been small villages or hamlets.

It seems likely that imports arrived up the river valleys, no doubt brought on foot since there was insufficient water for boats, since imported ceramics have been found at sites in two of the rivermouths - the Linta and the Manambovo. Exactly what was exchanged in return is not known - cattle, leather, semi-precious stones, or slaves are all possibilities. Whereas the urban centres of the Swahili coast and north Madagascar continued into the 15th century and even beyond, the manda of the south were abandoned in the 13th century. What followed was a substantial decline in population, combined with a move to more secure sites on hilltops or naturally defensive locations. Our survey results indicate sparse and dispersed population levels in the 14th-15th centuries. Yet some contact with the outside world was still maintained in the form of imported Chinese celadon. Exactly what caused such a major upheaval is unknown.

THE ARRIVAL OF THE EUROPEANS: CULTURAL ISOLATION AND RESISTANCE IN THE SOUTH

In Madagascar the period between the end of the 15th century and the beginning of the 18th century witnessed the rise of powerful kingdoms, particularly the Sakalava kingdom of the west which was eclipsed in the 18th century by the Merina kingdom of the central highlands (Kent 1970). This was also the period in which European powers entered the Indian Ocean although their impact on political processes within these kingdoms was not profound. In this era of state formation and European interference there were also profound social and ecological changes islandwide and also in southern Madagascar as agriculture intensified and populations grew.

At the end of the 15th century the Portuguese arrived in the Indian Ocean and their violent intervention was one of the causes of the decline of the Swahili states. They showed relatively little interest in southern Madagascar, but established a base in the southeast at Ranofotsy and supposedly constructed a stone building nearby at Trañovato. Most of the Portuguese who were left here or arrived as shipwrecked castaways along the reef-strewn south coast were massacred by the local people. Archaeologically, we may be able to see the impact of Portuguese arrival in these waters. Whereas the coastal dunes were inhabited until the 15th century, thereafter there was an almost complete absence of 16th-17th century sites along the coast or in the rivermouths. At the same time the river valleys of the south were emptied - an observation based on archaeological survey and on the account (1660) of a French colonial governor, Etienne de Flacourt, who commanded a French outpost at Fort Dauphin in the southeast of Madagascar in the mid 17th century.

Flacourt tells us that these watered and more fertile regions were deserted, 50 years before his arrival, for fortified villages on the dry sand plateau in between the rivers because of escalating warfare amongst rival leaders of the various polities of the south. In order to maintain supplies to Fort Dauphin, Flacourt sent out his men - armed with the only guns in the region - in military campaigns supporting one rival leader against another and bringing home the booty in the form principally of cattle. It seems very likely that the intervention of albeit small numbers of French troops had a catalystic effect on warfare. Fort Dauphin was the entry point for imports and these included guns, shot and powder which were exchanged for slaves and other goods. French, Dutch and English ship's crews who were unlucky enough to be marooned on the south coast in the late 17th and 18th century were confronted by armies of warriors increasingly armed with muskets and pistols, from whom they were unlikely to escape (Parker Pearson 1997b; Parker Pearson and Godden 2002).

It was only by the late 18th and 19th centuries that the Tandroy, the people of the extreme south, began to colonise back into the formerly deserted river valleys whilst still maintaining their waterless existence on the sand plateau. In the early 19th century the Merina empire directed armies against the Tandroy but lost two engagments and retreated from the south. In 1900 the French tried more successfully and brought towns and state control to the south, disarming the Tandroy of at least 12,000 muskets and rifles. During the colonial rule of the French - a brief 60 years - the Tandroy and other people of the semi-arid south experienced a sequence of droughts and famines. The most catastrophic, which killed thousands of cattle and perhaps hundreds of people, was famously an early example of biological warfare, caused by French introduction of the cochineal beetle which killed the cactus (Middleton 1999).

ACKNOWLEDGEMENTS

We wish to thank Jean-Aimé Rakotoarisoa and the staff of the Musée d'Art et d'Archéologie of the University of Antananarivo for their institutional support of the Androy project, which has been funded by the British Academy, British Institute in Eastern Africa, National Geographic, NERC, Nuffield Foundation and the Society of Antiquaries. Since 1991 the project has enjoyed the hospitality of communities throughout southern Madagascar, and special thanks go to Georges Heurtebize and all at Analamahery. Georges Heurtebize, John Mack, Chantal Radimilahy and Henry Wright have also provided valuable advice and help throughout. Figures 1 and 2 were drawn by MPP and figure 3 was drawn by Irene de Luis.

REFERENCES

BATTISTINI, R. AND VÉRIN, P.
1965. L'Age absolu de la disparition des grandes subfossiles dans l'extrême sud de Madagascar. Madagascar Revue de Géographie 7: 29.

BATTISTINI, R. AND VÉRIN, P.
1967. Ecologic changes in prehistoric Madagascar. In P.S. Martin and H.E. Wright (eds.) Pleistocene Extinctions: the search for a cause. New Haven: Yale University Press. 407-424.

BATTISTINI, R., VÉRIN, P. AND RASON, R.,
1963. Le site archéologique de Talaky: cadre géographique et géologique; premiers travaux de fouilles; notes ethnographiques sur le village actuel proche du site. Annales Malgaches 1: 111-153.

BROWN, M.,
1995. A History of Madagascar. London: Tunnacliffe.

BURNEY, D.A.,
1993. Late Holocene environmental changes in arid southwestern Madagascar. Quaternary Research 40: 98-106.

BURNEY, D.A. ,
1997. Theories and facts regarding Holocene environmental change before and after human colonization. In S.M. Goodman and B.D. Patterson (eds.) Natural Change and Human Impact in Madagascar. Washington DC: Smithsonian Institution Press. 75-92.

CHAMI, F.A.,
1994. The Tanzanian Coast in the First Millennium AD. Uppsala: Uppsaliensis SAA 7.

CHAMI, F.A.,
1998. A review of Swahili archaeology. African Archaeological Review 15: 199-218.

CLARK, C.D., GARROD, S.M. AND PARKER PEARSON, M.,
1998. Landscape archaeology and remote sensing in southern Madagascar. International Journal of Remote Sensing 19: 1461-77.

CROSSLAND, Z.,
2001. Time and the ancestors: landscape survey in the Andrantsay region of Madagascar. Antiquity 75: 825-36.

DESCHAMPS, H.,
1960 Histoire de Madagascar. Paris: Berger Levrault.

DEWAR, R.E.,
1984. Extinctions in Madagascar: the loss of the sub-fossil fauna. In P. Martin and R. Klein (eds.) Quaternary Extinctions, 574-93. Tucson: University of Arizona Press.

DEWAR, R.E.,
1996. The archaeology of the early settlement of Madagascar. In J. Reade (ed.) The Indian Ocean in Antiquity. London: Kegan Paul International. 471-86.

DEWAR, R.E.,
1997. Were people responsible for the extinction of Madagascar's subfossils, and how will we ever know? In S.M. Goodman and B.D. Patterson (eds.) Natural Change and Human Impact in Madagascar. Washington DC: Smithsonian Institution Press. 364-380.

DEWAR, R.E. AND WRIGHT, H.T.,
1993. The culture history of Madagascar. Journal of World Prehistory 7: 417-66.

EMPHOUX, J.-P.,
1979. Archaeology and migrations in northern Androy: a preliminary report. In R. Kent (ed.) Madagascar in History. Albany CA: Foundation for Malagasy Studies. pp. 32-41.

EMPHOUX, J.-P.,
1981. Archéologie de l'Androy: deux sites importants: Andranosoa et le manda de Ramananga. Omaly si Anio 13/14: 89-98.

FERRAND, G.,
1908. L'Origine Africaine des Malgaches. Journal Asiatique 10: 353-500.

FLACOURT, E DE.,
1661. Histoire de la Grande Isle Madagascar. Paris: Clouzier.

GREEN, G.M. AND SUSSMAN, R.W.,
1990. Deforestation history of the eastern rain forests of Madagascar from satellite images. Science 248: 212-15.

HEURTEBIZE, G.,
1986. Les anciennes cultures de l'Androy central. Taloha 10: 171-80.

KENT, R.,
1970. Early Kingdoms in Madagascar, 1500-1700. New York: Holt, Rinehart and Winston.

KUSIMBA, C.M.,
1999. The Rise and Fall of Swahili States. Walnut Creek, CA: Altamira.

MACPHEE, R.D.E. AND BURNEY, D.A.,
1991. Dating of modified femora of extinct dwarf Hippopotamus from southern Madagascar: implications for constraining human colonization and vertebrate extinction events. Journal of Archaeological Science 18: 695-706.

MACK, J.,
1986. Madagascar: island of the ancestors. London: British Museum Press.

MACKAL, R.P. ,
1983. Searching for Hidden Animals.

MIDDLETON, K.,
1999. Who killed "Malagasy cactus"? Science, environment and colonialism in southern Madagascar, 1924-30. Journal of Southern African Studies 25: 215-48.

PARKER PEARSON, M.,
1992. Tombs and monumentality in southern Madagascar: preliminary results of the central Androy survey. Antiquity 66: 941-48.

PARKER PEARSON, M.,
1997a. Qui étaient les Ampatois? L'archéologie des développements politiques du seizième au dix-huitième siècle en Androy. Etudes Océan Indien 23-24: 237-252.

PARKER PEARSON, M.,

1997b. Close encounters of the worst kind: Malagasy resistance and colonial disasters in southern Madagascar. World Archaeology 28: 393-417.
PARKER PEARSON, M. AND GODDEN, K.,
2002. In Search of the Red Slave: shipwreck and captivity in Madagascar. Stroud: Sutton.
Parker Pearson, M., Godden, K., Ramilisonina, Retsihisatse, SCHWENNINGER, J.-L. AND SMITH, H.,
1999a. Lost kingdoms: oral histories, travellers' tales and archaeology in southern Madagascar. In R. Funari, M. Hall and S. Jones (eds.) Back from the Edge: Archaeology in history. London: Routledge. 233-54.
PARKER PEARSON, M., RAMILISONINA AND RETSIHISATSE.,
1999b. Ancestors, forests and ancient settlements: Tandroy readings of the archaeological past. In P. Ucko and R. Layton (eds) Indigenous Landscapes. London: Routledge. 397-410.
POLO, M.,
1298-99 ,1908. The Travels of Marco Polo. London: Dent.
Preston-Mafham, K. 1991. Madagascar: a natural history. Oxford: Facts on File.
RADIMILAHY, C.,
1980. Archéologie de l'Androy - contribution à la connaissance des phases de peuplement. Mémoire de maîtrise soutenu au Département d'Histoire de l'EES Lettres. Antananarivo: Centre d'Art et d'Archéologie.
RADIMILAHY, C.,
1988. L'Ancienne Métallurgie du Fer à Madagascar. Oxford: BAR Supplementary Series 422.
RADIMILAHY, C.,
1998. Mahilaka: an archaeological investigation of an early town in northwestern Madagascar. Uppsala: Uppsaliensis.
RAHARIJAONA, V.,
1988. Etude de peuplement d'espace d'une vallée des Hautes Terres Centrales de Madagascar: archéologie de la Manandona (XVè-XIXè siècle). Etudes Océan Indien 10: 157-60.
RAKOTOARISOA, J.-A.,
1998. Mille Ans d'Occupation Humaine dans le Sud-est de Madagascar: Anosy, une île au milieu des terres. Paris: Harmattan.
RAMIANDRASOA, F.,
1967. L'oiseau "rok" et le "Madeigascar" dans le livre de Marco Polo. Revue de Madagascar 37: 55-68.
VÉRIN, P.,
1976. The African element in Madagascar. Azania 11: 135-51.
VÉRIN, P.,
1981. Madagascar. In G. Mokhtar (ed.) General History of Africa II: ancient civilisations of Africa. London: Heinemann & UNESCO. 693-717.
VÉRIN, P.,
1986. The History of Civilization in North Madagascar. Rotterdam: Balkema.
VÉRIN, P.,
1990. Madagascar. Paris: Karthala.
VÉRIN, P. AND HEURTEBIZE, G.,
1974. La tranovato de l'Anosy - première construction érigée par les Européens à Madagascar. Taloha 6: 117-42.
WRIGHT, H.T.,
1984. Early seafarers of the Comoro islands: the Dembeni phase of the IX-Xth centuries AD. Azania 19: 13-59.
WRIGHT, H.T.,
1986. Early communities on the island of Maore and the coasts of Madagascar. In K.P. Kottak, J.-A. Rakotoarisoa, A. Southall and P. Vérin (eds.) Madagascar: Society and History. Durham NC: Carolina Academic Press. 53-88.
WRIGHT, H.T.,
1993. Trade and politics on the eastern littoral of Africa, AD 800-1300. In T. Shaw, P. Sinclair, B. Andah and A. Okpoko (eds.) The Archaeology of Africa: food, metals and towns. London: Routledge. 658-72.
WRIGHT, H.T. AND FANONY, F.,
1992. L'Evolution des systemes d'occupation des sols dans la vallée de la rivière Mananara au nord-est de Madagascar. Taloha 11: 16-64.
WRIGHT, H.T. AND KUS, S.,
1976. Reconnaissances archéologiques dans le centre de l'Imerina. Taloha 7: 19-45.
WRIGHT, H.T., RAKOTOARISOA, J.-A., HEURTEBIZE, G. & VÉRIN, P.,
1993. Evolution of settlement systems in the Efaho river valley, Anosy: a preliminary report on archaeological reconnaissances of 1983-1986. Indo-Pacific Prehistory Association Bulletin 12: 2-20.
WRIGHT, H.T., VÉRIN, P., RAMILISONINA, BURNEY, D., PIGOTT BURNEY, L. AND MATSUMOTO, K.,
1996. The evolution of settlement systems in the Bay of Boeny and the Mahavavy River valley, north-western Madagascar. Azania 31: 37-73.

5

IMAGINING ISLANDS

TAMARA KOHN

University of Durham
Durham, England
EMAIL:

INTRODUCTION

Islands provide settings from which one can witness or partake in all sorts of splendid journeys. They are places to take off from and come home to with new riches. From the outside, they are often imagined to be romantic, remote, closed and insular social and physical spaces. The insider's imaginings of the island and its community are far less bounded, but are nonetheless variable and contested. It is the power of the imagination of island observers and island people that I will address in this short paper. I wish to critically examine a tiny portion of the great corpus of anthropological material on island communities to make a few observations that may be of use in our attempt to conceptualize 'islands' and 'island-ness'. I will draw a few examples from my research in the Scottish Inner Hebrides which examines the way that changing island identities of both 'islanders' and 'incomers' are expressed through social action in the present. The oft-expected rigidity of the island boundary gives way, through these examples, to ideas of fluidity, change, incorporation and innovation.

Islands provide settings from which one can witness or partake in all sorts of splendid journeys. They are places to take off from and come home to. The following excerpt from a poem by Constantinos Cavafy illustrates the splendour in this process:

ITHAKA

When you start on the way to Ithaka,
pray that the road be long,
full of adventure, full of knowledge.
Do not fear the Lestrygonians
and the Cyclopes and the angry Poseidon.
You will never meet such as these on your path,
if your thoughts remain lofty, if a fine
emotion touches your body and your spirit.
You will never meet the Lestrygonians,
The Cyclopes and the fierce Poseidon,
If you do not carry them within your soul,
If your soul does not raise them up before you.

Then pray that the road is long.
That the summer mornings are many,
that you will enter ports seen for the first time
with such pleasure, with such joy!
Stop at Phoenician markets,
and purchase fine merchandise,
mother-of-pearl and corals, amber and ebony,
and pleasurable perfumes of all kinds,
buy as many pleasurable perfumes as you can;
visit hosts of Egyptian cities,
to learn and learn from those who have knowledge.

Always bear Ithaka in mind.
Arriving there your journey's end.
But do not hurry your voyage at all.
The more years it lasts the better;

and anchoring at your isle an old man,
rich with all that you have gained on the way,
do not expect Ithaka to give you riches.
Ithaka has given you the beautiful journey.
Without her you would not have set your course.
Yet she has no more to give you.
And poor though you find her, Ithaka has not deceived you.
So wise you have become, so full of experience,
already you will have understood what Ithakas mean.

Out of these lines comes a zest for exploration - of gathering the riches and knowledge of the world and savouring the moments with a positive open soul - of returning to the island which has given you the gift of travel - the gift of the beautiful journey - but is always carried in your mind as a place to come home to. The island's imaginary is a focal point of a journey which has little to do with islandness per se. Likewise, one could suggest that an islander is one who carries the island in her mind wherever she goes and for whom the wisdom of experience and action on and off the island gives meaning to the island space.

It is the power of the imagination of island observers and island people that I will address in this short paper. I wish to critically examine a tiny portion of the great corpus of anthropological material on island communities to make a few observations that relate to a process of belonging found in my own research in the Scottish Inner Hebrides - namely the way that changing island identities are expressed through action in the present.

The most obvious general observation one might make from common knowledge and a comb through the literature is that not all islands are equal; some are tiny, others are huge, some are mountainous, others flat, some are friendly and welcoming, others are hostile and 'insular', some are linked in an archipelago, others are solitary and confined, some are multi-ethnic/multi-cultural, others are relatively homogenous, some are hot, others are cold.

Yet, they are all situated on margins and encircled by margins. They all have 'insides' and 'outsides' / 'insiders' and outsiders'. To the outside eye they give off illusions of boundedness - they appear self-contained and manageable, and this is perhaps one reason why anthropologists have flocked to them. Their readership finds islands alluring and mystical which perhaps explains the appearance of the word 'island' in so many volumes about island communities. It is interesting to see how little, however, the social space which islands provide us are analyzed vis-a-vis their 'islandness'. In social terms (even if not in political and economic ones) the island is often provided as a tasty wrapping for something else which we wish to observe. The island is a vessel in which all social life is contained.

This is, of course, evident in many earlier ethnographies, such as Radcliffe-Brown's study of the Andaman Islanders (1922) or Malinowski's study of the Trobriand islanders (1922, 1929), but such functional and structural-functional studies are bound to treat locales as vessels, whether they are islands, mountainsides or remote villages on a continent. That generation was looking for the social isolation and the lack of socio-cultural contamination that was seen to come with colonialism and other external contact.

The methodology of participant observation which remains a cornerstone of social anthropology is popularly understood to have been born on a Trobriand island, but the allure of islands can be traced further back in our discipline to Darwinian interests in islands as special places to study natural selection and limited numbers of species. When Haddon planned the Torres Straight expedition in the late 19th century, he thought it essential to go to islands since the protected habitats would provide as natural a laboratory for human cultural variation as it would for the biological evolution of plants and animals (Kuklick 1996). The island context was critical in Darwinian biogeography (ibid), and then later its invisibility in the early 19th century was explained by paradigms of the times, but I find it curious how more recent work carries on this tradition. Even though recent island ethnographies and other texts have challenged functional perspectives which disallow affects of change, contact and diversity, and even though they often celebrate the movement and change within and beyond island boundaries, 'islandness' is not often explicitly unpacked in the body of the work.

A book of poetry by the Jamaican writer Jean 'Binta' Breeze called 'On the edge of an Island' (1997) is full of people, voices, and the occasional mention of seafish and sand - the things of islands, but with no explicit talk of islands. Hogbin's well-known ethnography of Wogeo in Papua New Guinea called The Island of Menstruating Men (1970) begins with a brief description of the island's position and geological history, and then the island as a special space becomes invisible while chapters describe Wogeo social structure, myths, spirit world, taboo, initiation, menstruation, childbirth, illness, death, magic, and religion. Goodale's study of Tiwi Wives (1971) in N. Australia's Melville Island also begins with descriptions of the land, flora, fauna, and history of settlement, but thereafter the 'islandness' is insignificant in the rich descriptions of social structure, lifecycle, ceremony and value systems.

The great irony is that the more interesting, diverse and chaotic island cultures become, the more we unquestioningly observe them. This has everything to do with their conceptual remoteness in the Western imagination. The qualities of 'remote areas' (which need not all be on islands) give us some clues to their intrigue. Ardener's work on these sorts of spaces, drawing both from his Hebridean island work as well as his extensive fieldwork in Camaroon provides a useful guide to help us understand some possible qualities of islandness.

...the feature I describe of 'remoteness'... persists when it has lost its geographical correlates - that is, when the 'remote' area has been reached, and when it should now be meredly present. Thus people would visit the Cameroons, and ... stagger in to see us as if they had surmounted vast odds; as if there was an almost exaggerated contrary sense of the absence of any barrier to the world - a peculiar sense of excessive vulnerability, of ease of entry. With every improvement of communication over the decades, the more speedily did people appear to pour in uninvited; and yet the more they seemed to be on the last stages of an expedition to some Everest that terminated in the middle of your floor. That is a law of 'remote' areas - the basic paradox... The West still maintains ideals of such places.... You know you are 'remote' by the intense quality of the gaze of visitors, by a certain steely determination, by a slightly frenetic air, as if their clocks and yours move at different rates. Perhaps that is why the native of such an area sometimes feels strangely invisible - the visitors seem to blunder past, even through him (1987:215).

He goes on to suggest that, 'Western Islanders do not see themselves as resembling that artistic or textual remoteness. They are quite ordinary - as ordinary as anybody can be who has the regular experience of wild-eyed romantics tottering through his door' (ibid: 217-18). This enhancement of certain 'remote' physical and social spaces by visitors (including anthropologists) reminds us of Urry's descriptions of the tourist gaze on certain landscapes (1995) or Schneider's notions of enchantment (1993), both of which have been useful for understanding some of my own Scottish island material. Ardener's allusion to being in a different time frame is likewise significant - it's more than just the difference between rural and urban time. Remoteness, often accentuated by island edges and sea, encourages a different use or marking of time. Learning to belong in such places requires readjusting one's social body clock.

The romantic 'other' lives on islands. The other is understood in any way that will help keep the island a safe and special haven for study or holiday. A colleague recently said that she thought islanders tend to be sociable, outgoing and happy! 'Think of the Fijians,' she said. 'Maybe it's something to do with the protection they get from the sea - they can always see people coming and welcome diversity.' Perhaps she had recently read Christina Toren's work on mutual compassion, love and desire in Fijian identity (1999), but the first islanders I thought of were Scottish Hebrideans who are oft characterised as shy or wary of strangers. The second I thought of were the ex-cannibals of Dobu Island off Papua New Guinea (Benedict 1934) who are described by neighbouring islanders as dangerous magicians and warriors. The reality of islanders' diverse 'characters' externally described is of little import - more useful, certainly, is emic discourse on the things that make people belong to islands - internal and contested visions of what 'islandness' entails. The shifting boundaries between 'inside' and 'outside' is key not only in the study of islands, but in the constructions of landscape more generally (Hirsch 1995:16). The conceptual opposition between inside and outside defines the social and the wild (Hastrup 1990), the known and the unpredictable, the 'arrived' and the 'becoming'. The island is the journey between them.

My first fieldwork experience was on a small island in the Inner Hebrides of Scotland between 1984-1987. I wrote up my Oxford DPhil on the experience of incomers in that island community at the same time that Jackie Waldren was writing about 'insiders' and 'outsiders' in Deia, Mallorca. We have had so much in common – not least our interest in shifting island identities and the conceptual and experienced oppositions that are drawn and redrawn between the inside and outside (see Waldren 1996). I believe that we both demonstrate in our work some of the qualities of 'islandness' outlined generally above, starting with the romantic and processual directives in Cavafy's poem. Here, I cite an example from the Hebrides.

THE ISLANDER.

In the summer, Iain often sat in the sun outside of his cottage, fixing fishing nets or repainting his boat. Day-trippers would pass on their walk from the pier to the pub for lunch and they often stopped to chat. "Have you always lived here?" "Och aye I have, aye!" Then they would smile and once out of his earshot comment on how wonderful it is to see the true islanders and how sad it is to think about what will become of the place once they are gone. And yet if you know Iain, then you know that he spent most of his adolescence and adulthood away from the island – travelling the world and working in Glasgow – only returning to his home on the island in old age. Why did he lie? Or was he lying at all? Was he perhaps conflating time – pulling his childhood past towards his elderly present in order to deliver the desired externally demanded effect of islandness as being bounded in time and remote space? What truths about people's notions of island identity are rooted in this vignette? Iain's lived experience challenges the expected rigidity of islander mobility, but his presentation of himself to the outsider reify all their expectations. Island-ness is found both in his Ithakas-like adventure of the world as well as in the insular representations he provides to tourists and other outsiders.

Iain's friend and neighbour Robbie was born and raised on the island. He signed up to join the services at the very beginning of WW2 and didn't return until the end of the war, 6 years later. He escaped 4-5 times from prisoner of war camps in Germany, but kept getting caught and moved to a new one. He shared many detailed stories about his adventures during the war with me. These memories seemed to loom larger than most. He returned to his island home and never worked away from it again. He said he still knew a few German expressions and could understand quite a bit of the language. To illustrate he told me about how he had been working his way slowly up the plank onto the island-faring boat one day, and some Germans were coming up the plank behind him. They spoke in German about the 'Old Seaman', going on about how he really looks the part of the islander setting off to sea. Without missing a step or looking back, Robbie spoke back to them in German – something to let them know he understood all they had said (and probably quite heavily laced with the swear words he was apparently most acquainted with). So here we have another confirmation of outsiders' constructions of islandness, but unlike Iain who stepped outside the truth of his cosmopolitanism to support their expectations, in Robbie we see joy in their challenge.

'Islander' status would never be challenged for these two men because of their birth on the island and the deep geneological links they hold with the place. Many years spent elsewhere returns them to the island with only that much more to offer it. Islandness is not defined by insularity, but insularity is constructed out of dreams and imaginations.

THE INCOMER

The vast majority of residents on Iain and Robbie's island could not trace their roots to the place. It is considered by many to be an island of incomers. In my own work, I have been interested in "the way people unselfconsciously and subtly demonstrate through actions their procession along an incomer/islander continuum, and how these actions contrast with more consciously constructed identity markers and declarations of belonging which are often built upon idealised 'rememberings' of a general or local past" (Kohn, JRAI 2002 forthcoming). Ideas of 'becoming an islander' run contrary to the popular notion that many people from all walks of life (academics, historians, artists, poets, locals, tourists) take on unquestioningly about the decline of Scottish Culture; one that places the old Gaels who speak the Gaelic (like Iain and Robbie) as the very last of a kind, teetering on the brink of extinction. 'Becoming an islander' in Scotland appears to be a contradiction in terms. Such a process clashes with a popular model of Celtic decline, and the very word 'islander' is associated with an ascribed rather than an achieved status. The word implies geneological depth or at the very least, birthplace. Born in Leeds, London or Chicago, and you'll never be an 'islander'. You either have it or you do not. And yet, one must be able to look beyond the apparently rigid classifications of island belonging to observe the spaces of active movement between them. Through action we see people adapting ways of being and doing - sometimes highly selfconsciously, and yet often unselfconsciously and unreflexively, but always meaningfully. I have also tried to place the island 'tourist' into this potential continuum as well (Kohn 1997). The word 'potential' is significant – not all incomers or tourists can even imagine themselves to become rooted in any way to an island community in their lifetime, largely due to the romantic notions about islands and island populations that I introduced earlier. However, if such a shift does happen in the way individuals imagine themselves to belong to a place, then it only appears to occur in any lasting way through activity. 'Incomers' as well as 'islanders' have to act their islandness out in a socially engaged way. I will give another short illustration here to demonstrate this process.

The MacPhees were from Argyll, and they moved to the island shortly after I did. Billy had trained as a chef, and Mairi as a nurse. They had four children, and before they bought their big farmhouse in the west end of the island, he worked as the gamekeeper for a laird on a big estate. They lived in 'tied' cottages' on the mainland and became fed up with the lack of security. They bought the island farmhouse plus 250 acres of land for £50,000. Billy had only seen the island and farm once in the few hours between boats one day and decided to go for it. From the outset this family presented themselves as good proto-islander material. Two younger children went to the local primary school, the older ones to high school in Oban. Mairi picked up local work, joined in the knitting bee and other social gatherings open to women, and generally became an active member of the small island community. Billy very quietly applied himself to his farm and sheep. Once settled in, he hardly ever left the island in the past 15 years – tied to his land and disinterested in even getting away for a short holiday in the way that insular islanders are said to be. A few months ago, a university student of mine who had worked for a summer on the island, told me that he was now good friends with two 'islanders' – the youngest of Mairi and Billy's children who were now in their twenties. On a visit to Durham they all came over to my house for lunch and they shared stories that blended together their own histories of high school, travel abroad, University study and work on the island with many tales of island events and people and gossip. I thought of Cavafy's poem – these young folk carrying the island in their minds as they travelled distant lands – something they said they would never have done if they hadn't lived on the island – and then returning home with the riches, knowledge and education that the journey had offered them. It is because I now see the islander in the person who travels and returns that I can imagine these young incomers painting their boats on the old pier in 50 years' time.

In my own work in Scotland but also in the research I have conducted since in the hills of East Nepal, I have been interested in how, for incomers, dreamers, tourists, anthropologists, women marrying into new families, for people generally 'outside' home or on the periphery, senses may be enlivened (Kohn 1994, 1995, 1998). You can reflect more abstractly and significantly on your own identity and experience when you are outside of some perceived centre. Perhaps incomers who wish to make a Scottish island their 'home' are conscious of a need to actively involve themselves but are unable to articulate the real effect of that involvement (that they are becoming islanders). They are unable to do this, perhaps, because of the inordinate strength of imagined histories of nations and ideas about belonging to remote rural island communities which their minds are laden with. Many residents on the island where I lived do battle between two competing discourses; one which relies on action, social commitment, and an increased sense of belonging to the place, and another traditionalized one which suggests that people without the right names, linguistic backgrounds and histories cannot ever achieve islander status.

As Macdonald says of her work on Skye, 'Place is a place of return - an enduring location which outlasts all the sites in which a person may temporarily reside' (1987:144). From Ithaka to Mallorca to the Inner Hebrides, islands are places from which one may journey far and to which one eventually returns having demonstrated through a life of action who one is.

NOTES

I am very grateful to my many island friends in the Scottish Hebrides who made me feel like a proto-islander. A version of this paper was presented as a keynote lecture at the University of St. Andrews' 'Managing Island Life' symposium organised by Drs. J. Skinner and M. Hills (1998). I received much useful feedback from the participants. I am also grateful to Dr. Luisa Elvira Belaunde for finding me the beautiful translation of Cavafy's poem, Ithaka.

REFERENCES

ARDENER, EDWIN,
1987. 'Remote Areas': Some Theoretical Considerations. In Anthropology at Home (ed) A. Jackson, 38-54. London: Tavistock.
ARDENER, EDWIN,
1989. The Voice of Prophesy: Further Problems in the Analysis of Events. In Edwin Ardener: The Voice of Prophesy (ed) M. Chapman, 134-154. Oxford: Blackwell.
BENEDICT, RUTH, 1934. Patterns of Culture. Boston: Houghton Mifflin Company.
BREEZE, JEAN 'BINTA'
1997 On the Edge of an Island. Newcastle upon Tyne: Bloodaxe Books.
GOODALE, JANE,
1971. Tiwi Wives: a study of the women of Melville Island, North Australia.
HIRSCH, E. AND M. O'HANLON (eds),
1995. The Anthropology of Landscape. Oxford: Oxford University Press.
HOGBIN, IAN
1970. The Island of Menstruating Men: Religion in Wogeo, New Guinea. Scranton, London, Toronto: Chandler Publishing Co.
KOHN, TAMARA,
1988. Seasonality and Identity in a Changing Hebridean Community. Unpublished D.Phil. thesis, University of Oxford.
KOHN, TAMARA,
1994. Incomers and fieldworkers: a comparative study of social experience. In Social Experience and Anthropological Knowledge (eds) K. Hastrup & P. Hervik. London: Routledge.
KOHN, TAMARA,
1995. She came out of the field and into my home: reflections, dreams and a search for consciousness in anthropological method. In Questions of Consciousness (eds) A.P. Cohen, & N. Rapport. London: Routledge.
KOHN, TAMARA,
1997. Island Involvement and the Evolving Tourist. In Tourists and Tourism (eds) S. Abram, J.D. Waldren, & D.Macleod. Oxford: Berg Press.
KOHN, TAMARA,
2002. "Becoming an Islander Through Action in the Scottish Hebrides", JRAI…
KUKLICK, HENRIKA,
1996 "Islands in the Pacific: Darwinian biogeography and British Anthropology", American Ethnologist, Vol 23(3): 611-638.
MACDONALD, SHARON,
1987. Social and Linguistic Identity in the Scottish Gaidhealtachd: A Study of Staffin, Isle of Skye. Unpublished D.Phil. thesis, University of Oxford.
KOHN, TAMARA,
1997. Reimagining Culture: histories, identities and the Gaelic renaissance. Oxford: Berg Press.
MALINOWSKI, BRONISLAW,
1922 (1984) Argonauts of the Western Pacific. Prospect Heights, Waveland Press.
MALINOWSKI, BRONISLAW,
1929 The Sexual Life of Savages. NY & London: Harcourt Brace Jovanovich.
RADCLIFFE-BROWN, A.R.,
1922 (1964) The Andaman Islanders. New York: The Free Press.
TOREN, CHRISTINA,
1999 "Compassion for one another: constituting kinship as intentionality in Fiji", JRAI, Vol 5(2): 265-280.
URRY, JOHN ,
1995 Consuming Places. London: Routledge.
WALDREN, JACQUELINE,
1996 Insiders and Outsiders: Paradise and Reality in Mallorca, Oxford: Berghahn Books.

6

INNOVATION, CONTINUITY AND INSULAR DEVELOPMENT IN THE ISLE OF MAN

P. J. DAVEY (1) , **J. J. INNES** (2)

Centre for Manx Studies,
(1) School of Archaeology, Classics and Oriental Studies, Unversity of Liverpool, England
EMAIL: pjd1@liv.ac.uk
(2) Department of Geography, Queen Mary and Westerfield College, University of London
EMAIL: J.B.Innes@qmw.ac.uk

BACKGROUND

The Isle of Man lies in the centre of the northern Irish Sea, some 60km from the coasts of both north-east Ireland and north-west England, 70km from Anglesey and just 20km from Burrow Head in Galloway, south-west Scotland. The Island is just over 53km long, 21km wide, has an area of 588 sq km and comprises two upland areas that reach 621m at Snaefell separated by a low-lying central valley. There are further lowlands, based on Carboniferous limestone, in the southeast and an extensive northern plain composed of a range of glacial and post-glacial drift deposits (FIGURE 1).

That the Isle of Man was ice-free by about 16,000 years ago is attested by a radiocarbon date of 15150±350BP (Birm-754; Chiverrell et al 1999, 333) that was obtained from a kettle hole on the Jurby ridge in the northwest of the island. Opinion differs over the existence and dating of land bridges within the Irish Sea basin. One recent study has argued that the land-link with Ireland was broken by 9,500BP and that by around 9,000BP the link with Britain had been severed (Wingfield 1995, 233), but Lambeck (1996) presents models which indicate no connection with Ireland and the severance from England by 10,500BP.

Given its position almost equidistant from two much larger islands, and its visibility from different parts of the western coasts of Britain and eastern Ireland, it is hardly likely that Man would have escaped the attentions of the first Mesolithic hunter-gather arrivals into the region and have subsequently been open to cultural innovation from its near neighbours throughout later times. At the same time the Island is sufficiently distant from the other islands that, under certain conditions, indigenous developments might be expected to occur.

The purpose of this paper will be to review recent research into the prehistory, geomorphology and environmental history of the Isle of Man, especially its northern plain, and to focus on two related themes: the development, or otherwise, of insular cultural traits at different times in prehistory and the absolute chronology of its natural and human populations, particularly in relation to the neighbouring 'mainlands'.

RECENT RESEARCH

Since 1995 the Centre for Manx Studies has been carrying out a series of research programmes aimed at the production of a New History of the Isle of Man, in five volumes, to replace the last scholarly analysis published in 1900 by A W Moore. Specific research commissioned for `Volume Two – Prehistory' has involved some two dozen individuals at doctoral (PhD) and post-doctoral (D) level from a range of institutions in Britain and Ireland.

There are four main threads to the research: geomorphological, geoarchaeological, palaeo-environmental (including sea-level change) and archaeological.

The Geomorphological research has included new mapping and interpretation of the drift geology in the southern lowlands (D) and of the northern uplands (D) with special reference to Holocene features. Specific detailed study has been carried out of Holocene terrace formation in the Peel Embayment (D) and sub-surface radar study of the fossil storm beaches of the Ayres (D; Chiverrell et al forthcoming). The new information has contributed to the greatly enhanced drift coverage of the new 1:50,000 geological map of the Island (British Geological Survey 2001; Burnett and Quirk 2001).

Plain (D), together with palaeo-botanical studies of peat deposits in the Ballaugh Curraghs (D; Davey et al 2000), the northern uplands (D. Chiverrell et al 2001) and the central valley (D; Innes 1995) have allowed the vegetation history of the Island to be placed firmly into its British Isles context

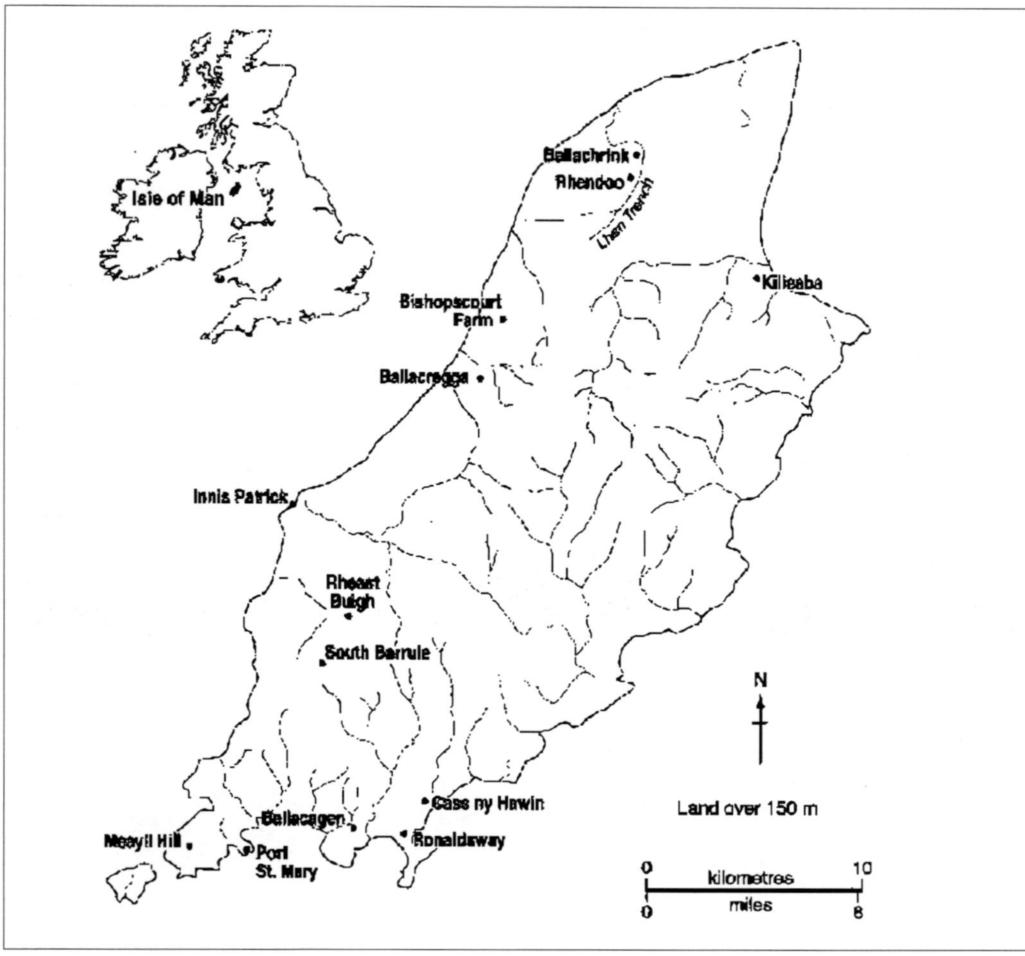

FIGURE 1 THE TOPOGRAPHY OF THE ISLE OF MAN

position in the Irish Sea and the location of the main sites referred to in the text.

(Chiverrell et al forthcoming).

The Geoarchaeological work has largely focused on the extreme north of the Island. Radiocarbon dated terrestrial and marine sediments, combined with good, discrete archaeological evidence has provided a convincing explanation of sea-level change at Port Cranstal, Bride (D), for the absolute chronology of the Ayres raised beaches (D) and for the sequence of human occupation of the area (Gonzales et al 2000). In addition, a study of the terrace systems in the northern uplands has revealed probable human impact on landscape evolution (D; Chiverrell et al 2001, 233-4).

In addition to new dating and ecological study of the great elk (Megaloceros megaloceros) that indicate its survival on Man into post-glacial times (D; Gonzales et al 2000), the main palaeo-environmental research has concentrated on providing a geographically comprehensive vegetation history. In particular, palaeo-ecological assessment of the kettle-holes in both the Peel Embayment (D) and the Northern

The Archaeological research has included a comprehensive study of Manx lithic industries (PhD; McCartan 1990, 1999), a new assessment of Neolithic pottery from the Island in relation to the Irish Sea area (PhD; Burrow 1997, 1999), a major Neolithic and Bronze Age excavation project at Billown, Malew (D; Darvill 1999), an account of the Manx Bronze Age and its Irish Sea relations (PhD; Woodcock 2001), a re-assessment of the Manx Iron Age and the problem of the Celts (D; Davey forthcoming), together with survey and excavation of sites in the Peel Embayment (D) and excavations of existing sites in order to obtain dating and environmental evidence (D; eg Woodcock; Johnson and Woodcock; McCartan and Woodcock; Woodcock and Tomlinson forthcoming).

The following short paper will highlight three recent projects that have changed perceptions of Manx prehistory.

The first deals with vegetation history and the evidence for human impact at the start of the Neolithic period. The second explores the relationship between the pollen evidence for early agriculture at Ballachrink and a later Mesolithic site at Rhendhoo in the Lhen Trench. The third looks at the emergence of a new type of middle Neolithic site that has been identified in the southern uplands of the Island. The paper will conclude with a brief re-assessment of period terminology applied to the Island's prehistory.

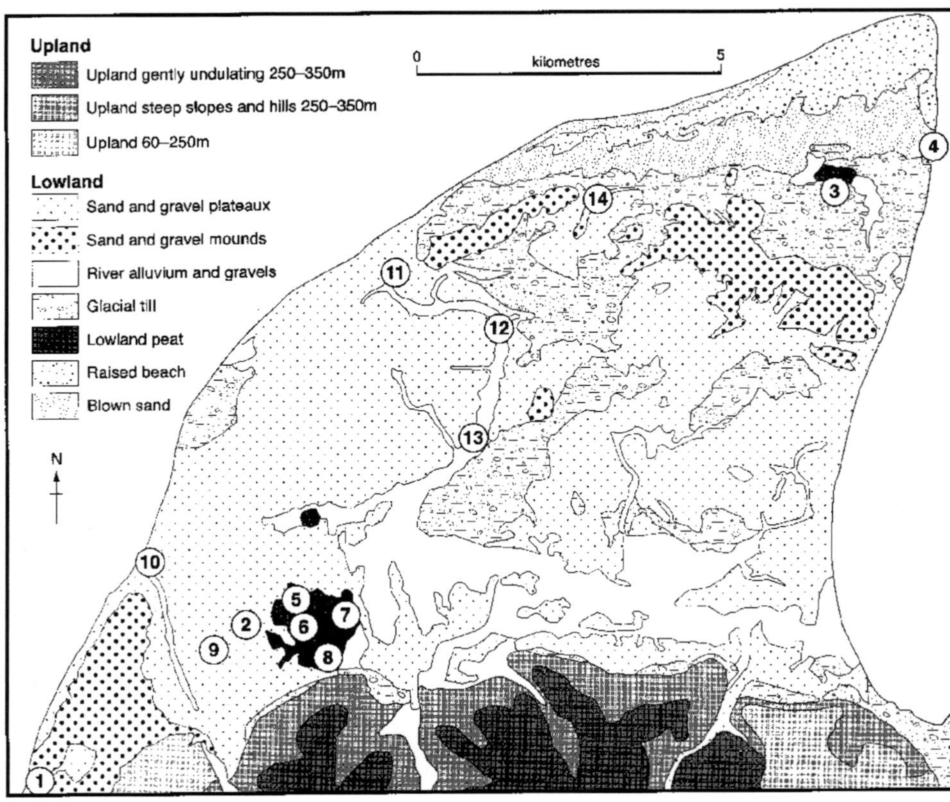

FIGURE 2 THE LOCATION OF RECENT POLLEN CORES FROM THE NORTH OF THE ISLE OF MAN.

VEGETATION HISTORY AND EVIDENCE FOR HUMAN IMPACT

Using the over-lapping chronologies of some 11 peat cores, it is now possible to provide a realistic overview of vegetation history in the lowlands of the north of the Isle of Man (FIGURE 2). Although Godwin's (1995) pollen Zones I to III are represented by a single core from one of the Kirk Michael kettle-holes, the remaining post-glacial phases from IV to VIIb are covered by a number of parallel cores.

In general the major arboreal taxa present and the dates of their rise and decline closely parallel other areas in highland Britain and Ireland. The decline of juniper appears to be late in the Ballaugh Curragh, as does alder at Phurt. Whilst some of these variations may be due to local, topographical factors, the identification of the elm-decline at the Cronk and in the Dhoo Valley at very similar dates, at 5,310BP (4,332 to 3,973 cal BC; AA 28397) and 5,313BP (4,245 to 3,999 cal BC; Chiverrell et al 1999, 331) respectively, does suggest that the general picture of natural vegetation history on the Island is normal for the region.

Virtually no part of the Island's vegetation has escaped the effects of human activity. The uplands have developed heather moorland, blanket bog and grassland maintained by sheep grazing and the lowlands are intensively farmed. On the cliffs just south of Lag ny Killey, in the Santon Gorge, at Traie ny Halsall, and at a few other locations, small groups of stunted oak trees can be found that may represent relict fragments of a 'primeval' forest cover.

HUMAN IMPACT ON THE VEGETATION

At two locations and periods, in the Lhen Trench and at Quarry Bends on the southern edge of the Ballaugh Curraghs, there is evidence for forest disturbance in the lowlands, conventionally ascribed to the activities of Mesolithic hunter-gatherers. After the elm decline such clearance episodes become more frequent and by the Bronze Age are ubiquitous. Pollen evidence from the upland locations at Mullagh Ouyr and Beinn-y-Phott also emphasizes the intensity of Bronze Age impact on the Manx environment (FIGURE 3).

THE EARLIEST AGRICULTURALISTS IN THE LHEN TRENCH

In 1998 Dr Jim Innes undertook a detailed palaeo-environmental survey of the Lhen Trench. Towards the centre of the Trench at Ballachrink the depositional sequence had been severely truncated by peat cutting and drainage. Despite this, sufficient peat survived, with good pollen preservation, to allow the production of a pollen diagram spanning the period c8,000BP to <6,000BP (FIGURE 4). Towards the top of the diagram cereal pollen, together with other clearance indicators was dated to 5,925±60BP [4,940 to 4,720 2 sigma Cal BC], a date which lies towards the earlier end of the dates from nearby Rhendhoo. At the time of writing, this date is as early as any acceptable for cereal pollen from anywhere in the British Isles. It forms parts of a cluster of early dates in the northern Irish Sea area that compare with the primary evidence from the North European Plain (Innes J B, Blackford J J, and Davey P J in press). At Billown in the south of the Island cereal grains have been identified in a number of contexts that predate the Neolithic enclosures. Two of these have been dated to 5684±40BP and 5780±40BP respectively (Darvill and Fairbairn1999, 14-16).

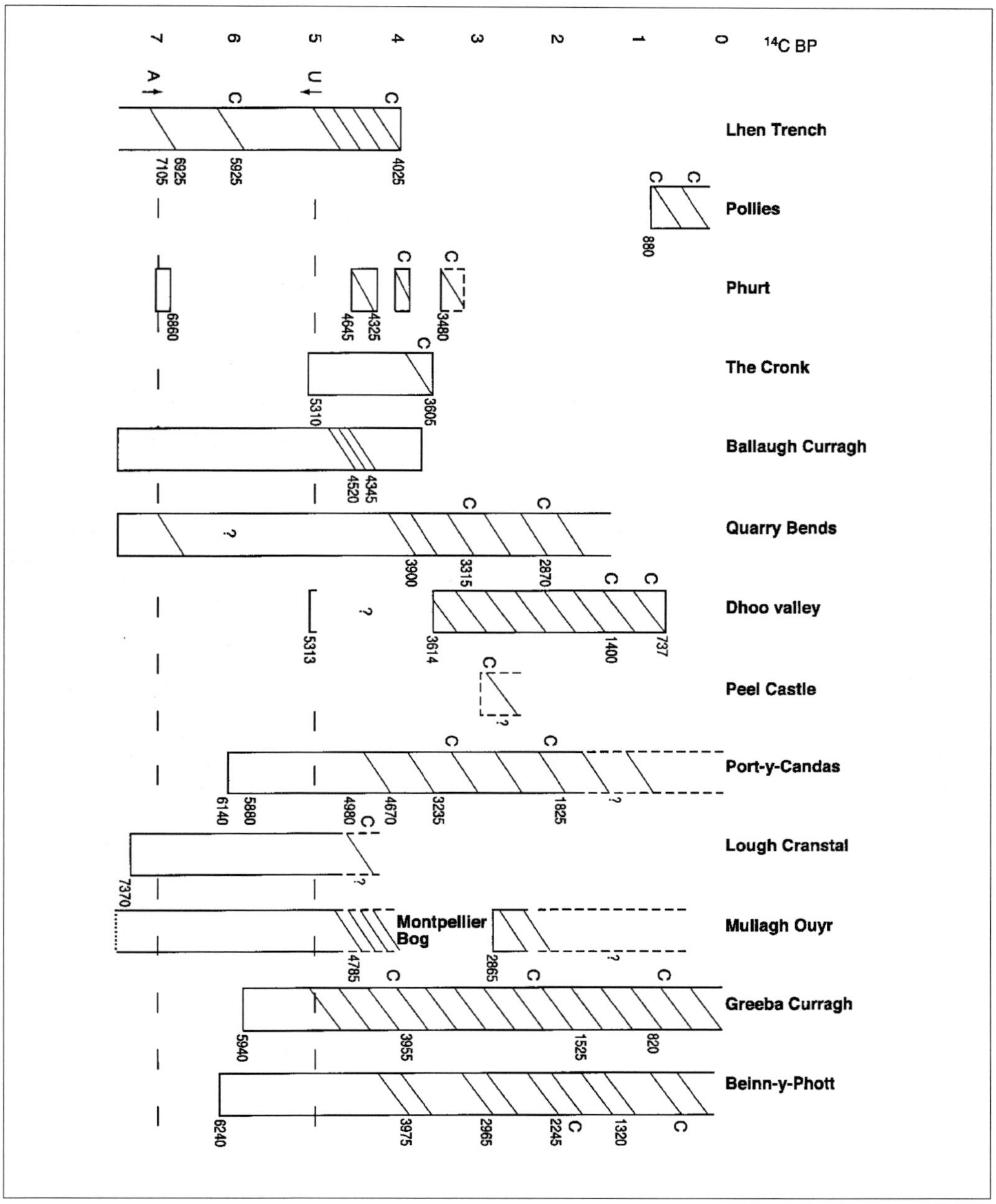

FIGURE 3 THE POLLEN EVIDENCE FOR HUMAN IMPACT
on the natural vegetation in the north of the Isle of Man [U_ = elm decline; A_ = alder rise; C = cereal pollen]

THE MANX MESOLITHIC AND EARLY NEOLITHIC

The Island is rich in Mesolithic sites, conventionally an 'earlier' microlithic industry and a 'later' heavy-bladed one (Woodman 1978). There is no sign of anything which could be assigned, on typological grounds, to the early Neolithic. A distinctive middle Neolithic ceramic type –

Meayll Hill ware (Piggott 1932), a round bottomed shouldered pottery equivalent to Lyles Hill ware (Sheridan 1995,6) - is associated with the earliest megalithic tombs, which are court cairns of Clyde-Carlingford Type. A reassessment of the earlier excavations in the Lhen Trench at both Rhendhoo ('heavy-bladed' Mesolithic) and of the stratigraphic position of the multi-period site at Ballachrink together with the environmental evidence is allowing a new synthesis to be proposed.

At Rhendhoo the site of a 'heavy-bladed' field-walking scatter was excavated in 1989 (cf FIGURE 5; McCartan 1994). The large butt-trimmed implements that are characteristic of this industry were named 'Bann Flakes' after the only other area in the British Isles, in northern Ireland, where they occur (cf Clarke 1935, 74-5). A major lithic assemblage was recovered, both from the plough soil and from stratified deposits that survived in places below. This excavation constitutes the Manx 'type site' for this industry. The report provides a detailed analysis of the technologies involved, including the refitting of a number of pebbles, and a discussion of raw material procurement strategies.

The four radiocarbon dates that were obtained, though widely spaced, give a very broad indication of the date range for heavy-bladed technology: -

Lab No	Conventional date	2 sigma Cal	Context
BM-2694	6100±50BP	5340-4900BC	Dump
BM-2693	5790±50BP	4780-4520BC	Dump
BM-2695	5170±50BP	4220-3820BC	Hearth
GU-2703	4770±230B	3940-2760BC	Testpit

Taken together with the new dating of the cereal pollen from nearby Ballachrink of 5925±60BP [4940-4720 cal BC] which lies at the earlier end of the Rhendhoo dates - the sites are only a few hundred metres apart - it would seem reasonable to infer that towards the end of the Mesolithic, heavy-blade using communities adopted agricultural practices, specifically cereal production, but continued to maintain a 'traditional' lithic technology for some time afterwards.

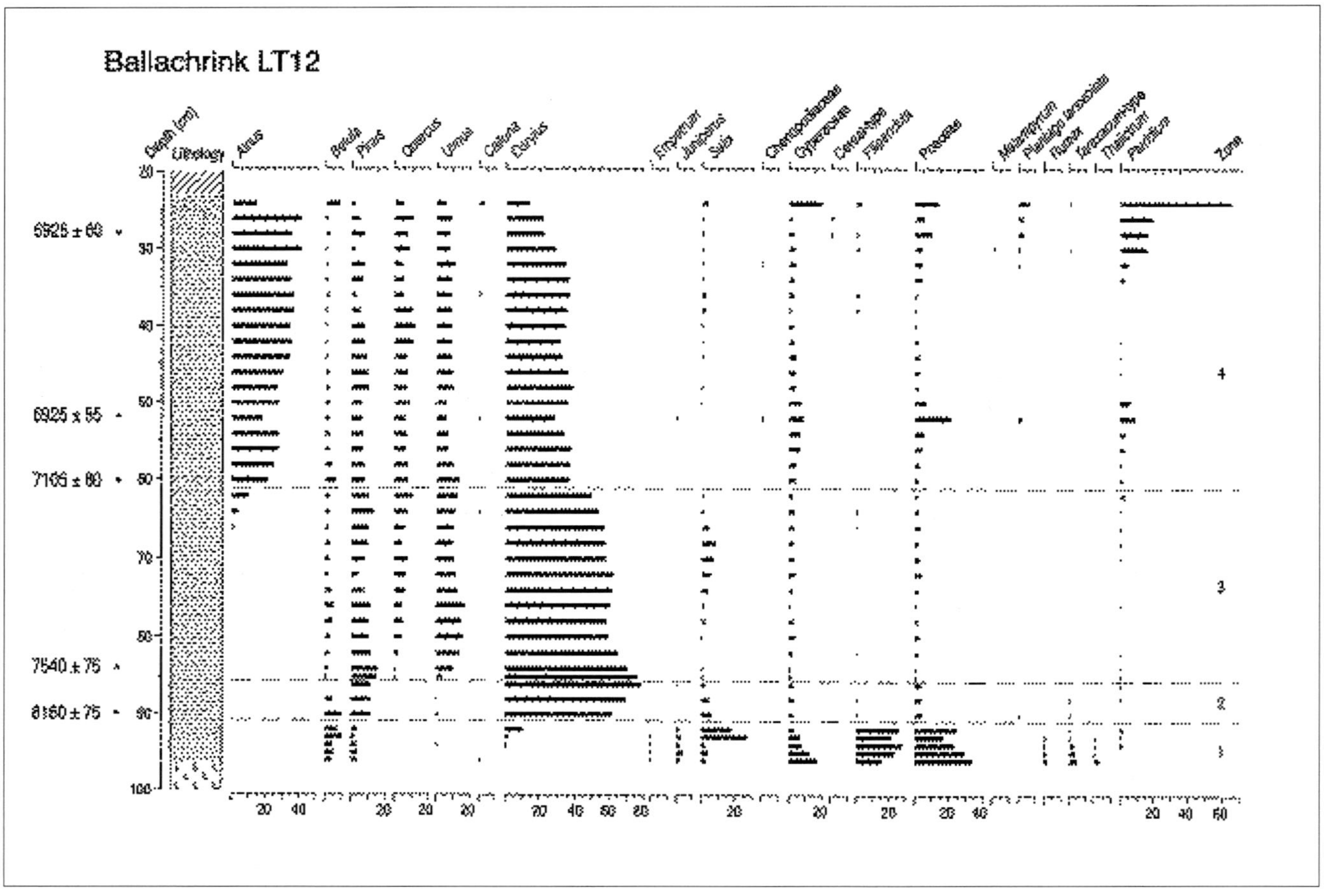

FIGURE 4 POLLEN DIAGRAM FROM BALLACHRINK, ANDREAS, ISLE OF MAN

SYNTHESIS?

Recent archaeological and palynological research has allowed a detailed picture of the evolution of the Manx environment and human impact upon it to be drawn. A number of further questions can now be posed: -

1. Does the abrupt typological break between the microlithic industries of the earlier Mesolithic and their heavy-bladed successors represent the change from hunter gathering to farming economies on the Island, or did both lifestyles coexist for a period?

2. Does the early cereal evidence from Ballachrink imply adaptation of an existing human population, or some kind of migration?
3. Whatever the answers to 1 and 2 above, why are the earliest cereal pollen dates from the British Isles concentrated in the north and west? Are they artefacts of past research distribution?
4. How can the apparently abrupt technological, behavioural and typological changes involved in the transition from the heavy-bladed lithic users to the tomb-building, ceramic and polished stone axe using societies of the middle Neolithic be explained?
5. What is the relationship between the earliest agricultural communities in the northern Irish Sea area and those in continental Europe that might explain the chronology that now seems to be emerging?

A similar range of questions can be posed for the earlier Mesolithic, the later Neolithic (The 'Ronaldsway Culture') and the Iron Age.

QUARTZ MOUNDS
During the period 1986 to 1997, M.B. Pitts a local amateur fieldworker identified a new class of site on the uplands in the south of the Island. These presented themselves as low mounds, only two or three metres across at most which, when ploughed, produced large quantities of freshly broken quartz associated with prehistoric lithic waste. Pitts surveyed a number of sites and recorded their disturbance, often when upland pasture was being created from heather moorland. A wide range of surface artifactual evidence was also collected. The results of his fieldwork were published in 1999 (Pitts 1999).

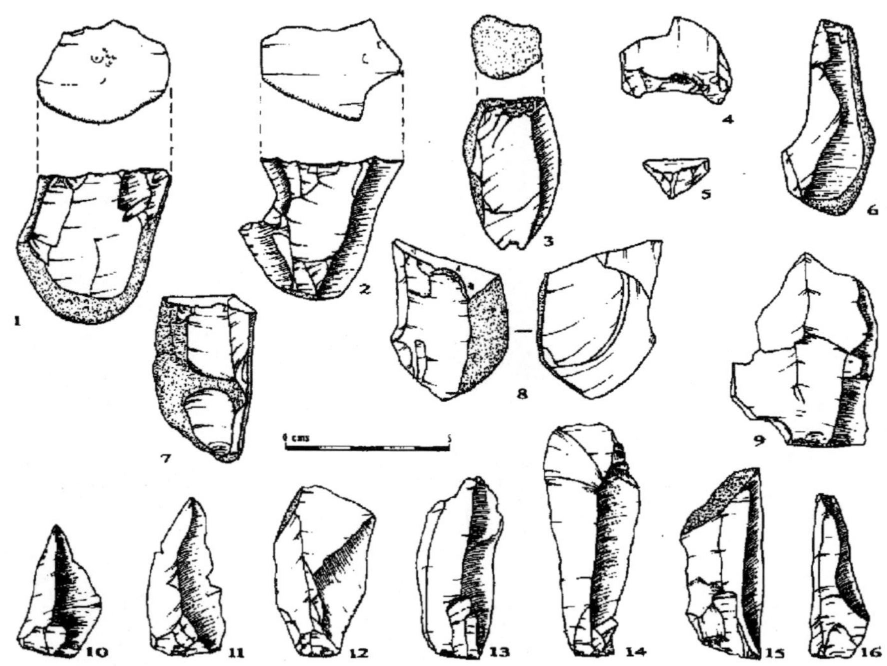

.FIGURE 5 PART OF THE LITHIC ASSEMBLAGE FROM RHENDHOO, JURBY, ISLE OF MAN.

EXCAVATIONS IN 1999 AND 2000
As part of an ongoing research programme directed towards the production of Volume 2 of A New History of the Isle of Man, Dr J J Woodcock of the Department of Archaeology, University of Liverpool carried out two short seasons of investigation of one of the mounds at Rheast Buigh which had escaped ploughing during recent 'improvement' by the Department of Agriculture (Davey and Woodcock in press; Woodcock forthcoming).

The mound proved to consist of freshly broken quartz with no evidence of carbon in or around the stone fragments or of in situ burning. Middle Neolithic pottery, of Manx 'Meayll Hill' type was recovered from within the cairn and from the old ground surface together with a range of flint tools and debitage. The mound partially overlay a post-built timber structure that is likely to have been rectilinear in form (FIGURE 6). Carbon dates from the old ground surface of 4640±100BP [3643-3041 cal BC] and from charcoal within the mound of 4695±55BP [3638-3358 cal BC] confirm a middle Neolithic date for the site. The possibility that some or all of the quartz had been burnt elsewhere and transported to the site is still under consideration.

This is the first evidence of human activity in this period from the Manx uplands and implies a much greater degree of impact at these altitudes than had previously been thought. It should be taken in context with pollen data from the northern uplands that show anthropogenic disturbance at 4785±55BP [3659-3378 cal BC] at Montpellier Bog and 3975±65BP [2828-2292 cal BC] at Beinn-y-Phott, both of which suggest some impact on the uplands during the Neolithic.

THE NATURE OF THE SITE

Initially it was thought that the mounds were likely to be the upland equivalent of burnt mounds which abound in the Manx lowlands and in the Irish Sea area generally (Garrad 1999). The majority of burnt mounds have proved to belong to the Bronze Age. The discovery that Rheast Buigh was of middle Neolithic date, that it did not represent the debris from burning and that it sealed a rectilinear timber structure was both surprising and intriguing. It remains to be seen whether other quartz mounds consist of similar structural elements or are of the same date.

It is quite unclear what this sequence of structures represents or what kind of activity was attracting its Neolithic inhabitants to the Manx uplands. So far no convincing parallels have been found elsewhere in the British Isles. What set of circumstances or forces prompted the evolution of this class of monument on Man, and not in the surrounding lands, in unclear. Suggestions and advice would be welcomed.

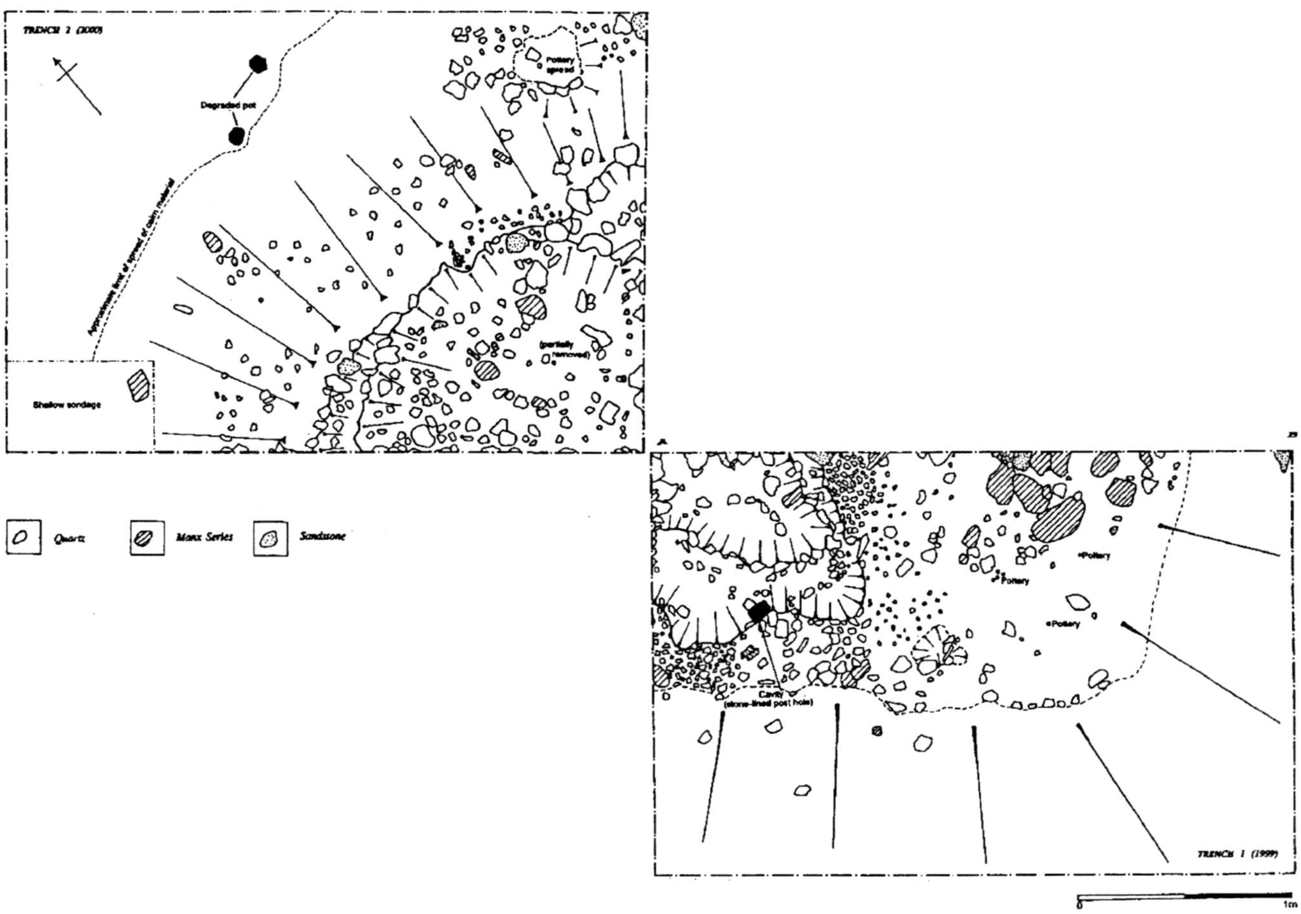

FIGURE 6 PLAN OF THE NEOLITHIC QUARTZ MOUND AT RHEAST BUIGH, GERMAN: TRENCHES 1 AND 2.

DISCUSSION

The two examples quoted above, that of Rhendhoo and Ballachrink in the Lhen Trench and of the quartz mounds in the southern uplands demonstrate the complex nature of Manx prehistory. Whilst the arrival of plants and people seems to parallel fairly closely the picture known from the surrounding lands, there is evidence for spasmodic local innovation. Cultural continuity between Mesolithic and Neolithic, with apparently more 'primitive' types persisting much longer than in Britain and Ireland, has to be assessed against the evidence for the earliest dating of cereal production in the British Isles. If the later Mesolithic occupies the period when early Neolithic cultural traits might be expected, why was arable agriculture adopted but none of its attendant attributes? What factors led both to the adoption of middle Neolithic cultural norms from overseas and to the insular creation of the quartz mounds? Such questions, in different forms, recur in the study of later Neolithic, Bronze Age and Iron Age societies in Man.

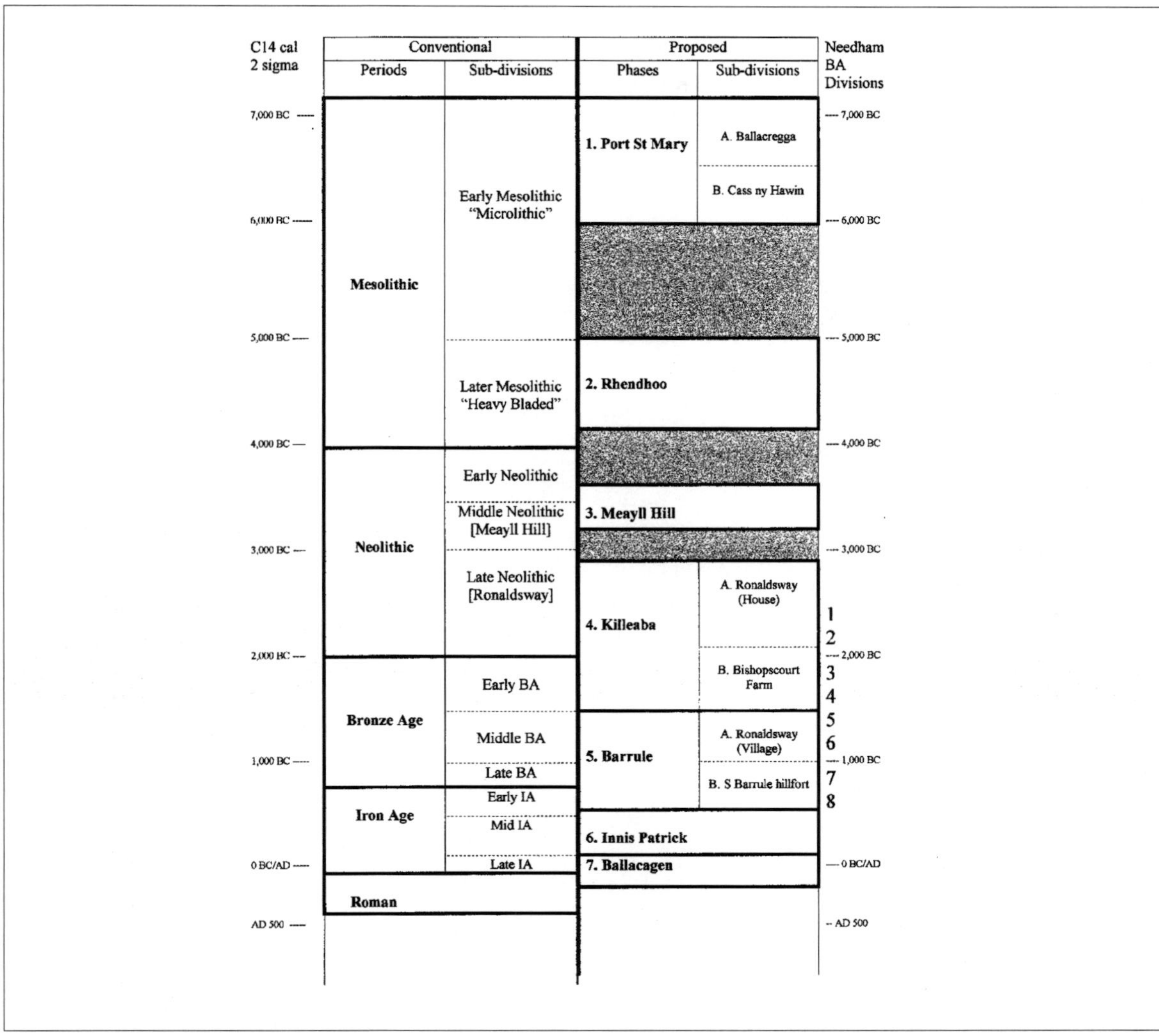

FIGURE 7 CONVENTIONAL AND PROPOSED PERIOD NAMES IN MANX PREHISTORY.

MANX PREHISTORY: TOWARDS A NEW CULTURAL SYNTHESIS

In view of the proximity of the adjacent islands - on a clear day it is possible to see neighbouring parts of Scotland, England, Wales and Ireland - Manx prehistory has normally been seen in terms of contemporary developments in those countries. Samuel Haining (1824, 91-94), for example, interpreted the megalithic tombs as Druidical temples, Swinnerton (1890, 1891) saw the Mesolithic sites at Glen Wyllin and Port St Mary as evidence of pre-Aryan settlement of the island. Some forty years later Clark, following continental European models, identified the same sites as Tardenoisian (Clark 1935, 71-74). More recent writers, whilst emphasizing the complexity of some of the insular evidence, and from time to time its individuality, have, nevertheless, sought to place the Manx post-glacial archaeological sequence firmly within a technological model framework that is accepted for the British Isles as a whole.

In the final part of this paper an attempt will be made to cast some doubt on the validity and appropriateness of this framework for the better understanding of Manx prehistory, and to propose a local terminology that better reflects the insular record.

THE CONVENTIONAL MODEL

Although there has been no scholarly attempt to write a new prehistory of the Island in the manner of Clark in 1935, individual specialists have conducted research and presented their results in terms of a four-age structure - Mesolithic, Neolithic, Bronze and Iron Age - with a range of dual or triple sub-divisions (Burrow 1997, McCartan 1999, Woodcock 2001). This conventional, 'technological model', chronological structure for post-glacial prehistory, as applied to the Isle of Man, is presented as the left-hand column of FIGURE 7 above.

A number of presuppositions lie behind this table. First, that the continuous horizontal lines represent major socio-economic and cultural change between blocks within which there is some degree of evolution and relatively little intrusive influence. Secondly, that the dotted horizontal lines divide each of the major 'culture-blocks' into two or three subdivisions, based on the apparent phasing of such evolution. Thirdly, it is assumed that the Manx sequence will normally follow the same general pattern as the rest of Britain and Ireland, any differences being minor and regarded as aberrations.

A brief assessment of the Manx evidence suggests that all three of these assumptions are mistaken. First, the continuous horizontal line between the late Neolithic and the Bronze Age cannot really be sustained. Detailed studies, from both sides of this line (Burrow and Darvill 1997 and Woodcock 2001) show that between c3000 and 1500 cal BC the Manx record exhibits a number of parallel continuums, in terms of pottery, lithics, burial rites and metallurgy. Similarly, with as yet, far less material evidence and fewer radiocarbon dates, the late Bronze Age to Iron Age transition appears extremely fluid. At the end of the Mesolithic, where the continuous line does appear to be justified, the distinction is between the heavy-bladed flint industry of 'Bann' type and nothing. To date no cultural material has been found that could reasonably be assigned to the early Neolithic, as known elsewhere in the British Isles.

Secondly, a number of the 'internal' period sub-divisions, indicated by dotted lines on the diagram, are much more abrupt and separate groups of material with far greater differences, than those visible between the Bronze Age and elements at either end beyond the continuous lines. In the case of the Mesolithic, for example, although the microlithic industries of the earliest phases do exhibit some signs of insular development, the transition to the use of heavy-bladed tools, involving a radical change in technology and, apparently, natural resource exploitation too, occurs with no sign of intermediate stages. Similarly, the transition from the middle to the later Neolithic is very much more sharply defined than that between the Neolithic and the early Bronze Age.

MANX PREHISTORY: A NEW NAMING PROPOSAL

If the Manx evidence is allowed to stand on its own, using the available radiocarbon dates and setting to one side the conventional British Isles sequence, a somewhat different structure may be suggested. Using existing type-site terminology, where appropriate, together with the most recent excavation evidence for the definition of new type phases, the right-hand column of FIGURE 7 also uses continuous horizontal lines to separate assemblages that are quite distinct from one another and dotted lines to indicate some degree on continuous evolution. Manx prehistory before AD 500 may now be divided into seven major phases with appropriate sub-divisions. The phases are described in summary below:-

PHASE 1 PORT ST MARY (c7200-6000 cal BC)[Swinnerton 1890]

Over 50 microlithic sites are known (McCartan 1999, 7). The earliest material, according to Woodman (1978, 126) is represented by Port St Mary and the latest by Swinnerton's finds from Glen Wyllin/Ballacregga (Swinnerton 1891; Woodman 1978, 127). More recent excavations by Woodman (1987) at Cass ny Hawin and a new site in Glen Wyllin/Ballacregga (McCartan in press), together with radiocarbon dates, allows a tentative two-fold division to be proposed; an earlier Phase A, typified by Ballacregga (McCartan in press) for which early 7th millennium or even late 8th millennium radiocarbon dates have been obtained, and a later Phase B, at Cass ny Hawin. Apart from some evidence for a shell midden at Port St Mary and hazel nuts from the recent Ballacregga excavations, the economic basis of the human populations represented by these microlithic industries is difficult to determine.

PHASE 2 RHENDHOO
(c5000 - 4000 cal BC) [McCartan 1994]

Over 60 major flint assemblages are known of Rhendhoo type (cf above p00; McCartan 1990, 523-31, Fig 3). They are more widely distributed than the Port St Mary-type finds and represent a fundamental change in lithic technology. The new pollen-analytical results from the north of the Island indicate forest opening phases of this period at a number of sites. Cereal pollen from Ballachrink (cf above p00) suggests that some of these openings may represent clearance for the earliest arable farming (Davey and Innes in press; Innes, Blackford and Davey in press). In conventional terms it seems likely that this phase spans the latest Mesolithic and earliest Neolithic.

PHASE 3 MEAYLL HILL
(c3300 - 3600 cal BC) [Herdman and Kermode 1894; Piggott 1932]

The Island boasts up to ten megalithic tombs, at least half of which are court cairns of Clyde-Carlingford type (Henshall 1978). The material culture represented on the sites, including local forms of western Neolithic pottery and developed lithics represents a major change from Rhendhoo. Although the megalith-building phase may be relatively short, it was long enough for the islanders to develop distinctive ideas of their own, especially as represented by the Meayll Circle. Piggott's remarks of 1932 are still apposite today:-

> The Mull Hill Circle is at present unique, and unless further examples of its type are recognized we must consider it as the tour de force of an insular community which, although not completely cut off from the main stream of events, was sufficiently isolated to develop its culture in certain directions along its own line, and so produce a structure which is in so many of its features perhaps even more puzzling than the average of those enigmatical structures, the megaliths (Piggott 1932, 157).

PHASE 4 KILLEABA

(c2900 - 1400 cal BC) [Cubbon 1978]

Since the definition of the Ronaldsway culture in 1946 (Bersu 1947; Bruce, Megaw and Megaw 1947) this phase of Manx prehistory has been seen as a distinctive, insular, manifestation of the latest Neolithic [Killeaba A], to be followed by a fairly conventional, 'intrusive' Early Bronze Age, best represented by the decorated bowl food-vessels of Bishopscourt Farm type (Woodcock 1999a, 105-6) and early copper and bronze flat axes of Irish type (Davey et al, 1999, 43-5) [Killeaba B]. The excavation of the cemetery at Killeaba, near Ramsey, by Marshall Cubbon at which 'Bronze Age' interments were set carefully between typical 'Ronaldsway' style burials and artifacts has generally been regarded as the product of, at best, a bizarre set of accidents. On the other hand, given the continuity of metal-working types from c2500 cal BC onwards, the overlaps between the Ronaldsway pottery dates, beakers and Irish Food Vessels (cf Woodcock 2001, Fig 6.41; Burrow and Darvill 1997), it seems reasonable to see the continuity apparent at Killeaba as the norm for this phase of human settlement on Man rather than as an abnormal deviation.

PHASE 5 BARRULE

(c1400 - 500 cal BC) [Gelling 1972]

Gelling excavated the hill-fort at South Barrule over a number of seasons and found it to consist of two major structural phases, the second associated with an outer rampart and stone chevaux de frise. Pottery from the fort was of Ronaldsway Village type dated by radiocarbon to the period 1500 to 1000 cal BC (Woodcock 1999b, 133) and suggests that the earliest occupation dates from this period [Barrule A]. A sample from the latest hearth in one of the buildings in the interior of the fort rendered a radiocarbon date of 2473±74BP [806-306 cal BC] implies that settlement continued well into the first millennium [Barrule B]. It is unclear whether the other Manx hillfort at Cronk Sumark, with its vitrified outer rampart, or any of the 21 promontory forts belong to this phase of occupation. The problem of the relationship between this and subsequent phases was well expressed by Clark:-

Evidence for the settlement of the Island either by the Iron Age invaders of Britain or by the Romans is conspicuously lacking. The survival of the Late Bronze Age people is, therefore, to be expected on a priori grounds (Clark 1935, 85).

PHASE 6 INNIS PATRICK

(c500 - 50 cal BC) [Freke in press]

Few Iron Age settlement sites have been excavated on Man, the most extensive lies on Innis Patrick beneath the medieval ecclesiastical and defensive structures of Peel Castle. A number of buildings and parts of buildings were excavated and a range of environmental evidence obtained (Freke 1990; Freke in press, 45-57). Recent survey and excavation on Slieu Curn by Dr J J Woodcock (Davey (ed) forthcoming) has identified an upland open settlement, broadly contemporary with the prehistoric structures at Peel.

PHASE 7 BALLACAGEN

(c50 cal BC - cal AD200) [Bersu 1977]

Bersu identified a group of large round houses, located mainly in the south of the Island and excavated three of them. They turned out to be multi-phased, undefended, complex structures with clear evidence of contact with Roman Britain during the first and second centuries AD. Carbon dates from two timbers of the primary phases of building at Ballacagen suggest a rather earlier dating, 2155±85BP [393 cal BC to cal AD76] and 2235±45BP [393-175 cal BC] (Chiverrell et al 1999, 327), though it is possible that this discrepancy is due to the age of the large structural timbers that were used for dating the site. At least one of the promontory forts, that at Close ny Chollagh, was also occupied during this period (Gelling 1958). It is possible that the development of these new house types, suggesting a significant increase in wealth and political stability, was a direct response to the arrival of the Romans in the northern Irish Sea.

CONTINUITY AND DISCONTINUITY IN MANX PREHISTORY

It remains to consider possible reasons for the number of abrupt changes in the Manx prehistoric record and for the gaps in time that seem likely to have occurred between them. The discussion is hampered by two further presuppositions, shibboleths of contemporary thinking. First, there is an a priori resistance to the interpretation of technological, stylistic or social change, as seen in the archaeological record, as the movement of peoples. Thus, there is an unstated assumption that once first colonized, the Island experienced a continuous human presence up to present. The general discrediting of the 'invasion hypothesis' in British prehistory has rendered any such possibility difficult to articulate. The obverse presupposition that cultural continuities are the norm has tended to see apparent breaks in the record as a lack of evidence rather than as evidence of possible discontinuities of occupation.

In the case of the Isle of Man, with the possible exception of the Iron Age, the evidence on the ground is prolific and has increased exponentially in the last decade, but the gaps are still there and show no signs of being filled. It might be argued that there are insufficient carbon dates for the proposed new structure to be sustained. In reply it should be noted that even should future absolute dating demonstrate a far greater degree of overlap than is in evidence at present, with the discrimination available it would not be possible to identify an occupation gap of even 100 years, let alone much shorter periods of human absence from the Island. The many new radiocarbon dates, if anything, have only served to emphasize further the episodic nature of much of earlier Manx prehistory. Against this, the evidence of copper and bronze metallurgy suggests that, from Needham's metal using Period 1 (2500-2300 cal BC) onwards the Island was in receipt of most of the common metalwork types current in the northern British Isles until his Period 7 (950-750 cal BC) (Davey et al 1999, 55). Is it possible that the Manx record is best interpreted as a combination of a number of periods of discontinuity, especially in the early part of the Holocene, and quite long periods of continuous settlement with varying degrees of internal evolution and external influence? Thus the Port St Mary, Rhendhoo and Meayll Hill complexes may best be understood as distinct and separated settlement phases, while the continuous occupation subsequent on the 'arrival' or 'development' of Killeaba A (Ronaldsway House) shows varying degrees of external influence - all of the Bronze Age pottery and metalwork, for example, can be paralleled in Britain and Ireland. The Manx hillforts and promontory forts share many characteristics with those in adjacent areas of the neighbouring islands. On the basis of their material culture, even the 'Bersu' roundhouses were constructed by a society in close contact with Roman Britain.

In many ways the Manx experience is the opposite of that described for the Balearics, where locally derived terminologies for describing insular prehistory are now perceived as creating serious problems for their integration into the wider social and economic history of the western Mediterranean (Walden this volume?). In contrast, on Man, prehistoric research needs to be freed from the straight-jacket of an alien technological model that has tended to disguise and distort appreciation of the unique character of its earliest inhabitants.

ACKNOWLEDGEMENTS

The writers are grateful to Sinéad McCartan of the Ulster Museum and Jenny Woodcock and Philippa Tomlinson of the Centre for Manx Studies, University of Liverpool for reading a first draft of this paper and making a number of helpful suggestions. They would also like to acknowledge and thank Ed Oliver for preparing FIGUREs 1 to 5 and Philippa Tomlinson for producing FIGURE 7.

REFERENCES

BERSU, G.,
1977, Three Iron Age round houses in the Isle of Man, Douglas, The Manx Museum and National Trust.
BURNETT, D. J, AND QUIRK, D. G.,
2001, Manx Geology - a guide to the rocks of the Isle of Man, Douglas, Manx Geological Survey.
BURROW, S.,
1997, The Neolithic culture of the Isle of Man,
Oxford, BAR British Series 263.
BURROW, S.,
1999a, 'The Ronaldsway pottery of the Isle of Man: a study of pro
BURROW, S.,
1999B. 'Neither east nor west: a social history of the Manx Neolithic', in: P J Davey (ed) Recent Archaeological Research on the Isle of Man, 27-38, Oxford, BAR British Series 278.
BURROW, S. AND D. ARVILL ,
1997, 'AMS dating of the Manx Ronaldsway Neolithic', Antiquity, 71, No 271, 412-9.
CHIVERRELL, R. C., DAVEY, P. J., GOWLETT, J. A.. J AND WOODCOCK, J. J.,
1999. Radiocarbon dates for the Isle of Man', in: P J Davey (ed) Recent Archaeological Research on the Isle of Man, 321-336, Oxford, BAR British Series 278.
CHIVERRELL, R.. C., THOMAS, G. S. P. AND HARVEY, A.. M. ,
2001. .Late Devensian and Holocene landscape change in the uplands of the Isle of Man, Geomorphology, 40, 219-236.
CHIVERELL, R. C., BLACKFORD, J. J.,, INNES J .B., ROBERTS,. D AND TOMLINSON, P. R.,
forthcoming. 'The Holocene' , Chapter 6 in: R C Chiverrell and G S P Thomas (eds) A New History of the Isle of Man, Volume 1, Liverpool, Liverpool University Press.
CLARK, G.,
1935. 'The prehistory of the Isle of Man', Proceedings of the Prehistoric Society, 1, 70-92.
CUBBON, A. M..,
1978. Excavation at Killeaba, Ramsey, Isle of Man', Proceedings of the Prehistoric Society, 44, 69-95.
DARVILL, T.,
1999. Billown Neolithic landscape project 1995-1997' in: P J Davey (ed) Recent Archaeological Research on the Isle of Man, 13-26, Oxford, BAR British Series 278.
DAVEY, P. J.,
forthcoming. 'The Iron Age', Chapter 5 in: P J Davey (ed) A New History of the Isle of Man, Volume 2, Liverpool, Liverpool University Press.
DAVEY, P. J., BLACKFORD, J. J., CHIVERRELL R, INNES, J. B. AND TOMLINSON, P. R.,
1999. 'Bronze Age metallurgy on the Isle of Man: a symposium', in: P J Davey (ed) Recent Archaeological Research on the Isle of Man, 39-62, Oxford, BAR British Series 278.
DAVEY, P. J., BLACKFORD, J. J., CHIVERRELL R, INNES, J. B. AND TOMLINSON, P. R.,
2000. 'The curragh wetlands of the Isle of Man' in: M Atherden (ed) Wetlands in the landscape: archaeology, conservation, heritage, 12-40, York, PLACE Research Centre.
DAVEY, P. J. AND INNES, J. B.,
in press, 'The early Neolithic and the Manx environment' in: Armit I, Murphy E, Nelis E and Simpson D (eds) Neolithic settlement in Ireland and Western Britain, Belfast, School of Archaeology and Palaeoecology, Queen's University and the Prehistoric Society.
DAVEY, P. J. AND WOODCOCK, J. J.
in press, 'Rheast Buigh, Patrick: middle Neolithic exploitation of the Manx uplands?' in: Armit I, Murphy E, Nelis E and Simpson D (eds) Neolithic settlement in Ireland and Western Britain, Belfast, School of Archaeology and Palaeoecology, Queen's University and the Prehistoric Society.
DAVEY P J, CHIVERRELL R, JOHNSON A A C, JOHNSON, N. C., TOMLINSON P R AND WOODCOCK, J. J., forthcoming 'The landscape of Upper Glendhoo, Ballaugh, Isle of Man', in: P J Davey (ed) Recent Archaeological Research on the Isle of Man, 1998 to 2000, Oxford, BAR British Series.
FREKE, D.,
1990. 'The prehistoric environmental evidence from the 1987 Peel castle excavations', Proceedings of the Isle of Man Natural History and Antiquarian Society, 1X, No 4, 491-500.
FREKE, D.,
in press, 'Excavations on St Patrick's Isle Peel, Isle of Man, 1982-88 prehistoric, Viking, medieval and later, Liverpool, Liverpool University Press.
GARRAD, L. S.

1999. `Field walkers' records of burnt mounds in the Isle of Man' in: P J Davey (ed) Recent archaeological research on the Isle of Man, 75-80, Oxford, BAR British Series 278.
GELLING, P. S.,
1958. 'Close ny chollagh: an Iron Age fort at Scarlett, Isle of Man', Proceedings of the Prehistoric Society, 24, 85-100.
GELLING, P. S.,
1972. 'The hill-fort on South Barrule and its position in the Manx Iron Age', in: F Lynch and C Burgess (eds) Prehistoric man in Wales and the West (essays in honour of Lily F Chitty), 285-292, Bath, Adams and Dart.
GELLING, P. S .,
1978. 'The Iron Age', in: P J Davey (ed) Man and environment in the Isle of Man, 233-243, Oxford, BAR British Series 54.
GONZALES, S., INNES J B, HUDDART. D., DAVEY, P. J. AND PLATER, A.,
2000. 'Holocene coastal change in the north of the Isle of Man: stratigraphy, palaeoenvironment and archaeological evidence', in: K Pye and J R L Allen (eds) Coastal and Estuarine environments: sedimentology, geomorphology and geoarchaeology, 343-363, London, The Geological Society, Special Publication No 175.
GONZALES, S., KITCHENER, A. C. AND LISTER, A. M.,
2000.'Survival of the Irish elk into the Holocene', Nature, 405, 753-754.
HAINING, S.,
1824. A historical sketch and descriptive view of the Isle of Man, London, Harris & Co.
HENSHALL, S.,
1978. 'Manx megaliths again: an attempt at structural analysis' in: P J Davey (ed) Man and environment in the Isle of Man, 171-176, Oxford, BAR British Series 54.
HERDMAN, W. A. AND KERMODE, P., M .C.,
1894. 'The excavations of the Neolithic stone circle near Port Erin, Isle of Man', Transactions of the Biological Society of Liverpool, 8, 1-14.
Hill, J. D.,
1995. 'The pre-Roman Iron Age in Britain and Ireland (ca. 800 B.C. to A.D.100): an overview', Journal of World Prehistory, 9, No 1, 47-98.
INNES, J. B.,
1995. The Dhoo Valley, Isle of Man: a palaeo-environmental assessment, Douglas, Centre for Manx Studies, Research Report 2.
INNES, J. B., BLACKFORD, J. .J, AND DAVEY, P. J.,
in press. 'Manx palaeoecological evidence for earliest cereal cultivation in the British Isles', J Quaternary Science.
JOHNSON, A. A. C. AND WOODCOCK, J. J,
forthcoming, 'Further excavations at Ballachrink, Andreas, 1999' in: P J Davey (ed) Recent Archaeological Research on the Isle of Man, 1998 to 2000, Oxford, BAR British Series.
MCCARTAN, S. B.,
1990. 'The early prehistoric colonisation of the Isle of Man: Mesolithic hunter-gathers', Proceedings of the Isle of Man Natural History and Antiquarian Society, 1X, No 4, 517-534.
MCCARTAN, S. B. AND JOHNSON, A. A. C.,
1992, 'A rescue excavation at Ballachrink, Jurby', Proceedings of the Isle of Man Natural History and Antiquarian Society, X, No 1, 105-122.
MCCARTAN, S. B.,
1994, 'A later Mesolithic site at Rhendhoo, Jurby, Isle of Man', Proceedings of the Isle of Man Natural History and Antiquarian Society, X, No 2, 87-117.
MCCARTAN, S. B.,
1999. 'The Manx early Mesolithic: a story in stone', in: P J Davey (ed) Recent Archaeological Research on the Isle of Man, 5-11, Oxford, BAR British Series 278.
MCCARTAN, S. B. AND WOODCOCK, J. J.,
forthcoming, 'Excavation of the early Mesolithic site at BalLA-CREGGA, GLEN WYLLIN,
2000', in: P J Davey (ed) Recent Archaeological Research on the Isle of Man, 1998 to 2000, Oxford, BAR British Series.
NEEDHAM, S.,
1996, 'Chronology and periodisation in the British Bronze Age', in: K Randsborg (ed) 'Absolute chronology: archaeological Europe 2500-500BC', Acta Archaeologica, 67, 121-140.
PIGGOTT, S.,
1932. 'The Mull Hill Circle, Isle of Man and its pottery', Antiquaries Journal, 12, 147-57.
PITTS, M. B.,
1999. `Quartz mounds: a preliminary assessment' in: P J Davey (ed), Recent Archaeological Research on the Isle of Man, 63-73, Oxford, BAR British Series 278.
SWINNERTON, F.,
1890, 'The early Neolithic cists and refuse heap at Port St Mary', Yn Lioar Manninagh, 1, No 5, 137-9.
SWINNERTON, F.,
1891, 'Pre-Aryan remains at Glen Wyllin, Isle of Man', Yn Lioar Manninagh, 1, No 9, 262-4.
WALDREN ,W.,
2001. 'Links in the Chain: evidence of sustained prehistoric contact and cultural interaction between the Balearic Islands andContinental Europe. this volume **
WALDREN ,W.,
2001. Evidence of ancient animal, water and land management, exploitation and depletion: a Case History, the Olezian Chalcolithic Old Settlement, Valldemossa, Mallorca.. this volume **
WINGFIELD, R. T. R.,
1995. 'A model of sea-levels in the Irish and Celtic seas during the end-Pleistocene to Holocene transition, in: R C Preece (ed) Island Britain: a Quaternary perspective, 209-242, London, The Geological Society.
WOODCOCK, J. J.,
1999A, 'The Bronze Age burials from Cottier's Field, Bishopscourt Farm, Ballaugh (SC 33289274)' in: P J Davey (ed) Recent Archaeological Research on the Isle of Man, 99-110, Oxford, BAR British Series 278.
WOODCOCK, J. J.,
1999B, 'A study of Bronze Age pottery from Ronaldsway Village, Malew' in: P J Davey (ed) Recent Archaeological Research on the Isle of Man, 121-137, Oxford, BAR British Series 278.
WOODCOCK, J. J.,
2001, In search of a cultural identity: a study of the Manx Bronze Age in its Irish Sea context, PhD, University of Liverpool.
WOODCOCK, J. J.,
forthcoming 'Excavation of a quartz mound at Rheast Buigh, Patrick, 1999 and 2000' in: P J Davey (ed) Recent Archaeological Research on the Isle of Man, 1998 to 2000, Oxford, BAR British Series.
WOODCOCK, J. J .AND TOMLINSON, P. R.,

forthcoming, 'Re-assessment of the archaeology and peat stratigraphy of the ditched platform at Port y Candas, German' in: P J Davey (ed) Recent Archaeological Research on the Isle of Man, 1998 to 2000, Oxford, BAR British Series.

WOODMAN, P. C.,

1978 'A re-appraisal of the Manx Mesolithic', in: P J Davey (ed) Man and environment in the Isle of Man, 119-139, Oxford, BAR British Series 54.

7

ISLAND ARCHAEOLOGY AND ISSUES OF POLITICAL CENTRALIZATION: MICRONESIAN EVIDENCE

WILLIAM S. AYRES

Department of Anthropology
University of Oregon
U.S.A.
EMAIL: wsayres@oregon.uoregon.adu

INTRODUCTION

This paper considers the archaeology of islands, ones in the tropical Pacific in particular, and addresses political change in small islands and on interrelated environmental, cultural and archaeological data on human adaptation. Ethnological studies of island peoples, such as Firth's in Tikopia and Mead's in Samoa in the Pacific, have long had a significant impact on the field of anthropology through comparative study of societal isolates. As well, archaeological and historical studies offer long-term perspectives on islands as cultural and environmental laboratories and provide new evidence with global implications (see, e.g., Kirch 1997). While insular land forms are diverse in geography and cultural traditions, focusing on them as a distinctive landscape form permits formulating an assessment of their special contribution to the study of human adaptation, ideological constructs, and culture change (Renfrew and Wagstaff 1982; Dewar 1997). Central among the issues are ones concerning: 1) islands as cultural and natural Alaboratories;@ 2) islands as contexts for modeling global environmental change; and 3) island settlement chaining, that is, population expansion through island chain colonization, as cultural and biogeographic phenomena; 4) the relationship of insularity and issues of scale to evolutionary processes; 5) island ecosystem transformation (especially as affected by technology, subsistence, and ideology); and 6) the significance of island coastal adaptations, that is, their fundamental aquatic-terrestrial component. I focus primarily on the fourth aspect here.

The means and patterns of colonization, item 3 above, is particularly significant for islands' study, often specifically linked to maritime transport issues, but also to general principles of human migration. The idea of beachhead colonization for islands is notable among five basic patterns for colonizing unpopulated regions (Moore 2001; Keast and Miller 1996). Island researchers have long been aware of and applied evidence of biogeographic processes, including the Sewall-Wright small population effects, to interpret human culture change and population viability, as well as the Malthusian demographic trends so often considered in island context (Kay 1999; Fitzhugh and Hunt 1997). Recent simulation studies indicate that while the "beachhead" process for Pacific Islands colonization is well supported, other more widely found patterns, the matrix and pulse models, are also characteristic of post-initial colonization demographic trends for islands (Moore 2001). These variants suggest that geographical or regional scale influences how we interpret human migration and its archaeological signatures. Moore (2001), for example, argues that a quick Malthusian takeoff in a colonizing population produces homogeneous culture and a cladistic expansion of human groups by fission, creating a population whose genetic structure indicates an early Abottleneck.@ Archaeological and genetic evidence for such bottlenecks has been sought in the Pacific Islands, especially for Polynesia (e.g., Kirch and Green 2001). While the colonization of Micronesian islands appears to be more complex in terms of multiplicity of events and sources of population than of Polynesia, it seems to have followed a similar process, that is, beachheads followed by pulses of migrants to form viable human colonies over the long term. These varying kinds of colonization models can have much to do with how politically complicated island societies became.

tral Carolines has had complex interrelationships with atolls in adjacent areas since prehistoric times (Takayama 1981). A different, independent atoll integration occurred in the Marshall and Gilbert atoll chains located at Micronesia's

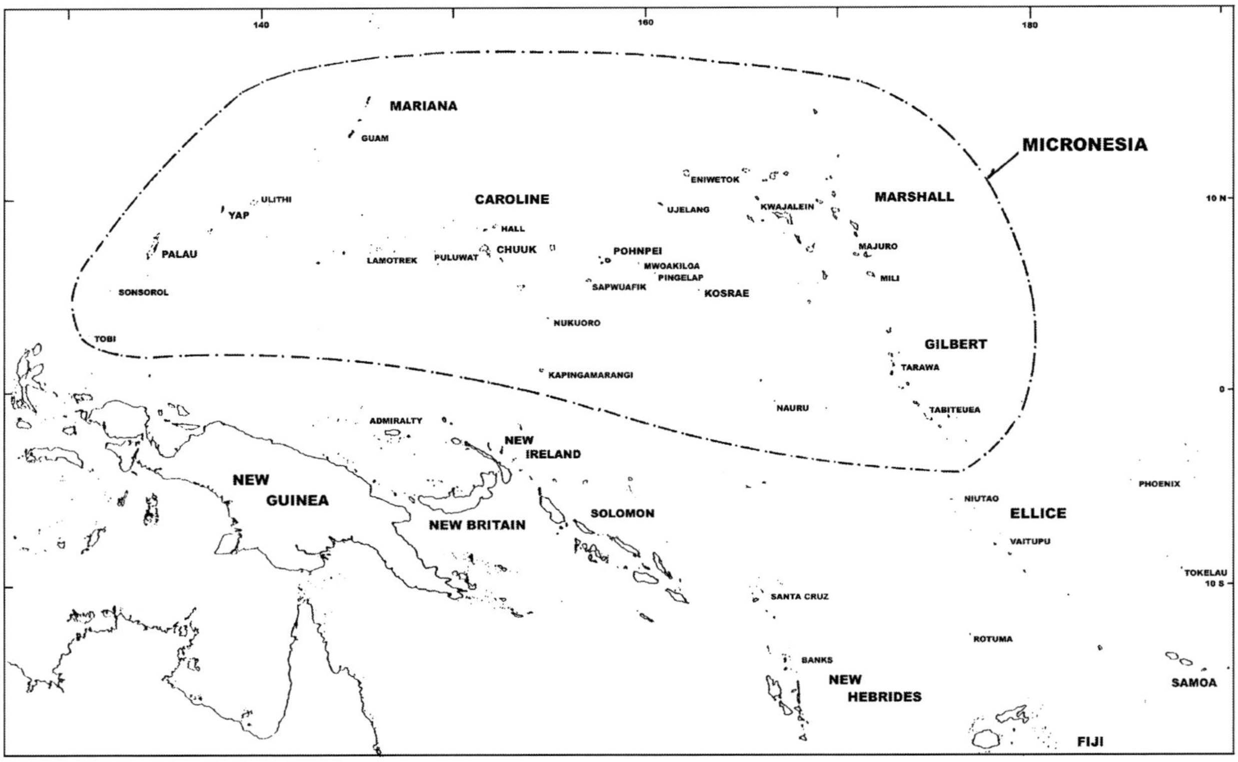

FIGURE 1. MAP OF MICRONESIA

eastern extreme where, for example, hierarchically ordered society consisting of three levels of chiefly territorial authority and two levels of chiefly titles existed among the atolls in the Ralik chain (Alkire 1978). Thus, the set of islands represented by Micronesia shows a remarkable range of island type and political complexity within which to consider the impact of scale on evolutionary processes.

MICRONESIAN ISLAND CONTEXTS

The world's islands provide several contrasting contexts, including those within continental interaction spheres, as in much of Southeast Asia, and those comprising the often smaller oceanic islands with no regular continental interaction, as in Polynesia. Of special interest are the relationships among sets of more remote, smaller islands, such as in Oceania, where they formed a settlement frontier. Two major contrastive island environments exist within Micronesia in the western Pacific, the low islands or atolls and the high islands (oceanic volcanic or continental rock forms) and this difference is reflected typically in the degree of political complexity. It may also signal differences in colonization. The most complex traditional political systems are known historically from the six major Micronesian high island groups; these include the Marianas, Palau, and Yap in the far west which are linguistically and culturally distinguishable from Chuuk, Pohnpei, and Kosrae in the central-eastern Caroline Islands (Fig 1). The atoll integration into high island political spheres is best known in the Yapese interaction with Caroline Island atolls extending approximately 1200 km to the east (Alkire 1978, 1980; Descantes 1999). As well, the almost-atoll of Chuuk in the central Carolines has had complex interrelationships with atolls in adjacent areas since prehistoric times (Takayama 1981).

RESEARCH ISSUES

In Remote Oceania, island adaptation and change often have been viewed in terms of a process of colonization, population growth, settlement and subsistence expansion and concomitant evolution of social and political complexity. One significant issue is the relationship of population size and density to political integration. Population density or nucleation, rather than an ill-defined population pressure, a major issue in any island context, should be recognized as a major factor promoting political domination (see, for example, Roscoe 1993). Efforts to link centralization and demography and to distinguish population density and nucleation from "pressure" represent a worthwhile endeavor. To develop this approach it is valuable to distinguish political centralization from sociopolitical complexity. While complexity relates to the number of sociopolitical components and their hierarchical arrangement, centralization should refer to the proportion of interaction or activity handled by a central place or individual (Feinman in Roscoe 1993). Thus, complexity may increase simultaneously with the degree of centralization, but not necessarily so. One advantage of distinguishing hierarchy from centralization is that provenance studies, for example, determining the sources of building stone and ceramics or other artifacts at Nan Madol, a major stone architectural site on Pohnpei, Micronesia, can be used effectively to indicate degree of centrality compared to other centers (Ayres et al. 1997).

Aims of this investigation are to clarify the role of public construction in chiefdoms and to assess its significance in differentiating levels of political complexity. The latter has been interpreted in many cases with reference to elaborate stone and earth architecture. The concept of "monumental" has figured significantly in anthropological investigation of cultural evolution and this has seemed appropriate because monumentality has traditionally been viewed in terms of massive, enduring structures set up to commemorate politically-significant persons and events. Monumental architecture and its associated site contexts represent critical features for marking social and political differences among Pacific Island groups and their leaders. This is particularly characteristic of chiefdoms, defined here as ranked societies in which redistribution is a major mechanism for economic exchange; redistribution for food distribution and transmitting status goods characterizes all Pacific Island chiefdoms. Hierarchies or ranked levels of chiefs fulfilling a variety of leadership roles and culminating in a paramount are also characteristic.

To assess the significance of architectural variation as a measure of sociopolitical complexity, it is necessary to consider monuments and monumental architecture in the broader context of settlement, subsistence, and demographic data.

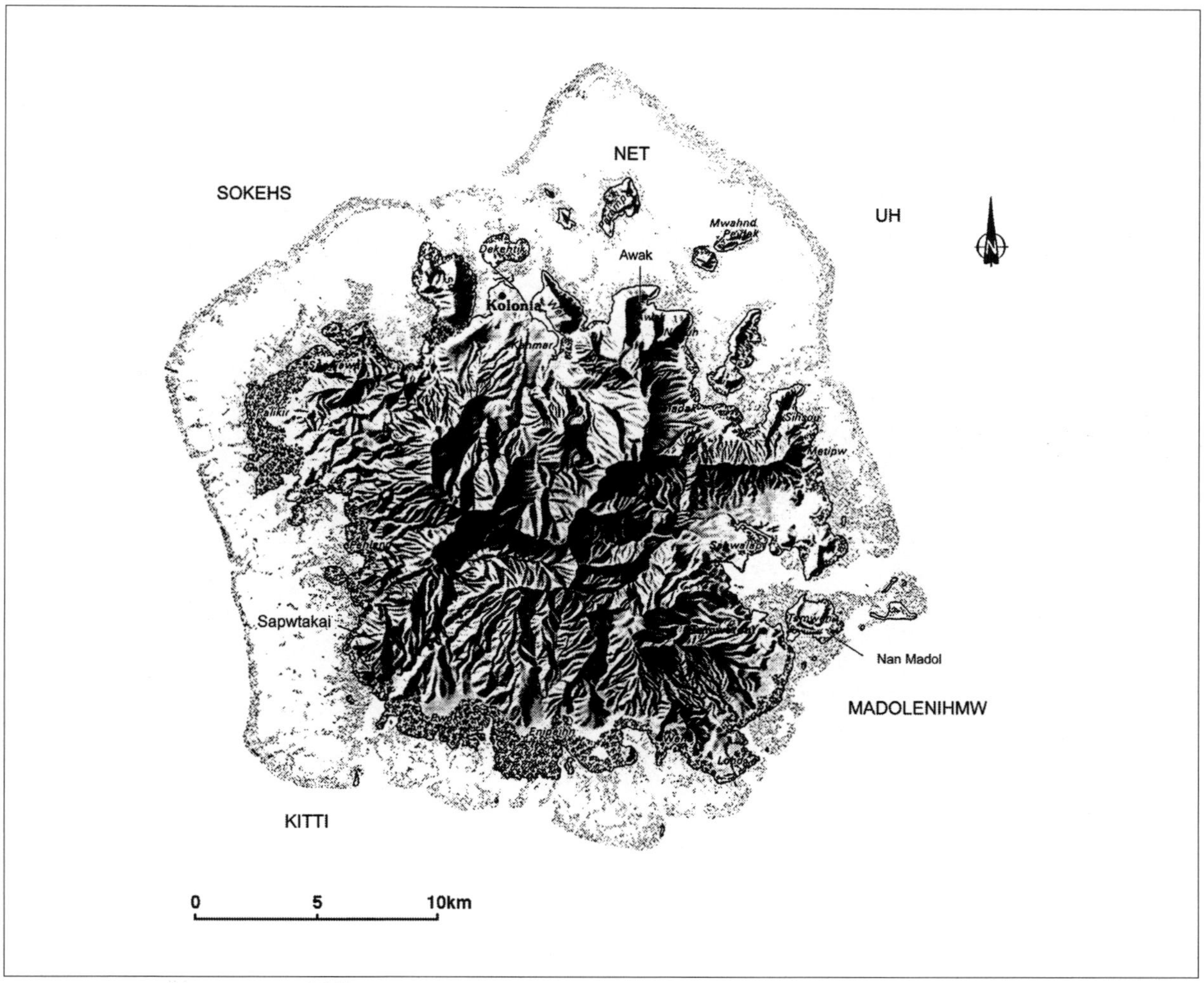

FIGURE 2. MAP OF PHONPEI ISLAND, CENTRAL CAROLINE ISLAND
showing research locations at Awak, Sapwtakai and Nan Madol.

A basic issue is whether architectural site size or scale can be interpreted as directly reflecting political centralization or hierarchies. Kolb (1994,1999) makes a case for increased political centralization in Hawai'i based on stone architecture that indicates a rise in the religious authority of the Hawai'ian chiefly hierarchy.

Pohnpei (FIGURE 2) and Kosrae are the two volcanic high islands in eastern Micronesia and here significant political centralization and social stratification appeared prehistorically. Kosrae at 109 sq km represents a highly stratified society on the smallest of the Micronesian volcanic high islands (other than the Chuuk island complex; 96 sq km; see Peoples 1990 for a comparison). Archaeological evidence suggests prehistoric connections between Pohnpei and Kosrae (Cordy 1983; Ayres 1990). Pohnpei's largest archaeological site complex, Nan Madol, is examined here from the standpoint of its administrative level relative to other island ritual centers. The island provides evidence pertinent for examining the relationship of insularity and scale to evolutionary processes; the research also points to the significance of coastal adaptations in island context and demonstrates the transformation of Pohnpei's ecosystem reflecting changes in subsistence, maritime technology, and ideology over a period of more than 2000 years.

Pohnpei is a classic chiefdom that continues to function today with redistribution supervised by more than 20 ranks of chiefs who provide access to status, including through the distribution of basic foodstuffs. Stone construction of varying scales on Pohnpei served in the past, and even today when it is less monumental, to define and differentiate social status and political rank differences. The island=s name, Apohnpei@ refers to Aupon a stone altar@ and exemplifies its hallmark characteristic. This also reflects the significant role that oral history and ethnographic models play in our interpretation of the rise of complexity in the Pacific Islands(Mauricio 1987; Ayres 1990).

Archaeologists must be fully appreciative of complexities and variation in political structures in extant chiefdoms in order to incorporate such information into analyses of prehistoric systems. Petersen (1991, 1992), a social anthropologist, claims that archaeologists tend to gloss over ethnographically-known variants in political organization and that they misunderstand the complex nature of the Pohnpei sociopolity. While it is debatable which aspects of variant political structures are crucial for forming models of pre-contact political form and interaction, Petersen emphasizes that: 1) there is a problem of one-to-one mapping of the Pohnpei chieftainship ideology onto the dynamics of Pohnpei socio-political life, and 2) archaeological data will be more useful if they are studied to instruct us about specific historical developments which are significant in their own right. The first point reminds us that ideology, although increasingly pointed to as a missing element in much archaeological explanation, is typically not directly linked through scientific methods to behavior that can be identified in archaeological context. Petersen's second point should en-

FIGURE 3. PREHISTORIC STONE ARCHITECTURE AT NAN MADOL
View of the main entry, Nan Douwas tomb, Nan Madol, Pohnpei.
The extensive use of columar lava rocks in wall construction is indicative of high status burial places.

courage more ethnoarchaeological study of links between behavior and ideology in historically placed cases.

THE POHNPEI ARCHAEOLOGICAL RECORD

Although the artifactual evidence available for the early prehistoric period on Pohnpei is limited, I have proposed an overall, but general, sequence of cultural phases: These range from a Settlement Phase (ca. 1000-0 BC), an early pottery producing culture; the Peinais Phase (AD 0-800), a later pottery producing culture showing the beginnings of stone architecture; the Nan Madol Phase (AD 800-1500), when stone construction escalated dramatically along with centralization of political function at Nan Madol; and the Nahnmwarki Phase (AD 1500-1825), representing the fragmentation of the earlier, more centralized polity. By this time pottery was no longer being produced. An Historic and Missionary Phase (AD 1825-1886), when Pohnpei first experienced intensive Euroamerican contact, and the Historic Government Phase (AD 1886-), when European governments began to form on the island a series of colonial administrations that increasingly transformed the traditional political system, follow. The time frame for the initial development of politically complex Pohnpeian systemsBcomparable to that known historically--on Pohnpei then falls into a relatively narrow time range of approximately 500-700 years (the Nan Madol Phase).

Hypotheses pertinent to the Nan Madol research have been 1) that the extension of hegemony by Nan Madol rulers over Pohnpei was based primarily on their role in religious systems; and 2) that the dissolution of the centralized polity after A.D. 1500 is associated with the rise of militarism and the formalization of a dual chiefly title system.

ARCHAEOLOGICAL ATTRIBUTES

The recognition of patterns in archaeological remains that documents the presence of particular complex sociopolitical forms is critical. A set of archaeological features that identifies past chiefdoms has not been completely or uniformly distinguished for Pohnpei, or for that matter for elsewhere, but available evidence permits a delineation of major hierarchical administrative centers. Interpretation of the archaeological settlement data in terms of individual, functionally-identified structures, the differentiation of various level centers, and agricultural and other subsistence related patterns are all important elements in archaeologically identifying prehistoric chiefdoms (Ayres and Haun 1985, 1992). Archaeological sites that reflect ritual activities include meeting houses, stone tombs, and complexes of these and various residential and subsistence units. A wide range of burial goods aids in differentiating status, but portable artifacts such as shell ornaments are typically not preserved outside of coral reef environments on Pohnpei and so architecture represents perhaps the most useful set of attributes to document. The intersection of evidence from food remains, portable artifacts, including pottery, and architecture has been useful in this study and is summarized here.

Where preserved, types and quantities of food remains indicate status feasting. In particular, the ethnographic model provides specific reference to several species of fish, turtles, and dog being reserved for high chiefs or for consump-

FIGURE 4. VIEW FROM THE NAN DOUWAS TOMB
rear wall area across three protective stone seawalls and towards the reef islet, Nahkapw, and the open ocean, Pohnpei.

tion in ritual feasting circumstances. One of the most distinctive of these is the hump-headed parrot fish, the kemeik (Bolbometopon m.), which has massive dentition that preserves well in a variety of site types. In larger specimens, the dentary, maxillary and pharyngeal bones are indicative of the size-grade, called tamwais, of the fish that were reserved traditionally for ritual use. Other status foods, including turtle and dog are found widely in Nan Madol structures, but food remains in main island sites are mostly found only in house structure firepits. Differential preservation makes use of food remains as a status feasting marker difficult to confirm everywhere on the island, but comparative study of prestige fish bones and their distribution does suggest a close association of the distinctive status foods and ritual areas marked by elaborate stone architecture. Environmental and resource changes at Nan Madol's coastal location span over 2500 years of mangrove fringe-coral reef exploitation. Sedimentation and reef species extraction shifted significantly during this period.

Portable artifacts, particularly grave goods, are often interpreted as indicators of social status differences and they can be used in this way for Pohnpei; however, again differential preservation of shell ornaments, because they are not preserved in main island sites with highly acidic soils, means that broad inter-site comparisons are not possible. Such ornaments do represent the main status-linked artifact form, but while we have tens of thousands of shell disk beads, rings or bands from conus (Conus sp.) shell, pendants from spondylus shell (Spondylus sp.), and pearl shell fishing lure blanks from the tombs at Nan Madol, we very rarely find these in main island sites. Thus, comparison is limited.

For the earlier prehistoric period, pottery might serve as a good marker of status if encountered as burial goods; however, although pottery was made on Pohnpei for more than 1500 years from the outset of colonization up until the middle (ca. AD 1200) of the Nan Madol Phase, none has been found in stone tombs or other burial contexts. The pottery brought into Nan Madol has been examined from the standpoint that its diversity of source or manufacturing location may serve as a good indicator of Nan Madol=s centrality as a place for ritual feasting. The diversity of pottery found at the site is suggestive of an island-wide ceramic Acatchment,@ but these studies are underway still; no pottery has been found with stone tombs or specialized ritual structures anywhere on the island (Descantes et al. 2001; Athens 1990)

Thus, stone architecture represents the best avenue at present for documenting the connections between ritual, status marking, and political authority. Nan Madol is the most elaborate of the sites of interest in this regard with its core area of approximately 100 artificial islets covering nearly 1.0 square km; one major tomb at Nan Douwas was built up on a 3000 square meter artificial island (FIGURE 3). Other artificial islets range up to 4,000 sq meters. Construction materials include massive boulders and other building materials that came from locations all over the island. Pohnpei, with only this one really large site complex and several lesser ones, represents a central place primate center settlement form (rather than rank-size) in terms of total site area, internal complexity, differentiation, and resident population. This site served as the focal point of a paramount chiefdom called a wehi in the traditional system.

DEMOGRAPHICS

Archaeological and ethnohistoric evidence suggests that a maximum population figure of 25,000 in prehistoric times for Pohnpei is a reasonable estimate (Haun 1984) and a resultant density of 80/sq km falls within the range for some Polynesian chiefdoms and also, for example, at the lower end of the range of 80-200/sq km for early states in the Valley of Mexico (see for example, Brumfiel 1983:268). Cordy (1983) has proposed a maximum population of ca. 6000 or a density of 55/sq km for Kosrae, the nearest high island east of Pohnpei; however, even though historically Kosrae was more politically centralized than Pohnpei, this density is considerably lower than the figure for Pohnpei. The range of Brumfiel's figures and Cordy's estimate suggest that there is no close relationship between population density and centralization of social organization; however, it appears that there is a closer relationship between overall population size and degree of social stratification on Pacific Islands (cf. Cordy 1986), probably because as islands they represent the circumscribed area that is a significant factor in developing social differentiation and of centralization of political processes. Circumscription may be either environmental or social (see, e.g., Kirch 1988). Roscoe (1993) comes to a similar conclusion but for reasons having to do with patterns of nucleation rather than total population size. However, population density figures may be misleading because the effective land area may be a relatively small portion of the total island area; that is, non-productive land may represent a very high percentage of the total island area. This reflects a difference in scale compared to continental areas.

If one compares the construction of the Nan Madol complex with efforts to build other centers such as those recorded in Awak and Wene, e.g., in terms of tomb construction, meeting houses, and residential complexes representing possible different centers and structural levels of sociopolitical organization, one is struck by the magnitude of the Nan Madol effort. Even Sapwtakai, a late prehistoric-early historic center on the island's south coastal Kitti area dating to ca. A.D. 1400 (Bath and Athens 1990) and perhaps the largest single complex of many small ones known outside of Nan Madol, is dwarfed by comparison (TABLE 1). Nan Madol has boulders weighing up to 60 tons as well as tens of thousands of columnar basalt pieces, some more than 5 meters long and 0.5 m across. An estimate of the total stone building materials brought into Nan Madol Central is 300,000 cu m and a weight of ca. 500,000 to 750,000 metric tons. While this total represents the culmination of several hundred years of construction activity, even one of the major artificial islets is substantial (Pahn Kadira, begun ca. A.D. 900-1000, ca. 40,000 metric tons; Pahnwi, begun ca. A.D. 1250, 40,000 metric tons). This center, at an estimated 1000 persons, is a very large settlement for the Pacific Islands. While not serving as the classic primate urban center for central place description, it certainly was a functional center serving a hinterland, here all of Pohnpei (and perhaps other islands), for which it provided essential services. One of these was its role in ritual related to marine deities and its coastal reef setting was crucial (FIGURE 4).

The largest site complex in Awak at the subdistrict or section (kousapw) level, by comparison, is approximately 0.6 ha (Site B7-116; this figure could be expanded to ca. 1.2 ha if site B7-117, B7-1, and 2 are added to the complex) which pales in comparison to Nan Madol Central at 75 ha. The comparisons below (TABLE 1) suggest that even though columnar rock was available in Awak, extra-local columns were brought in for construction of ceremonial architecture (especially lolong, stone walled tombs) where in one case (Site B7-2) one massive column 2.5 m long was the only large one used in the entire structure covering some 150 sq m. No local columnar rock is known in the vicinity of Salapwuk or of Nan Madol and this confirms the long-distance transport of this basic building material. The main Awak site complex is not that much smaller than the Sapwtakai complex at 1.1 ha (calculated from Bath 1984: FIGURE 2). Comparisons of individual structures mentioned above includes tombs, found in all of these complexes. One with limited numbers of large columns is illustrated in Figure 5. Further contrasts in stone work are outlined in TABLE 2.

Table 1. Comparison of Structural Sizes of Three Chiefly Centers on Pohnpei.

	Awak Local Chiefdom	Sapwtakai Regional Center	Nan Madol Central Paramount
Area of Complex	0.6 ha	1.1 ha	75 ha
Total Construction Volume	3,000 cu m	3,500 cu m (est)	300,000 cu m
Total Mass of Complex	8,000 m tons	10,000 m tons (est)	500,000 - 750,000 m tons
Large Single Tomb Feature (Area/Vol)	285 sq m/220 cu m	230 sq m/345 cu m	3,000 sq m/18,000 cu m
Largest Single Transported Stone	2 m tons	5 m tons	60 m tons

PROVENANCE STUDIES

Petrographic, elemental composition, and technological attributes of building stone and ceramics together aid in understanding prehistoric patterns of production and distribution (Ayres 1998, Ayres et al, 1997). Studies of stone building materials have provided evidence of at least 13 rock types used in Nan Madol construction. Among these, two main types of boulders that were used for basal wall anchors come from Temwen Island adjacent to Nan Madol. The distinctive columnar basalt materials come from multiple sources; this reflects needs for differing column sizes, uniformity of column shapes, and possibly structural strength as well as sources controlled by groups in differing geopolitical regions of the island.

The "micro-mapped" distribution of these rock types in wall constructions demonstrates building sequences in the case of some islands. In the site's major tomb islet (Nan Douwas) it was demonstrated that the front wall (60 m long by 8 m high) had been built either by people using different source materials or that sections were rebuilt. Most of the boulders used in construction are of types found on nearby Temwen Island as seen in the early building at Pahndipap Islet (ca AD 800) and the basal layer of the Nan Douwas front wall (AD 1200?). The columnar basalt pieces show a uniformity that requires more detailed study to differentiate them according to possible quarry areas. Present evidence of source location establishes that materials were brought to Nan Madol from some 30 kilometers away.

SUMMARY AND CONCLUSIONS:

The islands of Micronesia show a remarkable range of island type and political complexity within which to consider the impact of insularity and scale on evolutionary processes of political transformation. The island of Pohnpei in particular provides evidence useful for examining the relationship of monumentality to processes of centralization of administrative function. The research also points to the significance of coastal adaptations in island context and examines the transformation of Pohnpei's ecosystem, reflecting changes in subsistence, maritime technology, and ideology over a period of more than 2000 years (Ayres 1990).

The definition of complex society used in this paper relates to chiefdoms and states, and the former is appropriate when discussing Polynesian and Micronesian sociopolitical forms, which include marked ranking of social status, importance of redistribution, and multiple chiefly ranks. While the significance of redistribution has been brought into question as a defining characteristic of chiefdoms--due in part to the existence of varying forms of redistribution--collection and re-allocation of goods is an important function of the chiefs within complex ranked systems.

The model of Pohnpei=s early political system is based on historic and ethnographic data, now supplemented with archaeological data. Important elements are the territorial boundaries, defined social ranks, means of attaining social status integrated into the subsistence system, levels of administrative hierarchy, dual chiefly title lines, residential complexes or centers of various forms, and patterns of redistribution and feasting related to chiefly hierarchy. Present evidence supports the hypothesis that the Nan Madol rulers, Sau Deleurs, extended their control over Pohnpei The time frame for the initial development of a complex Pohnpeian polity comparable to that known historically falls into a relatively narrow time range of approximately 500-700 years (the Nan Madol Phase). This supports the conclusion that both the temporal and geographical scales of chiefdom development in oceanic islands are typically compressed.

Table 2. Comparison of Stone Sizes Per Type Transported for Construction in Three Levels of Chiefly Centers on Pohnpei.

CONSTRUCTION MATERIALS FOR THREE POHNPEI CHIEFLY LEVELS			
STONE TYPE	Awak - Local Chiefdom	Salapwuk - Regional Center	Nan Madol - Paramount
Largest Stone Type Transported	Volcanic Boulder/ Column	Volcanic Boulder	Volcanic Boulder
Type 1. Columnar Lava Rock: Size (Mass) Range of Natural Local Columns*	Range: 0.02 - 0.5 m ton Upper limit: 0.5	none known	none known
Type 1. Size (Mass) Range of Construction Columns Moved	Range: 0.02 - 2 m tons	Range: 0.05 - 1.5 m ton	Range: 0.04 -10 m tons
Type 2. Boulder Rock: Upper Mass Limit of Natural Local Boulders	45 m tons	60 m tons	130 m tons
Type 2. Largest Construction Boulder Moved	1 m ton	5 m tons (est.)	60 m tons
Type 3. Coral Boulder from Reef	Range: 1 -20 kg (Rare)	Range: Very Rare	Range: 5 kg -0.3 m ton

*note: mass is calculated at 2900 kg/cu m for the crystal-rich lava columns and 2700 kg/cu m for the flow-banded basalt boulders (the range is on the order of 2400 to 3100 kg/cu m for the basalt stone used in building). Columns range from nearly perfect prismatic forms that are neatly pentagonal or hexagonal for their entire length to ones with one end showing a hexagonal cross-section and then grading into an octagonal shape on the other, thus a close approximation of diameter is measured and this is used to calculate volume.

because of their role in religious systems. This study identifies: levels of administration or centralization comparable to historically described ones; networks reflected in nucleation; and degree of social ranking or stratification as represented by center status and function. A Pohnpeian administrative hierarchy formulated from the ethnographic model has been tested against several lines of pertinent data that permit an assessment of administrative levels from the archaeological record. Definition of social ranks or strata is much more difficult from archaeological evidence (see also, Kirch 1980, 1984 in this regard).

The intersection of evidence from food remains, portable artifacts, including pottery, and architecture has been useful in this study. Data on functionally-differentiated stone structures within settlements, subsistence related patterns, and the development of centers are all useful elements for archaeologically identifying prehistoric chiefdoms. Archaeological sites that reflect ritual activities include meeting houses, stone tombs, and complexes of these and residential and subsistence units.

Types and quantities of food remains indicate status feasting. In particular, the ethnographic model provides specific reference to several species of fish, turtles, and dog being reserved for high chiefs and ritual feasting. Archaeological evidence shows a close association of the distinctive status foods and ritual areas marked by elaborate stone architecture.

Portable artifacts, particularly grave goods, also indicate social status differences. Shell ornaments represent the main status-linked artifact form and these are known in a variety of forms from Nan Madol, but differential preservation on the main island makes comparative study difficult. The diversity of pottery found at Nan Madol prior to approximately AD 1200 indicates a wide set of ceramic source areas from around the island, thus highlighting Nan Madol=s central position in exchange and ritual during the time period from approximately AD 400 to 1200. The co-occurrence of portable artifacts, distinctive food remains and stone architecture aids in differentiating status; architecture shows the most useful set of attributes.

Stone architecture represents a basic way to document the connections among ritual, status marking, and political authority. Comparison of three archaeologically-well known site complexes, Sapwtakai, Awak, and Nan Madol shows differing levels of centrality in exchange, social status, and administrative extent. Nan Madol is the most elaborate of the sites in this regard with its extensive core area and connections--based on oral history as well as archaeological remains--to many other sites. Pohnpei, with one really large site complex and several lesser ones, represents a central place primate center settlement form in terms of total site area, internal complexity, differentiation, and resident population.

The Pohnpei centralization process is seen to culminate with the formation of the megalithic, monumental Nan Madol complex; its rapid dissolution can be seen as a systems collapse and perhaps because it is the most recent one of major scale on Pohnpei it may be the most archaeologically visible. The latest system collapse obscures earlier fluctuation and regional variation in complexity. While Nan Madol served as a major religious and political administrative center beginning as early as A.D. 1000, it became, at least by comparison, a non-center after ca. A.D. 1500. The shifting status of such centers in Pohnpeian sociopolitical life is seen to be paralleled in many other Pacific island cases, but the significance of Nan Madol as by far the most complex of such centers known on the island--and anywhere in Micronesia outside of Kosrae for that matter--is evident. Clearly, distinct geopolities on Pohnpei had varying relationships with the Nan Madol center and it appears

FIGURE 5. INNER CORNER OF THE ENCLOSING WALL AT A TOMB (LOLONG) IN AWAK, POHNPEI.
The low wall (approximately 1.5 m high) shows small boulders and limited use of columnar rock in this structure

that the Nan Madol chiefs were in the process of uniting the entire island--at least in terms of acceptance of their political and, especially, religious leadership--when they met their downfall through revolt.

The impact of Pohnpei=s island character on the political change process may be seen because of its small geographical scale and limited locales for settlement nucleation. Also, the rapidity of change may be distinctive. Significant technological change in one area contrasts with stability in another, for example, the elaboration of stone architecture at the same time pottery is declining in use. Finally, the emergence of a single, major center (central place) in both Pohnpei and Kosrae contrasts with most Polynesian situations such as Easter Island or Hawai=i. Two reasons for the relatively rapid emegence of centralized polities are the limited locales for nucleation of settlement, that is, compact environmental circumscription, and the presence of marked social status differentiation within the colonizing population or populations.

The Nan Madol case provides an interesting counterpoint to more widely known cases of political centralization in early civilizations. In addition to illuminating structural and processual change in microcosm, the geographically remote Pacific islands provide striking illustrations of human imagination and creativity in prehistory.

ACKNOWLEDGMENTS:
Mr. Emensio Eperiam, of the Pohnpei State Historic Preservation Office, and Dr. Rufino Mauricio, Federated States of Micronesia Historic Preservation Office, have facilitated the Pohnpei research over many years. The field studies were aided financially by the U.S. National Park Service, the National Endowment of the Humanities, the Wenner-Gren Foundation, and the Skaggs Foundation. I thank Gordon Goles, Robert Bryson, Joan Wozniak, Scott Fitzpatrick and Christophe Descantes for many fruitful discussions about Micronesian archaeology

REFERENCES

ALKIRE, W.,
1978. Coral islanders. Arlington Heights: AHM Publishing Co.

ALKIRE, W.,
1980. Technical knowledge and the evolution of political systems in the Central and Western Caroline Islands of Micronesia. Canadian Journal of Anthropology, 1(2), 229-237.

ATHENS, S.,
1990. Nan Madol pottery, Pohnpei. In R. Hunter-Anderson, ed., Recent Advances in Micronesian Archaeology, Supplement 2, Micronesica. Guam: University of Guam, 1990, 17-31.

AYRES, W. S.,
1990. Pohnpei's position in Eastern Micronesian prehistory. In R. Hunter-Anderson, ed. Recent Advances in Micronesian Archaeology, Supplement 2, Micronesica. Guam: University of Guam, 187-212.

AYRES, W.S.
1998 Lithic resources and uses of stone in Oceania. In Easter Island in Pacific Context: South Seas Symposium. Los Osos: Easter Island Foundation, 1998, 277-278.

AYRES, W. S. , G. GOLES, AND F. BEARDSLEY,
1997. Geochemical provenance study of lithic materials in Micronesia. In Prehistoric Long-Distance Interaction in Oceania, ed. by M. Weisler, New Zealand Archaeological Association Monograph 21. Auckland, 1997, 53-67.

AYRES, W.S., AND A. HAUN,
1985. Archaeological perspectives on food production in Eastern Micronesia. Proceedings of a Symposium on Prehistoric Intensive Agriculture in the Tropics. British Archaeological Reports, No.232, International Series, pp. 455_473.

AYRES W.S., AND A. HAUN,
1992. Prehistoric food production in Micronesia. In XVth Pacific Science Congress Symposium on Food Production, New Zealand 1983. Pacific Production Systems. Canberra: Research School of Pacific Studies, Australian National University.

BATH, J., AND S. ATHENS,
1990. Prehistoric social complexity on Pohnpei: The Saudeleur to Nahnmwarki Transformation. In R. Hunter-Anderson, ed., Recent Advances in Micronesian Archaeology, Supplement 2, Micronesica. Guam: University of Guam, 1990, 275-290.

BRUMFIEL, E.,
1983. Aztec state making: Ecology, structure, and the origin of the state. American Anthropologist, 85, 261-284.

CORDY, R.,
1983. Archaeological Survey of Innem, Okat and Loal, Kosrae Island. Micronesian Archaeological Survey Report Number 7. Saipan: Historic Preservation Office.

CORDY, R.,
1986. Relationships between the extent of social stratification and population in Micronesian polities at European contact. American Anthropologist, 88, 136-142.

DESCANTES, C.,
1999. Integrating Archaeology and Ethnohistory: The Development of Exchange Between Yap and Ulithi, Western Caroline Islands. Ph.D. Dissertation, Anthropology, University of Oregon, Eugene.

DESCANTES, C., H. NEFF, M. GLASCOCK, W. DICKINSON,
2001. Chemical characterization of Micronesian ceramics through Instrumental Neutron Activation Analysis: A preliminary provenance study. Journal of Archaeological Science, 28(11),1185-1190.

DEWAR, R.E.,
1997. Does it matter that Madagascar is an island? Human Ecology, 25(3),481-490.

FITZHUGH, B., AND T. L. HUNT,
1997. Introduction: Islands as laboratories: archaeological research in comparative perspective. Human Ecology 25(3), 379-383.

HAUN, A.,
1984. Prehistoric Subsistence, Population, and Sociopolitical Evolution on Ponape, Micronesia. Ph.D. Dissertation. University of Oregon.

KAY, E. A.,
1999. Biogeography. In The Pacific Islands: Environment and Society, ed. M. Rapaport. Honolulu: Bess Press, 1999, 79-92.

KEAST, ALLEN, AND SCOTT MILLER (EDS),
1996. The Origin and Evolution of Pacific Island Biotas: New Guinea to Eastern Polynesia. Amsterdam: SPB Academic Publications.

KIRCH, P. V.,
1980. Burial structures and ranking in Vava'u, Tonga. Journal of the Polynesian Society, 89(3), 291-308.
KIRCH, P.V.,
1988. Circumscription theory and sociopolitical evolution in Polynesia. American Behavioral Scientist, 31(4), 416-427.
KIRCH, P.V.,
1997. Microcosmic histories: Island perspectives on "Global" change. American Anthropologist, 99(1), 30-42.
KIRCH, P.V., AND R.C. GREEN,
2000. Hawaiki, Ancestral Polynesia: An essay in historical anthropology. New York: Cambridge University Press.
KOLB, M.,
1994. Monumentality and the rise of religious authority in precontact Hawai=i. Current Anthropology, 34(5), 521-547.
KOLB, M.,
1999. Monumental grandeur and political florescence in precontact Hawai=i: Excavations at Pi=ilanihale Heiau, Maui. Archaeology in Oceania, 34(2), 71-82.
MAURICIO, R.,
1987. Peopling of Pohnpei Island: Migrations, dispersal and settlement themes in clan narratives. MAN AND CULTURE IN OCEANIA, 1987: 47-72.
MOORE, J. H.,
2001. Evaluating five models of human colonization. American Anthropologist, 103(2), 395-408.
NUNN, P. G.,
1994. Oceanic Islands. Cambridge, Mass.: Blackwell.
PEOPLES, J. G.,
1990. The Evolution of complex stratification in Eastern Micronesia. In R. Hunter-Anderson, ed. Recent Advances in Micronesian Archaeology, Supplement 2, Micronesica. Guam: University of Guam, 1990, 291-302.
PETERSEN, G.,1990. Some overlooked complexities in the study of Pohnpei social complexity. In R. Hunter-Anderson, ed. Recent Advances in Micronesian Archaeology, Supplement 2, Micronesica. Guam: University of Guam, 1990, 137-152.
PETERSEN, G., 1992. Lost in the Weeds: Theme and variation in Pohnpei political mythology. Occasional Paper 35. Honolulu: Center for Pacific Islands Studies, University of Hawaii.
RENFREW, C., AND J.M. WAGSTAFF (EDS),
1982. An Island Polity: The archaeology of exploitation in Melos. Cambridge: Cambridge University Press.
ROSCOE, P.B.,
1993. Practice and political centralisation. Current Anthropology, 34(2), 111-140.
TAKAYAMA, J.,
1981. Early pottery and population movements in Micronesian prehistory. Asian Perspectives, 24(1),1-10.

8

NEW COALS ON OLD FIRES: THE QUESTION OF EARLY BALEARIC ISLAND SETTLEMENT

W.WALDREN (1) , **J. ENSENYAT ALCOVER** (2), **J. ORVAY** (3)

(1) Donald Baden Powell Quaternary Research Centre, Pitt Rivers Museum
University of Oxford, Oxford OX2 6PN England
(2-3) Deia Archaological Museum Research Center,
Deia de Mallorce, Baleares, Spain

INTRODUCTION

In the space alloted, this paper will attempt to bring into clearer perspective a particularly polemic period in the interpretation of Balearic Prehistory...... that refered to here as the Early Settlement Period, involving the initial settlement of the Balearics. Until recently, the earliest phases of this period have been represented by only a few radiocarbon dates, artefacts and contextual information. These factors have made its credence both tentative and its concept not always completely acceptable. This lack of so called solid evidence has never been more apparent within local response than during the last few years, in which it has fluctuated between acceptance, credulance and denial. Undoubtedly, the best assessment of the situation, regarding the pros and cons regarding early Balearic settlement, can be found in V. Lull's recent critique (Lull et al 1998). The source of recent interpretive arguments rightly centres around evidence supporting and clarifying this period, especially the initial phase of the island's earliest occupation. Until very recently, the Early Settlement Period, circa 5000 BC to circa 3000 BC, was supported by less than a half-dozen age determinations, which in the authors' opinion has never been at all satisfactory, although it was at the time and still is in the opinion of the present authors a basis and framework with which to work in a positive and constructive manner.

However, among certain recent, local investigators (Alcover et al 1999, 2001, 2002), these early dates for the initial human occupation have been challenged to the point of being invalid. In fact, it has been stated by these investigators that proof at present to the order of 95% probability exists that the island's were not occupied very much prior to 2000 cal BC. This conclusion in more than certain respects closes the door on or limits the possibility of earlier human presence in the Balearic, and of course has not taken into account, as it has claimed to have done, all of the data and evidence available. While recent solid evidence exists of climatic, change, hence, a diminishing of floral resources, is clearly evident during the V and IV millennia, the effects of these changes on the local environment are still very far from clear. As has been proven time and again, there are no real certainties in the study of prehistory, even with so called scientific proof, data changes, sometimes radically, almost from moment to moment.

The general scientific climate among the agnostics is, sadly enough, not one of adding to what has already been built up over the last four decades, but more a matter of tearing down what has been labourously built on available evidence. One might argue that this is the aim and purpose of scientific investigation, although it is but only one of the aims and purposes. However, the motives for this in the recent climate are somewhat more obscure, a division into two camps (pros and cons) and, perhaps, even personally motivated, otherwise it is a subconscious intention to return everything investigative to where it was forty years ago, to be personally rediscovered or invented again.

In any case, some of these agnostic arguments are addressed and challenged here as well as in another section of this publication (Waldren, this volume, Links in the Chain), where the question of initial settlement is argued on different levels and in more detail and for slightly different motives.

This particular paper deals with the chronometric issues in the three or four research sites most involved and will attempt to show that the latest estimate for Balearic early human contact is rather presumptuous and highly speculative and in some respects, flawed, incorrect and not based on all the empirical evidence. The recent assessments, apart from one or two, where the problems are meticulously analysed (eg. Lull et al 1999), are lacking in essential detail and for one reason or another have made the rather ambiguous decision to either ignore the strategies and purposes for what on the quantity and quality of investigation and research, and at times philosophical reasoning. (2) It will attempt to do this, in the space alloted, by presenting a number of large list of new as well as old recalibrated radiocarbon dates, all of which have 95.4% confidence probabilities, demonstrating human presence in Mallorca during 3000-5000 BC, considerably prior to recently made estimates. The inventory lists demonstrate within those probabilities unusual homogeneous and temporal continuity, as well as other interactive and correlative characteristics.

FIGURE 1. SAMPLE OF DATED CONSOLIDATED CAVE FLOOR CONTAINING CHARCOAL, BONE, CERAMIC AND FLINT. (stratum 27: 2675 cal BC)

has been long-term and on-going investigation and research, carried out over the last four decades. Or they have selectively used what best suits their arguments with the exclusion of other closely related and important evidence. They have also set aside the naturally high probability and essential truth that the strongest arguments can and do change with an unusual frequency. Hypothesis and interpretation are only temporal modes of fact. So, it would be perhaps more constructive and even more positive in the long run not to welcome, even encourage change at the least provocation, until more evidence becomes available.

THE OBJECTIVES

The present response to the problem of Balearic early settlement has two main objectives: (1) to defend, on the basis of past empirical data, the possibility and even high probability of early contact and occupation of Mallorca, based

However, before this process begins, it must be clearly understood that the present evidence, although it is believed to be of the highest quality and the most extensive on a site to site basis, although originating from a specific and limited geographic region, it is not to be considered definitive by any measure or for that matter as a signal of conclusive proof. Conclusive proof is conditional at most. It is a collection of modest but insistant indicators of what, regarding early settlement, in all possibility and reasonable probability did take place to an unknown degree somewhere along the time line, and for which we may possibily find further substantiation, hence verification, as time and future investigation and research goes forward, particularly within this specific but small geographical region.

The reader should take other factors into consideration. The empirical evidence presented in this paper is of three distinct categories: (1) chronometric, (2) stratigraphical and

(3) artefactual, including architectural evidence. It is also based on 40 years of extensive and inexhaustable investigation, excavation and research during that time and, most of all, a fund of collective experience, equally, observational, innovative and intuitive. All of it founded within the limited, although highly diversified, geographical and geological regional unit which forms basis of its strategy and planning.

FIGURE 2. RADIOCARBON SAMPLE COLLECTION IN THE EAST ENCLOSURE OF MATGE

Since the beginning of our discipline, the interpretation of prehistory has relied almost entirely on empirical evidence in the form of data and materials, but we do well to remind ourselves that it is also founded on other essential elements, both tangible and intangible as well. In the final analysis, no single piece of empirical evidence is ever sufficient to constitute fact. Fact or conclusive proof, conditional in its most productive mode, is the assemblage of numerous small pieces of empirical evidence of as many different categories as possible, collected over time. Prehistoric interpretation and hypothesis (abstractions to begin with) are therefore the assemblage of conditional facts, each in their own right made up of more and more small pieces of evidence.

In our attempts at gathering data and evidence, we have frequently relied heavily and often disasterously on sampling techniques for the establishment of fact, believing that one piece of data constitutes irreputable fact. While sampling techniques can at times be informative or indicative, the technique and its practice frequently has an ambiguous nature and result. Much of its value depends proportionately on the size and quality of the sampling, along with the complexity of the problems involved. One can use the following example as a useful hypothesis or guideline: the emergence of one or two samples, is indicative; ten or more samples, is reasonably evident; 20 or more samples, is serious evidence; a hundred or more, is solid evidence. However on the other hand, conclusive fact is much more allusive, as well as deceptive. It is unfortunately highly dependent on the eyes of the beholder, as well as self motivation.

An example of this sort of interpretive problem and its subsequent outcome, concerning the possible presence, sequence and development of early open-air settlement patterns in Mallorca took place about 20 years ago. This was in regard to the question of International Bell Beaker Cultural presence in the Balearic Islands (Waldren 1982; Waldren et al 1987, 1998), especially on Mallorca (the largest of the Balearics group and the island with most natural resources, making it hence highly condusive to early settle-

FIGURE 3 DEEPEST LEVELS ATTAINED IN CAMPAIGN- EARLY HEARTHS- FIRE LEVEL AND RED EARTH FLOOR
(strata 33, mm 42: charcoal radiocarbon date: 4565 cal BC)

ment or demographic receptical). At that time, there was an equally adamant opposition as to whether or not contact and Beaker settlement, circa 2500 BC, actually or ever had, taken place in the Balearics (Waldren, Ensenyat and Kennard 1987).

The fact is that, as a result of extensive investigation (sampling) and on-going research, we currently have nearly 100 radiocarbon dates with 95.4% confidence probabilities attesting to not only Beaker (Chalcolithic) contact, but actual settlement, all supported by open-air architectural achievement, abundant artefact and chronometric evidence. This is demonstrated in sites, like the Olezian Chalcolithic Old Settlement (SFO-OS) and illustrates just how sophisticated and well organized, social and economic conditions were on Mallorca during the whole of the III and II millennia BC.

The list of radiocarbon dates shown in Tables 1-4 represents not only this initial, open-air stage of urban settlement, but closely correlates with evidence in caves and rock shelters as well. In turn, these lists form an integeral part of the existing and on-going, radiocarbon master inventory of some 450 dates currently available for local prehistory (Waldren in preparation). As can be seen, part of the selected date lists presented in Tables 1-4 covers a temporal span not only for the whole of the III millennium but it also represents a similar argumentive situation and polemic period of time, over which there was a great deal of unproductive controversy, in that it, too, dealt at the time with a chronometric problem of verification, assessment and interpretation, regarding early urban settlement. One which eventually positively resolved itself

Some of that data and information is included here, for a number of reasons: (1) The time period involved represents a large body of dates that immediately relate to an unusually long continuity within an especially informative open-air situation (undoubtedly the forerunner of still earlier settlement), and not just in the 'closed' environment of a cave or rock shelter. (2) The data shows a closely, cross-correlative, chronometric sequence comparable with that of the Matge stratigraphy (Waldren 1982) and the other research sites discussed here. (3) In so doing, the sites and their data, collectively and individually, complement and support one another. In this particular instance, their time sequence and contexts represent a highly organized, open-air or rural settlement already in existence as early as 2500 calBC and continuing through to the sites abandonment circa 1300-1400 calBC, prior to the establisment of later Bronze Age Talayotic communities.

At the time, this evidence of earlier than Talayotic architectural settlement was unknown and therefore not easily or graceously accepted by most local investigators. Regardless, as will be seen, this is exactly the time period supported in the Matge stratigraphical sequence (Strata 14-23) (Waldren 1982), where similar and correlative dates can be found (TABLE 2). (4) Collectively, these dates and the corresponding artefacts and contexts also show that there is no bias in age to be had between dates and artefacts coming

FIGURE 4 COPROLITES BED , BUTCHERED AND BURNT MYOTRAGUS BONES CONTACT STRATA 32-33
(Strata 33 and 34 radiocarbon date:4635 cal BC- 4722 calBC, respectively)

from shallow, contextual environments and eroded situations, characteristic of open-air settlements and those from deep vertical deposits, usually found in caves and rock shelters. On the basis of difference in environment, one might well expect differences, however, none seem to exist. In actuality, close cross-correlation is possible on every level.

So, we can logically expect that these dates from the SFO-OS open-air situation, along with those from the rock shelter site of Matge, can be reliabily grouped together and, therefore, can be statistically counted on as a collective body of supporting data in the interpretation of their time representation. (4) Likewise, we can regard these 44 Chalcolithic SFO-OS urban dates and the 13 corresponding ones from the Matge rock shelter, 2523 calBC (QL-4191) to 1485 calBC (CSIC-178) (TABLE 2), as a singular, reliable body of data that can be used, not only, in the interpretation and assessment of the time period they represent, but also as solid evidence of III and II millennia BC open-air settlement of a highly sophisticated nature and character during that time.

Surely, we cannot logically expect that this very well organized and obviously, highly developed settlement appeared overnight, but must have undergone earlier stages of evolution and development, apparently, over-lapping and coexisting with that of an even earlier period of settlement, already established and carried on in rock shelters and caves, even as they occurred in mainland and continental sites.

As we will presently see, strong evidence of this earlier settlement does exist in rock shelters and caves (see Figure 6, also Waldren 1982), where dates exist that are well below the recent estimates of only just sliightly prior to 2030 cal BC (Alcover et al 2002).

FIGURE 5. ORIGINAL MATGE STRATIGRPHY SHOWING SUCCESSION OF CHARCOAL AND ASH STRATA: 1978

THE MATGE SITE

The Matge deposit consists of 35 cultural, verticallly successive levels (FIGURES 4 and 5). The site was excavated between 1968-1982 with teams in the field 6 months of the year, consequently, the Matge site now has 67 radiocarbon dates from its stratigraphy, with a prospective 30 new ones from recent re-excavations in the pipeline. So, there is no legitimate reason not to believe that the extensive list of dates from this site also represent as reliable a time sequence as that of the SFO-OS site, both of which, equally, conform within the high statistical probabilities and reliability of radiocarbon method. Also important is that two of the three sites (Matge and Muleta) conform to the probabilties and formalities of *stratigraphical succession* and all conform to artefactual correlation.

In this section, we will consider the entire range of the stratigraphical and chronometric sequence at Matge, circa 4635 calBC to 1485 calBC, by covering the time span referred to here as the Early Settlement Period (ESP), including the Neolithic Early Ceramic Phase (NECP) and the Chalcolithic Period, synonymous with the local Pretalayotic Period (PRT), with its Early and Late Beaker Phases (EBP and LBP) (TABLE 1). In turn, for presentation purposes, the individual date list from Matge can be compared with more than 100 dates representing other intersite inventories, related to and forming the overall chronometric sequence for this time as well as the periods preceding it.

In Table 1, the Matge dates begin with the first solid evidence of human occupation of the shelter, the very thick charcoal and ash level of the deposit, Stratum 33 dated at 4635 calBC (I-5516) and continuing through to Stratum 14, the top of the series of hearth fires extending through to Stratum 33, all of which form an impressive, successive series of compressed ash and charcoal layers up to three me-

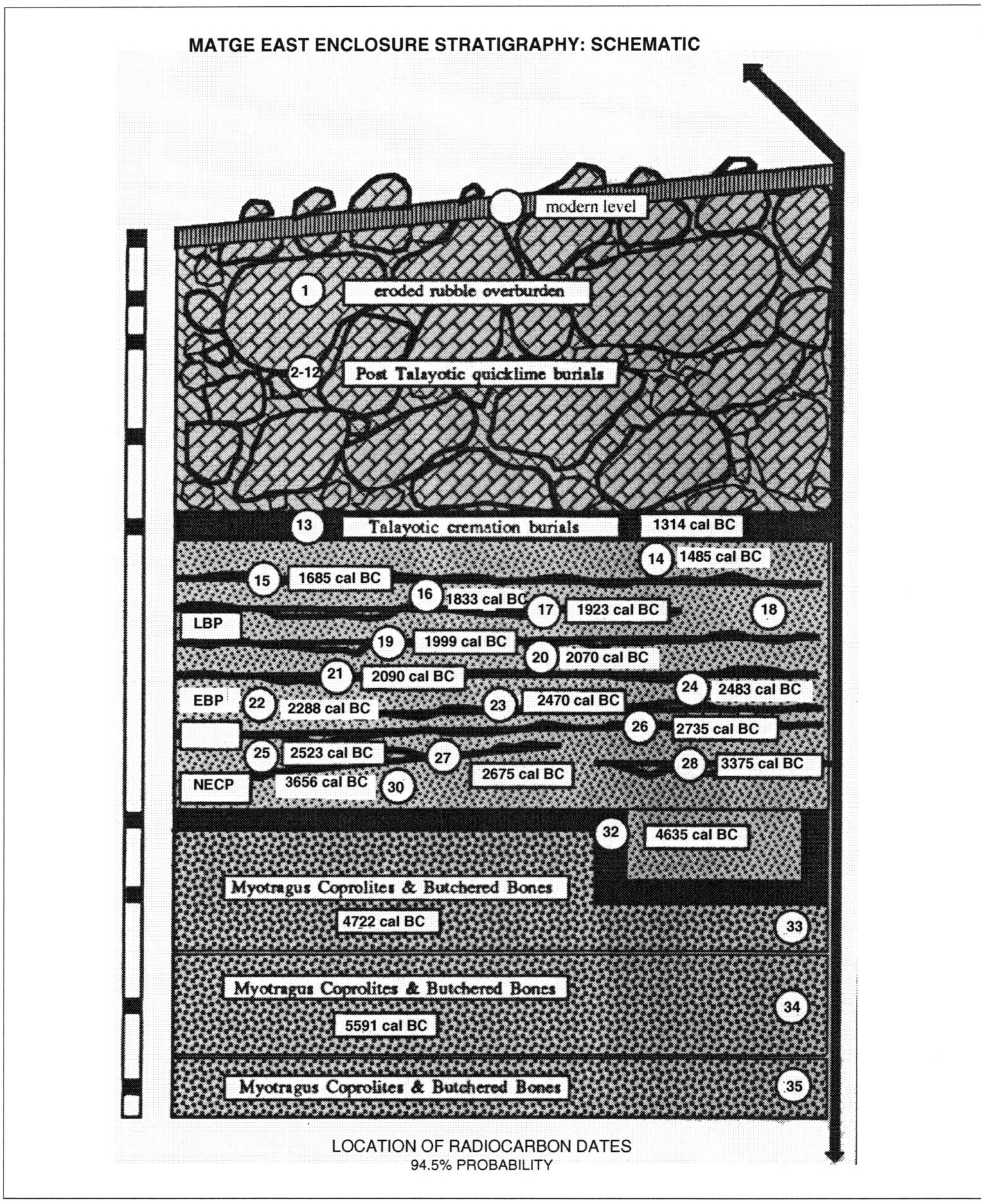

FIGURE 6.

ters thick (FIGURE 5). A succession of levels such as this, classical by every standard, is both rare and potentially very informative, when assessing and interpreting the question of human presence. A few local investigators have suggested , especially the first of these fire levels, to be the result of accidental agency, an unsatisfactory explanation particularly since the level like the others in the series contain cultural materials. On the basis of this, they are undeniably the result of kitchen fires, although their origin can always be argued, but not very convincingly.

We shall compare the sequence with new and old data from the cave of Muleta and Muleta Pocket Cave (SMPC) which have in their individual stratigraphical contexts, along with those of the SFO-OS site, comparative dates and other data representative of early open-air settlement in the III to V millennia BC.

Where necessary, a group of schematic stratigraphical drawings (FIGURES 8-9) have been provided the reader to clarify depositional contexts and demonstrate the stratigraphical position of the dating samples within the Matge and Muleta deposits. The authors' consider this as the best way in which to demonstrate, logically and visually, the bulk of the available chronometric and stratigraphic information present in the discussion. For those interested in more detailed descriptions other than the summation of the chronometric and sequencial order discussed here can trace a more detailed level by level account in Waldren's: Aspects in Balearic Prehistoric Ecology and Culture (Waldren 1982) and other related literature and reference.

The dates involved in the various intersite radiocarbon inventories originate from old and new sites investigated between 1962 and 2002 and consist of new data and information obtained from recent re-excavation and re-evaluation in the Rock Shelter of Son Matge (Waldren, Ensenyat and Van Strydonck, in preparation), as well as with new research relating to the Cave of Son Muleta and its subsidiary Son Muleta Pocket Cave (SMPC) site (Waldren, in preparation), as well as other dates and information which have not been already published elsewhere. Included also are some old data which have been ignored of late and which have a great deal of pertinence on the issue of earlier than V millennium settlement (eg. Kopper 1984). There is included still other very recent research, involving analytical materials originating from these and other sites, related to this polemic period of time (see Davis, this volume), as well as other important new projects in the pipeline involving recently discovered rock art (Ensenyat et al in preparation), the later two of which could, in the minds of the authors, eventually offer further evidence of human presence in the islands circa the IV and V millennia BC or even earlier.

Finally, the reader should keep in mind the fact that the present investigation and research, although it has a long history, is still on-going and that there are many aspects of the research still not resolved, for which further data are essential before we can reach any generally acceptable date for the initial settlement of the islands that will survive a test of time. However, the most of the material and evidence available must be considered as valid in any final reckoning.

THE EVIDENCE: MATGE REVISITED

After an absence of 28 years, it was decided to reopen excavations in the Matge rock shelter with the objective of rechecking the shelter's stratigraphical sequence, confirming former dates and collecting additional radiocarbon samples for further age determinations to be compared with and added to the present radiocarbon inventory of 67 dates, as well generally adding these to the fund of information already on hand.

The site, with its 35 cultural levels, circa 5591 cal BC to II century AD, offers so far the most complete cultural stratigraphy in the Balearics, as well as the islands longest term, on-going investigation and research (1968-1982).

The re-excavations were carried out with the help of students from the University of the Balearic Islands for a two week period in March 1999, under the co-directorship of the authors with the license of local authorities. Apart from the retrieval of further test samples and confirmation of the original stratigraphical sequence, another objective was one of the consolidation of the site, which unfortunately is being badly and irrepairably damaged by vandals and local rock climbers, who use the rock face of the shelter to practice rock climbing and mountainteering.

During the two week excavations some 30 new radiocarbon samples were collected, where deemed essential, from corresponding levels, as well as other levels where collection had not previously taken place.

One important sample from this new collection of dates, originating from a critical level (Stratum 27) in the stratigraphy, marking the beginning of the Neolithic Ceramic Phase and end of the Early Settlement Period, and consisting of charcoal, bone of M. balearicus and flint (FIGURE 1) has since been analysed. This was a large sample of a consolidated human occupational floor, corresponding in sequential order to the same critical group of levels found in the 1968-82 excavations (MM 41-42)(NECP, 400cms) and located between two former strata, Statum 26 and Stratum 28 (Waldren 1986), but slightly to the east of them. This particular set of strata mark two important moments within the deposit: (1) the earliest introduction of domesticated animals and (2) the oldest form of pottery. Only in the Rock Shelter of Muertos Gallard (AMG) does an older date exist for these two major events as 3988 cal BC (BM-1994R), a IV to V millennia age.

The dates recently received from the analysis of Stratum 27 are: 2675 calBC (UtC-9269) with a 95.4% probability. This date correlates well with the earlier dates of lower Stratum 28 of 3375 calBC (QL-998) and upper level 26 of 2735 calBC (BM-1408), both of which also have a 95.4% confidence probability (see TABLE 1). This series further correlates with Stratum 30 of 3656 calBC (BM-1994) and subsequently others below it. The close correlation of all these levels verifies and matches with very little doubt the sequential order and dating results of the earlier excavations. In the process, it confirms the III-IV millennia BC origin and human use of this particular area of the shelter and suggest an even earlier IV-Vmillennia origin in an o0ther rock shelter, that of Muertos-Gallard (Waldren 1982).

MALLORCA SETTLEMENT RADIOCARBON DATES FOR THE I-VTH MILLENNIA 2002

CAVE & ROCK SHELTER SETTLEMENT

ABSM

244 cal AD
25 cal BC

I

101 cal BC
196 cal BC
266 cal BC
278 cal BC
379 cal BC
379 cal BC
385 cal BC
391 cal BC
400 cal BC
400 cal BC
400 cal BC
402 cal BC
402 cal BC
408 cal BC
408 cal BC
641 cal BC
651 cal BC
651 cal BC
655 cal BC
687 cal BC
770 cal BC
779 cal BC
779 cal BC
779 cal BC
786 cal BC
786 cal BC
789 cal BC
786 cal BC
786 cal BC
791 cal BC
793 cal BC
795 cal BC
795 cal BC
801 cal BC
805 cal BC
809 cal BC
813 cal BC
826 cal BC
838 cal BC
897 cal BC
993 cal BC
993 cal BC

II

1144 cal BC
1314 cal BC
1485 cal BC
1673 cal BC
1685 cal BC
1710 cal BC
1740 cal BC
1833 cal BC
1923 cal BC
1999 cal BC

III

2075 cal BC
2090 cal BC
2288 cal BC
2470 cal BC
2483 cal BC
2523 cal BC
2675 cal BC

IV

2735 cal BC
3375 cal BC
3656 cal BC

V

4635 cal BC
4722 cal BC
5591 cal BC

SMPC

III

2370 cal BC
2480 cal BC
2524 cal BC
2524 cal BC
2577 cal BC
2780 cal BC

SM

V

272 cal BC
905 cal BC
2443 cal BC
4212 cal BC
4225 cal BC
7991 cal BC
7578 cal BC
8050 cal BC
7586 calBC
7893 cal BC
7893 cal BC
10,685
14,000
14,465
14,650
15,885
15,800
16,335
16,850
18,100
18,735
18,800
18,980
23,800
23,800
26,000
26,000
28,600
32,000
42,000

OPEN-AIR URBAN SETTLEMENT

SFO-YS

I

200 cal BC
200 cal BC
408 cal BC
408 cal BC
516 cal BC
646 cal BC
755 cal BC
757 cal BC
760 cal BC
685 cal BC
767 cal BC
770 cal BC
786 cal BC
786 cal BC
790 cal BC
793 cal BC
793 cal BC
801 cal BC
810 cal BC
818 cal BC
823 cal BC
868 cal BC
888 cal BC
897 cal BC
960 cal BC
975 calBC
998 cal BC

II

1100 cal BC
1100 cal BC
1240 cal BC

SFO-OS

1261 cal BC
1261 cal BC
1357 cal BC
1403 cal BC
1428 cal BC
1445 cal BC
1450 cal BC
1516 cal BC
1519 cal BC
1519 cal BC
1523 cal BC
1549 cal BC
1565 cal BC
1673 cal BC
1712 cal BC
1745 cal BC
1745 cal BC
1760 cal BC
1749 cal BC
1770 cal BC
1770 cal BC
1770 cal BC
1755 cal BC
1740 cal BC
1760 cal BC
1780 cal BC
1819 cal BC
1827 cal BC
1829 cal BC
1860 cal BC
1884 cal BC
1884 cal BC
1925 cal BC
1929 cal BC
1970 cal BC
1985 cal BC

III

1995 cal BC
1995 cal BC
2030 cal BC
2074 cal BC
2110 cal BC
2138 cal BC
2239 cal BC
2533 cal BC

SMSS

I

259 cal BC
271 cal BC
274 cal BC
278 cal BC
282 cal BC
282 cal BC
458 cal BC
464 cal BC
465 cal BC
467 cal BC
477 cal BC
477 cal BC
616 cal BC
617 cal BC
629 cal BC
629 cal BC
645 cal BC
666 cal BC
691 cal BC
716 cal BC
786 cal BC
786 cal BC
786 cal BC
787 cal BC
788 cal BC
788 cal BC
794 cal BC
799 cal BC
814 cal BC
832 cal BC
835 cal BC
852 cal BC
872 cal BC
915 cal BC
979 cal BC

II

1026 cal BC
1086 cal BC
1157 cal BC
1167 cal BC
1169 cal BC
1172 cal BC
1195 cal BC
1325 cal BC
1256 cal BC

III

1816 cal BC
1923 cal BC
1926 cal BC
2170 cal BC

TABLE 1.

RADIOCARBON DATES: ROCK SHELTER OF SON MATGE
Valldemossa, Mallorca, Baleares, Spain
2002

nos.	site designation	date cal BC	date BP	lab. inv. no.
1.	ABSMPT-L	244 cal AD	1706 BP± 30 yrs	IRPA-710L
2.	ABSM-PT-L	25 cal BC	1925 BP± 120 yrs	IRPA-710
3	ABSM-PT-L	101 cal BC	2050 BP± 65 yrs	QL-7a
4	ABSM-PT-L15	196 cal BC	2145 BP± 65 yrs	IRPA-695L
5	ABSM-PT-L	266 cal BC	2215 BP± 65 yrs	IRPA-778L
6	ABSM-LIA-T	278 cal BC	2215 BP± 100 yrs	QL-9
7	ABSM-PT-L	379 cal BC	2244 BP± 60 yrs	QL-22
8	ABSM-PT-GT	379 cal BC	2328 BP± 190 yrs	BM-1409
9	ABSM-PT-L	385 cal BC	2328 BP± 65 yrs	IRPA-789L
10	ABSM-PT-L	391 cal BC	2336 BP± 100 yrs	QL-5c
11.	ABSM-PT-L30	400 cal BC	2340 BP± 55yrs	IRPA-808L
12.	ABSM-PT-L	400 cal BC	2349 BP± 55 yrs	IRPA-710
13.	ABSM-PT-L	400 cal BC	2349 BP± 55 yrs	IRPA-789
14.	ABSM-PT-L	402 cal BC	2349 BP± 140 yrs	UtC-1003
15.	ABSM-PT-L	402 cal BC	2351 BP± 60 yrs	IRPA-803L
16	ABSM-PT-L45	408 cal BC	2352 BP± 55 yrs	IRPA-776
17.	ABSM-PT-L65	408 cal BC	2358 BP± 80 yrs	Y-2669
18.	ABSM-PT-L45	641 cal BC	2590 BP± 40 yrs	IRPA-790L
19.	ABSM-PT-L15	651 cal BC	2595 BP± 60 yrs	QL-10a
20.	ABSM-PT-L1	651 cal BC	2600 BP± 70 yrs	QL-10b
21.	ABSM-PT-MIA	655 cal BC	2604 BP± 70 yrs	QL-10
22.	ABSM-PT-L	687 cal BC	2635 BP± 60 yrs	IRPA-751
23.	ABSM-PT-L	770 cal BC	2719 BP± 80 yrs	QL-6
24.	ABSM-PT-MIA	779 cal BC	2728 BP± 80 yrs	QL-4
25.	ABSM-PT-EIA	779 cal BC	2728 BP± 80 yrs	QL-24
26	ABSM-PT-MIA	779 cal BC	2728 BP± 60 yrs	IRPA-752
27.	ABSM-PT-EIA	786 cal BC	2735 BP± 60 yrs	IRPA-676
28.	ABSM-PT-EIA	786 cal BC	2735 BP± 60 yrs	IRPA-751
29.	ABSM-PT-EIA	789 cal BC	2738 BP± 60 yrs	IRPA-803
30.	ABSM-PT-EIA	786 cal BC	2736 BP± 85 yrs	IRPA-811L
31.	ABSM-PT-L90	786 cal BC	2735 BP± 60 yrs	IRPA-752L
32	ABSM-PT-L	791 cal BC	2741 BP± 55 yrs	IRPA-676L
33.	ABSM-PT-L	793 cal BC	2743 BP± 60 yrs	IRPA-1012L
34.	ABSM-PT-L	795 cal BC	2744 BP± 100 yrs	IRPA-790L
35	ABSM-PT-L	795 cal BC	2744 BP± 100 yrs	QL-20
36.	ABSM-PT-L	801 cal BC	2750 BP± 160 yrs	IRPA-695
37.	ABSM-PT-L	805 cal BC	2754 BP± 60 yrs	IRPA-695B
38.	ABSM-PT	809 cal BC	2759 BP± 100 yrs	QL-27
39.	ABSM-PT	813 cal BC	2763 BP± 60 yrs	IRPA-811
40.	ABSM-PT-L10	826 cal BC	2775 BP± 60 yrs	IRPA-806L
41.	ABSM-PT-T	838 cal BC	2787 BP± 80 yrs	QL-11
42	ABSM-T	897 cal BC	2847 BP± 100 yrs	QL-7
43.	ABSM-T	993 cal BC	2943 BP± 50 yrs	QL-986
44.	ABSM-T	993 cal BC	2942 BP± 50 yrs	BM-2140N
45.	ABSM-T	1144 cal BC	3098 BP± 60 yrs	IRPA-750L
46.	ABSM-T	1314 cal BC	3263 BP± 110 yrs	BM-2140
47.	ABSM-PRT	1485 cal BC	3435 BP± 100 yrs	QL-4191
48.	ABSM-PRT	1673 cal BC	3622 BP± 60 yrs	QL-5
49.	ABSM-PRT	1685 cal BC	3634 BP± 50 yrs	BM-1995
50.	ABSM-PRT	1710 cal BC	3659 BP± 100 yrs	BM-2312R
51,	ABSM-PRT	1740 cal BC	3689 BP± 80 yrs	QL-5A
52.	ABSM-PRT	1833 cal BC	3774 BP± 80 yrs	CSIC-180
53.	ABSM-PRT	1923 cal BC	3872 BP± 80 yrs	CSIC-179
54.	ABSM-PRT	1999 cal BC	3948 BP± 80 yrs	UtC-902
55.	ABSM-PRT	2075 cal BC	4026 BP± 45 yrs	QL-24
56.	ABSM-PRT-CE	2090 cal BC	4049 BP± 60 yrs	IRPA-835
56.	ABSM-PRT	2288 cal BC	4237 BP± 120 yrs	Y-2359
58.	ABSM-PRT	2470 cal BC	4419 BP± 100 yrs	QL-23
59.	ABSM-PRT	2483 cal BC	4432 BP± 100 yrs	QL-5B
60.	ABSM-PRT	2523 cal BC	4472 BP± 120 yrs	CSIC-178
61.	ABSM-PRT-EE	2675 cal BC	4625 BP± 40 yrs	UtC-9269
62.	ABSM-PRT	2735 cal BC	4689 BP± 392 yrs	BM-1408
63.	ABSM-PRT-EE	3375 cal BC	5324 BP± 120 yrs	QL-988
64.	ABSM-PRT-EE	3656 cal BC	5520 BP± 50 yrs	BM-1994
65.	ABSM-PRT-EE	4635 cal BC	6592 BP± 115 yrs	I-5516
66.	ABSM-PRT-EE	4722 cal BC	6671 BP± 360 yrs	CSIC-176
67.	ABSM-PRT-EE	5591 cal BC	7538 BP± 120 yrs	QL-29

TABLE 2

What is more interesting is that thes particular levels are only three strata in the series of other hearth zones and human activity levels that are found above and below it, (9 hearths in all) (Waldren 1982) most of which have former

RADIOCARBON DATES
MULETA CAVE & SMPC
2002
Datelist includes recent analyses results from human remains in the SMPC deposit

COMBINED DATES AVAILABLE FROM MULETA

SON MULETA POCKET CAVE

no.	site/no.	sector	95.4% prob .	anal. mater.	lab. inv. no
1.	SMPC-1	Central Area	2780 cal BC	Human Tibia	KIA-14003
2.	SMPC-2	Central Area	2480 cal BC	Human Tibia	KIA-14008
3.	SMPC-3	Central Area	2577 cal BC	Femur Shaft	KIA-14026
4.	SMPC-4	Central Area	2370 cal BC	Human Bone	KIA-14004
5.	SMPC-5	Central Area	2524 cal BC	Human Tibia	KIA-14005
6.	SMPC-6	Central Area	2524 cal BC	Human Femur	KIA-13997

MULETA CAVE DEPOSIT

no	site/period	sector	conv.14c date	date bp	lab. inv. no
1.	SM-PT	C	272 cal BC	2180	SI-651a
2.	SM-T	C	905 cal BC	2760	SI-552
3.	SM-PRT	C-F 20	2443 cal BC	3910	Y-2359
4.	SM-PSP	X 65	4212 cal BC	5350	HAR-2968
5.	SM-150-PRT-	O 150	4225 cal BC	5435	KBN-640d
6.	SM175-PRT-	O 175	7991 cal BC	7135	KBN-640c
7.	SM200-PRT-SM-	O 200	7578 cal BC	7300	UCLA-1704j
8.	PSP-MB	O 175	8050 cal BC	8448	KBN-640b
9.	SM-PSP	O 150	7586 calBC	8570	UCLA-1740c
10.	SM-PSP	X IOO	7893 cal BC	8800	HAR-2932
11.	SM-PSP	X 145	7893 cal BC	8800	HAR-2909
12.	SM-PSP	O 175	12,050±350	10,685	KBN-640c
13.	SM-PSP	Z 350	12,515±315	14,000	SIO1a -RAA
14.	SM-PSP	Z350	12,700±850	14,465	SI-654
15.	SM-PSP	Z250	13,395±655	14,650	UCLA-1704b
16.	SM-PSP	D 250	13,850±600	15,885	SI-646
17.	SM-PSP	X 300	14,385±415	15,800	HAR-2933
18.	SM-PSP	X 300	14,900±200	16,335	SI6-48
19.	SM-PSP	E 350	16,150±600	16,850	UCLA-1704d
20.	SM-PSP	F 300	16,805±555	18,100	SI-649
21.	SM-PSP	X 350	16,850±500	18,735	SI-650
22.	SM-PSP	X 350	17,030±200	18,800	HAR-2971
23.	SM-PSP	Z 400	21,850±480	18,980	UCLA-1704e
24.	SM-PSP	E 400	21,850±500	23,800	SI-647
25.	SM-PSP	E 355	25,050±600	23,800	HAR-2912
26.	SM-PSP	Z 475	25,050±600	26,000	SIO-1cRAA
27.	SM-PSP	Z 525	26,650±600	26,000	HAR-2910
28.	SM-PSP	Z 600	30,050±715	28,600	UCLA-1704a
29.	SM-PSP	Z 750	40,050±1000	32,000	SIO1g -RAA
30.	SM-PSP	Z 650	42,450±400	48,000	HAR-2911
31.	SM-PSP	ST COL	48,000inf	48,000inf	QL-127
32.	SM-PSP	ST COL	58,000inf	58,000inf	QL-128
33	SM-PSP	FL ST FL	80,000pm	80,000pm	Y-1cf

** Most dates are uncalibrated beyond the range of calibration

TABLE 3.

radiocarbon dating. New samples were collected, where ever possible, from these lower strata for possible verification and re-evaluation of already existing dates and levels. The newly collected materials and eventual results will be studied and added to the already existing Matge inventory in order to further augment our knowledge and understanding of the site. These new samples are at present in the process of analysis and assessment.

Of particular importance and interest concerning the lower series is Stratum 32 consisting of a dense charcoal stratum about 10cm thick and compressed ash, running along the entire bottom of the level. This level was the first stratum to produce burnt Myotragus balearicus food bones and is undeniably representative of the first use of the rock shelter's east enclosure as a domestic area (FIGURE 6). This initial charcoal level has been dated at 4535 calBC (I-5516), with a 95.4 confidence probability, for a V millennium BC context.

The implications in this date are quite obvious and the dense fire level from which it originates strongly indicates, along with the other available dates above and below this dated level, of a IV-V millennium BC presence in the east enclosure of the rock shelter. The sequential correlation with those of the preceding levels, Stratum 28 of 3375 calBC (QL-998) and upper Stratum 26 of 2735 calBC (BM-1408) with the recent supporting date for Stratum 27, 2675 calBC (UtC-9269) completes the series or succession of fire and compressed ash strata of the shelter's East Enclosure, further tightening the argument of Balearic III and IV millennia settlement, but also further suggests a V millennium age.

This established succession of dates and strata is one of the strongest possible categories of evidence available in the present interpretation. If we cannot believe in a well dated vertical, stratigraphic succession, we have no tool of comparison or possibility of interpretation.

Beneath this dense charcoal layer (Stratum 32), lies Stratum 33 which is the first and uppermost of the three M. balearicus coprolite levels found under the fire level (Stratum 32). It has, in the overall sequence, a date of 4722 calBC (CSIC-176) and represents the use of the area of the east enclosure as an animal corral for M. Balearicus. Below this are found the second and third coprolite levels. The second of the three (Stratum 34) has a date of 5591 calBC (QL-29)

RADIOCARBON DATES: SON OLEZA CHALCOLITHIC OLD SETTLEMENT and DOLMEN AREA 2002

nos	site designation	years BP	95.4 probability range	date cal BC	lab.inv.no
1.	**SFO-OS-CH1**	2990 ± 160 yrs	1610- 820 (3559-2769)	***1261 cal BC***	QL-4044
2.	**SFO-OS-CH1**	2990 ± 160 yrs	1610- 820 (3559-2769)	***1261 cal BC***	IRPA-1053
3.	**SFO-OS-EXW**	3070 ± 50 yrs	1440-1250 (3389-3199)	***1357 cal BC***	HAR-3490
4.	**SFO-OS-C2**	3095 ± 70 yrs	1520-1250 (3469-3199)	***1403 cal BC***	BM-1698
5.	**SFO-OS-WC**	3150 ± 300 yrs	2192-2161 (4141-4110)	***1428 cal BC***	BM-1988
6.	**SFO-OS-SWF-A**	3180 ± 80 yrs	1670-1668 (3619-3617)	***1445 cal BC***	QL-4191
7.	**SFO-WOS-DOL1: 17-2M8 EX**	3130 ± 35 yrs	1520-1300 (3420-3250)	***1450 cal BC***	KIA-12274
8.	**SFO-OS-WC**	3230 ± 30 yrs	1670-1410 (3619-3359)	***1516 cal BC***	BM-1567
9.	**SFO-OS-CSS**	3240 ± 30 yrs	1611-1441 (3560-3390)	***1519 cal BC***	QL-1224
10.	**SFO-OS-C2**	3240 ± 30 yrs	1611-1441 (3560-3390)	***1519 cal BC***	IRPA-1054
11.	**SFO-OS-EXW-UW**	3260 ± 100 yrs	1854-1849 (3803-3798)	***1523 cal BC***	QL-4040
12.	**SFO-OS-WC-IN**	3280 ± 120 yrs	1880-1310 (3829-3259)	***1549 cal BC***	QL-1896
13.	**SFO-OS-H4**	3300 ± 60 yrs	1738-1708 (3687-3657)	***1565 cal BC***	UtC-3192
14.	**SFO-OS-WC**	3350 ± 330 yrs	2569-2538 (4518-4487)	***1673 cal BC***	BM-1988R
15.	**SFO-OS-WC**	3390 ± 80 yrs	1884-1682 (3833-3631)	***1712 cal BC***	BM-2312R
16.	**SFO-WOS-DOL1: 16-2M4**	3445 ± 40 yrs	1880-1610 (3830-3660)	***1745 cal BC***	KIA-12275
17.	**SFO-OS- EW-INT-B1**	3460 ± 45 yrs	2115-2085 (4064-4034)	***1745 cal BC***	UtC-9021
18.	**SFO-WOS-DOL3: 15-EEP**	3450 ± 30 yrs	1880-1680 (3880-3630)	***1760 cal BC***	KIA-12273
19.	**SFO-OS-CH1-INT4**	3450 ± 110 yrs	1878-1839 (3827-3788)	***1749 cal BC***	QL-4043
20.	**SFO-WOS-DOL1: 12- 4I3**	3440 ± 30 yrs	1880-1640 (3830-3590)	***1770 cal BC***	KIA-12267
21.	**SFO-WOS-DOL1: 11- 5L7**	3440 ± 35 yrs	1877-1840 (3826-3789)	***1770 cal BC***	KIA-10559
22.	**SFO-WOS-DOL1: 9- 5L8**	3445 ± 30 yrs	1930-1680 (3879-3629)	***1770 cal BC***	KIA-10558
23.	**SFO-OS-WCS**	3430 ± 30 yrs	1880-1630 (3830-3586)	***1755 cal BC***	KIA-16267
24.	**SFO-OS-SW-AG**	3405 ± 35yrs	1880-1600 (3830-3556)	***1740 cal BC***	KIA-16268
25.	**SFO-WOS-DOL3: 14-EWEX**	3445 ± 30 yrs	1880-1680 (3850-3590)	***1760 cal BC***	KIA-12270
26.	**SFO-WOS-DOL3: 13- NEX**	3460 ± 30 yrs	1880-1680 (3850-3590)	***1780 cal BC***	KIA-12268
27.	**SFO-OS-EW-UF-S**	3470 ± 50 yrs	1901-1742 (3850-3691)	***1819 cal BC***	QL-4100
28.	**SFO-OS-AT-QB**	3490 ± 30 yrs	1901-1742 (3850-3691)	***1827 cal BC***	QL-1859
29.	**SFO-OS-CH1-INF3**	3490 ± 30 yrs	1946-1768 (3895-3717)	***1829 cal BC***	QL-4042
30.	**SFO-WOS-DOL1: 2- 5M3**	3540 ± 30 yrs	2198-2159 (4147-4108)	***1860 cal BC***	KIA-10557
31.	**SFO-OS-EWF**	3550 ± 110 yrs	2007-2004 (3956-3953)	***1884 cal BC***	UtC-2742
32.	**SFO-WOS-DOL1: 12-5L7**	3550 ± 30 yrs	2030-1988 (3979-3937)	***1884 cal BC***	KIA-10560
33.	**SFO-WOS-DOL1: 1**	3565 ± 35 yrs	2030-1770 (3980-3720)	***1925 cal BC***	KIA-1228
34.	**SFO-OS-EXW**	3585 ± 35 yrs	2040-1880 (3990-3930)	***1929 cal BC***	UtC-8952
35.	**SFO-WOS-DOL1: 10-6M4**	3605 ± 35 yrs	2137-2076 (4086-4025)	***1970 cal BC***	UtC-10079
36.	**SFO-OS-EW-INT-B2**	3620 ± 50 yrs	2123-2097 (4072-4046)	***1985 cal BC***	UtC-9022
37.	**SFO-WOS-DOL1: 4-5M9**	3625 ± 30 yrs	2123-2097 (4072-4046)	***1995 cal BC***	KIA-10583
38.	**SFO-WOS-DOL1: 7-5M6**	3625 ± 30 yrs	2279-2231 (4229-4180)	***1995 cal BC***	KIA-10582
39.	**SFO-WOS-DOL1 J3-G13**	3640 ± 50 yrs	2199-2158 (4148-4107)	***2030 cal BC***	BM-1981R
40.	**SFO-OS-EW- INT-CC**	3680 ± 45 yrs	2199-2158 (4148-4107)	***2074 cal BC***	UtC-9023
41.	**SFO-WOS-DOL1: J9 W17**	3700 ± 30 yrs	2290-1930 (4240-3880)	***2110 cal BC***	QL-1592
42.	**SFO-WOS-DOL1: G13 FLT**	3720 ± 35 yrs	2275-2141 (4224-4198)	***2138 cal BC***	BM-1981
43.	**SFO-OS-NW-BB**	3790 ± 90 yrs	2480-1970 (4429-3919)	***2239 cal BC***	QL-1636
44.	**SFO-OS-EW-BB**	4030 ± 60 yrs	2585-2568 (4537-4512)	***2533 cal BC***	BM-1843R

TABLE 4 .

analysed on M. Balearicus bone, retrieved from the coprolite bed and, if accepted as part of a corral area, on the basis of the density the coprolite beds (Figure 4), extends human presence even as far back as the VI millennium, although admittedly this is not as convincing as the kitchen fire levels above it.

It has been one of the authors' opinions that the three dense coprolite beds, based on their depth (90 cms) and other considerations, represented evidence of the area having been used as a corral for M. balearicus, as well as strong indication of attempts at domestication of the animal. The argument being that M. balearicus, whose feeding habits were not only highly specialized, as seen by the development of its two ever-growing, chisel-like incisors in it lower jaw, but signal the caracteristics of a grubber and forager, rather than those of the normal browser or grazer. If so, in foraging, it probably never spent a long time in any given place, especially long enough for the coprolites to have accumulated to the depth they had in the shelter, despite the fact that animals of this sort do use the shelter of caves and rock shelters when sick or to get away from incumberrant weather.

These strata are also the levels in which the controversial 'v' trimmed horn specimens originated (Waldren 1982; Alcover et al 2000). It has been suggested that these are not the work of human intervention, but rather M. balearicus knawing on the bones, as a food suppliment, after the animals were dead, perhaps during a period when the animal was undergoing mass starvation. While this phenomenon is well known, such configurations, especially on two horn cores at once, and on all the heads retrieved from the deposit, are believed not always to be the case as actual culling of horns is quite common with horned domestic animals, to prevent them from doing damage to one another in corralled situations or to prevent them (for example goats: personal correspondence with local farmers) from becoming snarled in briar patches or wire fences, or irksome goats from butting. A method carried out as well as in the case of prize bulls doing damage to the cow during studding activities. (see Waldren 1982, pp. 594-602 and Plt. 27 for the complete argument).

In all events, it is not the intention here to examine all the possible arguments as to what does or does not constitute hard evidence, apart from vertically accumulated kitchen fires and kitchen debris, signs of human presence or the human use of M. balearicus as a possible food source. It is principally to examine the stratigraphical evidence, along with the closely related sequence of dates with their 95.4% confidence probabilities (TABLE 1), that are closely associate with levels involving the animal. In short, contexts which clearly show that human presence in all reasonable probability possibily took place in the Balearic Islands, as early as the IV and V millennim before the present epoch, along with the extra curriculum intention of off-setting some of the damage done in the recent claims that the Balearic Islands were only occupied slightly prior to 2030 cal BC. There are also numerous other logical arguments, too numerous to discuss here, for the likelihood that human settlement may possibly have been even earlier in the Balearics than the evidence suggests at present, evidence that will appear in the not too distant future.

In any case, the existing collective, chronometric evidence is the best we have for the strong support of such a presence in the islands earlier than assumed. This is further examined in the following sections where we will discuss the issues at hand in conjuction with the Muleta cave deposit.

THE EVIDENCE

MULETA REVISTED

To further the present evidence for a IV-V millennia settlement of Mallorca, a return to the research on existing material from the Cave of Muleta, where the question of early human settlement first arouse, was undertaken in 2000 to complement the Matge re-assessments and defend the original settlement hypothesis of an earlier than III millennium BC human contact.

It should also be understood that the authors do not consider the first mainland arrivals as a form of occupational invasion but initially the sporadic arrival of a small number of individuals who dispersed themselves, mainly, in the mountainous areas of the island, near the sea and in the protective cloisters of caves and rock shelters. Here, they survived until gradually others arrived with new cultural and technical innovations circa the 4th millennium. The intensity and frequency of those arrivals are by no means ascertainable at present, all attempts to do so would be sheer quess work and, perhaps, the only hints we have are what can can observe as having once transpired in adjacient mainland regions.

The present undertaking will review old data in the light of new findings from this important site and perhaps be a small but important issue toward the clarification of some of the ambiguity in the present local climate. However before this process can take place some background discussion is necessary as a matter of continuity and background.

BACKGROUND

The excavations first initiated in 1962 were carried without break until 1982, supported by Smithsonian Institute awards and National Geographic Society Grants for Research and Exploration during which time several major international survey programmes and analytical projects were initiated, using a number of new and established analytical techniques carried out on materials from the deposit (Waldren 1982; Waldren in press), with one end in mind to help clarify the origin and nature of the deposit.

Analytical results were obtained from such renown scientific organizations as the Smithsonian Institution, Scripps Institute of Oceanography, Oxford University, Yale University, Peabody Museum of Natural History, British Museum of Natural History and Antiquity, Harwell Atomic Laboratories, Geochronological Laboratories of the University of Washington and others, that have produced among other analytical surveys a radiocarbon inventory of some 31 dates for the stratigraphy (TABLES 1 and 3), along with other physical, biological and chemical analytical results (see Waldren 1982.

Enhancement of this intersite inventory list of radiocarbon

determinations and other dating results has since been possible, as well as the clarification and verification of the small number and range of dates available, especially those

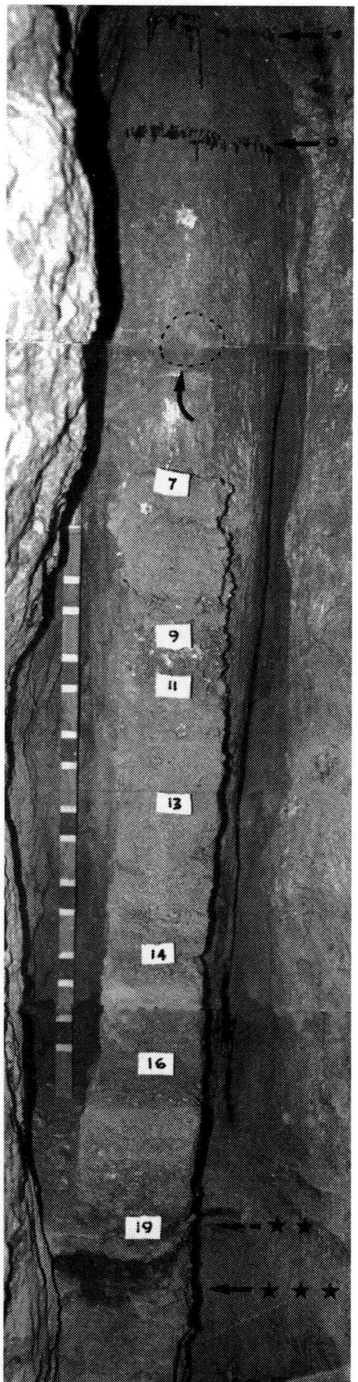

FIGURE 7. MULETA STRATIGRAPHIC TESTIMONIAL: 1972

involving human intervention. These are presented here in Tables 1 and 3, as a list of 6 new readings from a small subsidiary cave, along with the former dates from Muleta are only the first in subsequent analysis and research (eg Van Strydonck and Davis, two separate papers in this volume).

Since as early as 1970, human remains found in a small cave adjacent to the main Muleta system, referred to here as Son Muleta Pocket Cave (SMPC), have posed a problem in the interpretation of human presence in the cave proper. The SMPC is part of the same geological formation and overall cave sytsem, where the first collection of human bones and pottery were found. The inventory of human body components retrieved from Muleta proper and the SMPC site have shown finally that several individuals (3 or 4) were involved from the Muleta cave and not a single person as reported earlier. These consist of bone components of what was at first considered a single individual most of which was thought to have been removed from the Muleta cave proper by later cave inhabitants for reburial elsewhere in the SMPC site, the parts of which have since been recognised as being more than one body and not from the same single individual first assumed.

The recent remains analysed from the adjacent SMPC deposit were first considered part of the inventory of missing human remains encountered earlier and associated with M. balearicus in Muleta proper. These were subsequently thought to have been removed much later by occupants using the Muleta cave as a transcient camp, circa 2379 calBC (Y-2258) (dated kitchen debris) and reburied by them in the SMPC site. These human remains eventually deposited in the central Museum in Palma and have been a source of enquiry since their discovery in the early 1970s. As the Muleta proper remains were highly fragmentary, along with the fact that the body component inventory was far from complete, led to the idea that perhaps the bone components from the SMPC site had originally belonged to the remains from Muleta proper, which we now can regard as having not been the case, although one of these has been secretly dated and asessed as proof of the actual date of Mallorcan early settlement.. The human finds in the SMPC site consisted of badly broken, disartictulated bones of what was at first also considered part of the single individual that was removed from the Muleta cave proper by earlier occupants and hastily redeposited in the SMPC site, along with a single ceramic, with part of a badly broken and partial human skull in the pot. These human remains from the SMPC have since been recognised as being not only from more than one body but constitute materials from a collective burial cave used over a period of several hundred years, and no longer related directly to a single burial nor, as first thought, part of the Muleta proper human deposit.

Part of the earlier human bone components from the Muleta cave proper were dated in 1966 and found to be of the age range of 3930 bc (KBN-640D), a calibrated age of 4225 calBC (TABLE 3). While this tentative date was considered by one of the authors (Waldren 1982) as strong indication of its age and possible association with the islands endemic fauna M. balearicus, it was also realized that further evidence was necessary before it could be considered something more than an indication of such association and antiquity.

This supporting evidence was eventually forthcoming in

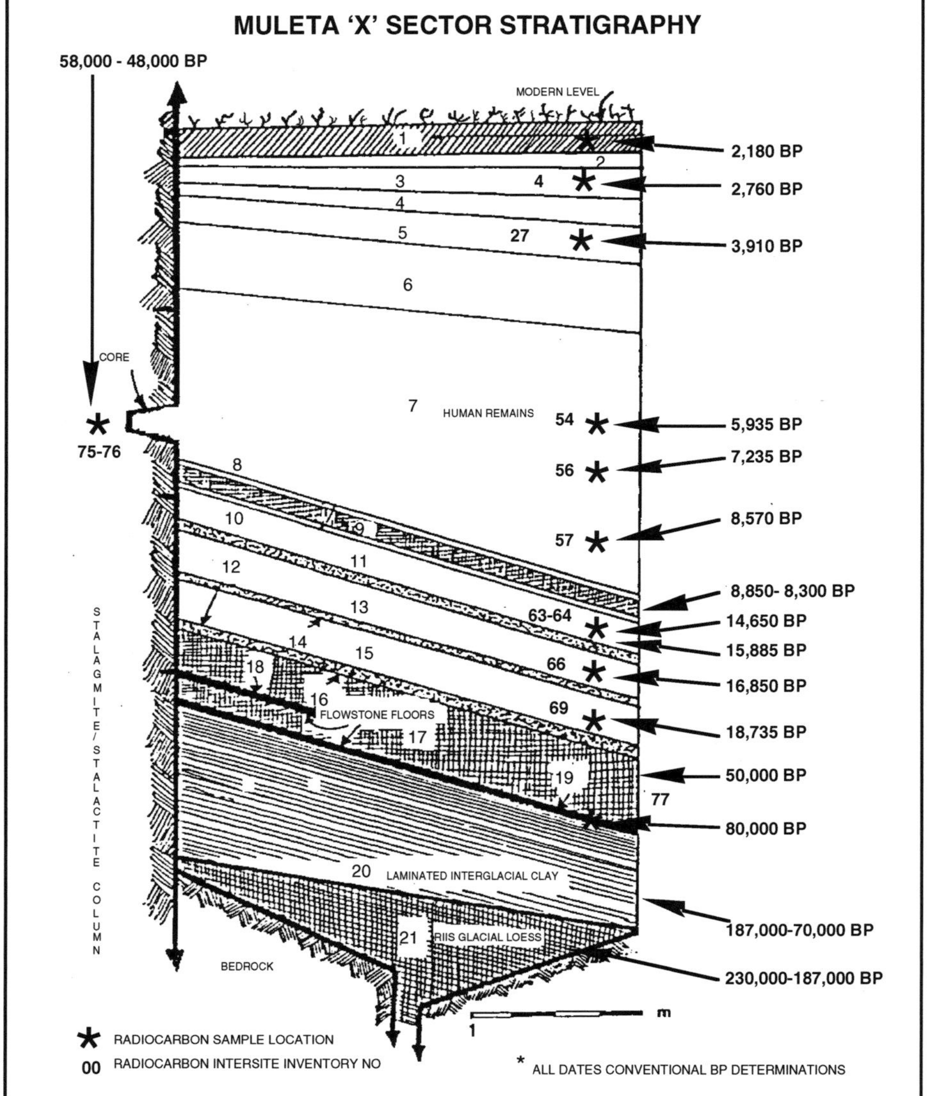

FIGURE 8. MULETA 'X' SECTOR STRATIGRAPHICAL SECTION SHOWING RADIOCARBON DATES IN YEARS BP
(250,000 years of sequential strata and the dates, representing the Riss-Wurmian-Holocene origin of this cave sector)

the Matge rock shelter which subsequently gave the affirmation discussed ealier in this text. Once this correlation with another site was made possible, the Muleta date was, although a singular example of early settlement, given additional credence and support. Unfortunately, at the time of the first dates large quantities (500grms) of human bone were necessary for a reliable date. However, not before smaller amounts were first analysed in the form of two 56 grms samples. The results of these samples gave huge statistical margins of error, but all the same gave a clue to their antiquity, despite the large discrepancies. The larger samples of 500 grms were subsequently submitted and analysed. The results from these larger samples gave the uncalibrated date of 3930 bc (KBN-640d) ('O' Sector, 150 cms) with a corresponding M. balearicus bone sample ('O' Sector, 175cms), taken from 25cms directly below the human sample gave a date of 5135 bc (KBN-640c). It has been subsequently suggested (Alcover et al 2001) that the sample used for dating included part of M. balearicus bone ; hence the old date. This was categorically not the case

Unfortunately, reaffirmation of these exact samples is not possible, because the bone used for analysis was destroyed in the analyses. However, despite this fact, there is no real reason to assume that these dates are totally unreliable, as recently suggested, because radiocarbon dating was not as accurate at that date as it is at present or by mixing of samples. It is an accepted fact that the same samples dated today might well give different answers, but how importantly or radically different we do not know, although the the motives for doing so may be personal rather than scientific. Admittedly events can go wrong at any point in time, despite the better technology, this is not always the case and old dates can certainly be as indicative as the new ones carried out at present.

For example, as pointed out earlier in the question of open-air Chalcolithic dates and any corresponding ones originating from the closed environments of caves and rock shelters samples, their recalibration, the time it takes for analysis, preparation and the equipment used, and of course our growing experience and knowledge of the analytical method. It is more a case of sour grapes to try and look for reasons to doubt older results when there are so many of them in use at present. Therefore, the argument of discarding old dates is not a reasonable or valid argument, but perhaps rather differently motivated. Samples processed at present and those processed in the early years, both uncalibrated

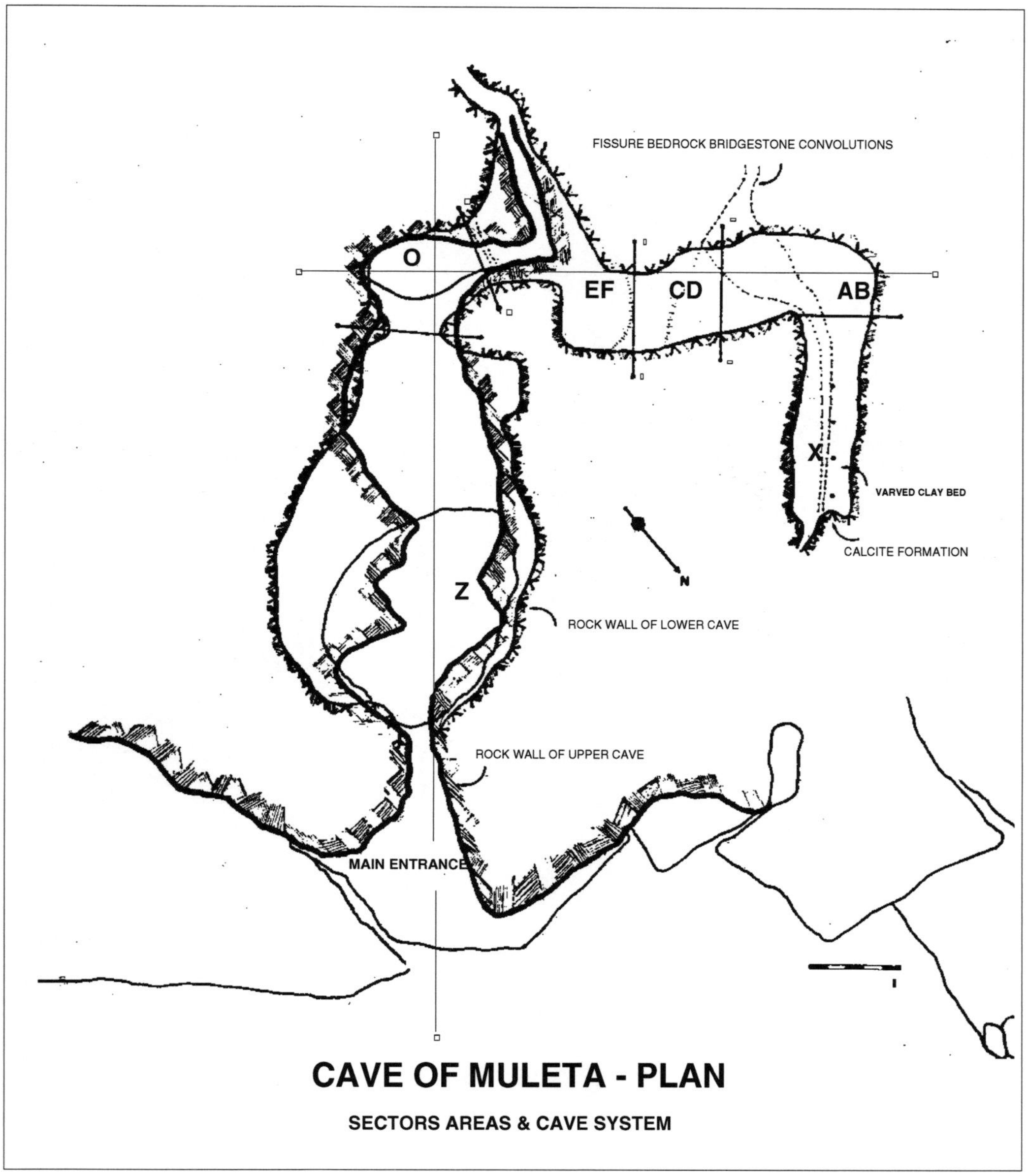

FIGURE 9.

do not apparently differ in the least from those conducted as early as 40 years ago from the same sort of contexts. Modern labs are still producing complementary dates comparable to those collected long ago, although there are large differences in staistical margins of error. Therefore, the principal differences lie in particulars like the amount of test material needed, the statistical margins of error of the and calibrated, do not for the most part significantly differ, apart from the statistical margin of lab error, statistical probabilities and the fluctuating calibration curves themselves. A margin of error of a hundred years or so one way or the other in distant antiquity is only as essential or critical as we tend to make it, and although we must seek to perfect the technique and its accuracy as well as our knowl-

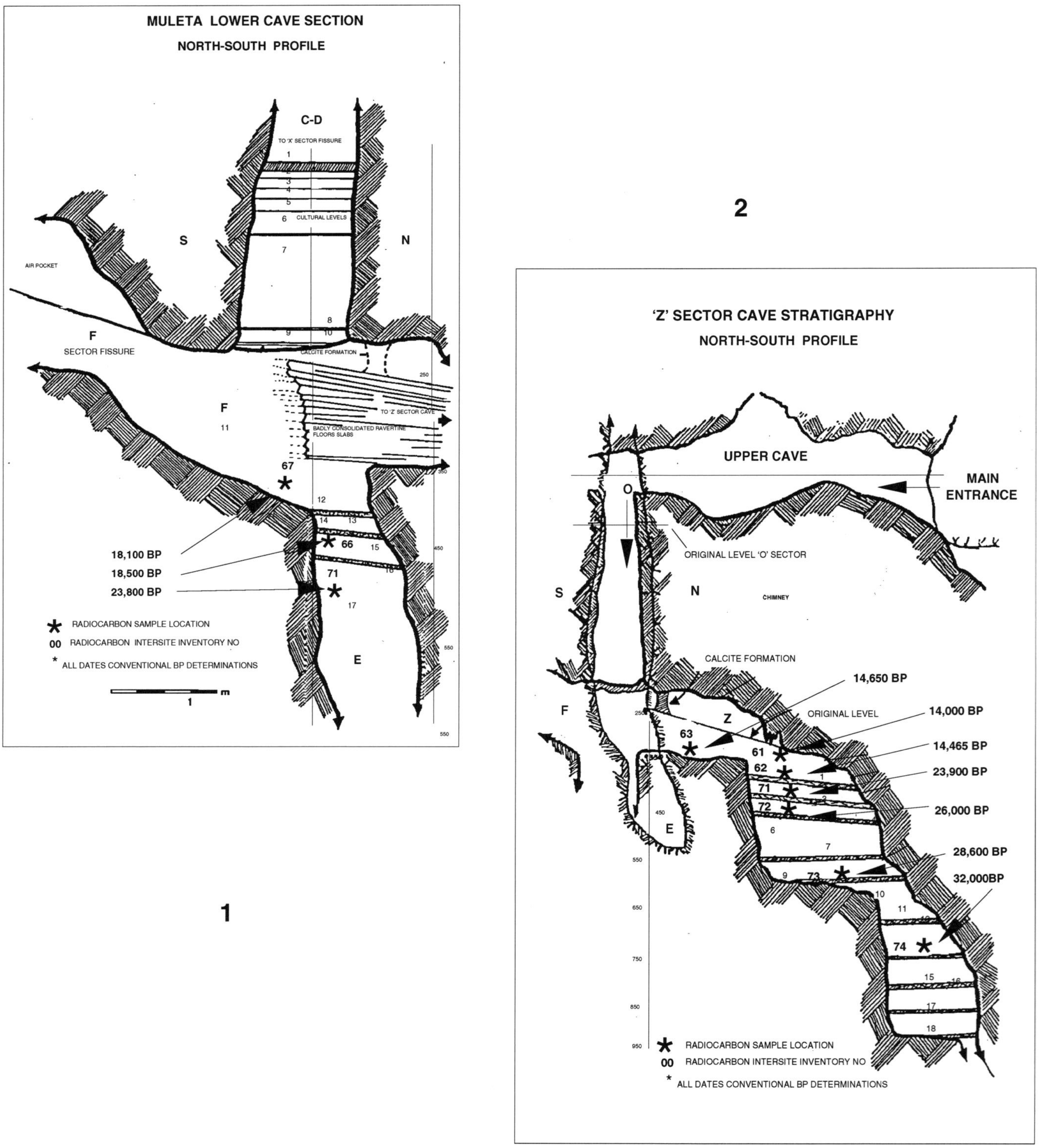

FIGURE 10. MULETA STRATIGRAPHICAL SECTIONS: NORTH-SOUTH PROFILES

edge of events, we never will be completely satisfied with results, either as archaeologists or scientists.

RADIOCARBON DATES: PREHISTORIC SANCTUARY OF SON MAS
Valldemosa, Mallorca, Baleares, Spain
2001

nos	site/sector	BPdate ±	calBC date	Lab Inv
1.	**SMSS-12W9**	3775 BP± 35 yrs	*2170 cal BC*	UtC-4676
2.	**SMSS-8O9**	3580 BP± 65 yrs	*1926 cal BC*	UtC-909
3.	**SMSS-9O8**	3570 BP± 65 yrs	*1923 cal BC*	UtC-908
4.	**SMSS-7O7**	3510 BP± 60 yrs	*1816 cal BC*	UtC-2020
5.	**SMSS-8J3**	3060 BP± 35 yrs	*1325 cal BC*	UtC-5892
6.	**SMSS-6N5**	3020 BP± 60 yrs	*1256 cal BC*	UtC-2756
7.	**SMSS-6L4**	2990 BP± 70 yrs	*1195 cal BC*	IRPA-1053
8.	**SMSS-12L5**	2970 BP± 40 yrs	*1172 cal BC*	UtC-2727
9.	**SMSS-6L7**	2960 BP± 60 yrs	*1169 cal BC*	IRPA-1049
10.	**SMSS-10W2**	2940 BP± 40 yrs	*1167 cal BC*	UtC-9017
11.	**SMSS-7N2**	2970 BP± 70 yrs	*1157 cal BC*	IRPA-976
12.	**SMSS-8L8**	2930 BP± 40 yrs	*1086 cal BC*	IRPA-984
13.	**SMSS-6K7**	2875 BP± 35 yrs	*1040 cal BC*	UtC-3044
14.	**SMSS-8J3**	2880 BP± 80 yrs	*1026 cal BC*	UtC-5891
15.	**SMSS-12J2**	2830 BP± 140 yrs	*979 cal BC*	UtC-2736
16.	**SMSS-12X5**	2795 BP± 35 yrs	*915 cal BC*	UtC-8954
17.	**SMSS-8K9**	2735 BP± 30 yrs	*872 cal BC*	UtC-5420
18.	**SMSS-8L5**	2690 BP± 70 yrs	*852 cal BC*	UtC-1256
19.	**SMSS-10S8**	2700 BP± 60 yrs	*835 cal BC*	IRPA-1055
20.	**SMSS-5M5**	2700 BP± 50 yrs	*832 cal BC*	UtC-2759
21.	**SMSS-8J8**	2655 BP± 35 yrs	*814 cal BC*	UtC-5675
22.	**SMSS-12X5**	2695 BP ± 70 yrs	*799 cal BC*	UtC-8953
23.	**SMSS-10L2**	2600 BP± 70 yrs	*794 cal BC*	UtC-1255
24.	**SMSS-8K9b**	2590 BP± 30 yrs	*788 cal BC*	UtC-5428
25.	**SMSS-18U9**	2590 BP± 50 yrs	*788 cal BC*	UtC-4170
26.	**SMSS-19V1**	2590 BP± 35 yrs	*787 cal BC*	UtC-4166
27.	**SMSS-7L7**	2580 BP± 25 yrs	*786 cal BC*	UtC-5421
28.	**SMSS-8J8b**	2565 BP± 30 yrs	*786 cal BC*	UtC-4857
29.	**SMSS-7L7**	2565 BP± 30 yrs	*786 cal BC*	UtC-5374
30.	**SMSS-5L3**	2470 BP± 25 yrs	*716 cal BC*	QL-4264
31.	**SMSS-16S2**	2480 BP± 50 yrs	*691 cal BC*	UtC-3190
32.	**SMSS-19V1**	2490 BP± 35 yrs	*666 cal BC*	UtC-4167
33.	**SMSS-12W5**	2480 BP± 30 yrs	*645 cal BC*	IRPA-1250
34.	**SMSS-7N8**	2520 BP± 80 yrs	*629 cal BC*	UtC-1002
35.	**SMSS-9K6**	2520 BP± 60 yrs	*629 cal BC*	UtC-1257
36.	**SMSS-8M6**	2500 BP± 40 yrs	*617 cal BC*	IRPA-836
37.	**SMSS-5L6**	2520 BP± 60 yrs	*616 cal BC*	IRPA-1025
38.	**SMSS-13U9**	2580 BP± 70 yrs	*613 cal BC*	IRPA-1051
39.	**SMSS-6K9b**	2440 BP± 90 yrs	*477 cal BC*	UtC-1258
40.	**SMSS-13U9b**	2430 BP± 40 yrs	*467 cal BC*	IRPA-1066
41.	**SMSS-13X4**	2430 BP± 25 yrs	*465 cal BC*	UtC-3188
42.	**SMSS-18Y1**	2400 BP± 70 yrs	*464 cal BC*	UtC-5426
43.	**SMSS-6V2**	2390 BP± 50 yrs	*458 cal BC*	UtC-3933
44.	**SMSS-10L2**	2360 BP± 140 yrs	*454 cal BC*	UtC-1003
45.	**SMSS-8M6**	2220 BP± 70 yrs	*282 cal BC*	UtC-1001
46.	**SMSS-12X5**	2200 BP± 40 yrs	*278 cal BC*	UtC-9018
47.	**SMSS-7O7**	2210 BP± 90 yrs	*274 cal BC*	QL-4200
48.	**SMSS-17Y4**	2210 BP± 110 yrs	*271 cal BC*	UtC-3045
49.	**SMSS-17Y7**	2140 BP± 50 yrs	*259 cal BC*	UtC-3189
50.	**SMSS-18Y6**	2140 BP± 70 yrs		UtC-3046
51.	**SMSS-7O7**	2050 BP± 150 yrs		QL-4201
52.	**SMSS-18Y2**	1960 BP± 40 yrs		IRPA-1026
53.	**SMSS-8L3**	530 BP± 50 yrs		IRPA-1024

TABLE 5.

Meanwhile, the Muleta and Matge dates have remained the unique representatives of the earliest human evidence in the Balearics, apart from one other even earlier date showing human association with the endemic fauna in the Canet Cave, Mallorca (Kopper 1984). These have been generally accepted by authorities both on the islands and elsewhere as reasonable assessment and estimates for early human presence in the islands, as unsatisfactory as their number may be. They have also given us the first accurate assessment for the survival of M. balearicus, which until then was thought to have become extinct during the last glaciation (40,000 to 20,000) years ago and have demonstrated its survival well into the Holocene (Clutton-Brock and Burleigh 1978), where there is every indication and reason to believe the animal co-existed with humans. Hence this first data can be regarded as a logical and practical starting point regarding IV -V the millennia early settlement.

The new series from the Muleta Pocket Cave should do much to fill in missing and essential data as to the question of early III to late IV millennium settlement and origin and age of the remains from the SMPC site, as well as further clarifying the Muleta proper series. All the dates show 95.4% confidence probabilities to support their age (TABLE 3). As all the past and present tests were conducted on human bone components and not on charcoal where the 'old wood' effect possibly could alter results, their scientific reliability must be seriously considered, along with the other pertinent radiocarbon dates presented in TABLES 1-4, as strong argument against an only slightly prior to 2030 calBC date for human settlement in the Balearics.

MULETA STRATIGRAPHY IN BRIEF

In this section a set of stratigraphical schematics are illustrated showing an analysis and synopysis of the Muleta sectors levels and their coinciding dates. This is done here as a matter of continuity and principally to show the sequential order of the cave strata and their related dates, and the manner in which they occurred throughout the different sectors of the deposit. In this way the readers may see the deposition in the different sectors for themselves It is also presented in this way to show the unusual sequential continuity of the dates in reasonable graphic detail. Most of all, they are presented graphically and statistically in order to relate them practically to the upper cultural levels and in turn to the questions and issues of early settlement.

Finally, as discussed above, this is the cave deposit from which the earliest dates for human presence first originated and that are currently in question and which despite recent argument are still believed to be valid, even though the original materials used for testing have been destroyed in analyses a long time ago. This way the readers may decide for themselves by graphically seeing the order in which samples were found, as well as their accompanying stratigraphical contexts. Should there be a need for still more detail, as was pointed out earlier, this can be obtained in Waldren 1982 and other pertinent literature.

THE DEPOSIT

The cave is far from a spectacular one. Briefly it is a relatively small solution cave system desolved out of a Jurrassic linestone,'karst' outcrop overlooking the sea. The cave has only one, small, keyhole shaped entrance, leading into an upper cave chamber, about 8 meters long, with a vertical chimney at the rear of it. This leads vertically down through an opening about 50 cms in diameter into the lower chamber in which the main deposit was located.

The cave deposit and the cave plan in general was divided into a series of cave sectors for excavation (FIGURE 9). The excavations were carried out on a six month a year basis from 1962 to 1982 and included an 8 year hydraulic microfauna extraction project in which 90% of the caves soils were washed through seven graduated screens. Muleta, because of its well preserved contents and stratified deposit, has been the test reservour for numerous dating projects and testing grounds for many other analytic techniques over the years (see Waldren 1982). It was the first Balearic cave to be systematically and extensively investigated and aspects of that research are still very much that of an on-going project to the present.

In FIGURES 8 and 10, the schematic illustrations clearly show the pertinent cave vertical stratigraphies and matching chronometric data retrieved. They also show, particularly in Figure 8, the extent and succession of levels which have enabled us to interpret and determine the age of the entire deposit in 'X'Sector of the cave, as an unbroken series of 21 consecutive levels extending from the Riis Glaciation, some 230,000-187,000 years ago, through the Riis/Wurmian Intergalacial Interim to the Wurmian Glaciation, some 80,000 years ago and into the recent Holocene Period.

There exists within this span of time, in this stratigraphical section alone, a datable succession of remains of the island's endemic macro and micro fauna as well as some human record. Other sectors of the cave are also represented in Figure 9, especially in 'Z'Sector, which has accumulated the oldest record of M. balearicus in the cave, circa 42,000 BP. Dates from these sectors and levels are recorded in Table 3.

One of the remarkable characteristics is the excellent preservation and condition of the bone remains which are osseous materials and not fossilized remains, hence, ideal for dating and biometric study surveys. This preservation is due to the almost constant temperature in the cave, which varies less than one degree C, maintaining an average temperature of 19.5 degrees centigrade, or 68.5 Farenheit, over the course of the year.

The deep, vertical stratification of the Muleta deposit, along with that from Matge can be regarded together as a prehistoric 'yardstick' covering the entire ecological and cultural pehistoric sequence in the Balearics, a factor that has made them extremely valuable as prehistoric test and reservoirs and reference sites, the investigation and research of which are still very much in progress.

CONCLUSIONS

What we can observe from the evidence at present is that probabilities favour a much earlier settlement than an early III millennium BC origin. There is still in the combined evidence in both the Muleta and Matge deposits considerable data supporting the past hypothesis of a IV-V millennia human arrival on the islands. In the authors' opinion recent assessments of only a little better than III millennium origin are just not adequate and by far have not taken all the evidence into account or been unbiasedly interpreted correctly. As an example elsewhere, island settlement is being increasingly shown to be earlier than thought and more and more examples emerge that demonstrate the fact that people moved around much earlier than believed, particularly in the Mediterranean (see Jacobsen 1976), where there is a long period of fine seasonal weather favourable to navigation, compared to other world regions where far less favourable conditions have shown people moved about very early snd over large distances (see Nash, this volume).

Is it possible that the Balearics and Pityussaes are the imbreded 'backwaters' they were believed to be thirty years ago, where nothing of any importance ever happened? The authors do not believe so. In prehistoric times, even as today, dramatic and universally important events happened in the rich prehistoric past of the islands. This has been demonstrated quite emphatically in the case of Chalcolithic settlement, where sophisticated and technologically advanced settlement existed as early as the middle III millennium BC, with occupational evidence in caves showing early and middle IV millennium BC origin and suggestion of even earlier contact in the Vth millennium. From this, one can, well deduce that the true date for this initial occupation is yet to be established and that the available dates are but a beginning. Having said this, there is no logical or probable reason that we may not eventually come to find evidence of an earlier than IV or V millennia arrival. After all such dates are only Neolithicin age. With such an endemic meat source as Myotragus balearicus a Mesolithic presence may eventually prove itself. It is a question of investigating and researching the right areas and having the good fortune and skills necessary to recognise the evidence when one sees it and procede from there. After all, the laws of differential preservation, as it has always done, makes evidence sparcer and more fragmentary the farther back we go in time.

If we look at the implications (see Links in the Chain, this volume), there is no valid reason to assume that the distances between the individual islands and between the islands and mainland presented any insurmountable obstacles to early voyagers or transcients. On the contrary, there never have been barriers anywhere that could not be overcome, given enough time and will to seek it out. In short, the evidence is all there and all we have to do is to look for it... in the right places. We cannot expect nor be so vain as to believe it will come easy and that we will find it in the first cave or rock shelter where we pass a season or two in sporadic prospection or frantic excavation. It has taken 4 decades to arrive at what we have discussed here and it is still only a beginning and far from what is lastingly essential.

It will be long-term commitment and dedication to a few potentially prolific sites and well designed on-going strategies and objectives that will be needed within a climate of cooperation and close communication to achieve the fruits of what our intelligence, investigation, research and the various disciplines are capable of realizing. It is certainly not any product of competitive behaviour, professional gain or egotistical motives.

THE RADIOCARBON INVENTORY DATES

In the lists of calibrated dates presented in TABLES 1-5, an interesting pattern emerges of a closely related series of dates that demonstrate an unusual continuity and homogeneity. It becomes evident from the comparison of the individual lists of dates from site to site that they are not only closely related but belong to particular blocks of time or periods, a factor which definitively unites and links them them convokingly as a body of dates. As such, among other things, they cover and support what are referred to here as the local Early Settlement Period (ESP), the Neolithic Early Ceramic Phase (NECP) and the Chalcolithic Early Beaker and Late Beaker Phases (EBP-LBP) of the local Pretalayotic Period (PRT), 5597 calBC-1400 calBC, a period of 4200 years.

What also makes these individual date lists of particular value and interest as a comparative body of dates is the fact that they come from sites that are in all probability interactively related and bonded together from a social and economic point of view throughout time and as such they form a workable and useful environmental model, admittedly as yet not as clear as desired.

The model itself includes two vertical stratigraphies and two horizontally distributed situations that demonstrate no major break in their continuity that cannot be logically and adequately explained, thus offering a classical example and reliable tool for archaeological interpretation. To date, regretably, there are no precedents for these vertical and horizontal ecological and cultural accumulations. It is equally interesting that the Muleta stratigraphy with its much older deposit and depth of over 10 meters reasonably shows a similar structural and chronological continuity as that of the cultural accumulation at Matge, although definitively older. Between the two sites, as briefly pointed out above, they form a single, overlapping stratigraphy representative of much of local prehistory.

Another interesting and reassuring aspect is found in the apparent close correlation made possible among the settlements and the caves and rock shelters, as well as recent ritual sites, such as the Prehistoric Sanctuary of Son Mas (SMSS), which, apart for an explainable break, has the same continuity (TABLE 5) (see Van Strydonck,Waldren, Ensenyat and Orvay, this volume). To be able to include rural settlement and other open-air activity areas in our assessments, where we do not have access to vertical stratigraphy for interpretation or verification, is an important advantage, one that did not exist before the discovery of these two deep stratigraphies. Balearic caves and rock shelters are not especially renown for their deep deposits.

The opposing roles normally played by these quite different environments and physical situations in interpretation have often been the crux of the matter involving settlement assessment. In the sites involved, paradoxically to a significant degree, what can be assertained as having occurred in one, can be seen often to be reflected in the other. The ability to do this type of comparison between the two, the authors believe, has been a great help in the interpretation and understanding of prehistoric time in the various sites investigated over the various periods.

It is only logical to assume, with the bulk of existing evidence and the knowledge of these comparative abilities between these two different environments, that it will take far more than the repudiation of a single date or a small array of dates by those unable see this or to use such comparisons for them to set back or disrupt what has been so far accomplished. If we cannot believe in the realities of vertical stratigraphical succession, controlled horizontal distribution and correlative data, as well as statistical probabilities, then our discipline is in a bad way and therefore has no solid basis for existence.

Fortunately, there are those (eg. Lull et al 1999) who are becoming increasingly aware that extensive, long-term and on-going research and investigation with clear strategical long-term objectives and innovated methodology and creative, impartial assessment are needed in the solution of prehistoric problems, untainted by personal motives and ambitions and the incapacitating awareness and pride of being 'insular' and an 'island to their own liking and thinking'.

The present paper, the subject of both long-term investigation and the chronometric issues related to early island settlement, does not deal adequately with the subject and certainly does not include the amount of detail necessary to do justice neither to the subject nor does it bring the necessary clarity and insight into the question as a whole. At best, it is merely a brief defense, to what the author's feel are some essential facts and details that have been unconsciously or consciously ignored, perhaps, even manipulated in the most recent assessment of early Balearic settlement (eg. Alcover et al 2001). It is also intended to offer and add further details to the astute and profound critique and assessment of initial Balearic settlement by Lull et al (2000). In all events, it would take a lot more space and time than allotted to respond more fully to the virtues or shortcomings of these recent assessments. For the moment, this is not possible, although the opportunity to do so is, perhaps, not too far off.

Finally, far more seriously, the objective here has been to present the new as well as some of the old data and reveal certain aspects of the on-going investigation and research so far not communicated to the outside or off-island scientific community. There should be nothing particularly controversial in the evidence offered here other than perhaps the brevity and manner in which it has been presented. It is believed however that the evidence, despite its brevity and manner, supports a much earlier than III millennium initial occupation of the Balearics and that this appears and offers, on the basis of the statistics and other evidence, both a reasonable, although at the same time tentative, argument and assessment of the situation and climateas a whole.

Therefore, it should be stressed once more that in the authors' view, it is difficult and almost impossible to assign any definitive date at present to the initial stage of local settlement. Nor was it the original intent to do so in the past assessments, which were always provisional, although they were accepted by some and rejected by others. The whole issue is at the moment ambiguous, our own argument included. There is at present too many missing factor and details for recent assessments to be convincing, as some would desire. At best, all we can do, as is the intention here, modestly to suggest possibilities and present probabilities, trying to add to the available data bit by bit and not agressively set about trying to removing it.

It is also logical and a matter of history to expect philosophically that the moment we close the door on possibilities by expressing statistical probability, we are at the same time opening the door for a disappointment, if not today, then tomorrow. It should be honestly expected and generally expected that probabilities are circumstantial evidence although they are tentatively the best we have, evidence, capable of changing at any moment. In real fact, no one will ever know with any real certainty what went on in prehistory and the lesson that nothing in the study of prehistory is ever definitive is often humanly hard to accept. The only certainty, that we can be sure of in life or the universe itself, is that change itself is the only certainty. All the rest is an ambiguity of possibility, probability and circumstantial evidence, with human curiousity being the impetus.

SYNOPSIS

What has been presented here, in its most positive sense, is a review of former research in the Muleta cave from 1962-2002, as well as in the Matge rock shelter from 1968- 2002 and the presentation for the first time of a combined number of over 120 absolute dates, in the form of intersite radiocarbon inventories. These intersite inventories are supported by artefacts evidence (briefly discussed, not represented here), all of which deal directly with settlement, the resulting stages of cultural development and human interaction in the Balearics.

The current combined number of radiocarbon readings for all periods and phases, along the entire range of local prehistory, stands at present close to 450 radiocarbon dates, with the overall inventory growing steadily and significantly, as more investigators begin to use it constructively (eg. Lull et al 1996, 1998; Plantalamor et al 2000; Van Strydonck et al 1999; Waldren et al 1962- 2000 and others). This undoubtedly qualifies the Balearic Islands as one of the best radiocarbon documented areas in Europe. It also gives a good idea of the researchers actively using the method as a means of interpreting the past and the dating of contexts and things in general.

Apart from the initial stages of settlement in caves and rock shelters, there are numerous other chronological areas, where the method has been used successfully. We do well also to remember that both caves and rock shelters, although the sources of the oldest evidence of human presence, were used for many different functions, constantly from the earliest record through to modern times. They were used for both living and as cemeteries, as well as other functions, such as workshop and animal corral areas. While serving these purposes, they continued in parallel development with the open-air establishments for many thousands of years. The Matge rock shelter, Muleta cave and many other sites on the islands are the best examples of this long use and variety of function.

For example, caves and rock shelters are universally excellent sources of data and information dealing with early settlement, in that they, usually, preserve evidence much better than outside sources and environments. They can also be highly polemic and have until recently been difficult to interrelateand correlate productively with other forms of settlement, such as those found in the open-air. Until recently, we related early Balearic settlement with the Culture of the Caves (the local Pretalayotic Period) and as a result its relationships with later Bronze Age Talayotic, open-air settlements remained ambiguously two quite different environments. At that time, there were none or very few indications of anything resembling socially and economically organized, sophisticated, architectural settlement prior to the I millennium building of the Talayots. What small indication that did exist was either not pursued or treated with distance.

We now know above all probability that by 2500 calBC, III-IVmillennia sites like the Ferrandell-Oleza-Mas Prehistoric Settlement Complex were flourishing and constructing permaent, sophisticated architectural establishments with highly developed technology in the form of metallurgy, ceramics and other skills, along with water/land and animal management.

The construction of sophisticated sanctuaries and ritual areas for astronomical observation and ceremony, such as the Prehistoric Sanctuary of Son Mas, with its intersite radiocarbon inventory of 53 radiocarbon dates (TABLE 1 and 5) (Hoskin 2000; Van Strydonck, Hoskin, Waldren, Ensenyat Alcover and Orvay 1999; 2000 and this volume) were also in operation as early as 2200 cal BC. Surely, this III millennium site too, like the SFO-OS site, did not appear over night, but underwent a period of jestation. Furthermore, this unusual site was in use as a ritual centre for the area, continually, from 2200 calBC to 200 BC and incipiently until 300AD, apart from an accountable hiatus of 300 years, c 1750-1450 calBC.

Dolmenic structures were constructed near the SFO-OS site and used for both living and for burial as family tombs from 2200 calBC to 1400 calBC (Waldren 1990 1994, 2000; Waldren an VanStrydonk 1999; Waldren, Ensenyat Alcover and Orvay, this volume) and were also in operation long before the Talayotic societies began. In fact, it is only recently that some of us have come slowly to terms with the fact that the Talayots are not as old as we formerly thought 40 years ago, as being in the range of 1800 BC forward, but now have sound estimates for construction closer to the range of 1100 calBC. Recent excavations in one Talayot (T1) of the Ferrandell-Oleza Younger Settlement SFO-YS site, however, has given strong evidence that the Talayot was constructed directly on top of a former`Pretalayotic structure containing Beaker pottery (Waldren, Ensenyat Alcover, Van Strydonck and Orvay in preparation)

Of course the dates for III-IV millennia Chalcolithic settlement, too, may eventually have to be reassessed somewhat, when we find more settlements of this age or when and if we find evidence of outdoor IV-V millennia Neolithic settlement, a possibility that is not out of the range of possibility.

In one site, that of the Olezian Chalcolithic Old Settlement (SFO-OS), sophisticated open-air architectural settlement in terms of long-term investment (Gilman 1986) is an archaeological fact, where 44 radiocarbon dates (TABLE 4) with 95.4% probability confidence, cover the occupational duration of nearly 1200 years from 2500 calBC to its abandonment circa 1400 calBC, possibly due to over exploitation and high erosion ofnearby agricultural fields. An occupation which in age and materials parallels Iberian and Continental Mainland settlements. There are some recent indications in the SFO-OS site that it too was built on former open-air structures that had the form of a cluster of possible Neolithic IV millennium huts.

The evidence and data presented above are, in the authors' minds sound indication of a much better than III millennium BC occupation of the islands. The sophisticated, organized, rural settlement pattern evidently, already, in operation and its contemporaneity with the occupation and use of caves for living and other purposes during still ealier periods, circa the IV and V millennia, increases the probability still further.

Artefact evidence and dating show that by 2000-2500 calBC, Mallorcan settlements had access to exotic items and material such as elephant ivory and metallurgical skills, with recent evidence of the inhabitants drawing on local, nearby copper sources for metal and local flint sources for a very sophistic lithic industry. Both aspects little known until very recently (Waldren, Sanders and Conesa Coll, 1984) There is sound indication that by 1800-1700 calBC tin bronze was in circulation (Waldren 1976; Northover 1984), something which further suggests access to outside tin sources, as the local geology has as yet no known resource of this element, one which would necessitate transport in either ingot form, or as finished products.

Evidence of access to such exotic and essential commodities, technologies and resources, in the authors' opinion, are the most direct indication of contact with areas that have these materials and are not too distant to make transport difficult. This fact alone would certainly be the best argument for those who would lead us to believe that the islands were ever too distant for early access (see Waldren, Links in the Chain, this volume).

So, we can logically expect that the Balearic are no different than other places, and did not stand isolated from other areas, where the earliest record for settlement is inevitably-found in rock shelters and caves and that contact may have taken place much earlier and more frequently than usually believed. Many of these considerations have been overlooked by some recent insular investigators, who would like to preserve or return to the 'status quo' of 40 years ago and whose experience is limited to a few years, in cave and rock shelter environments, or shorter experience in open-air excavation. Just how far the documentation and knowledge of local settlement patterns have progressed, how much our experience and understanding regarding these situations has grown, how the approach to and the strategies used in such sites have changed and advanced in the last 40 years and in final conclusion, the overwhelming realization of just how much there still remains to be done on both ends of the local prehistoric time scale, are and in the future will be, most salient to any clear and accurate attempt to evaluate Balearic Prehistory.

BIBLIOGRAPHY

ALCOVER, J. A, RAMIS, D, COLL, J. TRIAS, M.,
2001 'Sobre la cronologia de la primera arribada humana a les Balears, III Jornades del Medi Ambient de les Illes Balears, Societat d' Historia Natural de les Balears, Palma de Mallorca, Baleares, Spain

CHERRY, J.F.
1981. Pattern and Process in the Earliest Colonisation of the Mediterranean, Proceedings of the Prehistoric Society 47, pp.41-68.

CHERRY, J.F.
1984. Islands out of the Stream: Evolution and Interaction in Early Eastern Mediterranean Insular Prehistory', in A.B. Knapp and T. Slech (eds.), Prehistoric Patterns of Exchange in the Aegean and Eastern Mediterranean, UCLA Institute of Archaeology, Los Angeles.

CHERRY, J.F.
1984. The Initial Colonisation of the Western Mediterranean Islands in the Light of Island Biogeography and Palaeogeography, in W.H. Waldren, R.W. Chapman, J.W. Lewthwaite and R.C. Kennard (eds.), The Deya Conference of Prehistory: Early Settlement in the Western Mediterranean Islands and their Peripheral Areas', British Archaeological Reports (BAR) International Series

229 (i), Oxford , England, pp 7-28.
DAVIS, M.
2002 'Putting Meat On The Bone: An Investigation Into Palaeodiet In The Balearic Islands Using Carbon And Nitrogen Stable Isotope Analysis', this volume.
FERNANDEZ-MIRANDA, M. AND WALDREN, W.H.
1974. El Abrigo de Son Matge (Valldemosa) y la Periodización de la Prehistoria Mallorquina mediante las Análasis de Carbon 14', Trabajos de Prehistoria 31, Madrid, pp. 287-304.
FERNANDEZ-MIRANDA, M. AND WALDREN, W.H.
1979. Periodificación Cultural y Cronológia Absoluta en la Prehistoria de Mallorca, Trabajos de Prehistoria 36, Madrid, pp. 349-377.
GASULL, P.
1985. Arqueologîa, Enciclopedia España-Calpe, Suplemento 1983-1984, pp. 51-68.
GILMAN, A.,
1981. 'The Development of Social Stratification in Bronze Age Europe', Current Anthropology 22, pp. 169-224.
HOSKIN, M.
1990. 'The Menorcan Taulas', in Waldren, Ensenyat and Kennard (eds.) IInd Deya Conference of Prehistory: Archaeological Techniques, Technologyand Theory, BAR International Series 565, Oxford.
JACOBSEN, T.W.
1976. 'Excavations in the Franchthi Cave', 1969-1971, Hesperia 42, pp. 45-88 and 253-283
HOSKIN, M.,
2001. Tombs,Temples and Their Orientations: a New Perspective on Mediterrannean Prehistory, Ocarina Books, West Sussex, England
KOPPER, S.,
1984, 'Canet Cave', Esporlas, Mallorca, Early Settlement in the Western Mediterranean Islands and their Perpheral Areas, ed. Waldren, Chapman, Lewthwaite and
Kennard, I Deia Conference of Prehistory, BAR International Series, 229(i)
Oxford, pp.61-70
LULL,V.,MICO, R., RIHUETE HERRADA,C.,RISCH, R., 1999. La Cova des Carritx y la Cova des Mussol en Ideologîa y Sociedad en la Prehistoria de Menora, Consell Insular de Menorca, Mahon, Menorca.
Alcover
LULL, V., MICÓ, R., RIHUETE, C. and RISCH, R.
1999, La Cova des Càrritx y la Cova des Mussol. Ideología y Sociedad en la Prehistoria de Menorca, Consell Insular de Menorca, Barcelona.
LULL, V., MICÓ, R., RIHUETE, C. and RISCH, R.
2000a, "Rituales de vida y muerte en la prehistoria de Menorca", Revista de Arqueología, 225, pp. 18-33.
LULL, V., MICÓ, R., RIHUETE, C. and RISCH, R.
2000b, "Divinità nelle tenebre. Parlano le grotte delle Baleari", Archeologia Viva, 84, pp. 54-61.
NASH,G.
2002, Constructing Hypohetical Systems of Exchange: Doing Kula in the Danish Mesolithic, in V Deia International Conference of Prehistory, World Islands in Prehistory, eds Waldren, Ensenyat
WALDREN, W.H.
1967. 'Informe Preliminar Sobre Analisis de Radiocarbono en Mallorca'. X Congreso Nacional de Arqueologia. Mahon y Zaragossa, pp. 75-82.
WALDREN, W.H.
1979. 'A Beaker Workshop Area in the Rock Shelter of Son Matge, Valldemosa, Mallorca' World Archaeology, Routledge and Kegan Paul, London, Vol 11, No. 1, pp. 43-67.
WALDREN, W.H.
1982. 'Balearic Prehistoric Ecology and Culture: The Excavation and Study of Certain Caves, Rock Shelters and Settlements', British Archaeological Reports BAR International Series 149 (ii) Appendix 3A, pp. 675-711. Oxford.
WALDREN, W.H.
1984. ' Chalcolithic Settlement and Beaker Connections in the Balearic Islands'. The Deya Conference of Prehistory ...Early Settlement in the Western Mediterranean Islands and their Peripheral Areas. W.H. Waldren, R.W. Chapman, J.G. Lewthwaite and R.C. Kennard (eds), British Archaeological Reports, BAR International Series 229 , Oxford. vol. iii, pp.911-965.
WALDREN, W. H.,
1987. 'A Rose by any other Name...A Question of Balearic Beakers...the Evidence', Oxford International Western Mediterranean Bell Beaker Conference, St. Catherines College, University of Oxford.
WALDREN, W. H.
1990. 'Age Determination, Chronology and Radiocarbon Recalibration in the Balearic Islands', IInd Deya Conference of Prehistory...Archaeological Techniques, Technology and Theory, Waldren, Ensenyat and Kennard eds. British Archaeological Reports, BAR International Series 565, i-ii, Oxford.
WALDREN, W. H., ENSENYAT, J. AND CUBI GRIMALT, C.
1990. 'Son Mas: a New Mallorcan Prehistoric Sanctuary, circa 2000 BC-1750 BC', IInd Deya Conference of Prehistory: Archaeological Techniques, Technology and Theory, Waldren, Ensenyat and Kennard eds. British Archaeological Reports, BAR International Series 565, i-ii, Oxford.
WALDREN W.H., ENSENYAT, J. AND CUBI , C.
1994. 'Prehistoric Architectural Elements, Ferrandell-Oleza Chalcolithic Old Settlement: Radiocarbon Dating and Artefact Distributional Survey', DAMARC Series, 20, Deya, Mallorca.
WALDREN W.H.
1995. 'The Function of Bell Beaker Pottery as a Ceremonial and Votive Object', in Waldren, Ensenyat and Kennard (eds.), IIIrd Deya Conference of Prehistory, Ritual, Rite and Religion in Prehistory, Oxford, British Archaeological Reports, BAR International Series, 611 (i-ii), vol i, 238-263.
WALDREN W.H.
1997. 'The Definition and Duration of the Beaker Culture in the Balearic Islands: a Radiocarbon Survey'. Oxford Journal of Archaeology, Vol.16, No 1, Blackwells Publishers, Oxford, 25-48.
WALDREN W.H.
1998. 'The Beaker Culture of the Balearic Islands, An Inventory of Evidence from Caves, Rock Shelters, Settlements and Ritual Sites Western Mediterranean Series 1, British Archaeological Reports, BAR International Series, 709 , 267-312.
WALDREN, W.H.
1986. 'Balearic Pentapartite Division of Prehistory.. Radiocarbon and Other Age DeterminaMuseum, Oxford University, England.
WALDREN, W. H.,
1992b. ' Radiocarbon and Other Isotopic Age Determinations from the Balearic Islands: a Comprehensive Inventory', DAMARC 26, Donald Baden-Powell Quaternary Research Centre, Pitt Rivers Museum, Oxford University, England.

9

PACIFIC PREHISTORY: AN EXAMPLE OF CULTURAL EVOLUTION FROM THE ISLAND OF CIKOBIA (MACUATA, FIJI)

CHRISTOPHE SAND (1), **FRÉDÉRIQUE VALENTIN** (2)

(1)Département Archéologie, Service des Musées et du Patrimoine, Nouméa, New Caledonia
(2) CNRS UMR 7041, Maison de l'Archéologie et de l'Ethnologie, Nanterre, France
EMAIL: sand.smp@gouv.nc

SUMMARY

The reconstruction of the prehistory of the south-western Pacific has been the focus of much research during the past 20 years. Most of the programs have centred on the oldest periods of human settlement in the region, with lesser attention placed on later periods. This paper proposes to show the evolutions and transformations of an island society, from first discovery to European arrival, through the archaeological study of a small island of north-eastern Fiji, named Cikobia. This island at the border between the geographical regions of Melanesia and Polynesia has experienced during it's more than 2500 years of prehistory, different cultural influences and local developments, resulting in an original cultural evolution. The paper will mainly deal with questions of first settlement, landscape transformations, regional interactions, socio-political evolutions, warfare, to show how these different points enrich our understanding of the local chronology in a regional perspective.

INTRODUCTION

Studies on island's prehistories conduced throughout the world have shown over the decades the numerous specificities of these often smale-scale and sometimes isolated landscapes. Most researchers have discussed at length questions of isolation, outside contact, organized relations on long distances in an island context. Sometimes hotly debated interpretations of the archaeological data has shown the difficulty of reaching a consensus on these matters (ex: Renfrew and Bahn 1996). Largely neglected in these often Mediterranean-oriented debates, seen from a Pacific perspective, have been the questions of inland relations, contacts, tensions and conflicts on islands, that are the everyday life of islanders (Kirch 1984). As an example of what happened sometimes in the southern Pacific, this paper presents a case study from a small Melanesian island of Fiji called Cikobia (note 1), located at the border of Polynesia.

This island, about 10km long and 2km wide, is located 65 km off the nearest large island in the north-eastern part of the Republic of Fiji (FIGURE 1). Positioned on a basaltic outcrop appearing on its south-eastern part, the island is mostly covered by uplifted limestone formations (Sand and Valentin 1998). The present-day population, numbering around 150, lives in four villages located in the east of the island. This eastern end is partly protected by a reef, forming a small lagoon where as the rest of the coast is directly open to the ocean.

A joint program between New Caledonian, French, and Fijian archaeologists was conduced on Cikobia between 1997 and 1999, archaeologically unknown, with the purpose of defining the occupation pattern of the island and its changes during the prehistory, with a special focus on the inter-insular relationships in the south-west Pacific. The present paper exposes the major archaeological results of the program, focussing, after the presentation of the chronological background, on the late prehistoric data as can be defined though the 120 archaeological sites that have been recorded.

CHRONOLOGICAL BACKGROUND:
THE EARLY SETTLEMENT PHASE

The oldest known settlement of the Fijian archipelago

around 2950 years ago, is related to the rapid colonisation spread of Austronesian seafearing populations of the Lapita Cultural Complex into Remote Oceania (Anderson and Clark 1999; Kirch 1997, 2000). The geomorphology of Cikobia led us hypothesise that the first occupations had focused on the south-eastern part of the island, in areas protected by the lagoon and the coral reef and where sand dunes were present. Convincingly, the oldest signs of settlement have been identified in this region in several excavations (Sand et al. 2000). The results obtained in the Nakasaga and Naselala sites show that the chronology starts at the beginning of the first millennium B.C., with ceramics stemming from the Lapita tradition. The oldest date obtained is 2740+/-50 BP (Beta-109362), calibrated 990 (855) 805 BC, corresponding to the advent of the Late Eastern Lapita (Clark and Anderson 2001). As a demonstration of this cultural link, only a few sherds have a dentate stamped decoration and others have incised decorated rims. A set of other dates show slow increase in occupation during the whole first millennium BC (Sand and Valentin 1998).

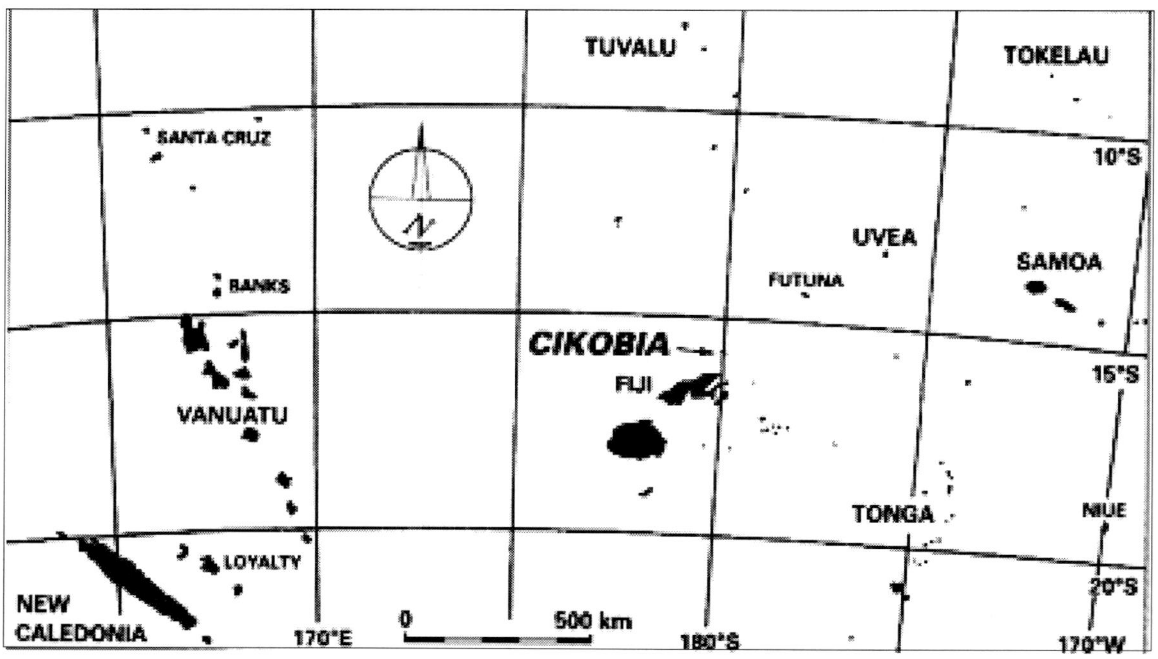

FIGURE 1. REGIONAL MAP OF THE SOUTH-WESTERN PACIFIC LOCATING CIKOBIA IN THE FIJIAN ARCHIPELAGO.

Today, Cikobia is characterized by restricted natural diversity, especially with few birds and few shells. The large island of Vanualevu, seen by good weather from Cikobia, having probably allowed a natural colonisation well before human settlement, this limited diversity reflects, at least partially, the human impact on the insular environment. This impact can be identified through various set of data. The excavations have clearly shown that today's lack in shells diversity does not reflect the situation at the beginning of human settlement. Some shell species like Anadara scrapha, not living in Cikobia's lagoon today, were still present 2000 years ago. The same situation apply for birds, probably due to forests destruction for horticultural cultivations and/or over-hunting.

Although fishing has probably always represented an important source of subsistence, as shown by the amount of fishbone in the excavations (Leach et al. 2000), the existence of a second-growth vegetation in a large part of the island indicates that cultivation was the central economic activity. Clearly, Pacific horticulturists had a major impact on the natural environment (Kirch and Hunt 1997). The identification of a second-growth forest, which involves the destruction and replacement of primary forest areas, and the important phenomenon of soil erosion, mostly resulted from fires. This can specially be observed on part of the volcanic soils of the south-eastern part of Cikobia, with the development of a talasiga-like dry vegetation, characterizing impoverished soils (Latham et al. 1983). This first major landscape transformation phase probably occurred during the first millennium of settlement.

The comparatively low density of sites related to the following millennium AD, seem to indicate that the island, after a first period of probable rapid demographic growth, was not densely occupied any more. This phenomenon is paralleled in other small islands of Fiji at the same period, like Lakeba (Best 1984) and Totoya (Clark et al. 1999). It may be related to the impoverishment of the soils and possible small-scale weather changes, leading to the departure and the resettling of groups on the two major islands of the archipelago. There, larger valleys like Sigatoka allowed for a better living, possibly with the first intensification of horticultural systems (Field 1998). This means that density fluctuations of island occupation can have occurred during a Neolithic-like chronology, with a demographic curve not necessarily directed upwards. During this period of the first millennium AD, the first Lapita-related ceramic traditions were progressively replaced by paddle impressed potteries (Sand et al. 2000).

FIGURE 2. ONE OF THE WELL-PRESERVED WALLS OF RUKUNIKORO FORT, OVER 4M HIGH.

LIVING ON AN ISOLATED ISLAND IN LATE FIJIAN PREHISTORY

Other kinds of potteries started to substitute the paddle impressed pots before the end of the first millennium A.D. (Sand and Valentin 1998). The paddle impressed pottery disappeared progressively, to be replaced by thicker and better fired pots, showing various incised, impressed, and relief decorations comparable to the other islands of the region (Best 1984). Special decorations are associated with special shapes. The two main reconstituted shapes are a large bowl with decorated rim and a jar with a neck and an out curved rim.

The data on an erosional process in Cikobia, with for example part of the soils of the Namasi hill eroding into the nearby Lobau swampy area, filling the site with over 1,5m of earth and giving it its present shape only in the last 500 years (Stevenson 1999), indicate a new development of horticultural processes in all the cultivable areas of the island during the late prehistoric period of the second millennium AD.

Large defensive stone arrangements appeared at the same period. Forts were constructed on all the hill-tops of Cikobia. The recorded data indicate that at the end of the prehistoric chronology, most of the villages were located in fortified areas. On this small island only 10km long, 11 fortified locations have been discovered (Sand et al. 1999). The two major sites, the forts of Namasi and Korotuku, covered the whole area of the two main hills of the south-eastern part of the island. The Namasi ridge-fort is located on top of the eastern-most hill of Cikobia. It is 150m long and completely surrounded by a defensive wall overhanging the cliffs. It was used as the last defence position of the Namasi village, located underneath on the first plateau about 350m long, mostly defended by a natural cliff up to 20m high. Two large broken adzes originating from Samoa about 700km to the east have been found in Namasi, along with one complete samoan adze in Vuninuku village and one in Rukunikoro fort. The huge Korotuku site covers a surface of over 15 hectares (650m by 250m). The fort is surrounded by different sets of walls, some over 3m high. In its south-eastern part, over 80 structures have been mapped, comprising house-mounds, burial mounds and ceremonial platforms. Clearly, this part of the site was once a populated village

(Sand et al. 1999).

The extent of the defensive arrangements and other structures suggests a rather numerous population during the last millennium of the prehistoric chronology, leading to conflicts over land. Oral traditions indicate that most of the conflicts resulted from local stress between the groups, and only one clear foreign trial at violent settlement appears in the local histories (Biggs and Veremalumu-Biggs 1975; Matararaba 1998). These conflicts can also be identified in the less fertile remote area of Caukaci, at the north-western point of Cikobia, with the presence of a large defensive ridge-fort and surrounding walls in the flat plain.

The conflicts on Cikobia were part of common life and continued well into the 19th century. Some where recorded by Europeans like Reverend John Hunt. In his daily Journal, this Methodist priest wrote on October 29th 1840 from his Somosomo Wesleyan mission on Taveuni island: " We have heard a sad account of a Tonga canoe, which was driven to Chikombea some time since by contrary winds. They were sailing in compagny with Josiah it appears from Vavau to Tonga (...) and were driven to Fiji as is often the case. After being six or seven days at sea they found Chikombea, a small island belonging to Macuata, but some distance from it. After coming near enough to the land some of them went ashore, leaving nine persons in the canoe to watch. It appears there are two towns on the island, one of which appeared friendly and received the Tonga people who went ashore, but the other determined to kill those who were on board the canoe and take their canoe from them. The people on board the canoe appear to have no suspicion, accordingly they were taken by surprise, killed and eaten, except one who was diseased, whom they buried. The rest were allow to live, but they have made slaves of them, and a few are in one place and a few in another that they may serve the Fijians. The canoe was taken to Macuata to Ziqi, the chief of the place, and one of the men on it who has come from Macuata here and given us this report. ... " (Methodist Overseas Mission, record 133).

FIGURE 3 : THE BURIAL GROUND OF KOROTUKU AT THE END OF THE EXCAVATIONS, BEFORE REBURIAL.

Beside classic fights over the control of land, the possession of the fresh water sources must have been a reason of competition between groups, in this island without streams. Population increase during the last part of the prehistoric chronology certainly led to stress on water resources, especially in drought periods (Sand and Valentin 1998). One of the main water points of Cikobia was and still is the swampy area of Lobau, in the eastern part of the island. Numerous abandoned walls can still be seen in this wet area today, probably indicating the strategic importance of the control of water, especially after the infilling of the former large swamp. It is also not surprising to find near to Lobau the Rukunikoro defensive fort. Located on top of an uplifted coral platform 9m high, this structure, reconstructed by our team in 1999, is more than 40m long, with some walls up to 4m high (FIGURE 2). The roots of the large trees had collapsed some of the major wall sections, but with the representatives of the different clans, we were able to clean the collapsed parts, easily finding the un-destroyed basis of the walls. The original platforms and protection parts where reconstructed on the in-situ level (Sand et al. in press). In another par of Cikobia, the Dreketi fort was constructed to protect the only water source of the area, the Waicinavi pool, located in the bottom of a cave. This fort, nearly 20m long, has a surrounding wall up to 2,8m high, with a well preserved fortified entrance.

The large majority of the burial sites surveyed are probably related to the late prehistoric period. The cemeteries linked with the oral traditions show arranged burials with coral and/or beachrock slabs. It appears that burial-grounds existed, forming - after use - large platforms with slab walls (Sand and Valentin 1998). One of these structures, located in the Korotuku fort, was excavated during the 1999 field-season. The burial mound, despite its small size (10 m by 7 m), results from the progressive combination of burials marked at the surface of the mound by seven stone arrangements (Valentin et al. 2001). The excavation revealed the remains of ten individuals, comprising two children and two females (FIGURE 3). Even though the remains of ten persons occupied the same funerary space, this burial complex cannot be assimilated to a mass burial, because each deceased had been buried in an individual pit. The mound was organised around two starting central burials surrounded with stones walls, progressively expanded through other burials surrounding this core. An individual had also been added between the two earliest interments. The central burials have independent enclosures, a large sepulchral space and a thick fill grave reaching 1 meter high, with a north/south orientation. They sheltered males interred with adornments, the hands sometimes brought on the shoulders, and the lower limb tightly wrapped in an envelope (probably mats or bark clothes). The ornaments consist in pig's tusk pendant, shell armbands, shark tooth, as well as tiny western glass trade beads. The lateral burials have enclosures resulting from the reuse of the central walls, but faced with beachrock slabs and with a much smaller internal space. They have received females, or males buried with glass beads ornaments, which have experienced another sepulchral treatment including a different envelope wrapping.

The presence of glass trade beads among four of the burials, indicates that the mound was built after first European contacts in this region of the Pacific Ocean. It was most probably erected during the middle or the end of the nineteen century, according to the glass beads typology and the glass composition (Francis 1994 ; Gratuze in press). The inhabitants of Cikobia could have received them by direct relations or through neighbouring islands. The complexity of the burial structure and treatment may reflect " relative socio-political status " of the interred individuals (Valentin et al. 2001). Despite some variations in the funerary treatment, in the architecture, the adornments and the energy expenditure involved in the construction, the Korotuku burial mound was certainly devoted to high-ranking persons. The variations could indicate gender distinctions, but the architectural changes could also reveal evolutions of the burial practices of the Cikobia people during the latest phase of the island's prehistory (note 2).

CONCLUSION

The research program carried out on Cikobia has allowed to identify the major characteristics of the prehistoric chronology and cultural evolution of this island located at the border of Polynesia. The chronology started at the beginning of the first millennium BC, with a settlement related to the Eastern Lapita Cultural Complex. The progressive human occupation led to transformations in the landscape and the natural environment, with the use of all the cultivable areas. The end of the prehistory was characterised by the occupation of fortified hill-tops as major habitation sites, as an answer to mostly local tensions, with fights within and between the island's groups.

These results bring new data and have thus a wider significance in the definition of the Fijian and regional cultural chronologies (Frost 1979; Best 1984; Clark 2000). One of the major results of the program has been to reconstruct a general ceramic chronology, which shows clear links with the rest of Fiji to the south-east and is different from the neighbouring West Polynesian typology after the first millennium of settlement. Clear signs of environment transformations have been observed, leading during the prehistoric chronology to variations of population densities over time. The first transformations can be termed as destructive, with the replacement of primary forests by second-growth forests. Erosional processes can be observed in the development of talasiga-like dry vegetation environments and the deposition of soils in small valley bottoms and swampy areas. The second set of transformations is clearly constructive, with the development of dry field systems, of pig-walls and of settlement areas including wide fortified constructions. Burial sites were located in strategic places, often in low-fertile areas, indicating the will at a proper use of all possible cultivable zones for economic purposes, in a densely populated island. Their typology clearly reflect the Fijian link of the whole late prehistoric cultural complex present in Cikobia, although the island was near to West Polynesia. The Korotuku fort, over 650m long, is one of the largest Fijian forts mapped to date and certainly the most densely settled in regard to habitation structures (Sand et al. 1999). The extent of fortified constructions on Cikobia is an archaeological confirmation to the first descriptions of the European sailors, who had a known fear of this badly renewed island populated by violent warriors. Their amount is a sign of regular tensions on this small island, a conclusion paralleled in most of Fiji, be it on the small or the big islands of the archipelago.

The results obtained from the archaeological research program on Cikobia confirm the potential of these type of small island environments to inform about the adaptive strategies used during prehistoric times in the settlement of the Pacific. The chronologies are complex, evolutionary, showing transformations in various fields and specially in settlement patterns (Kirch 2000). From first settlement to the protection on the hill-top fortifications, before the recent introduction of the Christian religion and the permanent settlement of the present-day villages, the nearly 3000 years of human history of an island like Cikobia are a synthesis of the complex history of the Pacific islands societies.

ACKNOWLEDGMENTS

The bulk of the funding obtained to work on Cikobia for three seasons was provided by the Ministère des Affaires Etrangères (Paris, France); the Département Archéologie, Service des Musées et du Patrimoine (Nouméa, New Caledonia); the French Embassy in Fiji (Suva, Fiji) and the GDR 1170 CNRS (Paris, France). Support and administrative assistance was provided by the Fiji Museum. This enterprise could not be performed without the agreement and the logistic support of the Fijian government and the Cikobia people, for which the team is very greatfull.

NOTES

(1) Cikobia is pronounced "Thikombia".

(2) As a classic attitude in Pacific Archaeology that we defend regarding the respect due to the deceased as well as the living community, the bones of the skeletons from the Korotuku burial mound were not removed from the graves. At the end of the excavation, after a Christian prayer by the local inhabitants and a new covering with traditional bark cloth, the in-situ remains were covered by the earth we had excavated.

REFERENCES

ANDERSON, A., and G. CLARK. ,
1999. The age of Lapita settlement in Fiji. Archaeology in Oceania 34 : 31-39.
BEST, S.,
1984. Lakeba : the Prehistory of a Fijian Island. Unpublished Ph.D dissertation, University of Auckland.
BIGGS, B., and M. VEREMALUMU-BIGGS.,
1975. Na ciri kalia, the oral traditions of Cikobia-I-ra northern-most island in the Fijian group. Working papers in Anthropology, Archaeology, Linguistics, Maori Studies. University of Auckland, Dep. Of Anthropology.
CLARK, G.,
2000. Post-Lapita Fiji: cultural transformation in the mid-sequence. Unpublished Ph.D. Thesis, Australian National University, Canberra.
CLARK, G., and A. ANDERSON. 2001. The Age of the Yanuca Lapita site, Viti Levu, Fiji. New Zealand Journal of Archaeology 22 (2000): 15-30.
CLARK, J., A.O. COLE, and P. NUNN. ,
1999. Environmental change and human prehistory on Totoya Island, Fiji. In J.C. Galipaud and I. Lilley (eds), The Pacific from 5000 to 2000 BP. Colonisation and transformations. Paris : IRD : 227-240.
FIELD, J. ,
1998. Natural and Constructed Defenses in Fijian Fortifications. Asian Perspectives 37 (1): 32-58.
FROST, E.L. ,
1979. Fiji. The Prehistory of Polynesia. Harvard University Press: 61-81.
FRANCIS, P. ,
1994. Beads at the crossroads of the continents. In Anthropology of the North Pacific Rim. W.W. Fitzhugh and V. Chaussonnet (eds.), Washington : Smithsonian Institution Press: 281 - 305.
GRATUZE, B.,
In press. Provenance analysis of glass artefacts. In Non destructive microanalysis of cultural heritage materials. K. Janssens, R Van Grieken, University of Antwerp Eds., Elsevier Science BV.
KIRCH, P.V. ,
1984. The Evolution of the Polynesian Chiefdom. New Studies in Archaeology. Cambridge: Cambridge University Press.
KIRCH, P.V.,
1997. The Lapita Peoples: Ancestors of the Oceanic World. London: Blackwell Publishers.
KIRCH, P.V.,
2000. On the Road of the Winds: An Archaeological History of the Pacific Islands Before European Contact. Berkeley: The University of California Press.
KIRCH, P.V., and T.L. HUNT. ,
1997. Historical Ecology in the Pacific Islands. New Haven, Con.: Yale University Press.
LATHAM, M., P.J. HUGHES, G. HOPE, and M. BROOKFIELD. 1983. Sédimentation dans les zones marécageuses de Lakeba: conséquences sur l'érosion et l'occupation humaine de l'île. In Iles Fidji orientales. M. Latham and H.Ç. Brookfields (eds). Travaux et Documents de l'ORSTOM 162: 103-119.
LEACH, F., J.M. DAVIDSON, K. FRASER, and A. BUDEC-PIRIC. ,
2000. Analysis of faunal material from several archaeological sites on Cikobia, Fiji. Museum of New Zealand Te Papa Tongarewa. Technical Report 28.
MATARARABA, S. ,
1998. Oral Stories from Cikobia and Naqelelevu. In Archaeological and Anthropological data on the Island of Cikobia (Fiji). 1998 field-season. C. Sand and F. Valentin (eds). Departement of Archaeology, New Caledonia Museum: appendix A.
RENFREW, C., and P. BAHN. ,
1996. Archaeology. Theories, Methods and Practice. London: Thames and Hudson Ltd.
SAND, C., J. BOLE, D. BARET, A. OUETCHO, and F. VALENTIN. Rukunikoro. Reconstruction of a late prehistoric fortification on Cikobia Island (Macuata), Fiji. Domodomo.
SAND, C., and F. VALENTIN. ,
1998. Cikobia : données archéologiques préliminaires sur une île fidjienne à la frontière entre Mélanésie et Polynésie. Journal de la Société des Océanistes106: 65-74.
SAND, C., F. VALENTIN, and T. SOROVI VUNIDILO. 2000. At the border of Polynesia: archaeological research in the East-Fijian Inslands of Cikobia and Naqelelevu. Indo-Pacific Prehistory: the Melaka papers. Bulletin of the Indo-Pacific Prehistory Association 20: 107-116.
SAND, C., F. VALENTIN, T. SOROVI VUNIDILO, J. BOLE, A. OUETCHO, S. MATARARABA, J. NAUCABALAVU, D. BARET, and L. LAGARDE. ,
1999. CIKOBIA-I-RA. Archéologie d'une île fidjienne; Archaeology of a Fijian Island; I Tukutuku Makawa. Nouméa : Les Cahiers de l'Archéologie en Nouvelle-Calédonie 9.
STEVENSON, J. ,
1999. Report on the pollen analysis from Lobau, Cikobia, Fiji. MS

on file at the Département Archéologie, New Caledonia Museum.
VALENTIN, F., C. SAND, I. LE GOFF, T. SOROVI VUNIDILO, S. MATARARABA, A. OUETCHO, J. BOLE, D. BARET , and J. NAUCABALAVU.
Burial practices at the end of the prehistoric period in Cikobia-i-ra (Macuata, Fiji). In The archaeology of Lapita dispersal in Oceania. G. Clark, A. Anderson and T. Vunidilo (eds). Canberra: Terra australis 17: 211-223.

10

AKANTHOU-ARKOSYKO (TATLıSU-ÇIFTLIKDÜZÜ): THE ANATOLIAN CONNECTIONS IN THE 9TH MILLENIUM BC

MÜGE SEVKETOGLU

Eastern Mediterranean University,
Faculty of Arts and Sciences, Department of Archaeology and Art History,
Gazimgusa, Turkish Republic of Northern Cyprus
EMAIL: muge.sevketoglu@emu.edu.tr

INTRODUCTION

Until recently, the earliest known human settlement on Cyprus dated to around 7000 BC, taking the form of an Aceramic Neolithic "Khirokitia Culture," which was thought to have arrived in a fully developed form from elsewhere and to have existed for some centuries before abruptly coming to an end in the 6th millennium BC. This period was seen as the first time people had colonised the island with their full economy and culture from an unknown origin and for unexplained reasons. The origins of the Khirokitia culture "colonists" have been questioned by some archaeologists (Watkins 1973, Stanley Price 1977, Le Brun 1989). The most likely areas of origin on grounds of proximity are the Syrian or Hatay coast or south Anatolia. The Anatolian plateau was not considered as a possibility by archaeologists because of the differences in the cultural assemblages, and the lack of relevant sites on the south Anatolian coast: although Anatolian obsidian has been found at several Khirokitia culture sites, it is in tiny quantities, and many archaeologists suggested that the small quantities could have arrived via the Levant rather than directly from Anatolia. Due to the location of Khirokitia in the south of the island, it was geographically more relevant that these people may have come from the neighbouring countries from southeastern direction. Although some resemblances to the Levantine Aceramic Neolithic existed, The Cypriot Aceramic Neolithic still represented an indigenous culture. In the mid-1980s, new discoveries at the coastal cave site of Akrotiri-Aetokremnos suggested that humans had visited the island as early as 9000 BC (Simmons et al. 1999). This site contained many pigmy hippopotamus and dwarf elephant bones, found in the same context with man-made tools and carbonized plant materials and remains. This controversial discovery has excited many archaeologists for the reason that here was a population arriving long before the Khirokitians, who could have been the ancestors who finally developed in their own way to Khirokitian culture. However, the radiocarbon chronology indicated a gap between Khirokitia and the pigmy hippo hunters that made this idea extremely hypothetical. It was suggested that the population of hunters must have caused the extinction of the pigmy hippos and elephants, and that they therefore left the island (or became extinct themselves). The discoveries during the excavations at Pareklisha-Shillourokambos have regained momentum for the search for the origins of the Khirokitian culture. The dates from Pareklisha-Shillourokambos begin in the middle of the gap between the Akrotirians and the Khirokitians, and then continue down to the Khirokitia period itself. Besides Pareklisha-Shillourokambos other Aceramic Neolithic sites predating Khirokitia excavated or now under excavation are Kalavassos-Tenta, Kissonerga- Mylouthkia, and Akanthou-Arkosyko (Tatlısu-Çiftlikdüzü).

Now the date of the earliest known Neolithic settlement has been revised. Recent archaeological excavations at Pareklisha-Shillourokambos and Kissonerga-Mylouthkia and also the re-evaluation of the radiocarbon dates and lithics from Kalavassos-Tenta, have pushed the date of settlement in Cyprus back by more than 1000 years. In Pareklisha-Shillourokambos in particular hundreds of obsidian blades with central Anatolia origin have been discovered (Briois,

1997), changing the importance of the arrival of obsidian from a single journey to a continuous supply. The discovery of cattle fallow deer, sheep, goat, einkorn and barley from Pareklisha-Shillourokambos, Kissonerga-Mylouthkia and Akanthou is also another dimension to the newly discovered early Aceramic Neolithic of Cyprus. With these new discoveries, the issue of where the colonists came from and when they arrived is once again being debated. Recent excavations at, Akanthou-Arkosyko (Tatlısu-Çiftlikdüzü), provide new clues which may help to resolve this debate.

in sight of the Anatolian Taurus mountains approximately 40 miles away; in addition, there are favourable currents that make the journey between the mainland and the island both short and easy sailing without going out of sight of land in clear days.

Akanthou is situated on top of a 10-15 metre-high extensive marine terrace terminating in limestone cliffs on the seaward side. Immediately below the cliff is a narrow belt of soil surrounding a small pebble beach. On the west end of this beach there is a small spring that could have been

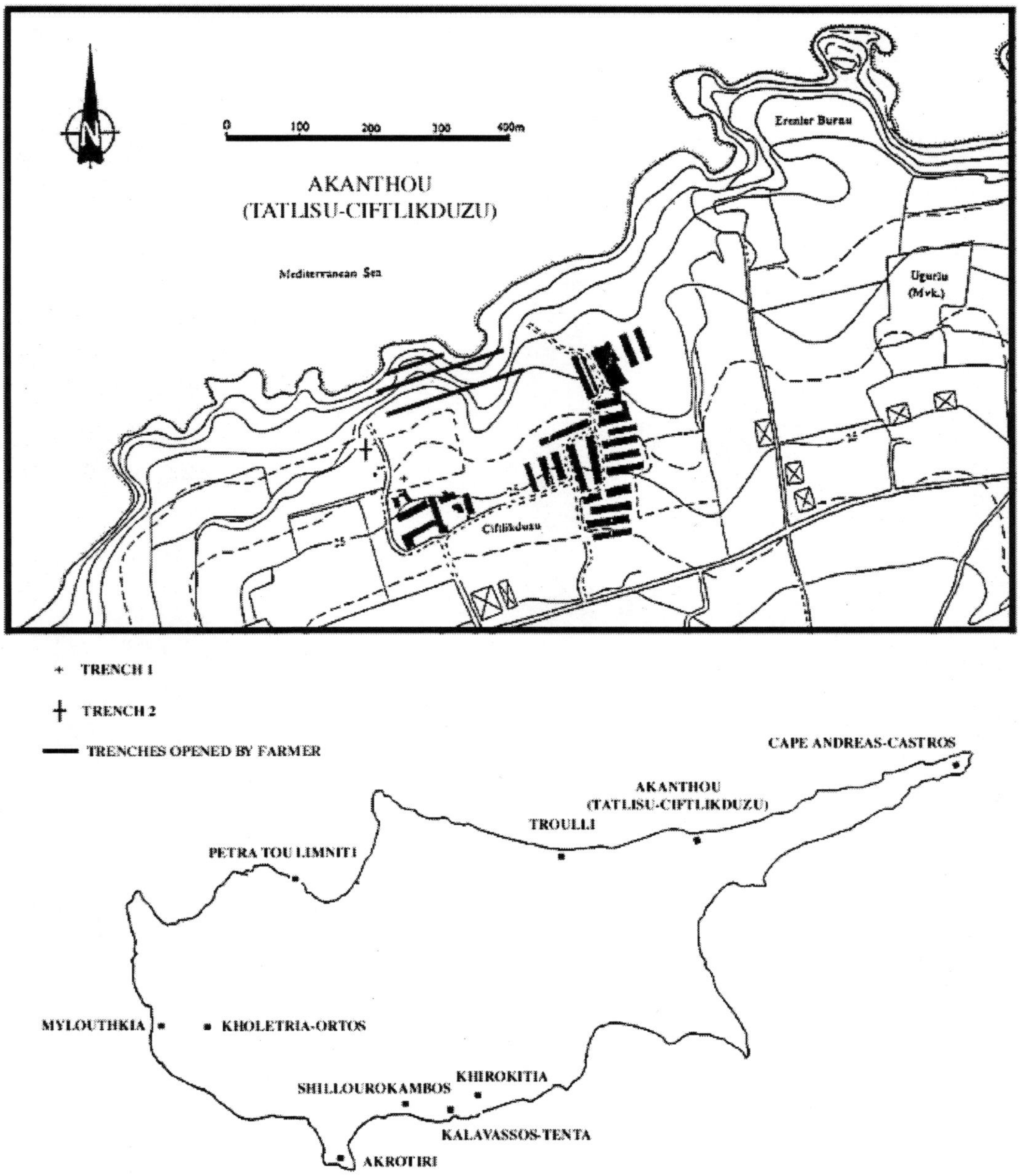

MAP 1: SHOWING SOME ACERAMIC NEOLITHIC SITES OF CYPRUS

LOCATION

Akanthou-Arkosyko (Tatlısu-Çiftlikdüzü) henceforth (Akanthou), a site on the north-east coast of Cyprus, is a new candidate to join the existing group of early Aceramic Neolithic sites. While the other sites are located inland in the south-west of the island, Akanthou is a coastal site with-

more active in the Aceramic Neolithic period, an important reason for attracting settlement. To the east of this bay is another sheltered bay, and to its west there are rocky inlets. The south of the site is bordered by the modern buildings of a chicken farm, extending east-west alongside the site, and perhaps even built on part of the settlement.

HISTORICAL BACKGROUND TO THE WORK AT AKANTHOU-ARKOSYKO (TATLıSU-ÇIFTLIKDÜZÜ)

Akanthou was first recorded by the Cyprus Survey in 1931. In 1945 it was reported by G. Anastasiou and in 1946 by P. Dikaios of the Cyprus Museum. In 1972-3 Stanley Price visited the site and added his own observations to these survey records (Stanley Price 1979). The surface finds reported from these surveys were: stone axe; hand-stone; stone vessels; chert flakes, blades, chunks, chips and cores; and other remains such as the bones of sheep, pig, pigmy hippo, fallow deer; a perforated shell; and 4 obsidian blades (Stanley-Price 1979:119). Stanley Price's surface observations during his visits to Akanthou in which he reported the obsidian blades attracted the attention of the present author. In 1996 a survey expedition to Akanthou under the NCASP (North Cyprus Archaeological Survey Project) was initiated. More archaeological work has been done at the site between 1999 and the present (Sevketoglu 2000).

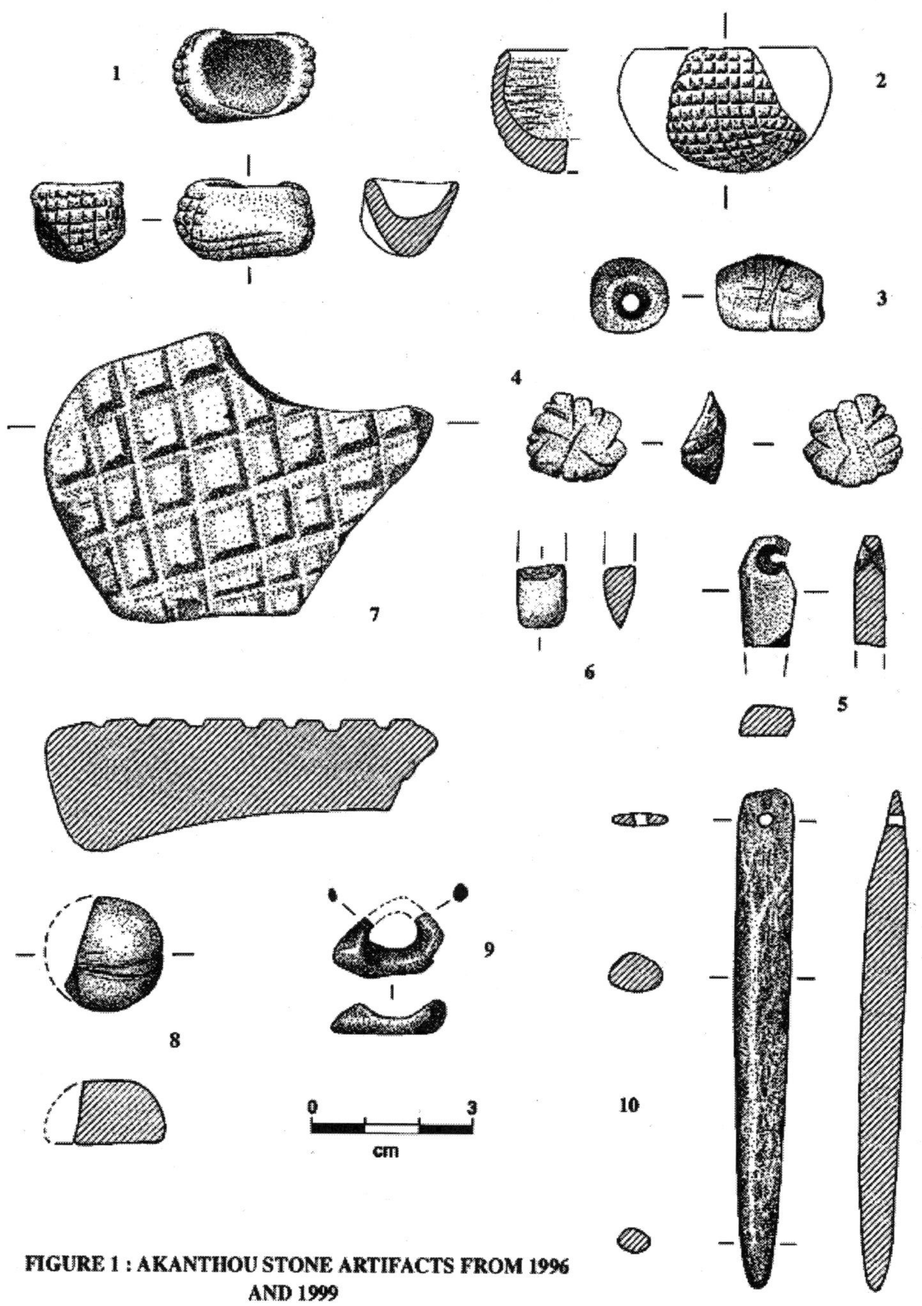

FIGURE 1 : AKANTHOU STONE ARTIFACTS FROM 1996 AND 1999

In 1996 on arrival, it was discovered that a series of regular trenches had been dug through the site by the farmer, to prepare for the burial of chicken entrails (MAP 1). A rescue operation based on sample sieving of the soil from the up cast of these trenches was carried out (Sevketoglu 2000). Many artefacts were recovered, including 20 obsidian blades and 65 chert tools; 39 ground stone implements; 9 axes; 55 animal bones identified as sheep/goat, pig, fallow deer, cattle; 2 worked bones; fallow deer antlers; 14 sea-shells, some of which were pierced as beads; stone beads; and a stone pendant (Sevketoglu 2000)(see FIGURE1, 10). to salvage as many artefacts as possible. Another aim of this was to provide a deterrent to the farmer's activities on the site, an aim that was achieved successfully at the end of

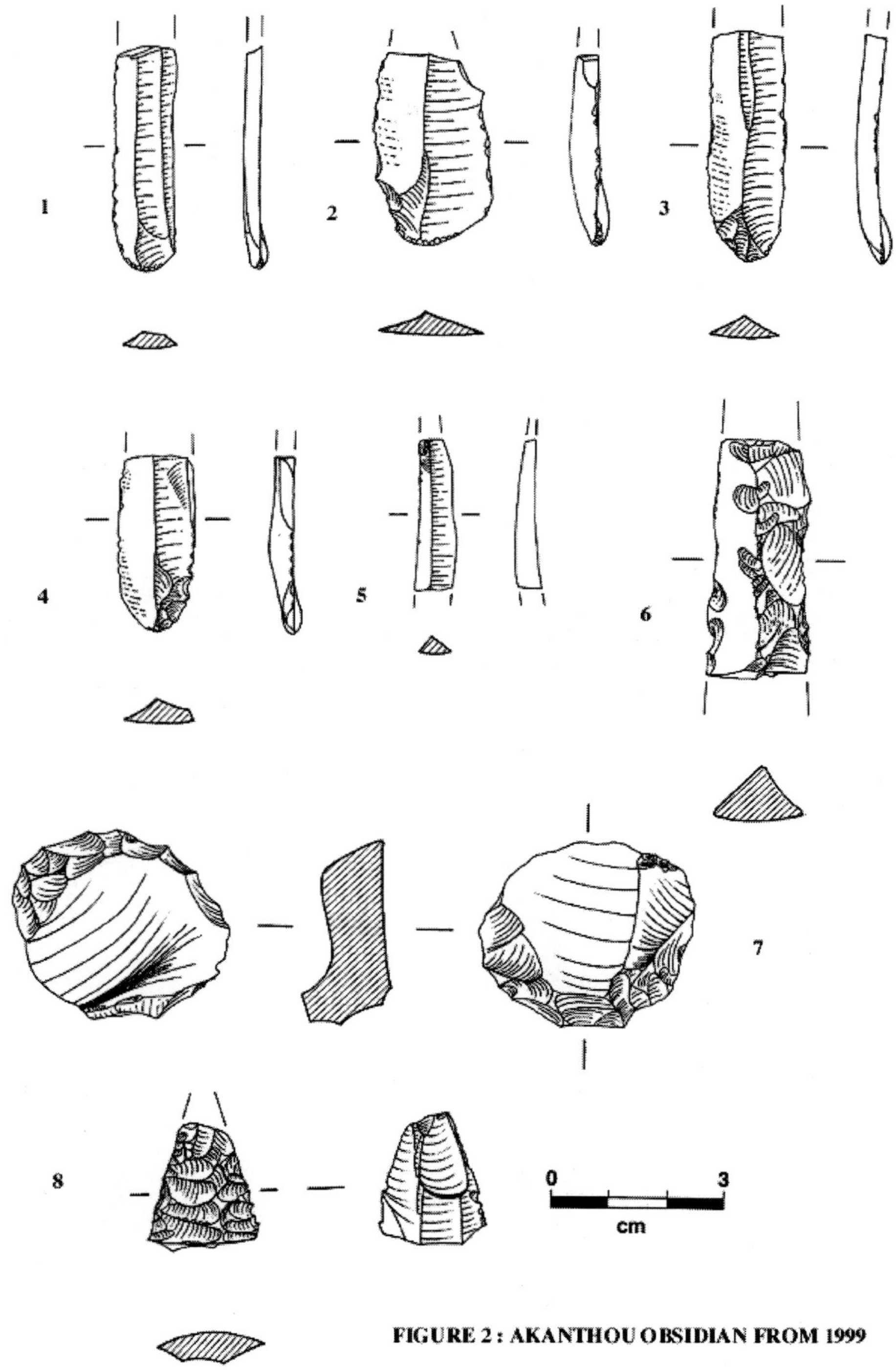

FIGURE 2 : AKANTHOU OBSIDIAN FROM 1999

Already, in this preliminary rescue operation, the number of obsidian pieces found at the site was second only to that found at Pareklisha-Shillourokambos. Identification of the two cattle bones in addition to the number of obsidian discovered at Akanthou was clearly a sign for the site's importance and its potentiality.

In 1999, more dry sieving was carried using a 4mm mesh this season. Each trench and soil mound was given a number, and a total of 44,550 litres of soil was sieved. The artefacts recovered from the mixed and re-deposited soil comprised: 453 complete and fragmentary ground stone bowls and other stone implements; 2861 chert; 422 obsidian some of which were 193 blades, 8 chunks, one core and 221 fragments of various blades and flakes; 4 picrolite objects, one of which was a bead (see FIGURE1, 3), one miscellaneous (see FIGURE1, 4) and two miniature cups (see FIGURE1, 1 and 2). The miniature cups are decorated by hatched incised motifs on the outer surfaces; animal bones, some worked; 76 fish vertebrae; 21 sea shells, some pierced as beads; 7 stone beads; chisel fragment (see FIGURE 1, 6); axes, and miscellaneous objects such as a fragment of a flat stone artefact with deep grooved hatchings on its surface (see FIGURE1, 7) (Sevketoglu 2000). However, the unstratified artefacts from sieving are difficult to interpret and date: we needed similar artefacts from secure, excavated contexts, and architectural/structural connections needed to be established for a more firm understanding of the site and its relation with the other contemporary sites in question.

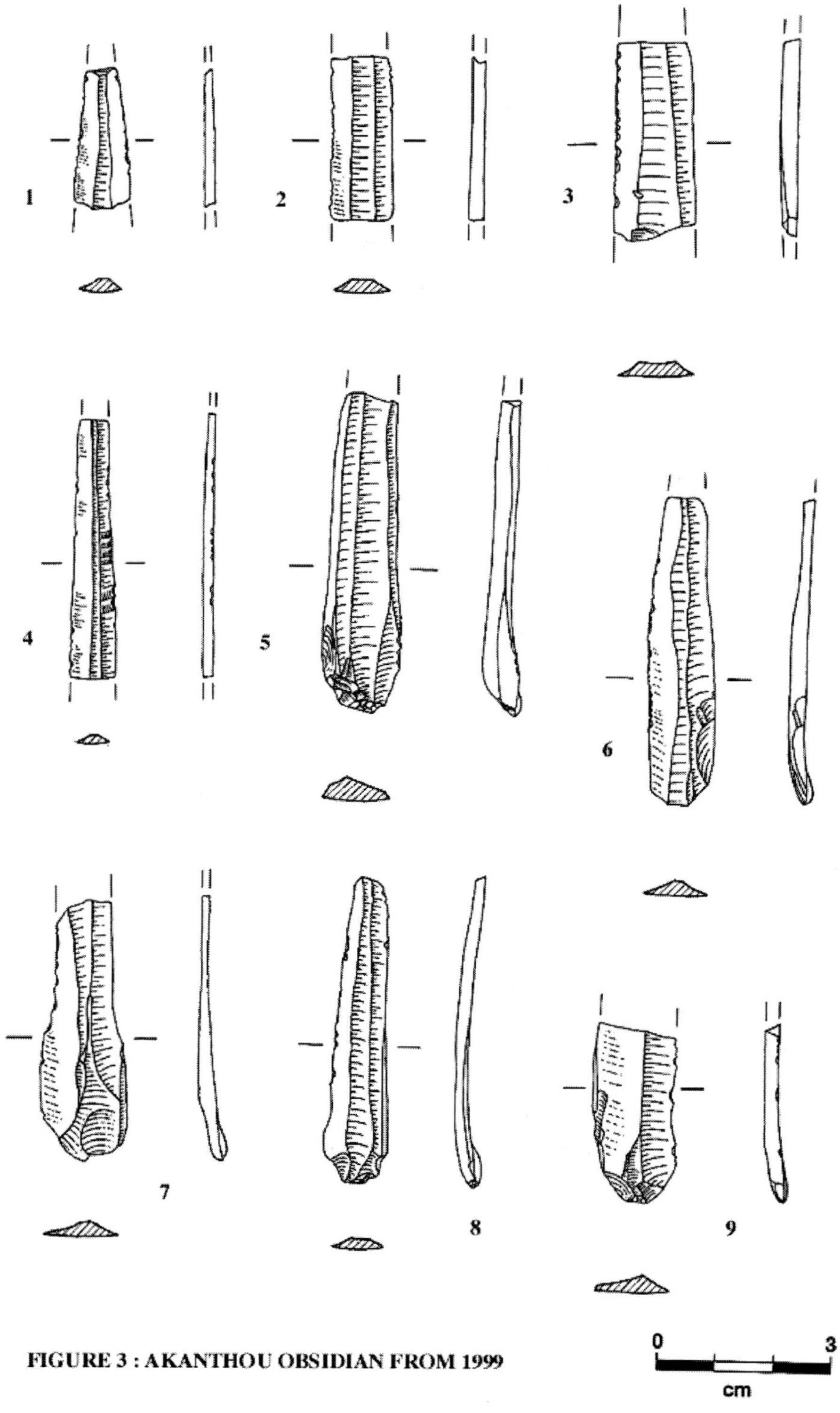

FIGURE 3 : AKANTHOU OBSIDIAN FROM 1999

THE EXCAVATION AT AKANTHOU

The first excavation season of fieldwork at the Aceramic site of Akanthou was conducted during 1 August–5 September 2000. The aim of the fieldwork was to discover architectural remains, establish stratigraphy and the extent of the site.

During this season, two small trenches were opened by hand, and all deposits removed were dry sieved through 4mm mesh. The most southerly of these, Trench 1, was sited away from known concentrations of surface finds, as an intentional trench to identify and evaluate the local surface geology and the extent of the site. The dimensions of the test trench were approximately 4 x 4 metres and 60cm in depth. In Trench 1, topsoil overlay a fine silty subsoil, which had probably been formed through long-term colluvial processes. Two discrete features were recorded cut into the natural soil of weathered limestone. These were pits—one was circular in plan, and the other was a larger and more irregular-shaped feature. Both pits contained animal bones, chert, and obsidian blades.

A larger trench was sited to the north of Trench 1, to explore an area of concentrated surface finds where the ground surface was discernibly raised. The dimensions of Trench 2 were initially 10m x 5m, and the trench was further extended to 15m in length, north to south. In Trench 2, topsoil was removed by hand to a depth of 0.35m. Beneath the topsoil was a disturbed soil, approximately 0.30m thick that overlay intact archaeological deposits. Archaeological deposits found included evidence for a stone walled structure, and fragmentary remains of beaten earth/plaster surfaces. In the trench extension to the south there was a more extensive and intact floor surface. A number of cut features were also investigated.

A second season of excavation in July–August 2001 enabled the further exploration of deposits revealed in 2000 (Plate 1). In particular, the aim was to reveal the full extent of the possible stone walled dwelling, as well as to increase the overall size of the trench. Trench 2 was reopened and extended 4.0m to the east and 4.0m to the south. Topsoil and disturbed underlying deposit were removed and dry sieved for finds. The eastern extent of stone walled structure and other structures were revealed.

As most things have not been fully excavated, results reported here are preliminary. However, there is definite evidence for 3 sub-square structures, and possibly one square structure, which at present all appear to be contemporary, as they are all constructed on the same hard limestone/plaster surface. In addition to this there are discrete features, including internal bins and pits. Several construction techniques were used, including mud-brick and stone finished with a plaster surface. All the houses have plaster floors, and there is clear evidence of replastering of the floors. There is further evidence that the walls were plastered as well as painted with red and brown paint.

In the buildings, and perhaps even outside, there are shallow plastered bins. Four such bins of various sizes were discovered. One of the bins excavated in 2000 contained a fill of large rounded flint pebbles, arranged around a grinder stone that was placed centrally. These could be interpreted as "potboilers"—stones heated in a fire and than transferred to the bin for boiling water. Since the other bins did not display the same fill, the bins could also be multi-functional. In 2001, another lined bin, relatively smaller than the bin found in 2000, was excavated: inside, there was a deliberately placed complete shed fallow deer antler. Two more plaster bins larger in size but very poor in condition were excavated. Besides the plastered bin with antlers, there was a stone-lined round structure with a grinding stone placed in the centre, as in one of the bins described earlier. This structure is identical to that discovered at Shillourokambos phase B and referred to as a hearth/fireplace dating to 7800-7500B.C. (Ministère des affaires étrangères [No date]. Shillourokambos Les premiers paysans de Chypre. from: hhtp://www.france.diplomatie.fr/culture/france/archaeologie chypre. [Accessed: 10.05.1999]. However, this comparison makes poor dating evidence as such structures could have been built and used over a large time period.

There are also postholes of various sizes, but their function has not yet been determined in relation to other structures.

ARTEFACTS AND REMAINS FOUND AT AKANTHOU

OCHRE

There are three colours evident from the excavations: red and yellow and brown. Evidence for red paint was discovered on large amounts of fallen wall plaster; red paint stains were also found, one on a quern and another on a stone mortar. A worked lump of brown/red ochre with rubbed sides was also found. The yellow ochre was discovered in small fragmentary quantities. Red paint and painting on walls is known from Kalavassos-Tenta (Todd 1981) and also at Khirokitia (Le Brun, 1997).

PUMICE

Several pieces of pumice were found on the site. One of them in particular shows signs of rubbing, for shining or smoothing purposes; Pumice could have easily used as equivalent to that of a sandpaper for smoothing bone tools or polishing obsidian. Pumice does not occur naturally in Cyprus, but is very common in central Anatolia, especially in the Cappadocia area where the origin of obsidian found in Cyprus is sourced.

OBSIDIAN

The number of obsidian blades recovered from Akanthou from rescue sieving of the mounds and from both seasons of excavation is now above 1000, including complete tools and débitage. This shows that it was regularly brought to the site in large quantities and worked on site. The size of the obsidian blades vary from 70 mm (the longest discovered at Akanthou) to 10 mm long. The initial preparation of the blades indicates a bi-directional blade industry. There are fragmentary pressure-flaked obsidian pieces (FIGURE2, 8), but they seem to be very few in number and fragmentary (FIGURE 2 and 3). The same blade technique is applied on the local fine-grained chert discovered at Akanthou. Trace element analysis work for the sourcing and dating work will be done in collaboration with Çukurova University.

PICROLITE

Tokens and miniature cups: The well known picrolite sources are in the Troodos mountains. One picrolite source nearer to Akanthou is marked on the geology map, but has not been recently investigated by any archaeologists or geologists. The three picrolite tokens are of various shapes and sizes—conical, rectangular cubical and round (FIGURE1, 8). The miniature picrolite cups or 'thimbles' are round or rectangular in shape (Fig1, 1and 2). All of these objects have incised hatching decoration on them.

PLATE 1 : AKANTHOU (TATLISU-CIFTLIKDUZU), AERIAL VIEW OF TRENCH 2

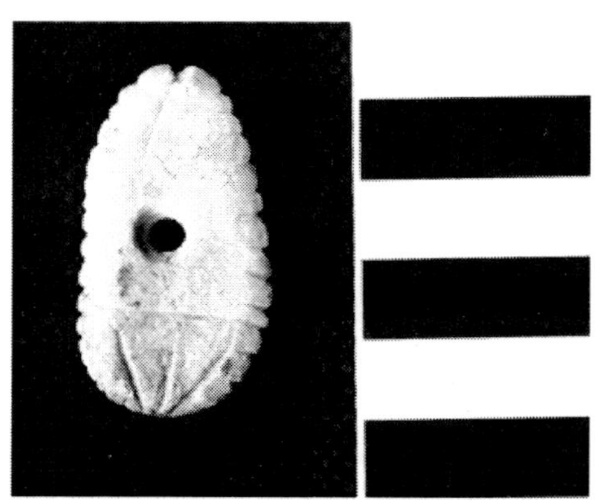

PLATE 2 : AKANTHOU (TATLISU-CIFTLIKDUZU) STONE PENDANT

CHERT Many varieties of chert tools were found, from large bulky blades of rough chert as large as 8cm long, to small blades of fine chert. These smaller blades are made of fine chert in the same technology as the obsidian blades.

WORKED BONES
The worked bone finds include fishhooks, needle with an eye which may indicate an Anatolian link (Cauvin, 2000), small spatula head, a long pin, and various bone points commonly known as awls. These indicate on site activities such as piercing and sewing possibly of animal skins. The two fishhooks are missing the bait end. One of them has incised parallel lines where the hook line was tied. Hooks fit with the results of certain fish types identified by Ruby Ceron-Carrasco, her results show some fish types from Akanthou such as hake, dogfishes and Atlantic bonito which could be caught with line and hook.

BEADS AND PENDANTS
Beads are made from various coloured stones, light green, white, black and turquoise (picrolite) (FIGURE 1,3). All the

stone except picrolite needs to be identified more precisely. Some may very well be imported and they may point to an origin of contact. Besides stone beads there are shell beads. These occur on two different types of shells, bivalves and cowries. String holes on them were made by using two different techniques. The first type was produced by rubbing the raised tip end of the bivalve until it was worn into a hole; the second was made by slicing a thin opening vertical to the length on each end of a cowrie shell. Dentalium and bone beads were also discovered.

White (FIGURE1, 5), black (FIGURE 1, 9) and orange (possibly carnelian) stone pendant fragments were discovered from Akanthou. Two complete pendants are both from an unstratified contexts. First one was discovered from 1996 seasons work, which is a green stone pendant (FIGURE 1, 10) (Sevketoglu 2000). The second one is a white stone pendant, discovered in 2001 season on top soil which may be one of the earliest representations of human figure found in Cyprus (Plate 2). This one may be one of the earliest representations in Cyprus of the human figure.

GROUND STONE ARTEFACTS

The Ground Stone vessels from the excavation are all fragmentary and rare. Excavated material is 100% sandstone and not igneous like the Khirokitian ground stone industry. There are, however, some igneous stone dishes and bowl fragments from the surface \t the east end of the site, where the later phase of Aceramic Neolithic occupation may shifted. Large basins, querns, and a few grinders, all of sandstone, were discovered from the surface and from the disturbed parts of the site from spoil heaps. All were possibly used for agricultural purposes; one of the querns however had residue marks of a red paint.

Several miscellaneous stone artefacts were discovered during 1999. One of the two most interesting is a heavily hatched flat stone discovered in the 1999 season during sieving (FIGURE1, 7) and a round pebble with flat base decorated with incisions (FIGURE1,8). Similar types of artefacts discovered at Shillourokambos were described by Guilaine as having connections with Mureybetian in Syria (Guilaine 2001).

ANIMAL REMAINS

The animal remains fall into two broad categories: fish, and terrestrial animal bones. Large amounts of well-preserved fish vertebrae were discovered during sieving and in the excavation. The preliminary work on the fish bones was carried out by Ruby Ceron-Carrasco, from the 1999 seasons collection. The sizes of these vertebrae vary belonging to a very large to small fishes. Her studies show that some identified fish are not common and may have been over-fished to extinction. Many of the fish are caught either by line or net and these methods require sailing in deep waters. Some of the types identified included the following: Hake (Merluccius merluccius), dogfishes and small sharks (Scyliorhinidae), nursehound (Scyliorhinus stellaris), long-finned tunny (Thunus alalunga), common tunny or blue-finned tunny (Thunnus thynnus) and Atlantic bonito (Sarda sarda).

The first identification of terrestrial animal bones was done in Edinburgh in 1996 by Dr Nicola Murray. She identified sheep/goat, pig, deer but also the cattle from Akanthou. In the same year Dr Louise Martin, from the Çatalhöyük project, confirmed the cattle identification. At present, the excavated animal bones are being studied by Dr. Sheelagh Frame, who has identified another more cattle, sheep/goat, pig, dog and a fox. (see Frame, this volume for further details)

PROVISIONAL DATING OF AKANTHOU

Although Akanthou is known to be an Aceramic Neolithic site, it is not clear to which phase within the new chronological framework it should be fitted. Several sealed contexts were excavated only last year (2001) from where animal bones are selected for radio carbon dating. The results of these are still pending. The work carried out in 1996 and 1999 produced enough artefact evidence for comparative dating with sites like Shillourokambos, Cape Andreas and Khirokitia. Here I will suggest a provisional date for Akanthou from the artefacts, relying especially on the large amounts of obsidian and the small amounts of cattle, and assigning the site to the Cypro-PPNB. This seems to be the upper levels of the excavation (30-65cm below the top soil) and the depth of occupation layer is at least 1.60cm calculated from the section taken from the chicken pit that cut through end of the Trench 2. As Peltenburg and Briois suggest that the large number of obsidian pieces may be evidence for Cypro-E/MPPNB (Brois et al. 1997; Peltenburg et al. 2001). Obsidian blades made from bi-directional cores are known from the Cypro-EPPNB (Fig2, 6).

CONCLUSION

In 1999 during the rescue operation, sieving work at Akanthou produced a large number of obsidian, and various kinds of artefacts and animal bones, reflecting only a small part of the large potential and importance of the Akanthou site. The promising results enabled the decision to begin excavation in the following year, in order to test the evidence for undamaged structures on the site. The first excavation season in 2000 produced obsidian and picrolite artefacts, and architectural remains including pits, plastered shallow bins, plastered floors and stone structures. The following season, in 2001, produced even more solid results for the Aceramic settlement, as explained above. More research needs to be done in C14 dating of the site; after this is done, we will be able to develop and consolidate Akanthou's role in the Aceramic Neolithic of Cyprus.

So far, these discoveries make Akanthou a clear candidate to join the group of existing early sites like Shillourokambos and Mylouthkia. With excellent preservation of organic and inorganic material starting 30 cm below the topsoil, and a deep stratigraphy spread over an area of 4 hectares, the site may prove to be the key that opens the door to an even earlier period of Cypriot prehistory than the Cypro-PPNB. It will also allow us to address questions such as whether this was a landfall site; whether the settlement served as an obsidian distribution centre for Cyprus and even for the Levant, and a centre of contact with Anatolia. Akanthou will be also interesting in other respects, for having continuing occupation throughout the Aceramic period. The type of surface materials as well as those from the chicken trenches seem to indicate a drifting of the settlement as time progressed.

The large amounts of obsidian from the excavated area of Akanthou, which must have an Anatolian source, suggests that the settlers of Akanthou retained strong links with mainland communities in central and south-east Anatolia, ensuring continued supplies of the material. At present, although there are early Aceramic Neolithic sites in Cyprus, there are no sites of that period known on the opposite coast of Turkey, or the coastlands of the Hatay, Syria, Lebanon or Israel. A survey along the coast of Ta_ucu, Adana and _skenderun Bay may reveal more information about this connection. Such a coastal survey is being carried out by TAY (Türkiye Arkeolojik Yüzeyara_tırmaları) of Istanbul University this year.

** In memory of Dr. William Dreghorn, friend and mentor

ACKNOWLEDGEMENTS

Acknowledgement is gratefully made to all my student who joined the excavation; field director Joe Partridge and supervisors Ian Hanson, Dann Eddisford and Miriam Clyne for their excellent conduct of the excavation; and the Tatlisu village Council and the inhabitants of the village; to the Department of Antiquities and Museums of the Turkish Republic of Northern Cyprus for the excavation permit, and to the sponsors of the excavation for all help and contributions: Eastern Mediterranean University, Shevketoglu Domestic, Cyprus Turkish Airlines, Onar Village, Ramiz Manyera Tesebbusleri, Astro Hypermarket, Ahmed Rashid Ltd. And Mustafa Emirzadeogullari. Special thanks to Professor T. Watkins for his valuable comments and suggestions on the paper and to Dr. J. Pillai for correcting the text.

BIBLIOGRAPHY

BRIOIS, F., B. GRATUZE, J. GUILANE,
1997. Obsidiannes du site Neolithique Preceramique de Shillourokambos (Chypre). Paleorient, vol. 23/1, p. 95-112.
CAUVIN, J.,
2000. The birth of the gods and the origins of agriculture. Cambridge University Press.
FRAME, S.
(This volume) Island Neolithics: Animal explotation in the Aceramic Neolithic of Cyprus.
HELD, S.,
1989. Colonisation and extinction on Early Prehistoric Cyprus, in Paul Astrom (ed.), Acta Cyprus: Acts of an International Congress on Cypriote Archaeology Held in Goteborg on 22-24 August 1991:104-164. Jonesered: Astrom. Studies in Mediterranean Archaeology Pocket Book 114.2.
HELD, S.O.,
1989. "Colonisation Cycles on Cyprus 1: The Biogeographic and Paleontological Foundations of Early Prehistoric Settlement". Report of The Department of Antiquities, 7-28.
LE BRUN, A.
1974. "Cap Andreas, Kastros. Rapport Preliminaire 1970-1972". Report of the Department of Antiquities of Cyprus.
LE BRUN, A.,
1989. La neolithisation de Chypre, in O. Aurenche and J. Cauvin (ed.): 95-107.
LE BRUN, A. ,
1997. Khirokitia A Neolithic Site. Bank of Cyprus Cultural Foundation. Nicosia.
MINISTTERE DES AFFAIRES ETRANGERES
[No date]. Shillourokambos Les premiers paysans de Chypre. from:hhtp://www.france.diplomatie.fr/culture/france/archaeologie/chypre. [Accessed: 10.05.1999].
PELTENBURG E., S. COLLEDGE, P. CROFT, A. JACKSON, C. MCCARTNEY AND M.A. MURRAY.
2000. "Agro-pastoralist colonisation of Cyprus in the 10th millennium BP: initial assessments". Antiquity, 74: 844-853.
PELTENBURG E., S. COLLEDGE, P. CROFT, A. JACKSON, C. MCCARTNEY AND M.A. MURRAY. ,
2001. "Neolithic dispersals from the Levantine Corridor: a Mediterranean perspective", Levant, 33:
35-64.
PELTENBURG E., S. COLLEDGE, P. CROFT, A. JACKSON, C. MCCARTNEY AND M.A. MURRAY.,
2001. "Well established colonists: Mylouthkia 1 and the Cypro-Pre-Pottery Neolithic B", in S. Swiny (ed.), The earliest prehistory of Cyprus: from colonisation to exploitation. Boston (MA): American Schools of Oriental Research. 61-93.
GUILAINE J., BRIOIS F.,
2001 "Parekklisha Shillourokambos: An Early Neolithic Site in Cyprus", in S. Swiny (ed.), The earliest prehistory of Cyprus: from colonisation to exploitation. Boston (MA): American Schools of Oriental Research. 37-53.
SEVKETOGLU, M.,
2000. Archaeological Field Survey of the Neolithic and Chalcolithic Settlement sites in Kyrenia District, North Cyprus. Systematic Surface collection and the interpretation of artefact scatters. BAR International Series 834. Oxford: Hadrian Books.
SIMMONS, A. AND ASSOCIATES.,
1999. Faunal Extinction in an island society: Pygmy hippopotamus hunters of Cyprus. New York (NY) : Plenium.
STANLEY PRICE, N.,
1977. "Khirokitia and the initial settlement of Cyprus", Levant 9: 66-89.
TODD, I.A.,
1981. "A Cypriote Neolithic Wall-painting" Antiquity, 55, 47-51.
TODD, I.A.,
1996. "The Vasilikos Valley: Its place in Cypriot and Near Eastern Prehistory". In Go to the Land I will show you Edited by J.E. Coleson and V.H. Matthews. 317-351.
WATKINS, T.,
1973. "Some problems of the Neolithic and Chalcolithic period in Cyprus", Report of the Department of Antiquities, Cyprus: 34-61.
WATKINS, T.,
1980. "The Economic Status of the Aceramic Neolithic Culture of Cyprus", Journal of Mediterranean Archaeology and Anthropology: 1, 139-149.

11

"THE ISLAND FACTOR": INSULARITY AS A VARIABLE IN THE ARCHAEOLOGICAL STUDY OF THE SOCIAL DYNAMICS OF HUNTER-GATHERERS

JORDI ESTÉVEZ, ASSUMPCIÓ VILA AND XAVIER TERRADES

Autonomous University of Barcelona and Archaeology Laboratory, CSIC, Barcelona,Spain
EMAIL: ilphg@blues.uab.es

ISLAND ARCHAEOLOGY AS EXPERIMENTAL ARCHAEOLOGY

As a social science, Archaeology aims to explain the essence of humankind as a social entity or, put another way, the reasons why processes of change have taken place through the interpretation of material remains (Argelés et al. 1995). Like other sciences, Archaeology has its experimental side. In fact, the excavation of any archaeological site may be seen as an experimental process through which data is obtained that makes it possible to contrast hypotheses and, eventually, to reformulate them.

The importance of studies in social dynamics with controlled variables is obvious: they allow us to gain greater insight into the mechanisms and laws of social interaction. In this case, living on an island means that the societies in question are relatively isolated as is the environment in which they go about their daily lives. This environment acts as a natural limit to a series of specific variables. Because of this, island societies, which are subject to a type of isolation which accentuates moments of stress, are an exceptional subject for social experimentation.

The importance of this fact (controlled or directed social experimentation) has been practically proven through the disguised realisation of a number of genuine social experiments in internal group dynamics. Anthropological ethics, which would oppose this type of experimentation on human subjects, has recently been cleverly circumvented by presenting these experiments as little more than a game. Television shows such as Survivors, Big Brother, and The Bus have proliferated. These shows are really no more than the selection of a group of individuals based on a number of variables so that they can be studied under conditions of controlled isolation and their reaction to a series of different inputs observed. In short, by using the same principle of insularity for a human group and inducing moments of stress, one succeeds in provoking reactions which under normal conditions are difficult to perceive or only occur over a much longer period of time.

The archaeological approach is by definition different to that of other social sciences. It is the study of social formations of the past, through the analysis of objective elements. There is not carried out through the direct analysis of the ideas or subjective experiences of the people who make up these human groups, but through the analysis of the material remains left by their productive and reproductive activities. The arrangement and nature of the majority of these remains is not the result of any desire to leave a legacy of subjective ideas for a later society to transform into a source of archaeological information. Therefore, the society of the past is unable to exert any subjective influence on the society which is studying it, nor can the latter, through any action of its own, affect the development of the former. So, as it is impossible to modify the social conditions of those under study, this ethical-social responsibility is transferred to the effect that this scientific-social knowledge has on present and future societies. Despite the fact that we analyse the society without affecting the subjects of the study themselves, this does not prevent the consequences of the study being of even greater relevance. In fact, the societies which may be influenced by these studies transcend the fact that they are or are not direct descendants of the societies under study.

In addition to the advantage that this characteristic of archaeological research offers, there is also the fact that in Archaeology we are able to study social change over a long period of time. The time scale of the object of an archaeological study is much greater than that of any of the other social sciences. Consequently, Archaeology is in a much better position to observe the phenomena of social change over this long period of time. On the other hand, the obvious disadvantage is the difficulty in detecting processes of rapid change, which is the critical level of influence for subjective factors, both social and ideological, and the specific decisions of small, discrete human groups. However, the possibility of making a dialectical analysis of the conjunction of archaeological and ethnographical information would clearly appear to compensate for the limitations inherent in the approaches of these two social sciences.

In short, the study of island societies increases the effectiveness of Archaeology as a social and experimental sci-

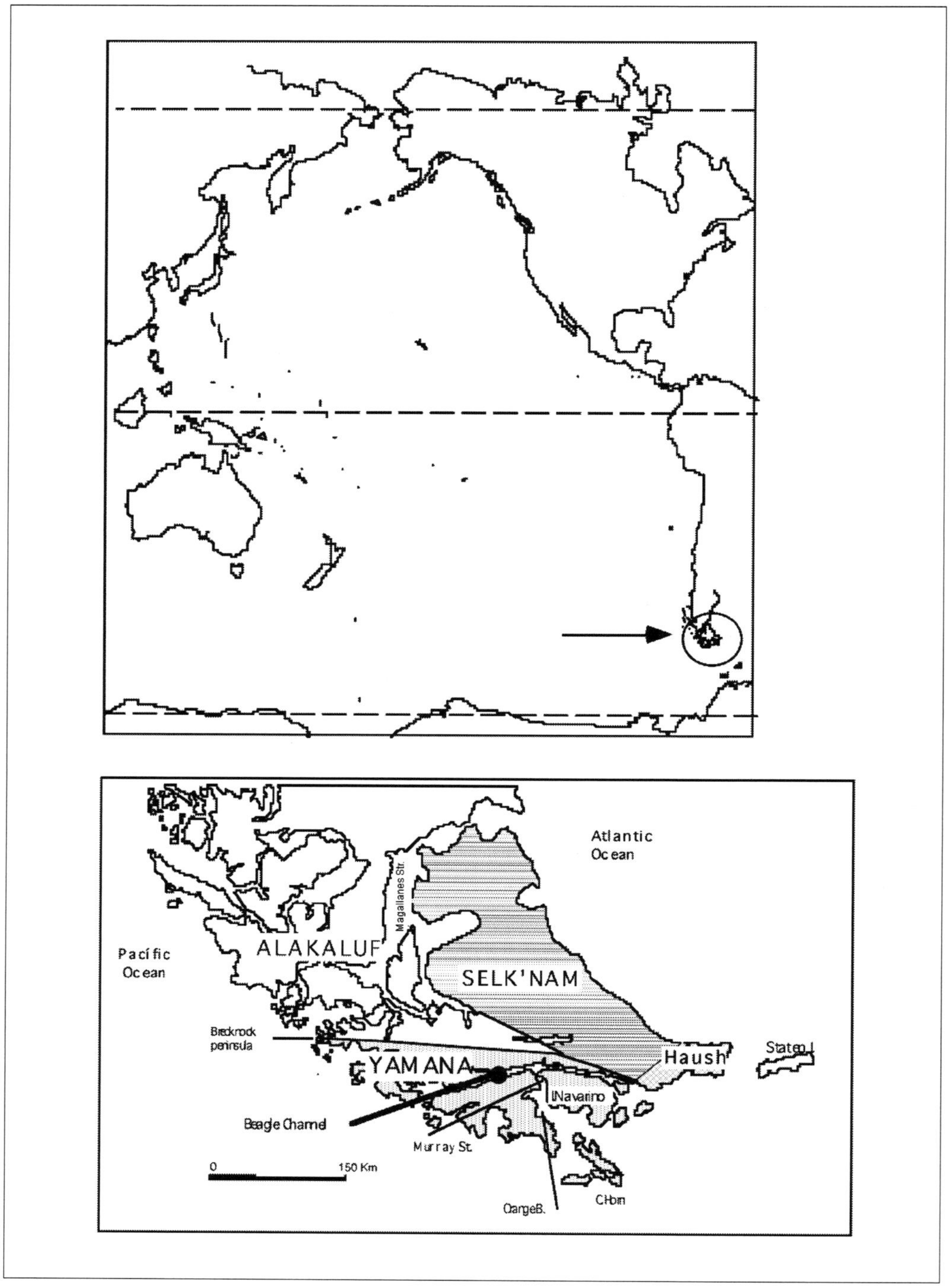

FIGURE1: POSITION OF TIERRA DEL FUEGO IN THE PACIFIC OCEAN: DISTRIBUTION OF THE FOUR GROUPS OF HUNTER-GATHERERS BASED ON ETHNOGRAPHICAL DIVISIONS.

ence both in terms of the analysis of long-term historical processes and, if an ethnoarchaeological perspective is added, in terms of the more subjective elements of social change. But, apart from this substantive objective, the ethnoarchaeological perspective makes it possible to contrast the methodological validity and capacity of the conceptual tools involved in archaeological and ethnographical analysis.

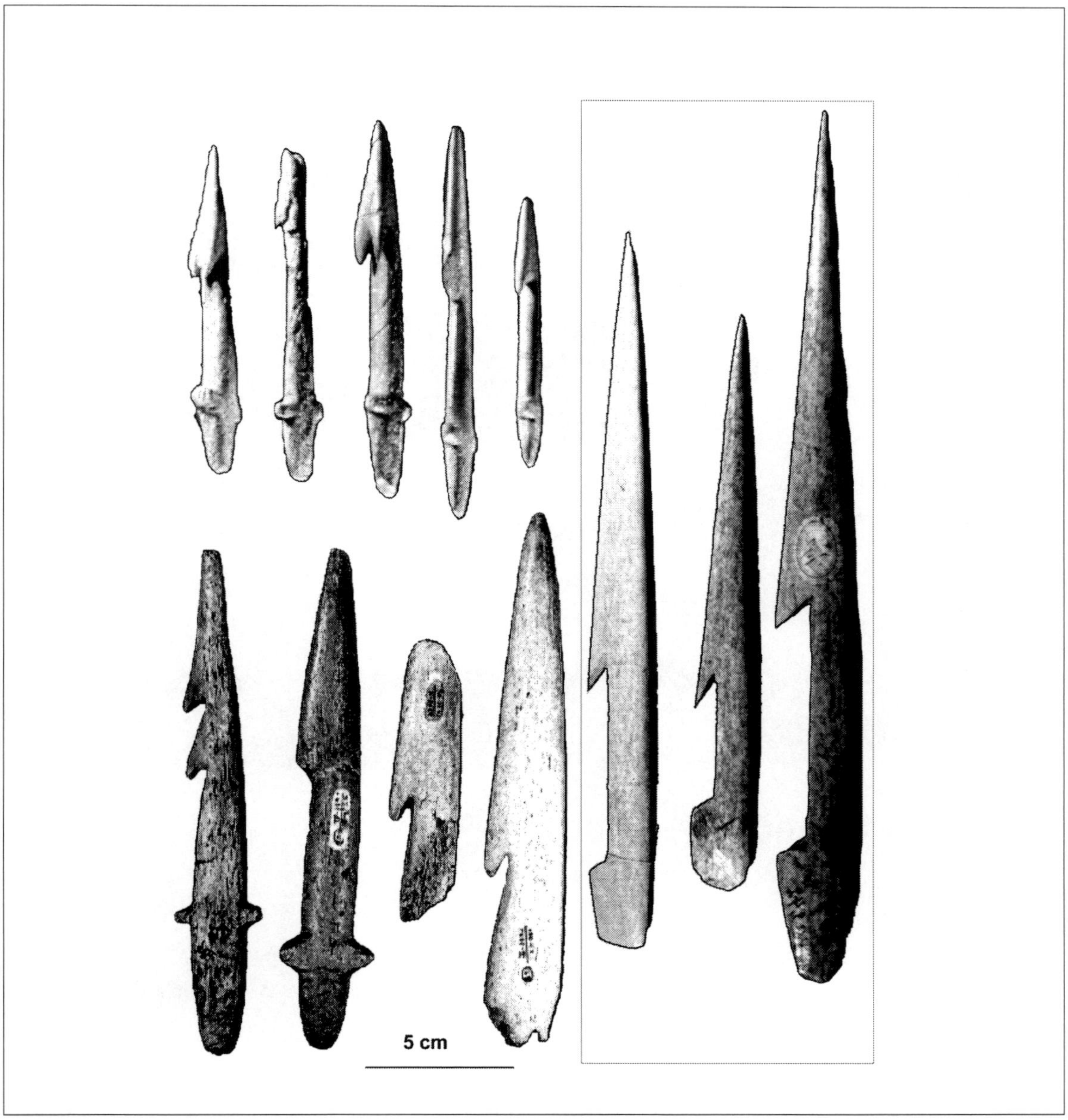

FIGURE 2. HARPOONS FROM TIERRA DEL FUEGO
an example of the formal development of this tool. On the left, the old archaeological examples. On the right, inset, modern examples from ethnographical collections.

TIERRA DEL FUEGO: AN EXCEPTIONAL STUDY LOCATION

The Isla Grande (Large Island) of Tierra del Fuego has a number of features which make it a highly suitable location for social-archaeological study. It has been separate from the southern end of the South American continent for 9000 years (Rabassa, et al. 1986). It is the largest island of the Tierra del Fuego archipelago with an area of 21,390 sq. km., and displays pronounced geomorphological and environmental gradients. This made it possible for different social groups to live there autonomously. From the time the island separated from the mainland (cf. Legoupil 1993-1994) and until their disappearance at the beginning of the 20th century, the indigenous groups that inhabited the island developed two fundamentally different strategies in a physically and environmentally limited area:

In the northern part of the island we find land-based groups that depended on the exploitation of land resources for their survival. They appear to have been isolated from other similar hunter-gatherer groups of continental Patagonia from the time of the opening of the Magellan Strait.

The southern and western parts of the same island were inhabited by other groups whose subsistence strategies depended mainly on the exploitation of the coast and sea channels. The availability of canoes facilitated contact with other littoral groups from the mainland and neighbouring archipelagos. (FIGURE 1).

At the end of their long history and having entered into contact with "white" people, these groups were classified into three or four "ethnic" groups: Selk'nam, Yaghan or Yamana, Alakaluf and Hausch (Orquera & Piana 1999b).

Temporal and social variables are therefore known. Until the arrival of European explorers, external influences were limited to possible contacts with canoe people with very similar strategies for exploiting resources whose origins were known. As a result, these societies did not receive a significant population input until the arrival of European colonists.

Unlike other societies that were affected earlier, these societies existed until the 20th century without coming into contact with neighbouring agricultural societies and were among the last to be affected by direct European impact. This has made it possible for researchers to make a wide and relatively complete ethnographical description, including photographs and collections of commodities. The varied and abundant ethnohistorical sources that are available in this case allow us to describe certain elements that would indicate the existence of an "island factor" in terms of the composition and survival of these hunting, gathering and fishing societies.

Well-preserved archaeological evidence, thanks to the relatively small amount of contemporary settlement that has taken place, climatic stability and the composition of the sites themselves (on the coast there are clearly visible shell middens, which accumulated very quickly) make it possible to obtain excellent and well-preserved archaeological remains.

Tierra del Fuego is important for the study of hunter-gatherer societies for another reason. Since Darwin, Fuegian native societies have become the paradigm for different theoretical-methodological approaches and have, as a result, been used as the main point of reference for descriptions and explanations of prehistorical hunter-gatherer societies. An analysis of these societies allows us to test the validity of certain approaches and hypotheses that have been taken for granted in the description of the prehistory of the Old World and have become axioms.

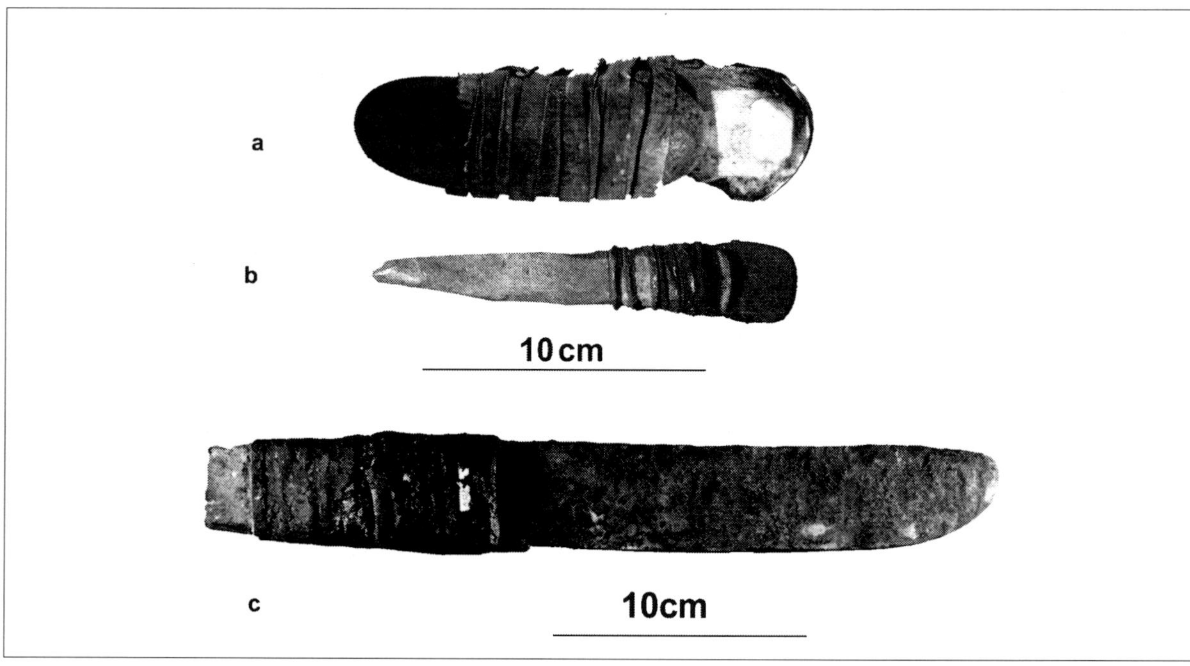

FIGURE 3. MODIFICATION OF KNIFE DESIGN IN THE ETHNOGRAPHICAL PERIOD.
Above, the traditional knife made from a mollusc valve. In the centre, a knife with a transversal metal blade and below, an imitation of the European knife with a longitudinal blade.

ETHNOARCHAEOLOGICAL PROJECTS: ADAPTATION OR SOCIAL ALTERNATIVES

Since 1975, a team from the National Council for Scientific and Technical Research (CONICET) of Argentina has been carrying out archaeological work in the south of the Isla Grande of Tierra del Fuego (Beagle Channel area). This work has focused on canoe societies and their diachronic change by verifying the significance of geographical factors that may be seen as independent causes of variability (Orquera & Piana 1999a). Since 1988, and in collaboration with the CSIC (Spanish High Research Council) and the Autonomous University of Barcelona, an ethnoarchaeologically-oriented research project has been carried out (Piana et al. 1992; Estévez & Vila 1996; Estévez & Vila 1998) with the intention of dialectically correlating ethnohistorical and archaeological sources and with both general theoretical-methodological and particular objectives regarding local development. In relation to the first objective, the main aim was the development of archaeological methods and theory as well as the verification of existing theories for prehistoric European hunter-gatherer societies. As far as the second objective is concerned, the aim was to follow the process of the last moments of a hunting, gathering and fishing society.

Through these projects we were able to evaluate to what extent societies which were clearly delimited by their relative insular isolation and living under relatively stable environmental conditions sustained what has been called a successful process of adaptation (cf. Piana 1984; Legoupil 1985-86).

In effect, studies carried out as part of a European Union project regarding the impact of industrial exploitation on marine resources which we carried out through the study of faunal remains and isotopic analysis of samples which were obtained through the exploration, sampling and dating of over 130 archaeological sites have led us to the conclusion that, despite evidence of minor oscillations since the Climatic Optimum, the insular-littoral ecosystem has been relatively stable for the last 6000 years (Vila et al. 1998).

Analysis carried out on the faunal remains shows great stability in this type of resource, which was exploited throughout this time frame. Almost from the beginning of human occupation, which took place at least 10,400 years ago (Massone 1987; Orquera & Piana 1999b), until the period of contact with Europeans (Estévez & Vila 1996) there is evidence of exploitation of marine resources: fishing, gathering of molluscs and the hunting of sea lions and birds. Despite the fact that there are minor fluctuations and a slight variability that may be attributed mainly to differences in the biotope, fishing and the gathering of molluscs were the critical variable in the effort invested in subsistence. Sea lions were always a crucial source of calories (Schiavini 1993).

If we contrast this subsistence base with the development of production tools, we may observe important variations that allow us to postulate the existence of mechanisms of change and social self-regulation.

Throughout the occupation period of the Channel (Orquera & Piana 1999a), it is possible to identify a substratum of tools which, despite remaining constant, develops in terms of their form (FIGURE 2). Such is the case of harpoons. Ancient forms of harpoon decoration are abandoned in later periods. Other minor changes are quantitative: for example, tools made of polished and smoothed stone are more common in ancient times (Mansur et al. 1989), and their presence decreases progressively until they all but disappear among archaeological remains in more recent sites. But there is also evidence of new developments or the adoption of new technologies. Thus arrowheads first appear in archaeological lithic remains around 2700 BP, after which their manufacture remains constant within increasingly standardised technical and morphological parameters. One also observes clearly differentiated tool groups, as far as their morphology is concerned, in sites such as Lancha Packewaia (around 4200 BP) that show moments of experimentation or instability in the means of production, despite the relative stability shown by the objects of production.

In short, it is not immovability, despite the constant presence of the objects of production (continuity in the exploitation of a wide range of marine resources) and the relative stability of the environment, but signs of social readjustment which are clearly visible. This is not only in terms of experimentation and changes in the instruments of production, but also in terms of the structure of the settlements themselves (ring-shaped shell middens are more common in recent periods). It seems clear that we are dealing with a dynamic effort to maintain a specific social system and not just a stable and successful adaptation.

By contrasting ethnography and archaeology we may observe the appearance of another set of extremely relevant phenomena which contrast with the ethnographical extrapolation that served as a basis for the construction of the archaeological representation of prehistoric hunter-gatherer societies (Estévez et al. 1999). It is true that the differences described between the groups of land-based hunters of the north of the island and the canoe people of the southern and western coast are not so apparent from their artefacts and commodities. Analysis of ethnographical materials and the extrapolation of those that were preserved and are visible in archaeological remains tells us that the majority are shared by all the social groups on the island. Any differences are quantitative or transitory rather than strictly qualitative (above all in relation to direct commodities), with the exception of production tools linked to exploitation of resources from canoes. This last aspect could also be wrongly interpreted, from an archaeological perspective, as a transitory fact (alternate cycles of coast-land exploitation) rather than as a genuine ethnical-cultural difference. Furthermore, the disparity between archaeological remains produced by the same ethnographically identified group are sometimes greater than that which can be found between the remains of different groups.

It would therefore be inappropriate to talk in terms of a simple correlation between archaeological remains and "culture" (whether one considers this to be an ethnic category or in terms of adaptation). But the most interesting question in this context and which is most relevant to the phenomenon of insularity is the fact that on the same island ethnographers have isolated four different groups. According to these ethnographers, this distinction was most marked in physical appearance and languages spoken, rather than in the strategies to exploit resources: there appear to have been two main groups of canoe people and two groups of land-based hunter-gatherers. In fact, both detailed and paleo-bioanthropological analysis show that there was a large degree of permeability among these supposed discrete groups, despite factors that should have favoured isolation such as exploitation strategies for disparate resources of coastal and land-based people, or insular dispersion and the difficulties of navigating around theTierra del Fuego archipelago

TIERRA DEL FUEGO ARCHIPELAGO. THE END OF THE FUEGIAN SOCIETIES

We can also link archaeological data with the analysis of social and ideological organisation in the last moments of these societies (Estévez & Vila 1997), which has been ethnographically documented.

The presence of European ships in the area signified the arrival of "new materials" to the coastal areas inhabited by these groups even before the direct physical presence of these explorers took place. This obviously increased the number of raw materials that these groups possessed and helped them to improve some of their activities. This did not, however, cause significant social changes, which only took place when the new arrivals started to settle.

In effect, 18th century sites along the coast of the Beagle Channel have revealed fragments of glass. This ability to rapidly assimilate new raw materials was not held back by the inertia of having used the same materials for millennia. On the contrary, we may assume that accumulated technical experience meant that the advantages of these new raw materials were quickly appreciated. In the case of glass, its superior and more predictable response to flaking (especially through pressure) was the feature that made it more popular for the manufacture of arrowheads or scrapers than local stone raw materials, which were much more difficult to shape through flaking. The use of metal, obtained from the Europeans, was also rapidly adopted as a (total or partial) substitute for other cutting tools. The greater resistance of this material, in comparison to the original knives, produced by hafting an Aulacomya valve as a blade, was also quickly perceived and generated intense demand for this material. At first, metal knives were produced by placing the edge in the same way, transversely, as the original (valve) models until the typical European longitudinal placement was adopted later (FIGURE 3). Metal was also adopted for the manufacture of awls to replace those made from bird bones, which were much more fragile. In this case fragments of nails were inserted directly into the diaphysis of the bones.

There is no doubt that the use of metal stimulated other work processes, such as woodwork. This shows that there were certain technical limitations, which were perceived as such, and which made it desirable to use and stimulated the adoption of this new material. Sure enough, with the introduction of chopping tools made from this material (axes, for example), the manufacture of canoes made from planks was begun which substituted the traditional ones made from tree bark. It is also probable that there was an increase in the construction of conical huts made from tree trunks as opposed to dome-shaped huts made from branches.

These technological changes took place quickly without the need for prolonged direct contact with European society and did not significantly affect general strategies of resource exploitation or social organisation, which remained unchanged.

The collapse of these societies came about with the definitive establishment of European settlers. These missionary-settlers brought about a catastrophic change in these societies, and their mechanisms and consequences can now be studied from a historical perspective. They imposed new forms of social organisation, property systems and ways of exploiting the resources of this territory and introduced agricultural and stockbreeding and even new species. They rapidly undermined and destabilised the hunter-gatherer societal organisations.

For the land-based groups that exploited the interior of the island the new strategy of the settlers was completely incompatible with their own. Livestock keeping limited the killing of the herbivores (sheep, in this case) which were the property of the new European holders of the land "concessions". For the canoe people, the collapse was brought about because the whites introduced new diseases which spread very quickly and were further stimulated by forced or encouraged sedentarisation. However, one must add another, internal, factor to this equation: the inability of the indigenous societies to react quickly enough to this catastrophic loss of individuals caused the population to drop below the critical level of recovery.

This phenomenon must be attributed to the strict and firmly entrenched social and ideological organisation (which is well documented by ethnohistorical sources) which maintained a restrictive strategy for reproduction (and limited production) at the moment of contact (Vila & Ruiz 2001). Only by accepting that at some moment during their long historical process these societies had succeeded in implementing an effective social strategy of this kind is it possible to put forward a hypothesis to explain the stability and continuity of the population throughout the 7000 years of human presence on the island.

It is probable that technological changes may offer a clue to understand how experiences were gradually accumulated and strategies implemented during moments of instability between the production of resources and reproduction in these societies. The example of Tierra del Fuego will also serve to illustrate how these technological changes are accepted more quickly than those which involve a restructuring of the social relations of reproduction and when this does finally take place it is the social and economic system itself that disappears.

ARQUEOLOGÍA DE LAS ISLAS COMO ARQUEOLOGÍA EXPERIMENTAL

La Arqueología como ciencia social pretende explicar el Ser social en su esencia o, lo que es lo mismo, las causas de sus procesos de cambio a través de sus manifestaciones materiales (Argelés, et al. 1995). Al igual que otras ciencias, la Arqueología tiene también un componente experimental. De hecho, la excavación de cualquier contexto arqueológico puede considerarse en sí misma un proceso experimental mediante el cual se obtienen datos que permiten contrastar implicaciones de hipótesis y, eventualmente, proceder a su reformulación.

La importancia de estudios sobre dinámicas sociales con variables controladas es muy evidente, puesto que permiten un mejor acercamiento a los mecanismos y a las leyes de interacción social. En este caso, el efecto insular propicia un aislamiento relativo de las sociedades sujetas a estudio y del medio ambiente en el que desarrollan sus actividades, que actúa como delimitante natural de determinadas variables. Y, en este sentido, las sociedades insulares, sometidas a un aislamiento que acentúa además los momentos de stress constituyen un campo de experimentación social privilegiado.

La importancia de este tema (experimentación social controlada o dirigida) ha sido evidenciada en la práctica con la realización más o menos encubierta de auténticos experimentos (de dinámica interna de grupos) sociales. La ética antropológica que se opondría a tal tipo de experimentación con sujetos humanos ha sido recientemente soslayada de forma muy hábil presentando esas prácticas de forma lúdica y restándoles trascendencia. Así, hemos asistido a la proliferación de programas televisivos tipo Supervivientes, Big Brother, el Bus, etc. que en definitiva no son otra cosa que la selección de un grupo definido de sujetos en función de determinadas variables para su observación en condiciones de aislamiento controladas y su reacción frente a distintos "inputs" que se van introduciendo. En definitiva, participando del mismo principio de insularización de un grupo humano y estimulando esos momentos de stress se consigue el efecto deseado de precipitar reacciones que en condiciones normales son difíciles de apreciar o que suceden en un plazo histórico más largo.

La aproximación arqueológica tiene por definición un carácter especial respecto a otras Ciencias Sociales. Se estudian las formaciones sociales de tiempos pasados, y específicamente a través de elementos objetivos; no se analizan directamente las ideas o subjetividades de las personas que componen esas agrupaciones humanas sino a través de los restos materiales, resultado de sus actividades de producción y reproducción. La disposición y caracteres de dichos restos no responden, en su gran mayoría, a ninguna intención de dejar un legado de ideas subjetivas a la sociedad contemporánea que los transforma en fuente de información arqueológica. Así pues ni la sociedad del pasado tiene ya la capacidad de influir subjetivamente en la sociedad desde la que se estudia ni ésta puede, a través de su acción, afectar el devenir de las sociedad objeto de estudio. Así pues al no poder modificar lo que fueron las condiciones sociales de los sujetos, esa responsabilidad ética-social queda transferida al efecto que tenga el conocimiento científico-social producido hacia las sociedades presentes y futuras. Aunque se analiza la sociedad sin afectar a los propios sujetos concretos objeto de estudio, ello no es óbice para que las consecuencias de los estudios puedan ser aún de mayor relevancia. En efecto, las sociedades sobre las que dichos estudios pueden tener influencia trascienden el hecho de que sean o no descendientes directas de las estudiadas.

A la ventaja que plantea esa característica de la investigación arqueológica se le añade el hecho que en Arqueología podemos considerar los cambios que se producen a largo plazo a partir de su manifestación diacrónica. La dimensión temporal del objeto de estudio arqueológico supera fácilmente la dimensión temporal de la de cualquiera de las otras Ciencias Sociales. Por ello la Arqueología está especialmente privilegiada para la observación de los fenómenos de cambio social a ese largo plazo. Por el contrario, la desventaja clara es la dificultad de acercarnos a los procesos de cambio temporal rápido, nivel en el que actúan prioritariamente los elementos subjetivos –sociales e ideológicos- y las decisiones concretas de grupos humanos reducidos y discretos. Sin embargo la posibilidad de analizar dialécticamente la conjunción de una información arqueológica y una etnográfica compensaría claramente las limitaciones de las aproximaciones de ambas Ciencias Sociales.

En definitiva, estudiar sociedades isleñas incrementa el poder de resolución de la Arqueología como Ciencia Social y Experimental, tanto para analizar los procesos históricos de largo alcance temporal como, si se le añade un enfoque etnoarqueológico, los elementos más subjetivos del cambio social. Pero más allá de este objetivo sustantivo el enfoque etnoarqueológico permite, a su vez, contrastar la validez y capacidad metodológica de los instrumentos conceptuales implicados tanto en el análisis arqueológico como en el etnográfico.

TIERRA DEL FUEGO: UN CASO DE ESTUDIO PRIVILEGIADO

La Isla Grande de Tierra del Fuego tiene unas características que la convierten en un lugar de experimentación social arqueológica idóneo. Ha estado separada del extremo meridional del continente americano desde hace unos 9.000 años (Rabassa, et al. 1986). Es la mayor isla del archipiélago fueguino, con una extensión de más de 21390 km2, y fisiográfica- y medio-ambientalmente tiene unos gradientes muy marcados, lo que permitió la existencia y sobrevivencia autónoma de diferentes formaciones sociales. Desde su separación del continente (cf. Legoupil 1993-94), las formaciones sociales aborígenes que habitaron la Isla fueron desarrollando, hasta su desaparición a principios del siglo XX, dos estrategias fundamentalmente distintas en un marco físico y ambiental limitado:

(1) en la parte septentrional de la Isla nos encontramos con unas agrupaciones humanas pedestres que basaban su subsistencia en la explotación de recursos desde tierra. Aparentemente, estuvieron aisladas desde la abertura del estrecho de Magallanes de otros grupos cazadores-recolectores similares de la Patagonia continental;

(2)en la parte meridional y occidental de la misma isla vivían otras agrupaciones humanas cuyas estrategias de subsistencia iban dirigidas principalmente hacia la explotación de recursos de la costa y canales marítimos. En este caso, la disponibilidad de canoas permitía el transporte y contacto con otros grupos litorales del continente y archipiélagos vecinos. (Figura 1)

En los momentos finales de su larga historia, ya en contacto con blancos, estos grupos fueron clasificados por algunos etnólogos en tres o cuatro etnias : selk'nam, yaghan o yámana, alakaluf, y haush. (Orquera & Piana 1999b).

Las variables temporal y social están pues controladas. Hasta la llegada de navegantes europeos, las influencias externas estuvieron limitadas a los posibles contactos con gente canoera con estrategias de explotación de recursos muy semejantes, y de procedencia conocida. Así, estas sociedades no recibieron un "input" de población significativo hasta la instalación de colonos europeos.

La perduración hasta el siglo veinte de estas sociedades sin contactos con sociedades agrícolas vecinas y el haber sido afectadas muy tardíamente por el impacto directo europeo ha posibilitado, a diferencia de otras sociedades contactadas antes, contar con una amplia y relativamente buena descripción etnográfica, con documentación fotográfica y con colecciones de objetos de consumo. Las variadas y abundantes fuentes etnohistóricas existentes en este caso concreto nos permiten describir ciertos elementos que remiten al efecto del factor insular en la constitución y la pervivencia de estas sociedades (cazadoras-recolectoras-pescadoras) isleñas.

La buena conservación de la evidencia arqueológica, gracias al relativo poco impacto urbanístico, a la estabilidad climática y a las propias características de formación de los yacimientos (en la costa son acumulaciones perfectamente visibles de valvas y caparazones de moluscos, concheros, de formación muy rápida) hace posible obtener un excelente registro arqueológico.

Tierra del Fuego tiene una importancia añadida para el estudio de sociedades cazadoras recolectoras. Desde Darwin ha servido como paradigma a los diferentes posicionamientos teórico-metodológicos y se ha utilizado, por tanto, como referente capital para la construcción de las representaciones y explicaciones de las sociedades cazadoras-recolectoras prehistóricas. Por ello, su análisis nos permite poner a prueba la validez de ciertos planteamientos e hipótesis explicativas que se han planteado como obviedades para la Prehistoria del Viejo Mundo, usándose finalmente como axiomas.

LOS PROYECTOS ETNOARQUEOLÓGICOS: ADAPTACIÓN O ALTERNATIVAS SOCIALES

Desde 1975, un equipo del Consejo Nacional de Investigaciones Científicas y Técnicas argentino (CONICET) viene desarrollando trabajos arqueológicos en el sur de la Isla Grande de Tierra del Fuego (zona del Canal Beagle), centrados en las sociedades canoeras y su cambio diacrónico, controlando además la significación de factores geográficos que pueden ser utilizados como variables independientes de variabilidad (Orquera & Piana 1999a). Y desde 1988, conjuntamente con el C.S.I.C. y la UAB se llevó a cabo una investigación con orientación etnoarqueológica (Piana et al. 1992; Estévez & Vila 1996; Estévez & Vila 1998) intentando correlacionar dialécticamente las fuentes etnohistóricas con las arqueológicas, y con objetivos tanto teórico-metodológicos generales como particulares acerca del desarrollo local. Entre los primeros, el principal era el desarrollo de métodos y teoría arqueológicos a la vez que verificar los modelos explicativos para sociedades cazadoras recolectoras prehistóricas europeas; entre los segundos, se trataba de seguir el proceso del final de una sociedad cazadora-recolectora-pescadora.

Con el desarrollo de estos Proyectos pudimos evaluar hasta qué punto unas sociedades con estas características, bien delimitadas por el relativo aislamiento insular y en condiciones ambientales relativamente estables, mantuvieron lo que se ha denominado un proceso de adaptación exitosa (cf. Piana 1984; Legoupil 1985-86).

Efectivamente, los estudios realizados en el marco de un Proyecto de la Unión Europea sobre el impacto de la explotación industrial en los recursos marinos que realizamos a partir del registro faunístico e isotópico conseguido en base a una prospección, muestreo y datación de más de 131 sitios arqueológicos, nos han llevado a la conclusión de que, a pesar de registrarse oscilaciones menores del clima desde el Óptimo Climático, existió una relativa estabilidad del ecosistema insular-litoral durante los últimos 6000 años (Vila et al. 1998).

Los análisis de fauna realizados muestran una gran estabilidad en este tipo de recurso, explotado a lo largo de todo este lapso de tiempo. Prácticamente desde la primera ocupación humana, que se produjo hace por lo menos unos 10400 años. (Massone 1987; Orquera & Piana 1999b), hasta la época de contacto con europeos (Estévez &Vila 1996) hay evidencias de explotación de los recursos marinos, pesca, recolección de moluscos y caza de lobos marinos y aves. Aunque existen fluctuaciones menores y una pequeña variabilidad que puede atribuirse mayormente a pequeñas variaciones en el biotopo, la pesca y la recolección de moluscos constituyeron la variable crítica en el esfuerzo invertido en la subsistencia. A su vez, los lobos marinos constituyeron siempre una base imprescindible en la consecución de calorías (Schiavini 1993).

Esta base subsistencial podemos confrontarla al desarrollo de los instrumentos de producción, y así vemos aparecer variaciones importantes que nos permiten plantear la existencia de mecanismos de cambio y de autoregulación social.

A lo largo de la secuencia de ocupación del Canal (Orquera & Piana 1999a) se observa un substrato en el instrumental que, aunque se mantiene, se va desarrollando formalmente (FIGURA 2). Tal es el caso de los arpones. En ellos la decoración que se realizaba en momentos antiguos se abandona en los momentos posteriores. Otros cambios menores son cuantitativos: por ejemplo, el instrumental en piedra pulida y alisada es más frecuente en los momentos antiguos (Mansur et al. 1989), reduciéndose progresivamente su presencia hasta su práctica desaparición del registro arqueológico en los asentamientos más recientes. Pero también hay constancia de nuevos desarrollos o adopción de nuevas tecnologías. Así, las puntas de flecha hacen su aparición en los conjuntos líticos arqueológicos hacia el 2700 BP, manteniéndose constante su manufactura bajo parámetros técnicos y morfológicos cada vez más estandartizados. Igualmente, se observan conjuntos instrumentales claramente diferenciados, en cuanto a su morfología, en yacimientos como Lancha Packewaia (hacia el 4200 BP) que nos ilustran momentos de experimentación o inestabilidad en cuanto a los medios de producción, al margen de la estabilidad relativa que se documenta en los objetos de trabajo.

En definitiva, lejos de la inmovilidad, a pesar de la persistencia en cuanto a los objetos de trabajo (una continuidad en la explotación de amplio espectro de los recursos marinos) y la estabilidad relativa del medio, se observan claramente fenómenos de reajuste social. Ello no sólo por esa experimentación y cambio en los instrumentos de trabajo sino también en cuanto a la propia estructuración de los asentamientos (las acumulaciones anulares de concheros parecen un patrón más frecuente en épocas recientes).

En definitiva, se debería hablar de un intento dinámico por mantener un determinado sistema social más allá de simplemente una adaptación estable y exitosa.

De la contrastación etnográfica con la arqueológica vemos aparecer otra serie de fenómenos de gran relevancia frente a la extrapolación etnográfica que sirvió de base a la construcción de la representación arqueológica de las sociedades cazadoras recolectoras prehistóricas (Estévez et al. 1999). Efectivamente las diferencias descritas entre las agrupaciones de cazadores terrestres del norte de la Isla y las de canoeros del litoral sur y occidental no son tan aparentes en los artefactos y en los bienes de consumo. El análisis de los materiales etnográficos y la extrapolación de aquellos que se conservarían y serían visibles en un registro arqueológico nos indica que la mayor parte son compartidos entre todos los grupos sociales insulares. Las diferencias serían en algún caso cuantitativas o coyunturales más que estrictamente cualitativas (especialmente en lo referido a los bienes de consumo directo), con la excepción de los instrumentos de producción ligados a la explotación de recursos desde las canoas. Este último aspecto también podría, en el caso arqueológico, ser interpretado equivocadamente como una cuestión coyuntural (ciclos de explotación alternante costa- interior) más que como una verdadera diferenciación étnico-cultural. Por otro lado, en ocasiones la distancia entre conjuntos arqueológicos producidos por el mismo grupo identificado etnográficamente son mayores de lo que se puede encontrar entre conjuntos de las distintas agrupaciones.

Así pues, no sería apropiado hablar en términos de una correlación simple entre conjuntos arqueológicos y "cultura" (tanto si se concibe ésta como una categoría étnica como en términos de adaptación). Pero la cuestión que resulta más interesante en este contexto y que es más relevante al fenómeno de la insularidad es el hecho de que en la misma isla los etnógrafos hubieran aislado esos cuatro grupos diferentes. Esa distinción tenía según esas fuentes su correlación más significativa con la apariencia física y con los idiomas hablados, más que con las estrategias de explotación de recursos: habría habido dos agrupaciones mayores de gente canoera y dos de cazadora-recolectora terrestre. Pero en realidad tanto del análisis detallado como del análisis paleo-bioantropológico se desprende que existió una gran permeabilidad entre esas pretendidas agrupaciones discretas, incluso a pesar de factores que deberían haber tendido a favorecer el aislamiento como son las estrategias de explotación de recursos divergentes entre la gente litoral y terrestre, o la dispersión insular y las dificultades de navegación en el archipiélago fueguino.

EL FINAL DE LAS SOCIEDADES FUEGUINAS

También podemos vincular los datos arqueológicos con el análisis de la organización social e ideológica correspondiente a los momentos finales de estas sociedades (Estévez & Vila 1997) documentada etnográficamente.

La presencia de los barcos europeos en la zona implicó la llegada de "nuevos materiales" a las costas habitadas por esos grupos incluso antes que se hiciera efectiva la presencia directa de los propios navegantes, lo cual evidentemente aumentó el número de materias primas de que disponían y mejoró la realización de algunas de sus actividades, sin que ello implicara en un principio cambios sociales significativos, que sólo tuvieron lugar con la efectiva instalación de las nuevas gentes.

En efecto, en yacimientos del siglo XVII sobre la costa del Canal Beagle se documentan fragmentos de vidrio. Esa rápida capacidad de asimilación de nuevas materias primas no fue frenada por la inercia de la utilización durante milenios de los mismos materiales. Al contrario, podemos pensar que la experiencia técnica acumulada hizo que se apreciaran rápidamente las ventajas de esas nuevas materias primas. En el caso del vidrio su mejor y más previsible reacción a la talla (especialmente frente a la presión) fue la característica que lo hizo preferible para la manufactura de puntas de flecha o raspadores frente a las materias primas líticas locales, mucho más difíciles de tallar. La utilización del metal, obtenido de los europeos, también se adoptó enseguida como substituto (total o parcial) de otros instrumentos de corte. La mayor tenacidad de este material frente a los cuchillos originales, fabricados con una valva de Aulacomya como parte cortante, también fue percibida rápidamente, y generó una acuciante demanda de este material. Al principio los cuchillos metálicos fueron fabricados siguiendo la misma colocación del filo, en forma transversal, que la del modelo tradicional (de valva) hasta que, después, se replicaron las formas europeas habituales, de filos longitudinales (FIGURA 3). También se adoptó el metal para la fabricación de punzones en substitución de los fabricados con huesos de ave, mucho más frágiles. En este caso se usaron fragmentos de clavos colocados directamente en el interior de diáfisis de los anteriores.

El uso del metal impulsó, sin duda, otros procesos de trabajo como el de la madera. Así, se demuestra que existían y eran percibidas como tales ciertas limitaciones técnicas que hicieron deseable e impulsaron la adopción de ese nuevo material. Efectivamente con la introducción de instrumentos de este material de corte contundente, hachas p.e., se empezaron a fabricar canoas de tablas que substituyeron a las tradicionales canoas de corteza, y también con toda probabilidad se construyeron más frecuentemente que antes cabañas cónicas de troncos más que las cabañas en forma de domo fabricadas con ramas.

Esos cambios tecnológicos se produjeron rápidamente sin necesidad de un contacto directo prolongado con la sociedad europea, y sin que llegaran a afectar sustantivamente a las estrategias generales de explotación de recursos ni a la organización social, que se mantuvieron estables.

El colapso de esas sociedades se produjo con el establecimiento definitivo de pobladores de origen europeo. Los misioneros-colonos impulsaron un cambio catastrófico para esas sociedades, cuyos mecanismos y consecuencias son ya estudiables históricamente. Impusieron nuevas formas de organización social y de propiedad y explotación de los recursos en este territorio e instauraron la agricultura/ ganadería, introduciendo incluso nuevas especies. Rápidamente desestructuraron y desequilibraron las organizaciones sociales cazadoras-recolectoras.

En el caso de los grupos pedestres que explotaban el interior de la isla la nueva estrategia colonizadora fue totalmente incompatible con la anterior. La explotación ganadera imponía la limitación de la matanza de los herbívoros (ovejas) sobre los que los nuevos propietarios europeos de las "concesiones" territoriales ejercían la propiedad. En el caso de las gentes canoeras tal colapso se precipitó al introducir los blancos nuevas enfermedades que se propagaron a gran velocidad favorecidas por la sedentarización estimulada o forzada. Pero a los factores anteriores hay que sumar un factor interno: la imposibilidad de reacción suficientemente rápida de las sociedades indígenas frente a esa disminución catastrófica de efectivos conllevó la disminución del poblamiento por debajo del nivel crítico de recuperación.

Tal fenómeno hay que atribuirlo sin duda a una organización social e ideológica estricta y ya muy bien afianzada (bien documentada en las fuentes etnohistóricas) que mantenía en esos momentos de contacto una estrategia restrictiva en la reproducción (y limitante en la producción) (Vila & Ruiz 2001). Sólo admitiendo que, en algún momento, durante su largo proceso histórico habían conseguido implementar una estrategia social efectiva de ese tipo es posible proponer una hipótesis explicativa para la estabilidad y continuidad de la población durante los 7000 años de historia en la Isla.

Es probable que los cambios tecnológicos sean una pista para llegar a comprender cómo se fueron acumulando esas experiencias e implementando las estrategias durante los momentos de desequilibrio entre la producción de recursos y la reproducción de esas sociedades. El ejemplo de Tierra del Fuego servirá también para ilustrar como esos cambios tecnológicos se aceptan con mayor rapidez que los que implican una reestructuración de las relaciones sociales de reproducción y que finalmente cuando se produce es la propia formación social y económica la que desaparece.

BIBLIOGRAPHY

ARGELÉS, T. ET AL.,
1995. Teoría para una praxis: "splendor realitatis". Trabalhos de Antropologia e Etnología, 35(1), 501-507.

ESTÉVEZ, J. & A. VILA COORDS.,
1996. Encuentros en los conchales fueguinos. Bellaterra: UAB-CSIC.

ESTÉVEZ, J. & A. VILA, (SCRIPTWRITERS),
1997. Humo en los Ojos. Video produced by the Institute of Education of the Autonomous University of Barcelona, Bellaterra.

ESTÉVEZ, J. & A. VILA,
1998. Tierra del Fuego, lugar de encuentros. Revista de Arqueología Americana, nº15, 187-219.

ESTÉVEZ, J. ET AL.,
1999. Cazar o no cazar, ¿es ésta la cuestión?. Boletín de Antropología americana, 33, 5-24.

ESTÉVEZ, J. ET AL.,
2001. Archaeological Analysis of Shell Middens in the Beagle Channel, Tierra del Fuego Island. International Journal of Osteoarchaeology, 11.1-2 , 24-33.

LEGOUPIL, D.
1985-86. Los indios de los archipiélagos de la Patagonia: un caso de adaptación a un ambiente adverso. Anales del Instituto de la Patagonia, vol.16, 45-52.

LEGOUPIL, D.
1993-94. El archipiélago del Cabo de Hornos y la costa sur de Isla Navarino: poblamiento y modelos económicos. Anales del Instituto de la Patagonia, vol.22 , 101-121.

MANSUR, E., L. ORQUERA Y E. PIANA,
1989. El alisamiento de la piedra entre cazadores-recolectores: el caso de Tierra del Fuego. Runa, XVII-XVIII, 111-205.

MASSONE, M.,
1987. Los cazadores paleoindios de Tres Arroyos, Tierra del Fuego. Anales del Instituto de la Patagonia, vol.22, 47-60.

ORQUERA, L. & E. PIANA,
1999a. Arqueología de la región del canal Beagle. Buenos Aires: Sociedad Argentina de Antropología.

ORQUERA, L. & E. PIANA,
1999b. La vida material y social de los Yámana. Buenos Aires: Eudeba.

PIANA, E.,
1984. Arrinconamiento o adaptación en Tierra del Fuego. Antropología Argentina, 1984 , 7-114.

PIANA, E., A. VILA, L. ORQUERA Y J. ESTÉVEZ,
1992. Chronicles of "Ona-Ashaga": archaeology in the Beagle Channel (Tierra del Fuego, Argentina). Antiquity, vol. 66, nº 252,771-783.

RABASSA, J., C. HEUSSER Y K. STUCKENRATH,
1986. New data on Holocene sea transgression in the Beagle Channel; Tierra del fuego, Argentina. Quaternary of South America and the Antartic Peninsula, IV. 291-309.

SCIAVINI, A.,
1993. Los lobos marinos como recurso para cazadores-recolectores marinos: el caso de Tierra del Fuego. Latin American Antiquity, 4(4), 346-366.

TERRADAS, X.,
1997. Lithic Raw Material Procurement Strategies of the Yamana People. In: R. Schild, & Z. Sulgostowska, eds. Man and Flint. Oxford, 1997, 123-126.

VILA, A.,
2000. Dadores de nombres/dadores de identidad. Secuencia para la Tierra del Fuego. In P.García, et al., coords. Estrategias de Poder en América Latina Barcelona: Universidad de Barcelona, 2000, 45-59.

VILA, A. & G. RUIZ,
2001. Información etnológica y análisis de la reproducción social: el caso yamana. Revista Española de Antropología Americana, nº31 , 275-291.

VILA, A. ET AL.,
1998. Marine Resources at the Beagle Channel prior to the Industrial exploitation: an Archaeological evaluation (Final Report). Barcelona: E. U., (project CI1*-CT93-0015).

12

SOCIAL AND IDEOLOGICAL CHANGES IN THE BALEARIC ISLANDS DURING LATER PREHISTORY

VICENTE LULL, RAFAEL MICÓ, CRISTINA RIHUETE HERRADA , ROBERTO RISCH

Departament d'Antropologia Social i Prehistòria, Universitat Autònoma de Barcelona
EMAIL: lubio@arrakis.es

INTRODUCTION

Recent discoveries on the island of Menorca have shed new light on social development in the Balearic islands during the second and first millennia cal BC. In 1995 and 1997 local speleologists discovered two caves, named the Cova des Carritx and the Cova des Mussol, in the western part of the island, which had remained unaffected by human activities since prehistoric times. Moreover, the atmospheric conditions in these caves favoured the exceptional preservation of organic materials, such as wood or human hair, which are unusual finds in the western Mediterranean. The archaeological remains registered in both sites showed that a wide range of ritual and funerary practices had taken place in different places in the cave system. Thanks to an extensive C14-dating programme it was possible to establish various phases of occupation of the caves between approximately 1600 and 800 cal BC. Several excavation campaigns have been carried out so far, and the systematic analyses of the palaeobotanical, faunal, anthropological and artefactual remains, as well as of their geological, geomorphological and chronological contexts has been carried out in recent years in the context of an interdisciplinary research project (Lull et al. 1999, 2000a, 2000b, 2001). This has not only implied the re-examination of the absolute and relative chronological sequence of the Balearic prehistory, but also we now begin to understand better the environmental conditions, the socioeconomic organisation and the ideological changes taking place on the islands during the second and first millennia BC. Thus, the discovery and investigation of the Cova des Carritx and the Cova des Mussol has resulted in a general shift in our view of the later prehistory of the Balearic islands.

The new evidence obtained from these two caves and the review of the archaeological record of Mallorca and Menorca allows us to raise a series of questions about the dynamics of the islands' societies in prehistoric times and their situation in the context of the Western Mediterranean. Frequently, it has been claimed, or implicitly assumed, that island communities are "special", in the sense that they develop specific forms of social or economic organisation, due to their circumscription on an area of land with defined natural limits. Yet, the specificity of this geographical situation can be understood in terms of isolation, or the opposite, i.e. exposed and open to constant interaction with other communities. These different ways of conceiving insularity have both operated in the explanation of islands in Prehistory, as can be seen, for example, comparing modern archaeological thinking in the eastern and the western Mediterranean. While in the east it is generally assumed that the sea facilitates communication and that islands mediate in social, economic and political interactions, archaeological models of the west emphasise local and gradual development and limited contact with other regions. These perceptions are not always implicit in the archaeological record, but seem to be influenced by a certain historical understanding of these areas. Since Medieval and, especially, in Modern times the islands of the eastern part of the Mediterranean basin have played a key role in the trading routes of the Byzantine empire, the Ottoman empire, the republics of Venice or Genoa, or under British rule. On the contrary, in the west, after the expulsion of the Islamic population, most of the islands remained under the rule of Spanish kings, who were soon looking towards the Atlantic Sea. Apart points of communication and exchange, which allowed the consumption of products coming from Africa, Iberia and Italy, as the ceramic evidence shows. Later again, from the

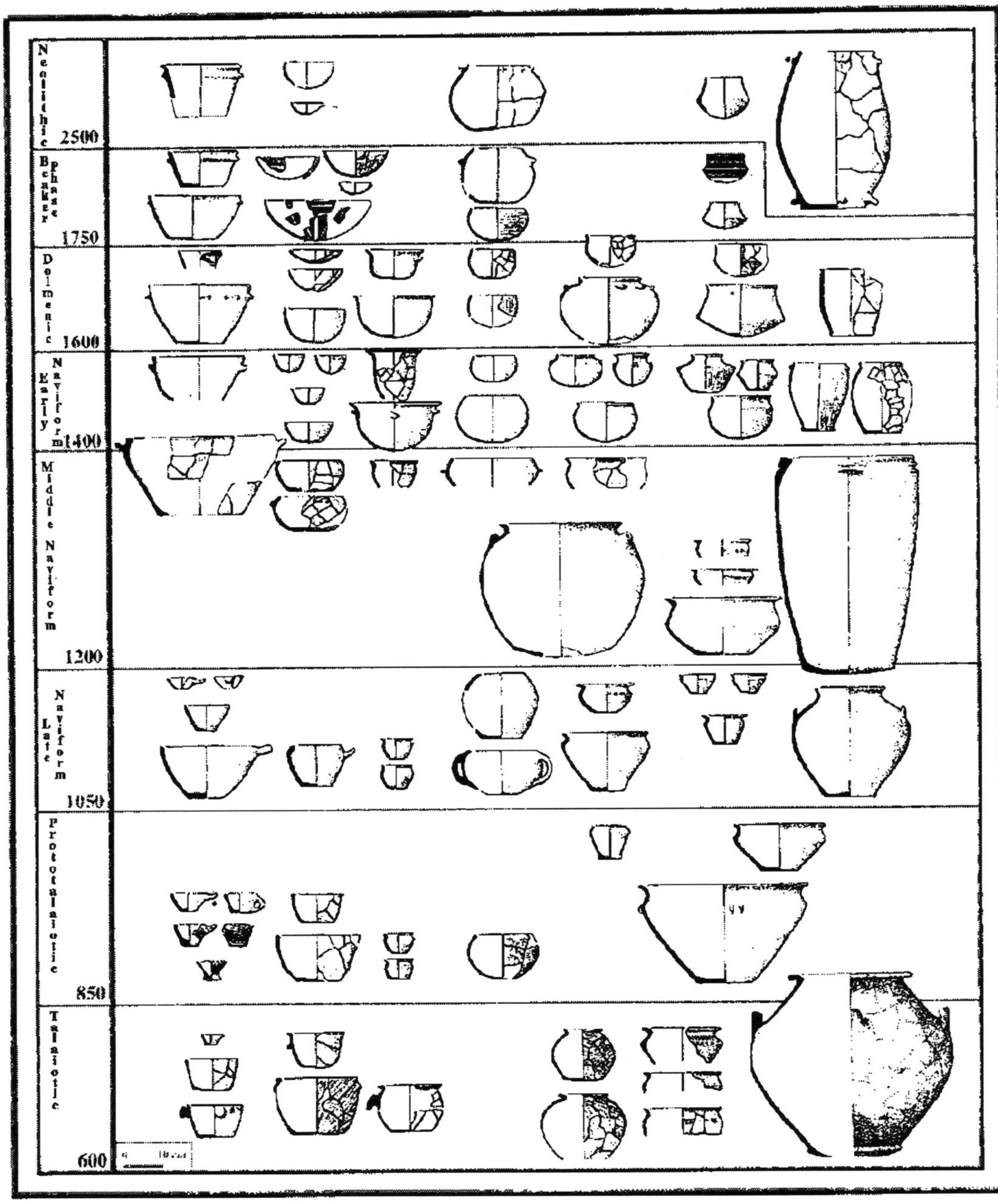

FIGURE 1 .DEVELOPMENT OF THE PREHISTORIC POTTERY IN THE BALEARIC ISLANDS THE NEOLITHIC UNTIL THE TALAIOTIC PHASE.

from particular situations, such as the development of Menorca under British occupation in the eighteenth century AD, islands in the western Mediterranean developed an economy based more on agriculture and husbandry, than on their role as trading ports. The archaeological record shows tha Èt this situation has not always been this way. During the 3rd-1st centuries BC, at a time when the whole of the western Mediterranean was involved in the Punic-Roman conflict, Sardinia and the Balearic islands became key 5th to early 7th centuries, when these islands formed part of the Vandal kingdom and the Byzantine Empire, their societies participated intensely in the influences and products coming from North Africa and the eastern Mediterranean.

This suggests that geographic isolation or long distance communications are not the consequence of an insular position, nor do such situations seem to depend exclusively on the development of seafaring technology or the distance to the nearest mainland. The social space of an island does not finish where its land ends, but integrates the surrounding sea as well. The perception of this space and the attitude of island societies towards the outside does not seem to be fixed but changes through time, due to different economic, social or political factors. Thus, an island situation can be perceived as isolated, networked, or both, depending on different aspects of the social organisation. Rather than insisting on the geographical constraints of islands and using purely Cartesian criteria for their spatial definition (e.g. Cherry 1981), it is the behaviour of "social insularity" which needs further discussion, if we want to understand the diversity of the islands trajectories and at the same time explain their overseas connections.

The recent archaeological discoveries from Cova des Carritx and Cova des Mussol on Menorca show that such changes in "social insularity" also occurred on the Balearic Islands during prehistoric times. Together with the extensive dating programmes carried out so far at different sites in Mallorca and Menorca (Waldren 1986) it is now possible to propose three moments of increased interaction between the late third and the early first millennia BC.

The first of these phases occurred during the Early Beaker phase, which started around 2500 and lasted until 2000 cal BC. It represented a moment when Mallorca especially was taking part in a much wider phenomenon, which, apart from certain ceramic decorative styles, involved the introduction of new funerary rituals, copper metalworking and the development of a varied stone and bone industry (Waldren 1998). The continued production of many of these new instruments and ornaments during the next one thousand or, some cases, even seventeen hundred years, underlines the importance of this moment of trans-European contacts and movements for the future development of the islands.

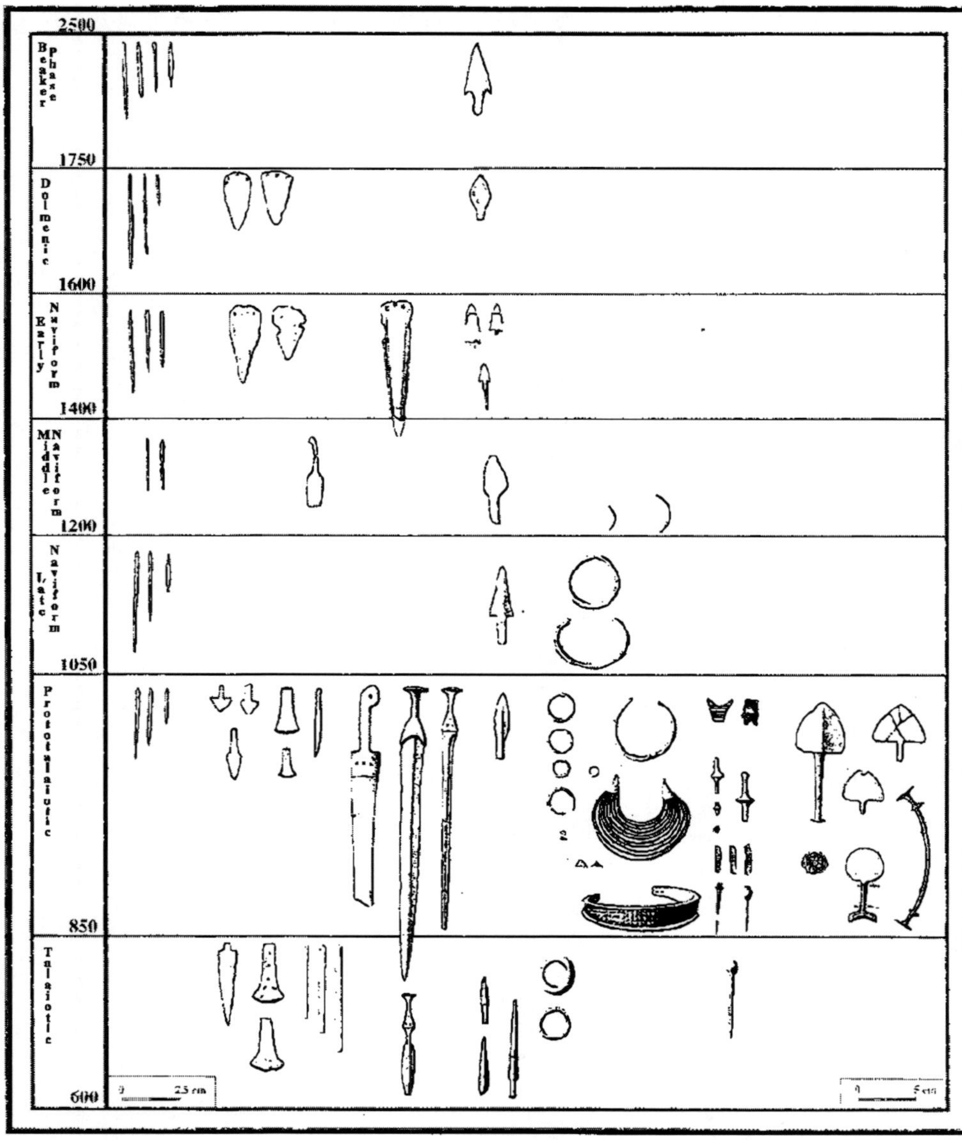

FIGURE 2. DEVELOPMENT OF METAL PRODUCTION 1N THE BALEARIC ISLANDS THE BEAKER PHASE UNTIL THE TALAIOTIC PHASE.

During the Beaker phase we also find the first evidence of occupation of the lowlands of Mallorca and of the islands of Menorca, Ibiza and Formentera. Such a sudden demographic and technological change suggests the arrival of new population groups from one or several points of the arc of immediate influence, as proposed by Waldren (1982). Moreover, it is important to note that these changes occur during a phase of intense social mobility and communication throughout large parts of Europe and the Mediterranean during the second half of the third millennium cal BC.

The archaeological record of the following centuries is rather sparse, but very little evidence exists for long distance contacts. The use of small rectangular megalithic chambers as funerary structures in Menorca and northeastern Mallorca could be understood as an indicator of the arrival of some new populations, but these megaliths represent an anachronistic type of construction around 1750-1550 cal

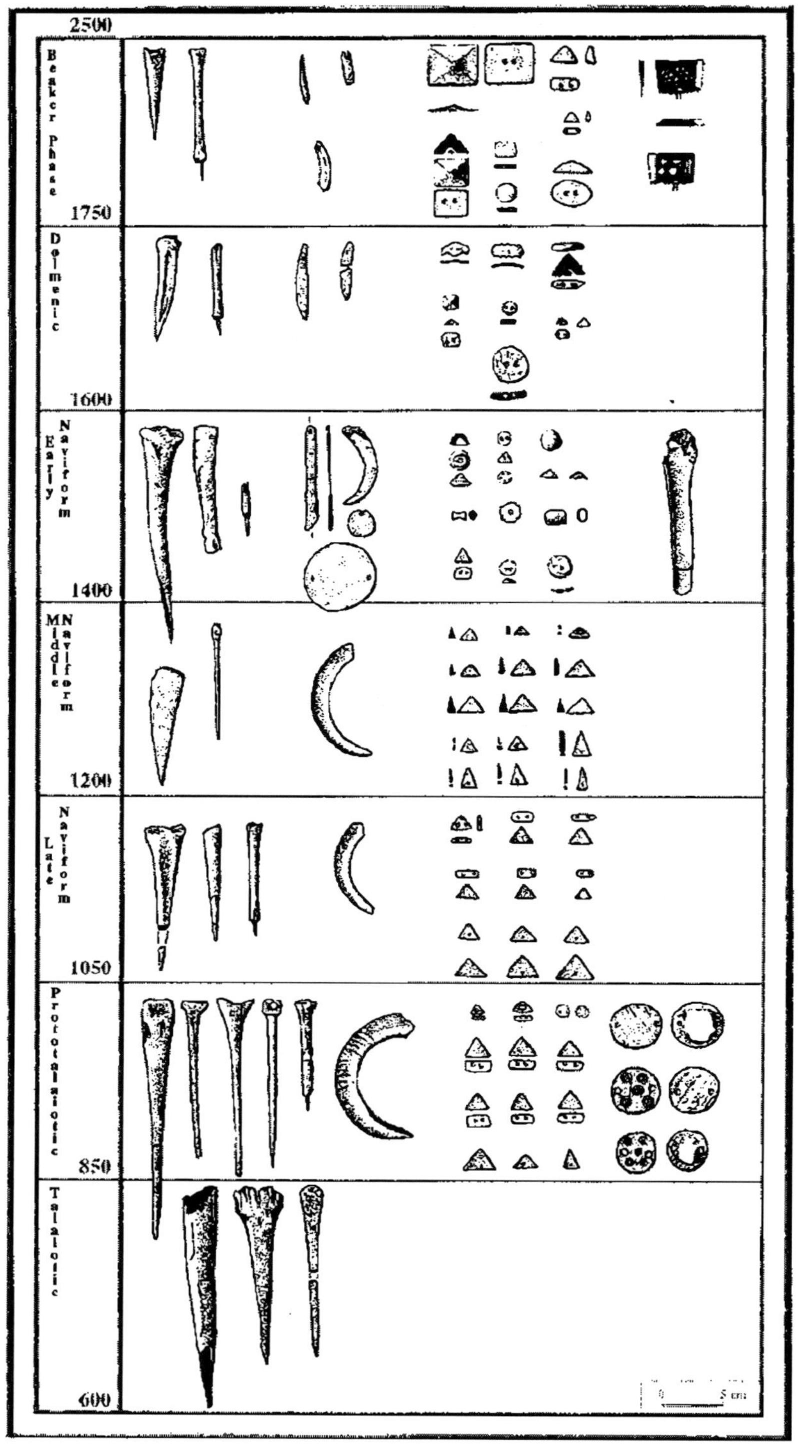

FIGURE 3. DEVELOPMENT OF BONE PRODUCTION IN THE BALEARIC ISLANDS THE BEAKER PHASE UNTIL THE TALAIOTIC PHASE.

BC on the continent, where they only continue appearing in marginal regions such as the Pyrenees. In general the Balearic Islands do not seem to have participated in the social and economic developments of the so-called European Early Bronze Age.

This situation changes around 1600 cal BC when troduction of tin-bronze and metal casting with moulds), and the presence of exotic materials (e.g. ivory) confirms the reactivation of long distance networks. Nevertheless, previous settlement and funerary forms continue to be used at an initial moment (Early Naviform), indicating that probably local, as well as new population groups arriving from

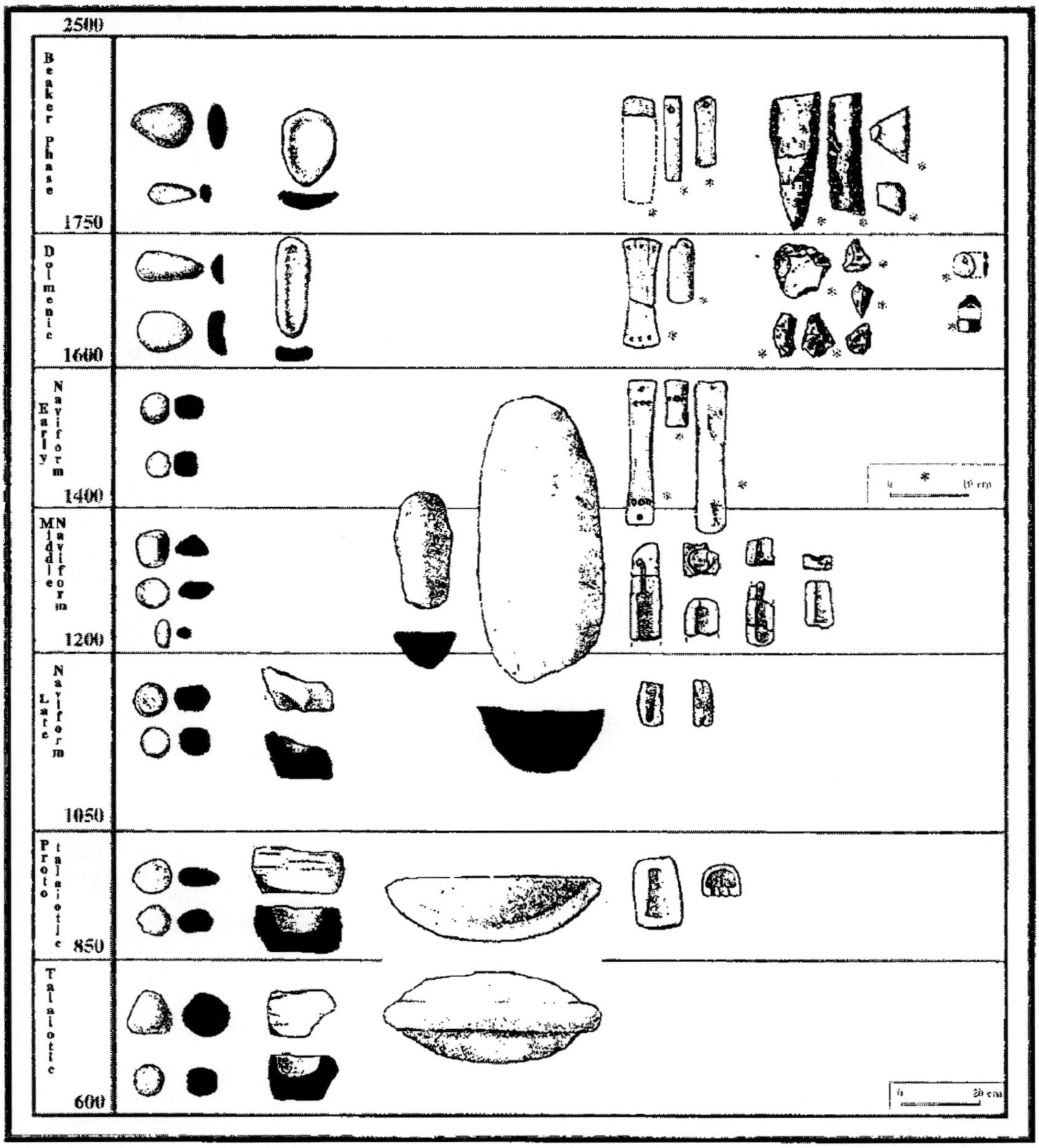

FIGURE 4. DEVELOPMENT OF STONE PRODUCTION IN THE BALEARIC ISLANDS THE BEAKER PHASE UNTIL THE TALAIOTIC PHASE.

new types of settlement and funerary structures appear in great numbers on the Balearics. A distinctive feature is the use of cyclopean construction techniques, as can be seen in the 14-18m long and 3-4m wide naviform (boat-shaped) buildings. These constructions become the characteristic habitation site of the period between 1600 and 1050 cal BC (TABLE 1). Also certain technological changes take place during this period, especially in the realm of metallurgy (in- outside were living side by side. The lack of weapons and fortified settlements implies that this coexistence was generally peaceful.

In this context the use of the Cova des Carritx and the Cova des Mussol as ritual spaces starts around 1600 cal BC. In both caves a series of human, faunal and artefactual remains have been found related to the worshipping of an anonymous and underground entity of feminine character

(Lull et al. 1999). Unfortunately the fact that such types of archaeological contexts are unknown in the western Mediterranean, hinders the possibility of establishing if these ritual practices were introduced by the new naviform communities or this process combi ”ned old as well as new elements and produced a particular social organisation, which continued to develop with little interference from the outside over the next few centuries.

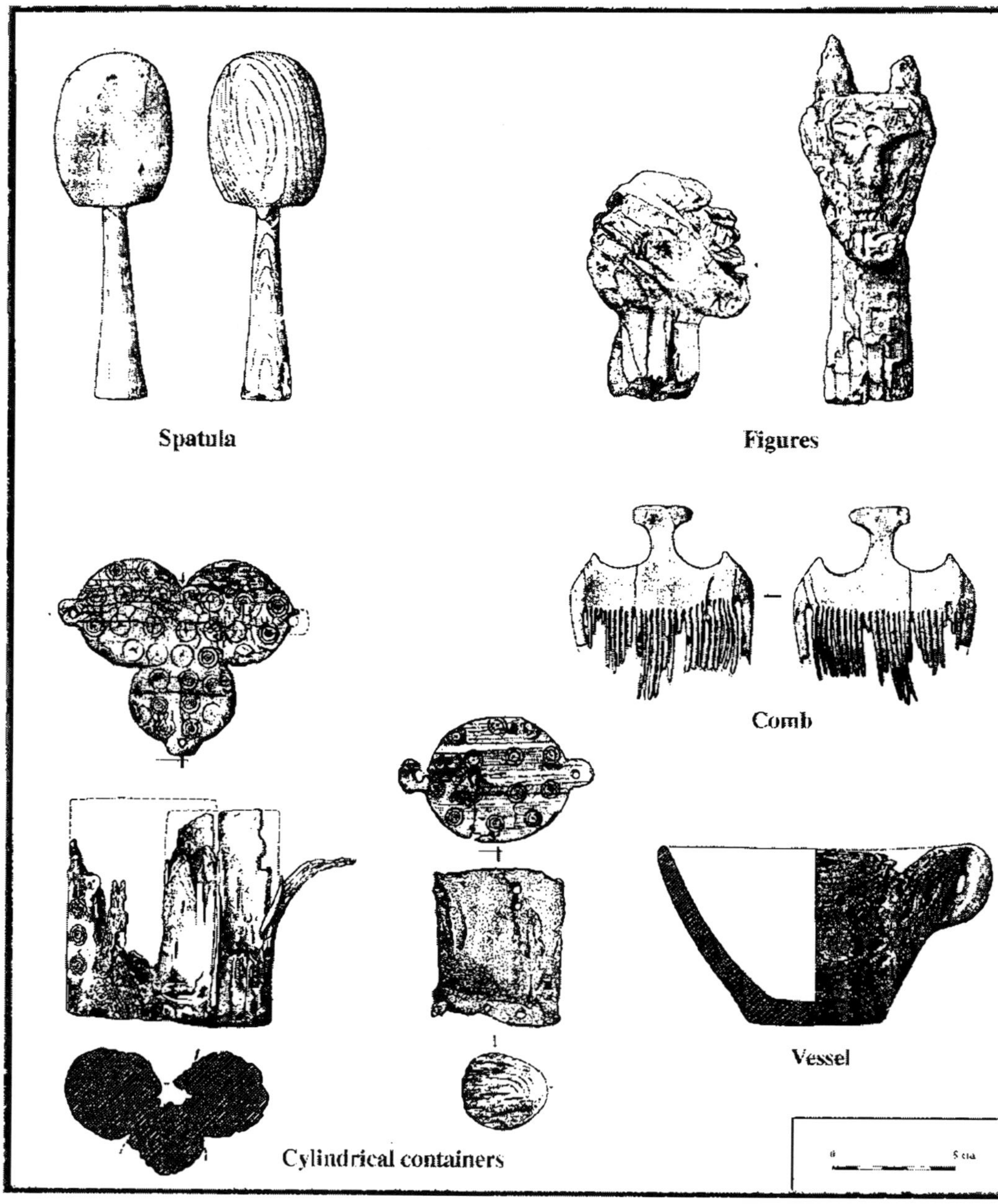

FIGURE 5. WOODEN, HORN AND BONE ARTEFACTS
THE COVA DES CÀRRITX AND COVA DES MUSSOL (CIUTADELLA, MENORCA).

formed part of earlier belief systems.

In any case, the patterns observed in the archaeological record of the sixteenth/fifteenth centuries cal BC show that, at this moment, the Balearic islands again became involved in wide ranging social and economic transformations which have been recognised in different parts of the Mediterranean and Central Europe. Yet, contrary to the situation during the Beaker phase,

It is not until the end of the second millennium cal BC that we find again signs of enhanced contacts with other parts of the western Mediterranean. After 1200 cal BC changes in the naviform settlements become evident through the appearance of larger population centres and more complex architectural structures. The different spheres of production seem to initiate a gradual process of intensification (metallurgy, bone industry), or a technological shift (pottery). Yet, the most important split occurs at the ideological level and was caused by extra-insular influences, as has been made evident by the recent discoveries of the Cova des Mussol. In one of its most inaccessible rooms a series of wooden carvings were found, two of which represent a zooanthropomorphic and an anthropomorphic figur øe. The first was placed in a dominant position in the rather small sanctuary part of the western Mediterranean.

The existence and intensity of these contacts, specially during the Prototalaiotic period (1050-850 cal BC), has been confirmed by a series of grave goods and ritual objects found in the Cova des Carritx and Cova des Mussol. Metal and faience artifacts, as well as a complex technology used in the production of small cylindrical containers made out of wood or horn, have their direct parallels in the Circumal-

Occasional arrivals and Neolithic occupation	c. 5000-2500
Early Beaker Phase	c. 2500-2000
Late Beaker Phase	c. 2000-1750
Dolmenic	c.1750-1600
Early Naviform	c.1600-1450/1400
Middle Naviform	c.1450/1400-1200
Late Naviform	c.1200-1050
Prototalaiotic	c.1050-850
Talaiotic	c.850-600/500
Posttalaiotic	c. 600/500-123

Table 1.Chronological scheme of the Prehistory of the balearic islands

TABLE 1.

and showed a very schematic carving with an impressive expression. On its head two horns appear, whose anatomical features correspond to antlers of a young deer, an animal that did not exist on the Balearic Islands. The detailed analysis of the complete archaeological context of the cave allows us to propose that here we are in a ritual space, where a small number of individuals participated in some kind of initiation practices, and where a supranatural creature, half human, half animal was worshipped (Lull et al. 1999). Such a ritual suggests a clear symbolic and ideological break on the islands. Surprisingly, the combination of human and deer features has its closest parallels in the Celtic symbolism of central Europe during the Iron Age (e.g. in the rock art of Val Camonica in northern Italy), which suggests the establishment of close contacts with the northern pine regions. Previous studies of the Late Bronze Age Balearic metallurgy have also pointed in this direction as the possible origin of certain types of weapons and ornaments (Delibes y Fenández-Miranda 1988). Yet, the two caves also provide evidence of contacts with northern Africa (ivory), southwest Spain (possibly copper ores) and, perhaps, Sardinia (bronze "mirrors"). Small iron objects of still unknown origin are also present in contexts before 1000 cal BC.

This evidence suggests that Minorca and Mallorca played an important role at the end of the second and the beginning of the first millennia cal BC in an extended network of social relations established between central Europe and the western Mediterranean. These contacts included the circulation of goods, technology and ideologies at a time when the

Balearic populations were especially open to external influences. The absence of any eastern Mediterranean products on the islands implies that the protagonists of these communication networks were some of the local populations of the west, and not Cypro-Phoenician traders, as has often been claimed (e.g. Sherratt and Sherratt 1993). Rather, it seems to have been the existence of a developed exchange system during the eleventh and tenth centuries cal BC that first attracted the mercantile interests of the eastern polities and then furthered the establishment of colonies in the west from the ninth century cal BC onwards.

This phase of early control of the trade routes by eastern sailors and merchants coincides with the classic Talaiotic period (c. 850-600/500 cal BC) on Mallorca and Menorca, and the foundation of a Phoenician trading post on Ibiza. Surprisingly, during this period, which is characterised on Mallorca by a significant demographic increase, nucleated settlement structures and large tower-like buildings (talaiots), the intense relations established specially with the north-eastern shores of the western Mediterranean broke down. The construction of talaiots on Mallorca and Menorca, together with the appearance of nuraghi on Sardinia and torres on Corsica, has been understood as an simultaneous socio-economic phenomenon of the western Mediterranean (Lethwait 1985). Yet, a growing number of C14 dates confirms that these cyclopean towers appear comparatively late on the Balearics, when similar constructions go through their final phase of development in the eastern islands. The Talaiotic population seems to have adopted this architecture in order to isolat itself from extra-insular contact. Apart from some sporadic findings in the sixth century, it is not until the fifth century that we again find evidence indicating some interaction with overseas, in this case with the Punic colony of Ibiza. This new phase of increasing circulation and contacts characterises the Post-Talaiotic period until the Roman conquest of the islands in 123 BC.

A discussion of the relations between the Balearic Islands and other Mediterranean regions in Prehistory shows that "insularity" was most of all a social construction. Internal as well as external factors seem to have played an important role in the behaviour or reaction of the local populations towards overseas. It can be noticed that while the major episodes of interaction and/or migration at European and Mediterranean scale, such as the Beaker phase (26th-22nd centuries), the transition from the Early to the Middle Bronze Age (17th/16th century) or the Final Bronze Age (12th-9th centuries), also affected Mallorca and Menorca, the response of the indigenous population and the consequences of this contact with other social organisations was different in each case. In some periods of extensive trading and seafaring, like early Phoenician and Greek colonisation, the Balearic societies even developed an isolationist attitude towards a deal which probably was conceived as disadvantageous to their own interests. This shows that the isolation of an island, at least in the Mediterranean, is not necessarily determined by its geographical particularities. The perception of the islands' landscapes changed over time, and differences must have existed between the perception of the local populations and that of the visitors. Probably it was more for these that the islands appeared as limited and isolated spaces, while the islanders were aware that their origins lay on the mainland and that the sea had been crossed before.

The archaeological record and the chronometric framework of the prehistory of the Balearic islands allows us to analyse more closely the factors implied in the construction of "social insularity". Three forms of interaction between the islands and the continent or other islands of the western Mediterranean can be distinguished. The first is the arrival or immigration of populations, which occupied the island in the beginning (Neolithic) and produced significant demographic increases in certain moments: early Beaker, early Naviform, early Roman and early Andalusian (Omeyan) periods. These immigration waves would explain the sudden development of very dense settlement patterns and are confirmed in later times by historical records. Their duration and intensity remains unknown, but they always triggered off a long lasting phase of social and economic development. New types of architectonical structures, burial rites and technical devices were introduced each time, most of which remained in use for many centuries, suggesting a common origin and a considerable number of new settlers. Probably the islands were also reached more or less constantly by smaller groups, but their impact on the existing communities, as well as their visibility in the archaeological record would be limited.

Equally important are population movements in the opposite direction, i.e. emigration away from the islands. Yet, no significant decrease in the number of archaeological sites can be identified during prehistoric times. An open question in Belearic archaeology is the development of Ibiza and Formentera from the end of the II millennium until the Phoenician control of these islands in the 7th century, which apparently prevented the formation of a Talaiotic society on these islands. Emigration coupled with depopulation, at least of certain parts of the islands, rather seems to be an historical phenomena and can be detected for the first time during the Roman Empire when most settlements of the central plain of Mallorca are abandoned. After rapid post-Roman recovery, the 'Muluk' period, when the islands practically gained back their independence from continental powers, sees again a strong demographic decrease, although, to a certain extend, this pattern can be the result of a lack of diagnostic archaeological materials for the 8th/9th centuries. In any case, the fertility and plentiful resources of the islands suggests that such depopulation phases were only possible through a forced military expulsion of the local agropastoral communities.

Communication of information and values represents a second form of interaction between societies and its material expression is the introduction and reproduction of technologies, styles and symbols. While all immigration phases supposed the introduction of a new technological and symbolic universe, the opposite does not hold true (table 2). At certain times, as at the transition from Naviform to Talaiotic, in Posttalaiotic times and Late Roman times, the communities of Mallorca and Menorca were open to technological innovations and new belief systems coming from outside. These must have been periods when a well developed communication network in the western Mediterranean coincided with important internal changes in the local socioeconomic structures. Archaeological and historical information supports such a view for the last centuries of the first millennium BC and the Vandal/Byzantine period, but we only start to grasp a very different form of organisation of the western Mediterranean 1000 years earlier.

Another way in which societies can interact is through the exchange of material goods. Unquestionable evidence for the introduction of foreign materials is always coupled to phases of enhanced communication, but not necessarily with significant immigration processes . Thus, communication with the islands' prehistoric societies seemed to depend more on their necessity or desire to acquire non local products, than on the arrival of new communities. We know very little about the organisation of the material exchange systems, if it resembled a gift giving network or if already a notion of commodities with established exchange values existed. Yet, an interesting observation is that only raw materials (ivory, copper, tin, iron and lead) were imported, but practically no finished products and specially no objects with a high exchange value and a low use value (so called 'prestige items'). Balearic metal and bone objects, as well as pottery and stone artefacts, might resemble certain continental prototypes, but follow their own stylistic and/or technical pattern (FIGURE 1-4). Until the 4th/3rd century BC, prehistoric communities show little interest in acquiring exotic objects, but rather attempted to obtain the necessary raw materials to maintain their local workshops. The absence of precious metals, such as silver or gold, and of imported symbols of power, agrees well with the idea of classless societies, with little political centralisation and economic asymmetry, which is based on the archaeological record previous to the Posttalaiotic period.

The lack of control over production and consumption by a state like organisation and the absence of a violent occupation of the Balearic islands during prehistoric times also seems to explain why many of products and technologies, once they had been introduced, continued being used for centuries and even much longer than on the continent where they originated. V-perforated buttons, flat axes, sharpening plaques (so called 'archers wristguards') or calcite tempered and hand made pottery are some of these long lasting productions which continued independently of the different moments of social interaction.

Another difficult question is what sort of products the islanders were giving in exchange for the received raw materials. As has been suggested for Sardinia, agricultural products, such as cattle and grain, could easily be produced beyond the local needs, but at least during the Prototalaiotic period a well developed craft production could also provide important exchange values.

So far we have proposed a tentative model of the forms of interaction developed by the Balearic communities and how 'geographical insularity' was socially modified and understood in very different ways over several millennia. In the future, new discoveries and the systematic analysis of the extremely large collections of prehistoric artefacts recovered during more than hundred years of 'archaeological' activities should help to improve this general scheme and to explain the social and economic situations that made Balearic societies more or less inclined to overcome or to enhance certain geographical limits. Moreover, it will be interesting to see if these phases of communication/isolation can also be observed on neighbouring islands.

ACKNOWLEDGEMENTS

The continuation of the research on the prehistory of the Balearic Islands is supported by the Dirección General de Enseñanza Superior (DGES, PB98-0889) and by the Comissió Interdepartamental per a la Recerca i la Tecnologia of the Generalitat de Catalunya (CIRIT, code 2000 SGR 00051 in the frame of the "Grup de Recerca d'Arqueoecologia Social Mediterrània"). Rafael Micó and Roberto Risch are researchers of the Ramón y Cajal programme of the Spanish Ministery of Science and Technology. We thank Sylvia Gili for the preparation of the illustrations, and Robert Chapman for valuable comments on the manuscript.

BIBLIOGRAPHY

CHERRY, J.F.

1981, "Pattern and process in the earliest colonisation of the Mediterranean islands", Proceedings of the Prehistoric Society, 47, pp. 41-68.

DELIBES, G. y FERNÁNDEZ-MIRANDA, M.

1988, Armas y utensilios de bronce en la Prehistoria de las Islas Baleares, Studia Archaeologica, 78, Universidad de Valladolid, Valladolid.

LEWTHWAITE, J.

1987, "Social factors and economic change in Balearic prehistory, 3000-1000 b.c.", in BAKER, G. and GAMBLE, C. (eds.), Beyond domestication in Prehistoric Europe, Academic Press, London.

LULL, V., MICÓ, R., RIHUETE, C. and RISCH, R.

1999, La Cova des Càrritx y la Cova des Mussol. Ideología y Sociedad en la Prehistoria de Menorca, Consell Insular de Menorca, Barcelona.

LULL, V., MICÓ, R., RIHUETE, C. and RISCH, R.

2000a, "Rituales de vida y muerte en la prehistoria de Menorca", Revista de Arqueología, 225, pp. 18-33.

LULL, V., MICÓ, R., RIHUETE, C. and RISCH, R.

2000b, "Divinità nelle tenebre. Parlano le grotte delle Baleari", Archeologia Viva, 84, pp. 54-61.

LULL, V., MICÓ, R., RIHUETE, C. and RISCH, R.

2001, "Neue Entdeckungen zur Vorgeschichte von Menorca. Beobaóhtungen zu gesellschaftlichen und ideologischen Verhältnissen auf den Balearen zwischen 1600 und 800 v.u.Z.", in Hispania Antiqua - Vorzeit, Deursches Archäeologisches Institut, Madrid, pp. 153-170.

SHERRATT, S. and SHERRATT, A.

1993, "The growth of the Mediterranean economy in the early first millenium BC", World Archaeology, 24,3, 361-378.

WALDREN, W. H.

1982, Balearic Prehistoric Ecology and Culture: The Excavation and Study of Certain Caves, Rock Shelters and Settlements, British Archaeological Reports, International Series, 149. Oxford.

WALDREN, W. H.
1986, The Balearic Pentapartite Division of Prehistory. Radiocarbon and other age determination inventories, British Archaeological Reports, International Series, 282, Oxford.
WALDREN, W. H.
1998, The Beaker Culture of the Balearic Islands. An inventory of evidence from caves, rock shelters, settlements and ritual sites, British Archaeological Reports, International Series, 709-Western Mediterranean Series, 1. Oxford.

13

ISLANDS OF INDIFFERENCE

ROSEMARY MCKECHNIE

Bath Spa University College
Newton Park
Bath BA2 9BN
EMAIL: r.mckechnie@bathspa.ac.uk

INTRODUCTION

ARE ISLANDS DIFFERENT?

What is so special about islands? The arguments in this paper concern peculiarities in the way islands are conceptualised. Thomas Hylland Eriksen (1993) points out that the island is ..'a powerful metaphor in everyday speech as well as in several academic disciplines'. Island metaphors have played a strong role in anthropological theorising and research, particularly in the heyday of functionalism and cultural relativism associated with a research focus on small-scale, self-sustaining social systems. The idea of the island connotes isolation and uniqueness. Islands are, it goes without saying, set apart and self-contained: they are different. My aim here is to look critically at the nature of the boundaries which set islands apart, and at the similarities in the ways that things old and new are classified when the cross this boundary. The most mundane, unmarked objects take on a life of their own as they move into island space, forcing us to look at them in a new light. I hope to show this is as true of dishwashers and levis, as microliths and megaliths.

The taken-for-granted nature of 'islandness' is important here. It is impossible to isolate some objective definition of what constitutes 'islandness' and then deconstruct it. When we think about islands, our imagination is coloured by organising assumptions, which are never fully articulated, and therefore not amenable to analysis. Yet we are employing some very basic categories about the way that spatial boundaries and socio-cultural boundaries relate to each other: how we think about space and people. I am going to argue here that 'islandness' conflates inchoate but central themes in Western thinking. The following paragraphs explore a few of those, I should emphasise that they concern islands as a concept, not real islands.

The first important point is that islands have such clear boundaries - actual topographical, incontrovertible cut off from other places boundaries. In a world where establishing boundaries is always difficult and often fraught with confrontation, islands appear blessedly free from ambiguity. They have a clear outside and inside. In the real world these spatial boundaries often contain disorder, and cultural conflict (as the example of Corsica below will show). But as an ideal, islands offer the possibility of closure and manageable scale. Islands are often associated with another overdetermined metaphor of closed social grouping: community (see Urry 2000 133). The 'island' and 'community' seems to fit naturally together, connoting not just specificity and cultural difference but cosy close social relations.

The marginality of islands is also important. Islands are often incorporated in larger political entities, part of nation-states. They are often distant from economic and political centres, and in a somewhat ambiguous position, officially just another region, but still thought of as different, set apart. As Edwin Ardener points out, the factors that distinguish 'real' centres and the 'real' peripheries do bear further investigation...

'There are certainly some topographical elements that are relevant. Mountains conventionally add to the 'remoteness' experience, but so very frequently do plains, forests and rivers - so much so that the inhabitants of unremote places

sometimes say that they do not have 'real' mountains, plains forests or rivers ... the actual geography is not the overriding feature - it is obviously necessary that 'remoteness' has a position in topographical space, but it is defined within a topological space whose features are expressed in a cultural vocabulary (Ardener 1987 41)

Ardener termed such regions 'remote areas', to capture the ambiguous nature of their relation to dominant social space. In Europe 'remoteness' is associated with a whole host of metaphorical imagery which sets peripheral places in opposition to dominant societies. Islands are archetypal 'remote areas' - romantic imagery depicts islands as less tainted by the ills of civilisation, as backwaters where life passes at a less frantic pace, where people know each other, live closer to nature etc. This imagery which maps the contours of tradition and modernity fits snugly in the closed conceptual space established for islands. And if the romantic literature of the nineteenth century has slipped out of fashion, travel brochure and media industries have taken their place in purveying island dreams to harassed commuters and city-dwellers. The stereotypes and assumptions that create imagined islands - the fishermen sitting outside cottages mending nets, palm trees swaying in a gentle breeze, or the postman who brings groceries and gossip – are the stuff of fantasy, island life and environment appropriated and dislocated from lived social relations. Real islands are often far from neat and not always marginal. The point here is that 'islandness' is a moveable feast - a conceptual picnic. Some islands are more 'island-like' than others - some islands are big and powerful, or industrial, or densely populated modern cities. Despite the apparently 'hard' objective nature of islands, conceptually 'islandness' is vague.

This indeterminate aspect of what makes an island an island is important. Many of the boundaries which divide populations are clearly politically and legally defined: they have obvious consequences for social relations. 'Islandness', on the other hand, hovers on the edge of consciousness, a fuzzy, semi-realised boundary. It is, I would argue, nevertheless, real for all that. Bourdieu describes how he developed the concept of habitus in order to make it possible to move away from describing the world in terms of formalised rules (1990 76-7). The habitus is a system of dispositions which give rise to certain practices and regularities in the way people behave. Not, however, the regularity of behaviour governed by explicit rules. Habitus goes hand in glove with vagueness and indeterminacy, improvisation and spontaneity (ibid). This kind of practical logic provides us with principles to impose order on action in everyday life; principles of classification, division, and hierarchisization - everything which enables us to distinguish between things the modes of behaviour (ibid 78). While formal rules confer uniformity and predictability even in diverse social contexts, we work with areas of life which have not been codified in this way through nothing more or less than a 'feel for the game' (ibid). The way we think about islands might be irrational, even ridiculous if looked at objectively. However, the issues involved are not frivolous: marginality, difference, the nature of modern and traditional societies. The order we impose on things, on the things we find on islands reflect our 'feel for the game' in relation to difficult issues. Islands provide a closed, manageable conceptual context, where some clarity can be established by mapping boundaries over a more messy reality. McDonald describes how identities are created and confirmed within European culture, how boundaries are drawn and reified: 'We can and do pack moral, political, economic linguistic, geographical biological historical and other social, cultural differences into the divide, and we both perceive the divide as these differences and perceive the differences in the divide (1987, 122).

Academic definitions of space, culture and difference do not have the luxury of remaining indeterminate. In fact establishing continuity and discontinuity in populations over time and space is a central problematic for both anthropology and archaeology. The formal categorisation of space and culture generated by academic theory and practice cuts across experiential concepts. Bauman, in discussing the nature of modern concepts of distance and difference (1998) starts from Leach's (1964) insight about similarities in categorisation of space, kinship and domestic/wild animals. He points out the 'near' and 'far' are crucial categories on which to base categories of distance "Near' is primarily what is usual, known to the point of obviousness; someone or something met, dealt with or interacted with daily, intertwined with habitual routine and day-to-day activities. As a cultural ideal type, 'near' is such a space, inside of which one can feel chez soi, at home...'(1998 110) On the other hand , to find oneself in a 'far away' space is an unnerving experience, venturing 'far away means being beyond one's ken, out of place, and out of one's element, inviting trouble and fearing harm. 'The crucial dimension is between certainty and uncertainty... 'far away' means learning rules... and mastering them through risky trials and often costly errors. These areas can be visualised as areas of condensation stretched between two poles along which all known beings and entities are plotted. (ibid) This opposition between what is transparent, familiar, and what is incomprehensible, is not a territorial measure - it might well correlate with measured distance but it is not inevitable.

The objective territorial distance which sets islands apart from mainlands may be small, but the subjective experience of that distance is coloured by expectation of strangeness: islands can be close yet different and therefore distant. Modern disciplinary measurement of distance by objective units depersonalises the subjective origins of the perception of space, removes those elements of danger, of unfamiliarity, of difference. Archaeology for example has developed a multitude of methods, concepts and technologies for precisely mapping objects in time and space. Archaeologists and anthropologists have created their own kind of islands. Metaphorically, islands of culture or material culture can exist anywhere, on the other hand objectively defined cultural continuities may well cross over contemporary topographical and topological boundaries, even island boundaries. Islands are places where processes of identification work overtime and the rational ordering of distance and difference confront other imaginative currents.

CORSICA – AN ARCHETYPAL ISLAND?

The 'islandness' of Corsica has certainly been central to its identification by outsiders and insiders alike. Corsica has always been peripheral to the dominant European powers. It has been defined as 'other' in opposition to a varied procession of dominant cultures, from the classical sea-faring powers to modern France. At the same time, because of its position, Corsica has been open to invasion, trade and intense population movements. People, things and ideas have crossed the boundaries to and from the island over the millenia. Corsica, like many islands (including the Mauritius described by Eriksen), has been anything but isolated. In the last quarter of the twentieth century, flows of Corsicans outwards towards the continent and beyond were matched by flows of continental French and North African workers, as well as global tourists towards the island. In addition the island was economically dependent on an influx of consumer goods, finance and mediated products from around the globe. Corsica is a prime example of the fundamental paradox about islands that though metaphors for specificity and isolation, islands are often intensely tied into larger social, cultural, political and economic networks. Corsica has gone through a series of catastrophic events, continually, it seems, on the brink of irreversible change. Yet the symbolic boundaries that surround it seem stubbornly impervious.

The symbolic boundary separating Corsica from France is woven into the geographical barrier of the Mediterranean Sea. The sea may be crossed in twenty minutes by plane, but conceptually it represents a boundary that not only cuts Corsica off from the mainland, but places it in another world (see McKechnie 1993). France is often described as the 'hexagon'. In many maps of France, Corsica is either missing or pasted on to the South Eastern coast somewhere. The concrete bounds of the island are enhanced in Corsica's case by its mountainous aspect. The mountain has dominated descriptions of Corsica from classical times. Virtually every modern work written about Corsica quotes Ratzel's description of a the island as 'a mountain in the sea'. Corsica might be said to possess all the hallmarks of a 'remote area', because of its position and geography we can readily imagine it as such. .

The fracture line which sets Corsica apart from France has many dimensions - historic, linguistic, economic, and so on. It has been created and recreated over time, and in this respect there are many parallels with other French regions, particularly Brittany (see Bertho 1980; McDonald 1989). Exiles, administrators and travellers have all brought back representations, whether bitter or ecstatic, of the strangeness of Corsica. From the inside nationalists and cultural activists have attempted to rescue and defend visions of Corsica from the encroaching dominant state. I will argue here that the 'islandness' of Corsica, the topological rather than the topographical boundary which both separates and encloses it is an important aspect of its continued singularity. This creates a particular sensitivity to movement across the boundary, shaping the way the comings and goings over the island's boundaries of both people and objects are conceptualised.

ARCHAEOLOGY AND THE MAPPING OF THE PAST

The lack of fit between visions of the past, between meticulously defined material cultures and, say heritage packaged for tourists, or authentic specificity marshalled by nationalists offers insight into contemporary processes of identification mapping social and cultural boundaries over territorial ones. The distant past of prehistory is in some ways, more open to inventive recreation in the present than historical texts. Dietler points out that archaeological evidence... 'is particularly labile, it can be appropriated by invented traditions and strategically used to validate the competing claims of 'imagined communities' (1995 584). He argues that it is important to look at how the ancient past is marshalled as a symbolic resource to establish cultural authenticity and continuity (ibid 585). As several commentators have pointed out contestation over the meaning of archaeological finds has been particularly heated where nationalism is involved (see Trigger 1984; Hroch 2001). Pomponi notes there have been two opposing tendencies in the interpretation of Corsican prehistory. The first ignores local cultural traits and focuses on signs of Mediterranean migrations and empires. The second 'inverted history' minimises links with external powers (1975 14). Whether remains are viewed as evidence of external cultural influences, or as examples of autochthonous development has important consequences for the conceptualisation the social and temporal boundaries that define contemporary populations (Smith 2001).

I became aware of the expectations archaeological evidence can be expected to fulfil, when attending a lecture to first year students, most of whom were nationalists. The careful presentation of the archaeologist M. Weiss was constantly interrupted by the disappointed audience. Wasn't it possible to translate European lithic traditions into Corsican and deliver the lecture in a more appropriate language? Surely, despite the fact that during the neolithic Corsica was not separated from Sardinia, and was only thirty km or so from the mainland, there was something that distinguished the material culture of early Corsicans from their neighbours? The interpretation of archaeological remains is certainly contested most stridently in tussles over the identification of origins rooted in the far past that can be used to establish and validate cultural boundaries. As Appadurai (1981) points out, the idea of the past as infinite symbolic resource has tended to gloss the structured processes which create the dominance of some voices over others in recreations of the past (see also Hodder 1986 150). Nationalist contestation which raises questions about the interests and hypothesis of archaeology is clearly an important aspect that needs to be explored (Champion 1996; Dias-Andreu 2001: 441). However, there are also more muted processes in action shaping contemporary relations to ancient things.

Why is it that some archaeological research is ignored, yet other findings garner so much popular attention, engender so much debate? Archaeology bears the mark as few disciplines do of the twin drives of the enlightenment: rationality and romanticism. The attraction and repulsion of these forces may go some way towards explaining mixed reception that archaeological findings receive. The strangeness and charm of attraction, the passion for lost worlds, for reaching into other lives that has also coloured the imaginative worlds of archaeologists and their audiences sits uncomfortably with the meticulous rigour of the archaeological quest. Is it 'dry as dust' or is it breathtakingly, passionately interesting? It is sometimes hard to predict, and the reason seems less to do with the past than with the present.

Careful, rigorous work mapping assemblages of material culture can be disappointingly ignored. Little public attention is paid to the assemblages of artefacts, painstakingly organised and dated, that document the Neolithic in Corsica and set finds there in relation to material from mainland Italy and Sardinia (see Grosjean 2000; Tossi and Weiss 2001). Another type of prehistoric remain has been more successful in capturing the attention of both the academic world and the public. Statue-menhirs clearly appeal to the imagination in a way that microliths do not. Interesting here is the description of Dorothy Carrington, who first instigated the interest of the archaeologist R Grosjean in the statue-menhirs of Filitosa:

'At this period of roaming the island in search of antiquities the occasional meetings with these stone figures were always exhilarating. The Italianate churches, altarpieces and saints' statues... invariably impressed me with their foreignness; I could never forget that they were subdued versions of prototypes created by and for much wealthier, urban societies. Only the statue-menhirs, rude, vigourous and archaic, spoke the authentic insular language.' (1984 27-28)

The unique character of some of the statue-menhirs has been seized on to represent the specificity of insular culture. For Dorothy Carrington, something essential in Corsican character which had endured for millennia was displayed visually on these stones:

'That Corsica - obscure, peripheral Corsica - should have brought forth the central concept of European art at so early a period is something not many people have have been willing to believe. Yet this monumental portrayal of the human being would seem a natural enough expression of the megalithic veneration of the dead, practised in an area isolated from other religious influences and there pushed to extremes. The artistic innovation issued from a spiritual development, one that implied the emergence of a new concept of man... an increased recognition of his own value. Confident in his power to stand up to forces seen and unknown, he assumed the vocation of the hero...'(1984:35)

'It is easier to degrade images than beliefs: there is evidence to show that the megalithic faith lived on in the deeper areas of the Corsican psyche, affecting customs, attitudes and values, almost to the present time... The hero-myth... became self-generating; the Corsicans never went back on their early realisation of the potential of the individual man. Napoleon is the lineal descendant of the warriors of Filitosa '(38)

These monuments can be seen as an expression of the character of an ancient people that somehow distils essential differences that can be used to make sense of the boundaries which divide modern populations. This is not uncommon in peripheral areas. Stonehenge is seldom used to explain any characteristic of the modern English population, or show that there is a strong mystical streak in the English psyche. However, the monuments of British Celtic areas, of Brittany, and more recently of Corsica have all been used to substantiate just this sort of claim. While history provides a vehicle for tracing the continuity of dominant populations through time, recent attention to 'peoples without history' identifies groups whose past never registered in dominant texts. Archaeological finds can provide an anchor in the past, but are disconnected from a chronological sequence of events that create movement through time to the present. Their past and present can be conflated. Further, in Europe, Megalithic tombs have been seen as providing a link between minority areas that other archaeological evidence has failed to show. Chiari (1960) has expanded on the links between the Corsicans and the Celts, and Carrington (1984) has drawn links between the megalithic culture that unites Corsica with Brittany, Wales and Scotland.

Megaliths have escaped from the objective disciplinary mapping of culture in space and time in a way that microliths have not. They have been appropriated by popular visions of Corsica., though this is a relatively recent phenomenon here. Twenty years ago I visited the site of Filitosa, guided by a cultural activist from the North of the island. Finding the site was not too difficult, but the paths were overgrown and we had to use maps to trace the relationship between different parts of the site. We were alone. Today, there is a hotel near the site, a visitor's centre, a craft shop and a wealth of beautifully produced information to guide the hundreds of thousands of tourists who visit each year around the stones. Many processes have contributed to this new found popularity. Nationalism has played its part, as has the tourist industry with its interest in the invention of heritage. From an archaeological point of view it marks a movement towards communicating living versions of the past, rather than fencing off findings from alternative readings (Hodder 1986 163-4). An internet search revealed 1 600 websites mentioning the site. Many of these carry pictures, and accounts of the experience of seeing the stones. This focus on the sensual experience of 'being there' is important. Urry draws attention to the importance of the 'materiality of place' and the role of objects play in contemporary populations' sense of place (2000 140). Megaliths are iconic objects par excellence, crystallising a whole host of impressions that can be 'felt' rather than intellectually learned. They sit in the contours of the maquis covered hillside, part of a sensual landscape that evokes an embodied sense of place. Urry uses Ingold's insights that 'landscape is constituted as an enduring record of - and testimony to - the lives and works of past generations who have dwelt within it, and in so doing, have left there something of themselves' (ibid 134) Insiders and outsiders alike in taking part in the 'Filitosa experience' are experiencing a sense of place which sets different kinds of time in relation to one another, different practices, different landscapes. Why did people living in this landscape create the megaliths? Just being there, and wondering, is a dramatic journey into the past that evokes emotional responses. This is very different from the kind of intellectual work involved in appreciating the story told by microliths about the dispersal of material culture over familiar boundaries.

Old things found on islands may well fire imaginations, but not all old things, and not all imaginations. Visiting Filitosa distils the experience of being on the island of Corsica now. The processes framing the past at work here are not only associated with islands, but they work particularly well in relation to islands, they are particularly seductive there. The key building blocks of the palette of experience/ knowledge that enables contemporary visitors and natives alike to use ancient stones to differentiate and order social groups in relation to difficult concepts such as tradition/ modernity; marginal/dominant, past /present can be constituted on islands with a clarity, that boundaries don't have elsewhere.

DISHWASHERS AND SHOPPING

An extension of the 'remoteness' paradoxes described by Edwin Ardener in his analysis of the way 'remote areas' are perceived from an outside standpoint (1987:54-50) might be that remote areas are visibly full of the latest technology and consumer goods. Things that disappear into the background of every-day life, that are unremarkable elsewhere, aquire a new significance in such areas; there they herald change and destruction. Modern technology and fashion are not just present, they are seen as symbols of the invasion of the modern state and its market economy heralding the demise of distinctive culture. The boundaries enclosing islands make this 'incursion' of the new into space viewed as traditional particularly visible.

Concern with cultural boundaries is common to most theorising of the intensification of flows of objects and meanings in contemporary times. The increasing reach of global consumer markets is associated with the destruction of indigenous culture, increasing homogeneity and the spread of superficial consumer culture. The grounded work of anthropologists is often at odds with the conclusions of more generalised theoretical schemas. Douglas, writing on goods as a system of communication, has criticised the way anthropologists have considered goods as objects and consumption in almost purely economic terms (1982 16-32). The political economy of consumer goods and their cultural significance cannot be collapsed together. Or, to put it another way consumption cannot be understood from the outside. As Miller (1998) has shown in another island context, Trinidad, even Coca-Cola, that most 'global' of products can have specific local meanings. Visual markers of difference may disappear but systems of local cultural meaning remain intact if fuzzy: the 'other' is still there if looked for beneath the surface.

Some Corsican nationalist discourses about consumption combine Romanticism with Marxism, excoriating the false values of consumer society, the manipulated identities of capitalism (see Bellone 1984). Even recent ideas concerning the kind of economic development which is suitable for the island which see new communications technologies as compatible with traditional economic strategies are self-consciously marking this technology as it moves across the island boundary into Corsican cultural space. One evening I was invited to the home of a Corsican Cultural activist to sample traditional Corsican fare of course, but also to view the way he had transformed his home, restoring the traditional Corsican features. After dinner I offered to help tidy up, as one does. He refused, not in line with traditional hospitality he elaborated at length. His wife was bored - 'Shit', she said, we're only going to stuff some dishes in the dish washer.' He blushed, for here in the kitchen, behind a salvaged cupboard door, was the dishwasher, and he was ashamed of it. The dishwasher had no place in traditional Corsican life.

There was a huge reservoir of indifference to this particular vein of cultural rationalising, Corsicans like shopping and they shop a lot. Dishwashers were very popular. While it might be tempting to smile at the paradoxes of island life no doubt we might also catch ourselves involved in similar activities closer to home if we care to think about it. As objects are transformed into heritage, the familiar European generational D-I-Y change around can be found everywhere: older people struggle to obliterate bread ovens, plaster over stone, install central heating, while the young are uncovering chimneys, sticking agricultural implements on the wall and raising eyebrows at P.V.C. double-glazing. The unwanted material reminders of poverty, and brutal labour of one generation becomes the 'must have' authenticity for the next. The transformation of the past, of things, the generational positioning, the paradoxical relationships (old, working Corsicans are least likely to valorise emblematic Corsican items) should at least warn us not to make too many assumptions about the meanings of material assemblages of the past or the present. Young Corsicans are transforming aspects of the architecture and material culture of the past and appropriating them to play a role in their everyday lives in the present , they are making their space both old and new. For most Corsicans there is no problem in setting their grandmothers traditional chest for kneading dough in a kitchen full of modern conveniences. They feel no need to carry the identification and organisation of their sense of place to such extremes.

Corsicans frequently pointed out to me that they were obsessed with acquiring things. This was identified as one of their 'Corsican' traits. I heard many ironic accounts of the sacrifices families made to buy expensive clothes and cars, the privations endured to ensure a good standard of appearance in public. There is a great deal of interest in the possessions of others and the ways in which they are displayed. Everyone knows that every individual and every family is going to do its utmost to achieve the best appearance possible, failing to do so would indicate that the family was incapable of providing for itself. Most conversations revolve around not failure to maintain decent presentation but the converse. There is a swings and roundabouts quality to the conversations concerning the expenses involved. The cost of consumption can be arbitrarily invoked as a cause for approval or disapproval. There are plausible reasons for finding someone's expensive clothes/car/home impressive, or a scandalous waste of money. The opinions expressed may be a ploy to find out allegiances and preferences, to threaten or to flatter. Gossip about the things people buy is an important activity, creating a certain conformity both in the level of consumption and in the styles of objects chosen.

Conformity in appearance and comportment has been noted as a 'traditional' Corsican trait. Ravis- Giordani has described the enforcement of conformity in Niolan villages:

'La comunauté ne règle pas seulement l'organisation du travail; elle contrôle régit et, eventuellement sanctionne le comportement social des individus et des groupes familaux qui la composent. L'habitat aggloméré du village oú, comme dit L. Cachon "tout se sait et tout se cache", rend possible une surveillance quasi permanente des faits et gestes de chacun. Les nouvelles et les reputations vont vite dans ces réseaux de vicinité que le moindre incident, la moindre nouveauté fait vibrer dans toute leur étendue. Il faut donc savoir tenir l'équilibre entre le conformisme, et la libre direction de sa vie; s'il est possible de manger ce que l'on veut - la nourriture quotidienne est du domaine de l'intimité familiale - en revanche le costume doit exclure l'excentricité: manghja a gustu toiu e vestitu a gustu d'altri (Eat to suit your own taste and dress to suit the tastes of others) (Ravis-Giordani, 1983)

This aspect of 'traditional' life seems to be relevant to contemporary Corsica. Reputation might be seen to have acquired (or been reduced to?) a commodified aesthetic indices, creating a general preoccupation with possessions and display. This could be compared with Herzfeld's description of changing values in Glendi. He discusses in a section entitled 'The passing of a style' (1985:269- 274) how displays of urbanity and wealth are becoming increasingly important in Glendi, superseding traditional aspects of masculine reputation.

At this point the narrative I am developing seems ready to collapse into the predetermined categories of Mediterranean Ethnography, interpreted in terms of stereotypical representations of the past and the present. Agonistic competition within a context of apparent egalitarianism could be seen as having lost moral substance, reduced to competitive consumption and the measurement of honour in purely material ways. However, rather than accepting this at face value I would like to discuss in more detail some of the complexity of the meaning of things in Corsica.

The discussion must necessarily be truncated here, but I wish to stress the need for close examination of the relationship between status, control and morality. What is as true in Corsica is that personal qualities are said to be hidden from all but family and perhaps very close friends. One is what one is seen to be. Scandals that might ruin a reputation are relatively rare; it is generally accepted that people hide their misdeeds. In the absence of 'hard facts' the slightest breaking with convention can be interpreted as a sign of moral failing. One woman in her thirties, married with two children, delighted in dressing originally (as she saw it) in expensive designer creations. Others, however, described her appearance in other terms. I was asked if I found her style 'bizarre'. 'Its a bit too bizarre in my opinion' added my questioner... 'people talk you know.' 'Well, I don't know', said a companion 'I mean I've never been married myself, but it seems to me that a woman should show her husband a little respect.'

The young display the status and wealth of their family. Their dress, their cars, their homes reveal the ability of the family to care for its own and reflect their own success; information that otherwise might remain hidden. They are also the vulnerable point in the family reputation. Their comportment could alter the family's standing. Since older members of the family generally have solidly established their reputations, scandal is most likely to visit a family through its younger members. Thus their appearance and comportment is particularly interesting. While consumption seems above all to be an indicator of economic status, it cannot be separated from observed comportment: aesthetic evaluation cannot be split from moral evaluation.

The difference that modern things - dishwashers, washing machines, cars, Levis and training shoes - had wrought on island life, is one of the more apparent indices of change and as such it was used by people to 'hang' comments about their view on the way life had altered. The opinions expressed do often contain an opposition of traditional life/ modern life. As Chapman has pointed out, however, the ethnographer must be wary of according more significance to this sort of native intellectual structure than any other (1981:15). One day, I was sitting outside with one woman in her sixties, for example, talking about her life before she married. As we talked, she watched her granddaughter wash her new car. The girl was dressed in skimpy shorts, a bikini top, Rayban sunglasses and high-heeled Western style boots. She was tanned, her non-naturalistic make-up was carefully applied, and her hair - style was impressive (long, layered and bouffant). Her gaze stayed on her granddaughter as she began to talk about the different ideas people had about beauty when she was young. She recounted that though she had to help her parents with agricultural work, she used to cover herself from head to foot, with sleeves, gloves and hat to protect her skin. She was very proud of her pale complexion, and had considered herself lucky to have red hair and green eyes, for dark colouring had been thought ugly then. She thought it was funny that the young who nowadays should loaf about in the sun half-naked with all the time they had. She had very few dresses or finery, and the work was terrible, she worked all day. She finished by saying: 'Aujourd'hui les jeunes femmes sont lès princesses, elles travaillent pas, elles sont toujours bien habillées... toutes comme des princesses.'

The remark was not a criticism of young women, but a recognition of the gulf between their lives. A comparison between two ways of being a young woman was set in train by my presence. Her meditation was triggered by her focus on the past as we sat in the present. This juxtaposition brought about by an outsider's eye may have made visible contrasting moral, aesthetic, and practical aspects of her grand-daughter's life and her own. However, it would, I think, be wrong to assume that this was a juxtaposition of incommensurable moral or aesthetic universes, or that one worldview had been replaced by an alien one. One of the distinguishing features of marginal populations in Western Europe, particularly those who live on islands, is that they have to explain the transition from their past to their present (ours is self-evident). Dominant society not only imposes the imperative, it constrains the responses which can be given. For this grandmother, the dress, the car, the 'immodesty' of her granddaughter were not alien or unimaginable. The difference which she singled out was that of work, the absence of exhausting toil and she did not appear to consider this a loss.

Corsicans stubbornly go against the grain of the neat oppositions of tradition and modernity, unless one can be traditional by looking more modern than populations that typify modernity. On Corsica, if the David of traditional culture is not slaying the Goliath of threatening modernity he is at least better dressed. Changes in the material conditions of life are accompanied by changes in value systems but the complex processes involved are all too often collapsed into a direct causal relationship between the two. The narrative of traditional 'authentic' culture being crushed by the homogenising forces of modernity continues to inform popular and some academic accounts. Yet, at the same time, difference continues to be experienced and boundaries continue to divide populations, though not necessarily the differences which are embroidered by dominant discourses. While the reproduction of ideas about difference are surprisingly static and repetitive, the construction and reproduction of specificity must be dynamic, particularly in periods of material change like the present. Minority groups may be 'the creatures of the structural context they inhabit' (McDonald 1993), however, the strange preoccupations of dominant and nationalist identity politics, in this case fixated on mapping oppositions of tradition and modernity over space and population, can only have a tenuous grasp of the lives within their organisational reach. I am arguing that social boundaries are being reproduced over time, through social change, and through the purchase of all global capitalism can produce and brand. Rather than pursuing authenticity Corsicans are shopping like other Europeans, but in a particularly Corsican kind of way. The quest for material prosperity, the dreams of dishwashers and the likes, have fuelled the contemporary migrations of people to and from the island. Not only those Corsicans whose families laboured and saved and sacrificed to send them away to the continent and further, but the North Africans who form the main body of manual labour on the island today. While the search for prosperity disperses families geographically, and changes the lifestyles of generations, the efforts involved in attaining a good standard of living also ties families together in mutual support and enlivens enjoyable everyday preoccupations like shopping and gossip in peculiarly Corsican ways.

CONCLUSION

Today there is enormous emphasis placed on defining, explaining and describing the differences between groups, and the knowledges they possess. In a world characterised by flux and uncertainty the pursuit of certainty seems to be increasing (James 1996). Anthropologists may now look quizzically at these processes of construction, and at the role of the discipline in formalising ideas of the exotic other. This will not necessarily be popular (see Hastrup 1995). As we start to disengage ourselves from processes of definition, as we stand outside and unpick the constructions of difference we will start to annoy (just as archaeologists do) those who have a stake in the definition. While archaeologists may wish to let the facts speak for themselves, and material cultures emerge where they will, the wilful fascination exerted by some things rather than others because of the story they could tell about boundaries stubbornly refuses to disappear. Observing the social life of things on Corsica certainly enables one to see how things old and new can acquire strange glamour or infamy on an island - can be 'of the place', or 'out of place'. We can examine how the self-conscious narratives of identity must englobe even the most mundane object and put them in their place. This might well give us food for thought on the role we might ourselves play in these.

Ardener (1989) claimed that in Western Europe we are space specialists, and one could add we're pretty good at time as well. Adept at tidying up boundaries, rolling them backwards and forwards chronologically and ironing out inconsistencies. Islands are perfect laboratories for this kind of activity -indeed they require it: for if processes of definition can't work there, if clarity, certainty can't be achieved there, where can it be achieved? Thus it is not surprising that contesting modern visions of Corsica should skirt around similar themes regarding the story that things tell about the populations who possess them whether in the past or the present.

As Hannerz (1999 45) points out the new concerns that globalisation has brought in its wake are disconcertingly similar to early twentieth century concerns with diffusionism. Anthropologists started with attempts to categorise and map patterns of culture and social organisation. They then moved into the different socio-cultural units, exploring different social structures and cultural boundaries from the inside. Now there seems to be a need to find ways of conceptualising difference, movement and change again, and the key categories of culture and society which had become naturalised stand again in need of attention. But there are some differences between the early and late twentieth century questioning. Anthropologists' concerns with the changes which accompany globalising processess have been accompanied by questions about how we can imagine and describe 'our' relations with 'others', as 'our' world impinges on 'theirs'. Anthropologists struggle to frame the knowledge they gain of others and ourselves is now complicated by reflection on prior ideological assumptions about progress, on political relations, on their own position as commentators on other people's lives. It is also complicated by natives own struggle with the same issues. Eriksen suggests that islands are empirically interesting to explore in relation to these problems. The extent to which cultural phenomena and human societies can be meaningfully conceptualised as discrete, autonomous and bounded, or alternatively seamlessly woven together can be explored in relation to the way island metaphors are developing in relation to the material changes experienced by island populations. He suggests that a modified diffusionism might well develop which could employ a new form of island metaphor which takes its substance from the construction of social and cultural boundaries rather than 'objective' territorial or biological properties. But then again the strange charm exerted by islands would also be missed both by those who live on them and those who yearn to visit.

BIBLIOGRAPHY

APPADURAI, A.,
1981) 'The Past as a Scarce Resource' in Man vol. 16(2) pp. 201-19.
APPADURAI, A.,
1986. 'Introduction: Commodities and the Politics of Value' in A. Appadurai (ed.) The Social Life of Things, Cambridge: Cambridge University Press.
ARDENER, E.,
1987. '"Remote areas" - some theoretical considerations', in A. Jackson (ed.), Anthropology at Home (ASA Monograph 25), London: Tavistock.
BAUMAN, Z.,
1998. 'Urban Space Wars: On Destructive Order and Creative Chaos' in Space and Culture vol. 3 pp 109-123
BELLONE, H.,
1984. 'La Crise d'Identite Culturelle dans la Corse D'Aujourdhui' in La Corse et La Folie, Bastia: CORS.
BERTHO, C.,
1980. 'L'invention de la Bretagne: genèse sociale d'un stereotype' in Actes de la Recherche en Sciences Sociales 35 pp 45-62.
BOURDIEU, P.,
1990. In Other Words. Essays Towards a Reflexive Sociology Cambridge: Polity Press.
CAISSON, M,
1984. 'L'Écriture des Pierres', in La Corse et la Folie: Rencontres Pluridisciplinaires, Bastia: Publication C.O.R.S.
CARRINGTON, D.,
1984. Granite Island: A Portrait of Corsica (2nd Edn), Harmondsworth: Penguin Books.
CHAMPION, T.,
1996. 'Three Nations or One? Britain and the National use of the Past' in , M. Diaz Andreu & T. Champion (eds) Nationalism and Archaeology in Europe, London: UCL Press
CHIARI, J. ,
1960. Corsica: Columbus's Isle, London: Barrie and Rockliff.
DIETLER, M.,1994. '"Our Ancestors the Gauls": Archaeology, Ethnic Nationalism, and the Manipulation of Celtic Identity in Modern Europe' in American Anthropologist 96:3, pp. 584-605.
ERIKSEN,T. H.,
1999. Do cultural Islands Exist? htttp://www.uio.no/~geirthe/Culturalislands.html
GROSJEAN, R.,
2000. La Corse avant l'Histoire Paris: Klincksieck
HANNERZ, U.,
1999. 'Epilogue: On some Reports from Free Space' in B. Meyer and P. Geschiere (eds) Globalization and Identity: Dialectics of Flow and Closure, Oxford: Blackwell.
HASTRUP, K.,
1995. A Passage to Anthropology: Between Experience and Theory London: Routledge
HERZFELD, M.,
1985. The Poetics of Manhood, Princeton: Princeton University Press.
HODDER, I.,
1986. Reading the Past: Current approaches to Interpretation in Archaeology Cambridge: Cambridge University Press
HROCH, M.,
1996. 'Epilogue' in M. Diaz Andreu & T. Champion (eds) , London: UCL Press
JAMES, W. (ed.) ,
1995. The Pursuit of Certainty; Religious and Cultural Formulations London: Routledge.
KUPER, A.,
1992b. 'Introduction', in A. Kuper (ed.) Conceptualising Society, London
MCDONALD M.,
1989. 'We Are Not French!' Language, Culture and Identity in Brittany, London and New York: Routledge.
MCDONALD M.,
1993. 'An anthropological Approach to Stereotypes' in S. MacDonald (ed.) Inside European Identities, Oxford: Berg
MILLER, D. (ed.),
1998. Material Cultures: Why Some Things Matter Chicago: University of Chicago Press..
POMPONI, F. ,
1979. Histoire de la Corse, Paris: Hachette.
RAVIS-GIORDANI, G.,
1983. Bergers Corses Les Communautés Villageoises du Niolu, Paris: Edisud.
SMITH, A.,
2001. 'Authenticity, antiquity and archaeology' in Nations and Nationalism vol.7(4) pp441-50.
TOSSI, C. WEISS, M.,
2001. Nouvelle DonnÈes sur le Neolithique Ancien de l'Aire Corso-Toscane' in Bulletin de la SocietÈ Prehistorique Francaise vol. 98(3) pp445-458.
TRIGGER, B.G.,
1984. Alternative archaeologies: nationalist, colonialist, imperialist Man 19(3) pp355-70
URRY, J.,
2000. Sociology Beyond Societies: Mobilities for the Twenty-First Century, London: Routledge

14

THE INSULARITY OF MALTA

D.H. AND B.A.V. TRUMP

Mac Donald Institute,
University of Cambridge, Cambridge, England
EMAIL: trump@defrave.freeserve.co.uk

INTRODUCTION

A theme which emerged early on in the Conference was the interplay of geographical and cultural factors in the development of insularity in prehistory. With gaps in the programme due to missing speakers, it was thought an extempore examination of the Maltese evidence on the issue would be useful. This paper is a more deeply considered appraisal of the same. It is a pity that John Robbís paper (2001), covering very much the same ground, was not known to the writers until afterwards.

The Maltese islandsí geography can be understood from any map of the Mediterranean, in which it occupies a central position. It lies 80km south of the southern tip of Sicily, with which, weather permitting, it is intervisible. However, it is comparatively low, rising to barely 250m above sea level, whereas Sicily has a vastly more prominent feature in the 3340m cone of Mount Etna. In other words, Sicily is much more easily seen from Malta than vice versa.

Another proviso is that this is the current situation. At periods of lowest sea level during the Pleistocene, at approximately -120m, the two were connectd by an isthmus. Even at -100m, a large island would have reduced the sea crossings to two narrow channels. Quite apart from the possibility of navigation by Homo erectus, as suggested by the evidence from Sardinia and Flores presented at this conference, they could at times have arrived in Malta without that help. But while access would have been much easier than usually envisaged, other geomorphological changes - soil erosion from the heights, deep alluviation in the valleys, and drowning of all coastlands - make the total absence of secure evidence for any colonisation before the Neolithic unsurprising.

Colonisability is primarily, but not solely, a matter of geography. Navigational competence is another factor, as is the willingness to venture overseas. We have no hesitation in accepting that both were present in the Neolithic, given the coastwise, and insular, distribution of Impressed Ware sites round the Central and West Mediterranean in the 6th millennium BC. The first proven settlers of Malta arrived from Sicily probably well before 5000BC, by which date radiocarbon dates from Skorba show them to have been well established. Their distinctive pottery, named after the cave site of Ghar Dalam, is identical with that of the Stentinello culture of Sicily (Evans 1953, 1959; Trump 1966 and forthcoming).

That the two areas remained in constant contact thereafter is amply demonstrated by the recovery from subsequent contexts in Malta of raw materials not found on the islands. These include obsidian from Lipari and Pantelleria, flint from the Monti Iblei, both for flaking into tools, red ochre for pottery decoration and ritual use, particularly in burials, and various hard stones, for polished axes and amulets, the nearest sources being the Monti Peloritani and Aspromonte, either side of the Straits of Messina. One amulet from the Xaghra Circle was even made from Piedmontese jadeite.

Despite that, the pottery styles diverge markedly, either by natural drift or by deliberate choice. Maltese Grey and Red Skorba are clearly distinguishable from contemporary wares in Sicily, although the latter shares with Sicilian Diana Ware a vivid red slip. This could possibly be due to influence of one on the other, though this is by no means certain, still less, in which direction the influence passed.

With the next phase, Zebbug, influence of Sicily on Malta is much stronger. Further research would be needed to show whether this implies a new influx of people from San Cono-Piano Notaro-Conzo, or merely copying of pottery styles. It is the last cultural break in the Maltese sequence for fifteen hundred years, though imports of stone continue.

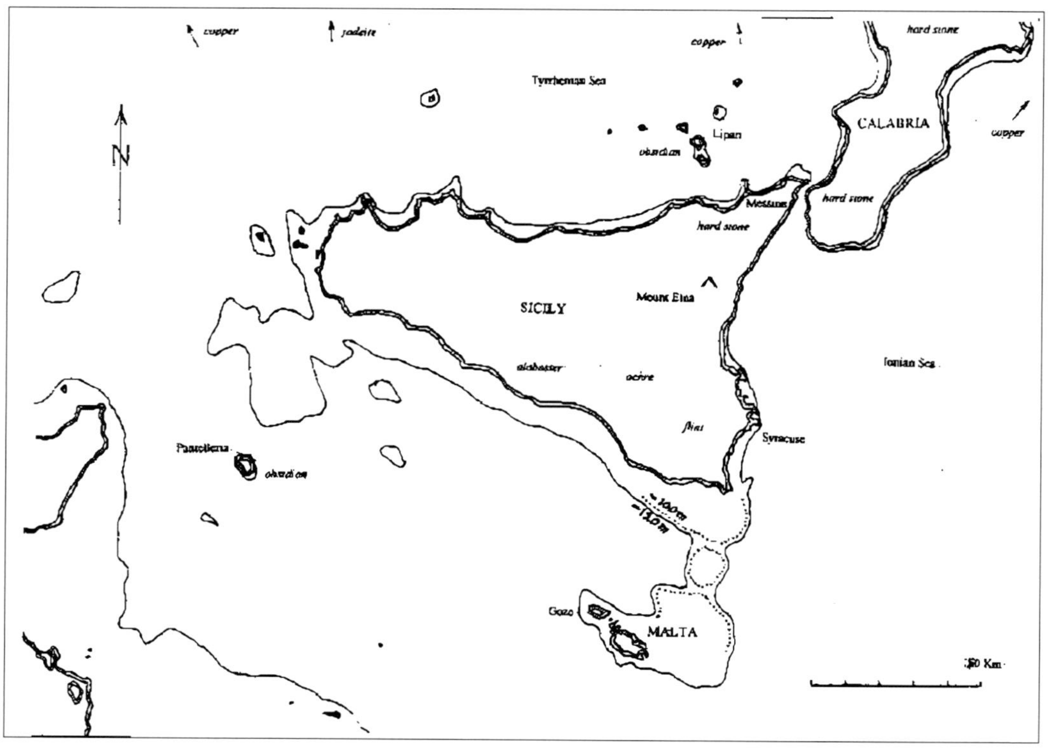

FIGURE 1. MALTA AND SICILY: OUTLINES AND RESOURCES.

There is even a handful of sherds of imported pottery from Skorba - four from Trefontane, one of painted Serra díAlto Ware and one of red-slipped Diana (Trump 1966). Contacts continued, then, but Maltese pottery developed along entirely different lines. Far more remarkable is the local development of architecture and statuary, in the Maltese temples and their ëfat ladyí figures. These are unparalleled outside the Maltese islands, and owe nothing to external influences.

A few foreign sherds continued to find their way to Malta, but it has to be stressed, extraordinarily few. From the Xaghra Circle were recovered single sherds of Sicilian Serraferlicchio and SantíIppolito type, and several, of only two vessels, of Piano Quartara, with an origin in Lipari (Trump, in Malone, Stoddart et al. forthcoming). More puzzling are a handful from Skorba and Xaghra, together with a complete vessel from Tarxien. They are more puzzling because, though very distinctive and patently not local, their oversea source has not been positively identified. They are of open tronco-conic bowls with a prominent thickened lip on the inner side. While there are also sporadic specimens from Castelluccio in Sicily and the Grotta delle Volpe near Cagliari in Sardinia, their area of origin would seem to be - one dare not put it more firmly - the northern Aegean, particularly Thermi on Lesbos (Lamb 1936) and Troy II (Blegen 1963).

In contrast to the few foreign pieces on Malta, remarkably no single Maltese sherd of this period has yet been recognised outside the islands. Although the imported materials already discussed prove contact, this would seem to have had no cultural effect whatsoever, in either direction, a clear case of insularity by choice.

There was then in c.2500 BC a break in cultural continuity so marked as to imply collapse of the temples and replacement of their culture, perhaps also their population (Evans 1956, 1959). A hiatus lasting several centuries would make it conceivable that the differences could be explained by local evolution, of which there is as yet no trace, but they are so marked in every cultural aspect that this is most unlikely. As well as the marked break in pottery development and the abandonment of worship in the temples, more significant are the introduction of metal and replacement of inhumation in rock-cut tombs and hypogea by urn cremation.

New immigration remains much the most economical explanation. That, however, begs a number of questions, above all, what happened to the previous inhabitants and their temples, and whence the newcomers came. Wherever that was, their Tarxien Cemetery culture in Malta very soon acquired a purely local flavour. This is particularly noticeable in their lavish pottery decoration and strange discoid figurines, neither, be it noted, showing the slightest similarity to those of the preceding temple culture.

The pattern of insularity in the Bronze Age remains much as before. Like the stone for flaking earlier, the copper, and bronze which later replaced it, had also to be imported, from even greater distances. The nearest sources were in Sardinia, Tuscany, the Balkans and Cyprus. By now, the markedly increased number of spindle whorls and loomweights found suggest that the Maltese were producing textiles with which to pay for their imports, although that leaves unanswered what fibres, animal or vegetable, they were spinning and weaving.

From this period, only a single sherd of imported pottery is known. It is from the lip of a Mycenaean drinking cup, painted with an octopus in red on a yellow ground and dating to the 13th century BC. It was found at the Borg in-Nadur temple site in the 1920s, though not recognised until 30 years later by John Evans (1953). It is unusual in that most Mycenaean exports were of storage jars, presumably as containers for desirable liquids, rather than drinking vessels.

On the other hand, exported pottery is more frequent, notably at two sites on the Sicilian coast, south and north of Syracuse. A number of thickened lip bowl sherds were found by L.Bernabo Brea (1966)at the Torre díOgnina, unusual locally and assumed to be brought from Malta. There are two lesser, and inadequately published sites, near Licata, namely Manfria and Casalicchio-Agnone, apparently with the same, though Tusa (1993) refers to the former as Tarxien Temple, which seems unlikely . If their source was indeed Malta, it is rather surprising that the more typical Tarxien Cemetery everted lip bowls and jars did not appear also. There is no question about the Borg in-Nadur vessels placed intact in rock-cut tombs at Thapsos. These tombs were also found to contain Mycenaean stirrup jars, confirming contacts, at least at second hand, between the Greek Argolid and Malta.

For the first time since Zebbug-San Cono two and a half millennia earlier, there are distinct resemblances in both vessel shape and decoration between Borg in-Nadur and Thapsos wares, though the former is nearly always red-slipped and the latter grey. Here, contacts did lead to some sharing of cultural traits, which is more what one would expect than the total dissimilarity noted at all other periods. There may even have been movement of people between the islands at this time.

With the Bahrija phase, we are faced with a different situation. Its most distinctive pottery, open cups with high loop handles and hemispheric bowls, both decorated with elegant cut-out square meanders incorporating panels of false-relief zigzags, all under an intense black slip, together indicate a close link with the Fossa Grave Culture of Calabria, in about the 10th century BC. However, much more frequent is a less elaborate ware with bands of parallel cut-out lines, often forming a dovetail design, especially below handles, and also black slipped. There is little doubt that this second element is of local evolution from Borg in-Nadur, where the red slip was being replaced by black, due to different, reducing, firing conditions. This ëlocalí Bahrija occurs in small quantities at sites throughout the islands, but the ëforeigní Bahrija virtually only at the type site. How do we explain this?

The locally derived ware demonstrates that we are not looking at an immigrant population, highy unlikely in any case at so remote a site as Il-Qlejgha ta Bahrija on the western cliffs. Though little more than a guess, one might envisage the Calabrian ëinfluenceí, first pointed out by John Evans (1953), as the result of the arrival from the Italian peninsula of a single family, or even a lone potter. Not finding employment in the more populous settlements of the lowland zone, he, she or they finally found acceptance at Bahrija. Perhaps their high quality wares were adopted to combat an inferiority complex of this large but remote and probably poor community, and so an ëisland within an islandí. This is highly speculative of course, but at least reconciles all the somewhat contradictory evidence.

In the 9th century occurred an even more profound change than that between the Temple Period and the Bronze Age, though this time without any implication of a displacement of population. The first effect of the arrival of Phoenican navigators/ traders was to convert Malta from a remote offshore island to the hub of a great commercial enterprise. Its geopolitical insularity had gone for good, though with no change to the physical map. So too the cultural insularity. Although there were still to be minor local variations, Maltese culture hereafter became little more than a regional variant of one or more foreign ones. What in later times became distinctively Maltese was not so much a local tradition as a unique blend of foreign ones.

The introduction of the potterís wheel brought with it a whole alien style of pottery, the local handmade tradition, as C. Sagona (2000) has pointed out, surviving only in inferior, and dwindling, products. For monumental architecture, at Tas Silg and Zurrieq, Malta had nothing to compete with Levantine skills, and though we have not the actual evidence, the same probably soon applied to domestic architecture also. Although writing was now introduced, we can be no more sure of the spoken language at the time than in the pre-literate periods. The script was the Phoenician alphabet, naturally used to write the Phoenician language, whatever the natives were speaking among themselves.

While Malta and Gozo were islands in the geographical sense for the whole of the period archaeologically documented, their idiosyncrasies (a word which does less than justice to their magnificent pottery, statuary and architecture, and presumably the religious beliefs and rituals which underpinned them) were not dictated by that but by cultural choice. Their inhabitants wished to emphasise the differences between ëusí and ëthemí, and did so spectacularly. Even when a foreign culture was adopted, it was not necessarily that of the next nearest land, Sicily.

One could compare this case with that of Zanzibar well into the 19th century. Although only 40km off the African coast, and with a substantial part of its population derived therefrom, its culture showed virtually no influence from the mainland. But instead of developing one entirely of its own, as in prehistoric Malta, it chose to share that of the

Yemen and Oman, nearly 3000km to the north. In this, it was more like Malta in the Phoenician period, when the latter more closely imitated the Levant, or at least Tunisia, than Sicily.

A similar point can be made from the later history of Malta. Its language, with its close connections to that of the Maghreb, was imposed, or at least adopted, in the period of North African rule, from AD 870 to 1090. Since the latter date, all cultural influences - Norman, Spanish, Italian and British - have come from the north, as shown by the vocabulary, though not the grammatical structure, of its language, and above all by its devout Roman Catholicism. This is even more apparent from its baroque architecture, though often with a distinctly Maltese flavour. And yet the islanders stoutly maintain their unique Semitic-based language, as the primary means of asserting their cultural identity today.

The Maltese islands, then, amply document the fact that, while geography undoubtedly plays a part, cultural insularity was always, and remains today, largely a matter of social choice. As John Robb (op. cit.) points out, ëislands are ideasí, or even more strongly, ëIn cultural terms, the Maltese didnít [just] live on an island, they built one.í

REFERENCES:

BERNABO BREA L.,
1966. Abitato neolitico ed insediamento maltese dellíeta del bronzo nellíisola di Ognina (SR) ed i rapporti fra la Sicilia e Malta dal XVI-XIII secoli a.C. Kokalos XII, 40 seq.
BLEGEN C.,
1963. Troy. London, Thames and Hudson.
EVANS J.D.,
1953. The Prehistoric Culture Sequence in the Maltese Archipelago, P.P.S. XIX (1), p.41-94.
EVANS J.D.,
1956. The ëDolmensí of Malta and the Origins of the Tarxien Cemetery Culture, P.P.S. XX, p.85-101.
EVANS J.D.,
1959. Malta. London, Thames and Hudson.
LAMB W. ,
1936. Excavations at Thermi in Lesbos
MALONE C., STODDART S. ET AL.
(forthcoming) Excavation of the Brochtorff Circle, Xaghra, Gozo.
ROBB J.F.,
2001. Ritual, travel and the creation of difference in Neolithic Malta. European Journal of Archaeology, 4(2), p.175.
TRUMP D. ,
1966. Skorba. Research Report XXII, Society of Antiquaries. Oxford University Press.
TRUMP D. ,
forthcoming. Malta: Prehistory and Temples. Malta, Midsea Publications.

15

THE ROLE OF RETURN MIGRANTS AND URBAN MIGRANTS ASSOCIATIONS IN THE REVIVAL AND REINVENTION OF GREEK ISLAND TRADITION: AN EXAMPLE FROM THE CYCLADES.

MARGARET E. KENNA

School of Social Sciences and International Development
University of Wales Swansea
EMAIL: m.e.kenna@swan.ac.uk

INTRODUCTION

In this paper I want to touch on some of the issues listed in the suggestions put forward by Jacqueline Waldren for the anthropologists who attended the conference. The brief was to present anthropological research on islands in a manner which might be useful and informative for the archaeologists and prehistorians who formed the majority of the audience. Topics included: concepts of identity, space and place, demographic changes, impacts of tourism and emigration/ immigration, geographical, geological, ecological and architectural environments, authenticity, museum exhibitions, 'cultural tourism', the use, control and politicisation of natural resources, (water, air, sea, sky), material culture, tradition, and history.

I will touch on some of these, by considering the role of return migrants and of migrants associations in Greece (see map, FIGURE 1). By return migrants I mean those who have come back to their place of origin, usually a village or small town, after living and working in one of the large cities of Greece such as Athens or Thessaloniki. Very few Anafiots have emigrated to Australia, New Zealand, the USA or Canada, although some have migrant experience of European, Mediterranean, and Middle Eastern countries (including Germany, Egypt, and Iraq). 'Place of origin' can also include the natal village of a parent or grandparent. Migrants Associations are organisations set up by internal and external migrants through their common tie to the same place of origin, village or region, usually to aid their own social interaction but also to act as channels for financial help to the place of origin and as a lobbying force to attract government aid to that place (Dubisch 1977, Sutton 1991).

Many return migrants become involved in local politics, and form local cultural and folkloristic associations (or local branches of the urban migrant associations to which they used to belong), through whose auspices museums are set up and local customs and traditional practices revived. Many are also involved in touristic enterprises, and return precisely because of opportunities offered for such initiatives in their place of origin. While many of these museums of local antiquities and folklife are ways of representing themselves to themselves, they are also ways for islanders to represent themselves to others (whether these others are fellow Greeks or European and international tourists), and their significance as tourist attractions is often one of the justifications used for setting them up.

In picking up some the themes suggested by Waldren, I am drawing on Hobsbawm and Ranger's seminal book The Invention of Tradition (1983), which showed how cultural elements may be consciously selected and developed to aid a nationalistic, or a romantic, agenda. Scottish tartans and bardic groups of Welsh druids are examples of such invented traditions. These ideas of constructing tradition through selection and innovation have been developed particularly in the field of tourism studies through concepts such as authenticity (MacCannell 1976) and commoditization (Greenwood 1989). MacCannell's discussion of authenticity claims that tourists are searching for the real essence of something in their holiday destination, something which they expect to be objectively true or real, whether it is an unspoilt landscape or genuine people, or the actual site of a battle or historical event, or non-westernised music or dancing. Greenwood developed the concept of the commoditization of culture (Greenwood 1989) to cover examples of indigenous traditions (authentic events in MacCannell's terms) being turned into commodities for tourist consumption. His example was of a local festival which regional officials wanted to be performed more than once on the day of its occurrence so that it could be witnessed by a larger number of tourists. In this case the connection between the two concepts, authenticity and commoditization, is that tourist demand for authentic spectacles, the desire for the opportunity to witness a real life celebration, turned the event from an expression of townspeople's identity and solidarity into a commodity out of which profit could be made.

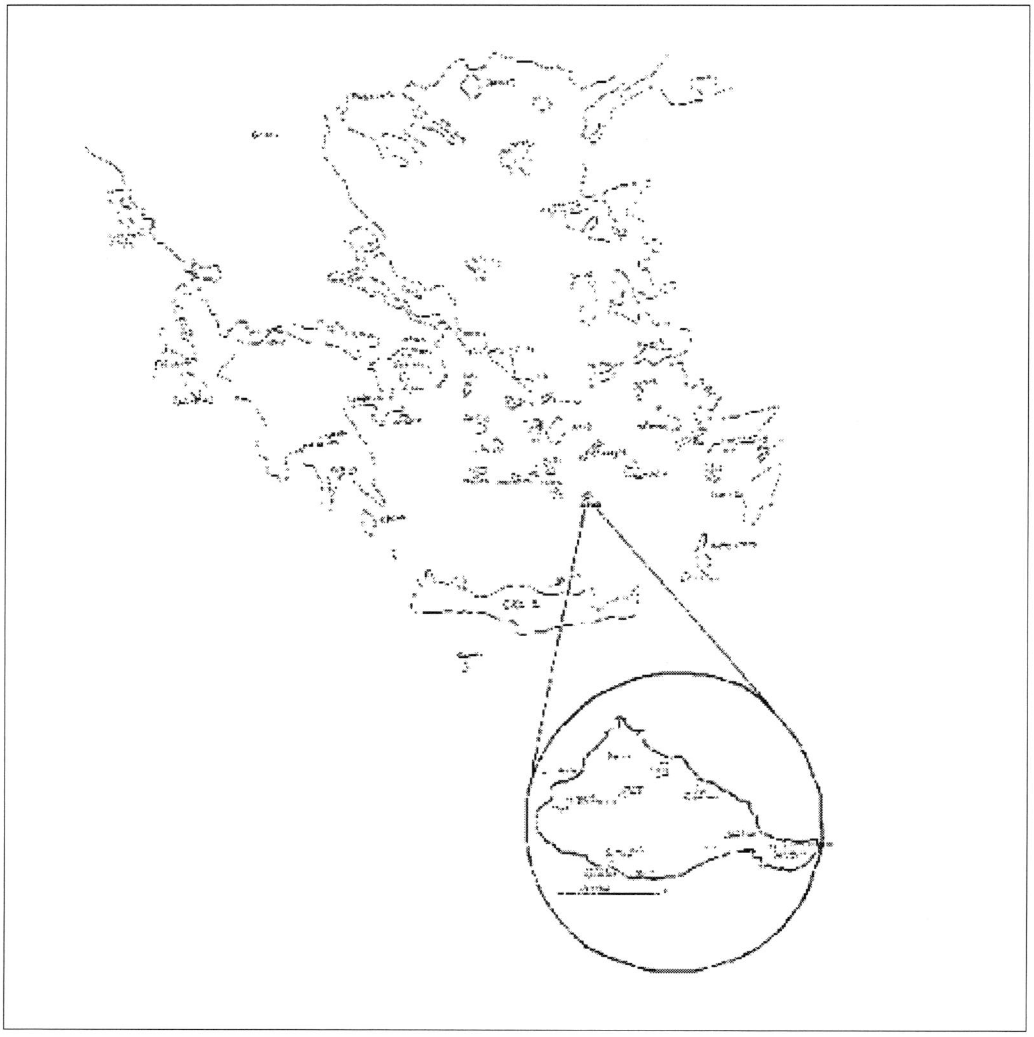

FIGURE 1. MAP OF GREECE, SHOWING ANAFI.

The invention, adaptation and packaging of authentic traditions for tourism are well- discussed and debated topics (see, for example, Taylor 2001). Some scholars have argued that for tourists it is the validity of their own experience which is important, not the actual objective authenticity of the place or event (see for example Wang on existential authenticity, Wang 1999, 365). Some scholars have argued that it is almost impossible for a tourist to go anywhere 'real' in this sense of 'authentic' because of the total construction of the experience of tourist travel both by travel companies and by host countries. (Whether solo travellers, or anthropologists, can ever go anywhere 'real' in this sense, is another question.) Others have argued that for the people of the host country, the conscious attention to aspects of their culture with a view to attracting tourists often results in the revival of crafts, the introduction of new techniques and practices, and the rediscovery of forgotten elements (and of course, there may well be contrary opinions and opposing factions on these issues among the hosts, as there are among scholars). Such outcomes may be beneficial or damaging, depending on one's perspective. How and why particular authentic items are selected for such commoditization and by whom are interesting questions. However, tradition is not only invented, revived, and commoditized for tourists, but also for members of a host country themselves. Modernisation and urbanisation often induce a sense of distance which is temporal as well as spatial, and connections or reconnections with conditions in the past as well as in other parts of the country are felt to be a kind of anchor in such situations of rootlessness. A rural past may be romanticised (as well as denigrated) even in a village itself, let alone in a city, and the impetus to record and preserve certain aspects of that past is as much for those who wish to claim that past as their own and represent it to themselves, particularly because they have moved beyond it, as it is for others.

FIGURE 2. VIEW OF THE NEW ROAD AND OF OLD AND NEW BUILDINGS,

In this paper I consider a particular example: the revival and reinvention of traditions on a small Greek island, partly a result of the return of island migrants, some of whom are involved in tourist enterprises (Kenna 1993, see also Kenna 2001a). Some traditions are perceived as of relevance for islanders only and as having no tourist dimension (for example, if they are events which occur outside the tourist season), others have emerged as a consequence of tourist demand rather than of the local creation of a product in advance. The key actors here are return migrants, locals, and the tourists themselves.

The clear-cut nature of the membership of each of these categories could be questioned and complexified: nearly all locals have been involved in migration at one time or another, usually seasonal or short-term migration to Athens, so that the distinction between resident islanders and returned migrants is in some respects an artificial one. Some tourists and holiday-makers are Greeks with some sort of prior connection to the island. Some of the return migrants have themselves been tourists elsewhere in Greece, or abroad, and bring to the island their knowledge and experience of tourist provision in these places. The members of each of these categories have different conceptions of tradition, history and heritage, conceptions acquired through socialisation, education, the media and experience.

Greeks. Greeks who before the time of the War of Independence had never heard of Socrates or Sappho, Aristotle or Antigone, soon began to give their children these names as signs of patriotism and of connection with their past. Tracing this line of descent for modern Hellenism, Ellenismós,

FIGURE 3. CONSTRUCTION OF ROOMS TO RENT AT THE EASTERN END OF THE VILLAGE

THE MYTH-HISTORY OF GREECE

In forming a view of themselves as a nation (ethnos) and as a state (kratos), the Greeks have to some extent adopted the views about their links with the ancient past which were held by many of the Europeans who had assisted in the struggle to detach Greece from the Ottoman Empire of which it had been part since 1453. For the Philhellenes, the Greeks of the late eighteenth and early nineteenth centuries, who had been 'under the Turkish yoke' for 'four hundred years of slavery', were the direct descendants of the ancient

predominated over, for example, tracing links with the Byzantine Empire, links which result in another form of Greekness, Romios_ni, 'Roman-ness', that is being the heirs of the Eastern Roman Empire (and being the subjects of the Ottoman Empire which was responsible for the fall of Byzantium). As far as the Greeks are concerned, the rest of Europe owes its civilisation and culture to the ancient Greeks, and this debt should be acknowledged. Who better to receive these thanks than the descendants of the ancient Greeks?

The sites of ancient temples and buildings, museum exhibits, and the physical landscape are heritage sites equally for Greeks and for visitors to Greece. The most common visual image which represents Greece in the media is without doubt the Acropolis rock on which the Parthenon stands, but the Parthenon is an ancient Greek building of the 5th century B.C. Is there one unequivocal image which stands for Greece today, modern Greece, as the Sydney Opera House stands for modern Australia while Uluru (Ayers Rock) represents natural and aboriginal Australia? as the Sagrada Familia church in Barcelona stands for modern Spain, post Gaudí, while the bull stands for Spanish tradition? Possibly the most recognisable image of modern Greece for tourists is the vernacular architecture of the countryside, particularly whitewashed churches, windmills, the white cube shapes of village houses and the barrel-vaulted houses of the volcanic Aegean island of Santorini. So while an ancient temple in the capital city stands for the ancient heritage of Greece (and of Europe), it is anonymous peasant buildings, particularly of the islands, which represent it as a contemporary reality as a glance at the images in advertisements published by the Greek National Tourist Organisation and in holiday travel brochures will demonstrate. (This generalisation may well have to be modified after the 2004 Olympics).

School-teachers and local historians start any account of a village or locality with its ancient heritage, often by reference to classical texts such as the Iliad and Odyssey, and confirm its Greekness with an account of the part played by local people in the War of Independence or with efforts to become united with the Greek state after its formation. This is a particular kind of history, or rather myth-history, which is conveyed through socialisation and education, and thus frequently represented in museums and other exhibits. Parts of this history become stressed or highlighted in the context of contemporary events (for example the controversy over the use of the name Macedonia for the former Macedonian Republic of Macedonia, see Sutton 1997). Other historical events are not mentioned, sometimes because they are not known about, or because they are thought not to be of interest (or worthy to be retold), or are too controversial (as were, until recently, the events of the Greek Civil War). Again, such events may be incorporated into the national or local story when they become significant to others: for example, stories of Greek Orthodox Christians as righteous gentiles who saved the lives of Jewish Greeks, which became salient particularly after the release of the film Schindler's List. Similarly, each island tries both to link itself to the national myth-history and to cultivate its own special version of this, linking itself to poets, philosophers, mythological characters and other figures whose names used to be well known to Europeans who had been educated in the Classics. Local people frequently remember these local myth-histories from their schooldays, and regional museums repeat them in placards explaining exhibits and in their brochures (see Damer 2001).

GREEK ISLANDS: SOME FACTS AND FIGURES

Depending on how an island is defined, there are between 3,000 and 1,000 Greek islands, and, again depending on definitional criteria, between 160 and 100 of them are inhabited (see Zenfell 1989, 26, 71; Kolodny 1974 vol.1, 42). A few islands (like Niarchos's Spetsopoula and Onassis's island of Skorpios, one of the Ionian Islands) are privately owned. The island population constitutes about ten per cent of the Greek population (currently numbering over 10 million people). Some Greek islands were part of the new nation-state from the beginning (i.e. from about 1832), but others joined at later points in the nation's history: the Ionian Islands in 1864, Crete in 1913, the islands of the Dodecanese (which had been under Italian rule since 1912) in 1947.

Many islanders have gained their living from the sea, whether by fishing, sponge-diving and boat-building or by sea-going trading and commerce. Successful mariners or traders acquired capital to invest. One of the items of conspicuous display they invested in was neo-classical houses which can still be seen on formerly prosperous islands such as Syros (where Greece's most important port was located before the port of Piraeus was constructed); other 'display items' included increased dowries for daughters and patrimony for sons, and the building or adorning of chapels and shrines. More recently, capital from city jobs, from merchant marine and naval occupations, and from overseas migration, has been invested in the building and running of restaurants, tavernas, hotels, and rooms to rent. The islands in the Saronic Gulf, that is, closest to the mainland where Athens is located, experienced early domestic tourism as city-dwellers escaped there from the city heat in summer, and the peninsula of Chalkidiki began to develop as a result of domestic tourism from Thessaloniki. In both areas, tavernas, restaurants and hotels were built to cater for such tourists (and for the Athenians or Thessalonikans who built holiday villas there), and thus the infrastructure for post-war international tourism was in place for travellers from elsewhere in Europe and further afield. It should be noted, however, that as the majority of urban Greeks retain ties with their (usually rural) places of origin, and frequently maintain voting rights there, many city-dwellers return to their own, their parents' or grandparents' natal village for summer holidays and on other occasions.

As well as looking outwards as seafarers, islanders also engage in agriculture and raising livestock, for both subsistence and export, and exploit local specialities such as marble or minerals. But island workforces have been depleted by urban migration and overseas migration for well over 150 years. Remittances from these internal and external migrants have been, and still are, a most important, even essential, element in the islands' and nation's economy. Certainly until the establishment of agricultural pension schemes, money sent home by migrant sons and daughters sustained the lives of many elderly islanders, and more successful migrants often transformed the lives of the whole community by financing the building of a hospital, the construction of a road, the repair of a village school, or the installation of a piped water supply.

RETURN MIGRANTS

Return migrants have never replaced this out-migration, but

have brought back with them knowledge of both urban and overseas ways of life. From the 'hello boys' coming back from America between the wars with their few words of English and tales of city life (Saloutos 1956), to postwar migrants to Athens returning to their place of origin for their retirement years, return migrants have played an important part in the development of their home communities. Many take an active part in the life of the community. And they are not all elderly retired people. Some return migrants are young or middle-aged adults who have left the city in order to exploit the economic opportunities offered on the islands by government initiatives and especially by tourism.

been extensive tourism development, particularly from the mid-sixties (Apostolopoulos and Sönmez 1999, Loukissas 1982, Tsartas 1992, 517).

As is usual in the Mediterranean area, the period of May to September is the main tourist season in Greece when three-quarters of the annual number of tourists arrive (i.e. about 8 million out of 11 million). July and August are the peak months in which half of the 'tourist season' tourists (nearly 4 million) flood into the country. Over the course of a year, almost every category of tourist identified by Cohen, and later by Smith (Cohen 1974, Smith 1989), visits Greece, from mass package tourists to those who would prefer to describe themselves as independent travellers.

FIGURE 4. POSTCARDS AND A SCHEMATIC ISLAND MAP (top left) ON SALE IN A VILLAGE GROCERY
Summer 1987.

TOURISM

Tourism is a very important element in the economy of Greece as a whole, with the total number of international tourists equalling or even exceeding the number of people in the population as a whole (11 million tourists in 1995). Domestic tourism is also very important, dating from the nineteenth century and the growth of the city of Athens, and now including such diverse elements as winter ski-ing holidays, rock-climbing and mountaineering weekend breaks, as well as the more predictable vacation trips from centres such as Thessaloniki and Athens to seaside areas of the mainland, to the mountains, and to the islands. International tourism, which dates back to the time of the European Grand Tour of the eighteenth and nineteenth centuries, has been of particular significance in the post-war era, since the late 1950s and 1960s. In the islands there has

Wickens (1994), somewhat tongue in cheek, typified five sorts of tourist to Greece: the Cultural Heritage types (interested in culture, history and the natural beauty of Greece); the Ravers, hardly concerned with being in Greece but more with the beach by day and clubbing by night; the Shirley Valentines, women who are looking for romance and sex; the Heliolatrous types (Sun-Worshippers) concerned with tanning on the beach; and the Lord Byrons, returning visitors who become familiar with particular places, return there regularly, and cultivate nostalgia about how things used to be. As Wickens points out, the search for pleasure - whether by relaxation, partying, intensified sexual experiences, or visiting ancient monuments - is the primary motivation of all tourists, and not the attempt to find authenticity as MacCannell maintains. While some of the tourists may incorporate a search for authenticity and tradition into their pursuit of pleasure (the Cultural Heritage types), few of them really want to undergo the physical hardships of traditional Greek culture. We leave that to the anthropologists.

TOURISM IN THE GREEK ISLANDS

Greek islands differ greatly in the extent of their reliance on tourism, depending on their other resources, but surveys have shown that as much as 70% of the different regions of islands (Ionian, Dodecanese, Cyclades, Sporades, etc.) derive their GNP from tourism, and that a similar percentage of income comes from the service sector, and thus by implication, from tourism (Tsartas 1992, 518). As we have already seen, on some islands the necessary infrastructure was already in place, partly because of domestic tourism (and with the entrepreneurial activity of returning migrants with savings to invest and urban investors), when the post-war tourism boom began, and package tourism developed (see Galani-Moutafi 1993, Kousis 1989, Stott 1985, Zarkia 1996). Some of these islands (Tsartas's Group A) became almost entirely dependent on tourism (Ios, Mykonos, Paros, Thera: see Tsartas 1992, 519, 524). Most of these tourists were Wickens's Ravers or Heliolaters; interested in recreational tourism, and with not much concern for the particular features or traditions of the place where they were holidaying. In these places, island traditions were much more a matter for locals and return migrants than for tourists. The individual characteristics of the islands which were marketed were those features which appealed to the tourists, rather than indigenous features: the gay club scene on Mykonos for example.

Other islands (Tsartas calls this category Group B and defines them as 'satellites of the region that consists of touristic islands', Tsartas 1992, 522) didn't have a similar infrastructure in place, nor such vigorous catalysts for tourism development. Inhabitants of such satellite islands were able to observe events on these other islands and had time to evaluate the effects and consequences of tourism on their neighbours. Some independent traveller tourists came to these satellite islands and as adaptations were made to them, so, slowly, a more regular incursion of more standard types of tourist began, although package tourism did not develop to the extent that it did on tourist-dependent islands. (Tsartas gives Serifos as an example of this type of satellite tourism development.) On these islands, the traditions and individual characteristics of the islands, important to their residents and migrants, were equally salient as attractions, given the kind of tourist who was interested in visiting them (i.e. not the Ravers and Sun-Seekers).

Still other islands (Tsartas's Group C) were opposed to tourism for reasons which Tsartas describes as a combination of 'religious feeling and tradition' (Tsartas 1992, 523) and also (and probably a more important factor) had other revenue-producing economic specialities (marble quarrying, agriculture). They were able to restrict tourism to the kinds and degree they wanted: for example by making sure that nudists were arrested, or by not giving licences to discos or tavernas with juke-boxes, etc. It has been reported that some influential (Greek) summer visitors were able to prevent the kind of tourism which they disliked, or rather, to prevent licences being granted to those kinds of enterprise, such as late-night bars and discos, which interfered with their own enjoyment of places towards which they had proprietorial feelings. It could be argued that on islands of this type, while tradition was something which return migrants and the influential summer visitors might be conscious of, to the islanders themselves, it was their way of life, and not something which could be envisaged as a commodity.

REVIVAL AND REINVENTION OF TRADITION ON GREEK ISLANDS

Migrants Associations in large Greek cities and abroad, and return migrants to the Greek islands (as well as to other parts of Greece), have been anxious not only to improve and modernise their places of origin but also to ensure that certain of their features are preserved. In the course of these efforts, some aspects were selected as worthy of preservation and others not. Factions within migrants associations and rivalries between families of return migrants, as well as similarly cross-cutting interest groups within members of the resident local communities, made these activities far from consensual. What counted as traditional, authentic, typically of the place, were all controversial. However, there are broad generalisations which can be made. Most Greeks would agree with nationalist categorisations of tradition as being associated with celebrations and (religious) festivals, with folksong and dance, with certain arts and crafts, with certain kinds of scenery, landscape, etc.

AN EXAMPLE FROM THE CYCLADES

I have been carrying out research on a particular Greek island since the 1960s. The island is Anafi, the most south-easterly of the Cyclades, just to the east of Santorini (see map, FIGURE 1). It is a small island, seven miles long and four miles wide at its widest point; it has one village, which was not electrified until the 1970s, a small harbour which did not have a deep-water jetty until the 1980s (the first steamer to dock there did so in September 1985), and sandy beaches all along its south coast. Construction work began on a motor road to connect the harbour, the nearest beach, and the village in 1988, and was completed in 1990. A bus was purchased from the neighbouring island of Santorini in the spring of 1993, used outside the tourist season only to transport passengers to and from the steamer once or twice a week, and more frequently during the summer months to link harbour, beach and village. At roughly the same time, a piped water system was supplied to village houses, although a sewerage system has still not been completed. The only people who visited the island in the 1960s were urban migrants on their summer holidays, a few pilgrims from neighbouring islands to the shrine of the island's patron saint, the occasional archaeologist interested in the ruins of the temple of Apollo at the eastern end of the island, and a very few tourists, both Greek and foreign. After an electricity generator was built in 1974, however, migrants began to return to the island to live there all year round; some to spend their retirement, but others to begin to cater for the very slowly developing tourist trade. These tourists were mostly back-packers, many of whom camped (illegally) on the beaches, but some of whom tried to find rooms to rent at the harbour, in the village, and at the beach nearest to the harbour where a few migrants had built themselves holiday cottages. Small boats from the neighbouring island of Santorini brought day-trippers over to experience 'a really unspoilt island', and some of these people stayed or came back later, and then passed on the name of the island to their friends and other travellers. It was almost immediately apparent that there was a greater demand for rooms than could possibly be fulfilled by the existing small stock of made-over store-houses, converted stables and custom-built accommodation, and that there was a potential clientele for a much larger number of cafés, tavernas, take-away fast-food outlets and bars than existed to cater for the islanders themselves and migrant holiday-makers. It was the returned migrants who had the capital and the know-how to invest in providing these facilities (see Kenna 1993).

Με πρωτοβουλία του συμπατριώτη μας Νεόφυτου Ρινάκη αναβίωσε το κάψιμο του Ιούδα. Ήταν μια ωραία κίνηση που έφερε στο νου μας παλιά ήθη και έθιμα του νησιού μας.

FIGURE 5 . A REVIVED OR INVENTED TRADITION? THE 'BURNING OF JUDAS' Easter 1996.

ARCHITECTURE

The local building style on Anafi is, or rather was, the barrel-vaulted shape familiar to anyone who has seen pictures of, or has visited, Santorini. While those who had migrated in the first half of the twentieth century brought back with them to the island the neo-classical style then popular in Athens, post-war Anafiot men who went to the city to work in the building trade copied on the island the modernist styles they were using in the suburban blocks of flats they were engaged in constructing (the kind of 'lagging emulation' which Friedl described in her village in mainland Greece, Friedl 1964). As well as houses for residents, the first sets of tourist rooms were built in this style, and there were no building regulations in force (as there were elsewhere, for example, on Santorini) to preserve the local traditional style (see FIGURE 2). However, some of those building rooms to rent began to copy the traditional styles of other Cycladic and Aegean Islands: what we might describe as the whitewashed cube style. It is only recently that arches and curved roofs have become part of the construction of rooms to rent and other buildings designed for tourist use. Rather than reviving the original vernacular architecture of Anafi itself, so I was told, which was impossible because 'there are no experts left who know those old skills' (see FIGURE 3), the builders were copying the style known to be used on Santorini and popular with tourists there (and now subject to laws governing the conservation of traditional architectural styles). These innovations in design and construction indicate that some Anafiots have begun to realise that the particular vernacular architecture of Santorini and Anafi, which marks them out as different within the Cyclades, has a particular appeal for tourists. In addition, Anafi has come to the attention of the Ministry of the Aegean as one of the smaller Cycladic islands which is deserving of development aid, and the earlier reluctance even to apply for building permits from the Ministry of the Environment, and the systematic flouting of building regulations, is increasingly difficult to get away with.

Partly anticipating, and partly responding to tourist demand, post-cards, a schematic map/poster, souvenir plates and cups, t-shirts, sun-hats, cigarette lighters with the name of the island on them, were ordered in Athens and brought to the island to sell (see FIGURE 4). Some of these items were subsidised by the Migrants Association (the postcards and the map) while others were ordered from suppliers by independent return-migrant entrepreneurs. As some guidebooks mentioned that the island produced thyme honey, tourists asked where they could get some. This honey is in scarce supply on the island, and mostly reserved for local people and their relatives in Athens. But to supply the tourist demand (I was told), Anafiots imported honey in bulk from elsewhere and decanted it into smaller containers to sell. The vendor was thus able to say (again, according to my informant), without falsehood: 'I put the honey in the jar myself'.

MUSEUM

It was return migrants, as well as the members of the Anafiot Migrants Association (based in Athens) and its committee, who agitated for the founding of a museum of antiquities on the island, worried that smaller pieces of carving and statuary might be stolen. Although an archaeological guard had been appointed, he could not supervise the unfenced site of a Hellenistic city on the south coast of the island as well as the ancient temple to Apollo right at its eastern end.

Issue 8 of the Migrants Association newspaper (July/ August 1986) had the headline 'there ought to be a museum on Anafi' with a photograph of a Hellenic sarcophagus located outside a chapel near the ancient city site. One of the points

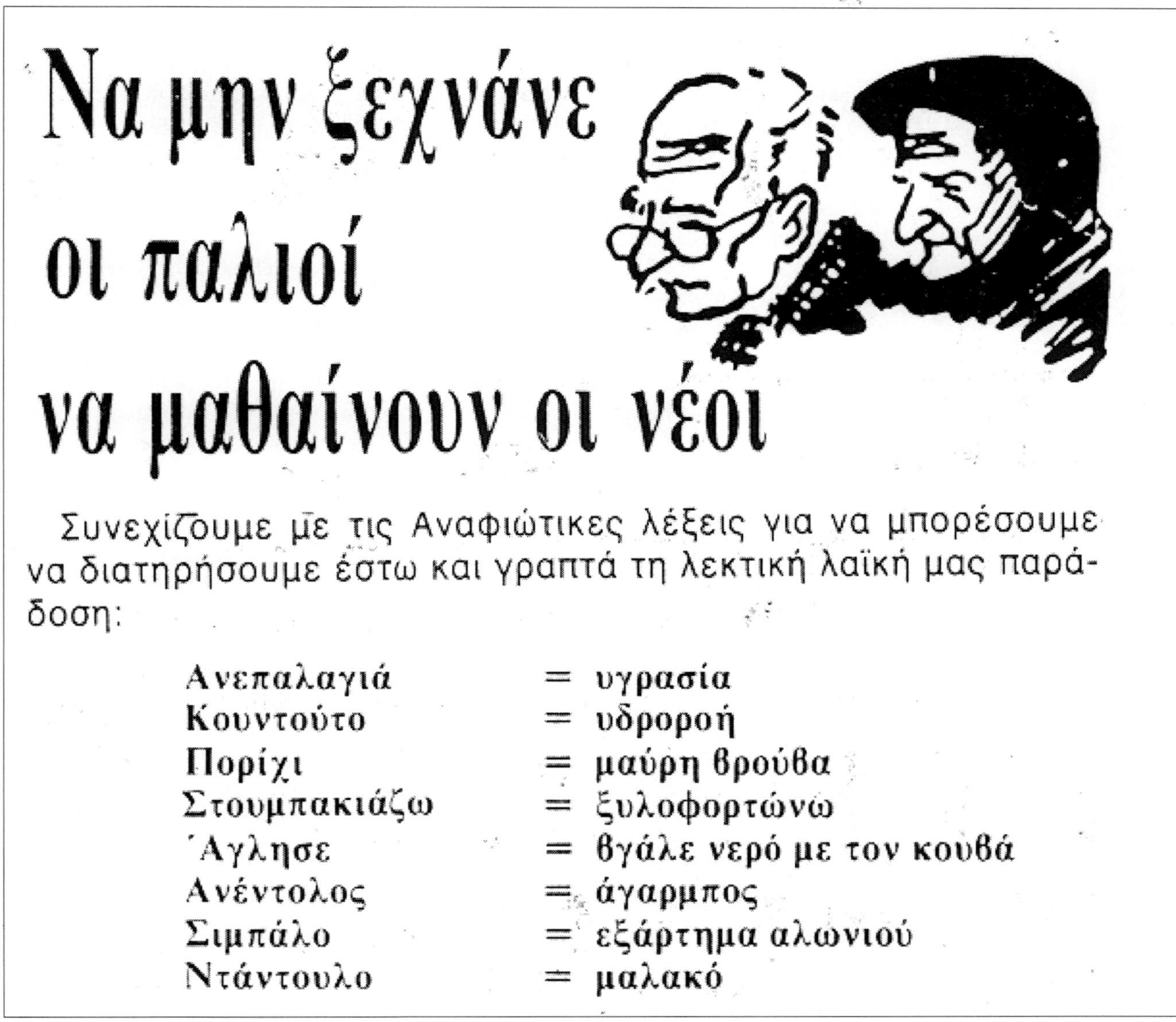

Να μην ξεχνάνε οι παλιοί να μαθαίνουν οι νέοι

Συνεχίζουμε με τις Αναφιώτικες λέξεις για να μπορέσουμε να διατηρήσουμε έστω και γραπτά τη λεκτική λαϊκή μας παράδοση:

Ανεπαλαγιά	=	υγρασία
Κουντούτο	=	υδρορoή
Πορίχι	=	μαύρη βρούβα
Στουμπακιάζω	=	ξυλοφορτώνω
Άγλησε	=	βγάλε νερό με τον κουβά
Ανέντολος	=	άγαρμπος
Σιμπάλο	=	εξάρτημα αλωνιού
Ντάντουλο	=	μαλακό

FIGURE 6. ANAFIOT DIALECT WORDS LISTED IN THE MIGRANTS ASSOCIATION NEWSPAPER 1987.

made in the article urging the Association to consider funding the building of a museum for the archaeological antiquities of the island was that it was important to emphasise that Anafi was not 'the island of exile , but above all the island of Civilisation/ Culture'. The final point made was that 'we should learn our History' as if the periods when the island was used as a place of exile were not part of its history.

Accident achieved what agitation could not. As a result of a fire (set off, according to local informants, by bee-hive fumigating equipment) which in July 1997 burnt one-third of the island and ravaged the site of a Hellenistic city, an empty house near the village council offices was, with the aid of the Ministry of the Aegean (whose Minister visited the island herself a few days after the fire) almost immediately rented and turned into a 'museum'. There moveable antiquities (mostly headless Hellenistic statues) from the ancient city could be displayed, and await eventual cleaning from fire damage.

FESTIVALS

Return migrants also established a Cultural Organisation for the preservation of local traditions and customs. A prominent member of this organisation was a return migrant who had initially made money in the building industry in Athens and now runs a business bringing supplies to the island. This man is claimed to have revived a custom of burning a stuffed effigy representing Judas (see FIGURE 5), an activity which seems in other parts of Greece to have had distinctly anti-Semitic overtones. The 'Burning of Judas', mentioned in the March/ April 1996 (issue 66) of the Migrants Association newspaper was not practised when I was carrying out fieldwork in 1966/67 and was never described to me as a traditional Anafiot practice. My impression is, therefore, that this custom has been invented, or introduced from elsewhere rather than revived, but the issue remains to be clarified.

The festival which is mentioned in those few guide books which include Anafi is that of the island's patron saint, the Virgin Mary in a particular manifestation, the Virgin of the Reed, Panayia Kalamiotissa, celebrated on 8 September (the Birth of the Virgin). Frequently the date given in guide-books for this festival is incorrect, probably because of confusion with a more widely celebrated, and peak summer, festival (the Dormition of the Virgin, 15 August). The festival on 8 September is recognised by locals and migrants to be attractive to tourists, and those who are on the island at the time are encouraged to attend, but it is not advertised or marketed as something to bring tourists to the island, rather as something for them to witness once they are there. The presence of tourists at the festival is regarded as prestigious for the island, in that they enlarge the size of the crowd and can admire the celebrations (which include dancing and a free meal using produce from the agricultural lands and livestock belonging to the monastery within which the saint's church is located) and can act as witnesses to the Anafiots' devotion to their saint. The presence of tourists also demonstrates to pilgrims from other islands that Anafi is known internationally. [Try the name of the island on any search engine (as Anafi, Anaphi, or Anaphe) and the majority of sites found will be set up by non-Greeks who came as tourists to the island. The sites mainly show their photographs and praise the island's natural beauty.]

Migrants in Athens who are not able to attend the festival on the island, partially recreate it in the city, at a chapel at the foot of the Acropolis in the area where the first Anafiot migrants used to live (the area of Anafiotika, just above the Plaka). The cult of the island patron saint, and its expression in this festival, is one of the major expressions of Anafiot identity. Indeed, although migrants and islanders are split into factions on almost every issue, any project which involves the renovation or improvement of the patron saint's shrines is wholeheartedly supported by all; though of course they argue about the details, such as whether or not the light bulbs around the door should be all one colour or variegated.

THE PRESERVATION OF LOCAL DIALECT

The Anafiot Migrants Association has, among other things, attempted to encourage the collection of dialect words through a column in its newspaper (see FIGURE 6). From 1981, for ten years, the newspaper published in twenty-six issues a regular column with a list of six to seven dialect words under the heading 'So that the old will not forget and so that the young will learn' . While other items in the newspaper occasioned letters and discussion, this column never once attracted letters offering additions, corrections, or alternative interpretations.

FOLK-DANCING AND FOOTBALL

I was interested to note that the newspaper did not suggest collecting and printing recipes for local specialities (although by the late 1980s the until then entirely male-dominated committee included a woman member). Nor did it mention compiling an account of local customs, folk-tales or folk-songs. The Association did, however, encourage those who played traditional island instruments such as the goat-skin bagpipes to perform at Association celebrations (such as the cutting of the New Year cake) and also tried to make a collection of recordings made on the island of folk-songs and music played on traditional instruments. However, people preferred to make and keep their own copies of these, particularly as the musicians were paid, and the host of the party was felt to have the rights to the music they performed. Questions of copyright did not arise then, but would probably do so now. Other Athens-based island migrants associations participated in dance and music displays and competitions at local, regional and national festivals, and in order to be able to participate in these (as well as to encourage younger migrants to become involved) the Anafiot Association formed a young persons' dance troupe to appear at these events and to perform at its own celebrations (see FIGURE 7). If there had been specifically Anafiot dances, knowledge of them had already been lost, and the dance troupe was instructed in folk dances of a more general type. Similarly, no-one seemed to have any idea what the traditional island costume had been (although I had read a description of a Anafiot woman's costume, said to have been 'entirely abandoned' in 'the last few years', Bent 1885, 91), and so, a century later, both men and women's costumes were invented along the lines of other Cycladic island dance troupes' apparel. The dance troupe seemed to be primarily of interest to young female members (teenagers still living at home, who thus had a legitimate opportunity to go out to rehearsals and to perform in public), while young men, and some older ones, were keen players in the Association's football team which played amateur matches with other similar teams all over the Athens suburbs every Sunday afternoon during the autumn and winter (see Kenna n.d.).

Interestingly enough, given that the Anafiot Association seemed to follow the lead of other Migrants Associations, they have not yet emulated those of other islands in setting up a museum of folk-life to preserve vanishing tools and practices of rural life (see Hart 1992, 82-83), a museum of the kind which might have 'living exhibits' (local people dressed in traditional costume and practising various crafts) and a gift and souvenir shop attached. Elsewhere such 'folklore and folklife' museums have proved particularly popular with foreign tourists. These displays, couched in the language of the preservation of heritage and tradition, are also ways in which return migrants can assess the extent to which they and their place of origin have progressed since the days of their parents and grandparents, and ways in which, implicitly, the extent of personal, economic and political struggles are recorded. By being 'museumised', a way of life is relegated to a past separate from the present, something to look back on with nostalgia and relief.

FIGURE 7. THE DANCE-TROUPE PERFORMS AT A DINNER-DANCE
March 1987.

Such museums can also be used to project a particular view of local history, however partisan, as Hart noted in the case of a folklore museum set up by repatriated migrants to the Peloponnesian village of Richia (Hart 1992, 84). We have already noted the view expressed in the Migrants Association newspaper that the use of Anafi as a place of exile should be counteracted by supposedly more positive images of it as an island with important ancient sites. Its past history as a place of deportation for common law criminals and political dissidents might possibly affect prospective Greek holiday-makers , but I doubt that foreign tourists' perception of it as a suitable place for a holiday would be adversely influenced by the knowledge of such a past. As far as explaining the lack of enthusiasm for setting up a museum of traditional local life, it should be noted that compared with other islands and other parts of Greece, Anafi was late receiving electricity, a deep-water jetty, running water, secondary school level education, a motor road, etc. So it is not surprising that a rural past, and the use of the island as a place of exile during the Junta, which are still within the recent memory of most middle-aged inhabitants, are regarded as not distant enough yet to look back on with nostalgia for those who were forced to leave school at the age of eleven and herd goats, or deal with the family's weekly wash in a wooden trough filled with rain water hauled up from a cistern with bucket and rope; or for those who feel that the state, and other Greeks, regard their island as a place where no one in their right mind would wish to live and therefore to be suitable as a place of punishment.

The cultivation of nostalgia which museums of folklife practise is, in Hart's view, a particularly Greek phenomenon with very specific origins, but in fact there are numerous examples of similar kinds of heritage creation for domestic and tourist consumption all over the world. Hence the tourism studies and business management courses which stress the importance of the creation of specialities for marketing purposes. Wherever they are put on, and whatever the national agenda for tourism development might be, these courses teach a common business truism: that a country should be associated with particular attractions, and regions and local areas should also cultivate their similar yet different products and use them in advertising themselves. So the graduates of these courses return to a local or national tourism office ready to search for, revive or develop a marketable local speciality, or to cultivate a particular type of niche marketing to attract particular types of tourist.

CONCLUSION

This paper has presented a case study of the revival and reinvention of tradition. The main players have been migrants and their Athens-based Migrants Association (also the key figures in the tourism development of Anafi over the past three decades), and the traditions were those of their island place of origin. The specific traditions which were resuscitated, some in Athens, others on the island, some in both places, allow them to construct a particular view of themselves and their place of origin. They have successfully distanced themselves, economically, socially and occupationally, from an agricultural background, poorly-remunerated hard physical labour and lack of opportunities for advancement, and have also linked their place of origin both to a nationalist agenda and to a regional identity which is gaining in salience as tourism focuses attention on distinctly local attractions. Many older local people and some migrants are happy to forget and forego many of the practices which are not so much traditions to them as unpleasant reminders of past rural poverty, and to gloss over those aspects of the recent past which they consider shameful to the island's reputation. In general, however, the cultivation of nostalgia has been a strongly motivating force in defining and reviving 'tradition', that is, in constructing items and occasions for the presentation and performance of Greek national, Cycladic regional, and specifically Anafiot identity for the consumption of some migrants, some islanders and most tourists.

ACKNOWLEDGEMENTS

I am grateful to Jackie Waldren for her invitation to make a paper proposal for the Deia conference, and to her and other anthropologists present at the conference for stimulating and fruitful discussions. I would also like to thank the School of Social Sciences and International Development of the University of Wales Swansea for funds to cover some of my travel costs to the conference. The research on which the main part of the paper is based was carried out in 1987/88 with a grant from the Social Affairs Committee of the Economic and Social Research Council (ref GOO 232341) which allowed me to take nine months' unsalaried leave after which University College of Swansea (as it was then) granted me three months' sabbatical leave. I am grateful to Sean Damer and Chris Stray who offered helpful comments on earlier drafts of this paper, although its final form is my own responsibility.

BIBLIOGRAPHY

APOSTOLOPOULOS, YIORGOS AND SEVIL F. SÖNMEZ,
1999. From farmers and shepherds to shopkeepers and hoteliers: constituency-differentiated experiences of endogenous tourism in the Greek island of Zakynthos. International Journal of Tourism Research, 1, 413-427.

BENT, JAMES THEODORE,
1885. The Cyclades, or, Life among the Insular Greeks. London: Longmans, Green.

COHEN, ERIC,
1974. Who is a tourist? A conceptual clarification. Sociological Review, 22, 527-555.

DAMER, SEAN,
2001. Signifying Symi: the social construction of authenticity. unpublished paper.

DUBISCH, JILL,
1977. The city as resource: migration from a Greek island village. Urban Anthropology, 6, 65-81.

FRIEDL, ERNESTINE,
1964. Lagging emulation in post-peasant society. American Anthropologist, 66 (3), 569-586.

GALANI-MOUTAFI, VASILIKI,
1993. From agriculture to tourism: property, labor, gender and kinship in a Greek island village (Part One). Journal of Modern Greek Studies, 11 (2), 241-270; From agriculture... (Part Two). Journal of Modern Greek Studies, 12 (1), 13-131.

GREENWOOD, DAVYDD,
1989. Culture by the pound: an anthropological perspective on tourism as cultural commoditization. In: Valene Smith, ed. Hosts and Guests: the Anthropology of Tourism. Philadelphia: University of Pennsylvania Press, 1989, 171-185.

HART, LAURIE KAIN,
1992. Time, Religion, and Social Experience in Rural Greece. Lanham, Maryland: Rowman & Littlefield.

HERZFELD, MICHAEL,
1982. Ours Once More: Folklore, Ideology, and the making of modern Greece. Austin: University of Texas Press.

HOBSBAWM, ERIC AND TERENCE RANGER,
1983. The Invention of Tradition. Cambridge: Cambridge University Press.

KENNA, MARGARET E.,
1990. Football and folkdancing: some concerns of a Migrants Association. Unpublished paper.

KENNA, MARGARET E.,
1993. Return migrants and tourism development: an example from the Cyclades, Journal of Modern Greek Studies, 11 (1), 60-74.

KENNA, MARGARET E.,
2001a. Greek Island Life: Fieldwork on Anafi. Amsterdam: Routledge Harwood.

KENNA, MARGARET E.,
2001b. The Social Organisation of Exile: Greek Political Detainees in the 1930s. Amsterdam: Routledge Harwood.

KOLODNY, EMILE Y.,
1974. La Population des Isles de la Grèce. 3 vols. Aix-en-Provence: Edisud.

KOUSIS, MARIA,
1989. Tourism and the family in a rural Cretan community. Annals of Tourism Research 16, 318-332.

LOUKISSAS, P. J.,

1982. Tourism's regional development impacts: a comparative analysis of the Greek Islands. Annals of Tourism Research 9, 523-543.
MACCANNELL, DEAN,
1976. The Tourist. New York: Schocken Books.
SALOUTOS, THEODORE,
1956. They remember America. Berkeley: University of California Press.
SMITH, VALENE. ED.,
1989 (revised edition). Hosts and Guests: the Anthropology of Tourism. Philadelphia: University of Pennsylvania Press.
STOTT, MARGARET,
1985. Property, labor and household economy: the transition to tourism in Mykonos, Greece. Journal of Modern Greek Studies, 3, 187-206.
SUTTON, DAVID E., 1997. Local names, foreign claims: family inheritance and national heritage on a Greek island. American Ethnologist 24, 837-852.
SUTTON, SUSAN BUCK,
1991. What is a 'village' in a nation of migrants? Journal of Modern Greek Studies 6, 187-216.
TAYLOR, JOHN P.,
2001. Authenticity and sincerity in tourism. Annals of Tourism Research 28, 7-26.
TSARTAS, PARIS,
1992. Socioeconomic impacts of tourism on two Greek isles. Annals of Tourism Research 19, 516-533.
WANG, NING,
1999. Rethinking authenticity in tourism experience. Annals of Tourism Research 26, 349-370.
WICKENS, E.,
1994. Consumption of the authentic: the hedonistic tourist in Greece. In: A.V. Seaton, ed., Tourism: the State of the Art. Chichester: Wiley, 1994, 818-824.
ZARKIA, CORNELIA,
1996. Philoxenia: receiving tourists - but not guests - on a Greek island. In: J. Boissevain, ed., Coping with Tourists: European Reactions to Mass Tourism place: publisher, 1996, 143-173.
ZENFELL, MARTHA ELLEN, ED.,
1989. Greek Islands. Hong Kong: APA Publications.
ZIPES, JACK,
1979. Breaking the magic spell: radical theories of folk and fairy tales. London: Heinemann Educational.

NOTES

(1) Kolodny uses a measure of an area of 5 square kilometres as the lower limit of what constitutes an island (1974 vol. 2, 693). He points out that some islands are or were inhabited seasonally (e.g. by shepherds, or curating staff) or by particular kinds of population (juvenile offenders and the staff of an open prison; monks; lepers) or visited on only a few days a year (a saint's chapel, for example). To complicate the definition there are normally uninhabited islands which have been used during dictatorships as prisons, like Yiaros and Makronissos.

(2) Studies comparing the originally collected and later published versions of folk-poems and folk-songs have shown that considerable bowdlerisation of both a political as well as a sexual kind took place in promoting these as part of a national cultural agenda (see Herzfeld 1982, Zipes 1979).

(3) Anafi was used as a place of exile (internal deportation) for those defined as dangers to the public from the 1920s to as recently as the Colonels' regime (see Kenna 2001b). Those considered to be public dangers included animal thieves, drug addicts, and political dissidents.

(4) The Migrants newspaper refers to this in the caption under a front-page photograph as 'old customs of our island' and uses the verb `to bring back to life, revive' in praising the initiative of a returned migrant. This same issue of the newspaper referred to the publication of a 128-page booklet for tourists to Santorini and Anafi in which 30 pages concerned Anafi. The Migrants Association had offered information to the publishers of this tourist guide.

(5) Those who were exiled during the Civil War and during the Junta period have annual commemorative trips to islands such as Makronissos which receive considerable media attention.

16

LINKS IN THE CHAIN: EVIDENCE OF SUSTAINED PREHISTORIC CONTACT AND CULTURAL INTERACTION BETWEEN THE BALEARIC ISLANDS AND CONTINENTAL EUROPE

W. WALDREN

Donald Baden-Powell Quaternary Research Centre, Pitt Rivers Museum, University of Oxford, Oxford, England.
Deia Archaeological Museum and Research Centre, Deia de Mallorca, Baleares, Spain.
EMAIL: wwaldrenconf@yahoo.com or william.waldren@linacre.ox.ac.uk

FOREWORD

As early as 1972, David Clarke in Models in Archaeology stressed the need for sound models in the interpretation of prehistory. Pointing out the heuristic nature of models, Clarke, emphasised the importance for strong empirical evidence as the foundation of any model or structure representing the distant past. At the same time, he particularly stressed the dangers of too theoretical postulations without proper data to support a hypothesis.

The current paper presents such an heuristic model. It presents the supporting empirical evidence for what is believed to have been sustained interaction in terms of cultural and commercial exchange between the Balearic Islands and Continental Mainland during most of local prehistory. In order to demonstrate this, it will use as a chronolgical framework a scheme that is in many respects different from that which has been more commonly or traditionally used. It will present as evidence a series of archaeological parallels accompanied by chronometric evidence and logical postulations to support the general hypotheses and to explain cultural and economic change as it appears alongside the traditional prehistoric sequence.

BACKGROUND

Over the last three decades, outstanding island research has been carried out in the Eastern Mediterranean by scholars and investigators such as J. Evans (1973 and 1977), J. Cherry (1984 and 1985), M. Patton (1996) and more recently by C. Broodbank (2001). Through their efforts, they have demonstrated in their geographic sectors the need, importance as well as productiveness of constructing workable models and practical frameworks by which to explain human interaction throughout prehistoric time.

Unfortunately, the Western Mediterranean islands are seriously lacking in similar focus and strategical regional design as that found in the eastern sectors. By comparison, in the Balearic Islands, there has been almost no modelling, very little systematic study in long range terms and even less explanation regarding the possible mechanisms and various kinds of interaction that may have taken place between the individual western islands as well as the reciprocal exchanges which possibly occurred between these islands and their peripheral mainland areas during prehistory.

In fact, it has been the general contention among most older and many of the younger local investigators that the Balearics and Pityussae Groups have had, compared to other islands, the rather unique geographic situation and otherwise 'privileged' status of having been particulary isolated and individual in their cultural and commercial development throughout prehistory. The author believes that in certain respects this has caused some investigators to turn in on themselves, believing in an 'island uniqueness' that becomes almost impossible to view retrospectively or as being constructive.

This sense of 'uniqueness' and, for the lack of a better word, 'insularity' has been one of the key obstacles responsible for the lack of more dedicated study and research regarding the probable relationships or other interaction that certainly transpired between these two island groups or between them and the mainland during prehistoric times. This is equally true concerning a knowledge of the possible mechanisms and processes that were responsible for what was certainly similar movements of people, here in the west, as that which took place in the eastern end of Mare Nostrom.

Although this sense of 'island uniqueness' and 'insularity' is far from a strictly Balearic phenomenon152, but one that is shared to varying degree by most islanders , it can manifest itself in different ways from island to island (see McKechnie, this volume). However, regardless of the manner in which it universally shows itself, it is in the author's opinion, one of the main stumbling blocks in coming to grips with many of the problems found in insular investigation, particularly in regard to the study of origins and contacts.

FIGURE 1. BALEARIC ARC OF IMMEDIATE INFLUENCE

In the Balearics and Pityussae island groups, the question of insularity and the belief in its importance as a matter of regional identity, especially in the minds of local investigators, are tantamount to 'trait' and 'tradition' and as such are difficult obstacles to overcome and equally difficult to work with because of the confining limits they impose on us in any quest to understand more accurately the course of prehistoric events.

Another strong influence in our thinking over the years has been the academic emphasis and research interest shown by mainland and other international researchers in the Eastern Mediterranean over that of the Western Mediterranean. Undoubtedly, a great part of this has been due to the eastern proximity to the wealth of classical, cultural presence in these areas, which has influenced and shaped academic priorities and centered them traditionally on Biblical and Classical studies, deeming these geographic areas more important and potentially fruitful than those of the western regions.

Apart from a few exceptions, all of this, until very recently, has been further complicated by the reluctance of most local investigators, despite the wealth of available evidence, chronolological and artefactual, to arrive at a wider than just localized perspective of prehistory. Todate, the main interest or focus has been to establish what seems to be an ever increasing number personal chronological frameworks and theories to explain local prehistory, where it has literally become a case of believing "according to one's own goth".

This apparent inability to arrive at a single workable chronological framework for local Prehistory, agreed on by all, has been especially perplexing since the Balearics are certainly one of the best radiocarbon documented areas in the western Mediterranean and more than equal to other areas elsewhere in Europe proper. In fact, the number of dates is increasing significantly on a yearly basis. With such a potential, one would think that this factor alone would produce some generally acceptable chronological format for the islands agreed on by a majority. Instead, there seems to be a general reluctance to use the available data and information more broadly and generally putting it to more personal use, which this author believes has resulted in a kind of insular disarray. One that has leads to further complications, not only, regarding chronology, but also others issues dealing with interpretation and methodology as well. It has seriously limited until now any significant broadening of archaeological perspective in terms of being able more productively to assimulate or correlate the existing data and information internally or for that matter further afield.

In addition to these shortcomings, until just recently, there has been an almost unprecedented lack of locally collaborated effort or interest among investigators, particularly concerning the discussion and solution of problems and questions regarding demographic movement or a better understanding of cultural and commercial exchange in this sector of the Mediterranean during prehistory. Such collaborated effort might possibly give us more detailed clues and insights into inquiries pertaining to cultural origins and a better understanding of the various mechanisms and processes behind socio-economic change and conditions, as well as give us better insight into details related to the religious organization within the islands' various groups during the different prehistoric periods and their phases. Such interrelated realtionships must have existed and influenced events, whether periodically or on a steady basis, throughout local prehistory, not only, between the individual islands, but also the Iberian mainland and other peripheral geographical areas as well. Nor for that matter have these possible relationships or events undergone better study regarding the various other sorts of interaction and communication that surely must have taken place between the smaller groups and communities. That is to say, interactions and contacts within individual sites and groups of sites in the different regions (Strategies and Use of Intersite Radiocarbon Inventories as Interpretative Tools, Waldren in preparation).

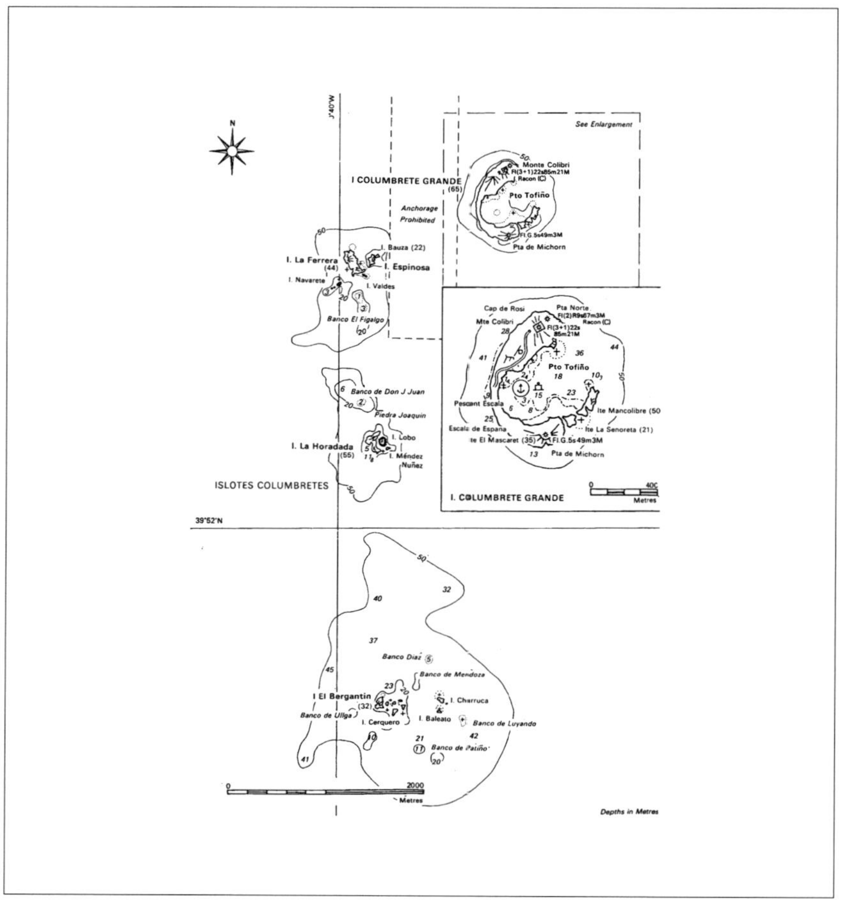

FIGURE 2. COLUMBRETE ISLANDS

In summary, the present author believes that in order better to understand insular problems both locally and in the Mediterranean as a whole, it will be necessary in the future to consider and apply more long range research objectives and strategies here in the west, before we can expect any significant improvement in the present situation. This equally applies to the use of more innovative and interdisciplinary thinking and even longer range exchange of ideas, such as has been attempted in the current international island conference, World Islands in Prehistory, where the large variety of interests in island research can be demonstrated. Thus, bringing local issues into a sharper focus and a more universal domaine than usually presented by an otherwise shorter ranged focus on localized issues that until now have been in such great interest, favour and demand.

ON THE CHRONOLOGICAL FRAMEWORK

For some time now, the use of traditional, local terminology to explain and define the Balearic archaeological sequence for the author has become increasingly unsatisfactory as well as inadequate in its chronological explanation of prehistoric events. It is hard to break with tradition and, therefore, that which has become traditionally entrenched in the local mind and vocabulary. However, for myself, I find that I have arrived at the point where it has become absolutely necessary for my understanding of local prehistory to break with the rigid framework and limited tripartite division of prehistory which we have inherited and with which we have become overly comfortable. For myself, the traditional framework has become equally self-defeating and unproductive, not only, as a tool for the interpretation of local prehistory itself, but also more importantly in its use and relationship with European Prehistory, as well as that of the rest of the Mediterranean.

In short, although it has been necessary in the past and to some degree still necessary at present to have a local framework to use internally, I have come to believe that the traditional local terminology has led over the years to more confusion and uncertainty than clarity, order and usefulness. It has fostered serious shortcomings and erected barriers leading toward and preventing any possible long-term solution or general agreement regarding a single, acceptable and workable framework for local prehistoric events. Consequently, the resulting chronological divisions and subdivisions proposed for the local prehistoric sequence by local investigators have become todate very much a personal affair, a matter of the individual picking and choosing what to use and what to believe, despite the common and acceptable use of the traditional terminology in terms of Pretalayotic, Talayotic and Post Talayotic. Each term of which, also has an individual connotation. As a result, even the local use of radiocarbon dating to support temporal hypotheses can be said to be selective rather than enlightening and productive and, except for one or two cases (eg.Lull et all 1998), the use of all the chronometric data has not been the ultimate or long-term objective in temporal interpretation, but mainly more often a means to an end.

Although the practice of personalizing the use of available data is a common one (the present author is no exception in this practice), the use of all the existing data would be preferable to selectively using dates to make one's point. In fact, it more or less boils down to the fact that each investigator is sure his sites and evidence are the most representative of the times or of the issues involved and often with the exclusion (other than as reference) of other data.

In certain respects, the present chronolgical proposal, and its arguements are part and parcel of a similar belief, although in this case the current chronological proposal can be supported by a radiocarbon inventory of over 400 dates (Waldren in press), well over half of which originate from the sites discussed and examined at present. The exception here being the fact that all the empirical evidence is supported and reinforced by the correlation of the chronometric data with diagnostic artefact evidence and comparative mainland parallels to reinforce the argument. With these

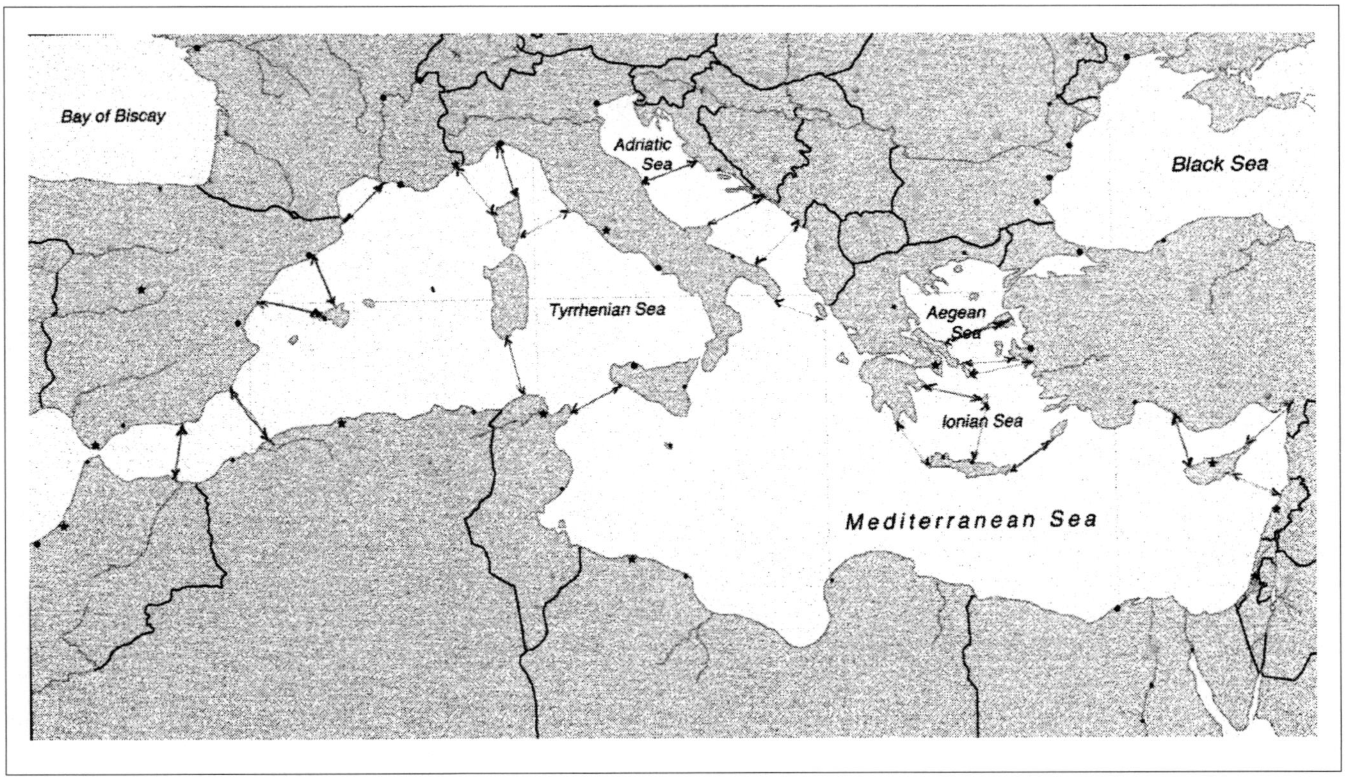

FIGURE 4. MAP OF THE MEDITERRRANEAN SHOWING 100 MILE DISTANCES

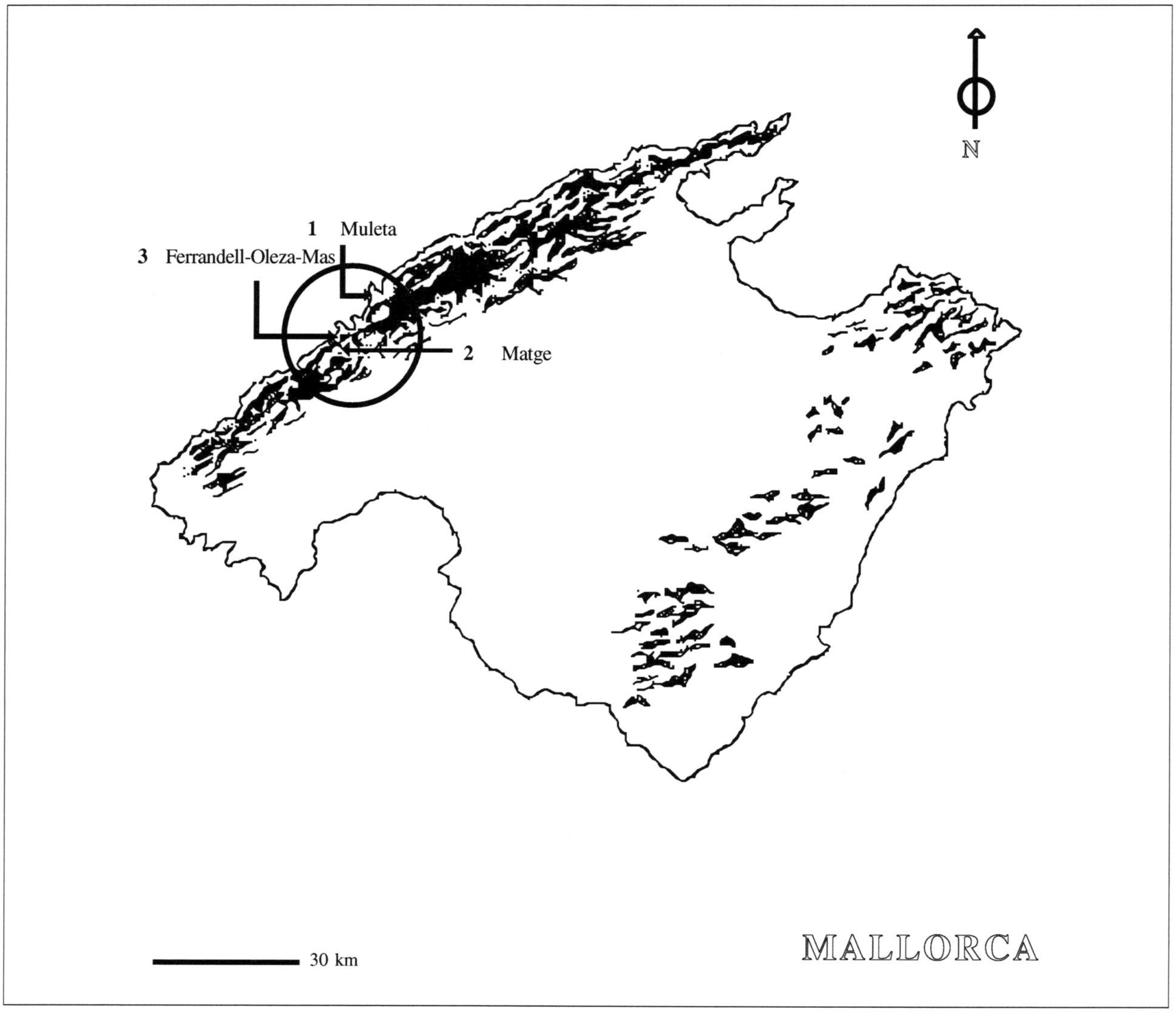

FIGURE 5. MAP OF MALLORCA SHOWING AREA OF THE SITE CATCHMENT AND NORTHERN SIERRAS

factors in mind, the author believes, the propositions outlined below are justifiable and that the use of different chronological nomenclature in the form of less restrictive and more universally acceptable termonology are equally within reason.

As a response to the situation and by way of demonstrating the empirical and other evidence for what the author believes to be strong indications of sustained contact and interaction with the mainland areas indicated, it will propose what is believed to be a more logical and, perhaps, a more practical comparative, chronological model or framework that will reflect the empirical evidence more clearly. This is to say, a framework that will use the most diagnostic evidence to build a larger more universal, external framework and wider perspective, rather than the usual detailed internal one, such as the one admirably and scholarly presented by Lull et al (this volume), where local socio-economic conditions and organization are closely examined.

In contrast, the current author has sought in the present paper basically to delocalize and disassociate his view of local environment and its internal developments from within the very restrictive limits of the traditional tripartite divisions and phasic subdivisions of the local sequence and viewing them in a different environment and perspective.

The present paper is therefore a response to this situation in that it attempts through somewhat different means to adopt a broader approach to the interpretation of Balearic prehistory than has been hitherto undertaken. One that will utilize a basically different approach to the interpretation and chronological division of local prehistory, bringing it into a European environment rather than a traditionally local one.

Although it will employ some of the conventional vocabulary and at times use and refer to a local pentapartite division of local prehistory proposed in 1982 and that has not changed in concept or in structure over the interim years, it will equally use other nomenclature in the construction of the chronological framework to follow. It will break with the traditional description and nomenclature for the local prehistoric sequence solely in terms of Pretalayotic, Talayotic and Post Talayotic periodizations. Instead, the author will try in this paper to avoid using these local, traditional terms as much as possible and use in their place continental

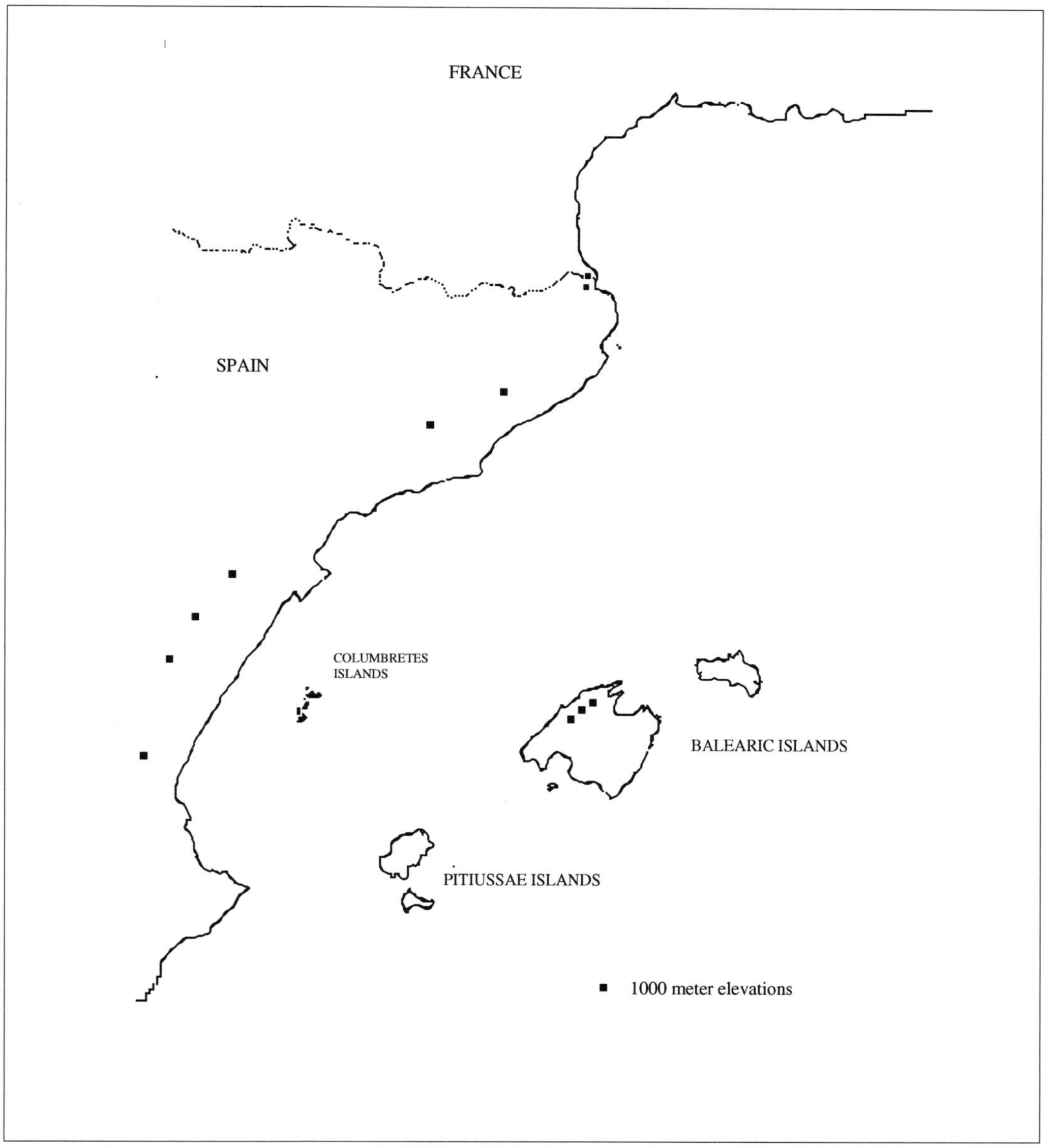

FIGURE 6. AREAS WITH MORE THAN 1000 m ELEVATIONS
Showing areas with over 1000 m elevations and location of the Columbrete Islands.

equivalents, such as Neolithic, Chalcolithic, Bronze and Iron Ages to describe the times.

It will use the generally acceptable local nomenclature only where absolutely necessary as reference in brackets, alongside the continental prehistoric periodization for the Balearic Islands, and graphically illustrate the strong 'parallelisms' that exists between the islands and mainland in terms of chronometric data and comparable artefact evidence that further supports the more theoretical propositions. Thus, perhaps, we shall arrive if only briefly at a broader and more detailed, as well as logical, perspective of prehistoric events as they transpired in the Balearics, as well as elsewhere, and in so doing, arrive at a framework that is more closely related to and simulates that of mainland Europe, and not a view seen only in the usual local, insular contexts and environment of the traditional scheme. In the final analysis, it might possibly provide us with a more productive and generally more informative overview and insight into what at present appears to lie just below the surface of consciousness and in this way possibly reveal what is closer to actual prehistoric occurrences than normally seen from the local perspective.

LOCAL CHRONOLOGICAL SCHEME

The following section is a review of the local Chronological Pentapartite Division of Balearic Prehistory introduced by the author in 1982 in which the traditional nomenclature and terminology are used. It is included here for no other reason than to familiarise those readers for whom the local prehistoric sequence is not familiar or for those that wish to equip themselves with a background and preparation for the break or divorsement the author is postualting with the more continental chronological framework. At the sametime, it may help to clarify some of the present chronological contentions, as well as give some detail and idea of the general archaeological atmosphere that has motivated the changes proposed in the perspective of local prehistory. Additionally, it present the more traditional framework for interpretative puproses and reference.

THE PENTAPARTITE SEQUENCE

EARLY SETTLEMENT PERIOD

This Early Settlement Period delineation, or Meso/Neolithic Period (c. 5600 calBC- 3000 calBC), was first used by the author in his Pentapartite Division of Balearic Prehistory (Waldren 1982). The logic behind this division was to define a period prior to the first open-air architectural settlement and to accommodate the earliest dates of human occupation of the Balearics. It conforms roughly with the period known as the 'culture of the caves' and is used with the understanding that cave dwelling and the utilization of rock shelters existed well into the Late Iron Age, until the islands' Classical Colonization, in 123 BC and not exclusively during any one period. It is further delineated as a period that is so far 'aceramic' and 'pre-domestic animal', although the endemic Myotragus balearicus coexisted with the island's first settlers as was the subsistence resource of the first settlers.

Before this period, a Presettlement Period, or Paleo/ Mesolithic Period, was suggested by the author to define the islands' fossil record (6-8 MYA to c. 5800 calBC) before the arrival of humans and deals extensively and exclusively with the arrival, evolution, survival and extinction of the early endemic fuana, Myotragus balearicus. While the author still adheres basically to this division of local prehistory, recent evidence has enlarged on the Early Settlement Period with the establishment of a series of dates well into the III-IV millennium calBC and potentially earlier evidence (see New Coals on Old Fires, this volume and Waldren, in preparation).

PRETALAYOTIC PERIOD

The term, Pretalayotic, first introduced into the archaeological vocabulary by the Spanish prehistorian, Colominas Roca during the first part of the last century and known as the culture of the caves. Initially, it was designed to describe the culture of the caves and to delineate the chronological period before the introduction of local megalithic (Cyclopian) Bronze Age Talayotic architectural building and organised, open-air settlement.

Since the early 1970s, with the discovery of the Olezian Chalcolithic Old Settlement and the advent and use of radiocarbon dating (Waldren 1982), this nomenclature of the culture of the caves is no longer correct or applicable in its former sense. In fact, the understanding of the Pretalayotic period (circa 3000calBC to circa 1300 calBC) is being extended in many areas. Its proposed sub-division into three phases (NECP, Neolithic Early Ceramic Phase, circa3000calBC - 2500calBC) and (EBP Early Beaker Phase, c. 2500 calBC -2000 calBC) and (LBP Late Beaker Phase, circa 2000 calBC and 1300 calBC) has been well established chronometrically as well as by artefact and architectural evidence.

In the Ferrandell-Oleza Chalcolithic Old Settlement, the principal representative of open-air settlement, there are close to 50 radiocarbon dates defining the origin, duration of occupation and eventually abondonment of the site (see Waldren, this volume).

In fact, the appearance and subsequent study of this Pretalayotic open-air settlement over the last two decades has brought about a re-assessment of the period as a whole and including the Bell Beaker question in particular (Waldren 1999). What was pointed out by Maluquer de Motes thirty years ago, when assessing the first meager evidence of Continental Bell Beaker Cultural contact found in caves during the first decades of the last century (Castillo 1925) and disreputed, along with the question of Pretalayotic open-air settlement generally, is today an unquestionable fact and not any longer a supposition.

Other areas and aspects of this period have also been enlarged upon in the last few decades, such as the discovery of dolmenic structures and the dating of these megalithic structures (Conesa Coll et al 1990; Lull et al 2000; Waldren 2001) where about 30 dates currently exist, c. 2200 calBC to circa 1500calBC. In one of these sites, the Olezian Chalcolithic Old Settlement, there appears to be a close relationship between the settlement and four of these dolmenic structures, making possible estimates of population for the settlement, as well as stimulating a rethinking concerning the function of these structures. In the Olezian dolmenic structures, here is some indication that they may have been used for living as well as burial.

IN RETROSPECT

An excerpt by J. Maluquer de Motes translated from his preface to the VI Symposium Internacional de Prehistoria Peninsular held in 1972 and published by the Institute of Archaeology and Prehistory, University of Barcelona, Barcelona, Spain summed up the situation at the time:

The discovery by José Colominas of an incised pottery fragment at the Cova dels Bous in Felanitx Mallorca, had been a unique find of its kind, and for that reason it was widely discussed and even refuted for a long time. Research carried out in the last years has provided not only the confirmation of the existence of an abundant and widespread incised decorated pottery horizon, but also its close relationship with the Bell Beaker Culture, although one with a personality of its own. Unfortunately, most of the finds come from burials, a factor which makes the evaluation of this cultural development difficult, especially when dealing with the assessment of its effects on the earlier coloniser population of the islands.

There is still a large chronological gap for which no evidence is available as yet. It will be necessary to direct our research towards the discovery of open-air and other habitational sites that could provide levels where the Incised Pottery Horizon is well documented, that is to say, levels with a chronology dating from around 2000 BC.

En los últimos años, aquella tímida aportación, por otra parte tan discustida, del fragmento inciso de la cueva dels

Bous, de Felanitx, hallado por José Colominas, se ha confirmado la realidad de la existencia no ya de cerámica incisa en abundancia, sino incluso con evidentes relaciones con el vaso campaniforme, bien que con una acusada personalidad local. Pero la major parte de esas cerámicas corresponden a hallazgos sepulcrales que no estadio de poblamiento isleño, y de modo concreto cuál sea su relaciones con la mas primativa población balear.

Queda un amplio espacio cronológico sin documentar. Será necessario extremar la búsqueda de yacimientos de habitación en los que puedan documentarse bien niveles que correspondan a esa cerámica incisa, es decir, que puedan situarse chronológicamente alrededor del 2000 a.C.

Over the last three decades, since those prophetic statements were made, the extensive use of radiocarbon dating and more systematical excavation have well fulfilled those requirements for proof of early open-air settlement. Today, there are over 100 radiocarbon dates supporting Continental Bell Beaker Cultural presence in association with local Pretalayotic settlement (TABLE 1) (also see Waldren, this paper) .

In the interim years since 1972, radiocarbon dating and field work has provided us with a better knowledge of the age and duration of the subsequent Talayotic (Bronze Age) as well as the ensuing Post Talayotic (Iron Age) periods. Part of that knowledge is the recognition that the construction of these later local megalithic architectural elements for which the local mother culture is named are neither the oldest nor first form of local prehistoric architectural achievement, as it was once believed to be.

Apart from increased and more methodically conducted excavations during these last decades, our understanding of the sequence of events during these later periods has been made potentially enlarged by an ever-growing inventory of radiocarbon dates, that have made it possible to delineated and defined key landmarks within the local chronological framework generally. So, making it further possible to compare them confidently with Mainland European evidence in most areas. With the evidence that exists today, with close to 400 radiocarbon dates for Balearic prehistory, we should have a much broader view of the chronological order of prehistoric events than we had say 20 years ago. Although when examined more closely, we really do not have a clearer or firmer grasp on events, basically because of the way in which the data has been used and is continuing to be used. Such an increased accumulation of data and information in the form of empirical evidence over the last two decades should have further broadened our prospective allowing us to relate that information farther afield. It should have permited us, by now, to compare and relate that data and evidence in a manner other than to the rather narrow insular perspective with a limited understanding of local prehistory, we have had until now.

With the present level of evidence, we should have been able to reinforced a strong local chronological framework (had there been one) and made it possible to differentiate, each of the local chronological periods by changes in burial customs, artifacts and technology and to find chronometric correlations and parallels that can be compared confidently with those of continental nomenclature: Neolithic, Chalcolithic, Bronze and Iron Ages.

In turn, these characteristics should clearly demonstrate comparative and corresponding social, economic, political and religious similarities, present in mainland cultures. For example, as well as in the local Chalcolithic period with the International Bell Beaker culture, the Talayotic Bronze Age and Post Talayotic Iron Age cultures can be seen to reflect strong influences within the mainland Bronze and Iron Age cultures of the European Hallstatt and La Tene. Such parallels can clearly be seen in the artefacts illustrated and radiocarbon evidence (TABLE 1-4), especially, those found in the deep vertical as well as horizontal stratigraphies of archaeological sites such as the rock shelter of Son Matge and the Ferrandell-Oleza-Mas Prehistoric Settlement Complex (Waldren 1996).

TALAYOTIC AND POST TALAYOTIC PERIODS

The diagnostic characteristic of this local period is the building of the prehistoric architectural elements for which the local mother culture get its name, a word derived from an arabic word signifying 'sentinal', hence ,'watch tower', the believed function of these identifying cyclopean structures and named by Colominas Roca and Cartaillac in the first decades of the last century. As pointed out below, these structures once believed to be the oldest local form of architectural expression, along with an assemblage of other large stone buildings, including Taula Sanctuaries, Navetas and other secular and sacred structures clearly, among other changes and differences, mark the Talayotic Period. All, of which to the author strongly indicate Mainland, but other possible origins toward the end of the period as well.

As stated by the author in his doctoral thesis (Waldren 1982), the local Talayotic culture contains many elements that are comparable to the European Halstatt Culture and Celtic Urn Field cultures. Furthermore, these comparisons are not only be supported by artefact parallels, as shown in the figures, but also by radiocarbon dating (TABLES 1-4). This is further manifested by the change of burial customs in the use of purification fires in the internment of the dead during the Talayotic period (circa 1000 calBC to 800 calBC) (discussed below) and a further change to inhumation in quicklime during most of the Post Talayotic period (circa 700 calBC to 123 BC), a burial method unique to the Balearics during the Celtic La Tene (also briefly discussed below).

It is generally thought and the author believes rightly so that a change in burial customs is a firm indication of the introduction and presence of another culture. Indications of cultural change can also occur with the introduction of such products as metals or other exotic commodities, such as amber, ivory and glass. As an example, the common circulation of bronze (especially tin bronze) and its subsequent and frequent appearance in burials during the Talayotic period is one of these indication. The common use of iron, lead and glass faience in Post Talayotic quicklime inhumations is another indication of possible cultural change and European and not eastern hertitage (Waldren 1996).

It is not only possible to associate these burial methods

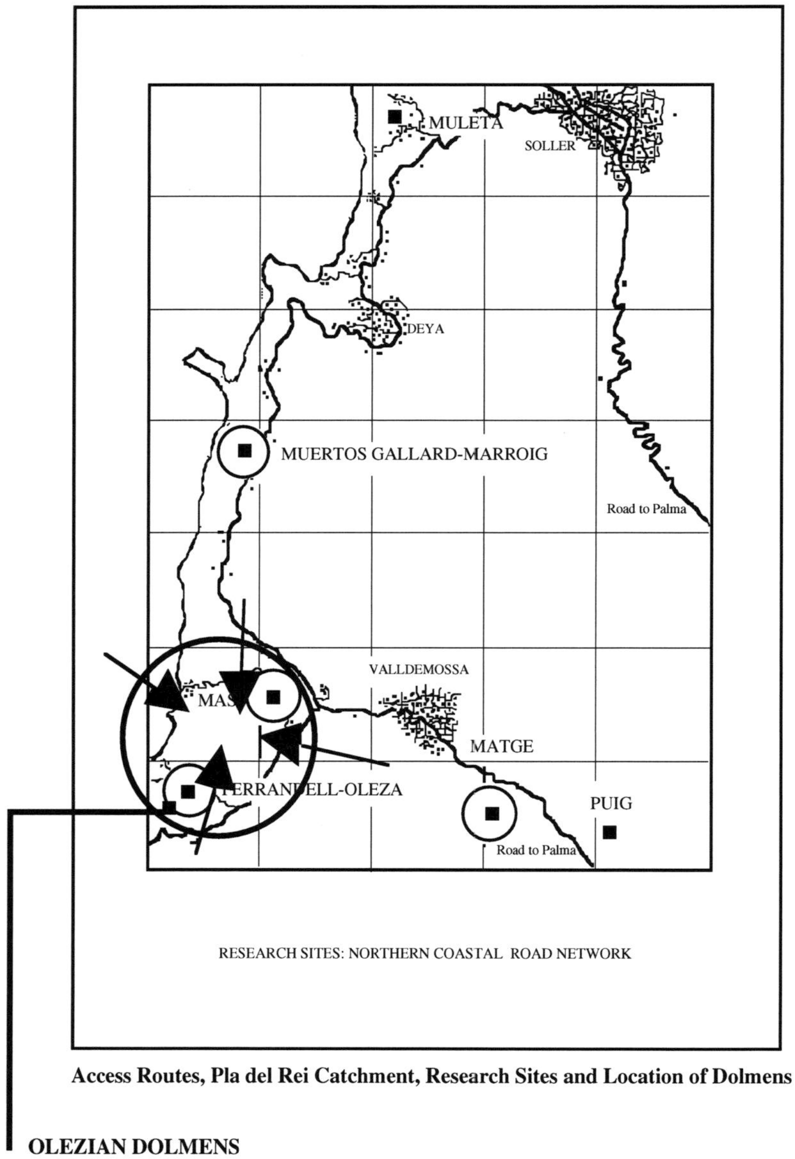

FIGURE 7. MAP OF MALLORCA'S NORTHERN COAST SHOWING RESEARCH SITES AND CATCHMENT AREA

and artifacts with those of the European Halstatt and La Tene, but the artefact evidence as well.

INTRODUCTION: THE QUESTION OF CONTACT

The following list of hypotheses and arguments will serve as both an introduction and conclusion to this paper. It will do this by presenting new and old empirical evidence in what is believed to be a slightly different way than normally. It will demonstrate, as Braudel pointed out as early as 1949, that the geographical maritime distances between the Balearic Islands and these regions have throughout prehistoric times never presented any great navigational obstacle, as believed by many. The evidence presented here strongly suggests that prolonged and relatively easy avenues of access in the form of maritime routes and commercial ties and other contact existed from the Vth millennium BC until the colonisation of the Balearic Islands in the second century BC. This is shown by presenting archaeological and cultural parallels in the form of artefact (FIGURES 5-14) and chronmetric evidence (TABLES 1-4).

On the basis of the empirical evidence supporting these hypotheses, it is possible to assemble the following list of contentions and other factors:

(**1**) This paper argues that, during most if not all of prehistory, there was a high frequency of both cultural and commercial contact and exchange that took place between the Balearic Islands, Iberian mainland and the south of France (FIGURE 1).

(**2**) It propounds that geographical and geological factors, such as proximity and visual orientation, are serious aspects to be considered as having been prime factors in the initial occupation of the islands, as well as in any subsequent contact.

(**3**) The fact that there was, to some extent, even as it is today, apparently no insurmountable geographical or other physical barriers between the Balearics and mainland regions, more than strongly suggests and reinforces this hypothesis of sustained cultural and commercial interaction.

(4) Furthermore, this can be demonstrated as manifesting itself in local artefact assemblages, architecture, chronological evidence and burial customs and other practices in the sites examined and which best demonstrate these ties.

(5) This apparent mutual sustained interaction further strongly suggests and indicates the probable cultural and geographic origins and influences for what until now was considered as largely, regional insular occurrences and developments of unclarified origins.

(6) It proposes that this artefact and other evidence demonstrates a great deal of parallelism between the geographical regions suggested within the arc of immediate influence, proposed by the author (Waldren 1982).

(7) It still further suggests, as well as indicates, that while some 'unique' insular and regional developments probably did take place and played active roles in the course of local prehistory, they did not quite exist to the degree nor have the importance considered before.

(8) The author believes that these concepts and possibilities, along with others to be examined, at present, open up opportunities to revise and enlarge our thinking concerning the frequency and nature of local interaction over time and space and the strategy used by which to approach local problems and solutions of interpretation.

(9) It proposes that on the basis of these aspects, it is possible to view these topics in a more global light, rather than in the usually limited perspective, and that although regional thinking can lead, paradoxically and eventually, to a more global view, it is not a logical course of action to pursue and if pursued then, only if the regional concepts do not in their wake overlook the origins and the influences in their wider meaning.

(10) As a pentultimate consideration, it is believed, we should bear in mind, more seriously and more frequently than we do, the universal factor that almost everything culturally as well as most things commercially, despite any independent or insular development or evolution, probably originated from elsewhere in the past, much the same way as they do today and that many of the mechanisms, forces and processes that at present shape current island events, in their manner shaped them in the past.

(11) That which is finally suggested and even sought afterhere lies between these two points of view (regional and global) regarding prehistoric interpretation and could and should be both valuable and productive in the future interpretation of both local and global prehistory.

THE HYPOTHESIS

This paper will illustrate that abundant, land-fall navigation points of orientation existed between the Balearic Islands and continental mainland regions, some of which have hitherto not been considered by other investigators. This evidence best illustrates the favorable navigational conditions and access routes that existed throughout prehistory, much knowledge of which still exists among the local fishing communities and that has been handed down from generation to generation.

Other than the short 50 mile distances that separate the islands that form a set of stepping-stones northward from the Valencian penisula, one of the little considered direct sea routes and one of the shortest between the island of Mallorca and the Iberian mainland and one which ancient Balearic Mariners would have certainly been familiar, lies between the Mallorca's northwestern coasts and the Columbretes Islands. These small but certainly useable and strategically situated islands are located approximately 30 sea miles off the Spanish coast and the port of Castellon de la Plana, north of the city of Valencia (FIGURE 5). This group of islands significantly shortens the distance between the mainland and the largest of the Balearic Islands. The island's are all well-known to local fishermen as well as those on the mainland. In fact, the seas around these small islands are renown today as one of the best fishing areas in the Mediterranean, something that we must regard as also known to prehistoric fishermen. They are also the haven for large numbers of sea and migrating land birds which use these islands as feeding and staging areas .

This paper considers access route factors such as these in support of the sustained traffic believed to to have existed between the islands and the mainland in prehistoric times. The author argues a 100 mile access distance as a common factor in the movement of prehistoric as well as protohistoric cultures throughout the Mediterranean, using as an example the access route between the Greek mainland and the Aegean island of Melos, where evident trade in obsidian and deep-sea fishing, occurred as early as the VIII Millennium BC (Jacobsen 1976). There is a growing belief and increasing empirical evidence to support it that people moved quite openly at earlier dates than presupposed, not only around the Mediterranean, which has long periods of calm weather for a good part of the year, but in other less favourable conditions (see Doing Kula, Nash, this volume), where fisherman and probably hunters braved rough icy sea in canoes in what appear to be regular trips as early as the Mesolithic Period.

Another serious aspect to be considered is the fact that geographically Mallorca has, in it's northern Jurassic Sierras, mountains which are well over 1000 metres in altitude. Mountain ranges of similar heights also exist along the coastal regions of the mainland, overlooking the areas of the Balearic Islands. The heights of these peaks on both sides of the sea area in question, along with small islands like the Columbretes (FIGURE 2) would have guaranteed extended visibility from the most ancient times and hence further supports the easy access these islands must have shared with the mainland. Such visibility factors should be seriously considered and brought into perspective along with the empirical and other evidence for sustained interaction from this area of the Western Mediterranean. Still another aspect, as briefly noted above, exists in the fact that the Balearic and Pityussae Islands Groups form a series of stepping-stones leading northward out from the Valencian peninsula into the Mediterranean with much shorter than 100 mile range sea distances between each landfall and from which the individual islands can be clearly seen from one another on any clear day (FIGURE 1). In fact it is claimed that under exceptional circumstance the Spanish-French Pyrennes can be seen from Mallorca's highest peak, the Puig Mayor.

Islands, apart from critical factors like geographic size, natural resources, environment and mainland proximity, islands are, primarily, potential recipients of most living entities and that without a full consideration of these basic island characteristics and potentials, along with the concsious recognition that nearly everything on islands has had to originate from somewhere else, we can never arrive at a productive understanding of either their true nature or the different forces and agencies that control and shape them throughout time. Regarding insular studies generally, it is absolutely necessary, the author believes, not only to recognise the special identity of islands, but at the same time to come to terms with the fact, that while there is certainly a basis of commonality among all islands regarding the elements and processes shaping them, that there is also a high degree of difference among them. It is also necessary to recognise the high probability of possible cultural affinities and close commercial links with their peripheral neighbors throughout time, taking into account the probable equally the two way reciprocal nature of these contacts and influences.

The reluctance by most local insular investigators and researchers to recognise the importance in the comparison and correlation of local artefact and chronometric evidence with those of mainland Europe and, particularly, with the more immediate peripheral areas to the islands, the author believes is an obstacle that has largely complicated and flavoured our understanding and interpretations of local events, as well as one that has influenced and flavoured chronological and other interpretative facets of research generally. As pointed out earlier, part of this has been also due to the tendency, until recently, to place too much importance on the possible local origin and interpretation of evidence and, although, one cannot completely overlook local or regional characteristics and development, the fact remains, the author believes, there has been too much emphasis placed on regionality, thus, resulting in an inabilty to attain the needed wider prespective of this particular area of the western Mediterranea that one would ideally wish.

THE RESEARCH EVIDENCE

The empirical evidence discussed in this paper is the result of nearly four decades of excavation and research in caves, rock shelters, open air settlement and ritual areas, situated along the coastal regions of the Northern Jurassic Sierras of the Balearic Island of Majorca (Figure 5).

It forms an environmental unit or catchment of sites in which an unusual interaction and homogeneity are demonstratable, along with a chronological and artefact continuity, capable of suggesting cultural origins.

CHRONOMETRY:
INTERSITE RADIOCARBON INVENTORIES

Over the last 30 years, a dating strategy has been underway in the formation of a series of *intersite* radiocarbon inventories to both demonstrate and explain social, economic and religious interaction between a selected number of archaeological stations within a limited although potentially rich, varied and productive geographic *catchment* or *environmental* model. This geographical catchment consists of the series of environmentally, geographically and geologically different sites, made up of rock shelters, caves, open-air settlement and ritual areas within a 10 and 20 kilometer radius of one another. The centre of this geographical catchment is a soil rich, Pliocene intramontaine basin forming an alluvial plain, known locally as the Plain of the King (Pla del Rei) and situated at an altitude of approxiamtely 400 meters along the northern coastal region of the Jurassic Sierras of Mallorca, between the town of Valldemossa and the small coastal Port of Valldemossa. Here, 28 prehistoric archaeological sites are located in and around as well as in the nearby hills of the alluvial plain, the later of which is made up of soils eroded down from the adjacent hills and mountain ranges, which surround the flatland on four sides (Waldren, this volume). There are four major access routes into the alluvial basin or plain and the approaches to the centre of the catchment (FIGURE 7), consisting of a northern and southern access route along the coast, a western inland access route from the central plains of the island and a final one eastward, inland from the sea. Along with the wealth of the land and water resouces as an alluvial mountain basin, along with its proximity to the sea have made it an ideal subject for long term and extensive investigation and study, a research project that has been underway since 1962 and still continuing at present (Waldren, et al in press).

Several of the intersite radiocarbon inventories refered to are presented in Tables 1-4 and form part of the chronological agrument, as well as give support to the hypothetical and nomenclature changes proposed and presented here. In turn they give substance to the correlation of artefacts, their assemblages and parallells. Together these, best demonstrate the close interaction that not only existed between the various research sites, but also between these insular examples and their mainland and continental counterparts as well.

ARC OF IMMEDIATE INFLUENCE:

Within and despite its limited geographical sphere, the site environmental catchment offers an unusual study opportunity for the correlation of artefacts, architecture and chronometric dating that is ultimately comparable farther afield, making it possible to place the evidence discussed firmly within what is known as a geographical arc of immediate influence, first suggested by the author in 1982.

This geographical arc of immediate influence extends from the southern regions of Almeria, through the Valencia peninsula, Catalonia and the French southern geographic regions of Languedoc, Provence and Bouche du Rhone. It is within these geographical areas that credible archaeological parallels and cultural influences within the Balearic Islands can be clearly demonstrated and their origins traced over a large part of local prehistory. Within the space available, this paper proposes a chronological as well as geographical model and framework (see FIGURE 8), one which forms part of a much larger and more detailed study to be published elsewhere (Waldren et al, in preparation).

MATERIAL PARALLELISM:
ARTEFACTS AND CHRONOMETRIC EVIDENCE OF INTERACTION

This section itemises the different examples of empirical evidence which strongly and best support the hypothesis of extensive and sustained cultural and commercial liasons between the islands and European mainland. It examines and discusses the different forms of evidence in chronological order by periods.

MESO/NEOLITHIC EARLY SETTLEMENT PERIOD (5600 calBC-3400 calBC)

Todate, the early settlement period is an aceramic phase and is diagnosticly identified by contexts in the stratigraphy of the Rock Shelter of Matge associated with subsistence use and attempts at domestication of Myotragus balearicus, the islands' major endemic species (Waldren 1982). This is the period in most need of more investigation and research. It is mainly supported by the the materials and dating of the earliest charcoal and bone levels at the Matge shelter, dated circa 5600 calBC, circa 4200calBC and circa 3400 calBC, respectively (see Waldren et al, New Coals and Old Fires, this volume).

NEOLITHIC NECP POTTERY
(3400 calBC-2500 calBC) (FIGURE 9)

The artefact parallels are best illustrated by the oldest pottery dated for this period. The pottery forms compare favourably to similar examples found in many areas of the European Mainland, particularly the peripheral areas of Catalonian and Almerian regions of the Iberian mainland. They consist of the islands' earliest dated pottery of coarse, heavily built, leathery, sack-like forms with lifting handles and perforated lugs for suspension, carinated and troncated shapes with cord-like rims. This period at Matge has also provided us with the earliest dates, so far, for introduction of domesticated species, sheep, goat, pigs and small cattle, circa 3400 calBC, which apart from evidence of earlier human presence marks the Mainland importation of domesticated animals, giving evidence of early navigation skills and further suggesting relatively easy access routes at this period (Waldren 1982).

CHALCOLITHIC BEAKER POTTERY
(2500 calBC- 1800 calBC) (FIGURES 10-13)

Little has to be said concerning this cultural item as a great deal of evidence supporting extensive and sustained cultural and commercial interaction during this local Pretalayotic cultural period exists and has been previously discussed in great detail (WALDREN 1996). One or more of the more diagnostic items however deserve some commentary here. This is in the form of items made of elephant ivory that can be dated to this period, circa 2000 calBC-1800 calBC, along with evidence of the introduction of the first metallurgical technology, as shown by beaker decorated and undecorated crucible sherds with encrusted copper oxides still present. Other luxury items consist of such objects as loom combs and 'V' perforated buttons made of elephant ivory that strongly substanciate some form trade in this exotic material during this period (Waldren 1982) and further builds up the case for sustained interaction between the Balearics and Iberian mainland to the south, otherwise this rare product would not be present in the islands at this early date. Other examples of Balearic Beaker parallelism can be seen in the included illustrations.

BRONZE AGE BOQUIQUE POTTERY
(1700 calBC-1400 calBC) (FIGURE 13)

Examples of this distinctive Bronze Age pottery found in the Balearic Islands is a recent development. On Mainland sites, the discovery and distribution of this distinctive pottery style has been limited until now to the Catalonian area and the Ebro Valley from Tarragona to Moncin and its spread accredited to transhumant shepherds. It's presence on Mallorca in the Ferrandell-Oleza Chalcolithic Old Settlement where it has been dated in contexts toward the end of the site's occupation is certainly conclusive evidence of exchange of some sort (marital, cultural or commercial) between the Balearics and the Mainland regions of Tarragona, within the arc of immediate influence. The pottery is distinguished by its decoration and micaceous clay fabric. Examples of both micaceous and non-micaceous samples exist (Mallorca at precent has no known clays containing mica) and strongly indicates transportation of the samples to the islands as a commercial item, gift or heirloom or sign of another form of alliance. Its presence on the island leaves little doubt of interaction or exchange of some sort during this period (WALDREN, in PRESS).

BRONZE AGE FRANCO-CANTABRIAN POTTERY SIGNS (1400 calBC- 800 calBC) (FIGURES 14-15)

Another indication of cultural contact and interaction during the early-late Bronze Age between the Balearics and Iberian mainland, as well as perhaps farther afield eastward, lies in the presence of Franco-Cantabrian pottery markings (FIGURE 14) (Waldren 1982), found on funerary pottery of the local early Talayotic period, circa 1400 calBC-1200 calBC. These markings first recorded by Schmidt (1903) and later by Leroi-Gourham (1967) are considered by scholars, because of their and wide and recurrent distribution throughout the entire Mediterranean, as evidence of some form of early linguistic abstraction and cultural interaction.

On pottery these symbols only appear in one place on the pot shoulder and because of this location, they are not considered decorations, as decorations would presumably occur all around a pot. The individual symbols or markings consist of incisions, punctates, digital navel-like impressions or sculptural breast-like protrusions. Schmidt (1903) recorded 17 different, Copper-Bronze Age signs throughout the Mediterranean of the time (FIGURE 14). Petrie (1912) records that there are no more than 50 such symbols that are present throughout the Mediterranean in ancient Mediterranean languages. The author has recorded over 50 such pottery markings on the funerary pottery from the rock shelter of Matge and these include more than and all but one of Schmidt's original inventory (Waldren 1982).

These pottery signs and their presence during a known period, circa 1400 calBC and 800 calBC on Mallorca (most of the European continental Bronze Age in the Balearic Islands and earlier in some parts) is further evidence of sustained cultural and economic contact and interaction between mainland areas as well as other areas of the Mediterranean during this period. This is further emphasised in a striking parallel from the south of France seen in Figure 15. Here the evidence implies cultural contact and perhaps other relationships not only with this specific area of the mainland, but as well as with the more northern Hallstatt and La Tene cultures, filtering down the Rhone Valley into the Mediterranean, through the Bouche du Rhone.

IRON AGE POTTERY AND IRON ARTEFACTS
(1000 calBC- 200 calBC) (FIGURES 17-18)

The earliest phase of this period is diagnostic and accredited to the first introduction of iron into the Balearics, circa 1000 calBC, a date that is proximal to the date iron was introduced into Northern Spain. It correlates therefore with the European Early, Middle and Late Iron Age and demonstrates great affinity to the Celtic Urn Field Invasion of Spain. The presense of Iron and its dating in the Balearics at 1000 BC strongly testifies to the speed, ease and frequency with which iron items that show such great correlation and affinity with mainland parallels (FIGURES 16-18) were capable of arriving in the islands.

Toward 654 BC, Carthaginian colonization of nearby Ibiza, began to open established trade routes and easy means of access of trade goods to the Balearics, some of which probably existed well before this time. Later in this period, there emerges strong and well established Celtic connections with the south of France and Catalunia and by 350 calBC as well as earlier these began to reap the benefits of classical eastern trade. During nearly the whole of this period, Classical, Carthaginian, Greek and Roman trade items appear in the local settlements and ritual sites in great abundance and variety. It is this period in which indisputable evidence can be found in the form of ever-increasing trade, population increase and other cultural developments, including military involvement, evidence of which is also in considerable abundance both materially and in early historical record in the accounts of Diodorus, Herodidus, Pliny and Strabo.

CHRONOLOGY :
FUNERARY PRACTICES AND BURIAL CUSTOMS

Changes in burial customs are strong indications of cultural, socio-religious and even economic conditions and other forms of identity, therefore, it is only logical that these should be excellent markers in the support of any proposed chronological framework or signal of cultural origin or change. There have been just such marked periodic changes in burial customs and methodology from one chronological period to another in the Balearic prehistoric burial sequence. In the research sites, used for the current proposal, there is datable evidence and other empirical support as to the probable order in which these occurred in the local burial sequence. At the same time, while there is specific ritual and methodology as well as precise and general formality as well as elaboration attached to human burial of any sort, these characteristics can be, equally, expedient, efficient, practical and at times economic in their execution. However, the key to their sequential order lies in large, radical changes, not only, in funerary architecture (receptacle), but mainly in the burial method (disposal and deposit), used in burial practices. For example, in the Balearic sequence, during the Pretalayotic period (circa 5000 calBC to circa 1200 calBC), primary and secondary earth burials appear to be the norm. By contrast, during the Talayotic period (c.1100 calBC to c. 800 calBC) in the majority of cases, there is a ceremonial and cleansing use of fire to dispose of the dead. Again, in contrast, during the Post Talayotic period (c.800 calBC to the Roman colonisation of the islands, there was the unique introduction of inhumation in quicklime, as well as the parallel continuation of former techniques.

In summary, the local tripartite division of prehistory is basically marked by three radical changes in funerary practices: (1) that of earth burials, either (a) primary or (b) secondary, (2) that of cremation or attempts at cremation by fire and (3) inhumation in quicklime. These differences in burial methodology alone would indicate radical cultural changes occurring, at least three times, if not more.

However, logical or even natural as this might seem at first glance, the situation is not that simple. It does not account for other factors, both dramatic and subtle. One of these involves the possible periodical arrival of new cultural influences or peoples, bringing their particular treatment of the dead, along with the introduction of new socio-economic conditions, brought about by trade and other sorts of more subtle interaction during these large chronological periods. One example of this is the use of sarcophagi, both stone or wood as in the burial cave of Baronat, Mallorca, where wooden sarcophagi of larch, a northern wood not found in the Balearics was used, circa 500-600 calBC or stone-cut children's sarcophagi in Son Olms on Mallorca circa 300- 200 calBC. All, of these burial variations as well as other details available through excavations and research should be taken into account when studying a funerary sequence as locally present in the islands. It seems to the author that the variation alone through time is a strong indication of sustained cultural interaction, especially since in the upper stages of the local chronological sequence many of these different methods appear to have coexisted and at times overlapped.

While such events can often reflect a reliable chronological order to events, other factors, such as possible changes in political climate, the emergence of elites, changes in fashion and even plain necessity and other factors are not always so clear or easily interpreted or arrived at, but must be seriously considered as perhaps having had effects on burial methods themselves, as well as on other more mundane or simple events.

Another factor related to the Balearic burial sequence is found in the the reality that the use of caves and rock shelters as burial sites in the Balearics is the most common form of cemetery throughout the whole of the prehistoric record, even continuing into the post Roman colonisation, where several different burial methods and cultures coexist. This simple factor can seriously and at times has complicated chronological assessments throughout the whole of the time sequence and therefore its existence must be perpetually kept in mind.

Then too, geography (terrain) along with other physical factors and charateristics, like the geology of the area, effect architectuyral burial design and practices. Such conditions as the lack or availablility of caves and rock shelters, shapeable stone resources, deep or shallow (eroded) top soil and other factors can be serious factors in the burial methods used as well as the funerary architecture itself (Waldren 2001).

The following rough chronometric divisions and temporal durations are proposed based on findings in the research sites. (1) Primary Earth Burials (circa 5600 calBC-3000 calBC); (2) Secondary Burials (circa 3000 calBC-1400 calBC); (3) Cremation Burials (circa 1400 calBC-700 calBC) and (4) Quicklime Burials (circa 700 calBC-100 calBC) (Waldren 1982).

(1) During the earliest part of the Early Settlement Period (Meso-Neolithic) (circa. 5600 calBC) and beginning of the local Pretalayotic period (Neo-Chalcolithic) (circa 3000 calBC), cave and rock shelter earth burials apparently were favoured, although in reality this is the least understood of the Balearic burial sequence and it is in this chronological area exactly where there is a need of more data, dedicated focus and investigation into cave burials. This period coincides with the initial settlement of the Balearics and was probably the result of accidental contact from the Mainland.

(2) During the beginning of the Neo-Chalcolithic Period (NECP-EBP of local Pretalayotic period) (c. 3000 calBC to c. 2000 calBC), earth burials in caves and rock shelters continued and secondary burials began in these locations toward the beginning of this period. During the beginning of the period perhaps slightly earlier, further arrivals of new mainland settlers brought with them not only pottery technology but also domestic animals and the construction of funerary architecture and more ceremonial and formalized burial practices. These include: reworked natural caves, rock-cut tombs, constructed dolmenic structures and fosa graves, all of which suggest and reflect discernable Mainland origins. Recent investigations have established during this period the arrival of Beaker cultural links, with their specialized burial practices and distinctive material culture, circa 2500 calBC.

Some time during the final stages of the EBP-LBP of the Chalco-Initial Bronze Age (local Middle and Late Pretalayotic) (circa 1800 calBC to 1400 calBC), disposal and burial of the dead changed from basically primary burials to that of assorted forms of predominantly secondary burials, not only, in caves but in especially prepared burial places, like dolmens, rock-cut tombs and in cases drying out places where the dead were exposed for specific periods of time to the natural elements prior to more formal secondary burial (Waldren 1982). Here, the dead were initially laid out prior to interment in an open-air situation with grave goods for ritual cleansing by insects and other scavengers, after which time the remains were placed along with grave goods in the various types of individual or family tombs.

(3) During the beginning of the Bronze Age, or local Talayotic period (EBA) (circa 1400 calBC to 1200 calBC), the arrival of still other peoples into the Balearics (Mallorca and Menorca) took place, bearing with them new technology and new burial customs. The subsequent increase in population during this period is evident in the reasonably rapid construction and distribution of complex urban settlements as well as large communal burial cemeteries in rock shelters or caves, therefore, resulting in the establishment of the first large Talayotic communities and the profuse vestiges of the cyclopian structures which we know and see today.

At this stage, the burial methods resort heavily to the use of fire, probably for ritual cleansing purposes and not bonified cremation, but certainly providing both practical and economic means for the disposal of the dead.

These ritual bone cleaning practices have a strong Celtic Hallstatt flavour and especially relate to the Urnfield cultures, circa 1400 cal BC- 1000 cal BC and reflect characteristics that are equally reflected in the grave goods and chronometric parallels of the Mainland regions of Catalunia and northern Spain, circa 1400 calB to 800 calBC. These newly arrived Bronze Age influences and contacts from the mainland are believed by the author to be responsible not only for the introduction of the Talayotic 'mother culture' but also for other social and economic as well as religious developments of the period as well as the radical changes appearing in the burial practices themselves.

While the method of construction of the Talayotic structures themselves bear great resemblance to similar structures that appear on Sardinia in the form of the Nuraghic culture, there is in the mind of the author some doubt as to whether or not the cultural influences inspiring these similar structure in the Balearics were the result of direct contact with Sardinia and hence the Nuraghic culture itself or another source, often referred to as the Sea People.

The open-sea route across the Bay of Biscay is not particulary condusive to navigation by small craft even today, much less in ancient times across the separating stormy, 360 kilometer Gulf of Lyon distance between Sardinia and closest point on Menorca. The cultural influences and contact, however, could have arrived via southern France and Catalonia in a round-about manner similar to that of Celtic influences arriving down the Rhone Valley and that evidently went along the coastal routes of the Pyrennes and Catalonia and into the Balearics by the proposed sea routes.

While primary and secondary burials were still practised in caves and rock shelters, during the whole of this time mostly in the islands' rural areas, eventually giving way to more elaborate funerary architecture of social importance, such as the especially constructed Chalcolithic dolmenic structures and tombs. These earlier family and individual burial places were to be also eventually replaced during the Talayotic Bronze Age by larger, communal, architectural burial places such as navetas (eg. Es Tudons, Menorca) and individual or closely clustered, family architectural types of sepulchers (eg. Necropolis of Son Real, Mallorca) during the Iron Age.

Such well organized, later, communal cemeteries were usually constructed near the larger Talayotic settlements and in these, later architectural types of sepulchers, the use of fire in burial became minimal, giving way to still another culture and burial method, inhumation in quicklime.

The cleansing by fire during this period may have been subsequently abandoned because one of the functions of these later, individual and family tombs, like those found in the Necropolis of Son Real and Navetas on Menorca, served dually as aeriating places and depository sites, where the decomposition process could take place under more controlled conditions and be less destructive.

Nevertheless, artefacts found in both types of burials have the same typology as more communial burial deposits, using either fire or quicklime in caves and rock shelters and therefore can be considered contemporary with the more prepared tombs. Also in these larger architectural funerary structures, such as those of the Navetas on Menorca, secondary burials seem to be the end result desired, where the bodies were first left to dry out in the aeriated upper chamber and later deposited in the lower chamber, after the bones were thoroughly cleaned. Although some later primary burials still occurred, for example on the Ille de los Porros, part of the Necropolis of Son Real, Mallorca, where individual crypt-like burials were found with iron artefacts, such as occur in warrior burials in northern Europe.

This important necropolis is now known to have served during nearly all the local Iron Age unitl the islands Roman Colonization (Hernandez-Gasch 1998) and give the impression that more than one culture is involved. This characteristic of independent tomb architecture, which shows architectural chronology in the tomb arrangement and typology spanning many hundreds of years, are highly suggestive as signs of the emergence of elites and a complex social distinction during this period.

The use of fire, both collective and individual, is carried out throughout most of the period as is the beginning of the use of quicklime from about 700 calBC, especially in the rural areas in caves and rock shelters, where its use marks differences once again and variation in the burial methods in practice at this time.

Toward the end of the Late Bronze Age, local Talayotic period (LBA-EIA) (circa 1100 calBC to 800 calBC), there appears to be a lessening not only of the use of fire but in architectural construction, but this phase is equally characterised by the first introduction of iron into the Balearics. The strange development here is found in the fact that little evidence for the age and construction of the Talayots in the lowlands of Mallorca or Menorca is older than 1000 calBC and no earlier than 1200 calBC-1100 cal BC in what the author believes are the older Talayots in the mountainous regions of Mallorca's north coast. If this turns out to be the case then the Talayotic Culture is minimally Bronze Age and predominantly Early Iron Age in origin and chronological duration.

(4) During the beginning of the Early Iron and Middle Iron Age (EIA-MIA) or Post Talayotic period (circa 1000calBC to 700 calBC), there is a distinctive changed once again in burial methods. This is demonstrated by the introduction of the unusual burial custom of inhumation in quicklime, around 700 calBC and known nowhere else in the prehistoric record, except in the Balearic Islands. This is manifested by the manufacture and the use of quicklime to dispose of the dead, a custom that while probably in essence a ritual practice, similar to the use of fire, was equally effective in disposing of the dead but exceptionally more destructive to the bodies and to any objects used as grave goods. Evidence shows that during the preparations the bodies were dismembered, probably to release body liquids for activation of the quicklime (Waldren 1982: Waldren 1996). The dismemberment, in contrast to the mostly entire burials of the Bronze-Iron Age Talayotic period, may have also had an alternative purpose.

The use of quicklime is a complex technological process of preparation that seems to have had two possible origins. It is a known fact that the Celts made and used quicklime by slaking it into the form of whitewash which they used to paint their bodies in preparation for battle. As producers of bronze and iron, one of the by-products would have been the quicklime. A second origin may have had eastern roots as lime making was known to the Phoenicians were they used to lime in the form of mortar to line the inside of their water cisterns (Stuiver and Waldren 1974; Van Strydonck and Waldren 1992).

The Carthaginian colonisation of the adjacent Pitiussse island of Ibiza in 654 BC marks the earliest historical recorded contact we have with the classical world. While there was contact and trade as well as subsequent military (mercenary) involvement in Carthaginian campaigns as early as 406 BC, there was no actual Punic colonisation in the Balearics, contrary to the belief of some local investigators (Guerrero Asuyo 1999). Actual contact with Carthaginian sources was limited to the establishment of trading posts and other market installations on several small off-shore islands of Mallorca (Guerrero Asuyo 1995).

The key to the question of funerary assessment during this later period, the author believes, lies in the fact that most of the artefacts surviving the corrosive action of the quicklime are made of iron and of Celtic La Tene typology and hence probable origin, as is the Bronze and Early Iron Age artefacts of Hallstatt typology. Whether or not these iron products arrived on the islands by the use of Carthaginian sailing vessels, hence as trade goods, is not known, although it is a strong possibility that cannot be overlooked.

All of the artefacts found in the quicklime burials were 'killed' and made unusable, by being either smashed or bent and twisted. This is also a Celtic custom believed done in order to render them forever unusable by anyone other than the owner in his life-time, and whose spirit may still have been attached after death. This also accentuates the importance and ability of being able to waste valuable material that might have otherwise been recycleable during this period. The abundance of iron objects in the graves also shows the common use and circulation of iron as a product and way of life. There is also the appearance in the artefact assemblages of the grave goods, apart from iron and bronze objects, for the first time also includes items of lead and glass fiaence.

Toward the end of the Post Talayotic period (LIA) (c. 500 calBC to the second century BC) evidence of an increasingly complex contact with Eastern Mediterranean classical cultures, beginning with the Carthaginians, Greeks and Romans is conspicuously evident. This can be seen in the presence in excavations and of a full range of classical imported items such as pottery, jewellery and other imported luxury goods in great quantities, which continues in an unbroken sequence to the Roman colonisation of the Balearic Islands in 123 BC and further to 300 AD.

Over the period from the 6th century BC to the 3rd century AD, the variety of burial methods used apparently auments in proportion with the population and cultural origins of the inhabitants. Large Punic and later Roman cemeteries appear on both groups of islands and stands as irreputable evidence of intense trade and contact, eventually leading to the disappearance of the Talayotic Mother Culture.

TECHNOLOGY:
QUESTION OF METALLURGY AND CHRONOLOGY

In any geographical region, without the physical presence of tin bronze, either as separate elements of tin and copper or as a composite import, it is impossible to postulate a Bronze Age with any certainty. The Balearics has no known tin resources such as would be necessary for Bronze production, although this usually amalgamated metal could well have been imported in the form of ingots, but as yet none have been found. Regarding copper resources, the region of this research does have at least 5 sources of copper known to the author, all along the Mallorca's northwest shoreline, near and accessible to the research sites discussed (Waldren 1999). Experimental smelting from two of these sources by the author (Waldren, in press) has produced copper ingots or modules and although the ore itself is not particulary high grade, consisting mainly of copper sulfites and oxides, it is completely usable. One of these geological exposures also contains veins of a poor quality coal in the copper bearing lode, which would seemingly faciltate copper ore removal.

The ore bearing deposit is located directly below the Ferrandell-Oleza Prehistoric Settlement Complex and easily accessible to the site. As mentioned, the grade of copper ore available is not what we can call rich, but is ample to produce moderate amounts of workable metal, certainly capable of supplying a modest Copper Age demand, subsequentally perhaps acting as a supplimentary source to later Bronze Age inhabitants of the area. It can therefore be concluded, until some tin sources are found, that tin would have had to be imported, at a later date, either in already worked form or incorporated as bronze ingot imports (see Waldren 1979)

For all intents and purposes, the local Chalcolithic (Pretalayotic) (Copper Age) extends until the arrival of tin bronze. While we cannot overlook the possibility of an intermittent stage of development in the production of bronze in the form of arsenic bronze being present as it does elsewhere, as a natural mixture of copper oxides and arsenic, it has not so far appeared in local ore sources. Altough in due course future analyses may determine its presence. Extensive study of the subject of bronze production in the Balearic Islands has done much in recent years to shed some light on Balearic metallurgical techniques (Delibes and Fernandez Miranda 1988).

On the basis of prior research, the earliest date for a true tin bronze in the Balearics originates from the rock shelter of Matge on Mallorca (Waldren 1979). It dates from levels directly associated with beaker pottery and established by mass spectrographic analyses and radiocarbon dating, at circa 1400 bc (1720 calBC). The date fits comfortably within mainland contexts. This coincides with the author's assessment for the Balearic Islands initial Bronze Age.

The analyses samples originate from a beaker workshop in the Matge rock shelter (Waldren 1979). The analyses were carried out on a the horde of copper awls discovered in the beaker workshop and represent the first of the series of metal analyses initiated by the author and P. Northover in Oxford University (Waldren 1979, appendix 1 and Waldren 1982).

Regarding other metal resources. lead ore is present not too distant from the research cites examined here. Lead artefacts are found quite frequently in funerary situations of the Post Talayotic Period, associated with quicklime inhumation in the form of elaborate open mould, stone-cast lead pectorals (Waldren and Cubi 1987b). In one site, the Prehistoric Sanctuary of Son Mas, there is evidence in the form of large lead ingots, gutter pieces from moulds, spill and scrap, lead pectoral miscasts and completed pectoral casts, indicating that the pectorals were made in the 'agora' area of the sanctuary for use in at least 3 nearby cemeteries. The local lead ore mines are found near the village of Orient, some 8 kilometers from the sanctuary and were exploited into modern times for the manufacture of lead tubing.

CONCLUSION

From the existing evidence presented here their is a solid argument supporting sustained interaction between the Balearics and the Spanish Iberian mainland and Southern France from the earliest of prehistoric record. The argument has a convincing quantity and variety, both in its chronological and artefact evidence to demonstrate in some detail the extent and origin of the cultural and economic influences and contacts involved. The subject matter and the existing prehistoric chronological data, the author believes, are deserving of more space and time than is afforded here. The final analysis and assessment of the current data and information is left up to the reader's judgement and discretion, especially, in regard to the arguments and hypotheses proposed here. The fact remains that the majority of the evidence reasonably supports the arguments and hypotheses, despite the available argumental time and space and limited geographic area, as well as number of research sites used in the propositions. The author's motives, opinions of local archaeological environment and political issues are, as they always should be personal and based on observation and long experience.

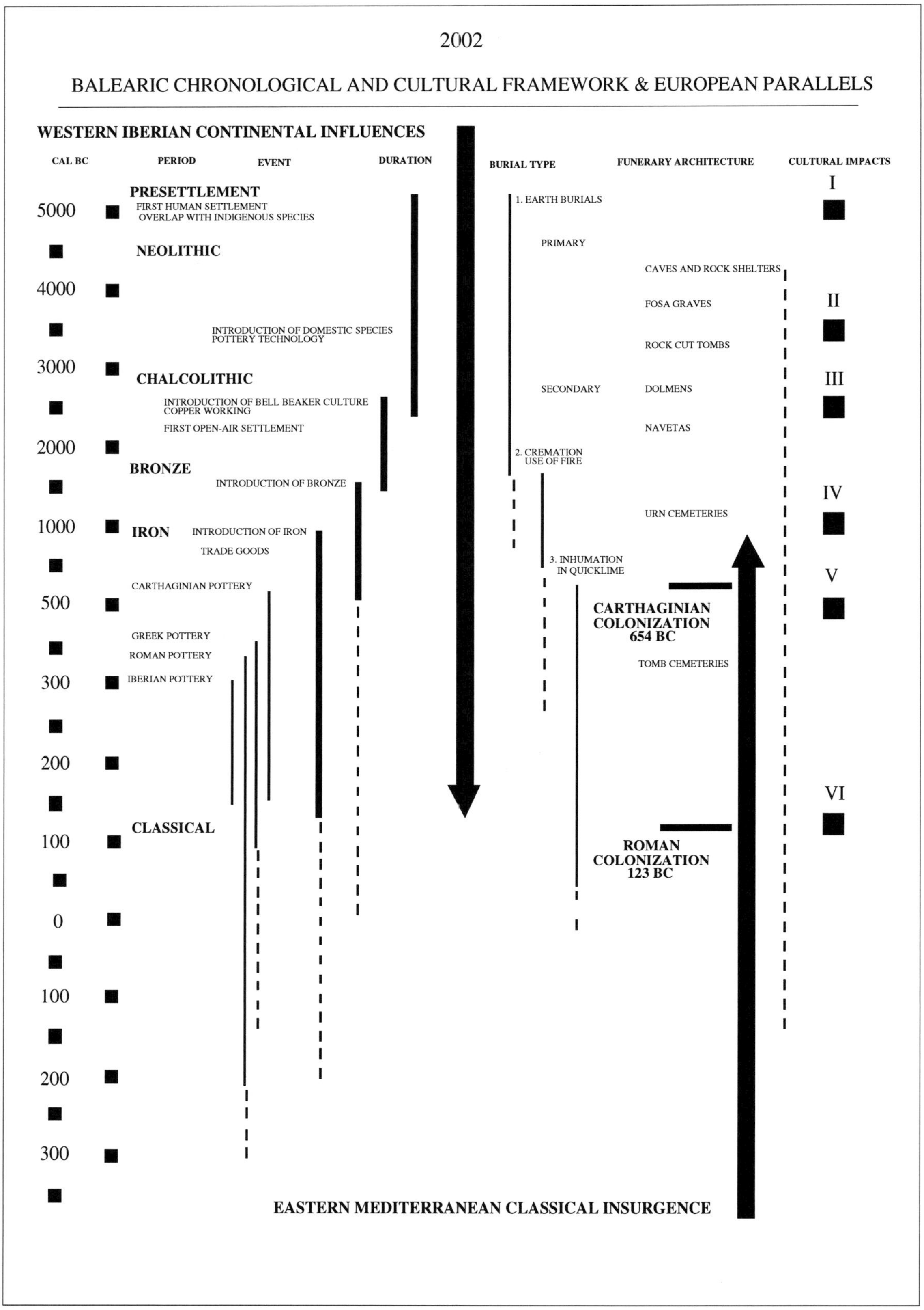

FIGURE 8. .BALEARIC CONTINENTAL CORRELATED PREHISTORIC CHRONOLOGICAL FRAMEWORK

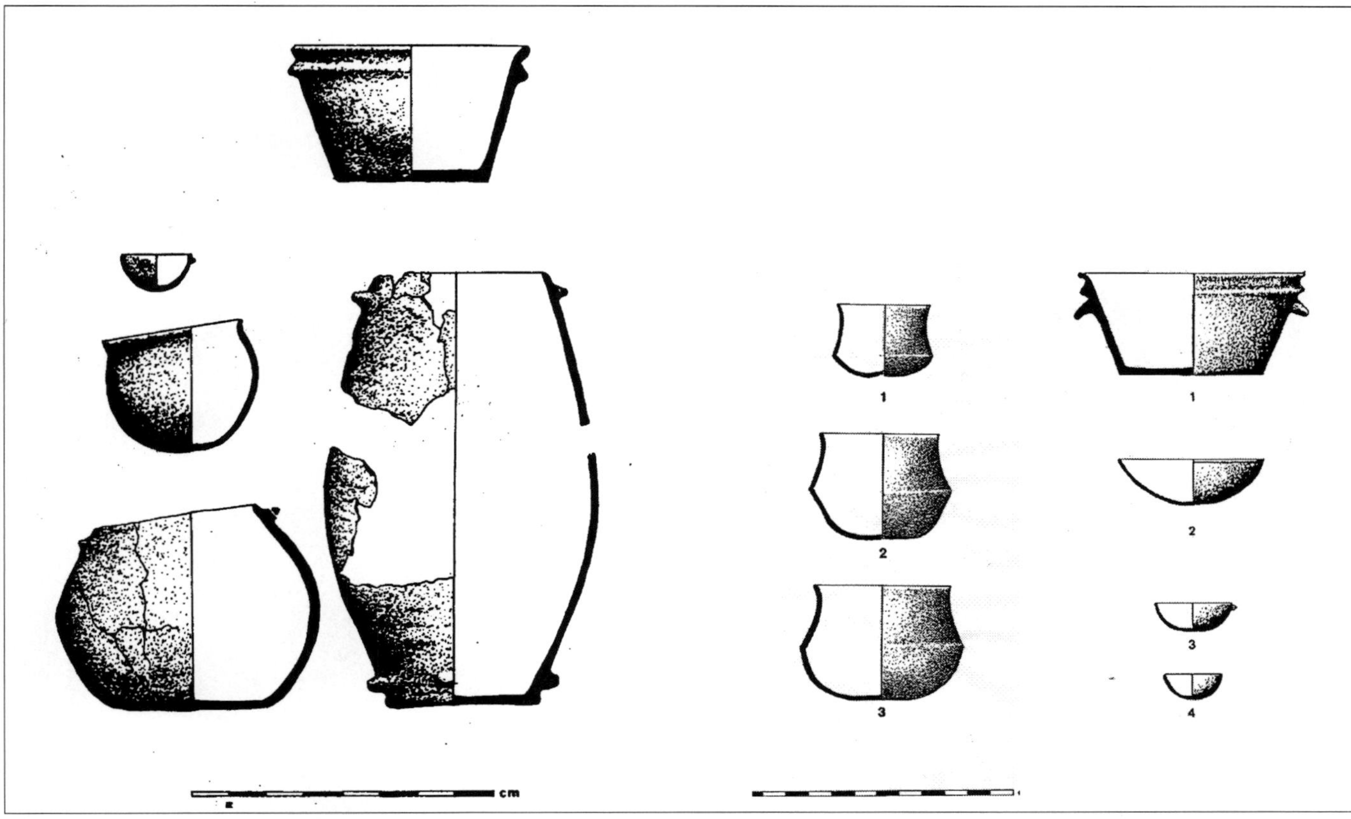

BALEARIC NEOLITHIC

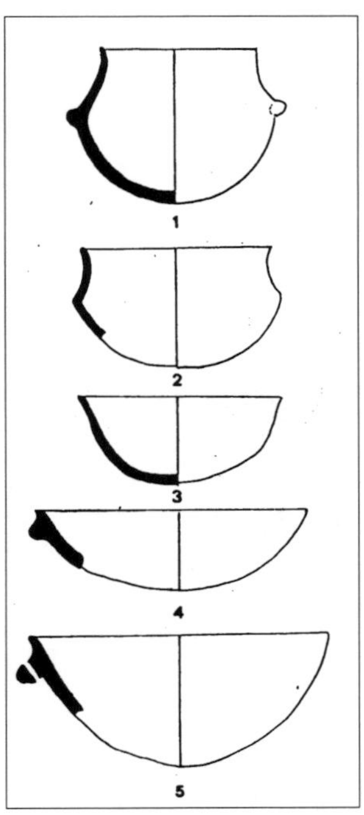

FRENCH NEOLITHIC

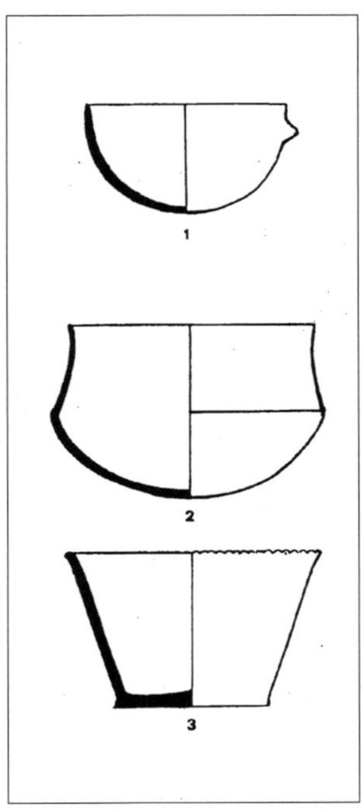

SPANISH NEOLITHIC

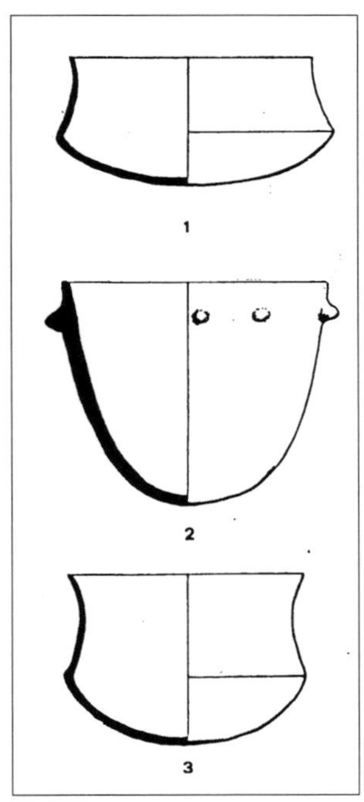

ITALIAN NEOLITHIC

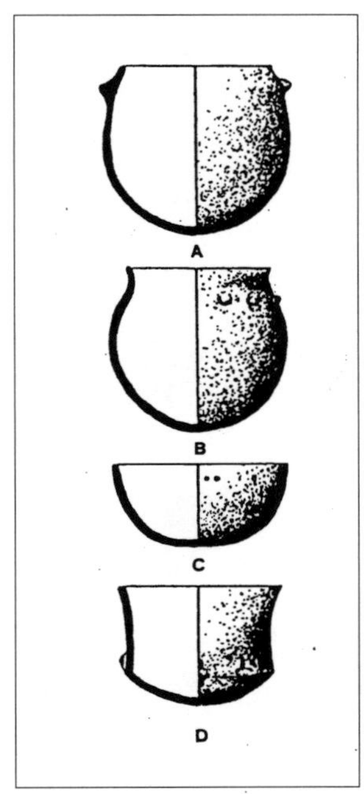

CORTAILLOD CULTURE

FIGURE 9. MAINLAND NEOLITHIC POTTERY PARALLELS: CIRCA 5000 cal BC- 3000 cal BC

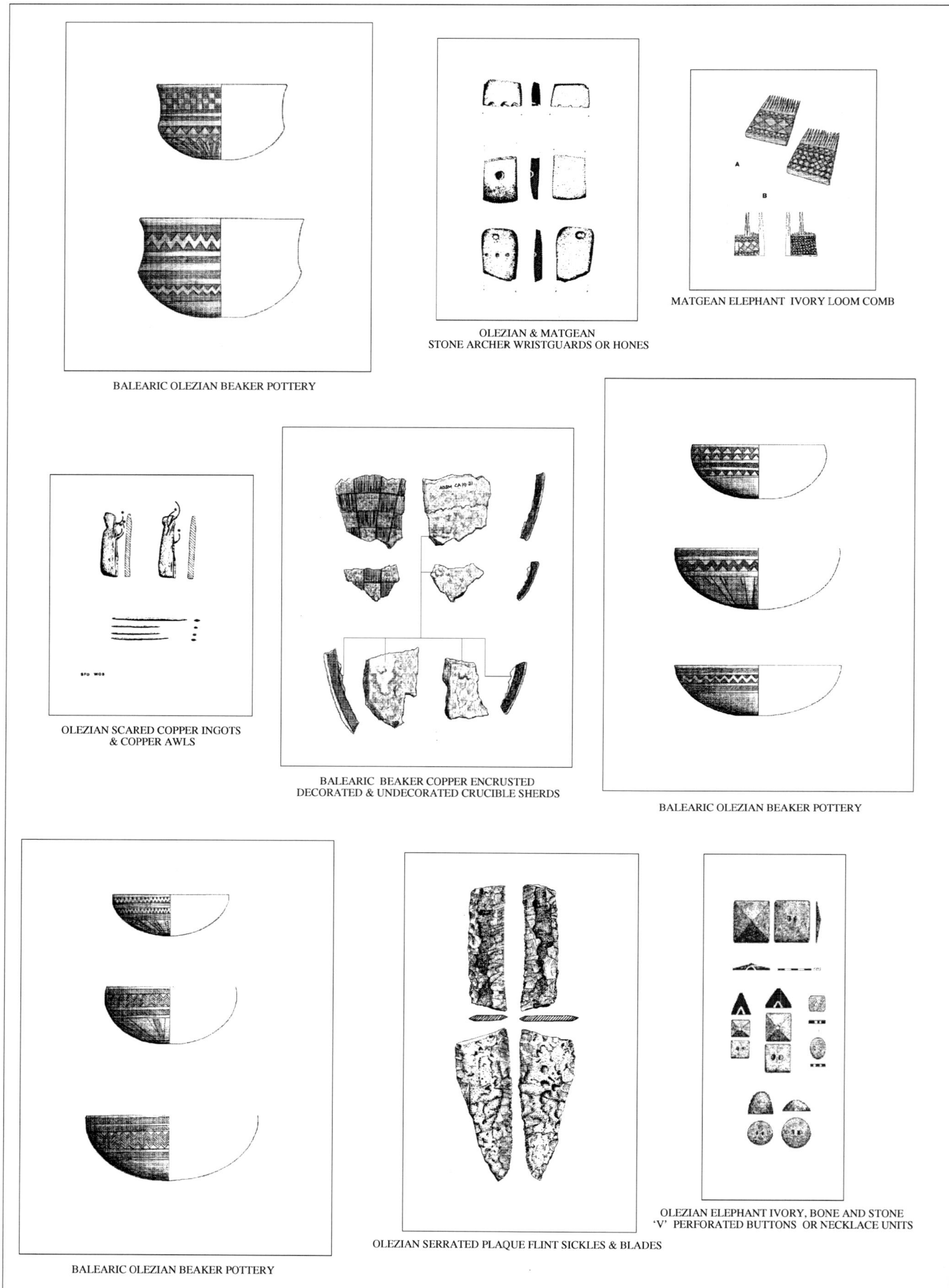

FIGURE 10. BALEARIC BEAKER ASSEMBLAGE: CIRCA 2500 cal BC to 1800 cal BC

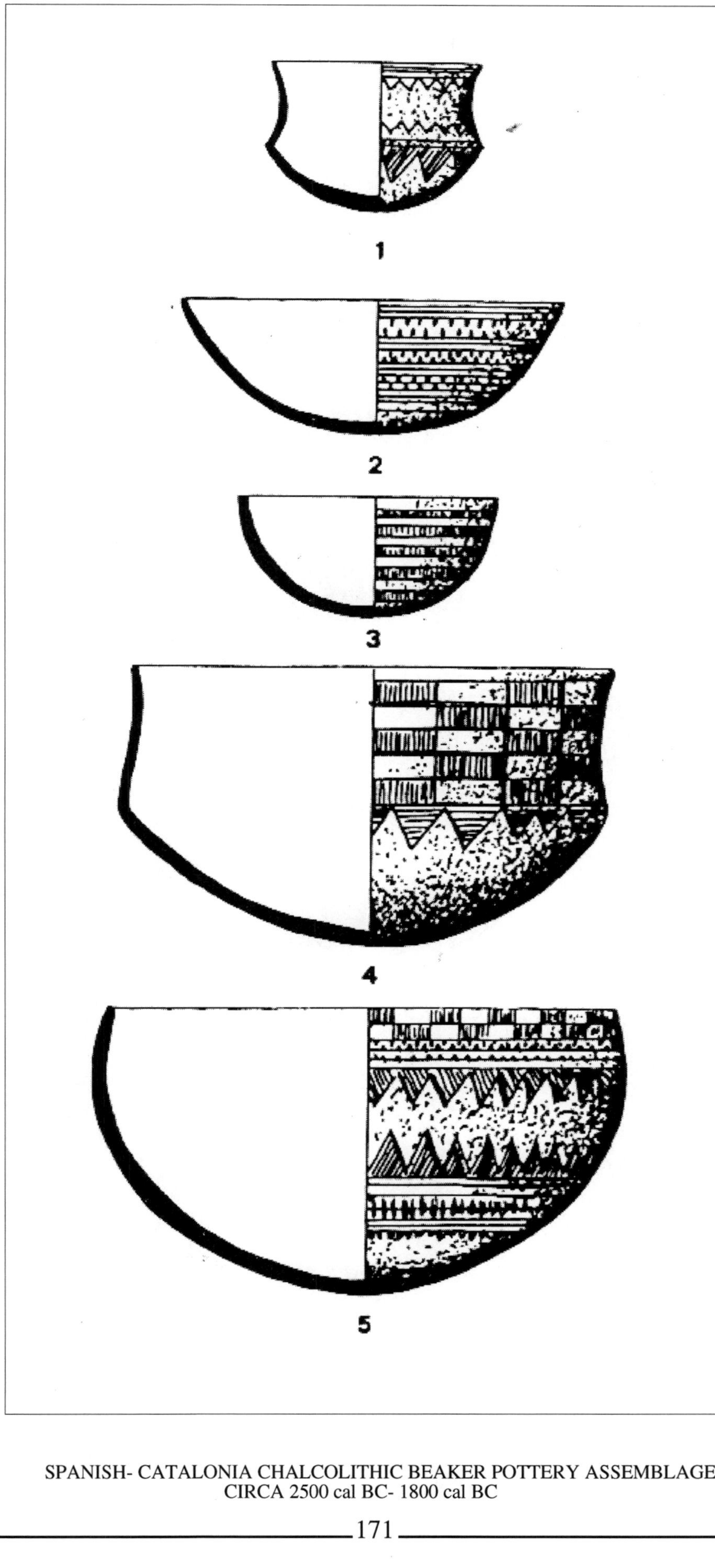

FIGURE 11. SPANISH- CATALONIA CHALCOLITHIC BEAKER POTTERY ASSEMBLAGE
CIRCA 2500 cal BC- 1800 cal BC

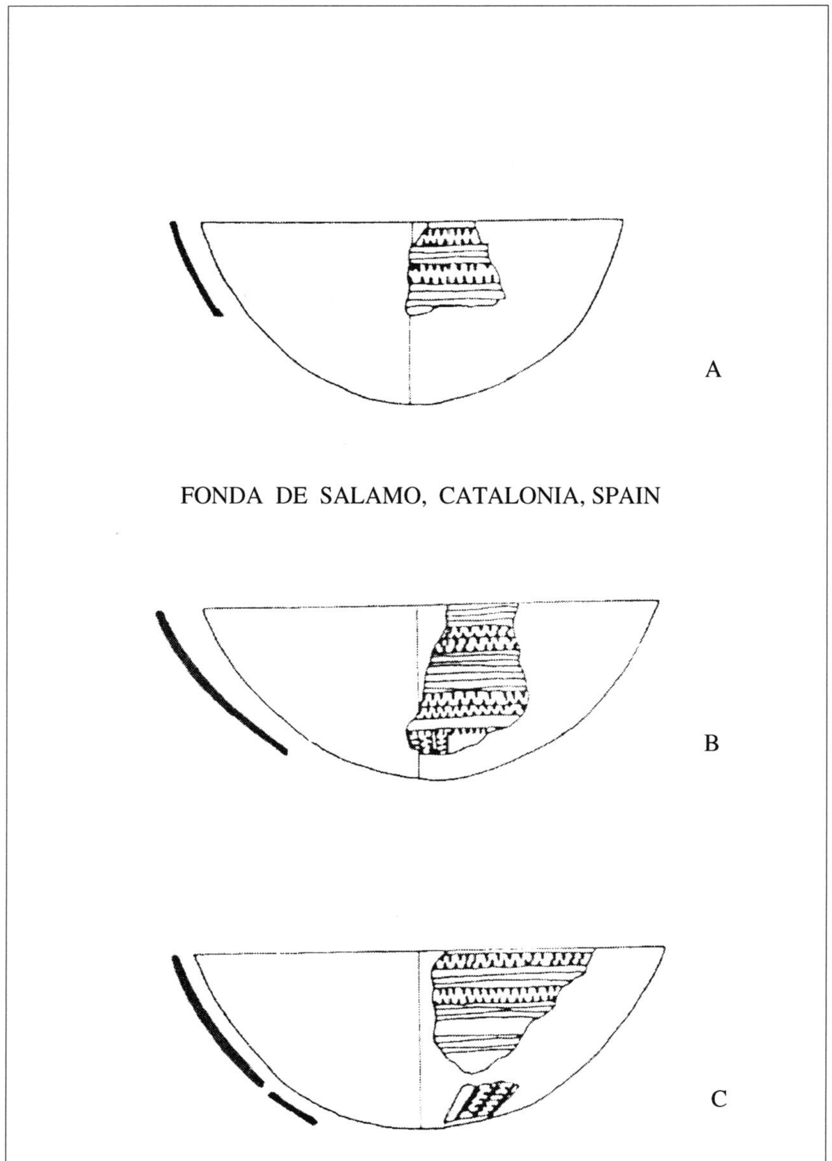

FIGURE 12. CATALONIA- MALLORCA COOPER AGE POTTERY PARALLELS: CIRCA 2100 cal BC-1800 cal BC

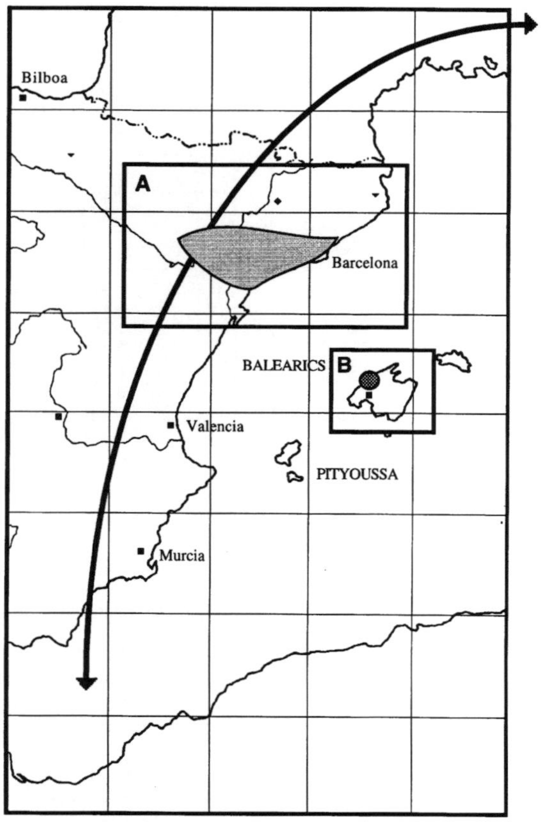

GENERAL GEOGRAPHIC SITUATION AND ARC OF IMMEDIATE INFLUENCE

MAINLAND BOQUIQUE AREA OF DISTRIBUTION B MALLORCAN FINDS

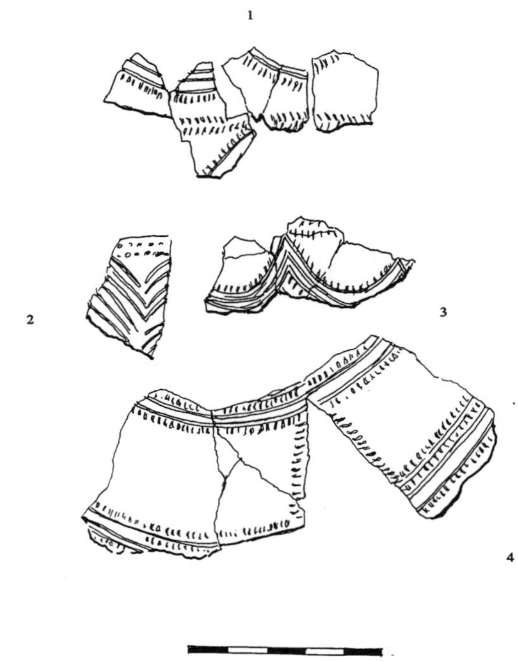

A SELECTION OF BOQUIQUE SHERDS
FROM SEVERAL MAINLAND CATALAN SITES

(1) Cova del Daniel, (2) Bobila Gonzales, (3)Corigo de Aigues Vives, (4) Cova de Porta Lloret

MAINLAND BOQUIQUE

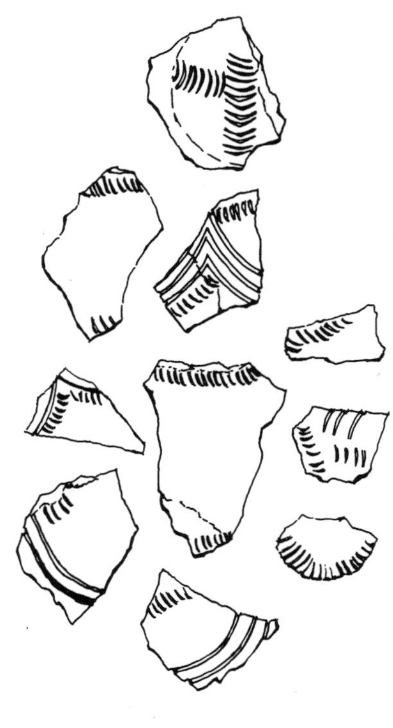

BALEARIC BOQUIQUE

FIGURE 13. BALEARIC AND MAINLAND INITIAL- EARLY BRONZE AGE BOQUIQUE PARALLELS, CIRCA 1700 cal BC-1400 cal BC

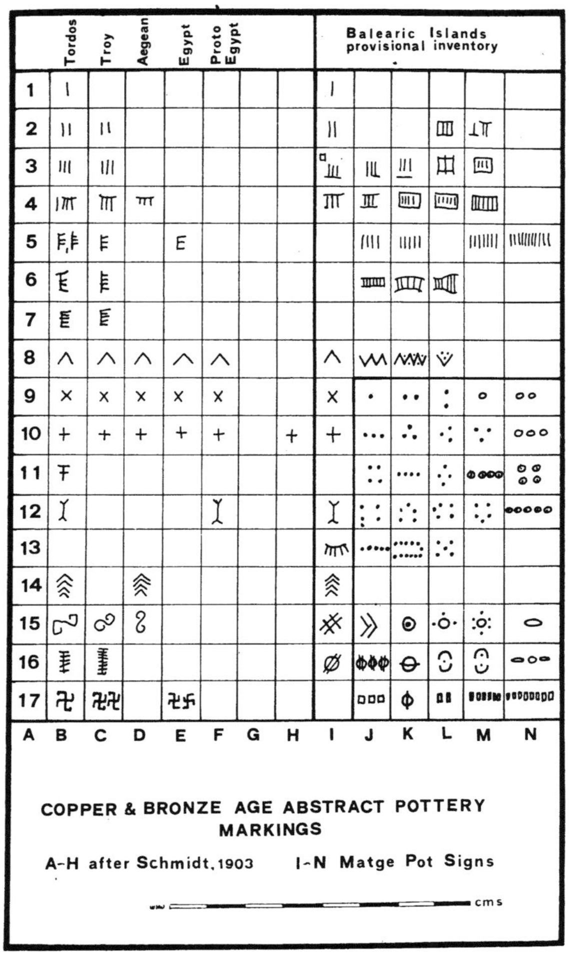

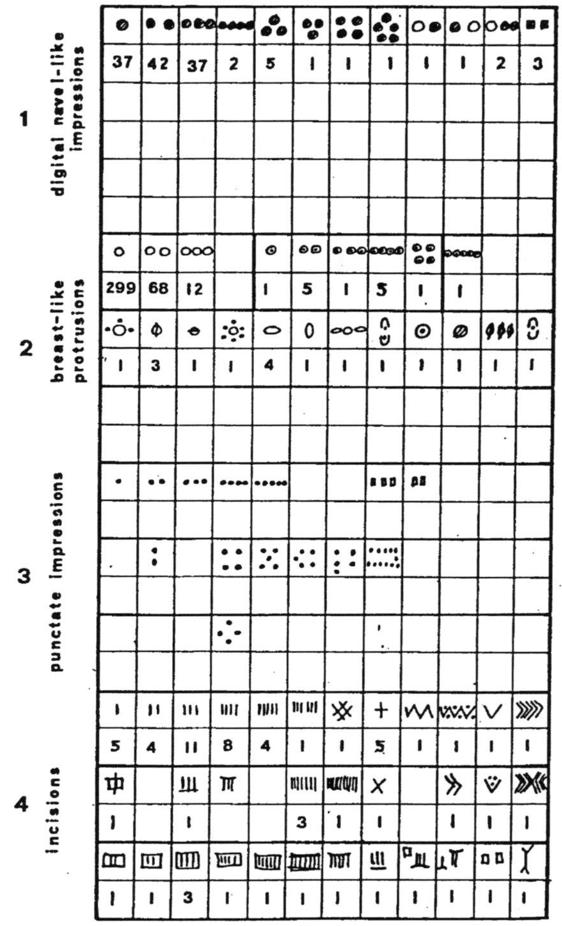

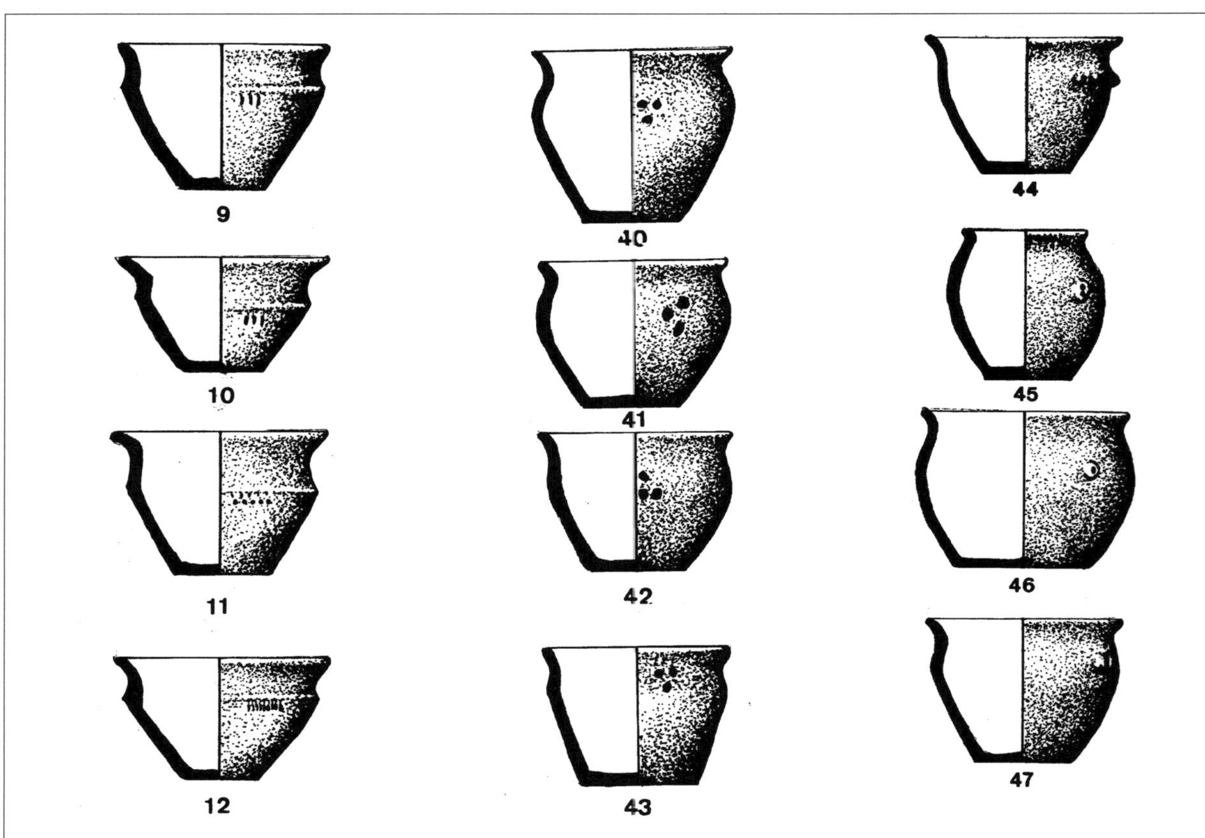

B

FIGURE 14. FRANCO-CANTABRIAN POTTERY MARKINGS: MATGE FUNERARY WARES
CIRCA 1400 calBC-1200 cal BC

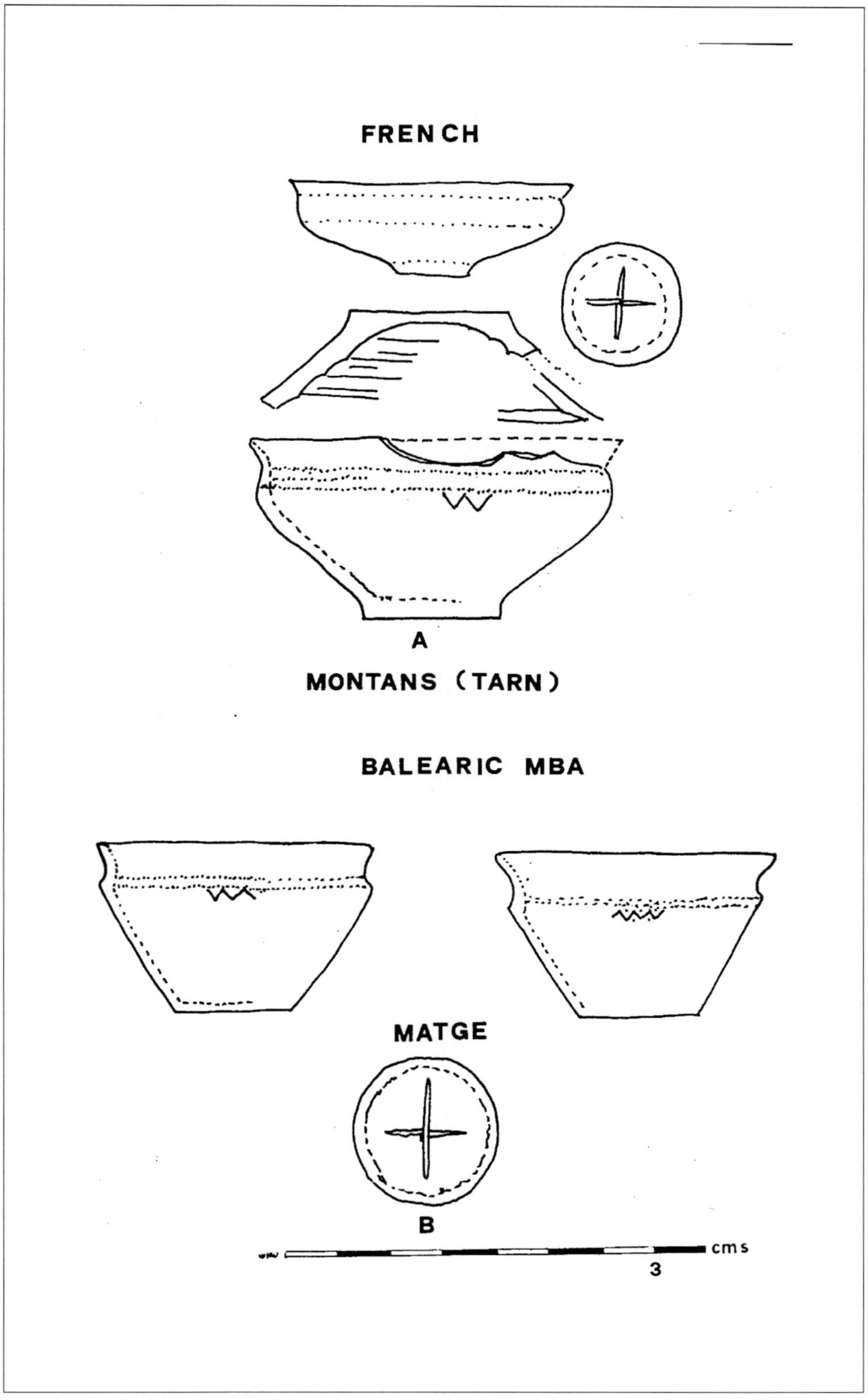

I B

FIGURE 15. BALEARIC AND FRENCH MIDDLE LATE BRONZE AGE PARALLELS : MATGE FUNERARY WARES CIRCA 1400 calBC-1200 cal BC

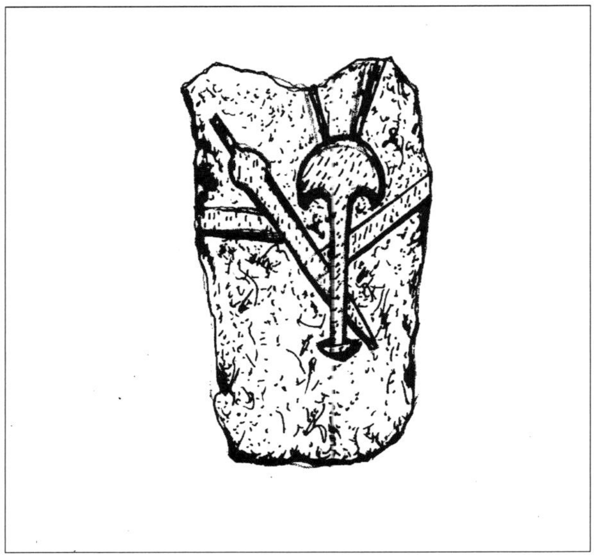

CARVED CIST SLAB, SETUBAL, PORTUGAL A

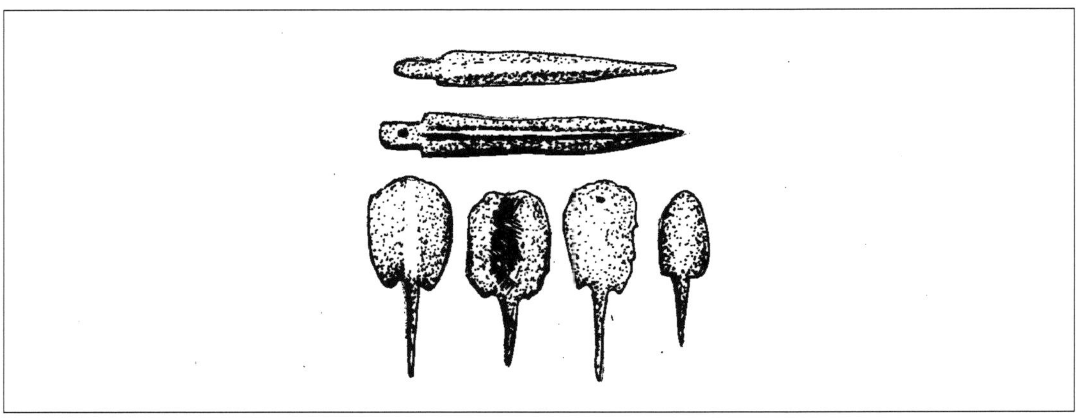

HUERTO DE ARRIBA, BURGOS, SPAIN
CIRCA 1200-1000 CAL BC B

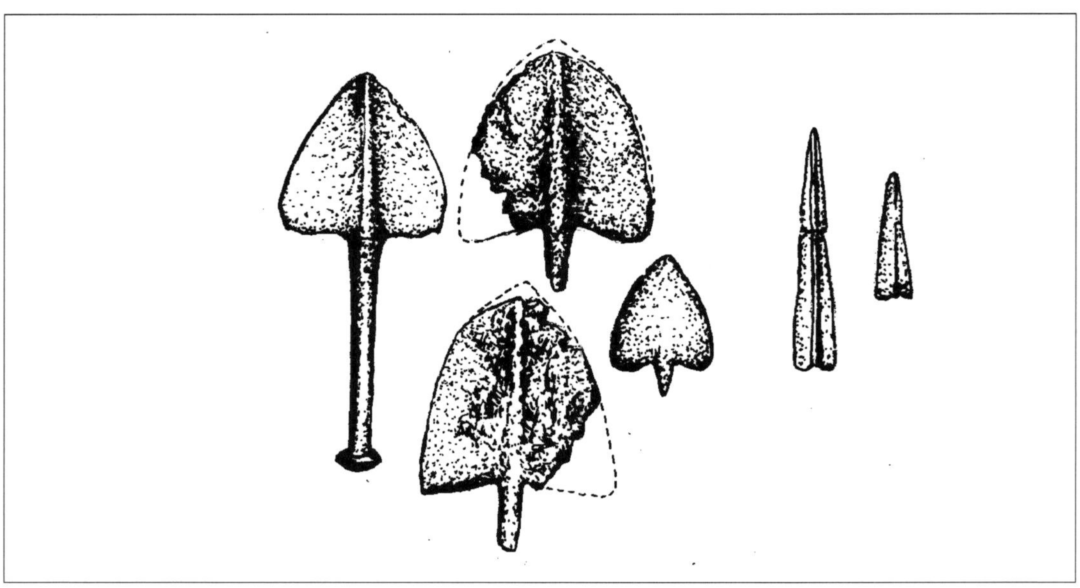

BALEARIC LATE BRONZE AGE RAZORS AND BLADES C

FIGURE 16. BALEARIC AND MAINLAND IBERIAN BRONZE AGE PARALLELS:
1400 calBC- 1000 cal BC

BALEARIC LATE IRON AGE (POTERIE GROSSIERE)

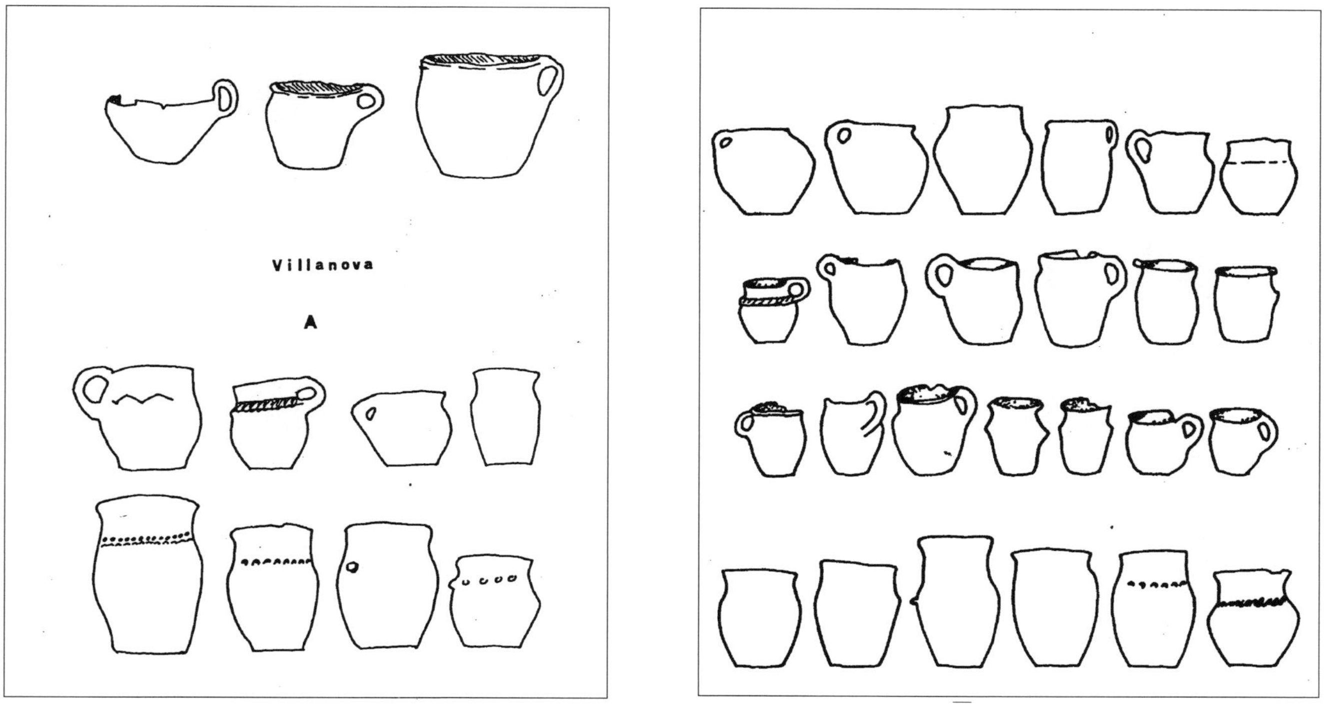

SOUTHERN FRENCH LATE IRON AGE (POTERIE GROSSIERE) (ENSERUNE & VILLANOVA)

FIGURE 17. SOUTHERN FRENCH-BALEARIC LATE IRON AGE POTTERY PARALLELS: CIRCA 400 calBC- 300 cal BC

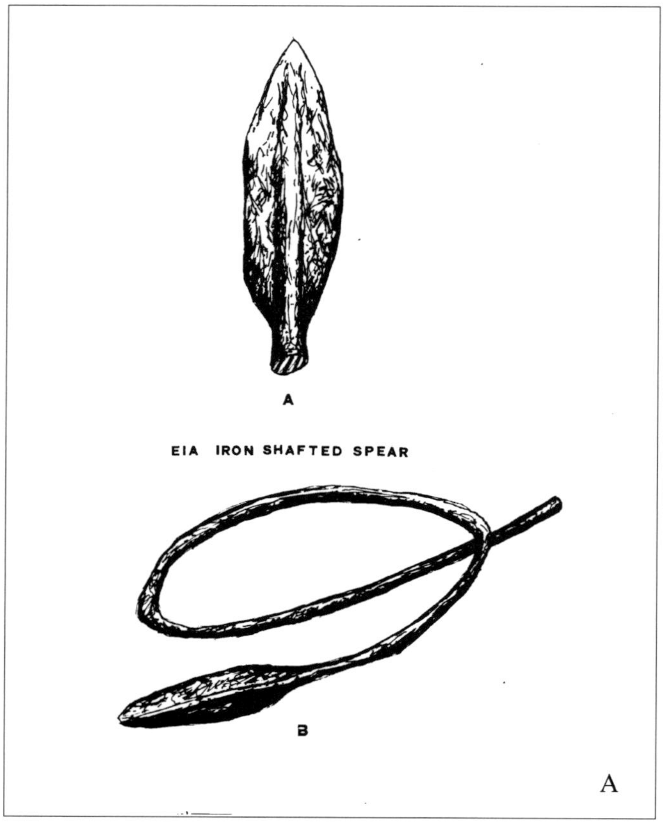

BALEARIC 'KILLED' IRON AGE CELTIC LANCE (MATGE)

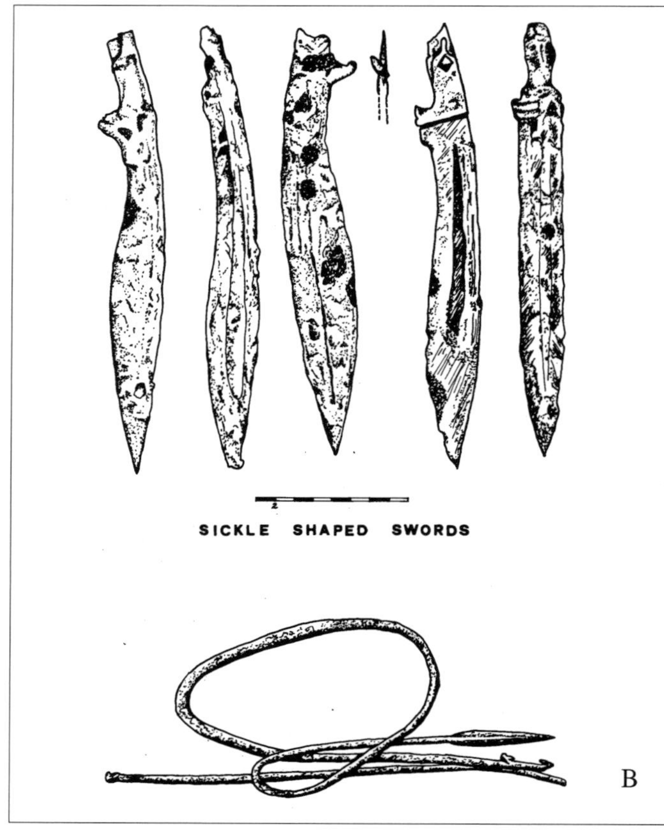

MAINLAND KILLED IRON CELTIC LANCE (BURGOS)

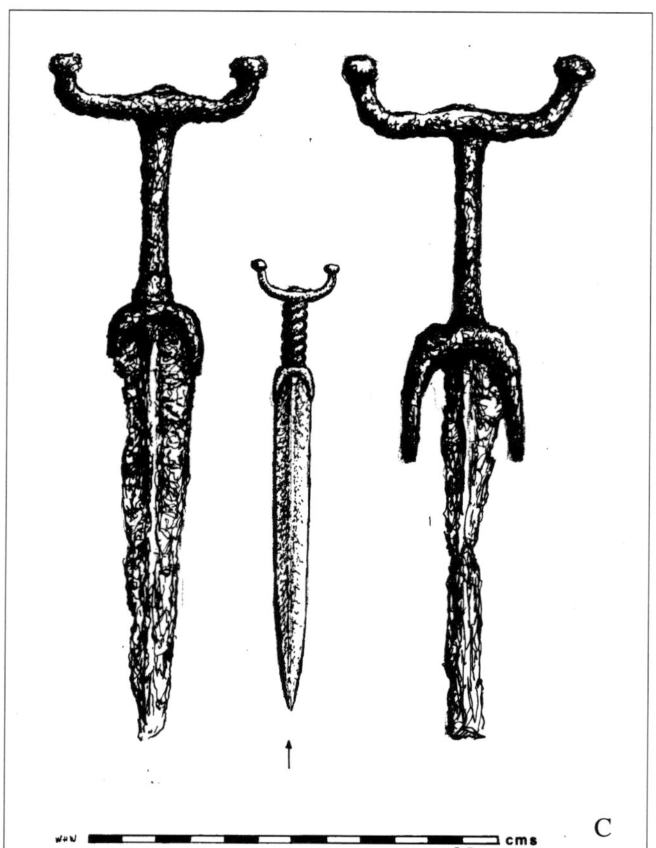

BALEARIC IRON ANTENNAE CELTIC SHORT SWORDS

MAINLAND IRON ANTENNAE CELTIC SHORT SWORDS

FIGURE 18. BALEARIC AND MAINLAND EARLY & LATE IRON AGE PARALLELS: CIRCA 1000 cal BC- 450 cal BC

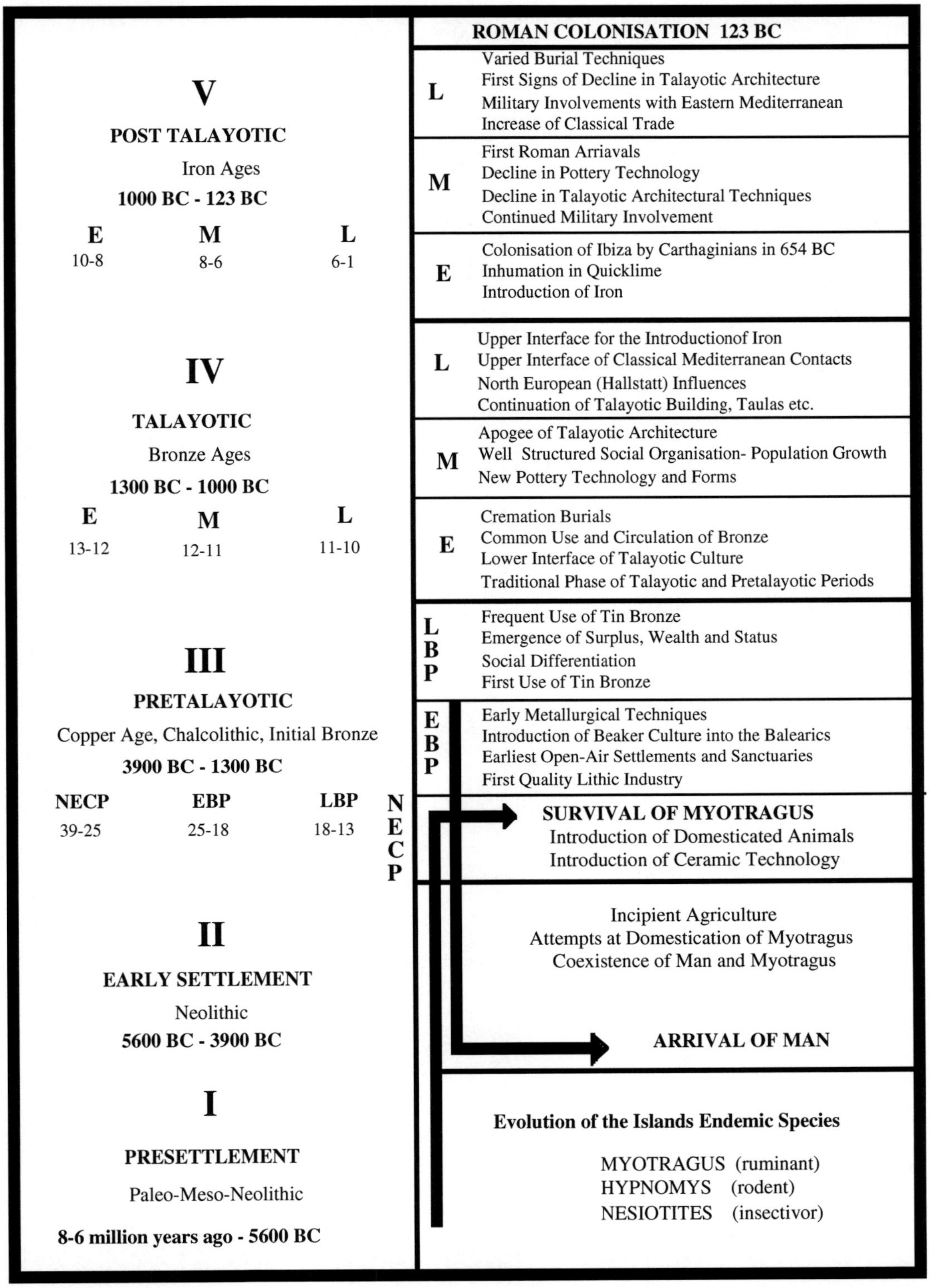

FIGURE 19. PROPOSED 1982 BALEARIC PENTAPARTITE DIVISION OF PREHISTORY USING TRADITIONAL NOMENCLATURE (after Waldren)

RADIOCARBON DATES: ROCK SHELTER OF SON MATGE
Valldemossa, Mallorca, Baleares, Spain
2002

nos.	site designation	date cal BC	date BP	lab. inv. no.
1.	**ABSMPT-L**	244 cal AD	1706 BP± 30 yrs	IRPA-710L
2.	**ABSM-PT-L**	25 cal BC	1925 BP± 120 yrs	IRPA-710
3	**ABSM-PT-L**	101 cal BC	2050 BP± 65 yrs	QL-7a
4	**ABSM-PT-L15**	196 cal BC	2145 BP± 65 yrs	IRPA-695L
5	**ABSM-PT-L**	266 cal BC	2215 BP± 65 yrs	IRPA-778L
6	**ABSM-LIA-T**	278 cal BC	2215 BP± 100 yrs	QL-9
7	**ABSM-PT-L**	379 cal BC	2244 BP± 60 yrs	QL-22
8	**ABSM-PT-GT**	379 cal BC	2328 BP± 190 yrs	BM-1409
9	**ABSM-PT-L**	385 cal BC	2328 BP± 65 yrs	IRPA-789L
10	**ABSM-PT-L**	391 cal BC	2336 BP± 100 yrs	QL-5c
11.	**ABSM-PT-L30**	400 cal BC	2340 BP± 55yrs	IRPA-808L
12.	**ABSM-PT-L**	400 cal BC	2349 BP± 55 yrs	IRPA-710
13.	**ABSM-PT-L**	400 cal BC	2349 BP± 55 yrs	IRPA-789
14.	**ABSM-PT-L**	402 cal BC	2349 BP± 140 yrs	UtC-1003
15.	**ABSM-PT-L**	402 cal BC	2351 BP± 60 yrs	IRPA-803L
16	**ABSM-PT-L45**	408 cal BC	2352 BP± 55 yrs	IRPA-776
17.	**ABSM-PT-L65**	408 cal BC	2358 BP± 80 yrs	Y-2669
18.	**ABSM-PT-L45**	641 cal BC	2590 BP± 40 yrs	IRPA-790L
19.	**ABSM-PT-L15**	651 cal BC	2595 BP± 60 yrs	QL-10a
20.	**ABSM-PT-L1**	651 cal BC	2600 BP± 70 yrs	QL-10b
21.	**ABSM-PT-MIA**	655 cal BC	2604 BP± 70 yrs	QL-10
22.	**ABSM-PT-L**	687 cal BC	2635 BP± 60 yrs	IRPA-751
23.	**ABSM-PT-L**	770 cal BC	2719 BP± 80 yrs	QL-6
24.	**ABSM-PT-MIA**	779 cal BC	2728 BP± 80 yrs	QL-4
25.	**ABSM-PT-EIA**	779 cal BC	2728 BP± 80 yrs	QL-24
26	**ABSM-PT-MIA**	779 cal BC	2728 BP± 60 yrs	IRPA-752
27.	**ABSM-PT-EIA**	786 cal BC	2735 BP± 60 yrs	IRPA-676
28.	**ABSM-PT-EIA**	786 cal BC	2735 BP± 60 yrs	IRPA-751
29.	**ABSM-PT-EIA**	789 cal BC	2738 BP± 60 yrs	IRPA-803
30.	**ABSM-PT-EIA**	786 cal BC	2736 BP± 85 yrs	IRPA-811L
31.	**ABSM-PT-L90**	786 cal BC	2735 BP± 60 yrs	IRPA-752L
32	**ABSM-PT-L**	791 cal BC	2741 BP± 55 yrs	IRPA-676L
33.	**ABSM-PT-L**	793 cal BC	2743 BP± 60 yrs	IRPA-1012L
34.	**ABSM-PT-L**	795 cal BC	2744 BP± 100 yrs	IRPA-790L
35	**ABSM-PT-L**	795 cal BC	2744 BP± 100 yrs	QL-20
36.	**ABSM-PT-L**	801 cal BC	2750 BP± 160 yrs	IRPA-695
37.	**ABSM-PT-L**	805 cal BC	2754 BP± 60 yrs	IRPA-695B
38.	**ABSM-PT**	809 cal BC	2759 BP± 100 yrs	QL-27
39.	**ABSM-PT**	813 cal BC	2763 BP± 60 yrs	IRPA-811
40.	**ABSM-PT-L10**	826 cal BC	2775 BP± 60 yrs	IRPA-806L
41.	**ABSM-PT-T**	838 cal BC	2787 BP± 80 yrs	QL-11
42	**ABSM-T**	897 cal BC	2847 BP± 100 yrs	QL-7
43.	**ABSM-T**	993 cal BC	2943 BP± 50 yrs	QL-986
44.	**ABSM-T**	993 cal BC	2942 BP± 50 yrs	BM-2140N
45.	**ABSM-T**	1144 cal BC	3098 BP± 60 yrs	IRPA-750L
46.	**ABSM-T**	1314 cal BC	3263 BP± 110 yrs	BM-2140
47.	**ABSM-PRT**	1485 cal BC	3435 BP± 100 yrs	QL-4191
48.	**ABSM-PRT**	1673 cal BC	3622 BP± 60 yrs	QL-5
49.	**ABSM-PRT**	1685 cal BC	3634 BP± 50 yrs	BM-1995
50.	**ABSM-PRT**	1710 cal BC	3659 BP± 100 yrs	BM-2312R
51,	**ABSM-PRT**	1740 cal BC	3689 BP± 80 yrs	QL-5A
52.	**ABSM-PRT**	1833 cal BC	3774 BP± 80 yrs	CSIC-180
53.	**ABSM-PRT**	1923 cal BC	3872 BP± 80 yrs	CSIC-179
54.	**ABSM-PRT**	1999 cal BC	3948 BP± 80 yrs	UtC-902
55.	**ABSM-PRT**	2075 cal BC	4026 BP± 45 yrs	QL-24
56.	**ABSM-PRT-CE**	2090 cal BC	4049 BP± 60 yrs	IRPA-835
56.	**ABSM-PRT**	2288 cal BC	4237 BP± 120 yrs	Y-2359
58.	**ABSM-PRT**	2470 cal BC	4419 BP± 100 yrs	QL-23
59.	**ABSM-PRT**	2483 cal BC	4432 BP± 100 yrs	QL-5B
60.	**ABSM-PRT**	2523 cal BC	4472 BP± 120 yrs	CSIC-178
61.	**ABSM-PRT-EE**	2675 cal BC	4625 BP± 40 yrs	UtC-9269
62.	**ABSM-PRT**	2735 cal BC	4689 BP± 392 yrs	BM-1408
63.	**ABSM-PRT-EE**	3375 cal BC	5324 BP± 120 yrs	QL-988
64.	**ABSM-PRT-EE**	3656 cal BC	5520 BP± 50 yrs	BM-1994
65.	**ABSM-PRT-EE**	4635 cal BC	6592 BP± 115 yrs	I-5516
66.	**ABSM-PRT-EE**	4722 cal BC	6671 BP± 360 yrs	CSIC-176
67.	**ABSM-PRT-EE**	5591 cal BC	7538 BP± 120 yrs	QL-29

TABLE 1

RADIOCARBON DATES: SON OLEZA CHALCOLITHIC OLD SETTLEMENT and DOLMEN AREA 2002

nos	site designation	years BP	95.4 probability range	date cal BC	lab.inv.no
1.	**SFO-OS-CH1**	2990 ± 160 yrs	1610- 820 (3559-2769)	***1261 cal BC***	QL-4044
2.	**SFO-OS-CH1**	2990 ± 160 yrs	1610- 820 (3559-2769)	***1261 cal BC***	IRPA-1053
3.	**SFO-OS-EXW**	3070 ± 50 yrs	1440-1250 (3389-3199)	***1357 cal BC***	HAR-3490
4.	**SFO-OS-C2**	3095 ± 70 yrs	1520-1250 (3469-3199)	***1403 cal BC***	BM-1698
5.	**SFO-OS-WC**	3150 ± 300 yrs	2192-2161 (4141-4110)	***1428 cal BC***	BM-1988
6.	**SFO-OS-SWF-A**	3180 ± 80 yrs	1670-1668 (3619-3617)	***1445 cal BC***	QL-4191
7.	**SFO-WOS-DOL1: 17-2M8 EX**	3130 ± 35 yrs	1520-1300 (3420-3250)	***1450 cal BC***	KIA-12274
8.	**SFO-OS-WC**	3230 ± 30 yrs	1670-1410 (3619-3359)	***1516 cal BC***	BM-1567
9.	**SFO-OS-CSS**	3240 ± 30 yrs	1611-1441 (3560-3390)	***1519 cal BC***	QL-1224
10.	**SFO-OS-C2**	3240 ± 30 yrs	1611-1441 (3560-3390)	***1519 cal BC***	IRPA-1054
11.	**SFO-OS-EXW-UW**	3260 ± 100 yrs	1854-1849 (3803-3798)	***1523 cal BC***	QL-4040
12.	**SFO-OS-WC-IN**	3280 ± 120 yrs	1880-1310 (3829-3259)	***1549 cal BC***	QL-1896
13.	**SFO-OS-H4**	3300 ± 60 yrs	1738-1708 (3687-3657)	***1565 cal BC***	UtC-3192
14.	**SFO-OS-WC**	3350 ± 330 yrs	2569-2538 (4518-4487)	***1673 cal BC***	BM-1988R
15.	**SFO-OS-WC**	3390 ± 80 yrs	1884-1682 (3833-3631)	***1712 cal BC***	BM-2312R
16.	**SFO-WOS-DOL1: 16-2M4**	3445 ± 40 yrs	1880-1610 (3830-3660)	***1745 cal BC***	KIA-12275
17.	**SFO-OS- EW-INT-B1**	3460 ± 45 yrs	2115-2085 (4064-4034)	***1745 cal BC***	UtC-9021
18.	**SFO-WOS-DOL3: 15-EEP**	3450 ± 30 yrs	1880-1680 (3880-3630)	***1760 cal BC***	KIA-12273
19.	**SFO-OS-CH1-INT4**	3450 ± 110 yrs	1878-1839 (3827-3788)	***1749 cal BC***	QL-4043
20.	**SFO-WOS-DOL1: 12- 4I3**	3440 ± 30 yrs	1880-1640 (3830-3590)	***1770 cal BC***	KIA-12267
21.	**SFO-WOS-DOL1: 11- 5L7**	3440 ± 35 yrs	1877-1840 (3826-3789)	***1770 cal BC***	KIA-10559
22.	**SFO-WOS-DOL1: 9- 5L8**	3445 ± 30 yrs	1930-1680 (3879-3629)	***1770 cal BC***	KIA-10558
23.	**SFO-OS-WCS**	3430 ± 30 yrs	1880-1630 (3830-3586)	***1755 cal BC***	KIA-16267
24.	**SFO-OS-SW-AG**	3405 ± 35yrs	1880-1600 (3830-3556)	***1740 cal BC***	KIA-16268
25.	**SFO-WOS-DOL3: 14-EWEX**	3445 ± 30 yrs	1880-1680 (3850-3590)	***1760 cal BC***	KIA-12270
26.	**SFO-WOS-DOL3: 13- NEX**	3460 ± 30 yrs	1880-1680 (3850-3590)	***1780 cal BC***	KIA-12268
27.	**SFO-OS-EW-UF-S**	3470 ± 50 yrs	1901-1742 (3850-3691)	***1819 cal BC***	QL-4100
28.	**SFO-OS-AT-QB**	3490 ± 30 yrs	1901-1742 (3850-3691)	***1827 cal BC***	QL-1859
29.	**SFO-OS-CH1-INF3**	3490 ± 30 yrs	1946-1768 (3895-3717)	***1829 cal BC***	QL-4042
30.	**SFO-WOS-DOL1: 2- 5M3**	3540 ± 30 yrs	2198-2159 (4147-4108)	***1860 cal BC***	KIA-10557
31.	**SFO-OS-EWF**	3550 ± 110 yrs	2007-2004 (3956-3953)	***1884 cal BC***	UtC-2742
32.	**SFO-WOS-DOL1: 12-5L7**	3550 ± 30 yrs	2030-1988 (3979-3937)	***1884 cal BC***	KIA-10560
33.	**SFO-WOS-DOL1: 1**	3565 ± 35 yrs	2030-1770 (3980-3720)	***1925 cal BC***	KIA-1228
34.	**SFO-OS-EXW**	3585 ± 35 yrs	2040-1880 (3990-3930)	***1929 cal BC***	UtC-8952
35.	**SFO-WOS-DOL1: 10-6M4**	3605 ± 35 yrs	2137-2076 (4086-4025)	***1970 cal BC***	UtC-10079
36.	**SFO-OS-EW-INT-B2**	3620 ± 50 yrs	2123-2097 (4072-4046)	***1985 cal BC***	UtC-9022
37.	**SFO-WOS-DOL1: 4-5M9**	3625 ± 30 yrs	2123-2097 (4072-4046)	***1995 cal BC***	KIA-10583
38.	**SFO-WOS-DOL1: 7-5M6**	3625 ± 30 yrs	2279-2231 (4229-4180)	***1995 cal BC***	KIA-10582
39.	**SFO-WOS-DOL1 J3-G13**	3640 ± 50 yrs	2199-2158 (4148-4107)	***2030 cal BC***	BM-1981R
40.	**SFO-OS-EW- INT-CC**	3680 ± 45 yrs	2199-2158 (4148-4107)	***2074 cal BC***	UtC-9023
41.	**SFO-WOS-DOL1: J9 W17**	3700 ± 30 yrs	2290-1930 (4240-3880)	***2110 cal BC***	QL-1592
42.	**SFO-WOS-DOL1: G13 FLT**	3720 ± 35 yrs	2275-2141 (4224-4198)	***2138 cal BC***	BM-1981
43.	**SFO-OS-NW-BB**	3790 ± 90 yrs	2480-1970 (4429-3919)	***2239 cal BC***	QL-1636
44.	**SFO-OS-EW-BB**	4030 ± 60 yrs	2585-2568 (4537-4512)	***2533 cal BC***	BM-1843R

TABLE 2 .

RADIOCARBON DATES: PREHISTORIC SANCTUARY OF SON MAS
Valldemosa, Mallorca, Baleares, Spain
2001

nos	site/sector	BPdate ±	calBC date	Lab Inv
1.	**SMSS-12W9**	3775 BP± 35 yrs	*2170 cal BC*	UtC-4676
2.	**SMSS-8O9**	3580 BP± 65 yrs	*1926 cal BC*	UtC-909
3.	**SMSS-9O8**	3570 BP± 65 yrs	*1923 cal BC*	UtC-908
4.	**SMSS-7O7**	3510 BP± 60 yrs	*1816 cal BC*	UtC-2020
5.	**SMSS-8J3**	3060 BP± 35 yrs	*1325 cal BC*	UtC-5892
6.	**SMSS-6N5**	3020 BP± 60 yrs	*1256 cal BC*	UtC-2756
7.	**SMSS-6L4**	2990 BP± 70 yrs	*1195 cal BC*	IRPA-1053
8.	**SMSS-12L5**	2970 BP± 40 yrs	*1172 cal BC*	UtC-2727
9.	**SMSS-6L7**	2960 BP± 60 yrs	*1169 cal BC*	IRPA-1049
10.	**SMSS-10W2**	2940 BP± 40 yrs	*1167 cal BC*	UtC-9017
11.	**SMSS-7N2**	2970 BP± 70 yrs	*1157 cal BC*	IRPA-976
12.	**SMSS-8L8**	2930 BP± 40 yrs	*1086 cal BC*	IRPA-984
13.	**SMSS-6K7**	2875 BP± 35 yrs	*1040 cal BC*	UtC-3044
14.	**SMSS-8J3**	2880 BP± 80 yrs	*1026 cal BC*	UtC-5891
15.	**SMSS-12J2**	2830 BP± 140 yrs	*979 cal BC*	UtC-2736
16.	**SMSS-12X5**	2795 BP± 35 yrs	*915 cal BC*	UtC-8954
17.	**SMSS-8K9**	2735 BP± 30 yrs	*872 cal BC*	UtC-5420
18.	**SMSS-8L5**	2690 BP± 70 yrs	*852 cal BC*	UtC-1256
19.	**SMSS-10S8**	2700 BP± 60 yrs	*835 cal BC*	IRPA-1055
20.	**SMSS-5M5**	2700 BP± 50 yrs	*832 cal BC*	UtC-2759
21.	**SMSS-8J8**	2655 BP± 35 yrs	*814 cal BC*	UtC-5675
22.	**SMSS-12X5**	2695 BP ± 70 yrs	*799 cal BC*	UtC-8953
23.	**SMSS-10L2**	2600 BP± 70 yrs	*794 cal BC*	UtC-1255
24.	**SMSS-8K9b**	2590 BP± 30 yrs	*788 cal BC*	UtC-5428
25.	**SMSS-18U9**	2590 BP± 50 yrs	*788 cal BC*	UtC-4170
26.	**SMSS-19V1**	2590 BP± 35 yrs	*787 cal BC*	UtC-4166
27.	**SMSS-7L7**	2580 BP± 25 yrs	*786 cal BC*	UtC-5421
28.	**SMSS-8J8b**	2565 BP± 30 yrs	*786 cal BC*	UtC-4857
29.	**SMSS-7L7**	2565 BP± 30 yrs	*786 cal BC*	UtC-5374
30.	**SMSS-5L3**	2470 BP± 25 yrs	*716 cal BC*	QL-4264
31.	**SMSS-16S2**	2480 BP± 50 yrs	*691 cal BC*	UtC-3190
32.	**SMSS-19V1**	2490 BP± 35 yrs	*666 cal BC*	UtC-4167
33.	**SMSS-12W5**	2480 BP± 30 yrs	*645 cal BC*	IRPA-1250
34.	**SMSS-7N8**	2520 BP± 80 yrs	*629 cal BC*	UtC-1002
35.	**SMSS-9K6**	2520 BP± 60 yrs	*629 cal BC*	UtC-1257
36.	**SMSS-8M6**	2500 BP± 40 yrs	*617 cal BC*	IRPA-836
37.	**SMSS-5L6**	2520 BP± 60 yrs	*616 cal BC*	IRPA-1025
38.	**SMSS-13U9**	2580 BP± 70 yrs	*613 cal BC*	IRPA-1051
39.	**SMSS-6K9b**	2440 BP± 90 yrs	*477 cal BC*	UtC-1258
40.	**SMSS-13U9b**	2430 BP± 40 yrs	*467 cal BC*	IRPA-1066
41.	**SMSS-13X4**	2430 BP± 25 yrs	*465 cal BC*	UtC-3188
42.	**SMSS-18Y1**	2400 BP± 70 yrs	*464 cal BC*	UtC-5426
43.	**SMSS-6V2**	2390 BP± 50 yrs	*458 cal BC*	UtC-3933
44.	**SMSS-10L2**	2360 BP± 140 yrs	*454 cal BC*	UtC-1003
45.	**SMSS-8M6**	2220 BP± 70 yrs	*282 cal BC*	UtC-1001
46.	**SMSS-12X5**	2200 BP± 40 yrs	*278 cal BC*	UtC-9018
47.	**SMSS-7O7**	2210 BP± 90 yrs	*274 cal BC*	QL-4200
48.	**SMSS-17Y4**	2210 BP± 110 yrs	*271 cal BC*	UtC-3045
49.	**SMSS-17Y7**	2140 BP± 50 yrs	*259 cal BC*	UtC-3189
50.	**SMSS-18Y6**	2140 BP± 70 yrs		UtC-3046
51.	**SMSS-7O7**	2050 BP± 150 yrs		QL-4201
52.	**SMSS-18Y2**	1960 BP± 40 yrs		IRPA-1026
53.	**SMSS-8L3**	530 BP± 50 yrs		IRPA-1024

TABLE 3.

RADIOCARBON DATES
MULETA CAVE & SMPC
2002

Datelist includes recent analyses results from human remains in the SMPC deposit

COMBINED DATES AVAILABLE FROM MULETA

SON MULETA POCKET CAVE

no.	site/no.	sector	95.4% prob .	anal. mater.	lab. inv. no
1.	SMPC-1	Central Area	2780 cal BC	Human Tibia	KIA-14003
2.	SMPC-2	Central Area	2480 cal BC	Human Tibia	KIA-14008
3.	SMPC-3	Central Area	2577 cal BC	Femur Shaft	KIA-14026
4.	SMPC-4	Central Area	2370 cal BC	Human Bone	KIA-14004
5.	SMPC-5	Central Area	2524 cal BC	Human Tibia	KIA-14005
6.	SMPC-6	Central Area	2524 cal BC	Human Femur	KIA-13997

MULETA CAVE DEPOSIT

no	site/period	sector	conv.14c date	date bp	lab. inv. no
1.	SM-PT	C	272 cal BC	2180	SI-651a
2.	SM-T	C	905 cal BC	2760	SI-552
3.	SM-PRT	C-F 20	2443 cal BC	3910	Y-2359
4.	SM-PSP	X 65	4212 cal BC	5350	HAR-2968
5.	SM-15O-PRT-	O 150	4225 cal BC	5435	KBN-640d
6.	SM175-PRT-	O 175	7991 cal BC	7135	KBN-640c
7.	SM200-PRT-SM-	O 200	7578 cal BC	7300	UCLA-1704j
8.	PSP-MB	O 175	8050 cal BC	8448	KBN-640b
9.	SM-PSP	O 150	7586 calBC	8570	UCLA-1740c
10.	SM-PSP	X IOO	7893 cal BC	8800	HAR-2932
11.	SM-PSP	X 145	7893 cal BC	8800	HAR-2909
12.	SM-PSP	O 175	12,050±350	10,685	KBN-640c
13.	SM-PSP	Z 350	12,515±315	14,000	SIO1a -RAA
14.	SM-PSP	Z350	12,700±850	14,465	SI-654
15.	SM-PSP	Z250	13,395±655	14,650	UCLA-1704b
16.	SM-PSP	D 250	13,850±600	15,885	SI-646
17.	SM-PSP	X 300	14,385±415	15,800	HAR-2933
18.	SM-PSP	X 300	14,900±200	16,335	SI6-48
19.	SM-PSP	E 350	16,150±600	16,850	UCLA-1704d
20.	SM-PSP	F 300	16,805±555	18,100	SI-649
21.	SM-PSP	X 350	16,850±500	18,735	SI-650
22.	SM-PSP	X 350	17,030±200	18,800	HAR-2971
23.	SM-PSP	Z 400	21,850±480	18,980	UCLA-1704e
24.	SM-PSP	E 400	21,850±500	23,800	SI-647
25.	SM-PSP	E 355	25,050±600	23,800	HAR-2912
26.	SM-PSP	Z 475	25,050±600	26,000	SIO-1cRAA
27.	SM-PSP	Z 525	26,650±600	26,000	HAR-2910
28.	SM-PSP	Z 600	30,050±715	28,600	UCLA-1704a
29.	SM-PSP	Z 750	40,050±1000	32,000	SIO1g -RAA
30.	SM-PSP	Z 650	42,450±400	48,000	HAR-2911
31.	SM-PSP	ST COL	48,000inf	48,000inf	QL-127
32.	SM-PSP	ST COL	58,000inf	58,000inf	QL-128
33	SM-PSP	FL ST FL	80,000pm	80,000pm	Y-1cf

** Most dates are uncalibrated beyond the range of calibration

TABLE 4.

BIBLIOGRAPHY-REFERENCES

BRAUDEL, F.
1949. The Mediterranean and the Mediterranean World in the Age of Philip II. Fontana? Collins, New York
BROODBANK, C.
2000. An Island Archaeology in the Early Cyclades, Cambridge University Press, Cambridge.
CLARKE, D.L.
1973. "Models and Paradigms in Contemporary Archaeology," en Renfrew (ed.), The Explanation of Culture Change: Models in Pre-
CHERRY, J.F.
1981. "Pattern and Process in the Earliest Colonisation of the Mediterranean Islands", Proceedings of the Prehistoric Society, 47, pp. 41-68, pp. 1-60. Londres: Dickworth.
CHERRY, J.
1984. "The Initial Colonization of the Western Mediterranean Islands in Light of Island Biogeography and Paleogeography," en Waldren, Chapman, Lewthwaite and Kennard (eds.), The Deia Conference of Prehistory: Early Settlement in the Western Mediterranean Islands and Their Peripheral Areas (British Archaeological Reports International Series 229), pp. 7-23, Oxford.
CHERRY, J.
1985. "Islands Out of the Stream: Isolation and Interaction in Early East Mediterranean Insular Prehistory," en Knapp and Stech (eds.), Prehistoric Production and Exchange: The Aegean and Eastern Mediterranean (UCLA Institute of Archaeology Monograph 25), pp. 12-29. Los Angeles.
DELIBES, G. y FERNÁNDEZ-MIRANDA, M.
1988. Armas y utensilios de bronce en la Prehistoria de las Islas Baleares, Studia Archaeologica, 78, Universidad de Valladolid, Valladolid.
EVANS, J.D.
1973. "Islands as Laboratories for the Study of Cultural process," en Renfrew (ed.), The Explanation of Culture Change: Models in Prehistory, pp. 517-520. Londres, Dickson.
EVANS, J.D.
1977. "Island Archaeology in the Mediterranean: Problems and Opportunities," en World Archaeology 9, pp.12-26.
FERNANDEZ-MIRANDA, M. AND WALDREN, W.H.
1974. El Abrigo de Son Matge (Valldemosa) y la Periodización de la Prehistoria Mallorquina mediante las Análasis de Carbon 14', Trabajos de Prehistoria 31, Madrid, pp. 287-304.
FERNANDEZ-MIRANDA, M. AND WALDREN, W.H.
1979. Periodificación Cultural y Cronológia Absoluta en la Prehistoria de Mallorca, Trabajos de Prehistoria 36, Madrid, pp. 349-377.
GASULL, P.
1985. Arqueologîa, Enciclopedia España-Calpe, Suplemento 1983-1984, pp. 51-68.
GILMAN, A.,
1981. 'The Development of Social Stratification in Bronze Age Europe', Current Anthropology 22, pp. 169-224.
GUERRERO, V.M.
1999. Arquitectura y poder en la prehistoria de Mallorca, "El Tall editorial" Palma.
GUERRERO, V. M.,
1995. 'Colonos, caciques y mercenarios. Una aproximación al contexto histórico del intercambio desigual en las Baleares», in: Jornades d'Arqueologia Fenicio-Púnica (X: Eivissa: 1995).
GUERRERO, V. M.,
1995b. 'Una sociedad en estado de jefatura (Chiefdoms). La cultura Talayótica Balear», in: Waldren, W.H. (ed.), Deya Conference of Prehistory. Ritual, Rites and Religion (III: 1993: Deià), Oxford: British Archaeological Reports, (International Series, 611, II), 295-313.
HERNÁNDEZ-GASCH, J.,
1998. 'Son Real. Necrópolis talayótica de la edad del hierro. Estudio arqueológico y análisis y análisis social». Arqueomediterrània 3, II (Universitat de Barcelona). Barcelona.
HOSKIN, M.
1990. 'The Menorcan Taulas', in Waldren, Ensenyat and Kennard (eds.) IInd Deya Conference of Prehistory: Archaeological Techniques, Technologyand Theory, BAR International Series 565, Oxford.
JACOBSEN, T.W.
1976. 'Excavations in the Franchthi Cave', 1969-1971, Hesperia 42, pp. 45-88 and 253-286
LEWTHWAITE, J.
1987. "Social factors and economic change in Balearic prehistory, 3000-1000 b.c.", in BAKER, G. and GAMBLE, C. (eds.), Beyond domestication in Prehistoric Europe, Academic Press, London.
LULL, V., MICÓ, R., RIHUETE, C. and RISCH, R.
1999. La Cova des Càrritx y la Cova des Mussol. Ideología y Sociedad en la Prehistoria de Menorca, Consell Insular de Menorca, Barcelona.
LULL, V., MICÓ, R., RIHUETE, C. and RISCH, R.
2000a, "Rituales de vida y muerte en la prehistoria de Menorca", Revista de Arqueología, 225, pp. 18-33.
LULL, V., MICÓ, R., RIHUETE, C. and RISCH, R.
2000b. "Divinità nelle tenebre. Parlano le grotte delle Baleari", Archeologia Viva, 84, pp. 54-61.
NASH,G.
2002. Constructing Hypohetical Systems of Exchange: Doing Kula in the Danish Mesolithic, in V Deia International Conference of Prehistory, World Islands in Prehistory, eds Waldren, Ensenyat
PATTON, M. 1996. Islands in Time: Island Sociogeography and Mediterranean Prehistory. Londres: Routledge.
STUIVER, M. and WALDREN, W.H. 1974. 14C Carbonate Dating and the Age of Post Talayotic Lime Burials in Mallorca. *Nature*, Vol. 255, June, pp. 475-476.
VAN STRYDONCK, M. and WALDREN, W. H.
1990. Radiocarbon dating of lime burials. PACT 29, pp. 403- 414
WALDREN, W.H.
1967. 'Informe Preliminar Sobre Analisis de Radiocarbono en Mallorca'. X Congreso Nacional de Arqueologia. Mahon y Zaragossa, pp. 75-82.
WALDREN,W.H.
1972. Age Determination by Radiocarbon Method, Sociedad de Historia Natural de las Baleares, Palma de Mallorca,pp 34-50.
WALDREN, W.H.
1979. 'A Beaker Workshop Area in the Rock Shelter of Son Matge, Valldemosa, Mallorca' World Archaeology, Routledge and Kegan Paul, London, Vol 11, No. 1, pp. 43-67.
WALDREN, W. H.
1982. Balearic Prehistoric Ecology and Culture: The Excavation and Study of Certain Caves, Rock Shelters and Settlements, British Archaeological Reports, International Series, 149. Oxford.
WALDREN, W.H.
1984b. The Lithic Industry of the Balearic Islands...Its Olezian Tradition of Tabular Flint Blades, in W.H. Waldren, R.W. Chapman, J.G. Lewthwaite and R.C. Kennard (eds.), Ist Deya Conference of Prehistory...Early Settlement in the Western Mediterranean Islands and their Peripheral Areas, Oxford, British Archaeological Reports, BAR International Series 229, Oxford, vol. iii, 859-909.
WALDREN, W.H. 1986: Balearic Pentapartite Division of Prehistory.. Radiocarbon and Other Age Determination Inventories, British Archaeological Reports, BAR International Series 282, Oxford.
WALDREN, W.H.
1984a. Chalcolithic Settlement and Beaker Connections in the Balearic Islands, inW.H.Waldren, R.W. Chapman, J.G. Lewthwaite and R.C. Kennard (eds.), Ist Deya Conference of Prehistory, Early Settlement in the Western Mediterranean Islands and their Peripheral Areas, Oxford, British Archaeological Reports, BAR Interna-

tional Series 229, vol. iii, 911-965.
WALDREN, W.H.
1987a. A Rose by Any Other Name...a Question of Balearic Beakers...the Evidence, Oxford International Conference, in W. Waldren and R.C. Kennard (eds.), Bell Beakers of the Western Mediterranean: Definition, Interpretation, Theory and New Site Data, Oxford, British Archaeological Reports, BAR International Series, 331 (i), vol. i, 369-482.
WALDREN, W. H., CUBÌ, C.
1987b. A Case History: the Function and Origin of Mallorcan Cast lead Votive Funerary Jewelry, Oxford International Conference, in W. Waldren and R.C. Kennard (eds.), Bell Beakers of the Western Mediterranean: Definition, Interpretation, Theory and New Site Data, Oxford, British Archaeological Reports, BAR International Series, 331 (i), vol. i.
WALDREN, W. H.
1986. The Balearic Pentapartite Division of Prehistory. Radiocarbon and other age determination inventories, British Archaeological Reports, International Series, 282, Oxford.
WALDREN, W. H., ENSENYAT, J. AND CUBÌ, C.
1991. Prehistoric Archaeological Elements, Ferrandell Oleza Chalcolithic Old Settlement, DAMARC Series 20, Donald Baden-Powell Quaternary Research Centre, Pitt Rivers Museum, Oxford University, England.
WALDREN W.H., ENSENYAT J. and CUBI GRIMALT, C.
1991. Son Mas: a New Mallorcan Prehistoric Sanctuary, circa 2000 BC-1750 BC', in Waldren, Ensenyat and Kennard (eds.), IInd Deya Conference of Prehistory...Archaeological Techniques, Technology and Theory, Oxford, British Archaeological Reports, BAR International Series, 574 (i-iii), vol ii, 267-312.
WALDREN W.H. and M. VAN STRYDONCK
1994. Prehistoric Sanctuary of Son Mas: a Radiocarbon Analysis Survey , Dating the Activity Sequences of the Santuary DAMARC Series, 24, Deya, Mallorca.
WALDREN W.H., ENSENYAT, J. AND CUBI , C.
1994. Prehistoric Architectural Elements, Ferrandell-Oleza Chalcolithic Old Settlement: Radiocarbon Dating and Artefact Distributional Survey, DAMARC Series, 22, Deya, Mallorca.
WALDREN W.H. 1995: The Function of Bell Beaker Pottery as a Ceremonial and Votive Object, in Waldren, Ensenyat and Kennard (eds.), IIIrd Deya Conference of Prehistory, Ritual, Rite and Religion in Prehistory, Oxford, British Archaeological Reports, BAR International Series, 611 (i-ii), vol i, 238-263.
WALDREN, W. H.
1997. 'The Definition and Duration of the Beaker Culture in the Balearic Islands: a Radiocarbon Survey' Oxford Journal of Archaeology, Blackwell Publishers,vol 16, no 1. pp 25-39.
WALDREN, W. H.
1998. The Beaker Culture of the Balearic Islands. An inventory of evidence from caves, rock shelters, settlements and ritual sites, British Archaeological Reports, International Series, 709-Western Mediterranean Series, 1.
WALDREN, W. H.
2001. 'A New Megalithic Dolmen from the Balearic Island of Mallorca: its Radiocarbon Dating and Artefacts', Oxford Journal of Archaeology, Blackwell Publishers,vol 20, no 3. pp 241-263.

17

NEGOTIATING ISLAND IDENTITIES: CERAMIC PRODUCTION IN THE CYCLADES DURING THE MIDDLE AND LATE BRONZE AGE

INA BERG

School of Art History and Archaeology, Oxford Rd, Manchester, M13 9PL, UK
EMAIL: ina.berg@man.ac.uk

SUMMARY

This paper is an attempt to write an island archaeology which perceives material culture as an active force in shaping one's identity. I wish to move away from the still predominantly passive view of material culture which portrays changes in ceramic production and style as determined by environmental changes or as a passive response to outside influences. This paper investigates the Minoan influence (Minoanisation) on the pottery production, repertoire and style at Phylakopi on Melos and Ayia Irini on Kea in the Middle and Late Bronze Age. It will be argued that these two sites which are frequently grouped together as one interpretative unit show great differences in their responses to Minoan influence. Each followed its own course. I will propose that the detected variations reflect the towns relationship with Minoan Crete; pottery production was used as a means to define the islanders identity vis á vis Cretan culture.

INTRODUCTION:

Investigating islands is different from investigating regions on the mainland. This is not to say that islands are the infamous 'laboratories' where a culture can be studied in almost total isolation. Islands, like other regions, have been exposed to varying degrees to outside influences and have, in the case of the Cyclades, interacted regularly with their surrounding world since the Neolithic. What distinguishes islands from regions is their proximity to the sea. The sea has frequently been regarded as either isolating or connecting. However, geographical units do not necessarily also constitute cultural ones. Insularity is culturally constructed and there is no intrinsic reason why islands should be the ultimate unit of analysis (Broodbank 2000). Working in prehistoric periods, material culture remains our only tool to investigate how islanders perceived themselves, what social strategies they adopted and how they negotiated their identities vis á vis other cultural units.

It is a common complaint by archaeologists working in the Cyclades that the available evidence is too limited to allow for conclusive answers. This attitude however hides the peculiarly down-played fact that the Bronze Age Cyclades are one of the most intensively researched and archaeologically data-rich islands in the world. For the two islands under investigation here, in particular, the evidence available is very good, as excavations of a multi-period settlement are complimented by one or more field surveys. Also, the local ceramic sequence and thus the chronology of those islands is comparatively well established and can be linked to other regional sequences in the Aegean. In contrast to scholars working in the Pacific or Oceania, Aegean scholars have been slow in fruitfully utilising this wealth of information for developing theoretical approaches to interaction, identity and island landscapes (notable exceptions are Cyprian Broodbank, John Cherry and Jack Davis). A tentative start in this new direction is attempted in this paper by investigating the changes in ceramic production at two Cycladic settlements; the observed changes are suggested to be an expression of the existence of different island identities.

Minoan influence (Minoanisation) in the ceramics production can investigated on several levels: Minoan shapes might be adopted and incorporated into the pottery repertoire. Minoan motifs and decorative styles might be adopted. Or, Minoan technologies might be taken over, such as the potter's wheel or the use of coarser clay for the foot or base of some vessels to provide a stronger support - a common practice on Crete. Claims for Minoanisation in the Cyclades are based on: a) Locally made Minoanising vessels become more popular over time; so do Minoanising motifs. b) The potter's wheel is quickly adopted by societies, and c) Minoan imports increase gradually over time, peaking in LM IA.

DISCUSSION OF THE EVIDENCE:PHYLAKOPI ON MELOS

One of the most apparent visual divisions in the pottery production at Phylakopi is that between the local and Minoanising styles and shapes. Although there were of course overlaps, every aspect of the pottery assemblage can be analysed in this way. The 'conical cup' fabric was exclusively used for Minoanising shapes, Cycladic White primarily for local ones. The local fabric was used for both. However, even in the local fabric, Minoanising vessels were regularly wheelmade while those of local tradition were made by hand. Motifs follow this trend: Minoanising motifs are commonly found on Minoanising shapes and traditional motifs on local shapes. In other words, there is pronounced division between Minoanising and local shapes. It appears that no complete technological change occurred in the Melian pottery production after it came in contact with Crete or the Mainland. Although the local potters certainly knew how to use the wheel, they only used it for Minoanising vessels. Their own local vessels continued to be made by hand.

This result is rather striking. It seems that the Melian potters did not adopt selected Minoan features but made a more or less exact copy of the original. This included the shape, the forming process and the decoration. Melian shapes, on the other hand continued to be manufactured according to the local tradition: local shapes, traditional motifs and the handmade mode. For whatever reason, Melian potters regarded their local tradition as separate from the Minoan one: what was appropriate for the production of Minoanising vessels, was not for the production of local shapes. These two groups were perceived as conceptually different.

Since we know that the Melian potters used the potter's wheel competently, we have to reject the hypothesis that this technology was only transferred incompletely. It is more likely that a change in the technology used for local shapes might have interfered with social taboos. We could hypothesise that it was in the handmade mode that the potters of Phylakopi expressed their group identity. To tamper with it might have resulted in unwanted changes to age, gender or class taboos. These taboos were not applicable to the Minoanising shapes which could therefore be wheelmade.

Finally, Minoan imports do not conform to the general pattern of a slow increase in the Middle Bronze Age period with a peak in LMIA. Although this trend is visible at many sites, at Phylakopi Minoan imports are most numerous in the late MC period and have started to decline already by early LCI. And by LCII, when there are still many LMIB imports into Ayia Irini, for example, Minoan imports into Phylakopi have almost ceased.

What was the reason for this non-conformist pattern? The following hypotheses can be put forward: a) Traders did not stop over at this settlement and the demand for Minoan vessels had to be saturated by the production of local imitations. b) A sufficiently high local supply cut down on the need for imports. c) Minoan pots were not fashionable at Phylakopi and Melian consumers explaining the fall of imports?

AYIA IRINI ON KEA

According to Overbeck (1989: 8) nearly all identifiable local products are hand-built in early MC. The use of the potter's wheel is rare. By the late MC many of the pots are wheelthrown, especially the smaller vessels (Davis 1986: 4). In LCI and LCII most of the vessels were wheelmade (Elizabeth Schofield, pers. communication).

The use of the wheel in the Keian local fabric set in earlier and was more complete than in the local fabric at Phylakopi. There does not appear to have been a divide between local and Minoanising vessels. Even local ones were overwhelmingly made with the wheel already in LCI. It can therefore be concluded that a conceptual division as at Phylakopi is not visible in the Keian pottery production - or at least not to the same degree.

The import pattern of Minoan pots also followed a different pattern from that of Phylakopi. Minoan imports were rare in the early MC period. Already by the end of this period they constituted a substantial proportion of the total bulk of pottery and had become the dominant import (Overbeck 1989: 11). Calculations from a representative pottery assemblage estimated that ca. 8% of the total assemblage were Minoan imports (Overbeck 1982: 40). This percentage seemed to increase between the early and late MC. In late MC Minoan imports had become abundant (Davis 1986: 6). Minoan influence is said to have been stronger than Mainland influence in this period. Minoan imports remain the most dominant import in LCI. Only in LCII did Minoan imports decrease and were overtaken by Mycenaean ones (Cummer & Schofield 1984: 146).

Finally, although no precise percentages are available, it appears that the local imitations of Minoan shapes became more and more popular over time (Davis 1986: 86-98; Cummer & Schofield 1984: 47-48).

DISCUSSION

As so often in archaeology, perfectly comparable data are not always available. However, the differences in pottery production between the two sites are marked and warrant at least speculating.

In Phylakopi there is a recognisable conceptual division between Minoanising and local shapes. Such a division does not appear to exist in the pottery production at Ayia Irini. At Phylakopi, Minoan imports already declined in LMIA and were virtually non-existent in LMIB. At Ayia Irini, on the other hand, Minoan imports peaked in LMIA and were only overtaken by Mycenaean imports in LMIB. Regarding local imitations, the two settlements reacted similarly. At both Phylakopi and Ayia Irini local imitations of Minoan shapes became more and more popular over time. And although no accurate percentages are available from Ayia Irini, the publications and the selection of pots studied by me give the impression that the local Minoanising production was greater than that at Phylakopi both in quantity and range of shapes. The potter's wheel was used primarily for Minoanising shapes at Phylakopi. At Ayia Irini, its use was more widespread and encompassing.

Thus, it appears that the pottery production at Phylakopi was relatively conservative and preserved their local manufacturing tradition. Potters at Ayia Irini, on the other hand, followed another strategy. The islanders fully embraced Minoan ceramic culture and incorporated it into their local pottery production, resulting in substantial technological changes. Their openness was not limited to ceramic production. Ayia Irini also adapted religious, economic, architectural, and artistic Minoan features to suit their own cultural needs.

To explain Ayia Irini's strategy accurately, we also have to take the potential exports into account. Finds of litharge and the attribution of analysed metal to the Lavrion field (Gale, Stos-Gale & Davis 1984: 400) point to Ayia Irini as a recasting and redistribution centre for metals; other industrial or craft activities, such as weaving, manufacturing aromatics, and stone bowls were spread across the settlement; the wide dispersal of these activities suggests part-time production which seemed to have involved every household. One could indeed claim that Ayia Irini was "one big 'workshop'" (Schofield 1990: 209). It is a distinct possibility that this town functioned as a meeting point and supply centre for Cretan and Mainland traders. By offering raw materials and finished products, coupled with its convenient location, the island reduced the need for the Cretans or Mainlanders to travel to the homeland of the respective trading partner. In demonstrating their openness toward outside influences the local may have actively encouraged such a development. The evidence of a wide range of diverse materials may be an indication that their strategy paid off, resulting in greater wealth and prosperity.

CONCLUSION

What does the above case study tell us about the process of acculturation in the Bronze Age Aegean? First and foremost, it has demonstrated that acculturation is not a one-way street; all societies involved can influence and negotiate the outcome of this interaction process. Minoan influence was not something that was forced upon the Cyclades, nor was it simply a natural strive by lower ranking societies to emulate the superior Cretan culture, but each society negotiated the depth of Minoan influence, based on internal social and cultural attitudes. By studying and comparing the changes in the local ceramic production archaeologists may indeed be able trace such attitudes.

REFERENCES

BROODBANK, C.,
2000. An Island Archaeology of the Early Cyclades. CUP. Cambridge.
CUMMER, W.W. AND SCHOFIELD, E.,
1984. Keos III. Ayia Irini: House A. von Zabern. Mainz.
DAVIS, J.L.,
1986. Keos V. Ayia Irini: Period V. von Zabern. Mainz.
GALE, N.H., STOS-GALE, Z.A. AND DAVIS, J.L.,
1984. The Provenance of Lead Used at Ayia Irini, Kea, Hesperia 53, 389-406
OVERBECK, J.C.,
1982. The Hub of Commerce in Keos and Middle Helladic Greece, Temple University Aegean Symposium 7, 38-49.
OVERBECK, J.C.,
1989. Keos VII. Ayia Irini: Period IV. von Zabern. Mainz
Rouse, I. 1977. Pattern and process in West Indian archaeology, World Archaeology 9, 1-11.
SCHOFIELD, E.,
1990. Evidence for Household Industries on Thera and Keos, in Hardy, D.A. et al (eds.), Thera and the Aegean World III. Proceedings of the Third International Congress, Santorini, Greece, 6-9 September, 1989. Thera Foundation. London, 201-211.
TERRELL, J.E.,
1977. Geographic systems and human diversity in the North Solomons, World Archaeology 9, 62-81.

18

STABLE ISOTOPES (^{13}C AND ^{15}N) AND DIET: ANIMAL AND HUMAN BONE COLLAGEN FROM PREHISTORIC SITES ON MALLORCA, MENORCA AND FORMENTERA (BALEARIC ISLANDS, SPAIN)

MARK VAN STRYDONCK (1) , **MATHIEU BOUDIN1 & ANTON ERVYNCK** (2).

(1) Royal Institute for Cultural Heritage, Jubelpark 1, B-1000 Brussels, Belgium.
(2) Institute for the Archaeological Heritage of the Flemish Community, Doornveld 1 box 30, B-1731 Zellik-Asse, Belgium
EMAIL: mark.vanstrydonck@kikirpa.be

INTRODUCTION

The Balearic Islands form an archipelago in the western Mediterranean Sea, a cluster of islands which lies 80 to 300 km east of the Spanish south-eastern coast. The Balearics have been colonised by humans since prehistory, with any meaningful habitation possibly originating in the period from 5000 to 3500 BC. A Neolithic way of life was introduced around 3500 BC, after which Neolithic, Chalcolithic and Early Bronze Age cultures developed. During this period, megalithic monuments were constructed. A following period, starting after 1600-1500 BC, witnessed the erection of different types of monuments, such as naviformes and navetas (stone-built structures with the form of an upturned boat), talaiots ('watchtowers' with their associated domestic structures) and taulas (T-shaped stone pillars) as late Bronze Age - early Iron Age successors of the former megalithic monuments. The prehistory of the islands lasted until their incorporation into the Roman empire (123 BC), although the period of 800-600 BC to 123 BC, often called 'protohistoric', can be distinguished by the growing influences from other Mediterranean cultures (Lull et al. 1999, Plantalamor Massanet & Benejam 1997, Waldren 1986).

Since the 1960s, and still repeated in recent publications, the hypothesis has been launched that the subsistence strategy of the late Bronze Age - early Iron Age culture focused on animal husbandry, while plant cultivation remained of minor significance (see the references in Hernandez-Gasch et al. 2001). During the following period cereal agriculture would have gained importance (Mayoral Franco 1984). On the other hand, the Neolithic to early Bronze Age period would have been characterised by a mixed farming economy (Lewthwaite 1985). These reconstructions, however, are based upon a set of evidence that is not proof to criticism. Especially the almost 'exclusive' animal husbandry system of the late Bronze Age - early Iron Age period has been questioned because it is clear that the cultural archaeological data could be biased due to preservation conditions, other taphonomic factors, excavation techniques, the choice of sites (being almost always ritual places), etc. (Hernandez-Gasch et al. 2001). Moreover, the environmental archaeological record for the Balearics is still very limited (see, e.g., the overview for animal remains in Chapman & Grant 1997).

The analysis of plant and animal remains from archaeological sites can reveal which species were used for consumption and how the exploitation strategies were organised. Generally, however, it is very difficult, if not impossible, to evaluate the relative importance of animal versus plant products. Not only in sites with a hypothetical 100% survival of organic remains this would still be a methodological puzzle, in areas with less favourable preservation conditions, this is completely hopeless. The poor record for plant remains from prehistoric sites can, for example, not be used in any interpretation, since the phenomenon could be the result of taphonomic factors, of problems with sampling and recovery (given the state of excavation methodology in the area) or of the (ritual) nature of the sites. The preliminary archaeozoological data indicate that, at least at Son Ferrandell, the rearing of animals for meat alone was not the major aim of the site's inhabitants, suggesting that vegetable food may have made the main contribution to the diet (Chapman & Grant 1997, 76). The same conclusion was reached by the analysis of tracer elements from 24 bone samples from Càrritx (Pérez Pérez et al. 1999). To gain importance, however, these interpretations should be corroborated by an independent line of evidence. Moreover, there are other important questions, such as whether the absence of fish remains from the sites is the result of preservation conditions, of the lack of sieving during excavation, or of the subsistence strategy of the prehistoric people?

An alternative approach for dietary reconstructions consists in the analysis of the isotopic fractionation of collagen from the bones found in human burials from the period under study. It is well known that the $\delta^{13}C$ (carbon) and $\delta^{15}N$ (nitrogen) values in the body reflect the diet of an organism, a phenomenon that is now with increasing frequency used for dietary reconstructions within archaeology (Sealy 2001). This study presents data recorded from human bones from prehistoric sites from Mallorca, Menorca and Formentera. The aim is to see whether information about the subsistence strategy of these people could be extracted from the isotopic signatures, and whether diachronic trends were observable from the material. Isotope measurements on animal bones, often from the same sites from which human bone was studied, were taken to serve as reference for the human data.

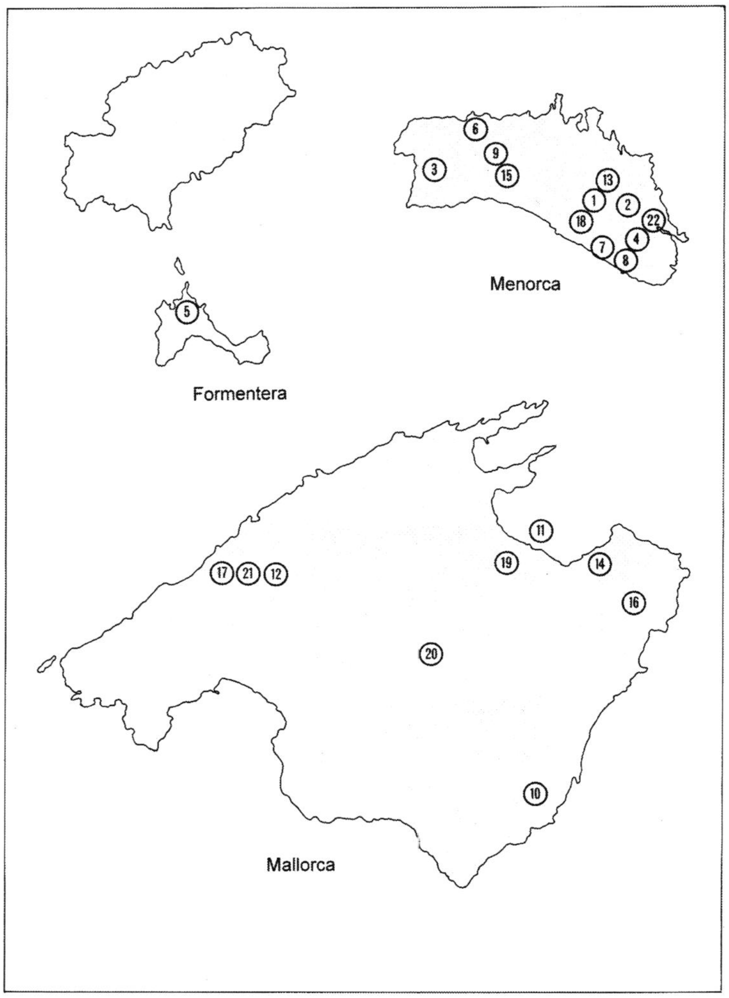

FIGURE 1. DISTRIBUTION MAPS

MATERIAL AND METHODS

The $\delta^{13}C$ and $\delta^{15}N$ data (always expressed as per mil, ‰) gathered for this paper are a by-product of the radiocarbon analyses on animal and human bone from different sites at the Balearic Islands, performed, since 1986, at the Royal institute of Cultural Heritage. For most of the samples only $\delta^{13}C$ data are available. Recently, also $\delta^{15}N$ data and C/N (carbon - nitrogen) ratios are measured as part of a Sample Quality Research Program (SQRP). The C/N values give an indication about the preservation condition of the bone. For gelatin derived from archaeological bones this ratio should be between 2.7 and 3.3 (De Niro 1985), while higher values indicate humic intrusion.

Table 1 lists the sites from which human and animal bones were analysed. Their location is depicted on Figure 1. For certain animal bones, a species identification was not available, because the specimens had been subjected to radiocarbon dating before they were studied by an archaeozoologist.

When an identification was made, we are always dealing with domestic mammals, i.e. sheep, goat, cattle or pig. As traditional in archaeozoology, the remains of sheep and goat could not always be separated. After calibration of the radiocarbon date (Stuiver et al. 1998)(TABLE 2), the human samples can be subdivided in three chronological groups, which cover roughly the periods of the end of the third to the end of the second millennium BC (phase 1), the end of the second to the beginning of the first millennium BC (phase 2), and the second quarter of the first millennium BC to the beginning of the first millennium AD (phase 3). Roughly, these time periods coincide with the megalithic period (phase 1), the late Bronze Age to early Iron Age (phase 2), and the protohistoric period (phase 3). Three samples were not included within this tripartite subdivision: the single samples from Binipati Nou and Rafal Rubi (they fall on the limit subdividing phase 2 and 3), and the human sample from Talati de Dalt (because of its medieval age).

The pre-analysis treatment of the bone samples, i.e., the extraction of the collagen, followed the methodology established by Longin (1971). The transformation of the collagen into graphite (for Accelerated Mass Spectrometry measuring) followed Van Strydonck & van der Borg (1991). All data obtained proved to come from bones with a good to very good C/N ratio, implying that no contamination with humic acid has to be taken into account. There is one exception, an animal bone from Biniparratx Petit (table 2), but the $\delta^{13}C$ and $\delta^{15}N$ values are not exceptional compared to those of the other samples from the same site.

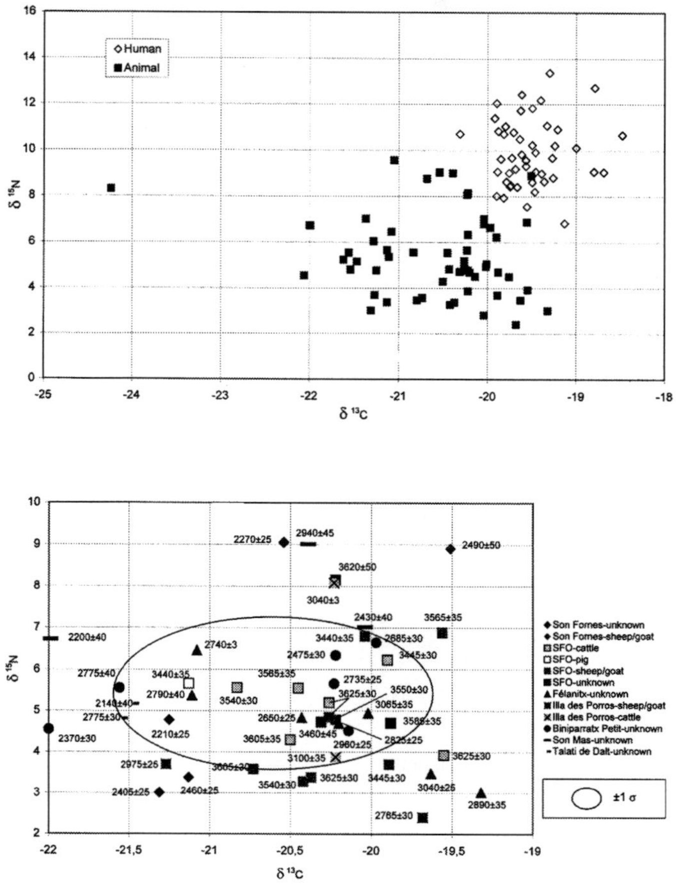

FIGURES 2 & 3.

RESULTS
TABLE 2 lists all data recorded from the sample set. The results will be discussed separately for the human and the animal sub-samples.

ANIMAL BONES
Figure 2 shows the isotope measurements for all material analysed. Except for one outlier, with an unusual low $\delta^{13}C$ value, the measurements for animals and humans group together in two clusters that only minimally overlap. The values for the domestic mammals show averages for $\delta^{13}C$ and $\delta^{15}N$ (FIGURE 3) that roughly coincide with those listed in the literature ($\delta^{13}C$= -21‰, $\delta^{15}N$= +5‰: Lanting and van der Plicht 1996, see also table 3). The slightly less negative $\delta^{13}C$ values (0.5‰) of the Balearic samples could perhaps be related to the geographic difference between the samples in the literature and the collection studied here. It has indeed already been noticed that there might be a climatological influence on the $\delta^{13}C$ values (Van Klinken et al. 1994).

All together, the data for the animals show a large variation, a trend that cannot be explained by inter-species differences (FIGURE. 3). However, there are many other factors that can have had an influence (see Van Strydonck & Wouters 2001). Most of the animal bones sampled come from herbivores, i.e., cattle, sheep or goat, except for the single skeletal element of pig (see further). Most of the unidentified animal samples will most probably also come from herbivores since, as far as we know, pigs were relatively rare in Balearic prehistory (Chapman & Grant 1997, 82, FIGURE 7.6). The variation of the isotope fractionation within the group of herbivores can have been influenced by differences in the plants they consumed. Not only different plant species can have different isotope characteristics (see the references in van Klinken et al. 2000, 43), but also the spot where they grow can be highly influential. Altitude, salinity, local humidity and amount of sunlight (canopy effect) can have an influence on the δ13C of plants (see the references in Bocherens 2000, 73) and therefore also on the $\delta^{13}C$ of the collagen of the herbivores. There are also differences in isotopic fractionation between different parts of the same plants (e.g., roots, leaves). The $\delta^{13}C$ of the bone collagen of an animal will thus also be significantly determined by its foraging habits (Heaton 1999). Furthermore, taphonomic factors can have been important: archaeological bones can, for example, suffer from an additional isotopic fractionation due to degradation processes (Sullivan and Krueger 1981). Finally, it should be stressed that isotopic fractionation is largely different between two groups of plant species, having a different physiology, i.e. C-3 and C-4 plants. The latter group, however, was not present on the Balearic Islands in prehistory.

It has been investigated whether there is a diachronic trend in the measurements for the animal bone samples. However, no trend could be observed: regression analyses between the uncalibrated radiocarbon dates and $\delta^{13}C$ or $\delta^{15}N$ values proved that were no relations between these variables (regression data not depicted here).

Table 1
Sites from which samples have been analysed (for more information see Van Strydonck *et al.* 1998, 2001 & in prep.). The numbers refer to the map (fig. 1)

N°	Site	Island	Community	Human bone	Animal bone
1	Alcaïdus	Menorca	Alaior	+	-
2	Biniai Nou	Menorca	Maó	+	-
3	Binipati Nou	Menorca	Ciutadella	+	-
4	Biniparratx Petit	Menorca	Maó	-	+
5	Ca Na Costa	Formentera	-	+	-
6	Cala Morell	Menorca	Ciutadella	+	-
7	Calas Coves	Menorca	Alaior	+	-
8	Cap de Forma Murada	Menorca	Maó	-	+
9	Es Tudons	Menorca	Ciutadella	+	-
10	Closos de Can Gaià	Mallorca	Felanitx	-	+
11	Illa des Porros	Mallorca	Santa Margalida	+	+
12	Naveta Baduia	Mallorca	Valldemosa	-	+
13	Rafal Rubi	Menorca	Alaior	+	-
14	S'Aigua Dolça	Mallorca	Colonia de San Pere	+	-
15	Ses Aritges	Menorca	Ciutadella	+	-
16	Ses Paises	Mallorca	Arta	-	+
17	Son Ferrandell Oleza	Mallorca	Valldemosa	-	+
18	So Na Caçana	Menorca	Alaior	-	+
19	Son Baulo	Mallorca	Santa Margalida	+	-
20	Son Fornes	Mallorca	Montuïri	+	+
21	Son Mas	Mallorca	Valldemosa	+	+
22	Talati de Dalt	Menorca	Maó	+	+

HUMAN BONES

The total set of measurements for the human bones shows somewhat less variation than is the case for the animals (FIGURE. 2). Could this be due to a more uniform diet within the human population? More important is the comparison between the animal and the human samples, because the differences observed could yield information about the human diet. It is known that, biochemically, a predator has a less negative value of $\delta^{13}C$ and a more positive value of $\delta^{15}N$ compared to its prey. Theoretical studies suggest that the differences, describing the shift or a trophic level within the food chain, ideally should be +5‰ for $\delta^{13}C$ (Ambrose 1993) and +2 to +4‰ for $\delta^{15}N$ (Ambrose 1991), meaning that, when the human diet would have consisted solely of the meat of herbivores, these differences should describe the variation between the mean values for the isotope fractionations of human and herbivore. When freshwater animals would have been added to the diet, more negative d $\delta^{13}C$ measurements would be obtained and more positive $\delta^{15}N$ values. When marine food products would have been consumed, a less negative $\delta^{13}C$ and a more positive $\delta^{15}N$ would be recorded (see table 3).

When a human population would have survived solely on plant products, their isotope fractionation values must have been identical to those of herbivore mammals. However, before any comparison between animals and humans can be made, it must first be established whether there are no significant differences within the human sample population. Most likely, diachronic trends could be expected and therefore the data were split up taking into account the three phases defined earlier (FIGURES 4 to 6).

The oldest human bones (phase 1, FIGURE 4) present a homogenous group in terms of isotopic values, except for two samples from Ca Na Costa, the only site from the small island of Formentera from which material was available for this study. Another outlier is formed by a sample from Biniai Nou; its different values must without doubt be related with the radiocarbon date which is too young compared to the other samples within this group. This individual must represent a younger burial, intrusive within a much older funeral monument, and the data will be transferred to phase 3 (see further). Remain the intriguing data for Ca Na Costa, indicating the possibility that diet was different on this small island. The isotope values could then perhaps indicate a subsistence strategy including a (limited) consumption of marine resources. Due to the limited data set from Formentera, however, this interpretation cannot yet be fully evaluated.

Table 2
Species identification, uncalibrated radiocarbon date, $\delta^{13}C$ and $\delta^{15}N$ measurements, and C/N ratio for all samples studied.

Site	**Species**	**^{14}C age (BP)**	**$\delta^{13}C$**	**$\delta^{15}N$**	**C/N**
Alcaïdus					
	human	3340±30	-19,64	+10,50	2,80
	human	3360±30	-19,82	+7,95	2,79
	human	3110±30	-19,90	+8,03	2,85
	human	3075±35	-19,75	+8,44	2,77
	human	3220±35	-19,79	+8,64	2,83
Biniai Nou					
	human	2270±35	-18,48	+10,66	2,79
	human	3745±35	-19,92	+11,39	2,88
	human	3290±40	-19,50	+8,63	2,99
	human	3635±35	-19,74	+8,49	3,05
	human	3200±35	-19,67	+8,42	2,87
	human	3605±30	-19,62	+9,83	2,84
	human	3660±25	-19,39	+9,00	2,80
Binipati Nou					
	human	2815±40	-19,50	+11,83	2,76
Biniparratx Petit					
	animal	2685±30	-19,97	+6,65	2,72
	animal	2735±25	-20,23	+5,66	5,32
	animal	2775±40	-21,56	+5,55	2,58
	animal	2825±25	-20,14	+4,52	2,77
	animal	2475±30	-20,22	+6,34	2,80
	animal	2370±30	-22,06	+4,55	2,99
Ca Na Costa					
	human	3595±35	-18,69	+9,06	2,82
	human	3535±40	-18,79	+12,74	2,73
Cala Morell					
	human	2875±45	-19,57	+9,33	2,98
Calas Coves					
	human	2475±25	-19,89	+9,08	2,86
	human	2470±35	-19,71	+10,79	2,76
	human	2595±30	-19,88	+10,83	2,76
	human	2480±25	-20,31	+10,71	2,86
	human	2415±30	-19,82	+10,71	2,81
	human	2525±35	-19,50	+10,23	2,91
Cap de Forma Murada					
	animal	2755±30	-21,28	+6,05	2,83
	animal	2930±35	-19,76	+4,51	2,84
	animal	2815±45	-20,79	+3,47	2,82
Es Tudons					
	human	2820±40	-19,46	+9,08	4,99
Felanitx					

The samples representing phase 2 (FIGURE. 5) are limited in number but show values roughly comparable to the human material from Mallorca and Menorca, dating to phase 1. Possibly, $\delta^{13}C$ values are somewhat less negative but statistically, this cannot be substantiated. The human bone samples from phase 3 (FIGURE 6) are more numerous and also more varied. One sample from Illa Des Porros shows rather low $\delta^{15}N$ values. Together with the sample from Biniai Nou (originally included within phase 1, but showing a date that is much younger), these specimens accentuate the larger variation in phase 3 compared to phases 1 and 2. It is even possible to postulate, on the basis of Figure 5, that the data from the sites of Calas Coves, Illa Des Porros (except one), Son Fornes and Biniai Nou could represent separate clusters located along a gradient of gradually less negative $\delta^{13}C$ values and more positive $\delta^{15}N$ values, perhaps indicating differences in diet, e.g., different extents of the addition of marine products. Remarkably enough, the named sites all show specific burial rituals, thus suggesting different cultural (sub-)groups, a pattern that also makes acceptable that variation in subsistence strategy existed. More data, however, will be needed to substantiate this suggestion. In total, phase 3 specimens show $\delta^{13}C$ values more or less comparable to those recorded from the older material (but more varied), and $\delta^{15}N$ measurements that are generally higher.

When the values observed for each phase are compared (FIGURE 7), it could be postulated that there are no diachronic trends in the δ13C values within the human material, but that phase 3 could be different from the preceding periods because of higher $\delta^{15}N$ measurements. Possibly, Figure 7 illustrates a diachronic trend towards more positive $\delta^{15}N$ values, from phase 1 (average $\delta^{15}N$= 9.0) over phase 2 (average $\delta^{15}N$= 9.2) to phase 3 (average $\delta^{15}N$= 10.9). These differences will be discussed further but, regarding the comparison between animal and human samples, the analysis will first be restricted to the animals and the humans from phase 1. The difference in the average $\delta^{15}N$ values between both data sets, around 3.7‰, well within the range mentioned in the literature (Ambrose 1991), could well be the result of the shift of one trophic level, meaning that the higher $\delta^{15}N$ value for the humans from phase 1 would be the result of the inclusion of the meat of herbivores in their diet. Of course, this interpretation is hardly shocking, given the mammal bones found as consumption refuse in the Balearic prehistoric sites (see Chapman & Grant 1997). But does this mean that the diet of the humans from phase 1 was predominantly carnivorous? This question cannot be answered, because the nitrogen within the collagen of the bones of a carnivorous or omnivorous species is almost exclusively derived from the animal part of the food consumed. Collagen, a protein product within the body, is mainly assembled from the protein taken in as food; only when there is a lack of protein intake will carbohydrates be used as building stones for collagen (Ambrose 1993). In a mixed diet, protein is dominantly found in animal products while the carbohydrates will mainly come from plants. Because nitrogen is hardly present in carbohydrates but abundant in protein, this explains why $\delta^{15}N$ only reflects the animal part of the diet. Of course, this reasoning starts from the assumption of an abundant mixed diet but what would be the effect of periods of meat shortage, or of strictly seasonal culling of the domestic herds? Perhaps then protein must be taken from plant sources, resulting in a lower value for the $d^{15}N$ trophic level shift? However, the present state of research does not yet allow to use the actual values of the shifts observed in case studies to make inferences about the relative proportions of plant and animal products within a diet. Ambrose (2000) lists values obtained from collagen in a number of studies, of +5.7‰ (Ambrose & De Niro 1986), +3.9 (Sealy et al. 1987), +3.2 (Schoeninger 1985) and +5.0 (Koch et al. 1991), but whether the possibility of obtaining higher values compared to the one observed in this study (+ 3.7) indicates that the diet of the people of phase 1 was sometimes in shortage of meat, therefore requiring the use of plant protein, is certainly too dangerous a statement.

The difference in $\delta^{13}C$ values is less pronounced (0.8‰ between the animals and the humans from phase 1), an observation more difficult to explain. Lanting and van der Plicht (1996) list a less negative $\delta^{13}C$ value for the consumers of the meat of herbivores (table 3) and in the literature a trophic level shift of +5‰ was proposed (Ambrose 1993). However, recent research has proven that the $\delta^{13}C$ value can differ widely between different parts of the body (Hubner 1985), making that the (selective) feeding behaviour of the predator will determine the $\delta^{13}C$ shift and that the average value for the prey as a whole is of less importance (van Klinken et al. 2000, 47). This pattern, which is not valid for the $\delta^{15}N$ distribution within the body (being almost equal between different tissues), can be responsible for a

Table 3:
Isotopic fractionation of carbon and nitrogen in the collagen of animals with different diets (Lanting and van der Plicht, 1996).

Bone collagen from animals having a 100% diet of:	$\delta^{13}C$ ‰	$\delta^{15}N$ ‰
C-3 plants	-21	+5
Meat of C-3 herbivores	-18	+8
C-4 plants	-7	+5
Marine food	-13	+18
Marine mollusks	-17	+11
River fish	-24	+16
Lake fish	-20	+16

minimal trophic level shift when the collagen of herbivores and carnivores is compared. Indeed, it must be realised that even when carnivores eat the collagen ('the bones') of their prey, most of the animal biomass consumed will consist of muscle or fat, which have $\delta^{13}C$ values that are 3 and 8‰ lower than collagen (data from Huebner 1985). The values for muscle and fat falling within the range of those for C3-plants (compare with Bocherens 2000, 68, FIGURE 4.1), and taking into account that humans often do not consume bones, this makes clear that $\delta^{13}C$ values can hardly be used as diet indicator within this context. Differences of 0.5 to 1‰ between herbivores and carnivores have often been recorded from modern and ancient samples (see the references in van Klinken et al. 2000, 47), and the Balearic material forms no exception to this pattern.

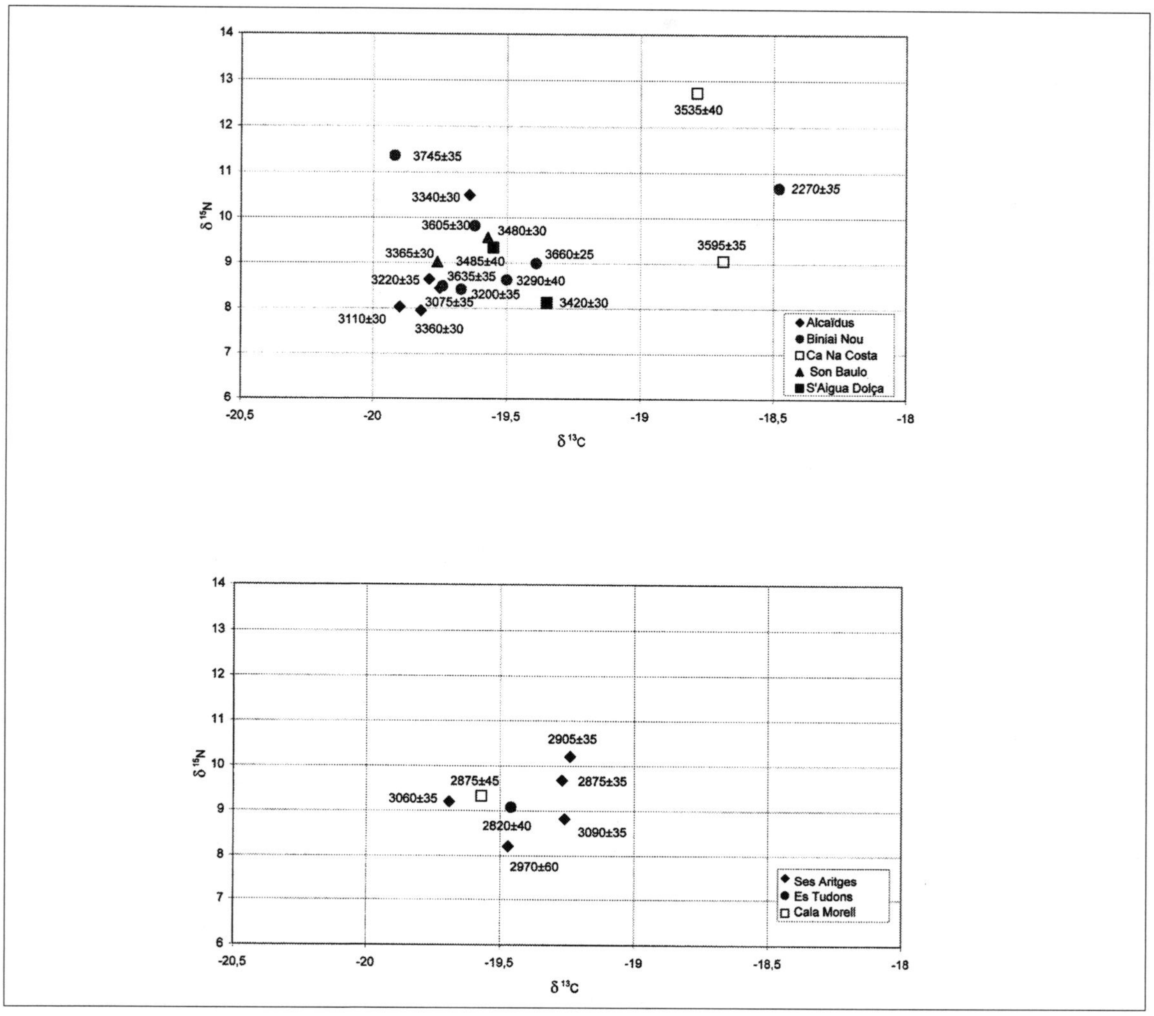

FIGURE 4 & 5.

Remain the differences in $\delta^{15}N$ values between the three human sub-samples (Fig. 7). Disregarding the possible differences between individual sites (see earlier), the slightly higher value in the youngest phase could generally be explained by a more varied diet compared to the earlier phases, for example including imported goods, such as salted marine products (Biniai Nou may be a good example for this). Such additions to the diet, however, must have been very limited because the differences of the $\delta^{13}C$ and $\delta^{15}N$ values observed are not dramatic. The fact that the Balearics came into full contact with other Mediterranean cultures during phase 3 is the most likely explanation for the (small) diet shift observed.

Generally, for the human samples of all three phases, it must be stressed that both N and C isotope fractionation values clearly indicate that the diet in Balearic prehistory was not based upon marine resources (implying that people living on an island were not fishing) and that people were also not concentrating upon the exploitation of freshwater animals (which can be explained by the islands' biogeography). Would one of these strategies have been incorporated within the subsistence system, clearly different $\delta^{13}C$ and $\delta^{15}N$ values would have been obtained (compare with table 3).

CONCLUSION

The foregoing analysis represents a first attempt to use stable isotope analysis in order to make inferences about the diet of the pre- and protohistoric people inhabiting the Balearic Islands. It has been impossible to evaluate the relative importance of animal versus plant products within the diet, but the hypothesis of a mixed diet, relying on the exploitation of both resources, has certainly not been contradicted. To make firmer statements, this research theme will have to be addressed in the future by new approaches, such as the comparison of isotopic fractionation in bone collagen and carbonate, and the evaluation of the spacing value of $\delta^{13}C$ between collagen and carbonate hydroxylapatite (see, e.g.,

993. Isotopic analysis of palaeodiets: methodological and interpretive considerations. In: M.K. Sandford, ed. Investigations of ancient human tissue. Langhorne: Gordon & Breach Science Publishers, 59-130.
AMBROSE, S.H.,
2000. Controlled diet and climate experiments on nitrogen isotope ratios of rats. In: S.H. Ambrose & M.A. Katzenberg, eds. Biogeo-

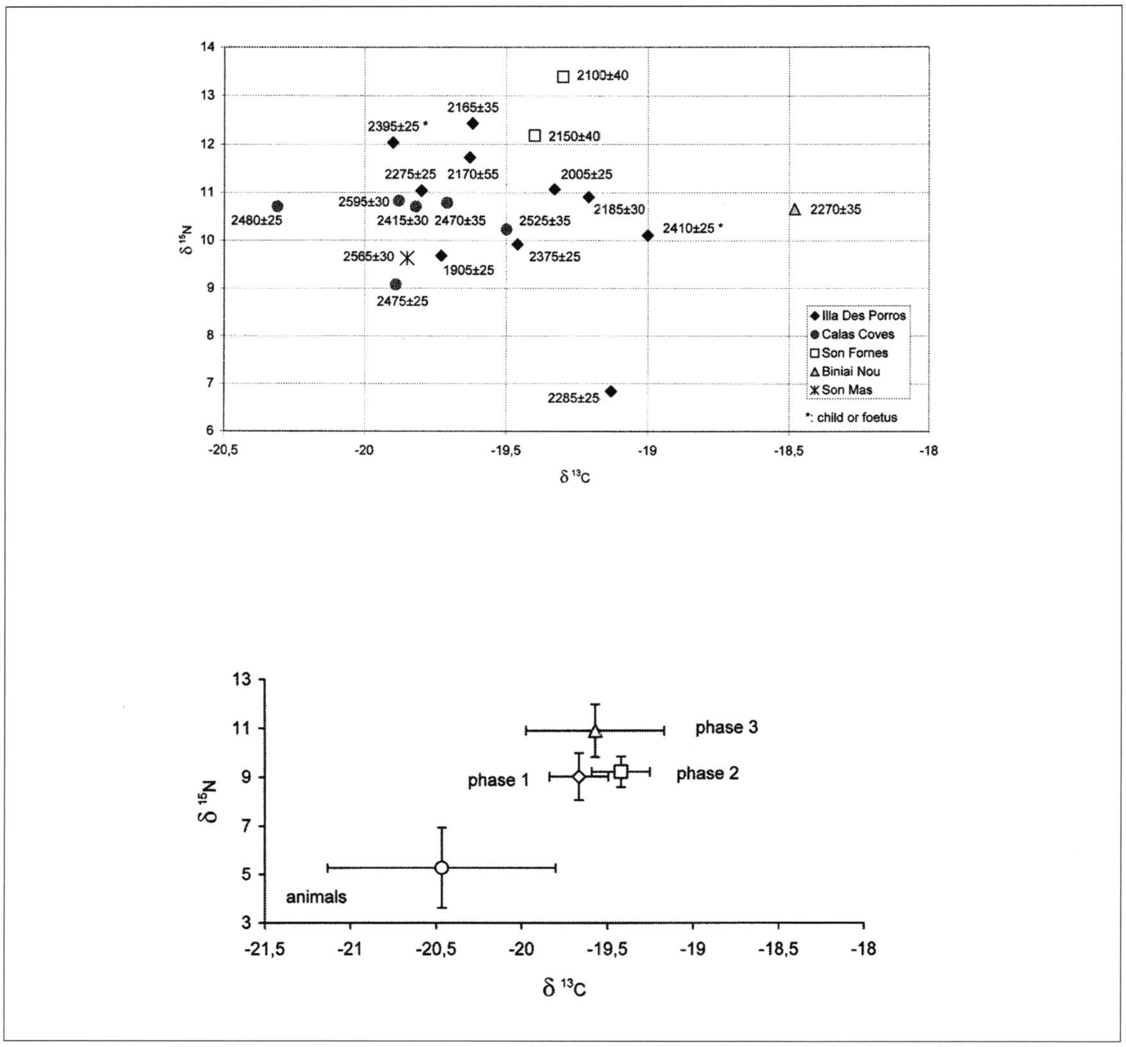

FIGURE 6 & 7.

Bocherens 2000). It is clear, however, that marine or freshwater resources were hardly used by the prehistoric communities, except for a slight change in the diet during phase 3. For the oldest phase, the difference between the data from the main islands and the small island of Formentera remains remarkable. During the youngest phase, slight differences in diet between sites may have been present.

chemical approaches to palaeodietary analysis. New York: Kluwer Academic / Plenum Publishers, 243-259.
AMBROSE, S.H. & DE NIRO, M.J.,
1986. The isotopic ecology of East Africal mammals. Oecologia, 69, 395-406.
BOCHERENS, H. ,
2000. Preservation of isotopic signals (13C, 15N) in Pleistocene mammals. In: S.H. Ambrose & M.A. Katzenberg, eds. Biogeochemical approaches to palaeodietary analysis. New York: Kluwer Academic / Plenum Publishers, 65-88.
CHAPMAN, R. & GRANT, A.,
1997. Prehistoric subsistence and monuments in Mallorca. In: M.S. Balmuth, A. Gilman & L. Prados-Torreira, eds. Encounters and transformations: the archaeology of Iberia in transition. Sheffield: Sheffield Academic Press, 69-87.

BIBLIOGRAPHY

AMBROSE, S.H. ,
1991. Effects of diet, climate and physiology on nitrogen isotope abundances in terrestrial foodwebs. Journal of Archaeological Science, 18, 293-317.
AMBROSE, S.H.,1

DE NIRO, M. J.,
1985. Postmortem preservation and alteration of in vivo bone collagen isotope ratios in relation to palaeodietary reconstruction. Nature, 317, 806-809.
HEATON, T.H.E.,,
1999. Spatial, species, and temporal variations in the δ13C/d12C ratios of C3 plants: implications for palaeodiet studies. Journal of Archaeological Science, 26, 637-649.
HERNANDEZ-GASCH, J., NADAL, J., MALGOSA, A. & ALES, A.,
2001. Economic strategies and limited resources in the Balearic insular ecosystem: the myth of an indigenous animal farming society in the first millenium BC. In: World islands in prehistory. International insular investigations. September 13-18, 2001. Conference resumes. Deia: Deia Archaeological Museum and Research Centre, chapter 22.
HUEBNER, H.,
1985. Geographische Variationen der Isotopen Zusammensetzung leichter Elemente im tierische, insbesondere menslichen Organismus. Mitteilungen Zentralinstitut für Isotopen- und Strahlenforschung, Leipzig, 106, 1-91.
KOCH, P., BEHRENSMEYER, A.P. & FOGEL, M.L.,
1991. The isotopic ecology of plants and animals in Amboseli National Park, Kenya. Annual Report of the Director, Geophysical Laboratory, Carnegie Institution, Washington D.C., 1990-1991, 163-171.
LANTING, J.N. & VAN DER PLICHT, J.,
1996. Wat hebben Floris V, skelet Swifterbant S2 en visotters gemeen? Palaeohistoria, 37-38, 491-519.
LEWTHWAITE, J.G.,
1985. Social factors and economic changes in Balearic prehistory, c. 3000-1000 bc. In: G. Barker & C. Gamble, eds. Beyond domestication in prehistoric Europe. London: Academic Press, 205-231.
LONGIN, R.,
1971. New method of collagen extraction for radiocarbon dating. Nature, 230, 241-242.
LULL, V., MICÓ, R., RIHUETTE HERRADA, C. & RISCH, R.,
1999. Ideología y sociedad en la prehistoria de Menorca. La cova de Càrritx y la cova des Mussol. Ciutadella: Consell Insular de Menorce, juntament de Ciutadella, Fundació Rubió Tudurí Andrómaco.
MAYORAL FRANCO, F.,
1984. La fase post-talayotica Mallorquina: periodizacion y dinamica economico-social. In: W.H. Waldren et al., eds. Early settlement in the Western Mediterranean islands and their peripheral areas. BAR International Series 229. Oxford: British Archaeological Reports, 1299-1313.
PÉREZ PÉREZ, A., FERNÁNDEZ, E. & TURBÓN, D.,
1999. Paleodieta. In: V. Lull, R. Micó, C. Rihuette Herrada & R. Risch, eds. Ideología y sociedad en la prehistoria de Menorca. La cova de Càrritx y la cova des Mussol. Ciutadella: Consell Insular de Menorce, juntament de Ciutadella, Fundació Rubió Tudurí Andrómaco, 197.
PLANTALAMOR MASSANET, L. & JUAN BENEJAM, G.,
1997. Entra a la Menorca prehistòrica. Maó: Museu de Menorca
SCHOENINGER, M.J.,
1985. Trophic level effects on 15N/14N and 13C/12C ratios in bone collagen and strontium levels in bone mineral. Journal of Human Evolution, 14, 515-525.
SEALY, J.,
2001. Body tissue chemistry and palaeodiet. In: D.R. Brothwell & A.M. Pollard, eds. Handbook of archaeological sciences. Chichester: John Wiley & Sons, 269-279.
SEALY, J.C., VAN DER MERWE, N.J., LEE-THORP, J.A. & LANHAM, J.,
1987. Nitrogen isotope ecology in southern Africa: implications for environmental and dietary tracing. Geochimica et Cosmochimica Acta, 51, 2707-2717.
STUIVER, M., REIMER, P.J., BARD, E., BECK, J.W., BURR, G.S., HUGHEN, K.A., KROMER, B., MC CORMAC, G., VAN DER PLICHT, J. & SPURK, M.,
1998. INTCAL 98 radiocarbon age calibration, 24,000-0 cal BP. Radiocarbon, 40(3), 1041-1084.
SULLIVAN, C.H. & KRUEGER, H.W.,
1981. Carbon isotope analysis of separate chemical phases in modern and fossil bone. Nature, 292, 333-335.
VAN KLINKEN, G.J., VAN DER PLICHT, J. & HEDGES, R.E.M.,
1994. Bone 13C/12C ratios reflect (paleo-)climatic variations. Geophysical Research Letters, 21, 445-448.
VAN KLINKEN, G.J., RICHARDS, M. & HEDGES, R.E.M.,
2000. An overview of causes for stable isotopic variations in past European human populations: environmental, ecophysiological, and cultural effects. In: S.H. Ambrose & M.A. Katzenberg, eds. Biogeochemical approaches to palaeodietary analysis. New York: Kluwer Academic / Plenum Publishers, 39-63.
VAN STRYDONCK, M., LANDRIE, M., HENDRIX, M., VAN DER BORG, K., DE JONG, A.F.M., ALDERLIESTEN, K. & KEPPENS, E.,
1998. Royal Institute for Cultural Heritage Radiocarbon dates XVI. Brussels: Royal Institute for Cultural Heritage.
VAN STRYDONCK, M., LANDRIE, M., MAES, A., VAN DER BORG, K., DE JONG, A.F.M., ALDERLIESTEN, K. & KEPPENS, E.,
2001. Royal Institute for Cultural Heritage Radiocarbon dates XVII. Brussels: Royal Institute for Cultural Heritage.
VAN STRYDONCK, M., LANDRIE, M., & BOUDIN, M.,
in prep. Royal Institute for Cultural Heritage Radiocarbon dates XVIII. Brussels: Royal Institute for Cultural Heritage.
VAN STRYDONCK, M. & VAN DER BORG, K.,
1991. The construction of a preparation line for AMS-targets at the Royal Institute for Cultural heritage Brussels. Bulletin de l'Institut Royal du Patrimoine Artistique/ Bulletin van het Koninklijk Instituut van het Kunstpatrimonium, 23, 228-234.
VAN STRYDONCK, M. & WOUTERS, J.,
2001. Pitfalls in radiocarbon dating of archaeological bone derived gelatine and specific gelatine fractions. In: Lodewijckx, M., ed. Belgian archaeology in a European setting 1. Acta Archaeologica Lovaniensia Monographiae 12. Leuven: Leuven University Press, 131-138.
WALDREN, W.H.,
1986. The Balearic pentapartite division of prehistory. BAR International Series 282. Oxford: British Archaeological Reports.

19

PUTTING MEAT ON THE BONE: AN INVESTIGATION INTO PALAEODIET IN THE BALEARIC ISLANDS USING CARBON AND NITROGEN STABLE ISOTOPE ANALYSIS

MICHAEL H.L.A. DAVIS
Research Laboratory for Archaeology and Art History,
University of Oxford, 6 Keble Road, Oxford, OX1 3QJ, England.
EMAIL:

INTRODUCTION

Bone collagen from a broad range of prehistoric human remains, extinct endemic fauna and domesticate animals, were employed in this research. Samples were taken from numerous sites on Mallorca (dates refer to associated age of human material used from these sites) including Son Muleta (5000-2000 BC), Son Matge (1400-200 BC), Son Mas (800-200 BC), Son Naveta (1500-1100 BC), Son Fornes (200-50 BC), Aigua Dolca (1800-1600 BC), Illa des Porros (700-100 BC), Son Real (700-100 BC), Covas des Carritz on Menorca and Es Forat de Ses Aritges (1600-800 BC). Few of these sites are discussed in this paper, as detailed reports appear in previous publications, including these British Archaeological Reports volumes. This paper is intended to give the reader some understanding into the theory of stable isotope research, and how such measurements are made (Part 1) before discussing some of the background to the research (Part 2) and then finally the results and conclusions resulting from this Balearic case study (Part 3).

DIET AND ISOTOPES

For the majority of human existence, we hunted, gathered and foraged for food. This continued for the most part until the Neolithic period when we began more sedentary lifestyles undertaking farming and animal husbandry. To understand past human subsistence, we have to gain evidence from a variety of sources. Paleoenvironmental evidence, through floral and faunal remains, allows us to build a picture of specific habitats that were settled in the past. How we adapted to these environments and subsisted, becomes clearer as we enter the Holocene, as more evidence survives from the post glacial period. To interpret what we ate, we have to look at archaeological remains and use a variety of techniques, to establish our dietary nutrition. Indirect or qualitative evidence of this comes from animal bones, plant remains, bone/stone tools and pottery (by type of vessel and residues, lipids or proteins that can be extracted). Unfortunately, the material record does not survive in its entirety and therefore we cannot rely on this evidence alone if we wish to build an accurate picture of human palaeodiet.

Only from the chemical analysis of human remains can we analyse quantitatively, what past populations ate throughout a large proportion of their lifetime. The early work in this field focussed on trace element analysis but this has gradually shifted towards light stable isotope measurements. Isotopes are atoms of the same element but with a different mass, in dietary studies we look at stable isotopes, as they can remain with a similar composition over time, if they are not altered diagenetically. The majority of research into stable isotopes focuses on bone collagen (Ambrose, 1993). Bone is a composite material made up from approximately 70% mineral (mainly hydroxyapatite) and 30% organic material, primarily collagen (Mays, 2000). Bone collagen is subject to continuous turnover throughout life, although this is slow as estimates suggest complete replacement takes 10-30 years in adults (Chisholm, 1989). Bone collagen therefore provides an excellent material for studying palaeodiet, as its composition contains average isotopic dietary signals for the last few decades of an individual's life (Mays, 2000).

Most elements of the natural environment, from which we

obtain our food, comprise of more than one stable isotope. These can be tracked from their source, the atmosphere through the food webs of the biosphere to their destination in a particular human tissue, like bone collagen or tooth enamel (DeNiro & Epstein, 1978, 1981). The reason we can use this to reconstruct diet is due to processes of isotopic fractionation. Mass differences between isotopes are sufficiently large for physical and chemical reactions to cause fractionations that are very small but measurable. Although the metabolic processes that create these fractionations are not yet entirely understood, we can see the proportion of isotopes in a starting material and the product of the reaction (Ambrose, 1993).

Stable isotopes of various elements may be used to infer palaeodiet from bone collagen namely carbon and nitrogen but possibly also sulphur, hydrogen and oxygen. In bone mineral or tooth enamel other isotopes can also be investigated, such as strontium, though this is less specific to diet and may be subject to greater diagenic alteration. In this study we are only looking at the stable isotopes of carbon (C) and nitrogen (N), which each exist in two stable isotopic forms. The stable isotopes of carbon are 12C (98.89%) and 13C (1.11%). The stable isotopes of nitrogen are 14N (99.634%) and 15N (0.366%). These isotopes are both measured in a isotope ratio monitoring mass spectrometer and their values are expressed as the ratio of the rarer heavy isotope to the light isotope, in units of parts per thousand or per mille (‰). As it is the change in this ratio that is of interest, and this is represented by the symbol d. Since δ is a change, it is defined relative to a standard. The more positive the ratio, the more enriched the sample is with the heavier isotope. The ratio of carbon is compared to the VDPB standard and nitrogen with the AIR standard, the former is calculated using the following equation:

Stable Isotope Ratios

All values expressed relative to a *standard* as δ values

e.g.

$$\delta^{13}C_{VPDB} = \left[\frac{\left({}^{13}C/{}^{12}C\right)_{sample} - \left({}^{13}C/{}^{12}C\right)_{VPDB}}{\left({}^{13}C/{}^{12}C\right)_{VPDB}}\right] \times 1000$$

Units are *permil* or

The more positive the δ value,
the more enriched the sample is in the heavier isotope

'YOU ARE WHAT YOU EAT... PLUS A SHIFT'
What we eat is reconstituted and incorporated in our body tissues. When we look at food chains, all organic tissue C and N is ultimately derived from atmospheric CO2 and N2. For animals especially, dietary C and N will have had many metabolic alterations from atmosphere to body tissue. Establishing a clear relationship between the isotopic composition of the dietary input and that of the body tissue being studied is essential to the process of palaeodietary reconstruction (Schoeninger & DeNiro, 1984; O'Connell & Hedges, 1999). The isotopic evidence that "you are what you eat" is seen from feeding experiments on a variety of animals from the 1970s to 1990s. In most well controlled studies a close relationship between diet and body isotope composition was observed (Ambrose, 1993).

CARBON STABLE ISOTOPES
The situation for carbon fractionation is quite complex. The isotopic food web begins with photosynthetic fractionation and plants are known to fractionate against 13C to varying degrees during this process (O'Leary, 1981, 1988). Photosynthesis is carried out in different types of plants by using different metabolic pathways in order to fix atmospheric CO2 from the atmosphere or marine/freshwater environments. The atmospheric value is approximately – 8‰ in the present day (in respect to the VDPB standard), but this has become more negative since the industrial revolution by around 1‰ due to the burning of fossil fuels (Friedli et al, 1986). Most plants photosynthesise by the Calvin-Benson or C3 pathway, and this leads to a depletion of the plant δ13C by approximately 19-20‰ relative to the atmosphere.

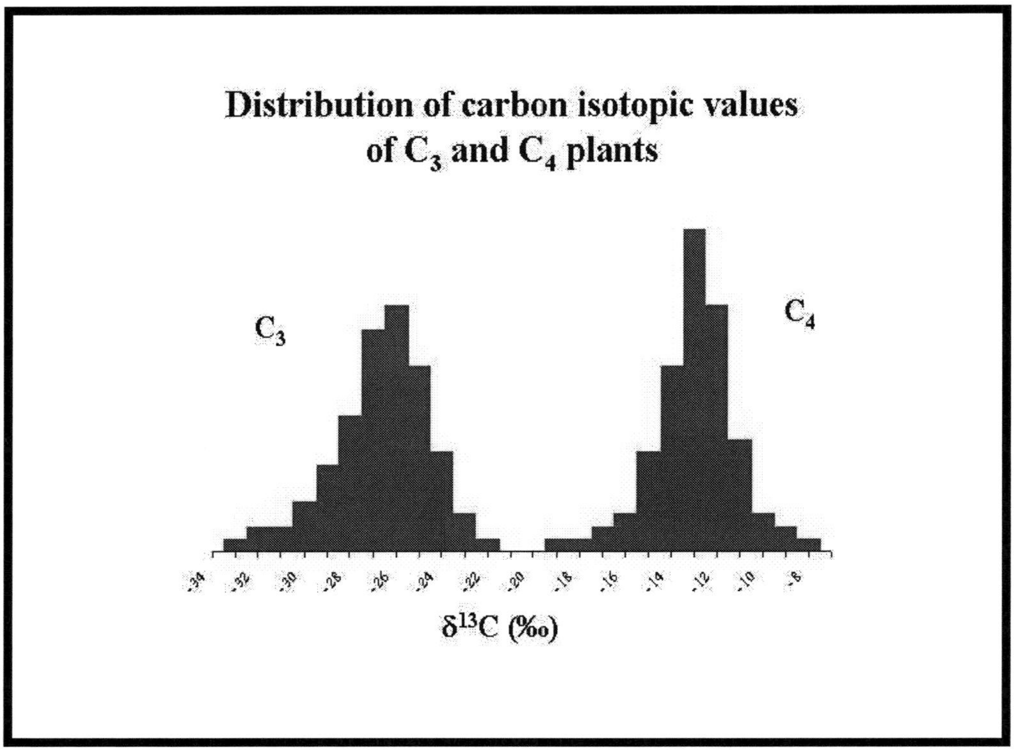

FIGURE 1. C3 AND C4 PLANT δ13C VALUES

Most terrestrial plants such as trees, wheat, barley, oats, rye, rice, beans and legumes are C3 and have a corresponding δ13C value based on the entire plant of approximately -26.5‰ (Hoefs, 1997; Pate, 1994).

Hatch-Slack or C4 plants include maize, millet, sorghum, tropical grasses and sugarcane. These plants generally occur in tropical environments with higher light levels and are thought to have been an evolutionary response to changing atmospheric or climatic conditions in the Late Miocene (Taiz & Zeiger, 1991). C4 plants have mechanisms for water conservation and use CO2 more efficiently. The C4 pathway discriminates less against 13C bringing the fractionation reaction closer to equilibrium. C4 plants are still slightly depleted relative to the atmosphere, there approximate value being around – 12.5‰. As we can see in Figure 1, C3 and C4 plants lead to well-differentiated δ13C values that do not overlap (O'Leary, 1998; Pate, 1994).CAM (Crassulacean Acid Metabolism) plants use photosynthetic processes common to C3 and C4 plants depending on their environmental situation (water availability, temperature and light). CAM photosynthesis is generally used by succulents as it is highly dynamic and most suited to desert environments (Taiz & Zeiger, 1991). Their δ13C values are variable between the ranges of C3 and C4 plants (O'Leary, 1981,1988) that could prove problematic for palaeodietary studies, but they are rarely important sources of food for humans.

The carbon isotope pattern from oceans is quite complex. There is a constant exchange between dissolved carbon dioxide, bicarbonate and carbonate ions, with associated isotopic fractionation. Enzymes associated with both C3 and C4 photosynthesis have been identified in marine algae. Thankfully, these variations tend to average out as we move up the food chain. Fish and marine mammals, which are important in human diet, seem to be nearer C4 values than C3 (Chisholm et al, 1983). The reason for this can be traced back to the difference in δ13C between atmospheric CO2 (~ -7 to –8 ‰) and the δ13C of dissolved CO2 (~ 0 to 1.5 ‰) in the ocean, which are both taken up at the base of the food chain (Hoefs, 1997; Pate, 1994). Therefore marine food webs are ~ 8 to 9 ‰ more enriched in δ13C than C3 terrestrial environments.

'PORQUOI FAIRE LES CHOSES SIMPLE QUAND EN PEUT FAIRE COMPLIQUÉ'

So we can see at the base of the foodweb, carbon fixing or photosynthetic organisms preferentially fractionate against the heavier 13C isotope, and these metabolic processes are well understood (see Figure 2). We can now move along the food chain and observe the situation becoming more complex. When we look at herbivores in the terrestrial environment and marine fish we see that isotopic fractionation turns in the opposite direction and 13C is preferentially retained over the lighter isotope in bone collagen (Ambrose, 1993; Koch et al, 1994). Here we can bring in the old collagen adage; 'You are what you eat… plus 5 ‰'. The shift for herbivores is seen to be around 5‰ heavier or more positive than diet (Ambrose, 1993) giving δ13C values of approximately between -22 to -19‰ (for C3 terrestrial herbivores). This does hold true for omnivores and carnivores but their diet is not based on the entire herbivore (in which case you would expect to observe a + 5‰ shift from the herbivore to carnivore bone collagen isotopic value). The carbon isotopic composition of different tissues within the same animal can vary by up to 7-8‰ (Lee-Thorp, 1989; Lajtha & Marshall, 1994; Hare et al, 1991; Keeling et al, 1999; Ambrose, 1993), but there is no significant variation between similar tissues of the same animal (Schoeninger et al, 1997; Lovell et al, 1986; DeNiro & Schoeninger, 1983). As omnivorous and carnivorous diets are generally based on the flesh and fat of the herbivore we would expect to see a corresponding + 5‰ from the tissue that they are consuming in their bone collagen.

While proteins are positive compared to whole body isotopic value (~ +3-5‰), fats are more negative (~ -2-3‰). This is complicated further by the unusual amino acid composition of bone collagen which unlike any other protein, it contains 33% glycine (isotopically heavy amino acid). There is not enough glycine in dietary protein so it must be partly synthesized from carbohydrates (which are more negative isotopically - Ambrose, 1993; see Figure 3). The resultant observed enrichment in 13C from herbivore bone collagen to omnivore/carnivore bone collagen is small, approximately 0–1 ‰ (Ambrose, 1993). To conclude we can state that although the routing of carbon in diet to bone collagen is not fully understood, studies have shown that carbon in bone collagen is preferentially derived from dietary protein but carbohydrate also has an input (Ambrose,

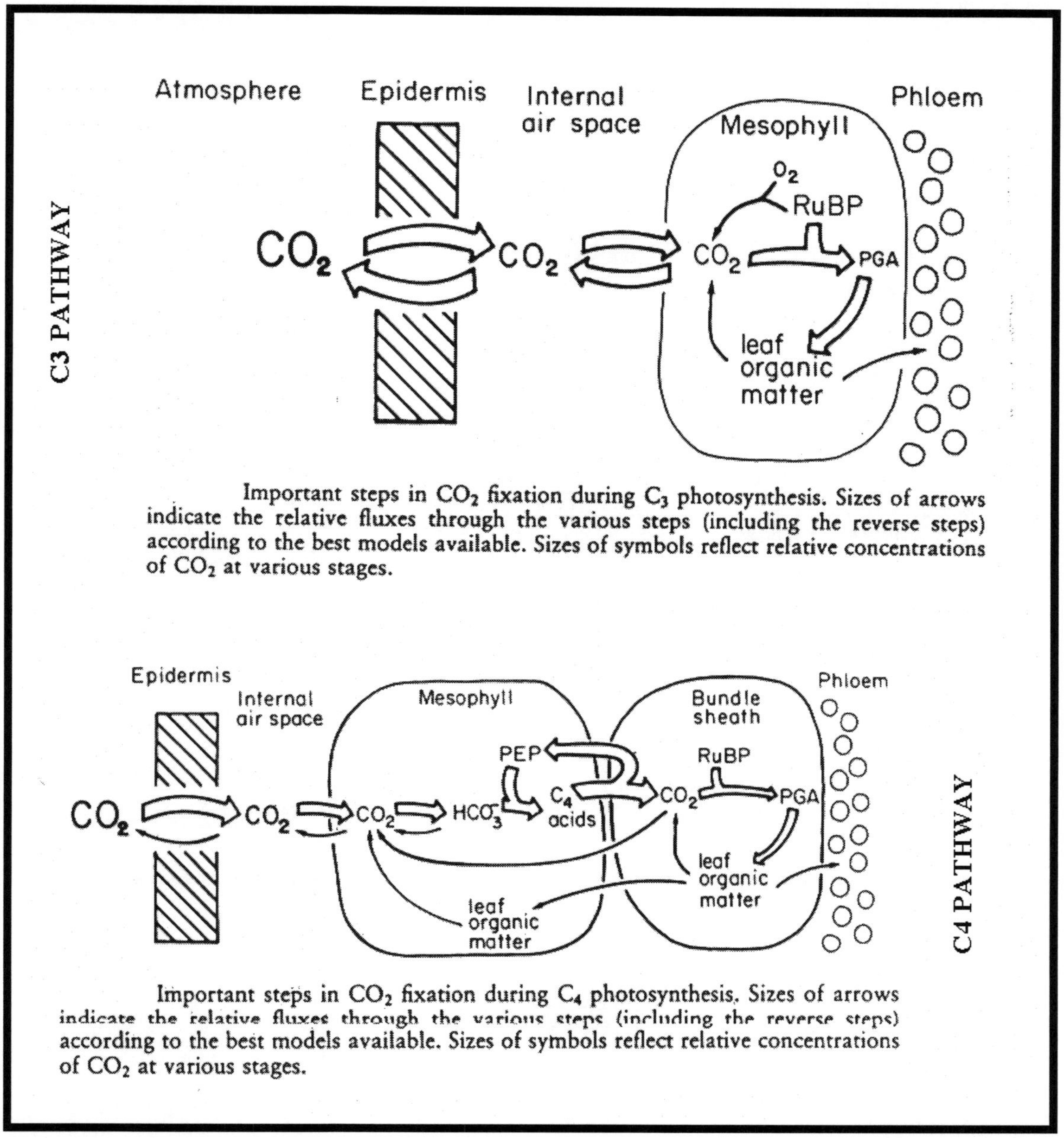

Important steps in CO_2 fixation during C_3 photosynthesis. Sizes of arrows indicate the relative fluxes through the various steps (including the reverse steps) according to the best models available. Sizes of symbols reflect relative concentrations of CO_2 at various stages.

Important steps in CO_2 fixation during C_4 photosynthesis. Sizes of arrows indicate the relative fluxes through the various steps (including the reverse steps) according to the best models available. Sizes of symbols reflect relative concentrations of CO_2 at various stages.

FIGURE 2. PROCESSES OF CO2 FIXATION IN THE C3 AND C4 PATHWAY

1993).

As was described earlier, the main variation seen in carbon stable isotope values is due to the photosynthetic pathway from which an organism fixes its atmospheric carbon. The resultant palaeodietary studies have capitalised on this by observing the introduction of maize agriculture into North America (Vogel & van der Merwe, 1977; van der Merwe and Vogel, 1978) and the Mesolithic-Neolithic transition in Atlantic Europe (Tauber, 1981; Richards, 1998). The former study observed the switch from C3 to C4 terrestrial diet over time and the latter observed the change from marine (similar to C4) to C3 terrestrial diet.

NITROGEN STABLE ISOTOPES

Atmospheric nitrogen (N2) is incorporated into biological systems through N2 fixation by bacteria and soil microorganisms. This is undertaken by legumes and blue-green algae without appreciable fractionation. These are depleted in 15N when compared with plants that utilise soil N2, as N2 containing compounds in the soil have a higher δ15N than atmospheric N2 (DeNiro, 1987). This is due to the cumulative effect of the fractionations associated with fixation and denitrification (Shearer & Kohl, 1986). N2 fixing plants have a typical δ15N range of –2 to +2‰, while non-nitrogen plants range between 0 - 6‰ (Virginia & Delwiche, 1982). These values generally apply to temperate regions with appreciable rainfall. Very arid regions can complicate matters but this will be discussed in more detail below. Marine phytoplankton utilises nitrate and ammonium ions in ocean water, as it does not fix atmospheric N2. As denitrification occurs mostly in oceans, we see more enriched δ15N values than those in terrestrial systems (Shearer & Kohl, 1986).

THE TROPHIC LEVEL EFFECt

The shift in δ13C from herbivore to omnivore/carnivore is not significant enough to determine trophic levels. The input of δ15N values into palaeodietary analysis allows this new parameter to be appreciated. For nitrogen, it appears that 'you are what you eat + 3-5 ‰' (Schoeninger, 1989; Ambrose, 1991). Each trophic level contains organisms whose average protein is roughly 4 ‰ enriched in 15N with respect to the trophic level below. The reason for the increase with trophic level is not yet fully understood, but clues can be derived from studies that have noted waste products from organisms are depleted in δ15N relative to diet (O'Connell, 2001; Ambrose et al, 1993).

Different tissues from the same organism have similar δ15N values (such as flesh, eggs, dairy products and blood). However, organisms of the same species vary in δ15N depending on their dietary input. A study by O'Connell and Hedges (1999) indicated that regular animal protein intake by humans elevated their δ15N values when compared to the δ15N level of vegans (based on hair keratin samples). This study showed that there is no distinction between vegetarians eating secondary animal products and meat eaters,

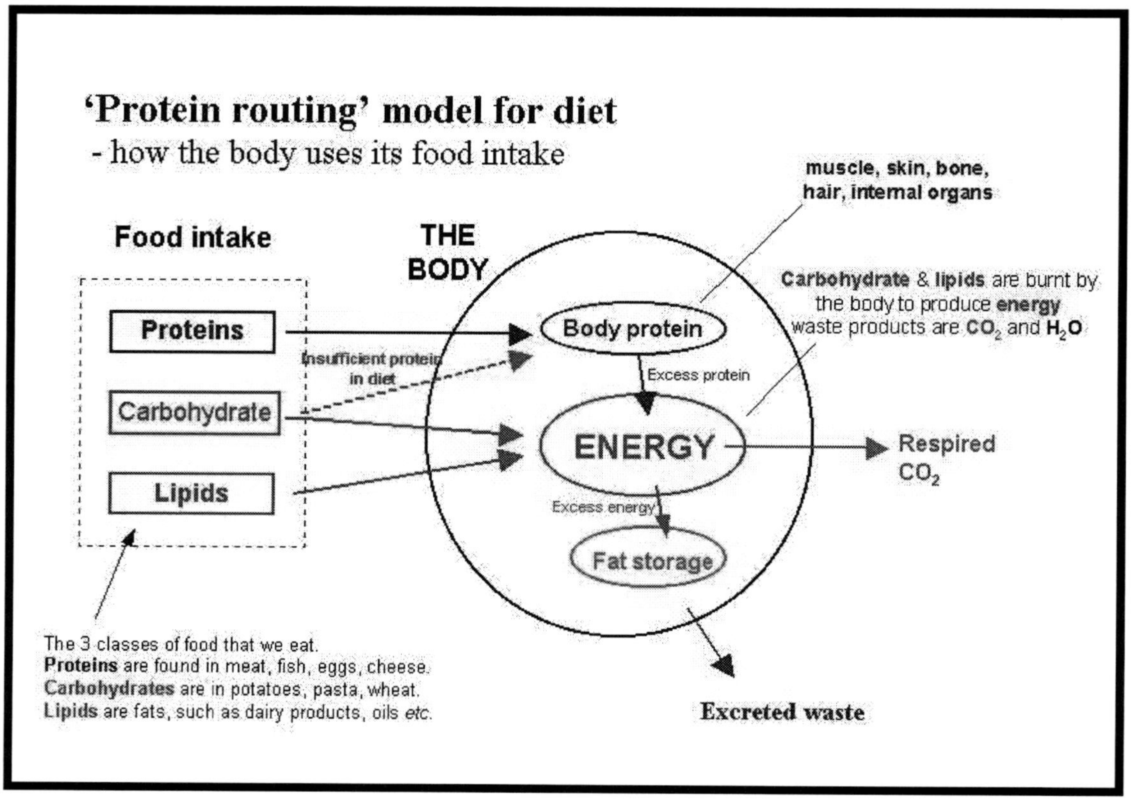

FIGURE 3. PROTEIN ROUTING MODEL FOR DIET

as we have already stated above that all animal tissues show similar δ15N values.

So we can see that nitrogen is relatively simple as only the though there is certainly overlap between categories this can be reduced in more specified geographical applications.

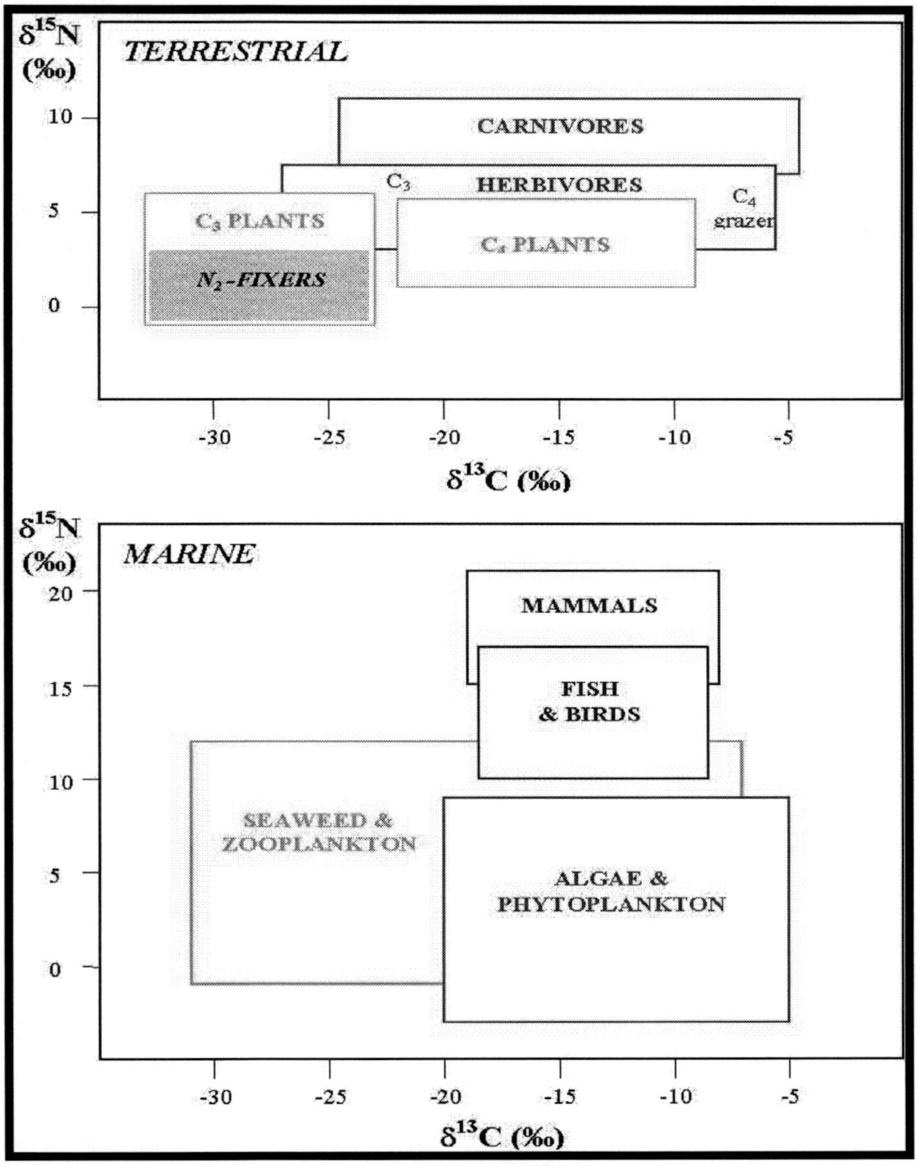

FIGURE 4. ISOTOPIC MAP OF MARINE AND TERRESTRIAL FOOD WEBS

The data obtained from the run has to configure to certain criteria before it can be used for palaeodietary analysis. The amount of nitrogen in the sample has to be greater than 200 micrograms. The final criteria for acceptable results is the C/N ratio which is calculated to give an atomic ratio of carbon to nitrogen atoms in the sample ([mass of carbon/mass of nitrogen] _ [14/12]). The C/N ratio of bone is deemed acceptable between 2.9-3.6 (theoretically 3.2). If the C/N ratios for the isotopic data fall outside this range it cannot be used. Diagenetic processes can significantly alter the C/N ratio to as low as 1.0 or as high as 15.0 (Koch et al, 1994). The final collagen yield of ~ 5% (by mass) from the original bone sample suggests good preservation (Ambrose, 1990, 1993), and samples less than 1% should not be included in analyses. Mass percentage yields were carried out for all samples and only those that conformed to this as well as all the above criteria were used in this analysis. Sample runs were conducted in triplicate where possible (i.e. if enough sample was available), to give an average isotopic value.

protein in the diet (as carbohydrates and fat do not contain nitrogen) and in the body composition need to be considered (see Figure 3). We can now combine the data from the δ15N and δ13C of different food webs to gain a clearer overall pattern in a bi-variate distribution (see Figure 4). Although there is certainly overlap between categories this can be reduced in more specified geographical applications.

CLIMATIC AND PHYSIOLOGICAL EFFECTS ON δ15N AND δ13C

In arid regions the δ15N values of humans and animals tends to increase with decreasing precipitation. This effect has also been noted with high temperatures and increasing salinity although all these environmental factors are connected (Heaton, 1987; Fogel et al, 1997). An explanation for this was isotopically lighter molecules evaporating from the soil in times of drought, therefore enriching the soils in 15N (Schwarz et al, 1999). The most likely model for this phenomenon centres on the function of urea in animals under drought stress. For most mammals the majority of excreted nitrogen is in the form of urea in urine. Urea is comparatively depleted in 15N compared to the body (O'Connell, 2001) and when animals are under heat stress they tend to increase their output of urea and thus retain more 15N in their bodies (Ambrose, 1991). This metabolic mechanism may be the reason for increased δ15N values in plants and animals in areas affected by drought (Heaton, 1986). Bacteria in digestive systems may also cause fractionations (O'Connell, pers.comm.) and other physiological factors can be seen. During growth, starvation, old age, pregnancy and lactation, tissue remodelling is either enhanced or slowed in rate (Bonsall et al, 1997). How or if these processes affect δ15N values is not yet fully understood.

MATERIALS

The human bone samples were collected from various sites mentioned above. No preference could be adopted for the type of bone to be analysed, as samples were rarely preserved as complete skeletons. Almost all of the animal bones were taken from the sites of Son Mas and Son Naveta. Most of the bones from these sites were fragmentary, so identifiable specimens were collected where possible. Bos, Capra, Ovis and Sus made up the domesticate group and the non-indigenous Eliomys (dormouse) was also collected. Bones of the endemic fauna, Myotragus and Hypnomys, were collected from Son Muleta and Son Matge. Very fragmentary, unidentifiable bones from recently discovered dolmens (local term for chambered tombs) located 70 meters west of Ferrandell-Oleza were also analysed in order to see if they could be placed within an identified isotopic group. Jacqui Mulville at the Oxford Natural History Museum verified all the bone identification.

SAMPLING AND MEASUREMENTS

Although bone is a material that 'survives' well in the archaeological record, it does not withstand alteration. In the context of palaeodiet research it is vital to understand diagenetic alteration to bone. Bone is made from an intertwined mineral and organic (protein) portion. We know that the mineral portion is altered to varying extents within differing burial environments and even within the same site, and therefore pre-treatment is essential before analysis. Although protocols exist for pre-treatment of bone for collagen extraction (Longin, 1971), it is rare for laboratories to use identical methods (Howland, 1999). Despite this variability, pre-treatments generally follow familiar stages of acid demineralisation, gelatinisation and lyophilisation based on the Longin method (Longin, 1971). The resultant 'collagen' fraction hopefully retains the bones isotopic composition, as it was when it was buried. The carbon and nitrogen isotopic differences we look for in bone 'collagen' are small, so if burial environments have transformed these, our results will be misleading. Therefore every precaution must be taken to ensure the data collected is acceptable. The sample integrity is judged after measurement in the isotope ratio monitoring mass spectrometer by conforming to certain qualifications similar to the ORAU (Oxford Radiocarbon Accelerator Unit) protocol (Hedges & van Klinken, 1992).

THE PREHISTORIC COLONISATION OF THE MEDITERRANEAN ISLANDS

The scarcity of pre-Neolithic finds in almost all of the Mediterranean islands have led some archaeologists to conclude that human colonisations did not occur before the beginning of the Neolithic (Cherry, 1981). This is probably a misconstrued judgement as the archaeological record by its very nature is invariably incomplete. We know humans have had sea-faring capabilities in the Mediterranean for over 10 000 years (Jacobsen, 1973; Payne, 1975), from the evidence of Melian obsidian and fish bones (from deep sea fish) found at Franchthi cave on mainland Greece. Earlier evidence in the archaeological record this can only be recognized for Cyprus and Sardinia. On Cyprus, man clearly exterminated the endemic population of dwarfed hippopotami during the early Lower Holocene. In Sardinia, it has been inferred from tool evidence that Homo erectus reached there by 0.8 mya (Sondaar et al, 1986).

Although direct human evidence of early island colonisation is limited, we must accept that much earlier settlement or even seasonal movements were entirely possible from the Palaeolithic. Pre-Neolithic use of seagoing craft from other parts of the world is evident. Hunter-gatherers from southern Asia reached Australia at least 40 000 years ago (Allen et al, 1977). In order to do so, even during phases of extremely low sea level, they had to cross stretches of open sea more than 100 km wide. They may have been aided in their enterprise by warm waters and favourable currents, but nevertheless, their travels were much more difficult to accomplish than any intentional or unintentional navigation in the Mediterranean (Schule, 1990).

It is of no coincidence that from the Palaeolithic onwards, the arrival of man to the islands in the Mediterranean has caused the extinction of all the endemic Pleistocene mammal species, flightless birds and giant tortoises that had previously survived for millions of years (Kurten, 1968; Kowalski, 1971). The fauna presently on the islands belong to the same species found on the continent, or can be classified as a subspecies of these.

It can be seen from palaeogeographic maps that vast tracts of coastal land at the end of the last Ice Age in Eastern Spain, Southern France and surrounding the Balearics were available for settlement. After the Younger Dryas (~ 10 000 BC), temperatures rose sharply and glacial melting caused sea levels to rise dramatically and subsequently coastal plains were flooded (around 1.5 million km2 were lost). The scenario suggests that post glacial hunter-forager communities would be feeling the pressure of losing many plant and animal resources from coastal areas (Cherry, 1981; Shackleton et al, 1984). Although an equivalent amount of land was freed from glaciations, it would have been tundra like environment less viable as a resource for hunter-gatherer subsistence. This may be interpreted in the archaeological record by seeing the marked increase in reliance in maritime resources, especially around the Atlantic coast (Richards, 1998). This resource pressure may have been the reason for the colonising the Mediterranean Islands, as fauna was abundant there.

THE ENDEMIC FAUNA OF THE BALEARIC ISLANDS

When at the beginning of the Pliocene the waters of the Atlantic returned through the newly formed Strait of Gibraltar, flooding the Mediterranean basin. Hardly any of the hypothetical animals of the Upper Miocene apparently survived. The only ungulate, which probably constituted a relict of this fauna, is the dwarfed Myotragus balearicus (Andrew, 1915 - see Figure 5). Myotragus remains are exclusively known from caves and rockshelters on Mallorca and Menorca (Bate, 1909).

At least three species of Myotragus existed in the Gymnesic Islands (Mallorca and Menorca – Ibiza and Formentor are referred to as the Pityusic Islands), Myotragus balearicus being the latest. The fossil evidence indicates at least 2 previous species, M.antiquus and M.batei (Waldren, 1982), the former dating to the Upper Pliocene (5-6 mya) and the latter to the Upper Pleistocene (2.5 mya). M.balearicus is thought to originate by the Middle Pleistocene before its Holocene extinction (~ 1 mya – 2000 BC). Very little evidence exists of the two earliest examples of the genus, but what can clearly be seen from the fossil evidence are the disappearance of the 2 pairs of normal incisors and 1 pair of pre-molars in the highly evolved M.balearicus (Waldren, 1982). Normally artiodactyls contain 3 pairs of normal incisors and M.balearicus replaced these with a single ever-growing incisor, similar to that of a rodent hence the name Myotragus, which means mouse goat. The evolution of the dentition of the Myotragus suggests environmental changes requiring the functional evolution of the animal (Alcover et al, 1981; Waldren, 1982). The advantage of this incisor is clear as it would not preclude the animal from grazing or browsing but also allow it to grub, excavating for roots, scraping bark from tough vegetation and even removing lichen from rocks.

FIGURE 5. RECONSTRUCTED SKELETAL REMAINS OF MYOTRAGUS BALEARICUS

Other morphological changes include its steep angled jaw that was supported by a powerful neck and shoulder muscles. It also had shortened fore and hind limbs. The extreme frontal position of the orbits would have given it a reduced visual field (~110°) but allowing stereoscopic vision. Its metacarpals and metatarsals were both very short and stout when compared to Cervus, Ovis and Capra (Waldren, 1982). This characteristic would result in slow locomotion and a low centre of gravity. These combined adaptations to the mountain dwelling creature would have made it more effective in its environment. Stereoscopic vision gave it better judgment of distance and a lower centre of gravity with strength and stability would allow easier climbing for scarce pasture and also allow jumping between rocks with more certainty (Waldren, 1982). Several of the Son Muleta individuals have been found with healed cranial scars or broken horns, and in some cases these were pathological. Waldren has also postulated this as evidence of rutting behaviour as a prelude to mating, as such behaviour can also be seen in Ovicaprids. This is emphasized by the sexual dimorphism in males where horns are longer and thicker and cranial thickening is also apparent (Waldren, 1982).

Another example of Balearic extinct endemic mammalian fauna is Hypnomys (a giant dormouse) of which there are two species, (Anderson, 1984; Kurten, 1968) H. morphaeus in Mallorca and H. mahonensis in Menorca. Both species of Hypnomys survived the Pleistocene but subsequently disappeared along with the last and smallest endemic mammal was Nesiotites hidalgo (a shrew). Reumer and Sanders (1984) stated that both species were present on Mallorca and Menorca when humans first arrived there and Vigne (1987) estimated the time of extinction of both species at around 2500 BC and was associated with ecological disturbances caused by the spread of people and their domestic livestock across the Balearic Islands (Martin, 1984; Vigne, 1987).

With the lack of carnivores on the Balearics, population

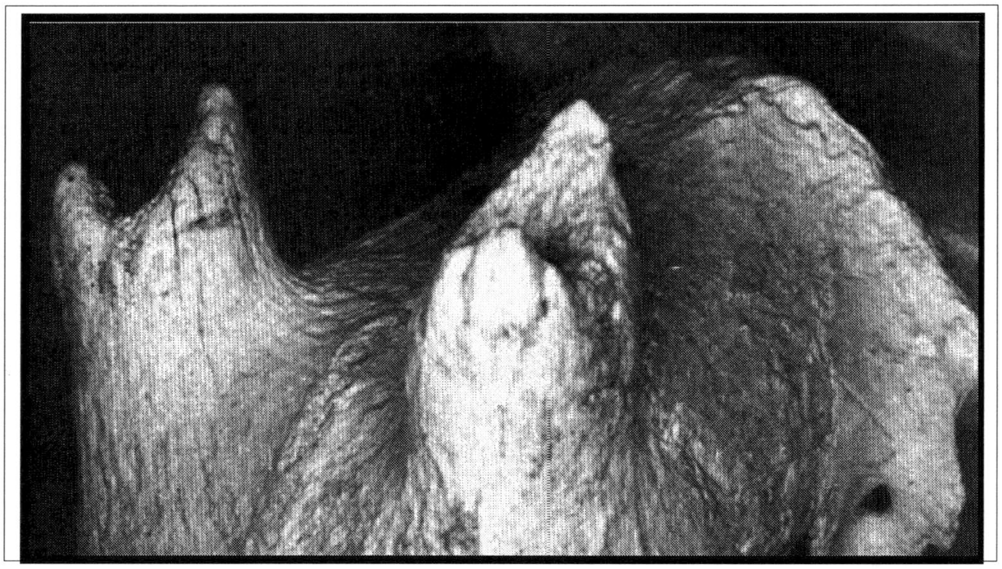

FIGURE 6. MYOTRAGUS CRANIUM, SHOWING SYMMETRICALLY 'V-TRIMMED HORNS'.

FIGURE 7. MYOTRAGUS HORN CORE FOUND WITH BUTCHERING MARKS.

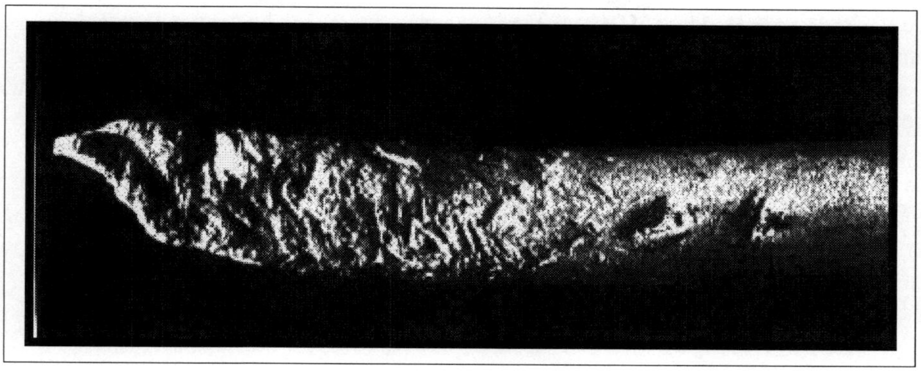

FIGURE 8. BONE CHEWED BY OVIS (SHEEP)

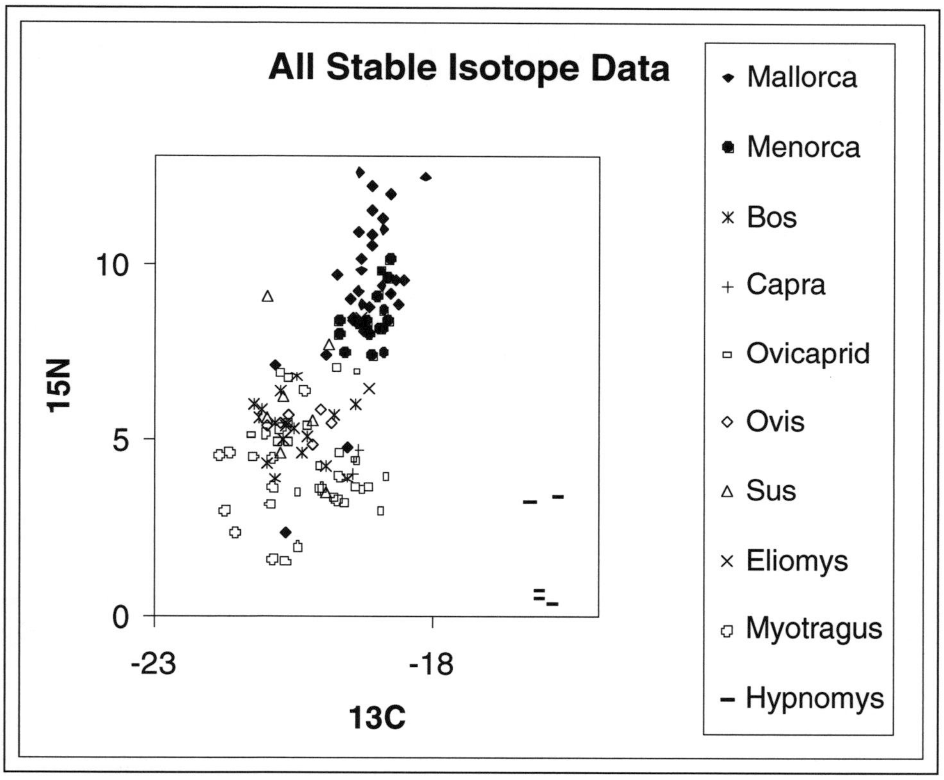

FIGURE 9. ALL THE ISOTOPIC DATA USED IN THIS RESEARCH
(MALLORCA AND MENORCA REFER TO HUMANS FROM THESE ISLANDS)

density would have been restricted mainly by trophic factors and the occasional natural disaster (Dewar, 1984). Carnivorous mammals need a much more extensive territory than ungulates. For this reason, islands of less than 20,000 km2 (Mallorca is just over 3 600 km2) will not sustain carnivorous mammals larger than a weasel. In the Gymnesics, Myotragus survived all climatic changes of the Pliocene and the Pleistocene. No terrestrial carnivore threatened its existence and no other animal competed for its pasture, until the arrival of man.

THE HUMAN COLONISATION OF THE BALEARIC ISLANDS

The evidence of the first human colonisation in the Balearics is on the island of Mallorca and therefore this is the main focus of this research. Early human occupation is currently limited to sites on the West of the island, near the Northern coast. The archaeological evidence suggests around 3500 years of co-habitation with man until Myotragus died out by around 2000 BC (Waldren, 1982). This is problematic for many archaeologists as endemic fauna in the Mediterranean islands is postulated to have died out within decades or centuries of the first colonisers (Schule, 1992). However, it seems that this is not the case in Mallorca and in my opinion unlikely to be the case on many of the other Islands of the Mediterranean. This is due to the compelling evidence showing the interaction of pre-Neolithic man with the extinct endemic fauna, Myotragus Balearicus. Archaeological evidence displays several thousand years of removed in a 'v-trimmed' manner (see Figure 6) with evidence of later healing. Other Myotragus bones were found with signs of butchering and burning in association with hearths. A horn core was also found with butchering marks (Waldren, 1982 - see Figure 7). Very few artefacts were present and this has led some archaeologists to claim that Myotragus caused the v-trimmed horns by chewing (Ramis & Bover, 2001). Osteophagic behaviour in artiodactyls does occur but mainly in Cervids, but it has also been noted in Bovids to a lesser extent (Sutcliffe, 1973; Brothwell, 1976). We can see that chewing marks by Ovis leave distinctive

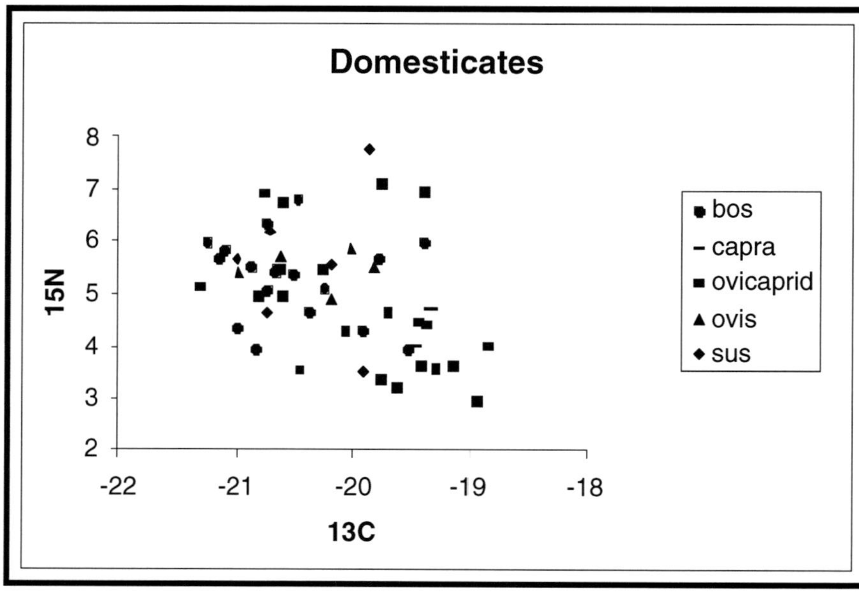

FIGURE 10. ISOTOPIC VALUES OF ALL MALLORCAN DOMESTICATES

coexistence with man before the extinction of the Myotragus (Waldren, 1982, also see Waldren, New Coals on Old Fires, this publication), providing a rare opportunity to investigate possible interaction.

The rock shelter of Son Matge is located near the town of Valledemossa, to the North of Palma. It has revealed the earliest inferred evidence (14C dated charcoal from hearths) of human occupation on the island (5691 BC). Situated in the narrowest section of the North coast road known as S'Estret (the narrow place), Son Matge would have controlled access to and from El Pla del Rei, as it is one of the only direct routes to the North coast of the island. This was possibly a migratory route for the Myotragus. Son Matge has provided the longest stratigraphical sequence of anthropogenic remains in the Balearic Islands (5600-200 BC) and the earliest deposits provide evidence for human interaction with the Myotragus (Waldren, 1982, also Waldren, New Coals on Old Fires this publication).

Coprolite beds were found over 1m in depth and represent the possible corralling of the animal in one sector of the rock shelter. Similar deposits in Mediterranean caves have also been inferred as such (Courty et al, 1991; Brochier et al, 1992). Myotragus remains were found with their horns

marks on the bone (see Figure 8) that cannot be seen on the 'v-trimmed' horns

If Myotragus required bone in their diet from mineral deficiencies, we may see this through stable isotope analysis. Bone is high in protein (~25-30%) and also positive in δ13C when compared to C3 terrestrial plants. Even if bone formed a small proportion of total energy in the diet (5-20%) it could produce a positive δ13C shift if the diet is low in protein (Ambrose, 1993).

Rodent chewing has also been postulated, but in over 2000 specimens at Son Muleta (a cave site 27km East of Son Matge), Myotragus bones or horns showed no signs of these distinctive gnawing marks, and this is in clear association with much rodent fauna. Experimental work has also been carried out in order to see how the horn was split in this manner (if we are to believe that it was human manipulation). It became apparent that by sawing straight across horn with a flint tool and then snapping the bone off caused splintering and this would clearly be of great pain to a living animal. However, when a bone was sawed with a flint tool in a v-shape and snapped it made a clean break. Although horns would have clearly been useful for making tools (needles etc), such care would not be necessary if the animal was only killed for meat. As mentioned previously, the Myotragus deposits from Son Muleta show convincing evidence of rutting, so we could therefore expect to see horn trimming in a corralled environment in order to prevent the animals injuring each other. This is postulated as an attempt to domesticate the Myotragus.

If this is the case and the animal was domesticated, it is the only known example in the world in which a domestication of a species has been followed by its complete extinction (Alcover et al, 1981). However, one could argue that what we are seeing is an attempt at human control rather than domestication of the Myotragus. If the animal were harvested for secondary products like milk or blood, it would offer a more reliable subsistence for human populations. This also explains the prolonged coexistence between man and Myotragus, as continual hunting of the animal would most likely have depleted their food source within several generations (Schule, 1992). Humans would most probably have subsisted on the endemic fauna until its numbers declined. This decline would have necessitated the greater use of domestic animals for meat, which would have to be brought from the continental mainland. The arrival of domesticates is dated to 3375 BC (2700 ± 120 bc uncalibrated – Waldren, 1984, 1997). This date of introduction of domesticates into Mallorca is extremely late when we consider their arrival into other major Mediterranean islands. In Cyprus, Crete, Sardinia and Corsica there is no significant time delay from the arrival of ovicaprines on the mainland to their appearance on nearby major islands, but in Mallorca it is over 2000 years (Vigne, J.D, 1999). This evidence adds support to the theory that humans on Mallorca controlled the Myotragus as a food resource for several millennia, rather than hunting it to extinction.

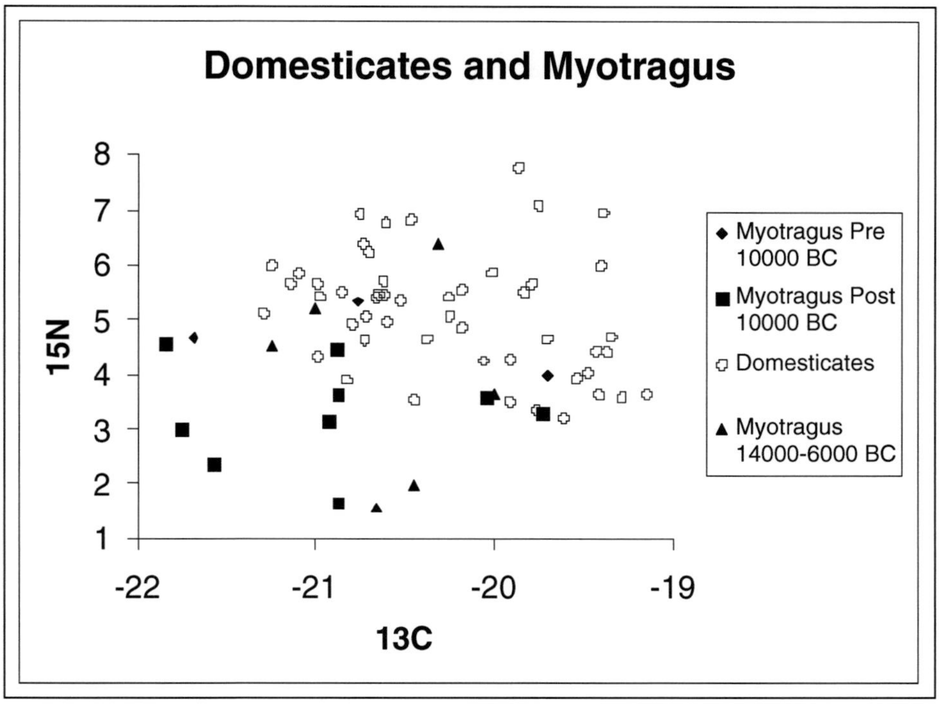

FIGURE 11. δ13C AND δ15N FOR INDIVIDUAL SAMPLES OF MYOTRAGUS (SHOWN AS BEFORE OR AFTER 100 000 BC OR WITH AN ASSOCIATED DATE RANGE) AND DOMESTICATES

ISOTOPE RESULTS FROM THIS RESEARCH

Successful collagen extraction was achieved from 40 humans, 52 domesticate animals, 18 Myotragus and 4 Hypnomys bones. This represents just over 70% of initial bone samples taken. The complete data set of all Balearic samples can be seen in Figure 9. Although this initial plot represents a rather complex picture, some clear distinctions can be seen between humans, domesticates and the endemic fauna. Initially we can see the familiar trophic level variation between humans (the 'diner') and animals (the 'dinner'), but the endemic fauna are clearly isotopically different to the domesticate group.

DOMESTICATED ANIMALS ISOTOPE DATA FROM MALLORCA

As can be seen in Figure 10, the domesticate animals from Mallorca form a fairly large spread in d13C (~ -21.5 to – 19‰) and δ15N (~ 3-7‰). There seems to be no clear distinctions between different species with possibly the exception of sheep (Ovis) and goat (Capra). The only nonruminant in the domesticate group is the pig (Sus). As the pig is an omnivorous animal we would expect to see slightly enriched δ15N values from the other domesticates but this is not really evident with the exception of one anomalous value.

topic data (post 10 000 BC), as climatic change after the last glaciation would cause isotopic variation (aridity and plant species available to eat would have differed and consequently had a knock on effect to isotopic values). Climatic effects may be evident from the elevated Myotragus δ15N ratios (See Figure 11), as they appear to have varied

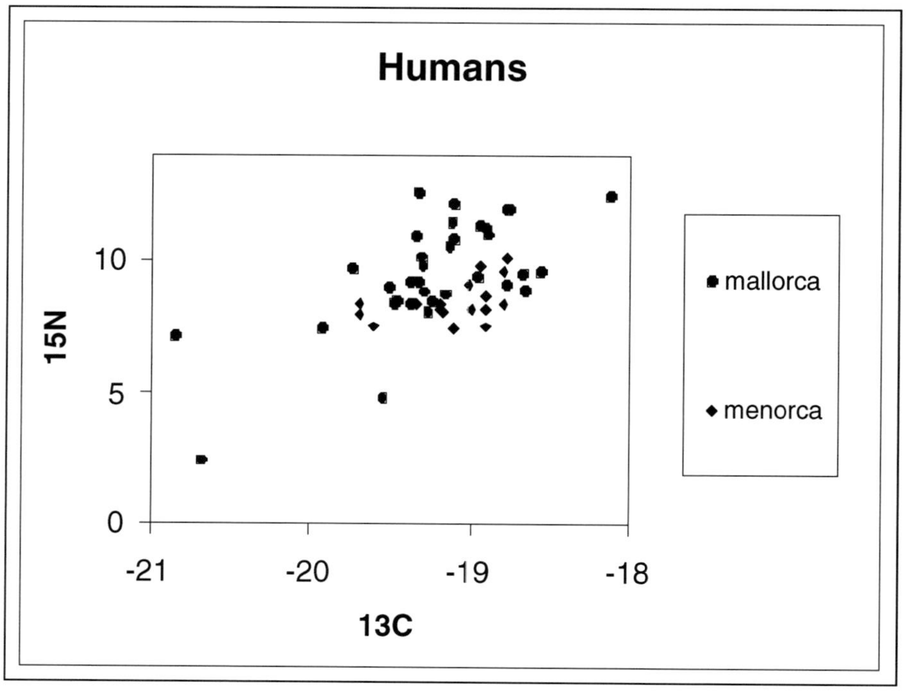

FIGURE 12. ALL HUMAN δ15N AND δ13C VALUES

through time. However, new radiocarbon dates with complimentary isotopic data would be necessary to prove the tentative correlation seen with the limited data taken so far.

MYOTRAGUS AND DOMESTICATED ANIMALS ISOTOPIC VALUES

As can be seen from Figure 11, the domesticate diet on Mallorca seems to be based on C3 terrestrial plants. The Myotragus has been commonly described as a grubbing animal basing its diet on roots and tubers (Waldren, 1982; Alcover, 1999). If a significant proportion of leguminous plants were constituent in the diet of Myotragus, we would expect to see a lower δ13C and δ15N ratio than seen in domesticatead animals, as the latter are generally grazers or browsers. The results of the stable isotope analysis seem to agree with this hypothesis, so it seems that the palaeodiet inferred from the palaeontological evidence could be fairly accurate.

The Myotragus samples tested ranged in date between 14-6000 BC. Some have a highly inaccurate age estimate as these bones were derived from between the 2 dated layers, one meter apart, with radiocarbon dates of approximately 6000 and 14 000 BC.

For this study all Myotragus data was used regardless of age, but ideally only the latest Myotragus deposits at Son Muleta should be used for comparison with the other iso-

HUMAN ISOTOPIC VALUES

All the human isotope values used in this research are shown in Figure 12. Human remains on the Balearics before 2000 BC are based solely on the material from Son Muleta Cave (See Figure 13). Human evidence on the island becomes more common after 2000 BC as expected in the Neolithic, when the population has grown and new burial customs are adopted. The human material taken for isotopic analysis in these later periods, generally dates between 1800-100 BC (See Figure 13)

The Mallorcan and Menorcan values suggest that marine protein was not a significant part of the diet, as d13C ratios are all lighter than -18‰ (Schoeninger & DeNiro, 1984). The Menorcan samples (1600-800 BC) forms a fairly tight isotopic cluster. The diet of these humans would seem to be based on terrestrial plants and domesticated animals. Some δ15N ratios are enhanced in the Mallorcan humans from later periods. In fact all values over 10‰ date after 800 BC (see Figure 13). This trophic level enhancement could correspond to freshwater fish being part of the diet, but this was unlikely to be available in the Balearics. A clearer explanation would be the consumption of animals with similar diets to humans (being fed on human food waste). Rats, mice or domestic dogs could possibly have formed part of the diet, which was not uncommon in Classical Rome and Greece (Brothwell et al, 1998). Cannibalism would also lead to such an enhancement as described above, but there is no evidence of butchery on human skeletal remains from the Balearic Islands.

Another possibility would be climatic effects causing the raised δ15N values. Increased aridity as mentioned earlier has been found to produce this phenomenon (Heaton, 1986). Climatic fluctuations and cooling due to changes in solar activity have been detected in at least three phases of the Holocene in Europe as well as in America and the Southern hemisphere. These climate changes have been related to the 2400 year solar cycle. The Homeric minimum is one of these periods (2750-2450 BC) and coincides with the change seen in palaeodiet in Mallorca. Higher precipitation rates and global cooling was the predicted climate change in this period (Van Strydonck, 1999). This rules out the aridity theory, but this phase is a time of cultural change and migration across Europe. This could be connected with the change in climate. On the Balearics we see major cultural changes as burial customs change (from cremations to lime burials) and the local Iron Age begins (Post Talayotic period) at around 850 BC (Lull et al, 1999). This may suggest that the change in diet was due to the cultural change that was occurring at this date, possibly due to the influx of new populations.

THE EXTINCTION OF THE MYOTRAGUS

The extinction of the Myotragus is often estimated to around 2000 BC (Waldren, 1982; Lalueza-Fox et al, 2000). The radiocarbon dating of burnt Myotragus bone (which was in association with domesticate bone) at Son Matge rock shelter gave the latest dated specimen of the animal (Waldren, 1984). This was dated by the British Museum to 2143 ± 392 bc (the calibrated age is 2735 BC with a 2_ error of approximately ± 700 years). However, Myotragus bone was thought to be found at the settlement of Son-Ferrandell Oleza, suggesting its survival into the Neolithic (Burleigh & Clutton-Brock, 1980). Over 40 radiocarbon dates for Son-Ferrandell Oleza placed the settlement in the period between 2500-1300 BC (but most relate to the period between 2100-1300 BC). Fragmentary unidentifiable bones from recently discovered 'dolmens' (local term for megalithic tombs) near Son Ferrandell-Oleza, were measured isotopically to see if they conformed to human or animal isotopic groups (see Figure 14). It appears that none of the bones were likely to be human, but were probably domesticates and some appeared to be Myotragus. This raises

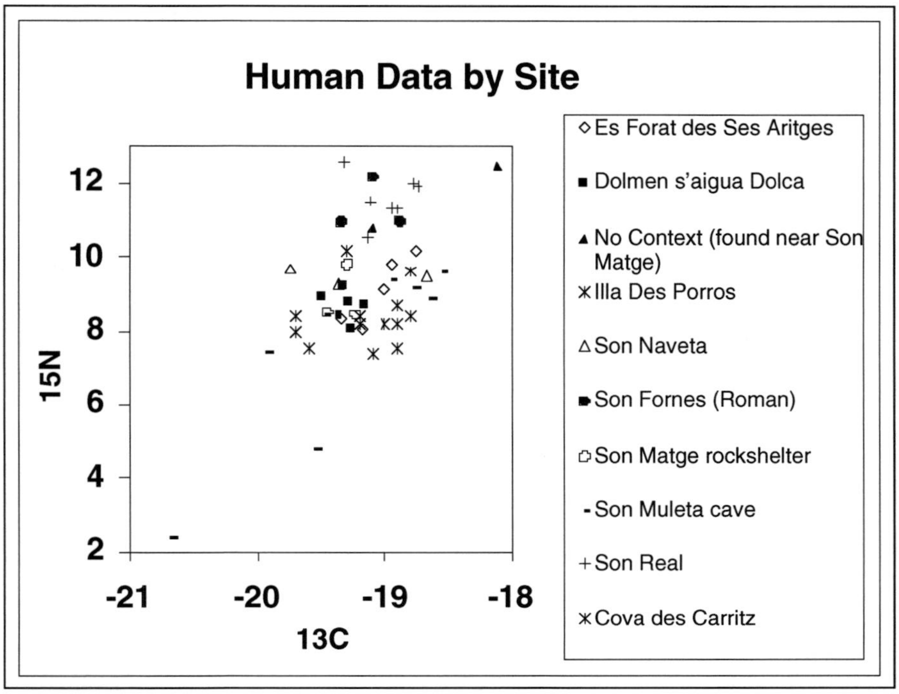

FIGURE 13. ALL ISOTOPIC HUMAN DATA DISPLAYED BY SITE WHERE REMAINS WERE FOUND: SON MULETA (5000-2000 BC), SON MATGE (1400-200 BC), SON MAS (800-200 BC), SON NAVETA (1500-1100 BC), SON FORNES (200-50 BC), AIGUA DOLCA (1800-1600 BC), ILLA DES PORROS (700-100 BC), SON REAL (700-100 BC), COVAS DES CARRITZ ON MENORCA AND ES FORAT DE SES ARITGES (1600-800 BC)

the query whether these structures were actually burial chambers. It is possible that they could be an early settlement contemporary with Son Ferrandell-Oleza. 12 radiocarbon dates have placed these dolmens between 2200–1800 BC. Therefore if Myotragus was part of the fauna here, the estimated time of extinction at 2000 BC does not seem unlikely, but the survival of the animal beyond this date is also possible.

IMPLICATIONS OF RESULTS ON PREVIOUS PALAEODIETARY RESEARCH

As mentioned earlier some current discussion has argued that Myotragus was osteophagic (Ramis & Bover, 2001). As we have already seen the Myotragus is isotopically lighter than any other animal so its diet would have probably been based partly on N2 fixing plants. If they required bone in their diet from mineral deficiencies, we might expect to see this by a significant increase in their δ13C and δ15N ratio. This is due to bone collagen being the isotopically heaviest tissue (in δ13C) in the body and also high in protein ~25-30 %. As was stated earlier, even if bone formed a small proportion of total energy (from all carbohydrates, fats and protein) in the diet (5-20%) we would expect to see a significant shift (+2-5 ‰) as the protein component of diet is routed mainly towards tissues like collagen (Ambrose, 1993). However, the Myotragus is lighter in 13C than any other animal, which strongly suggests that bone was not part of the diet. The giant dormouse (Hypnomys) isotope values have presented this possibility. Distinctly different from any other animal, we can see a significant positive shift in the d13C value in this animal (~ -16 ‰ - see Figure 9 & 14). In the absence of C4 plants being associated with diet of any animals on the island, and the likelihood that the mice were not fishing, it was mysterious as to why the giant dormouse had such a positive δ13C value. If it were eating bone this carbon shift might be explained. However, the dormouse is very light in 15N, and if it were gnawing Myotragus bone this would be more enhanced. It is possible that Hypnomys preyed on another small animal

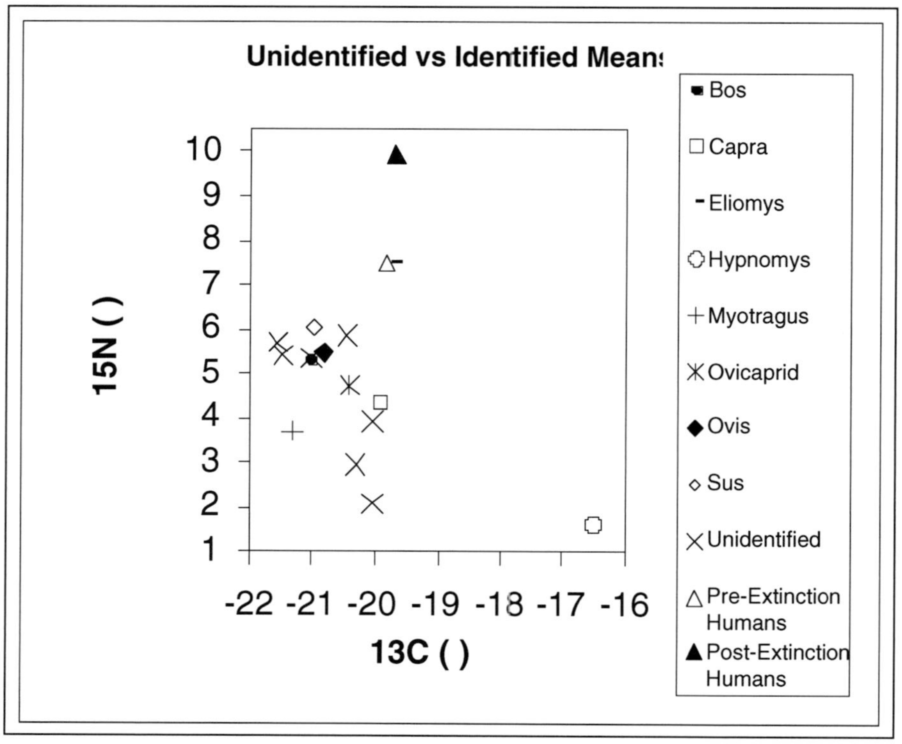

FIGURE 14. UNIDENTIFIED BONES PLOTTED AGAINST THE MEANS OF ALL SPECIES

CONNECTIONS BETWEEN HUMAN AND MYOTRAGUS ISOTOPIC VALUES

One of the main aims of this research was to determine whether or not human diet depended on the Myotragus before its extinction. The extinct animal is isotopically lighter than the domesticates that replaced it, so if humans from Son Muleta (contemporary with the Myotragus) exhibited lower δ13C and δ15N ratios than later populations, it would indicate that these early settlers had probably been deriving their meat/milk protein intake from the Myotragus. The isotope values of humans on the Balearics before 2000 BC are lower in carbon and nitrogen in comparison to Balearic human values after this date (pre and post extinction humans - See Figure 15). This possibly suggests that humans derived a significant proportion of their dietary protein intake from Myotragus before its extinction. In the period after its extinction the human diet has shifted to a higher trophic level appearing to link their protein intake to domesticate animals. The mean isotopic value of the Spanish humans from the mainland is almost identical to the mean isotopic value post extinction humans (see Figure 15), suggesting a similarity in diet.

light in 15N. By consuming and digesting the whole animal, a large δ13C shift (+ ~5‰) would be expected, and this would explain the significant positive shift seen in Hypnomys.

An investigation into the Myotragus diet by analysis of coprolite pollen (Alcover, 1999) concluded that Buxus balearicus was the extinct animals staple food, as it constituted 98% of the pollen (from a plant that only pollinates for about 3 weeks in every year – Richardson, pers.comm.) in the fossilized stool. Myotragus coprolites are currently being analysed at the Natural History Museum in Oxford, and as yet none of the highly poisonous boxwood has been identified from plant macrofossils. The anemophilous pollen may have contaminated the studied samples, but rigorous efforts were made to ensure samples were taken from the centre of the coprolites. The possibility remains that Myotragus could digest the highly poisonous boxwood (which can be fatal to sheep and goats), as it evolved alongside it. However, the isotopic results lend support to the argument that Buxus was probably not the staple food of Myotragus, as it does not appear to be a browser. It seems that from the palaeontological and isotopic evidence that the Myotragus is primarily a grubbing animal. This also suggests that the extinction of the animal was not caused by being out competed by the domesticate animals imported to the Balearics. In the knowledge that man and Myotragus survived several thousand years of coexistence, its extinction may well be due to disease borne from domesticate animals. The Myotragus, which evolved in an insular environment for millions of years, would have had little or no resistance to the bacterial and viral infections that the imported livestock no doubt carried.

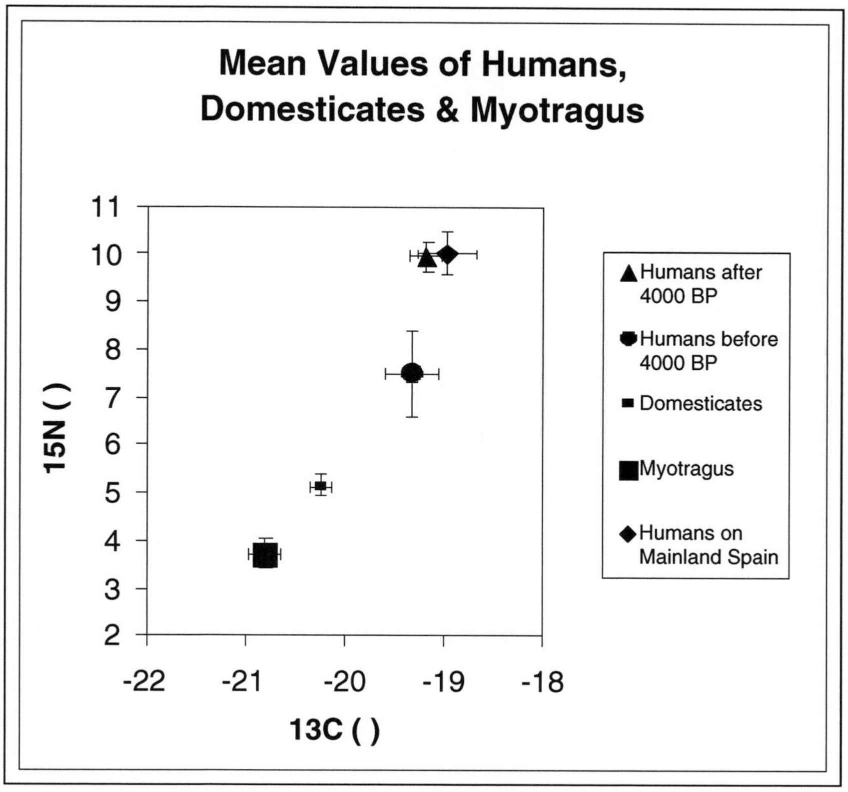

FIGURE 15. MEAN VALUES OF HUMANS (NEOLITHIC HUMAN MEAN FROM MAINLAND SPAIN TAKEN FROM PREVIOUS RESEARCH), DOMESTICATES AND MYOTRAGUS (ERROR BARS SHOW STANDARD ERROR ON EACH POPULATION)

SUMMARY

Carbon and nitrogen stable isotope analysis has been successful in separating the extinct endemic fauna from the Balearics, from human and domesticate groups. The diet of Myotragus and Hypnomys are still not certain, but the findings from this research suggest that the recent conclusions drawn by Alcover (1999) and Ramis & Bover (2001) are premature. The postulated human subsistence on Myotragus before its extinction is not proven but tentatively suggested from the isotopic data. The consumption of Myotragus by later humans (after 2000 BC) is unlikely, but this statement is rather obvious. The possibilities of future work based on some of the conclusions from this initial research seem boundless. Insight could be gained from several areas of study. Accurate dating of the extinctions of the endemic fauna would certainly be of the utmost importance for future investigations into palaeodiet in the Balearics. The unidentifiable bones from late contexts that seem isotopically similar to Myotragus could possibly be confirmed with ancient DNA analysis (Lalueza-Fox & Shapiro, 2001). Isotopic analysis of different light isotopes (sulphur, hydrogen and oxygen) and enamel carbonates could elucidate more dietary and climatic information. If dietary analysis can be taken to the level of amino acids in the future, a more quantitative reconstruction of diet would also become more realistic.

REFERENCES:

ADAMS, S.N.
1975. Sheep and cattle grazing in forests: a review, Journal of Applied Ecology 12, pp. 143–152.
ALCOVER, J.A. MOYA-SOLA, S. PONS-MOYA, J.
1981. Les quimeres del Passat. Els Vertebrats fossils del Plio-Quaternari de les Balears I Pitiusses, Palma: Editorial Moll
ALCOVER, J.A. SANS, A. PALMER, M.
1998. The extent of extinction of mammals on islands, Journal of Biogeography 25, pp. 913-918
ALCOVER, J.A. PEREZ-OBIOL, R.YLL, E.I. BOVER, P.
1999. The diet of Myotragus balearicus Bate, 1909 (Artiocdactyla: Caprinae), an extinct bovid from the Balearic Islands: evidence from coprolites, Biological Journal of the Linnean Society 66, pp. 57-74
ALLEN, J. GOLSON, J. JONES, R.
1977. Sunda and Sahul, Academic Press, London
AMBROSE, S.H.
1993. Isotopic analysis of palaeodiets: methodological and interpretive considerations. In Investigations of Ancient Human Tissue (ed. M. K. Sandford), Gordon & Breach Science Publishers, Langhorne, pp. 59-130
AMBROSE, S.H. & NORR, L.
1993. Experimental evidence for the relationship of the carbon isotope ratios of whole diet and dietary protein to those of bone collagen and carbonate. In Prehistoric human bone - archaeology at the molecular level, ed. Lambert, J.B. and Grupe, G., Springer Verlag, Berlin, pp. 1-37
AMBROSE S.H.
1991. Effects of diet, climate and physiology on nitrogen isotope abundances in terrestrial foodwebs, Journal of Archaeological Science, 18, pp. 293-317
ANDERSON, A.E. WALLMO, O.C.
1984. Odocoileus hemionus, Mammalian Species, no. 219, 9
ANDREWS, C.W.
1915. A description of the skull and skeleton of a peculary modified Rupicaprine Antelope (Myotragus balearicus) with a notice as a new variety (Myotragus volcaiens var. major). Philosophical Transactions of the Royal Society of London, 206, pp. 281–305
BATE, D.M.A.
1909. A new Artiodactyle from Majorca, Geological Magazine 5-6, pp. 385-388
BONSALL, J. C. et al.
1997 Mesolithic and Early Neolithic in the Iron Gates: a palaeodietary perspective, Journal of European Archaeology, 5, 50-92.
BROCHIER, J.E. et al.
1992. Shepherds and sediments: geo-ethnoarchaeology of pastoral sites. Journal of Anthropological Archaeology, 11, pp. 47-102.
BURLEIGH, R. CLUTTON-BROCK, J.
1980. The Survival of Myotragus balearicus, Bate 1909 into the Neolithic on Mallorca, Journal of Archaeological Science 7, pp. 385–388
BROTHWELL, D.
1976. Further Evidence of Bone Chewing by Ungulates: the Sheep of North Ronaldsay, Orkney, Journal of Archaeological Science 3, pp. 179-182
BROTHWELL, D.
1998, Food in Antiquity, Thames and Hudson, London.
CHERRY, J.F.
1981. Pattern and process in the earliest colonisation of the Mediterranean Islands, Proceedings of the Prehistoric Society, 47, pp. 41–68
CHERRY, J.F.
1984. The initial colonisation of the Western Mediterranean islands in the light of island biogeography and palaeogeography, In The Deya Conference of Prehistory Part 1, pp. 145-164
CHERRY, J.F.
1990. The first colonization of the Mediterranean islands: A review of recent research, Journal of Mediterranean Archaeology, 3/2, pp. 145-221
CHISHOLM, B.S,
1989. Variation in diet reconstructions based on stable carbon isotopic evidence, in Chemistry of prehistoric bone, ed. Price, T.D, CUP, pp. 10-37
CHISHOLM, B.S. NELSON, D.E. SCHWARZ, H.P.
1982. Stable-carbon isotope ratios as a measure of marine versus terrestrial protein in ancient diets, Science, 216, pp. 1131-1132
COURTY, M.A. MACPHAIL, R. WATTEZ, J.
1991. Soil micromorphological indicators of pastoralism; with special reference to Arene Candide, Finale Ligure, Italy. Rivistitudi Liguri, A. 57, pp. 127-150
DE NIRO, M.J.
1987. Stable isotopy and archaeology, American Scientist, 75, pp. 182-191
DE NIRO, M.J. EPSTEIN, S.
1978. Influence of diet on the distribution of carbon isotopes in animals, Geochimica et Cosmochimica Acta, 42, pp. 495-506
DE NIRO, M.J. EPSTEIN, S.
1981. Influence of diet on the distribution of nitrogen isotopes in animals, Geochimica et Cosmochimica Acta, 45, pp. 341-351
DEWAR, R.E.
1984. Extinctions in Madagascar: the loss of the subfossil fauna, In Martin, P.S. and Klein, R. ed. Quaternary Extinctions, University of Arizona Press, Tuscon, pp. 574–593
FOGEL, M.L. TUROSS, N. JOHNSON, B.J. MILLER, G.H.
1997. Biogeochemical record of ancient humans, Organic Geochemistry, 27 (5/6), pp. 275-287
FRIEDLI, H. LOTSCHER, H. OESCHEGER, H. SIEGENTHALER, U. STAUFFER, B.
1986. Ice core record of the 13C/12C ratio of atmospheric CO2 in the past two centuries, Nature, 324, 237-238
HARE, P.E. FOGEL, M.L. STAFFORD, JR. T.W. MITCHELL, A.D. HOERING T.C.
1991. The isotopic composition of carbon and nitrogen in individual amino acids isolated from fossil and modern proteins, Journal of Archaeological Science, 18, pp. 277-292
HEATON, T.H.E., VOGEL, J.C. VON LA CHEVALLERIE, G. COLLETT, G.
1986. Climatic influence on the isotopic composition of bone nitrogen, Nature, 322, pp. 822-823
HEATON, T.H.E. 1987 The 15N/14N ratios of plants in South Africa and Namibia: relationship to climate and coastal/saline en-

vironments, Oecologia, 74, 236-246
HEDGES, R.E.M. VAN KLINKEN, G.J.
1992. A review of current approaches in the pretreatment of bone for radiocarbon dating by AMS, Radiocarbon, 34 (3), pp. 279-291
HEDGES, R.E.M. MILLARD, A.R
1995. Bones and Groundwater: Towards the modelling of diagenetic Process, Journal of Archaeological Science, 22, pp. 155-164
HEDGES, R.E.M. MILLARD, A.R. PIKE, A.W.G.
1995. Measurements and relationships of diagenetic alteration of bone from three archaeological sites, Journal of Archaeological Science, 22, pp. 201-209
HOEFS, J.
1997. Stable isotope geochemistry, Berlin, Springer-Verlag
HOWLAND, M.R.
1999. Isotopic Fractionation of Nitrogen During Enzymatic deamination of Amino Acids Unpublished Thesis, School of Natural Sciences, University of Oxford
JACOBSEN, T.W.
1973 Excavations in the Franchthi Cave 1969-71, Hesperia, 42, 45-88, 253-83
JUNIPER, B.
1984. The natural flora of Mallorca, In The Deya Conference of Prehistory Part 1, pp. 145-164
KEELING, C.I. NELSON, D.E. SLESSOR K.N.
1999. Stable carbon isotope measurements of the carboxyl carbons in bone collagen, Archaeometry, 41, pp. 151-164
KOCH, P.L. FOGEL, M.L. TUROSS, N.
1994. Tracing the diets of fossil animals using stable isotopes, In Stable Isotopes and Environmental Sciences, Oxford, Blackwell Scientific, pp.63-92
KOWALSKI, K.
1971. European Upper Pliocene and Quaternary mammal faunas, stratigraphically arranged, with first and last appearances, In Flint, R.F. ed.
Glacial and Quaternary Geology, Wiley, New York, pp. 748–750
KRZANOWSKI,W.
1988. Principles of Multivariate Analysis, Oxford University Press
KURTEN, B.
1968. Pleistocene mammals of Europe, Weidenfeld, London.
LAJTHA, K. MARSHALL, J.D.
1994. Sources of variation in the stable isotopiccomposition of plants, In stable Isotopes in ecology and environmental science, Oxford, Blackwell Scientific, pp. 1-21
LALUEZA-FOX, C.L. SHAPIRO, B. BOVER, P. ALCOVER, J.A.
2001. Molecular Phylogeny of the extinct bovid Myotragus Balearicus, unpublished article
LEE-THORP, J.A. SEALY, J.C VAN DER MERWE, N.J.
1989. Stable carbon isotope ratio differences between bone collagen and bone apatite, and their relationship to diet, Journal of Archaeological Science, 16, pp. 585-599
LONGIN R,
1971. New method of collagen extraction for radiocarbon analysis, Nature, 230, pp. 241-242.
LOVELL, N.C. NELSON, D.E. SCHWARZ, H.P.
1986 Carbon isotope ratios in palaeodiet: lack of age or sex effect, Archaeometry, 28-1, pp. 51-55
LULL, V. MICO, R. RIHUETE-HERRADA, C. RISCH, R.
1999. Ideologia y Sociedad en la Prehistoria de Menorca: La Cova des Carritx y la Cova des Mussols, Consell Insular de Menorca, Ciutadella, 699
MARTIN, P.S.
1984. Prehistoric overkill: the global model, In Martin, P.S. and R. Klein, ed. Quaternary Extinctions, University of Arizona Press, Tucson, pp. 354–403
MARTIN, P.S. KLEIN, R.
1984. Quaternary Extinctions, University of Arizona Press, Tucson
MAYS, S.
2000. The archaeology of human bones, Routledge, London and New York
MERCER, D.
2001. Statistical Analysis of Carbon and Nitrogen Stable Isotope Data from the Balearic Islands, unpublished MSc thesis, Oxford University
O'CONNELL, T.C. HEDGES, R.E.M.
1999. Investigations into the effect of diet on modern human hair isotopic values, American Journal of Physical Anthropology 108, pp. 409-425
O'CONNELL, T.C. HEDGES, R.E.M.
2001. Isolation of urinary urea for natural abundance nitrogen isotopic analysis, poster presentation at SIMSUG Conference, Glasgow
O'LEARY, M.H.
1981. Carbon Isotope fractionation in plants, Phytochemistry, 20, pp. 553-567
O'LEARY, M.H.
1988. Carbon Isotopes in Photosynthesis, BioScience, 38, pp. 328-336
PATE, F.D.
1994. Bone chemistry and palaeodiet, Journal of Archaeological Method and Theory, 1-2, pp. 161-209
PAYNE, S.
1975. Faunal change at Franchthi Cave from 20000-3000BC, In Archaeological studies, 120-131, North-Holland Publishing Company, Amsterdam
PINNEGAR, J.K. POLUNIN, N.V.C.
2000. Contributions of Stable Isotope Data to elucidating food webs of Mediterranean rocky littoral fishes, Oecologia, 122, pp. 399-409
RAMIS, D. BOVER, P
2001. A Review of the Evidence for Domestication of Myotragus balearicus Bate, 1909 (Artiocdactyla, Caprinae) in the Balearic Islands, Journal of Archaeological Science 28, pp. 265-282
REUMER, J. W. F. SANDERS, E. A. C.
1984. Changes in the vertebrate fauna of Menorca in prehistoric and classical times, Z. Saugetierk, 49, pp. 321-25
RICHARDS, M.P. HEDGES, R.E.M. MOLLESON, T.I. VOGEL, J.C.
1998. Stable isotope analysis reveals variations in human diet at the Poundbury Camp cemetery site, Journal of Archaeological Science, 25, pp. 1247-1252
RICHARDS, M.P. MELLARS, P.A.
1998 Stable isotope evidence for similarities in the types of marine foods used by Late Mesolithic humans at sites along the Atlantic coast of Europe, Journal of Archaeological Science, 26, pp. 717-722

RICHARDS, M.P. HEDGES, R.E.M.
1999. A Neolithic revolution? New evidence of diet in the British Neolithic, Antiquity, 73, 891-897
SCHOENINGER, M.J.
1989. Reconstructing prehistoric human diet, In Chemistry of Prehistoric Human Bone, ed. Price, T.D., CUP, pp. 38-67.
SCHOENINGER, M.J. DE NIRO, M.J. TAUBER, H. 1983
Stable nitrogen isotope ratios of bone collagen reflect terrestrial and marine components of prehistoric human diet, Science, 220, pp. 1381-1383
SCHOENINGER, M.J. DE NIRO, M.J.
1984. Nitrogen and carbon isotopic composition of bone collagen from marine and terrestrial animals, Geochimica et Cosmochimica Acta, 48, pp. 625-639
SCHOENINGER, M.J. IWANIEC, U.T. GLANDER, K.E.
1997. Stable isotope ratios indicate diet and habitat use in New World Monkeys, American Journal of Physical Anthropology, 103, pp.69-83
SCHULE, W.
1990. Landscapes and climate in prehistory: interaction of wildlife, man and fire, In Goldammer, J.G. ed, Fire in the tropical biota, ecosystem processes
and global challenges, Ecological Studies 84. Springer, Heidelberg, pp. 273–318
SCHULE, W.
1990. Human evolution, animal behaviour, and Quaternary extinctions, Homo, 41, pp. 229–250
SCHULE, W.
1992. Vegetation, megaherbivores, man and climate in the Quaternary
and the genesis of closed rain forests In Goldammer, J.G. ed. Tropical forests in transition: ecology of natural and anthropogenic disturbance processes Birkhäuser, Basel, pp. 45–76
SHACKLETON, J.C. VAN ANDE, T.H. RUNNELLS, C.N.
1984. Coastal paleogeography of the Central and Western Mediterranean during the last 125,000 years and its archaeological implications, Journal of Field Archaeology, 11, pp. 307–314
SHEARER, G. KOHL, D.H.
1986. N2 fixation in field settings: estimations based on natural 15N abundance, Australian Journal of Plant Physiology, 13, pp. 699-756
SONDAAR, P.Y. SANGES, M. KOTSTAKIS T. DE BOER P.L.
1986. The Pleistocene deer hunters of Sardinia, Geobios, 19, pp. 17–25
SONDAAR, P.Y. SPOOR, C.F.
1986. Human fossils from the endemic island fauna of
Sardinia, Journal of Human Evolution, 15, pp. 399–408
SUTCLIFFE, A.
1973. Similarity of Bones and Antlers, gnawed by Deer to Human Artefacts, Nature, 246, pp. 428-431
TAIZ, L. ZEIGER, E.
1991. Plant physiology, Redwood City, CA, The Benjamin/ Cummings Publishing Company, Inc.
TAUBER, H.
1981. 13C evidence for dietary habits of prehistoric man in Denmark, Nature, 292, pp. 332-333
VAN DER MERWE, N.J. VOGEL, J.C
1978. 13C Content of human collagen as a measure of prehistoric diet in woodland North America, Science, 276, 815-816
VIGNE, J.D.
1987. L'exinction holocène du fond de peuplement mammalien indigène
des îles de Méditerranée occidentale, Memoires de la Societé Géologique de France, 150, pp. 167–177
VIRGINIA, R.A. DELWICHE, C.C.
1982. Natural 15N abundance of presumed N2 fixing and non N2-fixing plants from selected ecosystems, Oecologia, 54, pp. 317-325
VOGEL, J.C. VAN DER MERWE, N.J.
1977. Isotopic evidence for early maize cultivation in New York State, American Antiquity, 42, pp. 238-242
WALDREN, W.H.
1982. Balearic Prehistoric Ecology and Culture. The excavation and study of certain caves, rock shelters and settlements, British Archaeological Reports, International Series 149, Vols. I–III. British Archaeological Reports, Oxford
WALDREN, W.H.
1986. The Balearic Pentapartite Division of Prehistory: Radiocarbon and other age determination inventories British Archaeological Reports, International Series 282, British Archaeological Reports, Oxford
WALDREN, W.H.
1991. Age determination, chronology and radiocarbon recalibration in the Balearic islands, In Waldren, W.H. Ensenyat and Kennard ed. IInd Deya Conference of Prehistory, Archaeological Techniques, Technology and Theory, British Archaeological Reports, International Series 331, Vol. I. pp. 207-266
WALDREN, W.H.
1997. The definition and duration of the Beaker culture in the Balearic islands: A radiocarbon survey, Blackwell, Oxford
WALDREN, W.H.
2001. A new megalithic dolmen from the Balearic Island of Mallorca: Its Radiocarbon Dating and Artefacts, Oxford Journal of Archaeology, 20-3, pp. 241-262
YLL, E.I. PEREZ-OBIOL, R. PANTALEON-CANO, J. ROURE, J.M
1997. Palynological Evidence for Climatic Change and Human Activity during the Holocene on Menorca Islands, Balearic islands, Quaternary Research, 48, pp. 339-347

20

NEW CONTRIBUTIONS TO THE STUDY OF THE NEOLITHIC SEA/LANDSCAPE AND HUMAN INTERACTION ON THE AEOLIAN ISLANDS (SICILY, ITALY)

ELENA FLAVIA CASTAGNINO BERLINGHIERI,

Department of Archaeology,
University of Bristol, Bristol, England
EMAIL: emffidone2libero.it

INTRODUCTION

The wealth of information that can be found and examined about the Proto-historic human interaction on sea/ landscape is mostly connected to the impact that the natural system had on human beings or vice versa, especially when it's involved with natural calamities in a complicated volcanic environment such as that of the Aeolian Islands (FIGURE1). Although this places a huge limit to the information that can potentially be obtained, it seems to imply that the mutual interaction between man and his environment might lead to a better understanding of the benefits or the harmful changes affecting the whole system of living.

Moreover, thanks to a remarkable geological phenomenon, which has preserved an exceptional stratigraphy by the amassing of a fine volcanic ash carried by the West wind, the formation of an archaeological deposit of about ten metres depth (FIGURE 2), has revealed a significant pattern for the sequence of cultures and their interconnections with traders and mariners of the ancient world over several millennia.

In the picture of the Mediterranean maritime economy of Western Prehistory, where the Aeolian Islands have represented a remarkable focal point in ancient navigation - either because of the natural resources present in the islands or their strategic position at the centre of the South Tyrrhenian- the role of human interaction with the Neolithic sea/ landscape acquire a significant meaning towards the investigation of the early societies.

Evidence of archaeological remains which indicates that in the Early Neolithic age groups of people settled on the Aeolian Islands leads one to believe that these people might have found in these islands such a favourable landscape that they were induced to establish their own village. In the passage from seasonal nomadism to sedentariness, still characterised by hunting/gathering and an agricultural way of life, the choice of settlement would have to be dictated by practical criteria, observing the landscape in which the animal and plant populations live, the raw materials and the climate, which represent the main stimuli for selecting a particular site rather than another. The new data considered in this work seem to suggest that this choice was mainly connected to the evaluation of the paleo-environment system as well as to the exploitation of mining resources, which is notably obsidian; in fact, Lipari, one of the four obsidian sources in the Western Mediterranean, seems to have had greater activity and a wider level of diffusion in comparison with other producers (TABLE 1).

The question arises as to the landscape and environmental ecosystem that determined in primis the choice, and thus promoted the activity of exploitation of obsidian on the Aeolian Islands and Lipari in particular. It is not easy to answer these questions exhaustively, since many of them cannot be inferred from the study of the archaeological sites alone, but through examination of a wider scale of evidence, including paleo-environmental. Indeed, the study of organic remains in archaeological contexts, such as pollen, residues of fauna, and lumps of charcoal, if analysed with appropriate methodologies, may provide information on the flora-fauna ecosystem; it is also a vital clue regarding the diet, the strategies of exploitation of the resources, the tech-

niques of working as well as of collection and cultivation used by the inhabitants.
Unfortunately, the results of analysis carried out on the Aeolian Islands have been extremely compromised because of the acid nature reaction of the volcanic soil that does not (Climate, Natural resources), closely connected and complementary. The vector time indicates the chronological variation of the equilibrium established in the ecosystem according with the sub-system components. The analysis of the mutual relationships between the several components

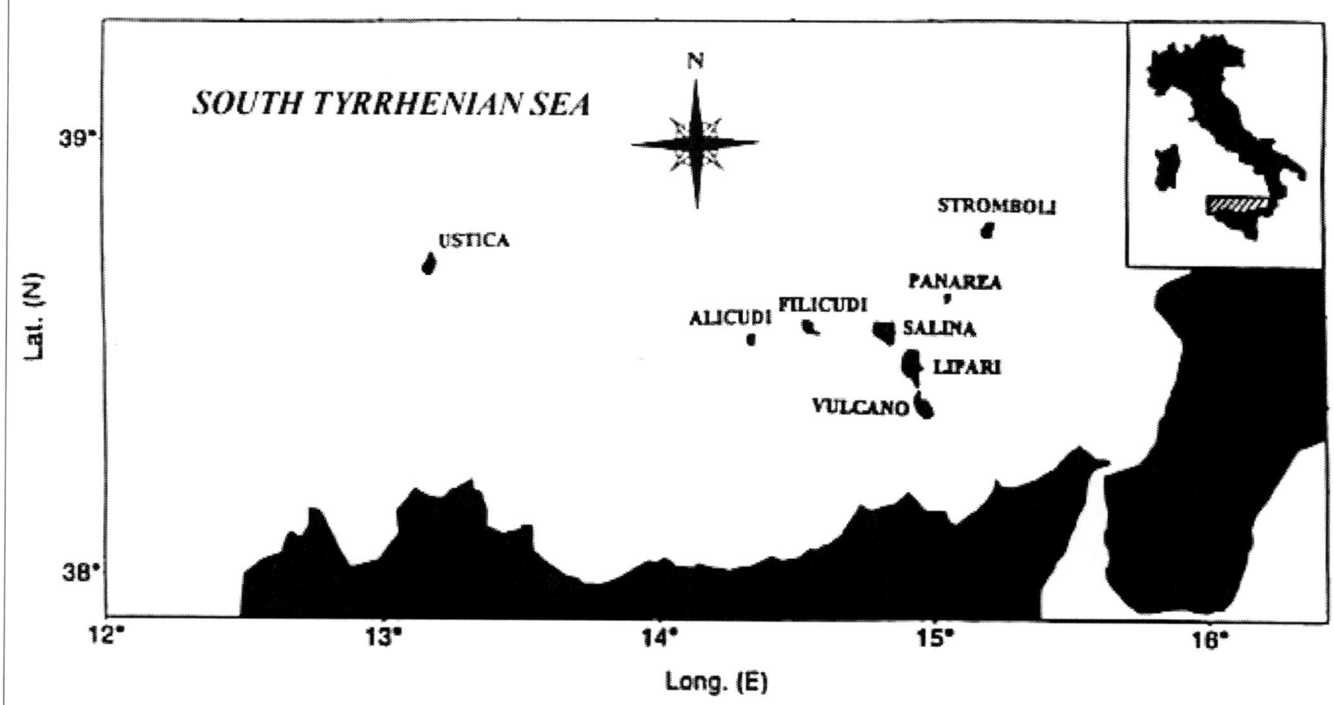

FIGURE 1. GEOGRAPHICAL LOCATION OF THE AEOLIAN ISLANDS.

permit monitoring of the pollinic and sporomorphic complexes characterising the vegetative ground of the Neolithic age. Given the lack of direct remains belonging to the flora/fauna eco-system in the form of organic substances found in a contextualised site, which had been wiped out by the high acid component of the volcanic soil, evidence of the paleo-environment is very restricted. However, taking into account the combination of all of these question marks, a proper analysis through a microeconomic scale, connected to the paleo-environmental situation of the islands, might thrown new light upon the whole question.

MICROECONOMICS SCALE OF EVALUATION

In an attempt to reach a reasonable answer through logical-deductive procedure in relation to the microeconomics approach, let us characterise the interaction man-landscape, imagining the Aeolian Islands located in the centre of the ecosystem synthesised in FIGURE3. Here, the several relations of interaction between human culture and natural environment are represented, articulated in two main subsystems of "biotic" type (Flora and fauna) and of "abiotic" type above reported is focused on the determination of following aims:

(a) the environment of flora and fauna, the islands on which the first settlement took place during the Neolithic age and the factors which have suggested this particular choice;

(b) the type of adaptation and cohabitation occurring between the newly arrived and the surrounding landscape.

Unfortunately, the several attempts aimed at the reconstruction of the archaeological sites on the Aeolian Islands in order to establish the paleo-environmental matrix by direct evidence (Bernabò Brea and Cavalier, 1966: 142-173; Picone Villari, 1985-86: 253-254) have failed.

Owing to the occurrence of oxidation as well as fermentation phenomena that have determined the sterility of those sediments frequently sampled (Villari, 1993: 307-322), we can only detect some information by looking at the indirect traces belonging to the biotic system (until a time comes when new advances provide appropriate typologies of analysis in volcanic contexts).

A connection with the flora environment came from the

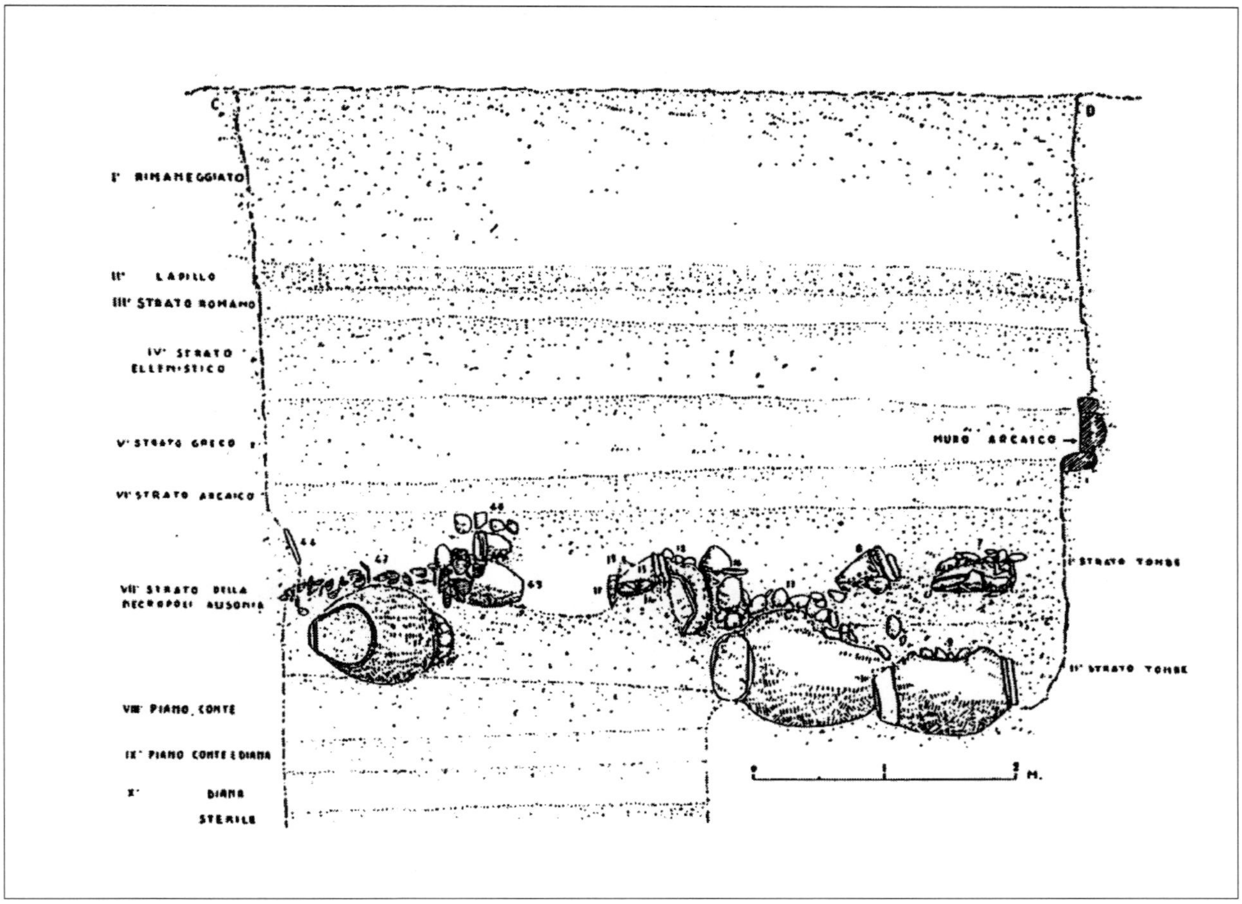

FIGURE 2. SAMPLE OF STRATIFICATION FROM THE NEOLITHIC AGE UNTIL THE LATE-ROMAN AGE, FROM "SAGGIO

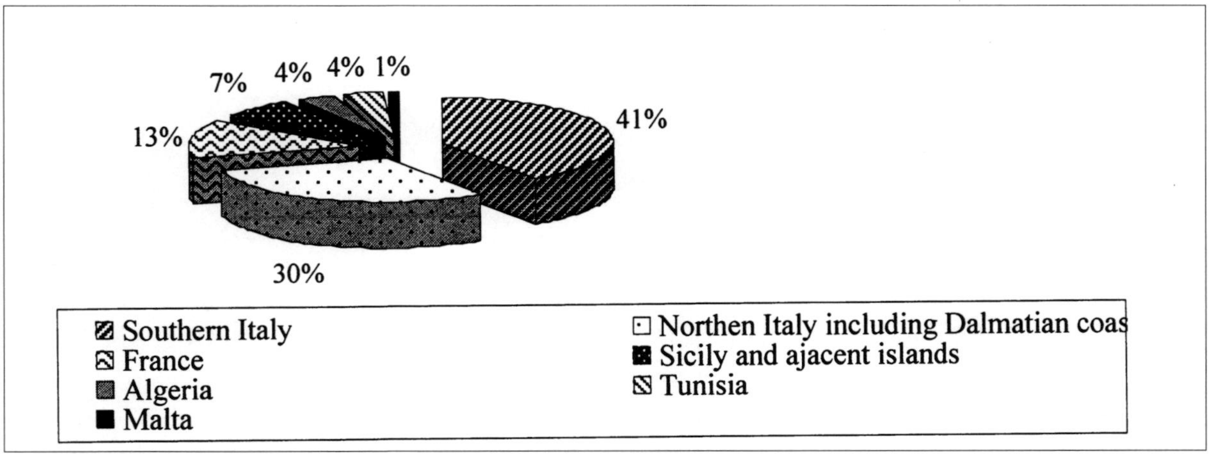

TABLE 1- Distribution of obsidian from Lipari attested to in the Central Mediterranean, including the consumer markets under the influence of Monte Arci in Sardinia and of Palmarola on the Pontine islands as well as of Pantelleria on the channel of Sicily (Data de- T he y

prints of branches which were big enough preserved in numerous cooked clay fragments of plaster of the huts (Bernabò Brea and Cavalier, 1995: 33) recovered in all the layers of the Neolithic hut, both inside and outside, in Lipari as well in Salina. Knowledge of this species - only the negative print of a wooden matrix of difficult interpretation - and the requirements of these in terms of temperature, humidity and light constitute a serious obstacle to specific deductions of the paleo-environment, climatic and ecological. represent, however, an indisputable, albeit generic, source of information. Indeed, these data suggest the presence of trees and perhaps more extensive undergrowth which must have also played an important role in supplying any kind of floating system in order for it to cross the sea.

The recent discovery of fossilised Palm (Chamaerops humilis) found in Filicudi (Lo Cascio and Pasta, in press), and also of scrubs characterised by the dominant presence of the Quercus Ilex or of oak wood with short-lived leaves like the Quercus pubescens (Lo Cascio and Navarra, 1997: 14;16;29), would seem to confirm the hypothesis of a moderately healthy paleo-environment in the islands. Moreover, the recent publication (Bonfiglio, 1998) of an 18th century collection of arboreal fossilised samples from Lipari, belonging to and written by the Sicilian baron Enrico Piraino di Mandralisca, shows the presence of several type of plants. The three plates, well-illustrated in colours by Charles Th. Gaudin, represent the following arboreal spices (FIGURE 4): Chamerops humilis, Hedera helix, Hedera Hibernica and Smilax mauritanica which belong to a species of stunted vegetation; Quercus ilex, Laurus canadiensis, Laurus nobilis belonging to a high trunk species.

Moreover the presence, in both villages, of grinding stones made of grey lava constitutes a clear indication of the presence of cereals, which is an obvious pointer to a rational agrarian activity. It also suggests that a kind of advanced organisation might have been in place, strictly linked with a productive cycle related to the exploitation of the subsystem source flora in terms of cultivation/processing/production to the aims of domestic subsistence, an aspect which presupposes a certain level of technological development.

Regarding the possible presence of animals both wild or domesticated, the bone remains of sheep, goats and cattle on the Aeolian Islands is not evidenced until the Middle Neolithic Age, while in Sicily it is already attested by the EN (TABLE 2). Despite the occasional presence of pig (Sus scrofa ferus) there is no evidence of local domestication and the bones belonging to sheep/goats were presumably introduced already domesticated.

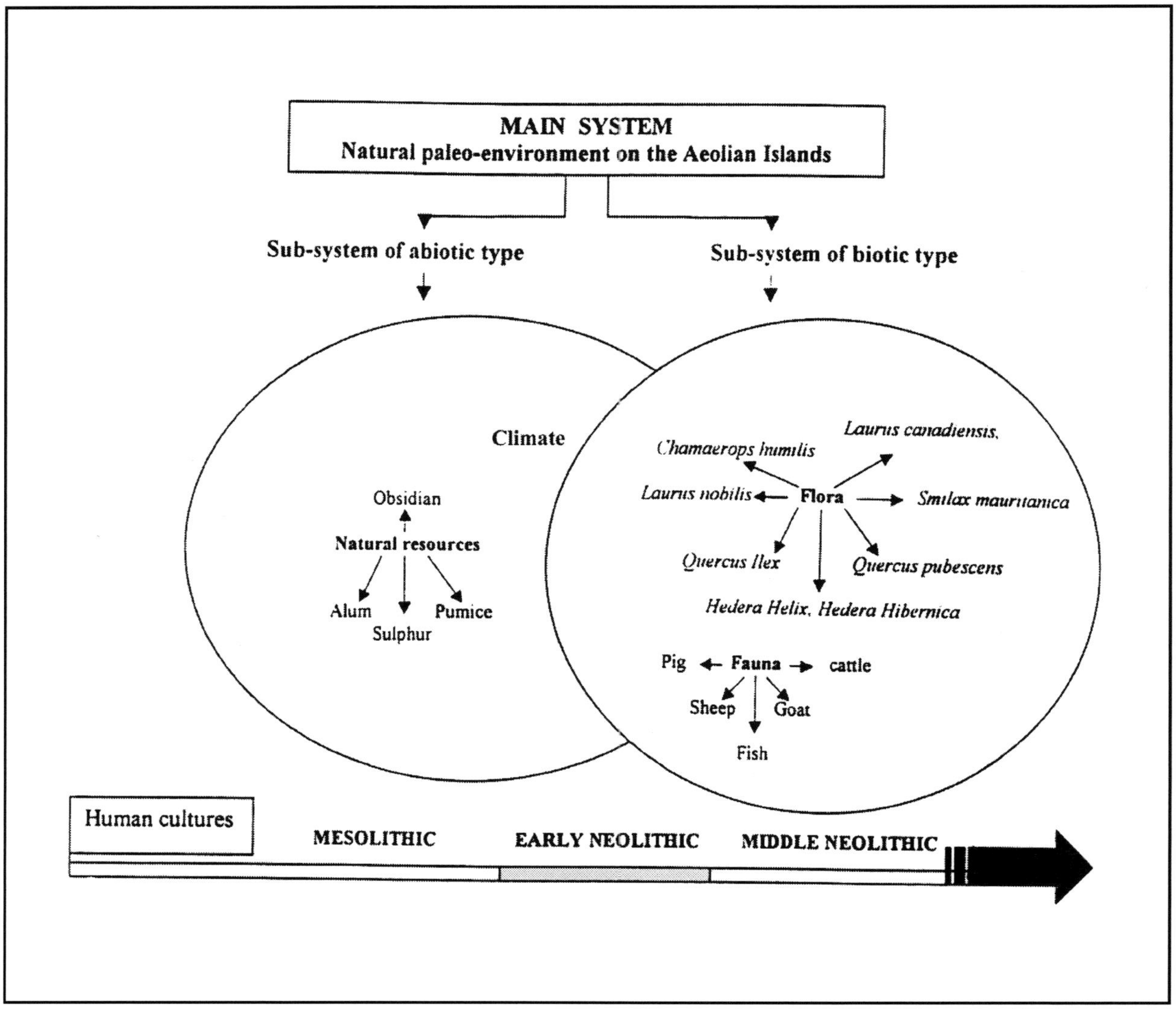

FIGURE 3. Synthesis of the paleo-ecosystem on the Aeolian Islands. Relation of interactions between human culture and natural environment. The vector time indicate the chronological variation of the equilibrium established in the ecosystem according with the sub-system components.

By comparison with the Sicilian context sheep and goat seem more abundant, cattle are quite common, pig is attested to, although not in great numbers. It is also interesting to observe, on the MN levels, the presence of worked instruments made of animal bones, mainly from the sheep species (FIGURE 5). In the case of Lipari-Acropolis, among the 16 bone instruments associated with trichrome pottery (MN), it is possible to recognise their offensive function, as they were conceived as a sort of weapon characterised by pointed end (TABLE 3).

Blades and pins were doubtless more common, mainly made using long sheep bones, even if cattle samples are also attested. Long fish bones made into very pointed pins are also attested. Those instruments which are not very big in size bring to mind implements used for slaughtering small animals, which seems to reflect both the persistence of hunting as well as the practice of domestication.

Even with the extreme lack of the paleo-environmental data as well as the scarcity of information which is also restricted to limited digging, it may be asserted that during the end of 5th millennium BC, the populations which settled on the islands were attracted by environmental factors like water or vegetation and possibly also animals.

TWO CASES OF STUDY: CASTELLARO VECCHIO AND CONTRADA RINICEDDA.

For detailed analysis two cases of study related to the Early Neolithic age have been selected: the Castellaro Vecchio near Quattropani on Lipari, and the Contrada Rinicedda in Salina. These examples are chosen because of their status as first settlements, the oldest inhabited villages recognised in an Aeolian context. It seems clear that the position of the first settlement of Lipari on the North West area of the island was deliberately chosen on the high top of the tableland. This choice suggests that the newcomers were more

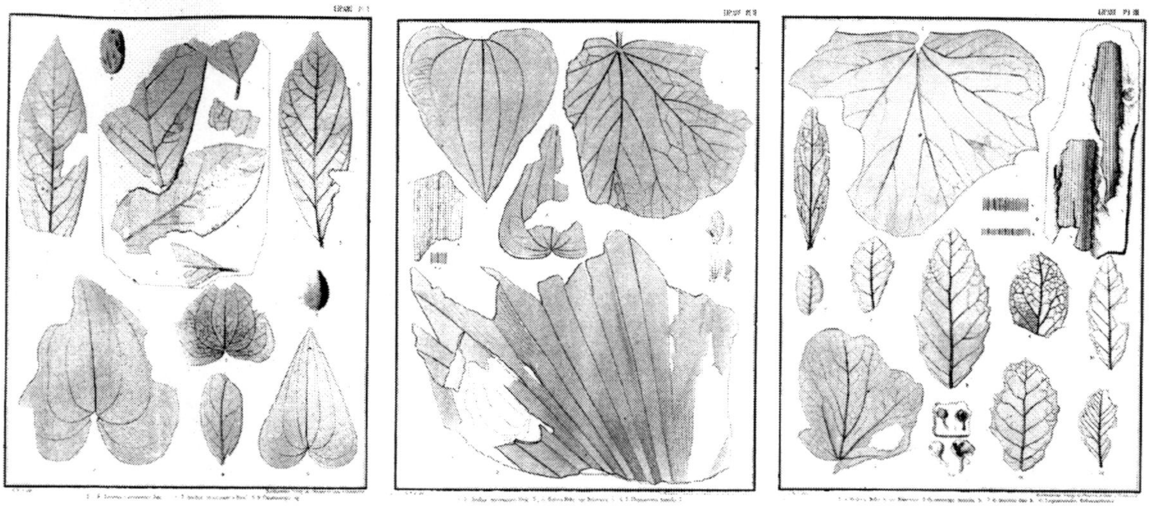

FIGURE 4. THREE TABLES BY CHARLES TH. GAUDIN (SOURCE: GAUDIN AND MANDRALISCA, 1860,

MIDDLE NEOLITHIC	**ANIMAL BONES**		
Lipari-Acropolis layers	**TOTAL** 207	**Instruments made of bone** 16	- 2 pins (fishbone) - 1 blade (unidentified animal) - 3 drifts (unidentified animal) - 2irregular sherds (unidentified animal) - 1 spatula (cattle) - 1 pointed long bone (cattle) - half pointed long bone (cattle) - 5 regular sherds (unidentified animal)
		Natural bones 191	Some of them show traces of use

TABLE 2. Comparison of faunal samples between Sicily and the Aeolian Islands. Number of animal bones recovered, arranged by period (Main sources: Bernabò Brea and Cavalier, 1980; Leighton, 1999).

bound up with agricultural tradition than a maritime involvement. Indeed, it is a strategic position placed between an old river basin with very fertile fields, the little Madoro spring, the obsidian source nearby: the first two factors related to reasons of subsistence, the second relating to production and exchange. The old river basin in the proximity of the Timpone Pataso, not far from the village, guarantees very rich soil, while the spring offers a natural water supply and a good environment for the vegetation. Abundant prints of plants remains, such as palms, olive leaves, and small fragments of arboreal species, are well documented in this place which is still the most fertile field of Lipari, nowadays prosperously planted with vineyard (FIGURE 6). The productive factor is connected to the cultivation of the fields allowing the inner maintenance, as well as to the extraction of obsidian on the eastern side of the obsidian flow of Lami-Pomiciazzo (Gabellotto), at that time also easily reachable approaching from the mainland (Castagnino Berlinghieri, 1997: 160-162). It is important to underline that, although this site is under cultivation, there is still an abundance of obsidian pieces spanning from small fragments of debitage to entire artefacts such as rasping tools or blades (FIGURE 7).

The geomorphologic situation of this area has been completely modified by the violent eruption occurring during the Medieval Age which produced the huge obsidian flow of Rocche Rosse together with the Forgia Vecchias' flow discharged from the Pirrera crater (see also Chapter 2). Indeed, the medieval flows covered most of the NE area of Lipari, resulting in a considerable transformation of its landscape. All of the possible Neolithic path-connections between the settlement of Castellaro Vecchio and the coastline, were buried under the medieval obsidian flow. The workshop-centres identified in Canneto-Lami (Keller,1970:75) and in Canneto-Acquacalda (Buchner, 1948) constitute obvious confirmations of this exploitation; the recovery of an obsidian deposit containing 26 obsidian cores in the first site is also witness to the safekeeping of the important utilitarian material or, more probably, it indicates a phase of gathering for the purposes of export.

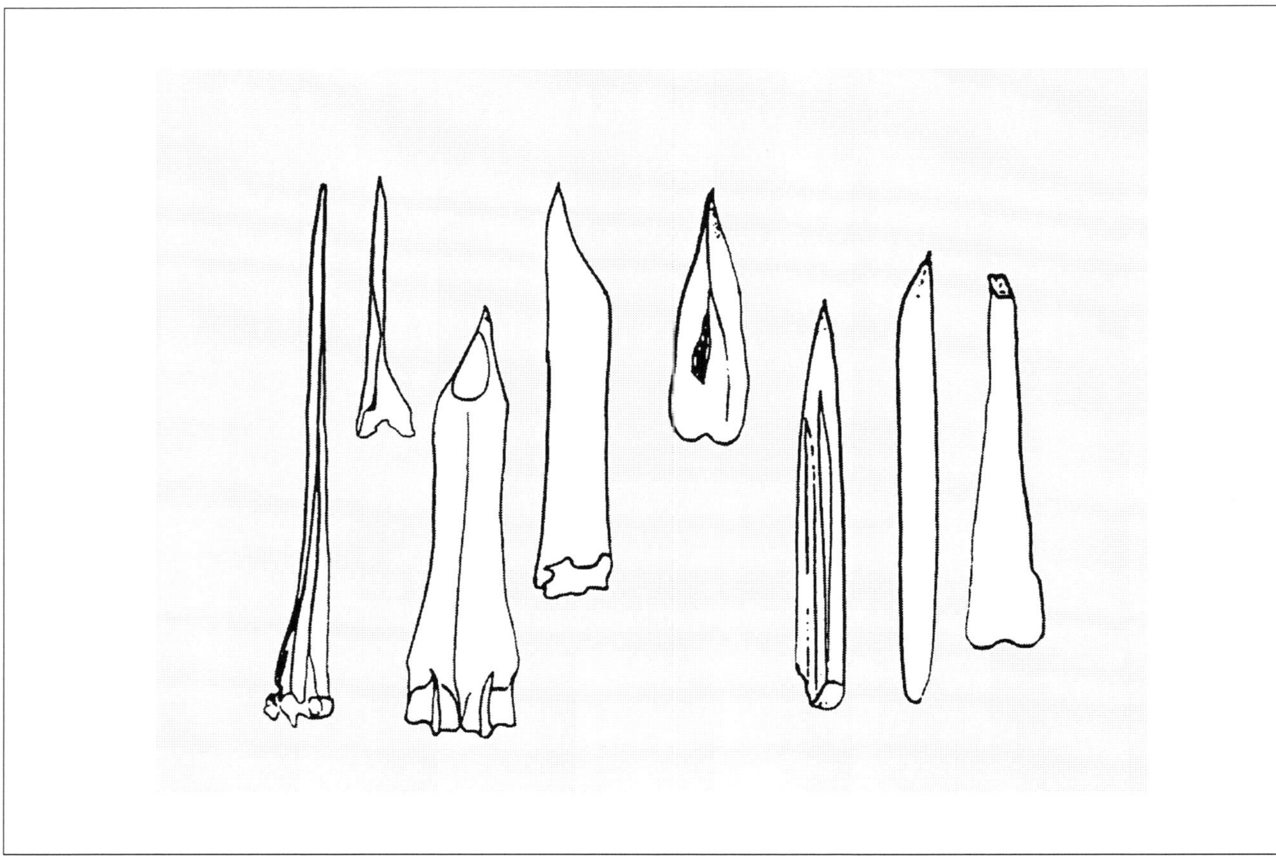

FIGURE 5. WORKED ANIMAL BONES RECOVERED FROM LIPARI-ACROPOLIS LAYERS
(Data derived from: Bernabò Brea and Cavalier, 1980).

As already discussed elsewhere (Castagnino Berlighieri, 1997:153-159), the maritime topography of the N/E coast of the island was characterised by a more favourable coastal situation during the Neolithic age. This coastal situation was very suited to supplying the activity of production and exportation of obsidian. A loading place for the cores or pre-cores (Ammerman, and Andrefsky, 1982) or even for instruments already fashioned and ready for export may be identified at the landing place the Spiaggia della Papesca. The shoreline profile between Punta di Sparanello and Faro di Porticello therefore simultaneously constituted the possible place of extraction, working, port of call and export of the main source of the island. Observing this place in economic terms, it can be seen that from the Spiaggia della Papesca it is very easy to establish operations of transport and loading, following the ratio of economy of means and distances that regulated the movements over the ages (Castagnino, 1997:153-159).

It is important to observe the topographical position of the Neolithic hut found in Salina, the only one systematically dug up till now, situated on the SW coast of the island. The particular location, characterised by a high flat back on Contrada Rinicedda - on which a wider village must have extended - seems to be the result of a considered choice. It is surrounded by two little streams Vallonazzo and Vallone di Rinella, a source of life and limit of natural defense, conveniently safeguarded and well sheltered from the prevailing winds, less than one mile of sight navigation from the NE slopes of Lipari, source of the supply of obsidian.

A contingency relationship between the Village of Salina (at the current state of the diggings limited only to a hut) and that on Lipari may be supposed in terms of material culture and perhaps also of domestic economy. This hypothesis is suggested by the typological homogenates which characterise the impressed as well as red-band painted ceramics associated with a lithic industry in obsidian in which blades and laminar flint are widespread. The Castellaro Vecchio on Lipari had probably constituted the first settled group of people, promptly followed by that of Salina. The Contrada Rinicedda Village on Salina is very easily reachable from Lipari sailing with coast in sight, crossing just a short section of sea. These two sites are respectively visible and easily distinguishable one from the other: the Salina hut is situated up to 40 meters above sea level, exposed to the South; Castellaro Vecchio is located up to 400 metres above sea level, situated to the NW. This would appear to suggest that Salina used to exploit the obsidian from Lipari, in a perspective of interchange between the two settlements (FIGURE 8).

FIGURE 6. VIEW OF THE FERTILE FIELD OF LIPARI, NOWADAYS PROSPEROUSLY PLANTED WITH VINEYARD AT CASTELLARO VECCHIO NEAR QUATTROPANI ON LIPARI

It seems that the first phase of adaptation man-landscape on Lipari was followed by a second phase of inhabited extension to Salina. It suggests a picture of balanced cohabitation of the two villages whereby, even if separated on two different islands, they were integrated with the surrounding landscape and probably interacted within the systems of material culture diffusion in order to support their daily maintenance. Only in such conditions could they assure a balanced standard of living, strongly stimulated by the exploitation of the geologic subsystem: the resource obsidian.

EXTRA EVIDENCE OF HUMAN INTERACTION

Analysing the first evidence of human interaction in the Central Mediterranean one can also recognise that some data clearly attest relationship between the Aeolian Islands and Sicily starting from the pre-Neolithic Age. The beginnings of sporadic extractive activity that presaged a systematic exploitation, a kind of preliminary to an economic system that was to become the most prosperous in the Prehistory of the Western Mediterranean, is testified to in at least two contexts of the pre-Neolithic age in Sicily. These were (FIGURE 9): Grotta dell' Uzzo in Western Sicily (Francavilla and Piperno, 1987: 31-39), dated to 6th millennium BC (Piperno, Tusa, and Valente,1980), and that one of Perriere Sottano in the mainland near Mount Etna (Aranguren and Revedin, 1996: 35) found in a Mesolithic context.

In both cases the radiocarbon dates show that the chronological context of the obsidian recovered marks a fundamental moment in the development of the economic process of interchange, underlining the relationship of mutual influence between Sicily and the Aeolian islands. It would be somewhat simplistic to think that in the context of the pre-Neolithic age only the site of Perriere Sottana on S/E Sicily and Grotta dell'Uzzo in Western Sicily had had contact with Lipari, even if indirect.

However it may be observed that the two mentioned sites, both quite far away from the original source, could represent just exempla of a wider picture of contacts. This is a situation which was certainly more extended either towards the East or towards the West, necessarily limited in interpretation owing to the current state of the research in the context of Sicily. What has become evident is that during the Mesolithic age, while the settlement of stable groups of populations in the archipelago had yet to occur, obsidian from Lipari had reached the rock-shelter of the hill of Perriere Sottano, on the left side of the Gornalunga river, in the Etna-Piedmont zone. The area of Etna and more extensively that located on the East and South/East of Sicily was highly receptive to the lithic industry generated by obsidian, which from the Neolithic age was to become one of the greater user areas in Sicily. It is also apparent that obsidian from Lipari, in the course of the 6th millennium BC, reached the coasts of the western Sicily, as is abundantly documented in the impressed pottery phase of the Grotta dell' Uzzo.

It is also important to recognise that the site of Perriere

FIGURE 7. CASTELLARO VECCHIO NEAR QUATTROPANI ON LIPARI:
abundance of obsidian pieces spanning from small fragments of debitage to entire artifacts such as rasping tools or blades (photo taken by the author in September 2000).

Sottano together with that of Grotta dell' Uzzo - the first in the Mesolithic period and the second in the Early Neolithic - documents at least two channels of interaction belonging to the same route of mixed exchange: one marine (Sicily-Lipari-Sicily), the other across the mainland. Indeed, the discovery of Liparian obsidian in Mesolithic levels of Pirrere Sottano has thrown new lights on the whole dispute regarding interchange and mobility during the Western Mediterranean Mesolithic age. Indeed, the settlement of Castellaro Vecchio near Quattropani marks the end of nomadism and the beginning of a stable phase of life on Lipari.

FIGURE 8. VIEW OF SALINA FROM CASTELLARO VECCHIO ON LIPARI (photo taken by the author in September 2000).

Earlier evidence of traded materials on the Aeolian Islands

It is generally accepted that coarse and impressed ware were manufactured in loco and did not travel very far away from the production site. The total absence of high quality clay deposits throughout the Aeolians Islands appears to lend support to the hypothesis that pottery was imported from outside and that the islanders were not so keen on this kind of production. By contrast, it would seem possible to look at this from a different perspective to demonstrate how the greatest proportion of the pottery was manufactured in Lipari using both local (raw clay and volcanic filler) and imported material (clays from outside) and how it is possible to analyse the different levels of development of this local industry. Among the Early Neolithic settlements spread throughout the Central Mediterranean communities, pottery was usually made locally and was decorated with some stamped and incised motifs known as impressed wares or Stentinello wares (so-called after the site near Syracuse, first found by Paolo Orsi and characterised by a large amount of these wares). Indeed, several tiny variations in style can be also distinguished by region within their area of distribution.

Within the Early Neolithic layers on the settlement of Castellaro Vecchio in Lipari (Bernabò Brea and Cavalier, 1980: 656-664) we can distinguish two main categories of those coarse wares: one heavier, plain or decorated with simpler incised lines, finger-pinched or using the edge of a shell (Cardium or Punctulus); another finer and much more elaborately decorated, clearly accomplished by using proper tools, such as stamps, combs or spatulas (FIGURE 10). The Castellaro Vecchio settlement shows that both categories of pottery were contemporary, although the first seems clearly manufactured locally using local raw materials, while the second seems to have been imported as whole vessels already made, as verified by petrological examination (Williams, 1980: 864 and 867). The occurrence of both categories (imported and locally made) within the same stratigraphic layer suggests that very little time elapsed between the importation of those vessels and the phase of experimentation aimed at producing pottery with local raw clay characterised by the volcanic clastic rocks and ash deposits of Lipari. It seems clear that the first category was produced in imitation of the second type of imported vessels, using local clay associated with a much simpler repertory of motifs (FIGURE 11); these vessels betray a level of artisanship which was still relatively undeveloped, and could be interpreted as an attempt to reproduce the models that had just begun to be imported.

Indeed, this situation provides a useful hint for the interpretation of the Early Neolithic age on the Aeolian Islands; it suggests that the islanders were first attracted by these exotic items such as pottery and then attempted to reproduce them using the available resources. Therefore, because they could not draw on any consolidated expertise or place reliance on the local clay, which indeed is not very amenable

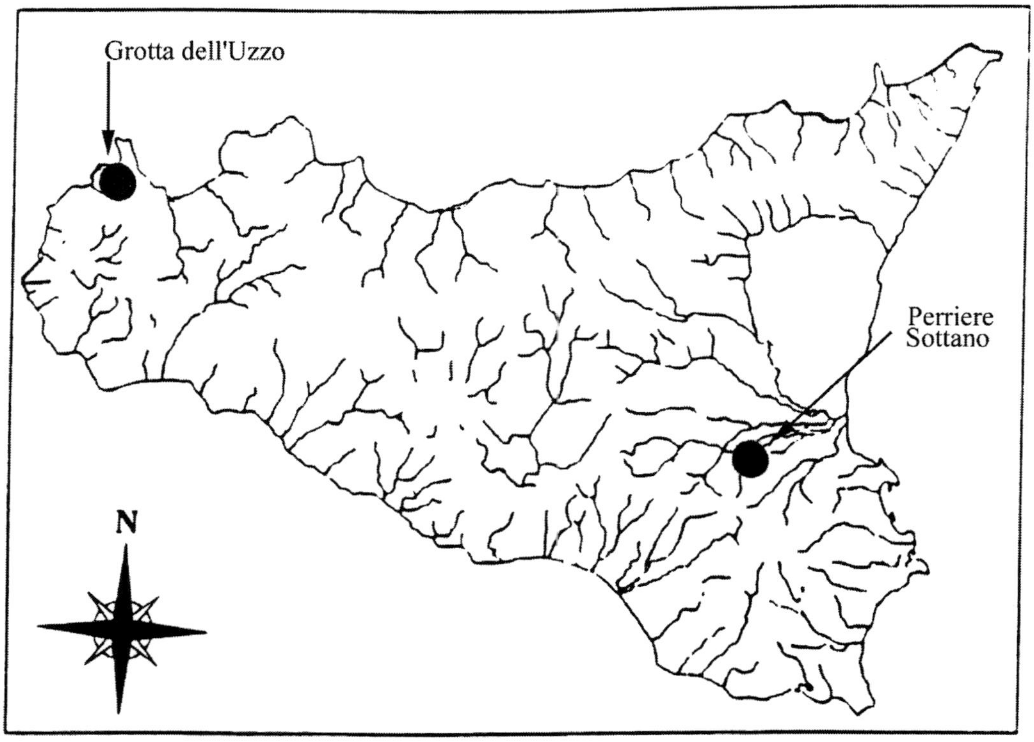

FIGURE 9. THE TWO PRE-NEOLITHIC SITES IN SICILY WITH OBSIDIAN FROM LIPARI: PERRIERE SOTTANO AND GROTTA DELL'UZZO.

DATE	SETTLEMENTS	SHEEP/GOAT	PIG	CATTLE	LOCATION
Early Neolithic	Stentinello	225	139	170	**Sicily**
	Megara	222	146	203	
	Matrenza	195	156	375	
Middle Neolithic	**Lipari-Acroplis**	225	139	170	**Aeolian Islands**
	Lipari-Serra d'Alto	222	146	203	

TABLE 3. Animal bones both natural and worked recovered in the Lipari-Acropolis layers, mainly associated with trichrome pottery (MN), classified according to common functional features. (Data based on: Bernabò Brea and Cavalier, 1980; Leighton, 1999).

to moulding, the result of this attempt produced a low quality pottery which manifested a clear distinction between the two categories.

Further evidence from Castellaro Vecchio suggests a progression in the production scale of pottery during the Middle Neolithic Age, were painted with three colours (so-called trichrome ware characterised by bands with red flames edged with black on a light background) and made of brown clay imported from outside and then mixed with the local volcanic filler.

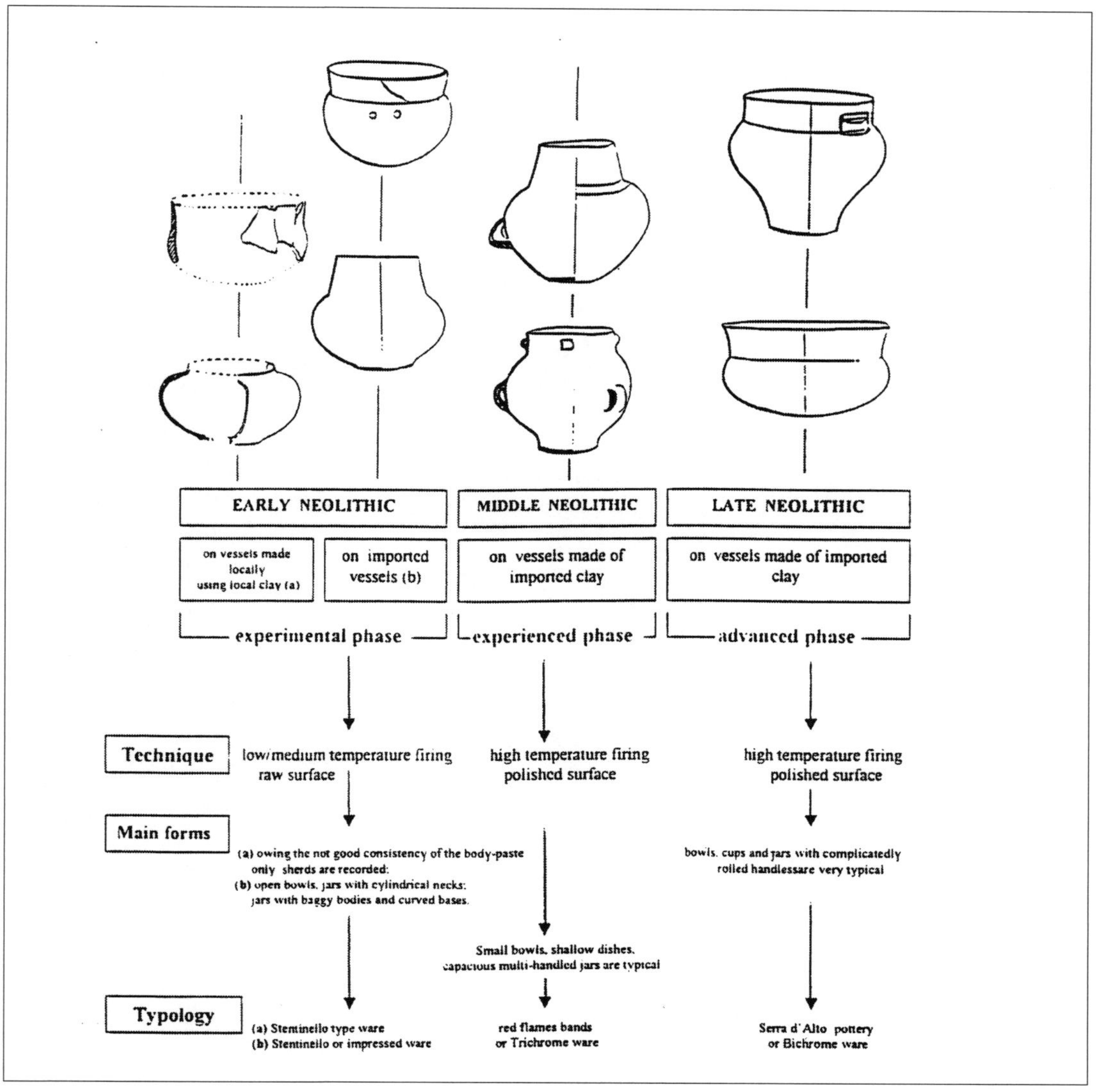

FIGURE 10. EVOLUTION OF THE POTTERY INDUSTRY OF LIPARI

(based on the Castellaro Vecchio and Lipari- Acropolis archaeological stratification),

dle Neolithic Age when the islanders realised that it was necessary for them to employ a better quality of clay from other sources. This situation seems to clearly indicate that, following the experimental phase occurring during the Early Neolithic Age which resulted in a very poor repertoire of motifs and forms moulded with an unsuitable clay, from the Middle Neolithic coarse wares were made locally and on a local scale; however, they used a kind of paste-body which is geologically distinct from the deposit which characterises Lipari's soil (compare Figg. 10, 11). Indeed, microscopic examination of thin section, which allows the evaluation of the mineralogical composition of the paste body of pottery, attests to the fact that coarse wares manufactured in Lipari are made locally from imported clays (Williams, 1980). These ceramics, very different from those of the previous

Moreover, during the Late Neolithic Age another typology of finely painted pottery appears on the Serra d'Alto layers of Lipari, still made with imported clay mixed with volcanic filler. This elegant painted style of ware, also imported as finished vessels, is decorated with two colors and is characterised by having thin walls similar to modern china and by meandering spiral motifs or simple geometric design such as hooks and small triangular shapes.

It appears that knowledge and experience in pottery production generated an increasingly specialised technology, which included the high-temperature firing of vessels by the Middle Neolithic Age. Indeed, it is easy to recognise an advanced development in the chain of pottery-production in Lipari in regard to which we can propose three different interpretative keys. They are as follows:

(a) people who settled on the Lipari-Acropolis coming from short-distance trade brought their own clay;

(b) people from Lipari going to Sicily to collect clay by direct procurement within one day's travel from the source;

(c) people from outside coming to Lipari as a central location for social relationship in order to exchange their goods (not only clay);

(d) people from far-off lands offering exotic fine clay as gifts to people settled in Lipari, by secondary reciprocal trade.

In order to take into consideration the aforementioned possible keys, it should be borne in mind that the EN layers of Lipari-Acropolis are characterised by both imported and locally made vessels, with clear features for each category. It seems highly unlikely that the same people who were able to bring clay to the Aeolian Islands should then prove unable to mould a good quality of pottery. Knowledge of clay also implies a proper understanding of its quality and geological components, as well as all the stages related to the firing phases and a common pattern of decoration.

The products of the local industry are not comparable with the imported fine wares which are well-fired, thin-walled and decorated with more complex designs typical of Stentinello's pottery phase. If those people had set up a workshop on Lipari, even with its own clay initially, they would surely have used at least part of their established repertory of motifs and forms on both categories of vessels, rather than the very simple ones attested to. From this, it can be seen that the fine ware was imported as whole vessels, while the other is just an earlier stage of imitation in the attempt to reproduce those imported. Stratigraphic evidence and radiocarbon dates (4211-3977 BC) suggest that most of the more developed forms classified as trichrome and later as bichrome vessels found on the MN and LN Lipari-Acropolis layers were made using clay from outside sources. Even if there may well be an overlap between the later trichrome and earlier bichrome, the association of imported and locally manufactured painted ware seems to represent a high point of craft production reached by Lipari by the MN.

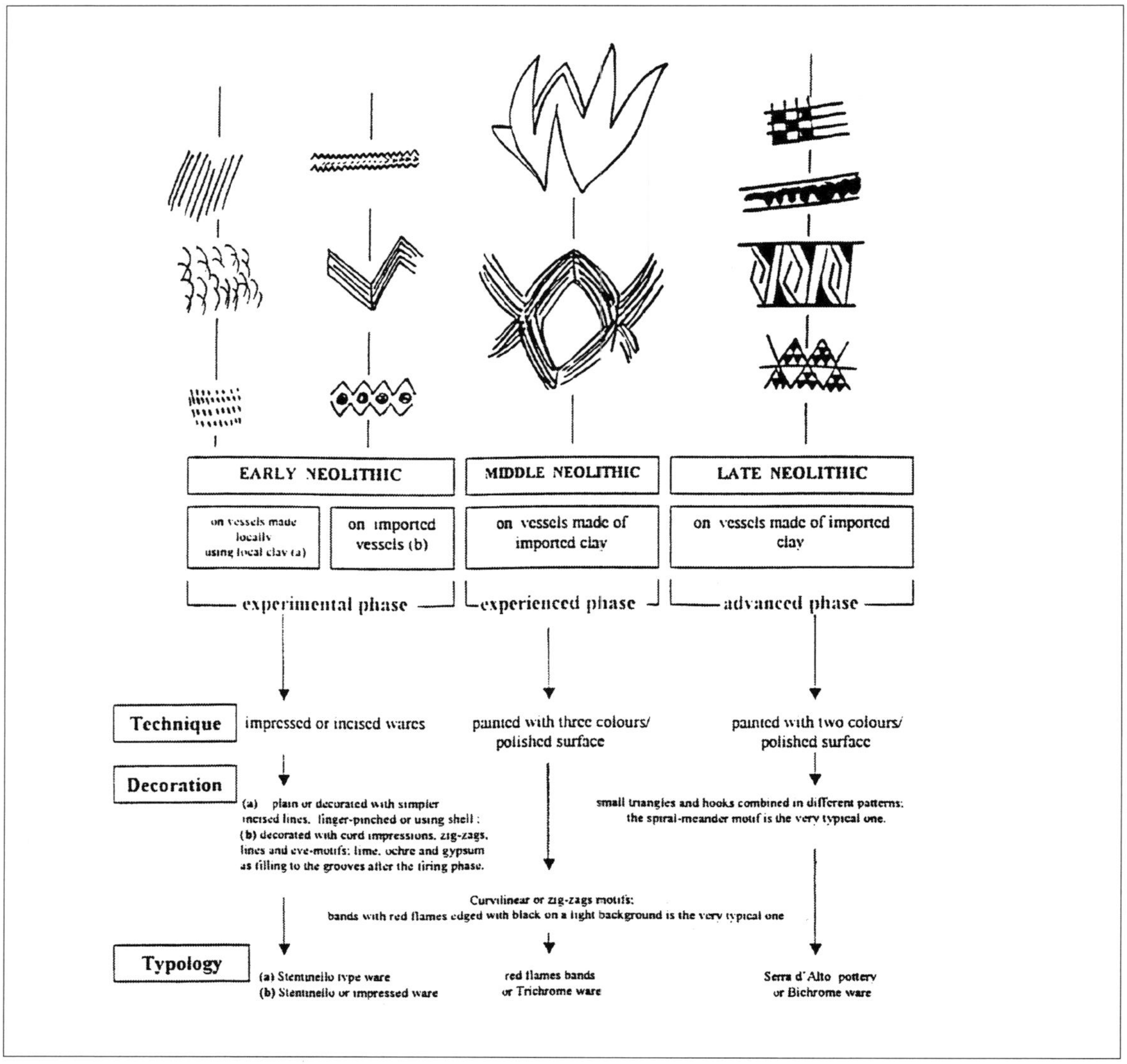

FIGURE 11. SAMPLES OF MOTIFS OF DECORATION
(Data derived from: Bernabò Brea and Cavalier, 1980)

It is quite likely that people from Lipari used to go to Sicily to collect clay to supply their own pottery industry, although it is also possible that Lipari itself played a role as an important network point of exchange of every kind of material, similar to a sort of emporium. Therefore, it should be pointed out that the first appearance of pottery together with the widespread use of obsidian coincided with the EN which highlights the reciprocal and complementary nature of exchange during the Neolithic Age. Summing up, all of these considerations point to the persuasiveness of a combination between hypotheses b) and c), which goes some way towards explaining the diversity of EN pottery, the changes which occurred during the MN and the supplying strategies adopted.

As far as the last hypothesis is concerned, the d), we have to recognise that there is also supporting evidence for the exchange of prestige fine wares, such as for example in the Biferno Valley (Barker, 1981:163) where bichromic pottery style has been found, which is clearly imported from further south. The discovery of the Ripoli trichrome sherd in the Dalmatian settlement of Danilo (Bray, 1966:100-106), on the other coast side of the Adriatic, comes as no surprise. Apart from pottery, the most persuasive example of this exchange of prestige items seems to be the shells of the Columbella rusticana species of Mediterranean mollusc, which have been recorded in a Mesolithic site in Southern Germany (Childe, 1973:39). The presence of this species of shell is also recorded, in the pre-Neolithic sites of Grotta dell'Uzzo and Perriere Sottano, by two interesting specimens, well decorated with delicate vertical cuts, which show a close similarity to each other.

The same problems arose for the cases of polished axes, flint tools and obsidian. It is evident that all those different kinds of material are characteristic of prehistoric cultures being of great value for their practical use in making sharp microlithic tools, before the discovery of metal techniques.

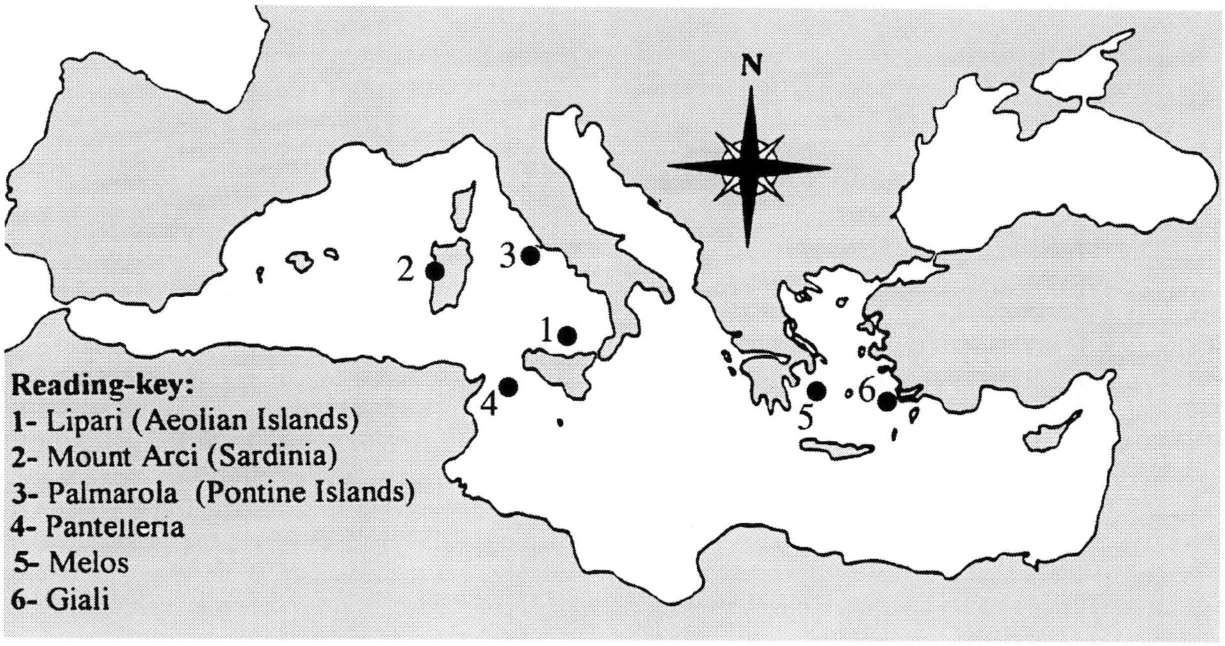

FIGURE 12. SOURCES OF OBSIDIAN IN THE MEDITERRANEAN.

DISCUSSION

The first Neolithic village established in Lipari marked the beginning of an intense maritime activity and began a process of extraction and working of obsidian that assured the Aeolian Islands a remarkable role which remained at least until the Late Neolithic age.

Evaluation of the range and frequency of Aeolian Prehistoric seafaring involves a series of interpretation problems, both relating to its organisation and the components of exchange involved. In archaeological terms, the resolution of these problems is closely connected as well as unavoidably limited to the remains of material culture, these being the only documentary source available for that age. The question which arises concerns the reasons for and manner in which Western Mediterranean society passed from its land-based hunting and gathering stage to maritime activity.

The widely accepted answer is that early maritime activity, attested by the distribution of obsidian, was connected with the catching of migratory fish (notably tunafish) as well as birds, as has been proposed by several scholars (Bintliff, 1977:117). Taking the example of the Franchthi Cave in the Argolid, it is clear that the use of Melian obsidian on the Greek mainland bears a witness to the earliest journeys in the Aegean, about which, however, it is not easy to establish anything related to market systems or anything of that sort. The picture that Lipari together with the Franchthi Cave gives us, is one of the great flair in maritime mobility which seems to have been quite ahead of its time in relation to the hypothesis of the attestation of stable settlements following fish migration. In fact, it should be stressed that in Franchthi Cave the obsidian from Melos has been found in earlier stratigraphic layers (Renfrew and Aspinall, 1990:257) than those in which evidences of off-shore fishing were first recognised.

In Lipari as well, the first direct evidence of sea-resources remains appears not earlier than the Middle Neolithic Age. Going into detail in Lipari's case, it is useful to observe that a fishbone of large dimensions has been recovered inside a pear-shaped pithos used as a burial place, stratified in the Middle Neolithic layer of Lipari's Acropolis (BernabòBrea and Cavalier,1960:113). From the size and typology analysis it has been assumed that the fishbone belongs to a species of tuna, which has been estimated as being rather big, perhaps a few kilos (Villari,1991:324). It should be emphasised that the fact of the tuna-fishbone's being recovered within a pithos, following the enchytrismos burial ceremony, seems to be of some significance, so much so that it became part of a grave-goods of the so-called Tomba 20.

Furthermore, the burial context itself seems to emphasis prestigious as well as remarkable rarity. It was not merely a part of the food remains left over from a ritual ceremony, but something more: an exotic item. Moreover, the Middle Neolithic settlements investigated on the Aeolian Islands show that no food remains involving large fishbones have been recorded. Indeed, there is no evidence here by which we can detect a food-chain including the tunafish resource either. In any case, the burial context together with the relatively scarcity of evidence attesting to such fish, or at least of such a size, tends to reduce the likelihood of the hypothesis regarding the migration of people following the migratory fish off-shore.

This view suggests that such a species of fish, which requires fishing equipment of a particular level of design for catching, as well as a boat capable of keeping its balance against the dynamism of the fish, had yet to be caught at that time. We can explain this particular occurrence of tuna-fishbone within a Middle Neolithic context only by admitting that it was an exceptional case. To sum up, this suggests that the people who arrived in the Aeolian Islands during the Early Neolithic period or even before were attracted by the islands themselves. The appeal of the islands was perhaps attributable to curiosity aroused by fire flows and clouds of smoke which lit up the archipelago, feeding the liveliest eschatological imaginations. As a nautical mark, resembling natural lighthouses in the maritime topography of the Southern Tyrrhenian, Vulcano towards the South and Stromboli towards the North East, are like 'fire islands', underlining the limits and defining the coastal spaces of the archipelago; they stand as huge luminous volcanic 'goals' across the Southern Tyrrhenian route. It may be hypothesised that people, possessing their rafts, were able to reach the coast of the Aeolian Islands, crossing the sea propelled by the currents and southerly winds.

However, in the absence of any direct evidence of shipwrecks or of any kind of water transport related to the West Mediterranean culture of this age, it is interesting to note that the obsidian from Lipari finds a wide market of distribution (TABLE 1).

It covered a fairly advanced level of distribution - in many cases also competitive – in comparison with the other producers of obsidian located on the western Mediterranean, such as that of Mount Arci in Sardinia, Palmarola on the Pontine Islands as well as on Pantelleria on the Channel of Sicily (FIGURE 12). Whilst archaeological evidence of any kind of water transport for seafaring in the Early Neolithic is largely indirect, the testimony in terms of material culture is both direct and concrete.

It is interesting also to observe that, while in Lipari the first settlement attested to is dated back to the Early Neolithic period, neither Melos island nor the other Cycladic islands show evidence of settlement until the Late Neolithic period. This indicates a situation where stable overseas settlements were already established by the Early Neolithic period, even if no significant indications of tuna-fish bones which could be related with early deep-sea seafaring are found in the East as well as in the West Mediterranean. Indeed, the presence of settlements in levels of Early Neolithic attested to in Lipari, antedating the human movement and their permanent community in the island, has thrown new light upon the whole question of trading contact and mobility in Western early seafaring.

CONCLUSION

Summing up, the main question arising in relation to the human interaction and the establishment of the Neolithic settlements on the Aeolian Islands relates to whether these were specialised seafaring communities or merely based on farming cultures which grew to exploit the fertile volcanic soil together with obsidian. How discussed in this paper, the first considerable pattern of extra-insular contacts occurs during the Early Neolithic and introduces a highly distinctive integration of the Aeolian Islands within the Central Mediterranean socio-economic systems, especially connected to the exploitation of the mining resources obsidian. It is evident that, even with very little knowledge of navigation, these communities got to Lipari probably by crossing the sea from Capo Milazzo, which is the easiest or shortest way of reaching the island. Even for people living close to the shore on the other side (Sicily) such a crossing would appear to implicate the quest for some advantage. If these people were skilled mariners possessing nautical knowledge of weather and currents, it is difficult to understand the reason why they should have settled on the highland of Quattropani instead of the coast.

In this regard it should be emphasised that no tools such as net-sinkers or simple fish-hooks have yet been identified that would confirm a predominance of maritime activity and that the first direct evidence of fish remains does not appear, even in a special circumstance (burial context), earlier than the Middle Neolithic Age (Bernabo' Brea and Cavalier, 1960: 113).

Moreover, the conclusion reached in the analysis of the pottery production indicates that Lipari was supported by its own pottery industry, despite an almost total absence of clay deposits throughout all the islands belonging to the archipelago. The stratigraphy of Castellaro Vecchio and the Lipari-Acropolis provides quite distinct evidence to establish that during the Early Neolithic the presence of several models of pottery with a wide repertory of complex decorations formed the basis for an indigenous pottery industry which could not have proceeded using the available clay. Secondly, it shows an evaluation of the paste composition and the recognition of and search for a more suitable source of clay. Thirdly, it highlights the beginning of a local pottery production that developed without interruption through the whole Prehistoric age from an exploratory phase (EN) until the establishment of a real chain of production.

Certainly, the human integration with the Aeolian Islands ecosystem give the birth to social and economic interactions either with the neighbouring mainlanders or the seafaring people from the western Mediterranean. Although the records analysed so far underline that the newcomers were associated with agricultural than maritime activity, the role played by the Aeolian Islands was clearly as a centre which operates over a long durée of at least two thousand years. It is now even more evident that the Aeolian Islands might be viewed as genuine meeting places where long-distance exchange of obsidian and other goods played a remarkable role in maintaining preferential ethnic or kin connections, and they thus point to the capability of Central Mediterranean seafaring people of making open water crossings for both ritual and exchange purposes.

In summary, the overall perception gained from the analysis of the evidence relating to the Neolithic sea/landscape and human interaction on the Aeolian Islands points to groups of people with land-based economies relying on their rafts being able to reach the coast of the Aeolian Islands and settling on the high plateau of Quattropani in order to take advantage of obsidian as well as the fertile volcanic soil while exploiting the landing places on the nearest beaches available.

REFERENCES

AMMERMAN, A. J. AND ANDREFSKY, JR.,
1982, Reduction sequences and the exchange of obsidian in Neolithic Calabria. In Ericson, J.E., Earle, T.K. (Eds.), Context for Prehistoric exchange, Academic Press, New York: 149-172.

AMMERMAN, A.J., TYKOT, R.H.,
1997, New Directions in Central Mediterranean Obsidian Studies, Antiquity, 71(274): 1000-1006, 1997

ARANGUREN, B. AND REVEDIN, A.,
1996, Problemi relativi all'insorgenza del mesolitico. In Leighton R., The Early Society in Sicily, London: 31-39.

Bernabò Brea, L. and Cavalier, M., 1980, L'acropoli di Lipari nella Preistoria. Meligunis-Lipara, IV, Palermo.

BERNABÒ BREA, L. AND CAVALIER, M.,
1968, Stazioni preistoriche delle Isole Eolie: Panarea, Salina Stromboli. Meligunìs Lipàra, III, Palermo.

BERNABÒ BREA AND CAVALIER, M., 1966, Ricerche paleontologiche nell'isola di Filicudi. In Bollettino di Paletnologia Italiana, N.S.XVII, 75: 142-173.

BERNABÒ BREA, L. AND CAVALIER, M.,
1995, Salina, Ricerche archeologiche (1989-1993). Meligunis-Lipara,VIII, Palermo.

BERNABÒ BREA, L., CAVALIER, M. A, VILLARD, F.,
1998, Topografia di Lipari in età greca e romana. Parte I: L'Acropoli. Meligunis-Lipara, IX, Palermo.

BONFIGLIO, L.,
1998, I resti fossili di vegetali dell'isola di Lipari conservati nelle collezioni Mandralisca. In Mastelloni, M.A. and Spigo, U.(Eds.), Agli albori della ricerca archeologica nelle Eolie, Museo Archeologico Regionale Eoliano, Messina: 32-34.

BUCHNER, G.,
1949, Ricerche sui giacimenti e sulle industrie ossidianiche in Italia. Rivista Italiana di Paleontologia e di Stratigrafia, 4: 162-168. / Buchner, 1948/ Controllare

CASTAGNINO BERLINGHIERI, E. F.,
1997, Isole Eolie: osservazioni sui siti sommersi. Scali marittimi, approdi e antiche linee di costa tra neolitico ed età del bronzo. Atti del I° Convegno Associazione Italiana Archeologi Subacquei, Edipuglia, Bari:153-168.

CASTAGNINO BERLINGHIERI, E.F.,
forthcoming, Attività umana e assetto costiero nella Protostoria eoliana: nuovi risultati di ricerca, Proceeding of the Convegno Internazionale Strumenti per la Protezione e la Valorizzazione del Patrimonio Culturale Marino nel Mediterraneo, Università di Milano Bicocca, Università di Palermo e Regione Siciliana, Palermo-Siracusa, 8-10 Marzo 2001.

CASTAGNINO BERLINGHIERI, E.F.,
unpublished, The Aeolian Islands: crossroads of the Mediterranean maritime routes of the South Tyrrhenian, Ph.D. Thesis, University of Bristol, 2000-2001.

Lo Cascio, P. and Paste, S., forthcoming, Phoinikodes, l'isola delle palme. Quaderni del Museo Archeologico Regionale Eoliano, 2, Messina.

LO CASCIO, P. NAVARRA, E.,
1997, Guida naturalistica alle aisole Eolie, L'Epos, Palermo, 1997: 14;16;29.

FRANCAVILLA, V. AND PIPERNO M.,
1987, La répartition et la provenance de l'obsidienne archéologique de la Grotta dell'Uzzo et de Monte Cofano (Sicilia). Revue d'Archéometrie, 11: 31-39.

GAUDIN C. TH. AND MANDRALISCA H.P. (DE),
1860, Contribution a la flore fossile italienne, V. Tufs volcanique de Lipari. In Memoires de la Societe helvetique des sciences naturelles, 18, Zurigo.

KELLER, J,
1980, The island of Vulcano. Rendiconti della Società Italiana di Mineralogia e Petrologia , 36: 369-414.

LEIGHTON, R.,
1999, Sicily before history: an archaeological survey from Paleolithic to the Iron age, Gerald Duckworth publisher, London.

PICONE VILLARI, R.,
1985-86, Acropoli di Lipari, scavo 1983: indagini paleobotaniche. Rivista di Studi Preistorici, 40,1-2: 253-254.

PIPERNO, M., TUSA, S. AND VALENTE, I., 1980, Campagne di scavo 1977 e 1978 alla grotta dell'Uzzo (Trapani). Relazione preliminare e datazione dei livelli mesolitici e neolitici. Sicilia Archeologica, 42: 49-64.

RICCI LUCCHI F., CALANCHI N., LANZAFAME G. ROSSI P.L.,

1988, Plant-rich pyroclastic deposits of Monte Sant'Angelo, Lipari, (Aeolian Islands). In Rendiconti della Società Italiana di Mineralogia e Petrologia, vol.43:1227-1251.
TYKOT, R.H.,
1996, Obsidian Procurement and Distribution in the Central and Western Mediterranean. In Journal of Mediterranean Archaeology 9(1): 39-82, 1996.
TYKOT, R.H.,
1996, Changes in Subsistence, Settlement, and Social Interaction during the Neolithisation of the Western Mediterranean. In XIII Congress of the International Union of Prehistoric and Protohistoric Sciences, Forlì (Italy), September 9, 1996.
TROÌA A., LA MANNA M., LO CASCIO P., PASTA S., PUGLIA A.M., QUATRINI P.,
2000, Conservation of rare endangered plant species in Aeolian Islands (Sicily). Abstract del 95° Congresso della Società Botanica Italiana, Messina, 28-30 settembre 2000; p. 114.
TROÌA A., CARDINALE M., LA MANNA M., LO CASCIO P., PASTA S., PUGLIA A.M., QUATRINI P., VOUTSINAS E.,
2001, Preliminary results of EOLIFE99, a project concerning the conservation of four endangered plant species of Aeolian Archipelago (South Tyrrhenian Sea, Italy). Abstract of the X "OPTIMA" Meeting, Palermo, 13-19 September 2001: p. 176.
VILLARI, P.,
1993, Le faune del villaggio preistorico di Capo Graziano nel contesto archeologico eoliano e siciliano dell'età del bronzo. In Bernabò Brea L. and Cavalier M. (Eds.), Meligunis Lipara, VI, Palermo: 307-322.
WILLIAMS, J. LL. W.,
1980, A petrographical Examination of the Prehistoric pottery from the excavation in the Castello and Diana plain of Lipari. In Meligunis-Lipara, IV, Appendix VII, Palermo: 845-868.
WILLIAMS, J. LL. W.,
1994, The petrological analysis of Capo Graziano pottery from Filicudi and Milazzese pottery from Panarea. In Meligunis-Lipara, VI, Appendix , Palermo: 239-259.
WILLIAMS, J. LL. W. AND LEVI, S.T.,
1995, The characterisation of Neolithic Stentinellian pottery fabrics from the Aeolian Islands and the type of Stentinello near Syracuse, Sicily. In Meligunis-Lipara, VIII, Appendix I, Palermo: 138-163.

21

ISLAND NEOLITHICS: ANIMAL EXPLOITATION IN THE ACERAMIC NEOLITHIC OF CYPRUS

SHEELAGH FRAME

Department of Archaeology and Art History
Eastern Mediterranean University, TRNC
EMAIL: sheelagh.frame@emu.edu.tr

INTRODUCTION

It has frequently been argued that islands are a kind of laboratory for both biological and archaeological studies. While this approach can sometimes overemphasize the isolation of island societies, it is true that islands can provide something like a series of alternative histories for archaeologists and biologists studying cultural and biological variation. For zooarchaeologists, who are concerned with the relationship between human societies and animal populations, Cypriot prehistory is a particularly interesting case. Humans have been implicated in the extinction of the island's endemic Pleistocene fauna (Simmons 1999); They are certainly responsible for the introduction of virtually all the historic and modern mammalian and reptilian fauna. The Cypriot Neolithic therefore offers a rare opportunity to study the choices these people made in turning this empty island into a Neolithic landscape, deliberately populating it with both wild and domestic animals as well as stowaway such as the house mouse.

The outlines of the animal economy of the early settlers has been well defined by archaeologists working at a variety of Neolithic sites (Croft 1982; Davis 1984, 1989, 1994; Watson and Stanley Price 1977; Fox 1953). One of the intriguing features of the emerging picture of the Aceramic Neolithic is that although the plant and animal components are certainly imported from somewhere on the mainland, it is not exactly like any mainland Neolithic economy currently known. Although these differences should not be exaggerated - they do share the general characteristics of early village life throughout the region - they should equally not be ignored. It is possible that these differences are due to our inadequate knowledge of the contemporaneous coastal sites, but it seems likely that it is also a reflection of decisions made by the aceramic settlers about the 'best' possible suite of plants and animals. And these choices are not always the ones a prehistorian would have predicted; sheep, goat, fallow deer, pig, dog, fox, cat and it now appears cattle (Peltenburg et. al. 2000; Guilaine et. al. 1995; Simmons 1998). From this list only sheep and goat are uncontroversially agreed to have been domesticated on the mainland at the time of the colonization of Cyprus, although claims have been made for the incipient domestication of pig at Hayaz Höyük (Buitenhuis 1985, but see Hongo and Meadow 1998 for the arguments against proto-domestication at Cayonu).

Cattle, fox and cat are all rare and the cattle are only found on the earliest sites and then disappear until they were reintroduced during the Chalcolithic/ Early Bronze Age. The other species appear on most Neolithic sites, but their relative proportions change considerably. It is not clear to what extent this variability is chronological, environmental or social (or perhaps in some cases methodological). It is clear that all of these animals were important and separate components of the Cypriot Neolithic culture. Each animal must have filled an economic or social niche that the others could not - a fallow deer was not a substitute for a pig at least under some circumstances.

These Cypriot peculiarities are interesting in themselves as alternative neolithics. They also present some interpretative problems that lead back to how we understand the Neolithic as a category and what exactly the process of domestication meant for the people who first engaged in it. The animals brought to Cyprus include a mix of domestic animals, animals that have the potential for domestication, but were perhaps not yet domesticated, and animals that have never been domesticated.

Ingold (1984) has argued that the difference between a wild and a domestic animal is not it's morphology or the way it is managed, but how it is articulated with the social relations of the human community. For instance animals that are pets, or that are wealth or bride price are objects or vehicles of social relationships and not solely, or even primarily calories. Wild animals on the other hand are outside societies boundaries – by definition they are 'wild'. Socially this is not a trivial distinction - food that is shared at a feast can have very different meanings if it is the product of a group hunt, rather than the gift of the feast giver. This does not imply that the 'wild' is necessarily a shared resource or that hunting is an arena for the statement of egalitarian tendencies. The wild can equally be the provenience of the elite and hunting can be an important feature in the creation of masculinity. The problem for zooarchaeologists is not only to determine the exploitation methods of the various species but also to examine the social relations and cultural categories that these animals were a vital part of. These categories may broadly correspond to our notions of wild and domestic, but have different social functions than we associate with these categories. Alternately, the division of animals into 'wild' and 'domestic' may have had no meaning at all for the aceramic farmers who began the process we now call domestication. This kind of analysis can only be done when equal weight is given to the consumption as well as the production of animal resources. This entails a detailed look at the different contexts in which the animal bones are found and comparisons between and within context types.

Recent excavations at the aceramic site of Akanthou-Arkosyko/Tatlisu-Çiftlikdüzü (hereafter Akanthou) provide an opportunity to explore these questions. The excavations at this site are discussed elsewhere in this volume (Sevketoglu, this vol.). This paper presents only the results from the field analysis of the faunal assemblage excavated in the 2001 season. Only a small portion of the total assemblage has been examined and the results may vary when the analysis is complete. Obviously a complete analysis is necessary to address the issues discussed in the previous paragraph, and most of these problems will not be addressed until a greater number of contexts have been fully recorded. However, even the preliminary work has produced some interesting patterns.

METHODS

Critically for the purposes of this analysis, the site of Akanthou has been excavated by archaeological context and all the deposits, including the top soil have been sieved through a 4mm mesh screen ensuring consistent and comprehensive recovery of all remains. Numerous studies have demonstrated that unsieved material seriously underestimates the relative proportion of smaller taxa and often completely misses the microfauna. Differential fragmentation rates between species is an important factor in the analysis of consumption and discard patterns and can only be reliably estimated when the complete set of fragments are collected from different context types. Small unidentifiable pieces are also vital for taphonomic purposes, since bone that has been burnt in a fire, digested or trampled tends to be present only in small pieces that can only be recovered through sieving. Currently none of the material from Akanthou has been wet sieved. Soil samples were taken from every archaeological context for flotation. When wet sieved these will provide a sample of the <4mm fraction where small bones including most of the fish and rodent bones can be recovered. In this paper I will only discuss the mammal bones from the dry sieve.

The ultimate goal is to study all the bone excavated in the 2000 and 2001 season, but at the moment, only a small sample have been analysed. The contexts in this paper are all securely Neolithic contexts from the final phase of occupation. They include pits, plaster bins and the fill found directly on top of the floors from the interior of the houses. I have also began the analysis of one of the top soil layers from 1999 in order to asses how mixed this layer was and how it related to the lower, undisturbed levels. The material from this top soil level appears to be largely Neolithic, but has been excluded from the totals discussed in this paper except where noted. This is to ensure that later material does not skew the figures.

The aims of this study are not limited to identifying the species of animals exploited by the Neolithic inhabitants of Akanthou. The recording system therefore does not only identify each specimen of bone to some level, but also records all attributes of each fragment including surface condition, modifications, weathering, fragmentation, burning, abrasion. This is labour intensive, but provides the detailed information necessary to determine the activities that created the deposits as well as the post-depositional taphonomic processes.

THE BONES

Seven mammalian species have been identified at Akanthou: sheep, goat, fallow deer, pig, dog, fox and cattle. This includes all of the species definitely known to have been found on Cyprus during the Neolithic, except for cat. The first 4 species make up 98.4% of the total (Table 1). Absolute counts are not reliable when trying to assess relative number of bones from different species since there are several well known sources of bias which are particularly acute when comparing pig to other ruminants. In order to correct for these biases I have used the 'diagnostic zone' method (Watson, 1979) to determine the relative amount of bone contributed by each taxa (Table 1).

While the list of taxa found at Akanthou is similar to the species list found at most other Neolithic sites, there seems to be considerable variability in the proportions of animals used at different Neolithic sites. The reason for this variation is puzzling, although it is possible it reflects different local environmental conditions. It may also indicate that the Neolithic was a period of experimentation and adaptation, as the new populations of both humans and animals modified their new environment in intended and unintended ways.

Overall the animal bone was moderately well preserved although with some calcareous deposits, which were difficult to remove without breaking the bones. These deposits were quite common in the upper level, but become rarer and softer below 30 cm. A total of 725 bones have been identified at least to genus. The percentage of identifiable pieces, which can be used as a rough guide to the degree of fragmentation (in that highly fragmented bone becomes unidentifiable) - is approximately 10%. This is not an unusual amount of fragmentation for a Neolithic site, especially one where the smaller bones have been routinely collected by sieving. The long bones appear to have been fractured for bone marrow extraction and the mandibles seem to have been toasted in the embers, a practice useful for extracting marrow from the mandible. Many of the bone epiphysis were deliberately smashed into small fragments, perhaps for bone grease extraction. The fragmentation is unlikely to be a post-depositional phenomenon, since the taphonomic evidence suggests most of the bones were buried relatively rapidly. The edges are sharp and do not appear to have been abraded or trampled and the bones surfaces are only slightly weathered. A preliminary interpretation suggests that most of the contexts so far examined were intensively processed and rapidly buried. Sheep and goat appear to be rather more highly fragmented than the fallow deer bone, although this may be due to the higher number of caprine metapodials. There is also variation in the fragmentation and surface condition of the bones between different context types, especially between house floors and the other contexts, but at the moment there is not enough data from each separate context types for these patterns to be clear.

One other rather surprising feature of the unidentified bone is the rarity of digested and gnawed bone. Although some small digested pieces which are characteristically left by scavenging dogs, may still show up in the wet sieve, the dry sieve material is full of fragments less than 1 cm and none are digested. Pigs usually completely digest small bone, but leave lot's of heavily gnawed large bone. There are no examples of this kind of destruction at Akanthou. While it is possible that the pigs are brought to the site as carcasses, the dogs must have been close to the village. The rarity of carnivore gnawing suggests that the dogs were not frequently found in the domestic areas of the site so far excavated. Dog bones themselves are extremely rare in the Neolithic deposits, suggesting that dogs were either rare or not in the same category of animals as the other species, at least in terms of consumption and discard patterns.

TABLE 1. TAXA

TAXON	DIA. ZONES	%	NISP
SHEEP/GOAT	74.	34.1	235 (264)
SHEEP	25.5	11.8	34 (44)
GOAT	7.	3.2	8 (11)
FALLOW DEER	79.0	36.5	169 (233)
PIG	27.5	12.8	113 (169)
DOG	2.	9.0	4 (4)
FOX	1.0.	.5	3 (3)
CATTLE			(2)

Figures in ( ) includes the specimens from top soil level

CATTLE

The cattle bones included in Table 1 were found in the top soil level. Visual inspection of other contexts has shown that a small number of cattle bone are scattered throughout the site, including the lowest levels so far excavated. However of the fully recorded contexts, cattle bones have only been found in the top soil. Cattle have been reported in small numbers from other aceramic Neolithic sites including Shillourokambos and Kritou Marottou - Ais Yiorkis (Peltenburg et. al. 2000; Vigne et. al. n.d.; Simmons 1998), but their presence on Cyprus at this early date is still extremely rare and they do not appear on any site after the 8th millennium B.P. until they were reintroduced during the Chalcolithic/Bronze Age period. The cattle appear to have been an unsuccessful import although they may have persited in very small quantities for as much as 1000 years. A sample of the cattle bones from Akanthou are currently being carbon-14 dated in order to try and clarify the issue at this site. None of the bones were complete enough for any measurements to be taken.

SHEEP/GOAT

Sheep and goat are the most common animals in the assemblage accounting for 49.1 % of the bones. Sheep are consistently more common than goats at a ratio of about 3.5 to 1. The two sheep horn cores so far identified are mouflon in type and the very tentative size range for the sheep bones that have been measured, suggests that the Akanthou sheep were somewhat larger than modern mouflons. Since the modern wild mouflon are feral animals either deliberately of accidentally released by prehistoric humans, it is not surprising that the Neolithic sheep are morphologically similar to the modern mouflons. The size reduction in the modern population may be fairly recent due to habitat pressure and overhunting or could be a result of the gradual dwarfing of large mammals that is common on islands. A long chronological sequence of mouflon bones from archaeological sites would be invaluable in clarifying this question. In contrast to the sheep, the goat appear quite small and are relatively few in number. No goat horn cores complete enough to identify to type have yet been found.

FALLOW DEER

It has been suggested that both roe deer and red deer were found on Cyprus during the Neolithic, but there is no evidence of either of these species at Akanthou. All of the deer bone at Akanthou that can be definitely identified to a particular species of deer are fallow deer and I have therefore considered all cervid bones to be fallow deer. Both shed and unshed antler are found but most of the antler is very fragmented. It is relatively rare in the pits compared to the amount of horn core even when it's fragility is taken into account. The post-cranial deer bones show an extremely marked sexual dimorphism. The size range suggests that the species is dama dama mesopotamica the larger of the two sub-species of fallow deer, although none of the antler fragments have been large enough to confirm this morphologically. If the size range is indeed due to sexual dimorphism, than female deer are twice as common as male deer.

Table 2: Age at Death

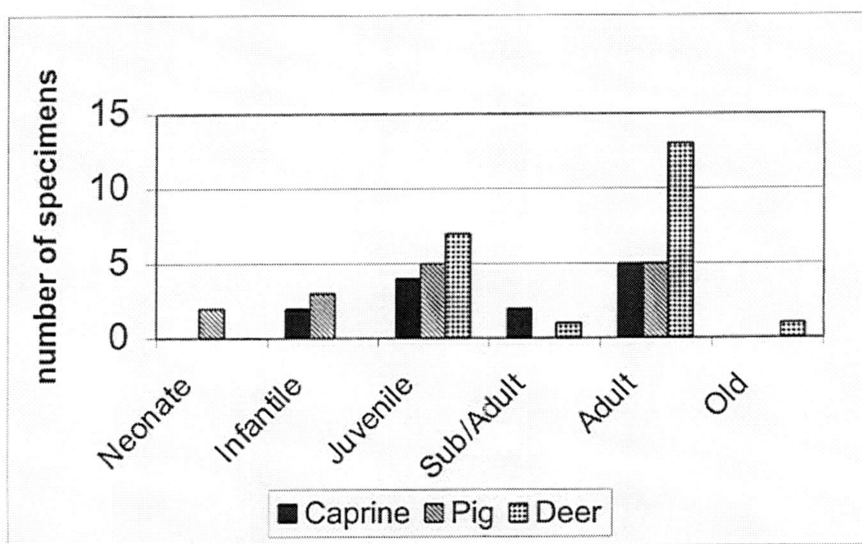

Some archaeologists have accounted for the importation of fallow deer to Cyprus by arguing that they were 'proto-domesticated (Jarman 1976; Zeuner 1958). However most recent analysts have argued that they were free ranging, hunted populations (Croft 1982; Davis 1984, 1989). The evidence from Akanthou including the age and element distribution data discussed below tends to support the idea that they were hunted. If the deer are hunted the high number of females may indicate the hunting of doe herds as a preferred strategy.

PIG

The nature of the pig exploitation at Akanthou is a particularly interesting problem due to the uncertainly about the nature and timing of early pig domestication. Presently there are not enough measurable pig bones to have a clear idea of their size. The small number that have so far been measured fall into a transitional range, although the uncertainty of what 'wild' population they come from makes this a very uncertain figure. They do appear larger than the pigs at Khirokitia (Davis 1989, 1994) but this could be a result of the tiny sample size. A perhaps more interesting pattern is the cull patterns which show a very high percent of infants and juveniles (Table 2). It is generally argued that a high percentage of juveniles in the animals culled is evidence that the animals were domesticated (Bökönyi 1969).

CARNIVORES

There are no carnivore species on Cyprus except those introduced by humans. It is interesting that among the earliest animals imported to Cyprus were a small number of fox and cat. It is probably that these predators had a dramatic effect on the local ecology, especially since most of their normal prey would not have been present on Cyprus. The element distribution of the fox bones at Akanthou, demonstrates that these were not just pelts imported to the site, but the whole animals.

Dog bones are very rare and are from small to medium size dogs. The scarcity of dog bone may have more to do with disposal practices rather than the number of dogs - most of the deposits so far examined appear to be post-butchery and post-consumption waste. However, the very low rate of digested and gnawed bones indicates that if there were many dogs they were not allowed to roam freely on site.

ELEMENT DISTRIBUTION

When adjusted for the different number of bones in a pig skeleton there are some interesting differences in the element distribution of the 3 major animals at the site (caprines have here been considered a single animal). All different species have been found in all of the contexts. However, the fallow deer are consistently underrepresented in foot bones and lower limb bones compared to the caprines and the pigs. Deer also have considerably more fore limbs than hind limbs. Together these suggest that perhaps the dismemberment of fallow deer carcasses is taking place farther away from the site than the pig or sheep/goat, and only the most meat bearing bones are being brought into the domestic area. When calculated using raw counts the deer seemed underrepresented in teeth as well, but both the diagnostic zone and MNI methods of quantification suggested a closer correlation between the figures for teeth and post-cranial elements. Closer inspection shows that there is a lack of deer maxillary teeth only. It is possible that the deer skull, with their useful antlers are not being processed and discarded with the rest of the carcasses. This is perhaps supported by the fact that although horn core fragments are fairly common in the rubbish dumps, antler fragments are rare as are pig canines, despite the large number of upper pig molars.

AGE AND SEX

I have not yet analysed the ageing information from the teeth. However, the age at death profiles from ephiphyseal fusion data show some interesting patterns (Table 2). As in the element distribution the fallow deer are distinct from the other common species. The majority of the fallow deer were slaughtered as adults or subadults. About one third of them were slaughtered as juveniles, between the age of 1.5 and 2.5 years of age and only a fewas infants less than 1.5 years old. This is almost the mirror image of the pig cull patterns. The only neonatal bones so far identified are from a pig. Two thirds of the pigs were slaughtered before they were adults, many of them as infants and young juveniles. The caprines are similarly skewed towards the younger age although almost 41 % reached at least subadult age, and the peak slaughter appears to be at the end of the juvenile stage when they have reached almost full size.

The difference between the pig and caprine ages is probably related to their different breeding patterns. Pig reproduce rapidly and slaughtering pigs young keeps the population young productive without risking overexploitation. Sheep and goat on the other hand usually produce only one offspring a year which are usually allowed to reach almost full size before slaughtering in order to maximize the amount of meat.

DISCUSSION

All of the animals brought to Cyprus except the carnivores, were probably used primarily for meat and not for secondary products such as milk and wool. All of the four main species are found mixed together in the deposits and seem to have been discarded in similar ways although there is a slightly higher rate of fragmentation in the sheep/goat bones which may indicate a more intensive use of the carcass. Of the four most common species there appears to be a distinction between fallow deer on the one hand and the caprines and pig on the other. The fallow deer appear to have been hunted, or at least butchered away from the houses with only the richest portions being brought back to the homes. The concentration on older individuals is consistent with the animals being hunted. The 2 to 1 ratio of females to males raises the possibility of managed herds, but could also be the result of hunting strategies that concentrated on doe herds, rather than the solitary males. The caprines and pigs on the other hand appear to have been slaughtered close to where they were deposited and the age distribution supports the idea that they were closely controlled populations.

In the aceramic Neolithic of Cyprus there seem to be at least three important categories of animals. There are the domestic or managed sheep, goat and pigs which make up the majority of the sample. Fallow deer which were hunted, but then prepared and eaten in a similar manner to the domestic animals make up the next largest group and then there are the fox and the cat, whose role is uncertain, but who were clearly important to the Neolithic inhabitants. At this stage of the analysis there is no apparent spatial separation between the hunted and managed animal remains and the treatment of the carcasses is similar, although not identical. The important difference may be that the 'wild' deer are a collective resource while the management and domestication of animals is based on the principle of divided access to both the living and dead animal resources. Apparently both of these principles were vital to aceramic social relations, perhaps allowing the same people to relate to each other in different ways, depending on the specific nature of the resource exploitation and consumption.

REFERENCES

BOKONYI, S.,
1969. Archaeological problems and methods of recognizing animal domestication. In: P. Ucko and G. Dimbleby, eds. The domestication and exploitation of plants and animals London: Duckworth, 219-29
BUITENHUIS, H.,
1985. Preliminary Report on the Faunal Remains of Hayaz höyük from the 1979-1983 Seasons. Anatolica 12, 61-74.
CROFT PAUL,
1982. Faunal remains from Tenta and Ayious in IA Todd Vasilikos Valley Project: Fourth Preliminary Report, 1979-80, Journal of Field Archaeology, 9, 35-77
DAVIS, S. J. M.,
1984. Khirokitia and its mammal remains. A Neolithic Noah's Ark. In: A. Le Brun, Fouilles recentes á Khirokitia (Chypre) 1977-1981. Paris: Editions Recherche sur les civilizatons 147-162
DAVIS, S. J. M.,
1989. Some more Animal Remains from the Aceramic Neolithic of Cyprus. In: A. Le Brun, Fouilles recentes á Khirokitia (Chypre) 1983-1986. Paris: Editions Recherche sur les civilizatons 189-221
DAVIS, S.J.M, 1994. Even more bones from Khirokitia: The 1988-1991 Excavations 1994 In Fouilles Recentes a Khirokitia (Cypre) 1988-1991. Alain Le Brun, Etudes Neolithiques Ediitons Recherche sur les civilisations Paris
GUILAINE, J. F. AND BRIOIS, J., COULAROU AND I. CARRERE, 1995. L'éstalishment néolithique de Shillourkambos (Parekklisha, Chypre). Premiers résultants. Report of the Department of Antiquities of Cyprus (1995) 11-32
HONGO, H AND MEADOW. R.H.,
1998. Pig Exploitation at Neolithic Çayönü Tepesi (Southeastern Anatolia). In: Ancestors for the Pigs: Pigs in Prehistory. MASCA Research Papers in Science and Archaeology 15, 77-95.
INGOLD, TIM,
1984. Time, Social Relationships and the exploitation of Animals: Anthropological Reflections on Prehistory. In: Juliet Clutton-Brock and Caroline Grigson, eds. Animals and Archaeology 3: Early Herders and their Flocks. BAR International Series 202
KING J.E.,
1953. Mammal bones from Khirokitia and Erimi pp 431-437. Appendix 3 in P. Dikaios, Khirokitia. Excavations on Behalf of the Department of Antiquities of the Government of Cyprus 1936-1946. Monographs of the Department of Antiquities of the Government of Cyprus no. 1. Oxford: Claredon.
PELTENBURG, EDGAR AND SUE COLLEDGE, PAUL CROFT, ADAM JACKSON, CAROLE MCCARTNEY AND MARY ANNE MURRAY,
2000. Agro-pastoralist colonization of Cyprus in the 10th millennium BP: initial assessments, Antiquity 74, 844-53
*EVKETO*LU, MUGE,2002, Akanthou-Arkosyko (Tatlisu-Çiftdüzü) and the Anatolian Connections in the 9th millennium B.C.E. in: W. Waldren, ed, Proceedings of the Islands In Prehistory Conference
SIMMONS, ALAN H..
1999. Faunal Extinction in an Island Society. Pygmy Hippopotamus Hunters of Cyprus. :New York:Plenum
SIMMONS, ALAN H., 1998. Hippos, Large Cows and Early Colonists in Cyprus. Journal of Mediterranean Archaeology, 11(2), 232-241
VIGNE, J.-D.
n.d. Statut des grands mammiferes de Shillourokambos, notamment Durant les phases anciennes (fin du 9e – debut du 8e millenaire. Paper at Le Neolithique de Chypre conference, Nicosia, 2001
WATSON, J.,
1979 The Estimation of the Relative Frequencies of Mammalian species: Khirokitia 1972. Journal of Archaeological Science 6, 127-137
WATSON, J.P.N. AND STANLEY-PRICE, N.P.,
1977, The vertebrate fauna from the 1972 sounding at Khirokitia Report of the Department of Antiquities, Cyprus:232-260
ZEUNER, F. E., 1958. Animal remains from a LateBronze Age sanctuary on Cyprus, and the problem of the domestication of fallow deer. Journal of the Palaeontological Society of India, 3, 131-135

22

EFFECTS OF ENVIRONMENTAL ALTERATION ON EASTER ISLAND (RAPA NUI) FROM THE MOAI ERA TO THE MODERN ERA, AD 1000-2000

JOAN A. WOZNIAK

Department of Anthropology, University of Oregon, Eugene, OR 97403-1218, U.S.A.
EMAIL: 'joanwoz@yahoo.com'

INTRODUCTION

Many researchers have concluded that deforestation on Rapa Nui (Easter Island) resulted from uncontrolled population growth and unsustainable resource management (Cliff et al., 2000); (Diamond, 1995); (Redman, 1999). In this paper I will present the perspective that in fact failure to limit natural resource use had little or nothing to do with the fall of the Rapanui society. I suggest, based on the results of my research on environmental alteration, that given the climatic limitations of the island environment, the Rapanui were successful in husbanding the resources they viewed as most beneficial to their society.

In order to acquire the resources they needed, the Rapanui modified the environment of Easter Island according to Polynesian tradition. They installed a system of introduced cultivable plants and domesticated animals--a "transported landscape" (Kirch, 1982) and sustainably managed valuable food crops, albeit at the expense of natural ecosystems.

The data that suggest changes in Easter Island flora are derived from study of pollen (Flenley and al., 1991) and wood charcoal remains (Orliac, 1998). Based on the results of these analyses, researchers concluded most of the native Rapanui forest, which had been in place for more than 30,000 years, had disappeared by AD 1000. Researchers also observed a corresponding increase in the remains of herbaceous and grass flora.

These changes in flora have been taken to signify environmental degradation. Some researchers have further stated that the deforestation led to widespread soil erosion and subsequently to the demise of the Rapanui people (Diamond, 1995).

However, I will provide evidence that the initial demise of the Rapanui people was not the result of shortsighted environmental degradation but the result of forces beyond their control. Unpredictable forces of Nature (short-term or long-term climatic changes) intensified human-induced alterations, which then affected the inhabitants' ability to produce sufficient food to maintain a large population. In a second phase of their virtual eradication the Rapanui population was reduced between 1864 and 1872 from several thousand individuals to a mere 111 persons as a result of kidnapping and introduced diseases.

CAUSES OF DEFORESTATION ON EASTER ISLAND

There is not argument that the native forest of Easter Island disappeared during human occupation. An examination of the reasons for the disappearance of the forest will show that the Rapanui did not consider environmental alteration degradation, but enhancement. At least three major factors accounted for the observed transformation of vegetation on Easter Island.

The primary factor is obvious. Given that Easter Island is a single island, only 160 square kilometers in size, and more than 3000 km distant from other islands or the American continent, the use of forest products to produce objects of material culture and to use as firewood is certainly one explanation that explains deforestation.

A second factor relates to the nature of the physical environment and the natural ecosystem that had evolved on the isolated island of Rapa Nui, which is surrounded only by the nutrient-poor waters of the eastern South Pacific Ocean.

A third and possibly the greatest cause of environmental change on the island is the Rapanui people's use of arable, that is forest, land for food cultivation.

In this paper I first summarize the known or suspected causes of environmental degradation on Rapa Nui. I then present my evidence regarding environmental alteration, which is based on observations I made while conducting a geoarchaeological survey of horticultural sites on the northwest coast of the island. Finally, I relate how my findings help explain the course of environmental changes on Easter Island, and the extent to which these changes might have affected the prehistoric (pre-European contact) Rapanui population.

generation of the palm forest (Flenley, 1993).

THE ENDEMIC FOREST OF EASTER ISLAND

The now extinct endemic Easter Island palm--classified as Paschalococus (Dransfield et al., 1984), appears to be the same species as Jubaea chiloensis--the Chilean wine palm, or a species within the same genus. This comparability is based on botanical remains (Grau, 1998) and geological evidence (casts of trunks and crowns in late Pleistocene lava flows). The outer skin of Jubaea trunks is known to be exceedingly hard and resistant to deformation (Gurley and Liller, 1997). However, the interior is fibrous. If the Rapanui palm is closely related to Jubaea palm and shared its morphology as evidence indicates, then the palm trunk with

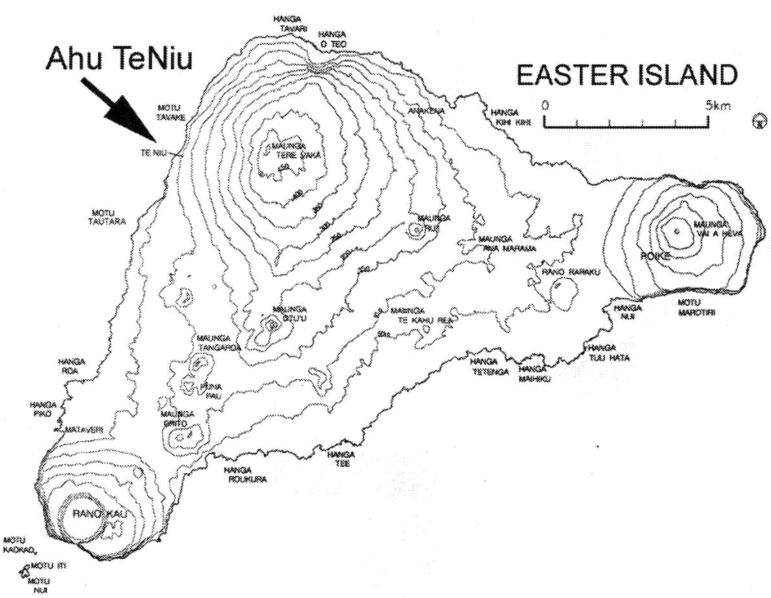

ENVIRONMENTAL ALTERATION

Many persons interested in environmental issues have portrayed a prehistoric Easter Island as a valuable example showing the negative effects of environmental alteration and the ultimate consequences of poor resource management (Diamond, 1995); (Redman, 1999).

Researchers have speculated that Rapanui people used palms for monument building, image carving, or making canoes for fishing (Cliff et al., 2000); (Diamond, 1995); (Grau, 1998). Whether the Rapanui used the actual palm trunks (stems is a more correct term) in canoe and raft construction, or as rollers to move the moai (the carved stone images) from the quarry to one of the 300 ahu monuments constructed around the island, is purely speculative. Moai transport had stopped by the time of the first European contact in AD 1722 (Roggeveen), but the last remains of Easter Island palm trees were observed during the mid-19th century (Palmer, 1870). Some researchers suggest that humans and their Polynesian rats ate all the palm seeds (as observed on a contemporary Lord Howe Island), thus precluding re-

its less than 1 cm thick skin and porous interior would make it difficult to manufacture a canoe form or to use the trunks as raft timbers, without them rapidly becoming waterlogged.

The diameter of Paschalococus trunks is such that the Rapanui workers in charge of moving the moai would have a difficult time manipulating them as rollers, or even as poles with which to construct a framework used to easily move the images to their ahu. Younger Jubaea tree trunks do not grow long and narrow like coconut palms. The trunks of even young Jubaea palms tend to be greater than 40 cm in diameter; a mature tree may be as thick as one meter. The trunk grows straight if supplied continuously with moisture, while the circumference of the new growth, at the top of the tree, increases or decreases as annual moisture supplies vary (Grau, 2000: 165).

The annual precipitation of Easter Island does vary greatly. Rainfall has fluctuated between 50 and 200 cm per year according to records maintained during the 20th century.

The Rapanui undoubtedly used the Paschalococus palms for

many purposes beyond those usually suggested --canoe construction or moai transport. The palm fronds likely served as thatch and bedding, and the nuts and sweet sap served as food for both humans and rats. Evidence supports both of these possibilities (Dransfield et al., 1984); (La Perouse, 1797).

The palm was not the only tree in the Rapanui forest. Many hardwood trees and shrubs were also present. These also were used for multiple purposes, including cooking fires, carving, and canoes (Orliac and Orliac, 1999).

BIOGEOGRAPHIC FACTORS

The composition of the native flora and fauna on Easter Island prior to the arrival of the Polynesians was more vulnerable to environmental and climatic alterations than those of many Pacific island ecosystems for several reasons. Easter Island is located in the subtropical dry zone of the Pacific, and so its vegetation was made up of slow-growing mesic varieties. Biodiversity was low and the total number of native species on the island were few because of the young age of the island and its biogeographic position within the Pacific (van Balgooy, 1996); (Zizka, 1990).

Plants on Easter Island are derived from biota of the western Pacific and from the Americas (Zizka, 1990). These were accidentally introduced, as seeds carried by water currents, or carried by birds blown off course during storms (Carson, 1996); (MacArthur and Wilson, 1967). Many of these introductions undoubtedly succumbed because the climatic conditions they could not reproduce in the dry and seasonally cooler environment. Others eventually evolved to become endemic specie.

The Rapanui forest evolved under dry conditions. Paschalococos palm made up the upper canopy foliage; the mid-story canopy vegetation was composed of at least ten species of shrubs and small hardwood trees (Orliac, 1998), and herbaceous plants and grasses formed the ground cover (Skottsberg, 1925); (Zizka, 1990). Such mesic forests are relatively slow growing. The inability of the ancient Rapa-

Figure 1. Stratigraphic profile from excavation 26-1-1 at ahu Te Niu (Site 26-1) on the northwest coast of Easter Island. Pedogenic A horizons are labeled A1 and A2. The burn layer represents the burning of the original forest at Te Niu, which was dated to AD 1300 (calibrated). Colluvial deposits occurred after AD 1300 to depths between 70 and 87 cm in front of the ahu.

nui forest to regenerate quickly likely played a role in its decline. Jubaea palms, for instance, need 30-50 years to reach maturity, at which time they begin to produce viable seeds (Grau, 2000: 165).

FOOD PRODUCTION ON EASTER ISLAND

A third cause of environmental alteration on Easter Island is rooted in the Polynesian tradition of food production. The Polynesians cultivated tubers and trees at the expense of native forest on all islands they colonized. Polynesians carried the biotic components of their horticultural system, a 'transported landscape' (Kirch, 1982) to provide a reliable source of food wherever they settled. In addition to sustenance, cultivation provided a means to fulfill social obligations through exchange rituals within a highly ranked society (Goldman, 1970).

In the centuries following colonization, the Rapanui replaced much of the native mesic palm forest with gardens of their imported tuber and tree cultigens. House gardens and fields contained bananas, sugar cane, taro, yams, sweet potatoes, gourds, and fiber-producing tree crops. The endemic palm and hardwood species diminished until the 19th century (Orliac and Orliac, 1999); (Palmer, 1870). Unlike many high volcanic islands of the Pacific, the low gently sloping mountains of Easter Island were easily accessible and therefore useable as garden space. The accessibility to humans of the land suitable for gardens may also explain why so much of the native forest disappeared relatively quickly

DATING ENVIRONMENTAL CHANGES ON EASTER ISLAND

Easter Island archaeologists have found evidence that monument building began prior to AD 1000. Early monuments were centered in a few locations near to the coast (Ayres, 1971); (Ayres, 1975); (Skjolsvold, 1994). The remains of settlements associated with the ahu indicate that prehistoric people chose to live near to the shore. Archaeological investigations at prehistorically cultivated garden and field sites within the settlements (Stevenson, 1999); (Wozniak et al., 2001) indicate there was significant forest clearing as early as the 12th century. By AD 1350 the Rapanui had established house gardens on most of the fertile forest soils in coastal areas. They also made seasonal use of garden sites on mountain slopes several kilometers inland of major south coast settlements after AD 1400 (Stevenson, 1999).
During the 15th and 16th centuries monument construction and the carving of large stone images--the moai--had become a major societal undertaking over much of the island (Heyerdahl and Ferdon, 1961). It is during this latter period that the Rapanui population may have approached the carrying capacity of the island resources, and according to some, crashed (Brander and Taylor, 1998); (MacIntyre, 1999).

ESTIMATING POPULATION SIZES

The timing of the 'fall' of Rapanui society—that period of final destruction of the environment and collapse of society--is often attributed to the late 17th century. According to oral tradition, this was a period of increased warfare, famine, and cannibalism. The cause of the warfare is attributed to environmental-induced (deforestation) or social-induced stress (revolt against excessive obligations to chiefs) (Diamond, 1995); (Redman, 1999). A decrease in population may have resulted from either intra-island warfare or famine, or both.

However such a decrease in Rapanui population is implied; since there is no empirical data to indicate the size of the population prior to the period of warfare, or even immediately afterwards. The earliest estimates of population size are from the early 18th century. These are provided by journals of Europeans and American, often viewing the island from on board their ships. The visitors from Holland, France, England, Spain, or the Americas, estimated that from several hundred to several thousand people lived on Easter Island between the early eighteenth and early nineteenth centuries (Forster, 1777); (Gonzalez y Haedo, 1908); (La Perouse, 1797); (Roggeveen, 1908). Considering that one cannot view the entire island from any one of the harbor sites used by the visitors, and each visit lasted a matter of hours or days, the numbers recorded for the early 18th century population are serious underestimation of actual population size. In the early 19th century when missionaries and more frequent visitors actually stayed on the island, "about 4000 persons" were there (Metraux, 1940).

CLIMATIC FACTORS

The timing of the supposed 'fall' of the Rapanui population (several decades prior to "contact" in AD 1722) also happens to coincide with a period of global cooling known as the Little Ice Age (Bradley, 1999); (Hunter-Anderson, 1998); (Jones, 1998); (McCall, 1993). Solar forcing (periodic changes in solar activity and irradiance) during the Maunder Minimum (mid-17th to early 18th century) caused regional temperature changes in the Northern Hemisphere to become 1-2oC colder than during the previous centuries (Shindell et al., 2001). While no empirical data or historic records are available for the eastern subtropical South Pacific for this time period, there is evidence that glaciers increased in size during the 18th century within New Zealand and South America as a result of cooler temperatures (Grove, 1996).

The period of the Little Ice Age (AD 1450-1850 in the Northern Hemisphere—possibly later in the Southern Hemisphere) was a long-term climatic change that had global consequences. Deleterious weather patterns during the Little Ice Age may have been exacerbated by short-term regional, or even global, climatic perturbations caused by episodic volcanic activity or ENSO (El Nino/Southern Oscillation) phenomena. Sulfur emissions, which are associated with volcanic activity of the 17th century in South America (Briffa et al., 1998) and elsewhere, affected global climate during this time period (Kondratyev and Galindo, 1997). The global impact of climatic events associated with the ENSO is also often associated with severe droughts interspersed with short periods of highly destructive rainfall in widely separated portions of the globe (Grove, 1998).

EVIDENCE OF ENVIRONMENTAL CHANGE RECORDED ON THE NORTHWEST COAST OF EASTER ISLAND

During the period of prehistoric environmental alteration resulting from human activities, there is no doubt that erosion occurred on Easter Island as forestland was converted to settlements and plantations. Some of these eroded sediments were likely lost to the sea, as they are today during exceptionally heavy rainfall.

During my study of marine resource use, horticulture, and environmental alteration in a coastal settlement, I surveyed a portion of the northwest coast of Easter Island. I collected stratigraphic data, soil samples, and charcoal samples and artifactual material used for dating from a systematic set of excavations, within residential and garden sites, and in locations having no obvious archaeological remains (as a control).

My data helped to elucidate the spatial and temporal character of environmental alteration. I was able to identify fertile soils and the geomorphic settings that had supported the endemic savanna forest vegetation in place when colonizing Polynesians arrived 1500 years ago. The stratigraphy of my excavation units in both garden and non-garden sites yielded evidence that included characteristic signs of deforestation, erosion, and sedimentation, over the period of the past millennium

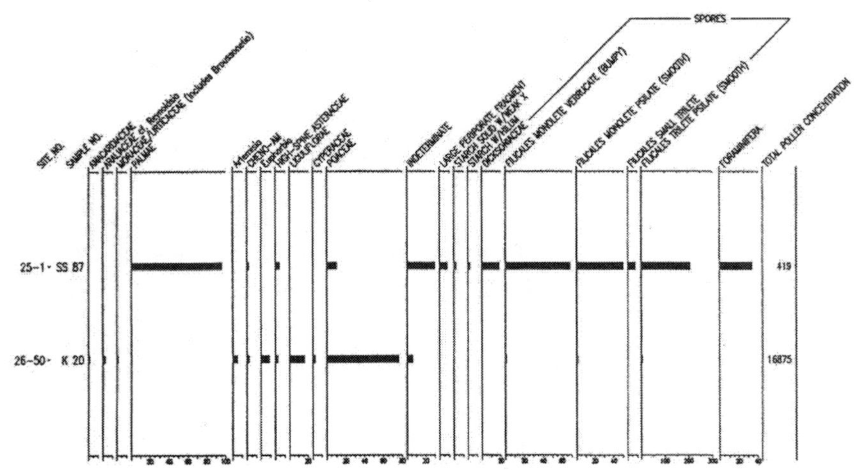

Pollen diagram from a cave and an *ahu* on Easter Island.

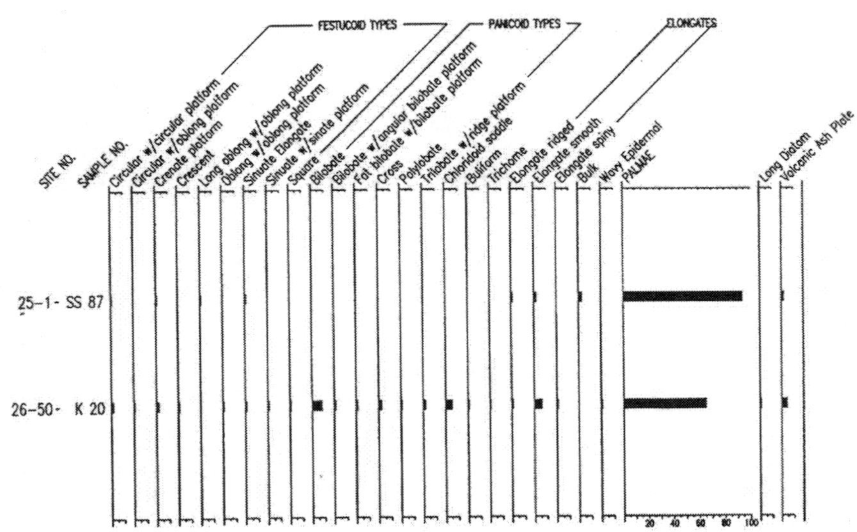

Phytolith diagram from a cave and an *ahu* on Easter Island.

Figure 2. Pollen and phytolith diagrams from the northwest coast of Easter Island. Soil samples for pollen analysis were taken from sediments under an ahu platform and inside a rock shelter; both sites obisidian hydrated dated to the 15th century.

The Rapanui living along the northwest coast had placed their settlement on a gentle mountain slope overlooking their central ceremonial location. At about AD 1300 the people of this settlement burned a large amount of vegetation on the ground surface where two ahu platforms were to be constructed. They did not build the platforms until the 15th to 16th centuries. Once built, the ahu structures served to preserve the history of forest clearing and subsequent sedimentation that occurred prior to monument building (Figure 1).

Under the ahu I found that a large fraction of the sediments eroded from the hill slopes were deposited at lower elevations shortly after AD 1300, again some time after AD 1650, and finally historically, after the introduction of sheep. More importantly, I found anthropogenic sediments located on top of still intact forest soils at many of the garden and residence sites. The fact that the forest soil was still intact indicates that erosion was localized, rather than widespread.

POLLEN AND PHYTOLITH ANALYSIS

Soil samples I collected from beneath this same ahu, from a rock shelter, and from rock covered gardens on the slope above the ahu, were submitted for pollen and phytolith (silica structures found in many monocots) analysis to Paleoresearch Laboratories, Golden CO (Figure 2). Pollen of Ipomoea (sweet potatoes), Moraceae (paper mulberry--used for tapa cloth), and cheno-ams was found along with palm pollen and fern spores under the ahu, while pollen of grasses and hardwood trees dominated the rock shelter deposits. Phytoliths were predominately those of palm in both settings. The rock shelter contained pollen of grasses and shrubs, but few palm pollen grains. Palm phytoliths dominated the rock shelter soil sample however. Consequently, the palm fronds were likely brought into the shelter for bedding or other use.

Soil sediments dating from the 15th to 17th centuries were extracted from prehistoric gardens for pollen analysis also. In addition to the palm, shrubs, and grass pollens, the pollen of Colocasia taro was found. Few phytoliths other than those of palms were found however

Unfortunately most Polynesian cultigens do not produce pollen or phytoliths. Those tuber or tree plants that do flower often produced fragile pollen and these did not survive in the Easter Island soil. Of the cultigens available to the Rapanui only banana plants and sugar cane produced phytoliths. The delicate Musa banana phytolith structures are easily destroyed; and those of sugar cane cannot be unequivocally distinguished from other Poaceae plants.

The phytological results support the geomorphological data indicating that the former savanna-palm forest ecosystem in place at Polynesian colonization had been modified to accommodate food cultivation (Cummings, 1998).

DATING THE SETTLEMENT ON THE NORTHWEST COAST OF EASTER ISLAND

The charcoal and obsidian debitage obtained from a rock shelter on the northwest coast (Site 26-50), from the ahu (Site 26-1) and from within several garden excavations, were used to date the settlement use and geomorphological changes (Figure 3). Forest vegetation growing on the edge of the coastal cliff at the ahu site was burned around AD 1300, as mentioned. The 1-cm thick charcoal lens of mixed vegetation remained at the site, preserved under the ahu structure at a depth of 90 cm below the present ground surface (an area approximately 10 by 30 meters). Some East Polynesian researchers believe the early ceremonial sites had no large platform structures (Green, 2001). This may explain why the two adjacent ahu platforms I examined were built almost a century after the site was cleared of vegetation. Remains within the charcoal lens indicate that humans were involved.

Within the rock shelter (Site 26-50), below sediments I collected for pollen and phytolith analysis are the remains of a small cooking fire, placed on bedrock. Charcoal from the cooking fire was radiocarbon dated to 550 BP (AD 1406 +/- 110 calibrated) and obsidian within the fire pit was hydration dated to between AD 1203 +/-90 and 1275 +/-83 (four flakes). In all thirty centimeters of sediment, combined with faunal remains and debitage from lithic tool manufacture, were deposited on top of the fire pit remains. Obsidian (10 flakes) retrieved from the material found within the rock shelter sediments provided hydration dates indicating a nearly continuous use of the rock shelter from the early prehistoric (AD 1200) to the late prehistoric period (AD 1715). The timing of the initial garden construction in the project area corresponds with the earliest phase of monument site selection--the burn event. Charcoal retrieved from garden excavations, was also dated (7 dates between AD 1285 and AD 1789, calibrated). The charcoal dates from seven gardens, spread over a distance of one kilometer, indicate widespread clearing took place between AD 1285 and 1450 (calibrated dates). The obsidian dates indicate that gardening practices continued until the19th century within the same gardens.

Based on the faunal analysis and the survey of garden sites it is clear that people living along the northwest coast of Easter Island used both marine resources and domesticated animals (chicken and Polynesian rat) and they grew predominately Polynesian cultigens for food.

SEDIMENTATION FOLLOWING ENVIRONMENTAL ALTERATION

Sedimentation, presumed to be the result of erosion on deforested slopes, occurred within concave landforms (U-shaped swales) once forested by palms and scrub hardwoods. These sediments and the original forest topsoil were used for gardening on Easter Island (Wozniak et al. MS in preparation). This phenomena has been documented on many inhabited islands of Polynesia and coincidentally during a similar timeframe (Athens and Ward, 1993); (Kirch and Ellison, 1994); (Leach and Leach, 1979); (Lepofsky, 1994).

Early prehistoric sedimentation was preserved under the ahu platform. The stratum laid down on top of the charcoal lens is twenty to thirty cm thick and composed of fine colluvial sediments. Once the ahu platform was built, around AD 1450, another 20 cm of sediments was deposited in front of the ahu structure. Within this colluvium stratum are thinner layers of sediment alternating with ahu construction materials, including a fine dust of volcanic tuff, produced as a result of moai rock carving at the site (Figure 1).

Evidence of high sedimentation rates is difficult to measure within the gardens of the northwest coast project area because the sediments were incorporated into the garden soil. The process of planting, weeding and harvesting mixes the garden sediments so that a thick, homogenous soil layer is formed. The anthropogenic soil layer is 30 to 50 cm thick, and in many cases I found the garden soil lying on top of the total thickness of the former forest topsoil and subsoil (A, AB, B, and C horizons). Many gardens and their underlying forest stratigraphy have been protected from historical erosion by a mulch of small lava rocks (Wozniak, 1999).

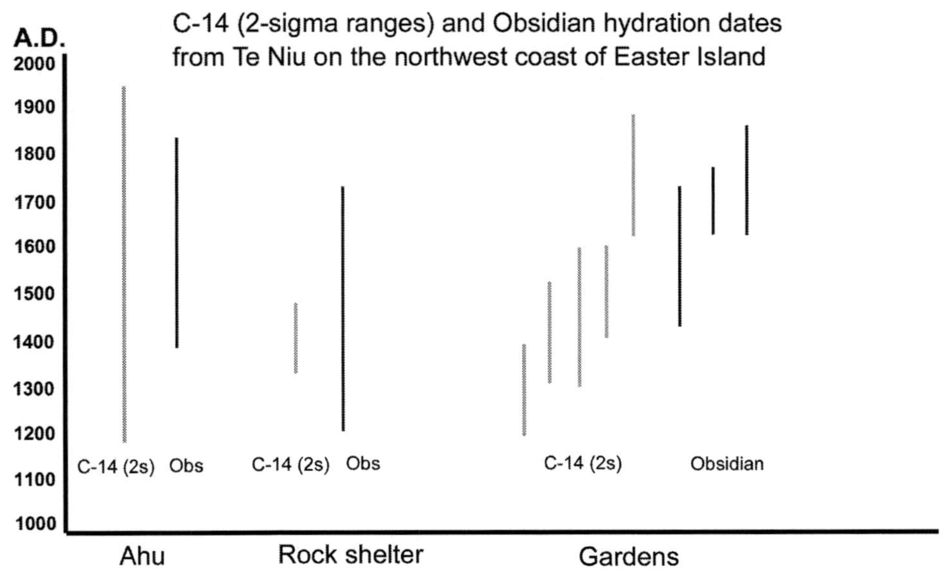

Figure 3. Radiocarbon, AMS, and obsidian hydration dates for sites at Te Niu on the northwest coast of Easter Island. Each bar represents the range of dates retrieved from one site. One charcoal sample from the rock shelter and charcoal from five garden sites was C-14 dated. Twelve samples of charcoal and bone from the ahu were dated. C-14 bars (2 sigma); Obs (1 sigma). Obsidian range bars represent 35 dates for the ahu, 10 for the rock shelter, 2 to 7 for each garden.

A third set of erosion events, while more difficult to date precisely, are the results of historical erosion from sheep and cattle ranching. These contemporary sedimentation events covered lithic tools, which had been abandoned at the ahu at the end of major ahu reconstruction events dated to the early 19th century (based on obsidian hydration dates). These tools were found under 20-30 cm of sediments. This contemporary surface topsoil is developing into topsoil typical of grassland vegetation (mollic epipedon).

The sedimentation events preserved under, and in front of, the ahu illustrate that erosion took place as several large events within a short time period rather than by slow accumulation over many centuries. On Easter Island, the most catastrophic of climatic events may have been periods of wide variations in mean annual precipitation, alternating periods of severe droughts and torrential rain. My data most closely supports this scenario.

ECONOMIC STRATEGY OF LAND USE ALTERATION

Based on the presence of garden soils on top of the former forest soils and phytological evidence of Polynesian and Ipomoea (prehistorically introduced American species) cultigens, I strongly believe that the environmental alteration occurring after the arrival of the Polynesians was part of a successful economic strategy. This strategy of Polynesian horticultural production was successful during the climatic conditions of socio-political and population growth.

CLIMATIC PERTURBATIONS

I further believe that it was the climatic conditions occurring after AD 1600 that led to, or increased the stress on food production. The climatic conditions of the 17th and 18th century are likely in part responsible for the purported famine, and the suspected reduction in population size that followed, and for an increase in social stress (inter-clan warfare). Droughts, exceedingly cool weather, or other aberrations from the "normal" weather conditions served to reduce the degree to which agricultural production during the 17th century on the island was successful.

THE NEAR DEMISE OF THE EASTER ISLANDERS

Neither environmental alteration (e.g. deforestation) per se, nor climatic conditions, were responsible for the demise of the Easter Island population described in the popular literature (Cliff, 2000), (Diamond, 1995); (Redman, 1999). Rather the reasons for the population reduction observed by visiting Westerners during the 18th century, were more likely related to the causes of the more serious reduction in population that occurred during the 19th century.

As documented by [McCall, 1990 #468], during the early 19th century kidnapping by "blackbirders" was a frequent occurrence, and during the 1860s over a thousand Rapanui were kidnapped for slave labor in Peru. The few Rapanui people who returned to their island from Peru brought with them Western diseases they had contracted while on the continent (Englert, 1974); (McCall, 1990); (Routledge, 1919). Smallpox and tuberculosis (pulmonary diseases) decimated the population, from "some 4000" (Metraux, 1940: 22) in the early 19th century to 111 persons by 1872.

There is no quantitative documentation for the changes in population reported during the 18th century. Speculative population estimates were provided (Roggeveen, 1908); (Forster, 1778); (La Perouse, 1797). However, given the apparent susceptibility of the Islanders to diseases introduced during the 19th century, it is reasonable to suppose that visits of foreigners during the 18th century had similar effects on the health of the Rapanui. Deaths from relatively minor diseases have been documented on other islands in the South Pacific (Cliff et al., 2000: 139-141). Unfortunately, census and epidemiological data on Rapa Nui prior to the 1860s, during the decades between Euro-American visits of the 18th century, are not available.

While the environment of Easter Island was altered to accommodate the needs for subsistence and rituals (ahu and moai images) during the prehistoric period, the extinction of native plants (deforestation) did not cause the fall of the Rapanui people. Population pressure likely did play a role in the decline of natural and produced resources. However, given the evidence of increased sedimentation during this time, climatic stress during the 17th century and its effect on the supply of cultivated food is the more likely explanation for the legend of famine (Metraux, 1940). The reduced cultivable resources in turn resulted in a decreased population, although the number who succumbed cannot be determined.

Loss of the natural ecosystem began under the Polynesians as a result of their introduction of the "transported landscape" to the Rapanui ecosystem. The native forest, and many of the Polynesian introduced plants, disappeared much later, between the late nineteenth and the twentieth century, as a result of sheep ranching introduced by Europeans.

Today, Rapanui people still live on Easter Island. The Rapanui environment is highly degraded and the culture greatly altered. Neither prehistoric climatic stress nor environmental degradation resulted in the near extinction of the Rapanui way of life as many researchers have concluded. The causes instead can be traced to contact with western civilization.

REFERENCES

ATHENS, J. S. AND WARD, J. V.
1993.
Asian Perspectives, 32, 205-223.
AYRES, W. S. ,
1971. Journal of the Polynesian Society, 80, 497-504.
AYRES, W. S. ,
1975. The University of South Carolina.
BRADLEY, R. S..
1999. Paleoclimatology, Harcourt-Academic Press, San Diego.
BRANDER, J. A. AND TAYLOR, M. S.,
1998.) The American Economic Review, 88, 119-139.
BRIFFA, K. R., JONES, P. D., SCHWEINGRUBER, F. H. AND OSBORN, T. J. ,
1998. Nature, 393, 450-458.
CARSON, H. L.,
1996. In The Origin and Evolution of Pacific Island Biotas, New Guinea to Eastern Polynesia: Patterns and Processes (Eds, Keast, A. and Miller, C. D.) SPB Academic Publishing, Inc., Amsterdam, pp. 7-17.
CLIFF, A. D., HAGGETT, P. AND SMALLMAN-RAYNOR, M. R.,
2000. Island Epidemics, Oxford University Press, Oxford.
CUMMINGS, L. S.,
1998. In Easter Island in Pacific Context (Eds, Stevenson, C. M., Lee, G. and Morin, F. J.) The Easter Island Foundation, Los Osos, CA, pp. 100-106.
DIAMOND, J.,
1995. Discovery: The World of Science, 16, 62-69.
DRANSFIELD, J., FLENLEY, J. R., KING, S. M., HARKNESS, D. D. AND RAPU, S.,
1984. Nature, 312, 750-752.
ENGLERT, S.,
1974. La Tierra de Hotu Matu'a, Editorial Universitaria, Santiago, Chile.
FLENLEY, J. R...,
1993. In Easter Island Studies, Vol. 32 (Ed, Fischer, S. R.) Oxbow Monograph, Oxford.
FLENLEY, J. R. AND AL., E.,
1991. Journal of Quaternary Science, 6, 85-115.
FORSTER, G.,
1777. A Voyage Round the World in his Britannic Majesty's Sloop, Resolution, commanded by Capt. James Cook during the Years 1772-5, London.
FORSTER, J. R.,
1778), London.
GOLDMAN, I.,
1970. Ancient Polynesian Society, The University of Chicago Press, Chicago.
GONZALEZ Y HAEDO, F.,
1908. Hakluyt Society, 13, 1-82.
GRAU, J.,
1998. In Easter Island in Pacific Context (Eds, Stevenson, C. M., Lee, G. and Morin, F. J.) The Easter Island Foundation, Los Osos, CA, pp. 120-124.
GRAU, J. ,
2000. Palmeras, OIKOS Ltda., Santiago de Chile.
GROVE, J.,
1996. In Time-Scales and Environmental Change (Eds, Driver, T. S. and Chapman, G. P.) Routledge, New York, pp. 39-87.
GROVE, R. H. ,
1998. Nature, 393, 318-319.
GURLEY, R. E. AND LILLER, W. ,
1997. Rapa Nui Journal, 11, 82-84.

HEYERDAHL, T. AND FERDON, E. N. (EDS.),
1961. Reports of the Norwegian Archaeological Expedition to Easter Island and the East Pacific: Archaeology of Easter Island, School of American Research, Santa Fe, NM.
JONES, P.,
1998. Science, 280, 544-545.
KIRCH, P. V. ,
198.) Pacific Science, 36, 1-14.
KIRCH, P. V. AND ELLISON, J. ,
1994. Antiquity, 68, 310-321.
KONDRATYEV, K. Y. AND GALINDO, I. ,
1997. Volcanic Activity and Climate, A.Deepak Publishing, Hampton, VA.
LA PEROUSE, J. F. D. G. ,
1797. A voyage Around the World Performed in the years 1785-1788 by the Boussole and Astrolabe, London.
LEACH, B. F. AND LEACH, H. (EDS.) ,
1979. Prehistoric Man in Palliser Bay, National Museum of New Zealand, Wellington.
LEPOFSKY, D. ,
1994. In Dept of Anthropology University of California at Berkeley, Berkeley, pp. 318.
MACARTHUR, R. H. AND WILSON, E. O.,
1967. The Theory of Island Biogeography, Princeton University Press, Princeton.
MACINTYRE, F. ,
1999. RNJ, 13, 35-41.
MCCALL, G.,
1990. In Circumpacifica, Band II: Ozeanien, Miszellen. Festschrift fur Thomas S. BarthelFrankfurt am Main: Peter Lang.
MCCALL, G.,
1993) Rapa Nui Journal, 7, 65-70.
METRAUX, A.,
1940. Ethnology of Easter Island, B.P.Bishop Museum Press, Honolulu.
ORLIAC, C.,
1998. Journal des Oceanistes, 107, 135-142.
ORLIAC, C. AND ORLIAC, M.,
1999. In Easter Island and East Polynesia Prehistory (Ed, Vargas Casanova, P.) Universidad de Chile Instituto de Estudios Isla de Pascua, Santiago, pp. 185-200.
PALMER, J. L.,
1870. Journal of the Royal Geographical Society (London), 40, 167-181.
REDMAN, C. L.,
1999. Human Impact on Ancient Environments, University of Arizona Press, Tucson.
Roggeveen, J. (1908) Works Issued by the Hakluyt Society, 13.
ROUTLEDGE, K.,
1919. The Mystery of Easter Island, Hazell, Watson & Viney, London.
SHINDELL, D. T., SCHMIDT, G. A., MANN, M. E., RIND, D. AND WAPLE, A.,
2001. Science, 294, 2149-2152.
SKJOLSVOLD, A. (ED.),
1994. Archaeological Investigations at Anakena, Easter Island, The Kon-Tiki Museum, Oslo.
SKOTTSBERG, C.,
1925. Bernice P. Bishop Museum Bulletin, 16, 47 pp.
STEVENSON, C. M.,
1999. Maunga Tari: an upland agricultural complex, Bearsville Press and Cloud Mountain Press, Los Osos, CA.
VAN BALGOOY, M. M. J.,
1996. In The Origin and Evolution of Pacific Island Biotas, New Guinea to Eastern Polynesia: Patterns and Processes (Eds, Keast, A. and Miller, C. D.) SPB Academic Publishing, Inc., Amsterdam, pp. 191-214.
WOZNIAK, J., HAOA, S. AND STEVENSON, C. M. (N.D.) Ms in preparation.
WOZNIAK, J. A.,
1999) Rapa Nui Journal, 13, 95-99.
ZIZKA, G.,
1990. In State and Perspectives of Scientific Research in Easter Island Culture, Vol. 3 (Ed, Esen-Baur, H. M.) Courier Forschunginstitut Senckenberg, Frankfurt, pp. 189-207.

23

SOCIAL INTERACTIONS ON CYPRUS IN THE PREHISTORIC PRODUCTION OF METALS : KISSONERGA AND A POSSIBLE STARTING POINT FOR A NEW RELIGION.

MARIA ROSARIA BELGIORNO

Institute for Technologies Applied to Cultural Heritage, National Council for Researches, Italy. CP. 10-00016. Monterotondo, St. (RM).
Italian Archaeological Mission at Pyrgos/Mavroraki, Cyprus: 7, Pallikaridi, Pyrgos Limassol.
EMAIL: belgiorno@pyrame.net

A NEW INTERPRETATION OF THE KISSONERGA CEREMONIAL AREA AND ITS DEPOSIT.

Is the Late Chalcolithic figurative repertoire of Cypriote religion linked to the rising of prehistoric metallurgical society in Cyprus? Between the Early Bronze Age and the Middle Bronze Age II, accepting that the religious remains of the metallurgical workshop of Enkomi and of Toumba tou Skourou go back to the Middle Bronze Age III, we have no other public architectural remains connected with religious items.

The suspected existence of rural "temenos" is related only to the coroplastic shrine models and to the vases decorated with representations of human activities. Public religious structures could have been connected with the production of copper as testified later by the industrial and religious contexts of Enkomi and Kition. The discovery of metallurgy probably changed still from the beginning the social system and the religious practises and credence of the Cypriote people.

The metallurgy reached the Mediterranean basin approximately at the half of V millenium BC. One of the most ancient country not far from the Mediterranean coast, where the copper was exploited and processed was the Egypt, most precisely the area of the North East desert at the shoulders of Alessandria.

In the same moment on the coast line of south west Cyprus, in front to the Egypt, started the copper metallurgy of the most rich in copper island of Mediterranean.

In turn the ancient tradition and the history gave different level of importance in producing copper to Egypt and Cyprus. Only recently the prehistoric capability and the natural consistence of the Egyptian mineral resources have been revalidated (Egyptian Congress on metallurgy). Many tentative to connect the most ancient Cypriote religion to the production of copper were not able to give a definition of the Copper Cypriote goddess far from Aphrodite, the traditional divinity of the Island. In later time Aphrodite was personified with the copper and even with the enormous massive of Troodos, which with its wealth of metals occupy the hearth of Cyprus.

The trade of the precious metal was one of the most important business of the antiquity since the half of second millennium BC. It was transported on shipwrecks as oxhide ingots, a famous Cypriote ingot shape of the Late Bronze Age probably linked with religious believes connected with the production of the copper, as demonstrate by the excavation at Kition and Enkomi. The prehistoric link of the Cypriote Aphrodite and the copper was trade by the Greek and Latin historians and enlarged by the archaeological discoveries and modern investigations. So today we could affirm that in the Greek and Roman world the divinity of copper was Aphrodite.

Meanwhile also in Egypt the copper had its own goddess, Hathor, whose symbol "Ka" became the symbol of copper.

On the southern slopes of the Troodos mountains, the existence of extensive mineral resources has been known since the Chalcolithic period. Its deposits of copper and iron have been wrought throughout the Roman period and its galleries reopened at the beginning of last century to exploit metals and to test the existence of the famous "Devil's Mud" of Cyprus, a "soft crumbly material -full of gold and silver, with the peculiar property of changing into liquid slime when pressed between the fingers" (Bear,1963, 185).

In antiquity Cyprus was considered one of the major centres for copper production in the Mediterranean basin. Modern studies have demonstrated that the historical tradition corresponds with the geological realty of the island even if, till today very little is known about its most ancient metallurgy.

The Massif of Troodos is mainly made of an enormous diabasic winding mass alternated with sedimentary rocks, from which volcanic scoriae of extrusive rocks and lava stand out. Igneous rocks in which cupriferous minerals are mingled with iron and precious minerals reach the slopes of Troodos seven-eight km from the coastline.

Copper archaeological findings from pre-Bronze Age Cypriot contexts are scarce, as not many extensive pre-Bronze age sites have been excavated in Cyprus so far. At the late Chalcolithic settlement of Kissonerga some of the most ancient metal objects of the island have been found together with fragments of crucibles and unalloyed smelted copper, testifying a small-scale copper processing .

With the exclusion of the ceremonial area of Kissonerga and some architectural enclosures found in Sotira (Dikaios, 1961) and Vrisy (Peltenburg, 1982), all belonging to the end of the Chalcolithic period, with some extension in the Early Bronze Age, no other remains of public or domestic shrines have been ever brought to light in other Early-Middle Bronze Age settlements.

The link of Cypriote metallurgy to religion was underlined first by V. Karageorghis (Karageorghis, 1982, 103-4, 116; 1985; 1990, 43) and then by other scholars, mainly B. Knapp who devoted most of his researches to the "Copper production and Divine Protection" (Knapp, 1986; 1988, 133-63; 1994, 377-452; 1996; 71-107).

Today, we know that almost all the Late Bronze Age excavated settlements of Cyprus were metallurgical sites. The method used to smelt and work copper was almost the same. But we have the chance to reconstitute some of the metalworking activity close to a sanctuary from the Late Bronze Age foundings of Enkomi (Courtois; 1982; 1986 8-13), Athienou (Dothan and Ben-Tor 1983), Kition (Karageorghis, 1985, Appendix V, VI, VII), and recently of Hala Sultan Tekke (Astrom, 1998, 2-6,).

They were organised public religious buildings (Muhly, 1991, 180-96; 1996, 45-61; Webb, 1999), metallurgical workshops whose production has been controlled by secular, not religious, elite as Knapp suggests (Knapp, 1996, 91).

Between the Early Bronze Age and the end of Middle Bronze Age II, accepting that the religious remains of the metallurgical workshop of Enkomi and of Toumba tou Skourou go back to the Middle Bronze Age III, we have no other public architectural remains connected with religious items.

The suspected existence of rural "temenos" for public ceremonies is related only to the famous acerbic debated coroplastic shrine models found in the tombs, and the vases decorated with the plastic representations of the symbolic human activities (Karageorghis, 1991,117-169). So we assume that the appearance of a public religious structures in Cyprus are deeply connected with the working and production of copper; an activity that, from the beginning changed markedly not only the social system of the Cypriote people, but also their religious practises and credence.

It would be no surprise to find that the "ceremonial area of Kissonerga" had been related to the most ancient evidences of copper working in Cyprus (Peltenburg, 1989, 195-197). We must not forget that the "ritual hoard", buried in the Mosphilia ceremonial area, "consisted of fire-cracked and ochre-stained stones, a stack of two complete and two inverted fragmentary Red-on-White bowls, nine pottery and ten stone figurines, a pottery stool, a triton shell, pestles and other ground stones and chipped tools" (Peltenburg 1989, 195). Among the different interpretations about the nature of this deposit (childbirth, ritual objects intentionally buried before the abandonment of the settlement) given by Peltenburg (1988, 47; 1989, 197), I would like to recall the observation of A Campo (1994, 183) "...the model found at Mosphilia supports to some extent the idea of special places. The radial floor partition and central rectangle especially seem to reflect an actual building type, of which two have been discovered thus far. What purpose these buildings served is more difficult to establish".... It is a place difficult to recognise in the two buildings close to the votive deposit. But it is clear that the whole is in relation and symbolises a ritual, which happened in another place.

In his conclusion, Elliott stresses that the ground stones tools of Kissonerga deposit... "do not represent tools from a foundation deposit. Neither is this cache of tools of outstanding value and quality such as may have been hidden before an impending attack or disaster for eventual retrireview-as is the case with several Late Bronze Age gold and bronze hoards... The connection of the Kissonerga deposit with childbirth rites, as interpreted from the figurine and building model evidence, underlines the suggestion that the tools are associated with the preparation of special foods ..."(Elliott 1991, 67).

But we know that the same tools were used for metallurgical purposes (Craddock 1995) and the discovery of evidence of Late Chalcolithic metal working, the most ancient in Cyprus (Peltenburg 1991, 28-9; Gale 1991, 37-62), in the Kissonerga/Mosphilia B3, move our attention towards a completely different interpretation of the deposit.

Why not postulate the hypothesis that the Kissonerga ritual deposit with its hoard of stone tools and fertility figurines was connected in some manner to the incipient metal working activity and used in rituals to consecrate places or actions to work the copper? It is commonly accepted that the smelting evolution has been based on two main factors: first, the technical culture treasure, resulting from experience of experiments passed on through generations, and second the magic religious substratum born to explain the mysteries connected with life and the growth of the metals inside the mother-earth and the possibility to smelt, to produce and to craft them. The very long period needed to reach a satisfactory technical level to work the metal, together with metallurgical knowledge coming from distant countries, contributed to create a background of legends, myths and magic rituals which formed a special religion of the metals, practically common all over the world.

This is a complex base of beliefs and techniques which remained practically untouched for centuries. Studying the traditions of African and Asiatic people who continued working the metals with ancient methods till to day, offering us a way to understand different aspects of primitive metallurgy, the contributions of Eliade and Forbes on these topics (Eliade 1962; 1977; Forbes 1942, 3-32; 1971; 1972) show us the magic world of the metallurgy.

If we analyse one by one the objects stored in the Kissonerga deposit we find strong connections with the portrait of the primitive metallurgy described by Eliade (1991, 64-66, 89-91) and Forbes (1972, 54-99). The magic power of metallurgical stone tools (for the use of grinding stones in the smelting process, Craddock 1995, 156-204), the fired and ochre-stained stones (symbolising the smelting and the propitiatory sacrifice), the stone figurines (symbolising the metals born like the men from the "terra mater, petra genitrix", Eliade 1991, 39-47) and the pottery figurines involved in a fertility ritual, which has its fulcrum in the parturient idol are all probably connected with a realistic or symbolic sacrifice of human foetus to assure the success of the smelting as described by the VII° cent. BC Babylonian tests from the Assurbanipal library (Meissner 1925, 382 sgg.; Rey 1930, 193; Berthelot 1938, 43 e sgg.; Eliade 1991, 64-65).

Forbes remarks: "Death and resurrection was their fate and the smith, who worked these "changed stones", performed a rite full of secret dangers. As he conjured the metal out of its ore with the help of the fire-god, his patron, he interfered with the harmonious growth of the metals in the earth. Perhaps the sacrifice of an embryo when building a furnace has the meaning of an expiatory offering, giving one life for the other, or should we read in it the "charging" of the metal with the budding life of the embryo?" (Forbes 1971, 78).

In Kissonerga (2° and 3° period) we are at the beginning of the transformation of the Cypriote social system: the last Neolithic traditions (funeral practises of Khirokitia and Kalavassos Tenta) were leading towards the new cultural overseas influences (Red on White Pottery, imported obsidians, anthropomorphic vases) and to the new beliefs linked to the miracle of the metal smelting. I am personally convinced that the Cypriote conception of the copper workshop sanctuary, as those of Kition and Enkomi, was born in the Chalcolithic Period together with the first attempts to produce copper. Even if it is too early to postulate a reconstruction of the religion and the public cult related to copper production for the paucity of Late Chalcolithic and Early Middle Bronze Age evidence, I think that with a diligent reexamination of the iconography material of these periods, found in tombs and settlements, we could find interesting religious links between the figurative repertoire and the prehistoric metallurgical society of Cyprus (Belgiorno 1997, pp. 119-146; 1998).

Some archaeological evidence, recently found at Pyrgos, in Limassol district, strengthen the hypothesis of the existence at Cyprus still in the Early and Middle Bronze Age of industrial and religious complexes similar to the Enkomi and Kition ones.

The prehistoric Early-Middle Bronze age site of Pyrgos/Mavroraki, situated in a central position of Limassol Gulf on the slopes of Troodos Mountain, close to the modern village of Pyrgos, has been investigated since 1996 by the Italian Archaeological Mission of the National Council for Researches of Rome. Regular excavations started in 1998. The settlement, as the context suggests, presents connotations strictly industrial to be referred to copper working which was probably extracted nearby.

The excavations have revealed that the industrial complex was probably made out of an entity of its own, facing a stream, which ran West-East. The orientation and the general lay out of the roofed building discovered in 2000, roughly reminds the Late Bronze Age pattern of Enkomi and Kition. The main room, which faces the open courtyard, measures 10 meters for 17 and has walls of 70-80 centimetres. A plastered channel, coming from the courtyard, enters in the building from the northern side.

The structures belong to a squared building, which apparently had different destinations.

It was opened on the external court and roofed. An assemblage of animal bones and fine pottery was found in the North Eastern corner. They were scattered in an area of two square meters around the location of a posthole, where a wood pillar had been accommodated in a bed of white plaster on a plastered low bench. The bones and the pottery were in some relation with this pillar and probably were suspended from it.

The pottery and bones assemblage displays characteristics that set it apart from the remains found at Pyrgos so far. The material was probably a result of ritual activities and the location of the choose of the mandibles as symbolic offers was probably significant in ritual terms.

The animal bones are of some interest in that they show an almost exclusive preference for the mandibles of cattle. This degree of selectivity in the assemblage is not so common at other Cypriote Bronze Age sites connected with copper and religion, such as Enkomi and Kition, where remains of cattle regarded only the bucrania. And it is possibly assumed to be a result of selection of bones of sacrificed animals according to precepts that were particular to the cult being worshipped.

For Pyrgos, cattle, pig and deer were apparently the species regarded as worthy of forming votive offerings. The parts of the carcass represented was also a noteworthy feature: for cattle (and pig) mandibles were most common, while horncores and phalanges were rare. In terms of purely dietary considerations, this represents the better-quality meat only for the pig. The lack of cattle horns is not easy to explain, but certainly represents a significant degree of selectivity in Pyrgos depositional practice.

The entire assemblage was positioned almost in front of a sort of niche created in the corner of the wall over the bench which runs inside along the walls of the main room. A black painted slab of coarse clay material was found enwrapped in the plaster wall of this niche. It has an almond shape and the rim gently rounded. Its external surface is smoothed and lustre, and painted originally with a red slip which turned in black for intentional or occasional rations.

The discovery of the painted slab well positioned against the wall of the bench behind the pillar suggests that the place was considered a sort of religious focus. The particular location of this slab and its existence can be interpreted as indicating the symbolic position of the divinity. Whether or not this is the most appropriate explanation, it is clear that pottery of high quality and specific animal bones were arranged there around a wooden pillar or left as votive offerings. These offerings appear to have been in situ at the moment of the wall collapsing and never disturbed by later occupation of the site.

Other ceramic material from the site includes a remarkable Red Polished III large Jug found not far from the bones. It has a plastic and incised decoration related with the snake, which is a peculiar motive of the Limassol and Larnaca regions in Early and Middle BronzeAge.

On the shoulders of this jug a snake is represented crossing the body of a human figure, flanked by other two snakes rising vertically. In antiquity the serpent was often considered a divine creature, related to magic and religious believes, and in the Greek Mythology there was a special link between snakes and metallurgy.

This regarded the myth of Erichthonios (Ovidio's Metamorphoses, Book 2), the son of Athena and Hephaistos, supposedly born half-man half serpent and guarded by divine serpents (FIGURE 1. Attic amphora). "Athena gave the daughters of Cecrops (the first king of Athens half-snake and half man) Aglauros, Herse, and Pandrosos a basket, strictly forbidding them ever to open it; but at night, when Athena was absent, curiosity got the better of them. Aglauros and Herse opened the container and saw Erichthonios, the mysterious child of Hephaistos, but at the same time one or two snakes darted out of the basket causing the girls to fall in terror over the north face of the Acropolis to their death" (Burkert, W. Greek Religion, Cambridge, 1985).

The Pyrgos jug holds the unique Cypriote representation of a human figure directly connected with snakes, and it could be significant that it has been found in the first discovered Cypriote copper workshop of the Early-Middle Bronze Age.

We have any literary documents which can confirm this hypothesis, but undoubtedly the start of representation of snakes on Cypriot pottery coincides with the diffusion of metallurgical technology. Probably it is not accidental that the east-southern regions of Cyprus, the most involved in the process of copper in Bronze Age, were so devotionals to the snake representations till the end of Late Bronze Age, as the pithos found in the industrial and religious complex of Enkomi testifies.

The special destination of the large room found at Pyrgos is suggested also by its unusual dimensions. Nothing of similar has been found at Cyprus in Early and Middle Bronze Age so far. The islands of the Aegean Sea present only few examples of "megaroid buildings" (using the "megaron" word in the Homeric sense of a main large hall) and the palatial building pattern is far from Pyrgos in terms of chronology and geographic distribution.

The connection with the open courtyard and the presence of secondary rooms around show the difference with the traditional prehistoric "megaron" shaped as a clear cut, two rooms "directional building", close in itself (K.Werner , Jonsered 1993). In turn, the storage factor presents in the western sector of the Pyrgos "megaron" with a group of almost three large store jars may indicates that the room had also a public nature (destination).

The dimensions of the building complex are waiting to be clarified, but the evidences found still demonstrate that it was not a domestic forge but a true metallurgical centre, specialised and exclusively created for processing bronze objects. Many aspects of the most ancient Cypriot metallurgical technology are for the first time coming to the light.

The end of Kissonerga 3° B and 4° periods roughly corresponds with the first occupation of Pyrgos/Mavroraki site. Even if most of the Pyrgos Early Bronze Age levels are disturbed, its existence may reflect a historic occurrence. The Late Chalcolithic sites seem to have had a rapid collapse with the consequent abandon and destruction of the settlements at the beginning of the Early Bronze Age (Peltenburg 1990, 18-19). But this event may be interpreted as a migration of people to the East in search of richer copper resources, to enlarge the production of the precious copper objects where the need was more urgent. The geological map of Cyprus shows us that the most important copper layers where located on the slopes of the Troodos mountain mainly inland. But we have to consider the possibility of recognising the metallic veins of the surface which was not as semi-desert as today. The inland coast between Episkopi and Larnaca was one of the richest in metal outcrops, easy to find and to exploit. The pine forest reached the sea coast and water brought from the numerous short rivers and streams was abundant. It is natural to find in this region the continuity of the Southern Cypriote culture of the Chalcolithic period.

The strategic position of Pyrgos in the middle of a large copper mining area is not casual. The site of Mavroraki, as the Kalavassos one, owns an enviable location amid the Troodos slopes, rich of copper minerals, woods and water, and near to the seacoast for the foreign trade.

Through these privileged occurrences the inhabitants of Pyrgos had a rich life style. The period of splendour of Pyrgos runs between the Early and Middle Bronze Age. In terms of copper objects recovered in the tombs, mainly on the Cypriote northern coast, it corresponds to the most productive copper period of "prehistoric" Cyprus and its decline is contemporary to the end of the Middle Bronze Age, a period which signs a deep transformation in the copper methodological processing and in terms of quality and quantity of metal products (Kassianidou, Phorades).

The structures of Pyrgos copper workshop were oriented in a strategic way on the slopes of Mavroraki hill in order to intercept the direction of sea winds. This position besides favouring the feeding of the fire for the furnaces removed the toxic smoke from the workers and their habitations.

The most impressive evidence of copper processing is constituted by the huge quantity of ashes, carbons and slags

(800) found everywhere in the filling soil with copper prills, fragments of crucibles and small bronze objects. The quality of slags full of pure copper prills demonstrates that the system used to smelt the minerals was still primitive. As the metallurgical analyses show, the Pyrgos metallurgists had difficulties to reach the temperature necessary to smelt the copper minerals and most of the copper drops remained imprisoned in the slag.

Many stone instruments found at Pyrgos, such as maces from mines, heavy ring hammers and mortars, should be connected with the mining activity that was carried out not far from the settlement. Many other stone tools of different shape and dimension belong to the processing of the metal objects. Among them there is a rare and unique anvil of andesite, which is probably the most ancient found in the Mediterranean area. It was found in a secondary room in the East sector close to the main room. It has a well smoothed surface, except below the base. It's upper "table" shows signs of long use of the object to hammer the metal: on one edge it has a deep regular imprint, which corresponds to the size and shape of same metal objects found in the area.

The courtyard brought to the light by the excavations of 1998 and 1999 at Pyrgos, is peppered by pits excavated, used, closed and reopened many times, which may be related to the intense activity and the long use of the place as copper workshop. Different patterns of pits were probably destined to different employment, and it is not easy to understand their real function. Some of them are simple pit realised in earth and plaster as depressions located in the centre of a rough white plastered floor. Their dimensions varied from 35 to 25 centimetres. A peculiar shape of two pits assembled together like Siamese twins, (one of the two has an hole of 5 centimetres in the bottom, found covered by a token of pottery worked on scale) could be a peculiar Cypriote arrangement to host together the bellow and the crucible. In one case a reused pottery spout was found still inserted from the rim in the inner wall of the pit found full of carbons and ashes and supposed destined to house the crucible.

In Kissonerga we are at the beginning of Cypriote metallurgy and at the start of its magic and religious beliefs when..."this mysterious trade, which was so different from the other Neolithic arts and crafts, formed the centre of a wealth of myths and legends and the smith grew to form a special social type encumbered with religious rites and taboos" (Forbes 1971, 54).

The religious aspect of the smithy is one of the most interesting but unknown topics of the history of metallurgy. All has been covered until very recently by a magic mantel of secrecy linked with sexual taboos, regenerating powers and complicated often bloody rituals. Of course Cyprus too had its own beliefs related to metallurgy and as copper processing had a prominent role in the Bronze Age Cypriote society we had to examine how this factor influenced all the prehistoric Cypriote religion from the discovery of metals.

The copper workshop of Pyrgos has opened up new possibilities to investigate the technological background of Cyprus. The certainty that Cyprus was perfectly organised for cultivating its richest mineral resources and for extracting copper, already from the end of the III millennium BC, imposes the re-examination of the role of Cyprus for the supplying and the trading of the copper in the Mediterranean area since the beginning of the Bronze Age. In that period also the Cypriote religion followed the social evolution and the cultural progress created from the evolutionary push coming from the discovery of the rich mineral resources of Cyprus. New believes and rituals formed the base of a new religion which still contained elements of the most ancient traditional Cypriote cults.

REFERENCES

ASTROM P.,
1998. "An Altar to an Unknown God at Hala Sultan Tekke?", JPR, XI-XII, 1998, 2-6.
BEAR L.,
1963. "The Mineral Resources and Mining Industry of Cyprus", Bulletin n. 1.1963, 185.
BELGIORNO M.R.,
1997a. "Preliminary report on Pyrgos. Survey 1996", SMEA, 38, 193-196.
BELGIORNO M.R.,
1997 "A Coppersmith Tomb of Early-Middle Bronze Age in Pyrgos (Limassol)", in RDAC, pp. 119-146;
BELGIORNO M.R.,
1999a. "Indagini archeologiche ed archeometallurgiche a Pyrgos: le fornaci del Bronzo Antico e Medio", in Simposio Italiano di Studi Egei, dedicato a L.B.Brea e G.P.Carratelli, S.A.I.A., 53-6.
BELGIORNO M.R.,
1999b. "Preliminary Report on Survey and Soundings at Pyrgos/ Mavroraki 1996-1997", R.D.A.C. 1999, 71-86.
BELGIORNO M.R.,
1999c. "Pyrame", Culture in Sustainable Development, (vol. II, Researches), Roma, 16-21.
BELGIORNO M.R.,
2000: "Project "Pyrame" 1998-1999: Archaeological, Metallurgical and Historical Evidence at Pyrgos (Limassol)" RDAC 2000, xx-xx.
BELGIORNO M.R., CAMPO A.L.,
1994: Anthropomorphic Representations in Prehistoric Cyprus: a Formal and Symbolic Analysis of Figurines, C.3500-1800 BC, Jonsered 1994.
BERTHELOT R.,
1938. La Pensée de l'Asie et l'astrobiologie, Parigi, 1938.
COURTOIS J.C.,
1982. "L'activité metallurique et les bronzes d'Enkomi au Bronze Récent (1650-1000 avant J.C.)", in J.D.Muhly, R.Maddin, and V.Karageorghis (eds.), Early Metallurgy in Cyprus, 4000-500 BC, pp.155-76 Nicosia 1982; 1986: Enkomi et le Bronze Récent à Chypre, Nicosia 1986.
CRADDOCK P.T.,
1995. Early Metal Mining and Production,Washinton 1995.
DOTHAN,T AND A.BEN-TOR
1983. Excavations at Athienou, Cyprus 1971-1972. Qedem 16.
DUELL, P.
1938.The Mastaba of Mereruka, Oriental Institute Publications

31&39, 1938.
ELIADE M.,
1968. The Forge and the Crucible, London 1962.
ELIADE M.,
1977: Forgerons et alchimistes, Paris 1977.
ELIADE M.,
1991: Arti del metallo e alchimia, (3° ed.agg.) Torino1991.
ELLIOTT C.,
1991. "The Ground Stone Artefacts", in Peltenburg E.J., A Ceremonial Area at Kissonerga, Lemba Archaeological Project v.II.2, SIMA LXX:3, Goteborg, 1991, 67.
FORBES R.J.,
1942: "Metaal en Metaalbewerking in mythe en maatschappij", Mensch en Maatschappij XVIII, 1942, 3-32.
FORBES R.J.,
1971: Studies in Ancient Technology VIII, Leiden 1971.
FORBES R.J.,
1972: Studies in Ancient Technology IX, Leiden 1972.
GALE, N.,
1991: "Metals and Metallurgy in the Chalcolithic Period", BASOR 1991 282/3, 37-62.
GIARDINO C.,
2000: "Prehistoric copper activity at Pyrgos", RDAC 2000, xx xx.
KARAGEORGHIS V.,
1978: The civilisation of Prehistoric Cyprus, Athens 1978.
KARAGEORGHIS V.,
1982: Cyprus from the Stone Age to the Romans, London 1982.
KARAGEORGHIS V.,
1985: Excavation at Kition,VI:2. The Pre-Phoenician Levels, Nicosia 1985.
KARAGEORGHIS V.,
1990 "The Late Bronze Age"in Footprint of Cyprus, D.Hunt ed. 1990.
KARAGEORGHIS V.,
1991: The Coroplastic Art of Ancient Cyprus, I, Nicosia 1991.
KNAPP B.,
1986: Copper Production and Divine Protection: Archaeology, Ideology and Social Complexity on Bronze Age Cyprus, Goteborg, 1986.
KNAPP B.,
1988: "Ideology, archaeology and polity", Man 23, 1988, 133-63.
KNAPP B.,
1994: "Problems and prospects in Cypriote prehistory" Journal of World Prehistory 8, 1994, 377-452.
KNAPP B.,
1996: "The Bronze Age Economy of Cyprus", in The Development of the Cypriot Economy, V.Karageorghis D.Michaelides ed., Nicosia. 1996, 71-107.
MEISSNER B.,
1925: Babylonien und Assyrien,vol.2, Heidelberg 1925.
MUHLY J.D.,
1991: "The Development of Copper Metallurgy in Late Bronze Age Cyprus" in Bronze Age Trade in the Mediterranean, N.H.Gale ed., Jonsered, 1991, 180-96.
MUHLY J.D.,
1996: "The significance of Metals in the Late Bronze Age Economy of Cyprus" in Development of the Cypriot Economy, V.Karageorghis D.Michaelides ed., Nicosia. 1996, 45-
PELTENBURG E.J.,
1982: Vrysi, a subterranean settlement in Cyprus. Warminster 1982.
PELTENBURG E.J.,
1988: "Kissonerga-Mosphilia 1987, ritual deposit,unit 1015", RDAC, 1988 43-52.
PELTENBURG E.J.,
1989: "Lemba Archaeological Project, Cyprus, 1987". Levant XXI, 1989, 195-197
PELTENBURG E.J.,
1990: "Chalcolithic Cyprus", in Cyprus before the Bronze Age: Art of the Chalcolithic period, (The P.Getty Museum) Malibu 1990, 18-19.
PELTENBURG E.J.,
1991: A Ceremonial Area at Kissonerga, Lemba Archaeological Project v.II.2, SIMA LXX:3, Goteborg, 1991.
PELTENBURG E.J., REY A., La Science orientale avant les Grecs, Parigi 1930.
PELTENBURG E.J., SWINY H. AND S.,
1930. "An anthropomorphic figurine from the Sotira area, RDAC 1983, 56-59.
PELTENBURG E.J., Tylecote R.F.,
1987 The early history of metallurgy in Europe, London and New York 1987.
WEBB J. M.,
1999 Ritual Architecture, Iconography and Practice in the Late Cypriot Bronze Age, (P. Aestrom's Forlag) Jonsered.

24

INTERPRETING DISTANCE AND DIFFERENCE: INTER-CULTURAL CONTACTS, SYMBOLIC MEANINGS, AND THE CHARACTER OF THE CYCLADIC "EXPANSION" IN THE AEGEAN DURING THE EARLY BRONZE AGE.

FANIS MAVRIDIS

EMAIL: metaaa@x-treme.gr

INTRODUCTION

Several classes of artifacts of EC character or origin have been seen as providing evidence of the presence of islanders or the influence of the Cycladic culture in several parts of the Aegean (FIGURE 1). On the basis of typological similarities and differences, several models of interaction and exchange have been proposed while, where the Cycladic presence is more profound, Cycladic colonies have been identified.

Beyond the application of economic based explanations, lie a number of questions concerning the role and meaning of material culture and the way that artifacts were used for the expression of cultural identity and the formation of cultural boundaries. These research topics are gradually considered as key factors in understanding mechanisms of exchange and distribution of material culture in the prehistoric Aegean, however, they have not been thoroughly considered in the Aegean Early Bronze Age.

Unfortunately, the available data are very few and usually not published in detail. Additionally, Cycladic items such as pottery and idols are known from surface finds and in most of the times have no specific context. Scientific analysis has relatively recently been applied for understanding aspects of production, consumption and distribution of "imported" artifacts.

Most researchers accept that the presence of certain classes of artifacts implies the activities of people with a specific origin and identity.

However, it is not yet clear what cultural elements need to be present in order human beings to be regarded as having a specific (ethnic) label. It is a matter of question if someone can measure the Cycladic-ness by totting up the number of features of certain material culture remains.

In the following pages, parameters that influence material culture distribution are discussed such as symbolism and identity on the basis of the analysis of material culture changing role and meaning.

THE CYCLADIC EXPANSION: DEFINITION AND CHRONOLOGY.

The Early Cycladic culture is primarily known from cemeteries, however, the excavation and publication of settlement sites the last few years, such as Markiani on Amorgos, Skarkos on Ios or Akrotiri on Thera, are expected to give information about unknown features of the life of the prehistoric islanders, especially everyday life, use of space, economy e.t.c. (Marthari 1997, Sotirakopoulou 1999, 2001, Doumas 2000, 29, 2001). Other sites such as the cave of Zas and the settlement of Grotta on Naxos are key sites for understanding the earliest phases of the EBA, however, few information is available (Zachos 1994, Kontoleon 1949, 112-122, 1963, 153, Chatzianastasiou 1985, 153-157).

The phases of the EC sequence are known from the horizontal stratigraphy of cemeteries and from the excavation of two settlement sites, Phylakopi on Melos and A.Irini on Kea (Renfrew and Evans 1984, Barber 1984, Wilson 1987, 1999, Doumas 2000). However, cemeteries comprise the most well known type of site in the EBA Cyclades.

During ECI, cemeteries contained small clusters of graves with an individual in each burial. The type of the cist grave with four slabs on each side was used and the dead were placed in a high-suspended position. It has been suggested that during this period cemeteries reflect a social organization based on the family or lineage structure (Doumas 2000, 21). These nuclear or extended families occupied individual houses or farmsteads, possibly not organized in settlements, and additionally because of the use of perishable materials for the construction of their houses, no architectural remains have been preserved (Doumas 1972, 153). On the basis of the evidence from the graves, metals were not widely exploited in opposition to marble for the manufacture of vessels, figurines and other objects (Doumas 1977, 16-17 2000, 23-26).

During the EBA II, cemeteries are more extended and graves are mostly used for multiple burials. At Chalandriani on the island of Syros, a large cemetery was excavated (Hekman 1994). Many distinctive characteristics were recognized such as the use of a corbelled type built tomb, the identification of the graves as organized in clusters e.t.c. (Doumas 2000, 27).

The EBA in the Aegean (c.3200-2000 BC), is marked by widespread changes in economy, society and material culture technology, production and distribution, the majority of which are best attested during EBAII (c.2700-2200 BC) (Heath-Wiencke 1989, Broodbank 2000a, 45).

There is evidence which supports the intensification of inter-regional exchange that is responsible, according to most researchers, for the distribution of characteristic high value items in the Aegean. The role of islanders is considered as vital in relation to the distribution of artifacts such as obsidian tools, various metal objects, marble figurines and vases, pottery e.t.c. Beyond artifacts that are characterized as "imported", several other objects or specific elements of objects (e.g. decoration) are considered as imitations or influenced by Cycladic types. In some cases, funerary architecture, burial practices and beliefs are also shared by the islands of the central Aegean and adjacent areas.

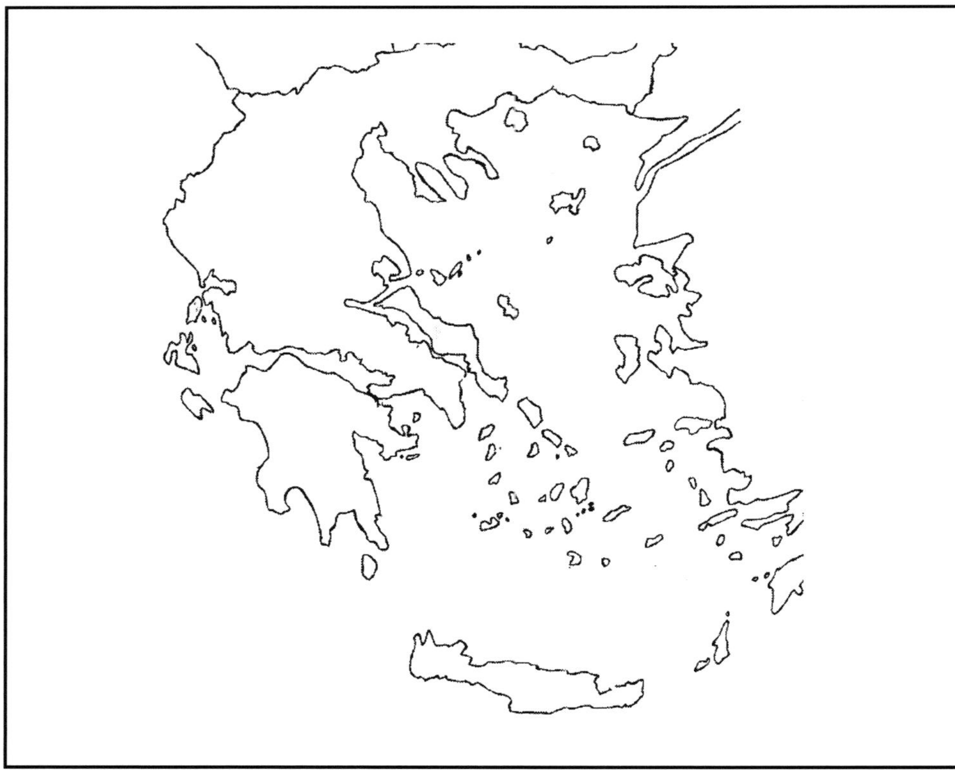

FIGURE 1. AREA MAP

In more detail, the connection of the islanders to the sea, made researchers identify relationships of these islands with several parts of the Mediterranean such as the Balearic islands, Mesopotamia, Syria, Palestine e.t.c. (see review in Cosmopoulos 1991, 156-162). However, the identification of general typological similarities is not proof of any direct relationship between these areas and the islands of the central Aegean.

In western Greece, the cemetery of Steno on the island of Leukas has produced many artifacts such as jewelry, bronze weapons or bone tubes, which have close parallels to the Cyclades, but also to Crete and the north east Aegean (Branigan 1975, 49), making the interpretation of the evidence rather problematic.

In the case of Attica, burial practices as well as several classes of artifacts such as marble vessels, figurines or pottery are also connected to the Cyclades. On the basis of this evidence, A.Kosmas (Mylonas 1959) was characterized as a Cycladic colony (Doumas 1976, 77, 1979). Important Cycladic elements are also known from the cemetery of Tsepi in the area of Marathon (Marinatos 1970, 350, Pantelidou-Gofa 1998, 41-50) as well as from other sites in Attica (Sampson 1988, 113-119). From the two latter sites as well as from the settlement site of Magoula-Kephalari in the Argolid (Dousougli 1987) and the cemetery of Manika in Euboea, elements of the late EBA I phase, the so called Kampos phase in the Cyclades are present. Lithares in Boetia has fewer Cycladic affinities; the presence of frying pans is significant (Tzavella-Evjen 1984, 176), however, they could be of local manufacture.

Important Cycladic finds are also known from Euboea (Calligas 1984, 89-93). The settlement and cemetery of Manika represent the best evidence from the area (Sampson 1985, 1988, Sapouna-Sakellaraki 1986a, 1987). Most of the artifacts from the cemetery indicate Cycladic influence as well as elements from the northeast Aegean, in contradiction to the contemporary settlement where the Helladic character of the finds is strong. In several occasions, finds of both traditions were found in the same graves. Manika has been interpreted as a trading post for the exchange of obsidian and metals (Sampson 1983, 516), however, many questions need to be further explored. Evidence of the presence of Cycladic traits comes from Asia Minor and sites such as the cemetery of Iasos or Miletos (Pecorella 1984, B.and W-D Niemeier 1997).

In the case of north-east Aegean, the evidence is sporadic and beyond obsidian several sherds from Troy or Poliochni could have been of a Cycladic origin. The absence of figurines or frying pans, widely used in other parts of the Aegean, is significant; however, no cemeteries are yet known from the area (Mavridis 1997, 113).

In relation to Crete, important Cycladic finds come from late Early Bronze Age I levels (Kampos phase), while during the EBA II the Cycladic influence has a different character. It is found in "Minoan" contexts (e.g. graves) and also in a larger number of sites.

Evidence of similarities between Cyclades and other Aegean EBA sites comes from other sites as well (Doumas 1976, 77-79, 2000, 51-53, Sakellarakis 1976, 150-158, Sapouna-Sakellaraki 1986b, Warren, Hankey 1989, 14-42, Cosmopoulos 1991, 162-167, Mavridis 1997).

The Cycladic influence attested in several areas of the Aegean created a debate among researchers for identifying Cycladic colonies. A.Kosmas in Attica and A.Fotia in Crete among other sites (Sakellarakis 1977, 97), have been considered as Cycladic colonies, however, the social organization and character of these societies as well as the overall "influence " of the Cyclades in the Aegean, remain almost unexplored. The Cycladic presence in the Aegean represents an ideal area of investigation for exploring the role of material culture in relation to aspects of identity, form and meaning of geographical and cultural boundaries.

THE CYCLADIC EXPANSION: MODELS OF EXPLANATION.

The numerous contacts in the Aegean during EBA II, led researchers to characterize the period as an international spirit phase (Renfrew 1972, 451-455) . The importance of exchange mechanisms and distribution of valuable artifacts is one of the distinctive features of the period. The increase of craft technology, specialized production and evidence for social differentiation, are some of the reasons responsible for this phenomenon.

The central role of the Cyclades in relation to the mechanisms of maritime exchange has been emphasized (e.g. Doumas 1977,62) as well as the small number of archaeologically visible imports in the Cyclades (picture which needs to be further explored after recent evidence), the new maritime technology (long boats), and the crucial role of metals as a fluid desirable medium and a means of acquiring status and power in generating this pattern of interregional linkage (Renfrew 1972, 455, Broodbank 2000a, 276). Islanders have been considered as middlemen of trade who reached several areas of the Aegean during trade expeditions, especially in relation to the distribution of metals, and on the basis of such activities they diffused their material culture, ideas and symbols (e.g. Karantzali, 1996, 48, Sapouna-Sakellaraki 1987, 262).

In this respect, the international spirit concept was used to explain the observations made in the archaeological record, regarded as a shared vocabulary of forms that overlaid continuing regional traditions (Broodbank 2000a, 249). Especially in the case of cemeteries, "imported or foreign" objects are related to the presence of traders or considered as indication of the presence of important members of the local communities who had access to foreign exotica (Renfrew 1972, 537-538, Doumas 1976, Koumouzelis 1981, 271, Barber 1987, 134-136, Sapouna-Sakellaraki 1991, 10, Manning 1994, 227). Another motivation adopted by most researchers for explaining the islanders maritime activities is the marginal agricultural land of most islands which motivated people to seek economic alternatives to basic subsistence farming (e.g. Wilson 1999, 233).

Moreover, according to some researchers, access to maritime activity has traditionally been related to major centers (Broodbank 2000a, 323) that occupy maritime nodal points or in relation to mainland areas, control their hinterland (Broodbank 1989, 326-327, 1993, 316). A characteristic example of this approach is the Early Minoan sites of the Messara plain in south Crete. Cycladic influence in this area is considered as filtered by intervening communities in northern Crete. In this respect, the variability observed in relation to the dispersion of Cycladic items is considered as a result of geographical position and access to exchange networks.

Most of the approaches presented above accept more or less, the presence of islanders as traders or colonists in several parts of the Aegean. That means that the presence of certain material culture remains has been considered as evidence of the activities of people with specific identity, suggesting a one to one relationship between material culture and ethnic identity. However, most models used for explaining this phenomenon were not able to explain the differentiation observed in relation to the presence of Cycladic objects in each context and identify specific patterns of distribution. According to the intensity and character of the Cycladic presence in each particular site, a model of access and trade is proposed (Cosmopoulos 1991, 162-167, Broodbank 2000a, 279). Where many Cycladic finds are identified, the presence of colonists or the activities of traders is suggested, while where the Cycladic presence is restricted to imitations of Cycladic artifacts and general typological similarities exist, these sites are considered as not being in direct contact with the islanders and their trade mechanisms (for a review see Warren, Hankey 1989, Coleman 1992, 266- 268, Broodbank 2000a, 302-303).

The need to focus on the study of parameters such as the role and meaning of material culture in relation to the expression of identity, power and prestige has been based on the acceptance by most archaeologists that access to resources, communication and exchange were non utilitarian in character (Hodder 1982, 199, Broodbank 1993, 315, Manning 1994, 226), indicating the importance of understanding the meanings of such processes. It is gradually becoming apparent that these complex symbolic parameters are important factors in understanding material culture distribution, beyond law-like models of economic behavior, usually applied to explain the distribution of artifacts in the Aegean EBA.

The cultural and symbolic aspects related to the distribution of material culture can be seen in several examples. The absence of Cycladic idols (however, cemeteries have not been excavated) from north-east Aegean sites cannot be easily explained in terms of access to resources or exchange networks since communication between the two areas is indicated by the presence of other classes of artifacts. Additionally, the north-east Aegean sites were important centers for the production and circulation of metal artifacts, the main reason, according to most researchers, for the intensification of exchange networks during EBA II. Moreover, the absence of Minoan pottery from major Cycladic sites such as Kavos or of golden objects from the Cyclades in general indicate the symbolic role and meaning of material culture and the effect that such parameters had in the distribution of material culture.

The role of exceptional trading centers in the EBA Cyclades needs to be further defined. This is because not much evidence is available, for example in relation to the settlement patterns in the Cyclades (Coleman 1992, 266, Manning 1994, 223) as well as to the control of certain sites over resources (Torrence 1986, 139-217, Wilson 1987, 44-46, Carter 1998, 73, Broodbank 2000a, 297). Moreover, sites on smaller islands such as Skarkos on the island of Ios do not seem to fit a model of "big trading centers and small fishing posts" being just an exception to the rule. According to this explanatory framework, the role of smaller sites remains unexplained as well as the way of dependence of these sites, if they were inhabited permanently e.t.c. In relation to the identification of exceptional sites, it is generally accepted that there is no one to one correlation between social complexity and people density (Soles 1983, 50, Manning 1994, 226).

Moreover, recent excavations at Akrotiri on Thera gave a picture of a very prosperous site during EBA II and III (Doumas 2001, 87-88). Ceramic finds showed a complicated picture of wares and imports present, a picture known from sites such as A.Irini on Kea and Kavos-Daskalio. Based on this evidence, it is a matter of question whether Akrotiri could be included or excluded from the catalogue of the few exceptional sites recognized in the Cyclades, responsible among other things for the exchange mechanisms

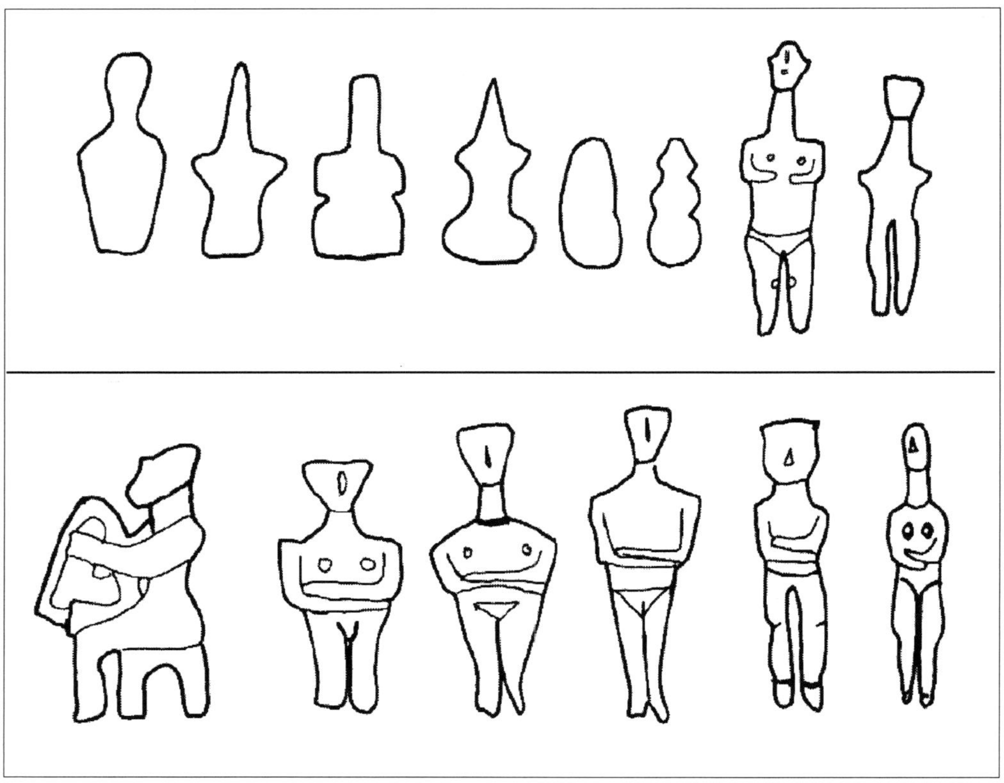

FIGURE 2. MARBLE FIGURINES AND VESSELS

in the Aegean. It is becoming apparent that as excavations continue to bring to light new sites in the Cyclades, the catalogue of the few "exceptional, trading" posts will be increased. Therefore, it is important to look into the social structure of each society, the symbolic role and meaning of maritime activities and other symbolic, social and economic parameters for understanding regional differences.

Moreover, opinions that support the absence of stratified, vertical social differentiation in the EBA Cyclades and accept a role of exchange networks and maritime activities as a communal enterprise (Doumas 1994, 117) need to be further clarified in relation to the presence and role of exceptional trading centers. The application of generalized models in explaining cultural phenomena does not usually take into consideration the fact that little information is available for settlements and patterns of habitation in the Cyclades as well as for other parameters such as social structures, symbolic meanings e.t.c.

The varying role and meaning of material culture in different contexts, human agency, symbolic power and how individuals and communities perceive and express their identity in relation to material culture are important for understanding patterns of artifact distribution.

BOUNDED CULTURES AND REGIONAL DIFFERENCES: THE CYCLADES.

The Cycladic EBA has traditionally been considered as having a homogenous material culture (e.g. Barber 1987, 25, Coleman 1992, 266, Birtacha 1999), however, regional differences have started to be identified (Broodbank 2000a, 53). Moreover, pottery analysis showed that 'foreign' influence and "imports" were not an occasional phenomenon in the EBA Cyclades.

This became apparent from the systematic analysis of pottery from two important sites, A. Irini and Kavos-Daskaleio. Most of the "foreign" wares at Kavos derive from other Cycladic islands and long range fine imports come from the Greek mainland. There is no clear indication for the presence of Minoan pottery from the site (Broodbank 2000b, 337). The pottery analysis indicated a large community of producers and consumers and that pottery data favour the picture for local agency; people at Kavos decided what to acquire from elsewhere (Broodbank 2000a, 245).

The site of A.Irini on the island of Keos is another characteristic example, indicating a complicated picture in relation to the production, distribution and consumption of pottery wares. The study of pottery from the EBA levels showed that there is no clear-cut cultural label to describe A. Irini II (EBA II). The local red brown wares show close similarities to contemporary settlements in the east coast of Attica, making possible to argue for a Keian/east Attica pottery style (Wilson 1987, 36-38, 1999, 231). Identifiable imports are not from mainland Greece but from the Cyclades and islands such as Syros, Siphnos and Melos (Vaughan and Wilson 1993, 179-182, Wilson 1999, 231). There are similarities in terms of fabric and surface treatment of urfirnis sauceboats between A.Irini, Knossos and Phylakopi, indicating a single potting center (Wilson 1994, 40); however, there is no sufficient evidence of the characterization of specific wares as imports or local products. Moreover, it is difficult to explain using traditional thought why A.Irini used Siphnian ores when it had an accessible source to Lavrion and why Naxos and Syros used Lavrion ores when

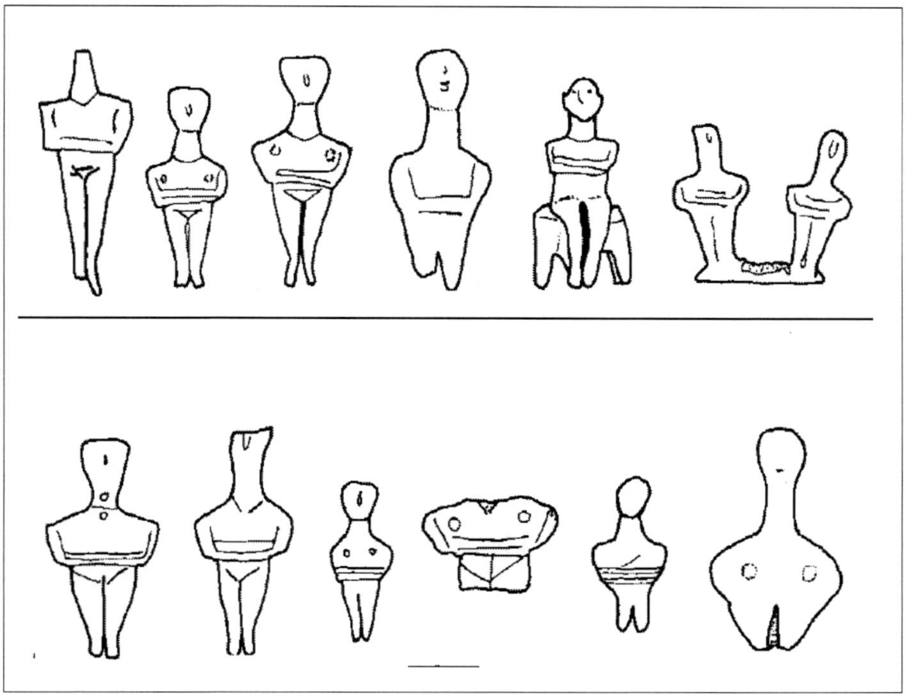

FIGURE 3. EXAMPLES OF IMPORTS, IMITATIONS OF CYCLADIC OBJECTS

Siphnos was closer (Wilson 1987, 44).

The pottery analysis from Kavos-Daskaleio and A.Irini showed a complicated picture of consumption and circulation of wares and shapes as well as the correlation of certain centers with the production of specific wares. Beyond such observations, there is yet no inadequate theory, which could connect patterns observed in the archaeological record and specific symbolic meanings. However, several observations indicate the existence of regional or symbolic traditions but the problem of the unknown context of most of the artifacts is a major obstacle. Such evidence is related to the marble figurines and vessels, the presence of specific traits in relation to burial architecture and customs or in relation to the painted or stamped designs of pottery (Broodbank 2000a, 199-201) (FIGURE .2). The production, circulation and consumption of artifacts seem to be influenced by the decisions of agents and symbolic aspects related to the expression of identity, power and prestige in each particular case. The identification of regional differences in the Cyclades is important for understanding the role and meaning of material culture and the character of the distribution of Cycladic material culture in other areas of the Aegean.

CYCLADIC EXPANSION: THE ROLE OF CONTEXTUAL VARIABILITY.

One important parameter in trying to understand material culture distribution is the existence of regional characteristics, which are gradually coming to light by recent research. There is evidence that symbolic expression influences the distribution of artifacts and in relation to pottery production, regional centers were responsible for the production of specific wares. The extent to which these regional differences are related to certain groups indicating specific identities, or the presence of "foreigners" and traders, is a matter of further investigation. These differences indicate the absence of direct correlation between geographical and cultural boundaries. In this respect, the Cycladic presence in the Aegean needs to be understood under the light of this evidence, indicating the importance of each particular context for the transformation and mediation of symbolic meanings. The discussion made so far indicates that economic approaches of material culture exchange, control of certain sites over resources or the control of foreign exotica need to be further investigated and re-analyzed under the concept of symbolic and ideological parameters.

In relation to Crete, some researchers have started to identify movement of people from outside the island during the Final Neolithic and the Early Bonze Age I that means the presence of groups with different origins and traditions (Soles 1983, 60, Wartrous 1994, 704, Nowicki 2001). These groups, whose cultural characteristics need to be further defined, support the idea of regional variability observed during the EBA.

In more detail, in the case of EBA Crete, regional differences are apparent in relation to the tomb architecture and burial practices among north Crete where chamber tombs were dug into bedrock (the "Cycladic" tombs of A.Fotia represent the best example), the tholos tombs of Messara and the so called house tombs of eastern Crete (Soles 1983, 1992, Branigan 1983, 1988, Wartrous 1994).

Moreover, the bronze artifacts from A.Fotia indicate in terms of style and isotope analysis a Cycladic origin (Gale 1990, 315-316, Stos-Gale 1985, 369-371, 1993, 125, 128). However, the provenance of the ore does not prove the place of manufacture of the artifacts and the identity of their manufacturers.

The pottery analysis from the same site showed that most of the assemblage is Cycladic and Minoan style pottery comprises a small amount (Day et al. 1998, 137-138). Similar pottery assemblages (with strong Cycladic affinities) are known from north- central Crete and sites such as Kyparisi and Pyrgos caves as well as from Poros. The pottery of Cycladic type comprises a homogenous, in terms of fabric, assemblage, also known from other contemporary sites such as Petras and Mochlos (Day et al. 1998, 136-137). It is of importance to note that Minoan pottery from A.Fotia originates in terms of fabric in central Crete while A.Onouphrios jugs originate in the area of Messara (Day et al. 1998, 138). Pottery from the same region has also been found in eastern Crete, while petrographic analysis of EM pottery from Myrtos indicated that half of the assemblage was imported (Whitelaw et al 1997). New information is expected from the analysis of EBA pottery from western Crete, especially in relation to wares that are usually considered imported (Nodarou pers. com.). This evidence indicates extensive exchange of pottery produced at a relatively restricted number of centers (Wilson, Day 1994, Day et al 1997), a shared phenomenon by both the Minoan and the Cycladic pottery production.

These regional differences were responsible for the differentiation observed in relation to the distribution of Cycladic traits in the Aegean. They are present not only in Crete but also in mainland Greece and other areas as well. In Attica for example, sites such as Palaia Kokinia, Keratsini, Rafina, Askitrario indicate a strong EH character, while other settlements such as A.Kosmas, Tsepi, A.Marina, Varkiza show a stronger Cycladic influence (Sampson 1988, 113-119, see also Wilson 1987, 1999).

The evidence of late Early Bronze Age I levels at Knossos and Poros- Katsambas represents the best example for understanding such processes as well as the active role of material culture in the expression of symbolic meanings.

It has already been referred that sites in the north part of the island, such as Poros-Katsambas, Pyrgos cave and A.Fotia showed close similarities in terms of material culture to sites of the so called Kampos group in the Cyclades. In the case of Pyrgos cave, over a third of the published pottery is comprised of wares of the Cycladic type. Some of these pottery types (e.g. conical shape pyxis, spherical pyxis with lugs at rim and deep bowls with perforated lugs below rim) have exact parallels to few sites of north Crete such as Eileithyia and Kyparisi caves and the Cyclades (Wilson, Day 2000, 56). New interesting evidence comes from the settlement site of Poros-Katsambas, considered as the harbor site of Knossos and gateway of the Cycladic affinities in Crete. The comparison of the pottery assemblages between Knossos and Poros during late Early Bronze Age I showed that there are important similarities in relation to the pottery groups, however, major differences occur when the Cycladic presence is considered (Wilson, Day 2000, 58). These differences have important implications for the character of the society as well as the way in which material culture was actively involved in the expression of identity and symbolism. The evidence of Knossos has been interpreted as the case of the ancestral site of the first settlers, while aspects of symbolism related to the pottery assemblage (food and drink ceremonies) and the evidence for craft activities identified at Poros are important for the negotiation of interpersonal and inter-group competition in the production and legitimation of power through mediation and transformation (Wilson, Day 2000, 61). What seems more apparent is that the role of gateways and exceptional sites need to be reconsidered as prime movers for explaining why Cycladic artifacts are found in specific sites and not in others. The study of each particular context is important for understanding the role and meaning of material culture (for example the presence of Cycladic objects in the Minoan context of Archanes or the presence of imported pottery in EBA II strata of Knossos).

What needs to be further explored is the way that material culture was tied to the expression of identity, power and prestige. There is evidence during EBA II of the important role of prestige objects in relation to the construction of power relations and individual identity. Also, the role of personal adornment, the widespread use of toilet items and weapons show that personal identity was an important aspect of the period (Broodbank 2000a, 249). The absence from sites in southern or western Crete (in opposition to sites situated in the north part of the island) of characteristic Cycladic affinities cannot be easily explained as being out of the Cycladic influence or exchange relations as the evidence from Katsambas and Knossos showed. There is evidence that pottery of the region of Messara as well as from other areas of Crete was in circulation among EM sites. This observation signifies the relations of these sites with the northern and eastern part of the island. For example, some of the finest A.Onouphrios jugs and bowls come from the deposits of Pyrgos and Kanli-Kasteli (Early Minoan I) in northern Crete (Branigan 1988, 25), as well as from A.Fotia which is considered as one of the best candidates for the existence of Cycladic colonies in the Aegean (Day et al. 1998, 138).The analysis of the metal objects from the Messara tholoi indicated sources of metals on Kythnos as it was the case of the metal artifacts from A.Fotia. Moreover, in contrast to approaches that support the role of gateways which control or filter the distribution of Cycladic items in Crete, the contents of the Messara tholoi such as obsidian, copper, marble vessels and jewelry indicate, according to some researchers, relations to other parts of Crete as well as to the Cyclades (Karantzali 1996, 238, 244). It seems that other parameters related to the symbolic role and meaning of material culture were important for the differences observed in relation to the intensity and character of the Cycladic presence in each particular site or region. Cycladic affinities cannot be simply explained as foreign exotica, because of their conscious expression of symbolic meaning, which could vary between different contexts. For example, the study of EH wares from mainland Greece, showed that every -day shapes, which could easily be produced locally, were exchanged. This evidence has been explained as an effort to promote stylistic uniformity and a sense of social cohesion (Attas, Fossey, Yaffe, 1987, Rutter, 1993).

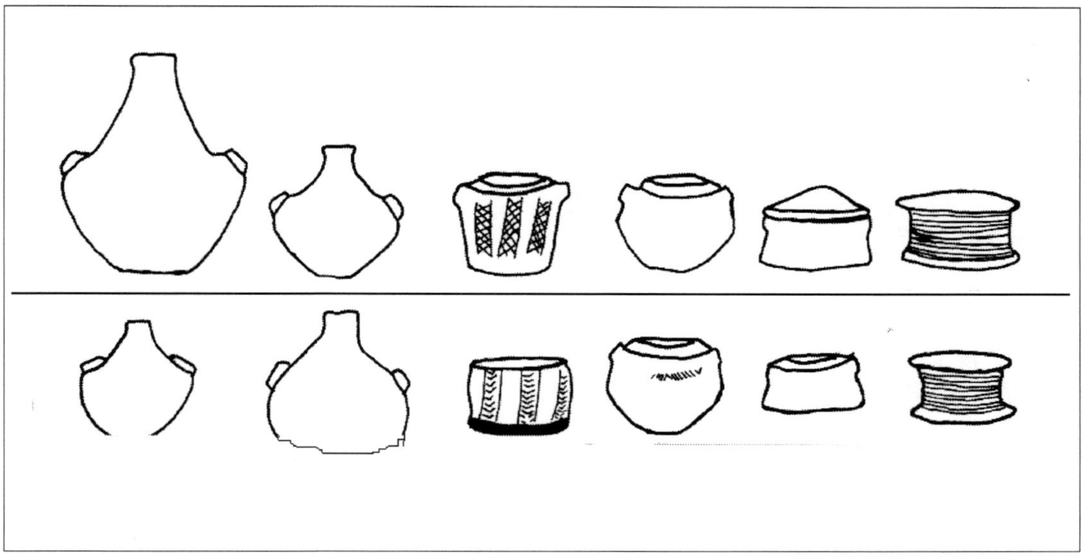

FIGURE 4. EXAMPLES OF IMPORTS, IMITATIONS OF CYCLADIC OBJECTS

The evidence of the Cycladic presence in the Aegean needs to be explored under the light of new evidence. In this way, the role of material culture, in all stages from production through deposition, the symbolic significance of their presence, associations and context of their discovery, all have a special importance of understanding the role of material culture and the character of "foreign" influence.

It is necessary to understand aspects that influence material culture distribution before applying generalized models that do not take into consideration each particular case and the active role of individuals.

IDENTITY, CULTURES AND SOCIAL BOUNDARIES.

There is growing evidence that supports the existence of regional variation in EM Crete in terms of the various symbolic statements produced by the presence, associations and context of the different classes of artifacts. The differences observed in relation to the production, consumption and distribution of material culture point against traditional approaches that treat cultures as homogenous and bounded entities (Eriksen 1993, 9). This is important for understanding spatial patterning of artifacts as well as the changing role and meaning of "imported" objects present in each particular context (Hodder 1986). What is also clear is that cultural differences are not the result of the lack of between group contacts since material culture boundaries are foci of interaction and not barriers (Jones 1996, 71). The example of Poros-Katsambas and Knossos is very indicative of this.

Most approaches related to aspects of exchange and contacts in the Aegean EBA have tried to explain the presence of Cycladic traits as evidence of the control of certain sites over exchange mechanisms. What has been stated so far is that the presence of certain artifacts does not correlate to ethnic groups directly and that complex processes of symbolism, ideology, power and identity were involved, parameters which have just started to be systematically studied in the Aegean EBA.

Culture-historical approach classifies spatio-temporal variation where discrete monolithic cultures constitute a mode of differentiation. Archaeological cultures are regarded as entities, the relationship between culture and past people is based on teleological reasoning in that culture is both representative of and constitutive of the nation of people concerned (Jones 1996, 65). The past is reconstructed, according to these approaches, in terms of the distribution of homogenous cultures whose history unfolds in an evolutionary narrative, prior to the formation of the first palace states in the case of the Aegean, measured in terms of objective events, such as contacts, migration e.t.c. The idea that ethnic groups are fixed, homogenous, bounded entities is a modern classificatory invention (Eriksen 1993). Cycladic or Minoan cultures were created more or less on the basis of this approach, representing entities with specific characteristics and geographical distributions.

The approach that takes as a fact the correlation of cultural similarities/differences and ethnic boundaries fails to resolve the relationship between agents, perception of ethnicity, cultural contexts and social relations in which they are embedded.

Regional material culture boundaries are not static; they are maintained and re-enacted through time. It is a continuous symbolic action that constitutes regional dichotomies where interaction is frequent (Cohen 1985). This may be the case of the changing picture between EB I and EB II in relation to the Cycladic presence on Crete, indicating changing patterns in terms of inter-group contact and expression of power and identity. The extent to which cultural similarity relates to interaction depends on the strategies and intentions of groups and on how they use, manipulate, negotiate material culture symbols as parts of those strategies (Shanks, Tilley 1987, 86). This indicates that there is no simple relation between degrees of interaction or scale of production and material culture patterning since other factors intervene such as intentions, strategies, ideologies (Jones 1996, 64). Culture attributes are shared between societies but they are integrated in different ways (Graves-Brown 1996, 91, Jones, Graves-Brown 1996, 7). It seems that symbolic parameters related to the role and meaning of material culture influenced the distribution of Cycladic traits in the Aegean. It is reasonable to assess that the understanding of their meaning and significance will help to investigate the differences observed in the archaeological record.

The above discussion indicated that archaeologists cannot analyze groups solely in terms of degrees of interaction; internal organizations as well as symbolic meanings need also to be taken into consideration (Hodder 1982). Discontinuities or boundedness in cultural distribution result from the nature of between group relations and within group organization and structure, in the context of ongoing social interaction and not in terms of social or physical distance (Hodder 1982, Jones 1996, 71). Regional cultural discontinuities are linked to aspects of society and economy by local concepts, principles of symbolic meanings and by the way meanings are used and maintained in each context (Jones 1996, 76). Different societies can share similar symbols but not necessarily the meaning. This observation needs to be taken into consideration in relation to the Cycladic presence in the Aegean. Cycladic idols in Cretan sites for example, do not necessarily seem to relate their meaning to a Cycladic context, in each particular case.

Moreover, the study of technology and production and the transformation of raw materials are equally important. The presence of a finished object alone is not determinant for sharing symbolic meanings between societies.

In more detail, the character and intensity of Cycladic presence in the Aegean has been based on typological affinities of artifacts which are characterized as imports, imitations of Cycladic objects e.t.c (FIGURES 3,4). The criteria used by archaeologists vary and only relatively recently provenance and characterization studies have tried to investigate aspects of production, distribution, and consumption of various classes of artifacts.

The stylistic variation concept, used by archaeologists to reach conclusions about similarities and differences of artifacts in order to reveal aspects of social relationships, especially as indicators of patterns of communication and spatial distribution, has been debated in terms of how deposits have been formed, how to analyze stylistic variation of different classes of artifacts and above all what constitutes socially meaningful variation in artifacts, whether technological and decorative practices play an active role in the way by which people adjust social relationships e.t.c. (see for example Plog 1995, Brown 1995)

Ethnographic work has shown that material culture was actively used in efforts of self-definition, constructing identities based on perceived cultural or social differences. In order to reveal such information, the presence or absence of certain objects, but also their associations, social context, symbolic meaning and use need to be examined (Hodder

1986, Jones 1997, 27).

The Cycladic influence in the Aegean varies quantitatively as well as qualitatively in different contexts. In relation to obsidian blades for example, it has been proposed that they represent a developed and restricted technology whose symbolic meaning has been taken as both a social and ethnic marker, tied to technological knowledge and power (Carter 1998). In this respect, it has been observed that obsidian technology follows Cycladic rather than Minoan tradition in the case of A.Fotia while in Messara the situation seems to be different (Carter 1998). That means that all the procedure from production to distribution and consumption was important in both the fields of symbolism and identity. There is also evidence of figurines imported from the Cyclades, others regarded as imitations of the Cycladic types and finally the "Cretan" variety of folded arm figurine, the so called Koumasa variety. It seems that these objects have different life-cycles since the raw materials used can be transformed into something with different value and meaning. Obsidian or metal ores can be procured by "local" craftsmen and can be physically transformed (Nakou 1995, 13-14, Carter 1998, 73).

Such evidence needs to be further explored in relation to the formation of regional variability and the meaning that this variability has in explaining cultural characteristics and differences in material culture. So, specific context, technological know-how, form, use and meaning of material culture need to be taken into consideration for understanding the differences observed in the archaeological record.

In the face of this variability of meaning, it is gradually becoming apparent that the labels archaeologists put on certain classes of artifacts in order to classify ethnic groups and cultures undervalue the complex nature of symbolic meanings related to such processes as well as to the changing role and meaning of material culture. Most traditional approaches accept a one to one relationship between cultural similarities and ethnic group identity (Jones, Graves-Brown 1996, 7). However, under the light of new evidence, one cannot easily answer the question, using traditional thought, whether a Minoan label fits better Messara sites, Mochlos or Pseira or whether A.Fotia is Cycladic or Cycladic and Minoan. It is gradually becoming apparent that there were important differences between sites in the same region in relation to the meaning and role of the "imported" objects, however, symbolic meanings cannot yet be correlated with the expression of specific identities and groups.

DISCUSSION AND CONCLUSIONS.

As it has already been proposed, a single analytical framework is inadequate for identifying patterns related to the Cycladic data. Symbolic and ideological parameters influence the distribution, role and meaning of material culture.

Before the so-called international spirit phase (EBA II), there is evidence for the existence of an earlier phase of the Cycladic expansion during the Kampos group phase (Warren, Hankey 1989, Broodbank 2000a, Wilson, Day 2000). This presence or influence is not restricted to Crete but it is found in other areas as well (Mavridis 1997). It also seems that it characterizes specific sites, showing a contradiction in relation to the way that symbolic expression and material culture were tied during the next phase (EBA II).

The EBA II Keros-Syros culture comprises common forms of prestigious symbolic objects of wider Aegean distributions such as painted or stamped pots, the frying pans, metal weapons, jewelry and figurines (Broodbank 2000a, 54). During EBA II it seems that social boundaries are more open and a common language exists between sites in the negotiation of power, prestige and identity. However, the study of these objects in different contexts, their spatial distributions and contextual associations, aspects of their production and consumption, use and meaning may guide to different conclusions in relation to the international spirit concept and the way that social identities were expressed.

Finally, a note is necessary concerning the interpretation of the character of the Cycladic expansion and the distribution of the Cycladic traits in the Aegean. There is first of all no indication that the Cyclades operated as a cultural and geographical entity, and that they represent a bounded and naturally defined area of investigation (Broodbank 2000a, 68). The Cycladic material culture has a strong anthropogenic, symbolic character with an active role in the formation of insular social strategies, social production and reproduction. Maritime expansion and exchange mechanisms operated as a form of social identity, providing agents with meaning and organizational channels, considered therefore as a form of social organization and reproduction (see also Broodbank 2000a). However, the way that these aspects operated in each particular site or region remains a problem.

Foreign influence in the Cyclades cannot be disputed; it seems however that it had a different meaning and expression in relation to the Cycladic one in other parts of the Aegean. This is possibly due to the varying expressions of identity, internal organization and how production and reproduction of different social systems were achieved and expressed in the symbolic sphere and the way that material culture was used in expressing such ideas.

Until recently, terms such as Early Cycladic or Minoan were considered as homogenous related to bounded cultures that situated in specific geographical areas. Minoan and Cycladic used as terms implying bounded and homogenous ethnic groups, restrict the understanding of the heterogeneous processes involved in the negotiation of power and identity. The diversification observed within the limits of these areas in terms of material culture, burial practices, beliefs, ideas, technology, production and consumption of artifacts, strengthen the idea of the character, role and meaning that these differences had in relation to the expression of social identity.

These observations indicate a complex pattern of overlapping material culture distributions related to the repeated realization of social identity in different contexts (Jones 1996, 72). What we see in the archaeological record is that patterns in the production and consumption of material culture involved in the communication of the same identity vary in quality and quantity in different contexts. Items of material culture which are widely distributed and used in a variety of social contexts were consumed in different ways and become implicated in the signification of a variety of expressions of social identity (Shanks, Tilley 1987, 97). Even if a culture copies artifacts from another, they are produced in terms of different conceptions of appropriateness, something which expresses different technical approaches, chaines operatoires (Graves-Brown 1996, 91). What appears in the surface to be a widely cultural identity in terms of artifact form may turn out to be something more complex, if one considers how an artifact is situated within the chaines operatoires of different groups. In this respect, the international spirit concept takes other meaning and significance. Such differences have been observed in sites with Cycladic presence and it seems that the systematic collection and study of such evidence in each particular context will help to understand the meaning of the variability observed in relation to the distribution and meaning of material culture.

It is apparent in the Aegean that the many dimensions of identity are unexplored as well as symbols that signify it. What is missing is an adequate theory of the relationship between identity and culture, including culturally inscribed relations of production and reproduction (Jones 1997, 69). Ethnic groups are rarely a reflection of the total of similarities and differences in objective cultural traits. Meaning of material culture will alter in different contexts, consumed in different ways, appropriated and incorporated into various symbolic structures according to historical traditions and social context. In this respect, the Cycladic presence in the Aegean needs to be understood in the context of heterogenous and often conflicting constructions of cultural identity. This is a possible reason why there is no easily identifiable pattern in relation to the distribution of Cycladic objects in the Aegean since there is not a single meaning of identity or association to be extracted from a particular style of material culture.

REFERENCES

ATTAS, M., M. FOSSEY, L. YAFFE,
1987. An archaeometric study of Early Bronze Age pottery production and exchange in Argolis and Korinthia, Greece. Journal of Field Archaeology, 44, 77-90.
BARBER, R.,
1984. The pottery of Phylakopi, first city (phase ii, i-ii). In: J. MacGillivray,R. Barber, eds.. The prehistoric Cyclades, 88-94. Edinburgh: Edinburgh University Press.
BARBER, R.,
1987. The Cyclades in the Bronze Age. London: Duckworth.
BIRTACHA, K.,
1999. Evidence for social differentiation in the Cyclades in the 3rd millennium. In: Eliten in der Bronzezeit. Mainz: Monographien des Romisch-Germanischen Zentralmuseums 43, 1-7.
BRANIGAN, K.,
1975. The round graves of Leukas reconsidered. The Annual of the British School at Athens, 70, 37-49.
BRANIGAN, K.,
1983. Some observations on state formation in Crete. InE. French, K. Wardle, eds. Problems in Greek Prehistory. Bristol: Bristol Classical Press, 63-68.
BRANIGAN, K.,
1988. Pre-palarial. The foundations of palatial Crete. A surrvey of Crete in the Early Bronze Age. Amsterdam: Adolf M. Hakkert.
BROODBANK, C.,
1989. The longboat and society in the Cyclades in the Keros-Syros culture. American Journal of Archaeology, 93, 319-337.
BROODBANK, C.,
1993. Ulysses without sails: Trade, distance, Knowledge and power in the early Cyclades. World Archaeology, 24, (3), 315-331.
BROODBANK, C.,
2000a. An island archaeology of the early Cyclades. Cambridge: Cambridge University Press.
BROODBANK, C.,
2000b. Perspectives on an EBA island center: An analysis of pottery from Daskaleio-Kavos (Keros) in the Cyclades. Oxford Journal of Archaeology, 19, (4), 323-342.
BROWN, D.P.,
1995. Style, selection and historicity. In: Carr, Neitzel, eds. Style, society and person. New York: Plenum Press, 123-141
CALLIGAS, P.,
1984. Euboea and the Cyclades. In: Fitton, ed. Cycladica. London: British Museum, 89-94.
CARTER, C.,
1998. Reverberations of the "International spirit". Thoughts upon Cycladica in the Messara. In: Branigan, ed. Cemetery andsociety in the Aegean Bronze Age. Sheffield: Sheffield Academic Press, 59-77.
CHATZIANASTASIOU, O.,
1985. Oikopedo Dimitrokali. Praktika tis Archaeologikis Etairias, 153-157.
Cohen, A.,
1985. The symbolic construction of community. New York: Ellis Horwood Ltd.
COLEMAN, J.,
1992. Greece, the Aegean and Cyprus. In: Ehrich, ed. Chronologies in old world Archaeology. Chicago: The University of Chicago Press, 247-288.
COSMOPOULOS, M.,
1991. Exchange networks in prehistory: The Aegean and the Mediterranean in the third millennium BC In: R. Laffineur, L. Basch, eds. Thalassa. Aegeum 7. Universite de Liege, 155-168.
DAY, P., D. WILSON, E., KIRIATZI,
1997. Reassessing specialization in pre-palatial Cretan ceramic production. In: R. Laffineur, Ph. Betancourt, eds. TEXNH. Aegeum 16. Universite de Liege, 257-290.
DAY, P. W. WILSON, E. KIRIATZI, 1
998. Pots, labels, and people: Burying ethnicity in the cemetery of A. Fotia Siteias. In: Branigan, ed. Cemetery and society in the Aegean Bronze Age. Sheffield: Sheffield Academic Press, 133-149.

DOUMAS, C.,
1972. Notes on Cycladic architecture. Archaeologisher Anzeiger, 87, 151-170.
DOUMAS, C.,
1976. Proistorikoi Kykladites stin Kriti. Archaeologika Analekta ex Athinon, 9, 69-80.
DOUMAS, C.,
1977. EBA burial habits in the Cyclades. SIMA 48.Goteborg: Paul Astroms Forlag.

DOUMAS, C.,
1979. Proistorikoi Kykladites stin Kriti. Archaeologika Analekta ex Athinon, 12, 104- 109.

DOUMAS, C.,
1994. I Protokykladiki koinonia. Mia prospatheia prosegisis tou kosmou tis noisis. In: I. Probonas, S. Psaras, eds. I Naxos diamesou ton aionon. Koinotita Philotiou : Athens, 115-122.

DOUMAS, C.,
2000. Protokykladikos politismos. Syllogi N.P. Goulandri. Athens : Museum of Cycladic Art.

DOUMAS, C.,
2001. I Thera tis epochis tou Chalkou: Polotismiko stavrodromi stin anatoliki Mesogeio. In: I. Danezis, ed. Santorini. Athens: Pergamos Ekdotiki, 87-96.

DOUSOUGLI, A.,
1987. Makrovouni-Kephalari, Magoula-Talioti: Bemerkungen zu den Stufen FH I and FH II, IN der Argolis.Praehistorische Zeitschrift, 62, 164-220.

ERIKSEN, T.H.,
1993. Ethnicity and nationalism. Anthropological perspectives. London: Pluto Press.

GALE, N.,
1990. The provenance of metals for EBA Crete-local or Cycladic? Acts of the sixth International Cretological Congress, _. Chania, 299-316.

GRAVES-BROWN, P.,
1996. All things bright and beautiful. In: P. Graves- Brown, S. Jones, C. Gamble, eds. Cultural identity and Archaeology. London: Routledge, 81-95.

HEATH-WIENCKE, M.,
1989. Change in early Helladic II. American Journal of Archaeology, 93, 495-509.

HEKMAN, J.,
1994. Chalandriani on Syros:Am EBA cemetery in the Cyclades.Archaeologiki Ephimeris, 133, 47-74.

HODDER, I.,
1982. Symbols in action. Cambridge: Cambridge University Press.

HODDER, I.,
1986. Reading the past. Cambridge: Cambridge University Press

JONES, S.,
1996. Discourses of identity in the interpretation of the past. In: P. Graves-Brown, S. Jones, C. Gamble, eds. Cultural identity and Archaeology. London: Routledge, 62-80.

JONES, S., P. GRAVES-BROWN,
1996. Introduction. In: P. Graves-Brown, S. Jones, C. Gamble, eds. Cultural identity and Archaeology. London: Routledge, 1-24.

Jones, S., 1997. The Archaeology of ethnicity. London: Routledge.

KARANTZALI, E.,
1996. Le Bronze Ancien dans les Cyclades et en Crete. Oxford : B.A.R., Int.S., 631.

KONTOLEON, N.,
1949. Anaskaphai en Naxo. Praktika tis Archaeologikis Etaireias, 112-122.

KONTOLEON, N.,
1963. Anaskaphai en Naxo. Praktika tis Archaeologikis Etaireias, 152-153.

KOUMOUZELI, M.,
1981. La tombe prehistorique d' Elis. Ses relations avec le sud-est Europeen et les Cyclades. Archaeologika Analekta ex Athinon, 14, (2), 265-272.

MANNING, S.,
1994. The emergence of divergence : Development and decline on Bronze Age Crete and the Cyclades. In: C. Mathers, S. Stoddart, eds. Development and decline in the Mediterranean Bronze Age. Sheffield: J. Collis publications, 221-270.

MARTHARI, M.,
1997. Apo ton Skarko stin Poliochni. In: Doumas, La Rosa (eds.), I Poliochni kai I proimi epochi tou Chalkou sto voreio Aigaio. Athens: Panepistimio Athinon-Italiki Archaeologik Scholi, 362-382.

MARINATOS, S.,
1970. Further discoveries at Marathon. Archaeologika Analekta ex Athinon, 349-362.

MAVRIDIS, F.,
1997. Monodiastati Archaeologia kai polydiastati anthropini symperifora. H archaeologia ton antallagon kai to Protokykladiko "emporio". M.Phil. diss., University of Athens.

MYLONAS, G.,
1959. Aghios Kosmas. An Early Bronze Age settlement and cemetery in Attica. Princeton.

NAKOU, G.,
1995. The cutting edge: A new look at early Aegean = metallurgy. Journal of Mediterranean Archaeology, 8,(2), 1-32.

NIEMEIER, B. AND W.D., A. GRAVES, A. RAYMOND, 1997. Most recent Evidence for the EBA, Chalcolithic and Neolithic periods at prehistoric Melitus and its Aegean connections. In: Erkanal, ed. The Aegean in the Neolithic, Chalcolithic and Early Bronze Age. 13-19

NOWICKI, K.,
2001. New evidence for the Final Neolithic in Crete. Acts of he eighth Cretological Congress. Abstracts.

PANTELIDOU-GOFA, M.,
1998. _ekrotapheio Tsepi Marathonos. Praktika tis Archaeologikis Etaireias, 41-50.

PECORELLA, P.,
1984. La cultura preistorica di Iasos a Caria. Rome.

PLOG, S.,
1995. Approaches to style. Complements and contrasts. In: C. Carr,J. Neitzel, eds. Style, society and person. New York : Plenum Press, 369-387.

RENFREW, C.,
1972. The emergence of civilization. The Cyclades and the Aegean in the third millennium BC London: Methuen.

RENFREW, C., R. EVANS,
1984. The earliest Bronze Age at Phylakopi. In: R. Barber, S. MacGillivray, eds. Prehistoric Cyclades. Edinburgh: Edinburgh University Press, 63-69.

RUTTER, J.,
1993. Review of Aegean Prehistory II: The prepalatial Bronze Age of the southern and central Greek mainland. American Journal of Archaeology, 97, 745-797.

SAKELLARAKIS, J.,
1976. The Cyclades and Crete. In: Thimme, ed. Kunst und Kultur der Kykladeninseln im 3. Jahrtausend v. Chr. Karlsruhe: Badisches Landesmuseum, 149-158.

SAKELLARAKIS, J., 1977. _a Kykladika stoicheia ton Archanon. _rchaeologika _nalekta ex _thinon, 10, 93-115.

Sampson, A., 1983. Oi palaeoeladikoi tafoi tis Manikas kai oikoinonikooikonomikes synthikes tis Proimis Epochis tou Chalkou stin Euvoia. Anthropos, 5,(10), 513-528.
SAMPSON, A.,
1985. Manika I. Mia protoelladiki polis tin Chalkida. Athens.
SAMPSON, A.,
1988. Manika II. O protoelladikos oikismos kai to nekrotapheio. Athens.
SAPOUNA-SAKELLARAKI, E.,
1986a. Manika. Stromatographiki ereuna ston oikismo, oikopedo Zousi. Archaeologikon Deltion, 41,(A), Meletai, 101-270.
SAPOUNA-SAKELLARAKI, E.,
1986b. Kykladika tis Skyrou. In: Filia Epi Mylona, I Athens, 193-299.
SAPOUNA-SAKELLARAKI, E.,
1987. New evidence from the Early Bronze Age cemetery at Manika, Chalkis. The annual of the British School at Athens, 82, 233-264.
SAPOUNA-SAKELLARAKI, E.,
1991. Nouvelles figurines Cycladiques et petite glyptique du BA d' Eubee. Antike Kunst, 34, 3-12.
SHANKS, M., C., TILLEY,
1987. Social theory and Archaeology. Cambridge: Polity Press.
SOLES, J.,
1983. Social ranking in prepalatial cemeteries. In: E. French, K. Wardle, eds. Problems in Greek Prehistory. Bristol: Bristol Classical Press, 49-61.
SOLES, J.,
1992. The pre-palatial cemetery at Mochlos and Gournia and the house tombs of Bronze Age Crete. Hesperia suppl. 24. Princeton.
SOTIRAKOPOULOU, P.,
1999. Akrotiri Thiras. H Neolithikn kai Proimi Epochi tou Chalkou epi ti basei tis keramikns. Athens: Archaeologiki Etaireia.
SOTIRAKOPOULOU, P.,
2001. Archaeologikes martyries gia tin proimi katoikisi tis Theras. In: Danezis, ed. Santorini. Pergamos Ekdotiki: Athens, 97-104.
STOS-GALE, Z.A.,
1985. Lead And silver sources for Bronze Age Crete. Acts of the fifth Cretological Congress, A,(I) , 365-371.
STOS-GALE, Z.A.,
1993. The origins of metals used for making weapons in early and middle Minoan Crete. In: Scarre, Healy, eds. Trade and exchange in prehistoric Europe. Oxbow: Oxford, 115-129.
TORRENCE, R.,
1986. Production and exchange of stone tools. Cambridge: Cambridge University Press.
Tzavella-Evjen, H., 1984. Lithares. Athens: TAP.
VAUGHAN, S., D. WILSON,
1993. Interregional contacts in the Aegean in Early Bronze Age II: The talc ware connection. In: C. Zerner, (eds.). Wace and Blegen: Trade as evidence for trade in the Aegean in the Aegean Bronze Age. J. C. Gieben: Amsterdam, 169-186.
WARREN, P., H. HANKEY,
1989. Aegean Bronze Age Chronology. BrIstol: Bristol Classical Press.
WARTRUS, L.V.,
1994. Review of Aegean prehistory III: Crete from the earliest prehistory through the proto-palatial period. American Journal of Archaeology, 98, 695-753.
WHITELAW, T., P. DAY, E. KIRIATZI, V. KYLIKOGLOU, D. WILSON,
1997. Ceramic traditions at EM IIa Myrtos, Fournou Korifi. In: R. Laffineur, Ph. Betancourt, eds. TEXNH. Aegeum 16. Universite de Liege, 165-274.
WILSON, D., 1987. Kea and East Attica in EBA II. Beyond pottery typology. In: J. Fossey, ed. McGill. J. C. Gieben: Amsterdam, 35-49
WILSON, D.,
1994. Knossos before the palaces:An overview if EBA (EM I-III). In: D. Evely, H. Hughes-Brock, N. Momigliano, eds. Knossos: A labyrinth of history. Papers in honor of C. Hood. British School at Athens: Oxford, 23-44.
WILSON, D.,
1999. Keos IX: Ayia Irini I-III. The late Neolithic and Early Bronze Age settlements. Part I. The pottery and small finds. Mainz on Rheine: Philipp von Zabern.
WILSON, D., P. DAY,
1994. Ceramic regionalism in prepalatial Crete. The Messara imports from Early Minoan I to Early Minoan IIaKnossos. The annual of the British School at Athens, 89, 1-87.
WILSON, D., P. DAY,
2000. Early Minoan I chronology and social practice: Pottery from the early palace tests at Knossos. The annual of the British School at Athens, 95, 21-63.
ZACHOS, K.,
1994. Prchaeologikes ereunes sto spilaio Za Naxou. In: I. Probonas, S.Psaras, eds. I Naxos diamesou ton aionon. Koinotita Philotiou : Athens, 99-113.

25

SAFFRON CROCUS DOMESTICATION IN BRONZE AGE CRETE

MOSHE AND ORA NEGBI

Agriculture Botany Department, The Hebrew University of Jerusalem, Rehovot and Institute of Archaeology, Tel Aviv University, Tel Aviv, Israel
EMAIL:onegbi@tau.ac.il

INTRODUCTION

"...throughout the whole of the palace period, one of the sources of wealth to the Lords of Knossos was to be found in the culture of saffron."

Sir Arthur Evans, 1935.

The dried orange-red style-branches of the saffron crocuses, which constitute the precious spice, were also used in antiquity as panacea, perfume and pigment. Evidence from modern botanical studies; Classical authors; Aegean Bronze Age art and scripts, is presented here in an attempt to trace the location and date of saffron crocus domestication.

BOTANICAL EVIDENCE

There are two saffron taxa: Crocus cartwrightianus, a wild and fertile diploid (2n=16chromosomes), endemic to Attica and 14 Greek Islands (Mathew, 1977: 95; 1999: 22; FIGURE 1A; Map), and its sterile mutant, the cultivated triploid (3n=24), Crocus sativus (FIGURE 1B). *"The most likely explanation of the origin of our present day C. sativus,"* wrote B. Mathew (1977), *"is that it was a naturally occurring triploid which was selected because of its vigour and length of style. Vegetative reproduction of the clone is very rapid so that its subsequent dispersal by man throughout Europe and Asia is not very remarkable."* Cytological analyses of eleven C. sativus populations showed high affinities and strengthen the idea that they derived from a common source (Brighton, 1977; Mathew, 1999: 22-23). The establishment of a new crop-plant like C. sativus, from a sterile parent, is a function of the number of daughter corms developed yearly (Negbi et al., 1989). A single crocus plant produces, annually, 2-3 flowering size corms (Skrubis, 1990). Hence, the offspring of a single plant by the 10th year would be 210=1,024 or 310= 59,049 propagating corms, sufficient for the initiation of a crop plant (92-100 thousands corms per acre are needed for good plantation [Goliaris, 1999: 76]).

SAFFRON IN CLASSICAL LITERATURE

Referring to its sterility Theophrastus (371-287 BC) points out that in Greece and Cyrenaica saffron crocus is propagated by "acorn-shaped" roots [corms], and discusses its autumnal flowering, either before or after leaf appearance (Theophrastus, 1916, 1926: Enquiry into Plants [=HP]: Greene, 1983: 128-134; Negbi, 1989: 19-20). He described the importance of drying the style-branches, in order to enhance the saffron`s scent, an obligate step still performed (Basker, 1999; Goliaris, 1999):

"...of the flowers some, like rose, have their potencies from the time when they are fresh, but others have their potencies after drying, as saffron crocus..." (Theophrastus, 1926: Concerning Odours 7).

By Roman times C. sativus reached Asia Minor, Italy and Sicily. Pliny (23/24-79 AD) states that:

"Wild saffron [C. cartwrightianus?] is better than any other. To grow it in Italy is most unprofitable, as a whole bed of saffron yields only a scruple of essence. It is propagated from a bulb of a root [corm]. The cultivated saffron is broader, larger and more handsome, but much less potent; it has degenerated everywhere, and it is not prolific even at Cyrene, where grows a saffron whose flowers have always been very famous. But the prime favourite is that of Cilicia... Some have given second place to the saffron of Thera. Nothing is adulterated as much as saffron…" (Pliny, 1969, XXI 31-2).

Dioscorides (mid 1st century AD) provided a recipe for the crocus plants painted on a vase (1972: 30-32; pl. 61a) from Akrotiri, he argued that its decoration might attest to its contents (cf. Porter, 1999: 613). Other crocus decorated vases from Thera (Marinatos, 1974: 32-33) may attest to their contents and or function (FIGURES 3, 4).

Yet, the wall paintings from the Palace of Minos at Knossos and Xeste 3 at Akrotiri provided the most detailed evi-

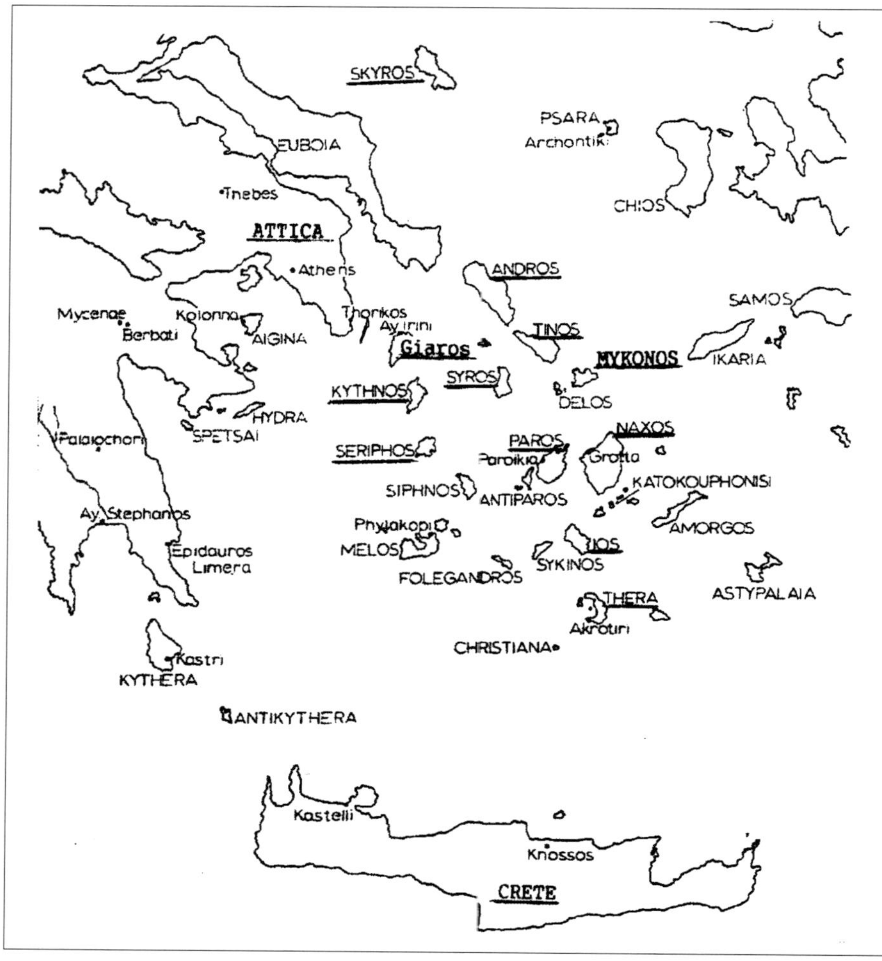

MAP: DISTRIBUTION AREA OF THE WILD SAFFRON CROCUS: CROCUS CARTWRIGHTIANUS.

medicinal unguent crocomagma, in which saffron is the principal component. Like Pliny, Dioscorides recommends Asia Minor's saffron to that of Cyrene (Materia Medica I.25-6).

SAFFRON IN AEGEAN BRONZE AGE ART

Saffron flowers were portrayed in various media of Minoan and Cycladic art during the second millennium BC (Amigues, 1988; Betancourt, 1985: FIGURES 82-83; Doumas, 1992; Immerwhar, 1990; Walberg. 1992). The earliest depiction of saffron flowers dates back to the initial phases of the Old Palatial Period (MM IB/IIA, 19th–18th centuries BC), including the elaborate painting crocus flowers on a Kamares ware vessel, where the style-branches and stamens are painted in their natural colour (FIGURE 2).

Marinatos (1972: 30-32) raised an interesting point regarding the large number of post- Kamares vessels with painted saffron flowers found on Thera, in New Palatial context (LM IA, circa. 1550 BC). In reference to flowering dence for the popularity of saffron crocus in the Aegean. Wall paintings from both sites describe scenes of saffron flower harvest in rocky settings. In the first, which probably dates back to MM III (17 century BC), the "Saffron Gatherer" (FIGURE 5) is a blue monkey (Smith, 1965; Shaw, 1993). In the Theran scene (FIGURE 6), assigned to the LM IA (16th century BC), the harvesters are young women wearing saffron-coloured (dyed?) ceremonial dresses decorated with chains of saffron style-branches on their sleeves (Amigues, 1988; Doumas, 1992; Sarpaki, 2000). M. Shaw (1993) considered the possibility that crocuses, among other wild plants, were cultivated in a rock garden of the Late Minoan Palace at Phaistos. This leads us to regard the harvest scenes on Knossos and Akrotiri as taking place in crocus rockeries rather than in a natural habitat. Wall paintings from the House of the Frescoes at Knossos, the villas at Amnissos and Hagia Triada indicate that similar gardens probably existed elsewhere on Crete at the same date (Shaw, 1993). Moreover, note that in the saffron harvest

FIGURE 1. A. crocus cartwrightianus (Porter, 2000). B: c. Sativus (Greenberg et al., 1983). The flowers of these species differ in length and orientation of their style-branches: short, upright or slightly curved in the former, long, heavier and drooping in the latter.

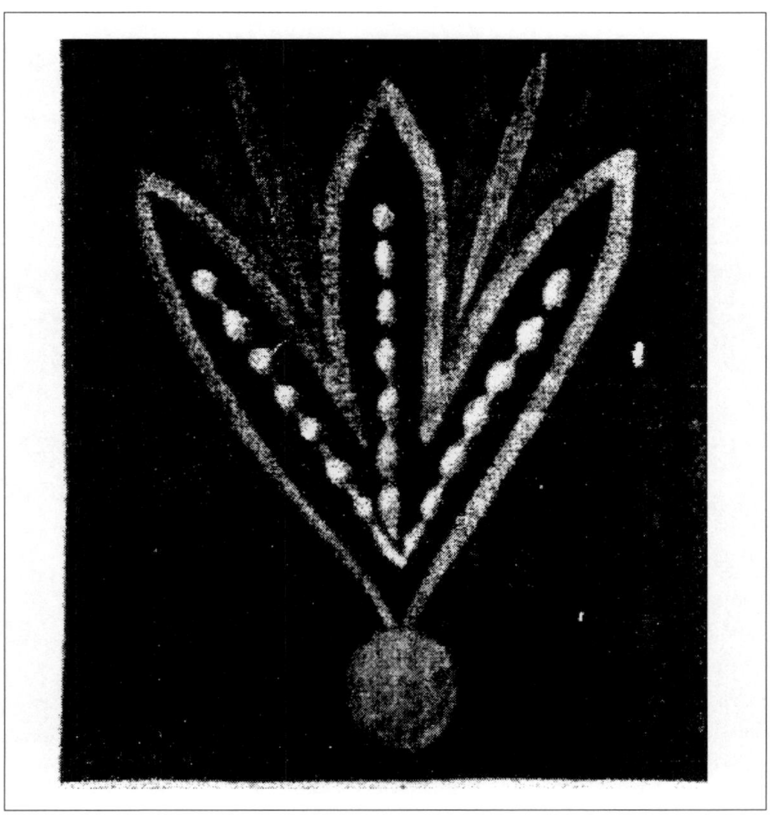

FIGURE 2. A stylised saffron flower painted on a Kamares cup from the Palace of Knossos.

FIGURE 3. LM IA strainer vase decorated with saffron crocus flowers on the base and on the lid (Marintos, 1974: coloured pl. 10).

FIGURE 4. LM IA elongated "kymbe" (boat) decorated with roe-deer and ibex grazing in crocus field (Marintos, 1974; coloured pl. 11).

FIGURE 5. Detail from the Saffron Gatherer Fresco at Knossos (Smith, 1965, fig. 103; drawn by Judith Dekel)

FIGURE 6. Crocus gatherers on a fresco from Xeste 3, Akrotiri (Doumas, 1992: 152).

frescoes in Knossos and Thera, there is only one kind of plant in the scene. In Xeste 3 fresco the crocuses are evenly dispersed as cultivated plants. It is in contrast with the habitat of Crocus cartwrightianus, which is "open rocky hillsides, sometimes in short turf or in scrub or sparse pine woods..." (Mathew, 1999: 22; cf. Sarpaki, 2000: FIGURE 9 in which C. cartwrightianus grows among thyme bushes).

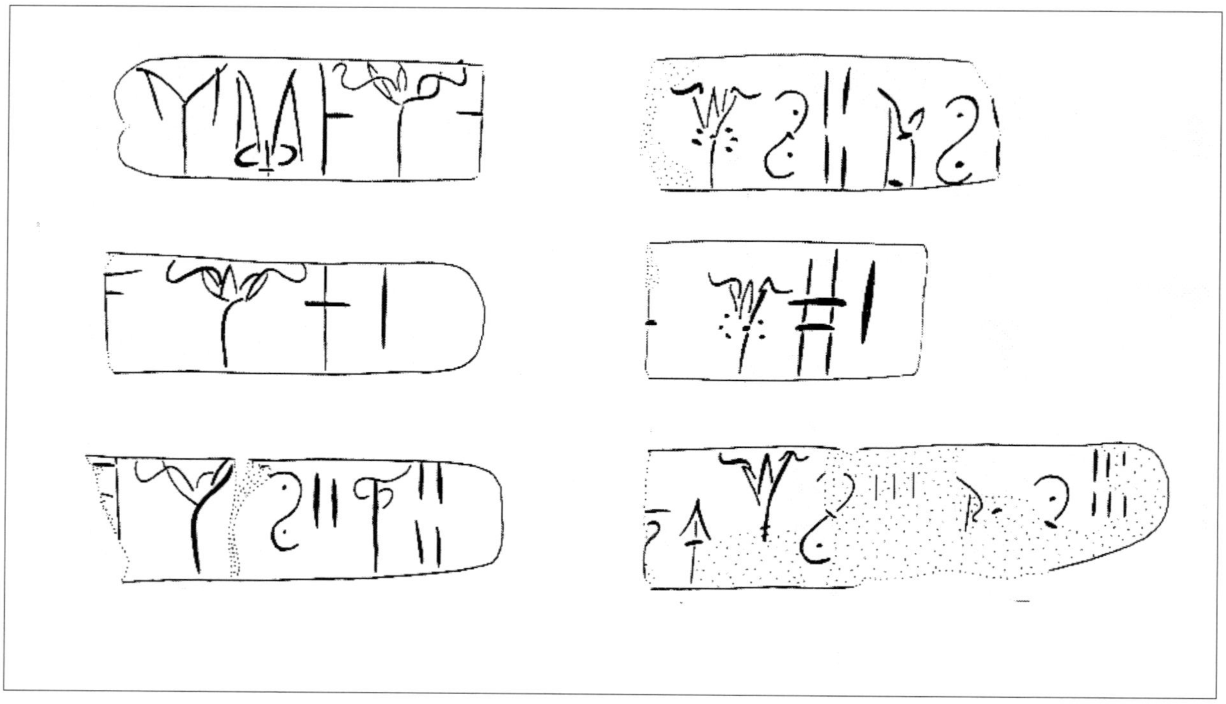

FIGURE 1. CROC tablets of the domesticated Crocus sativus from Knossos (Chadwick et al, 1986: 112–114; 344–346).
Note that in Np (1) series, left, the style–branches originate at the base of the flower, whereas in Np (2) series, right, they appear on its upper part. Such crocus flowers, with long style-branches, do not appear on pottery and wall.

The saffron flowers portrayed in Minoan and Cycladic pictorial art have straight or slightly curved style-branches, projecting out of three-petaled corolla. Based on morphological considerations (FIGURE 1) it is, generally accepted, that these saffron flowers represents, in spite of its variability, the wild C. Cartwrightianus (Mathew, 1977; Amigues, 1989; n. 18 to Théophraste VI, 6; Porter, 2000: 614-623; Sarpaki, 2000: 660-663).

The Minoan and Cycladic supremacy of saffron crocus portrayal was followed with a few examples of pottery depictions in Crete and mainland Greece in LM I (Vatrous, 1992: 105, FIGURE 66, Pl. 46) and LH I (Mountjoy, 1986: 10; 1993: FIGURE 32).

THE EVIDENCE FROM LINEAR B SCRIPT

By contrast to the wealth of Minoan pictorial evidence from Crete and Thera the written evidence are quite limited. Two series of inscribed tablets found in the Palace of Minos at Knossos are crucial in reconstructing saffron domestication on Crete. The tablets originated in a well-defined deposit known as "The Room of the Chariot Tablets," since its bulk enlists chariot's paraphernalia. Yet, a series of tablets from the same deposit bear CROC ideograms, which have distinctive morphological features diagnostic to the domesticated Crocus sativus (Killen and Olivier, 1989: 259-264). Moreover, the CROC tablets refer to "very small quantities, measured by weight in the units used for gold..." (Ventris and Chadwick, 1973: 130), which attest to the preciousness of the domesticated saffron crocus at Knossos (FIGURE 7). There is a growing consensus, supported by stratigraphic data that the Room of the Chariot's Tablets dates to LM II (15th century BC). As such, its contents are to be regarded as the earliest Linear B deposit found at Knossos (Driessen, 1990; 1992: 200-203; Bennet, 1993; Olivier, 1994: 166). The unique features of the Linear B CROC Tablets with long and curved style-branches lead us to the assumption that saffron crocus was already domesticated on Crete in the 15th century BC, when the last Minoan Palace became the residence of a new Mycenaean administration (Olivier, 1994: 167-1

CONCLUSIONS:

SAFFRON CULTIVATION AND DOMESTICATION IN AEGEAN GARDENS

Saffron crocus domestication could have taken place in three discrete stages:

(1) Saffron was collected from the wild C. cartwrightianus in its native habitat, as it is still done today in Greece: Attica, Eoboea, Skyros, Crete, and 12 Cycladic islands, including Thera (Amigues 1988: 228-230; Mathew, 1977: 95, 1982; 1999: 21; Sarpaki, 2000: n. 21).

(2) C. cartwrightianus plants transplanted from the wild and cultivated in gardens.

(3) A sterile mutant of the wild saffron crocus, having larger style-branches, was selected and propagated vegetatively by corms. This led to the domestication of the mutant (C. sativus).

Assuming that mutation frequency is equal in all wild populations, the likelihood of locating a mutated wild saffron crocus in a garden is greater than in its natural habitat (Sarpaki, 2000. FIGURE 9). Hence, an attentive gardener may have observed the mutant and barren C. sativus, and propagated it vegetatively. The Cretan and Theran frescoes showing saffron harvest, and the CROC ideogram in Linear B script from Knossos, allow us to speculate that an observed mutated saffron crocus was selected for further cultivation and domesticated by the close of LM II (15th century BC).

The gap in pictorial and written evidences concerning the saffron crocus lasts seven centuries from LM II (15th century BC) when the Linear B CROC tablets were inscribed in Knossos and the Archaic period when Homer mentioned crocus in the Iliad (XIV, 348; XIX, 1). Later, in the Classical period crocus is mentioned again by the dramatists (Skrubis, 1990: 172) and in the Hippocratic medical corpus (5th/4th centuries BC). Later Theophrastus (4th/3rd centuries) gave the saffron crocus an accurate botanical description and considered it as an important crop in Greece and in Cyrenaica (see above).

BIBLIOGRAPHY

AMIGUES, S.,
1988. Le crocus et le safran sur une fresque de Théra. Revue Archéologique, 1988 (2), 227–242.
BASKER, D.,
1999. Saffron technology. In: M. Negbi, ed. Saffron, Crocus sativus L. Amsterdam: Harwood, 1999, 95–101.
BENNET, J.,
1993. Book review: "An early destruction in the Mycenaean Palace at Knossos. A new interpretation of the field-notes of the south-east area of the west wing, by Jan Driessen."American Journal of Archaeology, 97(1), 172–174.
BETANCOURT, P. P.,
1985. The History of Minoan Pottery. Princeton: Princeton University Press.
BRIGHTON, C. A.,
1977. Cytology of Crocus sativus L. and its allies (Iridaceae). Plant Systematics and Evolution, 128 (3-4), 137–157.
DIOSCORIDES.,
1902. Des Pedanius Dioskurides aus Anazarbos Arzneimittellehre in fünf Bucheren (Translated and edited by J. Berndes). Stuttgart: F. Enke.
DOUMAS, C. G.,
1992. The wall paintings of Thera. Athens: The Thera Foundation.
DRIESSEN, J.
1990. An early destruction in the Mycenaean palace at Knossos. Acta Archaeologica Lovaniensia Monographiae 2. DRIESSEN, J.,
1992. "Collector's items." Observations sur l'élite mycénienne de Cnossos. In: J.-P. Olivier, ed. Mykenaïka, BCH Supplément, 25: 196–214.
EVANS, A. J.,
1935. The Palace of Minos IV: 718. New York: Biblo and Tannen.
GOLIARIS, A.,
1999. Saffron cultivation in Greece. In: M. Negbi, ed. Saffron, Crocus sativus L. Amsterdam: Harwood, 1999, 73-85.
GREENBERG, S. AND E. LAMBERT ORTIZ,
1983. The spice of life. London: Michael Joseph/Rainbird, 67-84.
GREENE, E. L.
1983. Landmarks of botanical history, ed. F. N. Egerton. Stanford: Stanford University Press.
KILLEN, J. T. AND J–P. OLIVIER,
1989. The Knossos Tablets. 5th ed. Salamanca: Ediciones Universidad de Salamanca.
MARINATOS, S.,
1972 & 1974. Excavations at Thera V (1971 season) & VI (1972 season). Athens: Bibliotheke tes en Athenais Arckaiologikes Hetaireias.
MATHEW, B.,
1977. Crocus sativus and its allies (Iridaceae). Plant Systematics and Evolution,
128 (1-2), 89-103.
MATHEW, B.,
1982. The Crocus: a revision of the genus Crocus (Iridaceae). London: Batsford.
MATHEW, B.,
1999. Botany, taxonomy, and cytology of C. sativus L. and its allies. In: M. Negbi, ed. Saffron, Crocus sativus L. Amsterdam: Harwood, 1999, 19-30.
MOUNTJOY, P. A.,
1986. Mycenaean decorated pottery: a guide to identification. Göteborg: P. Åströms Vorlag.
MOUNTJOY, P. A.,
1993. Mycenaean pottery, an introduction. Oxford: Oxford University Press.
NEGBI, M.,
1989. Theophrastus on geophytes. Botanical Journal of the Linnean Society, 100 (1), 15-43.
NEGBI, M. AND O. NEGBI,
In Press. The painted plaster floor of tel Kabri palace: Reflections on saffron domestication in the Aegean Bronze Age.
NEGBI, M. ET AL.,
1989. Growth, flowering, vegetative reproduction and dormancy in the saffron crocus (Crocus sativus L.). Israel Journal of Botany, 38 (1), 95–113.
OLIVIER, J.- P.,
1994. The inscribed documents at Bronze Age Knossos. In: D. Evely et al. eds. Knossos: A Labyrinth of History. Oxford: British School of Athens, 1994, 157-170.
Pliny.,
1969. Natural History. Cambridge, MA: Harvard University Press.
PORTER, R.,
2000. The flora of the Theran wall painting: living plants and motifs – sea lily, crocus, iris and ivy. In: S. Sherratt, ed. The wall paintings of Thera. Athens: Thera Foundation, 2000, II, 603-630.
SARPAKI, A.,
2000. Plants chosen to be depicted on Theran wall paintings: Tentative interpretations. In: S. Sherratt, ed. The Wall paintings of

Thera. Athens: Thera Foundation, 2000, II, 657-680.
SHAW, M. C.,
1993. The Aegean garden. American Journal of Archaeology 97 (4) 661–685.
SHAW, M. C.,
1993. The Aegean garden. American Journal of Archaeology, 97 (4), 661–685.
SKRUBIS, B.,
1990. The cultivation in Greece of Crocus sativus L. In: F. Tammaro and L. Marra, eds. Lo Zafferano. L'Aquila: Università degli studi L`Aqila e Accademia Italiana della cucina, 1990, 171-182.
SMITH, W.S.S.,
1965. Interconnection in the Ancient Near East. New Haven and London: Yale University Press.
THÉOPHRASTE,
1989. Recherchers sur les plantes. Établi et traduit par S. Amigues. Tome III. Paris: Les belles lettres.
THEOPHRASTUS,
1916, 1926. Enquiry into plants and minor works on odours and weather signs, English by A. Hort, I–II. London: Heinemann.
Vatrous, L. V.,
1992. Kommos III. The Late Bronze Age pottery. Princeton: Princeton University Press.
VENTRIS, M. AND J. CHADWICK,
1973. Documents in Mycenaean Greek, 2nd edition. Cambridge: Cambridge University Press.
WALBERG, G.,
1992. Minoan floral iconography. Aegeum 8: 241–246.

26

ECONOMIC STRATEGIES AND LIMITED RESOURCES IN THE BALEARIC INSULAR ECOSYSTEM: THE MYTH OF AN INDIGENOUS ANIMAL FARMING SOCIETY IN THE FIRST MILLENNIUM BC.

JORDI HERNÁNDEZ-GASCH , JORDI NADAL, ASSUMPCIÓ MALGOSA , ALÍCIA ALESÁN , JORDI JUAN

University of Barcelona,
Barcelona, Spain.
EMAIL: gasch@trivium.gh.ub.es

INTRODUCTION.

HOW THE ARCHAEOLOGISTS BUILT A MYTH.

The idea of a Late Bronze and Iron Age talayotic society (in general from the first millennium BC), based on an animal farming economy, has been largely assumed and defended by several authors. This topos, which we qualify as a myth, has been built over several years, using all sorts of evidence, even negative or ex silentio.

Thus, in the bibliography, thirty years ago, we can find researchers such as Cerdà and Rosselló-Bordoy , who, despite references to agriculture, clearly focused on animal husbandry, partly because of the amount of data recorded from this later, partly because of the absence of data related to agriculture .

Since systematic collections of food remains of an animal origin were kept by archaeologists, and since they produced early studies from such remains, the tendency to exaggerate the role of animal husbandry was intensified . Interpretation of stratigraphic records from a single site, partially excavated , and interpreting the whole of the talayotic society in this way, provided this point of view with a scientific justification beyond doubt .

The same writers have logically recognised the existence of agriculture, among other economic practices to provide food, in terms of subordination to the animal farming economic model . They have also proposed that this trend came about in the pretalayotic (at present naviform) period .

This assumption has even produced different explanations of the social, political and symbolic structure of the talayotic culture, when compared with cultural and anthropological data . Some of them have proposed that the exchange of goods was redistributive and controlled by a tribal chiefdom which assured flows of group surplus, following strict rules of generalised reciprocation (Guerrero, 1995a: 20; 1995b: 299). Some others proposed that it was a mainly egalitarian and self-sufficient system, "cuyo objetivo es un suministro regular y diario de comida para mantener a un máximo de la población" (Gasull, Lull, Sanahuja, 1984: 83), or even, as asserted by another author, that perhaps the predominance of this economic basis would explain those "delays" in the formation of complex societies. (Fernández-Miranda, 1991: 49). Indirectly, interpretation of some other data from the archaeological record using a animal farming society key (e.g. the monumental trends of the talayotic public buildings) has brought a reinforcement of this hypothesis, which now, used as a starting point, had already became a thesis.

Going back to the origins, we saw how Cerdà (and also Rosselló) linked the talayotic economic reality to biogeographic data. Later on, Ferrés went back to this topic pointing out the environmental impact of exploitation practices . Relationship between the developed economy and the environment, as conditioning strategies available for people under demographic pressure, was explored by Fernández-Miranda. Aramburu-Zabala has been one of the latest researchers to compile in detail paleoeconomic information related to the talayotic society . He went further in his goal to find new data in the landscape analysis and within studies about the productive capacity since medieval times to present days (grazing surfaces, herding load, productivity of by-products, caloric power, etc.). He calculated in detail herds which could have been supported by each village, use of meat and by-products, daily contribution to the diet from each community, and overall results for the island of Majorca (Aramburu-Zabala, 1998: 93-100). Nevertheless, his hypotheses come from relative data (faunal remains preserved), which are taken as absolute ones. They are constructed following logical proposals and accumulative conditionings, added from one proposal to the following one. The results are an unverified hypothesis often expressed as an absolute truth.

Research history in the Balearic Islands shows a predominant orientation towards excavations in necropolis and sanctuaries. These sites had traditionally provided museums and private collections of objects and, in the case of the Taula's sanctuaries, they had been per se attractive monuments. In recent years, plunders at these types of site had led to rescue diggings. Altogether, this created an important bias in research, in the way archaeologists recover and interpret data.

As a result, excavation in settlements has not received much attention. An early example comes from the fieldwork of Josep Colominas early in the twentieth century in Es Pedregar and Capocorb Vell, although a lack of extensive publication meant a poor contribution to the following research. Something similar occurred in Ses Païsses, excavated by G. Lilliu and very poorly published, except for the first season. On Minorca, quite often the only thing released from settlements largely excavated has been a plan.

Apart from not publishing the results, the data collecting system was quite or severely deficient, not always due to the stage of development of archaeology as a science. Thus, it's not rare to find in some Minorcan sites, even the ones excavated in recent years, many grounding stones (amolons in Catalan) not recovered by the archaeologists, or just abandoned in place after the digs and plundered later on (as in the case of the dozens of amolons found in Sant Vicenç d'Alcaidús, according to the published pictures).

In the late eighties, the diggings in Son Fornés and Son Ferrandell started to make use of flotation techniques, recovering samples and analysing them in order to set the economic basis of the talayotic society. In the nineties, new diggings in Puig de Sa Morisca (Majorca) and Biniparratx Petit (Minorca) and the reprisal of excavations in Ses Païsses (Majorca) and Talatí de Dalt (Minorca) pursued these goals from the beginning, although their results still have not been published. Palinologic analyses in the Balearic islands are quite recent too, just a few of them can be referred to the period we are discussing now.

The data used to sustain the animal farming hypothesis have been:

(1) The appearance of big faunal assemblages, although they come very often from sanctuaries and some other buildings, which could have had a ritual function.

(2) The spread of taurolatric votive, assuming a roughly direct translation can be made from the ritual sphere to the economic reality, representing a central role in the basis of subsistence of the indigenous people (Aramburu-Zabala 1998: 99). Nevertheless, this wrong interpretation of bull symbolism has no tradition in the previous bibliography, as Llompart (1970: 256), thirty years ago correctly interpreted it within the "agrarian cultural horizon"

(3) The characterization of some sites as "ceremonial centres" which would hold a political function; ritual, but basically redistributive, interpreting archaeological data from sanctuaries and other ritual buildings as evidence of the distribution of meat within an animal farming society (Guerrero, 1995a: 29) (Aramburu-Zabala, 1998).

Amongst the data ex silentio:

(4) In Majorca, a small number of grounding mills. As a result of the simple tipology of the majorcan grounding mills, and due to the lack of extensive diggings in settlements, these have often been neglected . However, in methodologically controlled digs, such as the one in Turó de Ses Beies, grounding mills appeared in almost every sector excavated (Camps, Vallespir, 1985: 337), and they were also found in the contexts of Son Fornés (Gasull, Lull, Sanahuja, 1984: 84). Recent excavations at Son Ferrandell and Ses Païsses have recovered more specimens than was average at that date.

(5)On the same island, a lack of storage pits has also been shown, although there are many other ways of storing food, such as big ceramic containers. These are very frequent amongst the archaeological finds, 23% in Son Fornés's Houses (Gasull, Lull, Sanahuja, 1984) or even in perishable structures not detected. In fact, we know of the discovery of big pots filled with carbonizate wheat in Son Oms (Rosselló, 1972: 126) and carbonizate seeds in some other sites. Oat was recovered in Sa Marina de Sa Punta (Amorós 1952: 13, Mascaró, 1968: 677), and wheat, oats and carob bean were found in the Ca N'Amer talayot (Llompart 1965: 30; Mascaró 1968: 677). Amoros's findings were referred to in detail by Cerdá (1971). He asserted that they appeared in a room, which might be dated in the 7th century BC and they were found carbonizated in a pot. According to Cerdá, the species recorded were Triticum vulgare, Hordeum exasticum, and Vicia sativa L. In a late context from the talayotic village of Ses Païsses, remains of acorns were found (Mascaró, 1968: 677). In the Turó de Ses Beies site the discovery of some italic amphorae with cut necks, sometimes polished (and, in one case, repaired with lead clamps) leaning against the walls, was taken as evidence of storing pots for grain or flour. This was suggested, since it was impossible to dig storage pits in the ground (as there is just clay close to the surface) and because of the repairs done to a pot with clamps as, being broken, they couldn't contain liquids, but solids. Basins made of clay, and small slabs containing sediments with high proportion of organic material are also related to food processing (Camps, Vallespir,1985: 336-337).

Strangely enough, on Minorca, where the characteristic grounding mills (amolons) are very common and there is an important amount of storage pits, the animal husbandry hypothesis was adopted without further discussion. Although there is no analysis to show which products were ground or stored, they are, at least, data to take into account. These boatshaped grounding stones are large and heavy, and they often have a moulding at the back, considered an element used to grasp it. They have been seen as active parts, the passive ones being large blocks with a flat surface, such as the ones located in the settlement of Trepucó. There is no proportion between the objects and even the smallest one (the amoló) seems too big to be considered an active part moved by hand. These considerations shade the interpretation with doubt. Some of the items (of the ones with a boat shape and moulding) examined by us, from talayotic domestic levels show a transversal concavity to the longitudinal axis, which allows us to consider them passive parts, never active on a flat surface (as there is no contact point and, as a result, no friction surface). Therefore, the moulding at the back matches the "grounding body position": kneeling, the upper body slightly bent forward, the mill leaning one part on the floor, the other part on the lap . Finally, one of these was put outwards in a small pit, where it was fixed.

(6)The lack of a set of tools for agricultural purposes faces the same issues: little or limited excavations in settlements and poor preservation of iron in those sites. Further back in time, in Son Oleza, excavated in modern times, and dated in the chalcolithic period, many flint blades and sickles were recovered from what are regarded as the site fields (Waldren, 1988). During Early and Middle Bronze Age they are found inside a boatshaped house (naviform) in naveta Alemany (Ensenyat, 1971). In some Iron Age settlements, well recorded, iron tools were found. (In Turó de Ses Beies a couple of hoes or adzes appeared in a context interpreted as a forge, and an iron point which might have been used as a spade "para remover la tierra; o sea, que también se puede admitir alguna actividad relacionada con la agricultura" (Camps, Vallespir 1985: 340). In recent excavations, such as the one in the talayotic habitat of Ses Païsses, amongst the few metallic objects recovered, a well preserved adze stands out. The context where it was found can be dated quite possibly in the first century BC. Its use must be related to wood carving, but it is also possible it was used to prune trees).

(7) The lack of archaeobotanic studies is a major factor in the ignorance of the agricultural activities in the late prehistory of the Balearic Islands. Poor results in the few carpological analyses carried out and published to date, are reduced to a couple of possible seeds in Talayot 4 of Son Ferrandell (Chapman, Grant, 1995), since there was no presence of cereals in the Son Fornes samples (Yll in Gasull, Lull, Sanahuja, 1984: 134). This lack of direct evidence of domesticated vegetal food has lead mostly to the proposal of economic models based on herding and foraging activities. The same researchers proposing these models recognise, however, that the absence of such evidence is not enough to deny an agricultural practice (Guerrero, 1997; Guerrero, Gornés, 2000, Lull et alii, 1999).

Thus, amongst the defenders of a basically herding economy, lately we can find important nuances. Lull et alii (1999: 67) defend that "una economía mixta con predominio de la ganaderia parecía ser el modelo practicado en época talayótica", pointing out that "encara que els indicadors agrícoles fossin pobres no es descarta la práctica d'estrategies de conreu o de recol.lecció per proveir els complements dietètics necessaris" (Gasull et alii, 1984: 83-90).

Criticism of the animal farming hypothesis

In the sixties, Lilliu (1962-1963: 22-48) defended the idea of a mixed economy, consisting of cereal agriculture, sheep and goat husbandry and pig rearing. Game hunting and shellfish gathering were an indispensable complement to diet. Despite this, it is only in recent years that some voices have been raised against the established animal farming theory. Chapman and Grant (1995: 29 and f.), whilst studying animal remains from Son Ferrandell, conclude that vegetables had to have made up the majority of the diet, since the animals were sacrified at the end of their productive cycle. These authors also claimed for the implement of more analysis . Recently, since the analysis of trace elements and other in individuals from naviform contexts in Cova des Càrritx (Menorca) some have insisted (Fornés, Salvà 2001) in the existence of a mix diet.

In this respect, it is obvious that we cannot put food production and diet on the same level, comparing what a significant part of the people daily consume to the economic structure or productive basis, since a large part of this production could be to exchange and to obtain goods for consumption, food or other, and to sustain people who work in secondary and third activities. However, some authors refer either to diet or to the economic production, or both things, ignoring this relation or just assuming their identity. As a result, when following their thoughts on the composition of the talayotic economy, we can find many references to diet.

Some of the first studies carried out in the Balearic Islands showed a settlement pattern according to different conditions, such as proximity to seasonal streams and fields suitable for crops (extensive crop of dry regions). For these authors, animal husbandry was just an economic complement (Coll, Mazaira, Riutort 1984: 120; Da Mota 1988, 1995, Rodríguez 1995, Salvà 1992: 417 and f., Vallespir et alii 1987: 14).

Finally, palinological studies detected alterations in the vegetal cover, due to climatic and anthropic causes. These could be related to agriculture as well as to animal farming practices (Yll, 1984: 134). Increases have been shown in some taxons, such as Cerealia, Olea, Vitis, and Juglans, in mid first millenium BC (Burjachs et alii 1994: 206, 208).

Archaeozoological data: Methodological criticism of their interpretation

Intellectual processes, such as finding animal bones and inferring from these the strategic economic importance of animal farming, have led to misunderstanding and misinterpretation of archaeological data. Animal remains are often overestimated compared to vegetal remains, due to better preservation during diagenetic processes. This is an independent data, which has nothing to do with their cultural importance. Post-depositional processes (acidity/alcalinity of the soil, carbonation...) and human activities (cleaning of certain areas of the site) can also alter samples. All of these must be assessed before establishing the real importance of animal origin resources.

Something similar occurs within the faunal remains themselves. Some species leave important remains in the archaeological sites, some others, few or none at all. Thus, big mammals and shell molluscs are overestimated, whilst molluscs without shells, crustaceans, fish and birds are underestimated, whose remains are disturbed by cultural behaviour, fossildiagenetic processes, and even missed in field research works (especially in old excavations).

In some studies, quantitative results (the number of remains, minimum number of individuals, biomass, in absolute figures or percentages) have been confused with the real number of individuals existing (usually underestimated).

Finally, projections of figures, and economic activities can be wrong, when the raw data come from sites other than settlements (cemeteries, sanctuaries) or those just partially excavated. Although a link exists between economic orientation and ritual behaviour, it cannot be interpreted in a lineal way. Domestic animals play a big role in societies based on agriculture (celebrations, imagery).

Cultural criticism of the paleozoological data

When animal husbandry is not a complementary strategy, balanced with other economic activities such as agriculture, gathering, fishing, but the main activity for the community as a whole or just a section (age, gender) within it, ethnological studies in contemporary societies (with different degrees of technology and social complexity) show the following trends:

(a) Vast territories are required as pastures to breed animals. Although they are used in a seasonal way, they are regarded as their own fields: Nuer and Massai with cattle, central Asian tribes with horses, camels, sheep and goats, Lapps and Northern Siberians with reindeers, European populations since the Middle Ages with cattle and sheep, and so on. These territories need to have boundaries that are not well-defined, able to be extended in critical times (drought) by different means (violent or not). This makes a sensible explanation for specialised animal farming more difficult, as a basic strategy for food and economic resources, in insular ecosystems, which are always precisely limited. So, it is easier to explain complementary herding with horticultural practices, which is generally more important in economic terms, although not necessarily more valued. In this respect, it has to be borne in mind that the Balearic Islands are small islands.

(b) Animal farming societies do not set up big settlements due to their extreme mobility. As well as having big territories, it is necessary to consider their organisation and, as result, a clear hierarchy of sites is developed, with metropolis and other satellite settlements. Obviously this does not match up at all with the archaeological data on Majorca, where settlements of similar extension (76 % of the sites) are located at an average distance of 2.47 km (in a range from 0.5 km up to 5.8) (Aramburu-Zabala, 1998: 179 and 190). Needless to say, this high density of similar sized sites sets small catchment areas for each of them.

(c) A distinctive orientation towards specialisation, breeding one (or two) species, which are the basis of economic and material culture. Commonly, they avoid species which hinder mobility. That might result in their being banned from the diet, which in turn is converted into taboo. Thus, the presence of the pig, an animal not adapted to moving too far, determines, at some extent, the existence of settled communities, based on agriculture. Diversity in the faunal spectrum, as it occurs in the Balearic Islands, points to a lack of specialisation and, furthermore, the absence of an animal farming society.

(d) The lack of sufficient vegetables for their consumption means these communities make large use of dairy products, blood, and other goods (wool...) to trade for the desired vegetables. Thus, they avoid killing animals until they finish their productive cycle. That leaves strong sacrifice patterns (especially regarding female specimens). However, this pattern is not exclusive to animal farming societies, being adapted to the orientation that each society gives to a specific species.

(e) Cults of the bull, according to many religion historians, are widespread in essentially agrarian societies, rather than in animal farming societies, as in many societies in the Mediterranean basin during Prehistory and Antiquity. In iconography, the bull's image appears as a masculine divinity secondary to the Mother God (Eliade, 1978, 1981; Gimbutas, 1989). These cults are not significant for herding societies and they are well recorded on both Majorca and Minorca.

AVAILABLE AND NEW DATA

Our paper intends to provide a summary of the paleofaunal, paleoenvironmental (carpology, analysis of residues, palinology) and anthropological (trace elements, diet, pathologies, which can show diets: caries, raquitism, etc.) applied to the talayotic world, both the published up to date and preliminary data still unpublished from the paleofaunal and vegetal macroremains and phytoliths studies carried out by some of the authors.

As seen, archaeofaunal studies has had a relatively long tradition, and they provide most of the data in discussion. We will try to assess the value and quality of these data (only references to faunal findings, preliminary records and lists, describing species, or complete faunal studies), taking into account the bias which has affected the samples. We will need to contextualize their origin (also according to their origin: from settlements, sanctuaries or cemeteries) and establish similarities and differences to be able to interpret them. Nevertheless, we can suggest that a high lack of proportion exists between the number of talayotic settlements known and/or excavated and the faunal studies published, up to the extent that we can only compare within the island of Majorca.

Further discussion will enable us to assess which species were represented, finding out whether or not specialisation existed. Looking at ages of death and sex patterns will allow us, following more classical interpretations, to examine the real importance of animal farming, establishing whether or not it was basic for the economy and subsistence, and eventually, more important than agriculture.

A profuse use of our own data (from Illa des Porros and Punta des Patró, Majorca) will be made in this part, as well as other comparable data, coming from bibliographic infor-

Settlements	Sheep and Goat %	Pig %	Cattle %
S Illot Pre-roman (Uerpmann, 1971)	72,2	9,5	18,2
S Illot transition[i] (Uerpmann, 1971)	74,5	9,3	15,2
Son Forn s (Est vez, 1984)	45	30	24
Son Ferrandell Oleza. Phase 3 (Chapman, Grant, 1995)	86	7	7
Son Ferrandell Oleza. Phase 7 (Chapman, Grant, 1995)	80	9	11
S Illa des Porros. Whole (Nadal, 2000)	59,8	10,5	29,5
Punta des Patr . Preliminary (Nadal, unpublished)	75	7,14	17,85
Son Real (Nadal,1998)	14,3	9,5	66,6
Almallutx. E6+E7 (Romero, 1971)	65,4	18,2	16,2

[i] The name used by Uerpmann to refer to the assemblages *that might be mixed up with remains of following occupations.* We added it because of its similarity to the good preroman (talayotic) assemblage. Possibly these similarities are due to the fact that the transition samples are basically from a talayotic origin.

TABLE 1.

mation.

To some extent, all data are comparable. In any case, we have chosen to contrast these data through proportional figures allowed by the number of remains (NR) from each report. We have done so, since, despite its simplicity, we consider the NR the most effective system to weigh the importance of diverse taxons (above the Minimal Number of Individuals (MNI), weight, or some other calculations around biomass). Thus, using relative figures and not the absolute ones, allows us to balance differences of register between different sites. In any case, absolute values can be seen in the original works of reference (Uerpmann, 1971 for S'Illot, Estévez, 1984 for Son Fornés, Chapman and Grant, 1995 for Son Ferrandell Oleza, and Nadal, 1998 for Son Real). We just need to point out that, in the case of our unpublished works for S'Illa des Porros, we have dealt with 2793 determined bone remains (which belong to three big chronological periods, according to the Stratigraphic Units –SU or contexts- dated). In addition, the data from Punta des Patró are utterly provisional, with less than 200 determined studied remains (a first list with the species represented can be found in Hernández-Gasch, Sanmartí, 1999, without quantitative data). In Almallut's faunal study (Romero, 1971), we have only used information from E6 and E7 structures, because, according to the monograph (Fernández-Miranda, Enseñat, Enseñat, 1971), this couple of structures are the only ones without a mixture of archaeological material with other periods later than the talayotic (roman and medieval times). In these cases, the remains are also extremely scarce.

In this study, we just have taken into account the outcome corresponding to the three main taxonomic groups related to the subsistence production system: sheep and goats, cattle and pigs. As a result, the total of figures presented is often not 100 %. The rest of the osteologic assemblages, always of a lower importance, is composed by horses and dogs, amongst the domestic fauna, and rabbit, deer, gnawing and pinnipeds, amongst the wild fauna (sometimes, clearly intrusive and some others of difficult cultural interpretation by now). Perhaps the clearer data, and at the same time easier to assume, is the presence of Mediterranean seal in S'Illot and S'Illa des Porros). So, now, we will focus on the comparison of data from sheep and goats, pigs and cattle, and, only at the end, will we mention the rest of the archaeofaunal record.

The results of the relative figures obtained from the NR of sheep and goats, pigs and cattle are asseen above:

The first thing we have to say is that the archaeozoologic focus cannot prove or deny by itself a given specialisation or economic strategy. Animal remains, as they are generally found in archaeological contexts, as mingled and disconnected debris, do not allow an estimate of the real number of individuals, which, at some point and in a determined space, once existed. Any sort of calculation always gives very theoretic results. Indeed, with the application of ethnoarchaeological techniques, we cannot see, throughout deposition of faunal remains, a relationship between the number of remains and a higher or lower dependence on resources of animal or vegetal origin. Besides, societies which depend to a higher degree on produced animal resources (herding societies), being more mobile, tend to leave a poorer archaeological record. Agricultural societies, however, being more settled, are susceptible to leave a bigger concentration of archaeological remains, and so bone remains as well, which could lead, in a simplistic interpretation of data, to consider a strong dependence on faunal resources, being in fact only a repeated activity of rejecting remains.

Finally, the herding specialisation would involve a series of very standarized behaviour patterns that would be reflected to some extent in the tanatomass. Firstly, there is a specialisation in a very concrete taxonomic group; secondly, a taxonomic homogeneity between different groups in the same cultural area, and thirdly, determined slaughter pattern, homogeneous between different groups, according to herd orientation.

	First moment	Second moment
Sheep and goats	588	750
Ovis	15	13
Capra	20	31
Sheep and goats + ovis + capra	623 (61,57%)	794 (57,70%)
Bos	286 (28,26%)	431 (31,32%)
Sus	103 (10,17%)	151 (10,97%)
Total	1012 remains	1376 remains

TABLE 2.

WHAT CAN WE LEARN, IN THIS RESPECT, FROM THE DIFFERENT FAUNAL REPORTS?

Undoubtedly, there is a tendency towards a predominance of sheep and goats, at the front of the couple of taxonomic groups remaining. Thus, we could argue a certain degree of specialisation in sheep and goats. Nevertheless, the reflected proportions are more similar to those we can find in other Mediterranean areas at similar dates (e.g., those observed in the Iberian period in the Catalan area –Albizuri, Nadal, 1999: 48-49); and so, rather than specialisation, it seems a basic adaptation to any kind of herding in dry Mediterranean contexts . Moreover, in the case of the sites studied by us, when a discrimination exists between sheep (Ovis aries) and goats (Capra hircus), proportions between them are more balanced than in other mainland sites, where there is usually a predominance of sheep over goats (e.g., global results from S'Illot, Son Ferrandell Oleza and our unpublished study from S'Illa des Porros). We argue that a specialised herding, in this case in sheep and goats, undoubtedly would have clearly favoured one of the two species.

We can not dismiss, however, those cases where this tendency towards a mediterraneanisation of animal farming with sheep predominance seems to be broken. On one hand, we have the Son Fornés case. Sheep and goats there make up no more than 50 % of total number of remains. On the other hand, we have the Son Real case with a spectacular predominance of cattle.

In our first case, we have to take into account the existence of a high concentration of cattle and pig within the talayots, largely bigger than the percentage documented within the houses (Estévez, 1984). Mean figures, as we expressed them in the graph, reflect this bias. The fact is that these differences pose some questions as whether these deposits are comparable or not, whether they have the same domestic character or they are exceptional or common. Several interpretations proposed about the function of the talayot itself as a ritual and communal space might reflect some special deposits of a different, non-domestic character. Researchers in this site propose an explanation based on an adaptative mechanism to the surrounding environment, as Son Fornés is located in a place which would have been rich in marshes until the drainage realised in the Arabic period. So, it is an exceptionally rich area in terms of quality grass, favouring cattle breeding. In the Son Real case, several factors clearly act in what is an obvious bias of the constitution of herds in the talayotic period. First of all, the remains come from a funerary context, which by itself can distort the economic reality of these populations. This very same funerary context and the fact that we are considering con-

Sheep and goats	Without epiphysis	With epiphysis	Whole sample
First moment	12 (21,42%)	44 (78,58%)	56
Second moment	27 (44,26%)	55,74 (55,74%)	88

Cattle	Without epiphysis	With epiphysis	Whole sample
First moment	6 (13,95%)	37 (86,05%)	43
Second moment	6 (17,14%)	29 (82,86%)	35

Pigs	Without epiphysis	With epiphysis	Whole sample
First moment	11 (45,83%)	13 (54,17%)	24
Second moment	14 (46,66%)	16 (53,34%)	30

TABLE 3.

texts from an 'old' excavation have limited the number of remains (only 21), which is statistically not significant. Moreover, cattle is represented exclusively by modified femur heads (called "stopper", "tap" in Catalan by the archaeologists). Being a cultural artefact, they are useless to analyse the proportions of cattle representation in a study about subsistence. Finally, from the excavation's monograph, we can easily see that they did not exhaustively collect every single bone, but mainly those culturally modified ("taps", punch) or rarer (hen egg) (Tarradell, Hernández, 1998). There is no need to go as far as the Son Real extreme in order to consider whether a site's functionality can determine or not proportions of representation of macromammals in the S'Illa des Porros and Punta des Patró cases. Despite the totally preliminary character of the Punta des Patró data, they coincide sufficiently with the ones proceeding from several levels in S'Illot and even more with the ones from Son Ferrandell Oleza. In the S'Illa des Porros case, an increase in the number of cattle at the expense of sheep and goats might be a result of the funerary use of the site. In any case, during our study, the similarity of this assemblage with one expected in an assemblage fully-oriented to subsistence was quite surprising.

Discrimination of fauna according to SU (stratigraphic units), gathered in two long periods does not seem to produce a variation in the results.

All those SU, dated by different means (some of them through radiocarbon method) in the second half of the second millennium up to the beginning of the first millennium, belong to the first moment. Culturally, in recent times archaeologists refer to this moment as three different cultural periods: Naviform II, transitional period, and beginning of talayotic culture.

The second period lasts from the fourth century BC to the second century BC., which is in LIA, at the end of the talayotic culture.

Some others SU can correspond to a third moment, in between the previous two, centred in the 5th century BC, but it has been dismissed because of volume of remains was not comparable to the others.

We have compared it in two senses, according to the represented species and according to the slaughter time for each specie in every length of time considered. For this calculation, we established two different groups: epiphysated remains and non epiphysated ones, whatever the bone examined. It is obvious that different bones from a same animal fuse in different times. However, an animal that has not completely finished its bone fusion in its extremities can still grow. Thus, it is young, whatever its age, and what is more important for our purposes, it is in the meat optimum which means it has been oriented towards meat production, not towards by-products. In this sense, provisory classification of remains according epiphysis is right, in spite of the existence of a problematic margin between the first and last bones fused.

The proportions in slaughter age do not seem to vary either. Perhaps there is a slightly orientation towards life products (animals slaughtered in adult ages) amongst sheep and goat in the first moment, whereas in the second one proportions between bones with and without epiphysis are very similar to the ones for pigs in both periods. And pigs are only oriented, in terms of subsistence, towards meat consumption. Being a complex calculation, we have only used postcranial data, but not teeth.

Another element which might reveal some data about the degree of specialisation of the talayotic herding is the slaughter age patterns of animals. In this respect, all the sites analysed seem to coincide greatly: sheep and goat might have been a mixed function, with some animals slaughtered as young as subadults oriented to meat production, and with some others, in similar proportion, slaughtered in adult age, in order to make a good use of their products when alive. Pigs, as we might expect, are slaughtered mainly in sudadult age, as there are not many products, if any at all, exploited during the animal's life. Finally, cattle are allowed to remain alive well into adult age, which means good use is made of resources while alive. Labour force would have been one of these, and so it could be associated with an economy, not exclusively, but at least clearly completed with agriculture. Amongst the specialised herding economies, it is very common for almost all animals to pass meat optimum, because, as well as products they provide during lifetime, animals become a status symbol, maybe the most appreciated aspect within the community.

Archaeobotanic evidence of agricultural activities in the Ba learic Islands: a critical discussion of the available data. The introduction of new analysis techniques is basic in order to study the agriculture of the Balearic Islands, specially archaeobotanic techniques, which are necessary in order to find a less biased explanation from the available data. Research carried out to date about vegetal food exploitation shows it is absolutely not advisable to globalise and speculate about the agricultural strategies of people from the islands exclusively from these data.

We have to consider evidence which can add some information about agricultural development, both direct and indirect, in order to discover the existence of one or maybe several types of agricultural practices. At present, data available are focused on scarce archaeobotanic studies (basically, palinological analysis from Majorca and Minorca) and some documentation from stone tools (sickle blades and grounding stones).

a) Palinology

Palinology was the first of the archaeobotanic specialisations used for research into direct evidence of crops. Thus, it has been used profusely in paleoenvironmental reconstruction, especially in the past few years, according to the data from soundings carried out in Majorca (Cova Estreta, Alcúdia, Abric de Son Matge, Cova des Moro and S'Arenal) and Minorca (South coast: Cala Galdana, Algendar, Son Bou, Cala'n Porter, Hort Timoner; North coast: KF14) (Burjachs et alii, 1994; Pantaleón-Cano et alii, 1996 a, 1996b; Pérez-Obiol et alii, 1996, 2000; Yll et alii, 1994, 1995, 1997, 1999). These data are complemented with palinological studies, carried out in archaeological sites in Majorca: Cova de Sa Moleta (Waldren, 1982), Son Fornés (Yll, 1984), and in several sites in the north and east of Minorca (Mariscal, 1996). Regarding agriculture, some outstanding data come from the diagram of the Albufera of Muro (Majorca), where we can find a significant presence of the taxon Cerealia since mid first millennium (Burjachs et alii, 1994). In the upper levels of Son Bou, in Minorca, there are also important figures of Cerealia.

Some interesting data revealed using palinology in the Balearic Islands shows that it can register phenomena of climatic change and landscape transformation. From 6000 to 5000 BP, there is a transition period from the dominance of a vegetation formed by Buxus, Corylus and Juniperus, characteristic from the previous period, to a period of more xerophilus type, dominated by Olea, accompanied of Juniperus, Ericaceae and Quercus. This involves the extinction of some animal and vegetal species. Amongst the most significant, there is the practical disappearance of Buxus both in Majorca and Minorca (although, it still exists at present in Cabrera, the smallest of the islands, in more dry conditions than in Minorca, which is the most humid one, which suggests that there are some other factors involved), the extinction of Corylus in Minorca and its almost complete disappearance in Majorca, the extinction of Hypnomus morpheus, Nesiotes hidalgo and, above all, Myotragus balearicus. A landscape dominated by Olea and ericacies appears; the last one is the main specie amongst the scrub that colonises this landscape. There is an increase in Pseudoschizaeae, a palinomorf that probably comes from seaweed, generally interpreted as an indicator of erosive conditions in the environment (Pantaleon-Cano et alii, 1996 a). A significant increment in the curves of Plantago, often indicating deforestation practices, is produced by 6000 BP in Cala Galdana, around 5000 BP in Cala'n Porter and about 4000 BP in Algendar. These increases are associated with the documentation of pollen of Cerealia, and in the case of Cala'n Porter, with Asphodelus, associated with pastures. This phase was previously related to the discussed human colonisation of Majorca. From 4000 BP and 2000 BP, the Mediterranean trends of vegetation increase, there is a predominance of Olea, and this, together with Quercus, represents almost the whole population of trees. Grasses, scrub and bushes indicate open landscapes and a human action over territory. It is formed by Asteraceae, Poaceae, Plantago, Fabaceae, Rumex and Polygonum. The presence of Pseudoschizaea must be related to eroding processes.

Despite the available data for a paleoenvironmental reconstruction, pollen studies are not very appropriate for identifying and defining specific horticultural practices (Zapata, 1996). As a limiting factor in terms of analysis and methodology, we have to underline how difficult it is to identify domesticated species, such as cereals and legumes, especially when dealing with fossilised pollen. Moreover, we have to take into account that pollinisation of cereals and weight of these pollens make their capacity of dispersal quite limited, with a mean reach of 2 km (Iriarte & Arrizabalaga, 1995).

In short, with some exception in extreme cases, palinology does not give solid, direct evidence of the presence of domesticated species and crops in the archaeological record in the recent prehistory from the Balearic Islands. Nevertheless, the presence of pollen of Cerealia in the records studied has been a common evidence used as a proof of agricultural activities.

This attitude has favoured, at least, the development of the idea that only through archaeobotanic studies are we going to be able to answer questions such as what they grew, where, how and since when they did it.

CARPOLOGY

Up to the very moment before starting new archaeobotanic research, references to the appearance of seeds and fruits in prehistoric contexts in the Balearic Islands were no less than exotic. This is the case of some documentation about olives (Olea) in Talayot 4 of Son Ferrandell Oleza Chapman, Grant 1995: 29), or seeds of cereals without determination in the sanctuary of Sa Marina de Sa Punta (Amorós, 1952), naveta of Son Oms (Rosselló, 1979) and surrounding circular talayot (Guerrero, 1997).

In front of this research for alternative models (quite conformists, as they have been based on 'no presence' of evidences), the application of new techniques of information recovering (contained in archaeological deposits) means a new advance in the knowledge of agricultural practices, as in the case of sites such as Cova des Càrritx (Lull et alii, 1999) and Biniparratxet in Minorca, and S'Illa des Porros, Punta des Patró and Son Real in Majorca (Juan-Tresserras, in study).

In the Cova des Carritx case, they documented the presence of barley (Hordeum vulgare) oats (Avena sp.), figs (Ficus carica), olives (Olea sp.), blackberry (Rubus fruticosus), raspberry (Rubus ideaeus) and beetroot (Beta vulgaris) (Lull et alii, 1999).

In levels dated in the 4th century up to the 2nd century BC in the cemetery of S'Illa des Porros, preliminary data show the presence of cariopsides of common wheat (Triticum aestivum/durum); the same than in levels from the 2nd century BC in Punta des Patró, where, apart from cariopsides of common wheat (Triticum aestivum/durum), we have found a juniper berry (Juniperus sp.), grape seeds (Vitis vinifera var. sativa DC), a fragment of fig (Ficus carica) and cotyledons of acorns (Quercus sp.) (Juan-Tresseras, in study).

To sum up, carpological studies offer good prospects for the investigation of agricultural activities and their development during the recent Prehistory in the Balearic Islands.

PHYTOLITS AND STARCH

Studies of Phytolits (vegetal remnants silicified and crystals of oxalate of calcium) and particles of starch can give us significant data about agriculture in the Balearic Islands. To date, research in these islands is restricted to studies of dental calculus from individuals found in S'Aigua Dolça 5 dolmen, in the talayotic cemetery of S'illa des Porros (Pla et alii, 2001) and in the punic chamber of Sant Antoni de Portany (Juan-Tresserras, 1997). However, the last case, being from the same period we are studying in this paper, belongs to a complete different culture, where agriculture, as a basic productive source, had been largely established. At present, we are studying sedimentological deposits and contents of some pots that come from several sites in both islands: S'Illa des Porros, Punta des Patró, and Son Real, in Majorca, and Biniparratxet,. in Minorca (Juan-Tresserras, in study).

We need to avoid making wrong interpretations after the analysis of dental calculus in S'Aigua Dolça 5 (Juan-Tresserras, 1997), as there are some authors who interpret our results as associated with the consumption of "wild cereals" (Guerrero, in press). There is no evidence of wild cereals in the Western Mediterranean. In the study of dental calculus from the same individuals, the presence of siliceous is scarce and most of the morfotypes identified can be ascribed to the family of Poaceae, especially to the subfamily of Festucoideae in which cereals such as wheats and barleys are included. This can also be seen in S'Illa des Porros (Pla et alii, 2001) except for in some peninsular sites studied (Juan-Tresserras, 1997). Well-sieved flour from cereals hardly shows any remnants of covers such as phytoliths. Starch identified in S'Aigua Dolça 5 belongs basically to the family of Fagaceae, represented in this area of Majorca by different species of the Quercus gender which can produce edible nuts (Juan-Tresserras, 1997). Thus, we need to retain the real data to avoid general interpretational problems.

In the case of calculus from individuals studied from S'Illa des Porros, samples of phytoliths are not found. By contrast, starch is quite significant. There is a dominance of morfotypes of starch such as Triticeae, characteristics from wheat and barley, above identification of some other morfotypes of starch such as Vicea and Quercus (Juan-Tresserras, Malgosa & Pla, in study).

It would be highly important to study dental calculus, excrements, and stalls areas of domesticated animals to see whether or not there are subproducts of agriculture activities in their diet. This methodology has been useful to clarify strategies of the farming cycle (Juan-Tresserras et alii, 1993; Matsumoto & Sugawara, 1997), especially in iron age sites in the lower part of the river Ebre (Juan-Tresserras, unpublished). In the Majorcan case, research into Myotragus balearicus excrements has allowed the proof of the importance of Buxus in the diet of this mammal. At least, that is what occurs in the analysed samples from the Majorcan highlands, although, as the researchers point out, we need to have more analyses of samples from other areas. If Myotragus balearicus was able to assimilate box toxins, it could possibly have eaten anything (Alcover et alii, 1999).

INDIRECT EVIDENCE

One of the most traditional forms of evidence used in order to defend the presence of agricultural activities related to cereals is what is known in the bibliography as "sickle pieces". We are not aware of any in depth study of these pieces found in several Balearic archaeological sites. Cut traces of cereals are unmistakable at microscopic level (Ibáñez, González, 1996). However, agriculture of cereals does not require the use of sickle or knives for harvesting purposes (Zapata, Ibáñez, González, 1997), but the use of other techniques are difficult to record archaeologically. Nevertheless, we have to bear in mind the close relation between the harvesting techniques of cereals and morphologic and physiologic characteristics of each variety. In the Euroasian zone, almost ten different models of harvesting have been described (Sigaut, 1991).

Grinding tools have also traditionally been considered as indirect archaeological evidence of farming practices. However, the direct relation between these tools and the processed cereal must be considered. There is a need to develop a typology of these artefacts in the Balearic Islands and study their functionality carefully. Grinding stones and mortars can be both related to the preparation of flours and foods based on cereals, legumes and dried fruits, as well as to the processing of mineral materials (Juan-Tresserras et alii, 1996). There is also a lack of data about storage pits and large storage pots. Studies on them have produced quite positive results in peninsular sites (Juan-Tresserras 1997, 1998a, 1998b, 2000 a, 2000b, 2000c; Juan-Tresserras et alii, 1999).

Methodological advances and a new attitude developed towards the archaeological accounts imply a reinforcement of this mixed economic model. Nonetheless, at present, and despite it being too soon to give a categorical insight, according to the real archaeological evidence about the role of agriculture in the exploitation of talayotic systems, some authors tend to consider that it did not have a major role in the subsistence strategies (Guerrero 1997, Guerrero & Gornés 2000; Llull et al. 1999a).

PALEOANTHROPOLOGICAL CONTRIBUTION. PALEODIET.

Alimentation and resources used by a society can be analysed from the same human remains. As the skeleton reacts to changes suffered by the body and the environment, it reflects the food which reaches it, in particular, the amount and its composition. In order to reconstruct a diet from the study of human remains we count on the chemical composition of bones through analysis of trace elements and stable isotopes, and on the analysis of dental pathologies and dental wear (Malgosa, Subirà, 1996). Macroscopic study also informs us about diet through indicators of stress and malnutrition problems, due to scarcity or excess of some types of food (Cohen, Armelagos, 1984). Finally, the study of demographic patterns, growth, and bone pathologies can also give us some data about the nutritional state and overall health of population. Thus, it is possible to determine from the study of human remains, the diet of the individuals, and, to some extent, the alimentary resources used by the society, at least the ones directed to the alimentation of their own population, since the ones produced to trade escape this analysis.

In the Balearic Islands, many human remains have been recovered over time, most of them useable for paleodietary examination, but it is in Majorca, in the large talayotic cemeteries of Son Real and S'Illa des Porros, where it has been possible to develop population studies. Although some of these studies were published a decade ago, surprisingly they have not yet entered archaeological discussion about the basis of the talayotic economy.

From the analyses of all these anthropological evidences, we can conclude that there is not a dependence on animal farming resources. This is highly significant, since, at least, Son Real is not considered to represent the whole numbers of a talayotic society, but only the elite (Hernández, 1998). As a result, we can expect a strong bias towards a diet more oriented to animal resources (dairy products and meat), as it could be possible that a privileged diet existed, amongst other items of status, for a sector of the society. At the same time, these populations did not have nutritional problems resulting from shortage of food and wide use was made of local resources such as fish, seafood, fruit and nuts.

TRACE ELEMENTS

In the site of S'Illa des Porros, we quantified trace elements of human bones and compared them with figures obtained from animal bones found in the same site and hence brought by the same people who created and used the cemetery. Both kinds of remains have been affected by the same diagenetic and post-depositional processes. Comparison between the values obtained in each set of samples shows on the one hand:

(1)Not very high consumption of vegetables (low levels of Ba), maybe compensated by nuts (high levels of Mg) (acorns, for instance).

(2)Moderate consumption of meat (medium levels of Zn), probably from a marine origin (high levels of Sr) (Subirà 1989; Subirà, Malgosa 1991,1992a; Subirà et alii 1989). However, this last aspect can be confirmed after isotopic analysis we have underway.

On the other hand, we did not find any difference in relation to age and sex (Subirà, Malgosa 1992b), or distinguished by burial ritual (inhumation or cremation) (Subirà, Malgosa 1993, 1994).

In Son Real, with a sample of 32 individuals, we got similar results, a mixed diet with a poor composition in animal proteins (Pérez-Pérez 1990).

TEETH MARKS AND DENTAL PATHOLOGY

Teeth and their support are the part of the skeleton most directly related to diet, as they are in direct contact with food. This produces a wide range of reactions, variable according to dietary compounds and metabolism, which can be seen easily (Malgosa, Subirà 1996).

From the study of teeth in S'Illa des Porros we can conclude that macroerosion, in general, is moderate. Microerosion is abrasive, with very coarse surfaces, in anterior teeth, and moderate, with polished surfaces, in posterior teeth. There is also a high frequency of microflakes. This evidence can be interpreted as consumption of abrasive food, maybe some kind of flour with coarse residues, and acid compounds, such as the ones acorns provide (Pérez 1990).

DENTAL PATHOLOGIES

Dental pathologies found (Carrasco, Malgosa 1990; Carrasco, Malgosa, Campillo 1991) have been compared to other contemporary populations and they can reveal a balanced diet in which fibrous vegetables and fruits would be moderate (low frequency of decay). This does not exclude cereal and nut consumption, and meat would be present as well (high frequency of calculus). We can conclude that the population buried in S'Illa des Porros had good oral health in comparison to some other prehistoric or even modern populations (low frequency of caries, periodontitis, and dental enamel hypoplasia).

At present, some analysis of phytoliths and microremains of dental calculus is underway, which will give us more conclusive data about the sort of vegetal food consumed. Preliminary microscopic analysis shows that samples are not very rich in phytoliths. This fact can be interpreted as a product of a careful manipulation of cereals (extraction of covers), as result of a consumption of plants with a low amount of this sort of biominerals, or due to some diagenetic processes.

What we are finding is large quantities of starch, mainly cereals, which are found fashioning a paste (characteristic state of starch of some food after grounding it), also fragmented (with shows some kind of cooking or heating). This fact matches with presence of vegetal fibres, which in many cases contained starch in the inner part, which also indicates consumption of grounded food, such as, for instance, semolina. There also appears to be a high level of microcharcoal, interpreted as remnants of burning.

PALEODEMOGRAPHY

Demographic data can indicate pressures on the population due to the shortage of food resources. Infant and child mortality in S'Illa des Porros (calculated from JI: Bocquet-Appel and Masset, 1996) is 29 % for the 0-1 age-group, and 46.4 % for the 0-5 age-group (Alesan et alii, 1999). This level of infant mortality is in the range of hunter-gatherer societies, as well as agricultural and animal farming ones (Hewlett, 1991).

The relation m0-1 > m1-5 suggests a child mortality pattern with a lactation period that might not extend beyond 12 months. This matches the development of a more intensive agriculture, as it allows a reduction in the suckling period when introducing substitute and supplementary food at the time of weaning (Bentley et alii, 1993). The first growth spurt (infancy), according to the general pattern of skeleton growth, would take place in the second-third year of life (Alesan, 1990), which means weaning would have taken place long before, since the inherent difficulties of this process would not have allowed this growth spurt. Thus, an early weaning is possible, not much later than 12 months.

The total fecundity rate (TFR) would be of about 6 children per woman during her reproductive life (Malgosa et alii 2000). These fecundity figures are quite high, and more likely to appear in populations with a subsistence pattern based on intensive agriculture, or also on intensive gathering, but not in populations which are not dependent on agriculture. However, there is a high variability in these populations according to the fecundity values they have (Bentley et alii 1993, Sellen and Mace 1997),

GROWTH AND BONE PATHOLOGIES

Growth and development parameters (Alesan, 1990) are adjusted to the skeletal growth in some other ancient populations, and they do not show a slowdown. This also shows sufficient food supplies in this talayotic population.

Bone pathologies can be related, in some cases, to malnutrition, either by nutritional deficiencies or excess of some foods, according to diet, and eventually to the subsistence economy (Cohen, Armelagos 1984). Most of these pathologies are non-specific and the result of different stress factors, which end up as malnutrition by qualitative or quantitative means (Martin et alii 1985).

There are a large number of skulls which show cribra orbitalia, and some of them have hyperostosic osteoporosis. As the trace elements do not show any difference between individuals with and without cribra (Subirà, Alesán, Malgosa 1992), it does not seem to be related to a population nutritional deficiency. No problems such as rickets or hypovitaminosis have been found in S'Illa des Porros.

CONCLUSIONS

The global analysis of this zooarchaeological, paleobotanic, and anthropological evidence, apart from the historic and cultural evidence, allows us to conclude that in the prehistoric Balearic Islands a society based or specialised in herding as a basic engine of the economy and main supplement of diet did not exist. But what the same data show is that a mixed horticultural and herding economy existed, complemented with hunting, fishing and foraging. Their subsistence pattern would depend on an intensive use of close resources, such as marine resources and wild food from the island, together with maybe not intensive but undoubtedly extended agriculture and diversified animal farming, which would ensure the nutritional needs of these populations.

Moreover, anthropology provides us with exceptional data: those talayotic populations analysed did not have problems of subnutrition derived from resource scarcity.

From the archaeobotanic point of view, the first data provided through carpology are highly informative, although at present they refer to the last period of the talayotic culture (LIA). Current studies about starch from dental calculus also provide new data about the consumption of grounded cereals, probably cooked. We hope that phytoliths and trace studies in grounding stones, planned for current research projects, soon will inform us of their use. The gathering of wild vegetal food such as berries and acorns is also possible, as the ones attested in Punta des Patró (juniper berry, acorn, and Chamaerops humilis) (Juan-Tresseras, in study) or pistacia lentiscus, as reported in Ses Païsses, whose oil might be used (Allué, 2001).

Finally, adding some considerations not directly learned from bone analysis, we consider that the environmental limitations in an island the size of Majorca, and a Mediterranean context, make it more difficult to accept a specialised herding economy. We understand that, not only in the sense of mainland moving herders, who favoured seasonally diversified pastures and could establish cultural and trade contacts with basically agricultural people, but as an outstanding basis of the economic production on which subsistence and diet is founded. It seems that a subsistence economy of a mixed agriculture/herding type would have been the most efficient model. We can not totally forget the complement provided by foraging. In this sense, hunting game would have been largely limited by scarce if any faunal diversity. The little data about wild mammals would have to be studied carefully and prudently, as the importing of big wild mammals before romanization is still highly problematic, despite the scattered data from the talayotic period (some of it new, and unpublished). In this respect, we have to bear in mind that some products could reach the islands cured and not necessarily alive. Ultimately, we cannot forget that they probably made good use of birds, and, especially in coastal and subcoastal sites of marine resources; fishing, seafood gathering, marine mammals and birds. In the coastal sites, some remains of Monachus monachus have been found, and large amounts of diversified malacology (in spite of this, in many cases we would need to have previous taphonomic studies in order to see whether we are dealing with human contribution or not). Fish remnants are scarce in the studied sites, but that might be due more to a problem of conservation and recovery rather than lack of use in prehistoric times. However, whether the floated soils from S'Illa des Porros and Punta des Patró didn't contribute with significant data, high levels of strontium in the analysed human remains show possibly an important contribution of food of marine origin.

To conclude, an evolution and intensification of some economic strategies over almost the whole millennium considered possibly existed (and also at regional level). However, we would never be able to talk of an economic specialisation in any of the sectors and, on the contrary, the mixed character of the economy was maintained. Undoubtedly this occurred in the final period of the talayotic culture. Thus, around the second century BC, in a moment when the islands intensified trade relations with the Italic and Iberian Peninsulas, in some sites such as Turó de Ses Beies and Son Fornés (Lull et alii, 2001: 104) the rotation mill was introduced. One of these was discovered in a small chamber of Turó de Ses Beies, together with some amphorae, probably reused, and some other wares. This indicates a certain specialisation of this room and grounding activity on a large scale, bigger than the family scale documented by the grounding stones found in high numbers in the domestic rooms (Camps, Vallespir, 1998: 128-130). In this village, a furnace, probably a cooking one, was also located (Camps, Vallespir, 1985: 337). These are early signs of a major change in the insular society and economy, which led to the exploitation of new resources (such as vine and olive trees) in a probably less diversified and more intensive system, both in vegetal and in animal resources.

NOTES

(1) Donald Baden-Powell Quaternary Research Centre, Pitt Rivers Museum, University of Oxford. 60 Banbury Rd OX2 6PN Oxford (England, U.K.). Postdoctoral research grant from the Ministerio de Educación y Cultura E-mail: gasch@trivium.gh.ub.es

(2)SERP. Departament de Prehistòria, Història Antiga i arqueologia. Universitat de Barcelona. c/ Baldiri Reixac s/n 08028 Barcelona (Catalonia, Spain). E-mail: jnadal@campus.uoc.es

(3)Unitat dAntropologia. Departament de Biologia Animal, Biologia Vegetal i Ecologia. Facultat de Ciències. Universitat Autònoma de Barcelona. 08193 Bellaterra (Barcelona, Catalonia, Spain) E-mail: Assumpcio.Malgosa@uab.es

(4) Idem. E-mail: Alicia.Alesan@uab.es

(5) Universitat de Barcelona (Barcelona, Catalonia, Spain). E.mail: juan@trivium.gh.ub.es

(6)"La población talaiótica en nuestra isla vivía la vida primitiva de los pueblos pastores y agricultores. La cría de cerdos, ovejas y cabras eran sus principales ocupaciones. En las zonas fértiles de nuestra geografía (...) los poblados talaióticos se suceden a una distancia del uno del otro de unos tres kilómetros. Esa ordenación parece que formaba parte de una estructura económica, pues la situación de cada poblado dominaba una zona de pastos que podía estar separada de otra, ya perteneciente a la siguiente comunidad (...). Esa zona de pastos y sus rebaños suelen pertenecer a todos los habitantes del poblado, pues la posesión de las tierras y ganado es rara entre pueblos primitivos." (Cerdá 1971: 424).

Forzosamente su sistema económico, basado en la ganadería, tuvo que tener bastante importancia en aquellos conjuntos humanos. Las excavaciones, además, dentro de su pobreza, han dejado vislumbrar la existencia de ganado lanar y cabrío, junto con el porcino (...)" (Rosselló Bordoy 1968: 26).

However, in the same article we find:

"El problema de la agricultura talayótica carece de planteamiento por el momento. (...) Es preciso pensar en que la sociedad talayótica (...) vivió en una etapa de civilización en la cual la agricultura era sobradamente conocida (...) Conviene, pues, pensar que en el apogeo talayótico la agricultura era una fuente de riqueza de escasa importancia; sin embargo, la dispersión del hábitat prehistórico demuestra que la vida mallorquina en la antigüedad fue eminentemente rural (...) con una rudimentaria agricultura de tipo exhaustivo (?) (sic), hasta llegar al agotamiento de los campos, ganadería para el propio consumo de la tribu (...)" (Rosselló Bordoy 1968: 27-30)

(7) "Todo rastro de técnica agrícola queda en la más completa oscuridad." (Rosselló 1965: 148).

(8) "A partir del estudio de Uerpmann se deduce que la ganadería en SIllot fue la más importante fuente de recursos alimenticios, desempeñando un papel primordial el ganado ovino y el caprino (...) Los cerdos desarrollarían un papel complementario." (Rosselló, 1979: 146).

(9) "La ganadería parece ser, según las evidencias que poseemos, la actividad económica fundamental de los habitantes de Son Fornés." (Gasull, Lull, Sanahuja, 1984: 83).

"El análisis de los restos faunísticos recuperados en Son Fornés sugiere una economía agraria tipo granja a partir de un rebaño mixto que garantiza estabilidad alimentaria, así como una diversidad funcional de las especies (Gasull, Lull, Sanahuja, 1984: 83) (...) el talaiot más que la casa del jefe parece ser la torre de control del espacio destinado a los rebaños de la comunidad (Fernández-Miranda, 1991: 44).

"Evidentemente en la época talayótica el paisaje estaría ya alterado, pues la economía era básicamente ganadera (...) Como hemos mencionado, la utilización de este paisaje por una cultura ganadera implica una alteración del mismo. " (Ferrés, 1984: 132).

"Podemos definir el substrato talayótico como fundamentalmente ganadero, siendo abundantes los restos óseos tanto a nivel cuantitativo como cualitativo(...) En la fase postalayótica se produce un desarrollo importante de la agricultura, a la cual se supedita el mantenimiento de la base ganadera por la influencia del substrato talayótico." (Mayoral 1984: 1303).

"Ahora bien, en Son Fornés, y seguramente en muchas comunidades talayóticas, el recurso principal de subsistencia era básicamente la ganadería" (Guerrero 1995a: 20, 1995b: 298)

"Sólo apuntar que mantenemos la misma línea argumental básica de considerar a la cultura talayótica como una organización social de rasgos tribales y una subsistencia fundamentalmente ganadera (...)" (Guerrero, 1999: 54).

(11) "Aunque los elementos arqueológicos son escasos en relación a otros sistemas primarios de producción debemos dejar abierta la posibilidad de que la dieta alimenticia se completase, en Son Fornés, con los productos de una agricultura subsidiaria y de la recolección. La caza, hasta este momento, no está registrada." (Gasull, Lull, Sanahuja, 1984: 83).

"En resum, la comunitat talaiòtica que va viure a Son Fornés mantenia una economia de subsistencia, en la qual la ramaderia exercia un paper fonamental. Aquesta activitat era complementada amb la collita de plantes silvestres i, possiblement, amb algun tipus de cultiu a petita escala." (Lull et alii 2001: 54).

"No cabe descartar tampoco la posibilidad de que se cultivasen terrenos llanos, que son los que en la actualidad se emplean para estos fines (...) La utilización de todas estas posibilidades y la presencia entre los restos encontrados no sólo de varias especies de animales domésticos (cerdo, cabra, buey, caballo) sino incluso de dos razas para algunas las especies, nos sugiere un panorama de recursos aunque no excesivamente abundante, sí al menos muy diversificado, el cual, añadiéndoles la recolección (frutos, tubérculos), podría asegurar el suministro de alimentos durante todo el año." (Ferrés, 1984: 132).

"En un espacio geográfico donde la caza resulta irrelevante y no se tienen indicios de recolección habitual de productos de origen marino, no resulta descabellado pensar en una economía mixta pero con predominio evidente de la ganadería. Igual ocurre como se ha visto, en Mallorca, al menos en el único caso bien documentado, el poblado talayótico de Son Fornés." (Fernández-Miranda, 1991: 48).

"Cuantitativamente los restos óseos expresan una ganadería limitada, cuya relación con la agricultura es estrecha por cuanto el acondicionamiento y extensión de terrenos para dedicarlos a aprovechamiento agrícola reduce la zonas de pasto." (Mayoral 1984: 1304).

"La actividad económica, básicamente ganadera, debía complementarse con una agricultura subsidiaria." (Guerrero, 1995a: 19).

(12) "Un modelo subsistencial pastoril, en un medio poco apto para una agricultura primitiva como es el de Menorca, pudo constituir la base del desarrollo económico a la largo de los tiempos pretalayóticos e incluso durante una buena parte del talayótico antiguo." (Fernández-Miranda, 1991: 48).

"Esta especialización económica en el pastoreo continúa y con seguridad se intensifica a lo lrgo del Bronce final". (Guerrero, 1997: 112-113).

(13) "Por ello, aunque los cerdos sean los animales básicos en cuanto a ingresos cárnicos, se observa que los pueblos que priman su domesticación deben recurrir al sacrificio del segmento excedentario de la manada realizando grandes matanzas en acontecimientos especiales que cobran por lo general caracteres de manifestación social (Watson, 1977)." (Gasull, Lull, Sanahuja, 1984: 83).

"Suponemos que la tierra y las zonas de pastos eran comunales y quizás una de las funciones del o de los "jefes" era la de controlar las tierras de pastos en relación a los poblados vecinos." (Gasull, Lull, Sanahuja, 1984: 89).

"¿Cabe pensar en la introducción de algún tipo de arado a lo largo del talayótico antiguo y con él la ruptura de un sistema económico mucho más igualitario basado preferentemente en la ganadería ovicaprina?" (Fernández-Miranda, 1991: 48).

"Si utilizamos el paradigma de los pueblos ganaderos del Este de África, expuesto por Lincoln (1991: 52), toda esta iconografía pudo estar ligada a la práctica de ofrendas sacrificiales, cuyas normas litúrgicas quedan fijadas en el mito del primer sacrificio y la estatuaria tauromorfa directamente a la teofanía del soberano celeste o dios supremo con quien la casta sacerdotal estaría en más directa conexión." (Guerrero, 1999: 103).

(14) "Su clima que actúa de modo singular en los recursos económicos." (Rosselló Bordoy 1968).

(15)"Como resumen podemos pensar que los hombres talayóticos vivieron dentro de un marco agrícola-pastoril impuestos por el medio geográfico." (Rosselló, 1979: 148).

(15) "Como hemos mencionado, la utilización de este paisaje por una cultura ganadera implica una alteración del mismo. Probablemente los talayóticos combinaban la utilización directa de algunas comunidades para pastoreo (bosques de olmos, encinares, vegetación halófila), como se hace aún en la actualidad en muchas zonas del Mediterráneo, con la destrucción de parte de la vegetación por medio de incendios para obtener pastos, práctica muy extendida en todas las culturas ganaderas." (Ferrés, 1984: 132).

(16) "En un medio de este tipo se desenvuelven bien comunidades de pocos individuos (..) en época pretalayótica, a partir del ganado, la recolección de los frutos silvestres y tal vez una rudimentaria agricultura cerealista (...)No parece que el incremento de la agricultura pueda asegurar la alimentación a un número de individuos cada vez mayor en cada unidad de poblamiento (...) sólo la ganadería parece ser capaz de garantizar su alimentación en particular la de ovejas y cabras (...) los primeros recintos talayóticos como Trebalúger o talaiots aislados con habitaciones adosadas pudieron surgir a consecuencia de un apreciable incremento del ganado que hizo posible garantizar la subsistencia a un grupo mayor de personas (...) el incremento de la población no dependería de poseer tierras cultivables amplias sino extensiones para el ganado. La situación de grandes poblados como Torre den Gaumés (...) corrobora la vigencia de un modelo económico basado en el ganado como principal estrategia alimentaria, al margen de que otras aglomeraciones igualmente relevantes como puede ser Torralba den Salort, elijan espacios donde esa fórmula subsistencial se pueda combinar mejor con la agricultura y quizá incluso el regadío." (Fernández-Miranda, 1991: 48).

"Además, como ya he comentado, la situación de algunos poblados relevantes de la fase final de la cultura talaiótica no refleja una especial elección de terrenos aptos para el cultivo y sí para la ganadería." (Fernández-Miranda, 1991: 49).

(17) This author includes himself in the proposal that says " la mayoría de los autores defienden que se trataba de la ganadería, básicamente de oveja y cabra." (Aramburu-Zabala, 1998: 90).

(18) Mascaró (1968: 678) pointed out this when writing: "sabemos cierto, por los numerosos hallazgos efectuados, principalmente en excavaciones arqueológicas, que el hombre prehistórico balear se aprovechaba del cerdo (...), de la oveja y de la cabra".

(19) "Los símbolos taurinos en líneas generales han aludido siempre, ayer como hoy, dentro de la cultura agrícola y ganadera inaugurada en el Neolítico al poder, a la fertilidad, a la fecundidad (...) la humana, que asegura la pervivencia y la afirmación de los clanes; la animal, que proporciona medios de subsistencia y la fertilidad de los campos que está a la base de las dos anteriores". (Llompart, 1970: 257-258).

(20)footnote* Despite the fact that they are mentioned in the bibliography (Rosselló, 1979: 147), when they have been recovered, they have usually not been published, nor studied under any aspect (technological, petrologic, use marks, residues, and phytolits).

(21) Experienced in Biniparratxt Petit excavation, 2001 season.

(22)Antes de reconstruir una economía de subsistencia basada exclusiva o predominantemente en la ganadería, debemos admitir la necesidad de muestreos a gran escala para la flotación sobre contextos domésticos bien conservados." (Chapman, Grant 1995: 29).

(23) "L'anàlisi combinada de les hipoplàsies de l'esmalt i lhiperostosi osteoporòtica com indicadors de pressió ambiental, la reconstrucció de la dieta estimada a partir danàlisis químiques i patologies dentals i els paràmetres demogràfics establerts (...), permeten proposar que la població dEs Càrritx disposava dun règim alimentari equilibrat basat en una dieta variada i de tipus mixte" (Rihuete,2000, 383)

(24) "La cría de animales exclusivamente para el aprovechamiento cárnico no parece haber sido el principal objetivo de la ganadería practicada en Son Ferrandell, lo que sugiere que los alimentos vegetales debieron aportar la mayor parte de la dieta." (Chapman i Grant 1995: 36).

"Tan sólo los relativamente abundantes restos de marisco proporcionan alguna precisión (...) más de la mitad de los moluscos pertenecen a dos especies comestibles, Cerastoderma Glaucum y Monodonta Turbinata (...) No dudamos que una de las bases de su alimentación sería el pescado." (Camps, Vallespir 1985: 341).

(25) The name used by Uerpmann to refer to the assemblages that might be mixed up with remains of following occupations. We added it because of its similarity to the good preroman (talayotic) assemblage. Possibly these similarities are due to the fact that the transition samples are basically from a talayotic origin.

ACKNOWLEDGEMENTS

We would like to express our gratitude to:

Dr. W.H. Waldren who invited us to participate in the 4th International Conference in Deià, and who has always shown a real interest in the several works developed around the cemeteries of Son Real and S'Illa des Porros.

Damià Ramis and Tomeu Salvà for their bibliographic advice, and valuable corrections and comments on the first version of this text.

Naomi Barbour, for the revision of the English version.

Needless to say, only the authors, not they, are responsible for the mistakes that still might be found.

REFERENCES

ALBIZURI, S.; NADAL, J. ,
1999. 'Aprovechamiento y producción animal en época ibérica. Consideraciones generales económicas», Limes, 6-7: 41-52.

ALCOVER, J.A.; PÉREZ-OBIOL, R.; YLL, E.I.; BOVER, P. ,
1999. 'The diet of Myotragus balearicus Bate 1909 (Artiodactyla: Caprinae), an extinct bovid from the Balearic Islands: evidence from coprolites», Biological Journal of the Linnean Society: 57-74.

ALESAN, A. ,
1990. 'Estudi d'una població subadulta de l'Edat del Ferro: Demografia, Antropometria i Creixement». Master's dissertation. Universitat Autònoma de Barcelona. Unpublished.

ALESAN, A.; MALGOSA, A.; SIMÓ, C. ,
1999. 'Looking Into the Demography of an Iron Age Population in the Western Mediterranean. I. Mortality», American Journal of Physical Anthropology, 110: 285-301.

ALLUÉ, E.,
2001. 'Informe de l'anàlisi antracològica del jaciment de Ses Païsses (Mallorca)". Unpublished.

AMORÓS, L. R. (1952): "La Edad del Bronce en Mallorca». Panorama Balear, 23 (Palma).

ARAMBURU-ZABALA, J. ,
1998. El patrón de asentamiento de la cultura talayótica de Mallorca. Palma

BENTLEY, G.R.; GOLDBERG, T.; JASIENSKA, G. (1993. 'The Fertility of Agricultural and Non-Agricultural Traditional Societies», Population Studies, 47: 269-281.

BOCQUET-APPEL, J.P.; MASSET, C.,
1996. 'Paleodemography: expectancy and phalse hop', American Journal of Physical Anthropology, 99: 571-583.

BURJACHS, F.; PÉREZ-OBIOL, R.; ROURE, J.M.; JULIÀ, R.,
1994. 'Dinámica de la vegetación durante el Holoceno en la isla de Mallorca», in: Mateu, I.; Dupré, M.; Güemes, J.; Burgaz, M.E. (eds.): Trabajos de palinología básica y aplicada. X Simposio de Palinología (X: 1994: València),199-210.

CALVO, M.; COLL CONESA, J. GUERRERO, V.M. (1997. 'El dolmen de s'Aigua Dolça (Mallorca)», Revista de Arqueología, 191:18-29.

CAMPS, J.; VALLESPIR, A.,
1974. 'La estación del Turó de Ses Beies' (Calviá)", in: Prehistoria y arqueología de las Islas Baleares. Symposium de Prehistoria Peninsular (VI: Barcelona: 1974), 101-114.

CAMPS, J.; VALLESPIR, A.,
1976. 'Excavaciones en Es Turó de Ses Beies", Noticiario Arqueológico Hispánico, 5 (Madrid): 399-404.

CAMPS, J.; VALLESPIR, A.,
1985. 'La vida cotidiana en una comunidad prehistórica mallorquin'", in: Jornada d'estudis històrics locals (3: Palma: 1985). Palma, 331-343.

CARRASCO, T.; MALGOSA, A.,
1990. 'Patología oral y dieta», Dynamis, 10: 17-37.

CARRASCO, T.; MALGOSA, A.; CAMPILLO, D.,
1991.'Patologías dentarias de la Necrópolis Talayótica de "S'Illot des Porros" (Mallorca)», in: Actas del Congreso Nacional de Historia de la Medicina (IX: Zaragoza: 1989), 1189-1199.

CERDÁ, D.,
1971. 'Economía antigua de Mallorca», a: Mascaró, J. (ed.), Historia de Mallorca, Palma, 417-448.

CHAPMAN, R.; GRANT, A.,
1995. 'Talayot 4, Son Ferrandell Oleza: Problemas de los procesos de formación, función y subsistencia», Revista d'Arqueologia de Ponent, 5 (Lleida): 7-52.

COHEN, M.N.; ARMELAGOS, G.J. (Eds),
1984. Paleopathology at the Origins of Agriculture. Orlando: Academic Press Inc.

COLL, J.; MAZAIRA, L.; RIUTORT, S. (1984. 'Evolución del hábitat durante la prehistoria y la antigüedad en el término municipal de Alcúdia', Arqueología Espacial, 2 (Teruel): 11-129.

DA MOTA, P.J..
1988. Torres and Torrentes: a new approach to the settement pattern of Mayorcan talayots. Master's dissertation. University of Reading, Unpublished.

DA MOTA, P.J.,
1988. 'Torres and Torrentes: a Spacial Analysis of Mallorcan Talayots", in: K. LILLIOS (ed.), The Origins of Complex Societies in Late Prehistoric Iberia. Michigan (International Monographs in Prehistory: Archaeological Serie, 8), 121-139.

ELIADE,M.,
1978. Historia de las creencias y de las ideas religiosas (vol 1: de

la Prehistoria a los misterios de Eleusis). Ed. Cristiandad. Madrid. Cap. 2.
ELIADE,M.,
1981. Tratado de Historia de las religiones. Ed. cristiandad. Madrid.
ENSEÑAT, C. ,
1971. 'Excavaciones en el navetiforme alemany, Magalluf, calvià", Noticiario Arqueológico Hispánico, 15. Madrid
ESTÉVEZ, J. (1984. 'Restos alimentarios e industria ósea de Son Fornés (Mallorca)», in: Gasull, P.; Lull, V.; Sanahuja, M.E.: Son Fornés: La fase talayótica. Ensayo de reconstrucción socio-económica de una comunidad prehistórica de la isla de Mallorca. Oxford: British Archaeological Reports (International Series, 209), 138-178.
FERNÁNDEZ MIRANDA, M.,
1978. Secuencia cultural de la Prehistoria de Mallorca. Madrid: C.S.I.C, Biblioteca Arqueológica Hispánica, XV.
FERNÁNDEZ MIRANDA, M. ,
1991. 'La transición hacia la Cultura Talayótica en Menorca», Trabajos de Prehistoria, 48, 37-50.
FERNÁNDEZ-MIRANDA, M.; ENSEÑAT, C.; ENSEÑAT, B.,
1971. 'El Poblado de Almallutx (Escorca, Baleares)». Excavaciones Arqueológicas en España, 73.
FORNÉS, J.; SALVÀ, B.,
2001. 'La cultura naviforme de Mallorca y su contexto cronocultural en el Mediterráneo", in: Actas del Congreso Español de Estudios del Próximo Oriente. Oriente y Occidente. De las primeras sociedades productoras a los comienzos de la romanización (II:Cádiz-Puerto de Santa María: 2001).
GASULL, P.; LULL, V.; SANAHUJA, M.E.,
1984. Son Fornés: La fase talayótica. Ensayo de reconstrucción socio-económica de una comunidad prehistórica de la isla de Mallorca. Oxford: British Archaeological Reports (International Series, 209).
GIMBUTAS, M.,
1989. The language of the Goddess. Tames and Hudson. Ed. Londres.
GUERRERO, V. M.,
1995. 'Colonos, caciques y mercenarios. Una aproximación al contexto histórico del intercambio desigual en las Baleares», in: Jornades d'Arqueologia Fenicio-Púnica (X: Eivissa: 1995).
GUERRERO, V. M.,
1995b. 'Una sociedad en estado de jefatura (Chiefdoms). La cultura Talayótica Balear», in: Waldren, W.H. (ed.), Deya Conference of Prehistory. Ritual, Rites and Religion (III: 1993: Deià), Oxford: British Archaeological Reports, (International Series, 611, II), 295-313.
GUERRERO, V.M.,
1997. Cazadores y pastores en la Mallorca Prehistórica. El Tall. Palma.
GUERRERO, V. M.,
1999. Arquitectura y poder en la prehistoria de Mallorca. El Tall. Palma.
GUERRERO, V. M. (coord) ,
in press. El dolmen de S'Aigua Dolça.
GUERRERO, V.M.; GORNÉS, S. (coors.),
2000. Colonización humana en ambientes insulares. Interacción con el medio y adaptación cultural, Palma: Universitat de les Illes Balears.
HERNÁNDEZ, J.,
1998. 'Son Real. Necrópolis talayótica de la edad del hierro. Estudio arqueológico y análisis y análisis social». Arqueomediterrània 3, II (Universitat de Barcelona). Barcelona.
HERNÁNDEZ, J.; SANMARTÍ, J. ,
1999.'El santuari de Sa Punta des Patró a l'àrea cultual i funerària de Son Real (Santa Margalida, Mallorca). Avenç dels resultats», Mayurqa, 25: 113-138.
HEWLETT, B.S.,
1991. Demography and childcare in preindustrial societies, Journal of Anthropological Research, 47, 1:1-37.
IBÁÑEZ, J.J.; GONZÁLEZ, J.E.,
1996. 'El uso de útiles en sílex de los niveles neolíticos de la cueva de «Los Murciélagos» (Zuheros, Córdoba). Primeros resultados», in: Congrés del Neolític a la Península Ibèrica. Formació i implantació de les comunitats agrícoles (I: Gavà: 1996), Rubricatum, 1 (Museu de Gavà), 169-175.
IRIARTE, M.J.; ARRIZABALAGA, A.,
1995. 'Aportación de la palinología al conocimiento de la primera economía de producción en Euskal Herria", in: Primeros agricultores y ganaderos en el Cantábrico y alto Ebro, Cuadernos de Sección. Prehistoria-Arqueología, 6 (Eusko Ikaskuntza, San Sebastián):141-153.
JUAN-TRESSERRAS, J.,
1997. Preparación y procesado de productos vegetales. Aportaciones del estudio de fitolitos, almidones y lípidos en yacimientos arqueológicos prehistóricos y protohistóricos del cuadrante NE de la Península Ibérica. Doctoral Dissertation. Universidad de Barcelona. Unpublished.
JUAN TRESSERRAS, J.,
1998a. 'La cerveza prehistórica: investigaciones arqueobotánicas y experimentales", in: Maya, J.L.; Cuesta F.; López-Cachero, J.L. (eds.), Genó: un poblado del Bronce Final en el Bajo Segre (Lleida). Barcelona: Publicacions de la Universitat de Barcelona-SERP-San Miguel, 239-252.
JUAN-TRESSERRAS, J.,
1998b. 'El cultivo de la vid y la elaboración del vino en la Península Ibérica en la Antigüedad. Aportaciones de los análisis de residuos», in: Col.loqui Internacional d'Arqueologia Romana. El Vi a l'antiguitat: economia, producció i comerç al Mediterrani Occidental. Actes (II: Badalona:1998), Monografies Badalonines, 14 (Museu de Badalona), 87-92.
JUAN-TRESSERRAS, J.,
2000a. 'La cerveza: un producto de consumo básico entre las comunidades ibéricas del N.E. Peninsular», in: III Reunió d'Economia en el Món Ibèric (III: València: 2000), Saguntum-Plav, Extra 3: 139-145.
JUAN-TRESSERRAS, J.,
2000b. 'Estudio de residuos vegetales conservados en recipientes y asociados a material de molienda en yacimientos de la Edad del Hierro del NE de la Península Ibérica», in: Buxó, R.; Pons E. (dirs.), Els productes alimentaris d'origen vegetal a l'Edat del Ferro de l'Europa Occidental: de la producció al consum. Actes del XXII Col.loqui Internacional per a l'Estudi de l'Edat del Ferro, Monografia del MAC-Girona, 18: 371-377.
JUAN-TRESSERRAS, J.,
2000c. 'La arqueología de las drogas en la Península Ibérica. Una síntesis de las recientes investigaciones arqueobotánicas», Complutum, 11: 261-274.

JUAN-TRESSERRAS, J.,
2000d. 'Estudio de contenidos en cerámicas ibéricas del Torrelló de Almazora (Castellón)», AespA, 73: 103-104.
JUAN-TRESSERRAS, J.; LALUEZA C.; NADAL, J.,
1993. 'Estudio de fitolitos en dientes como indicadores de patrones de dieta y su aplicación en arqueologia», Trabalhos de Antropologia e Etnologia, 33,1-2: 471-485.
JUAN-TRESSERRAS, J.; ECHAVE, C.; ALBERT, R.M.,
1996. 'El procesado de vegetales y la interpretación funcional del utillaje neolítico de molido y triturado en la Península Ibérica», in: Congrés del Neolític a la Península Ibèrica. Formació i implantació de les comunitats agrícoles, (I: Gavà: 1996), Rubricatum, 1 (Museu de Gavá), 201-206.
JUAN-TRESSERRAS, J.; MAYA, J.L.; LÓPEZ-CACHERO, X.,
1999.'Primeros análisis de contenidos en recipientes cerámicos de Genó (Aitona, Segrià, Lleida)», in: European Meeting on Ancient Ceramics. Archaeological and archaeometric studies (IV: Andorra, 1997), 182-199.
LILLIU; G.,
1962-63. 'Cenno sui piu recenti scavi del villagio talaiotico di Ses Païsses ad Artà-Maiorca (Baleari)", Studi Sardi, XVIII, 22-52.
LLOMPART, G.,
1965. 'Prehistòria i protohistòria de Mallorca». Monografies de l'Obra Cultural Balear, 3 (Palma).
LLOMPART, G.
1971. 'La religión del hombre prehistórico en Mallorca», in: Mascaró Pasarius, J. (coord.), Historia de Mallorca, Palma, 225-288.
LULL, V.; MICÓ, R.; RIHUETE, C.; RISCH, R.,
1999a. La Cova des Càrritx y la Cova des Mussol. Ideología y sociedad en la prehistoria de Menorca. Consell Insular de Menorca.
LULL, V.; MICÓ, R.; RIHUETE, C.; RISCH, R.,
2001. ' La prehistòria de les illes Balears i el jaciment arqueològic de Son Fornés (Montuïri, Mallorca). Montuïri.
MALGOSA, A. (1992): La població talaiòtica de Mallorca. Barcelona: Institut d'Estudis Catalans, Arxius de les Seccions de Ciències, Secció de Ciències Biològiques.
MALGOSA, A.; SUBIRÀ, M.E.,
1996. 'Antropologia i dieta: metodologies per a la reconstrucció de l'alimentació de les poblacions antigues», Cota-Zero, 12 (Vic): 15-27.
MALGOSA, A.; ALESAN, A.; SIMÓ, C.,
2000. 'Insights into the mortality and fertility of Majorca prehistory», in: Congress of the European Anthropological Association (XII: Cambridge: 2000), Poster.
MARTIN, D.L.; GOODMAN, A.H.; ARMELAGOS, G.J. ,
1985. 'Skeletal Pathologies as Indicators of Quality and Quantity of Diet», in: Gilbert, R.I.Jr; Mielke, J.H., The analysis of prehistoric diets, Orlando: Academic Press Inc., 227-279.
MARTÍNEZ PÉREZ-PÉREZ, A.,
1990. Evolución de la dieta en Cataluña y Baleares desde el Paleolítico hasta la Edad Media a partir de restos esqueléticos. Doctoral Dissertation. Universitat de Barcelona. Unpublished.
MASCARÓ PASARIUS, J.,
1968. Corpus de toponimia de Mallorca. Palma.
MATSUMOTO, H.; K. SUGAWARA,
1997. 'Use of opal phytoliths as a marker for the identification of plant species and herbage intake ingested by grazing animals. 1. Identification of plant species by opal phytoliths and changes in opal phytolith composition at growth stages», Grassland Science, 43, 3 (Tohoku, Japan): 249-257.
MAYORAL, F.,
1984. 'La fase posttalayótic mallorquina: periodización y dinámica económica y social», in: Waldren, W.H., R. W. CHAPMAN, J.G. LEWTHWAITE, R.C. KENNARD (ed.), The Deya Conference of Prehistory. Early settlement in the Western Mediterranean Islands and their peripheral areas (I: Deià: 1982), Oxford: British Archaeological Reports (International Series, 229, IV), 1299-1312.
NADAL, J.,
1998. 'Informe sobre los restos faunísticos correspondientes al yacimiento de Son Real», in: Hernández, J., Son Real. Necrópolis talayótica de la edad del hierro. Estudio arqueológico y análisis y análisis social, Arqueomediterrània 3, II (Universitat de Barcelona): 219-222.
NADAL, J.,
2000. Informe per a l'Estudi de les restes faunístiques procedents del jaciment S'Illa des Porros (Santa Margalida, Mallorca). Unpublished.
PANTALEÓN-CANO, J.; PÉREZ-OBIOL, R.; YLL, E.I.; ROURE, J.M.,
1996a. 'Significado de Pseudoschizaea en las secuencias sedimentarias de la vertiente mediterránea de la Península Ibérica e Islas Baleares», in: Estudios palinológicos, (Universidad de Alcalá de Henares): 101-105.
PANTALEÓN-CANO, J.; ROURE, J.M.; YLL E.I.; PÉREZ-OBIOL R. ,
1996b. 'Dinámica del paisaje vegetal durante el Neolítico en la vertiente mediterránea de la Península Ibérica e Islas Baleares», Rubricatum, 1:29-34.
PÉREZ, V.,
1990. Técnica de análisis de los patrones de microdesgaste dentario (M.O.) como indicadores de dieta. Master's dissertation, Universitat Autònoma de Barcelona, Unpublished.
PÉREZ-OBIOL, R.; YLL, E.I.; PANTALEÓN-CANO, J.; ROURE, J.M.,
1996. 'Historia de Buxus y Corylus en las Islas Baleares durante el Holoceno», in: Ramil-Rego, P.; Fernández Rodríguez, C.; Rodríguez Guitián, M. (eds.), Biogeografía Pleistocena-Holocena de la Península Ibérica, Xunta de Galicia, 87-97.
PÉREZ-OBIOL, R.; YLL, E.I.; PANTALEÓN-CANO, J.; ROURE, J.M.,
2000. 'Evaluación de los impactos antrópicos y los cambios climáticos en el paisaje vegetal de las Islas Baleares durante los últimos 8000 años», in: Guerrero, V.; & Gornés, S. (eds.), Prehistoria de Baleares, 73-89.
PLA, A.; JUAN-TRESSERRAS, J.; PINILLA, A.; MALGOSA, A.,
2001. 'Análisis de microresiduos indicadores de la dieta en poblaciones antiguas: La necrópolis de S'Illot des Porros», in: Congreso de la SEAB (XII: Bellaterra: 2001), Poster.
RIHUETE HERRADA, C.,
2000. 'Dimensiones bioarqueológicas de los contextos funerarios. Estudio de los restos humanos de la necrópolis prehistórica de la Cova des Càrritx (Ciutadella, Menorca). Tesi Doctoral. Universitat Autònoma de Barcelona.
RODRÍGUEZ, A. L.,
1995. 'Aproximación estadística al paisaje humano en la prehistoria de Mallorca», Complutum, 6:167-192.

ROMERO, R.,
1971. 'Estudio de los restos de fauna de las habitaciones significativas. Apéndice II», in: Fernández-Miranda, M.; Enseñat, C.; Enseñat, B., El Poblado de Almallutx (Escorca, Baleares), Excavaciones Arqueológicas en España, 73, 131-133.
ROSSELLÓ, G.,
1965. 'Una aproximación a la prehistoria de Mallorca», Ampurias, 25 (Barcelona).
ROSSELLÓ BORBOY, G. ,
968. 'Aspectos económicos de la cultura talayótica», in: Reunión de Historia de la Economía Antigua de la Península Ibérica, (I: Valencia: 1968), 23-30.
ROSSELLÓ BORBOY, G.,
1972. 'La Prehistoria de Mallorca. Rectificaciones y nuevos enfoques al problema", Mayurqa, 22, VII (Palma): 115-143.
ROSSELLÓ BORBOY, G.,
1979. La cultura talayótica en Mallorca. Bases para el estudio de sus fases inciales. Palma (second edition).
SALVÀ, T.,
1992. 'El patró d'assentament de la prehistòria a la romanització de Felanitx", in: La prehistòria de les Illes de la Mediterrània Occidental: Jornades d'estudis històrics locals (X: Palma: 1991), Palma: Institut d'estudis Baleàrics, 407-418.
SELLEN, D.W.; MACE, R.,
1997. 'Fertility and Mode of Subsistence: A Phylogenetic Analysis», Current Anthropology, 38, 5: 878-889.
SIGAUT, F.,
1991. 'Les techniques de récolte des grains. Identification, localisation, problèmes d'interprétation», in: Cauvin, M.C., Rites et rythmes agraires, Travaux de la Maisson de l'Orient, 20 (Maisson de l'Orient Médieterranéen, Lyon), 31-43.
SUBIRÀ, M.E.,
1989. Estudi d'elements traça en la població talaiòtica de «S'Illot des Porros» (Mallorca). Doctoral Dissertation. Universitat Autònoma de Barcelona. Unpublished.
SUBIRÀ, M.E.; MALGOSA, A.,
1991. 'Some alkaline earth metal content of human bones at the site of S'Illot des Porros (Iron Age, Mallorca, Spain)», International Journal of Anthropology, 6,4 (Firenze): 225-231.
SUBIRÀ, M.E.; MALGOSA, A.,
1992. 'Multi-element analysis for dietary reconstruction at a Balearic Iron Age Site», International Journal of Osteoarchaeology, 2: 199-204.
SUBIRÀ, M.E.; MALGOSA, A.,
1992b. 'Trace element contents in the bone as an age and sex indicator. A case of the necropolis of S'Illot des Porros (Spain)», International Journal of Anthropology, 7, 2 (Firenze): 65-70.
SUBIRÀ, M.E.; MALGOSA, A.,
1993. 'The effect of cremation on the study of trace elements», International Journal of Osteoarchaeology, 3: 115-118.
SUBIRÀ, M.E.; MALGOSA, A.,
1994. 'Incineración e inhumación en el estudio de elementos traza», in: Bernis, C. (ed), Biología de Poblaciones humanas. Problemas metodológicos e interpretación ecológica, Madrid, 89-97.
SUBIRÀ, M.E.; ALESAN, A.; MALGOSA, A.,
1992. 'Cribra orbitalia y déficit nutricional. Estudios de elementos traza», Munibe, supl. 8 (San Sebastián): 153-158
TARRADELL, M.; HERNÁNDEZ, J.,
1998. 'Son Real. Necrópolis talayótica de la edad del hierro. Catálogo e inventarios». Arqueomediterrània 3, I (Universitat de Barcelona). Barcelona
UERPMANN, H.P.,
1971. 'Dies tierknochenfunde aus der Talayot-siedlung von S'Illot (San Lorenzo, Mallorca)», Studien über frühe tierknochenfunde von del Iberrischen Halbinsel, 2: 110.
VALLESPIR, A.; PROHENS, J.M.; ORFILA, M.; MERINO, J.,
1987. 'Los yacimientos de Santa Ponça (Calvià)", Mayurqa, 21 (Palma): 1-30.
WALDREN, W.H.,
1982. Balearic Prehistoric Ecology and Culture: The excavation and study of Certain Caves, Rock Shelters ans Settlements. Oxford: British Archaeological Reports (International Series, 149).
WALDREN, W. H. ,
1998. The Beaker Culture of the Balearic Islands. An inventory of evidences from caves, rock shelters, settlements and ritual sites. Oxford: British Archaeological Reports (International Series, 709-Western Mediterranean Series, 1).
YLL, E.I.,
1984. 'Análisis polínico de Son Fornés», in: Lull, V.; Sanahuja, E.; Gasull, J., Son Fornés I: La Fase Talayótica. Ensayo de reconstrucción socio-económica de una comunidad prehistórica de la isla de mallorca. Oxford: British Archaeological Reports (International Series, S209), 133-135.
Yll, E.I.,
1995. Estudi de l'evolució de la vegetació i el clima durant el Tardiglacial i el Postglacial a partir d'anàlisis pol.líniques del delta de l'Ebre i de Menorca. Publicacions de la Universitat Autònoma de Barcelona.
YLL, E.I.; PANTALEÓN-CANO, J.; ROURE, J.M.,
1999. 'Cambio climático y transformaciones del medio durante el Holoceno en las Islas Baleares», Saguntum, extra 2 : 45-51.
YLL, E. I.; PÉREZ-OBIOL, R & JULIÀ, R.,
1994. Vegetational change in the Balearic Islands (Spain) during the Holocene. Historical Biology, Vol. 9 :83-89.
YLL, E. I.; PÉREZ-OBIOL, R.; PANTALEÓN-CANO, J.; ROURE, J. M.,
1995, 'Dinámica del paisaje vegetal en la vertiente mediterránea de la Península Ibérica e Islas Baleares desde el Tardiglaciar hasta el presente», in: Aleixandre, T.; Pérez-González, A. (eds.), Reconstrucción de paleoambientes y cambios climáticos durante el Cuaternario, Monografías del Centro de Ciencias Medioambientales-CSIC, 3 (Madrid), 319-328.
YLL, E. I.; PÉREZ-OBIOL, R.; PANTALEÓN-CANO, J.; ROURE, J.M.,
1997. 'Palynological Evidence for Climatic Change and Human Activity during the Holocene on Minorca (Balearic Islands)», Quaternary Research, 48: 339-347.
ZAPATA, L.,
1996. 'Modos de subsistencia en el Cantábrico oriental durante el cuarto milenio B.C.», in: Congrés del Neolític a la Península Ibèrica. Formació i implantació de les comunitats agrícoles (I: Gavà: 1996), Rubricatum, 1 (Museu de Gavà), 101-108.
ZAPATA, L; IBÁÑEZ, J.J.; GONZÁLEZ, J.E.,
1997. 'El yacimiento de la cueva de Kobaederra (Oma, Kortezubi, Bizkaia). Resultados preliminares de las campañas de excavación 1995-97», Munibe, 49 (San Sebastián): 51-63.

27

LANDSCAPE AND MANAGEMENT OF FOREST RESOURCES IN THE BALEARIC ISLANDS DURING THE II -I MILLENNIUM BCE

RAQUEL PIQUÉ AND MARC NOGUERA

Divisió de Prehistoria, Universitat Autònoma de Barcelona
EMAIL: raquel.pique@uab.es

INTRODUCTION

The forest environment provides raw materials that through simple appropriation techniques can be used to satisfy many basic necessities. Fruits, tubers, fungi, grass and leaves can be gathered and used for feeding humans and livestock. Likewise wood is used as raw material in construction and production of artefacts. Other materials such as resins, tannin, bark, etc. can also be obtained in the forest. No less important is the collecting of vegetal fuel, the main source of energy for prehistoric societies. However, the relationship between human groups and its forest environment is not exclusively focused on the appropriation of products. The forest is also the center of other basic economic activities of subsistence that are generated by a whole series of complex relations between human societies and their environment. Historically, agriculture and livestock farming have developed to the detriment of the forest. The forest has been felled to obtain agricultural land and pasture whereas the use of forest resources guarantees the viability of animal husbandry practices.

The practice of forest exploitation has often had dramatic consequences on vegetal coverage: deforestation and erosion, modification of the structure of the forest, substitution and the disappearance of species are phenomena that have taken place and continue taking place as a consequence of the relationship between human societies and their environment. This relationship is dialectic in so far as the appropriation or collection of resources implies a transformation of the environment that has consequences on the availability of resources and therefore on society itself.

Traditionally, research has not given much importance to forest resource management during prehistory. Undoubtedly one of the causes is the very nature of vegetal remains, which do not preserve very well in certain atmospheres. Few archaeological records document proof on the collection, transformation, redistribution and consumption of forest products. The explanatory limitation of archaeobotanical studies is also due, in part, to the theoretical frameworks of archeological research that have not generated the questions and the methodologies adequate to solve them.

Our interest is focused on those trees and shrubs that were deliberately introduced into a settlement and are therefore the product of a whole series of working processes that have transformed them from natural materials into a resource and have continued transforming them until they have been converted into working means or goods that are distributed and consumed. Changes in economic and social practices that developed at every historical time implied also the transformation of forest resource management practices themselves. This could jeopardize or favour the capacity of the forest for regeneration, up to the point of provoking irreversible changes in its very structure. The demand for wood, pastures, cropland and other forest products lead to constant pressure on the environment since these are basic needs for human subsistence. Because of this analysis of the economic practices of a society cannot be reduced to the study of practices for obtaining foodstuffs inferred from faunal and charcoal remains or to production activities that generate inorganic waste (metallurgy, pottery, stone industry). It is also necessary to look at how society organized the exploitation of the forest.

Charcoal from archaeological contexts is usually present in practically all prehistoric sites. This is because the presence of combustion waste is tied to the use of wood in the energy production process, which has been the normal practice of human societies since the Upper Pleistocene. Even though this waste is testimony to the working processes involved in the combustion of wood it is not easy to understand from studying it how the production, distribution and consumption of vegetal fuel was organized. However, by studying these remains we can reach an approximation of how these resources were collected and used (Piqué, 1999a). Energy is undoubtedly one of the basic requirements for every society. Energy makes most productive processes possible, it is necessary for the transformation of foodstuffs and raw materials and in general for the development of social activities that allow social reproduction. Technological development has gone hand in hand with an increase in energy demand; vegetal fuel has met this demand throughout history, and continues to do so. Firewood has constituted the main source of energy, jointly in some cases with excrement or other vegetal residual products, especially when firewood has been scarce.

The study of present day communities that depend on firewood to solve their energetic needs provides revealing data on the role of vegetal fuel in the economic plan (Reddy, 1983; Lucia, 1983; Morgan, 1983). In present day agricultural subsistence communities it is the women who are usually responsible for obtaining firewood to meet the basic energy needs of the domestic unit, each domestic unit being self-sufficient and gathering what it needs for its own use. Firewood is very rarely purchased from specialized fuel production people. The situation changes, however, when dealing with urban populations that depend mainly or exclusively on vegetal fuel or else when demand comes from the needs of craftsmen, in these cases the collection and distribution of fuel often falls to intermediaries. This is different in the case of public institutions in small rural communities whose energy needs can be satisfied by contributions made by all members of the community. We see then that not only can the demand for energy vary according to the type of society but also present substantial differences in the organization of production. We cannot consider that these models of organization were the same as those used in prehistoric societies, but they worthwhile keeping in mind when interpreting charcoal analysis records. Population concentration, ownership relations and political control of the territory undoubtedly determined important changes in the exploitation of forest territories

Evidence of consumption of forest products for other ends, for example the production of goods is more difficult to find. Uncarbonized wood is only preserved in anaerobic water or atmospheres where the activity of bacteria and fungi has been inhibited by different causes. Conservation of organic remains is therefore unusual in the majority of archaeological sites, where it is only accidental or intentional

	MARTORELLET	MUSSOL	CARRITX		FORNÉS	FERRAGUT	FORNÉS	
FUNCTION	**FUNEREAL**	**RITUAL**	**RITUAL**	**FUNEREAL**	**DOMESTIC/CONSTRUCION**			
CHRONOLOGY	**1620 cal BCE**	**1550 cal BCE**	**1580 cal BCE**	**1450-1400 y 800 cal BCE**	**850-c 550 cal BCE**	**750/700-475 cal BCE**	**400-250/200 BCE**	**s. II-I BCE**
Arbutus unedo				**3,07**				
Cistus sp		**1,57**		**0,38**	**0,08**		**3,02**	**3,86**
Clematis cirrhosa				**0,38**				
Erica sp.			**95,12**	**5,36**				
Ficus carica		**3,14**		**0,19**				**0,62**
Juniperus sp	**18,18**			**0,57**				
Leguminosae	**1,81**							
Monocotyledonous	**29,09**			**4,98**				
Olea europaea	**25,45**	**93,19**		**22,61**	**99,82**	**93,71**	**41,38**	**29,36**
Pinus halepensis		**0,52**	**2,44**	**9,77**	**0,08**	**6,29**	**54,07**	**62,13**
Pistacia lentiscus		**1,57**	**2,44**	**16,28**			**1,51**	**3,86**
Rhamnus-Phillyrea	**25,45**			**18,58**				**0,15**
Rosmarinus officinalis				**3,45**				
Minimum number of taxa	5	5	3	12	3	2	5	5
Number of fragments	55	195	41	522	1166	1463	331	647

Table1. Taxa represented in charcoal from different sites and time phases. Cova des Càrritx shows results of charcoal from rooms 1, 2 and 3. Son Ferragut shows results from assemblages III, V y VI.

carbonization of artefacts that provide any type of evidence. In general only reduced samples are available and these do not allow us to observe any significant recurrences in supply patterns of raw materials. Without doubt, wood usage has been conditioned by the properties of the wood and its availability in the environment, but it must not be forgotten that resource collecting strategies are socially determined and that factors such as technological capacity, production organization and demand have taken part in the development of these strategies for using forest resources.

This paper sets out to evaluate forest resource management practices implemented during the second and first millennium BCE in the Balearic Islands, its relationship to social and economic strategies and repercussions on the forest environment. Diverse studies carried out on charcoal and wood proceeding from archaeological sites on the islands of Mallorca and Menorca are presented, reflecting variations in the exploitation of tree and shrub resources, and the causes of the observed variability are discussed. All the changes observed in archaeobotanical records cannot be considered to be an exclusive response to a change in supply, i.e. the environment. Other socially determined factors such as demand or production organization have taken part in the formation of archaeobotanical assemblages and can therefore also be a basic cause of the variability observed. Management strategies of vegetal resources that develop human communities cannot be interpreted apart from the historical context in which they take place.

EXPLOITATION OF TREE AND SHRUB RESOURCES ON THE BALEARIC ISLANDS

A number of different sites studied on the Islands of Mallorca and Menorca have allowed us to compare the use of woody raw materials at sites that were occupied or used during different historical periods on the Balearic Islands: Can Martorellet, Cova des Mussol, Cova des Càrritx, Son Fornés and Son Ferragut. The charcoal and wood studied come from sites that had different functions i.e.: funereal and habitation. These wood remains document different uses at the same sites: remains of carbonized construction implements, combustion waste for energy production and wooden furniture artefacts, in some cases carbonized and in others not. In spite of this the sample is very reduced, data is lacking for many historical phases and for most there is only data proceeding from one site. This panorama does not allow conclusions to be drawn on forest resource management practices; it is only possible to make some hypotheses that can then be contrasted in further studies.

The data on exploitation of forest resources for the oldest period come from funereal contexts (Can Martorellet, Cova des Càrritx) and ritual uses (Cova des Mussol).

The cave at Can Martorellet (Pollença, Mallorca) was used for funereal purposes during the initial Naviforme settlement phase (1600-1400 cal BCE) or a previous moment, the dating obtained situates the use of the sepulchral cave at 1620 cal BCE (Pons Homar, 1997, quoted in Lull et alii 1999). Preliminary study of charcoal remains and uncarbonized wood, highlighted by the presence of various combs, has allowed differences to be seen between the taxa used for the manufacture of wooden artefacts and those used for combustion (TABLEs 1 and 2). The range of taxa used for manufacture is limited to Buxus cf balearica, however, between the coal and unworked wood greater diversity has been documented: Olea europaea, Juniperus sp., Pistacia lentiscus, Rhamnus/Phillyrea, monocotyledonous and leguminous plants. It must be stressed that the studied sample was very reduced and it is necessary to contrast the results with the study of a larger population.

Two periods are documented for the Cova des Mussol (Ciutadella, Menorca) the oldest assemblage is the one that is formed by recovered charcoal in one area of combustion, located in room 1 of the cave. This assemblage provided a dating of 1550 cal BCE, which also corresponds to the initial Naviforme phase; at this time the cave was used in for ritual purposes (Lull et alii, 1999). The charcoal analysis obtained (Piqué, 1999b) in the combustion area is composed of taxa that can be found in contemporary Balearic wild olive woods: Olea europaea, Pinus halepensis, Pistacia lentiscus, Cistus SP, (TABLE 1). The other taxon present, Ficus carica, is typically Mediterranean although it was probably introduced into the Balearic Islands. The best-represented taxa are Olea europaea (93.5% of the identified fragments).

The Mussol cave wasn't used again until several centuries later, again for ritual purposes, although these were of a very different nature, thus revealing an ideological change in the communities that lived on the island of Menorca (Lull et alii, 1999). The studied assemblage was formed by different anthropomorphic and zoomorphic figures, undetermined and formless fragments of wood dated between 1200-1000 cal BCE. The woods used for the manufacture this items provided an assemblage taxonomically poorer than the one obtained from the charcoal since only two taxa (TABLE 2) have been identified. Most of the figures were made from Olea europaea, which again shows that the supply of wild olive woods was used. The only exception was an item manufactured from Buxus balearica.

The sepulchral cave at Càrritx (Ciutadella, Menorca) provided a wide assemblage of charcoal and wood remains, among them furniture artefacts, which were deposited at different times. The oldest assemblage comes from room 6, dated at 1580 cal BCE, where it seems that an activity of ritual character was carried out (Lull et alii, 1999). The charcoal remains came from a fireplace, where practically the only represented taxon is Erica sp., a kind of hardwood, which makes excellent fuel (TABLE 1). The dating obtained for the remains of room 6 put it at some time in the initial Naviforme (1600-1400 cal BCE).

Undoubtedly the most important assemblage of remains from the Càrritx cave is that from the collective burial necropolis that was functioning between 1450-1400 and 800 BCE. Dating shows that the necropolis was functioning during most of the Naviforme phase, the Proto-Talayótica and beginnings of the Talayótica phase. The taxa assemblage identified from charcoal analysis (TABLE 3) is dominated by shrubs that can be found in the present day Menorcan wild olive community (Olea europaea, Pistacia lentiscus, Rhamnus/Phillyrea, Liliaceae and Clematis); and in the secondary vegetal communities that owe their origin to the degradation of the potential forest communities: the wild olive (Pinus halepensis, Cistus, Erica, Rosmarinus) or holm-oak woods (Arbutus, Pinus halepensis). Charcoal analysis shows differences in the pattern of fuel usage in the different sites studied, probably due to specific processes that led to the incorporation of the remains in archaeological records (Piqué, 1999c). In Room 1 hardwood taxa Olea, Pistacia and Rhamnus-Phillyrea dominate, although a minimum of 15 taxa is documented. In Room 2, which was a thoroughfare, easily inflammable taxa dominate (Liliaceae and Pinus) and in general a lesser taxonomical diversity (a minimum of 8 taxa) has been documented.

	MARTORELLET	MUSSOL		CARRITX						FERRAGUT	CARRITX	COVA MURADA
FUNCTION	**FUNEREAL**	**RITUAL**		**FUNEREAL**						**DOMESTIC**	**FUNEREAL**	**FUNEREAL**
CHRONOLOGY	**1620 cal BCE**	**1550 cal BCE**		**1450-1400 y 800 cal BCE**						**750/700-475 cal BCE**	**s. V cal BCE**	**s. V BCE**
ARTEFACT	**COMB**	**ANTHROPOMORPHIC FIGURE**	**NO DETERMINED**	**CUP**	**COMB**	**SPOON**	**STICK**	**TUBE**	**FORMLESS**	**COMB**	**STRETCHER BED**	**COMB**
Buxus cf *balearica*	X		X	X	X			X				
Erica sp.						X		X		X		X
Ficus carica									X			
Olea europaea		X					X	X			X	
Pinus halepensis											X	
Pistacia lentiscus								X				
Rhamnus-Phillyrea											X	
Minimum number of taxa	1	1	1	5						1	3	1

Table 2. Taxa represented in wood used for artifact at different sites and in different phases. Results from room 1 and room 5 (deposit) at Cova des Càrritx. Remains from Son Ferragut were carbonized.

The study of wood remains with traces of having been worked from the Cova des Càrritx reveals that the needs for raw material for manufacturing were covered with a lesser number of taxa than those used for energy production (TABLE 2). The artefacts come mainly from a deposit in room 5, dated towards the end of the period when the necropolis was used. Olea europaea, Pistacia lentiscus, Buxus cf. balearica, and Erica sp were used in the manufacture of these objects. The taxon Ficus carica should also be added, most of its remains were located in the same deposit of room 5 where some of the artefacts were found, and therefore, it is possible that it belonged to a manufactured item that has not been preserved. In general, the raw material supply pattern for manufacturing combines the collection of widely distributed local raw material (Olea, Pistacia, Erica) with the selection of material less abundant in the area (Buxus cf. balearica and Ficus carica), which implies as much a knowledge of the physical and chemical properties of these as of its distribution range.

These sites were in use during the Naviforme phase (1600-1050 cal BCE). Using the historical time sequencing of Balearic society proposed by Lull, Micó, Risch and Rihuete (1999, 2001) some of the significant features of this phase are now detailed. From 1600 BCE onwards the construction of Naviforme enclosures, buildings of considerable size with strong walls of large stone blocks, is documented in practically all the archipelago. These buildings would have probably been introduced by populations from outside the islands who would have lived with the local communities settled in villages with precarious structures in the open air or in shelters. The function of the Naviforme enclosures is related to productive activities and living spaces, and the lowland location not very far from the coast, is near to flatlands suitable for agriculture and livestock farming. They are frequently found in concentrated form. Funereal expressions in artificial and natural caves are contemporary to these occupations of the initial Naviforme. During the middle Naviforme (1450/1400-1200 cal BCE) habitat forms and burial practices were homogenised, this is when the Naviforme agglomerates and funereal practices in natural caves with cyclopean entry walls predominate. There is no indication of political centralization in this phase and society seems to have been organized in basically autonomous subsistence units. Between 1200-1050 cal BCE (final Naviforme) an urbanistic change is observed with the appearance of new architectonic structures along with the survival of settlements of the previous phase, funereal practices continue the tradition of the previous phase. The changes in settlement patterns suggest that a population concentration in denser units was now taking place. Empirical data on the strategies of subsistence during the Naviforme phase are generally scarce. The presence of faunal remains allows the basic role of animal husbandry in the subsistence of island communities to be considered. However, settlement in areas suitable for agriculture, the presence of storage containers of great dimensions and the presence of manual mills, along with the presence of cereals at the sites of Son Oms (Mallorca) and the Cabaña Torralba d'en Salord (Menorca) already during the final Naviforme period, could lead us to think that agriculture had acquired considerable weight in the whole range of subsistence strategies.

The exploitation of forest resources in the Talayótica phase and onwards is documented especially for the island of Mallorca, from the sites at Son Fornés and Son Ferragut. The Cova des Càrritx (Menorca) was also frequented during the Proto-Talayótica phase.

The Son Fornés site (Montuïri, Mallorca) was occupied between 850 cal BCE and the Roman period (Lull et al, 2001). Analysis of charcoal from this site allows to make evident an exploitation pattern of tree and shrub resources limited to a few types (Noguera and Piqué, e.p) During the Talayótica phase (850- circa 550 cal BCE) most of the remains belonged to Olea europaea (wild olive tree), the presence of Pinus halepensis and Cistus is testimonial (1 single fragment respectively) which allows us to think that they did not play any role in the economic plan (table 1). Olea is in evidence as much in fireplaces as at floor and collapsed levels, which would seem to indicate that it was used as much as fuel as building material

The results obtained from charcoal from Puig Morter de Son Ferragut (Sineu, Mallorca); dated between 750/700-475 cal BCE (Castro, Escoriza and Sanahuja, in this volume) allows a pattern to be made of how the inhabitants of the settlement exploited forest resources. This is focused also on a low number of taxa (Piqué and Noguera, e.p.), only three taxa have been identified: Erica sp. Olea europaea, Pinus halepensis (TABLE 1). The studied charcoals came from the floor, collapsed and abandoned levels of the Alfa building. The formation characteristics of the deposits, as well as the morphology and distribution of the remains observed during digging, let us consider that part of these came from wooden building structures, these could be mixed with remains of artefacts and consumed firewood. The species most used was Olea europaea; this seems to be indicated as much by the volume of remains as by the distribution since it is represented in all the sectors of the building. The use of Pinus halepensis is restricted to certain sectors and was used less intensively. The presence of a comb manufactured in Erica sp, allows us to suppose that other furniture artefacts could have been carbonized up to a point that makes it impossible to recognize their morphology.

The Talayótica phase following the historical time sequencing suggested by Lull et al (1999, 2001) covers the period between 1050-600 cal BCE. The initial Proto-Talayótica (1050-850 cal BCE) phase is characterized by a population concentration phenomenon and by the maintenance of an important ceremonial component expressed in funereal rituals. In Mallorca the first attempts at construction of talayots are documented. The Talayótica phase (850-600 cal BCE) is characterized by the proliferation of talayots on both islands. These constructions would have had, among others, a political function for controlling the territory. Funereal rituals were less important. A greater furnishing difference between Mallorca and Menorca is observed from 1050 cal BCE onwards, which would indicate a decrease in communication between the islands. In Mallorca walled villages and talayots are extensive but any number of necropolises in relation to the high density of population of the villages has hardly been documented. In Menorca, funereal expressions in natural caves and navetas are usual between 1000 and 800 cal BCE, whereas talayots were not constructed before 850 cal BCE. Economic strategies data proceed mainly from the Son Fornés site, according to this subsistence was based on a mixed economy with animal husbandry predominating. Meat was supplied fundamentally by cattle and pigs; these were used more than sheep and goats. Agriculture is little documented but the location of the settlements in areas suitable for crops favour the existence of a mixed agricultural stock farming economy.

The site at Son Fornés has provided evidence on the exploitation of woody resources during the Post-Talayótica phase (400-250/200 BCE). A greater quantitative importance of Pinus halepensis is observed and in general the use of a greater diversity of taxa as much in combustion waste as well as in construction (TABLE 1). This tendency is maintained in the Roman phase of occupation documented at Son Fornés, here again the quantitative importance of Pinus halepensis in relation to other taxa and a greater taxonomical diversity is observed (use of a minimum number of 6 taxa is documented).

The Càrritx cave was also occupied again at the end of the c. V cal BCE. Various wooden artefacts were deposited inside the cave at this time. They comprise some pieces from a stretcher bed made of Pinus halepensis and Rhamnus-Phillyrea, wooden nails made of Olea europaea, a T-form artefact, another one horn-shaped and some formless fragments of Pinus halepensis. The Erica SP wooden comb at Cova Murada (Menorca) is another of the remains documented from the Post-Talayótica phase.

In the middle of the s. VI a series of conflicts took place on the islands that gave rise to the end of the Talayótic society. During the Post-Talayótica phase (c. 550-123 BCE) contact with the colonial powers of the Mediterranean is also documented (Lull et al, 2001). Funereal practices experienced an apogee in regard to previous centuries, necropolis in natural caves, partially modified or totally artificial are usual. Inhabited areas correspond to walled villages, where use is made of parts of previous constructions and isolated buildings, a highlight is the construction of new plan as underground hypostyle chambers, with a possible function as storage space or communal meeting centres, and shrines of different architectonic types. It is considered that the Post-Talayótic economic structure was more diversified than in the previous phase, an increase in the number of mills in settlements, written sources and polinical data seem to indicate the importance of agricultural practices at this time, especially cereal farming. Livestock farming is documented in archaeological data as well as in written sources. The data provided by the Post-Talayótic constructions at Son Fornés indicate a greater importance of sheep and goats for meat consumption to the detriment of cattle and pigs. The Romanization of the Balearic islands began in the year 123, when, according to written sources, Roman troops occupied the islands militarily, however this official conquest did little more than consolidate a situation which had started at the end of the c. III and the beginning of the c. II BCE. Romanization meant the appearance of urban planning, a change in trade circuits, the adoption of technological innovations (rotation mills, iron instruments) and a change in economic policies and structures. However, it can be considered that there was certain continuity in population, ideology and ritual planning characteristic of the Post-Talayótica society up to the c. I CE (Lull et al, 2001). Urban rearrangement is already documented from beginnings of the c. II CE at Son Fornés; there is a change in function of the buildings, with a specialization of space. Cereal and livestock farming denote continuity from the previous phase, meat supply came from cattle and goats followed at a distance by pigs, cattle were also exploited for secondary products and as draught animals.

LANDSCAPE AND FOREST RESOURCE MANAGEMENT IN THE BALEARIC ISLANDS: CONTINUITY AND BREACH.

Overall, the results obtained from studying different prehistoric sites on the Balearic Islands present some differences, although we can also confirm recurrences in the use of certain taxa that should be considered. The taxa assemblage identified in the studied sites is characteristic of maquis and secondary communities. These were the forest formations that were exploited, independently of chronological phases and geographical locations. Olea europaea is well represented at all sites and at all phases. This taxon is part of the wild olive maquis that makes up the potential vegetation of most of the Balearic archipelago at 600 m below sea level in the present day. Pistacia lentiscus, Rhamnus-Phillyrea, Erica sp, Juniperus sp, Pinus halepensis also evoke the exploitation of maquis vegetation, although they appear sporadically and usually at lesser frequency. The presence of small shrubs like Cistus sp and Rosmarinus officinalis would perhaps indicate more open conditions; they may be a product of the degradation of the dominant shrub formations. A feature common to practically all the sites is the absence of taxa characteristic of the holm-oak woods, the other great potential forest formation of the archipelago that flourishes in the rainiest areas of the islands of Mallorca and Menorca.

CHRONOLOGY	1580 cal BCE	1450-1400 y 800 cal BCE								
ROOM	RH	DEP	HI	HI	R1	R1	R2	R2	R3	R3
Material	C	W	C	W	C	W	C	W	C	W
Arbutus unedo	0	0	0,93	1,07	4,37	1,88	0	0	0	0
Buxus cf. *balearica*	0	57,72	0	0	0	1,65	0	0	0	0
Cistus sp.	0	0	3,16	1,07	0,54	1,41	0	0	0	0
Clematis cirrhosa	0	0	0	0	0,54	0,23	0	0	0	0
Erica sp.	95,12	10,56	1,11	2,15	6,28	3,30	8,92	0	0	0
Ficus carica	0	4,06	0	0	0,27	0	0	0	0	0
Juniperus sp.	0	0	0,18	0	0	0,70	1,78	0	8	0
Leguminosae	0	0	0,18	0	0	0	0	0	0	0
Liliaceae	0	0	2,42	0	2,18	2,12	28,57	0	8	0
Olea europaea	0	26,82	54,19	36,56	27,32	20,99	14,28	0	40	0
Pinus halepensis	2,43	0	0,74	3,22	9,28	29,48	30,35	88,88	0	100
Pistacia lentiscus	2,43	0,81	19,74	17,2	20,76	12,97	7,14	0	20	0
Quercus ilex-coccifera	0	0	0,74	0	0	0	0	0	0	0
Quercus sp caducifolio	0	0	0,37	0	0	0	0	0	0	0
Rhamnus/Phillyrea	0	0	16,01	38,71	24,31	21,46	5,35	0	20	0
Rosmarinus officinalis	0	0	0,18	0	4,09	3,77	3,57	11,11	4	0
Number of fragments	41	127	586	111	427	478	65	10	30	5

Table 3.-Relative frequencies (based on the number of fragments)where identified taxa appearing each of the rooms studied.. C= Coal, W=Wood, T= Total, DEP= Deposit, **HI= Hoyo Interior**, R1= Room1, R2= Room 2, R3= Room 3, RH= Room 6. Quercus remains come from an intrusion level

Although some of the taxa identified can also accompany the Balearic holm oak woods, the absence of Quercus ilex-coccifera in the studied assemblage (only present at an intrusion level at Cova des Càrritx) seems to indicate that the shrubby plant supply was preferably taken from maquis, perhaps dominant in the landscape environment. The presence of Buxus balearica would seem to indicate that raw material collection could also have been carried out in other communities not so near the settlements.

According to polinical analyses carried out in the Balearic archipelago, the establishment of vegetation dominated by Olea, Pinus, Quercus perennifolio, Erica, etc. would have been produced towards 4000 BP (Burjachs et al, 1994; Yll et al, 1995, 1997), probably due to environmental conditions of greater aridity. Buxus practically disappears from the polínical sequences of Menorca from 4000 BP onwards, in Mallorca it is relegated to certain areas. Despite this, the survival of Buxus in Menorca posterior to this date is documented as much in polinical (Mariscal, 1995) as in archaeological records (Piqué, 1999b) at least until the I millennium BCE.

Another common feature in some of the assemblages studied is a low taxonomical diversity. That could indicate a selective use of environmental raw materials restricted to using only a few kinds of the forest formations that could have flourished on the island. Differences in geographical site locations are not reflected in the supply pattern of different woody raw materials; however, at some of the sites (Can Martorellet, Càrritx, Mussol) the use of raw materials for the manufacture of goods proceeding from more inaccessible places (Buxus) is documented.

Differences observed between the assemblages could be due to various factors. It has to be remembered that the sites reflect different functions (ritual, funereal, domestic) and within them different uses of wood (construction, combustion, furniture artefact manufacture). They are a reflection of different historical stages when changes occurred in settlement patterns as well as in economic activities. The summary table of the results obtained from charcoal analysis (TABLE 1) allows differences between sites and historical phases to be seen.

A very heterogeneous pattern of fuel usage is observed in the oldest sites (Martorellet, Mussol, Càrritx), a different floral assemblage is found in each case. The fireplace in room 6 at Càrritx shows a decreased diversity, most of the remains are of Erica SP. However, the remains proceeding from the other rooms at Càrritx are those that have provided greater taxonomical diversity, perhaps due to the long interval of time involved. The other two sites present few similarities between each other, either qualitatively or quantitatively. At Mussol the best taxon represented was Olea europaea, whereas at Can Martorellet monocotyledonous plants and Rhamnus-Phillyrea are present at high frequency.

A very low taxonomical diversity is found between the Talayótic sites (Son Fornés and Son Ferragut), fuel and building material needs were solved with Olea europaea. A lesser difference is observed between these sites, the wood usage pattern is more homogeneous and there is a greater recurrence in specie usage.

The taxonomical diversity increases in the Post-Talayótica and Roman phases at Son Fornés, moreover, Olea europaea ceases to be the best represented taxa; now the one that provides the most remains is Pinus halepensis. This change in the nature of exploitation of woody resources within contexts of similar function (domestic fuel and construction) could be explained by the general changes that took place in Balearic society from this point onward. A more intense use of Pinus halepensis from the Post-Talayótica occupation onward could indicate a change in forest management practices. Possible causes could be a greater need for woody raw materials or an increasing degradation of the maquis that would have favoured an expansion of the use of pine, however, it is also necessary to relate this change to global productive strategies developed by the societies that lived on the islands.

There could be many reasons for the causes of the differences observed throughout the sequence. On the one hand, the chronological range represented in the charcoal and wood analysis record includes different historical times, when big changes in subsistence strategies and social organization took place. On the other hand, the studied sites reflect different functions. The greatest heterogeneity between the oldest assemblages could be related to the fact that every site had its own particular function (funereal and/or ritual practices). These sites were also used over differing intervals of time that could have been longer or shorter. However, at the most recent sites, the charcoal comes from domestic contexts. Therefore, although there is a chronological difference between the assemblages studied, in an appraisal of fuel collection strategies we cannot fail to take into account the function that the wood was destined for, which as we see is different in every period.

To understand the forest management practices as a whole it is not enough to contemplate them only as source of woody resources. According to archaeological evidence the importance of agricultural practice on the islands seems to have increased from the Post-Talayótica period onwards, a trend that continues until the phase of Roman occupation (Lull et al 1999, 2001). Balearic society during the Naviforme and Talayótica phase based its subsistence mainly on animal husbandry, while agriculture would have taken second place. The change from an animal husbandry society to one mostly agricultural undoubtedly had repercussions on forest management practices.

The feasibility of livestock farming in Mediterranean biomes is generally based on the exploitation of forest products for fodder since Mediterranean herbaceous formations have very low productivity. The herbaceous layer of the Mediterranean forests is scarce and poor and livestock feed on leaves and fine branches. Although these have little food value they are a guaranteed source of food throughout the whole year as they are less sensitive to cold or drought than herbaceous plants. The total or partial felling of the tree stratum is a usual practice for increasing herbaceous communities. The abusive practice of felling the tree stratum for obtaining pastures would lead to the degradation and erosion of the forest and, therefore, in the long-term, a decrease in its productivity. Because of this the maintenance of one part of the tree stratum guarantees the feasibility of animal husbandry as well as satisfying the need for other forest products like firewood.

The concurrence of a greater importance to agricultural practices and the increase in consumption of Pinus halepensis from the Post-Talayótic period onwards could be the result of a different relationship of Balearic society to its forest environment. Livestock farming practices could have had a lesser impact on the forest environment than agricultural practices. The first would not have jeopardized the regeneration capacity of the forest. However, the intensification of agricultural activity would have implied a greater pressure on the surroundings, in the first place because fields are obtained at the expense of the forest masses, secondly, because that implies a reduction of pasturelands and a greater pressure on these with respect to the previous phases. The degradation of maquis and holm oak woods would have favored the secondary expansion of pine formation too.

Changes in settlement patterns throughout the documented phases could also have brought about an increasing fixed pressure on the environment with consequences for tree coverage. The increasing use of the pine to the detriment of other exponents of the Balearic maquis could be a consequence of this continuous pressure put on local resources by sedentary societies. The degradation of the maquis would have favored the expansion of secondary formations of pine. However, a more sedentary and concentrated population is documented from the Naviforme phase onwards, therefore they are phenomena previous to the exploitation of pinewoods. It seems that the phenomenon of more concentrated population entailed a more homogeneous forest management practice, focused on exploitation of the tree stratum of the wild olive tree maquis. For centuries it was not necessary to go to other woody resources to satisfy the needs for fuel or timber for construction, the wild olive tree maquis could be used for this end without restrictions. Forest management practices implemented by the Talayótica society don't seem to have jeopardized the regeneration of the existing forest communities.

Changes observed from the Post-Talayótica phase onwards could be the consequence of continued pressure on the maquis throughout hundreds of years. However, the greater importance of agricultural practices in subsistence strategies and the change that that meant in the relationship between the population and its most immediate environment has to be considered.

The function of the studied sites was undoubtedly a key factor in the taxonomical composition of the assemblages. It seems that the production of firewood was organized in a very different way depending if it was orientated to domestic or ritual-funereal purposes, the testing of both cases seems to show a more standardized supply of resources in the domestic area than in the funereal. Combustion waste proceeding from funereal contexts shows that Olea is always the dominant taxon, although the shrubby component is also well represented at Martorellet, Càrritx and Mussol, which could indicate more random fuel collecting. However, in the waste from domestic contexts the shrubby component is less represented, which indicates that collection was orientated to specific types, Olea during the Talayótica occupations and Pinus halepensis during the Post-Talayótic and Roman.

A restricted number of species was used for furniture production (TABLE 2), although this could be due to the characteristics of the sample studied, which was very reduced in some cases. The site that revealed a greater diversity of artefacts (Càrritx) also provided greater taxonomical diversity. Once again the best represented taxon at any of the sites studied is Olea. The main difference in relation to other uses of wood is that the use of less accessible taxa is documented, perhaps transported from greater distances. This is the case of Buxus cf. Balearica. It is also worth noting that the woods used for artefact and furniture production (Buxus, Erica, Olea, Pistacia) are notable for their qualities as very hard and dense woods, much appreciated in the present day in cabinet making and joinery. The studied sample is very reduced in some cases and does not allow a comparative study between sites. The study of raw materials used for the manufacture of furniture artefacts at Càrritx showed a careful selection of wood according to the type of artefact, the cups were made of Buxus, as was the comb, the spoons were manufactured in Erica sp, the sticks in Olea, which proves a certain standardization of production (Piqué, 1999b). The combs at Can Martorellet were manufactured of Buxus.

A change in the raw materials used for artefact production is also observed between the different historical phases studied. The combs made of Buxus wood during the Naviforme and Talayótic period are made of Erica wood during the later periods. The use of Pinus halepensis and Rhamnus-Phillyrea for artefact production is also documented during the Post-Talayótica period although in this case it is a type of artefact not previously documented. Pinus halepensis and Erica sp increase when wild olive and holm oak communities are destroyed, which again may indicate a more intensive exploitation of secondary communities than in previous periods. This could have been due to a greater importance attached to these in the environment.

Finally, it should be noted that exploitation of holm oak woods for fuel or artefact construction has not been made evident in any of the studied sites, in spite of it being, jointly with the wild olive maquis, the potential vegetation in the present day on the islands of Mallorca and Menorca. The cause of this absence could be due to the proximity of the sites to areas of dominant maquis. However, given that displacement to obtain specific woody raw materials was a usual practice, it is surprising that holm-oak wood was not exploited. Perhaps new studies can provide data on the exploitation of the holm-oak woods; in the meantime it is fitting to pose the hypothesis that this forest formation was reserved for other uses. A historically documented practice of Balearic herdsmen was to leave herds of pigs to graze freely through holm-oak woods feeding on acorns and tubers.

BIBLIOGRAPHY

BURJACHS, F.; PÉREZ-OBIOL, R.; MOURE, J.M. y JULIÀ. R. 1994. "Dinámica de la vegetación durante el Holoceno en la Isla de Mallorca". In: MATEU, I.; DUPRÉ, M. GÜEMES, J.; BURGAZ, M.E. (eds.) Trabajos de Palinología Básica y Aplicada. X Symposio de Palinologia (A.P.L.E.). Universitat de València, València, pp: 199-210

LULL, V.; MICÓ, R.; RIHUETE, C.; RISCH, R. (1999) Ideologia y Sociedad en la Prehistòria de Menorca.La Cova des Càrritx y la Cova des Mussol. Consell Insular de Menorca, Universitat Autònoma de Barcelona.

LULL, V.; MICÓ, R.; RIHUETE, C.; RISCH, R. ,
2001. La prehistòria de les Illes Balears i el jaciment arqueològic de Son Fornés (Montuïri, Mallorca). Montuïri: Fundación Son Fornés

MARISCAL, B. ,
1996). "Variación de la vegetación de Menorca en los últimos 4000 años" Revista de Menorca,1:197-297

DELUCIA, R. ,
1983. "Defining the scope of wood fuel surveys" Wood fuel surveys. Forestry for Local Community Development Programme. Roma: FAO

MORGAN, W.B. ,
1983. "Urban Demand:Studying the Commercial Organization of Wood Fuel Supplies" Wood Fuel Surveys. Forestry for Local Community Development Programme. Roma: FAONOGUERA, M; PIQUÉ, R (e.p.) "Prácticas de explotación forestal y uso de las maderas en el poblado prehistórico de Son Fornés"
PIQUÉ, R. ,
1999a. Producción y uso del combustible vegetal: una evaluación arqueológica. Treballs d'Etnoarqueologia, 3. Madrid: Universitat Autònoma de Barcelona, Consejo Superior de Investigaciones Científicas.
PIQUÉ, R. ,
1999b. "Análisis de las maderas y carbones del yacimiento de la Cova des Mussol". In: LULL, V.; MICÓ, R.; RIHUETE, C.; RISCH, R. Ideologia y Sociedad en la Prehistòria de Menorca. La Cova des Càrritx y la Cova des Mussol. Barcelona: Consell Insular de Menorca, Universitat Autònoma de Barcelona.
PIQUÉ, R. ,
1999.) "La gestión de los recursos leñosos en la cueva des Càrritx". In : LULL, V.; MICÓ, R.; RIHUETE, C.; RISCH, R. Ideologia y Sociedad en la Prehistòria de Menorca. La Cova des Càrritx y la Cova des Mussol. Barcelona: Consell Insular de Menorca, Universitat Autònoma de Barcelona.
PIQUE, R.,
in press. "Anàlisi de les fustes i carbons de Can Martorellet (Pollença, Mallorca)".
PIQUÉ, R.; NOGUERA, M.,
in press. "La gestión de los recursos forestales durante la prehistoria de las Baleares: el yacimiento del Puig Morter de Son Ferragut"
YLL, E-I; PEREZ-OBIOL, R.; PANTALEON-CANO, J.; ROURE; J.M. ,
1997. "Palynological Evidence for Climatic Change and Human Activity during the Holocene on Minorca (Balearic Island)". Quaternary Research, 48:339-347
YLL, E-I; PEREZ-OBIOL, R.; PANTALEON-CANO, J.; ROURE; J.M. ,
1995. "Dinámica del paisaje vegetal en la vertiente mediterránea de la Península Ibérica e Islas Baleares desde el Tardiglaciar hasta el presente". In: T. ALEIXANDRE y A. PÉREZ GONZÁLEZ (Eds.) Reconstrucción de paleoambientes y cambios climáticos durante el Cuaternario. Centro de Ciencias Ambientales, CSIC, Madrid, pp: 319-328
REDDY, A.K.,
1983 "Rural Fuelwood: Significant Relationships" Wood Fuel Surveys. Forestry for Local Community Development Programme. Roma: FAO

28

A CASE HISTORY: EVIDENCE OF ANCIENT ANIMAL, WATER AND LAND MANAGEMENT, EXPLOITATION AND DEPLETION, SON OLEZA CHALCOLITHIC OLD SETTLEMENT, VALLEDEMOSSA, MALLORCA, BALEARES, SPAIN.

W. WALDREN

Donald Baden-Powell Quaternary Research Centre, Pitt Rivers Museum, University of Oxford, Oxford, England
Deia Archaeological Museum and Research Centre, Mallorca, Baleares, Spain
EMAIL: lina0142@excite.com or william.waldren.prm.ox.ac.uk

SUMMARY

Of all natural resources, water is with little doubt the most essential to human existence and prosperity. Its availability and use are core and pivot point around which individuals and communities have thrived and revolved throughout time and around which they will continue to subsist in the future. Its distribution, abundance, management, exploitation and lack determines such things as economic growth and survival itself, as well as the further establishment of social status and other forms of power structure and display. Those individuals or communities that own and control water sources are potentially and understandably rich. They become in certain respects masters of their survival as well as that of others. The difficulties and efforts by which it is acquired and coveted are well recorded and shown to be diverse in motive. Its control is perfectly capable of changing terratorial borders, as well as social and economical regimes. Among even the smallest of communities, water as a resource predetermines and often directs the outcome of successful land and animal management, as well as food and much other production.

In prehistoric contexts, it is rare to be able to demonstrate by empirical and other evidence, the importance and way local water resources were used within a specific geographical environment. This is especially so regarding the way water resources were arranged and applied to such essential everyday subsistence activities as agricultural practices, land and animal management within a small insular community and its larger environmental setting. It is further rare to be able to observe within such a community the results and effects of the over-exploitation of resources during a specific period of prehistoric time.

In this particular instance, that of the Son Oleza Chalcolithic Old Settlement, the example occurs in a period between 2500 BC and 1400 BC. This is a period recently defined by the author as the Age of the Accumulation of Surplus (Waldren 1998), or the period more commonly known as the European Copper and Initial Bronze Ages.

It is equally unusual to be able to mark the various cultural, economic as well as demographic changes, as they transpired and continued into the ensuing Bronze and Iron Ages, leading up to the Classical Period of Roman Colonization of the island and the final decline of the local Talayotic culture. It is equally more unusual to be able to delineate these different stages chronometrically, not just relatively, demonstrating the strong continuity and various sorts of interaction that existed during these various stages by using extensive radiocarbon dating, and not only by the artefact evidence. All of these are aspects that heighten its value, pertinence and relevance still further.

In the space alotted, this paper discusses, the aspects of subsistence within a small community both objectively and subjectively and forms an additional part of the larger project of modelling and correlation of data and information to some extent already outlined and discussed earlier in this publication, where it does this in a wider geographical and cultural perspective. The difference is that, there it was seen from a more continental context and not as a regional one, as it is presented here.

In the present paper I will attempt to construct and demonstrate, using empirical evidence supported by radiocarbon documentation, a model of a small regional prehistoric community and its responses and solutions to water, land and livestock management. Also, it will present evidence of the different kinds of use, exploitation and depletion of these essential resources over that period of time, along with the causes and effects behind them. These are events and conditions, all of which resulted in the site´s final abandonment and probable movement to another locale.

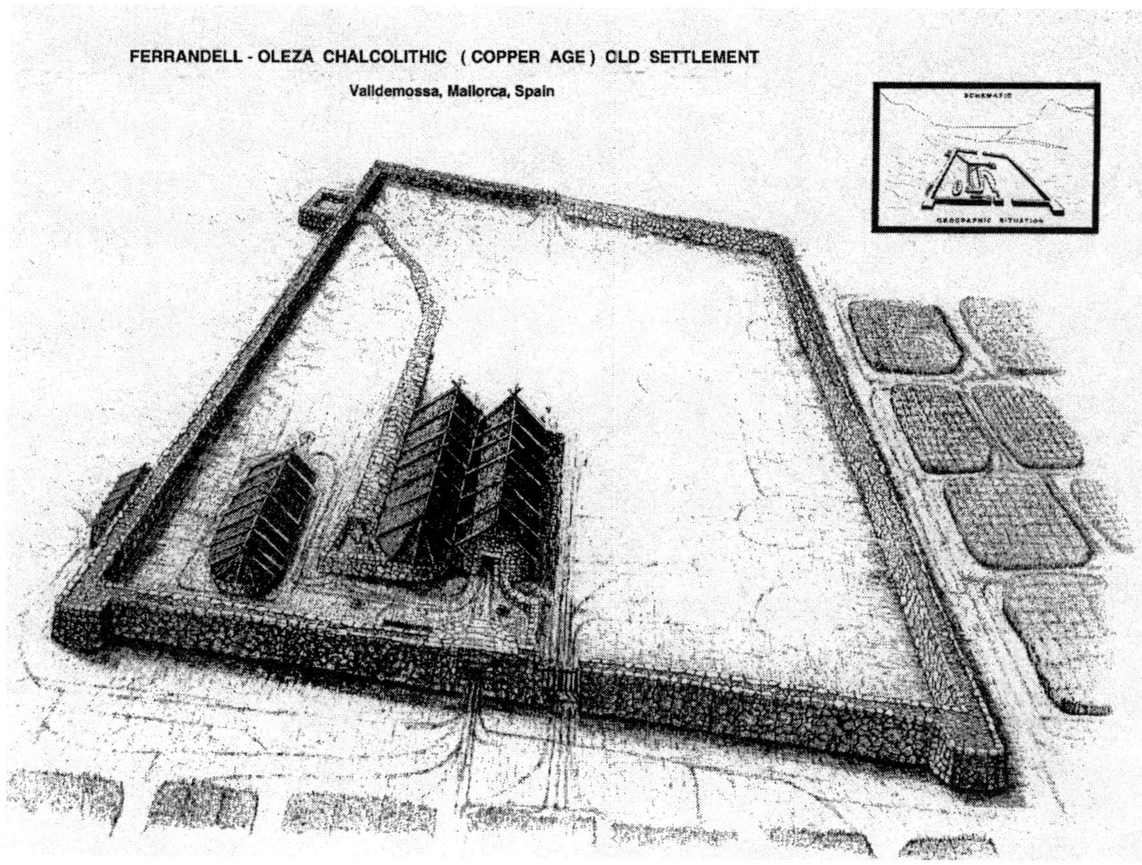

FIGURE 1. RECONSTRUCTION OF THE OLEZIAN CHALCOLITHIC OLD SETTLEMENT

INTRODUCTION

Circa 2500 BC, a small group of settlers, probably an extended family, clan or tribe of an estimated 12 to 16 individuals, cleared a dense, quercus oak forested area at the extreme western end of a soil-rich, intremontane, alluvial basin or plain . The plain is known as the Pla del Rei or Plain of the King. Here, after slash-burning an area of approximately 40,000 m2 (200 by 200 m), the settlers first constructed a series or cluster of huts, circa 2500 BC. Then later, circa 2200 BC, they enlarged the compound by stages into a permanent, high stone compound, enclosing an inner area of approximately 3,600 m2. They tilled the fields on the east and south sides of the walled compound, grew wheat and probably barley, which they harvested with finely made, serrated flint sickles and blade (Figure 3), storing the grains they reaped in large home crafted pots and possible grain storage pits. They raised sheep, goats, pig and small cattle, along with dogs, imported some 1000 years earlier, circa 3500 BC, as demonstrated from nearby cave and rock shelter deposits. As well as growing grain and tending domestic animals, they also produced wool, leather and dairy products, such as cheese, as seen in the artefact evidence of clay, curd strainers and spindle-whorles. They had access to and a knowledge of metallurgy along with the ability to extract local copper ore , which they mined from open face sources near the sea and from the extracted metal fabricated simple and modest items, like copper awls and needles. They had access to and a source of such exotic raw-material as elephant ivory and fabricated items like prismatic, pyrimidial, conical and flat v perforated necklace elements and buttons (FIGURE 4), as well as items like decorated ivory loom combs (FIGURE 5). Most impressive of all ,were their pottery skills, which consisted of finely made, highly burnished and intricately executed, geometrically decorated pottery accredited to the European Bell Beaker Culture (FIGURES 6-8).

Between 2000 BC and 1800 BC, the architectural features of the settlement were to grow to result in an impressive, large stonewall enclosed compound, raising to an estimate height of three meters and divided into distinctive areas of activity: the domestic area, in the west of the inner compound and the work area or animal corral in the eastern sectors (FIGURE 1). The domestic half of the walled compound to the west consisted of two opposing, apsidal, U-shaped, stone constructed long-houses, together forming a single, linked, S-shaped structure, the entrance of one facing south, the other to the north. The northern facing, absidal long-house, in which the western flank or outer wall was shared by a 60m long, elevated stone-lined water channel that transected the walled compound in a northwestern to southern direction, leading from a water reservoir outside the northwest corner of the compound (FIGURE 6). The stone-lined water channel was also covered with stone slabs to prevent evaporation and contamination and carried water to a catch basin at the southern end of the channel, making water available to both the apsidal long-houses.

The water channel was clay lined to filter and purify the water as it ran along the length of the channel's clay bed and contained a series of gradiated levels and a series of stone locks to check the flow of water at stages along its passage to the catch basin. The catch basin was constructed to have two levels, thus, controlling the supply of water held in it. When the larger of the two became empty, the lower one still contained a supply of water and hence signalled the time for the refilling of the catch basin. Its location in a sort of forecourt made it easily accessable to the long-houses and its placement close to the main entrance, built into the south compound wall, thus formed the whole arrangement into a major activity area, within the south sectors.

Main access into the compound was made through a covered passage in a tower-like reinforced area found in the middle of the south compound wall (Figure 1). The stone covered passage through the reinforced area of the south wall was paved with flat stone slabs or flagging like the forecourt. It also contained a recessed area large enough to house a dog, forming an impressive architectural element in the wall, known as the guard´s niche. Combined these architectural features, their overall appearance and organisation demonstrate considerable wealth and power, along with indicating social differentiation.

FIGURE 2. VIEW OF THE PLA DEL REI :
SURROUNDING SOUTHERN HILLS WITH OAK FOREST ABOVE THE MODERN AGRICULTURAL TERRACES

CONCERNING THE MATERIAL EVIDENCE.

By way of the combined evidence, it has been possible to trace the sequence of events that occurred in the Olezian settlement over the span of twelve hundred years of occupation and exploitation. It has been possible to suggest a logical scenario of events for these occurrences over chronometric time, particularly, pertaining to the sequence of probable effects and results of this long duration of occupation and exploitation of the landscape. Thus, arriving the probable causes of the final demise and abandonment of the site.:

Beginning with the period in which the first settlers moved

FIGURE 3. VIEW OF THE HIGHLY ERODED AGRICULTURAL FIELDS TO THE SOUTH AND ADJACENT TO THE WALLED COMPOUND WHERE THE FLINT SICKLES AND BLADES WERE FOUND

into the area, slashed and burned out the dense, endemic quercus oak forest, which still today forms the predominent floral coverage in the higher elevations above the modern agricultural terraces, surrounding the plain. They cleared a large land area on the northwestern, seaward slopes of the plain opening it to intense yearly soil erosion, which over time was to become extremely serious and essentially problematic.

The question of erosion of the surrounding hillsides and slopes has been solved in relatively modern times by the construction of stepped, stone terraces, which effectively regulate erosion. However, this terrace system was not adopted before medieval times and until then the slopes enclosing the alluvial basin relied on the erosional deterent of the floral root system to maintain soils and prevent severe soil displacement. The quercus oak is a tree which has a particularly intricate, tenacious and complicated root system capable of maintaining itself on steep slopes and quickly building rich top-soil from leaves and acorns. Apart from a few endemic pines, it was also the basis of essential wood on Mallorca, long before the introduction the olive and other productive trees. However once that root system is destroyed, except for sage and sedge-like vegitation, there is little to hold the top soil in place.

In areas surrounding the plain, there are today large areas where terracing has not taken place in modern times. In these areas, erosion has been especially severe and is evident even today. Despite the fact that the tenaciouness of the quercus oak still prevails in higher regions and shows strong signs of reincroachment in places where it has been seriously cut back, it is still tenacious enough to become helpful to the preservation of the general landscape. Although at the same time, it is detrimental to the maintennance of the prehistoric archaeological and architectural remains, present in such areas. In locales, such as the immediate area of the Olezian settlement and other sectors where prehistoric architectural remains are found, and modern terracing has not been carried out, these situations, despite the presistent quercus regrowth, the landscape still bears highly visible evidence of the damage done over time from over exploitation of the soils and quercus clearance, especially on the more sloped terrain of the basin.

The location chosen by the first settlers of the alluvial basin was undoubtedly selected for the particular advantage of its shallow gradient, while still having the advantage to command a full view of the major sectors of the alluvial flatlands to the east. It was also selected, undoubtedly, for its equally strategical situation and security afforded by a steep escarpment and ridge which drops precipitous to the sea in the west, directly behind the settlement, as well as for the protection from the prevailing, strong north winds sweeping in from the Mainland, across the Gulf of Lyon.

Recent evidence demonstrates that early Chalcolithic settlers to the alluvial basin originated from the Balearic arc of immediate influence, argued earlier in this conference, in this instance from the areas of southern France and the Catalan regions. It is also reasonably certain that when they first arrived in the area, still earlier settlers were present, dwelling in rock shelters and caves, as there is strong empirical evidence that occupation was in progress in these sites as early as 6th to 4th millennia (Waldren 1982) with other even stronger evidence in the 4th and 3rd millennia (Waldren 2001, in Press). This latter 4th-3rd millennia period, as shown by recent and older radiocarbon documentation from nearby caves and rock shelters, is highlighted by the earliest introduction of domesticated species and Neolithic pottery technology. This was followed by the initial 3rd and early 2nd millennia Chalcolithic settlers arriving from the Mainland. In the author's opinion, this Chalcolithic arrival was markedly different in every way from the island's former 4th to 3rd millennia settlers, who probably introduced the first domestic species and pottery technology during that period, following the islands very first and earliest of human settlement in the 6th to 4th millennia.

During the initial, 3rd millennia period of Chalcolithic/Initial Bronze open-air settlement, the first, small family groups of agriculturists and herders, who prior to moving into the open,

FIGURE 4. AN EXAMPLE:
FINELY MADE SERRATEDTABULAR FLINT SICKLE
FROM THE ' OLEZIAN' SETTLEMENT SITE

evidently, first occupied nearby caves and rock shelters for a time.. Eventually moving into the open, they began clearing a vast area in which they first built simple huts and farmed the cleared land. They gradually became increasingly successful and efficient, managing important subsistence activities, such as crop cultivation, animal and water control, they built by the middle part of this period, circa 2500 BC, an increasing more complex, self-sustaining, profitable and demonstrative expression of their efforts, that of a well delineated and highly organized, permanent settlement, which was to survive until circa 1400 BC. At that time, due to the gradual, over-exploitation and severe erosion of the surrounding agricultural fields, the site was finally abandoned by1300 BC .

LAND EXPLOITATION AND DEPLETION

An intersite radiocarbon inventory of 44 dates supports the evidence of the duration of use of the Olezian Chalcolithic Old Settlement as that of approximately 1200 years (Table 1). During this period, the adjacent fields immediately to the east and south of the walled compound were agriculturally exploited. Today, these areas as well as the areas running down into the alluvial basin are highly eroded to the point that great areas of limestone bedrock are exposed, rendering these zones useless for growing crops, except in the support of olive grove cultivation (FIGURE 3).

This severe erosion of the slopes of the basin and the areas, particularly those involving the settlement compound, are believed to be the result of the first and subsequent intense agricultural activities of the region. The first settlers slashed and burned a large area of the western slope, destroying the complicated root system of the Quercus oak and laying open the soil to erosion for more than a millennium, during which time the denuding of the tops soils did the first stages of the severe damaged seen at present. Today, there is an unbroken line of erosion with a gradient of 10 meters, The gradient is open and visible from the northwestern corner of the settlement compound and runs to the southeastern corner of the settlement. This line of erosion continues with a further unbroken line in a southeast direction until it reaches the flat soils of the alluvial basin or plain. In fact, this unbroken line of erosion has been the subject of extensive study by the Department of Geography of Lampeter University in Wales, during a 4 year period as an unusual and classic example of soil erosion involving an early human settlement (Bell et al, in press).

The seasonal exploitation of this region by early settlers by plowing and harvest would have over a 1200 year period severely depleted the soils immediately surrounding the walled compound, bringing about the sites abandonment. The site's date of abandonment has been ascertained from the analysis of organic garbage (animal bone debris) excavated from inside the settlement site's, stone sided, clay-lined water channel. This organic garbage would have certainly not been present in the water channel if the site was occupied. The excavated evidence of over 149,000 pottery sherds not one of which is accreditable any other age than the Copper/Initial Bronze Age, also testifies to the site's abandoment by 1400 calBC, well before the Talayotic Bronze and Iron Ages.

ANIMAL MANAGEMENT:
STRATIGICAL ORGANIZATION

The situational strategy and manner in which the Olezian Old settlement was planned and organized are additional considerations in any final assessment or interpretation of the site. While these have a certain degree of sophistication, uncommon to our knowledge of Chalcolithic settlement, not only in the management of the agricultural lands themselves, but they also demonstrate an equally important role as well as direct bearing on the kind of specialized animal and water management carried out. All of which is highly discernable from the existing evidence.

The site's location on the sloping western side of the hills of the basin was not in the true sense of the meaning a defensive element, that of protection from an enemy. While it did have a commanding view of the whole of the alluvial plain, its situation more importantly afforded protection from the strong north prevailing winds during winters. Furthermore, the natural topographical contours of the sloping hills was not steep enough so that the construction of the compound walls would have been obligated to follow the undulations of the land itself. Instead the settlement planner's were able to build a more functional, although unconventional, rectangular walled compound in place of the irregular- shaped, curved citadel walls and bastions, normally diagnostic of Copper Age settlement.

The construction of other architectural elements also support the non-defensive nature of the site. Despite the defensive appearence of the 3-4 meter high walls, the site's inventory of artefacts does not contain a single item that could be interpreted as a weapon. Another sign of the non-defensive nature of the compound walls lies in the fact that the site's water storage reservoir is built on the exterior junction of the north and west compound walls, where it would be easily accessible to an enemy. Furthermore, the wide undefended rear entrance, opposite the impressive, reinforced, tower-like main entrance is completely discerna-

FIGURE 5. STONE SLAB-CLAY LINED WATER CHANNEL

ble and would have probably had only a stout wooden gate, serving as an entrance into the corral area of the compound. Therefore the compound walls can be logically interpreted not as defensive in the accepted sense, but rather as a delineation of proprietorial domaine, as well as a statement of social differentiation or class distinction in regard to other settlers of the plain.

Most of all, the construction of such a well built stone wall compound had a more logical and practical reason, that of forming a permanent animal corral. As shown by abundant animal remains of sheep, goat, pig and cattle, the occupants of the compound had a wealth of domesticated livestock, the management of which needed particular attention. Such a variety of herbivors would have necessitate a permanent area, such as a corral, to house, and tend them, but most of all to keep them out of the adjacent grain fields for the longest part of the year until the crops were ready for harvest. What better place than a high, stone walled compound, that would have equally served as an efficient animal corral as well as delineation of proprietorial domaine and class status.

WATER MANAGEMENT

A knowledge of the management of water is also acutely evident in the Olezian Chalcolithic Old Settlement. This is illustrated by the clay-lined, stone-slab, elevated water channel that brought water literally into the outside of two main apsidal-shaped long houses from an uphill water reservoir outside the northwest corner of the west compound wall. This was immediately accessible to the houses (FIGURE). The flow of the water from the reservoir was controled by a sophisticated series of step and lock-like sections filling and flowing over a stone slab divider and into the subsequent section, gradually arriving at the end by stages .

The water filling the reservoir was apparently obtained from rain gathered during the wet season and channeled from the bedrock gulleys of the escarpment, against which the reservoir and northwest corner of the walled compound was built and stored for use in the dry summers. Water was probably also obtained from a water font which was found higher up on the escarpment and which was also channeled down a system of bedrock fissures and into the stone-slab aligned channel during the wet season, leaving the reservoir water in reserve.

Evidence of a few remaining stones demonstrates s the water channel was roofed over by flat slabs to protect against evaporation and contamination. The trickling of the water over the channel's clay lining and filtering also help to assure the quality of the water on arrival at the catch-basin. Such practical features gives us insight into how such hydraulic systems operated and must have similarly functioned elsewhere during the Chalcolithic Period.

HABITATIONS AND DEMOGRAPHICS

There is architectural evidence of two well constructed, stone apsidal (naviform) long houses, each approximately 8-9m long by 3-4m wide, with available floor space of approximately 24 -36 m2, the entrances of which are opposing., one facing south and one facing north. These would have had beamed or rafter roofs with thatching (FIGURE 1). The two apsidal habitations shared a common central wall which divided the two structures. There was another smaller, less well built, stone -sided, thatched building or work shed on the west side of the forecourt and from the evidence of large quantities of pottery fragments was probably a storage area or involved pottery making, although the latter function has no evidence of a firing area, but under the circumstance and preservation this evidence of firing may well have eroded away.

Based on the size of the two dwellings, the general areas and water supply, suggest an occupation of probably 4-5 nuclear families (16-20 individuals), probably an extended family or clan.

This theory is further substantiated by the recent discovery of 4 dolmenic structures to the south of the settlement compound which if considered, as they are in most places as individual family tombs, would give a realistic confirmation of the population estimates suggested . A radiocarbon series of 20 dates for these recent dolmen-like structures covers a period of circa 2200 cal BC to 1400 cal BC., The remaining 24 dates for the settlement's constructional /occupational activity areas and abandonment cover a period of circa 2500 cal BC to 1300 cal BC, therefore, one confirms the other.

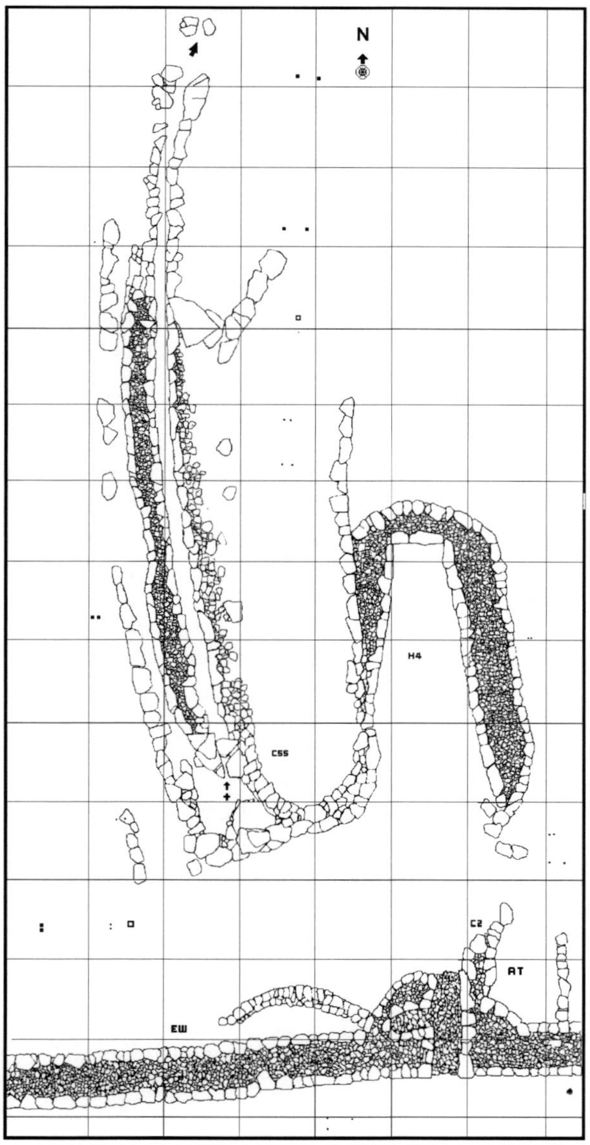

FIGURE 6. WATER CHANNEL TWO APSIDAL HOUSES, SOUTH COMPOUND WALL AND MAIN ENTRANCE

ARTIFACTUAL INTERPRETATION

The sophisticated and exceptional wealth of artefacts and other evidence available, leaves no doubt as to the economic success and social status of the occupants. This is especially evident not only in the architectural achievement, geographic situation, wealth in livestock but also in the possession and availability of the items represented by the artefacts. Excluding the presence of large quantities of skilled and beautifully decorated Beaker pottery, even the more utilitarian common ware show the wealth of pottery the occupant possessed. There are numerous pottery cheese strainers, spindle whorls that testify to wool and dairy production. Well made bone tools and presence of beach cobbles demonstrates leather making. The numerous elephant ivory, bone, stone and shell buttons or necklace elements also show the wealth of smaller personal items. All of which cannot be otherwise interpreted in any other manner apart from that as a strong indication of personal wealth and prosperity. Such evidence is also the sign of a particular sophisticated refinement, technological skill and a high socal status of the occupants of the Olezian settlement as well as a social diffentiation regarding the other occupants of the Pla del Rei. More importantly, however, it is evidence concerning and enrichening, not only, our knowledge and understanding of the local Chalacolithic Period, but also that of the Continental European Copper Age, as well as that of the International Bell Beaker Culture itself.

THE EVIDENCE AS CONCLUSION

The empirical evidence for this scenario in the Olezian Chalcolithic Old Settlement can be summarized in the following:

(1) the current list of 44 radiocarbon analyses from circa 2500 BC to circa 1300 BC demonstrates the 1200 year duration of the site's occupation. These individual dates also assist us in our knowledge of the different stages of development of the site from its inception to its abandonment, an aspect that is not so clear in other sites of this period elsewhere.

(2) This long duration is further reflected in the site's architectural evidence, where we see its long-term and sophisticated development into a permanent establishment, one which displays both proprietorial domaine and capital investment (Gilman 1981), as well as the evidence of the technological skills possessed by the occupants. This in turn gives us insight into its social and economic structure and organization, showing the aspects of social differentiation and apparent economic stability over that period. There is also a display of the kinds of interaction, both economic and social as well as religious, that took place over time between the different sites in the region, until its eventual decline, due to over exploitation and land depletion.

In the site's architectural features alone, we can observe

the occupants'various responses to everyday problems such as water, field and livestock management. These are particularly visible in the inhabitant's construction of a permanent corral and work area sector and in the sophisticated and effective hydraulic system.

We see reflected the use of the adjacent fields for crop production, along with the development of the necessary lithic industry essential for the harvesting of the crops and other tasks.

The proximity of the fields used for agriculture and their close relationship to the architectural wall features of the compound, tells us something of the organization of both food production and management of the animals, which would have had for much of the year to have been very closely monitored, particularly while the nearby crops were maturing. This additionally shows a desire to control both in order to secure the maximum production and to assure and attain their invested interests.

The cleverly engineered water system consisting of reservoir, channel and catch basin further enhances our understanding of their exceptional inventive skills and the general importance, response and respect toward the problem of water management.

(3) The extensively dated series of four recently discovered dolmenic structures, contemporarily datable with the settlement itself, give substance to the argument regarding the demographic assessments of the site. This is an estimate of which is equally based on the available water and number of dwellings and living space. The number of 16 to 20 individuals or four to five nuclear families give us clues as to the probable demography of the site as that of a tribe or extended family. At the same time, these estimates give us some insight as to the possible population of the region itself.

(4) The obvious wealth of the artefact assemblages is evident in the nature and character of the various objects recovered from the site, some of which are exotic and strong evidence of long distant trade links, such as imported elephant ivory and other obviously off-island items. The extraordinary quantity and diversity of luxuriously, decorated Beaker wares and apparently later imported Boquique pottery gives substance to the arguments regarding the duration of occupation the social status and state of the economy of this small community group. The presence of a quality lithic and bone tool industry and simple copper tools reflect the technologies present. Other items such as the tools and vessels obviously used for the production of dairy products and wool, still further emphasise and enhance our knowledge of the occupant's possessions and responses to daily subsistence practices.

(5) One of the major physical factors and the single major cause leading to the final abandonment of the site is the question of over exploitation and depletion. This is found in the unbroken line of erosion, extending from the northwest corner of the settlement to the southeast and into the alluvial basin of the Pla del Rei. There has been no modern terracing carried out along this line of erosion that would have deterred or interrupted the depletion of the top soil, particularly once the indigenous quercus oak was removed.

Consequently, the present-day landscape clearly demonstrates the result of this prehistoric and present-day continual alteration of the soils over the last 4500 year period. This has left vast areas of denuded bedrock exposed, where the fields cannot sustain normal agricultural practices, apart from the olive groves. Besides, recent research has demonstrated that olive cultivation is extremely detrimental to soil preservation and even excellerates erosion. Therefore, there is little doubt that this unbroken erosion line, along with the seasonal and increasing exploitation of the areas to the east and south of the settlement for agriculture during the prehistoric period played the major factor in the site's final abandonment, circa 1400 BC. At this time, we can assume that the surrounding agricultural fields became no longer viable or productive. As there is none of the general signs of a catastrophic abandonment of the site, such as would be found in the form of burning and distruction of the site's architectural elements, conditions that were very much present in the nearby Son Ferrandell Late Bronze and Iron Age Younger Settlement. Contrarily, in the Chalcolithic Old Settlement, there is only evidence of an apparent organized and orderly abandoment, where the inhabitants took with them whatever could be salvaged and resettled closer to the more fertile and abundant soils of the alluvial basin. There is strong recent evidence that this was the case.

The date of the abandonment of the Chalcolithic Old Settlement is well established as circa 1300- 1400 cal BC. This evidence is represented by the presence of organic matter in the form of animal bone and other debris found during excavations in the water channel and its radiocarbon dating accreditable to that date. This logically indicates that no occupation of the site would have taken place after this essential architectural element became disfunctional.

In conclusion, the present example of the use, management and final depletion of natural and man-made resources by a small but highly informative and unusual community, offers us an unique opportunity in the study of urban prehistoric settlement and environment. It is also one that is accreditable to a particularly problematic area in prehistory and one that involves the increasingly, socially and economically structured stages of the local as well as European Copper Age, a period when societies were first emerging from the former Neolithic Period. It also offers us an opportunity to observe the effects and results of land use, management of water and livestock, over an extensive and little known period of prehistoric time, along with the combination of geographic factors, natural and human, responsible for the culmination of what appears, otherwise, to have been an exceptional prehistoric continuity.

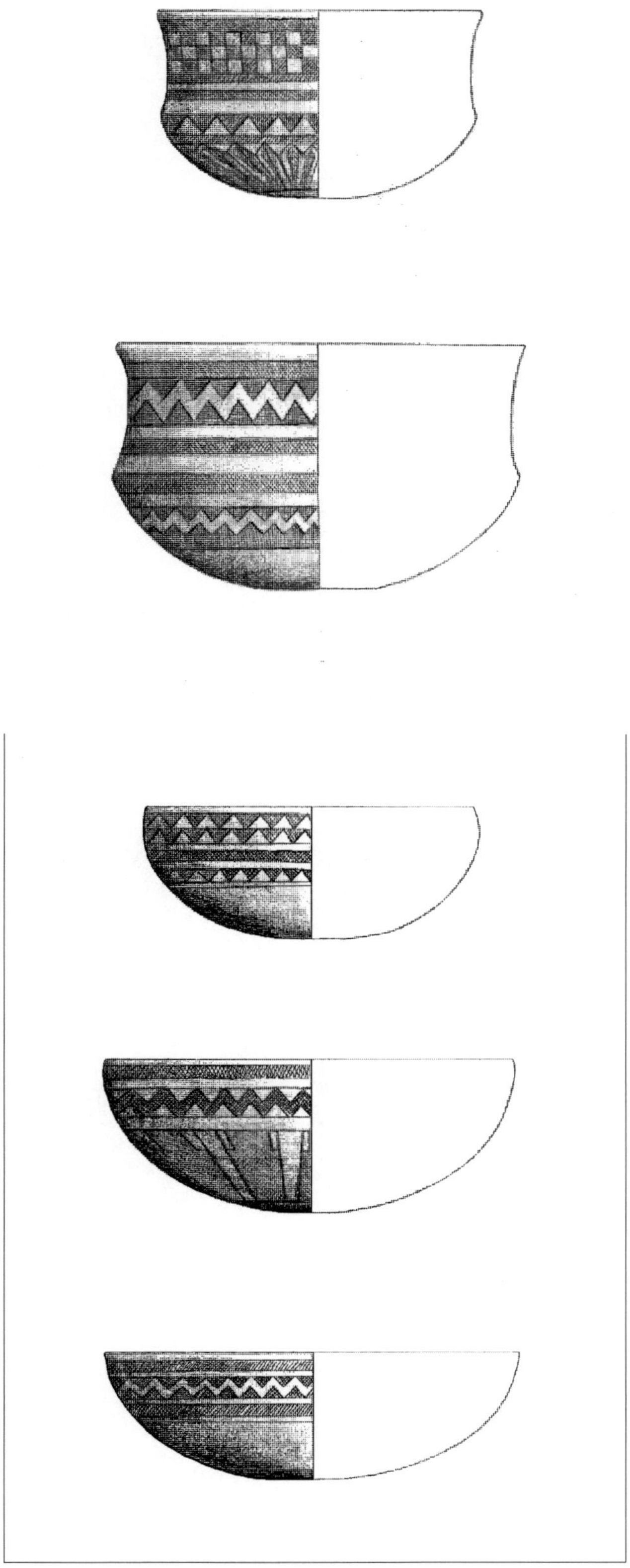

FIGURE 7. EXAMPLES OF BALEARIC BEAKER POTTERY

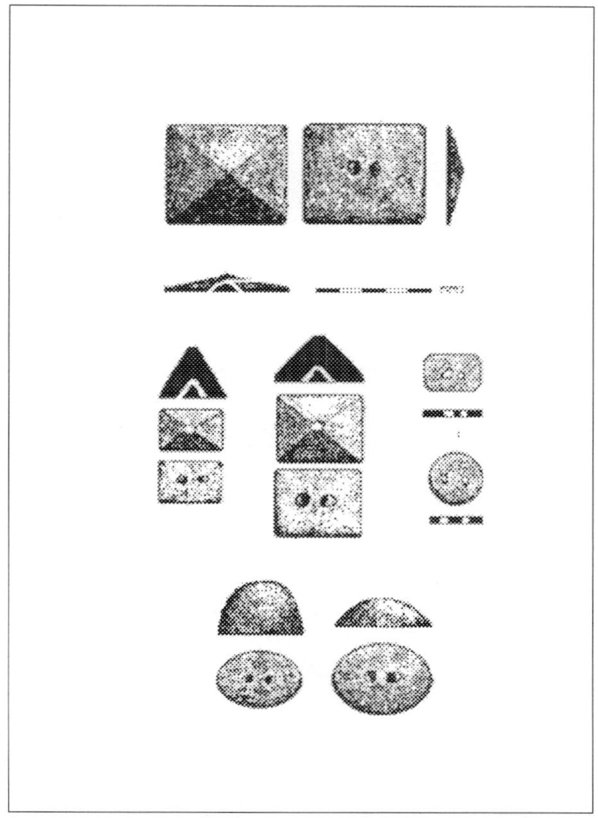

FIGURE 8. EXAMPLES OF 'V' PERFORATED BUTTONS

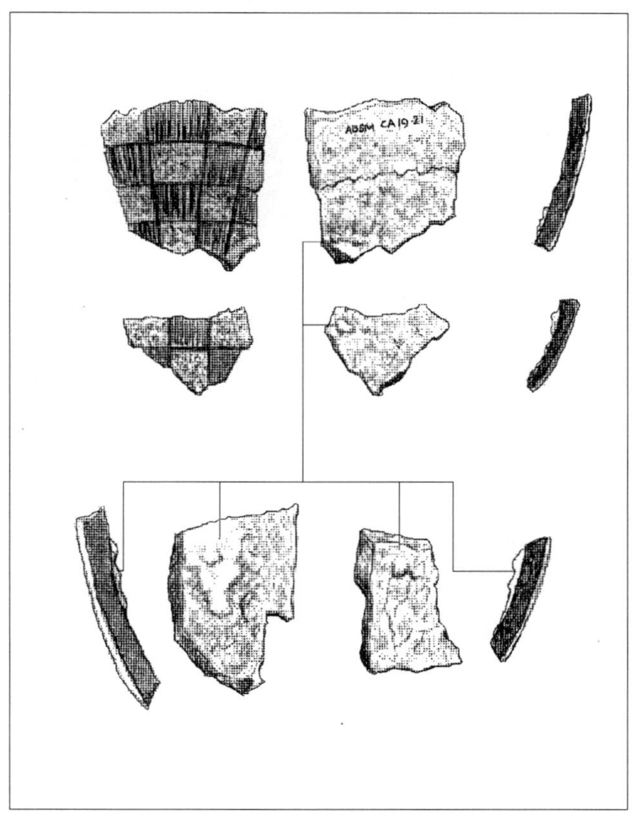

FIGURE 9 EXAMPLES OF BEAKER CRUCIBLE SHERDS

RADIOCARBON DATES: SON OLEZA CHALCOLITHIC OLD SETTLEMENT and DOLMEN AREA 2002

nos	site designation	years BP	95.4 probability range	date cal BC	lab.inv.no
1.	**SFO-OS-CH1**	2990 ± 160 yrs	1610- 820 (3559-2769)	***1261 cal BC***	QL-4044
2.	**SFO-OS-CH1**	2990 ± 160 yrs	1610- 820 (3559-2769)	***1261 cal BC***	IRPA-1053
3.	**SFO-OS-EXW**	3070 ± 50 yrs	1440-1250 (3389-3199)	***1357 cal BC***	HAR-3490
4.	**SFO-OS-C2**	3095 ± 70 yrs	1520-1250 (3469-3199)	***1403 cal BC***	BM-1698
5.	**SFO-OS-WC**	3150 ± 300 yrs	2192-2161 (4141-4110)	***1428 cal BC***	BM-1988
6.	**SFO-OS-SWF-A**	3180 ± 80 yrs	1670-1668 (3619-3617)	***1445 cal BC***	QL-4191
7.	**SFO-WOS-DOL1: 17-2M8 EX**	3130 ± 35 yrs	1520-1300 (3420-3250)	***1450 cal BC***	KIA-12274
8.	**SFO-OS-WC**	3230 ± 30 yrs	1670-1410 (3619-3359)	***1516 cal BC***	BM-1567
9.	**SFO-OS-CSS**	3240 ± 30 yrs	1611-1441 (3560-3390)	***1519 cal BC***	QL-1224
10.	**SFO-OS-C2**	3240 ± 30 yrs	1611-1441 (3560-3390)	***1519 cal BC***	IRPA-1054
11.	**SFO-OS-EXW-UW**	3260 ± 100 yrs	1854-1849 (3803-3798)	***1523 cal BC***	QL-4040
12.	**SFO-OS-WC-IN**	3280 ± 120 yrs	1880-1310 (3829-3259)	***1549 cal BC***	QL-1896
13.	**SFO-OS-H4**	3300 ± 60 yrs	1738-1708 (3687-3657)	***1565 cal BC***	UtC-3192
14.	**SFO-OS-WC**	3350 ± 330 yrs	2569-2538 (4518-4487)	***1673 cal BC***	BM-1988R
15.	**SFO-OS-WC**	3390 ± 80 yrs	1884-1682 (3833-3631)	***1712 cal BC***	BM-2312R
16.	**SFO-WOS-DOL1: 16-2M4**	3445 ± 40 yrs	1880-1610 (3830-3660)	***1745 cal BC***	KIA-12275
17.	**SFO-OS- EW-INT-B1**	3460 ± 45 yrs	2115-2085 (4064-4034)	***1745 cal BC***	UtC-9021
18.	**SFO-WOS-DOL3: 15-EEP**	3450 ± 30 yrs	1880-1680 (3880-3630)	***1760 cal BC***	KIA-12273
19.	**SFO-OS-CH1-INT4**	3450 ± 110 yrs	1878-1839 (3827-3788)	***1749 cal BC***	QL-4043
20.	**SFO-WOS-DOL1: 12- 4I3**	3440 ± 30 yrs	1880-1640 (3830-3590)	***1770 cal BC***	KIA-12267
21.	**SFO-WOS-DOL1: 11- 5L7**	3440 ± 35 yrs	1877-1840 (3826-3789)	***1770 cal BC***	KIA-10559
22.	**SFO-WOS-DOL1: 9- 5L8**	3445 ± 30 yrs	1930-1680 (3879-3629)	***1770 cal BC***	KIA-10558
23.	**SFO-OS-WCS**	3430 ± 30 yrs	1880-1630 (3830-3586)	***1755 cal BC***	KIA-16267
24.	**SFO-OS-SW-AG**	3405 ± 35yrs	1880-1600 (3830-3556)	***1740 cal BC***	KIA-16268
25.	**SFO-WOS-DOL3: 14-EWEX**	3445 ± 30 yrs	1880-1680 (3850-3590)	***1760 cal BC***	KIA-12270
26.	**SFO-WOS-DOL3: 13- NEX**	3460 ± 30 yrs	1880-1680 (3850-3590)	***1780 cal BC***	KIA-12268
27.	**SFO-OS-EW-UF-S**	3470 ± 50 yrs	1901-1742 (3850-3691)	***1819 cal BC***	QL-4100
28.	**SFO-OS-AT-QB**	3490 ± 30 yrs	1901-1742 (3850-3691)	***1827 cal BC***	QL-1859
29.	**SFO-OS-CH1-INF3**	3490 ± 30 yrs	1946-1768 (3895-3717)	***1829 cal BC***	QL-4042
30.	**SFO-WOS-DOL1: 2- 5M3**	3540 ± 30 yrs	2198-2159 (4147-4108)	***1860 cal BC***	KIA-10557
31.	**SFO-OS-EWF**	3550 ± 110 yrs	2007-2004 (3956-3953)	***1884 cal BC***	UtC-2742
32.	**SFO-WOS-DOL1: 12-5L7**	3550 ± 30 yrs	2030-1988 (3979-3937)	***1884 cal BC***	KIA-10560
33.	**SFO-WOS-DOL1: 1**	3565 ± 35 yrs	2030-1770 (3980-3720)	***1925 cal BC***	KIA-1228
34.	**SFO-OS-EXW**	3585 ± 35 yrs	2040-1880 (3990-3930)	***1929 cal BC***	UtC-8952
35.	**SFO-WOS-DOL1: 10-6M4**	3605 ± 35 yrs	2137-2076 (4086-4025)	***1970 cal BC***	UtC-10079
36.	**SFO-OS-EW-INT-B2**	3620 ± 50 yrs	2123-2097 (4072-4046)	***1985 cal BC***	UtC-9022
37.	**SFO-WOS-DOL1: 4-5M9**	3625 ± 30 yrs	2123-2097 (4072-4046)	***1995 cal BC***	KIA-10583
38.	**SFO-WOS-DOL1: 7-5M6**	3625 ± 30 yrs	2279-2231 (4229-4180)	***1995 cal BC***	KIA-10582
39.	**SFO-WOS-DOL1 J3-G13**	3640 ± 50 yrs	2199-2158 (4148-4107)	***2030 cal BC***	BM-1981R
40.	**SFO-OS-EW- INT-CC**	3680 ± 45 yrs	2199-2158 (4148-4107)	***2074 cal BC***	UtC-9023
41.	**SFO-WOS-DOL1: J9 W17**	3700 ± 30 yrs	2290-1930 (4240-3880)	***2110 cal BC***	QL-1592
42.	**SFO-WOS-DOL1: G13 FLT**	3720 ± 35 yrs	2275-2141 (4224-4198)	***2138 cal BC***	BM-1981
43.	**SFO-OS-NW-BB**	3790 ± 90 yrs	2480-1970 (4429-3919)	***2239 cal BC***	QL-1636
44.	**SFO-OS-EW-BB**	4030 ± 60 yrs	2585-2568 (4537-4512)	***2533 cal BC***	BM-1843R

TABLE 1 .

BIBLIOGRAPHY

GILMAN, A.,
1981. 'The Development of Social Stratification in Bronze Age Europe', Current Anthropology 22, pp. 169-224.
WALDREN, W.H.,
1979. A Beaker Workshop Area in the Rock Shelter of Son Matge, Valldemosa, Mallorca, World Archaeology, Routledge and Kegan Paul, London, vol. 11, No. 1, 43-67.
WALDREN, W.H.,
1980. Radiocarbon Determination in the Balearic Islands: an Inventory. Donald Baden-Powell Quaternary Research Centre, Oxford University, Oxford.
WALDREN, W.H.,
1982. Balearic Prehistoric Ecology and Culture: The Excavation and Study of Certain Caves, Rock Shelters and Settlements, British Archaeological Reports, BAR International Series 149 (ii), Appendix 3A, 675-711.
WALDREN, W.H.,
1984a. Chalcolithic Settlement and Beaker Connections in the Balearic Islands, in W.H. Waldren, R.W. Chapman, J.G. Lewthwaite and R.C. Kennard (eds.), Ist Deya Conference of Prehistory, Early Settlement in the Western Mediterranean Islands and their Peripheral Areas, Oxford, British Archaeological Reports, BAR International Series 229, vol. iii, 911-965.
WALDREN, W.H.,
1984b. The Lithic Industry of the Balearic Islands...Its Olezian Tradition of Tabular Flint Blades, in W.H. Waldren, R.W. Chapman, J.G. Lewthwaite and R.C. Kennard (eds.), Ist Deya Conference of Prehistory...Early Settlement in the Western Mediterranean Islands and their Peripheral Areas, Oxford, British Archaeological Reports, BAR International Series 229, Oxford, vol. iii, 859-909.
WALDREN, W.H.,
1986. Balearic Pentapartite Division of Prehistory.. Radiocarbon and Other Age Determination Inventories, British Archaeological Reports, BAR International Series 282, Oxford.
WALDREN, W.H.,
1987a. A Rose by Any Other Name...a Question of Balearic Beakers...the Evidence, Oxford International Conference, in W. Waldren and R.C. Kennard (eds.), Bell Beakers of the Western Mediterranean: Definition, Interpretation, Theory and New Site Data, Oxford, British Archaeological Reports, BAR International Series, 331 (i), vol. i, 369-482.
WALDREN, W.H.,
1987b. A Balearic Beaker Model, Ferrandell-Oleza, Valldemosa, Mallorca, Oxford International Conference, in W. Waldren and R.C. Kennard (eds.), Bell Beakers of the Western Mediterranean: Definition, Interpretation, Theory and New Site Data, Oxford, British Archaeological Reports, BAR International Series, 331 (i), vol. i, 207-266.
WALDREN, W. H., CUBÌ, C.
1987b. A Case History: the Function and Origin of Mallorcan Cast lead Votive Funerary Jewelry, Oxford International Conference, in W. Waldren and R.C. Kennard (eds.), Bell Beakers of the Western Mediterranean: Definition, Interpretation, Theory and New Site Data, Oxford, British Archaeological Reports, BAR International Series, 331 (i), vol. i.
WALDREN, W. H.
1986. The Balearic Pentapartite Division of Prehistory. Radiocarbon and other age determination inventories, British Archaeological Reports, International Series, 282, Oxford.
WALDREN, W. H., ENSENYAT, J. AND CUBÌ, C.
1991. Prehistoric Archaeological Elements, Ferrandell Oleza Chalcolithic Old Settlement, DAMARC Series 20, Donald Baden-Powell Quaternary Research Centre, Pitt Rivers Museum, Oxford University, England.
WALDREN W.H., ENSENYAT J. and CUBI GRIMALT, C.
1991. Son Mas: a New Mallorcan Prehistoric Sanctuary, circa 2000 BC-1750 BC', in Waldren, Ensenyat and Kennard (eds.), IInd Deya Conference of Prehistory...Archaeological Techniques, Technology and Theory, Oxford, British Archaeological Reports, BAR International Series, 574 (i-iii), vol ii, 267-312.
WALDREN W.H. and M. VAN STRYDONCK
1994. Prehistoric Sanctuary of Son Mas: a Radiocarbon Analysis Survey , Dating the Activity Sequences of the Santuary DAMARC Series, 24, Deya, Mallorca.
WALDREN W.H., ENSENYAT, J. AND CUBI , C.
1994. Prehistoric Architectural Elements, Ferrandell-Oleza Chalcolithic Old Settlement: Radiocarbon Dating and Artefact Distributional Survey, DAMARC Series, 22, Deya, Mallorca.
WALDREN W.H. 1995: The Function of Bell Beaker Pottery as a Ceremonial and Votive Object, in Waldren, Ensenyat and Kennard (eds.), IIIrd Deya Conference of Prehistory, Ritual, Rite and Religion in Prehistory, Oxford, British Archaeological Reports, BAR International Series, 611 (i-ii), vol i, 238-263.
WALDREN, W. H.
1997. 'The Definition and Duration of the Beaker Culture in the Balearic Islands: a Radiocarbon Survey' Oxford Journal of Archaeology, Blackwell Publishers,vol 16, no 1. pp 25-39.
WALDREN, W. H.
1998. The Beaker Culture of the Balearic Islands. An inventory of evidence from caves, rock shelters, settlements and ritual sites, British Archaeological Reports, International Series, 709-Western Mediterranean Series, 1.
WALDREN, W. H.
2001. 'A New Megalithic Dolmen from the Balearic Island of Mallorca: its Radiocarbon Dating and Artefacts', Oxford Journal of Archaeology, Blackwell Publishers,vol 20, no 3. pp 241-263.

29

SOURCES OF WATER ANCIENT AND MODERN : LINKING ARCHAEOLOGY AND HYDROGEOLOGY IN SOUTH-WEST VITI LEVU, FIJI.

JEFFREY DAVIES (1), **JANETTE DAVIES** (2)

(1) British Geological Survey, Wallingford, UK
EMAIL: jdav@bgs.ac.uk
(2) Centre for Cross-Cultural Research on Women, University of Oxford, UK
EMAIL: janette.davies@qeh.ox.ac.uk

SUMMARY

The Sigatoka coastal dune complex, Sigatoka Valley and Navua Delta in south-western Viti Levu are important sites of archaeological study in Fiji. Artefacts including Lapita pottery found in the dune complex indicate one of the earliest sites of human settlement in Fiji. In the other two areas, pre-European contact Fijian mataqali (village) groups constructed defensive hill- top and ring ditch forts with irrigated gardens and terraces for the cultivation of root crops. These areas now produce staple food such as rice, cassava and dalo, export crops such as sugar, and good quality fruit and vegetables for the expanding tourist industry. Along the Sigatoka valley river water is drawn with low-lift pumps to irrigate crops. Groundwater, from the underlying Holocene age alluvial aquifers, is used to irrigate flood plain areas more remote from the river. Late Quaternary sea level has impacted upon the hydrogeological characteristics of these alluvial aquifers whose groundwaters have been important to both past and present communities in the areas.

INTRODUCTION

Fiji lies at the junction of Melanesia and Polynesia within the South Pacific. The Sigatoka Valley and coastal dune complex, and Navua Delta on the main island of Viti Levu are important sites of archaeological study (Birks, 1973, Parry, 1981 and Parry, 1987). Following the presence of Lapita culture peoples between about 3500 and 2400 yrs BP, several waves of Melanesian and Polynesian peoples migrated into south-western Viti Levu where they developed hill forts, defensive ring ditch villages and irrigated gardens for root crop cultivation. Modern Fijian history dates from the mid 19th century period of European contact when the Fijian language was first set down in written form.

Viti Levu, the largest island in the Fiji group, is located between longitude 177°20' E and 178°50' E and latitude 17°20' S and 18°20' S (FIGURE 1). The island is oval in shape with a land area of about 10 400 km2. Much of the rugged mountainous centre of the island lies at elevations higher than 1000 metres above mean sea level (msl). The climate experienced by the island is controlled by prevailing south-east trade winds, the leeward north-western half frequently experiencing drought conditions during the May to October dry season. An average annual rainfall of 2000 mm is received by the south eastern half of the island during the wet ‘cyclone season’ between November and April, when intense ‘wet season’ storms produce flash flood conditions. These waters, heavily laden with suspended silts, discharge along the main rivers to flow into the sea turning the off-shore waters red. The Sigatoka River is the longest river in Viti Levu, rising in the central highlands, to flow to the southwest within an incised valley to the coast at Sigatoka. The Navua River drains the Namosi Highlands, flowing through a deep gorge and the Navua Delta to the sea. The distribution of landforms, settlement patterns and water sources along the Sigatoka Valley and within the Navua Delta were mapped using the interpretation of aerial photography and ground surveys by Parry (1981 and 1987).

During the 1980s and 1990s, tourism expanded with the

development of new resorts in the drier south-western part of Viti Levu. The resultant increased demand for potable water and fresh agricultural produce prompted investigation and development of sources of sediment free potable water (Asian Development Bank 1986 and Fawcett Mouchell 1988). South-western Viti Levu includes the coastal towns of Nadi, Sigatoka and Navua, the inland village of Keiyasi and some of the most scenically attractive and agriculturally productive land in Fiji.

The Sigatoka area, referred to as the 'Salad Bowl of Fiji', is a main source of local food staples, such as rice, cassava, yams and dalo, as well as sugar, the main cash crop produced for export. The cultivation of good quality fruits and vegetables to supply the demands of the expanding tourist industry is being developed using low-lift pumps to provide irrigation water from the Sigatoka River (Asian Development Bank 1986). This development is limited to areas within 400 m of the riverbank by the capacity of the spray irrigation systems used. The irrigation of areas further from the river requires the use of groundwater from Late Quaternary age alluvial sediments underlying the Sigatoka Valley (Davies 1992).

Study of these resources requires collection of data relating to the geology, climate, topography, agriculture and geomorphology of the areas of interest. As a basis maps prepared by Parry (1981 and 1987) were used to identify areas underlain by Quaternary age sediments and the location of water sources in the Sigatoka Valley and Navua Delta areas. This paper presents the results of joint studies undertaken by the British Geological Survey (BGS) and Mineral Resources Department (MRD) of Fiji, outlining the possible impact of Late Quaternary changes in sea level upon alluvial sediment deposition, and describing the importance of these aquifers to the current development of Fiji.

ARCHAEOLOGY

Parry (1981 and 1987) studied the pre-history of the Sigatoka Valley and Navua Delta areas using a combination of ethno-history and aerial photographic interpretation to identify pre-European contact period settlement patterns, water sources and irrigated areas. The impact of environmental changes, included change in water levels, rates of flooding that caused river channel migrations, and rates of coastal dune accumulation and erosion, on settlement patterns were assessed. Parry (1987) argues that such changes may be responsible for gaps in archaeological records that may otherwise be interpreted as abandonment features. These factors were collated with cultural aspects recorded by oral tradition and ethno-history to produce a cultural sequence showing the development of traditions of village fortifications, specialised irrigation and pond-field cultivation practices. These studies are good examples of how the integration of ceramic, oral tradition, ethno-historic and environmental data, can produce a conceptual model of cultural activity that can be used, with settlement location patterns, as starting points for further archaeological studies.

Parry (1987) outlined the importance of the limited ceramic evidence available to define the chronology of Fijian history between 3500 yrs BP and 1650 AD, identifying 3 phases of ceramic tradition in Fiji.

LAPITA TRADITION

The Austronesian peoples who moved eastwards through Melanesia into western Polynesia to become established in western Fiji between 3000 – 3500 yrs BP brought the distinctively patterned Lapita ceramic ware into the region (Spriggs, 1997). They were coastal people who relied on fish and shellfish for food supply and made use of stone and shell implements. 14C dates of 2460-3540±100 yrs BP have been recorded from charcoal recovered from sites at Natunuku, Yanuca and Sigatoka (Birkes, 1973).

IMPRESSED CERAMIC TRADITION

The Natavu phase peoples, apparently of Melanesian origin, moved into the Viti Levu area to become dominant during the 100 BC – 1000 AD period. As well as having a distinctive ceramic ware with impressed patterns there is evidence that they hunted wild pig, kept domestic animals such as dogs and chickens, and engaged in root crop horticulture.

INCISED CERAMIC TRADITION

The Vuda or Ra phase peoples, that appear to date from 1000 AD, may have been Polynesian immigrants from the east. This immigration into the Sigatoka area coincides with the construction of the earliest earthwork fortifications and fortified caves, such as those found in the Qalamari Limestones near Dubalevu. This event may have initiated a period of continuous warfare and hostility that remained as features of Fijian life until the mid 19th century.

Birks (1973) recorded these three levels of cultural occupation in his study of the Sigatoka sand dune complex. The levels were dated using 14C radiocarbon of charcoal fragments, at:

(1) Level one 2460±90 yrs BP.
(2) Level two 1720±80 yrs BP.
(3) Level three approx 1000 yrs BP (based upon dune sand accumulation)

Shards of Lapita style pottery were found in the seasonally occupied Level one. This level was separated from later occupation levels by periods of abandonment caused by changes in the course of the Sigatoka River and prevailing wind directions that would have impacted upon the rate of dune accumulation.

Between 1880 and 1965, the Fiji Native Lands Commission collected the General Histories of the Native Land Enquiries (Parry 1987). These volumes of oral tradition relate to the origins and migrations of different groups of people and the circumstances leading to their occupation of particular areas of land. These traditions indicate that originally a Melanesian people migrated through the Sigatoka Valley to south-west Viti Levu from the north of the island. Latterly the migration of people of Tongan origin into the lower reaches of the Sigatoka valley led to a period of upheaval and warfare between 1450-1650 AD, as preserved in local Meke chants. It was during this period that the present day settlement pattern, with hill forts, ring ditch fortifications and irrigation systems and gardens in southern and western Viti Levu were established. Turmoil occurred again during the 18th century by peoples forced westward by invasions of Tongans from the east, and by the arrival in about 1750 AD of Tongans who settled at Korotogo on eastern side of the lower Sigatoka valley. From then until the arrival of the Europeans in the mid 19th century there followed a period of occasional violent warfare with short periods of peace. European influence began with the arrival of European traders from Australia dealing in sandalwood, and Americans dealing in Bech de Mer during1840-1850, who also introduced guns to Fiji.

Following European contact, missionaries introduced Christianity to the lower Sigatoka valley in 1850-1860 AD. The hill tribes did not accept Christianity until after the disastrous measles epidemics of 1875-1876. Von Hugel, reported that the disease, introduced from Sydney by the crew of the warship HMS Dido, resulted in the death of at least 25% of the rural population of Viti Levu (Roth and Hooper 1990). The death of their village chiefs due to measles caused an uprising of hill peoples, which was put down by British led local troops during the 1880s. During this campaign Naval officers drew the first maps of the Sigatoka Valley, while the troops systematically destroyed the hill forts of the area.

The Navua River and its distributaries dominate the geomorphology of the Navua Delta (Parry 1981). Communities were established on levees between the distributary channels. Food crops such as dalo were cultivated in low-lying well-watered gardens established in the abandoned channels. Sea fish provided an important source of protein. The delta area was isolated from the rest of Viti Levu by the Namosi Highlands to the north, breeched by the deep Navua gorge, and the sea to the south. Parties of hill tribes later entered the area from the north by way of the gorge. In response to these raids the local communities erected fortified villages surrounded by moats or ring ditches. A number of these abandoned ring ditch settlements are identifiable from the air and on aerial photographs of the delta and are described by Parry (1981).

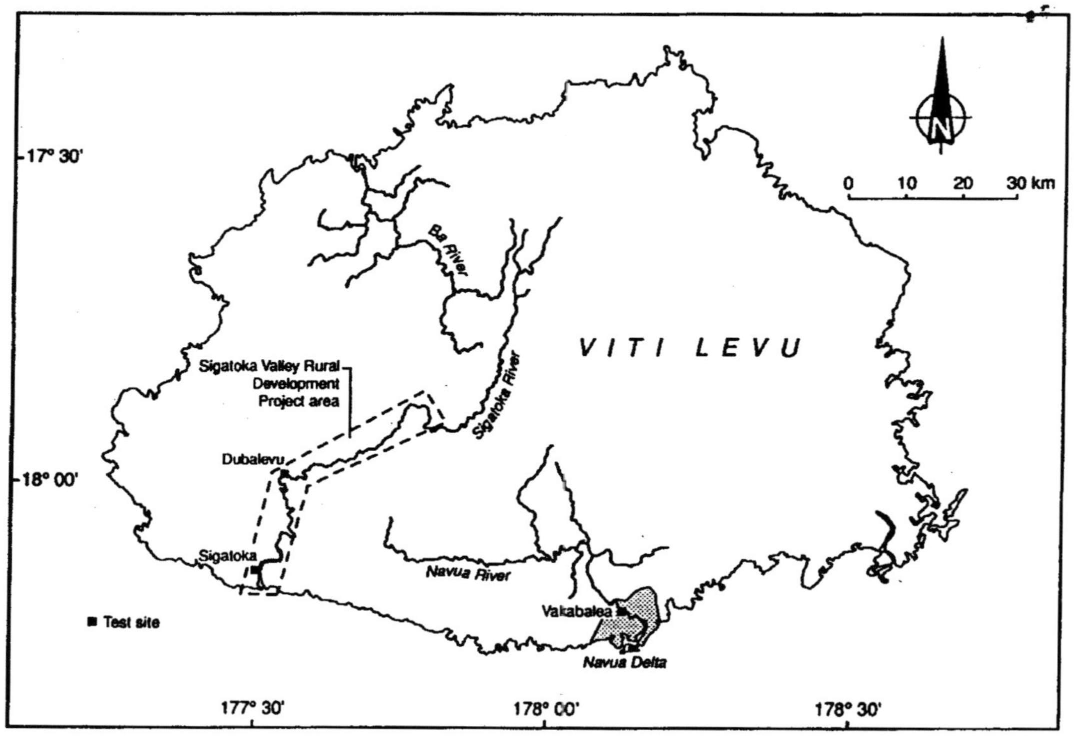

FIGURE 1. MAP OF VITI LEVU
SHOWING THE LOCATIONS OF PROJECT AREAS AND MAIN TEST SITES.

CULTURAL ASPECTS

Rural society in Fiji is formerly structured and requires certain ceremonies of welcome to be enacted on first contact, during official discussions and at special events. These ceremonies and social structure were developed during the pre-historic period and are still very much part of everyday life (Tora 1986).

Special ceremonies, as enacted for the visit of the President of Fiji to the Sigatoka Valley to discuss the progress of the Sigatoka Valley Rural Development Project (SVRDP) with local communities at Keiyasi, require the presentation of several tabua (whale tooth), a yaqona bush, decorated mats, several pigs and heaps of dalo as a sevusevu (obeisance). The kava ceremony is elaborate involves local men in traditional dress, arranged according to rank and custom. Following prolonged discussions and drinking of kava there is a feast, with foods cooked in a traditional lovo earth oven, accompanied by groups of men and women dancing and chanting meke. The meke themes symbolise major aspects of cultural life such as seasonal farming and fishing activities as well important historical events such as battles. Meke are usually accompanied by the beating of several lali or wooden drums made from hollowed out tree trunks.

Typically, local communities are involved in hydrogeological survey field work activities. Before field work can begin a sevusevu (obeisance) must be undertaken with the presentation of yaqona root to the local mataqali ratu or village chief. The yaqona is then crushed and infused in water to produce a fluid, kava, to be drunk with due dignity during discussions relating to the nature of the field work to be undertaken. The dispensation of kava is ceremonial with various parties taking traditional positions according to their standing in the community, usually within the chief's bure or meeting house. The proceedings can be quite lengthy, the participants sitting cross-legged on mats with the kava served from a tanoa (large wooden bowl) periodically, using a bilo (half a coconut shell).

Survey teams must be made aware of the locations of sacred sites such as burial grounds, former villages and hill forts. They also need to be aware of sites of water sources used traditionally for the cultivation of root crops such as dalo and yams.

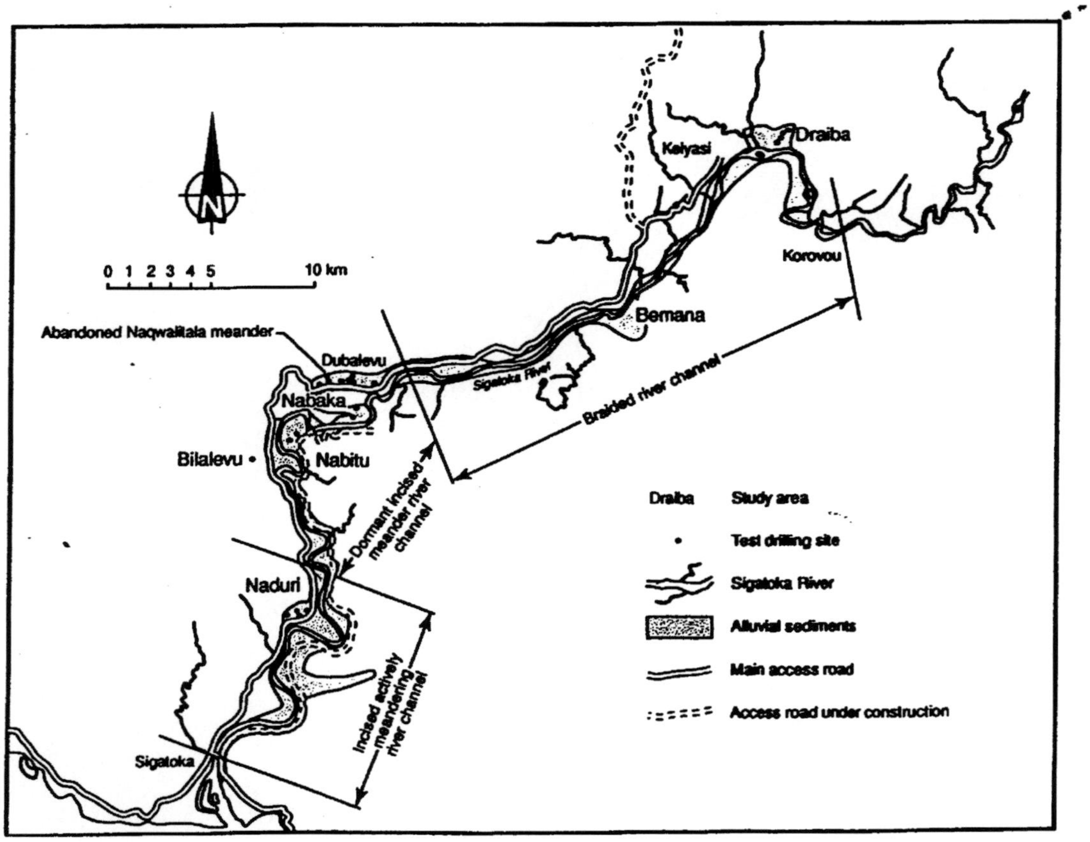

FIGURE 2. THE DISTRIBUTION OF LAND-FORMS, FLOOD-PLAINS, TEST SITES AND BOREHOLES ALONG THE SIGATOKA VALLEY BETWEEN DRAIBA

GEOLOGICAL SETTING

The bedrock formations present in SW Viti Levu are as follows:

Age Group Formation Main Lithologies
Late Quaternary Alluvial silts, sands and gravels
Upper Miocene onwards Medrausucu Group Serua Conglomerate
Vesari Sandstone Sandstones and conglomerates, various volcanic episodes
Navosa Group
Middle to Upper MioceneCuva Group Breccia, conglomerate and sandstone
Middle-Upper Miocene ColoPlutonic Group Plutonic rocks such as Gabbro and Tonalite
Upper Oligocene-Middle Miocene Wainimala Group Qalimare Limestone
Nabu Formation

Dakadaka Basalt
Tari Formation Limestones
Volcaniclastics, sandstones, mudstones
Volcanics,
Volcaniclastics, sandstones, mudstones
Upper Eocene - Lower Oligocene Yavuna Group
Basaltic pillow lavas, intrusive rocks with minor limestones and volcaniclastic rudites

The Yavuna Group is part of a volcanic arc that represents the beginning of the tectonic evolution of Fiji soon after the convergance of the Pacific and Indo-Australian plates in the Mid-Eocene. The Eocene to Lower Oligocene-age Yavuna Group, that includes the oldest known rocks in Fiji, crops out in western Viti Levu. The Upper Oligocene - Middle Miocene-age Wainimala Group is separated from the Yavuna Group by an angular unconformity. In southern Viti Levu the Wainimala rocks form an east-north-east-trending anticlinorium which is intruded by rocks of the Middle – Upper Miocene-age Colo Plutonic suite. On Viti Levu, th Upper Miocene sediments and interbedded volcanics are separated from the Wainimala Group by a depositional hiatus. Since the Late Miocene the main islands of the Fiji group have formed part of a micro-plate within the complex zone between the still tectonically active Vanuatu and Tonga arcs (Hathway 1994).

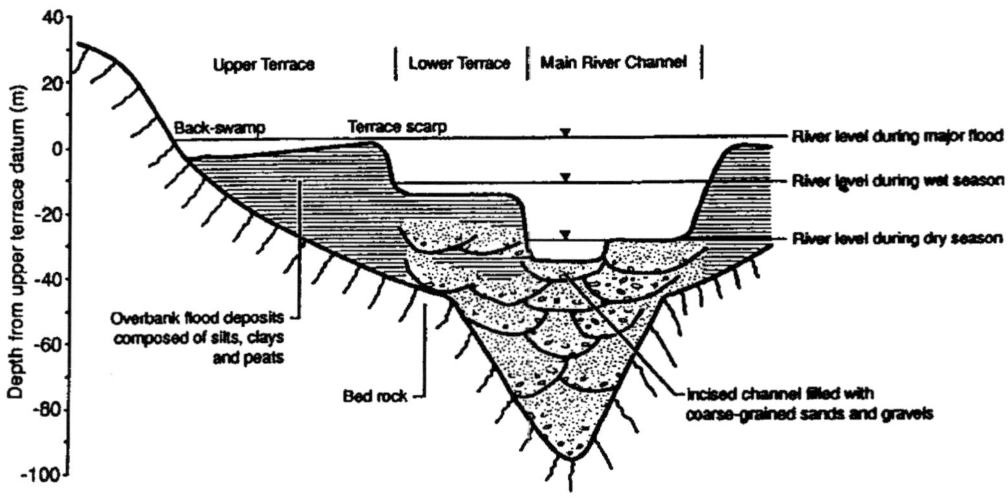

FIGURE 3. SUMMARY DISTRIBUTION OF SEDIMENT FACIES ALONG THE SIGATOKA VALLEY BETWEEN DAIBA AND

Smith (1989) and Smith and Saphore (1987) used marine seismic surveys to determine the depths of river channels in the estuaries of the Sigatoka and Navua rivers. These proved the existence of buried channels to depths of about 100 m below msl, incised to this level by 18 ka BP at the end of the last ice age maximum. Parry (1981 and 1987) identifies and maps the extent of alluvial terraces and other geomorphologic features, when studying the archaeology of the Navua Delta and Sigatoka Valley areas.

Houtz (1960, 1963) mapped the bedrock geology of the Sigatoka Valley recognising Miocene age volcanic and sedimentary deposits that he assigned to the Wainimala and Sigatoka groups. Hathway (1988) remapped the Keiyasi and northern Sigatoka areas reassigning the formations describeded by Houtz to the Wainimala, Tuva and Navosa groups. Carr (1973) noted that moderately thick Quaternary to Recent age unconsolidated alluvial sediments in excess of 36 m at Dubalevu and 68 m and Naduri occur along the Sigatoka Valley deposited as alluvial terraces and gravel and sand bars.

Band (1968) described the bedrock geology of the Navua area, noting that the Navua Delta area is underlain by Miocene - Pilocene age Serua Conglomerate and Vesari Sandstone of the Medrausucu Group that unconformably overlies Lower Tertiary volcanic conglomerates and tuffs of the Wainimala Group. The delta, some 50 km2 in area, is composed of Late Pleistocene and Holocene-age alluvial silts, sands and gravels deposited during the sea-level rise after the last glacial maximum at 21 ka BP. A few small hills composed of Vesari Sandstone crop out in the delta.

HYDROGEOLOGY

The unconsolidated Quaternary to Recent alluvial sediments, found along the lower reaches of the larger rivers, form the best potential aquifers in SW Viti Levu (Gale and Booth 1991). Joint studies of the hydrogeological development potential of these alluvial aquifers, at sites located on FIGURE 1, were undertaken by BGS and MRD:

(a) As part of the Sigatoka Valley Rural Development Project (1988-1990)
(b)In the Dubalevu area, Sigatoka (1991)
(c) In the Vakabalea area of the Navua delta (1990 and 1992)

These hydrogeological surveys of the study sites included:

(1) Aerial photograph and map interpretation
(2) Community consultation
(3) Geophysical surveys of possible irrigation areas
(4) Borehole drilling and construction
(5) Borehole test pumping
(6) Hydrochemical analysis of groundwater samples

(7) Lithological and grain size analysis of formation samples
(8) 14C dating of peat samples
(9) Water level, rainfall and river flow monitoring

The detailed results of these studies are presented in Davies et al (1991), Davies (1992) and Davies et al (1993). Summaries of the results obtained from the sites studied are presented below.

THE HYDROGEOLOGY OF THE SIGATOKA VALLEY

The SVRDP planned to provide irrigation to about 500 ha of farmland located on river terraces thereby increasing the capacity of farmers to better utilise the dry season growing period through the use of irrigation methods (Asian Development Bank 1986). The mobile low-lift pump method of abstracting water from the Sigatoka River for irrigation purposes was the preferred system but its effectiveness was limited to areas within 400 m of the riverbank abstraction point. A proportion of land identified as suitable for irrigation under the SVRDP was located at more than 400 m from the Sigatoka River. The use of groundwater to irrigate these lands, abstracted from boreholes, was proposed. This required study of the geology of the underlying alluvium and bedrock formations and their hydrogeological development potential. Limited hydrogeological data were available from:

(1) An inconclusive study of the Naduri Meander area that included a seismic-refraction survey and test drilling to depths of 60 to 68 m to prove the presence of mixed sediments containing large quantities of good quality water, derived from the Sigatoka River (Simpson 1973).

(2) Exploration drilling proved a sequence of transmissive unconsolidated, coarse-grained alluvial sediments in the former Meander at Dubalevu. Similar sediments were thought to underlie the Sigatoka River downstream of this abandoned meander site (Carr 1973).

(3) Fourteen test/exploration and four production boreholes were drilled into Sigatoka River Gravels to depths of 10-15 m at Nacacolevu, several kilometres upstream of Sigatoka Town. Water pumped from the four production wells to supply the Sigatoka Township and nearby Coral Coast tourist centres, is naturally filtered by the river gravels, especially during flood events (Peach 1988).

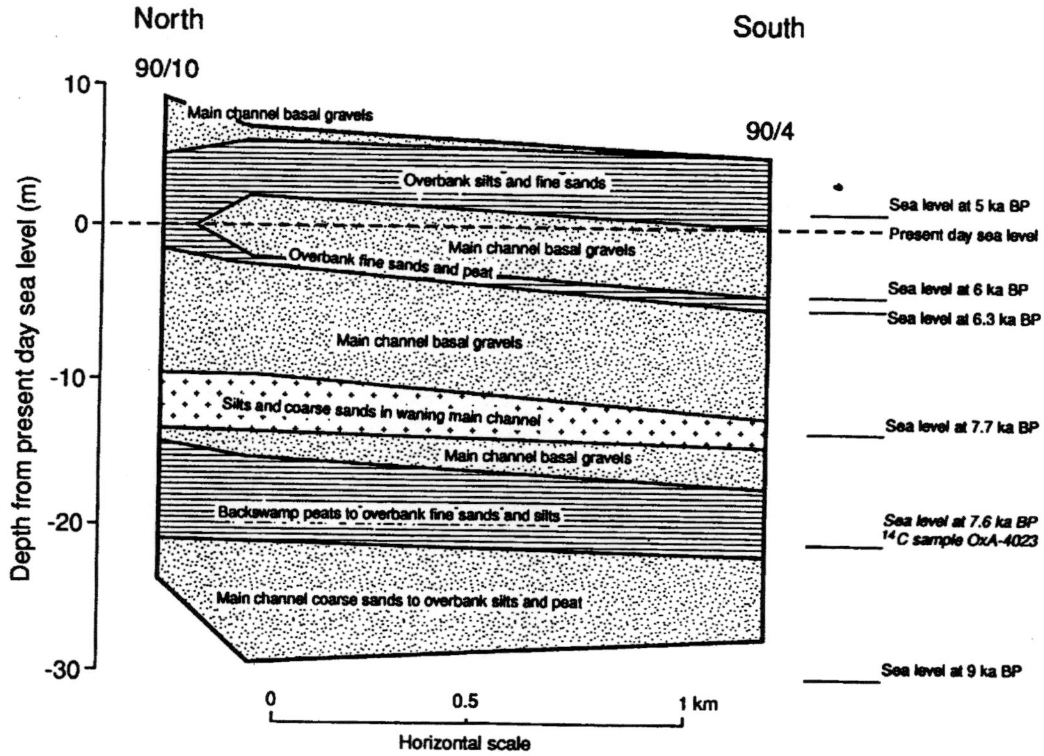

FIGURE 4. A CROSS-SECTION OF THE INCISED SIGATOKA VALLEY, BETWEEN NADURI AND DUBALEVU,

The SVRDP recommended construction of 30 boreholes, each to be completed at 150 mm diameter to an average depth of 50 m providing groundwater at 8 l sec-1 for the irrigation of 16-20 hectares.

Information drawn from maps of geomorphologic features of the Sigatoka Valley produced by Parry (1987) and previous hydrogeological and geological surveys were used to delineate flood plain areas suitable for irrigation with groundwater. Investigation of the hydrogeological development potential of alluvial sediments and underlying bedrock formations were undertaken at Naduri, Dubalevu, Nabitu, Nabaka, Bilalevu and Draiba (FIGURE 2). At the time of the survey local communities were using the river and springs as sources of water, the same sources as used in pre-historic times according to the results of the surveys of Parry (1987). This included geophysical surveys; electrical resistivity depth probes were used to determine the thickness and nature of alluvial sediments present and electromagnetic traversing using EM34 equipment was used to location near vertical fracture zones. During 1988-1990, twenty-one test/production boreholes were drilled; nineteen using a direct circulation rotary mud flush GEMCO drilling rig and two using a Hydromaster 2000 cable tool percussion drilling rig.

ties of 40-650 m2 day-1 were determined from these boreholes when test pumped at yields up to 11 l/sec. Water quality is poor and of NaHCO3 type, too poor for irrigation use. The results obtained supported the concept that the upper terrace areas are underlain by up to 60 m of fine grained silts above basement rocks, whereas earlier test drilling had proved the existence of 60-70 m of coarse sand and gravels within the incised mid channel below the present river course.

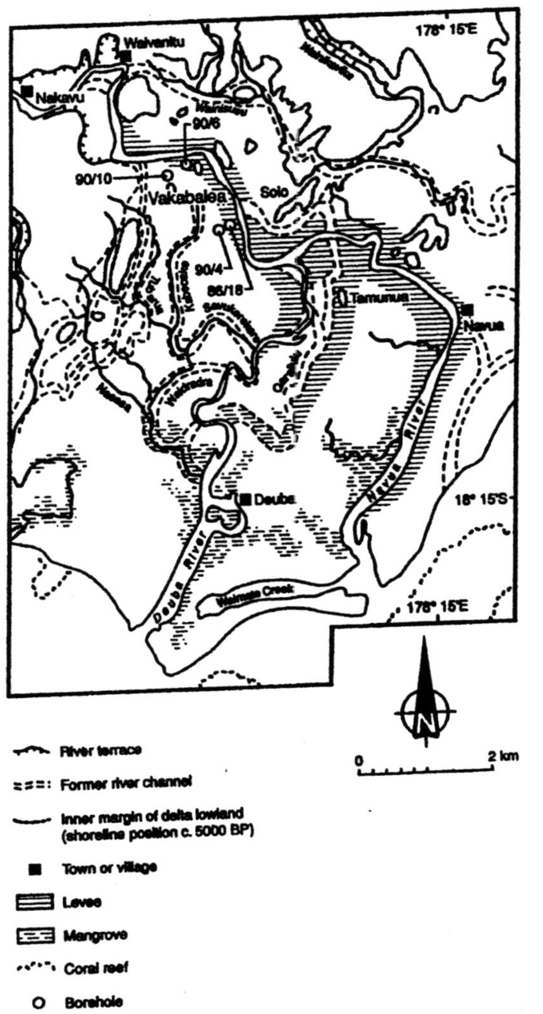

FIGURE 5. A MAP OF THE LAND FORMS OF THE NAVUA DELTA SHOWING THE LOCATION OF THE VAKABALEA TEST SITE AND EXPLORATION BOREHOLES.

NADURI

A poorly drained 3 km2 section of the upper terrace, along the valley side and more than 400 m from the river in the south west corner of Naduri was geophysically surveyed using EM34 traverses and resistivity depth probes. These surveys indicated that alluvial silts, which lie above Early Miocene basalts and turbidite rocks, underlie the survey area. Of the four exploration boreholes drilled, through up to 60 m of silts at sites selected using geophysical survey results, two were completed as production boreholes. Transmissivi-

BILALEVU

At this site the groundwater potential of fractured basement rock was assessed. Aerial photograph interpretation and geophysical surveys were used to locate a suitable drilling site where a former river channel overlay a fracture zone. The borehole passed through 21 m of alluvium and peat before entering basalts to 60 m. The borehole was tested at a yield of 3 l sec-1 for a drawdown of 25 m producing a transmissivity of 3-16 m2 day-1. Good quality Ca-HCO3 type water was produced. Although this yield was insufficient for irrigation it was enough to provide a local domestic water supply. However a local dispute resulted in local farmers backfilling the borehole.

NABITU
Geophysical surveys were used to determine the apparent thickness of fine-grained silty alluvium upon Upper Miocene conglomerates and faulted basalts. Two low yielding (2 l sec-1) boreholes were drilled through 7-20 m of silt into underlying siltstones. The water produced was of good quality and sufficient to supply the domestic needs of the local community.

NABAKA
Geophysical surveys undertaken in the southern limb of the Dubalevu Meander indicated the presence of silty alluvium above Miocene siltstones and greywackies. Two very low yielding (<2 l sec-1) were drilled through 26-40 m of upper terrace silts clays into siltstones and basalts. Although the water produced is of good quality the community is likely to continue to use its traditional source of a spring fed pond located centrally in the village.

DUBALEVU
The Dubalevu meander is a former part of the incised Sigatoka River cut off by the modern river. The results of a series of geophysical surveys indicated the presence of 10-40 m of alluvial sediments along the course of the northern limb of the Dubalevu meander. Seven test boreholes were drilled and test pumped to investigate the distribution and hydraulic properties of the alluvial sediment infill. On the basis of data obtained three high yielding (25 l sec-1) production boreholes were installed into water bearing gravels were found at 15-45 m depth. Several boreholes were test pumped at discharge rates up to 30 l sec-1, transmissivities in the region of 500-2000 m2 day-1 being determined. Monitoring of water table levels showed that the coarse sand and gravel aquifer is unconfined and in hydraulic continuity with the Sigatoka River. Problems were experienced with the blockage of borehole screens with fine-grained sediments on test pumping. The Dubalevu gravels form a good aquifer that can be developed as a source of groundwater for crop irrigation.

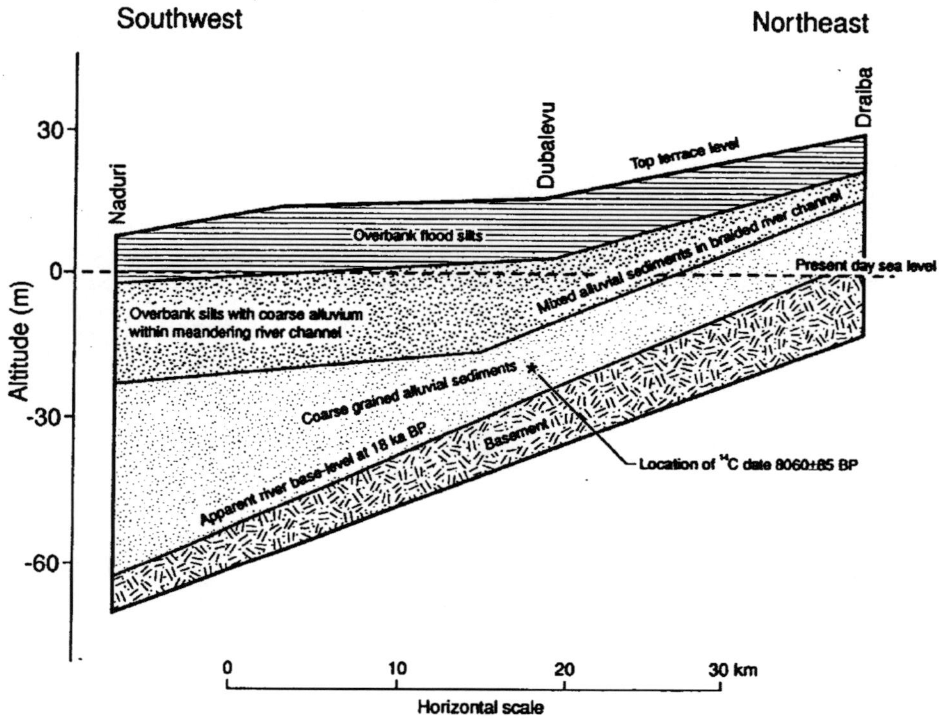

FIGURE 6. A GEOLOGICAL CROSS-SECTION THROUGH SEDIMENTS BELOW THE VAKABALEA TEST SITE: APPROXIMATES DATES OF THE LEVELS OF DEPOSITION AS INTERPRETED FROM CLARK AND LINGLE 1979.

DRAIBA
The northernmost study site, located east of Draiba and 3 km northeast of Keiyasi, encompasses the Namada Flats at the confluence of the Namada and Sigatoka rivers. The area is underlain by Late Miocene mudstones blanketed by coarse-grained fluvial river channel infill and fanglomerate gravels. Heavy, poorly drained gley type soils mantle the area. The distribution of these sediments and bedrock was mapped using electrical resistivity surveying. Of the four test boreholes drilled, two were completed as production boreholes yielding 9-29 l sec-1 from a 4 m thick transmissive (1400 m2 day-1) coarse sand and gravel aquifer. Good quality water is found in this formation that is in hydraulic continuity with the Sigatoka River gravels. The underlying mudstone formation is dry.

The following hydrometric monitoring data were collected:

Weekly water levels were collected at 12 boreholes and 17 hand dug wells during 1989-90,

(1)Daily rainfall records were obtained at 3 sites, Nacocolevu, Keiyasi and Mavua during 1988-90,
(2) Monthly river flow levels were recorded at 19 sites during 1989.

These data demonstrated that the basal coarse sand and gravel alluvial aquifer systems studied along the Sigatoka Valley are unconfined and in hydraulic continuity with the Sigatoka River.

Geological data collected from the boreholes, geophysical surveys and surface observations were used to produce conceptual models of the distribution of sediments within and along the Sigatoka valley. These can be related to distribution of river terraces and the geomorphologic features along the valley. The hydraulic and hydrochemical parameters determined can be used to indicate patterns of groundwater flow through the alluvial sediments.

A cross-section across the valley showing the distribution of river and flood levels and sediment infill within the incised valley has been constructed (FIGURE 3). This shows the range of river rise in response to storm events that renders the lower terraces of the valley uninhabitable. The deepest part of the incised valley is infilled with very permeable gravels, the best borehole drilling sites. Away from the present river channel the upper terrace areas pass into poorly drained back swamp areas underlain by low permeability clays and silts. These are the preferred sites used for dalo cultivation as shown on Parry's maps. They are areas of slow drainage and valley side spring zones.

A conceptual longitudinal section of the Sigatoka Valley between Draiba and Sigatoka explains the distribution of basal coarse sands and gravels and finer over-bank flood silts deposited along the incised valley in response to sea level rise (FIGURE 4). This pattern of sediment distribution can be related to the present day geomorphologic components of the river course:

(1) incised and meandering channels down stream of Dubalevu,
(2) the cut-off channel at Dubalevu,
(3) the middle braided river section between Dubalevu and Draiba,
(4) the nick point at Draiba,
(5) up stream of Draiba the river channel is still cutting down in response to sea level shift.

Using this longitudinal section it is possible to predict where water bearing river gravels exist in each part of the river valley.

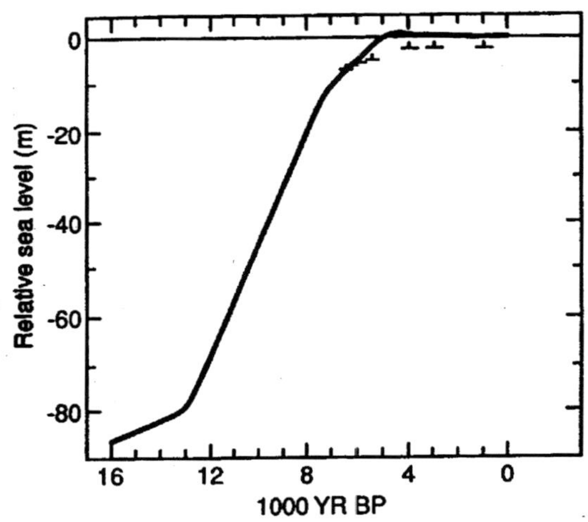

FIGURE 7. CURVE OF SEA-LEVEL CHANGE IN MICRONESIA (AFTER CLARK AND LINGLE 1979)

Additional test and observation boreholes at Dubalevu, jointly funded by the Government of Fiji and the UK Overseas Development Administration, were drilled and tested (Davies, Kumar and Herbert 1991). There, the nature of groundwater flow to boreholes from mixed fluvial sediment sequences composed of interbedded fine- to coarse-grained sand and gravel layers were studied in detail. The results were used for the better construction of high yielding boreholes installed there and within similar sediment sequences in the Navua Delta.

THE HYDROGEOLOGY OF THE NAVUA DELTA.

During 1986 the Hydrogeology Section of MRD drilled and tested two hydrogeological investigation boreholes within the Vakabalea area. The drill sites were located using data from geophysical surveys of the area. Good quality groundwater was found to be present within the coarse alluvial sediments beneath the Vakabalea area. However consultants employed by the Public Works Department concluded, from the limited hydrogeological data available, that neither the potential yield nor the groundwater resource was large enough to warrant development even though the provision of a water supply from a groundwater resource was considered to be the cheapest option. The consultants estimated that the water requirements of the Navua-Deuba area could reach 4010 m3 by the year 2010 (Fawcett Mouchel 1988)

Geological, geomorphologic and seismological survey results suggested that a structurally controlled, north-west to south-east trending river channel system, an extension of the deep Navua gorge, lies buried beneath the Navua delta sequence (Smith and Saphore 1987). These surveys proved fluvio-deltaic sediments above Vesari sandstone and Serua conglomerate. This palaeo-surface lies at –35 m below msl upstream of the Navua road bridge, falling to –110 m below msl some 3.8 km downstream.

Gale and Booth (1991) described the 1986 hydrogeological investigations undertaken within the Navua delta. Data from resistivity depth soundings indicated the distribution of the coarser sands and gravels containing fresh groundwater, argillaceous delta deposits, bedrock and areas of saline groundwater. Drilling south-east of Vakabalea proved a 22 m thick sand and gravel aquifer, above a 15m thick aquiclude of organic rich clays, with peat at 32.5 to 37.5 m below msl. Saline groundwater occurs in the underlying Seru conglomerate. Test pumping of the upper sand and gravel aquifer produced a transmissivity of 500 m2 d-1.

In 1990, the Water Supply and Drainage Department of Public Works Department requested the MRD to make a further assessment of the groundwater resource potential of Holocene-age alluvial sediments at Vakabalea in the Navua Delta for the Navua Water Supply Project. Methods of borehole drilling, construction and aquifer testing used during the SVRDP were utilised by a joint MRD/BGS team to investigate the deltaic alluvial sediments at Navua. Four test/production and seven observation boreholes were installed within coarse-grained sands and gravels and tested during 1990 and 1992. The test site was located at a gravel abstraction quarry adjacent to the Navua River at Vakabalea (FIGURE 5). The map of the Navua Delta drawn by Parry (1981) informed the team of geomorphologic, cultural and archaeological sites of interest. Much of the delta area has been given over to commercial farming, especially the cultivation of rice.

A 43 m deep borehole penetrated a 20 m thick upward fining sequence of gravels and sands before entering a 10m thick layer of fine sands and dark brown peat. Below this peaty layer coarse sands and gravels with saline water were found. The base of the peaty layer, at 30 m bgl, was dated by 14C dating at 7570±90 yrs BP (OxA-4033). Detailed grain size analyses and assessment of modes of deposition were made of the sediments investigated (FIGURE 6).

Test/production boreholes, drilled in 1992 and test pumped at a discharge of 18 l/sec, produced specific capacities of 8.5 to 13.4 l sec-1 m-1 and transmissivities of 2500-2750 m2 day-1. Two boreholes drilled earlier in 1990, tested at discharges of 18 and 29 l sec-1, produced specific capacities of 18 and 11 l sec-1 m-1 and transmissivities greater than 3000 m2 day-1. These results indicate that sufficient water can be obtained at Vakabalea to meet the supply needs of Navu township and adjacent tourist resort developments.

LATE QUATERNARY SEA LEVEL MOVEMENTS/ PALAEOCLIMATE

Clark, Farrell and Peltier (1978) produced a theoretical model of sea-level change since the last glacial maximum at 18 ka BP. Their results indicate that the oceans can be divided into six zones, each of which is characterised by a specific form of relative sea-level curve. In four of these zones emerged beaches are predicted, and these may form at considerable distance from the ice sheets themselves. In the remaining zones submergence is dominant, no emerged beaches being expected. Fiji is located within the latter type. The close agreement of these predictions with available data suggests that no net change in ocean volume has occurred during the past 5 000 years. The Pleistocene-Holocene boundary at approx 10 ka BP is marked by climate change from arid to moist environments within the tropics as indicated by a marine transgression, as at Navua. (Gates 1976 and Fairbridge 1982).

Nunn (1997) discusses possible environmental changes experienced by the southern Pacific region during the last interglacial/glacial cycle of the Pleistocene and the following Holocene periods. Clark and Lingle (1979) report that in the south Pacific sea levels declined to about 120 m below msl by the last glacial maximum. Between 21-18 ka BP the region may have experienced a period of aridity. This was followed by increased rainfall during the Late Pleistocene as indicated by the deposition of coarse rained gravels as the sea levels rose from a minimum of 130 m below msl to about 35-40 m below msl by the start of the Holocene at 10ka BP. There then followed a cooler period, several thousand years in duration, when sea levels may have declined several metres or remained static. This period is marked by the deposition of a layer of fine-grained over-bank, commonly peaty deposits, as found in Navua between 20-30m depth and in Dubalevu at 18-20m depth. This pattern has been noted elsewhere in Micronesia (Chappell and Polach 1976), world-wide by Pirazolli (1991) and in similar deltaic and fluvial environments in Bangladesh (Davies 2001).

Recent groundwater resource studies of Late Quaternary age alluvial aquifers in south-western Viti Levu indicate that these sea level changes affected valley erosion and sediment deposition patterns in that area. Marine seismic studies of the Sigatoka and Navua estuaries indicate valley incision into basement to -115 m below msl at the coast. Between the coast and Dubalevu the Sigatoka River occupies a series of incised meanders, the effect of incision decreased inland but still apparent at Draiba, more than 50 km inland. Infilling of the lower valley with fluvial sediments took place as the sea level rose between 18-5 ka BP. Coarse sands and gravels were first deposited in the lower reaches and fluvial gravel deposits in the high reaches. Following the hiatus at the end of the Late Pleistocene, fine- to medium-grained sand and silt overbank flood sediments were deposited as valley side terraces and coarse-grained sands and gravels were deposited within the meandering main channel downstream of Draiba. The Dubalevu Meander is a former part of the incised main river channel cut off by the modern river. This abandoned meander contains flood plain lands, suitable for crop cultivation, underlain by a gravel and coarse-grained sand aquifer. A sample of peat from 38m bgl in these gravels produced a 14C date of 8060±85 yrs BP (OxA-3832). Similar alluvial sediments were deposited by meandering channels within the Navua Delta. There a peat layer at 30m below ground level has produced a 14C date of 7570±90 years BP (OxA-4033), about 8 m below the theoretical equivalent level of Clark and Lingle (1979). Parry (1987) mentions that studies of Holocene sea level changes at Navua have demonstrated a sharp rise in sea level from 5 ka to 3 ka BP to 1.6 m above msl (HIPAC Team, 1984). This is also at variance with the rates of sea level rise and highest levels proposed by Clark and Lingle (1979) (FIGURE 7). It is possible that these differences in levels are indicative of a downward shift in the basement due to tectonic activity in the Navua area.

DISCUSSION

This paper demonstrates areas of overlap between studies of the hydrogeology of the Late Pleistocene/Holocene age sediments and spatial archaeology. This overlap occurs at several levels:

(1) Cultural experience during fieldwork,
(2) Field mapping techniques, especially the use of aerial photography,
(3) Impact of sea-level shift, tectonic movement and associated sediment distribution patterns,
(40 14C dating of sediments,

When studying Late Pleistocene/Holocene age unconsolidated alluvial aquifer systems information about aquifer sediment composition, patterns of deposition and hydraulic characteristics can only be obtained during borehole drilling and testing phases. The drilling of boreholes is expensive; therefore every effort should be made to ensure the accurate collection of samples and measurements with depth at such point sources. Such data and samples should be of interest to archaeologists. A detailed knowledge of sediment composition, grain size distribution and colour change with depth can be used to tentatively identify the sedimentary facies of deposition and the nature of the weathering process subsequently experienced.

As well as studying the hydrogeological characteristics of these alluvial sediments, the impact of sea level change during the Holocene period upon their distribution can be assessed. A good depth sequence of 14C dates need to be obtained from the Navua Delta and specific sites along the Sigatoka Valley if the effects of sea level change are to be differentiated from those of tectonic movement.

The uses of water by past and present communities have been compared. In the Sigatoka Valley vuci or traditional wetland gardens used for the cultivation of dalo are mainly found on the upper river terraces, especially between Dubalevu and Draiba in the area of Bemana, away from the effects of the present day annual floods. Such areas are fed by valley side springs and seepages and are underlain by poorly drained silty and clayey soils rich in peat. Test pumping and monitoring of the interaction of rainfall, water levels and river flows indicate that the abstraction of water from river gravels adjacent to the main channels of the Sigatoka and Navua rivers will have little or no effect upon water flow to traditional gardens. Small yields obtained from boreholes drilled into faulted bedrock, although too small for irrigation, could provide important sources of water for small village communities, if the waters produced conform to World Health Organisation chemical quality standards.

CONCLUSIONS

Available archaeological and ethnohistorical information indicate that Fijian pre-history starts at about 3500 yrs BP. Written Fijian history first began with European contact during the mid 19th century. Many of the cultural practices present in pre-historic times are still evident amongst Fijian rural communities today.

Survey methods used for archaeological study in Fiji have much in common with those used for hydrogeological surveys i.e. production of maps showing landforms, settlement patterns and water sources and agricultural practices as shown in this paper. Hydrogeological surveys require interaction with local communities, especially the observance of local customs and traditional water source, village, burial and agricultural sites, all of which were probably in use during pre-historic times.

Hydrogeological studies require a full understanding of the nature of aquifer systems to enable proper design of production boreholes. Therefore, the distribution of sediments types with depth and their hydraulic properties need to be defined. The data obtained from these studies can be used to indicate environmental conditions during the Late Pleistocene and Holocene, including rates of sediment accumulation in relation to rates of sea level change. Such data can be used to explain why no archaeological remains older than 3500 yrs BP have been found to date at Sigatoka. The absence of remains on the flood plain terraces of the Sigatoka valley may be due to their erosion by or burial by deposition of sediments from annual river flood.

Together the results generated by these studies and those from earlier archaeological surveys enable better understanding of the nature of Late Quaternary climate change and its impact upon past and present patterns of settlement, agriculture and water use.

REFERENCES

ASIAN DEVELOPMENT BANK,
1986. Appraisal of the Sigatoka Valley Rural Development Project. (Report FIJ:AP-9).
BAND, R.B.,
1968. The geology of southern Viti Levu and Mbengga. (Bulletin 15). Geological Survey of Fiji.
BIRKS, L.,
1973. Archaeological excavations at Sigatoka dune site, Fiji. (Bulletin 1). Suva, Fiji: The Fiji Museum.
CARR, P.A.,
1973. Survey of the groundwater resources of Fiji. Suva: United Nations Development Programme.
CHAPPELL, J. AND POLACH, H.A.,
1976. Holocene sea-level change and coral reef growth at Huon

Peninsula, Papua New Guinea. Geological Society of America Bulletin, 87, 235-240.
CLARK, A.C., FARRELL, W.E. AND PELTIER, W.R.,
1978. Global changes in postglacial sea level: a numerical calculation. Quaternary Research, 9, 265-287.
CLARK, A.C. AND LINGLE, C.S.,
1979. Predicted relative sea-level changes (18,000 years BP to present) caused by late-glacial retreat of the Antarctic ice sheet. Quaternary Research, 11, 279-298.
DAVIES, J., KUMAR, P. AND HERBERT, R.,
1991. Shallow aquifers characterised by poor sediment sorting and mixed textures. (Technical Report WD/91/79). British Geological Survey.
DAVIES, J.,
1992. Sigatoka Valley Rural Development Project: Groundwater for Irrigation. (Hydrogeological Report 3). Suva, Fiji: Mineral Resources Department.
DAVIES, J., HERBERT, R. AND KUMAR, P.,
1993. Shallow aquifers characterized by poor sediment sorting and mixed textures, Final Report. (Technical Report WD/93/17). British Geological Survey.
DAVIES, J.,
2001. Geology and Sedimentology. In: Kinniburgh, D.G. and Smedley, P.L. eds, Arsenic contamination of groundwater in Bangladesh, Final Report. (Technical Report WC/00/19). Keyworth: British Geological Survey and Department of Public Health Engineering, Dhaka, Bangladesh, 2001, 2, 17-46.
FAIRBRIDGE, W.F.,
1982. The Pleistocene - Holocene Boundary. Quaternary Science Reviews,1, 215-244.
Fawcett Mouchel Ltd., 1988. Navua - Deuba Regional Water Supply Study. (Final Report). Suva, Fiji: Public Works Department.
GALE, I.N. AND BOOTH, S.K.,
1991. Hydrogeology of Fiji. (Technical Report WD/91/36). British Geological Survey.
GATES, W.L.,
1976. Modelling the ice-age climate. Science, 191, 1138-1144.
HATHWAY, B.,
1988. The Eocene to Miocene geology of south-west Viti Levu. (PhD thesis). Oxford, England: Oxford Polytechnic.
HATHWAY, B.,
1994. Sedimentation and volcanism in an Oligocene-Miocene intra-oceanic arc and fore-arc, southwestern Viti Levu, Fiji. The Journal of the Geological Society, 151(3), 499-514.
HIPAC TEAM,
1984. Sea level changes and tectonics in the middle Pacific. (Report of the HIPAC Project in 1981-83). Nada, Japan: Department of Earth Science, Kobe University.
HOUTZ, R.E.,
1960. Geology of Sigatoka area, Viti Levu. (Bulletin 6). Geological Survey of Fiji.
HOUTZ, R.E.,
1963. Regional geology - Keiyasi area. (Bulletin 10). Geological Survey of Fiji.
NUNN, P.,
1997. Late Quaternary environmental changes on Pacific islands: controversy, certainty and conjecture. Journal of Quaternary Science, 12, 443-450.
PARRY, J.T.,
1981. Ring-ditch fortifications in the Navua Delta, Fiji: air photo interpretation and analysis. (Bulletin 7). Suva: the Fiji Museum.
PARRY, J.T.,
1987. The Sigatoka Valley - Pathway into Prehistory. (Bulletin 9). Suva: the Fiji Museum.
PEACH, D.W.,
1988. Groundwater Investigation and Development in Fiji. (Hydrogeological Report 1). Suva, Fiji: Mineral Resources Department.
PIRAZZOLI, P.A.,
1991. World Atlas of the Holocene sea-level changes. Elsevier oceanography series, 58. London: elsevier.
ROTH, J. AND HOOPER, S.,
1990. The Fiji Journals of Baron Anatole Von Hugel. Suva, Fiji: The Fiji Museum.
SIMPSON, A., 1973. Naduri Project. (Preliminary report). Suva, Fiji: Mineral Resources Department.
SMITH, R., 1989. Seismic reflection survey of the lower reaches of the Sigatoka River. (Technical Report 96). Suva, Fiji: CCOP/SOPAC.
SMITH, R. AND SAPHORE, E., 1987. Offshore survey techniques for a hydrogeologial application, the Navua River. (Note BP35/10). Suva, Fiji: Mineral Resources Department.
SPRIGGS, M., 1997. The Island Melanesians. Oxford: Blackwell.
TORA, V., 1986. Yaqona in Fiji. In: Deverell, G. and Deverell, B., eds. Pacific Rituals: Living or Dying. Suva, Fiji: Institute of Pacific Studies of the University of the South Pacific, 1986, 25-35.

30

L'ARCHIPEL DE MOLÈNE (FINISTÈRE, FRANCE) MISE AU POINT D'UN INVENTAIRE DES SITES PRÉHISTORIQUES

Y. PAILLER, Y. SPARFEL, S. CASSEN, P. GOULETQUER, M. LE GOFFIC, A. LEROY, G. MARCHAND, A. TRESSET ET E. YVEN

UMR 6038 du CNRS), Université de Bretagne Occidentale, Collaborateur UMR 6566 CNRS, BP 814, 29285 Brest
EMAIL: yvan.pailler@club-internet.fr

PRÉSENTATION GÉOGRAPHIQUE

Situé à l'extrémité nord-ouest de la péninsule armoricaine, l'archipel de Molène constitue une sorte de chaussée entre les côtes du Léon et celles de l'île d'Ouessant. Il s'agit d'une ancienne plate-forme littorale actuellement immergée s'étendant du Conquet au Chenal du Fromveur et composée d'une dizaine d'îles et îlots annexes, appelés ledenez, et d'une multitude d'écueils rocheux (Hallégouët, 1982 ; Gourmelon (dir.), 1995). La plupart de ces îles ne dépassent pas dix mètres d'altitude au-dessus des plus hautes mers, mis à part Molène qui dans sa partie centrale atteint 21 mètres. La superficie totale des terres émergées représente environ 300 hectares, avec une grande variabilité entre les îles (de 7 hectares pour Litiri à 127 pour Molène ; FIGURE1).

La question de l'insularité de ces territoires durant les périodes préhistoriques se heurte à l'absence de recherches spécifiques concernant l'évolution des niveaux marins autour du plateau molénais. Néanmoins, on peut considérer qu'au Mésolithique final, les plus hautes mers avoisinent la courbe bathymétrique des –6 mètres. Dès cette période, le plateau molénais pourrait donc former une vaste île séparée du continent par le chenal du Four (Morzadec-Kerfourn, 1974 ; Giot, 1990).

D'un point de vue géologique, il n'existe pas de rupture entre l'archipel et le continent. On note une réelle homogénéité entre Beniget et Trielen. Le substrat est composé essentiellement de gneiss et de granodiorite de la Pointe des Renards, très riche en enclaves. Sur Molène, on assiste à la réapparition du granite de Saint-Renan présent sur le continent. De même, le granite de l'Aber Ildut se prolonge jusqu'à Balaneg et le leucogranite de Ploudalmézeau se retrouve sur Banneg (Chauris & Hallégouët, 1989).

HISTORIQUE DES RECHERCHES

Les recherches menées au début du XXe siècle par Paul du Chatellier (1907, p. 116-125) et Alfred Devoir ont mis en évidence une densité importante de monuments mégalithiques. Il est nécessaire de rappeler qu'à cette époque et ce jusqu'au moins la moitié du siècle dernier, les îles de l'archipel de Molène étaient habitées, pour la plupart à l'année, par des fermiers qui en cultivaient les terres et, saisonnièrement, par des goémoniers venus du continent. Ces conditions ont largement favorisé, notamment par le biais du pâturage, les observations pour les monuments en élévation et encore plus pour les zones de forte concentration de pierres taillées.

Depuis les années 50, l'archipel a été le cadre de quelques interventions archéologiques assurées par les membres du Laboratoire d'Anthropologie Préhistorique (Université de Rennes I) et d'observations régulières qui ont fait l'objet d'une publication cosignée par Pierre-Roland Giot et Bernard Hallégouët en 1980. Plus récemment, Michel Le Goffic (1994) et Bertrand Grall ont effectué des relevés des principaux vestiges mégalithiques de l'île de Beniget.

STATUT ADMINISTRATIF

La désertion progressive des îles et le classement en réserve d'une partie de celles-ci font que l'archipel connaît aujourd'hui une pression anthropique modérée et très réglementée.

Trois îles (Trielen, Balaneg et Banneg) sont classées en réserve naturelle, deux autres (Kemenez, Litiri) sont privées, Beniget est quant à elle gérée par l'Office National de la Chasse. Molène, la dernière île habitée de l'archipel, et ses ledenez constituent une commune à part.

plus particulièrement notre attention. Le Mésolithique est-il représenté sur l'archipel? Quelles sont les relations entre Ouessant, le plateau molénais et le continent au cours de la Préhistoire récente? Existe-t-il une spécificité insulaire dès le Néolithique et quels sont les moyens de l'appréhender?

Au fil des investigations, nos problématiques se sont en-

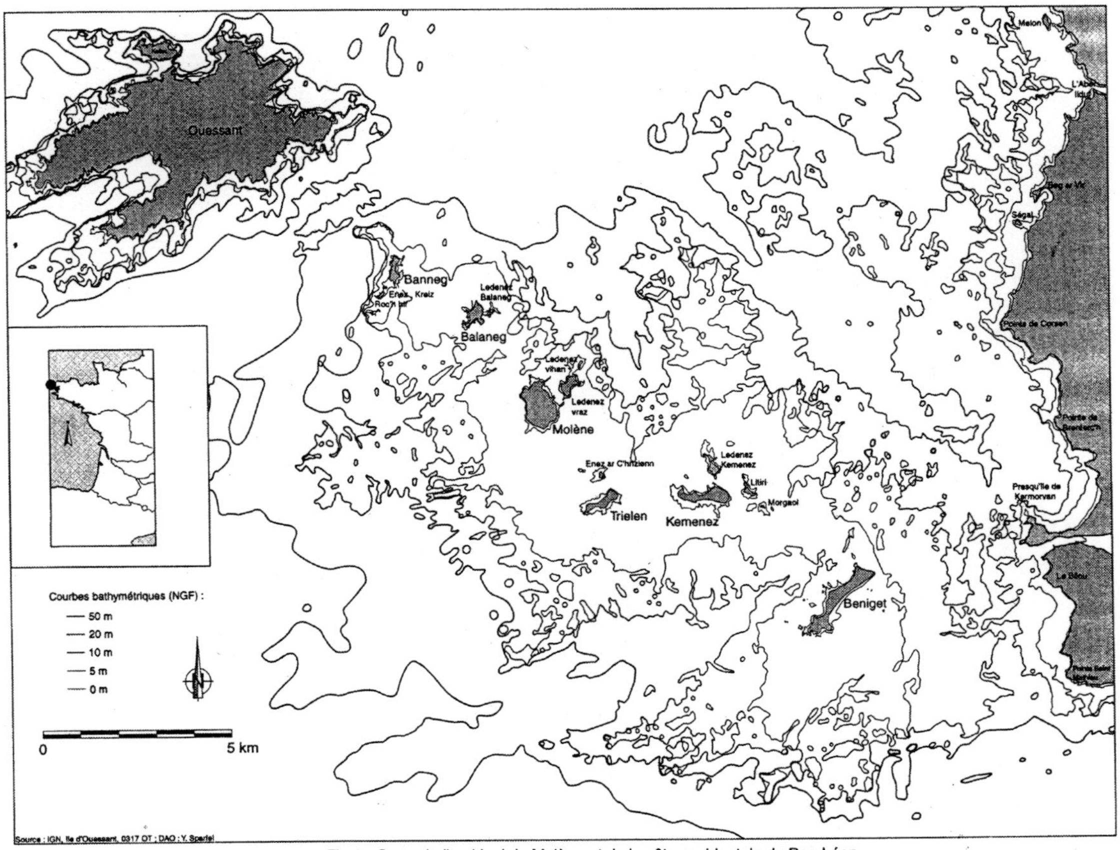

Fig.1 - Carte de l'archipel de Molène et de la côte occidentale du Bas-Léon

L'éloignement du continent pour certains de ces îlots et les difficultés d'atterrage constituent des freins naturels à l'arrivée en masse des plaisanciers durant la période estivale. Ces éléments font qu'en plus d'être une réserve naturelle, l'archipel constitue un véritable conservatoire du patrimoine archéologique, légèrement perturbé par des travaux agricoles et les besoins en pierres des goémoniers.

PROBLÉMATIQUE

Dans le cadre de nos recherches sur la néolithisation et le Néolithique en Basse-Bretagne (Y. Pailler) et le phénomène mégalithique qui l'accompagne (Y. Sparfel), il était logique que nous intégrions l'archipel de Molène à notre réflexion.

Le but de notre travail n'est pas de remettre en question l'exhaustivité ni la qualité des inventaires de nos prédécesseurs, mais de les préciser en comblant les zones non inspectées, en réalisant des relevés systématiques de toutes les structures archéologiques reconnues, en décrivant ces dernières et le cas échéant en les interprétant.

Une prospection spécifique a été mise en place concernant le "patrimoine archéologique diffus" . Après une année de prospection sur l'archipel, plusieurs questions retiennent richies de nouveaux thèmes aidés en cela par différents spécialistes. Les principaux domaines abordés sont le mégalithisme, l'étude du mobilier lithique, l'étude de la céramique, l'archéozoologie, l'étude des niveaux coquilliers au sens large et de la malacofaune. Un volet ethnologique vient s'ajouter à ces différentes spécialités archéologiques.

MÉTHODOLOGIE

La méthode que nous avons mise en place comprend deux phases qui comportent des interactions.

La première débute par le rassemblement des sources (bibliographie, enquêtes orales). Sur le terrain, un état des lieux est réalisé en confrontant la situation actuelle aux anciens documents.

La prospection de surface permet de couvrir l'intégralité du territoire concerné mais est fortement limitée par l'état de la couverture végétale. Plusieurs concentrations de matériel lithique signalées par nos prédécesseurs dans des secteurs piétinés, érodés ou cultivés sont actuellement invisibles en raison de la recolonisation de ces zones par la végétation. Au moment des interprétations, il conviendra d'être extrêmement prudent et de conserver en mémoire que sur l'île d'Ouessant, toute proche, des décapages de grande envergure effectués par Jean-Paul Le Bihan et son équipe ont mis en évidence des structures d'habitat qui n'étaient pas décelables en surface (Le Bihan, 2001).

La seconde phase consiste à effectuer des plans et élévations de mégalithes, d'amas coquilliers, d'abris temporaires modernes et de fours à soude. Ces monuments sont relevés selon une méthode traditionnelle de dessin manuel à partir d'un axe fixe et localisés à l'aide d'un GPS différentiel. La haute précision de ce dernier appareil offre également la possibilité de réaliser des modèles numériques de terrain (MNT) des cairns et tumulus inventoriés.

PRÉSENTATION DE L'INVENTAIRE

Cette présentation est le reflet de nos prospections effectuées entre le 02 novembre 2000 et le 14 décembre 2001 sur l'archipel de Molène. Elle vise à dégager les traits généraux du potentiel archéologique que renferment ces îlots.

LE PALÉOLITHIQUE

Dans l'état actuel de nos connaissances, les indices du Paléolithique sont rares. Sur Molène, un chopper découvert en contexte remanié a déjà fait l'objet d'une note. L'auteur le rattache, sur des critères typologiques, à l'industrie colombanienne datée du Paléolithique Inférieur (Molines, 1992).

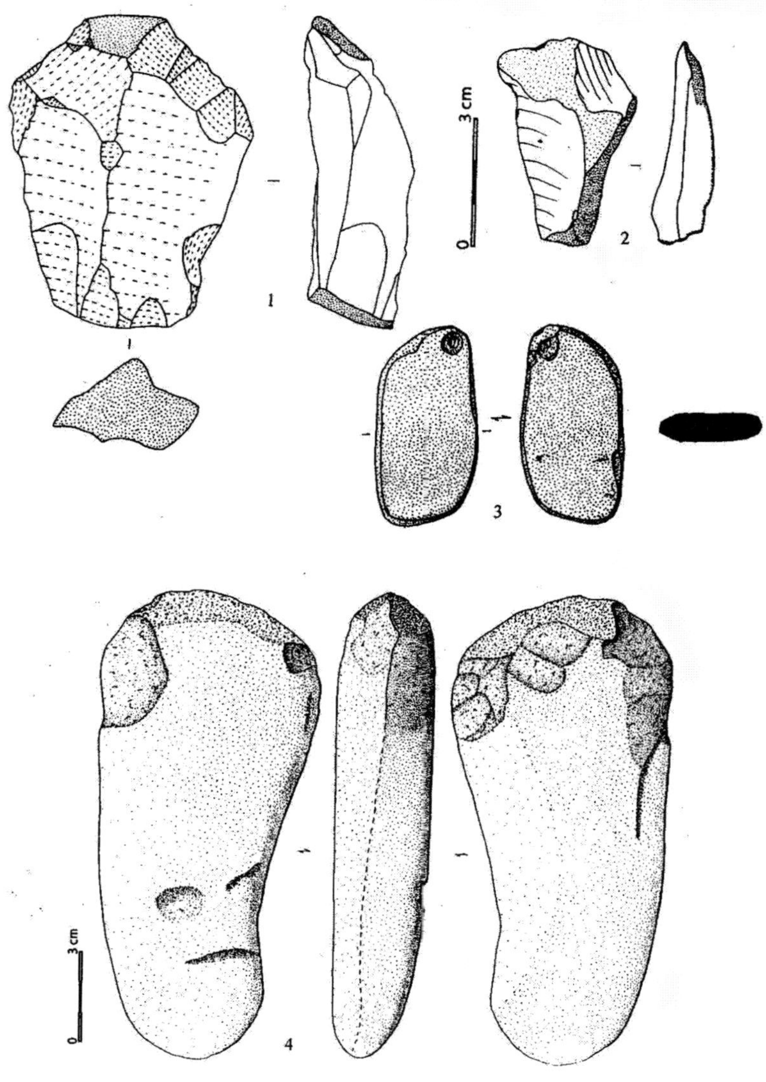

Fig.2 – Indice de site n° 8 (Balaneg) : 1 : éclat en quartz – 2 : éclat en silex.
Indice de site n° 22 (Balaneg) : 3 : pendeloque polie en hématite.
Indice de site n° 23 (Balaneg) : 4 : galet biseauté.

Dans une coupe de micro-falaise à Balaneg, deux éclats, l'un en quartz, l'autre en silex (n° 1 et 2, FIGURE 2) ont été récoltés. Malgré le faible caractère diagnostique de ces deux éléments, leur localisation stratigraphique, dans un paléosol qui correspondrait à un niveau périglaciaire Wiechsélien (com. pers. Bernard Hallégouët), plaiderait en faveur d'une attribution au Paléolithique moyen.

LE MÉSOLITHIQUE

L'occupation mésolithique de l'archipel n'est guère mieux représentée. Seuls trois nucléus à débitage lamellaire, découverts sur Balaneg et Molène, témoigneraient d'une fréquentation à l'Holocène ancien.

L'existence de plusieurs amas coquilliers pouvait laisser penser à une occupation plus dense à cette période, cependant tous ceux observés à ce jour présentent des indices attribuables au Néolithique final voire aux périodes postérieures.

LE NÉOLITHIQUE
Le Néolithique est la période la plus représentée à la fois par la densité de vestiges mégalithiques et la forte présence d'indices témoignant d'installations temporaires ou prolongées.

de cercle et un monument au plan complexe.

CERTAINS DE CES MONUMENTS ONT FAIT L'OBJET DE RELEVÉS PRÉCIS ET APPELLENT DE CE FAIT QUELQUES COMMENTAIRES.
Situé au sud du parcellaire moderne, l'ensemble n° 9 est constitué de deux monuments bouleversés distant d'environ 3,50 mètres (FIGURE 4).

A l'ouest, le mieux conservé est composé de huit dalles de chant formant un hémicycle de 3,50 mètres de diamètre. Ces dalles pourraient correspondre au parement interne

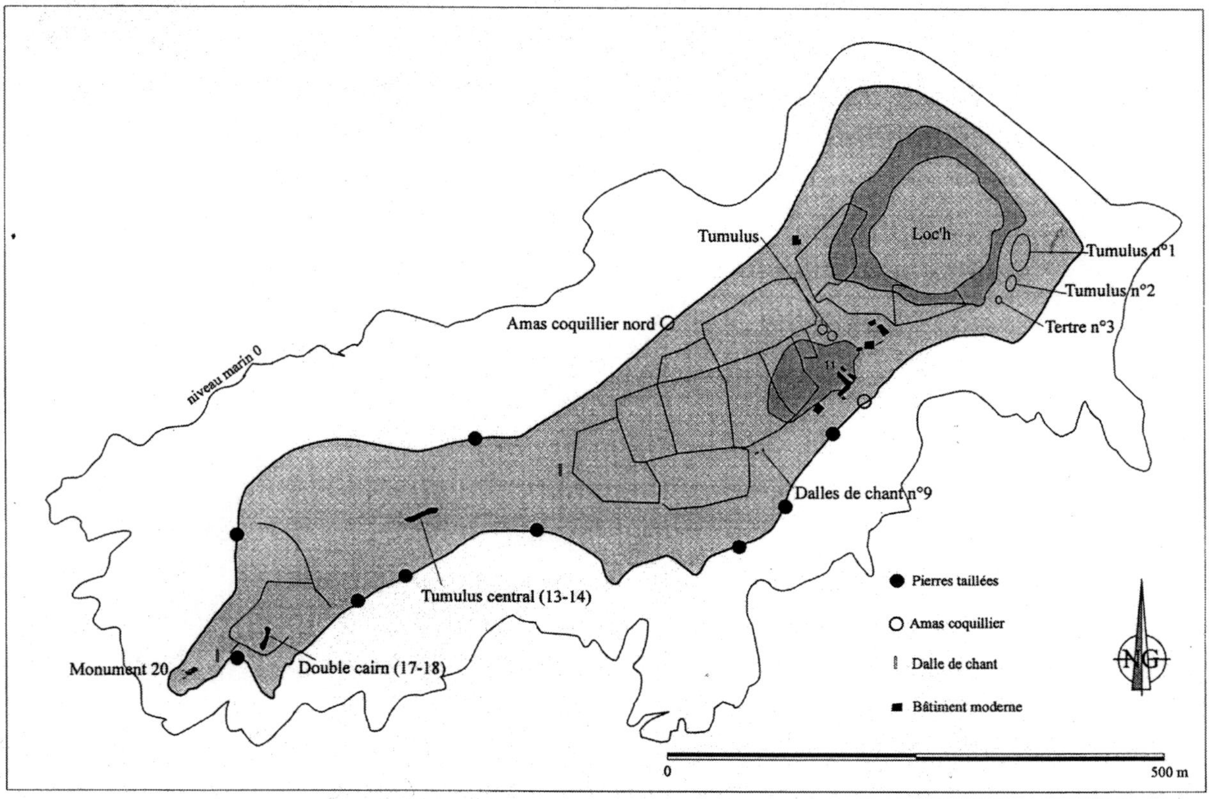

Fig.3 - Carte de répartition des vestiges archéologiques de Trielen

LE MÉGALITHISME
L'état de conservation des monuments étant très variable, nous avons choisi de ne décrire que quelques ensembles caractéristiques du mégalithisme de l'archipel.

L'île Trielen possède un grand nombre de structures mégalithiques plus ou moins bien conservées (FIGURE 3). Les concentrations principales se répartissent en trois secteurs.
L'une sur le pourtour du loc'h se compose d'un premier ensemble de deux tumulus et d'un tertre bas en enfilade puis de deux tumulus partiellement démembrés par une voie charretière qui traverse le hameau.
De part et d'autre du parcellaire moderne, s'égrènent les vestiges de monuments alignés le long de la ligne de crête. L'ensemble le mieux conservé est un vaste tumulus d'environ 70 mètres de longueur.
A l'extrémité sud-ouest de l'île, parmi un certain nombre de structures bouleversées, sont reconnaissables deux cairns reliés par un talus, quelques dalles de chant formant un arc d'une chambre dont les murs en pierre sèche auraient été détruits. Ce détail architectural est connu pour les chambres A, B et D du cairn secondaire de Barnenez et les trois chambres du cairn II de Guennoc pour ne citer que quelques exemples proches (Giot, 1987).

Le second ne comporte que quatre dalles de chant qui ne présentent aucune disposition particulière. Ces deux monuments mégalithiques semblent insérés dans un reste de tumulus d'où affleurent ici ou là des pierres. La présence d'un talus, venu perturber l'ensemble, nous a empêchés de relever l'enveloppe monumentale. Le plan de la structure occidentale semble correspondre au fond d'une chambre circulaire. Il est donc probable qu'il s'agisse là des vestiges très perturbés de dolmens à couloir incorporés dans un tumulus.
A l'échelle de l'archipel, les structures internes visibles montrent une prédominance de tombes à couloir, certaines sont circulaires, d'autres compartimentées.

Le tumulus référencé par les numéros 13-14 (FIGURE 5)

fait partie de cette série. A l'extrémité oriental, son versant est occupé par un groupe de cinq dalles de chant autour desquelles sont présents d'autres blocs, certains affleurant du sol, d'autres simplement posés à la surface de celui-ci. Deux blocs allongés, excentrés par rapport à cet ensemble, sont situés en contrebas. Parmi les éléments en place, quatre dalles plantées affectent la forme schématique d'un arc de cercle qui pourrait constituer le reste d'une chambre funéraire.

Dans la partie centrale du monument, une dalle plate d'une hauteur de 1,25 mètre hors sol se dresse. Perpendiculairement à la face nord de cette pierre dressée et depuis son arête orientale, part une ligne composée de deux dalles, dépassant de quelques centimètres du sol, et se terminant à angle droit par une troisième pierre. Parallèlement aux dalles alignées et commençant cette fois-ci à l'extrémité occidentale, est plantée une quatrième dalle de plus petite taille (FIGURE 6).

A l'ouest, un examen minutieux a permis de repérer un second ensemble de cailloux fichés en terre dont seuls les sommets sont visibles (inférieurs ou égaux à 5 centimètres). Des espaces de quelques dizaines de centimètres séparent ces pierres. On distingue malgré tout, trois lignes formant un trapèze ouvert sur son bord oriental. Deux interprétations sont possibles. Nous pouvons avoir affaire à deux structures distinctes, ce qui expliquerait des différences de gabarits des dalles et la différence d'orientation des axes des deux ensembles. Il peut aussi s'agir d'un seul monument asymétrique dont l'organisation est difficile à appréhender dans son état actuel. Quoiqu'il en soit, il apparaît assez clairement que nous sommes en face d'au moins une structure semi-enterrée de type "coffre" avec présence d'une stèle pouvant faire office de dalle de chevet ou de marqueur de sépulture. Avec prudence, ce type de monument peut être rattaché à une phase précoce du Néolithique. Cette attribution est étayée par des relations de fouilles anciennes comme celles du Tevenn au Conquet (Devoir, 1917) et de Kerbois à Carnac (Le Rouzic, 1921 ; Boujot, Cassen, in Cassen (dir.), 2000, p. 185-186).

Distantes de 14 m vers l'ouest, quatre dalles couchées au sol et dégagées sur tous leurs côtés sont alignées sur une longueur de 4,50 m au sommet du tumulus. Un interstice entre les deux blocs situés au nord-ouest permet d'observer deux murets parallèles édifiés en pierres sèches. A 1,50 m au nord-ouest, un dernier bloc ne présente que sa surface émergeante. Deux groupes de cailloux qui paraissent se dégager de la masse du tumulus se distinguent : l'un se trouve à proximité des dalles, l'autre forme un arc de cercle sur le versant nord-ouest de l'enveloppe.

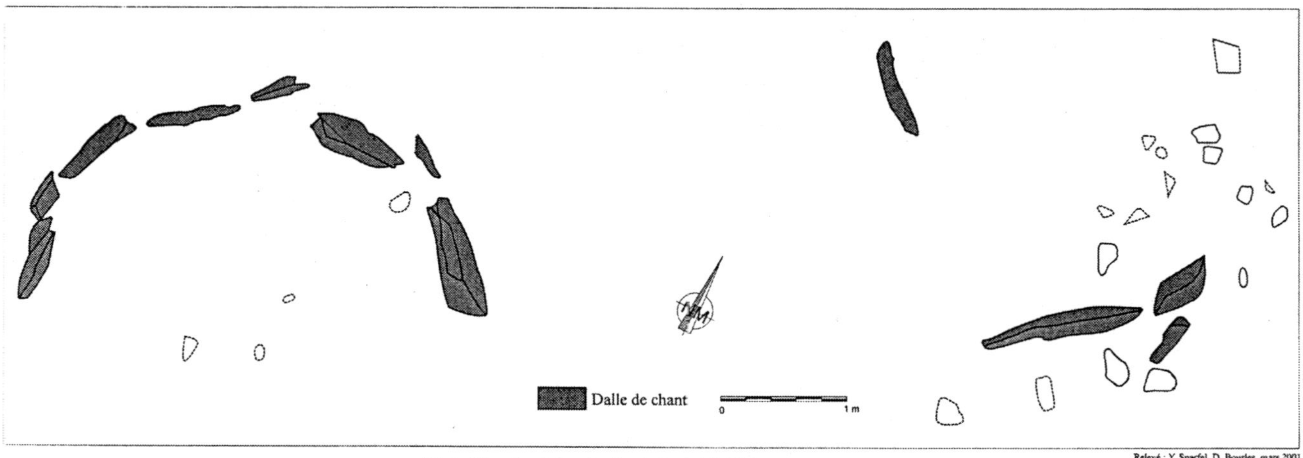

Fig.4 - Plan des dalles de chant n°9 de l'île Trielen

L'excellente conservation du couloir suffit pour reconnaître dans ce monument une sépulture à couloir typique du Néolithique moyen II, dont plusieurs autres exemplaires bien conservés existent en d'autres points du Léon (îles de Carn, de Guennoc). Il est difficile de savoir si le dernier bloc fait partie du couloir ou de la chambre.

Ce monument, difficile à comprendre d'un point de vue volumétrique sur le terrain, est d'interprétation plus aisée grâce aux relevés GPS. Schématiquement, au moins deux sous-ensembles peuvent être isolés dans la masse du tumulus. A l'est, une première enveloppe débute quelques mètres en avant de ce qui peut être le reliquat d'une chambre funéraire et s'étend à l'ouest jusqu'à englober le coffre, placé sur un versant. Autour de ce dernier, un tumulus hypothétique semble se dessiner, à moins qu'il s'agisse de la continuité du précédent. Le deuxième sous-ensemble, aux dimensions plus importantes et aux limites plus nettes, entoure la sépulture à couloir occidentale, située au sommet de l'enveloppe. En définitive, la lecture des courbes de niveau semble suggérer que les enveloppes des deux monuments extrêmes sont venus s'appuyer sur celle de la sépulture centrale dont la construction serait alors antérieure aux autres structures.

Implanté à l'ouest de l'île, l'ensemble n°17-18 (FIGURE 7), orienté NNE-SSO, s'étend sur une longueur d'environ 40 mètres. La levée de terre faisant le lien entre les deux élévations rend difficile la reconstitution de leurs formes, les limites se faisant face étant ennoyées dans le talus. Le monument septentrional présente à son sommet une dépression décentrée laissant apparaître quelques blocs basculés sans organisation apparente. Cette excavation fait penser à une fouille ancienne ou à l'effondrement d'une chambre en encorbellement, le remplissage terreux ne facilitant pas l'interprétation. La façade sud du monument méridional, plus grand et plus élevé que le précédent, est assez rectiligne bien qu'entaillée en son milieu par un décrochement. L'extrémité se présente comme un front abrupt ; un appareillage de pierres sèches le long de celui-ci est visible sur quelques mètres de longueur. Cet agencement de moellons n'est pas visible sur la façade nord du premier monument à cause de la végétation. Néanmoins le relevé montre clairement que ces deux extrémités sont de morphologie identique. Ceci laisse à penser que nous avons bien affaire aux limites de parement de ces monuments. Le talus, haut de 0,60 mètre, en faisant jonction, se fond dans la masse des cairns, ce qui plaide en faveur de sa grande antiquité ou même de la contemporanéité de l'ensemble.

Un second élément architectural vient conforter cette dernière hypothèse. Il s'agit de la présence de galets de gros gabarit (jusqu'à un mètre de diamètre) de quartz blanc teintés de rose-orange (FIGURE 8) dans le talus et à ses abords, mais aussi sertis dans les cairns. Cet ajout lithique lisse et blanchâtre rompt nettement avec l'aspect anguleux des dalles en granodiorite ou en gneiss et leur monochromie grisâtre. La valeur esthétique et/ou symbolique de cette disposition ne fait aucun doute. L'adjonction de mégalithes en quartz n'est pas inédite en Bretagne, elle se retrouve dans différents ensembles mégalithiques comme au tertre de la Croix-Saint-Pierre à Saint-Just (Ille et Vilaine ; Giot & L'Helgouac'h, 1955), dans l'enveloppe du dolmen à couloir de Coëby (Trédion, Morbihan ; Gouezin, 1992).

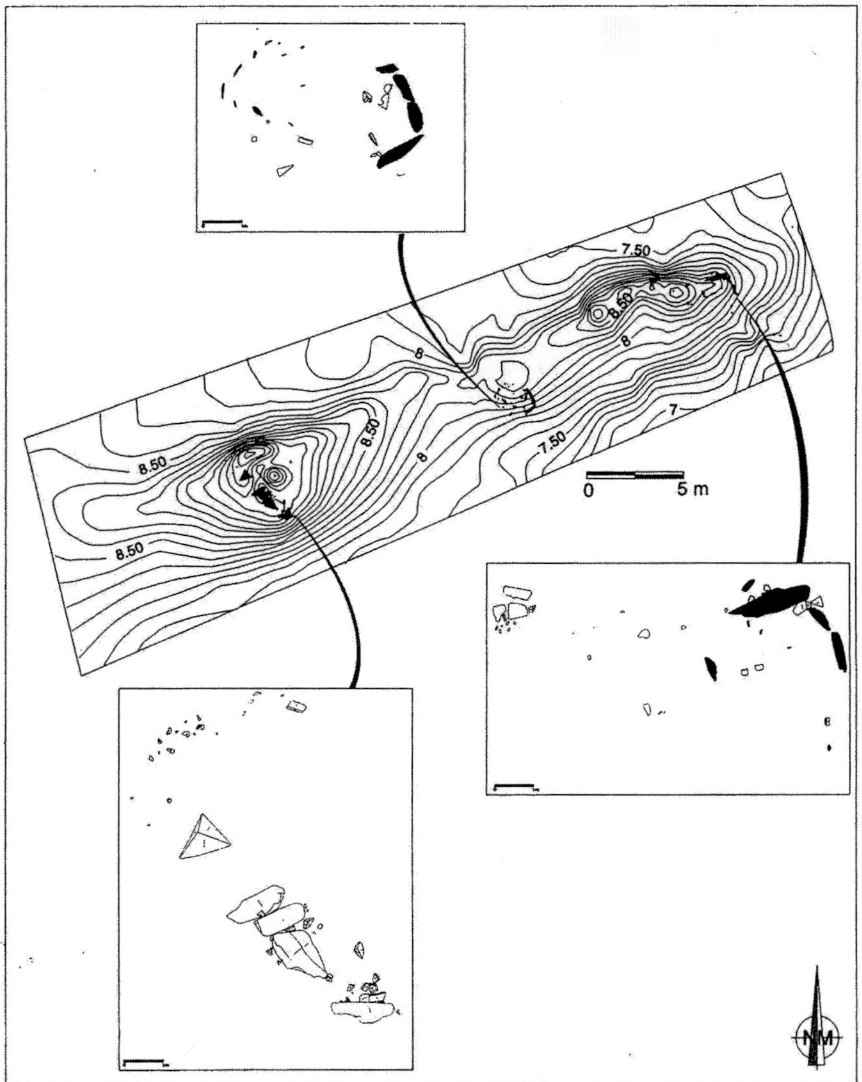

Relevé archéologique : Y. Sparfel, D. Bourles, A. Lourdeau, avril 2001 ; Relevé GPS : S. Pacaud ; DAO : Y. Sparfel

Fig. 5 - Plan du tumulus central n°13-14 et de ses structures internes sur l'île Trielen

Sur la pointe rocheuse située à l'extrémité ouest de l'île Trielen est visible un ensemble monumental complexe de 13 mètres de longueur sur environ 5 mètres de largeur (FIGURE 9). L'aspect compliqué de cette structure est accentué par l'enchevêtrement de nombreuses dalles couchées et d'autres de chant. La partie orientale semble compartimentée par de petites loges (au moins trois). Il est difficile de savoir si les nombreux blocs épars sont les vestiges du cairn ou des éléments de la structure interne. D'après S. Cassen, ce monument pourrait être une sépulture à couloir à chambres compartimentées. Si cette proposition était confirmée, nous serions en présence de la sépulture de ce type la plus septentrionale (Le Goffic, 2000).

Bon nombre d'autres vestiges sont trop perturbés pour être rattachés à un type de monument particulier. Ce constat vaut essentiellement pour Molène, où plusieurs ensembles composés de deux ou trois dalles de chant ont été observés.

A l'inverse, des monuments, peu endommagés par les agents naturel ou anthropique, ne se distinguent du sol environnant que par une élévation plus ou moins marquée. Dans ce cas, l'attribution typo-chronologique se révèle difficile, même si les matériaux constituant l'enveloppe et la morphologie de celle-ci peuvent donner quelques indications (Sparfel, 2001). Les monuments les mieux préservés sont dans bon nombre de cas les plus discrets comme le petit tertre bas n°3 de Trielen ou encore ceux implantés sur des îlots (Ledenez Vraz et Ledenez Vihan) ou dans des secteurs destinés au pâturage des bêtes. Ainsi, dans la zone appelée Le Tevenn, au sud-ouest de Molène, nous avons mis au jour sous un roncier un ensemble de plusieurs cairns alignés délimités par des parements conservés sur plusieurs assises.

Fig.6 - Cliché du coffre implanté au centre du tumulus n°13-14 sur l'île Trielen

Fig.8 - Cliché d'un galet de quartz incorporé dans le talus joignant les cairns n°17-18 sur l'île Trielen

Les pierres dressées ont le plus souffert des déprédations. En 1835, M. Hesse signale plusieurs files de menhirs sur l'île de Beniget qui n'existent déjà plus lors du passage de P. du Chatellier au début du XXe siècle (1907).

Actuellement, sur Kemenez, un grand menhir isolé de 3,25

mètres se dresse à l'extrémité occidentale, à l'angle d'un mur. Deux, encore debout au centre, devaient faire partie d'un ensemble plus imposant si l'on en juge par les dalles couchées avoisinantes.

Quant à Molène, au moins six autres menhirs sont encore debout, deux paires étant situées dans la partie méridionale (FIGURE 10).

En 1909, A. Devoir faisait déjà remarquer que "les tumulus surmontés de menhirs se trouvent presque uniquement dans l'ouest du Finistère sur les îles (Béniguet, Lédénès de Molène, Sein)". Dans le sud-est de Beniget, un tertre est surmonté de deux menhirs, élevés de 2,50 mètres. Sur la face de l'un d'eux, M. Le Goffic (1997) a remarqué la présence de dix cupules. A ces exemples, nous pouvons ajouter le monument 13-14 de Trielen et celui de la presqu'île de Kermorvan (Le Conquet ; Le Goffic, 1994).

Un autre type de structure est représenté par les talus au sein desquels sont parfois incorporés ou érigés des blocs de pierre. A l'extrémité ouest de Trielen sont visibles des élévations linéaires qui ne peuvent être confondues avec le parcellaire moderne constitué par des murets de pierres sèches : c'est le cas du talus reliant les cairns n°17-18.

LE MATÉRIEL DIFFUS

Les micro-falaises des îles de l'archipel offrent les principales sources d'information concernant la vie quotidienne des hommes néolithiques.

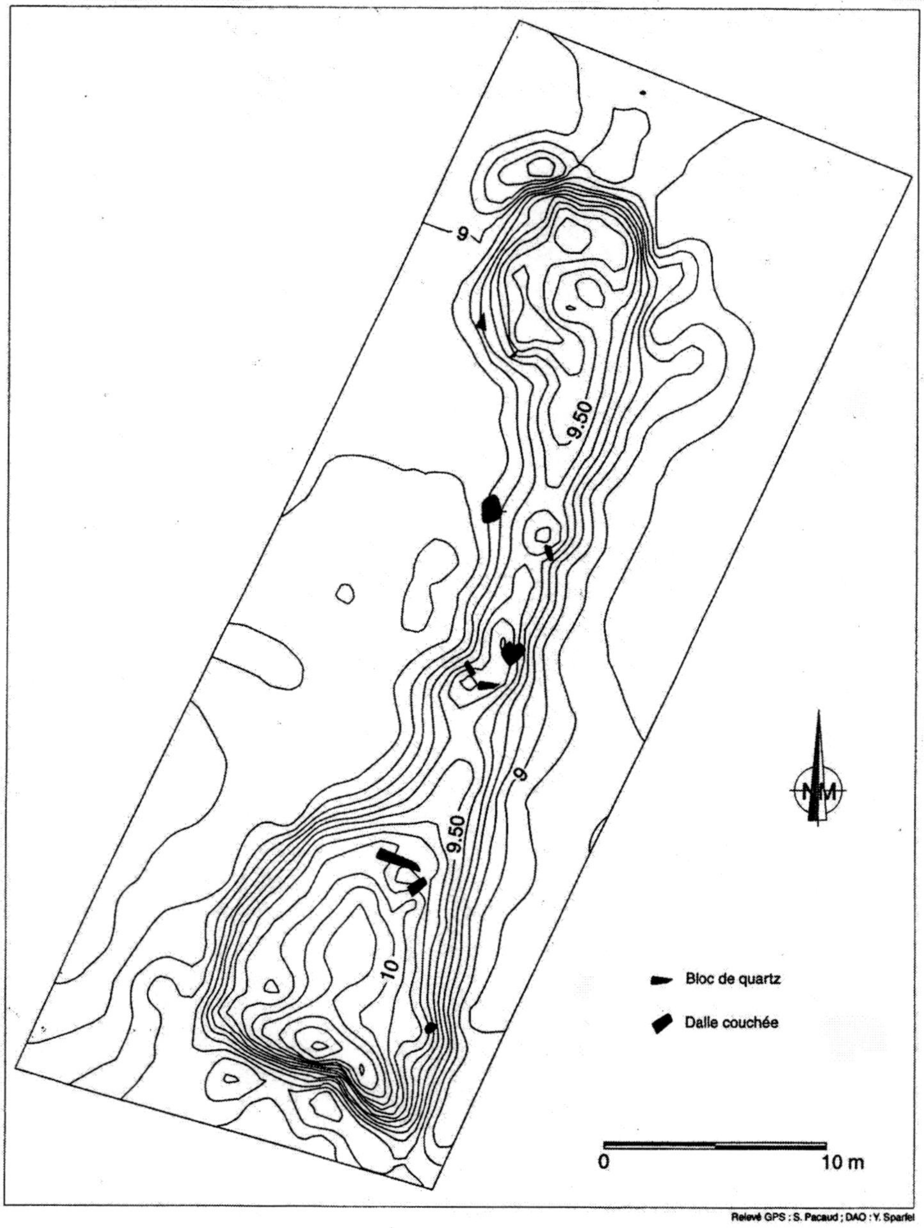

Fig. 7 - Plan du double cairn n° 17-18 sur l'île Trielen

A côté d'un semis d'indices caractérisés par quelques silex taillés ou tessons dispersés, d'autres stations semblent plus pérennes, comme le laissent penser des structures en creux (fosses, trous de poteaux), des aménagements (dallages) et des niveaux coquilliers riches en patelles. Le dénominateur commun de tous ces sites est la présence de la percussion sur enclume, mode opératoire ayant permis l'ouverture des galets comme leur plein débitage.

Les micro-falaises des îles Trielen, Molène et Balaneg ont livré des niveaux coquilliers de quelques décimètres d'épaisseur.

Sur Trielen, le niveau coquillier nord se situe au fond d'une grande anse. Il est épais de 0,20 à 0,30 mètre et s'étend sur deux mètres de longueur. Il comprend des tessons de céramique et des produits de débitage obtenus par percussion sur enclume. Un aménagement de petites dalles posées à plat est visible en coupe.

L'île de Balaneg présente la particularité de posséder plusieurs chaos granitiques, certains formant des abris sous roche, ayant été occupés durant la Préhistoire et le XXe siècle.

Le niveau coquillier de Balaneg, sous un abri sous roche, est composé d'une poche de patelles très dense. Un tesson recueilli a été analysé par S. Cassen qui y a vu une impression réalisée à l'aide d'un poinçon bifide. Le dégraissant est épais. Il s'agit peut-être d'une carène, que l'on pourrait attribuer, sous toutes réserves, au Néolithique ancien ou à l'Age du Bronze.

En général, ces structures ne comportent que peu d'éléments de datation, les tessons étant trop petits et les techniques de débitage du silex peu caractéristiques. Le débitage sur enclume des galets signerait plutôt une période néolithique que mésolithique, mais en la matière, des déterminismes locaux sont forts possibles.

A la pointe de Beg ar Loued sur l'île Molène, un niveau coquillier, dans une anse bien protégée des vents dominants, est soumis à une érosion rapide sous l'assaut des vagues. La couche archéologique, visible en micro-falaise, est constituée de nombreux vestiges (coquilles, silex, tessons de poterie, os, pierres rubéfiées et macro-outillage) et s'étend sur six mètres de long et 0,40 mètre d'épaisseur. A son pied, plusieurs structures en creux ont été repérées dans les restes d'un paléosol.

Les matières premières exploitées pour l'outillage lithique sont exclusivement locales. Les supports du mobilier sont toujours des galets côtiers, qu'ils soient en silex, en grès, en quartz ou en granite.

Bien que nous ayons repéré lors de nos prospections des galets de silex de belles dimensions sur l'estran situé au sud de l'île, les hommes ayant séjourné à Beg ar Loued se sont approvisionnés en galets de silex de petits modules (5 à 6 cm maximum). A contrario, ceux de grès sont de grandes dimensions et ont permis l'obtention de larges supports.

Deux techniques de débitage sont présentes sur le site. La plus fréquente est la percussion sur enclume caractérisée par l'obtention de produits de débitage spécifiques : galets fendus, "quartiers", pièces esquillées. Cette fracture en split est reconnaissable par différents stigmates, tels que l'absence de talon et de bulbe, des esquillements au point de percussion, des faces d'arrachement planes et des points d'impact sur la partie distale de l'objet. Cette technique de débitage est connue dès le Mésolithique final (Beg an Dorchenn à Plomeur, Finistère ; Rault, 1992), mais c'est surtout sur les sites du Néolithique qu'on la retrouve, tel l'atelier de débitage de Guernic à Saint-Pierre-Quiberon (Morbihan ; Guyodo, 1998) et l'habitat d'Er-Yoh à Houat (Morbihan, Guyodo, 1997).

La seconde technique employée est la percussion directe au percuteur dur. L'exploitation des nucléus a été le plus souvent réalisée à partir d'un seul plan de frappe. Les produits ainsi obtenus sont caractérisés par des talons lisses et larges et des bulbes bien marqués. Les accidents de taille, type siret ou réfléchissement, sont fréquents.

Le matériel lithique est toujours en cours d'étude, mais il apparaît, d'après les décomptes effectués, que la proportion d'outils est assez faible. Plusieurs pièces esquillées ont été récoltées. Il s'agit, avec les éclats retouchés, des outils les

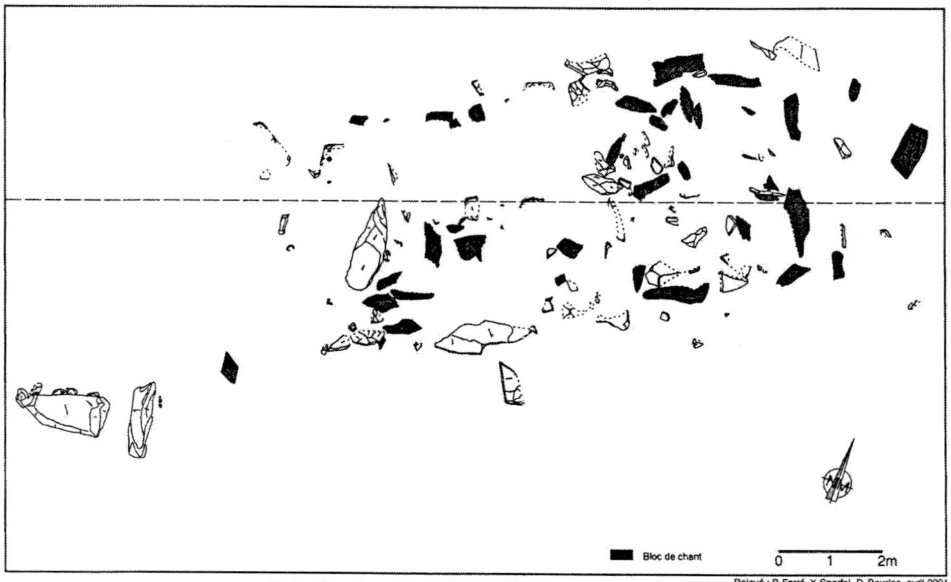

Fig.9 - Plan du monument 20 sur l'île Trielen

Relevé : P. Forré, Y. Sparfel, D. Bourles, avril 2001

plus représentés dans la série. Le débitage lamellaire est anecdotique. Sur les cinq lamelles étudiées, une est tronquée et deux autres portent des traces d'utilisation. La découverte d'une dizaine de mèches de foret pourrait indiquer l'existence d'une production spécifique, peut-être de perles en test de coquillage, comme Luc Laporte l'a démontré sur plusieurs sites de l'île d'Oléron (Charente) se rattachant aux groupes céramiques de l'Artenacien (Laporte & al., 1998). Plus proches géographiquement, l'habitat du Néolithique final/récent d'Er Yoh a également livré à ses fouilleurs une série de perçoirs fusiformes (Le Rouzic, 1930). Des grattoirs sur entame viennent compléter l'outillage. Le macro-outillage comprend une molette en granite rubéfiée dont une des faces est polie par usure et un galet allongé à l'extrémité esquillée.

Un fait important de cette série concerne l'absence d'armatures et de matériel poli. S'agit-il d'une lacune liée au ramassage ou d'une caractéristique du site ? On rappellera que le site d'Er-Yoh, malgré un mobilier nombreux, possède une proportion assez faible de ces types d'outils.

Les autres séries récoltées sur l'archipel sont plus modestes. On peut néanmoins considérer l'outillage de Beg ar Loued comme représentatif de la majorité du mobilier qui a été trouvé jusqu'à présent. La seule exception concerne une pendeloque polie en hématite (n° 3, FIGURE 2) recueilli in extremis à la faveur d'un affaissement de terrain à Balaneg.

Sur Balaneg et Trielen, trois galets biseautés (n° 4, FIGURE 2) ont été découverts en micro-falaise. Des manipulations récentes nous ont montré qu'un galet brut utilisé en percussion glissée se transformait en quelques minutes en galet biseauté. Sans vouloir déflorer un sujet qui fera ailleurs l'objet d'un travail plus développé (Pailler, Dupont, en cours), nous proposons, suivant en cela plusieurs auteurs anglo-saxons (Lacaille, 1954 ; Woodman et al., 1999), que certains de ces galets puissent être des "limpet hammers", autrement dit des "décolle berniques".

La plupart des tessons de céramique trouvés sur l'archipel offrent peu de critères diagnostiques. En revanche, l'échantillonnage de Beg ar Loued a permis de recueillir plus d'une vingtaine de tessons issus de céramique commune du Néolithique final. Malgré un état général très fragmentaire et érodé, certains ont été suffisamment bien conservés pour faire l'objet d'un examen par S. Cassen et A. Leroy. La couleur de leur pâte varie du noir au gris rouge avec un dégraissant moyen essentiellement constitué de petits cristaux de quartz. L'épaisseur des tessons varie de 0,60 à 1,10 centimètres. La cuisson des objets a été totalement réductrice (surface brune et cœur noir), elle est représentative de la plupart des sites finistériens du Campaniforme (Salanova, 2000, p.36).

DEUX PETITS TESSONS COMPORTENT DES DÉCORS.

Le premier tesson est un bord comportant un décor fin, érodé sur un tiers de sa surface mais qui est encore lisible. Il se caractérise par des lignes parallèles plus ou moins régulières formant des triangles incisés emboîtés, répartis en général sur le haut de la céramique. Il est attribuable au groupe Conguel.

Le second possède sur sa surface externe un décor typique du groupe Croh-Collé (défini autour de la baie de Quiberon). Les traits, finement cannelés, sont horizontaux et subparallèles entre eux. Ce genre de décor est l'une des variantes du décor originel des poteries de type Conguel.

Ces deux cultures sont proches thématiquement et géographiquement. Leurs principales différences se situent dans la forme et le décor (Cassen, 1993). Dans le groupe Conguel, les récipients sont biconiques et les décors dessinent des triangles.

Un autre tesson présente un cordon rapporté qui se situe la plupart du temps sur le haut de la poterie. L'extérieur du fragment a été lissé à la spatule ; des traces sont visibles à l'œil nu. La couleur de la pâte est gris foncé sur la partie externe du tesson et beige sur sa face interne. L'épaisseur du fragment varie de 1 à 1,50 centimètre. Sa datation s'étire du Néolithique récent/final au Campaniforme. Un dernier fragment de céramique est un fond ombiliqué attribuable au Campaniforme. La pâte est de couleur rouge foncé et le lissage fin ; ce sont deux éléments caractéristiques du Chalcolithique.

Ces trouvailles au sein d'une couche riche en coquilles viennent compléter celles du niveau d'habitat fouillé à Pe-

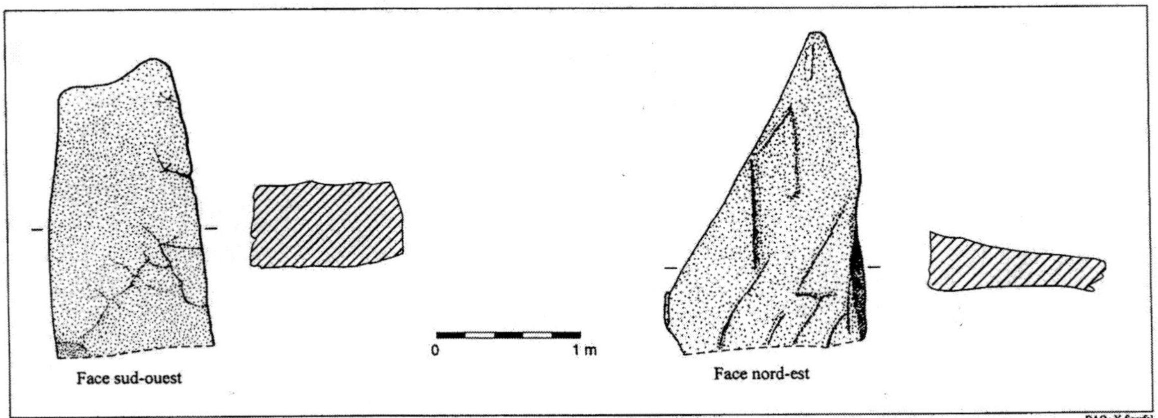

Fig.10 - Paire de menhirs au sud-ouest de l'île Molène

nancreac'h (Quimper ; Le Bihan & al., 1994) et permettent de combler partiellement un vide relatif concernant les découvertes du Campaniforme faites en contexte domestique. En effet, le Campaniforme bas breton est essentiellement connu à travers les fouilles de monuments mégalithiques, qu'il s'agisse de dolmens à couloir ou de monuments plus tardifs. Autre fait remarquable, la reconnaissance du groupe céramique Conguel/Groh-Collé parmi ce lot modeste de tessons trouvé sur l'archipel de Molène en fait la découverte la plus occidentale à ce jour.

La forte densité de coquillages, essentiellement des patelles mais aussi des bigorneaux et des ormeaux, a permis la conservation d'ossements, parmi lesquels Anne Tresset a identifié divers restes renvoyant essentiellement aux Caprinés (Mouton et peut-être Chèvre) et au Bœuf. Il est à noter que certains des restes de ces deux taxons appartiennent à de très jeunes individus, âgés de quelques semaines tout au plus.

L'association Bœuf/Mouton n'a rien de surprenant. Elle est déjà bien représentée dans le site du Néolithique final d'Er Yoh et est fréquente au Néolithique sur les îles de l'Atlantique (Tresset, soumis). Un ramassage récent sur le site a également permis de mettre en évidence le Porc.

En contrebas sur l'estran de Beg ar Loued, cinq structures en creux ont été repérées, dont quatre sont manifestement liées au site. Une autre au nord nous semble douteuse car ce secteur a été perturbé par une décharge moderne. Leur remplissage limoneux brun se lit facilement par contraste sur l'encaissant brun clair, plus argileux. De formes circulaires ou ovales, elles mesurent entre 0,60 et 1,80 mètre. Plusieurs blocs, certains ayant subi l'action du feu, sont incorporés dans chacune de ces fosses (n° 12, FIGURE11). Dans la structure C, quelques éclats de silex, des micro-charbons de bois, des boulettes d'argile et des esquilles osseuses ont été recueillis. La présence de ces structures en creux, probables trous de poteaux, ainsi qu'une couche archéologique riche en déchets culinaires et domestiques, interprétée comme un dépotoir, sont de bons indices en faveur de l'interprétation du site de Beg ar Loued comme habitat.

CONCLUSION

Si la moisson de sites et d'indices de sites découverts au cours de la campagne 2000-2001 a été fructueuse, elle n'apporte pas les réponses à toutes les questions que nous nous étions posées. La principale concerne l'occupation de l'archipel au Mésolithique, la dernière possibilité de la découvrir réside, pour ainsi dire, dans la prospection de l'île Beniget, la plus proche du continent, sur laquelle ont été repérés d'importants amas coquilliers. Pour le Néolithique ancien et moyen, nous nous trouvons dans le même cas de figure, les indices d'habitats sont inexistants et pourraient, par l'effet de la remontée du niveau marin, se trouver sous les eaux. Nous savons pourtant que les hommes de ces époques étaient bien implantés dans la région, au vu des nombreux tertres et dolmens à couloir édifiés. A contrario, le Néolithique récent/final, très bien représenté par le mobilier recueilli en surface, l'est très peu par les sépultures. Cette situation pourrait s'expliquer par la destruction des sépultures de cette période comme cela a été probablement le cas sur l'île Beniget (Du Chatellier, 1907), par l'éloignement relatif des monuments funéraires qui seraient implantés sur le continent, comme l'allée couverte de la presqu'île de Kermorvan (Du Chatellier, 1903) ou par l'ennoiement de ceux-ci, à l'instar de l'allée couverte de la plage des Blancs-Sablons (Le Conquet ; Sparfel, 2001). Une autre raison, plusieurs fois observée en d'autres lieux, serait la perduration ou la réutilisation de monuments plus anciens.

Il semblerait que les hommes du Néolithique final aient entretenu des relations avec le continent, comme l'atteste l'existence des styles céramiques Croh-Collé, Conguel et Campaniforme. D'une manière générale, l'outillage lithique, peu diversifié, témoigne d'une certaine spécialisation. L'absence de matériel poli sur l'archipel étonne d'autant plus du fait de la proximité de l'atelier de fabrication de lames polies en fibrolite connu sur la presqu'île de Kermorvan (Pailler, 1999) et par la présence de plusieurs outils réalisés dans ce matériau sur l'île d'Ouessant. Cela amène à se demander si, dès cette époque, ces territoires insulaires n'étaient pas dépourvus d'arbres.

Le fait que certaines questions soient encore sans réponse est un bon stimulant pour retourner sur le terrain avec un œil critique et tenter de les résoudre. Nonobstant cela, nous espérons que cette première contribution aura permis de réactualiser et de compléter de manière substantielle les inventaires préexistant. La deuxième campagne développera plusieurs axes : prospection-inventaire des îles Beniget, Kemenez et Litiri, systématisation des relevés au GPS, relevés complémentaires des monuments sur Molène, Trielen et Balaneg, rééchantillonages de sites repérés précédemment.

REMERCIEMENTS

Les autorisations de débarquer sur les îles nous ont été accordées pour la Réserve Naturelle d'Iroise par son comité consultatif et son conservateur Louis Brigand, pour l'île de Beniget par l'Office National de la Chasse et son conservateur Pierre Yesou, pour Kemenez par sa propriétaire Mme Tassin-Darcque, pour l'île de Litiri par son propriétaire de Mr de Kergariou.

Cette recherche a également bénéficié du soutien logistique de la réserve d'Iroise, de l'aide constante sur le terrain de Jean-Yves Le Gall, garde de la Réserve, et de son adjoint David Bourles, et de la participation de Sandrine Pacaud qui a réalisé les relevés au GPS.

BIBLIOGRAPHIE

BOUJOT C., CASSEN S.,
2000. Tertres et pierres dressées, in Cassen S. (dir.), Boujot C., Vaquero R., 2000. Eléments d'architecture, mémoire XIX, Association des Publications Chauvinoises, Chauvigny, p. 181-206.

CASSEN S.,
1993. Le Néolithique récent sur la façade atlantique de la France. La différentiation stylistique des groupes céramiques, Zephyrus, 44-45 (1991-1992), Salamanca, Editiones Universidad Salamanca, p. 176-182.

CHAURIS L., HALLÉGOUËT B.,
1989. Le Conquet, carte géologique de la France au 1/50 000, édi-

tions du BRGM, Orléans.
DEVOIR A.,
1909. Lettre au Sous-Préfet, Proposition de classement de monuments mégalithiques, commune du Conquet, Archipel de Molène, Archives du Laboratoire d'Anthropologie de Rennes I, inédit.
DEVOIR A.,
1917. Note sur la stèle du Téven de Kermorvan, Bulletin de la Société Archéologique du Finistère, t. 44, p. 1-7 et p. 1-19.
DU CHATELLIER P.,
1903. La pointe de Kermorvan, en Ploumoguer, ses monuments, pierres à cupules, Bulletin de la Société Archéologique du Finistère, t. 30, p. 90-95.
DU CHATELLIER P.,
1907 (2e éd.). Les époques préhistoriques et gauloises dans le Finistère, Lib. Plihon et Hommay, Imp. Leprince, Paris, Rennes, Quimper.
GOUEZIN P.,
1992. La néolithisation du Morbihan intérieur : la nécropole de Coëby, commune de Trédion. Premiers résultats, in Le Roux C.-T. (éd.), Paysans et Bâtisseurs, Actes du 17e colloque interrégional sur le Néolithique, Revue Archéologique de l'Ouest, suppl. n°5.
GOULETQUER P., PAILLER, Y., SPARFEL Y., 2000. Projet de prospection archéologique sur l'archipel de Molène, multigraphié.
GOURMELON F. (DIR.), 1995. Atlas de la Réserve de biosphère de la Mer d'Iroise, Parc Naturel Régional d'Armorique, Briec.
HALLÉGOUËT B.,
1982. L'archipel de Molène, Penn ar Bed, tome 110, p.79-82.
Giot P.-R.,
1987. Barnenez, Carn, Guennoc, Travaux du Laboratoire "Anthropologie-Préhistoire-Protohistoire Quaternaire armoricains", Rennes, 2 vol.
GIOT P.-R.,
1990. Le niveau de la mer : changeant, fluctuant, mouvant…, Bulletin de l'Association Manche-Atlantique pour la Recherche Archéologique dans les Iles, n°3, p. 5-16.
GIOT P.-R., HALLÉGOUËT B.,
1980. Les réserves naturelles de Bretagne : Intérêt Archéologique, Penn ar Bed, t. 101, p. 285-296.
GIOT P.-R., L'HELGOUAC'H J.,
1955. Le tertre tumulaire de la Croix-Saint-Pierre en Saint-Just, Annales de Bretagne, t. 62, p. 282-292.
GUYODO J.-N.,
1997. L'industrie lithique du site d'Er-Yoh (île de Houat, Morbihan), mémoire de DEA, université de Paris I, multigraphié.
GUYODO J.-N., 1998. Un atelier de débitage de silex : le site de Guernic à Saint-Pierre-Quiberon (Morbihan), rapport de sondage, multigraphié.
LACAILLE A.D.,
1954. The stone age in Scotland, Oxford University Press, London, New-York, Toronto.
LAPORTE L., DESSE-BERSET N., GRUET Y., TRESSET A.,
1998. Un lieu de production de parure au Néolithique final et son économie de subsistance : le site de Ponthezières à Saint-Georges-d'Oléron (Charente-Maritime), in Gutherz X., Joussaume R. (dir.), La néolithisation du Centre-Ouest de la France, actes du XXIe colloque inter-régional sur le Néolithique, mémoire XIV, Association des Publications Chauvinoises, Chauvigny, p. 237-255.
LE BIHAN J.-P., ROBIC J.-Y. ET TINEVEZ J.-Y., 1994. Quimper "Pennancreac'h", Bilan scientifique de Bretagne (1993), p. 43.
Le Bihan J.-P. (dir.), Villard J.-F., 2001. Archéologie d'une île à la pointe de l'Europe : Ouessant, tome 1, "Le site archéologique de Mez Notariou et le village du premier Age du Fer", Centre de recherche archéologique du Finistère, RAO, Quimper.
LE GOFFIC M.,
1994. Documents de la Zone de Protection du Patrimoine Architectural Urbain et Paysager du Conquet, Etude des sites archéologiques, multigraphié.
LE GOFFIC M.,
1997. Les cupules en relation avec les monuments mégalithiques du Finistère (Bretagne, France), Brigantum, t. 10, p. 355-375.
LE GOFFIC M.,
2000. Rapport de fouille : la nécropole mégalithique de la pointe du Souc'h en Plouhinec (Finistère), multigraphié.
LE ROUZIC Z., 1921. Fouilles faites dans le Morbihan, Bulletin de la Société Polymathique du Morbihan, p. 1-11.
LE ROUZIC,
1930. Carnac, fouilles faites dans la région, îlot d'Er-Yoh (Le Mulon), commune de Houat, 1924-1925, impr. Lafolaye et J. de Lamarzelle, Vannes.
MORZADEC-KERFOURN M.-T.,
1974. Variations de la ligne de rivage armoricaine au Quaternaire. Thèse de doctorat ès sciences, Université de Rennes I, Mémoire de la Société géologique et minéralogique de Bretagne, t. 17.
MOLINES N.,
1992. Le chopper de l'île Molène et le chopper de l'île Ilur, Bulletin de l'Association Manche-Atlantique pour la Recherche Archéologique dans les Iles, n°5, p. 19-23.
PAILLER Y.,
1999. Un site du Néolithique au Conquet : étude du matériel en fibrolite provenant de Kermorvan, Bulletin de la Société Archéologique du Finistère, t. 128, p. 89-98.
PAILLER Y., SPARFEL Y. Avec la participation de Marchand G., 2001. Prospection archéologique de l'archipel de Molène, deuxième rapport, multigraphié.
PAILLER Y., SPARFEL Y. A.,avec la collaboration de Cassen S., Arzel P., Tresset A., Le Prospection archéologique de l'archipel de Molène, troisième rapport, multigraphié.
PAILLER Y., DUPONT C.,
en cours. Analyse fonctionnelle des galets biseautés du Mésolithique à la fin du Néolithique dans l'ouest la France, l'ouest des îles britanniques et l'Irlande.
RAULT O.,
1992. L'économie du débitage sur le site mésolithique récent/final de Beg-an- Dorchenn (Plomeur, Finistère), mémoire de maîtrise, université de Paris I, multigraphié.
SALANOVA L.,
2000. La question du Campaniforme en France et dans les îles anglo-normandes, Comité des travaux historiques et scientifiques, Société préhistorique française, Paris.
SPARFEL Y.,
2001. Prospection et inventaire des mégalithes, tumulus, cairns et tertres dans l'ouest du Léon (Finistère-Bretagne), 2 vol., mémoire de maîtrise, Université de Bretagne Occidentale.
TRESSET A.,
soumis. French connections II : of Cows and Men (the introduction of domesticates and husbandry in north-western Europe – 5th –3rd millennia BC). in : Neolithic settlement in Ireland and western Britain. Proceedings of the Conference organised by the Pre-

historic Society and the department of Archaelogy and Palaeoecology, University of Belfast, 20-22nd April 2001.
WOODMAN P.C., ANDERSON E. AND FINLAY N.,
1999. Excavations at Ferriter's Cove, 1983-1995 : last foragers, first farmers in the Dingle Peninsula, Wordwell, Bray (Eire).

LISTE DES AUTEURS :

Yvan PAILLER : Doctorant, Centre de Recherche Bretonne et Celtique (UMR 6038 du CNRS), Université de Bretagne Occidentale, Collaborateur UMR 6566 CNRS, BP 814, 29285 Brest Cedex.
Yohann SPARFEL : Etudiant en DEA, Centre de Recherche Bretonne et Celtique (UMR 6038 du CNRS) Université de Bretagne Occidentale, 29285 Brest Cedex.
Serge CASSEN : Chargé de Recherche au CNRS, Civilisations Atlantiques et Archéosciences, UMR 6566 du CNRS, Laboratoire de Préhistoire, Université de Nantes, BP 81227, 44312 Nantes Cedex 3.
Pierre GOULETQUER : Chargé de Recherche au CNRS, Centre de Recherche Bretonne et Celtique (UMR 6038 du CNRS), Université de Bretagne Occidentale, Chercheur associé UMR 6566 CNRS, BP 814, 29285 Brest Cedex.
Michel LE GOFFIC : Archéologue Départemental du Finistère, Service Départemental de l'Archéologie, 1, rue du Général de Gaulle, 29590 Le Faou.
Aude LEROY : Etudiante en DEA, Université de Rennes II.
Grégor MARCHAND : Chargé de Recherche au CNRS, Civilisations Atlantiques et Archéosciences, UMR 6566 du CNRS, Laboratoire de Préhistoire, Université de Nantes, BP 81227, 44312 Nantes Cedex 3.
Anne TRESSET : Chargé de Recherche au CNRS, Laboratoire d'anatomie comparée, ESA 8045 du CNRS.
Estelle YVEN : Doctorante, Centre de Recherche Bretonne et Celtique (UMR 6038 du CNRS) Université de Bretagne Occidentale, Collaborateur UMR 6566 CNRS, BP 814, 29285 Brest Cedex.

31

LA COLONIZACIÓN DEL ARCHIPIÉLAGO CANARIO: UN PROCESO MEDITERRÁNEO?

PABLO ATOCHE PEÑA

Universidad de Las Palmas de Gran Canaria, Spain
EMAIL: patoche@sinf.ulpgc.es

SUMMARY

Despite more than a hundred years having gone by since scientific interest in the first inhabitants of the canarian archipelago was aroused, there are still considerable gaps in knowledge as regards fundamental aspects of protohistorical canarian cultures, especially with reference to the process of discovery and colonisation of the islands. However, in our opinion it is obvious that in order to become familiar with and to understand the cultural manifestations of canarian protohistorical societies, it is necessary to determine how the colonisation phenomenon was produced.

In recent years, research has provided us wisth an increasing number of cultural elements of western Mediterranean origin which are contextualised within the realm of colonial expansion of Phoenician-Punic civilisation. Thus, the Phoenician-Punic culture plays an important role in the discovery an colonisation of the islands. The aims of this paper are to determine this role as well as to draw attention to some of the resulting cultural and environmental consequences of human settlement on the islands.

A pesar de haber transcurrido más de cien años desde que se inició el interés científico por los primeros habitantes del archipiélago canario, aún subsisten profundas lagunas en torno al conocimiento de aspectos fundamentales de las culturas canarias protohistóricas, en especial por lo que se refiere al proceso de descubrimiento y colonización de las islas. Sin embargo, desde nuestra perspectiva resulta incuestionable que para llegar a conocer y comprender en toda su complejidad las manifestaciones culturales de las sociedades protohistóricas canarias, resulta prioritario determinar cómo se produjo el fenómeno colonizador.

En los últimos años la investigación nos ha ido proporcionando un número cada vez mayor de elementos culturales originarios del Mediterráneo occidental contextualizados en el ámbito de la expansión colonial fenicio-púnica. De esa manera, la cultura fenicio-púnica se nos presenta jugando un papel destacado en el descubrimiento y colonización de las islas. Determinar ese papel es el objetivo de esta comunicación, al igual que apuntar algunas de las consecuencias culturales y medioambientales resultantes del establecimiento humano en las islas.

INTRODUCCIÓN

Frente al tradicional interés por hacer de la realidad histórica canaria anterior a la conquista normando-castellana un proceso aislado y radicalmente distinto de lo que ocurría en su entorno geográfico y cultural más cercano, o por anclar a los primeros pobladores de las islas en el Neolítico, algunos de los trabajos arqueológicos dados a conocer durante la última década obligan a concebir el mundo aborigen canario en un contexto protohistórico, marcado por los prolongados contactos que las sociedades establecidas en las islas mantuvieron con las culturas mediterráneas, los cuales dieron lugar a interesantes fenómenos de cambio social y cultural.

Indiscutiblemente, las significativas evidencias materiales proporcionadas por la arqueología canaria a lo largo de los últimos años de la pasada centuria, con la constatación de la presencia de elementos materiales e infraestructuras púnicas y romanas, han contribuido a establecer una nueva visión histórica, caracterizada por la incorporación del archipiélago canario al proceso de expansión comercial y colonización llevado a cabo por navegantes mediterráneos a lo largo de la fachada atlántica africana durante el primer milenio antes de la Era. De esa manera, Canarias constituyó un curioso ejemplo de la forma en que se produjo la adaptación de un modelo de colonización ampliamente ensayado en el Mediterráneo por los marinos fenicios a unas islas oceánicas deshabitadas. Las Islas Canarias se convierten así en el último archipiélago mediterráneo en colonizarse, vinculándose el inicial establecimiento humano en las islas al fenómeno de expansión del comercio internacional que se produjo durante el Bronce final y la Primera Edad del Hierro en el Mediterráneo occidental.

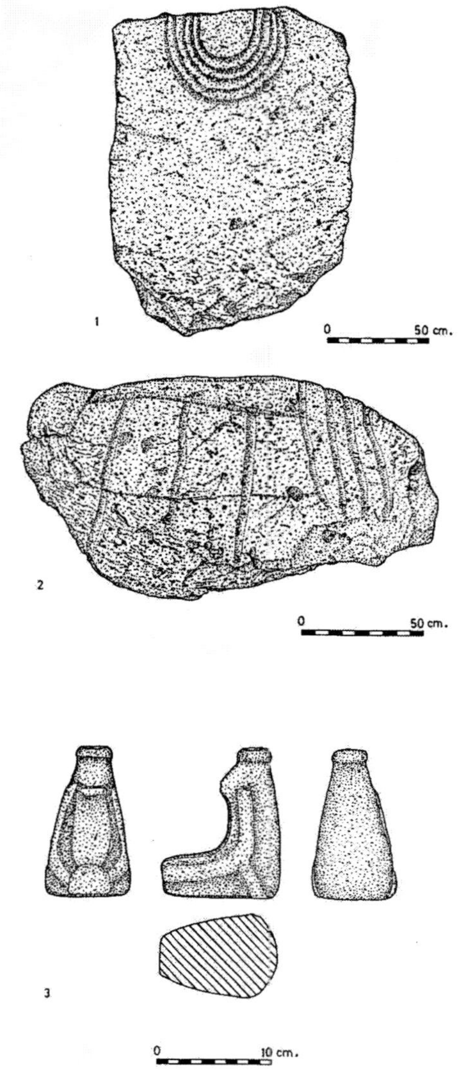

FIGURA 1: Estela lítica relacionada con el culto a Baal Hammón/Tanit (1). Escultura zoomorfa representando a un carnero identificado

CÓMO SE HA INTERPRETADO EL DESCUBRIMIENTO Y COLONIZACIÓN DEL ARCHIPIÉLAGO CANARIO?

Los vínculos que desde el Mediterráneo occidental se establecieron con las Islas Canarias durante el Bronce final/Hierro antiguo han estado durante mucho tiempo envueltos por el mito y el misterio. Una situación en parte propiciada por el desconocimiento que de las islas y sus habitantes tuvieron los conquistadores bajomedievales o, más recientemente, por la interpretación que se hace en el siglo XIX de los textos clásicos greco-latinos, claramente mediatizada por los intereses de la entonces naciente industria turística canaria y que lleva a identificar las islas como el lugar real donde se ubican mitos y leyendas de la antigüedad tales como la Atlántida o los Campos Elíseos (Bory de Saint-Vincent, 1805; Stone, 1887).

Cuando la Europa renacentista de los siglos XIII al XVI

entró en contacto con las islas y sus culturas se despertó el interés humanista por conocer el origen de sus pobladores. Como fue común en esa época, la incógnita se resolvió para algunos recurriendo a los textos bíblicos mientras que otros acudieron a las fuentes documentales clásicas; el resultado fue una visión más próxima a la mitología o la utopía que a la realidad de los hechos. Sin embargo, frente a la actitud más generalizada, encontramos que algunos eruditos de los siglos XVI al XIX abordaron la cuestión de una manera más objetiva, como fue el caso de las interpretaciones propuestas por L. Torriani (1978 [c. 1592]) o Fr. J. de Abreu Galindo (1977 [c. 1602]), quienes relacionaron las culturas insulares con pueblos y culturas mediterráneas de la antigüedad (etruscos, griegos, cartagineses y romanos).

En cualquier caso, no debe extrañar que la imagen que han proyectado las islas hasta la última década del siglo XX, hija en gran medida de los planteamientos esbozados por los historiadores románticos de finales del siglo XIX, nos haya mostrado unas sociedades inmutables, a las que los más de dos milenios transcurridos desde su arribada a las islas hasta la conquista europea bajomedieval no parecían haber supuesto ningún proceso de cambio cultural realmente significativo. Llegamos así al último tercio del siglo XX, momento en que desde ámbitos académicos se producen diversas tentativas dirigidas a dar respuesta a la incógnita del poblamiento, caracterizadas por incorporar al menos dos planteamientos comunes: considerar que el archipiélago se hallaba deshabitado hasta fechas muy tardías, cercanas al cambio de Era, e identificar a los protagonistas de la colonización con bereberes prehistóricos procedentes del continente africano. Mientras el primer argumento poseía una cierta apoyatura arqueológica, el segundo incurría en claras contradicciones, en especial cuando se intentaba caracterizar culturalmente a esos bereberes prehistóricos, ya que se acudía indefectiblemente al rastreo de concomitancias con las actuales poblaciones bereberes magrebíes. Se inicia así una infructuosa búsqueda africana de la cultura madre en la que debían reflejarse fielmente las culturas insulares; una tentativa cuyo fracaso hizo preciso reconsiderar la maternidad única, y es entonces cuando se recurre a la maternidad compartida, que en todo caso seguía siendo africana si bien circunscrita al noroeste del Magreb y al Sahara.

Esa búsqueda externa influyó de tal manera en la investigación, que todo cambio observado en el ámbito de la cultura material se identificó inexorablemente como resultado de la arribada de nuevos grupos humanos, negándose implícitamente a las poblaciones insulares ya establecidas cualquier posibilidad de cambio social y cultural que no estuvieran inducidos por factores externos. Sin embargo, lo anterior no impedía que paralelamente se defendiera la singularidad de las culturas insulares. Es en ese estrecho y contradictorio ámbito interpretativo en el que las producciones cerámicas pasan a convertirse en el caballo de batalla de la arqueología canaria, de tal manera que la menor variación morfotécnica o decorativa observada en las producciones insulares servirá como prueba de la superposición de sucesivas oleadas de poblamiento, de las que se han llegado a identificar hasta cuatro en La Palma, dos en La Gomera, etc... Asombrosamente esas oleadas estaban protagonizadas en todos los casos por grupos humanos cuyo mayor nivel de desarrollo cultural les permitía imponerse a las sociedades insulares previamente establecidas. En esencia, un modelo interpretativo basado en un fenómeno continuo de arribada de grupos de población, cuya supuesta procedencia de distintas áreas norteafricanas aportaba el argumento que explicaba las variaciones observadas en los contextos materiales a nivel de isla y la diversidad cultural a nivel del archipiélago.

De las muchas contradicciones que encerraba la interpretación señalada quizás la más significativa radicaba en considerar que el fenómeno colonizador afectó a unas poblaciones caracterizadas, desde la arribada a las islas, por su aislamiento, por su primitivismo cultural y su carácter tardío, además de por su incapacidad para desarrollar cualquier otro tipo de navegación que no pasara por el empleo de medios extremadamente arcaicos. Navegantes oceánicos que desconocían el arte de navegar, pero con la suficiente osadía/necesidad como para adentrarse en un océano ignoto valiéndose de unos medios de transporte rudimentarios fabricados con madera, piel u odres, pequeñas arcas de Noé con las que, arrastrados por las corrientes, alcanzaron milagrosamente el archipiélago canario en auténticas Anavegaciones de fortuna (Navarro, 1983: 93).

La consecuencia científica de todo lo anterior fue que la arqueología canaria se adentró en la última década del siglo XX careciendo de estudios que abordaran la colonización humana del archipiélago dentro de una hipótesis articulada y global, alejada de los tópicos históricos. Un problema estructural que había terminado por manifestarse en el generalizado desinterés científico por toda línea de trabajo que se alejara de lo puramente indígena, de lo puramente Aguanche.

Sin embargo, la segunda mitad de la década de los años 90 del pasado siglo ve surgir un punto de inflexión en la situación anterior coincidiendo con la publicación de diferentes evidencias materiales demostrativas de que las islas estuvieron abiertas a los procesos culturales, políticos y económicos que se desarrollaron en el Mediterráneo occidental, al menos desde el siglo VI a.n.e. al siglo IV d.n.e. (Muñoz, 1994; Balbín et alii, 1995; Atoche et alii., 1995; González et alii, 1995; Atoche et alii, 1999; Atoche, Martín et Ramírez 1999; Atoche et Martín, 1999; Atoche et Paz, 1999; Atoche et Ramírez, 2001). En concreto, la constatación de que los primeros hombres y mujeres que colonizaron las islas poseían un bagaje cultural en el que abundaban los elementos asimilados de la civilización fenicio-púnica establecida en el norte de África desde finales del segundo milenio antes de la Era, está permitiendo hallar explicaciones a fenómenos culturales de las comunidades canarias dentro del universo semita y de los resultados de la aculturación de las poblaciones paleobereberes norteafricanas. La consecuencia más inmediata de todo ello está siendo un saludable rejuvenecimiento de las perspectivas, la relectura de numerosos elementos culturales inadecuadamente interpretados y la reconstrucción de los primeros capítulos del devenir histórico de las islas sobre la base de su incorporación al contexto de la expansión comercial desarrollada a partir del primer milenio antes de la Era en la fachada atlántica africana.

COLONIZACIÓN: LA HIPÓTESIS MEDITERRÁNEA

La colonización de las islas mediterráneas, iniciada al finalizar el Pleistoceno, se acelera a partir del VII milenio espoleada por fenómenos de cambio social y cultural originados en el desarrollo de economías de producción de alimentos, las únicas capaces de incrementar la capacidad de carga de la mayor parte de las islas aún deshabitadas y permitir la supervivencia de grupos humanos. Ese será un proceso que, en última instancia, dará lugar a la emergencia de amplias redes de intercambio a larga distancia.

Con posterioridad, durante el IV y III milenios, se produce la colonización de otro importante grupo de islas mediterráneas; en esa ocasión el motor es en parte la denominada Arevolución de los productos secundarios@ (Sherratt, 1979), un fenómeno basado en la cría de ovicápridos, un ganado capaz de prosperar en áreas con escasos recursos. Esta nueva oleada colonizadora discurre paralela al desarrollo de una cada vez mayor diferenciación social, materializada a lo largo del Mediterráneo en la aparición de estructuras civiles y religiosas, las cuales requirieron incrementar la producción de excedentes, una de las razones que animarían la revolución de los productos secundarios. También estuvo alentada por las cada vez más competitivas relaciones sociales, sirviendo de válvula de escape a las ambiciones de rivales derrotados o segundones desheredados (Alvar, 1998).

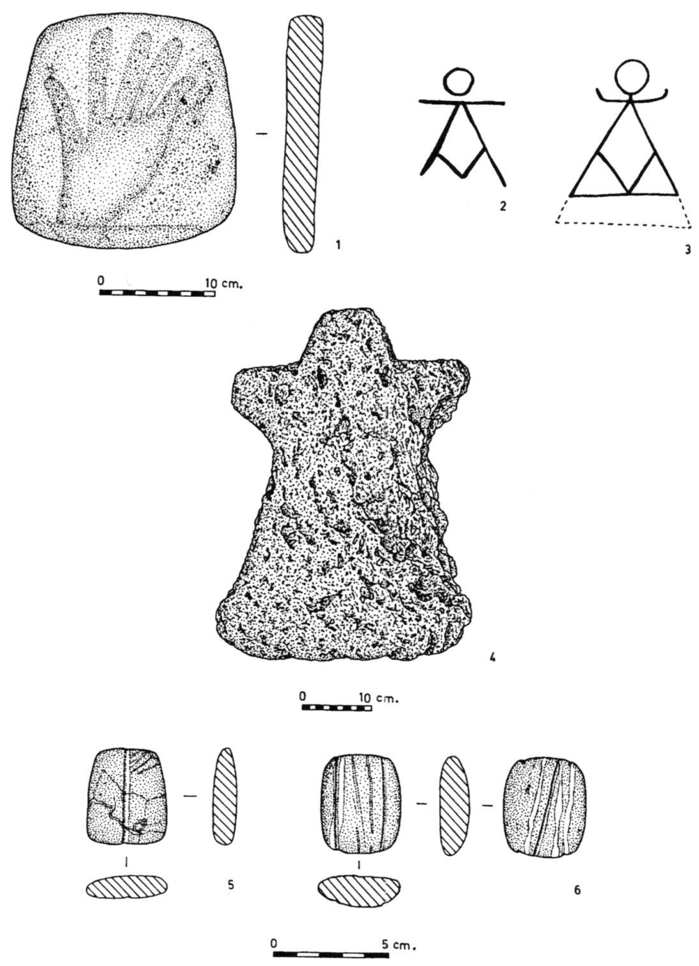

FIGURA 2: Betilo con representación grabada de mano derecha, símbolo de Tanit (Zonzamas, Lanzarote) (1). Signo grabado de Tanit: según sus descubridores (2) y en base a nuestras observaciones (3) (Pozo de la Cruz, Rubicón. Lanzarote). Estela con morfología del signo de Tanit (La Aldea, Gran Canaria) (4). Amuletos egiptizantes de ascendencia fenicio-púnica (Lanzarote) (5 y 6). (Dib. Mª A. Ramírez).

En el primer milenio antes de la Era se inició la última fase de colonización de las islas mediterráneas, ocupándose entonces la práctica totalidad de las que aún permanecían deshabitadas. El acicate de esa nueva corriente colonizadora lo constituyó la expansión del comercio internacional, que entre otras consecuencias dio lugar a la integración de las economías de extensas áreas a lo largo del Mediterráneo, un hecho que se produjo al mismo tiempo que se desarrollaban sociedades estatales en el Próximo Oriente, Grecia e Italia. Es de esa manera como, durante la primera mitad del primer milenio antes de la Era, se incrementó la influencia fenicia en el Mediterráneo central y occidental, contextualizada en un programa de fundación de colonias en el norte de África y la Península Ibérica, afrontando durante los siglos VII y VI a.n.e. la expansión de la influencia griega en el Mediterráneo occidental.

A ese fenómeno colonizador del primer milenio debemos asociar la colonización del archipiélago canario. Un proceso que, aunque localizado en la fachada atlántica africana, siguió el modelo puesto en práctica por los fenicios en el Mediterráneo occidental, inevitablemente con adaptaciones que alteraron el patrón inicial, pero que en esencia convirtieron a Canarias en el último archipiélago mediterráneo en colonizarse.

La que hemos denominado hipótesis mediterránea se viene estructurando, como modelo para explicar la colonización humana de las Islas Canarias, a lo largo de la última década. Se asienta sobre la aseveración de que existen suficientes argumentos científicos como para sostener una postura en la que los protagonistas del primer establecimiento humano en el archipiélago canario fueron navegantes fenicio-púnicos procedentes del Círculo del Estrecho, quienes en realidad tomaron el relevo a tradiciones anteriores que habían propiciado, cuando menos a partir del Bronce final, la frecuentación de las aguas insulares, todo ello con unos fines económicos muy concretos. Analicemos algunos de esos argumentos.

La existencia en las culturas canarias de influjos fenicio-púnicos ha constituido una posibilidad que ya fue apuntada, entre otros, por L. Torriani en las postrimerías del siglo XVI (1978 [c. 1592]); no obstante, habrá que esperar a la segunda mitad del siglo XX para que se plantee como posibilidad científica por investigadores como A. Gaudio (1958), quien sugirió el desarrollo de navegaciones tartésicas, fenicias, púnicas y romanas por las aguas canarias apoyándose, ante la ausencia de pruebas arqueológicas, en el estudio de las corrientes y los vientos de la zona, en las rutas de los bancos de túnidos y en la cita de textos clásicos. Sin embargo, no ha sido hasta los últimos años cuando se han comenzado a detectar en numerosas manifestaciones de las culturas canarias elementos asimilados de la cultura fenicio-púnica; unos elementos que, en las islas, al igual que el resto de los componentes culturales originariamente norteafricanos, aparecen filtrados por el tamiz de la interpretación y la elaboración de las poblaciones insulares, de ahí que hayan sido plasmados según los esquemas, los gustos y las posibilidades de los primeros pobladores.

Es precisamente en el ámbito de la cultura material donde hallamos un amplio conjunto de artefactos que señalan el tipo de actividades económicas que debieron propiciar inicialmente el descubrimiento y con posterioridad la colonización de las islas. Es el caso de los anforoides modelados a mano registrados en Canarias para fechas muy cercanas a los momentos iniciales del establecimiento humano y que reproducen la morfología de ánforas pre-romanas gaditano-ibicencas (Ramón, 1995, figs. 136 y 235) y tunecino-sicilianas (Op. cit., figs. 139 y 248) destinadas a contener salazones y otros productos cuya vigencia en los establecimientos del Círculo del Estrecho se prolongó entre los siglos VI y III a.n.e. (González et alii, 1995: 170). En la misma dirección apuntan las infraestructuras costeras establecidas en varios puntos de las islas, sin duda con el objetivo de facilitar la explotación de los recursos insulares y de las que nos ocuparemos más extensamente en otro apartado de este trabajo.

Por lo que se refiere a los elementos pertenecientes a la superestructura ideológica, y que por esa razón constituyen la mejor referencia acerca del grado de asimilación de la cultura fenicio-púnica que se dio entre los pobladores insulares, encontramos que las culturas canarias incorporaron dos sistemas de escritura, el líbico-bereber y un segundo sustentado en una lengua semítica, púnica, y que correspondería a la norteafricana escritura libio-fenicia (López Castro, 1992: 54), utilizada en las islas para expresar teóforos o teónimos propios de la cultura púnica . Un elevado índice de aculturación lo reflejan también las creencias que observaron y los ritos que practicaron los primeros canarios, en los que están presentes divinidades fenicias, dioses y amuletos egipcios o ritos de ascendencia semítica que muestran con nitidez que las comunidades protohistóricas canarias rindieron culto a deidades de origen mediterráneo-oriental, entre las que R. Muñoz (1994) encontró citadas en inscripciones bilingües de grafía libio-fenicia y líbico-bereber a divinidades como Amón, Us, Yahweh o Iuppiter, todas las cuales formaron parte del panteón fenicio-púnico y reflejan un hecho común entre los paleobereberes en contacto con la cultura fenicio-púnica del primer milenio antes de la Era.

Los canarios protohistóricos tuvieron una estructura religiosa politeísta en la que existió una divinidad principal, conocida bajo diferentes denominaciones (Acorán, Achamán,...) pero con características comunes para todo el archipiélago. A ella se le rendía un culto prioritario, con ritos propiciatorios destinados a obtener su protección, en el que son notables las semejanzas con respecto a las formas rituales semitas, tanto en lo referido a los lugares de celebración como al ritual desarrollado:AAdoraban a un Dios, levantando las manos al cielo. Hacíanle sacrificios en las montañas, derramando leche de cabras con vasos que llaman gánigos, hechos de barro@ (Abreu Galindo,1977 [c. 1602]: 57).

El panteón de los primeros canarios era complejo, como lo demuestra la existencia de grabados de toros, un animal ausente en los contextos arqueológicos canarios, representación del dios Baal Hammón (Balbín et alii, 1995), la escultura de carnero representando a Ammón-Baal Hammón (FIGURA 1, nl 2) o las frecuentes figuraciones de Tanit, deidad que aparece bajo varias de las formas en que solía evocarse en el norte de África (signo antropomorfo grabado o esculpido a modo de estela, rombo con cruz inscrita grabado, mano abierta en señal de bendición,...) (Atoche et Ramírez, 2001) (fig, 2, nl 1, 2, 3 y 4). Junto a las anteriores se hallan otras divinidades mediterráneas de gran aceptación en Cartago, como la diosa egipcia Tueris (González et alii, 1995), o los amuletos de calcedonia que reproducen escaraboides egipcios o pseudo-egipcios (Atoche, Martín et Ramírez, 1999), un elemento apotropaico muy frecuente en los ambientes fenicios del Mediterráneo, por donde los extendió precisamente la actividad comercial fenicio-púnica (FIGURA 2, nl 5 y 6).

Esa complejidad de la que hablamos se refleja igualmente en la diversidad de ritos funerarios que se practicaron. El más común consistió en depositar el cadáver en el interior de una cueva dispuesto en posición decúbito supino sobre tablones de madera u otros tipos de soportes , reproduciendo una ceremonia que hizo su aparición en el norte de África a raíz de la llegada de los colonizadores fenicios (Lancel, 1994: 61). Menos frecuente que el rito anterior, si bien manteniendo en común la conexión norteafricana, también hallamos documentada la práctica de la cremación en las islas de La Palma, El Hierro y Tenerife, en la primera de ellas al menos durante los instantes inmediatos al inicio del establecimiento humano. La cremación fue introducida en el norte de África por los colonizadores fenicios, un espacio geo-cultural en el que igualmente eran habituales las inhumaciones infantiles en recipientes cerámicos, ritual funerario que se ha constatado en las islas de Gran Canaria y Tenerife, donde guarda una gran semejanza con el que se conoce en los contextos fenicios de Ibiza, Huelva o la necrópolis tunecina de Kerkouane (Fantar, 1988: 59).

Los primeros canarios depositaban a sus difuntos en su morada eterna de forma individual o colectiva y sin que la disposición del cadáver siguiese una orientación fija (Arco, 1976: 21), utilizando oquedades naturales o artificiales, que en algunos casos atestiguados en Gran Canaria (Cueva Pintada de Gáldar, Cenobio de Valerón o Cuatro Puertas) se identifican como ejemplos de haouanet norteafricanos realizados al modo de los hipogeos de las colonias fenicias del occidente mediterráneo (Balbín et alii., 1995: 28), un hecho que nos vuelve a conducir por el camino de esa simbiosis cultural libio-fenicia de la que venimos hablando. Los cadáveres se dirigían a la otra vida vestidos con prendas semejantes a las que llevaban en vida y envueltos en un sudario elaborado con varias capas de piel de ovicápridos, ocasionalmente teñidas de rojo, o con mortajas elaboradas con tejidos vegetales, habiéndose empleado en Tenerife y Gran Canaria sarcófagos fabricados mediante el vaciado de troncos de árboles.

FIGURA 3: Mapa del N.O. de África en época fenicio-púnica, con indicación de los dos itinerarios que pudieron utilizarse para alcanzar

El culto funerario se sustentaba en creencias en la supervivencia tras la muerte similares a las que conocemos para la cultura fenicio-púnica. Los difuntos eran acompañados por un número limitado de ofrendas de alimentos y objetos, como adornos personales/elementos apotropaicos, recipientes cerámicos completos o más comúnmente fragmentados, punzones de hueso, lascas de obsidiana, armas de madera o, excepcionalmente (Tenerife), un perro con huellas de momificación. La ceremonia fúnebre aparejaba además la realización de libaciones y la celebración de comidas rituales.

En la mayor parte de los casos los cadáveres han llegado hasta nosotros esqueletizados, excepcionalmente (necrópolis de Ucazme, Tenerife) como resultado de un proceso de descarnamiento debido a la acción antrópica, una práctica poco habitual en las islas pero para la que de nuevo hallamos precedentes entre los ritos escatológicos fenicio-púnicos. Frente a esa norma general, en islas como Tenerife, Gran Canaria o La Palma se conocen ejemplos de momificación o mirlado, un complejo tratamiento funerario que implicó un ritual muy específico al que creemos relacionado en origen con Egipto, si bien readaptado en el Círculo de Cartago desde donde pasaría al archipiélago canario, ámbito en el que fue objeto de una postrera adaptación.

Como vemos, es muy numeroso el conjunto de evidencias arqueológicas, manifestaciones de la cultura material y espiritual, que atestiguan la vinculación de los paleobereberes que poblaron Canarias con el Mediterráneo y en concreto con el mundo fenicio-púnico del Círculo de Cartago: dioses egipcios, divinidades fenicio-cartaginesas (Tanit,...), ritos de ascendencia semítica,... Todo ello sin entrar en el campo de la toponimia insular o en el origen de los nesónimos, ámbitos en los que son extremadamente frecuentes los términos de origen semita . Pero sigamos analizando otros argumentos que apoyan nuestra hipótesis.

EL DESCUBRIMIENTO DE LAS ISLAS

Frente a la visión mítica surgida en las culturas griega y romana (Plutarco, Estrabón, Pomponio Mela, Salustio, Diodoro, Plinio el Viejo, Ptolomeo,...) que reflejaron el conocimiento de unas islas atlánticas, las Islas de los Bienaventurados, en la actualidad sabemos que el archipiélago canario conoció a lo largo del primer milenio antes de la Era sucesivos descubrimientos, si bien no será sino a partir del siglo I a.n.e. cuando aparezca en las fuentes literarias greco-latinas como una realidad geográfica reconocible.

El claro desfase cronológico existente entre el descubrimiento inicial y la aparición de las primeras noticias escritas se debió a diferentes factores responsables de que, durante un amplio espacio de tiempo, las islas se hallaran sometidas a un proceso de consecuencias comparables al que se ha descrito para el Mediterráneo occidental durante la precolonización fenicia: implantación de elementos culturales hoy muy difíciles de delimitar en toda su amplitud al contextualizarse en un periodo oscuro, aparentemente desprovisto de suficientes argumentos para la investigación, y por la propia actitud de los protagonistas, celosos por defender sus territorios comerciales. Ese es un proceso que en Canarias se agudiza debido a factores propios, tales como la lejanía de los focos colonizadores mediterráneos o el despoblamiento inicial, además de por desarrollarse la colonización en una etapa, los siglos VII y VI a.n.e., de gran confusión para las colonias fenicias de Occidente. Todo ello explica la escasez de referentes literarios fenicio-púnicos sobre las islas y su casi nula presencia en la tradición literaria anterior a Iuba-Plinio. En consecuencia, si bien los fenicios abrieron al conocimiento el Mediterráneo occidental y en particular el sur de la Península Ibérica, da la impresión de que también hicieron todo lo posible por mantener en la nebulosa de lo mítico o en el campo de lo desconocido el mundo existente más allá de las Columnas de Hércules. De hecho, tal y como reflejan o se intuye a través de algunas fuentes literarias posteriores a la desaparición de Cartago, el conocimiento que tuvieron de las regiones atlánticas fue muy superior al que hoy ni siquiera podemos imaginar (Santana et alii., e.p.).

Por tanto, y a falta de documentación escrita precisa y de datos arqueológicos que nos informen del instante en que se produjo el descubrimiento del archipiélago canario, un hecho por demás que apenas dejaría evidencias arqueológicas, habremos de intentar conjeturar acerca de cómo y cuándo se produjo. En ese sentido, debemos recordar que durante los últimos cinco siglos se ha propuesto la presencia en aguas canarias de marinos procedentes del Círculo de Cádiz, afirmándose la existencia de una navegación tartésica anterior a la fenicia (García y Bellido, 1942: 177), a la que se debería el descubrimiento de las islas. Sin descartar esa posibilidad, los vestigios arqueológicos y las dataciones disponibles obligan por el momento a señalar como protagonistas del descubrimiento a navegantes fenicio-púnicos en algún instante de la primera mitad del primer milenio antes del cambio de Era, espacio temporal en el que el intervalo comprendido entre los siglos VIII y VI a.n.e. es el que cuenta por ahora con mayores posibilidades de haber visto la arribada de los primeros pobladores. Esta aseveración se sustenta en el hecho constatado de que los marinos fenicios navegaron con normalidad más allá del Estrecho al menos desde el siglo VIII a.n.e.

Técnicamente, fenicios y cartagineses eran perfectamente capaces de navegar hasta las islas y entre las islas, y de adentrarse en el Atlántico si aceptamos el hallazgo producido en 1749 en el archipiélago de las Azores de un conjunto de monedas púnicas de los siglos IV y II a.n.e. (Blázquez, 1977). Sus travesías estaban motivadas por razones económicas: la búsqueda de recursos agrícolas y pesqueros para sus colonias costeras, materias que transformar en las factorías y mercancías potencialmente intercambiables en los mercados del Mediterráneo. Necesariamente, en esa labor de exploración no les debió pasar desapercibido el archipiélago canario, el cual habría sido visitado con anterioridad al siglo VI a.n.e., la fecha más antigua que se posee para un asentamiento humano en las islas (Arco et alii, 1997), por navegantes y mercaderes en busca de fondeaderos seguros y recursos de todo tipo procedentes tanto del medio terrestre como del medio marino. Lo anterior obliga a considerar la existencia, tras el descubrimiento, de una etapa previa de valoración o precolonización, que debió permitir calibrar las posibilidades estratégicas y económicas de cada una de las islas mediante sucesivas visitas e incluso el establecimiento temporal de colonos en puntos estratégicos.

Si acudimos al Periplo de Hannón, las rutas africanas debieron haber sido conocidas y frecuentadas durante largo tiempo sólo por gaditas y lixitas, quienes las mantuvieron en secreto y con ellas la existencia de islas en el Océano al menos hasta el desarrollo de la citada expedición si tenemos en cuenta que aquélla requirió embarcar pilotos conocedores de esos itinerarios en Gadir y/o Lixus. De hecho, el celo fenicio-púnico por mantener ocultas las fuentes de origen de sus riquezas explicaría en última instancia el por qué las primeras noticias literarias acerca de Canarias las proporciona Iuba II, heredero de los Libri Punici, cuando la realidad de los hechos nos muestra que las rutas marítimas al occidente de Gibraltar, al menos hasta Mogador, eran frecuentadas como muy tarde desde el siglo VIII a.n.e., lo que mente en el archipiélago. De esa manera, el establecimiento definitivo de grupos humanos debió responder a una aceleración del proceso de explotación de los recursos insulares, en un momento coincidente con profundos cambios económicos, políticos y culturales en el Mediterráneo occidental.

POR DÓNDE LLEGARON? LAS RUTAS DEL ATLÁNTICO ORIENTAL AFRICANO

Quienes han estudiado las rutas marítimas que conducían

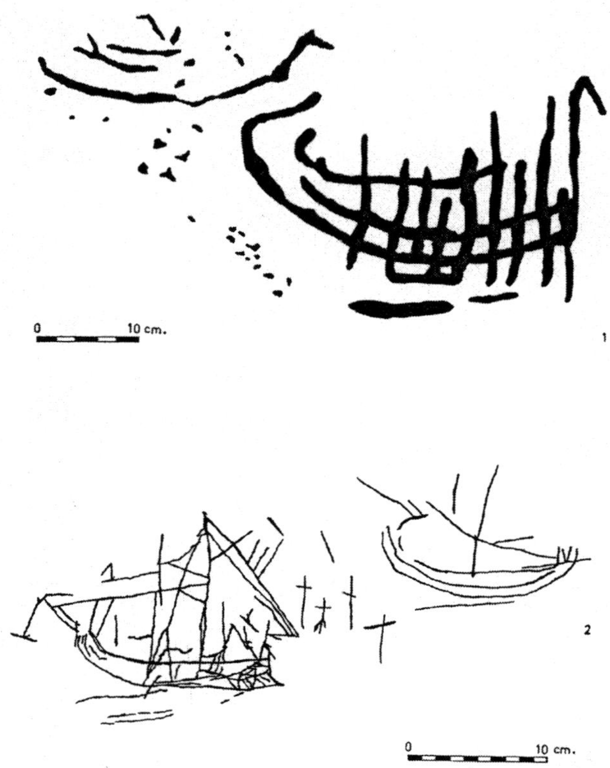

FIGURA 4: Grabados de naves localizados en las Islas Canarias: El Cercado (La Palma) (1) y Barranco Hondo (Tenerife) (2). (Dib. Mª A. Ramírez).

evidencia un profundo y muy antiguo conocimiento de la "mar abierta" (Fernández-Miranda, 1988: 460).

Lo anterior plantea la hipótesis de un descubrimiento temprano si bien una colonización tardía ya que tras su descubrimiento el archipiélago canario no parece haberse poblado inmediatamente, quizás porque no era ese el objetivo inicialmente perseguido; lo más probable es que sólo se ocuparan algunos puntos costeros estratégicamente situados para permitir la explotación de los recursos insulares, en especial los pesqueros. Se trataría de puestos avanzados de recalada y repostaje en los que se practicarían unas actividades semejantes a las que llevaron a cabo en época romana aquellos temporeros que, según M. Ponsich (1976: 70), seguían el recorrido migratorio de los túnidos, pescándolos al inicio del verano en la costa africana y al final del mismo en la costa de la Baetica, generando un nomadeo estacional de población que, por esa razón, sólo recalaría temporal- en la antigüedad hasta Canarias han prestado quizás una excesiva importancia a los riesgos que encierra navegar a lo largo de la costa noroccidental africana, de la que se ha resaltado (Aubet,1987) su carácter inhóspito y los escasos fondeaderos existentes desde Tingis hasta Lixus. Unos inconvenientes naturales que se vieron incrementados en diferentes épocas por acciones piráticas (Gozalbes, 1988). Sin embargo, esas dificultades no parecen haber constituido un obstáculo que impidiera a los marinos fenicios, por lo general valiéndose de pequeñas embarcaciones, alcanzar desde fechas muy tempranas la factoría de Mogador A...un relais pour les pêcheurs lixites@ (Jodin, 1967: 261- 262), o que a partir del siglo I a.n.e. y hasta el siglo IV d.n.e. (Atoche et alii., 1995) se mantuviera vigente un itinerario marítimo hacia Canarias, contextualizado en un fenómeno de mayor amplitud espacial y temporal que aseguró la presencia romana en extensas áreas de la costa occidental africana hasta más allá del sur del actual Marruecos (Rebuffat, 1987; Euzennat, 1989a).

En la factoría fenicio-púnica y romana de Mogador los vestigios materiales recuperados (Jodin, 1967) indican un prolongado asentamiento que arranca en el siglo VIII a.n.e., en época fenicia, y que perdura hasta el siglo V d.n.e., ya en época tardorromana (Jodin, 1967; Amadasi, 1992), cuando el sitio se abandona por razones que tuvieron su origen en la crisis político-económica del siglo III. Esa amplia ocupación, unida a su posición extrema y a su cercanía a Canarias la convierten en un magnífico punto de referencia de cara al conocimiento de las rutas marítimas utilizadas a lo largo de la costa africana durante algo más de un milenio . A lo anterior podemos sumar las noticias derivadas de las fuentes literarias latinas, en especial las que sobre la expedición enviada por Iuba II de Mauritania (25 a.n.e.-23 d.n.e.) a las islas recoge en su Historia Natural Plinio el Viejo. Esa información no sólo nos muestra la ruta elegida para alcanzar el archipiélago sino que sirve además para confirmar la incuestionable capacidad de la técnica naval de la época para arribar hasta él, al tiempo que testimonia la presencia de gentes romanas y/o romanizadas en los mares de Canarias desde unas fechas tempranas que, grosso modo, coinciden con el punto de partida del establecimiento romano en la Mauritania occidental (siglo I a.n.e.) y con las cronologías proporcionadas por los vestigios romanos recuperados en El Bebedero (Atoche et alii., 1995).

Pero aquélla no fue la primera ocasión en que gentes procedentes del Círculo del Estrecho recalan en las islas, como lo demuestra el mismo texto de Plinio con el amplio conocimiento que posee acerca de ellas, adquirible sólo empíricamente como resultado de una dilatada presencia de navegantes a lo largo de sus costas. De hecho, es muy probable que para el instante en que debió producirse su colonización, en la primera mitad del primer milenio antes de la Era, los itinerarios marítimos que permitieron el acceso al archipiélago y que éste mantuviera unos contactos fluidos con las culturas mediterráneas hasta los inicios del siglo IV d.n.e. sin duda debieron consolidarse sobre la base de las rutas establecidas por los marinos fenicios en esta parte del Atlántico a partir de su presencia en las Columnas de Hércules.

Por otro lado, resulta indudable que Gadir/Gades fue el punto de origen y de término de los itinerarios marítimos que discurrían por la costa occidental africana tanto en época fenicio-púnica como romana. Su importancia en el extremo Occidente a partir del Bronce final es indiscutible (Schubart et Arteaga, 1986), máxime si tenemos en cuenta que su puerto centralizó gran parte del comercio generado en los circuitos comerciales del Mediterráneo centro-occidental, convirtiéndose en la statio aduanera que controlaba el tráfico naval hacia el Mare Cantabricum, el Mare Britannicum (Martínez et Carreras, 1993: 102) y, con toda seguridad, hacia el Atlántico africano. Pues bien, si a lo anterior unimos el hecho de que en Gadir residiera el punto de partida de la ruta hacia el Atlántico sur, como nos muestra el Periplo de Hannón, se hace inevitable que debamos asignarle a ese centro urbano un papel protagonista en el proceso de descubrimiento y colonización del archipiélago canario. En esa línea no hay que perder de vista la importancia que se le ha asignado a Gadir en el surgimiento de centros coloniales en el norte de Marruecos (López,1988: 742), alcanzando con toda probabilidad su actividad comercial A... un lugar tan lejano como es el valle del Sus, a partir del islote de Mogador, donde se han encontrado contenedores, seguramente signados en Gades, en una fecha muy temprana, siglos VII y VI a.C. Habiéndose constatado el comercio con los indígenas del Cabo Ghir. (...) la presencia gaditana sigue siendo viva en la Mauritania Occidental aún en el siglo I a.C. y I d.C.@ (op. cit.: 742-743). M. Ponsich (1969: 234) no duda en afirmar que la industria de las salazones de pescado creó claros vínculos A... entre el sur de la Península Ibérica y la región de Tánger desde la época púnica ...@, una actividad de la que sabemos, a través de Estrabón, que al menos desde finales del siglo II a.n.e. los pescadores gaditanos faenaban en la costa mauritana e incluso más hacia el sur (Ponsich et Tarradell, 1965; García, 1973; Blázquez, 1977; González et alii., 1995), utilizando embarcaciones de reducido tonelaje, los hippoi.

En consecuencia, no resulta aventurado afirmar que las razones que dieron lugar al descubrimiento y colonización de las Islas Canarias debieron estar muy cercanas a las que estimularon a marinos y mercaderes fenicios a poner en marcha su expansión atlántica: la posibilidad de obtener ininterrumpidamente y con un escaso esfuerzo en infraestructuras, ingentes cantidades de la materia prima necesaria para la producción de salazones de pescado y garum. De ese modo, la beneficiosa demanda de salazones impulsaría entre los pescadores del Círculo del Estrecho el interés por abrir en el Atlántico una ruta sur, oceánica, que denominamos la ruta de los atunes, la cual seguiría un trayecto superpuesto al que hasta hoy siguen utilizando esos peces en sus migraciones anuales desde el Golfo de las Guineas hasta el Mediterráneo. La ruta partiría de Gadir con dirección oeste hasta el Cabo de San Vicente, desde donde se orientaría hacia el sur utilizando a partir de entonces como referencias las sucesivas bajas que, casi en línea recta, conducen a las Islas Salvajes, archipiélago cuya cercanía a Canarias hace factible su visualización y, en consecuencia, poder alcanzarlas teniéndolas a la vista (FIGURA 3).

Además de la ruta de los atunes existió un segundo itinerario de acceso a las islas, la ruta de las factorías, el cual seguiría un derrotero costero propio de la navegación de cabotaje y del que tenemos una buena referencia a través del relato del Periplo de Hannón y de la expedición organizada por Iuba II al archipiélago canario. Esta segunda ruta comparte con la primera su origen en el puerto de Gadir, si bien a partir de ahí realiza un recorrido paralelo a la costa africana hasta alcanzar Lixus (Larache), lugar considerado (Aubet,1987) como el punto de arranque de la vía meridional que conducía desde la colonización fenicia hasta las Islas Purpurarias (Essauira) (Desjacques et Koeberlé, 1955), archipiélago del que Plinio hace arrancar la ruta A... hacia las islas Afortunadas...@. Desde ese lugar la navegación podía dirigirse con dirección oeste hacia mar abierto buscando enlazar en las Islas Salvajes con la ruta de los atunes, y de ahí alcanzar por el norte las Afortunadas (La Palma, Gomera, Hierro, Tenerife y Gran Canaria), o bien podía seguir paralela a la costa hasta el Cabo de Iuba (Cabo Juby), punto desde donde se adentra en alta mar con dirección oeste para arribar a las islas de las Hespérides (Lanzarote, Fuerteventura y La Graciosa) (Santana et alii, e.p.).

El itinerario de las factorías necesariamente debió establecerse con anterioridad al siglo VII a.n.e. si tenemos en cuenta que existe constancia de su pleno funcionamiento en la segunda mitad de ese siglo, momento en el que mercaderes fenicios ya están plenamente establecidos en Mogador, una colonia integrada junto con otras factorías en una amplia red dedicada a explotar los recursos de la costa noroccidental africana. A partir del siglo VI a.n.e. los cartagineses trabajarán por reforzar la presencia semita en esa vertiente del continente africano, fortaleciéndose como consecuencia la ruta de las factorías mediante la fundación de nuevas colonias, el objetivo más evidente encomendado por Cartago al almirante Hannón. Un periplo que también debió perseguir otros fines menos confesables en un documento que debía quedar expuesto a la curiosidad de todo el mundo en el templo de Baal Hammón, con toda probabilidad relacionados con el comercio del oro africano, una de las razones que para W. Huss (1993: 20) motivaría gran parte de las actividades fenicias en el actual Marruecos y que permitieron durante mucho tiempo que Mogador se mantuviera en ese itinerario como baluarte en el comercio del oro procedente de África occidental . Con posterioridad Iuba II aprovechó las infraestructuras fenicio-púnicas de los islotes de Mogador para instalar una floreciente industria tintorera.

Una vez en aguas canarias, el trayecto que seguirían las dos rutas señaladas estaría avalado por la existencia de una serie de jalones, representados por diferentes vestigios arqueológicos, entre los que sin duda destacan las representaciones grabadas de barcos del norte de La Palma (Garafía), norte de El Hierro, sureste de Tenerife (Barranco Hondo), norte de Gran Canaria, sureste de Fuerteventura (Barranco de Tinojay) o sur de Lanzarote (Rubicón). En el archipiélago son frecuentes y variadas las representaciones de naves , localizadas por lo general en estaciones rupestres cercanas a la costa, las cuales ponen de manifiesto la continuidad temporal de este tipo de iconografía, que parece arrancar desde los inicios mismos de la colonización de las islas.

De las representaciones conocidas llama la atención sin duda el conjunto de grabados de El Cercado (La Palma, Garafía), localizado en una zona de antigua y amplia ocupación humana en el vértice noroeste de la isla, a escasa distancia de la línea de costa en la que se hallan dos buenos puertos naturales, en un punto por el que es obligado el tránsito si se pretende efectuar un recorrido del archipiélago de oeste a este accediendo desde el norte, tal y como lo refleja el trayecto seguido por la flota de Iuba II. Pues bien, de los diferentes elementos elaborados en El Cercado nos interesa la figuración de dos embarcaciones piqueteadas que se han representado fondeadas muy cerca la una de la otra; en la que aparece en primer plano se observa una nave vista por su borda de estribor, de casco asimétrico, con el puntal de proa recto y elevado, rematado en un prótomo zoomorfo, quizás una cabeza de caballo o león, y popa curvada y sobreelevada, en la que no aparece timón o remos timoneles. Los mástiles y las velas están ausentes, pero en cambio aparece una estructura central que podría corresponder a una superficie cubierta con una especie de toldilla de las que servirían para proteger a los marineros de los rigores del clima. El barco también presenta ocho remos que descienden por la borda de estribor (FIGURA 4, nl 1). En consecuencia, responde a un tipo de iconografía ampliamente conocida en el sur de la Península Ibérica (Almagro-Gorbea, 1988) y otras regiones del Mediterráneo, como es el caso de Túnez (Longerstay, 1990: Pl. VI), de morfología asimétrica, con alta popa sobreelevada, que reproduce modelos de naves rápidas del tipo galera como las representadas en algunas monedas de Cartago. De hecho estas figuraciones son totalmente comparables con las que se conocen en Cádiz (Laja Alta, Jimena de la Frontera) y, por tanto, con los paralelos señalados para aquéllas. Su presencia en el noroeste de La Palma, en un entorno donde existen otros grabados de naves correspondientes a periodos más recientes, adquiere significado a la luz de los datos que venimos barajando, adjudicando a ese extremo de la isla el papel de puerta de entrada occidental del archipiélago desde una etapa muy temprana de su desarrollo histórico.

La otra estación rupestre significativa se halla en el sureste de Tenerife (La Negrita, Barranco Hondo), en un punto desde el que se domina gran parte del canal que separa las islas de Tenerife y Gran Canaria. Ahí se han grabado dos embarcaciones de diferente morfología, fondeadas muy cerca la una de la otra (FIGURA 4, nl 2). En un caso, se trata de una nave de casco asimétrico, con el puntal de proa recto y elevado, rematado en espolón, y popa curvada y sobreelevada, en la que no aparece timón o remos timoneles. Presenta dos mástiles y aparejo sin velas. Sus paralelos son los mismos que hemos señalado para las naves de El Cercado, hallándonos en este caso ante otra galera morfológicamente muy próxima al tipo de la pentecontera. La segunda embarcación posee un casco curvo, simétrico, aparentemente sin remos y con un mástil para vela cuadrada, la cual se asemeja a las naves del tipo hippoi, derivadas de los pequeños cargueros fenicios de uso muy antiguo en el Egeo (Luzón, 1988: 458), donde a pesar de sus escasas posibilidades de carga fueron de gran utilidad tanto para el transporte local como para la pesca. Su uso se extendió en el Círculo del Estrecho a partir de la colonización fenicia con destino a actividades pesqueras y al tráfico de cabotaje.

Finalmente, en el extremo oriental del archipiélago, en las islas de las Hespérides, las navegaciones antiguas han dejado también importantes evidencias, entre ellas sin duda la más destacada es el asentamiento de Rubicón, en el sur de Lanzarote. El sitio ha conocido una prolongada reutilización desde la antigüedad, dado que ocupa una posición estratégica en el itinerario más idóneo para entrar o salir por el extremo oriental del archipiélago desde o hacia la cercana costa sahariana, jugando de hecho un papel semejante al que señalamos para Garafía, en este caso como puerta oriental del archipiélago . El yacimiento domina una amplia rada bien protegida de los vientos reinantes, con varias playas que facilitan el fondeo y/o el varado de navíos, en la que se concentran amplios recursos pesqueros, ya que se halla en la vía de paso de las rutas migratorias de túnidos y escómbridos, y cuenta con un gran potencial de cara a la producción de sal.

Situado en la Playa de los Pozos, a escasos metros de la línea de costa, el asentamiento consta de una pequeña estructura habitacional rectangular, levantada con muros de piedra y mortero de barro en el extremo más bajo de una colina (15 m.s.n.m.) que domina el cauce derecho de un barranco de escaso desarrollo, en donde se abren dos pozos con cámara que fueron construidos siguiendo en un caso los patrones arquitectónicos empleados en el norte de África para la apertura de pozos y cisternas en los ambientes punicizados, y en el otro los propios de los contextos romanizados, donde ese tipo de obras hidráulicas estuvo vinculado a factorías de garum y salazones (Ponsich, 1988) reestablecidas o erigidas ex novo por Iuba II a partir del último tercio del siglo I a.n.e. En consecuencia, dos soluciones arquitectónicas extensamente empleadas por las civilizaciones mediterráneas del primer milenio antes del cambio de Era para la captación y almacenamiento de agua, las cuales acreditan que nos hallamos ante los restos de un asentamiento utilizado a lo largo de un amplio espacio temporal para facilitar la navegación y el aprovechamiento de los numerosos recursos del entorno. De hecho, los límites cronológicos fijados para la presencia en Lanzarote de gentes romanas o romanizadas en base a las excavaciones en El Bebedero permiten establecer para Rubicón un desarrollo paralelo al de las factorías púnico-romanas de la costa marroquí y su incorporación al ámbito económico, comercial y cultural que abarcó, con diferente intensidad, todo el noroeste africano y sur de la Península Ibérica. Su fundación se debería inicialmente a marinos fenicio-púnicos, acrecentándose en época romana sus capacidades por medio de la construcción de nuevas infraestructuras, justo en un momento que corrió paralelo al resurgimiento y potenciación de las antiguas factorías pesqueras africanas por parte de Iuba II.

Nos encontramos pues ante un sitio cuyas características han sido tipificadas en el Mediterráneo como paisaje fenicio, las cuales podemos reconocer por ejemplo en la relativamente cercana factoría de Mogador, cuyos islotes ofrecían una rada protegida, excelente situación, proximidad a la costa del continente, disponibilidad de buenos materiales de construcción (rocas calizas, madera,...), y en donde la ausencia de agua se suplió almacenando la de lluvia en cisternas. Como hemos visto, son precisamente dos estructuras hidráulicas las que caracterizan al sitio de Rubicón (Atoche et alii, 1999).

LA ENTRADA A LAS ISLAS

Sobre la base de lo señalado hasta ahora, resulta plausible considerar que en el entorno del Estrecho de Gibraltar confluyen a partir del inicio del primer milenio antes de la Era las capacidades técnicas, el conocimiento del océano y los condicionantes económicos que propiciaron el descubrimiento y la colonización del archipiélago canario (Schubart et Arteaga, 1986). Ese sería un proceso que inicialmente afectaría a aquellas áreas insulares que mejor se adaptaban a los propósitos perseguidos por los colonizadores; indudablemente, y en base a las características geomorfológicas que presentan las islas, debieron ser las vertientes meridionales las que más facilidades proporcionaron, con zonas de calma ricas en recursos pesqueros, y costas bajas con playas de arena.

Sin embargo, la arqueología canaria aún no ha podido aportar un conjunto lo suficientemente preciso de pruebas que permitan alcanzar conclusiones definitivas en relación con las áreas que sirvieron inicialmente en cada isla como cabezas de puente desde las que arrancó la colonización, por lo que debemos guiarnos por algunos indicios que pueden arrojar luz a la cuestión. Así, si bien las fuentes etnohistóricas recogen, al menos para una de las islas (Tenerife) y casi un siglo después de finalizada la conquista normando-castellana, la explicación que los guanches daban a su presencia en la isla , señalando un lugar de la costa norte como área elegida para el establecimiento, lo cierto es que las mismas fuentes aportan también indicios que permiten reconocer la posibilidad de que en las dos islas centrales (Tenerife y Gran Canaria) la colonización haya centrado inicialmente sus esfuerzos en la costas que miran al suroeste, áreas donde los protagonistas del poblamiento hallarían, en base a lo que hemos señalado, las condiciones que inicialmente perseguirían: playas bajas, zonas de calma, puertos naturales, ingentes cantidades de túnidos y escómbridos, elevada insolación media anual que proporciona la energía necesaria para la producción de sal y una alta potencialidad agrícola y ganadera de los entornos terrestres. En ese sentido resulta sintomático que, tanto para Gran Canaria (Morales, 1978: 375, 381 y 433) como para Tenerife (Espinosa, 1967: 40 [1594]; Morales, 1978: 378), los cronistas de la conquista recojan noticias que revelan la existencia en los inicios de la colonización de estructuras políticas unitarias gobernadas por jefaturas con residencia en las regiones meridionales, Adeje para Tenerife y Arguineguín para Gran Canaria, las cuales coinciden precisamente con las áreas insulares que poseían las mejores condiciones para convertirse en las cabezas de playa desde las que proceder a la apropiación de los territorios insulares.

COLONIZACIÓN Y ESTABLECIMIENTO

El descubrimiento de las islas alentaría en sus descubridores el interés por conocer sus potencialidades que, una vez valoradas, debieron resultar lo suficientemente rentables como para poner en marcha la colonización, un fenómeno que sin embargo no parece haber afectado de manera sincrónica ni con la misma intensidad a todo el archipiélago. El momento en que se inicia, las razones, la forma y los protagonistas son cuestiones que trataremos seguidamente.

En cuanto al momento, y si tomamos como referencia las dataciones radiocarbónicas disponibles (González et alii, 1995; Arco et alii, 1997), que sitúan la más antigua presencia humana en las islas en el siglo VI a.n.e., tendría sentido afirmar que la colonización debió iniciarse con anterioridad a esa fecha, quizás en algún momento del siglo VII a.n.e. o incluso antes, máxime si tenemos en cuenta que esa datación indica que el descubrimiento se produjo previamente. El proceso arrancó con toda probabilidad a partir de alguna de las dos islas centrales del archipiélago, al ser éstas las que reúnen al menos tres condiciones básicas para haberse constituido en las iniciales protagonistas: una posición central estratégica, mayor superficie y unos recursos lo suficientemente amplios y variados.

Por lo que a las razones se refiere, un proceso tal como la colonización y definitivo establecimiento de grupos humanos en el archipiélago canario y, en general, cualquier proceso cultural humano puede y, en muchos casos, debe explicarse a partir de argumentos económicos. El descubrimiento y posterior colonización de las Islas Canarias, a semejanza de la ulterior presencia romana o el abandono y aislamiento que sufre el archipiélago a partir del siglo IV d.n.e., estuvieron motivados por similares condicionantes económicos: el interés generado en los mercados mediterráneos por los recursos insulares y el posterior cese de ese interés tras la crisis de las estructuras económicas mauritanas.

En consecuencia, las causas que incitaron la colonización de Canarias hay que rastrearlas en su entorno más inmediato, allí donde los recursos canarios podían resultar rentables: el Mediterráneo occidental y la costa atlántica norteafricana. En ese espacio geocultural se sucedieron, a partir del primer milenio antes del cambio de Era, una serie de acontecimientos que reflejan sucesos originados en la nueva situación político-económica que se impone en el Próximo Oriente y que pone en marcha en el Mediterráneo un proceso de fundación de asentamientos comerciales destinados a la captación de metales y productos agrícolas, además de servir de vía de control demográfico (Aubet, 1987: 55). De esa manera, en el Mediterráneo occidental, ya en el siglo VIII a.n.e., Gadir y Lixus poseen una estructura urbana desarrollada que permite fijar su fundación y el reconocimiento de su entorno y sus potencialidades económicas con bastante anterioridad, explicando por qué a partir del siglo VII a.n.e. se suceden las fundaciones fenicias en la costa marroquí que supondrán el control de rutas y recursos económicos atlánticos muy valorados en el Mediterráneo y la aparición de periodos orientalizantes en las culturas de los pueblos con los que toman contacto los navegantes y colonos semitas; pero sobre todo, constituye un fenómeno de suma importancia para nuestras islas al proveer las condiciones básicas para que se produzca su colonización.

A partir del siglo VI a.n.e. Cartago acomete su propio fenómeno colonizador en el Mediterráneo occidental, mediante la fundación de establecimientos o ejerciendo su influencia sobre los preexistentes en pugna con las colonias griegas del Mediterráneo central o la naciente Roma. Es en ese contexto en el que se organizan los viajes de Himilkón a las fuentes del estaño en el Atlántico norte (las Cassitérides) y de Hannón hacia el África atlántica, expedición que nos sirve de referencia para conocer la manera en que Cartago actúa: intervención estatal, fundamentada sobre argumentos políticos, estratégicos y económicos, que supone el desplazamiento de un numeroso contingente de población norteafricana perfectamente pertrechado para colonizar. La presencia púnica más allá de las Columnas de Hércules, en zonas de interés económico para Gadir y Lixus, pudo hacerse, como afirma F. López (1994: 67), para defender los intereses de los enclaves del Círculo del Estrecho, o bien para usurpar territorios y hacerse con un lugar en el control de las riquezas al margen de dichos enclaves. Desconocemos cual sería la razón concreta, pero en cualquiera de los casos ello no es sino un aspecto más de toda una política que condujo al enfrentamiento de Cartago con sus competidores en un intento por alejarlos del extremo occidental del Mediterráneo y del Océano, tal y como lo recogen Diodoro y el Pseudo Aristóteles (Blázquez, 1977: 38-39).

Tras las Columnas existe una fuente de riquezas clara: la pesca que surte a las factorías de salazón y de fabricación de garum. Junto a ella hay otras, como el comercio con las poblaciones de la zona -que es citado por Herodoto en el siglo V a.n.e.-, que proveería a los enclaves fenicio-púnicos del actual Marruecos de metales, tintes y sal, entre otros productos. Pues bien, Canarias está en el centro de todo ese espacio económico y posee recursos propios, algunos escasos o inexistentes en el litoral africano, que hacen que su colonización tenga una importancia geoestratégica y económica de primer orden. Por tanto, si bien los recursos derivados de la pesca debieron concentrar inicialmente una mayor atracción, a tenor de las óptimas condiciones existentes en las aguas canarias para la captura de túnidos y escómbridos (Ponsich et Tarradell, 1965; González et alii, 1995), las islas también ofrecían la oportunidad de obtener otros recursos de sumo interés comercial durante la antigüedad, como la sal, los elementos tintóreos (múrex u orchilla), la sangre de drago o la grasa y el ámbar gris de las diversas especies de cetáceos que residen en las aguas canarias aún en la actualidad. Tampoco hay que desdeñar el papel que el archipiélago pudo jugar como plataforma desde la que obtener productos del cercano continente, tales como el oro africano (Jáuregui, 1954; Mahjoubi, 1983), sobre todo si tenemos en cuenta que constituyen la base ideal para acceder a los yacimientos auríferos del África occidental y paso obligado en el viaje de retorno.

En consecuencia, se imponía el control efectivo de un entorno de alto valor estratégico, rico en recursos explotables, mediante el establecimiento de un contingente poblador garante de la hegemonía fenicio-púnica en la zona y la explotación de sus riquezas, hasta el punto de que, como ha señalado E. Gozalbes (1988: 773), la actitud de Cartago ante los transgresores de su monopolio era la propia de los piratas A... una piratería organizada como función colectiva, el ataque a todo tipo de barcos ...@.

Evidentemente, las ventajas derivadas del establecimiento de comunidades en estas islas serían elevadas: desde la colocación de excedentes demográficos, con posibilidades de llevar a cabo una colonización agraria y controlar con gentes afines una zona de gran importancia económica, hasta la obtención de bastimentos y mano de obra con la que transformar los recursos existentes como paso previo a su transferencia a las colonias y factorías africanas del Atlántico y de ahí a los mercados donde fuesen demandados. Esas circunstancias hacen de las ciudades fenicias del Círculo del Estrecho y del Círculo de Cartago (Bernardini, 1996: 35) las candidatas favoritas de cara a la adjudicación de la autoría del hecho colonizador desarrollado en Canarias.

En cuanto a la forma en que se produjo la llegada de los

primeros seres humanos, habría que destacar el hecho de que las profundidades existentes entre las islas y entre éstas y el continente africano hacen inviable el planteamiento de cualquier hipótesis que intente explicar la arribada de gentes sobre la base de la existencia de puentes terrestres que conectaran tierra firme con el archipiélago y a su vez las islas entre sí. La comunicación terrestre interinsular sólo fue posible en determinados momentos del Cuaternario para las dos islas orientales, Lanzarote y Fuerteventura. En consecuencia, los grupos humanos que se establecieron en el archipiélago utilizaron sin duda como vía de llegada el mar. A partir de este punto las hipótesis propuestas para explicar la forma en que se produjo la arribada han argumentado las ideas más peregrinas, desde afirmar que se debió a un fenómeno de deportación en época romana, hasta el resultado de una navegación de fortuna en pequeñas embarcaciones. Lo primero se basa en la denominada leyenda de las lenguas cortadas, una elaboración erudita moderna citada ya en Le Canarien a comienzos del siglo XV, según la cual una revuelta en el África romana motivó el abandono en barcos sin remos de sus responsables, acompañados de sus familias y de bastimentos, quienes alcanzarían las islas tras haberles cortado las lenguas de antemano para evitar que difundieran tal suceso. Por otro lado, la navegación de fortuna, ni se puede demostrar ni sirve para explicar la colonización, con todos los medios que son imprescindibles para ello, de siete islas rodeadas de corrientes y calmas difíciles de sobrellevar por ese tipo de embarcaciones, descritas en la mayor parte de los casos como simples canoas de madera y/o pieles.

En consecuencia, los protagonistas del poblamiento humano de las islas dominaban la navegación oceánica o fueron trasladados por gentes que la dominaban, que para el momento en que debió producirse sólo pudo tratarse de marinos fenicio-púnicos, los únicos que en el entorno del archipiélago canario poseían las razones y la necesaria tecnología naval para llevarlo a cabo. Por otro lado, el hecho de que el archipiélago se encontrara en el centro de una rica zona económica convertida en objetivo de la expansión púnica es indicativo de que en un primer momento su colonización debió tratarse de una empresa estatal, cartaginesa, con claros objetivos económicos y geoestratégicos, en la que se emplearon contingentes de población norteafricana afín, paleobereberes punicizados. Sin embargo, el establecimiento definitivo sólo se produciría como resultado de una aceleración del proceso de explotación de los recursos insulares, coincidente con intensas transformaciones en las estructuras socioeconómicas del Mediterráneo occidental, convirtiéndose las islas en el escenario donde era posible la obtención de ingentes beneficios que enriquecerán a personas e instituciones localizadas en el Mediterráneo y que en las islas ponen en marcha una profunda transformación medioambiental.

LOS PROTAGONISTAS: CÍRCULO DEL ESTRECHO VERSUS CÍRCULO DE CARTAGO

La procedencia norteafricana de los primeros habitantes de las islas es una hipótesis que ya apuntaron incluso los eruditos del siglo XVI y en la que hoy parece existir un amplio consenso entre los investigadores, si bien la cuestión se plantea a la hora de concretar de qué área o áreas geográficas y culturales del norte de África partieron. Para resolver esa incógnita resultan de una gran ayuda los numerosos elementos de ascendencia semita que hallamos en las culturas canarias, los cuales orientan la búsqueda hacia el ámbito norteafricano de la colonización fenicio-púnica.

El establecimiento de las primeras colonias fenicias en la costa magrebí puso en marcha un proceso de contacto y difusión de la cultura fenicia entre las poblaciones autóctonas que, a partir del siglo VI a.n.e., se acelera creando entre semitas y libios profundos lazos personales y étnicos; el resultado étnico-cultural sería la aparición de los libiofenicios, un colectivo que contribuyó a engrosar los contingentes humanos que fueron trasladados a zonas que debían quedar bajo el control político y económico de Cartago, como Sicilia, Cerdeña, el sureste de la Península Ibérica o los propios territorios africanos, cuya explotación se produjo mediante el empleo de población autóctona sometida. Al sur de la Península Ibérica se trasladaron colonos libiofenicios durante los siglos V y IV a.n.e., A... sometidos a un régimen de servidumbre, y (cuyo) asentamiento se contextualiza en un programa colonizador dirigido por Cartago a todas las áreas hegemonizadas por ella ...@, protagonizando una auténtica colonización agraria (López Castro, 1992: 47). El propio Periplo de Hannón constituye un buen ejemplo de empresa colonizadora utilizando población norteafricana en una época relativamente temprana de la expansión púnica. En definitiva, Cartago utilizó a libio-fenicios, africanos que habían adoptado elementos de la cultura fenicia, con lazos de parentesco con los semitas y organizados a su modo, para servir como mercenarios en el ejército de la metrópoli o como elementos transportables a zonas que debían quedar bajo el control político y económico de Cartago, como hacen los bárquidas en la Península Ibérica en el siglo III a.n.e. (Frutos, 1991).

Pues bien, ese mestizaje fenicio-africano resulta esencial para entender tanto el modelo de colonización implantado en el archipiélago canario como la especificidad de las culturas que se desarrollarán en él; se trata de dos fenómenos cuyos protagonistas fueron gentes procedentes de aquel contexto geo-cultural, africanos punicizados o libio-fenicios portadores de una cultura mestiza en la que, como hemos visto, no resulta extraño hallar facetas intensamente vinculadas al modelo social, político y económico libio-fenicio magrebí, en especial del entorno tunecino de Cartago. Ahí radica la razón de que tanto las evidencias arqueológicas como las fuentes etnohistóricas canarias revelen que las sociedades que se desarrollaron en las islas desde la arribada de las primeras poblaciones humanas hasta el inicio y culminación de la conquista normando-castellana poseyeran unas formas sociales complejas y una cultura mestiza en origen, resultado de la simbiosis de antiguas tradiciones mediterráneo-africanas con elementos fenicio-púnicos. No se trata de unas poblaciones africanas contaminadas por elementos de una cultura más pujante sino de poblaciones que, tras varios siglos de convivencia continuada con la cultura fenicio-púnica, acabaron por asumirla como propia.

Por tanto, si resulta evidente el protagonismo del Círculo

del Estrecho en el descubrimiento de las islas, el establecimiento de población en ellas se nos presenta más como un fenómeno promovido y llevado a cabo desde el Círculo de Cartago, circunstancia que acabaría por explicar el que numerosas manifestaciones de las culturas insulares presenten sus mejores paralelos en el marco del Bronce medio y final del Mediterráneo central.

LA SECUENCIA CULTURAL DE LA PROTOHISTORIA CANARIA

Atendiendo a la hipótesis expuesta y basándonos en los innumerables aportes realizados por la arqueología, es posible establecer un marco cronológico y secuencial de la protohistoria canaria, la cual se prolongó a lo largo de algo más de dos milenios, espacio temporal en el que se pueden diferenciar dos fases separadas por una época de ruptura: la primera fase fue de integración en los procesos económicos, sociales y políticos del Mediterráneo occidental, mientras que la segunda lo fue de aislamiento a causa de una amplia etapa de decadencia que afecta al occidente africano, iniciada durante el Bajo Imperio Romano y que alcanzó hasta la Alta Edad Media. De una manera más pormenorizada, la protohistoria canaria discurriría por las siguientes etapas:

PRECOLONIZACIÓN (CIRCA SIGLO X AL SIGLO VI A.N.E.

Constituye una etapa de unos cuatro siglos durante la cual el proceso de exploración y colonización fenicio centrado al otro lado de las Columnas de Hércules hace que entren en contacto con las islas atlánticas (Azores, Canarias,...). A partir de ese hecho, el archipiélago canario sería sometido a un proceso de evaluación de sus recursos y posibilidades mediante sucesivas visitas o incluso con la creación temporal de alguna colonia, fenómeno que culminaría con el establecimiento de población en un momento anterior al siglo VI a.n.e., a tenor de las fechas absolutas disponibles, que nos aseguran la presencia de comunidades humanas plenamente asentadas durante ese siglo en el norte y posiblemente el sur de Tenerife. Un temprano comienzo que, aunque no se ha atestiguado en otras islas, debió producirse igualmente al menos para Gran Canaria, máxime si tenemos en cuenta que, tanto Tenerife como esta última, se caracterizan por contener unos amplios y variados recursos y una estratégica posición central que favorece tanto el acceso al resto del archipiélago como a las rutas de entrada y salida de éste, aspectos que en conjunto sin duda debieron captar la atención de los navegantes fenicio-púnicos.

COLONIZACIÓN Y ESTABLECIMIENTOS DEFINITIVOS (SIGLOS VI A.N.E. AL IV D.N.E.

Comprende una fase de intensos contactos entre los grupos que protagonizan la colonización del archipiélago y quienes la deciden y diseñan. La presencia fenicio-púnica debió ser muy regular durante los momentos iniciales y más críticos del establecimiento humano. Más tarde, cuando menos a partir del siglo I a.n.e. en base a los datos suministrados por la arqueología, los contactos con las islas y sus habitantes se prolongan, de nuevo a través de navegantes, ahora romanizados, pero como antes procedentes del Círculo del Estrecho, quienes transitan regularmente las aguas canarias hasta finales del siglo III o comienzos del IV d.n.e. Durante el periodo romano los contactos debieron ser muy fluidos, si bien se espaciarían progresivamente hasta su interrupción a raíz de la crisis político-económica que afecta al Imperio en el siglo III d.n.e., circunstancia que reduce el interés romano por la costa atlántica africana y por sus producciones. A partir de ese instante carecemos de noticias sobre la presencia de gentes ajenas al archipiélago hasta aproximadamente el siglo XI d.n.e., momento en el que las islas reaparecen en la historiografía árabe.

De la situación de abandono que se instala, pero sobre todo de la esperanza de un futuro restablecimiento de los contactos, podrían ser muestra algunos textos recogidos en las fuentes etnohistóricas (Torriani, 1978 [c. 1592]: 75 y 204; Espinosa, 1967 [c. 1594]: 58-59; Abreu, 1977 [c.1602]: 68, 92-93), que reflejan mitos existentes entre las poblaciones de El Hierro, Fuerteventura, La Gomera y Tenerife en el momento de la conquista normando-castellana referidos a la esperada llegada por oriente de sus señores a bordo de negras naves.

DESARROLLO DE LAS CULTURAS CANARIAS (SIGLOS IV AL XV D.N.E.):

Esta fase arranca tras la interrupción de los contactos con el exterior. La desconexión con los centros que dieron origen y sentido a la colonización obliga a las poblaciones insulares a desarrollarse en un relativo aislamiento, generándose unos procesos culturales caracterizados por hallarse inmersos en un estadio tecnológico Neolítico forzado. Esa será una situación a la que comienza a poner fin el redescubrimiento del archipiélago, producido en el siglo XIV, y el posterior proceso de conquista normando-castellana que se desarrolla a lo largo del siglo XV. Durante los casi mil años a lo largo de los cuales discurre esta fase, las diferentes culturas insulares adquirieron sus características definitivas, aquellas de las que hoy poseemos una mayor cantidad de información al llegarnos tanto a través de las fuentes etnohistóricas como de la mayor parte de las intervenciones arqueológicas efectuadas en el último siglo.

CONSECUENCIAS DE LA COLONIZACIÓN

Los efectos ecológicos y culturales derivados de la insularidad constituyen temas de estudio que se han tratado con relativa frecuencia en la literatura científica, especialmente tras considerarse a las islas laboratorios mensurables con mayor claridad que cualquier otro espacio ecológico variable (Patton, 1996).

Como no podía ser menos, la colonización del archipiélago canario puso en marcha fenómenos culturales que, mediatizados por el factor isla, no siempre transcurrieron como cabría esperar que lo hicieran en las regiones continentales de origen de los protagonistas, lo que en última instancia contribuye a explicar algunas de las divergencias observables entre las diferentes culturas canarias una vez consolidadas. Es indudable que el traslado de unas poblaciones de un espacio continental a otro insular y su asentamiento definitivo ejerce sobre aquéllas la necesidad de adaptar sus estructuras socio-económicas. Reconocer las transformaciones que afectaron al sistema cultural original o las transformaciones medioambientales que se produjeron tras el establecimiento humano en el archipiélago canario resultan cuestiones aún muy difíciles de precisar, aunque intentaremos al menos perfilar algunas propuestas.

TRANSFORMACIONES CULTURALES: CULTURAS ENDÉMICAS

Si algo caracteriza a toda forma viviente que participa en procesos de colonización insular, es el carácter diferencial y específico que adquiere con el paso del tiempo. Ejemplos de este tipo de procesos son muy frecuentes en el archipiélago canario, no en vano se trata de uno de los territorios con mayor número de endemismos animales y vegetales de todo el estado español. Plantas y animales han debido adaptarse a las peculiaridades de las islas, a las formas geomorfológicas en las que se insertan y al resto de las especies con las que protagonizaron el fenómeno colonizador. El hombre, por supuesto, no podía ser una excepción, y debió adaptarse a las características del nuevo y limitado espacio en el que pretendía subsistir. Ello da lugar, de manera inmediata, a modificaciones en el ámbito de su cultura que circunstancialmente nos llegan reflejadas bajo la forma de cambios en el subsistema tecnológico de los recién llegados. Si a lo anterior añadimos que un grupo que se desgaja de su cultura madre no trasmite la totalidad de los caracteres culturales de la población a la que pertenece, resulta sencillo explicar por qué se establecieron diferencias desde el primer momento, tanto entre las poblaciones de cada isla como entre aquéllas y las poblaciones continentales de origen. Se trató por tanto de fenómenos dinámicos, con capacidad para producir soluciones particulares que implicaron incorporaciones, tales como la adopción de procesos tecnológicos ya por entonces marginales en los ambientes culturales norteafricanos, o exclusiones y con ellas la inevitable pérdida de ancestrales rasgos culturales.

La interacción hombre/medio ambiente condujo a los colonos que se establecieron en las islas a una espiral de adaptaciones que afectaron al sistema cultural original, seleccionando aquellos elementos que mejor se adecuaban a las nuevas condiciones o incluso revisando las estrategias de subsistencia. Con ello se asiste a un cambio en la identidad cultural de esas primeras poblaciones isleñas, a la aparición de particularismos y en esencia al arranque de un proceso de elaboración de nuevos sistemas culturales, último capítulo de la transformación de una cultura libio-fenicia en origen que se prolongó a lo largo de algo más de dos milenios. En consecuencia, no es extraño que encontremos en las culturas canarias facetas vinculadas al modelo social, político y económico fenicio-púnico magrebí y del sur de la Península Ibérica, y que sus producciones materiales, sin menospreciar la capacidad creadora propia de todo grupo humano, respondieran al menos a tres condicionantes: la tradición cultural inicial que alcanzó las islas con los colonos libio-fenicios; la evolución adaptativa que surge como respuesta a unos espacios insulares con recursos limitados y las influencias ejercidas, al menos durante el primer milenio de ocupación de las islas, por navegantes procedentes del Círculo del Estrecho, responsables en unos casos de la intensificación del proceso de colonización (mediante la aportación de nuevos contingentes pobladores, el asentamiento de población en las islas aún deshabitadas,...) y en otros del mantenimiento de las interacciones culturales entre el archipiélago y el Mediterráneo occidental.

Entre los aspectos culturales que debieron reorientarse se encontraba sin duda el subsistema económico, de tal manera que durante la etapa protohistórica debieron sucederse al menos dos modelos de subsistencia, pasándose de un modelo de dependencia externa e intercambio desigual a otro autárquico basado en una economía agraria de amplio espectro. El primero está en el origen de la propia colonización de las islas, reproduciéndose mientras se dieron las razones que la alentaron. El segundo se desarrolla a partir del momento en que se produce el aislamiento de las poblaciones insulares con respecto al exterior y será el causante de un proceso cultural que conduce a un estadio tecnológico que hemos denominado Neolítico forzado (Atoche et Martín, 1999). Ese aislamiento, resultado de la desaparición de las motivaciones político-económicas que propiciaron la frecuentación y posterior colonización de las Islas Canarias, favoreció el desarrollo singular de las poblaciones canarias, reforzando así las especificidades que imponía de por sí la insularidad. En términos biológicos pasaron a convertirse en Aculturas endémicas@ (Atoche et Ramírez, 2001). Es entonces cuando la obligada adaptación a los ecosistemas insulares propiciará la activación de soluciones tecnológicas extremas, de modo que la desconexión con el exterior unido a las limitadas posibilidades del medio insular modelan las características de las sociedades insulares al impedir el normal desarrollo de tecnologías bien conocidas en el ámbito de origen norteafricano de los primeros colonos, como la metalurgia . Esa circunstancia dará lugar a un curioso fenómeno que lleva a los libio-fenicios asentados en Canarias a rejuvenecer o, incluso, reinventar viejas tecnologías, como la talla de rocas, y a poner en marcha unas formas de subsistencia en las que la apropiación del territorio insular se basó en la ganadería extensiva (cabras y ovejas), complementada por la agricultura y actividades pesqueras y de recolección de elementos marinos y terrestres. A partir de entonces se hacen más acusadas las diferencias culturales entre islas o incluso en el seno de algunas de ellas debido al distinto grado de implantación y desarrollo que tuvo en cada una el mismo modelo económico . Algo que también se evidencia en otros aspectos culturales, como la organización social o la estructura política.

Las distintas comunidades insulares evolucionaron de manera particular en razón de sus propias experiencias y expectativas, desarrollando diferentes estrategias que es posible observar a través de los registros arqueológicos. Así se explica por qué existen disparidades en los componentes animal y vegetal de las dietas de las comunidades del norte y sur de Tenerife, o que en Gran Canaria se considere a la actividad agrícola la principal estrategia de subsistencia, con importantes excedentes almacenados en silos colectivos.

TRANSFORMACIONES MEDIOAMBIENTALES

La necesidad de conocer cuáles fueron las relaciones que se establecieron entre los grupos humanos asentados en el archipiélago canario y el medio natural constituye un tema de análisis que surge recientemente (Martín Oval et alii., 1999) y al que nos hemos acercado impulsados por la necesidad de conocer las razones que dieron lugar a la colonización de un archipiélago oceánico periférico en el primer milenio antes de la Era.

La primera evidencia que se nos presenta a nuestros ojos es que las islas eran de una manera muy diferente a como las vemos en la actualidad, de tal forma que hacia el siglo VII a.n.e. debían mostrar a los navegantes una cubierta vegetal mucho más extensa y frondosa de la que hoy conocemos, las selvas que aún pudieron observar en algunas islas los cronistas de la conquista normando-castellana dos milenios más tarde. Los recursos freáticos se hallarían en niveles óptimos, siendo frecuentes los cursos de agua estacionales e incluso los permanentes; tampoco estaban ausentes las amplias vegas de tierra fértil aptas para el cultivo, así como los puertos naturales y fondeaderos seguros donde sustentar actividades pesqueras y/o viajes más al sur. Sobre esa realidad natural las consecuencias que supo la llegada de grupos humanos aún no se han perfilado; de hecho sólo contamos con datos parciales para Lanzarote, la isla más oriental del archipiélago, donde el reconocimiento de procesos de destrucción de paleosuelos y de formación de potentes paquetes sedimentarios nos está permitiendo determinar algunas de las causas que motivaron la transformación medioambiental.

En Lanzarote, las peculiaridades de las prácticas agrícolas actuales nos proveen un elevado número de sitios en valles y fondos de cubetas donde la extracción de tierras para la construcción artificial de huertos, los denominados Aenarenados@, ha dejado al descubierto potentes suelos sedimentarios, en algunos lugares superiores a los 10 metros, formados por la erosión de las laderas circundantes. En esos rellenos es posible seguir la historia geológica más reciente de la isla, demostrando que la destrucción de la cobertura vegetal insular comienza a ser especialmente intensa a partir del siglo I a.n.e., coincidiendo con una explotación intensiva de ganado menor (cabras y ovejas) y con la presencia de navegantes romanos o romanizados en las islas, como ha demostrado el asentamiento de El Bebedero y otros yacimientos de Lanzarote. En consecuencia, entre los siglos I a.n.e. y IV d.n.e. la destrucción de suelos parece estar relacionada con el desarrollo de procesos productivos intensivos orientados a satisfacer la demanda de mercados exteriores, a donde se envían carnes en salazón, cueros curtidos,... A partir del siglo IV d.n.e., la interrupción de los contactos con el exterior y la consiguiente necesidad de reorientar las actividades subsistenciales hacia un modelo autárquico ralentizaron el proceso de destrucción de suelos.

CONCLUSIONES

La reciente constatación arqueológica de la presencia de elementos materiales e infraestructuras púnicas y romanas en diferentes puntos del archipiélago canario unido al amplio conjunto de elementos asimilados de la cultura libio-fenica norteafricana identificados en las culturas canarias ha traído como consecuencia más inmediata la apertura de la visión histórica establecida y la posibilidad de explicar muchos fenómenos culturales de los primeros canarios.

En Canarias se adaptan las estrategias colonizadoras ensayadas en las islas mediterráneas a lo largo de los tres milenios anteriores al cambio de Era, en especial las desarrolladas durante el primer milenio. En consecuencia, tras su descubrimiento, la colonización sólo debió depender de que se consideraran rentables sus numerosos y variados recursos, comparables con los que hallaron en muchas islas a lo largo del Mediterráneo. De esa manera, en las islas se limitarían a reproducir un patrón de comportamiento ancestral, dando continuidad en el Atlántico a un proceso que practicaron de manera habitual, siendo por ello posible reconocer en el procedimiento de colonización del archipiélago canario un fenómeno resultado de la adaptación a sus peculiaridades del modelo ensayado por griegos y sobre todo por los fenicios durante el primer milenio antes de la Era. Las islas se convierten así en el último archipiélago mediterráneo en colonizarse, la postrera región a la que alcanzó la expansión comercial del siglo VI a.n.e., amparada en un fenómeno más amplio, la explosión del comercio internacional que condujo a la integración de las islas en la economía mediterránea como una zona productora de valiosas materias primas total o parcialmente elaboradas.

BIBLIOGRAFÍA

ABREU, FR. J. DE,
1977 [c. 1602]. Historia de la conquista de las siete islas de Canaria. Santa Cruz de Tenerife: Goya Ed.

ALMAGRO-GORBEA, M.,
1988. Representaciones de barcos en el arte rupestre de la Península Ibérica. Aportación a la navegación precolonial desde el Mediterráneo oriental. Congreso Internacional AEl Estrecho de Gibraltar@, t. I, 389-398.

ALVAR, J.,
1998. Comunidad de navegantes: aspectos sociales de la navegación fenicia. En: Rutas, navíos y puertos fenicio-púnicos. XI Jornadas de Arqueología Fenicio-Púnica. (Eivissa, 1996), 49-59.

AMADASI, M.,
1992. Notes sur les graffitis phéniciens de Mogador. En: Lixus. Actes du colloque organisé par l=Institut des Sciences de l=Archéologie et du Patrimoine de Rabat avec le concurs de l=École Française de Rome, (Larache, 1989), 691-704.

ARCO, M0.C. DEL, 1976. El enterramiento canario prehispánico. Anuario de Estudios Atlánticos, 22, 13-124.

ARCO, M0.C. DEL, ARCO, M. DEL, ATIÉNZAR, E., ATOCHE, P., MARTÍN, M., RODRÍGUEZ, C. ET ROSARIO, M0.C.,
1997. Dataciones absolutas en la Prehistoria de Tenerife. En: A. Millares, P. Atoche et M. Lobo, coords. Homenaje a Celso Martín de Guzmán (1946-1994), 65-77.

ATOCHE, P., RODRÍGUEZ, M0.D. ET RAMÍREZ, M0.A.,
1989. El yacimiento arqueológico de AEl Bebedero@ (Teguise, Lanzarote).Resultados de la primera campaña de excavaciones. Madrid: Secretariado de Publicaciones de la Universidad de La

Laguna.
ATOCHE, P.,
1993. Excavaciones arqueológicas en El Bebedero (Teguise, Lanzarote). Segunda Campaña, 1987. Eres (Arqueología), 4, 7-19.
ATOCHE, P., PAZ, J.A., RAMÍREZ, M0.A. ET ORTIZ, M0.E.,
1995. Evidencias arqueológicas del mundo romano en Lanzarote (Islas Canarias). Arrecife: Cabildo Insular de Lanzarote.
ATOCHE, P. ET MARTÍN, J., 1999. Canarias en la expansión fenicio-púnica por el África Atlántica. II Congreso de Arqueología Peninsular (Zamora, 1996), t. III, 485-500.
ATOCHE, P. ET PAZ, J.A., 1999. Canarias y la costa Atlántica del N.O. africano: difusión de la cultura romana. II Congreso de Arqueología Peninsular (Zamora, 1996), t. IV, 365-375.
ATOCHE, P., MARTÍN, J., RAMÍREZ, M0.A., GONZÁLEZ, R., ARCO, M0.C. DEL, SANTANA, A. ET MENDIETA, C.,
1999. Pozos con cámara de factura antigua en Rubicón (Lanzarote). VIII Jornadas de Estudios sobre Lanzarote y Fuerteventura, (Arrecife, 1997), t. II, 365-419.
ATOCHE, P., MARTÍN, J. ET RAMÍREZ, M0.A.,
1999. Amuletos de ascendencia fenicio-púnica entre los mahos de Lanzarote: ensayo de interpretación de una realidad conocida. VIII Jornadas de Estudio sobre Lanzarote y Fuerteventura (Arrecife, 1997), t. II, 421-458.
ATOCHE, P. ET RAMÍREZ, M0.A.,
2001. Canarias en la etapa anterior a la conquista bajomedieval (circa s. VI a.C. al s. XV d.C.): colonización y manifestaciones culturales. En: Arte en Canarias: siglos XV-XIX. Una mirada retrospectiva. Madrid: Gobierno de Canarias. Dirección General de Cultura, t. I, 43-95 y t. II, 475-479.
AUBET, M0.E.,
1987. Tiro y las colonias fenicias de Occidente. Barcelona: Bellaterra.
BALBÍN, R., BUENO, P., GONZÁLEZ, R. ET ARCO, M0.C. DEL,
1995. Datos sobre la colonización púnica de las Islas Canarias. Eres (Arqueología), 6 (1), 7-28.
BALIL, A.,
1990. Comunicaciones marítimas. El comercio marítimo romano en el Mediterráneo occidental. P.A.C.T., 27, 63-72.
BERNARDINI, P., 1996. Giustino, Cartagine e il tofet. Rivista di Studi Fenici, XXIV (1), 27-45.
BLÁZQUEZ, J.M0., 1977. Las Islas Canarias en la Antigüedad. Anuario de Estudios Atlánticos, 23, 35-50.
BORY DE SAINT-VICENT, J.B.G.M.,
1988 [1803]. Ensayos sobre las Islas Afortunadas y la antigua Atlántida o Compendio de la Historia general del Archipiélago Canario. La Orotava, Tenerife: J.A.D.L.
CAMPS, G.,
1980. Berbères aux marges de l'Histoire. Paris: Éditions des Hesperides.
DESJACQUES, J. ET KOEBERLÉ, P.,
1955. Mogador et les Îles Purpuraires. Hespèris, XLII, 193-202.
EUZENNAT, M.,
1989a. Le limes de Tingitane. La frontière méridionale. Paris: C.N.R.S. Études d'Antiquités Africaines.
EUZENNAT, M.,
1989b. Remarques sur la description de la Maurétanie Tingitane dans Pline, H.N. V, 2-18. Antiquités Africaines, 25, 95-109.
ESPINOSA, FR. A. DE,
1967 [c. 1594]. Historia de Nuestra Señora de Candelaria. Santa Cruz de Tenerife: Goya Ed.
FANTAR, M.,
1988. Carthage: archetypes et specificité. Carthage et son territoire dans l'antiquité. Actes du IV Colloque International sur l'Histoire et l'Archéologie de l'Afrique du Nord, I, 53-65.
FERNÁNDEZ-MIRANDA, M.,
1988. La navegación fenicia hacia el lejano Occidente y el Estrecho de Gibraltar. Congreso Internacional AEl Estrecho de Gibraltar@, t. I, 459-472.
FRUTOS, G. DE,
1991. Cartago y la política colonial. Los casos norteafricano e hispano. Écija: Gráficas Sol.
GARCÍA, R.,
1973. Túnidos de las pesquerías canario-africanas. Homenaje a Elías Serra Ràfols, IV, 61-76.
GARCÍA Y BELLIDO, A., 1942. Fenicios y cartagineses en Occidente. Madrid: C.S.I.C.
GAUDIO, A.,
1958. Sur l'origine des canariens prehispaniques (étude comparée). Anuario de Estudios Atlánticos, 4, 115-167.
GONZÁLEZ, R., BALBÍN, R., BUENO, P. ET ARCO, M0.C. DEL,
1995. La Piedra Zanata. Museo Arqueológico de Tenerife. La Laguna: Organismo Autónomo de Museos y Centros. Cabildo Insular de Tenerife
GOZALBES, E.,
1988. La piratería en el Estrecho de Gibraltar en la Antigüedad. Congreso Internacional AEl Estrecho de Gibraltar@, t. I, 769-778.
GOZALBES, E.,
1993. Observaciones acerca del comercio de época romana entre Hispania y el Norte de África. Antiquités Africaines, 29, 163-176.
GOZALBES, E.,
1997. Economía de la Mauritania Tingitana (Siglos I a.C-II d.C.). Ceuta: Instituto de Estudios Ceutíes.
HUSS, W.,
1993. Los Cartagineses. Madrid: Ed. Gredos.
JÁUREGUI, J.J., 1954. Las Islas Canarias y la carrera del oro y la púrpura en el Periplo de Hannón. I Congreso Arqueológico del Marruecos Español, 271-276.
JODIN, A., 1967. Les établissements du roi Juba II aux Iles Purpuraires (Mogador). Tanger: Éditions Marocaines et Internationales.
LANCEL, S., 1994. Cartago. Barcelona: Crítica.
LENOIR, M.,
1992. Lixus à l'époque romaine. En: Lixus. Actes du colloque organisé par l=Institut des Sciences de l=Archéologie et du Patrimoine de Rabat avec le concurs de l=École Française de Rome, (Larache, 1989), 271-287.
LONGERSTAY, M.,
1990. Représentations de navires archaïques en Tunisie du Nord. Contribution à la chronologie des haouanet. Karthago. Revue d=Archeologie Mediterranéenne, XXII, 33-44.
LÓPEZ, F.,
1988. Apuntes sobre la intervención hispana en el desarrollo de las estructuras económicas coloniales en Mauritania Tingitana. Congreso Internacional AEl Estrecho de Gibraltar@, t. I, 741-748.
LÓPEZ, F.,
1994. El Periplo de Hannón y la expansión cartaginesa en el

África occidental. V Jornadas de Arqueología fenicio-púnica (Eivissa, 1990), 59-70.
LÓPEZ CASTRO, J.L.,
1992. Los libiofenicios: una colonización agrícola cartaginesa en el sur de la Península Ibérica. Rivista di Studi Fenici, XX (1), 47-65.
LUZÓN, J.M., 1988. Los hippoi gaditanos. Congreso Internacional AEl Estrecho de Gibraltar@, t. I, 445-458.
MAHJOUBI, A.,
1983. El período romano y post-romano en África del Norte. I. El período romano. Historia General de Africa. Salamanca: Tecnos/ Unesco, vol. II, 473-507.
MARTÍN, M., ATOCHE, P., CASTILLO, C. ET CRIADO, C.,
1998. La microfauna del yacimiento de El Bebedero (Teguise, Lanzarote): Implicaciones paleobiológicas, históricas y medioambientales. XIV Jornadas de Paleontología (Tenerife, 1998). Paleontología y Medioambiente, 121-124.
MARTÍNEZ, M., 1995. Canaria. Gran Enciclopedia Canaria, t. III, 755.
MARTÍNEZ, J. ET CARRERAS, C.,
1993. Ánforas de salazón y navegación comercial vía atlántica en época romana. La conexión Baetica-Britannia. XXII Congreso Nacional de Arqueología, 101-110.
MORALES, F.,
1978. Canarias: crónicas de su conquista. Sevilla: Ayuntamiento de Las Palmas/El Museo Canario.
MUÑOZ, R., 1994. La piedra Zanata y el mundo mágico de los guanches. Santa Cruz de Tenerife: Organismo Autónomo de Museos y Centros. Cabildo Insular de Tenerife.
NAVARRO, J.F.,
1983. Poblamiento humano de Canarias. En: Canarias origen y poblamiento. Madrid: Círculo de Estudios Sociales de Canarias, 85-96.
OSUNA, J.,
1988. Lixus,)El Gadir Magrebí en el ACírculo del Estrecho@?. Congreso Internacional AEl Estrecho de Gibraltar@, t. I, 563-576.
PAIS, F.J.,
1996. La economía de producción en la prehistoria de la isla de La Palma. La ganadería. Estudios Prehispánicos, 3. Santa Cruz de Tenerife: Dirección General de Patrimonio Histórico.
PATTON, M., 1996. Islands in Time. Island Sociogeography and Mediterranean Prehistory. London and New York: Routledge.
PEACOCK, D.P.S. ET WILLIAMS, D.F., 1986. Amphorae and the Roman Economy. An introductory guide. London and New York: Longman Archaeology Series.
Picard, G.C., 1978. Les romains en Côte d'Ivoire. Archéologie, 116, 22-27.
PONSICH, M. ET TARRADELL, M.,
1965. Garum et industries antiques de salaison dans la Méditerranée Occidentale. Paris: P.U.F.
PONSICH, M.,
1969. Nouvel aspect de l'industrie préromaine en Tingitane. Bulletin Archéologique, fasc. 4, 225-235.
PONSICH, M.,
1976. A propos d'une usine antique de salaisons à Belo (Bolonia-Cadix). Mélanges de la Casa de Velázquez, XII, 69-79.
PONSICH, M.,
1988. Aceite de oliva y salazones de pescado. Factores geoeconómicos de Bética y Tingitana. Madrid: Universidad Complutense.
RAMÓN TORRES, J.,
1995. Las ánforas fenicio-púnicas del Mediterráneo central y occidental. Barcelona: Publicacions Universitat de Barcelona. Col. Instrumenta 2.
REBUFFAT, R.,
1987. L'implantation militaire romaine en Maurétanie Tingitane. Atti del IV Convegno di studio sur L'Africa romana, 31-78.
REBUFFAT, R.,
1992. La frontière du Loukos au Bas-Empire. Lixus. Collection de L'École Française de Rome, 166, 365-377.
SANTANA, A., ATOCHE, P., MARTÍN, J. ET ARCOS, T.,
e.p. El conocimiento geográfico de la costa noroccidental africana en Plinio: La posición de las Canarias. Hildesheim-New York: Georg Olms Verlag. Colección Spudasmata.
SCHUBART, H. ET ARTEAGA, O.,
1986. El mundo de las colonias fenicias occidentales. Homenaje a Luis Siret (1934-1984). Cuevas del Almanzora (junio, 1984), 499-525.
SHERRATT, A.G.,
1979. Plough and pastoralism: Aspects of the Secondary Products Revolution. En: I. Hodder, N. Hammond et G. Isaac (eds.). Patterns in the Past, 261-305. Cambridge University Press.
STONE, O.M.,
1995 [1887]. Tenerife y sus seis satélites. Las Palmas de Gran Canaria: Cabildo Insular de Gran Canaria.
TORRIANI, L.,
1978 [c. 1592]. Descripción e Historia del Reino de las Islas Canarias antes Afortunadas, con el parecer de sus fortificaciones. Santa Cruz de Tenerife: Goya Ed.

32

EL CONTROL DEL TERRITORIO DURANTE LA EDAD DEL BRONCE EN EL ÁREA DE DORGALI (NUORO, CERDEÑA)

LILIANA SPANEDDA, TRINIDAD NÁJERA COLINO, JUAN ANTONIO CÁMARA SERRANO

Departamento de Prehistoria de la Universidad de Granada
Grenada, Spain
EMAIL: spanedda@ugr.es

RESUMEN

En este artículo se presentan los resultados de un análisis del patrón de asentamiento utilizado durante la Edad del Bronce en el término municipal de Dorgali (Nuoro, Cerdeña), a través del uso de diferentes índices que tienen en cuenta la posición topográfica relativa de cada uno de los asentamientos. Después se discute la organización territorial de la zona y la articulación entre los diversos tipos de yacimientos en el marco general de la Cultura Nurágica.

RIASSUNTO
In questo articolo presentiamo i resultati di un'analisi del modello di insediamento realizzata per i siti abitativi dell'Età del Bronzo appartenenti al territorio di Dorgali (Nuoro, Sardegna). Per la realizzazione di questa analisi si sono utilizzati differenti indici che considerano la posizione topografica di ogni singolo insediamento. In seguito vengono trattati l'organizzazione territoriale della zona e l'articolazione tra i diversi tipi di giacimenti nel contesto generale della Cultura Nuragica.

ABSTRACT
In this paper we present the results from an analysis of the settlement pattern used during the Bronze Age in the local territory of Dorgali (Nuoro, Sardinia). We use different index which take into account the relative topographical situation of every settlement. Afterwards we talk about the territorial organization of this area in Prehistory and the relation among the different kinds of sites inside the Nuragic Culture global frame.

INTRODUCCIÓN
Hemos usado diversas técnicas estadísticas y, en particular, el Análisis de Componentes Principales, a través de determinados índices elaborados por el Grupo de Estudios de la Prehistoria Reciente de Andalucía (Departamento de Prehistoria de la Universidad de Granada), para estudiar los asentamientos nurágicos de una zona concreta de Cerdeña, el municipio de Dorgali, en el que trabajos de campo recientes y revisiones bibliográficas han puesto de manifiesto la existencia de una gran variedad de asentamientos nurágicos (Lo Schiavo, 1980; Manunza, 1980, 1985, 1988,1995; Spanedda, 1994-95; Moravetti, 1998) cuya articulación espacio-temporal se pretende indagar, y sobre la que profundizaremos en el futuro en el marco de un análisis global sobre los asentamientos de la Edad del Bronce del Golfo de Orosei (Nuoro, Cerdeña, Italia).

En lo que respecta al uso del método que aquí probamos hay que decir que tras unos primeros estudios en que se procedió a contrastar los resultados del Análisis de Componentes Principales (de ordenación) con los del Análisis Cluster (o de agrupación) (Nocete, 1989, 1994), los trabajos más recientes, sobre todo en el caso del patrón de asentamiento, han procedido a desarrollar únicamente el primero de ellos aunque analizando pormenorizadamente los resultados numéricos de las variables en cada uno de los grupos obtenidos (Moreno, 1993; Lizcano, 1995; Lizcano et al., 1996; Moreno et al., 1997). La validez de este método para articular las diferentes componentes en que se basa el

Análisis de Componentes Principales, debido a las dificultades que origina la reducción a dos dimensiones de los gráficos obtenidos, ha sido probada en el estudio de la distribución de las tumbas megalíticas del pasillo de Tabernas (Maldonado et al., 1997; Cámara, 1998, 2001) y también en los análisis morfométricos del material cerámico de diversos yacimientos (Contreras et al., 1992; Moreno, 1993; Lizcano, 1995; Lizcano et al., 1997; Contreras y Cámara, 2000a, 2000b), para los que también se contaba con precedentes basados en la articulación de diversas técnicas (Contreras, 1986; Contreras et al., 1987-88; Nocete, 1989; 1994).

Los análisis sobre patrones de asentamiento prehistóricos desarrollados en el marco de las investigaciones del Grupo de Estudios de la Prehistoria Reciente de Andalucía (Departamento de Prehistoria de la Universidad de Granada) se han centrado en la articulación a través de diversos análisis multivariantes (Contreras, 1984) de diferentes variables que ponen en conexión el yacimiento con el entorno concreto en que se ubica (lo que se ha denominado Unidad Geomorfológica de Asentamiento, UGA, -Nocete, 1989, 1994-) y con el área en que se sitúa ésta (Area Geomorfológica, definida hasta hoy por límites métricos convencionales de 1, 3 ó 5 kms. en torno al asentamiento).

Para el análisis del patrón de asentamiento se han utilizado en los trabajos anteriormente referidos (Nocete, 1989, 1994; Moreno, 1993, etc.) diversos índices que se pueden agrupar en tres conjuntos:

(1) Conjunto de índices referidos a la articulación del asentamiento con el área que lo circunda y en la que sus habitantes teóricamente desarrollaron la mayor parte de sus actividades:

(a) YCAIP (Índice de pendiente del área geomorfológica). Busca relacionar el yacimiento con un determinado tipo de condicionante natural en cuanto a recursos subsistenciales, obstáculos para el control y capacidades estratégicas.

(b) YCAI1 (Índice de dominio visual 1). Relaciona la situación del yacimiento con la máxima altura del área buscando desentrañar hasta qué punto la elección estuvo motivada por objetivos estratégicos, lo que viene complementado por el siguiente índice.

(c) YCAI2 (Índice de dominio visual 2). Pone en relación la situación del yacimiento en cuestión con la mínima altura del Area Geomorfológica, lo que puede tener especial interés en la determinación de yacimientos dependientes.

(2) Índices referidos a la Unidad Geomorfológica de Asentamiento, el elemento concreto del paisaje, más o menos individualizado, donde se sitúa éste.

(d) YCUIC (Índice de compacidad de la Unidad Geomorfológica). De particular trascedencia para mostrar las capacidades defensivas del asentamiento y su aislamiento relativo del entorno.

(e) YCUIT (Índice de pendiente teórica de la Unidad Geomorfológica). Se trata de la primera aproximación a las características internas del lugar concreto en que se ubica el asentamiento y que condicionan tanto el hábitat, que en algunos casos conducen al aterrazamiento, como la accesibilidad general.

(f) YCUIR (Índice de pendiente real de la Unidad Geomorfológica). Ayuda a discernir la posibilidad de la existencia de áreas concretas de alta pendiente en la UGA, sea ésta llana o no.

(g) YCUIS (Índice de compacidad de la sección de la UGA). Busca la individualización de los rasgos propios de la zona del asentamiento en la UGA, intentando mostrar si existió un reforzamiento de sus potencialidades.

(h) YCUIA (Índice de amesetamiento). Distingue p. ej. los grandes poblados fortificados en unidades amesetadas de los espolones o aquellos aislados por barranqueras.

(3) Por último, y en relación tanto con la problemática anterior como con los condicionantes del Area geomorfológica, interesa estudiar

(i) el Índice de pendiente teórica del asentamiento en sí (YCYIT), que es otro modo de discernir la elección de una dificultad de acceso aun en zonas relativamente llanas.

Los análisis en curso sobre los asentamientos prehistóricos del Pasillo de Tabernas (Almería) han mostrado los problemas del denominado Índice de amesetamiento, dado que la presentación de la fórmula para obtener éste por parte de F. Nocete (1989, 1994) al presentar como dividendo la longitud de la sección y como divisor la diferencia de alturas de la parte que superaba la máxima pendiente conducía a valores extremos con respecto a los otros índices, por lo que hemos sugerido la alternancia del dividendo y el divisor, aun obviando el problema (YCUIA), conducía al efecto contrario con bajos resultados respecto a los otros índices, por lo que nuestro trabajo se ha centrado en la experimentación a partir de la inclusión o no de este índice, que en cualquier caso con esta transformación creemos que refleja mejor la inaccesibilidad de las mesetas, cuanto más alto es.

De la misma forma a los problemas de escasez de datos sobre la localización exacta de los asentamientos nurágicos debemos sumar la entidad arquitectónica de los monumentos nurágicos, aspectos ambos que nos sugieren que debemos prescindir de el YCYIT en nuestro análisis, para que la comparación entre los yacimientos sardos y españoles sea factible.

ANÁLISIS DEL PATRÓN DE ASENTAMIENTO EN DORGALI EN ÉPOCA NURÁGICA

INTRODUCCIÓN

Hemos realizado el Análisis de Componentes Principales sobre los 8 índices anteriormente referidos, excluyendo el YCYIT, y sus resultados muestran que la Varianza Acumulada en las dos primeras componentes es sólo del 52,63 %, mientras alcanza el 64,88 % si atendemos a las tres primeras componentes.

Es fundamental, por lo tanto, a la hora de establecer una tipología de los yacimientos nurágicos del municipio de Dorgali tener en cuenta las tres componentes y contrastar los resultados atendiendo a los valores reales de los índices establecidos.

Las correlaciones entre las variables son bastante bajas, a excepción del 0,69, que alcanza la correlación entre el Índice de Compacidad de la Sección (YCUIS) y el Índice de Amesetamiento (YCUIA), variables que, sin embargo, sólo adquieren significado en un número mínimo de los yacimientos estudiados.

En torno al 40 % rondan otras correlaciones. Entre el Índice de la Pendiente Real más pronunciada de la Unidad Geomorfológica (YCUIR) y el Índice de Pendiente Teórica de la Unidad Geomorfológica (YCUIT) la correlación alcanza el 0,47 y entre esta última y el Índice de Amesetamiento el 0,45. Más bajas son las correlaciones del Índice de Pendiente Teórica del Área Geomorfológica (YCAIP) con el Índice de Altura Relativa 1 (YCAI1) (0,40), con el Índice de Altura Relativa 2 (YCAI2) (0,35) y con el Índice de Pendiente Teórica de la Unidad Geomorfológica (0,39).

En cuanto al peso de las variables en los diferentes componentes, debemos decir que en la 1ª Componente priman de forma positiva el YCUIT (0,77), el YCUIS (0,73), el YCUIA (0,79), y, en menor medida, el YCUIR (0,61). En la 2ª Componente prima negativamente el YCAIP (-0,69), es decir aumenta cuando descendemos en el gráfico, y, positivamente, el YCAI1 (0,73). En la 3ª Componente prima negativamente el YCUIC (-0,72). Menos relevancia tiene el YCAI2 que alcanza un máximo negativo en la 2ª Componente con -0,53.

Desde esta distribución de los índices en las Componentes ha partido la división en tipos, subtipos y variedades. La división en tipos (denominados con números romanos) ha atendido fundamentalmente a la distribución en la 1ª Componente, y, por tanto, a la diferenciación en YCUIT, YCUIS e YCUIA, quedando a la derecha de los gráficos los yacimientos que presentan YCUIS e YCUIA (tipo I) y a la izquierda los que presentan un YCUIT más bajo (tipo IV) (FIGURE 1).

La división en subtipos (denominados por letras minúsculas) ha atendido a las variaciones en las componentes 1, 2 y 3 al interior de los tipos, con especial relevancia de la componente 3 en lo que respecta a la división de los tipos I, II y III. Es decir se han añadido como criterios de subdivisión el YCUIR, el YCAI2 y el YCUIC.

También la Componente 3 ha sido básica en la definición de las variedades, definidas por números arábigos, a partir del YCUIC. A la hora de definir éstas se ha procurado también tener en cuenta el carácter formal-funcional del yacimiento (nuraghe, villaggio, etc.), aspecto implícito también en las divisiones anteriores.

	COMPONENTES		
	1ª	**2ª**	**3ª**
Valores propios	2,41	1,80	0,98
% Varianza	30,13	22,50	12,25
% Varianza Acumulada	30,13	52,63	64,88

TABLA 1. Valores propios y varianza del Análisis de Componentes Principales realizado sobre los asentamientos de la Edad del Bronce del municipio de Dorgali

DESCRIPCIÓN DE LOS TIPOS

El análisis realizado sobre los nuraghi y poblados del municipio de Dorgali, en base a los datos de ubicación disponibles (Spanedda, 1994-95), tras el estudio de la tipología obtenida a partir del Análisis de Componentes Principales realizado a partir de los 8 índices antes referidos (YCAIP, YCAI1, YCAI2, YCUIC, YCUIT, YCUIR, YCUIS e YCUIA) ha ofrecido interesantes resultados a nivel global, sin embargo debemos realizar previamente una serie de consideraciones.

En primer lugar la baja definición de los mapas 1:25.000 del I.G.M. italiano ha conducido a que la obtención del YCUIS e YCUIA, como definidores de la existencia de un área especialmente relevante (y defendible) al interior de la Unidad Geomorfológica de Asentamiento, sólo fuera posible en los casos en que este área era más destacada.

En segundo lugar, y comparando con los análisis realizados anteriormente sobre el poblamiento prehistórico de diferentes áreas del sur de la Península Ibérica, llama la atención la excepcional variabilidad del YCAI2, uno de los índices que revelan el control visual general sobre el entorno. Este fenómeno es resultado en muchos casos de la cercanía al mar de zonas relativamente elevadas donde, en muchos casos, se sitúan los yacimientos.

En tercer lugar es de lamentar la escasez de datos sobre la cronología de los yacimientos, establecida, en los pocos casos en que ha sido posible, a partir de los restos materiales localizados (Campus y Leonelli, 2000). Este problema se une a la larga utilización de los monumentos y a las escasas estimaciones sobre la extensión de los yacimientos.

Comentando sucientamente los valores de cada uno de los tipos y subtipos, antes de abordar, también de forma resumida, su significado en términos de posición y control del territorio, debemos señalara la diferencia fundamental que existe entre el tipo I por un lado y el tipo IV por otro, re- (0,600-2,000), alto YCAI2 (3,275-41,500) y relativamente bajo YCAI1 (0,435-0,844). Los subtipos IIa y IIb se definen por su altísimo YCAI2 (12,000-41,500) dada su cercanía al mar. El subtipo IIb se diferencia por la mayor pendiente del Área Geomorfológica (0,700) aunque dentro de ésta se ha elegido una Unidad de menor pendiente (YCUIT e YCUIR=0,600). Los subtipos IIa y IIb incluyen sobre todo

	YCAIP	YCAI1	YCAI2	YCUIC	YCUIT	YCUIR	YCUIS	YCUIA
YCAIP	1							
YCAI1	-0,4	1						
YCAI2	0,35	-0,12	1					
YCUIC	-0,14	0,14	-0,05	1				
YCUIT	0,39	-0,10	0,26	-0,06	1			
YCUIR	0,12	0,13	0,05	-0,01	0,47	1		
YCUIS	0,09	0,24	-0,01	0,13	0,30	0,29	1	
YCUIA	0,16	0,22	0,02	0,16	0,45	0,25	0,69	1

TABLA 2. Correlaciones entre las variables utilizadas

specto a los tipos que ocupan el centro de ambos gráficos (FIGURE 1).

El tipo I, situado a la izquierda de ambos gráficos, presenta valores altos en todos los índices, especialmente el YCAI1 (0,692-1,000) y el YCAI2 (1,424-6,900). Si exceptuamos el Subtipo Id el YCUIR se sitúa entre 0,880 y 10,000. YCUIS e YCUIA son, sin embargo, los índices centrales para la definición del tipo, y oscilan entre 0,260-0,798 y 0,060-0,293 respectivamente.

En nuestro tipo I, con altas pendientes y alta visibilidad y con yacimientos cercanos a los puntos más altos dentro del Área Geomorfológica de 1 Km. de radio, predominan los nuraghi, aunque encontramos también nuraghi con poblado y los poblados encastillados de Tiscali y Tilimba, siendo el primero de ellos el único del que podemos sugerir su uso entre el Bronce Final y el Hierro inicial (Lilliu, 1988).

La variedad Ia1 se caracteriza por su alto YCUIT (1,260-1,342) y la Ia2 por su alta compacidad de la Unidad Geomorfológica (0,991) y su relativamente menor YCUIR (0,880). El subtipo Ib cuenta con un elevadísimo YCUIR (10.000), resultado de la existencia de una verdadera pared vertical en uno de los límites de la Unidad Geomorfológica. La variedad Ic1 presenta mayor YCAIP, YCUIR e YCUIS que la variedad Ic2. La variedad Ic3 presenta menor YCAI2 (1,407-2.044). Al subtipo Id queda definido por sus altas compacidades y, sobre todo, por sus bajas pendientes de la Unidad Geomorfológica (YCUIT=0,153, YCUIR=0,166). Está integrado por el poblado de Sos Mucarzos, tal vez nos encontremos ante un poblado fuertemente amurallado o de un gran desarrollo temporal (tipo tell).

El tipo II presenta altos YCUIT (0,600-1,000) e YCUIR nuraghi y nuraghi con poblado, pero también el poblado de Fruncunieddu. Sólo del nuraghe Mannu se tienen datos cronológicos que lo sitúan entre el Bronce Medio y la época romana (Fadda y Pruneti, 1997:40; Campus y Leonelli, 2000). El subtipo IIc se sitúa en áreas de menor pendiente y presenta menor YCAI2, e incluye únicamente el nuraghe de Inghirai que ha sido atribuido al Bronce Medio y Reciente (Campus y Leonelli, 2000).

El tipo III presenta todavía un YCAI2 mayor de 1,300 y, sobre todo, un YCUIR normalmente mayor de 0,200. El YCUIC también es relativamente alto. El subtipo IIIa se caracteriza por las altas pendientes de la Unidad Geomorfológica (YCUIT entre 0,600 y 1,000 e YCUIR entre 1,000 y 1,250) y sólo incluye poblados con una visibilidad alta o muy alta. Sólo el de Ziu Santuru ha sido atribuido al Bronce Final e Hierro Inicial (Campus y Leonelli, 2000). El subtipo IIIb sólo presenta YCUIR alto (0,480-1,000), ya que el YCUIT desciende, sobre todo en la variedad IIIb2. presentan visibilidad alta (IIIb1) o muy alta (IIIb2) y pendientes de la Unidad Geomorfológica moderada. La compacidad de la Unidad Geomorfológica es alta en todo el subtipo. Por otra parte el subtipo IIIc presenta YCUIR similar (0,400-1,250), en un Área Geomorfológica de menor pendiente (YCAIP=0,241-0,393) y una compacidad de la Unidad Geomorfológica menor en la variedad IIIc1, e incluye sólo nuraghi. El subtipo IIId presenta YCUIT e YCUIR entre 0, 160 y 0.,500. Las variedades se definen, sobre todo, en función del YCAI1. El subtipo IIId incluye sobre todo poblados con alta visibilidad y pendientes moderadas-altas. Los nuraghi de S. Giorgio y Mannu de S. Anna caracterizan el tipo IIId3, con el poblado de Sidda 'e Josso. La muy alta visibilidad del tipo IIId1(Nuraghe Arvu) y de Tinniperargiu está en relación con su cercanía al mar (Spanedda, 1994-95). Sólo Nuraghe Arvu (Bronce Medio) y Su Tintinnau (Bronce Medio/Reciente) han podido ser adscritos cronológicamente (Campus y Leonelli, 2000), y del primero de ellos se han referido también materiales de superficie adscribibles a la cultura de Bonnanaro del Bronce Antiguo (Moravetti, 1998:23).

Más altas son las pendientes de la Unidad Geomorfológica del Subtipo IIIe (YCUIT entre 0,333 y 0,666 e YCUIR entre 0,500 y 2,000) en un Área de menor pendiente (0,136-0,263). Incluye sólo poblados con visibilidad relativamente alta, especialmente Pranus. El subtipo IIIf presenta YCUIS e YCUIA en un Área de baja pendiente, aspecto compartido con el subtipo IIIg de mayor YCUIR (0,666-3,200). Ambos subtipos comparten también un YCAI1 ALTO (0,610-1,000). El subtipo IIIf incluye un nuraghe con poblado (Sa Pramma) de muy alta visibilidad y compacidad pero bajo amesetamiento.

El tipo IV presenta un YCUIT más bajo (0,010-0,520) y un YCUIR especialmente bajo desde el subtipo IVb. El subtipo IVa es así aquel donde encontramos pendientes de la Unidad Geomorfológica mayores. Las variedades IVa1, IVa2 y IVa3 presentan menor YCAI1 (0,480-0,753), pero la IVa2 presenta un alto YCUIR (2,000) y la IVa1 el menor. La variedad IVa4 presenta mayor YCAI1 (0,390-0,996) y tiene fuertes pendientes de la Unidad Geomorfológica en lo que respecta al YCUIR.

Poblados y nuraghi se encuentran en las variedades IVa1 y IVa2 de pendientes moderadas y visibilidad alta pese a situarse lejos del punto más alto del Área Geomorfológica. La variedad IVa2 (Golunie) presenta un mayor YCUIR y corresponde al único nuraghe simple (sin poblado y cercano al mar) (Spanedda, 1994-95). En este subtipo contamos excepcionalmente con datos cronológicos sobre 3 yacimientos: Zorza I (Bronce Medio al Hierro Inicial), Sas Perdas Ladas (Edad del Hierro) y Balu Virde (Bronce Medio/Reciente) (Campus y Leonelli, 2000).

La variedad IVa4 presenta normalmente el YCAI2 bajo, con pendientes bajas del Área Geomorfológica, moderado YCUIT y moderado-alto YCUIR, situándose los yacimientos siempre cerca del punto más alto del Área Geomorfológica. Aquí contamos con dos yacimientos con datos cronológicos: S. Basilio (Bronce Medio al Hierro Inicial) y Toloi II (Bronce Medio) (Campus y Leonelli, 2000). En la visibilidad destacan Sa Icu (nuraghe), Corallinu (poblado) y Muristene (poblado).

El subtipo IVb presenta pendientes reales más pronunciadas en torno a 0,500 y, en general, un YCAI2 alto. La variedades se definen, en este caso, básicamente por la compacidad de la Unidad Geomorfológica. La variedad IVb1 incluye poblados y un nuraghe complejo con poblado (Luargiu) en áreas de pendientes moderadas y con visibilidad alta pese a situarse lejos del punto más alto del Área Geomorfológica de 1 Km. de radio. Sorgolita ha sido atribuido al Bronce Medio (Campus y Leonelli, 2000). La variedad IVb2 incluye poblados y un nuraghe simple (La Favorita). La visibilidad es normalmente baja como también la pendiente teórica de la Unidad Geomorfológica, mientras son moderadas las otras. Sólo el nuraghe La Favorita y el poblado Thomes presentan cierto control visual. La variedad IVb3 de mayor YCAI1 y compacidad incluye el poblado Isportana atribuido al Bronce Medio/Hierro Inicial (Campus y Leonelli, 2000).

La variedad IVb4 presenta pendientes bajas excepto la más pronunciada, se sitúan cerca del punto más alto y presentan visibilidad alta sea en los poblados o en los nuraghi. El poblado de Serra Orrios ha sido atribuido al Bronce Medio/Hierro Inicial (Ferrarese Ceruti, 1980; Fadda, 1990, 1996a; Campus y Leonelli, 2000) y de él ha sido destacada su extensión (Manunza, 1995:106, 118; Moravetti, 1998:28), lo que se podría relacionar con su misma an-

ÍNDICES	**COMPONENTES**		
	1ª	**2ª**	**3ª**
YCAIP	0,43	-0,69	-0,20
YCAI1	0,11	0,73	0,20
YCAI2	0,26	-0,53	-0,28
YCUIC	0,08	0,42	-0,72
YCUIT	0,77	-0,30	0,15
YCUIR	0,61	0,03	0,49
YCUIS	0,73	0,38	-0,11
YCUIA	0,79	0,31	-0,18

TABLA 3. Incidencia de los índices utilizados en cada una de las componentes

tigüedad si tenemos en cuenta los materiales adscribibles a Bonnanaro recogidos y atribuidos a una tumba de gigante destruida (Moravetti, 1998:23). La variedad IVb5 presenta pendientes bajas y moderadas con algunas excepciones en la YCUIR y alta visibilidad situándose cerca del punto más alto del Área Geomorfológica. La más baja visibilidad es paradójicamente la del nuraghe Lottoniddo, vinculado posiblemente a la defensa directa del poblado y auxiliado en el control territorial del nuraghe simple homónimo (variedad IIIg1). La compacidad de esta variedad IVb5 es todavía más alta y encontramos aquí sobre todo nuraghi y 3 poblados de los que se ha atribuido al Bronce Medio/Reciente el caso de Locu Secau y al Bronce Medio/Hierro Inicial el caso de Toloi I (Campus y Leonelli, 2000).

El sutipo IVc se caracteriza también por su alto YCAI2, mayor de 1,900, a excepción de las variedades IVc5 y IVc6, de menor pendiente del Área Geomorfológica, que comparten con la IVc4 y la IVc7, así como la mayor compacidad de la Unidad Geomorfológica. Estas variedades presentan también pendientes muy bajas y mayor YCAI1 (0,662-1,000). Las variedades IVc1, IVc2 y IVc3 sólo incluyen poblados con bajas y moderadas pendientes y alta visibilidad pese a situarse lejos del punto más alto del Área Geomorfológica. Las variedades IVc5 y IVc6 presentan respecto a las anteriores una visibilidad normalmente más baja que contrasta con los casos de Muristene (nuraghe simple) (variedad IVc4) y Poddinosa (nuraghe complejo) (variedad IVc7). Dentro de estas variedades Predu 'e Ponte (poblado) ha sido atribuido al Bronce Medio/Hierro Inicial, Sini (poblado) al Bronce Medio/Reciente y Giorgi Poddighe (nuraghe simple con poblado) al Bronce Medio/Reciente (Campus y Leonelli, 2000).

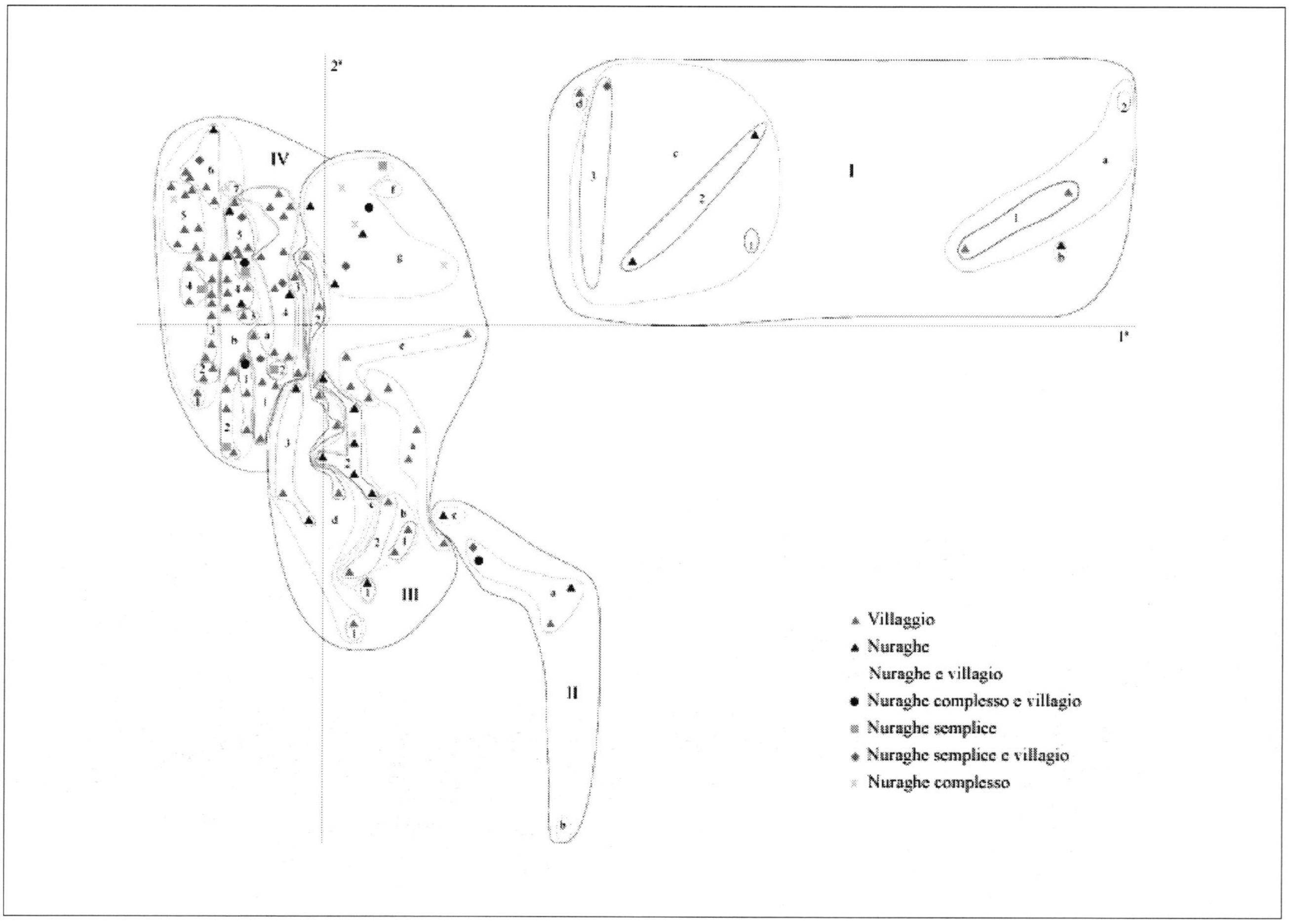

FIGURA 1. ACP De Los Yacimientos Nurágicos Del Municipio De Dorgali. Gráfico De La 1ª Y 2ª Componentes

CONCLUSIONES GENERALES

En lo que respecta a la distribución general de los tipos por el territorio de Dorgali (FIGURE 2) debemos señalar:

(1) Que dentro del tipo I los poblados del subtipo Ia (Tiscali y Tilimba) se sitúan al sur y en la parte más alta de las cuencas fluviales pero en zonas de muy alta pendiente. Lo mismo puede decirse de S. Diliga (subtipo Ib) y del nuraghe con poblado de Suttaterra también al sur (variedad Ic3). Su Casteddu (variedad Ic3) se asocia a un poblado y a tumbas de gigante (Manunza, 1995:131-133) y puede considerarse el culmen del control territorial en la zona en que se sitúa. El resto de los yacimientos inscritos en el tipo I se sitúan en el centro del territorio de Dorgali y el único poblado sin nuraghe es Sos Mucarzos (subtipo Id). En otras zonas se conocen también poblados de montaña de este tipo, que llegan hasta los 1000 mts. de altitud, como S'Urbale (Teti, Nuoro), Dovilineò (Orgosolo, Nuoro) y Mereu (Orgosolo, Nuoro)

(Fadda, 1990:102).

(2) Casi la totalidad de los yacimientos incluidos en el tipo II corresponden a lugares cercanos al mar, de ahí el altísimo YCAI2 del subtipo IIa que incluye nuraghi con poblado, un nurghe sin poblado (Toddeitto) y un poblado (Fruncunieddu) al sur. En el subtipo IIc se incluye un nuraghe de la misma zona ligeramente más interior.

(3) En el tipo III ya hemos dicho que encontramos nuraghi y poblados. En ambos casos parecen alinearse con los cursos fluviales principales. Los nuraghi se sitúan generalmente en los puntos más altos de la cuenca o en las cabeceras de los afluentes. Esto incide en su mayor visibilidad (subtipos IIIc, IIIf y IIIg y variedad IIId3, sólo superada por los yacimientos cercanos al mar (variedad IIIb1) entre los que se debe incluir Nuraghe Arvu (variedad IIId1), poblado de gran extensión (Manunza, 1995:106), y con materiales superficiales que se podrían remontar al Bronce Antiguo (Moravetti, 1998:23), y Tinniperargiu (variedad IIId2 por su mayor YCAI1 pese a estar en un área de menor pendiente). El nuraghe complejo Ruju situado controlando, desde el interior de un meandro, el río Cedrino, adquiere especial relevancia en una zona de gran concentración de monumentos (Spanedda, 1994-95; Manunza, 1995, Moravetti, 1998), pero es interesante también la articulación entre Su Marrone, S. Giorgio y Santa Diliga, este último nuraghe de la variedad Ia2, respecto al control del valle del Isalle (Manunza, 1995:107).

(4) En el tipo IV han quedado incluidos poblados y nuraghi situados fundamentalmente en las llanuras interfluviales. Los nuraghi situados en este tipo suelen ser complejos y en la mayoría de los casos están acompañados de poblados o se sitúan cerca de ellos. A veces presentan incluso tumbas de gigante asociadas. Destaca en este sentido Biristeddi, incluido en la variedad IVb5 del Análisis de Componentes Principales, y situado en un meandro con alta visibilidad pero que muestra bajas pendientes de la Unidad Geomorfológica, incluso respecto al resto de los monumentos de la variedad en que se inscribe.

Otros nuraghi como Golunie (variedad IVa2) y La Favorita (variedad IVb2) se sitúan en relación al control costero ya referido con respecto a otros tipos de mayores pendientes. Relativamente cerca de la costa se sitúa también Su Barcu, relacionado con su poblado y tumbas de gigante (variedad IVb4).

Son así todos ellos nuraghi más directamente vinculados al control directo de los pasos, de los cursos fluviales, aunque a veces en la cabecera de algunos afluentes (Zorza I, Sortei y Lotteniddo) pero siempre en zonas de menor pendiente y, por tanto, más vinculados a las zonas de hábitat. No en vano ya G. Lilliu (1962:14) había señalado que aunque si normalmente los nuraghi se sitúan sobre elevaciones estratégicas, formando una red de intervisibilidad entre las torres (...). Existen también nuraghi intencionalmente ocultos, o aislados, en plena llanura, que responden a estrategias defensivas, económicos u otras diversas de los precedentes. No importa en cualquier caso que el número de nuraghi fuese inferior al de los poblados (Fadda, 1990) si tenemos en cuenta que el sistema debió funcionar como un conjunto (Moravetti, 1998:28) estatal, como después discutiremos.

Entre las zonas de hábitat algunos poblados como Serra Orrios, Corallinu, Muristene, Thomes, Sa Paule Dorrisolo y, en general, los incluidos en las variedades IVb4 y IVb5, presentan importante visibilidad. Mª.R. Manunza ha referido la gran abundancia de poblados en relación al número de nuraghi en nuestra zona de estudio, donde además éstos tienden a situarse en los puntos estratégicos de las vías de tránsito en lugar de asociarse a los poblados (Manunza, 1995:105, 112) . Los nuraghi, en general, cumplen así la función primordial de los nuraghi simples en otras áreas pero esto no excluye, ni minusvalora como se había pretendido (Fadda, 1990:102), la existencia aquí de nuraghi complejos y simples, como hemos visto. La asociación de poblados a determinados nuraghi, en los grupos III y IV, como muestran, Neulè (variedad IVb5) o S'Ulumu (variedad IIIg1), éste último complejo, y asociado a tumbas de gigante (Manunza, 1995:133-138), y Su Barcu, de tipología formal no definida, también asociado a tumbas de gigante, incluido en nuestra variedad IVb4 y relativamente alejado del poblado asociado, nos refieren la importancia de la defensa en todos los casos. Por otra parte los poblados sin nuraghe no carecen de defensas ya estén conformadas sobre todo por el agrupamiento de las casas como en Serra Orrios (variedad IVb4) (Manunza, 1995:119 FIGURE 157; Moravetti, 1998:34 FIGURE 25), tal y como hemos referido también para la Edad del Bronce de la Península Ibérica (Cámara, 1998a), aunque sea un aspecto rechazado por determinados autores (Ferrarese Ceruti, 1980:110-111), o por murallas como la que rodea Arvu (variedad IIId1), o aquellas asociadas al nuraghe y poblado Mannu (subtipo IIa de control costero) (Manunza, 1995:157, 161-162) y que hallan sus raíces en las estructuras calcolíticas.

LOS "NURAGHI": POBLADOS FORTIFICADOS Y TORRES DE CONTROL

La articulación entre nuraghi simples y complejos y la relación de los poblados con éstos no parece homogénea en todo el territorio sardo ni tampoco lo es, como hemos referido anteriormente, en todo el periodo cultural que cubre la denominada civilización nurágica.

El aislamiento relativo de los pequeños nuraghi, más que en el Sureste de la Península Ibérica, aunque existieran también poblados asociados a determinados nuraghi con campesinos libres estabilizadores de la sociedad clasista , favorece el desarrollo de las clientelas de estas élites, si bien se daban relaciones entre ellas demostradas en el control ordenado del territorio que continuará en el Bronce Final (Moravetti, 1990:65-72) cuando según G. Lilliu se desarrollen casi todos los nuraghi complejos (Lilliu, 1982:62), aunque hemos visto que los datos sobre la ocupación del Bronce Medio no están ausentes del municipio de Dorgali, y que sólo el tipo I no cuenta con yacimientos que hayan proporcionado, hasta ahora, materiales atribuibles a ese periodo. Las diferencias se intuyen, con los pocos datos disponibles, entre los subtipos, con algunos posiblemente tardíos, como

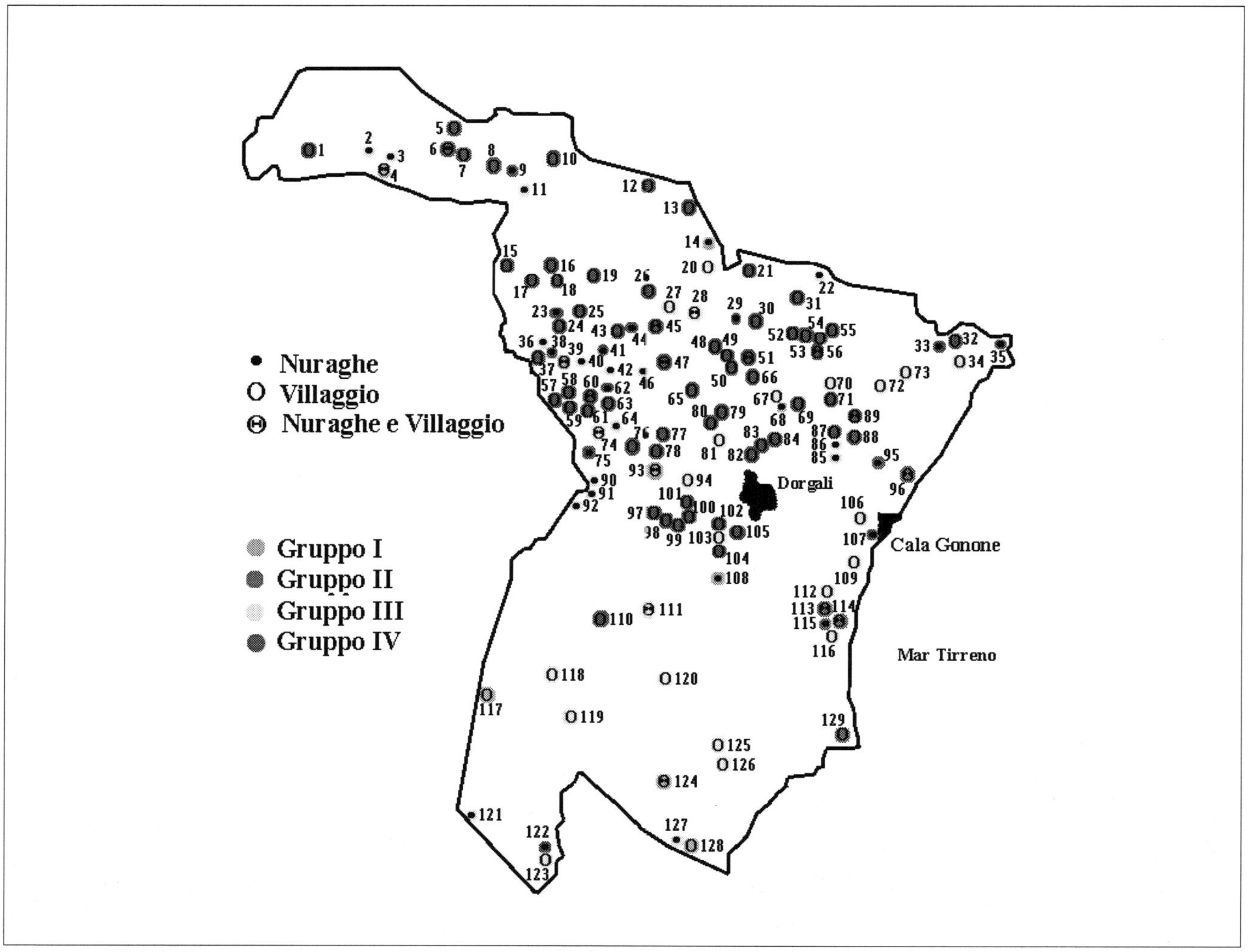

1. Villaggio Sorgolitta. 2. Nuraghe Su Marrone. 3. Nuraghe S. Giorgio. 4. Nuraghe e villaggio S. Diliga. 5. Villaggio Lottoni. 6. Nuraghe complesso e villaggio Luargiu. 7. Villaggio Predas Ruias. 8. Villaggio Sa Pira. 9. Protonuraghe Orrule. 10. Villaggio Thomes. 11. Nuraghe Bias 'Ebbas. 12. Villaggio Coscone. 13. Villaggio Sa Paule Dorrisolo. 14. Nuraghe Casteddu 'e Ghistala. 15. Villaggio S. Basilio. 16. Villaggio Fologhe. 17. Villaggio S'Arcu 'e su Linu. 18. Villaggio S. Jacu. 19. Villaggio Fruncudunue. 20. Villaggio S'Aliderru. 21. Villaggio Su Saucu. 22. Nuraghe Concas de Janas. 23. Nuraghe Paule Marras. 24. Villaggio Marras. 25. Villaggio Mattanosa. 26. Villaggio Biriddo. 27. Villaggio Chidera. 28. Nuraghe e villaggio Sa Pramma. 29. Nuraghe Picchio. 30. Villaggio Predu 'e Ponte. 31. Villaggio Biascotta. 32. Villaggio Su Barcu. 33. Nuraghe Su Barcu. 34. Villaggio Gutturu 'e Jacas. 35. Nuraghe semplice Golunie. 36. Nuraghe semplice Lottoniddu. 37. Villaggio Lottoniddo. 38. Nuraghe Lottoniddo. 39. Nuraghe semplice e villaggio Su Casteddu. 40. Nuraghe complesso Abba Noa. 41. Nuraghe complesso Oveni. 42. Nuraghe complesso Noriolo. 43. Villaggio Serra Orrios. 44. Nuraghe Sa Icu. 45. Nuraghe complesso e villaggio Biristeddi. 46. Nuraghe complesso Ruju. 47. Nuraghe semplice e villaggio Neulè. 48. Villaggio Caschiri II. 49. Villaggio Caschiri I. 50. Villaggio Caschiri III. 51. Nuraghe semplice e villaggio Giorgi Poddighe. 52. Villaggio Sa Carruba. 53. Villaggio Cubida Mou. 54. Villaggio Sas Luas. 55. Villaggio Su Tiresu. 56. Nuraghe e villaggio Sortei. 57. Villaggio Ollei Buscai. 58. Villaggio Sas Tuculas. 59. Villaggio Nastallai. 60. Nuraghe semplice e villaggio S. Nicola. 61. Villaggio Dughine. 62. Nuraghe semplice Muristene. 63. Villaggio Muristene. 64. Nuraghe complesso (?) Isili. 65. Villaggio Dugulana. 66. Villaggio Tillai. 67. Villaggio Ziu Santoru. 68. Nuraghe Sos Pruvereris. 69. Villaggio Sos Pruvereris. 70. Villaggio S. Giovanni Su Anzu. 71. Villaggio Zorza II. 72. Villaggio Siddai 'e Susu. 73. Villaggio Siddai 'e Josso. 74. Nuraghe complesso e villaggio S'Ulumu. 75. Nuraghe complesso Poddinosa. 76. Villaggio Iriai II. 77. Villaggio Iriai I. 78. Villaggio Mariscai. 79. Villaggio Nicola Mesina. 80. Villaggio Sini. 81. Villaggio Su Tintinnau. 82. Villaggio Locu Secau. 83. Villaggio Sas Perdas Ladas. 84. Villaggio Balu Virde. 85. Nuraghe Su Tuppedie. 86. Nuraghe S'Istrumpu. 87. Villaggio Cascozza. 88. Villaggio Littu. 89. Nuraghe semplice e villaggio Zorza I. 90. Nuraghe S. Pantaleo. 91. Nuraghe N. S. degli Angeli. 92. Nuraghe Iscra Duacore. 93. Nuraghe e villaggio Coazza. 94. Villaggio Toddoschi. 95. Nuraghe Inghirai. 96. Nuraghe e villaggio Codula Manna. 97. Villaggio Corallinu. 98. Villaggio Istipporo. 99. Villaggio Toloi I. 100. Villaggio Toloi II. 101. Villaggio Campumannu. 102. Villaggio Su Calavreri. 103. Villaggio Iscupidana. 104. Villaggio Sos Mucarzos. 105. Villaggio Isportana. 106. Villaggio Nuraghe Arvu. 107. Nuraghe semplice La Favorita. 108. Nuraghe S. Elene. 109. Villaggio Tinniperargiu. 110. Villaggio Finiodde. 111. Nuraghe semplice e villaggio Biriculi. 112. Villaggio Pranus. 113. Nuraghe complesso e villaggio Nuragheddu. 114. Nuraghe semplice e villaggio Nuraghe Mannu. 115. Nuraghe Toddeitto. 116. Villaggio Toddeitto. 117. Villaggio Tiscali. 118. Villaggio Monte Tundu. 119. Villaggio Maidreu. 120. Villaggio Marchesi. 121. Nuraghe Nuragheddu. 122. Nuraghe Mannu de S. Anna. 123. Villaggio S. Anna. 124. Nuraghe e villaggio Suttaterra. 125. Villaggio Ghivine. 126. Villaggio S'Ungrone 'e sa Mesa. 127. Nuraghe Punta Ghirudorgia. 128. Villaggio Tilimba. 129. Villaggio Fruncunieddu.

FIGURA 2. DISTRIBUCIÓN DE YACIMIENTOS EN EL MUNICIPIO DE DORGALI.

el Ia. Otro problema es valorar la evolución de los monumentos y su continuidad.

A partir del estudio de las relaciones sincrónicas entre los nuraghi hay que tener en cuenta que incluso se ha llegado a postular una tipología de las diferentes torres sin cubierta cupuliforme (Manca Demurtas y Demurtas, 1984a:635, 1984b:173-174, 180), incluyéndolas bajo el término protonuraghi (Manca Demurtas y Demurtas, 1984a:631, 1984b:167; Moravetti, 1992b:188) que cubrirían en su evolución formal desde el Bronce Antiguo con lo que la jerarquización en esos momentos (y antes) era evidente, respondiendo a la necesidad de ejercer el dominio sobre el territorio circundante (Manca Demurtas y Demurtas, 1984b:168). Se confirma así la antigua hipótesis de E. Contu (1959:110), que fue fuertemente criticada en su día en base al uso sólo de los materiales exóticos y metálicos de estratigrafía dudosa para establecer la cronología de algunos protonuraghi complejos como el rectangular de Albucciu (Arzachena, Sassari) (Ferrarese Ceruti, 1963:194-202) .

De hecho aunque G. Lilliu (1982:17) ha señalado la antigüedad de las estructuras protonurágicas en base a asociaciones a dólmenes y galerías cubiertas como en el caso de Santa Caterina (Macomer, Nuoro), y considera el nuraghe Albucciu un compromiso entre el nuraghe a tholos y el de galería , en una época tardía (de revival) posterior al 1200 a.c., también coloca en esta época el de Peppe Gallu (Uri,

Sassari) que había estudiado E. Contu (Lilliu, 1982:78) . Esto ha llevado también a otros autores a señalar la contemporaneidad de ambos tipos de nuraghi (L. Usai, 1992:14) y a rechazar el término protonuraghi (Ferrarese Ceruti y Lo Schiavo, 1992:125) pero L. Manca Demurtas y S. Demurtas conscientes de esta problemática (Manca Demurtas y Demurtas, 1984b:184) han estudiado los materiales de superficie de las protonuraghi de Fruscos (Paulilatino, Oristano) y Friarosu (Mogorella, Oristano) adscribiéndolos a Bonnanaro, por lo que A nivel cronológico, por tanto, parece probable que el inicio de la evolución del protonuraghe y el tránsito sucesivo hacia el tholos se inserten en un periodo que va desde el Bronce antiguo al Bronce medio sin una verdadera ruptura radical de raíz cultural entre un tipo de nuraghe y otro (Manca Demurtas y Demurtas, 1984b:187) en un proceso progresivo de mejora técnica y consecución de espacio interno (Contu, 1985:18) si bien con continuidad tal y como parecen mostrar las dataciones por hidratación de la obsidiana (Manca Demurtas y Demurtas, 1987:497, 499 n. 33).

En el territorio objeto de estudio sólo contamos con un monumento definido como protonuraghe e inscrito en la variedad IVb5, junto a la mayoría de los poblados (villaggi) pero ello no oculta que el sistema de control territorial debió asentar sus raíces en las épocas precedentes y que debió sufrir modificaciones como muestra el subtipo Ia, posiblemente tardío como hemos referido.

Al menos en el Bronce Reciente (Nurágico III, 1600/1450-900 cal A.C.) Integradas en un sistema de torres menores oportunamente distribuidas en el territorio y colocadas en puntos estratégicos estas fortalezas que dominan desde una posición más alta el hábitat inmediato y los campesinos y pastores de los inmediatos alrededores, y representan el centro del poder en pequeñas ciudades-capitales, albergaban monarcas militares que extendían su dominio a estados limitados, sin alcanzar el grado de un fuerte señorío supracantonal (Lilliu, 1982:70). El problema es establecer el grado de las relaciones entre estas élites para definir los verdaderos estados en que se integran, en proceso de consolidación en estos momentos del Bronce Final sobre unas bases aristocrático-gentilicias que, sin embargo, creemos, también aquí se vieron parcialmente superadas por el modelo oligárquico más o menos acusado y la importancia de la "ciudad". Para G. Lilliu los nuraghi complejos supusieron pasar de una función de simple control de bienes materiales a un sistema de defensa interna y externa (Lilliu, 1995:33).

Como hemos sugerido en nuestro análisis para otras zonas se ha indicado además que los poblados que no están en relación directa con un nuraghe se sitúan siempre en zonas donde otros nuraghi podían facilitar el control del territorio (Bafico y Rossi, 1993:47). En cualquier caso no todos los nuraghi complejos tienen la misma función y si bien no se puede generalizar a todos los nuraghi de zonas altas P. Bartolini (1993:29) ha resaltado que la presencia de estructuras complejas en las vías de tránsito y en áreas de interés estratégico estaba destinada a impedir el avance fácil a un eventual agresor al ser capaces de reunir fuerzas considerables que no se podían dejar a las espaldas. Esta hipótesis revela la importancia de estas verdaderas "colonias", en terrenos donde antes no había ocupación permanente, tal y como las hemos referido para la Península Ibérica (Cámara et al., 1995, 1996; Cámara, 1998) donde, sin embargo, en el Sureste no se asocian a esas torres aisladas presentes en Cerdeña y que tal vez pueden paralelizarse, como veremos, con modelos del Bajo Guadalquivir o la Andalucía Central poco estudiados.

Las objeciones (Trump, 1992:198) respecto a la imposibilidad de defender todos los nuraghi monotorre son absurdas dado que también en época medieval las torres aisladas servían sólo de puntos de avistamiento cuya defensa era imposible ante ejércitos vastos (o rebeliones), por lo que cumplida su función tendían a ser abandonadas . Aunque D. H. Trump (1992:199) niega que cada torre sea una entidad independiente, ha mantenido en su análisis (Trump, 1992:200), como G. Ugas (1990), prácticamente cada nuraghe complejo como el centro de una unidad política. Este último porque cree que es improbable que verificaran las condiciones para el nacimiento de un sistema político organizado sobre bases jerárquicas análogas a aquellas "feudales" (Ugas, 1990:24, énfasis en original). Las condiciones naturalmente no son idénticas dado que no se desarrollan sistemas tan complejos de dependencia entre los nobles y el esclavismo está en la Prehistoria en ascenso y no en decadencia con lo que las otras formas de servidumbre (tributarias) están menos agudizadas y se basan más en impuestos y servicios militares generalizados que en labores en las "tierras del señor" (Cámara, 1998a). Sin embargo las relaciones "familiares" entre las élites sí existen pues los estados aristocráticos derivan de situaciones centralizadas previas, utilizan elementos exóticos en la misma justificación de su existencia y de sus élites , y conducen a nuevas formas de centralización (por conquista, revueltas, política matrimonial, desarrollo de la adscripción servil y clientelar, etc.) a nuevas formas de organización centralizadas que hemos denominado gentilicias (Cámara, 1998a). Realmente un sistema de tan pequeños estados sería tan coyuntural como para ser considerado imposible dado que las ambiciones conducirían a agregaciones por conquista hasta conducir a territorios más estables, además qué sentido tendría el control por cada estado de una pequeña parte de una vía si en un sistema tan anárquico ninguno podría circular . Este panorama no niega luchas internas (guerras civiles, revueltas, etc.) de las que no están libres tampoco los estados posteriores.

En realidad el estudio de muchas zonas de Cerdeña, como se puede ver en nuestro análisis (FIGURE 1) y en la Conca di Isili (Navarra, 1998:336 FIGURE 1), nos muestra cómo los nuraghi monotorre tienden a circundar las áreas de valle donde se sitúan diversos nuraghi complejos, e incluso otros simples destinados a funciones específicas o poblados fortificados o no (en este último caso mostrando una seguridad mayor al menos en determinadas etapas de desarrollo). Pero este sistema no implica que cada uno de estos valles fuera una unidad autónoma ya que incluso, sobre todo en determinados momentos, debieron darse procesos de integración/conquista.

En los últimos años se ha iniciado a prestar atención a los sistemas de control territorial de índole comarcal (Alba,

1998:73; Foddai, 1998:85-87), siendo especialmente interesante la variabilidad del Logoduro-Meilou, donde se aprecian sistemas de alineación de torres en torno a estructuras complejas, la alineación de éstas en relación a los valles principales y la vinculación de ciertas torres a nuraghi especialmente relevantes como Santu Antine (Torralba, Sassari) (Foddai, 1998:85-87), sin embargo cuando se marcan los polígonos Thiessen sobre el territorio no se tiene en cuenta ni la tipología, ni la cronología ni las agrupaciones de nuraghi.

En este contexto el nuraghe era el símbolo del poder (Lilliu, 1982:40), especialmente cuando en los nuraghi complejos, la torre central, el bastión y las torres que en él se inscriben, el antemural y sus torres y el poblado conforman una estructura descendente vinculada al culmen de la torre nurágica (Lilliu, 1982:64). Este poder iba más allá del estadio tribal configurando verdaderos estados, por lo que expuesta claramente por G. Lilliu la distinción social sorprende que se atribuyan las fortificaciones sobre todo a defensa contra grupos invasores (Lilliu, 1982:64) . Como hemos visto, y aun con los problemas cronológicos referidos para el inicio de los nuraghi complejos, se debió dar un sistema articulado entre éstos (o los poblados "abiertos" de la zona de Dorgali) y las torres de llanura o altura. De hecho, frente a lo que señala R. M. Bonzani (1992:214), no todos los nuraghi simples de llanura se asocian a estructuras de hábitat y no todas estas son de la misma entidad ni tienen el mismo control visual.

Otros autores como R. Zucca refiriendo las destrucciones de nuraghi, entre ellos Su Nuraxi (Barumini, Cagliari), ha señalado que los estudiosos admiten que tales destrucciones deban imputarse preferiblemente a los procesos de expansión de algunos potentados indígenas a expensas de otros, no exluyéndose en tales empresas, aun así, la influencia político-militar (o la alianza) de las primeras comunidades urbanas fenicias (Carales, Nora, Bithia, Sulci, Othoca, Tharros, Bosa) establecidas sobre las costas suroccidentales sardas entre el 750 y el 650 a.C. (Lilliu y Zucca, 1988:100).

EL "NURAGHE" COMO FORTALEZA Y LAS DEFENSAS DE LOS YACIMIENTOS NURÁGICOS

La función intimidatoria de los nuraghi está clara en el caso de las saeteras de S. Antine (Torralba, Sassari) o Palmavera (Alguero, Sassari) (Moravetti, 1988) o Su Nuraxi (Barumini, Cagliari) incluso en las torres del antemuro (Contu, 1985:43), la posición en alto de los accesos a la torre central o al corredor que conduce a la terraza (Contu, 1985:15-17, 78) y los pozos de que constan muchos de los nuraghi al interior de los bastiones, en el patio frecuentemente, (Lilliu, 1982:69, 70; Lilliu y Zucca, 1988:39; Moravetti, 1988b:46, 51-52; 1992a:26, 58; Contu, 1985:36, 38, 78, 1990:48, 51, 1995b:108; Badas, 1995:163; Lo Schiavo y Sanges, 1994:36) e incluso al interior de la cámara central como en Is Paras (Isili, Nuoro) (Lo Schiavo y Sanges, 1994:11), Cabu Abbas (Olbia, Sassari) y Flumenelongu (Alghero, Sassari) (Contu, 1985:20, 1990:39). En el caso del nuraghe Arrubiu (Orroli, Nuoro) se ha demostrado el carácter de cisterna de la estructura con una canalización para la recogida del agua (Lo Schiavo y Sanges, 1994:42). En Su Nuraxi (Barumini, Cagliari) la importancia de estas estructuras se revela en la construcción en la Fase C de un segundo pozo al interior de la torre E, cuando con la reestructuración se habían sellado sus saeteras (Lilliu y Zucca, 1988:95-96). En el nuraghe Losa (Abbasanta, Oristano) una posible cisterna o silo al exterior del bastión pero dentro del recinto del antemuro más amplio es defendido por dos torres conectadas entre sí con un muro (Contu, 1985:41, 1990:57) que forma parte de un antemuro parcialmente destruido que conectaría con otra torre situada frente a la entrada principal del bastión. El antemuro interior y las torres con su disposición frente a las zonas de los accesos y obligando a giros para alcanzar las partes más internas del complejo nurágico.

Estos pozos se pueden relacionar en lo que respecta a la Península Ibérica no sólo con estructuras destinadas a resistir un asedio, en forma de cisternas en Fuente Álamo (Cuevas del Almanzora, Almería) (Schubart y Arteaga, 1986; Schubart et al., 1985) y posiblemente Peñalosa (Baños de la Encina, Jaén) (Contreras y Cámara, 2000b), galerías en Gatas (Turre, Almería) (Chapman et al., 1987), o inclusión de arroyos y manantiales entre las diversas partes de un poblado como en el Cerro de las Casas (Vilches, Jaén) (Pérez et al., 1992), sino que también se han documentado verdaderos pozos al interior de las Motillas presentes en La Mancha como la del Azuer (Daimiel, Ciudad Real) (Molina y Nájera, 1987) que recientemente han sido interpretadas en función de las necesidades no sólo de las personas sino de los rebaños (Martínez y Afonso, 1998), aunque no debe ser el caso de aquellos pozos situados en las zonas más fortificadas, excepción hecha de los animales (caballos) usados en la defensa. La defensa de los pozos puede ser también bastante antigua en Cerdeña si se confirman los resultados de Su Coddu (Selargius) donde el posible pozo más antiguo, de fase Ozieri, se sitúa al centro del asentamiento, en el que todavía no se conocen trazas de cierre, si bien los autores prefieren una interpretación sacra de este hecho en función del hallazgo de un cráneo de bóvido en la estructura vecina (Ugas et al., 1989:262).

También existen los "santuarios" de pozo, de los que los más clásicos son los de Santa Cristina (Paulilatino, Oristano) (Sebis, 1995:139) y S. Vittoria (Serri, Nuoro) (Lo Schiavo, 1995b:187), pero a veces creemos que como en el caso de las aldeas-santuario calcolíticas se ha abusado del término pues se aplicado a poblados que se asocian a una estructura nurágica y donde la ausencia de excavación no ha permitido definir la conexión del momento de construcción de ésta con las ofrendas como en S. Antonio de Siligo (Sassari) (Lo Schiavo, 1990a:27-29). En cualquier caso como hemos intentado destacar en estas páginas una función ritual sacra, conectada con la afirmación de una unidad (Lilliu, 1987:27; Zucca, 1988:8, 15) o unas relaciones de dependencia entre núcleos aristocrático-oligárquicos centrífugos, no excluye una función militar, como parece reflejarse en determinadas estructuras del santuario de S. Vittoria (Serri, Nuoro) (Lo Schiavo, 1995a:187) especialmente las que se adosan al primitivo nuraghe del Bronce Medio (Zucca, 1988:15) como aquélla que presenta saeteras (Zucca, 1988:45-47) y el mismo muro que cierra el espolón (Contu, 1985:103-105), incluyendo en su perímetro el templo de pozo y lo que se ha denominado "cabaña de la guardia" (Zucca, 1988:21), si bien sin duda hay modificaciones espaciales a lo largo del tiempo incluso en lo que es el mismo Bronce Medio (Puddu, 1995:193-197). Más claro aun es el caso del pozo con cubierta a tholos de Cuccuru Nuraxi (Settimo San Pietro, Cagliari) que sitúa su acceso al interior de una de las torres de un nuraghe considerado coetáneo, mientras una poza de ofrendas se sitúa en el patio (Atzeni, 1987:280-281, 283, 289-FIGURE 1) con lo que el acceso a la estructura quedaba especialmente restringido como sucede en otras estructuras (Fadda, 1992:28), si es que aceptamos que se dieran también ofrendas al interior. Similar debe ser la configuración de Nurdole (Oroni, Nuoro) (Lo Schiavo, 1990b:260).

La misma restricción que se da en los accesos a estructuras especiales como el taller de fundición de Gremanu (Fonni, Nuoro) (Fadda, 1997:75) que ilustra cómo no se puede englobar todas las estructuras especiales con el término genérico de templos sin explicar la función concreta que ocupan dentro del esquema general, aspecto que, desgraciadamente con escasos datos, persiguen las denominaciones de Taramelli sobre Santa Vittoria (Serri, Nuoro). Como hemos dicho la producción de determinados elementos por parte de "especialistas" vinculados a los santuarios supone una forma de enmascaramiento de la apropiación de riqueza, una apropiación que aun haciendo intervenir el valor de cambio se inscribiría posiblemente otras veces en una presunta redistribución de favores de raíz marcadamente tributaria.

De la misma forma habría que explicar las estructuras especiales que aparecen en determinados poblados como los templos de megaron en Serra Orrios (Dorgali, Nuoro) (Lo Schiavo, 1980:15; Ferrarese Ceruti, 1980:109; Fadda, 1996a:168; Moravetti, 1998:55-69) o en el mismo santuario de Gremanu (Fonni, Nuoro) (Fadda, 1997:75) que se han querido relacionar con la influencia micénica (Ferrarese Ceruti, 1980d:112), pese a que hay estructuras similares más antiguas como las de Malchittu (Arzachena, Sassari) (Ferrarese Ceruti, 1980:112; L. Usai, 1992:13, 51-53; Bafico y Rossi, 1993:47).

EL CARÁCTER DE LA SOCIEDAD NURÁGICA Y EL CONTROL TERRITORIAL MILITARIZADO

D. H. Trump ha indicado que los primeros nuraghi no tienen por qué interpretarse como fortalezas militares (Trump, 1992:198), sino que en el caso de los protonuraghi habría que hablar de plataformas elevadas para el hábitat, cuyo interior serviría como almacén, dado por ejemplo que Bonu Ighinu (Mara, Sassari), no se sitúa en el punto mejor defendido del área. Habría que preguntarse por qué razón se situaban entonces las casas en posición elevada. En el caso de las torres aun señalando que determinados rasgos como la preeminencia geográfica, la única entrada, la intervisibilidad y los parapetos de la terraza superior sugieren un uso militar, D.H. Trump prefiere señalar que son monumentos para ser admirados de lejos (Trump, 1992:198-199). Esto, sin embargo, no excluye, naturalmente, la función militar, en el sentido de que los castillos también eran la expresión monumental de la clase en el poder, y que, además los nuraghi, y otras construcciones de la Edad del Bronce europea, suponen la sustitución de otra forma de delimitación y control del territorio que había acompañado el desarrollo del nacimiento de las sociedades estatales (los megalitos y otras estructuras coetáneas) (Cámara, 1998a).

El problema es que el autor, como muchos otros (Trump, 1992; Bonzani, 1992; Bernabò Brea, 1994; Mathers y Stoddart, 1994:16; Barfield, 1994:137, 141; Navarra, 1998:307, 334), situado en un marco teórico evolucionista es incapaz de admitir la dualidad básica comunidad-estado y prefiere situar términos intermedios como "jefatura", tan oportunamente criticados como construcciones irreales cuyas presuntas características se pueden aplicar a un gran número de estados (de tipos y cronologías diversas) (Nocete, 1984, 1994; Lull y Risch, 1995; Cámara, 1998a).

Otro ejemplo nos puede mostrar como la aceptación acrítica de ciertos modelos antropológicos discutibles conduce, con una teórica iluminación, la interpretación de la organización social en la Prehistoria Reciente, bajo el presupuesto de que las sociedades occidentales, al carecer de escritura, no podrían haber alcanzado el nivel social al que llegaron las sociedades del Mediterráneo Oriental. L. Navarra (1998) ha intentado ver cómo la evolución del poblamiento sardo en la Prehistoria se adecúa al modelo evolucionista propuesto por R.L. Carneiro, basado realmente en un determinismo demográfico-ambiental en lo que respecta a la exposición de las causas que condujeron a la explotación del hombre por el hombre, y descriptivo en lo que respecta a las características del estado que se define en función del poder de enrolar, someter a tributo y legislar (Navarra, 1998:309). El problema es el considerar si una clase que disponen de clientes y séquito ha enrolado personas, si los regalos a los jefes, enmascarados o no en la ficción comunitaria, son considerados tributo, y si la legislación debe ser escrita o no, puesto que en la práctica todos esos rasgos son compartidos por casi todas las sociedades que se incluyen bajo su concepto "jefatura". Por otro lado la estratificación social está siempre unida indisolublemente al estado y la presentación de la articulación social compleja que presenta R.L. Carneiro dentro de sus jefaturas en tres clases (Navarra, 1998:313) lo que nos revela es la articulación de diversos modos de producción dentro de los estados iniciales.

En otro orden de cosas es indudable que el estado implica coerción (Navarra, 1998:310) pero ésta no siempre está relacionada con un proceso de conquista (Navarra, 1998:312) ya que la coerción no es únicamente el resultado de la guerra ya que es la acumulación desigual uno de los condicionantes que dan lugar al desarrollo de las clases, junto a la explotación previa de las mujeres y de hombres procedentes del exterior, en este caso sí en relación a la rapiña, o identificados ideológicamente con éste (Cámara y Afonso, en prensa; Cámara, 1998a). Por otra parte frente a R.L. Carneiro (Navarra, 1998:312) hemos de decir que si el énfasis se coloca en la oposición exterior, el desarrollo del estado tendrá lugar más por la vía esclavista que por la tributaria.

Por el contrario nosotros hemos preferido abordar la evolución en Europa occidental, no desde modelos universales, sino desde la cadena de tipos de "formaciones sociales" que darán origen al mundo capitalista, lo que no supone una sucesión de modos de producción, dado que el "tributario" será el dominante en toda la Prehistoria Reciente (y de nuevo en la Edad Media) aun cuando en la Edad del Bronce la importancia del esclavismo y de una servidumbre estricta sea más importante (Contreras et al., 1995; Cámara et al., 1996; Cámara, 1998a).

Es en este marco teórico donde apreciamos que los nuraghi, y otras fortificaciones, como hemos referido, no son tanto una defensa externa como un medio de control interno y de exhibición del poder, y, en este contexto, las fiestas que propone el autor (Trump, 1992:199), tendrían cabida como expresión del tributo y de la ideología de emulación, como se ha destacado para el caso del Cerro de la Encina (Cámara, 1998a; Martínez y Afonso, 1998). De hecho se debe destacar que la presunta "redistribución" (Mathers y Stoddart, 1994:15) es sólo una forma de restituir una parte de lo sustraído, de canalizar el tributo de forma centralizada , y, por tanto, la élite es siempre explotadora.

Teniendo en cuenta que entre los "bienes" a controlar en los primeros momentos también estarían los mismos hombres debemos señalar que el proceso simplemente muestra la agudización de las contradicciones.

En Cerdeña el sistema de la Edad del Bronce parecería mostrar ya una aristocracia gentilicia que hace pasar por ella cualquier vinculación con la comunidad en forma de servidumbre y clientela (Cámara, 1998b). De ahí la proliferación de pequeños pero ya verdaderos nuraghi desde el Bronce Medio (Lilliu, 1982:32; Moravetti, 1990:67-71, 1988:50, 56) no asociados a conjuntos residenciales amplios y también el modelo de poblados con estructuras relativamente aisladas entre sí aunque formando unidades en torno a un patio central , más complejas desde el siglo VIII (Contu, 1985:86, 102-103) .

Así, con respecto a las diferencias que hemos marcado respecto a la Península Ibérica en relación a la escasez de grandes núcleos de poblamiento en las primeras fases de la Edad del Bronce hay que decir que en los últimos años se han indicado construcciones contemporáneas de la torre central y el bastión desde el siglo XIV a.C como en Arrubiu (Orroli, Nuoro) (Lo Schiavo, 1995b:195; Lo Schiavo y Sanges, 1994:43, 55) y tal vez en otros casos como Losa (Abbasanta) y Madonna del Remedio (Oristano) (Santoni, 1995:172) y se ha sugerido una planificación conjunta de todo el nuraghe añadiéndose el bastión tras un corto periodo de tiempo (L. Usai, 1992:14; Ferrarese Ceruti y Lo Schiavo, 1992:124) como en Belveghile (Olbia, Sassari), realmente un protonuraghe (Sanciu, 1990:257; L. Usai, 1992:81), si bien otros nuraghi parecen tener una evolución lenta como Is Paras (Isili, Nuoro) (Moravetti, 1985e:28; Melis, 1995:180) o media como Genna Maria (Villanovaforru, Cagliari) (Badas et al., 1988:189), pero es que además existen casos como el Nurague Nolza (Meana Sardo, Nuoro) donde el bastión cuatrilobulado se sitúa sobre estructuras precedentes (Cossu y Perra, 1998:97). En cualquier caso la ordenación del territorio parece, al menos en Dorgali, haberse establecido ya desde el Bronce Medio.

Sin embargo, aun admitiendo esta cautela ante las fechas más antiguas propuestas que, como hemos visto, nos parecen aceptables e incluso cortas por lo que respecta a la construcción de las estructuras, que no tal vez a los primeros estratos de desecho, pervive un tercer problema que es el de la correlación de las diversas estructuras que debe hacerse en base a la estratigrafía continua, con lo que la limpieza del exterior de los muros en las excavaciones antiguas se revela una técnica totalmente errónea, y de los materiales en su conjunto.

A la espera de nuevos datos, y de la publicación de los perfiles estratigráficos y sus correlaciones, parece así más seguro retener la cautelosa afirmación de R. Zucca que señala que Aun así no negándose la existencia de complejos fortificados nurágicos de varias torres desde el Bronce Medio y Reciente es preferible admitir que los nuraghi de adición concéntrica se desarrollan entre el Bronce Reciente y el Bronce Final, probablemente en relación a una mejor organización del control territorial de la isla y eventualmente en relación a peligros externos crecientes (Lilliu y Zucca, 1988:85), simbolizando además el mayor control sobre los hombres de la misma forma que se ejerce un mayor control sobre el territorio (Santoni, 1980:174) .

A. Moravetti ha hablado de "revolución" (Moravetti, 1992b:195) porque los verdaderos nuraghi tienden a ocupar totalmente el territorio (Santoni, 1980:156; Moravetti, 1992b:192), pero tal concepción debería afirmar que todos los nuraghi de falsa cúpula se construyen al mismo tiempo y además supondría primar el desenmascaramiento de la sociedad aristocrática que culmina, verdaderamente, en el Mediterráneo Occidental hacia el 1800 cal A.C., sobre la génesis de ésta, al menos precediendo al Campaniforme (Shennan, 1982) y dependiente del desarrollo de los mecanismos de control de los estados anteriores y de sus intentos de expansión (Cámara, 1998a) hasta el punto que los especialistas que se han referido para determinados fortines de las Campiñas del Alto Guadalquivir y los dirigentes de los poblados dependientes son el germen de una nobleza periférica y centrífuga (Nocete et al., 1986; Nocete, 1994), que remarcó de una nueva forma la propiedad del territorio como el mismo A. Moravetti ha destacado (Moravetti, 1992b:192, 1993:100) .

Sin embargo, aunque no se diese el desarrollo de nuraghi complejos a partir de las torres aisladas precedentes en momentos tempranos correspondientes al Bronce Medio se podría pensar en un proceso bastante más similar al de la Península Ibérica, a partir de la dispersión de poblados (con o sin nuraghe) desde el Bronce Medio en determinadas zonas como el Dorgalese. Contaríamos así en la misma Cerdeña con modelos de asentamiento diferentes y, al menos en algunas zonas, complementarios, desde la delimitación del territorio a base de fortines a la presencia de verdaderas colonias. Aun así la continuidad con el Bronce Final, la proliferación de torres aisladas y la concentración en poblados de relativa entidad frente a otros pequeños mantendría los mayores paralelismos con el Occidente de Andalucía (Cámara, 1998b), debiéndose destacar de nuevo que, en ningún caso, se puede plantear un mayor nomadismo en las fases antiguas (A. Usai, 1988:62), simplemente no se habrían delimitado de forma estricta determinadas zonas del territorio, antes sacralizadas muy a menudo con dólmenes, menhires y domus de janas.

En este contexto de aprovechamiento integral del territorio la valoración de la implantación de los nuraghi en relación al tipo de tierras no debe hacerse solamente en función de las cualidades de estas para la agricultura (Moravetti, 1990:69-70), dado, además, que la existencia de una condición natural de la producción no implica su utilización por una sociedad. El problema es que cuando se valora la ganadería tiende a pensarse en una dieta únicamente cárnica y en comunidades trashumantes, extremos ambos que hay que rechazar.

El sistema debió implicar durante mucho tiempo una trashumancia a pequeña escala dentro de un contexto agropecuario sedentario (Webster y Michels, 1986:228), hasta el punto de que en algunos casos las típicas tumbas de gigante de la Edad del Bronce están ausentes o son escasas y se siguen utilizando las domus de janas, modificadas o no (Moravetti, 1992a:29-34). De hecho una especialización pastoril no tiene sentido si no es en contextos donde el intercambio haya adquirido un importante papel en la reproducción social (Hole, 1968; Lewthwaite, 1984:256-257).

Pero, frente a lo señalado por distintos autores y, como ya hemos sugerido, la ganadería tiene especial relevancia en lo que respecta a la acumulación de riqueza (Cámara, 1998a), sin descartar que en determinadas zonas el metal también fuera un objetivo prioritario que condujera al control de las vías de penetración internas (Usai, 1991:123). No existe en nuestra opinión ningún contexto mediterráneo, pasado o presente, en el que la reproducción de los rebaños fuera difícil por motivos ecológicos (Lewthwaite, 1984:259). En cualquier caso los límites a esa reproducción de la propiedad familiar vendrían dados por una distribución global y desigual de la tierra, y hemos planteado, en otro lugar (Cámara y Afonso, en prensa) que la división de la tierra efectiva (que no su propiedad eminente que es anterior) (Cámara, 1998a) sucede (y no precede) a la de los rebaños, tal y como se puede apreciar en determinados modelos etnográficos (Lindström, 1988; Jacobson-Widding, 1988). Por contra el modelo que presenta la incapacidad de autorreproducción a largo plazo del sistema de pequeñas propiedades (Lewthwaite, 1984.259) sin interrogarse sobre el por qué son éstas tan pequeñas lo que tiende es a justificar el desarrollo del latifundismo y, por tanto, una forma específica de la explotación del hombre por el hombre.

NOTAS

(1) Este trabajo ha sido realizado gracias a la Beca Posdoctoral concedida a Juan Antonio Cámara Serrano por la Universidad de Granada para realizar un trabajo de investigación bajo la dirección del profesor Alberto Moravetti del Dipartamento di Storia della Università degli Studi di Sassari entre septiembre de 1997 y agosto de 1998, y a la Beca concedida a Liliana Spanedda por los Ministerios de Asuntos Exteriores italiano y español para realizar un trabajo de investigación en el Departamento de Prehistoria y Arqueología de la Universidad de Granada entre enero y mayo del 2001, bajo la dirección de la Profesora Trinidad Nájera Colino. En la realización de las figuras queremos agradecer la ayuda prestada por Rafael Turatti Guerrero.

(2)Es interesante que de Serra Orrios y de Sa Paule Dorrisolo se haya destacado la abundante producción metalúrgica (Manunza, 1995:115. 118).

(3)Si bien como otros autores enfatiza el tráfico metalúrgico por estas vías en base a los hallazgos de Serra Orrios (Manunza, 1995:106).

(4)Incluso en las primeras fases como muestra Sa Turrìcula (Muros, Sassari) adscrito al Nurágico I con materiales Bonnanaro (Lilliu, 1982:24).

(5)Esta evolución ya había sido propuesta por V. Santoni (1980:141, 156, 158, 164), si bien hemos de rechazar aquí uno de los argumentos utilizados por este autor para definir la antigüedad de algunas de estas estructuras, ya que la asociación a dólmenes y galerías cubiertas (Santoni, 1980:150, 156, 158, 170) no implica una contemporaneidad sino la sustitución de una forma de definir el territorio por otra (Cámara et al., 1996), siendo este el contexto en el que se pueden explicar las diferentes estructuras de Ortachis (Bolotana, Nuoro) (Santoni, 1980:170) o la superposición en Sant'Efis (Orune, Nuoro) (Fadda, 1996b:173).

(6)Algunos de estos materiales se interpretan hoy como resultado de una reocupación de los siglos VIII-VII a. C.. mientras que la construcción del nuraghe debió tener lugar antes del 1600 A.C. (1300 a.c.) (L. Usai, 1992:50).

(7)El que prevalezca la masa sobre los espacios y la articulación interna de los ambientes hacen adscribir Albucciu a la categoría de los nuraghi "de corredor", pero las cubiertas tronco-ojivales, el corte trapezoidal de las puertas y la adición de las torres laterales lo resitúan en experiencias consolidadas de los nuraghi de tholos (Antona, 1995:68).

(8)El mismo E. Contu ha señalado que De cualquier forma no se cuenta con suficientes elementos para afirmar que (...) [los pseudonuraghi] deban considerarse sólo fruto de una mayor antigüedad (respecto a los nuraghi de tholos) y no también una consecuencia de situaciones y objetivos particulares (Contu, 1990:63 y 65), aunque parece atestiguarse esa mayor antigüedad por la forma de la construcción, con poco espacio útil respecto al volumen construido, y determinadas asociaciones materiales en monumentos como Sa Corona (Villagreca, Cagliari) y el recinto de Monte Baranta (Olmedo, Sassari) (Contu, 1990:71).

(9) Independientemente de las raíces en las fortificaciones calcolíticas reveladas por las asociaciones materiales de Sa Corona (Villagreca, Cagliari) y Bruncu Màdugui (Gésturi, Cagliari) (Lilliu, 1988:176-178).

(10)Se trata así sólo del cambio de una manifestación no verdaderamente trascendente.

(11)Le la cronología ha tendido a envejecerse, incluso por parte de Lilliu (1987:15), cuando no a ponerse en duda que algunos nuraghi no fueran ya en su origen complejos (Lo Schiavo, 1995b:195; Lo Schiavo y Sanges, 1994:43, 55; Santoni, 1995:172).

(12)Un ejército en marcha tampoco se podría permitir sitiar todas y cada una de las torres.

(13) Realmente el desenmascaramiento, que no el surgimiento como ya hemos discutido (Cámara, 1998a), de la propiedad privada está aún en sus inicios.

(14)Y las dotes nupciales son una forma de acceder a éstos así como una forma de impedir (cuando son sólo mobiliarias) o favorecer la concentración de la propiedad.

(15)Objeciones que se pueden hacer extensivas a los modelos ofrecidos para el sur de la Península Italiana (Peroni, 1993:120), presentados pese a que la evolución y las diferencias de tamaño entre los asentamientos muestran que no todos tuvieron la misma función (Peroni, 1993:107).

(16)Se trata de una simbología que se ha destacado también en la Península Ibérica (Schubart y Arteaga, 1986; Chapman, 1991; Contreras et al., 1995; Cámara, 1998a), aunque en el caso sardo al no incluirse las sepulturas la exhibición del poder adquiría, en el contexto del poblado, un carácter menos sacro, quizás más agudo si se confirmara que pudieron haber existido precedentes funerarios de la arquitectura nurágica en la misma isla (Lilliu, 1982:19).

(17)Aunque tal hipótesis se inscribe en un contexto claramente nacionalista (Lilliu, 1982:9). Por el contrario R. Zucca refiriendo las destrucciones de nuraghi, entre ellos Su Nuraxi (Barumini, Cagliari), ha señalado que los estudiosos admiten que tales destrucciones deban imputarse preferiblemente a los procesos de expansión de algunos potentados indígenas a expensas de otros, no exluyéndose en tales empresas, aun así, la influencia político-militar (o la alianza) de las primeras comunidades urbanas fenicias (Carales, Nora, Bithia, Sulci, Othoca, Tharros, Bosa) establecidas sobre las costas suroccidentales sardas entre el 750 y el 650 a.C.

(Lilliu y Zucca, 1988:100).

(18)Si bien la torre B donde se refieren éstas está atribuida sólo al siglo IX (Moravetti, 1992a:120-121).

(19)E. Contu (1985:78) las ha considerado más bien ventanucos para destruir las torres de asedio en base al hecho de que la parte más estrecha estaba hacia el exterior. Sin embargo si bien este hecho reducía la visibilidad impedía también que el arquero del interno fuera presa fácil de las flechas exteriores, como, frente a las troneras, puede apreciarse en cualquier castillo medieval.

(20)Es el proceso que siguen todos los sistemas impositivos.

(21)En nuestra opinión el conjunto es la vivienda.

(22) Esta disociación no puede llevar a negar el carácter de "ciudad" para muchas de las agrupaciones mayores, que debe establecerse en función de su papel en el orden político (Cámara et al., en preparac.), y mucho menos se puede negar la jerarquización como hacen algunos autores (p. ej. Contu, 1985:103).

(23)ampoco son grandes los asentamientos destinados al control del territorio fuera de los valles centrales en la Península Ibérica, especialmente si dejamos fuera el caso argárico (Cámara, 1998a).

(24) Este nuraghe muestra la complejidad del sistema defensivo no sólo en la estructura del bastión sino en los añadidos antemurales y las divisiones que forman de los espacios intermedios. De hecho aparte de las zonas en que queda dividido el espacio interno del primer antemuro (en realidad una estructura doble) con siete torres, se produce también una segunda adición con cuatro o cinco torres (Lo Schiavo y Sanges, 1994:23).

(25)El problema es que se sigue hablando de propiedad comunitaria de la tierra (Santoni, 1980:180), sin señalar qué partes podían mantenerse aún en común y sin destacar la falacia de la propiedad en el poder (ver críticas en Cámara, 1998a).

(26) Esta mayor amplitud planimétrica de los protonuraghi, más que a una arquitectura todavía incierta y la búsqueda de un canon arquitectónico - es más fácil, ciertamente, construir en extensión que en altura - se justifica sobre todo con la utilidad como habitación que la terraza parece tener en los protonuraghi: la necesidad, por tanto, de un espacio mayor que de hecho no era indispensable para la terraza de los nuraghi cuya función era exclusivamente de vigilancia y defensa (Moravetti, 1992b:195).

(27)En este contexto creemos que se ha exagerado la oposición del mundo nurágico a un mundo "urbano" presuntamente homogéneo en Oriente (Santoni, 1980:174, 178), cuando realmente ni la civilización micénica ni la estructura de oikos, están caracterizadas por ciudades en el mismo sentido que la civilización clásica grecorromana y, particularmente, en pocas zonas merecerían los agregados de la Edad del Hierro griegos el calificativo de ciudades aun en el mismo sentido formal (disposición urbanística) o de tamaño. La definición de la ciudad como residencia principal de la clase dominante (Benevolo, 1977:6, 18) más allá de la especialización o no en determinadas funciones, como su símbolo, debe tener en cuenta las diferentes formas de extraer el excedente de ésta (Ste. Croix, 1988) y, por tanto, atender a la jerarquización entre los asentamientos, más que a su forma, aunque teniendo también en cuenta elementos que hablan de planificación y exhibición (murallas complejas, lugares de culto restringido, grandes necrópolis, etc.) (Pérez y Cámara, 1999).

BIBLIOGRAFÍA

ALBA, E.,

1998. The distribution of Nuraghi in the Nurra in relation to the geomorphologic aspects of the territory, Papers from the EAA Third Annual Meeting at Ravenna 1997. Volume III: Sardinia (A. Moravetti, M. Pearce, M. Tosi, Eds.), British Archaeological Reports. International Series 719, Oxford, 1998, pp. 72-83.

ANTONA, A.,

1995. Nuraghe Albucciu (Arzachena, Sassari), Sardegna (A. Moravetti, C. Tozzi, Cur.), Guide Archeologiche. Preistoria e Protostoria in Italia 2. XIII Congresso Internazionale dell Scienze Preistoriche e Protostoriche, U.I.S.P.P./Ministero per i Beni Culturali e Ambientali, A.B.A.C.O. Edizioni, Forlì, 1995, pp. 66-73.

ATZENI, E., 1987. Il tempio a pozzo di Cuccuru Nuraxi. Settimo San Pietro - Cagliari (Nota preliminare), La Sardegna nel Mediterraneo tra il secondo e il Primo Millennio a.c. Atti del II Convegno di studi "Un millennio di relazioni fra la Sardegna e i Paesi del Mediterraneo" (Selargius-Cagliari 27-30 novembre 1986), (G. Lilliu, G. Ugas y G. Loi, Cur.), Credito Industriale Sardo, Cagliari, 1987, pp. 279-297.

BADAS, U. ,

1995. Nuraghe Genna Maria (Villanovaforru, Cagliari), Sardegna (A. Moravetti, C. Tozzi, Cur.), Guide Archeologiche. Preistoria e Protostoria in Italia 2. XIII Congresso Internazionale dell Scienze Preistoriche e Protostoriche, U.I.S.P.P./Ministero per i Beni Culturali e Ambientali, A.B.A.C.O. Edizioni, Forlì, 1995, pp. 162-169.

BADAS, U., ATZENI, E., COMELLA, A., LILLIU, C.,

1988. Villanovaforru, L'Antiquarium Arborense e i civici musei archeologici della Sardegna (G. Lilliu, Cur.), Banco di Sardegna, Sassari, 1988, pp. 181-198.

BAFICO, S., ROSSI, G.,

1993. La società dell'Età del bronzo, Sardegna. Civiltà di una isola mediterranea. Genova, Palazzo Ducale, Logia degli Abati (19 dicembre 1993 - 20 febbraio 1994), (G. Rossi, Cur.), Nuova Alfa Editoriale, Bologna, 1993, pp. 47-48.

BARFIELD, L.H.,

1994. The Bronze Age of Northern Italy: Recent Work and Social Interpretation, Development and Decline in the Mediterranean Bronze Age (C. Mathers, S. Stoddart, Eds.), Sheffield Archaeological Monographs 8, J.R. Collis Publications, Sheffield, 1994, pp. 129-144.

BARTOLINI, P.,

1993. Il complesso nuragico di Tzirimagus, Sardegna Antica. Culture Mediterranee 4, Nuoro, 1993, pp. 28-30.

BENÉVOLO, L.,

1977. El arte y la ciudad antigua. (Diseño de la ciudad 2), Gustavo Gili, Barcelona, 1977.

BERNABÒ BREA, Mª. (1994): Le terremare, queste sconosciute: alla ricerca del tempo perduto, "... Le terremare si scavano per concimare i prati..." La nascita dell'archeologia preistoria a Parma nel dibattito culturale della seconda metà dell'Ottocento. Catalogo della Mostra (Parma, Museo Archeologico Nazionale 12 maggio - 30 novembre 1994), (M. Bernabò Brea, A. Mutti, Cur.), Ministero per I Beni Culturali e Ambientali. Soprintendenza Archeologica per l'Emilia-Romagna/Museo Archeologico Nazionale di Parma/ Biblioteca Palatina di Parma/Soprintendenza Speciale al Museo Nazionale Preistorico Etnografico L. Pigorini di Roma/Archivio di Stato di Parma, Silva Editore, Parma, 1994, pp. 150-166.

BONZANI, R.M.,

1992. Territorial boundaries, buffer zones and sociopolitical complexity: a case study of the Nuraghi on Sardinia, Sardinia in the Mediterranean: a footprint in the sea. Studies in Sardinian Archeology presented to Miriam S. Balmuth, (R.H. Tykot, T.K. Andrews, Eds.), Monographs in Mediterranean Archaeology 3, Sheffield Academic Press, Sheffield, 1992, pp. 210-220.

CÁMARA, J.A.,

1998a. Bases metodológicas para el estudio del ritual funerario utilizado durante la Prehistoria Reciente en el sur de la Península Ibérica, Tesis Doctoral Microfilmada, Universidad de Granada, 1998.

CÁMARA, J.A.,

1998b. El control del territorio en la Edad del Bronce. Una comparación entre las situaciones sarda y andaluza, Papers from the EAA Third Annual Meeting at Ravenna 1997. Volume III: Sardi-

nia (A. Moravetti, M. Pearce, M. Tosi, Eds.), British Archaeological Reports. International Series 719, Oxford, 1998.
CÁMARA, J.A.,
2001. El ritual funerario en la Prehistoria Reciente en el sur de la Península Ibérica, British Archaeological Reports. International Series 913, Oxford, 2001.
CÁMARA, J.A., AFONSO, J.A. ,
en prensa. Una propuesta sobre el desarrollo de la desigualdad y las clases sociales en la Prehistoria Reciente de Andalucía, 1er Congreso Iberoamericano de Arqueología Social (La Rábida, Junio, 1996). (En prensa).
CÁMARA, J.A., CASADO, P.J., SÁNCHEZ, R., SUTIL, M.,
1995. El Cerro de la Horca: una aproximación a la sociedad del piedemonte subbético y zonas aledañas en la primera mitad del II Milenio, Sumuntán. Revista de estudios sobre Sierra Mágina V, Jódar, Carchelejo, 1995, pp. 205-212.
CÁMARA, J.A., CONTRERAS, F., PÉREZ, C., LIZCANO, R.,
1996. Enterramientos y diferenciación social II. La problemática del Alto Guadalquivir durante la Edad del Bronce, Trabajos de Prehistoria 53:1, Madrid, 1996, pp. 91-108.
CAMPUS, F., LEONELLI, V.,
2000. La tipologia della ceramica nuragica. Il materiale edito, Soprintendenza Archeologica per le province di Sassari e Nuoro, BetaGamma Editrice, Sassari, 2000.
CHAPMAN, R.W. ,
991. La formación de las sociedades complejas. El sureste de la Península Ibérica en el marco del Mediterráneo Occidental, Crítica, Barcelona, 1991.
CHAPMAN, R.W., LULL, V., PICAZO, M., SANAHUJA, Mª.E.,
1987. El yacimiento de Gatas. Antecedentes, Proyecto Gatas. Sociedad y economía en el Sudeste de España, c. 2500-800 a.n.e. 1. La prospección arqueoecológica (R.W. Chapman, V. Lull, M. Picazo, Mª.E. Sanahuja, Eds.), British Archaoelogical Reports. International Series 348, Oxford, 1987, pp.107-118.
CONTRERAS, F.,
1984. Clasificación y tipología en Arqueología: el camino hacia la cuantificación, Cuadernos de Prehistoria de la Universidad de Granada 9, Granada, 1984, pp. 327-385.
CONTRERAS, F. (1986): Aplicación de métodos y análisis estadísticos a los complejos cerámicos de la Cuesta del Negro (Purullena, Granada), Tesis Doctoral, Univ. Granada, 1986.
CONTRERAS, F., CÁMARA, J. A.,
2000a. El poblado de la Edad del Bronce de Peñalosa (Baños de la Encina, Jaén). La cerámica, Análisis Histórico de las Comunidades de la Edad del Bronce del piedemonte meridional de Sierra Morena y Depresión Linares-Bailen. Proyecto Peñalosa. (F. Contreras, Coord.), Arqueología. Monografías 10, Consejería de Cultura. Dirección General de Bienes Culturales. Sevilla, 2000, pp. 77-128 incluye 91/2-91/46 y 109/2-109/5 en CD-ROM.
CONTRERAS, F., CÁMARA, J. A.,
2000b. El poblado de la Edad del Bronce de Peñalosa (Baños de la Encina, Jaén). Análisis espacial y estructural del poblado de Peñalosa, Análisis Histórico de las Comunidades de la Edad del Bronce del piedemonte meridional de Sierra Morena y Depresión Linares-Bailen. Proyecto Peñalosa. (F. Contreras, Coord.), Arqueología. Monografías 10, Consejería de Cultura. Dirección General de Bienes Culturales. Sevilla, 2000, pp. 273-286 incluye 274/1-274/304 en CD-ROM.
CONTRERAS, F., CAPEL, J., ESQUIVEL, J.A., MOLINA, F., TORRE, F. de la,
1987-88. Los ajuares cerámicos de la necrópolis argárica de la Cuesta del Negro (Purullena, Granada). Avance al estudio analítico y estadístico, Cuadernos de Prehistoria de la Universidad de Granada 12-13, Granada, 1987-88, pp. 135-156.
CONTRERAS, F., CÁMARA, J.A., MOYA, S., SÁNCHEZ, R.,
1992. Primer avance metodológico del estudio de la cultura material del poblado de Peñalosa (Baños de la Encina, Jaén), Anuario Arqueológico de Andalucía 1990:II, Sevilla, 1992, pp. 281-290.
CONTRERAS, F., CÁMARA, J.A., LIZCANO, R., PÉREZ, C., ROBLEDO, B., TRANCHO, G. (1995): Enterramientos y diferenciación social I. El registro funerario del yacimiento de la Edad del Bronce de Peñalosa (Baños de la Encina, Jaén), Trabajos de Prehistoria 52:1, Madrid, 1995, pp. 87-108.
CONTU, E.,
1959, I più antichi nuraghi e l'explorazione del Nuraghe Peppe Gallu(Uri, Sassari), Rivista di Scienze Preistoriche XIV (1959), Firenze, 1959, pp. 59-121.
CONTU, E. (1985): L'architettura nuragica, Ichnussa. La Sardegna dalle origini all'età classica (E. Atzeni, F. Barreca, Mª.L. Ferrarese Ceruti, E. Contu, G. Lilliu, F. Lo Schiavo, F. Nicosia, E. Equini Schneider), Garzonti/Scheiwiller, Milano, 1985 (2ª Ed.), pp. 3-176.
CONTU, E.,1990. Il nuraghe, La civiltà nuragica (E. Atzeni, F. Barreca, P. Bernardini, E. Contu, Mª.A. Fadda, Mª.L. Ferrarese Ceruti, F. Lo Schiavo, A. Moravetti, M. Sanges, V. Santoni, C. Tronchetti, G. Ugas), Electa, Milano, 1990, pp. 35-99.
CONTU, E.,
1996.:Il nuraghe Santu Antine (Torralba, Sassari), Sardegna (A. Moravetti, C. Tozzi, Cur.), Guide Archeologiche. Preistoria e Protostoria in Italia 2. XIII Congresso Internazionale dell Scienze Preistoriche e Protostoriche, U.I.S.P.P./Ministero per i Beni Culturali e Ambientali, A.B.A.C.O. Edizioni, Forlì, 1995, pp. 106-113.
COSSU, T., PERRA, M.,
1998. Two contexts of the Bronze Age in the Nuraghe Nolza of Meane Sardo (Nuoro), Papers from the EAA Third Annual Meeting at Ravenna 1997. Volume III: Sardinia (A. Moravetti, M. Pearce, M. Tosi, Eds.), British Archaeological Reports. International Series 719, Oxford, 1998, pp. 97-109.
FADDA, Mª.A.,
1990. Il villaggio, La civiltà nuragica (E. Atzeni, F. Barreca, P. Bernardini, E. Contu, Mª.A. Fadda, Mª.L. Ferrarese Ceruti, F. Lo Schiavo, A. Moravetti, M. Sanges, V. Santoni, C. Tronchetti, G. Ugas), Electa, Milano, 1990, pp. 101-119.
FADDA, Mª.A.,
1992. Lo scavo di Su Tempiesu e la sua architettura, Su Tempiesu di Orune. Fonte sacra nuragica (M.A. Fadda, F. Lo Schiavo), Quaderni della Soprintendenza ai Beni Archeologici per le province di Sassari e Nuoro 18, Ozieri, 1992, pp. 17-59.
FADDA, Mª.A.,
1996a. Dorgali (Nuoro). Località Serra Orrios. Villaggio nuragico, Bollettino di Archeologia 19-20-21 (1993), Roma, 1996, pp. 168-169.
FADDA, Mª.A.,
1996. Orune (Nuoro). Località Sant'Efisio. Complesso archeologico di Sant'Efis, Bollettino di Archeologia 19-20-21,
1993. Roma, 1996, pp. 173-174.
FADDA, Mª.A.,
1997. Gli architetti nuragici di Gremanu, Archeologia viva 63,

Maggio-Giugno 1997, pp. 70-75.
FADDA, Mª.A., PRUNETI, P.,
1997. Nuraghe Mannu, Archeologia viva 61, Gennaio-Febbraio 1997, pp. 38-43.
FERRARESE CERUTI, Mª.L.,
1963. Nota preliminare alla I e alla II campagna di scavo nel Nuraghe Albucciu (Arzachena, Sassari), Rivista di Scienze Preistoriche XVII (1962), Firenze, 1963, pp. 161-204.
FERRARESE CERUTI, Mª.L.,
1980. Il villaggio nuragico di Serra Orrios, Dorgali. Documenti Archeologici (AA.VV.), Soprintendenza Archeologica per le province di Sassari e Nuoro/Chiarella, Sassari, 1980, pp. 109-113.
FERRARESE CERUTI, Mª.L., LO SCHIAVO, F.,
1992. La Sardegna, L'Età del Bronzo in Italia nei secoli dal XVI al XIV a.c. (Viareggio 26-30 Ottobre, 1989) (D. Cocchi Gerick, Coord.), Rassegna di Archeologia 10 (1991-92), Firenze, 1992, pp. 123-141.
FERRARI, A., PESSINA, A.,
1996a. La neolitizzazione: un'introduzione, Sammardenchia e i primi agricoltori del Friuli (A. Ferrari, A. Pessina, Cur.), Banca di Credito Cooperativo di Basiliano, Tovagnacco, 1996, pp. 11-18.
FODDAI, L.,
1998. The distribution of Nuraghi in "Logoduro-Meilogu" in relation to geomorphologic aspects of the territory, Papers from the EAA Third Annual Meeting at Ravenna 1997. Volume III: Sardinia (A. Moravetti, M. Pearce, M. Tosi, Eds.), British Archaeological Reports. International Series 719, Oxford, 1998, pp. 84-96.
HOLE, F.,
1968. Evidence of social organisation from Western Iran, 8000-4000 B.C. New perspectives in Archaeology. (S.R. Binford y L.R. Binford, Eds.), Chicago, 1968, pp. 245-265.
JACOBSON-WIDDING, A.,
1988. Death rituals as inversions of Life Structures. A comparison os Swedish an African Funerals, On the meaning of Death. Essays on Mortuary practices and eschatological beliefs (S. Cederroth, C. Corlin, J. Lindström, Eds.), Uppsala Studies in Cultural Anthropology 8, Uppsala, 1988, pp. 137-154.
LEWTHWAITE, J.M. ,
1984. Pastore, padrone: the social dimensions of pastoralism in prenuragic Sardinia, The Deya Conference of Prehistory. Early settlement in the Western Mediterranean Islands and their Peripheral Areas. Part I (W.H. Waldren, R. Chapman, J. Lewthwaite y R.-C. Kennard, Eds.), British Archaeological Reports. International Series 229:1, Oxford, 1984, pp. 251-268.
LILLIU, G.,
1962. I nuraghi. Torri preistoriche della Sardegna, La Zattera, 1962.
LILLIU, G.,
1982. La civiltà nuragica, Sardegna Archeologica. Studi e Monumenti 1, Carlo Delfino Editore, Sassari, 1982.
LILLIU, G.,
1987. La Sardegna tra il II e il I MIllennio a.c., La Sardegna nel Mediterraneo tra il secondo e il Primo Millennio a.c. Atti del II Convegno di studi "Un millennio di relazioni fra la Sardegna e i Paesi del Mediterraneo" (Selargius-Cagliari 27-30 novembre 1986), (G. Lilliu, G. Ugas y G. Loi, Cur.), Credito Industriale Sardo, Cagliari, 1987, pp. 13-32
LILLIU, G.,
1988. La civiltà dei Sardi dal Paleolitico all'età dei nuraghi, Nuova Eri Edizioni RAI, Torino, 1988 (3ª Ed. rev. y amp.).
LILLIU, G. ,
1995. Preistoria e protostoria del Sulcis, Carbonia e il Sulcis. Archeologia e territorio (V. Santoni, Cur.), Soprintendenza Archeologica per le province di Cagliari e Oristano/Comune di Carbonia, Oristano, 1995, pp. 11-50.
LILLIU, G., ZUCCA, R.,
1988. Su Nuraxi di Barumini, Sardegna Archeologica. Guide e Itinerari 9, Carlo Delfino Editore, Sassari, 1988.
LINDSTRÖM, J.,
1988. The monopolization of a spirit. Livestocks prestations during an Iramba funeral, On the meaning of death. Essays on mortuary practices and eschatological beliefs, (S. Cederroth, C. Coslin, J. Lindström, Eds.), Uppsala Studies in Cultural Anthropology 8, Uppsala, 1988, pp. 169-183.
LIZCANO, R. ,
1995. Las comunidades del Neolítico Final en el Alto Guadalquivir, Tesis Doctoral, Univ. Granada, 1995.
LIZCANO, R., PÉREZ, C., NOCETE, F., CÁMARA, J.A., CONTRERAS, F., CASADO, P.J. MOYA, S. ,
1996. La organización del territorio en el Alto Guadalquivir entre el IV y el III milenios (3300-2800 a.c.), I Congrés del Neolític a la Península Ibérica. Formació e implantació de les comunitats agrícoles (Gavá-Bellaterra, 1995). Actes. Vol. 1. (J. Bosch, M. Molist, Orgs.), Rubricatum 1:1, Gavà, 1996, pp. 305-312.
LIZCANO, R., CÁMARA, J.A., RIQUELME, J.A., CAÑABATE, Mª.L., SÁNCHEZ, A., AFONSO, J.A.
1996. El Polideportivo de Martos. Estrategias económicas y símbolos de cohesión en un asentamiento del Neolítico Final del Alto Guadalquivir, Cuadernos de Prehistoria de la Universidad de Granada 16-17 (1991-92), Granada, 1997, pp. 5-101.
LO SCHIAVO, F.,
1980. Dorgali e il suo teritorio nella Preistoria e Protostoria: osservazioni e problemi, Dorgali. Documenti Archeologici (AA.VV.), Soprintendenza Archeologica per le province di Sassari e Nuoro/Chiarella, Sassari, 1980, pp. 9-16.
LO SCHIAVO, F. (1990a): Il santuario nuragico di S. Antonio di Siligo (Sassari), Nuovo Bullettino Archeologico Sardo 3 (1986), Sassari, 1990, pp. 27-36.
LO SCHIAVO, F.,
1990b. Oroni (Nuoro). Loc. Nurdole, Bollettino di Archeologia 1-2 (1990), Roma, 1990, pp. 260.
LO SCHIAVO, F. ,
1991. Orroli (Nuoro). Nuraghe Arrubiu, Bollettino di Archeologia 4 (1990), Roma, 1991, pp. 117-118.
LO SCHIAVO, F.,
1993. Il nuraghe Arrubiu - Orroli (NU), Sardegna. Civiltà di una isola mediterranea. Genova, Palazzo Ducale, Logia degli Abati (19 dicembre 1993 - 20 febbraio 1994), (G. Rossi, Cur.), Nuova Alfa Editoriale, Bologna, 1993, pp. 53.
LO SCHIAVO, F.,
1995a. Santuario nuragico di S. Vittoria (Serri, Nuoro), Sardegna (A. Moravetti, C. Tozzi, Cur.), Guide Archeologiche. Preistoria e Protostoria in Italia 2. XIII Congresso Internazionale dell Scienze Preistoriche e Protostoriche, U.I.S.P.P./Ministero per i Beni Culturali e Ambientali, A.B.A.C.O. Edizioni, Forlì, 1995, pp. 186-193.
LO SCHIAVO, F.,
1995b. Nurague Arrubiu (Orroli, Nuoro), Sardegna (A. Moravetti, C. Tozzi, Cur.), Guide Archeologiche. Preistoria e Protostoria in

Italia 2. XIII Congresso Internazionale dell Scienze Preistoriche e Protostoriche, U.I.S.P.P./Ministero per i Beni Culturali e Ambientali, A.B.A.C.O. Edizioni, Forlì, 1995, pp. 194-201.
LO SCHIAVO, F., SANGES, M.,
1994. Il Nuraghe Arrubiu di Orroli, Sardegna Archeologica. Guide e Itinerari 22, Carlo Delfino Editore, Sassari, 1994.
LULL, V., RISCH. R.,
1995. El Estado Argárico, Homenaje a la Dra. Dª Ana Mª Muñoz Amilibia (AA.VV.), Verdolay 7, Murcia, 1995, pp. 97-109.
MALDONADO, Mª.G., MOLINA, F., ALCARAZ, F.M., CÁMARA, J.A., MÉRIDA, V., RUIZ, V.,
1997. El papel social del megalitismo en el Sureste de la Península Ibérica. Las comunidades megalíticas del Pasillo de Tabernas, Cuadernos de Prehistoria de la Universidad de Granada 16-17 (1991-92), Granada, 1997, pp. 167-190.
MANCA DEMURTAS, L., DEMURTAS, S.,
1984a. I protonuraghi... (Nuovi dati per l'Oristanese), The Deya Conference of Prehistory. Early settlement in the Western Mediterranean Islands and their Peripheral Areas. Part II (W.H. Waldren, R. Chapman, J. Lewthwaite y R.-C. Kennard, Eds.), British Archaeological Reports. International Series 229:2, Oxford, 1984, pp. 629-669.
MANCA DEMURTAS, L., DEMURTAS, S.,
1984b. Observaciones sobre los protonuraghes de Cerdeña, Trabajos de Prehistoria 41, Madrid, 1984, pp. 165-204.
MANCA DEMURTAS, L., DEMURTAS, S.,
1987. Di un tipo architettonico mediterraneo (talaiot Rafal Roig Mercadal - Minorca, La Sardegna nel Mediterraneo tra il secondo e il Primo Millennio a.c. Atti del II Convegno di studi "Un millennio di relazioni fra la Sardegna e i Paesi del Mediterraneo" (Selargius-Cagliari 27-30 novembre 1986), (G. Lilliu, G. Ugas y G. Loi, Cur.), Credito Industriale Sardo, Cagliari, 1987, pp. 493-503.
MANCA DEMURTAS, L., DEMURTAS, S.,
1992. Tipologie Nuragiche. I protonuraghi con Corridoio Passante, Sardinia in the Mediterranean: a footprint in the sea. Studies in Sardinian Archeology presented to Miriam S. Balmuth, (R.H. Tykot, T.K. Andrews, Eds.), Monographs in Mediterranean Archaeology 3, Sheffield Academic Press, Sheffield, 1992, pp. 176-184.
MANUNZA, Mª.R.,
1980. Siti archeologici del Dorgalese, Dorgali. Documenti Archeologici (AA.VV.), Soprintendenza Archeologica per le province di Sassari e Nuoro/Chiarella, Sassari, 1980, pp. 179-201.
MANUNZA, Mª.R.,
1985. Il patrimonio archeologico del comune di Dorgali (Nu), 10 anni di attività nel territorio della provincia di Nuoro, Soprintendenza Archeologica per le province di Sassari e Nuoro, Nuoro, 1985, pp. 14-16.
MANUNZA, Mª.R.,
1988. Dorgali, L'Antiquarium Arborense e i civici musei archeologici della Sardegna (G. Lilliu, Cur.), Banco di Sardegna, Sassari, 1988, pp. 147-156.
MANUNZA, Mª.R.,
1995. Dorgali. Monumenti antichi, Soprintendenza Archeologica per le province di Sassari e Nuoro, Oristano, 1995.
MARTÍNEZ, G., AFONSO, J.A.,
1998. Las sociedades prehistóricas: de la Comunidad al Estado, De Ilurco a Pinos Puente. Poblamiento, economía y sociedad de un pueblo de la Vega de Granada (R. Peinado, Ed.), Diputación Provincial de Granada, Granada, 1998, pp. 21-68.
MATHERS, C., STODDART, S.,
1994. Introduction, Development and Decline in the Mediterranean Bronze Age (C. Mathers, S. Stoddart, Eds.), Sheffield Archaeological Monographs 8, J.R. Collis Publications, Sheffield, 1994, pp. 13-20.
MELIS, P.,
1995. Nuraghe Is Paras (Isili, Nuoro), Sardegna (A. Moravetti, C. Tozzi, Cur.), Guide Archeologiche. Preistoria e Protostoria in Italia 2. XIII Congresso Internazionale dell Scienze Preistoriche e Protostoriche, U.I.S.P.P./Ministero per i Beni Culturali e Ambientali, A.B.A.C.O. Edizioni, Forlì, 1995, pp. 178-185.
MOLINA, F., NÁJERA, T.,
1987. Motilla del Azuer, Arqueología en Castilla-La Mancha. Excavaciones 1985 (F. Masa, D. de Paz, A. de Juan, Coords.), Junta de Comunidades de Castilla-La Mancha, Toledo, 1987, pp. 33-34.
MORAVETTI, A.,
1985. Nuraghe Is Paras - Isili, 10 anni di attività nel territorio della provincia di Nuoro, Soprintendenza Archeologica per le province di Sassari e Nuoro, Nuoro, 1985, pp. 28-29.
MORAVETTI, A.,
1988. Architettura del nuraghe S.Antine di Torralba, Il nuraghe S. Antine nel Logudoro-Meilogu, Carlo Delfino Editore, Sassari, 1988, pp. 45-60.
MORAVETTI, A.,
1990. Nota preliminare agli scavi del Nuraghe S. Barbara di Macomer, Nuovo Bullettino Archeologico Sardo 3 (1986), Sassari, 1990, pp. 49-113.
MORAVETTI, A.,
1992a. Il Complesso nuragico di Palmavera, Sardegna Acheologica. Guide e Itinerari 20, Carlo Delfino editore, Sassari, 1992.
MORAVETTI, A.,
1992b. Sui Protonuraghi del Marghine e della Planargia, Sardinia in the Mediterranean. A footprint in the sea. Studies in Sardinian Archaeology Presented to Miriam S. Balmuth, (R.H. Tykot, T.K. Andrews, Eds.), Monographs in Mediterranean Archaeology 3, Sheffield, Academic Press, Sheffield, 1992, pp. 185-197.
MORAVETTI, A.,
1993. Protostoria nel Marghine e nella Planargia, Archeologie et ambiente naturale. Prospettive di cooperazione tra le autonomie locali nel sud dell'Europa, Amministrazione Provinciale di Nuoro, Nuoro, 1993, pp. 91-103.
MORAVETTI, A.,
1998. Serra Orrios e i monumenti archeologici di Dorgali, Sardegna Archeologica. Guide e Itenerari 26, Carlo Delfino Editore, Sassari, 1998.
MORENO, Mª.A.,
1993. El Malagón: un asentamiento de la Edad del Cobre en el Altiplano de Cúllar-Chirivel. Tesis Doctoral. Univ. Granada. 1993.
MORENO, Mª.A., CONTRERAS, F., CÁMARA, J.A.,
1997. Patrones de asentamiento, poblamiento y dinámica cultural. Las tierras altas del sureste peninsular. El pasillo de Cúllar-Chirivel durante la Prehistoria Reciente, Cuadernos de Prehistoria de la Universidad de Granada 16-17 (1991-92), Granada, 1997, PP. 191-245.
NAVARRA, L.,
1998. Chiefdoms nella Sardegna dell'Età nuragica? Un'applicazione della circumscription theory di Robert L. Carnei-

ro, Origini. Preistoria e Protostoria delle civiltà antiche XXI (1997), Roma, 1998, pp. 307-353.
NOCETE, F. ,
1989. El espacio de la coerción. La transición al Estado en las Campiñas del Alto Guadalquivir (España). 3000-1500 A.C., British Archaeological Reports. International Series 492, Oxford, 1989.
NOCETE, F.,
1994. La formación del Estado en Las Campiñas del Alto Guadalquivir (3000-1500 a.n.e.), Monográfica Arte y Arqueología 23, Univ. de Granada, Granada, 1994.
NOCETE, F., RUIZ, A., MOLINOS, M., CASTRO, M.,
1986. Productos, lugares de actividad y estructuras en el asentamiento del Cobre Final del Cerro de La Coronilla (Cazalilla, Jaén). Coloquio sobre el Microespacio 2 (Teruel, 1986). Arqueología Espacial 8. Del Paleolítico al Bronce Medio, Teruel, 1986, pp. 203-218.
PÉREZ, C., CÁMARA, J.A.,
1999. Intervención arqueológica en Marroquíes Bajos (Jaén). Sector Urbanístico RP-4. Parcela G-3, Anuario Arqueológico de Andalucía 1995:III, Sevilla, 1999, pp. 256-270.
PERONI, R.,
1993. La Sibaritide prima di Sibari, Sibari e la Sibaritide. Atti del Trentaduesimo Convegno di Studi sulla Magna Grecia (Taranto-Sibari, 7-12 Ottobre, 1992), (A. Stazio y S. Ceccoli, Cur.), Istituto per la Storia e l'Archeologia della Magna Grecia, Taranto, 1993, pp. 103-136.
PUDDU, Mª.G.,
1995. Serri (Nuoro). Località Santa Vittoria. Santuario federale nuragico. Scavi nella zona occidentale, Bollettino di Archeologia 13-15 (1992), Roma, 1995, pp. 187-197.
SANCIU, A.,
1990. Olbia (Sassari). Nuraghe Belveghile, Bollettino di Archeologia 1-2 (1990), Roma, 1990, pp. 255-257.SANTONI, V. (1980): Il segno del potere, Nur. La misteriosa civiltà dei Sardi (AA.VV.), Cariplo, Milano, 1980, pp. 141-187.
SANTONI, V.,
1995. Il complesso nuragico Su Nuraxi (Barumini, Cagliari), Sardegna (A. Moravetti, C. Tozzi, Cur.), Guide Archeologiche. Preistoria e Protostoria in Italia 2. XIII Congresso Internazionale dell Scienze Preistoriche e Protostoriche, U.I.S.P.P./Ministero per i Beni Culturali e Ambientali, A.B.A.C.O. Edizioni, Forlì, 1995, pp. 170-177
SCHUBART, H., ARTEAGA, O. ,
1986. Fundamentos arqueológicos para el estudio socioeconómico y cultural del área de El Argar, Homenaje a Luis Siret (1934-1984), Consejería de Cultura, Sevilla, 1986, pp. 289-307.
SCHUBART, H., ARTEAGA, O., PINGEL, V.,
1985. Fuente Álamo. Informe preliminar sobre la excavación de 1985 en el poblado de la Edad del Bronce, Ampurias 47, Bar. lona, 1985, pp. 70-107.
SEBIS, S. (1995. Il santuario nuragico di Santa Cristina (Paulilatino, Oristano), Sardegna (A. Moravetti, C. Tozzi, Cur.), Guide Archeologiche. Preistoria e Protostoria in Italia 2. XIII Congresso Internazionale dell Scienze Preistoriche e Protostoriche, U.I.S.P.P./ Ministero per i Beni Culturali e Ambientali, A.B.A.C.O. Edizioni, Forlì, 1995, pp. 138-149.
SHENNAN, S.,
1982. Ideology, change and the European Bronze Age. Symbolic and structural archaeology, (I. Hodder, Ed), New Directions in Archaeology, Cambridge University Press, Cambridge, pp. 155-161.
SPANEDDA, L.,
1994-95. Archeologia del territorio. Emergenze archeologiche dal Paleolitico alla tarda età romana nei Fogli 195 e 208 dell'I.G.M.., Tesi di Laurea, Sassari, 1994-95.
STE. CROIX, G.E.M. de,
1988. La lucha de clases en el Mundo Griego Antiguo. De la Edad Arcaica a las conquistas árabes, Crítica, Barcelona, 1988.
TRUMP, D.,
1996. Militarism in Nuragic Sardinia, Sardinia in the Mediterranean: a footprint in the sea. Studies in Sardinian Archeology presented to Miriam S. Balmuth, (R.H. Tykot, T.K. Andrews, Eds.), Monographs in Mediterranean Archaeology 3, Sheffield Academic Press, Sheffield, 1992, pp. 198-203.
UGAS, G.,
1990, La tomba dei guerrieri di Decimoputzu, Norax. Collana di studi archeologici 1, Edizioni della Torre, Cagliari, 1990.
UGAS, G., USAI, L., NUVOLI, M.P., LAI, L., MARRAS, M.G. ,
1989. Nuovi dati sull'insediamento di Su Coddu - Selargius, La Cultura di Ozieri. Problematiche e nuove acquisizioni. Atti del I Convegno di Studio (Ozieri, gennaio 1986 - aprile 1987), (L. Dettori Campus, Cur.), Il Torchieto, Ozieri, 1989, pp. 239-278.
USAI, A.,
1988. Tomba megalítica in località "Mitza 'e Fidi". Donori (Cagliari), Quaderni della Soprintendenza Archeologica per le province di Cagliari e Oristano 4:I (1987), Cagliari, 1988, pp. 59-66.
USAI, D.,
1991. Modelli d'insediamento nel Sarrabus dal Neolitico all'Età del Bronzo, Quaderni della Sopraintendenza Archeologica per le province di Cagliari e Oristano 7 (1990), Cagliari, 1991, pp. 117-134.
USAI, L.,
1992. Gallura. Dolmen, nuraghi, castelli, Itinerari di Archeologica nella provincia di Sassari, Editrice Democratica Sarda, Sassari, 1992.
VALENTI, F.,
1996. Insediamenti dell'età del bronzo nel territorio dell'antica Leontini, L'antica età del bronzo. Atti del Congresso di Viareggio (9-12 Gennaio 1995), (D. Cocchi Genick, Cur.), Comune di Viareggio/Civico Museo Preistorico e Archeologico "Alberto Carlo Blanc"/Conseil d'Europe, Octavo. Franco Contini Editore, Firenze, 1996, pp. 600-601.
WEBSTER, G.S.,
1991. Monuments, mobilization and Nuragic organization, Antiquity 65, pp. 840-856.
WEBSTER, G.S., MICHELS, J.W.,
1986. Paleoeconomy in West-Central Sardinia, Antiquity 60, pp. 226-229.
ZUCCA, R.,
1988. Il santuario nuragico di S. Vittoria di Serri, Sardegna Archeologica. Guide e Itinerari 7, Carlo Delfino Editore, Sassari, 1988.

33

DECORACIÓN, REPRESENTACIONES FIGURADAS Y ÁREAS RITUALES EN LA PREHISTORIA RECIENTE SARDA: ACUMULACIÓN, CONTROL DEL TERRITORIO Y JERARQUIZACIÓN

JUAN ANTONIO CÁMARA SERRANO y **LILIANA SPANEDDA**

Departamento de Prehistoria y Arqueología de la Universidad de Granada
Granada, Spain
EMAIL: jacamara@ugr.es

RESUMEN

Este trabajo pretende señalar que los denominados ídolos o representaciones escultóricas exentas, las figuras presentes en determinadas sepulturas (pintadas, incisas o en relieve) y los mismos contenedores de estas representaciones tienen una profunda conexión con los rituales destinados a la defensa de la organización social, ya sea como justificación de la propiedad de determinados elementos o del territorio en su conjunto, o de la relación entre hombres y mujeres o de la relación entre los distintos grupos. Esto se manifiesta a partir de la comparación con otros tipos de rituales estudiados en el sur de la Península Ibérica y presentes en gran parte de Europa y que implican la inhumación de animales y personas, a veces en asociación. Por otra parte la dispersión de los monumentos o su concentración junto a determinados poblados se relaciona con el control del territorio y la afirmación de la capitalidad de determinados poblados. Por ello todos estos aspectos sólo pueden incluirse en el marco de una "religiosidad" prehistórica si definimos cuál era el objetivo de esas manifestaciones rituales y como, con el tiempo, cada uno de los elementos implicados pudo convertirse en la representación, más o menos directa, de una divinidad, o de alguno de sus atributos.

RIASSUNTO

L'argomento trattato in questo lavoro riguarda la relazione che gli idoletti, le sepolture e le figure (dipinte, incise o in relievo) presenti in queste ultime, hanno con i rituali utilizzati per la difesa dell'organizzazione sociale, sia come giustificazione della proprietà di certi elementi o del territorio stesso, sia della relazione tra uomini e donne o tra diversi gruppi. Questo viene dimostrato, anche, attraverso la comparazione con altri tipi di rituali studiati nel sud della Penisola Iberica e presenti in gran parte d'Europa, nei quali, a volte, s'incontrano inumazioni di animali e persone . La dispersione delle sepolture o la loro concentrazione presso i villaggi si relaziona con il controllo del territorio e dimostra il ruolo (primario o secondario) che ricoprono gli insediamenti ai quali si associano. Perciò tutti questi aspetti non possono essere inclusi in una "religiosità" preistorica se non si definisce l'obiettivo di queste manifestazioni rituali e, come, con il tempo, ognuno di questi elementi è diventato la rappresentazione, più o meno reale, di una divinità, o di qualche suo attributo.

ABSTRACT

This paper tries to show that the elements which have been considered idols and other sculptoric representations situated over the wall of many tombs, joined to these same tombs, have a deep connection with the ritual phenomenons directed to legitimize the social situation. These figures and monuments act by exhibiting the ownership of several elements and animals or the territory, or by showing and maskint the relation among men and women and among the different groups. This process can also be seen comparing this situation with other rituals, used in the southern Iberian Peninsola and other European areas, which present burials of animals and people, sometimes mixed. On the other way the dispersion of the monuments or their concentration next to the main villages must be relationed to the territorial control and the demarcation of a capital. So, all these aspects can only be included into a prehistoric religiousness if we define what was the aim of these ritual acts and how every used element could be converted in the representation, better or worse, of a god or any of its characteristics.

INTRODUCCIÓN

Los elementos culturales son usados ideológicamente en las manifestaciones rituales, que según P. Scarduelli (1988:32) son sistemas de comunicación repetitivos y estructurados destinado a la reproducción de la sociedad.. Además a los ritos (...) se otorga la tarea de transmitir los conocimientos indispensables para la supervivencia (Scarduelli, 1988:55). El ritual no sólo es utilizado por los dominadores para imponerse sino que, como aspecto ideológico, también es utilizado en la lucha social general (Thomas, 1990) y reinterpretado en diversas formas (Scarduelli, 1988).

La denominación como "arte" de las representaciones figurativas de la Prehistoria debe tener en cuenta que el objetivo principal de ellas no era satisfacer una demanda estética precisa sino que adquieren una más que precisa función en el marco de la sociedad que las originó, y por otra parte no debió existir una separación entre el "artista" y cualquier otro artesano (Carandini, 1984). En cualquier caso estos planteamientos suponen que estas representaciones no se separan mucho de la mayoría de las que hoy consideramos "artísticas", sobre todo aquéllas anteriores al Renacimiento desde un doble punto de vista: en primer lugar, en las representaciones históricas está muy presente también esa función práctica (didáctica, simbólica, etc.) y, en segundo lugar, el afán decorativo no está ausente de muchas representaciones prehistóricas como se puede apreciar por ejemplo en la cerámica (Abarquero, 1997), sin perjuicio de que en determinados momentos estas representaciones adquirieran un significado específico.

LA DECORACIÓN ARQUITECTÓNICA DE LAS TUMBAS

En las domus de janas, presentes en Cerdeña desde principios del IV Milenio cal A.C., existen diversos elementos que reproducen las casas de los vivos. En primer lugar debemos referir la imitación de la estructura general de la vivienda (Tanda, 1984:II-25).

Dentro de esta debemos citar en primer lugar los techos, por ejemplo en los magníficos ejemplos de la tumba III de Mandra Antine (Thiesi, Sassari) (Contu, 1965:234; Solinas, 1999:103, fig. 104), o los diferentes ejemplos de la necrópolis de Sant'Andrea Priu (Bonorva, Sassari), donde encontramos la representación del techo cónico de una cabaña circular en la tumba V, del techo a doble vertiente de una cabaña rectangular, parte central de una tumba compleja, en la tumba VIII (Lám. Ia), o el techo cónico de una semicircular, como ábside de una vivienda más amplia, en la tumba VI (Solinas, 1999:103, fig. 103, 2000:30, 33, 35). Dos ejemplos de cabañas rectangulares con techo a doble vertiente y sostenido por pilares se conocen también en la necrópolis de Sa Pala Larga (Bonorva, Sassari) (Solinas, 1999:104, fig. 105, 106, fig. 108 y 107, fig. 109). En ciertas manifestaciones se ha conservado la representación de los techos con las vigas centrales y los travesaños perpendiculares en más de una cámara, como apreciamos en Tanca Bullitas (Alghero, Sassari) (Moravetti, 1996:149, fig. 114) en la tumba XVI de Mesu 'e Montes (Ossi, Sassari) (Derudas, 2000:28, 115, fig. 116) o en la Tomba dell'Ovile di S'Adde 'e Asile (Ossi, Sassari) (Derudas, 2000:28, 139). Representaciones claras de techos radiales en ambientes subcirculares se encuentran también en la Tumba XIII de Mesu 'e Montes (Ossi, Sassari) (Derudas, 2000:109, figs. 110, 111) y en la Domus della Casa de Noeddale (Ossi, Sassari) (Derudas, 2000:163, figs. 164, 165) donde, en el mismo eje de los dos vanos con techo a doble vertiente, se abre un tercer ambiente semicircular con techo cónico y travesaños en estrella (Derudas, 2000:29). En la tumba 6 de S'Acqua Salida 'e Corongiu (Pimentel, Cagliari) las paredes y el techo se hallan divididos en paneles excavados o en relieve para simular la curvatura de las paredes y el techo de las viviendas (Usai, 1989:218, 226 fig. 4).

Muy a menudo las dimensiones de la cámara funeraria, especialmente su altura, y el tamaño limitado de los accesos, como en el caso de la tumba VIII de Puttu Codinu (Villanova Monteleone, Sassati) (Demartis, 1991a, Lám. 27, 28, 30) (Lám. Ib), nos recuerdan que estamos ante manifestaciones funerarias, relacionadas también con la clasificación de las personas y con el carácter secreto y restringido de determinadas ceremonias, aunque también haya ejemplos monumentales como las denominadas Tumbas-santuario de Montessu (Villaperùccio, Cagliari) (Atzeni, 1995b:219).

El análisis de los techos representados permite además a G. Tanda precisar que los planos y los convexos, según las representaciones de vigas y la curvatura del techo en casos como Mandra Antine (Thiesi, Sassari) o Sant'Andrea Priu (Bonorva, Sassari), son esquematizaciones de los techos a doble vertiente y de sus deformaciones en la realidad como consecuencia del peso (Tanda, 1984:II-26-29, 56; Caprara, 1986:17). Esta hipótesis encuentra apoyo en las representaciones intermedias como las de la tumba II de Mesu 'e Montes (Ossi, Sassari) (Derudas, 2000:29, 86) y la Tomba Maggiore de S'Adde 'e Asile (Ossi, Sassari) (Derudas, 2000:29, 136, fig. 137, 138). Es, sin embargo, muy discutible el rechazo por parte de G. Tanda de la posibilidad de viviendas con techo plano en el Mediterráneo en la Prehistoria en base a la reconstrucción climática cuando las viviendas argáricas disponían en la Península techumbres planas o con una ligerísima inclinación para facilitar los desplazamientos entre las terrazas (Contreras y Cámara, 2000).

En las domus de janas sardas ambién encontramos pilares (Lám. Ia), pilastras y columnas (Demartis y Canalis, 1989:57; Canalis, 1995:56; Contu, 2000:11). Los primeros son los más frecuentes, como prueba su concentración en alguna de las necrópolis de la zona de Ossi con ejemplos como la tumba II de Mesu 'e Montes (Ossi, Sassari), con basas y decoraciones de bóvidos a las que después nos referiremos (Derudas, 2000:33, 86, fig. 87, 88), la Tomba delle Spirali de Noeddale (Ossi Sassari), etc. (Derudas, 2000:33, 165, fig. 166, 167). También en Santu Pedru (Alghero, Sassari) encontramos pilares en las tumbas I y VI (Contu, 1995:20, fig.; Demartis, 1999c:14, fig. 14, 18, fig.

18 respectivamente). Las pilastras suelen representarse en los ángulos pero a veces también aparecen en el centro de una pared como en la tumba II de Mesu 'e Montes (Ossi, Sassari) (Derudas, 2000:30, fig. 31).

Las columnas no son tan frecuentes como los pilares o las pilastras pero existen en necrópolis renombradas como demuestran la tumba II (ó IV según las numeraciones) de Santu Pedru (Alghero, Sassari) (Contu, 1964:2; Moravetti, 1991:112; Demartis, 1999c:17, fig. 17), la tumba de Silittari (Bosa, Nuoro) (Moravetti, 1993:93), la tumba XXbis de Anghelu Ruju (Alghero, Sassari) (Contu, 1962:626; Demartis, 1995b:31, Lám. XVIII, 1999b:28) o la tumba II de S. Andrea Priu (Bonorva, Sassari), un monumento excepcional por su número de cámaras (diecisiete) (Contu, 1964:68) y por presentar dos cámaras rectangulares en el mismo eje, ambas con columnas (Contu, 1964:71), si bien esta tumba, conocida ahora como la número VI, sufrió importantes modificaciones tardorromanas y bizantinas, que según algunos autores no sólo incluyeron la adición de la segunda de las cámaras rectangulares, inexistente en época prehistórica sino la modificación de los pilares en columnas (Caprara, 1986:15, 39 y ss.). La imitación de las viviendas en la misma necrópolis está atestiguada sin embargo también en la tumba VIII con pilares y techo a doble vertiente curvado (Lám. Ia) para sugerir mejor la realidad de las presiones en las cámaras reales (Tanda, 1984:II-26-29, 56; Caprara, 1986:17-18; Solinas, 2000:33), en el techo cónico con representación de vigas en el hipogeo V (Caprara, 1986:13; Solinas, 2000:30) y en el hogar de la antecámara del hipogeo XIII (Caprara, 1986:24, fig. 15), habiéndose referido ya algunos de estos aspectos precedentemente. Las columnas, que también pueden ser la representación en piedra de postes de madera más o menos transformados, están presentes en otras sepulturas como la 3 y la 4 de Monte Ruggiu (Padria, Sassari) (Galli, 1991:26).

LÁM. IA. Techo A Doble Vertiente Y Pilares De La Tumba Viii De Sant'andrea Priu (Bonorva, Sassari).

Se aprecian incluso los zócalos a partir de los cuales el alzado de las viviendas sería de materiales orgánicos o, tal vez, como sucede a veces en la Península Ibérica de adobe o tapial, estando representada en las domus también la decoración pictórica (Contu, 1965:234, 239; Demartis,

1990:251, 1991b:107, 1992:1; Moravetti, 1995:156), y los relieves figurativos que después referiremos, aunque determinados investigadores sardos prefieren enfatizar en el primer caso la conexión sacra a partir del color predominante, el rojo, si bien se ha resaltado también el carácter arquitectónico, y en el segundo caso el presunto carácter divino de los animales representados.

La imitación de las viviendas puede incluir también a las divisiones del espacio, incluso con pequeños muros (Contu, 1964:75, 1995:19) y, sobre todo en el sur, la presencia de bancos adosados a las paredes en torno a un hogar central como en el hipogeo I de Morobu (Carbonia, Cagliari) (Lilliu, 1995a:18). Se ha referido incluso la reproducción de elementos del mobiliario como camas en la Tomba del Capo de Monte d'Accoddi (Sassari) y escabeles en la Tumba III de Li Curuneddi y en la IV de Ponte Secco (Sassari) (Demartis, 1999a:11).

Tal vez los elementos más frecuente sean los hogares con anillo delimitador resaltado presentes por ejemplo en Filigosa (Macomer, Nuoro) (Lilliu, 1988:221; Moravetti, 1993:93), Puttu Codinu (Villanova Monteleone, Sassari) (Demartis, 1991a:15, 46), Sant'Andria Priu (Bonorva, Sassari) (Caprara, 1986:24, fig. 15) (Lám. IIa), etc . La misma intención imitadora debían tener los hogares situados al exterior de las domus y que se han considerado áreas sacras, como por ejemplo en Montessu (Villaperùccio, Cagliari) (Atzeni, 1995b:219) o en S'Acqua Salida 'e Corongiu (Pimentel, Cagliari) (Usai, 1980:33, 1989:217).

Además a veces podemos pensar que los denominados "hoyos" para ofrendas, cuando forman círculos, también debieron representar los hogares y los agujeros realizados para incluir junto a ellos o dentro de ellos los recipientes (Lám. IIb), tal y como también hemos constatado en contextos de hábitat en Los Castillejos de Las Peñas de los Gitanos (Montefrío, Granada) (Afonso et al., 1996). Éste sería el caso de las representaciones de las tumbas VI, VII y IX de Sant'Andrea Priu (Bonorva, Sassari) (Solinas, 2000:25, 31), pero incluso cuando el anillo del hogar aparece perfectamente definido podemos encontrar hoyos al interior, como se aprecia en las tumbas I y XIII de Sant'Andrea Priu (Solinas, 2000:25, 29, 44). Además se ha planteado que determinados relieves interiores dispuestos en ángulo como los de la tumba II de Mesu 'e Montes (Ossi, Sassari) podrían representar las piedras destinadas a apoyar los recipientes sobre el hogar (Derudas, 2000:31, figs. 32, 88, 89). En esta línea otros hoyos podrían interpretarse como la imitación de agujeros para encajar los recipientes en las viviendas con lo que la vinculación de las ofrendas a la realidad cotidiana no haría sino afirmarse.

Como se aprecia en todos estos casos la vida de ultratumba está muy conectada a la reproducción de la vida terrena con todas sus comodidades, o incluso algunas más si tenemos en cuenta las características de la mayoría de las viviendas de la cultura Ozieri y posteriores en la isla de Cerdeña. De hecho sólo los hallazgos recientes del poblado de Serra Linta (Sedilo, Oristano) muestran una perfecta relación en época Ozieri con la arquitectura representada en las domus de janas (Meloni, 1998:121), aunque las viviendas absidales de Monte Baranta (Olmedo, Sassari) (Moravetti, 2000) y Biriai (Oliena, Nuoro) (Castaldi, 1983b) en época Monte Claro también se relacionan perfectamente con la tipología de muchas tumbas (Hayden, 1999:114). En realidad, si pensamos en la compleja estructura, pese a lo que se ha dicho, de determinados poblados calcolíticos dependiendo de la función, y de los residentes (familias relevantes, solteros, grupos corporativos, etc.) las estructuras de habitación en el mismo poblado podrían ser diversas (Castaldi, 1983b:169-170), del mismo modo que lo son las domus de janas en la misma necrópolis, imitando éstas tal vez sobre todo las viviendas de más alto nivel (Caprara, 1986:9; Lilliu, 1988:83; Demartis, 1991a:43). Por otra parte el crecimiento de los ambientes de las domus de janas, hasta llegar a las 21 cámaras de la Tomba Maggiore de S'ÇAdde 'e Asile (Ossi, Sassari) (Derudas, 1000:23, 125, 136, fig. 24), podría relacionarse con la multiplicación de los ambientes en las viviendas familiares o clánicas de los poblados y la formación de "agregados".

De tal forma la representación "artística" aúna "decoración" e intención ritual, ya que también nosotros resaltamos la importancia simbólica de la representación ideal de la vivienda en la muerte, como una forma de vinculación de los vivos a los muertos, a la tradición, y una forma de remarcar la continuidad no sólo en el territorio como a menudo se ha destacado (Renfrew, 1975) sino en la misma residencia, como muestran los enterramientos junto o bajo las viviendas del Neolítico en determinadas zonas europeas hasta el punto de que en Martos (Jaén, España) los muertos se incluyen en una estructura de hábitat que se sella pero a la que se superpone una nueva residencia (Lizcano et al., 1997; Lizcano, 1999). La imitación así es una forma de conexión y no tanto de trascendencia y si bien la idealización de la vivienda y los posibles procesos de descarnación indicarían que lo que pervive y renace es la comunidad en sí, o un grupo clánico (Bloch, 1981, 1982), la exhibición de la decoración y la individualización, así como la restricción y el alejamiento del público de las ceremonias, tal vez algo posteriores, marcarían la importancia del individuo y de su grupo de descendencia específico..

LA REPRESENTACIÓN DE ANIMALES EN LA DECORACIÓN DE LAS DOMUS DE JANAS

La presencia de representaciones de bóvidos que incluyen la cabeza, en las domus de janas, por supuesto no anula las hipótesis de trasposición de la decoración de las casas (de algunas viviendas) a las tumbas (Contu, 1965; Canalis, 1995; Lilliu, 1995), dado que los trofeos y otras demostraciones de la capacidad de movilizar más riqueza en fiestas y de hacer contraer nuevas deudas a los invitados (Cámara, 1998a), se exhibirían en éstas también en la forma de complementos o estructuras arquitectónicas, al menos en determinadas áreas del Mediterráneo, produciéndose imitaciones, una combinación de elementos procedentes de los animales con estructuras arquitectónicas o una evolución a formas meramente decorativas, lo que explicaría la convivencia de elementos naturalísticos y esquemáticos (Ferrarese Ceruti, 1992:12-13). Es absurda así la división entre elementos arquitectónicos o de uso doméstico y elementos rituales (Demartis, 1995a:39, 2001:25-26), dado que como muestran incluso las figuras de algunos libros (Demartis, 2001: figs. 28, 29, 80) todas representan imitaciones de elementos susceptibles de ser encontrados en las viviendas, lo que no excluye la función ritual de todos ellos que no es exclusiva, sin embargo, de las representaciones figuradas. De nuevo indicamos que esto no supone negar la simbología de estos motivos tal y como I. Hodder (1987) ha destacado al analizar la distribución de motivos similares en las viviendas de Çatal Huyuk, o como hemos destacado nosotros en relación a rituales que afectan a estructuras de hábitat en Martos (Jaén) (Cámara y Lizcano, 1996; Cámara, 1998a).

la Tomba del Capo de Monte d'Accoddi (Sassari) (Demartis, 2001:52, fig. 53) y, sobre todo, en los ejemplos de la Tomba delle Spirali de Sa Pala Larga (Bonorva, Sassari), donde la cabeza naturalística de bóvido se adosa al techo y se asocia a numerosas espirales, y de la Tomba Dipinta de la misma necrópolis (Solinas, 1999:106, fig. 108, 107, fig. 109) para la que se ha llegado a decir incluso que la cabeza de toro parece constituir un capitel del pilar (Solinas, 1999:107). En la tumba V de Montalé (Sassari) las representaciones aparecen sobre pilastras y sobre el arquitrabe de una puerta de la cámara central (Basoli y Foschi, 1988:323, fig. 5; Demartis, 1999a:13) pero en la pared occidental de

LÁM. IB. Techo A Doble Vertiente De La Tumba Viii De Puttu Codinu (Villanova Monteleone, Sassari)

Naturalmente, es posible que no todos los trofeos, incluso en las viviendas, decoraran elementos de sostén o sus representaciones, pero incluso las superposiciones de estos motivos bovinos, como en la Tumba IIA de Ponte Secco (Sassari) (Contu, 2000:20; Demartis, 2001:51) podrían interpretarse en esta línea de imitación (Cámara, en prensa a). Tal sería el caso de Tomba Maggiore (Ossi, Sassari) donde además las representaciones también están presentes en el pilar (Tanda, 1977a, fig. 8), como sucede también en

ésta se sitúa además bajo una cornisa en relieve (Basoli y Foschi, 1988:318, 323 fig. 5). Muy significativo en esta línea de interpretación es el alargamiento de los cuernos a lo largo de toda una pared y su aproximación al techo, que parecen sostener, especialmente cuando estas representaciones se asocian a las de vigas y travesaños en el techo, como se aprecia en Tanca Bullitas (Alghero, Sassari) y Tanca Calva (Alghero, Sassari,) (Moravetti, 1996:149, fig. 113, 152, fig. 122, 150, fig. 118), en la Tumba II de Mesu 'e

Montes (Ossi, Sassari) (Derudas, 2000: figs. 34, 87, 88, 89), en la tumba S'Incantu (Putifigari, Sassari) (Demartis, 2001: figs. 88, 89, 90) o en la Domus delle Doppie Spirali de Oredda (Sassari) donde se asocian a cabezas de carnero (Demartis, 2001:65, fig. 65).

En este contexto el que la cabeza del bóvido quede configurada en un determinado momento por una puerta de las domus de janas (o una falsa puerta) (Galli, 1984:270, 274) puede ser por un lado el resultado de la reducción o sustitución de los trofeos, lo que también explicaría la esquematización de los cuernos (Tanda, 1977a:10, 23; Galli, 1984:274) y, por otro, sugerir a partir del símbolo de la riqueza el cambio de posición social del finado (convertido en un ancestro de relevancia posiblemente comunal) y la posición de sus familiares (según la riqueza que representan y que posiblemente sirven en los funerales) en el punto de tránsito hacia el enterramiento (puerta de la cámara) o más allá (falsa puerta) . La posición de los bóvidos en estas puertas sería así más que una mera protección divina para los vivos y para los muertos (Contu, 1965:252, 1967:199 n.3), una protección del orden social a través de la deposición de los muertos, sin descartar, tampoco en este caso su función arquitectónico-decorativa dado que relieves totalmente lineares (travesaños) sobre las puertas pueden apreciarse en los accesos a domus de Dorgali (Nuoro) como la de Canudedda y la de Isportana (Manunza, 1995:45, 53, 50 fig. 58, 58 fig. 72).

El carácter decorativo de estas representaciones queda definido también por la presencia de cuernos recilíneos sobre las puertas a modo de cornisas y dinteles (Moravetti, 1989:83-84, 86 95-97 figs. 1-3, 102 Lám I), la asociación a decoraciones geométricas como en las tumbas II y XIX de Montessu (Villaperùccio, Cagliari) (Atzeni, 1995b:220, 225 fig.) e incluso la presencia de recuadros que enmarcan las representaciones como en la tumba 2 de S'Acqua Salida 'e Corongiu (Pimentel, Cagliari) (Usai, 1980:31, 231 fig. 1, 1989:217, 225 fig. 3).

En otros casos la posición del "trofeo" sobre el dintel adquiriría una mera función decorativa-simbólica, previa a su uso en el elemento arquitectónico mismo, como parece mostrar la representación de la tumba VI de Sos Furrighesos (Anela, Sassari) (Tanda, 1977a, fig. 14). Así en algunos ejemplos es su mayor cercanía a los modelos naturalísticos (Tanda, 1977a:10-11; Contu, 2000, fig. 4) lo que los aleja de la mera función arquitectónica pero no de la simbología que hemos propuesto en relación a su función decorativa.

Por otra parte en relación a la representación de pilares y pilastras en las domus de janas sardas nos gustaría llamar la atención aquí sobre las espirales grabadas con prolongación rectilínea (Lo Schiavo, 1984:138) cuya simbología creemos que va más allá de la interpretación de divinidad femenina (Tanda, 1977b:191-192; Lilliu, 1988:85, 218) o zoomorfa que se le quiera atribuir (Lo Schiavo, 1984:178-179; Demartis, 1990:251), sugiriendo de nuevo (Cámara, en prensa a) la trasposición de un esquema arquitectónico, el pilar, pilastra, poste o columna, con capitel y, a veces, basa, a las cuevas artificiales. Esto vendría mostrado por la situación de esta decoración en muchas de las tumbas sobre los pilares y columnas y por la frecuencia con que aparece la doble espiral con el motivo vertical central bien remarcado como en el caso de la tumba IV de S'Eligue Entosu (Cargestre, Sassari) (Lo Schiavo, 1984;172-173), aun cuando a veces aparezca bajo los cuernos de bóvidos y relacionada con una de las puertas como en Tanca dell'Uliveto (Sassari) (Lo Schiavo, 1984:173), o la Domus delle Doppie Spirali (Oredda, Sassari) donde se dan los dos casos (Antona Ruju y Lo Schiavo, 1989:58, 69 fig. 5, 72 Lám I). Incluso cuando se sitúan bajo las representaciones de cuernos de bóvidos, la asociación no tendría nada de extraño ya que para éstos se podría pensar en la representación de parte de los travesaños y en ménsulas a través de los cuernos de los bóvidos que se superponen, si bien existirían representaciones de doble espiral más difíciles de explicar como las del techo de la tumba III de Mandra Antine (Thiesi, Sassari) (Contu, 1965:233). También claro parece el ejemplo de la tumba 4 de Giorrè-S'Elighe Entosu (Cargeghe - Murros, Sassari) en relación a líneas verticales perpendiculares a otra horizontal paralela al pavimento (Tanda, 1977b:176-177) aunque realmente no presenten ningún relieve en el colmo de la esquematización.

Resulta particularmente interesante que se haya referido que las representaciones que muestran los cuernos hacia abajo (Castaldi, 1979:397) y las de la doble espiral (Lo Schiavo, 1984:175-176), pretenden representar a un muflón o carnero a partir del ejemplo de la Domus dell'Ariete (Perfugas, Sassari). Interpretación rechazada por otros autores que prefieren seguir hablando de Diosa-madre a partir de una presunta representación de sus ojos (Tanda, 1977b:191-192; Lilliu, 1988:85, 218). Esta última identificación en base meramente a su asociación a la presunta divinidad bovina no nos parece muy plausible en el contexto de las interpretaciones aquí vertidas. Por otra parte cuando se han buscado paralelos en la Península Ibérica para esta presunta Diosa-madre (Lilliu, 1995b:440, 460) se olvida que las representaciones oculadas no implican indudablemente una mujer, salvo en determinados casos, cuando las líneas incisas que los acompañan se leen en términos de un tatuaje (Molina y Martínez, 1995), como el mismo autor reconoce (Lilliu, 1988:252), y no en términos religiosos.

Además la presencia de representaciones de carnero en edificios nurágicos considerados sacros (Fadda, 1997:73,75) mostraría no sólo la posibilidad de representaciones más antiguas sino la importancia de este animal que se sigue vinculando a ofrendas religiosas, una forma de justificación de la sociedad y, muchas veces, una forma enmascarada de tributo no sólo a través del consumo diferencial de las ofrendas o del diferente carácter de éstas sino de la producción de determinados elementos aptos para ellas por parte de la élite sacerdotal. Esto en nuestra opinión pondría en relación estas figuras no con una oposición entre comunidades agrícolas y comunidades ganaderas (Castaldi, 1979:397 n. 15) sino con la importancia de esta última actividad en lo que respecta a la justificación social.

Así esto no excluye la simbología social de estos elementos pero sí una lectura meramente "divina" sin explicar el proceso que había conducido a la importancia de tales animales y de sus representaciones (Cámara, en prensa a) . En cualquier caso la interpretación zoomorfa se puede combinar perfectamente con la representación arquitectónica, sobre todo si tenemos en cuenta que las viviendas, o algunas de ellas, podrían estar decoradas con estas mismas, o parecidas representaciones .

Así E. Castaldi (1978:439) ha señalado la importancia de los sacrificios de los bóvidos en las ceremonias relacionadas con el ritual funerario y ha enfatizado el cambio de significado de sus representaciones y A. Usai habiendo señalado las diferentes funciones y significados (A. Usai, 1996:382) que se esconden tras el presunto culto al toro mediterráneo se ha mostrado más propenso a imaginar la creación local y la circulación de mitos y prácticas rituales concernientes a los bóvidos como un efecto secundario del desarrollo de las técnicas pastoriles en las sociedades neolíticas, no sin la persistencia de concepciones conectadas con las actividades cazadoras propias del estadio paleolítico (A. Usai, 1996:383).

LÁM. IIA. Hogar Con Anillo Perimetral Resaltado De La Tumba Xiii De Sant'andrea Priu (Bonorva, Sassari).

ANIMALES Y RITUAL EN LA PREHISTORIA RECIENTE

Sin embargo en nuestra opinión el argumento no está completo pues sigue sin explicarse por qué es tan importante la representación de estos animales. Hemos planteado en otros lugares (Cámara y Lizcano, 1996; Cámara, 1998a; Cámara y Afonso, en prensa) que desde el Neolítico Reciente, al menos, (fines del V milenio cal A.C), lo importante era la posesión de los animales, lo que explicaría el afán por sustituirlos por estatuillas en las ofrendas o la menor valoración de otros animales (Castaldi, 1978:449-451), especialmente los no domésticos o de menor porte cuya acumulación tenía menos trascendencia en términos cuantitativos y de prestigio social . Tal sustitución se inscribe además en el cese de la movilización masiva de riqueza en los funerales y otras ceremonias (Kristiansen, 1984; Sorensen, 1987) a partir sobre todo del Bronce Final en determinadas zonas europeas ya que no debemos olvidar la enorme importancia que como elemento de acumulación de riqueza, como símbolo y como medio de producción agrícola habían adquirido sobre todo los bóvidos y los équidos que reúnen en mayor medida (y en mayor masa) ambos aspectos (Martínez y Afonso, 1998; Cámara, 1998a; Cámara y Afonso, en prensa).

Podemos recordar aquí lo que hemos referido en torno a

los rituales en los que están implicados estos animales en el sur de la Península Ibérica, en una aproximación que creemos deja justamente de lado la problemática de la divinización o no de los animales (Castaldi, 1978) y de su identificación mítica con una comunidad o con determinados representantes (Castaldi, 1986b) . La representación frecuente de los animales, especialmente los bóvidos y los carneros, en determinados momentos de la Prehistoria Reciente se relaciona en nuestra opinión con la importancia que van adquiriendo estos animales como símbolos de riqueza y como elementos a partir de los cuáles la devolución de presuntos favores (correspondientes a fiestas, a dotes desiguales, a la indicación del origen, etc.) se convierten en un tributo enmascarado (Cámara, 1998a; Cámara y Afonso, en prensa) .

En este ámbito la presencia de representaciones de carnero (o muflón) en el contexto sardo con menor frecuencia a las representaciones de bóvidos podría relacionarse no con una menor importancia de una presunta divinidad ovina (Lo Schiavo, 1984:178-179) sino con la menor relevancia de estos animales en la representación de la posición social, dado además que no se emplean como fuerza de tracción y su papel como medio de producción es menor (Cámara y Afonso, en prensa). También así se podría leer el mayor énfasis en las representaciones de bóvidos en las asas de la cerámica de Serra d'Alto en el sur de Italia desde el VI Milenio cal A.C. en momentos en que los ovicápridos parecen haber desplazado en la importancia subsistencial a los bóvidos (Cassano, 1994:227, 229) , vinculándose también a un mayor énfasis en el control de zonas complementarias desde asentamientos más grandes donde se ha concentrado la población (Cassano, 1994:224, 228-229). En cualquier caso la jerarquización de las presuntas divinidades estaría en función de este factor y del papel adquirido por la religión en la justificación de un orden social que prima la propiedad de determinados elementos.

Por el contrario G. Tanda (1977c:207 y ss.) se centra en buscar el significado de cada elemento particular y la seriación entre ellos, interpretando la oposición de triángulos con el vértice hacia arriba y hacia abajo (técnica común en frisos decorativos de la cerámica por ejemplo) como una esquematización de la figura femenina (también presente en vasos Ozieri). En cualquier caso tales representaciones se inscribirían en el gusto decorativo-narrativo de la pintura rupestre neolítico-calcolítica del Mediterráneo Occidental adquiriendo su asociación a bóvidos nuevos significados que algunos de nuestros colegas (Martínez y Afonso, 1998) leerían en relación a la circulación de riqueza con las dotes nupciales, y que, en definitiva, habría que relacionar con el control de la fuerza de trabajo y de los medios de producción (Cámara, 1998a), y no simplemente como representaciones de divinidades femeninas (Tanda, 1977c:209) en una lectura fácil que cierra las posibilidades de investigación de la sociedad en cuestión si no indagamos cómo se produjo esta presunta identificación divina y a qué responde. En este sentido no se comprende por qué le parece a G. Lilliu (1988:85) poco plausible la hipótesis que ve en esta multiplicación el deseo de multiplicar los rebaños. En nuestra opinión, por el contrario este deseo se inscribiría en la necesidad de acumular riqueza pecuaria, en la exhibición de la que se tiene, la que se moviliza por los funerales, la que se desea alcanzar, o la que se sustituye por su representación una vez que no es necesaria la movilización de los recursos.

Otra línea de argumentación en relación a los enterramientos de bóvidos presentes en Centroeuropa desde el 3500 A.C. los ha puesto en relación con la aparición del carro y de la pareja de bueyes destinada a arrastrarlo. Aunque, como ha puesto de manifiesto A. Pollex (1999:545-546) existen diversos problemas en relación a tal argumento, no siendo el menor de los cuales el hecho de que cuando aparecen parejas de bóvidos y se ha podido determinar su carácter no siempre nos encontramos con bueyes más dóciles, aunque haya algunos ejemplos como Oscherleben, y también la escasez de evidencias de carros reales, aunque si haya representaciones de éstos y de partes (discos), incluso en la cerámica, que se han relacionado con el sol y el "carro solar" (Pollex, 1999.548).

Aunque es cierto que el significado de los enterramientos puede ser diferente, la importancia económico-social admitida para los bóvidos, y otros animales y la variabilidad de las deposiciones que había llevado a H. Behrens a presentar diferentes hipótesis: ofrendas a individuos de alto status, animales como compañeros (y propiedad en el otro mundo), ofrendas, totemismo, respeto por animales especiales, etc. (Pollex, 1999:544), obliga por un lado a buscar el verdadero significado de estas manifestaciones en la propiedad y en la mitificación de la riqueza, de los que la poseen y son representados (totem) por ella, y, por otro lado, a conectar más directamente los bóvidos con otras actividades de las sociedades agropecuarias como la tracción en los arados, que a veces representaban también la domesticación de la naturaleza incluyendo la santificación de los asentamientos y de la propiedad. Todo ello explicaría y conduciría a la deificación de los animales o su conversión en símbolos de la divinidad como proponemos en nuestra interpretación de las representaciones en las domus de janas sardas (Cámara, en prensa a). En cualquiera de los casos y aunque el sacrificio implicar el cierre del círculo nacimiento/muerte/renacimiento con la muerte del mismo dios o su representación, o simplemente una ofrenda (Pollex, 1999:543), como en todos los rituales lo que se persigue en origen es el mantenimiento del status quo.

Además tales argumentos no hacen sino reforzarse si tenemos en cuenta en primer lugar que en las delimitaciones rituales se usaban un toro y una vaca en lugar de bueyes y que el arado pudo convertirse en la forma de acceder a parcelas más extensas, y por tanto más riqueza, que era capaz de ser explotada (Cámara y Afonso, en prensa).

REPRESENTACIONES ESCULTÓRICAS EN BULTO REDONDO, "DIVINIDADES" Y DIVISIÓN SEXUAL

Una diferencia entre los enterramientos más antiguos de Cerdeña, entre finales del V y principios del IV milenio cal A.C., y los documentados en el sur de la Península Ibérica (Lizcano et al., 1997; Cámara, 1998a), o en el sur de Italia (Pitti y Tozzi, 1978:94; Cipolloni Sampó, 1983:227), es la presencia en los enterramientos sardos de elementos de ajuar y especialmente de las figurillas femeninas (Santoni,

1982:71) que, en los casos publicados, se asocian a enterramientos masculinos (Germanà y Santoni, 1993:6, 12-15; Santoni, 1995b:131) .

Representaciones de este tipo se conocen, aunque escasas y normalmente de contextos problemáticos, también en el sur de Francia (Guilaine, 1996). De confirmarse en futuros trabajos tal asociación exclusiva habría que pensar que el proceso de minusvaloración de la mujer a través de la apropiación de su capacidad productiva y reproductiva era relativamente más fuerte en el caso sardo, aunque, también, puede tratarse de una forma de manifestar el menor enmascaramiento de una situación que se venía gestando incluso desde el Paleolítico Superior (Conkey, 1980; Bender, 1989; Estévez et al., 1999). No cabe pensar como hacen F. Germanà y V. Santoni en una valoración del papel de la mujer en estas sociedades (Santoni, 1982:80; Germanà, 1996:452) pues en ese caso habría que preguntarse en primer lugar dónde se hallan sus tumbas y, en segundo lugar, cuál fue el momento en que se produjo el salto hacia su sumisión posterior.

No se trata, por tanto, meramente de un ritual doméstico, como se ha planteado (Hayden, 1999:120), dado que, aunque hay algunas adscribibles a poblado (Lilliu, 1999:12) con problemas ya que muchas proceden de superficie (Lilliu, 1999:186, 195, 198, etc.), otras están claramente asociadas a tumbas pre-ozieri, y otras representaciones más esquematizadas adscritas a fases más tardías entre mediados del IV y finales del III Milenio A.C. (Lilliu, 1999:68) proceden, a menudo, de cuevas artificiales (Ferrarese Ceruti, 1989:39; Moravetti, 1989:87-88, 90; Lilliu, 1999:217-222), y en otros casos de grutas naturales (Lilliu, 1999:209-210,.224-226, etc.), aunque algunas procedan de estructuras posiblemente domésticas (Lilliu, 1999:235, 253, 259).

LÁM. IIB. Posible Hogar Con Agujeros Para Recipientes Y Palos De Sostén De La Tumba Ix De Sant'andrea Priu (Bonorva, Sassari)

Realmente la minusvaloración de la mujer, condicionada desde el plano ideológico tiene lugar desde antes y las mismas estatuillas "paleolíticas" se han reinterpretado en términos de dominio masculino expresado en la subordinación de las formas femeninas a un esquema fálico (Martínez y Afonso, 1998), aspecto que no parece lejano a las representaciones del tipo "cicládico" (Lilliu, 1988:241) y que los mismos autores sardos insconscientemente resaltan al llamar la atención sobre el aspecto destacado de estas representaciones femeninas (los atributos de reproducción o, al menos, dedicados a la maternidad) (Lilliu, 1988:227, 241, 256), una minusvaloración de la contribución femenina a la actividad productiva que no se aprecia al considerar las estatuas sólo expresiones de la divinidad. Una divinidad que, en cualquier caso, es siempre una trasposición (y una justificación) de las sociedades en que se vive y no el resultado de un "inconsciente colectivo".

En este sentido, además, la identificación de esta fertilidad humana, destinada a la reproducción de la fuerza de trabajo, con la fertilidad de la tierra, en la forma de la Diosa Madre, tal y como se ha planteado por algunos autores (Lilliu, 1989:21) adquiere sentido, cuando la tierra es un verdadero medio de producción extensivo permanente, y esto sólo sucede tras la sedentarización plena y el inicio de la jerarquización, tal y como hemos referido en otros trabajos (Cámara, 1998a; Cámara y Afonso, en prensa). La tierra no es en sí medio de producción, por lo que su apropiación simple no conduce a ningún resultado, sino que es una condición natural de la producción (Gutelman, 1981). Sólo se convierte en medio de producción por el trabajo humano y, por tanto, lo será a menor escala en las sociedades cazadoras-recolectoras, en las que sin duda el territorio tiende a ser delimitado pero en otras formas (Godelier, 1989) . Lo que sí podría darse es una mayor importancia de la tierra como soporte de otros bienes, particularmente en las sociedades agropecuarias los rebaños, por lo que no sería extraña la vinculación de la fertilidad humana a la fertilidad animal, las representaciones de la tierra a las de animales, pero también, frecuentemente como veremos, las representaciones de éstos últimos parece que representan un orden social no igualitario al menos en lo que respecta a los sexos (Martínez y Afonso, 1998).

Por esto puede ser significativa la preponderancia en este sentido de las representaciones taurinas en las domus de janas, cuya simbología, no exclusa por su papel decorativo en las tumbas y posiblemente en las viviendas, aparte de relacionarse como todo el ganado con las formas más rápidas y reproducibles de acumular riqueza, como hemos visto, se vincula también al trabajo agrícola, con lo que de la misma forma que en la reproducción de los hombres, de la fuerza de trabajo, adquiere más fuerza la parte masculina a nivel ideológico (Lerner, 1990) en estrategias económicas globalmente desarrolladas por ambos sexos. Así en la producción de alimentos adquieren más importancia las labores desarrolladas con los animales, o gracias a ellos, y estos mismos (Cámara y Afonso, en prensa), lo que sería del mayor interés de poder probarse un cambio en las actividades masculinas y femeninas a partir del uso del arado de tracción animal en estas labores.

LA POSICIÓN DE LAS DOMUS DE JANAS Y LOS MEGALITOS

La observación de la posición de las necrópolis de domus de janas en algunas publicaciones (Usai, 1991) nos muestra que, también éstas, después del Neolítico Reciente, debieron ceder el papel de demarcación del territorio, en determinadas zonas, a otras manifestaciones, por ejemplo a los menhires, a los que a continuación nos referimos, y a los dólmenes y otro tipo de tumbas megalíticas, que arrancan sin embargo de momentos antiguos (Guilaine, 1996), de ahí la oposición en densidad de ambos elementos entre la Planargia (domus de janas) y el Marghine, al interior (sepulturas megalíticas) (Moravetti, 1985:8-9, 1993:94). Al mismo tiempo el papel ideológico de las domus de janas se ciñe en estos casos a la clasificación de las personas en las ceremonias y en el acceso al enterramiento y a la demarcación, posiblemente de los asentamientos más destacados, como posteriormente discutimos.

No es conveniente, por tanto, aislar la diferente fenomenología funeraria de la Prehistoria Reciente especialmente si tenemos en cuenta la presencia de materiales de la Cultura Ozieri en el sepulcro de corredor de Motorra (Dorgali, Nuoro) y, sorprendentemente, en la galería cubierta de Sculacacca (Oniferi, Nuoro) (Ferrarese Ceruti, 1980:63; Lilliu, 1988:189; Demartis, 1993:31; Moravetti, 1998:25), o en los círculos megalíticos, normalmente sin corredor, de Cerdeña (Lilliu, 1988:66; Antona, 1995:83; Atzeni, 1985:XL, 1988a:449-450, 1989:202) y Córcega (Lanfranchi, 1980:450, 458-459, 1992:122-124; Lanfranchi y Weiss, 1980:479-480). Las mismas domus de janas también se asocian a menhires (Lilliu, 1988:66, 68-69; Cesari, 1992:105).

Más grave es separar estos últimos del fenómeno megalítico a la hora de estudiar la distribución de éste en Cerdeña (Moravetti, 1998b), especialmente cuando J. Guilaine (1996) ha destacado la generalidad de este tipo de sepulturas de cámara simple y normalmente individuales o destinadas a pocos individuos en la génesis del Megalitismo en el Mediterráneo Occidental. En Córcega, de hecho, están destinados, al menos los tipos más antiguos (en coffre o bancali, es decir con cámara subterránea) (Lanfranchi, 1980:450, 458) a enterramientos individuales (Lanfranchi y Weiss, 1980:469-470) .

No es suficiente así atribuir estas diferencias entre domus de janas y megalitos al tipo de roca y si en algunos casos la distribución de un tipo u otro de sepultura puede indicarnos una diferencia cultural regional correspondiente a una o varias formaciones sociales como podría ser el caso de la escasez de los hipogeos artificiales en la Gallura (al noreste de Cerdeña) (L. Usai, 1992:10; Salis, 1996:39) en otros casos se debe pensar en diferentes formas de remarcar el espacio de explotación. Así debemos resaltar el hecho de que mientras en la Planargia extensas necrópolis parecen sugerir la existencia de vastos poblados, estos dólmenes [del Marghine], más bien modestos bajo criterios arquitectónicos y además aislados, no nos sugieren que estén relacionados con núcleos de habitación amplios; al contrario nos llevan a pensar más en un uso periódico del territorio que en una presencia estable, como por otra parte conviene a una economía preferentemente pastoril, la única posible en terrenos poco adaptados a una extensiva actividad agrícola (Moravetti, 1993:94, ver también 1998:25) mas debemos decir que los desplazamientos se darían desde zonas vecinas donde estarían los asentamientos permanentes correspondientes a estas comunidades, posiblemente, en parte, desde la Planargia.

Sí podría ser cierto que para remarcar áreas de desplazamiento se eligieran monumentos de alta visibilidad mientras junto a los poblados se prefiriera el secreto y la ocultación a través de cuevas artificiales como hemos sugerido para el valle del Guadalquivir (Cámara, 1998a) y como se ha planteado para otras áreas (Barnatt, 1998:96). En cualquier caso habría megalitos más cercanos al área de hábitat permanente y es también esta vecindad entre las áreas de los monumentos y las de los poblados la que se está viendo en Córcega situándose los megalitos en el borde del espacio agrícola (Cesari, 1992:112). En cualquier caso en Cerdeña existen pequeñas concentraciones de dólmenes (Moravetti, 1998:26) que pueden relacionarse con los puntos de descanso o fin del trayecto.

En otros casos todas las funciones las desempeña un mismo tipo de tumba, y así a una diferencia en forma, escasez de decoración, densidad y agrupación de las domus de janas, como sucede, con respecto a las zonas septentrionales, en la zona de Dorgali, y en general en la costa centro-occidental (Lo Schiavo, 1980:10; Fadda, 1980:47; Ferrarese Ceruti, 1980:57) corresponde también una diversidad de función con necrópolis muy dispersas como la de Concas de Janas (Fadda, 1980:47; Manunza, 1995:33-35) destinadas a remarcar las zonas de desplazamiento, y no simplemente la zona de llanura, aunque, como hemos dicho, también se pueden señalar agrupaciones como la de Sirilo (Orgosolo, Nuoro) con 13 tumbas (Fadda, 1985:61-63). Se explica así la asociación en esta región de las domus de janas a los dólmenes y menhires (Ferrarese Ceruti, 1980:57, 61; Atzeni, 1982:18 fig. 3, 19; 1988b:524; Manunza, 1995:72), debiendo destacarse también que como en el Pasillo de Tabernas (Almería) (Maldonado et al., 1997) se dan asociaciones entre distintos tipos de dólmenes, rectangulares y poligonales (Ferrarese Ceruti, 1988:67-68; Manunza, 1995:66).

Sin embargo también aquí, pese a la escasez de datos sobre hábitats prenurágicos (Spanedda, 1994-95; Manunza, 1995:22, 82) se puede pensar en base a la mayor densidad de las tumbas hipogeicas en torno a Dorgali (Manunza, 1995:21) y la dispersión de los megalitos en el altiplano cercano (Manunza, 1995:70) que el principal papel en la demarcación de éste lo adquirían los dólmenes, sin excluir la función de determinadas domus de janas, las excavadas en bloques aislados (Manunza, 1995:35, 43, p. ej.) especialmente cuando también aquí se dan estructuras mixtas con corredores ortostáticos añadidos en Mariughia I (Dorgali, Nuoro) (Manunza, 1995:42) o círculos externos megalíticos en Campu Marinu I (Dorgali, Nuoro) (Manunza, 1995:43).

Debemos tener presente también así que en los momentos calcolíticos las variaciones de las tumbas megalíticas y de las domus de janas, no se deben meramente al tamaño de los grupos familiares que las utilizan (Foschi, 1984:535), o a la zona geográfica, sino que deben tener relación con la jerarquización social. De la misma forma importantes necrópolis parecen situarse cerca de los poblados como también hemos sugerido en la Península Ibérica en relación a su capitalidad (Cámara, 1998a). Sin embargo es un problema el desconocimiento de los hábitats en momentos del Neolítico Reciente (cultura de Ozieri), excepto en lo que respecta a los "campos de silos" del sur.

Debemos destacar aquí como, pese a la escasez de datos, parece que, como en el sur de la Península Ibérica hacen las cuevas artificiales, las necrópolis de domus de janas se sitúan cerca de los poblados más importantes o aquellos de más tradición, como puede ejemplificar Monte Noe, existente, al menos, desde un momento avanzado del Neolítico Medio (Trump, 1984:517), argumento que, con abundantes reservas, dado lo exiguo de la información, se puede extender a Sorralia (Norbello, Oristano) (Depalmas, 1989:8) y a Cuccuru S'Arriu (Cabras) (Santoni, 1992; Germanà y Santoni, 1993), siendo más problemáticos en su evaluación los hallazgos de dispersiones de sílex junto a otras necrópolis (Demartis y Canalis, 1989:41 n. 5; Locci, 1991; Pitzalis, 1996), aunque a veces hay referencias a cerámica y alteraciones debidas al cultivo (Sanges, 1980:182).

Además es en estas concentraciones donde se sitúan las tumbas más complejas y las mayores agrupaciones como aquélla de Cannas di Sotto (Carbonia, Cagliari) junto a un hábitat Ozieri (Santoni y Usai, 1995:53), lo que nos muestra también la importancia que pueden tener las variaciones en el número de tumbas incluido en cada necrópolis hasta llegar a los máximos de necrópolis como Su Cruciffissu Mannu (Porto Torres, Sassari), Anghelu Ruju (Alghero, Sassari), Sas Concas (Oniferi, Nuoro), Santu Pedru (Alghero, Sassari) y Montessu (Villaperuccio, Cagliari) (Atzeni, 1985:XXXII; Ferrarese Ceruti, 1989:37; Moravetti, 1992:15-16; Demartis, 1995a:39; Atzeni, 1995:219; Santoni, 1995b:115) .

En el caso de Santu Pedru (Alghero, Sassari) si bien el número de tumbas no es muy amplio la relación con Monte Baranta (Olmedo, Sassari) parece clara (Moravetti, 1992:20), aun teniendo en cuenta la existencia de otros poblados de la época como Tanca Baranca (Olmedo, Sassari) y Padru Salari (Olmedo, Sassari) de menor entidad (Moravetti, 2000:14). Sin embargo esta necrópolis quedaría oculta a la vista desde el poblado como hemos referido en relación a las necrópolis de Las Peñas de los Gitanos (Montefrío, Granada, España) (Cámara, 1998a) . En el caso de Montessu (Villaperuccio, Cagliari) la necrópolis se asocia a 14 menhires en la zona de Terrazzu, que parecen configurar una vía hacia ella, y cercana al poblado de S'Arriorgiu (Villaperuccio, Cagliari) (Atzeni, 1995:219; Lilliu, 1989:22, 1995a:24), habiéndose hablado de estructuras monumentales (tumbas-santuario, especialmente en relación a las tumbas V y XVII) con entradas de dos metros de alto y ancho, hogares, bancos perimetrales y alineaciones de piedras formando recintos en toda el área de la necrópolis que incluye también una galería cubierta, etc. (Atzeni, 1995c:219; Cicilloni, 1995:54). Su continuidad viene reflejada no sólo en la aparición de corredores ortostáticos sino en la presencia de una tumba a galería (Atzeni, 1995c:219-220).

La dispersión o concentración de las tumbas debe explicarse en función de la organización social compleja con distinta función de la sepulturas, por un lado una afirmación sacra del control del territorio, por otro una afirmación y justificación de la capitalidad y de las diferencias sociales (Cámara, 1998a). Es por ello por lo que las agrupaciones se dan en el llano como hemos sugerido para el sur de la Península Ibérica (Cámara, 1998a; Maldonado et al., 1997).

PETROGLIFOS, MENHIRES Y ESTELAS. DIOSES, PROPIEDAD Y ARISTOCRACIA

No sólo los megalitos sino también otros monumentos prehistóricos más lejanos del carácter arquitectónico y vecinos a lo que tradicionalmente se engloba bajo el término "arte" en los estudios prehistóricos , parecen usarse en la delimitación del territorio y las áreas de desplazamiento como se ha sugerido para los menhires en Galicia (Villoch, 1998:110, 113-114), en Cantabria (Serna, 1997), o en Cerdeña (Cámara, en prensa b), o los petroglifos en Galicia, aun cuando cada elemento puede ocupar una posición en el espacio y desempeñar una función específica diferente (Villoch, 1995:49-52), existiendo cromlech y menhires en Portugal (Cardoso et al., 1997:209; Silva, 1997:217-218; Rocha, 1997:221) para los que se puede plantear la misma función. Además recientemente se ha demostrado la vinculación de los abrigos con pinturas rupestres a los mismos intereses, sin descartar también la exhibición del orden social y el parentesco (Martínez García, 1998).

Aunque los menhires sardos se asocian en numerosos casos a estructuras hipogeicas (Sanges, 1980:182; Lilliu, 1981:10, 123; Atzeni, 1982:26-27; Pitzalis, 1996:169-191; Cossu, 1996:301; Frau, 1996:294-298) que, como hemos visto, muestran una mayor exhibición que las de la Península Ibérica, e incluso se sitúan dentro de ellas y dentro de cuevas naturales (Sanges, 1980:193), se asocian también a sepulcros megalíticos (Atzeni, 1982:19, 25), como a la tumba I de Masone Perdu (Làconi) (Atzeni, 1988a:450) o Monimentos (Boddusò, Sassari) (Basoli, 1987). Sucede el mismo fenómeno ya comentado de la asociación de diversos tipos de monumentos funerarios como dólmenes , galerías cubiertas y tumbas de gigantes, existiendo en estos casos, muchas veces una relativa cercanía a nuraghi que no son tanto los indicios de la erección tardía de algún monumento (Lilliu, 1982:20-21) sino que son la expresión de la culminación de la delimitación sacra del territorio por adición en la forma de un control más militarizado (Cámara, 1998b), ya que en cualquier caso la asociación del dolmen de Sa Coveccada (Mores, Sassari) a hipogeos (Atzeni, 1968:135-137) pone en relación las diversas fases de la inhumación colectiva sarda, habiéndose señalado que los primeros dólmenes hacia finales del IV Milenio a. C.. estarían relacionados con el sur de Francia y Cataluña (Atzeni, 1988a:450).

En Cerdeña y Córcega la presencia de menhires (Atzeni, 1988a, 1988b; Pitzalis, 1996), incluye estatuas-menhir con representaciones femeninas , tal vez de dimensiones más modestas (Saba, 2000:136) y, sobre todo, masculinas (Atzeni, 1982), utilizadas en la demarcación del territorio, a veces en asociación a las tumbas (Cesari, 1992:105; Lanfranchi, 1992:122), sobre todo en el caso de los que no son antropomorfos (Frau, 1996) que presumiblemente parecen tener un carácter más antiguo a partir de su presencia en Pranu Mutteddu (Atzeni, 1982:34) . Esta función delimitadora y marcadora de rutas la hemos sugerido para determinadas distribuciones megalíticas de la Península Ibérica (Cámara, 1998a). Así en el área de Perda Longa (Tortolì) la presencia a pocos metros del menhir 23 (el Limba 'e Boe, aunque removido) de un importante camino, atestiguado al menos a partir de época romana, hace posible conjeturar sobre la existencia de una relación entre el camino y las alineaciones (Frau, 1996:293), y aunque la relación no tiene por qué ser directa sino de visibilidad (Perra, 1994:18), las referencias, aun en el contexto de autores que intentan primar sobre todo un impulso sacro, son constantes, y G. Lilliu nos dice que Están en la proximidad cuando no en el margen de antiguas vías que aún hoy se recorren (Lilliu, 1981:66) y que Es plausible que todos estos monumentos hubieran sido pensados y realizados, además de con la función local de objetos de culto por parte de los habitantes de los poblados del área en la cual se inscribían, también como puntos de referencia, señales y lugares de descanso de los viandantes, y tuvieran así un interés más general y diríamos también público (Lilliu, 1981:66-67; ver también Catte y Salis, 1985:81; Lilliu, 1988:87, Saba, 2000:130-132).

Existen significativas asociaciones de menhires protoantropomorfos y estatuas-menhir como en Corte Noa-Perda Iddocca o en Genna Arrele (Laconi, Nuoro) y en el primer caso incluso asociaciones a estructuras megalíticas (Atzeni, 1982:11-19). La alineación de Corte Noa-Perda Iddocca es particularmente relevante (Atzeni, 1982:18 fig. 3, 19; 1988b:524) y nos muestra la progresiva formación de un paisaje ritual, en el que los monumentos funerarios se sitúan en lugares elevados y se atribuyen al Cobre Pleno (Atzeni, 1988b:524).

Esta demarcación sería aun más relevante de confirmarse el hecho de que diferentes tipos de decoración de los menhires antropomorfos (estatuas-menhir) se localizan en determinadas regiones (Perra, 1994:37-38; Cossu, 1996:322-324), por lo que desde esa perspectiva las estatuas-menhir de la isla deberían interpretare como el producto de grupos humanos que, aun en el interior de un marco ideológico común, estarían fuertemente asentados en un territorio determinado, reconocido como propio, diferenciándose de los grupos vecinos también gracias a la elaboración de una específica tradición de mitos y de símbolos (Perra, 1994:37-38).

Al menos en la Edad del Bronce también se utilizarán en la justificación del papel de determinadas personas que en ellos se representan o que se apropian de ellos y asociadas, en algunos casos a tombe di giganti (Moravetti, 1985a) o a tumbas de otra tipología (Saba, 2000:116), fenomenología que no aparece con claridad en nuestra península hasta el Bronce Final, aunque determinados ídolos y placas decoradas incluso con representaciones antropomorfas pudieron servir a similar fin.

En otro sentido la asociación de elementos guerreros, como los puñales (Atzeni, 1982:13, 14, 15, 22, 23 fig. 5, 44 Lám. II, etc., Saba, 2000:114, 120, 123, 126, 147, 151) a estas estelas antropomorfas, al menos desde el Calcolítico, podría entroncar con el desarrollo de la jerarquización social y con los ajuares de armas, especialmente cuando tenemos posibles representaciones de cascos y otros elementos (Ambrosi, 1985:367). En este sentido determinadas representaciones capovolte o simplemente triangulares en la parte superior de las estatuas-menhir (Saba, 2000:160-164, fig. 2-6) definidas como cejas y nariz incluso en los casos hiperdesarrollados (Saba, 2000:128, 142-144), en nuestra opinión pueden ser interpretadas como parte de un casco. Si ello es así la parte superior del casco estaría representada por las presuntas cejas y lo que se interpreta como nariz sería una prolongación sobre la frente. Bajo este elemento frecuente aparecería, a veces, el llamado capovolto (Atzeni, 1982:13, 20, 22, 37), la protección de la nariz y las mejillas tal vez, con lo que además, al cambiar la escala, determinados elementos considerados senos podrían verse en otras figuras como los ojos (veánse por ejemplo la lámina II:3 en Atzeni, 1982:44, o las figuras en Saba, 2000:161-163). Este aspecto conduciría también a una reinterpretación de las representaciones presentes en determinadas domus de janas , por ejemplo las de la tumba V de Ponte Secco (Sassari) (Ferrarese Ceruti, 1989:39) o las de la Tomba dell'Emiciclo o la Tomba Nuova Ovest de Sas Concas (Oniferi, Nuoro) (Santoni, 1995a:115, fig. pág. 116, 118, 119, 120, 121).

No tiene por qué tratarse además de un culto separado del de los muertos (Mezzena, 1997:98) dado que las representaciones de las estelas más que a un panteón divino ideal podrían referirse a los ancestros (o a antepasados más inmediatos) más o menos divinizados y convertidos en héroes (Strahm, 1998:31 y 36), de ahí la inmediatez de las representaciones, cada vez más agudizadas, de armas y vestuario, con el fin de identificarlos con una élite ya consolidada, que no tiene por qué relacionarse con culturas pastoriles patriarcales (Santoni, 1982:80), ni referirse una subordinación de la divinidad femenina (Lilliu, 1988:237), que, en todo caso, no sería sino un resultado de un cambio social producido bastante antes (Lerner, 1990), como hemos referido. En este contexto además el abatimiento de las estelas y su reutilización, bastante frecuente (Saba, 2000:152-153) se debe considerar en el marco de las continuas transformaciones sociales que tuvieron lugar en la Prehistoria Reciente, como forma de negar la justificación ideológica del sistema sustituido, en un fenómeno que se aprecia en otras zonas de Europa Occidental (Gomes et al., 1985:87-88; Boujot y Cassen, 1990-91; Kirk, 1993; Mezzena, 1997:70, 72, 92, 99-100, 104-106; Soares y Silva, 2000:220).

Como bien han resaltado A. Catte y G. Salis (1985:81) en relación a la concentración de menhires en el yacimiento de Biriai (Oliena, Nuoro) la función de éstos sería diversa a la demarcación sacra del territorio antes referida, aspecto que también hemos reseñado anteriormente en relación a las necrópolis dolménicas o hipogeicas concentradas en torno a determinados poblados (Cámara, 1998a).

La presencia de menhires en áreas de poblado merece una discusión más detallada en cuanto al control del territorio y de los hombres pues junto a las sepulturas de diverso tipo y los menhires en la Edad del Cobre aparecen asentamientos fortificados como Monte Baranta (Olmedo, Sassari) y Monte Ossoni (Castelsardo, Sassari) adscritos a la cultura Monte Claro (Moravetti, 1988:528, 2000:29; Contu, 1988:448; Atzeni, 1988a:452). Habría que pensar en realidad que la distinción entre poblados y santuarios llega a ser muy difusa como muestran los hallazgos de Pranu Muttedu (Goni) con menhires en alineamientos largos, círculos megalíticos y tumbas con cámaras excavadas en bloques de roca (Atzeni, 1982:33-34, 1988a:449-450, 1989:202) y relacionado con un poblado de fondos de cabaña (Atzeni, 1989:200). Lo mismo se puede aplicar al yacimiento de Monte d'Accoddi (Sassari) con ziggurat que incluye menhires y estela con figura femenina, piedras altar (S. Tiné et al., 1989:19; S. Tinè, 1992:VII-VIII; Traverso, 1992:XXII) y otros elementos junto al poblado al menos desde época Filigosa, aunque recientemente se ha atribuido la primera área sacra (menhir y una de las piedra-altar) a una fase del Neolítico Medio (Lo Schiavo, 1995:11).

La relación entre poblado y santuario es aún más compleja en Biriai (Oliena) por su posición estratégica (Castaldi, 1992:81). Erigido en época Monte Claro, muestra 13 menhires y una estructura de culto junto al poblado en la parte de mayor visibilidad a la que se accedería por medio de terrazas (Castaldi, 1983b:154, 162, 1987:122, 128, 1992:77), el poblado de Monte Baranta (Olmedo, Sassari) donde menhires, círculos y muros megalíticos y edificio fortificado en herradura dan especial relieve al poblado, aunque situándose al exterior (Moravetti, 1992:20, 2000:53-57) en una zona que, sin embargo, también es ocupada por estructuras de hábitat (Moravetti, 2000:35). No debemos olvidar tampoco las cistas con túmulo asociadas a menhires y a dispersiones líticas y cerámicas que pueden corresponder a poblados en Li Muri, Malchittu, Monte 'Ncappiddattu, Li Tauli (Arzachena, Gallura), o la de Tracucu e Genna Tramonti (Lotzorai, Nuoro) con 25 domus de janas, un menhir fragmentado, un posible altar, diez estructuras megalíticas y dos fuertes concentraciones de restos líticos y cerámicos (Frau, 1996:294-298) . Son evidencias que hay que relacionar con las que ya hemos referido respecto a la posible asociación de las necrópolis más importantes de domus de janas a los poblados de más entidad o, al menos, de mayor tradición, siendo de destacar que frente a la relación posible entre Santu Pedru y Monte Baranta (Olmedo, Sassari) (Moravetti, 1992a) no se ha localizado la necrópolis de Biriai (Castaldi, 1986a:29).

A veces, sin embargo, habría que indagar en la contemporaneidad de cada uno de los fenómenos (poblados, menhires, tumbas de diverso tipo, etc.) que se encuentran en un yacimiento concreto, pues por ejemplo la continuidad de los poblados no está clara en ausencia de sondeos arqueológicos y cuando la mayoría del material recogido en superficie se sitúa en una única fase como sucede en las áreas de S. Michel-chiesa (Fonni), Pedras Fitas (Ovodda), etc. (Lilliu, 1981:12-13, 39-40, 51, 105, 118, 126). La interpretación puede ser diversa. Primero si las estructuras son, al menos, parcialmente contemporáneas cabe pensar en santuarios dentro de los poblados, mas no se podría aceptar en base a los datos presentados, con pocos materiales posteriores a Ozieri y ninguno calcolítico (Lilliu, 1981:105-115) una perduración de los poblados hasta una época que incluyera las galerías cubiertas y las tumbas de gigantes. Segundo si las estructuras se superponen a un hábitat previo o se asocian a puntos de descanso de escasa entidad se está produciendo un fenómeno de sacralización de las áreas de ocupación tradicional y de las líneas de desplazamiento que iniciándose en el Neolítico culminaría con las tumbas de gigante, siendo el carácter esporádico del descanso el que explicaría los escasos materiales tardíos .

EPÍLOGO

A lo largo de este trabajo hemos pretendido mostrar cómo los fenómenos rituales documentados en la Prehistoria Reciente sarda tienen una profunda relación con la organización socioeconómica de las comunidades que los realizaron y, por tanto, con la justificación y/o la subversión del orden establecido. En este sentido creemos que referir únicamente la relación de una manifestación arqueológica con la religiosidad o con divinidades presuntamente universales no genera ninguna explicación real de la organización social prehistórica y, por tanto, es un hábito simplificador que debe, absolutamente, ser rechazado.

NOTAS

(1) Este trabajo ha sido realizado gracias a la Beca Posdoctoral concedida a Juan Antonio Cámara Serrano por la Universidad de Granada para realizar un trabajo de investigación bajo la dirección del profesor Alberto Moravetti del Dipartamento di Storia della Università degli Studi di Sassari entre septiembre de 1997 y agosto de 1998, y a la Beca concedida a Liliana Spanedda por los Ministerios de Asuntos Exteriores italiano y español para realizar un trabajo de investigación en el Departamento de Prehistoria y Arqueología de la Universidad de Granada entre enero y mayo del 2001, bajo la dirección de la Profesora Trinidad Nájera Colino.

(2) Las referencias a nuestras láminas se encuentran en el texto en cursiva, mientras que las referencias a figuras y láminas de otras publicaciones se presentan en redonda.

(3) Estas estructuras son similares a las que se constatan en el sur de la Península Ibérica en las viviendas neolíticas y calcolíticas, aunque con cierta evolución y diferenciación según la función (Molina et al., 1986; Afonso et al., 1996; Ramos et al., 1997)

(4) La autora (Castaldi, 1983b:169-170) ha relacionado incluso la variabilidad de las viviendas de la cultura Monte Claro de Biriai (Oliena, Nuoro) con las representaciones esquemáticas en el abrigo de Frattale (Moravetti, 1980:204, 204 n.13).

(5) Como también había destacado E. Contu (1962:634) en relación a la importancia de la ganadería en base al alto número de representaciones en una misma domus de janas.

(6) En este último caso de confirmarse las fechas avanzadas (Tanda, 1984:II-71 y ss.) de la trasposición podría sugerirse el alejamiento del mundo de ultratumba de la Naturaleza y la confección de una verdadera religión sobrenatural y enmascaradora.

(7) Con lo que la suposición de que los pilares representan el cuerpo fálico presentada por determinados autores (Lilliu, 1988:253, 256) es innecesaria, y si bien es compatible con nuestra interpretación no lo es en absoluto con la del mismo G. Lilliu que considera las espirales la representación de una divinidad femenina (Lilliu, 1988:229) en la que el falo estaría fuera de lugar.

(8) Si bien no conocemos ninguna aseveración tajante de esta función decorativo-arquitectónica sí conocemos una opinión totalmente contraria de E. Contu (1964:81-82).

(9) Especialmente cuando bóvidos y équidos se comienzan a usar también en las labores agrícolas (Martínez y Afonso, 1998; Cámara y Afonso, en prensa).

(10) En esta obra creemos además que E. Castaldi (1986b) no ha sabido deslindar lo que es el agradecimiento a los fenómenos naturales, y su escenificación, del desarrollo de religiones sobrenaturales, y su conexión, a través de la representación animal antropomorfa, con la creación de unidades sociales y con la justificación del poder.

(11) A veces, pese a señalarse el papel de bien de prestigio de los rebaños y su importancia para la producción se valoran los símbolos (del tributo o de la dote) en términos de "intercambio" (Cassano, 1994:223, 227, 229). El carácter perfectamente reglado de éstos, su concentración al interior de la formación social, el diferente carácter de los bienes "intercambiados", y el papel de esta circulación en la reproducción de la organización social desigual, en su acentuamiento, nos han llevado a caracterizarlos como un "tributo enmascarado" (Cámara, 1998a).

(12) Representaciones de bóvidos y carneros en arcilla se conocen también en el sur de la Península Ibérica (Arribas y Molina, 1979a, 1979b).

(13) Si bien en la Península Ibérica también se ha sugerido que los animales de mayor talla tendrían más relevancia en cuanto a la exhibición de la riqueza (Martínez y Afonso, 1998).

(14) El uso de los animales en los rituales de fundación y delimitación se aprecia en el área megalítica de Saint-Martin-de Corleans en la periferia de Aosta, con cenizas y restos de cráneo de carnero o bóvidos, según las últimas publicaciones, en la base de la primera alineación de postes (Mezzena, 1982:123, 1997:85), interpretados también como trofeos exhibidos sobre los postes (Mezzena y Mollo Mezzena, 1992:18). Mientras también aquí la utilización de los restos humanos (en este caso dientes) se da en un momento ligeramente posterior y asociados a marcas de arado previas y a la erección de nuevos postes, estelas y estructuras en plataforma e incluso, tal vez, una tumba megalítica (la número II) (Mezzena, 1982:124, 1997:88-89, 103).

(15) Hay que señalar la presencia de una figurilla, de sexo indeterminado, junto a las manos de un individuo joven también en una cueva artificial neolítica de Apulia, concretamente en Arnesano (Lecce) (Lo Porto, 1973:357) adscrita a un momento ligeramente posterior a la de Serra d'Alto (Matera) y contemporánea alas de Scoglio del Tonno y Masseria Bellavista (Taranto).

(16) Por otra parte el considerar un dominio femenino en base a la importancia de la agricultura (Lilliu, 1989:21; Atzeni y Santoni, 1989:34) (por demostrar) supone también un determinismo económico, de raíz contraria al que B. Bender (1989) critica a C. Meillassoux (1987) al mantener que las mujeres deben ser reconocidas por el trabajo que realizan (¿acaso lo son hoy?) y basado además en una equivalencia burda de Economía, con estrategia económica (agricultura).

(17) Hemos destacado también el caso del caballo que no se reduce a la labor agrícola y el transporte general sino que se convierte en medio de transporte personal y símbolo de riqueza (Molina, 1983; Cámara, 1998a).

(18) Recientemente se ha hablado de enterramientos de "galería" en fechas tempranas (Lanfranchi, 1992:122) pero por sus dimensiones se pueden considerar cistas largas.

(19) Respecto a la agrupación de las domus de janas hay que decir que, al menos en el norte de la isla, en los últimos años las referencias a tumbas aisladas escasean (Demartis, 1990) salvo en los casos de necrópolis dispersas referidas para el Dorgalese (Fadda, 1980:47), predominando las menciones de tumbas en grupos relativamente dispersos (Melis, 1990; Pulacchini, 1994a, 1994b), que llegan a alcanzar hasta las trece tumbas en Oreharva (Orgosolo, Nuoro) (Fadda, 1987), hasta el punto que la definición de las necrópolis es problemática y se rige por la toponimia (Ferrarese Ceruti, 1989:37). En la necrópolis de Partulesi (Ittireddu, Sassari) se sitúan 30 tumbas en 1 km. (Galli, 1991:31), mientras al menos 27 incluye la necrópolis de San Pantaleo (Ozieri, Sassari) (Basoli et al., 1988:72), e incluso en zonas cercanas a la Gallura como el municipio de Oschiri (Sassari)se han citado veinticinco hipogeos en la necrópolis de Malghesi (Basoli, 1996:226). En algunos casos incluso se han podido recuperar referencias a tumbas hoy no visibles (Moravetti, 1989:83, 91 n. 4).

(20) El paralelo, sin embargo, no es exacto desde el momento que, en primer lugar el poblado de Monte Baranta (Olmedo, Sassari) muestra una potencia estratigráfica escasa y reducida a los aspectos Monte Claro, que existen otros yacimientos en el área muy cercanos (Moravetti, 2000:14) y, por último, que el uso de la necrópolis es, por el contrario mucho más duradero y antiguo (desde época Ozieri) (Moravetti, 1988, 1991, 1992).

(21) Es curioso como la arquitectura no suele recibir este tratamiento en los “manuales” prehistóricos, aun cuando, a veces, como en las tumbas, su función ideológica sea clara. Parece por ello que uno de los rasgos “clave” para definir un elemento como “artístico” en nuestra disciplina sea el de “significado oscuro”.

(22) El hecho de que no existan tales asociaciones en algunos casos (Frau, 1996:291) podría indicarnos que a la necesidad de demarcación cumplida por los menhires no se había asociado una exhibición de determinados clanes o familias a través de sus tumbas, lo que no quiere decir que la desigualdad no se manifestara en el desarrollo de las domus de janas y de centros ceremoniales como Tracucu e Genna Tramonti (Lotzorai) (Frau, 1996:294).

(23) Como la de Genna Arrele III (Laconi, Nuoro) (Atzeni, 1978:48, 1982:14, 23, fig. 5:3, 44 Lám. II:3) o, tal vez, algunas de las de Serra Monti Arcu como la I y la III (Isili, Nuoro) (Saba, 2000:118-120), claro que en muchos casos la interpretación es dudosa, al faltar la parte superior de la estela, y, sobre todo, si tenemos en cuenta la técnica de realización de los presuntos senos (incisión) y la posición de éstos en la estela, como veremos.

(24) Se han relacionado con representaciones antropomorfas esquemáticas presuntamente masculinas y asociadas a las coppelle, presuntamente femeninas, presentes en algunas necrópolis de domus de janas del área y que se han situado en el Calcolítico (Santoni, 1995a:115-116).

(25) Tal vez lo mismo se podría decir de las estatuas-menhir de la Liguria concentradas, sin embargo, en el área Lunigiana al este (Maggi y Lucchese, 1988:332).

(26) La secuencia propuesta por J. Cesari para las estatuas-estela corsas creemos, sin embargo, que peca por exceso al buscar la perduración hasta el Bronce Final (Cesari, 1992), aunque debe recordarse que es una fenomelogía presente en diversas zonas europeas, entre ellas el suroeste de la Península Ibérica (Galán, 1993) en esa época.

(27) Sin perjuicio de que, al menos algunos, pudieran tener otra utilidad sacralizada y Pitzalis (1996:204, 212) ha propuesto el uso de algunos como molinos múltiples.

(28) En este último caso interpretada como tumba de un personaje relevante al incluir también pequeños monolitos y un área circular, posiblemente ceremonial, rodeada de ortostatos (Lilliu, 1988:199).

(29) Áreas "sacras" se refieren también para momentos Ozieri, como los casos de Pedras Fittas (Ovodada, Nuoro) relacionada con un hábitat, Orrui (Fonni, Nuoro) y Ghortene (Orgosolo, Nuoro) (Castaldi, 1983b:203-204).

(30) Independientemente de la posible continuidad de la gran muralla de Monte Baranta que podría cerrar el acceso a la estructura en "herradura" hay que señalar la existencia de otras referencias a zonas cerradas tal vez aisladas entre sí en poblados fortificados sardos como Sa Punta 'e Sa Zittade, también llamado Monte Nieddu (Ottana, Nuoro) (Tanda, 1990:48; Depalmas, 1990:219) y la situación especial de cada uno de estos puntos para complementar el control del territorio, lo que es evidente en el recinto de Monte Baranta. De la misma forma y en relación a las áreas sacras la necrópolis de Los Millares se sitúa al exterior de las líneas de defensa, salvo una tumba que queda englobada en la muralla más reciente, habiéndose interpretado en los últimos años como una primera línea de defensa y delimitación sagrada (Jorge, 1994) y también como un área de regulación de la circulación de personas y animales (Martínez y Afonso,1998).

(31) Se señala un hábitat Ozieri a 600 metros de Li Muri (Arzachena, Sassari) (Antona, 1995:84).

(32) La presencia de tal variedad de monumentos, todos conectados al fenómeno megalítico, en un área de valor sacro-funerario y sobre todo de aquello que en las primeras investigaciones de superficie aparece como un poblado-santuario realizado con una técnica megalítica, la convierte cuanto menos en un área digna de interés y de estudio (Frau, 1996:294).

(33) Otras noticias como las de Lasasai (Oroni, Nuoro) con una muralla de 27 metros, aterrazamientos y 16 menhires (Manunza, 1987:401) que se ha querido comparar con Biriai (Oliena), o las de de Su Siddadu (Montrasta, Nuoro) donde dos círculos de menhires y un muro de cierre no se han asociado a materiales (Foschi, 1995:163), son todavía más difíciles de evaluar, aunque supondrían un paso más en la documentación de la generalización de las fortificaciones en el Calcolítico sardo.

(34) Aunque esto podría ser un problema de definición del material típico de esta región oriental para esta fase.

(35) Así hemos explicado los pequeños yacimientos de Velefique (Almería) para los que sin embargo, desgraciadamente, también carecemos de datos precisos de excavación (Cámara, 1998a; Maldonado et al., 1997)..

BIBLIOGRAFÍA

ABARQUERO, F.J.,
1997. El significado de la cerámica decorada de Cogotas I, Boletín del Seminario de Arte y Arqueología LXIII, Valladolid, 1997, pp. 71-96.

AFONSO, J.A., MOLINA, F., CÁMARA, J.A., MORENO, M., RAMOS, R., RODRÍGUEZ, Mª.O.,
1996. Espacio y tiempo. La secuencia en Los Castillejos de Las Peñas de Los Gitanos (Montefrío, Granada), I Congrés del Neolític a la Península Ibérica. Formació e implantació de les comunitats agrícoles (Gavà-Bellaterra, 1995). Actes. Vol. 1. (J. Bosch, M. Molist, Orgs.), Rubricatum 1:1, Gavà, 1996, pp. 297-304.

AMBROSI, A.C.,
1985. Quercia di Reusa (com. di Casola Lunigiana, prov. di Massa Carrara), L'Età dei metalli nella Toscana Nord-Occidentale (Mostra, Viareggio, Maggio-Dicembre 1985) (D. Cocchi Genick, R. Grifoni Cremonesi, Cur.), Pacini Editore, Pisa, 1985, pp. 367-369.

ANTONA, A.,
1995. La necropoli di Li Muri (Arzachena, Sassari), Sardegna (A. Moravetti, C. Tozzi, Cur.), Guide Archeologiche. Preistoria e Protostoria in Italia 2. XIII Congresso Internazionale dell Scienze Preistoriche e Protostoriche, U.I.S.P.P./Ministero per i Beni Culturali e Ambientali, A.B.A.C.O. Edizioni, Forlì, 1995, pp. 82-89.

ANTONA RUJU, A., LO SCHIAVO, F. ,
1989. Oredda - Sassari, la Domus delle Doppie Spirali, La Cultura di Ozieri. Problematiche e nuove acquisizioni. Atti del I Convegno di Studio (Ozieri, gennaio 1986 - aprile 1987), (L. Dettori Campus, Cur.), Il Torchieto, Ozieri, 1989, pp. 49-74.

ARRIBAS, A., MOLINA, F.,
1979a. El poblado de "Los Castillejos" en Las Peñas de los Gitanos (Montefrío, Granada). Campaña de excavaciones de 1971. El corte número 1, Cuadernos de Prehistoria de la Universidad de Granada. Serie Monográfica 3, Granada, 1979.

ARRIBAS, A., MOLINA, F.,
1979b. Nuevas aportaciones al inicio de la metalurgia en la Península Ibérica. El poblado de Los Castillejos de Montefrío, (Granada), Proceedings of the fifth Atlantic Colloquium, (M. Ryan, Ed.), Dublin 1979, pp. 7-34.

ATZENI, E.,
1982. Menhirs antropomorfi e statue-menhirs della Sardegna, Annali del Museo Civico "U. Formentini" della Spezia II (1979-80), 1982, pp. 9-64.

ATZENI, E. ,
985. Aspetti e sviluppi culturali del Neolitico e della prima età dei metalli in Sardegna, Ichnussa. La Sardegna dalle origini all'età classica (E. Atzeni, F. Barreca, Mª.L. Ferrarese Ceruti, E. Contu, G. Lilliu, F. Lo Schiavo, F. Nicosia, E. Equini Schneider), Garzonti/Scheiwiller, Milano, 1985 (2ª Ed.), pp. XIX-LI.

ATZENI, E.,
1988a.Megalitismo e Arte, [L'Età del Rame nell'Italia insulare: la Sardegna (E. Atzeni, E. Contu, M.L. Ferrarese Ceruti)], Rassegna di Archeologia 7, Congresso Internazionale L'Età del Rame in Europa (Viareggio, 1987), Firenze, 1988, pp. 449-456.

ATZENI, E. ,
1988. Statue-menhir di Laconi (Nuoro), Rassegna di Archeologia 7, Congresso Internazionale L'Età del Rame in Europa (Viareggio, 1987), Firenze, 1988, pp. 524-525.

ATZENI, E. ,
1989. Nota sulla necropoli megalitica di Pranu Mutteddu (Goni), La Cultura di Ozieri. Problematiche e nuove acquisizioni. Atti del I Convegno di Studio (Ozieri, gennaio 1986 - aprile 1987), (L. Dettori Campus, Cur.), Il Torchieto, Ozieri, 1989, pp. 201-209.

ATZENI, E. ,
1995a. La "cultura del vaso campaniforme" nella necropoli di Locci-Santus (S. Giovanni Suergiu), Carbonia e il Sulcis. Archeologia e territorio (V. Santoni, Cur.), Soprintendenza Archeologica per le province di Cagliari e Oristano/Comune di Carbonia, Oristano, 1995, pp. 117-143.

ATZENI, E.,
1995b. Necropoli ipogeica di Montessu (Villaperuccio, Cagliari), Sardegna (A. Moravetti, C. Tozzi, Cur.), Guide Archeologiche. Preistoria e Protostoria in Italia 2. XIII Congresso Internazionale dell Scienze Preistoriche e Protostoriche, U.I.S.P.P./Ministero per i Beni Culturali e Ambientali, A.B.A.C.O. Edizioni, Forlì, 1995, pp. 218-225.

ATZENI, E., SANTONI, V.,
1989. L'età prenuragica. Il Neolitico. L'Eneolitico, Il museo archeologico nazionale di Cagliari (V. Santoni, Cur.), Banco di Sardegna, Sassari, 1989, pp. 31-56.

BARNATT, J.,
1998. Monuments in the landscape: Thoughts from the Peak, Prehistoric ritual and religion (A. Gibson, D. Simpson, Eds.), Sutton Publishing, Phoenix, 1998, pp. 92-105.
BASOLI, P.,
1996. Oschiri (Sassari). Censimento territoriale, Bollettino di Archeologia 19-20-21 (1993), Roma, 1996, pp. 226-227.
BASOLI, P., FOSCHI NIEDDU, A. (1988): La Tomba della Protomi Taurine in località Montalé (Sassari), Rivista di Scienze Preistoriche XL (1985-86), Firenze, 1988, pp. 317-325.
BASOLI, P., LO SCHIAVO, F., DETTORI CAMPUS, L., GUIDO, F.,
1988. Ozieri, L'Antiquarium Arborense e i civici musei archeologici della Sardegna (G. Lilliu, Cur.), Banco di Sardegna, Sassari, 1988, pp. 71-92.
BENDER, B.,
1989. The roots of inequality, Domination and resistance (D. Miller, M. Rowlands, C. Tilley, Eds.), One World Archaeology 3, Unwin Hyman, London, 1989, pp. 83-95.
BLOCH, M. (1981): Tombs and states. Mortality and Inmortality, the anthropology and archeology of death, (B.C. Humphreys, H. King, Eds.), Academic Press, New York, 1981, pp. 137-147.
BLOCH, M.,
1982. Death, women and power. Death and regeneration of life. (M. Bloch, J. Parry, Eds.), Cambridge University Press, Cambridge, 1982, pp. 211-230.
BOUJOT, C., CASSEN, S.,
1990-91. Los problemas cronológicos del megalitismo en la fachada atlántica de Francia, Cuadernos de Prehistoria y Arqueología Castellonenses 15, Castellón, 1990-91, pp. 127-141.
CÁMARA, J.A.,
1998a. Bases teóricas y metodológicas para el estudio del ritual funerario utilizado en la Prehistoria Reciente del sur de la Península Ibérica, Tesis Doctoral, Universidad de Granada.
CÁMARA, J.A.,
1998b. El control del territorio en la Edad del Bronce. Una comparación entre las situaciones sarda y andaluza, Papers from the EAA Third Annual Meeting at Ravenna 1997. Volume III: Sardinia (A. Moravetti, M. Pearce, M. Tosi, Eds.), British Archaeological Reports. International Series 719, Oxford, 1998, pp. 67-71.
CÁMARA, J.A.
(en prensa a): Acumulación de riqueza pecuaria y representación ideológica. Algunas hipótesis sobre el significado de las representaciones animales en las "domus de janas", Nuovo Bollettino Archeologico Sardo, Sassari.
CÁMARA, J.A.
(en prensa b): Monumentalidad, variabilidad y exhibición del ajuar como diferentes formas de exhibición de la lucha social. Los ejemplos sardos e ibéricos en la Edad del Bronce, Nuovo Bollettino Archeologico Sardo, Sassari.
CÁMARA, J.A., AFONSO, J.A.
(en prensa): Una propuesta sobre el desarrollo de la desigualdad y las clases sociales en la Prehistoria Reciente de Andalucía, 1er Congreso Iberoamericano de Arqueología Social (La Rábida, Junio, 1996).
CÁMARA, J.A., LIZCANO, R. ,
1996. Ritual y sedentarización en el yacimiento del Polideportivo de Martos (Jaén), I Congrés del Neolític a la Península Ibérica. Formació e implantació de les comunitats agrícoles (Gavá-Bellaterra, 1995). Actes. Vol. 1. (J. Bosch, M. Molist, Orgs.), Rubricatum 1:1, Gavà, 1996, pp. 313-322.
CANALIS, V.,
1995. Mesu 'e Montes: sintesi dell'arte ipogeica, [Una casa per le fate (B. D'Arragon, Cur.)], Archeologia Viva XIV:54, Novembre-Dicembre 1995, pp. 56.
CAPRARA, R.,
1986. La necropoli di S. Andrea Priu, Sardegna Archeologica. Guide e Itinerari 3, Carlo Delfino Editore, Sassari, 1986.
CARANDINI, A.,
1984. Arqueología y Cultura Material, Textos de Antropología, Mitre, Barcelona, 1984.
CARDOSO, J.L., CANINAS, J.C., HENRIQUES, F.,
1997. Contributos para o conhecimento do megalitismo na Beira Interior (Portugal): A regiâo do Tejo Internacional, II Congreso de Arqueología Peninsular (Zamora, 24-27 de Septiembre de 1996). T. II. Neolítico. Calcolítico y Bronce, (R. de Balbín, P. Bueno, Eds.), Fundación Rei Alfonso Henriques. Serie Actas, Zamora, 1997, pp. 207-215.
CASSANO, S.M.,
1994. La facies Serra d'Alto:intensificazione delle attività podutive e aspetti del rituale, Origini. Preistoria e Protostoria delle civiltà antiche XVII (1993), Roma, 1994, pp. 221-253.
CASTALDI, E.,
1978. Il "culto del toro" nella preistoria della Sardegna ed il problema delle tre cavità sull'alto dei prospetti della tombe di giganti, Archivio per la Antropologia e la Etnologia CVI (1976), Firenze, 1978, pp. 439-458.
CASTALDI, E.,
1979. Una particolare rappresentazione zoomorfa in ipogei sardi, Rivista di Scienze Preistoriche XXXIII (1978), Firenze, 1979, pp. 393-398.
CASTALDI, E.,
1983a. Ancora sulla stelle delle tombe di giganti, Bullettino di Paletnologia Italiana. Nuova Serie XXIV:82 (1975-80), Roma, 1983, pp. 267-285.
CASTALDI, E.,
1983b. Villaggio con santuario a Biriai (Oliena, Nuoro). (Il Relazione preliminare), Rivista di Scienze Preistoriche XXXVI (1981), Firenze, 1983, pp. 153-221.
CASTALDI, E.,
1986a. Biriai e le fortezze pre-nuragiche: per una volutazione socio-culturale, Studi Urbinati B3, LVIII (1985), Urbino, 1986, pp. 29-54.
CASTALDI, E.,
1986b. Totemismo in preistoria, Seminario di Scienze Antropologiche VII (1985), Urbino, 1986, pp. 139-156.
CASTALDI, E.,
1987. L'architettura di Biriai (Oliena, Nuoro), Rivista di Scienze Preistoriche XXXIX (1984), Firenze, 1987, pp. 119-153.
CASTALDI, E.,
1992. Il santuario di Biriai, Monte d'Accoddi. 10 anni di nuovi scavi. Relazione Preliminare e Atti del Colloquio (Sassari, 18-20 ottobre 1990), (S. Tinè, A. Traverso, Cur.), Istituto Italiano Archeologia Sperimentale/Soprintendenza Archeologica di Sassari e Nuoro, Genova, 1992, pp. 77-81.
CATTE, A., SALIS, G.,
1985. La comunità montana. Itinerari archeologici, Oliena. Territorio, ambiente, architettura, Comune di Oliena/Università degli

Studi di Cagliari/Comunità Montana del Nuorese, Sassari, 1985, pp. 81-87.
CESARI, J.,
1992. New contributions to the study of the Megalithic in Corsica, Sardinia in the Mediterranean: a footprint in the sea. Studies in Sardinian Archeology presented to Miriam S. Balmuth, (R.H. Tykot, T.K. Andrews, Eds.), Monographs in Mediterranean Archaeology 3, Sheffield Academic Press, Sheffield, 1992, pp. 105-117.
CIPOLLONI SAMPÓ, M.,
1983. Scavi nel villaggio neolitico di Rendina (1970-76). Relazione preliminare, Origini. Preistoria e Protostoria delle civiltà antiche XI (1977-82), Roma, 1983, pp. 183-323.
CONKEY, M.,
1980. The identification of prehistoric Hunter-gatherer aggregation sites: the case of Altamira, Current Anthropology 21:5, 1980, pp. 609-620.
CONTRERAS, F., CÁMARA, J. A.,
2000. El poblado de la Edad del Bronce de Peñalosa (Baños de la Encina, Jaén). Análisis espacial y estructural del poblado de Peñalosa, Análisis Histórico de las Comunidades de la Edad del Bronce del piedemonte meridional de Sierra Morena y Depresión Linares-Bailen. Proyecto Peñalosa. (F. Contreras, Coord.), Arqueología. Monografías 10, Consejería de Cultura. Dirección General de Bienes Culturales. Sevilla, 2000, pp. 273-286 incluye 274/1-274/304 en CD-ROM.
CONTU, E.,
1953. Costruzione megalitica in località Monte d'Accoddi (Sassari), Rivista di Scienze Preistoriche VIII (1953), Firenze, 1953, pp. 199-202.
CONTU, E.,
1958. Argomenti di cronologia a proposito delle tombe a poliandro di Ena 'e Muros (Ossi, Sassari) e Motrox 'e Bois (Usellus, Cagliari), Studi Sardi XIV-XV:I (1955-57), Cagliari, 1958, pp. 129-196.
CONTU, E.,
1959. I più antichi nuraghi e l'explorazione del Nuraghe Peppe Gallu(Uri, Sassari), Rivista di Scienze Preistoriche XIV (1959), Firenze, 1959, pp. 59-121.
CONTU, E.,
1962. Alcune osservazioni su "domus de janas" edite ed inedite di Alghero e Sassari, Studi Sardi XVII (1959-61), Cagliari, 1962, pp. 626-635.
CONTU, E.,
1964. La tomba dei vasi tetrapodi in località Santu Pedru (Alghero, Sassari), Estratto di Monumenti Antichi XLVII, Accademia Nazionale dei Lincei, Roma, 1964.
CONTU, E.,
1965. Tombe preistoriche dipinte e scolpite di Thiesi e Bessude (Sassari), Rivista di Scienze Preistoriche XIX (1964), Firenze, 1965, pp. 233-263.
CONTU, E.,
1967. Ipogei con "corna sacrificali" plurime di Bròdu (Oniferi, Nuoro), Rivista di Scienze Preistoriche XXI (1966), Firenze, 1967, pp. 195-200.
CONTU, E.,
1988. Problemática ed inquadramento culturale, [L'Età del Rame nell'Italia insulare: la Sardegna (E. Atzeni, E. Contu, M.L. Ferrarese Ceruti)], Rassegna di Archeologia 7, Congresso Internazionale L'Età del Rame in Europa (Viareggio, 1987), Firenze, 1988, pp. 441-447.
CONTU, E.,
1995. Necropoli ipogeica di Santu Pedru (Alghero, Sassari), Sardegna. Guide Archeologiche. Preistoria e Protostoria in Italia 2 (A. Moravetti, C. Tozzi, Cur.), A.B.A.C.O. Edizioni, Forlì, 1995, pp. 18-25.
CONTU, E.,
2000. L'altare preistorico di Monte d'Accoddi, Sardegna Archeologica. Guide e Itinerari 29, Sassari, 2000.
COSSU, A.Mª. (1996): Nuove statue menhirs ed un inedito petroglifo nel territorio di Allai (Oristano), Studi Sardi XXX (1992-93), Cagliari, 1996, pp. 299-328.
DEMARTIS, G.Mª.,
1990. Sassari. Loc. Oredda. La Domus delle Doppie Spirali, Bollettino di Archeologia 1-2 (1990), Roma, 1990, pp. 250-251.
DEMARTIS, G.Mª.,
1991a. La necropoli di Puttu Codinu. Villanova Monteleone (SS), Sardegna Archeologica. Guide e Itinerari 13, Carlo Delfino Editore, Sassari, 1991.
DEMARTIS, G.Mª.,
1991b. Putifigari (Sassari). Località Monte Siseri. Tomba dell'Architettura Dipinta o di S'Incantu, Bollettino di Archeologia 4 (1990), Roma, 1991, pp. 107-110.
DEMARTIS, G.Mª.,
1992. La tomba dell'Architettura Dipinta. Un ipogeo neolitico di Putifigari, Bollettino di Archeologia 7 (1991), Roma, 1992, pp. 1-21.
DEMARTIS, G.Mª.,
1993. L'architettura funeraria, Sardegna. Civiltà di una isola mediterranea. Genova, Palazzo Ducale, Logia degli Abati (19 dicembre 1993 - 20 febbraio 1994), (G. Rossi, Cur.), Nuova Alfa Editoriale, Bologna, 1993, pp. 31.
DEMARTIS, G.Mª.,
1995a. Necropoli ipogeica di Anghelu Ruju (Alghero, Sassari), Sardegna (A. Moravetti, C. Tozzi, Cur.), Guide Archeologiche. Preistoria e Protostoria in Italia 2. XIII Congresso Internazionale dell Scienze Preistoriche e Protostoriche, U.I.S.P.P./Ministero per i Beni Culturali e Ambientali, A.B.A.C.O. Edizioni, Forlì, 1995, pp. 38-45.
DEMARTIS, G.Mª.,
1995b. La necropoli di Anghelu Ruju. Guida al percorso, Soprintendenza Archeologica per le province di Sassari e Nuoro, Chiarella, Sassari, 1995.
DEMARTIS, G.Mª.,
1999a. Tumba V de Montalé, Sassari. Necrópolis de Su Crucifissu Mannu, Porto Torres, El Triángulo de la Nurra 2-4, Soprintendenza Archeologica per le Province di Sassari e Nuoro, Sassari, 1999.
DEMARTIS, G.Mª.,
1999b. Necrópolis hipogea de Anghelu Ruju, Alghero, El Triángulo de la Nurra 10, Soprintendenza Archeologica per le Province di Sassari e Nuoro, Sassari, 1999.
DEMARTIS, G.Mª.,
1999c. Necrópolis hipogea de Santu Pedru, Alghero, El Triángulo de la Nurra 11, Soprintendenza Archeologica per le Province di Sassari e Nuoro, Sassari, 1999.
DEMARTIS, G.Mª.,
2001. Le Domus de Janas della Nurra, Il Triangolo della Nurra. Il circuito archeologico della Sardegna nord-occidentale, Soprintendenza Archeologica per le Province di Sassari e Nuroro, Imago

Media Editrice, Sassari, 2001.
DEMARTIS, G.Mª., CANALIS, V.,
1989. La Tomba II di Mesu 'e Montes (Ossi, Sassari), Nuovo Bullettino Archeologico Sardo 2 (1985), Sassari, 1989, pp. 41-74.
DEPALMAS, A.,
1989. L'insediamento preistorico di Sorralia (Norbello, Oristano), Quaderni della Soprintendenza Archeologica per le province di Cagliari e Oristano 5 (1988), Cagliari, 1989, pp. 7-20.
DEPALMAS, A.,
1990. Schede dei monumenti, Ottana. Archeologia e territorio (G. Tanda, Cur.), Amministrazione Comunale di Ottana, 1990, pp. 173-230.
DEPALMAS, A.,
1998. The Burial as an Hypogeum 2 in Iloi-Ispiluncas (Sedilo, Oristano), Papers from the EAA Third Annual Meeting at Ravenna 1997. Volume III: Sardinia (A. Moravetti, M. Pearce, M. Tosi, Eds.), British Archaeological Reports. International Series 719, Oxford, 1998, pp. 20-24.
DERUDAS, P.Mª.,
2000. Archeologica del territorio di Ossi, Il Triangolo della Nurra. Il circuito archeologico della Sardegna nord-occidentale, Soprintendenza Archeologica per le Province di Sassari e Nuroro, Imago Media Editrice, Sassari, 2000.
FADDA, Mª.A.,
1980. Domus de janas: aspetti di architettura ipogeica, Dorgali. Documenti Archeologici (AA.VV.), Soprintendenza Archeologica per le province di Sassari e Nuoro/Chiarella, Sassari, 1980, pp. 47-55.
FADDA, Mª.A.,
1987.Orgosolo (Prov. di Nuoro), Rivista di Scienze Preistoriche XXXIX (1984), Firenze, 1987, pp. 409-410.
FADDA, Mª.A.,
1997. Gli architetti nuragici di Gremanu, Archeologia viva 63, Maggio-Giugno 1997, pp. 70-75.
FERRARESE CERUTI, Mª.L. (1980): Le domus de janas di Mariughia e Canudedda e il dolmen di Motorra, Dorgali. Documenti Archeologici (AA.VV.), Soprintendenza Archeologica per le province di Sassari e Nuoro/Chiarella, Sassari, 1980, pp. 57-65.
FERRARESE CERUTI, Mª.L.,
1988. Il campaniforme in Sardegna, [L'Età del Rame nell'Italia insulare: la Sardegna (E. Atzeni, E. Contu, M.L. Ferrarese Ceruti)], Rassegna di Archeologia 7, Congresso Internazionale L'Età del Rame in Europa (Viareggio, 1987), Firenze, 1988, pp. 456-460.
FERRARESE CERUTI, Mª.L.,
1989. La Necropoli di Su Crucifissu Mannu - Porto Torres e di Ponte Secco - Sassari, La Cultura di Ozieri. Problematiche e nuove acquisizioni. Atti del I Convegno di Studio (Ozieri, gennaio 1986 - aprile 1987), (L. Dettori Campus, Cur.), Il Torchieto, Ozieri, 1989, pp. 37-47.
FERRARESE CERUTI, Mª.L.,
1992. Le necropoli a domus de janas nel territorio di Monte d'Accoddi, Monte d'Accoddi. 10 anni di nuovi scavi. Relazione Preliminare e Atti del Colloquio (Sassari, 18-20 ottobre 1990), (S. Tiné, A. Traverso, Cur.), Istituto Italiano Archeologia Sperimentale/ Soprintendenza Archeologica di Sassari e Nuoro, Genova, 1992, pp. 10-20.
FOSCHI, A.,
1984. I resultati degli scavi 1981 nella necropoli prenuragica di Serra Crabiles, Sennori (Sassari), The Deya Conference of Prehistory. Early settlement in the Western Mediterranean Islands and their Peripheral Areas. Part II (W.H. Waldren, R. Chapman, J. Lewthwaite y R.-C. Kennard, Eds.), British Archaeological Reports. International Series 229:2, Oxford, 1984, pp. 533-552.
FRAU, M.,
1991: Aspetti del megalitismo prenuragico nella Sardegna centro-orientale, Studi Sardi XXX (1992-93), Cagliari, 1996, pp. 219-298.
GALÁN, E.,
1993. Estelas, paisaje y territorio en el Bronce Final del Suroeste de la Península Ibérica, Complutum Extra 3, Universidad Complutense, Madrid, 1993.
GALLI, F.,
1984. La domus de janas di Butule (Ozieri, Sassari), Ricerche e scoperte nella Sardegna centro-settentrionale 1980-1982 (P. Desantis, M.A. Fadda, F. Galli, F. Lo Schiavo, M.R. Manunza, G. Pitzalis), Rivista di Scienze Preistoriche XXXVII (1982), Firenze, 1984, pp. 267-276.
GALLI, F.,
1990. Padria. Il Museo e il territorio, Sardegna Archeologica. Guide e Itinerari 18, Carlo Delfino Editore, Sassari, 1991.
GERMANÀ, F.,
1996: La morte nella Sardegna preistorica e protostorica (Annotazioni e commenti di un antropologo), Studi Sardi XXX (1992-93), Cagliari, 1996, pp. 441-478.
GERMANÀ, F., SANTONI, V.,
1993. La necropoli di Cuccuru S'Arriu (Cabras) e I paleosardi medioneolitici, Quaderni della Soprintendenza Archeologica per le province di Cagliari e Oristano 9 (1992), Cagliari, 1993, pp. 5-30.
GODELIER, M.,
1989. Lo ideal y lo material. Pensamiento, economías, sociedades, Taurus, Barcelona, 1989.
GOMES, R.V., GOMES, M.V., SANTOS, M.F. dos,
1985. Santuárico exterior e povoado calcolítico do Escoural, Clio/ Arqueología. Revista da UNIARCH 1 (1983-84), Lisboa, 1985, pp. 77-78.
GUILAINE, J.,
1996. La Neolithisation de la Méediterranée Occidentale, The Neolithic in the Near East and Europe. The Colloquia of the XIII International Congress of Prehistoric and Protostoric Sciences (Forlì, Italia, 8-14 September 1996). Vol 9 (R. Grifoni Cremonesi, J. Guilaine, J.L'Helgouach, Eds.), Colloquium XVII. The Processes of Neolithisation and the Development of Neolithic Cultures in Large Geographical Areas: Anatolia and the Near East; the Aegean and the Balkans; the Central Danubian Area and the Black Lands; the Loess Central Danubian Area and the Black Lands; the Loess Areas and Northern Europe; the Western Mediterranean, A.B.A.C.O., Forlì, 1996, pp. 53-68.
GUTELMAN, M.,
1981. Estructuras y reformas agrarias. Los problemas agrarios y los métodos para su estudio, Fontamara, Barcelona, 1981 (2ª Edic.).
HAYDEN, C.,
1999. Houses and monuments: two aspects of settlements in Neolithic and Copper Age Sardinia, Making Places in the Prehistoric World. Themes in Settlement Archaeology (J. Brück, M. Goodman), UCL Press, London, 1999, pp. 112-128.
HODDER, I.,
1987. Contextual Archaeology: an interpretation of Çatal Hüyuk

and a discussion of the origins of agriculture, Bulletin of the Institute of Archaeology 24, London, 1987, pp. 43-56.

JORGE, S. de O.,
1994. Colónias, fortificaçôes, lugares monumentalizados. Trajectória das concepçôes sobre um tema do Calcolítico peninsular, Revista da Faculdade de Letras II Série XI, Porto, 1994, pp. 447-546.

KIRK, T.,
1993. Space, subjectivity, power and hegemony: megaliths and long mounds in Earlier Neolithic Brittany, Interpretative Archaeology (C. Tilley, Ed.), Explorations in Anthropology Series, Berg, Exeter, 1993, pp. 181-223.

KRISTIANSEN, K.,
1984. Ideology and material culture: an archaeological perspective, Marxist perspectives in Archaeology, (M. Spriggs. Ed.), New Directions in Archaeology, Cambridge University Press, Cambridge, 1984, pp. 72-100.

LANFRANCHI, F. de,
1980. Les mégalithes corses et leur contexte. Problèmes et associations remarquables, Atti della XXII Riunione Scientifica dell'Istituto Italiano di Preistoria e Protostoria nella Sardegna centro-settentrionale (21-27 Ottobre 1978), (AA.VV.), Istituto Italiano di Preistoria e Protostoria, Firenze, 1980, pp. 449-467.

LANFRANCHI, F. de,
1992. The megalithic monuments of Corsica and Sardinia: a comparative study, Sardinia in the Mediterranean: a footprint in the sea. Studies in Sardinian Archeology presented to Miriam S. Balmuth, (R.H. Tykot, T.K. Andrews, Eds.), Monographs in Mediterranean Archaeology 3, Sheffield Academic Press, Sheffield, 1992, pp. 118-127.LANFRANCHI, F. de; WEISS, M.C.,
1980. Les chambres funéraires mégalithiques incluses dans un tumulus limité par des cercles de pierres ou dalles, Atti della XXII Riunione Scientifica dell'Istituto Italiano di Preistoria e Protostoria nella Sardegna centro-settentrionale (21-27 Ottobre 1978), (AA.VV.), Istituto Italiano di Preistoria e Protostoria, Firenze, 1980, pp. 469-473.

LERNER, G.,
1990. La creación del patriarcado, Crítica,
Barcelona, 1990.

LILLIU, G.,
1981. Monumenti antichi barbaricini, Quaderni della Soprintendenza ai Beni Archeologici per le province di Sassari e Nuoro, Sassari, 1981.

LILLIU, G.,
1982. La civiltà nuragica, Sardegna Archeologica. Studi e Monumenti 1, Carlo Delfino Editore, Sassari, 1982.

LILLIU, G.,
1988. La civiltà dei Sardi dal Paleolitico all'età dei nuraghi, Nuova Eri Edizioni RAI, Torino, 1988 (3ª Ed. rev. y amp.).

LILLIU, G.,
1989. Lineamenti di cultura materiale dal Neolitico all'Alto Medioevo, Il museo archeologico nazionale di Cagliari (V. Santoni, Cur.), Banco di Sardegna, Sassari, 1989, pp. 21-30.

LILLIU, G.,
1995a. Preistoria e protostoria del Sulcis, Carbonia e il Sulcis. Archeologia e territorio (V. Santoni, Cur.), Soprintendenza Archeologica per le province di Cagliari e Oristano/Comune di Carbonia, Oristano, 1995, pp. 11-50.

LILLIU, G.,
1995b. Betili e betilini nelle tombe di giganti della Sardegna, Atti della Accademia Nazionale dei Lincei. Memorie Serie IX, Vol. VI, Fasc. IV, Roma, 1995, pp. 421-507.

LILLIU, G.,
1999. Arte e religione della Sardegna prenuragica. Idoletti, ceramiche, oggetti d'ornamento, Sardegna Archeologica. Studi e Monumenti 4, Sassari, 1999.

LIZCANO, R.,
1999. El Polideportivo de Martos (Jaén): un yacimiento neolítico del IV Milenio A.C., Obra Social y Cultural Cajasur, Córdoba, 1999.

LIZCANO, R., CÁMARA, J.A., RIQUELME, J.A., CAÑABATE, Mª.L., SÁNCHEZ, A., AFONSO, J.A.,
1997. El Polideportivo de Martos. Estrategias económicas y símbolos de cohesión en un asentamiento del Neolítico Final del Alto Guadalquivir, Cuadernos de Prehistoria de la Universidad de Granada 16-17 (1991-92), Granada, 1997, pp. 5-101.

LOCCI, Mª.C,
1991. Necropoli ipogeica in località Fossada (Escalaplano, Nuoro), Studi Sardi XXIX (1990-91), Sassari, 1991, pp. 11-54.

LO PORTO, F.G.,
1973. La toma neolitica con idolo in pietra di Arnesano (Lecce), Rivista di Scienze Preistoriche XXVII:1 (1972), Firenze, 1973, pp. 357-372.

LO SCHIAVO, F.,
1980. Dorgali e il suo teritorio nella Preistoria e Protostoria: osservazioni e problemi, Dorgali. Documenti Archeologici (AA.VV.), Soprintendenza Archeologica per le province di Sassari e Nuoro/ Chiarella, Sassari, 1980, pp. 9-16.

LO SCHIAVO, F.,
1984. La domus dell'Ariete (Perfugas, Sassari), Rivista di Scienze Preistoriche XXXVII (1982), Firenze, 1984, pp. 135-186.

LO SCHIAVO, F.,
1995. Santuario prenuragico di Monte d'Accoddi (Sassari), Sardegna (A. Moravetti, C. Tozzi, Cur.), Guide Archeologiche. Preistoria e Protostoria in Italia 2. XIII Congresso Internazionale dell Scienze Preistoriche e Protostoriche, U.I.S.P.P./Ministero per i Beni Culturali e Ambientali, A.B.A.C.O. Edizioni, Forlì, 1995, pp. 10-17.

MALDONADO, Mª.G., MOLINA, F., ALCARAZ, F.M., CÁMARA, J.A., MÉRIDA, V., RUIZ, V.,
1997. El papel social del megalitismo en el Sureste de la Península Ibérica. Las comunidades megalíticas del Pasillo de Tabernas, Cuadernos de Prehistoria de la Universidad de Granada 16-17 (1991-92), Granada, 1997, pp. 167-190.

MANUNZA, Mª.R.,
1995. Dorgali. Monumenti antichi, Soprintendenza Archeologica per le province di Sassari e Nuoro, Oristano, 1995.

MARTÍNEZ, G., AFONSO, J.A. (1998): Las sociedades prehistóricas: de la Comunidad al Estado, De Ilurco a Pinos Puente. Poblamiento, economía y sociedad de un pueblo de la Vega de Granada (R. Peinado, Ed.), Diputación Provincial de Granada, Granada, 1998, pp. 21-68.

MARTÍNEZ, J.,
1998. Abrigos y accidentes geográficos como categorías de análisis en el paisaje de la pintura rupestre esquemática. El Sudeste como marco, Arqueología Espacial 19-20. Arqueología del Paisaje, Teruel, 1998, pp. 543-561.

MEILLASSOUX, C.,1987. Mujeres, graneros y capitales. Eco-

nomía doméstica y capitalismo, S. XXI, México, 1987, (8ª edición).
MELIS, P.,
1990. Sedini (Sassari). La domu di Li Algasa, Nuovo Bullettino Archeologico Sardo 3 (1986), Sassari, 1990, pp. 295-300.
MELONI, G.Mª.,
1998. The relationship between Funerary and Civil Architecture in Prenuragic Sardinia, Papers from the EAA Third Annual Meeting at Ravenna 1997. Volume III: Sardinia (A. Moravetti, M. Pearce, M. Tosi, Eds.), British Archaeological Reports. International Series 719, Oxford, 1998, pp. 121-126.
MEZZENA, F.,
1982. Ricerche preistoriche e protostoriche in Valle d'Aosta. Risultati e prospettive, Rivista di Studi Liguri XLI-XLII:1-4 (1975-76), Bordighera, 1982, pp. 92-146.
MEZZENA, F.,
1997. La Valle d'Aosta nel Neolitico e nell'Eneolitico, Atti della XXXI Riunione Scientifica dell'Istituto Italiano di Preistoria e Protostoria in valle d'Aosta (Courmayer, 2-5 Giugno 1994), (F. Mezzena, A. Palma di Cesnola, R. Peroni, Orgs.), Istituto Italiano di Preistoria e Protostoria, Firenze, 1997, pp. 17-133.
MEZZENA, F., MOLLO, R.,
1992. L'Archeologia, Dai dolmen alla città. Progetti per l'area megalitica di Saint-Martin-de-Corléans (Aosta Centro Saint-Benin 15 febbraio 5 aprile 1992), (AA.VV.), Regione Autonoma Valle d'Aosta. Assessorato al Turismo, Sport e Beni Culturali. Soprintendenza peri i Beni Culturali, Mesumeci Editore, Quart, 1992, pp. 11-37.
MOLINA, F.,
1983. La Prehistoria, Historia de Granada I. De las primeras culturas al Islam, (F. Molina, J.M. Roldán), Granada 1983, pp. 11-131.
MOLINA, F., MARTÍNEZ, G.,
1995. Estudio preliminar, Religiones neolíticas de Iberia (L. Siret), Col. Luis Siret Arqueología 2, Almería, 1995, pp. 9-29.
MOLINA, F., CONTRERAS, F., RAMOS, A., MÉRIDA, V., ORTIZ, F., RUIZ, V.,
1986. Programa de recuperación del registro arqueológico del Fortín 1 de Los Millares. Análisis preliminar de la organización del espacio, Coloquio sobre el Microespacio 3 (Teruel, 1986). Arqueología Espacial 8. Del Paleolítico al Bronce Medio, Teruel, 1986, pp. 175-201.
MOLLO MEZZENA, R.,
1997. L'età del Bronzo e l'età del Ferro in valle d'Aosta, Atti della XXXI Riunione Scientifica dell'Istituto Italiano di Preistoria e Protostoria in valle d'Aosta (Courmayer, 2-5 Giugno 1994), (F. Mezzena, A. Palma di Cesnola, R. Peroni, Orgs.), Istituto Italiano di Preistoria e Protostoria, Firenze, 1997, pp. 139-223.
MORAVETTI, A.,
1980. Riparo sotto roccia con petroglifi in località Frattale (Oliena, Nuoro), Atti della XXII Riunione Scientifica dell'Istituto Italiano di Preistoria e Protostoria nella Sardegna centro-settentrionale (21-27 Ottobre 1978), (AA.VV.), Istituto Italiano di Preistoria e Protostoria, Firenze, 1980, pp. 199-226.
MORAVETTI, A.,
1985a. Statue-menhirs in una tomba di giganti del Marghine, Nuovo Bullettino Archeologico Sardo 1 (1984), Sassari, 1985, pp. 41-67.
MORAVETTI, A.,
1985b. Il patrimonio archeologico del comune di Birori. Piano particolareggiato di censimento, valorizzazione e fruizione, Comune di Birori, Cagliari, 1985.
MORAVETTI, A.,
1988. La cultura di Monte Claro nella Sardegna settentrionale, Rassegna di Archeologia 7, Congresso Internazionale L'Età del Rame in Europa (Viareggio, 1987), Firenze, 1988, pp. 528-529.
MORAVETTI, A.,
1989. La tomba ipogeica di Littoslongos, Ossi, La Cultura di Ozieri. Problematiche e nuove acquisizioni. Atti del I Convegno di Studio (Ozieri, gennaio 1986 - aprile 1987), (L. Dettori Campus, Cur.), Il Torchieto, Ozieri, 1989, pp. 83-102.
MORAVETTI, A.,
1991. Alghero (Sassari). Necropoli di Santu Pedru: le tombe II e III, Bollettino di Archeologia 4 (1990), Roma, 1991, pp. 112.
MORAVETTI, A.,
1992. Il Complesso nuragico di Palmavera, Sardegna Acheologica. Guide e Itinerari 20, Carlo Delfino editore, Sassari, 1992.
MORAVETTI, A.,
1993. Protostoria nel Marghine e nella Planargia, Archeologie et ambiente naturale. Prospettive di cooperazione tra le autonomie locali nel sud dell'Europa, Amministrazione Provinciale di Nuoro, Nuoro, 1993, pp. 91-103.
MORAVETTI, A.,
1995. Località Santu Pedru. Scavi archeologici nella necropoli ipogeica, Bollettino di Archeologia 13-15 (1992), Roma, 1995, pp. 156.
MORAVETTI, A.,
1996. Il territorio dal Neolitico all'età romana, Alghero e il suo volto. Vol. 1 (AA.VV.), Carlo Delfino Editore, Sassari, pp. 139-165.
MORAVETTI, A.,
1998. On the dolmens of Pre-Nuragic Sardinia, Papers from the EAA Third Annual Meeting at Ravenna 1997. Volume III: Sardinia (A. Moravetti, M. Pearce, M. Tosi, Eds.), British Archaeological Reports. International Series 719, Oxford, 1998, pp. 25-45.
MORAVETTI, A.,
2000. Il Complesso prenuragico di Monte Baranta, Sardegna Archeologica. Guide e Itinerari 28, Sassari, 2000.
PERRA, M. (1994): Statue-menhir in territorio di Samugheo (Oristano), Nuovo Bullettino Archeologico Sardo 4 (1987-92), Sassari, 1994, pp. 17-42.
PITTI, L, TOZZI, C.,
1978. Gli scavi nel villaggio neolitico di Catignano (Pescara). Nota preliminare, Rivista di Scienze Preistoriche XXXI:1-2 (1976), Firenze, 1978, pp. 87-107.
PITZALIS, G.,
1996. L'Ogliastra preistorica tra ipogeismo e megalitismo. Le "domus de janas" con corridoio dolmenico d'accesso e i "menhirs" peritafici, Studi Sardi XXX (1992-92), Cagliari, 1996, pp. 169-218.
POLLEX, A.,
1999. Comments on the interpretation of the so-called cattle burials of Neolithic Central Europe, Antiquity 73, Cambridge, 1999, ppl 542-550.
PULACCHINI, D.,
1994a. Florinas (Sassari). Domu de janas Su Cannuju, Nuovo Bullettino Archeologico Sardo 4 (1987-92), Sassari, 1994, pp. 261-262.
PULACCHINI, D.,

1994b. Florinas (Sassari). Domu de janas Alzola Trigale, Nuovo Bullettino Archeologico Sardo 4 (1987-92), Sassari, 1994, pp. 262-263.
RAMOS, U., AFONSO, J.A., CÁMARA, J.A., MOLINA, F., MORENO, M.,
1997. Trabajos de acondicionamiento y estudio científico en el yacimiento de Los Castillejos en Las Peñas de los Gitanos (Montefrío, Granada), Anuario Arqueológico de Andalucía 1993:III, Sevilla, 1997, pp. 246-252.
RENFREW, C.,
1975. Megaliths, territories and populations, Acculturation and continuity in Atlantic Europe mainly during the Neolithic Period and the Bronze Age (Papers presented at the IV Atlantic colloquium, Ghent, 1975), (S.J. de Laet, Ed.), Dissertations Archaeologicae Gandenses, De Tempel, Brugge, 1976, pp. 198-220.
ROCHA, L.,
1997. Os menires de Pavia, Mora (Portugal), II Congreso de Arqueología Peninsular (Zamora, 24-27 de Septiembre de 1996). T. II. Neolítico. Calcolítico y Bronce, (R. de Balbín, P. Bueno, Eds.), Fundación Rei Alfonso Henriques. Serie Actas, Zamora, 1997, pp. 221-228.
SABA, A.,
2000. Le statue-menhir di Isili (NU), Studi Sardi XXXII (1999), Cagliari, 2000, pp. 111-164.
SALIS, Mª.L.,
1996. Aspetti della Gallura nell'età prenuragica, Archeologia del territorio. Territorio dell'Archeologica. Un sistema informativo territoriale orientato sull'archeologica della regione ambientale Gallura (R. Caprara, A. Lucian, G. Maciocco, Cur.), Carlo Delfino Editore/SIPIA, Cagliari, 1996, pp. 39-44.
SANGES, M.,
1980. Due "domus de janas" nel territorio di Oliena (Nuoro), Atti della XXII Riunione Scientifica dell'Istituto Italiano di Preistoria e Protostoria nella Sardegna centro-settentrionale (21-27 Ottobre 1978), (AA.VV.), Istituto Italiano di Preistoria e Protostoria, Firenze, 1980, pp. 181-197.
SANTONI, V.,
1982. Il mondo del sacro in età neolitica, La Scienze 170, Ottobre 1982, pp. 70-80.
SANTONI, V. ,
1992. Cabras - Cuccuru S'Arriu. L'orizzonte eneolitico sub-Ozieri, Quaderni della Soprintendenza Archeologica per le province di Cagliari e Oristano 8 (1991), Cagliari, 1992, pp. 15-47.
SANTONI, V.,
1995a. La necropoli ipogeica di Sas Concas (Onifori, Nuoro), Sardegna (A. Moravetti, C. Tozzi, Cur.), Guide Archeologiche. Preistoria e Protostoria in Italia 2. XIII Congresso Internazionale dell Scienze Preistoriche e Protostoriche, U.I.S.P.P./Ministero per i Beni Culturali e Ambientali, A.B.A.C.O. Edizioni, Forlì, 1995, pp. 114-121.
SANTONI, V.,
1995b. Il sito preistorico di Cuccuru S'Arriu (Cabras, Oristano), Sardegna. Guide Archeologiche. Preistoria e Protostoria in Italia 2 (A. Moravetti, C. Tozzi, Cur.), A.B.A.C.O. Edizioni, Forlì, 1995, pp. 130-137.
SANTONI, V., USAI, L.,
1995. Domus de janas in località Cannas di Sotto (Carbonia), Carbonia e il Sulcis. Archeologia e territorio (V. Santoni, Cur.), Soprintendenza Archeologica per le province di Cagliari e Oristano/ Comune di Carbonia, Oristano, 1995, pp. 51-82.
SERNA, Mª.R.,
1997. Neolitización y megalitismo en la cornisa cantábrica: el yacimiento de Guriero-Hayas, II Congreso de Arqueología Peninsular (Zamora, 24-27 de Septiembre de 1996). T. II. Neolítico. Calcolítico y Bronce, (R. de Balbín, P. Bueno, Eds.), Fundación Rei Alfonso Henriques. Serie Actas, Zamora, 1997, pp. 199-206.
SILVA, E.J.L. da,
1997. O Recinto Megalítico de S. Cristóvâo (Resende) - Primeira noticia, II Congreso de Arqueología Peninsular (Zamora, 24-27 de Septiembre de 1996). T. II. Neolítico. Calcolítico y Bronce, (R. de Balbín, P. Bueno, Eds.), Fundación Rei Alfonso Henriques. Serie Actas, Zamora, 1997, pp. 217-220.
SOARES, J., SILVA, C.T. da,
2000. Capturar a mudança na Pré-História Recente do sul de Portugal, 3º Congresso de Arqueologia Peninsular (UTAD, Vila Real, Portugal, Setembro de 1999). Actas. Vol. 4. Pré-História Recente da Península Ibérica (P. Bueno, J.L. Cardoso, M. Díaz-Andreu, V. Hurtado, S.O. Jorge, V.O. Jorge, Coord.), Porto, Adecap, 2000, pp. 213-224.
SOLINAS, Mª.,
1999. L'Età Neolitica, Bonorva. Museo Archeologico (Mª. Solinas, Coord.), Comune di Bonorva, Bonorva, 1999, pp. 95-112.
SOLINAS, Mª.,
2000. La necropoli di Sant'Andrea Priu, Guida: La necropoli di Sant'Andrea Priu (Mª. Solinas, A. Boninu, Cur.), Comune di Bonorva, Bonorva, 2000 (2ª Edic.), pp. 23-46.
SORENSEN, M.L.S.,
1987. Material order and cultural classification: the role of bronze objects in the transition from Bronze Age to Iron Age in Scandinavia. The Archaeology of contextual meanings, (I. Hodder, Ed.), New Directions in Archaeology, Cambridge University Press, Cambridge, 1987, pp. 90-101.
SPANEDDA, L.,
1994-95. Archeologia del territorio. Emergenze archeologiche dal Paleolitico alla tarda età romana nei Fogli 195 e 208 dell'I.G.M.., Tesi di Laurea, Sassari, 1994-95.
STRAHM, C.,
1998. Il bicchiere campaniforme: fenomeno e cultura, Simbolo ed enigma. Il bicchiere campaniforme e l'Italia nella Preistoria europea del III milenio a. C. (La Rocca di Riva del Garda, 12 maggio - 30 settembre 1998), (F. Nicolis, E. Mottes, Cur.), Provincia Autonoma di Trento. Servizio Beni Culturali. Uficcio Beni Archeologici, Trento, 1998, pp. 21-44.
TANDA, G.,
1977a. Arte preistorica in Sardegna. Le figurazioni taurina scolpite dell'Algherese nel quadro delle rappresentazioni figurate degli ipogei sardi a domus de janas, Quaderni della Soprintendenza ai beni archeologici per le province di Sassari e Nuoro 5, Sassari, 1977.
TANDA, G.,
1977b. Una "domu de janas" con motivi a spirali di Cargeghe - Muros (SS), Archivo Storico Sardo 3, Sassari, 1977, pp. 175-192.
TANDA, G.,
1977c. Le incisioi della "domu de janas" di Tisiènnari Bortigiadas, Archivo Storico Sardo 3, Sassari, 1977, pp. 199-211.
TANDA, G.,
1984. Arte e religione della Sardegna Preistorica nella necropoli di Sos Furrighesos, Chiarella, Sassari, 1984 (2 vols.).
TANDA, G.,

1990. Sequenza culturale dal Neolitico all'età medievale, Ottana. Archeologia e territorio (G. Tanda, Cur.), Amministrazione Comunale di Ottana, 1990, pp. 35-85.
THOMAS, J..,
1990. Monuments from the inside: the case of Irish megalithic tombs. World Archaeology 22:2 Monuments and the Monumental. London, 1990, pp. 168-178.
TINÈ, S.,
1992. Dati di scavo 1979-1987, Monte d'Accoddi. 10 anni di nuovi scavi. Relazione Preliminare e Atti del Colloquio (Sassari, 18-20 ottobre 1990), (S. Tinè, A. Traverso, Cur.), Istituto Italiano Archeologia Sperimentale/Soprintendenza Archeologica di Sassari e Nuoro, Genova, 1992, pp. IV-XV,
TRAVERSO, A.,
1992. L'area intorno al santuario. Lo scavo dell'area del menhir, Monte d'Accoddi. 10 anni di nuovi scavi. Relazione Preliminare e Atti del Colloquio (Sassari, 18-20 ottobre 1990), (S. Tinè, A. Traverso, Cur.), Istituto Italiano Archeologia Sperimentale/ Soprintendenza Archeologica di Sassari e Nuoro, Genova, 1992, pp. XX-XXII.
TRUMP, D.,
1984. The Bonu Ighinu project... Results and prospects, The Deya Conference of Prehistory. Early settlement in the Western Mediterranean Islands and their Peripheral Areas. Part II (W.H. Waldren, R. Chapman, J. Lewthwaite y R.-C. Kennard, Eds.), British Archaeological Reports. International Series 229:2, Oxford, 1984, pp. 511-532.
USAI, A.,
1996. Considerazioni sulle relazioni tra la Sardegna e L'Egeo durante il Neolitico e il Calcolitico, Studi Sardi XXX (1992-93), Cagliari, 1996, pp. 329-439.
USAI, D.,
1991. Modelli d'insediamento nel Sarrabus dal Neolitico all'Età del Bronzo, Quaderni della Sopraintendenza Archeologica per le province di Cagliari e Oristano 7 (1990), Cagliari, 1991, pp. 117-134.
USAI, E.,
1980. Una domu de janas dipinta della necropoli di S'Acqua Salida di Pimentel - Cagliari, Interaction and Acculturation in the Mediterranean (J.G.P. Best, N.M.W. de Vries, Eds.), B.R. Güner Publishing Co., Amsterdam, 1980, pp. I:31-35, II:231-238.
USAI, E.,
1989. La cultura Ozieri a Pimentel e a Siddi, La Cultura di Ozieri. Problematiche e nuove acquisizioni. Atti del I Convegno di Studio (Ozieri, gennaio 1986 - aprile 1987), (L. Dettori Campus, Cur.), Il Torchieto, Ozieri, 1989, pp. 217-230.
USAI, L,
1992. Gallura. Dolmen, nuraghi, castelli, Itinerari di Archeologica nella provincia di Sassari, Editrice Democratica Sarda, Sassari, 1992.
VILLOCH, V.,
1995. Monumentos y petroglifos: la construcción del espacio en las sociedades constructoras de túmulos del noroeste peninsular, Trabajos de Prehistoria 52:1, Madrid, 1995, pp. 39-55.
VILLOCH, V. (1998): Un nuevo menhir en Cristal, Gallaecia 17, Santiago de Compostela, 1998, pp. 107-119.

34

MAJOR PATTERNS AND PROCESSES IN BIODIVERSITY: TAXONOMIC DIVERSITY ON ISLANDS EXPLAINED IN TERMS OF SYMPATRIC SPECIATION

JOHN DE VOS (1), **ALEXANDRA VAN DER GEER** (2)

(1) Naturalis, Nationaal Natuurhistorisch Museum, Postbus 9517, 2300 RA Leiden, The Netherlands,
EMAIL: vos@naturalis.nnm.nl
(2) Instituut Kern, Nonnensteeg 1-3, Postbus 9515, 2300 RA Leiden, The Netherlands,
EMAIL:archimedesfoundation@altavista.com

SUMMARY

The history of life on earth is characterized by biodiversity, which is not limited to taxa, geological ages nor regions. Pandemic biodiversity is due to evolutionary processes in which every species is supposed to have evolved from another species through natural selection in a broad sense. Selection is mainly initiated and directed by availability and character of ecological niches. This is true for all taxonomical and phylogenetical levels, and is time independent.
As a rule, radiation within a taxon takes place as soon as a new, free ecological niche is entered successfully, in other words, after an innovation of this taxon or after a mass extinction of other, competitive taxa. For example, the development of jaws (innovation), and later fins, in the agnatha was followed by a large-scale, long-term radiation into numerous groups of fishes during the Devonian. A similar pattern is seen in amphibians in the Carbonian (innovation: land dwelling), the amniota in the Trias (innovation: egg with membranes), sauriers in the Mesozoic (survivors of mass extinction), and mammals in the Kenozoic (survivors of mass extinction; innovation: versatile dentition). The same is seen in lower phylogenetic levels, resulting in medium-scale, medium-term radiations, e.g. ungulates (innovation: stilt-like limbs to run the plains), mainly in the Miocene.

Each newly evolved taxon displays a minor change from the innovative ancestor, and as a specialist occupies a part of the total newly entered niche. Thus, after each innovation, sympatric speciation events take place, later followed by dispersals and allopatric speciation events. The speciation leads to homochronuous specialists, each adapted to a subniche of the main niche that was originally occupied by the innovative pioneer or colonizer.

The above observation leads automatically to the solution of the small-scale, short term radiation seen in endemic insular taxa. On islands, biodiversity within one taxon is the rule, and the most parsimonious explanation is to consider this diversity, too, as determined by availability and character of ecological niches, as on the mainland. The colonizers have no competitors, so have the unique possibility to radiate beyond the degree seen on the mainland, where close niches are always already occupied. Island species, depending on plasticity of functional structures, will enter all possible niches, and adapt to all available niches. In a way, the colonization on itself can be considered equal to the situation met with after a mass extinction on the mainland, which is also usually followed by a radiation within the surviving taxon. Clear examples are the large continental islands in sensu lato: the carnivorous marsupials and herbivorous mammals of South America during the Tertiair, and the marsupials of Australia during the Kaenozoic were able to radiate into a huge variety of taxa after colonization. But also small, true islands can be considered examples of sympatric radiation, as shown by the deer of Crete in the Pleistocene, and Darwin's finches of the Galapagos, and the haplochromine cichlids of East African Great Lakes of modern times. The biodiversity as seen in endemic insular monophyletic genera in the form of adaptive radiation is based upon sympatric speciation, in complete accordance with the major patterns and processes that gave rise to the long-term, large-scale and the medium-term, medium-scale biodiversity on the mainland.

INTRODUCTION

The history of life on earth is characterized by biodiversity through all taxonomic levels. Since Darwin's publication (1859), there is a general agreement that this biodiversity is caused by evolution through natural selection, although there is a discussion if this evolution is gradualistic (Makurath & Anderson, 1973; Makurath, 1974; Gingerich, 1974, 1976; Bookstein et al., 1978; Chaline & Laurin, 1986), punctuated (Eldredge & Gould, 1972, 1988; Eldredge, 1974; Stanley, 1975, 1978; Gould & Eldredge, 1977, 1986), or both, depending on the functional structure studied (Spaan et al., 1994; Sondaar, 1994). The end result, however, is the same: new taxa.

On earth there are several ecosystems, which comprise communities of several organic species and their environment (biotic and abiotic factors). These ecosystems are recognized on different levels, such as micro- (e.g. a tree trunk), meso- (e.g. a forest), and macro-ecosystems (e.g. an ocean or continent). In each ecosystem a species will occupy a particular niche. When one of the biotic and/or abiotic factors changes, the habitat may change, impelling the organisms to adapt to this new environment. If this environment contains free ecological niches, taxa appear to occupy as much niches as possible. As a result, natural selection on different changes in the locomotory and digestive mechanisms will take plac, resulting in an adaptive radiation, and thus a sympatric speciation into a vast range of species or morphotypes. If the variety is large, and the genetic distance small, the term flock is sometimes used (e.g. Greenwood, 1974; Echelle & Kornfield, 1984, and references therein).

Depending on the level of taxonomical, or phylogenetical, organisation, the observed biodiversity is a large-scale (continent), medium-scale (mainland) or small-scale (island) adaptive radiation, and depending on the end result, the radiation is of long-term (resulting in families), medium-term (resulting in new genera) or short-term (resulting in species or morphotypes). Large-scale, long-term radiations are the most conspicuous of all macroevolutionary phenomena (Carroll, 1997) and here the adaptive radiation is strong: the radiation of metazoan phyla during the Cambrian, that of primitive fish and early vascular plants in the Palaeozoic, of dinosaurs and flowering plants in the Mesozoic, and of mammals and birds in the Cenozoic. In small-scale, short-term evolution the radiation will be weaker. The most famous examples of this is the adaptive radiation of the Darwin's Finches.

The radiation as observed in endemic insular taxa is usually explained in terms of allopatric speciation. In actual fact, this implies the existence of an archipelago, e.g. Darwin's Finches on the Galápagos Islands. The Pliocene island Monte Gargano (Italy) has been considered part of a larger archipelago (Apulo-Dalmatic Realm) in relation to the micromammals (De Giuli & Torre, 1984; De Giuli et al., 1985, 1987) and the ochotonid Prolagus (Mazza, 1986a, 1986b) to explain the presence of several sister taxa, evolved in a relatively short time. Another explanation is the occurrence of more than one invasions, e.g. as suggested for the deer Candiacervus on Crete (Kuss, 1975; De Vos, 1984, however cf. De Vos, 1996, 2000; Capasso Barbato, 1992; Caloi & Palombo, 1996) and the rodents on Monte Gargano (Freudenthal, 1976). In the latter case, it is no explanation as to where and why this new sister taxon evolved. In the former case, it does not explain sufficiently why sister taxa are found together on one and the same island. In both cases it is not clear why the sister taxa wait with migration untill full speciation has taken place.

The end result of island evolution is always a range of taxa, closely related, which in a way can be compared to the medium scale, medium term radiations, and on its turn with the large scale, long term radiations as well. In this paper we compare these three types of radiations for the major vertebrate classes, to demonstrate that the processes of biodiversity are the same for all scales and all terms. The main driving force in all cases is the urge to occupy free ecological niches. In our opinion, the resulting speciation can be explained best in terms of sympatric speciation.

CLASS PISCES

LARGE SCALE, LONG TERM RADIATION

During the Ordovician-Silurian, a rapid dispersal of Ostracodermi took place, after a long period of relative stability in the omnipresent Agnatha. Later, at the end of the Silurian, the Placodermi with their primitive jaws underwent a major adaptive radiation, and the first predators arose. The more derived Actinopterygii and Sarcopterygii also radiated in many taxa, each of them in a different ecological niche. In addition, the Actinopterygii with their swimming bladder entered the deeper waters of the ocean, and occupied the available deep sea niches. Gradually, all modern osteichtyes evolved from this order, and took over all available aquatic niches in salt, sweet and brackish waters.The Sarcopterygii with their primitive lungs never radiated to the degree seen in the other order. Nowadays, the only descendants are the coelacanth, and the Dipnoi, who are restricted to tropical regions with periodical dry seasons.

Small scale, short term radiation

In the East African Great Lakes, the haplochromine cichlid fishes went through a rapid speciation (adaptive radiation), resulting in a wide range of trophotypes (Fryer & Iles, 1972; Greenwood, 1974; Barel et al., 1977; Witte, 1981; Keenleyside, 1991). Initially they differed little from their immediate reverine ancestors, and there is no evidence of significant new morphological changes that facilitated their differentiation into many trophic levels; rather they capitalised on a biological versatility already present (Liem & Osse, 1975). The cichlids were obviously capable of a much higher rate of speciation than were other fish in the East African Great Lakes, and were able to differentiate into many different trophic levels with a minimum of morphological change (Carroll, 1997). It resulted mainly in differences in the mouth, which gradually became adapted to different types of food: detritus, fishes, shells, crabs, insects, phytoplankton, zooplankton. For taxonomy, such radiations are a disaster, as taxonomy deals with fixed, clearly defined subunits of the observable world, whereas in reality such a species flock approaches a continuum. The taxonomical problems become evident through the many revisions and reconsiderations of the classification of the haplochromine cichlids (e.g. Greenwood, 1981; Witte & Witte-Maas, 1981; Hoogerhoud, 1984; Van Oijen, 1991).

MECHANISM

In all cases, the fishes encountered free ecological niches. In the Ordovician Ostracodermi the first pair of gills had been transformed to an opening at the head side between the eyes, through which opening water could be let in, without mud coming between the gills. The result was a rapid dispersal of Ostracodermi. Later, the Silurian Placodermi had developed innovative structures: jaws and fins. The fins made these fishes independent of the mud, so that they could enter a vast scala of free ecological niches, and the jaws gave the possibility to prey upon other fish. The innovations gave the Placodermi the possibility to occupy a new range of free ecological niches. Again later, the placoderm-type triangular fin evolved into two main types, the stray fin and the brush fin (Actinopterygii and Sarcopterygii, resp.). Both types of fins increased the swimming ability enormeously, and with it the possible range of habitats and size of territory. Within the Actinopterygii another innovation, which further increased their swimming ability, especially in deeper waters. Primitive osteichthyes had besides gills also lungs, to be able to survive periodical dry periods, and some had even external gills during the larval stage to cope with these dryness. The Actinopterygii, however, transformed their lungs to a swimming bladder. Now that they could enter the deep waters, i.e. the ocean, a new range of free ecological niches became available to them. The subsequent adaptive radiation resulted into the vast spectrum of living modern fishes.

The Holocene cichlids entered a new and still unoccupied lake: Lake Victoria was filled about 14,000 years ago, due to the creation of the Rif Valley, which started to arise from 750,000 years ago. Immediately after the formation of the lake, a host of adaptive zones became available. The entering of a zone with free niches gave the cichlids the possibility to radiate beyond the degree seen in related cichlids. The haplochromine species flock of Lake Victoria is a good example of recent speciation, that took place in less than 200.000 years (Meyer et al., 1990).

CLASS AMPHIBIA

LARGE SCALE, LONG TERM RADIATION

At the end of the Devonian, the Rhipidistia with their choanae were the forerunners of the amphibians. As soon as a tongue and a strong central axis had developed, an adaptive radiaion took place on the land amongst the amphibians, starting with the Devonian Ichthyostega. During the Carbone, two amphibian orders arose: the frog-like Aspidospondyla and the salamander-like Urodelomorpha. Both orders were preserved into modern times and its as a rule small sized representatives occupied especially insectivorous and carnivorous niches.

MECHANISM

During the Devonian enough oxygen had developed to reduce the ultraviolet light of the sun, but the land was still unoccupied by vertebrates. At the end of this period, the first amphibians appeared. Due to a combination of innovations (the development of choanae and lungs in the adult, crawling ability, tongue to swallow food), they were able to explore the available niches on land. The Labyrintodontia had complicated enamel patterns on their molars, which gave them the possiblity to occupy different trophic niches. The adaptive radiation of the amphibians, however, never reached the extent of that of the fishes, as amphibians are dependent on sweet water for their reproduction and dry up easily. This might also be the explanation why amphibians hardly reach islands (salt water barrier) or isolated areas (dry land barrier).

CLASS REPTILIA

LARGE SCALE, LONG TERM

At the end of the Carbone, the first amniota, representatives of the Aspidospondyla, laid their eggs on land. This was followed, during the Permian, by a quick radiation of these reptiles over the vast continental regions, which were still relatively empty after the drying up of marshes.

Later, at the end of the Permian (250 Ma), a mass extinction occurred (Sepkoski, 1989, 1990; Raup, 1991), the nature of which is still unclear, though most likely it had to do with relatively suddden major changes in the positions of the continents. After this mass extinction, the sea and the continents were "empty" once more. During the Mesozoic (era of the reptiles), an explosive radiation of sauriers took place, which gave rise to many orders: dinosaurs, ichthyosaurs, plesiosaurs, mesosaurs, and the flying pterodactiles. Within the Upper Triassic, the dinosaurs diverged further into three major groups: the bipedal carnivorous theropods and herbivorous ornithischians, and the quadrupedal sauropodomorphs.

MEDIUM SCALE, MEDIUM TERM

Still during the Mesozoic a further rapid radiation took place among secondarily aquatic groups of sauriers: nothosaurs, ichthyosaurs, and mosasaurs. They all occupy complete different ecological niches. Later, the same occurred within the major dinosaur groups. During the Jurassic and Cretaceous, the theropods, the ornithischians and the sauropodomorphs each evolved into a succession of divergent families.

MECHANISM

At the end of the Carbon (280 Ma), the marshes disappeared. Simultaneously, most of the amphibians became extinct, due to their dependentness on water. This was the change for the reptiles, with their innovative amniotic eggs and protective skin, to explode into a radiation. Their eggs were independent of the water: a large yolk to provide food, an amnion around the yolk to prevent dehydration, an allantois to respirate, a chorion to hold germs and chemicals outside, and a calciferous shell for mechanical stability. The fact that the amphibians had disappeared gave the reptiles the possibility to enter free ecological niches that were once occupied by amphibians. The innovation in reproduction, however, made them able to explore even more ecological niches than the amphibians could, as the distance to water became a minor factor. The keratinisation of the skin, the development of complicated lungs, and the homeothermy, agian other innovations, made them even much more adapted to a terrestrial life than the amphibians.

But also the sea was available, as there were no fish as large as a saurier. The secondarily aquatic forms had no aqautic enemies either, because of their size, thick skin and advanced dentition. After the innovation of a hydrodynamic body, the aquatic sauriers were able to occupy all available ecological niches.

The most remarkable innovation is seen in the pterodactiles: the development of wings. The air was almost free, except for some, much smaller sized early birds.

CLASS AVES

LARGE SCALE, LONG TERM

The modern bird orders are almost entirely restricted to the Cenozoic. Numerous specimens from the late Mesozoic have been described as possible ancestors, but none of them with conclusive evidence. Three orders can be distinguished during the Cretaceous (Hesperornithiformes, Ichthyornithiformes, Enantioornithes), all three not related to any of the now living birds (Chiappe, 1995). Feduccia (1994, 1995) has recently suggested that nearly all of the Cretaceous bird lineages became extinct at the end of the Mesozoic, at the same time as the dinosaurs. The almost complete absence of ancestors of the modern bird orders in the Cretaceous is not the result of an incomplete fossil record, but because they had not differentiated until the early Cenozoic. Feduccia (1994, 1995) argued that nearly every type of modern bird evolved within about five to ten million years, most of them from an assemblage termed the transitional shore birds. All modern bird orders had diverged by the Eocene, including the passerines (Boles, 1995).

MEDIUM SCALE, MEDIUM TERM

Radiations within bird families are also known. The result of these radiations is a number of monophyletic genera or species that differ in ecological niche. They share the same habitat, but occupy a different niche. For example the flamingoes in the brackish Nakuru Lake in East Africa, where the small flamingo (Phoeniconaias minor) and the pink flamingo (Phoenicopterus roseus) live together, each with their own trophic niche, respectively alges and salt-water crustacea (Grzimek, 1968).

SMALL SCALE, SHORT TERM

The finches from the Galápagos archipelago are remarkably similar in plumage, but differ in beak shape and size (Lack, 1947), whereas in continental passerine birds closely related species tend to differ from each other in plumage, whereas they are usually similar in beak and other structural charac-ters. Differences in beak more usually characterize the broad-er units, the different genera. In Darwin's finches the opposite seems to be the case. A further unusual feature in Darwin's finches is that some species are highly variable (Lack 1947: 12).

The differences in beaks indicates an adaptation to different habitats. It became clear that islands had species which were different from the mainland and were adapted to the different habitats of islands.

Six genera and about fourteen species are recognized on the basis of bill size and shape and body size (Grant, 1986). They are mostly sparrow-sized, and are all dull in colour. All these species are adapted to different habitats and occupy different niches. Patterson (1978: 112) recognizes several trophic groups: the seeds and insects eating ground finches (coastal zone, lowlands) with one cactus flower eating exception, the seeds and insects eating tree finches (forest zone), the insects eating warbler finch, the fruits, buds and soft seeds eating tree finch (forest zone), and the insects eating woodpecker-like finches (forest zone, including mangroves).

The variation within one species can be large. For example, the Geospiza fortis of Santa Cruz shows a clear biomodality in the bill depth of the male medium ground finches, measured at the Bahia Academia. As in this area there is a wide range of floral taxa (Abbott et al., 1977), it is suggested that G. fortis is in the middle of a process of fission, due to disruptive selection in a heterogenuous environment (Ford et al., 1973), but others (Grant, 1986: 276) have a different explanation.

Congeneric species each occupy a different niche, for example Cactospiza on Isabela. The woodpecker finch lives in dry territories and mesic woods, the mangrove finch in the mangrove woods at the coast, strictly separated from the rest of the island by a zone of bare lava (Grant, 1986: 274), although the woodpecker finch is found brooding in the mangroves at the east side of the island (Gifford, 1919).

Besides the Darwin's Finches there are more such cases of passerine birds known: the honeycreeper finches of Hawaii (Drepanididae), and the birds of paradise of New Guinea (Paradisaeidae).There is even more differentiation in morphology in the Hawaiian finches (nine genera, twenty two species) than in Darwin's Finches. Diets consist of honey and small seeds (e.g. Palmeria), hard seeds and nuts (e.g. Psittirostra, Chloridopa, Pseudonestor), and insects (e.g. the woodpecker-like Hemignathus), and the types of beak range from short and slender to the robust parrot-like beak of Pseudonestor. However, although these differences in diet are huge, other differences are much less. For example, they are all medium-sized (10-20 cm), and have a monochrome plumage (olive green, yellow, red or black), without distinct pattern, except for the tufted honeycreeper Palmeria dolei, who is grey with red spots and stripes, and has a short tuft above the nasal openings (Burton & Friedhoff, 1975).

The radiation is so extreme that it is not possible anymore to tell which group is the ancestor: a finch, a honeycreeper or a tangare. The fact, however, that the variation in colour and size, in type of nest, and in number and colour of eggs, is so small is highly indicative for a monophyletic origin of the Hawaiian honey-creepers. The time range of their isolation might be long, as they are all highly specialized. Major changes in their environment, mainly of a human nature, nowadays leads to their rapid extinction, as they are not able to adapt themselves easily.

MECHANISM

After the extinction of the pterosaurs, the air was free for the birds. They were more advanced than the pterosaurs as they had developed feathers, an aerodynamic beak, and a crop plus stones in their first stomach to postpone mastication. However, they could take their change only when the dominant pterosaurs disappeared.

The extreme adaptive radiation as seen in the passerine island birds is caused by the total lack of competition by other land birds, and led to the occupation of all imaginable trophic niches. Sometimes the principle of fissions might apply. Fissions occur essentially when the habitat becomes overpopulated, absolutely or relatively. An example is provided by the cactus finch Geospiza conirostris on Genovesa (Grant, 1986). Two types of males occur in one and the same population, with a different song and a different bill. In dry seasons they have a different diet and different eating methods, but in good seasons there is no such dietary differences. Adult males prefer to sit next to a neighbour with the other type of song, which results in a mosaic pattern of songs. In good seasons, females seem to have no preference, although they choose a mating partner of the song type opposite to that of their previous partner when their 'broedsel' failed (Grant & Grant, 1983). In dry seasons with food shortage on the contrary, fission occurs as a result of disruptive selection by the females (Grant, 1986:279; B.R. Grant, 1985). Now the population is instable and oscillates between fission and fusion (Grant & Grant, 1983), but it can be easily imagined that a series of dry seasons will induce a complete, irreversible fission yielding two different species. In Grant's opinion (1986: 279), only an extreme reduction in population size, giving change a role to play, could induce such a reproductive isolation, and thus speciation, but in our opinion, it is the relative overpopulation that is the drive behind the selective mating behaviour. Positive assortative mating is the only strategy to cope with a shortage on food in order to enter new ecological niches. Reduction in population size is not a strategy: it is a symptom of being unable to cope with the unfavourable situation.

CLASS MAMMALIA

LARGE SCALE, LONG TERM

During the Triassic, the cynodonts, a group of advanced, mammal-like reptiles, developed into several families of small and medium-sized carnivores and two families of herbivores. Some of the later members of the group are almost indistinguishable from the first mammals, which can be distinguished from reptiles by the fact that the jaw joint shifted from one position to another, and that the old jaw joint was still present in the form of three middle ear bones (malleus, incus, stapes) inside the ear (Benton, 1991). The best known is the adaptive radiation of placental mammals in the Early Cenozoic. Within ten million years, from 65 to 55 million years ago, nearly all the modern placental orders had appeared (Carroll, 1997; MacFadden, 1992; Szalay, Novacek & MacKenna, 1993; Prothero & Schoch, 1994).

Marsupials also radiated rapidly at the beginning of the Cenozoic. They were moderately diverse in North America at the end of the Cretaceous but were reduced to a single lineage following the end-Cretaceous extinction. Later, they also dispersed to the Palaearctic, but these lineages all became extinct with little further radiation. On the other hand, the lineage that entered South America, radiated extensively in that continent, spread to Antarctica, and from there extended its range into Australia as these continents first came in contact with one another and then drifted apart between the end of the Cretaceous and the beginning of the Eocene. They radiated rapidly in both South America and Australia (Woodburne & Case, 1996).

MEDIUM SCALE, MEDIUM TERM

The first horses (Hyracotherium, Mesohippus) were forest animals, with lophodont molars and subungulate, tridactyl, padded foot. In the Early Miocene, a new group enters the fossil record in North America: the grazing horses. They are recognized by an elongated first phalanx, and a true unguligrade tridactyl foot. Soon after their occurrence, a radiation took place, starting with Parahippus, who had still lophodont molars, though more complex, and later the rather hypsodont Merychippus. The lateral toes were still functional, and touched the ground during maximal dorsiflexion, resulting in a more allround locomotion (Renders, 1984; Renders & Sondaar, 1987). Its diversity is at its top in the Late Miocene, at which time about thirteen different genera of grazing horses can be distinguished, amongst others the tridactyl Hipparion and the first monodactyl Dinohippus, ancestor of Equus, both hypsodont grazers living at the same time, but in different niches. In the forests, the padded tridactyl horses such as Anchitherium and Hypohippus remained occupying the ancestral niche, without further major innovation and without remarkable radiation. On the plains, the degree of hypsodonty increased progressively and gradually in all lines, whereas the change in locomotion occured punctuated and restricted (Spaan et al., 1994).

The artiodactyls also radiated, and to much larger extent than perissodactyls: giraffids, deer, bovids, tylopods. Within these groups, smaller radiations occurred, for example within the antlered deer. As a result, in habitats suitable for deer, two or three species or genera live in the same habitat but differ considerably in size. As a rule, in each deer-suitable habitat we find a small, medium and large deer-like species together, providing the habitat is rich enough to support more than one species. There are exceptions to this rule, for example Rucervus duvauceli and Rusa unicolor in some parts of India, who have the same size, but in this case the species occupy different ecological niches or a slightly different habitat (Van Bemmel, 1973: 295). Quite probably, this is true for all similar cases.

SMALL SCALE, SHORT TERM

The Pleistocene deer Candiacervus of Crete shows a large variety. No matter how the final genera and species are placed into a taxonomical phramework (for a complete overview, see Dermitzakis & De Vos, 1987 and De Vos, 2000), six size groups (De Vos, 1979), and in one of them three morphotypes of skulls and antlers (De Vos, 1984), can be dinstinguished, based on biometrics of cranial and postcranial materials from different sites (De Vos, 1979, 1984). These eight taxonomic units most likely all belong to the monophyletic genus Candiacervus (De Vos, 2000), or alternatively, to two paraphyletic genera, be it Megaloceros and Cervus (Capasso Barbato, 1989) or Megaceroides and ?Pseudodama (Caloi & Palombo, 1996). Irrespective of the taxonomical problems, the eight Cretan deer types in any case differ clearly too much to assume a similar ecological niche; more likely is the hypothesis of different niches. On the ground of body proportions, molar morphology and wear pattern, the specialist trophic niches occupied by the eight taxa might, tentatively, be summarized as follows: grassy food or prickly bushes on a rocky hill (C. ropalophorus, Candiacervus spp. II), grasses on a steppe-like plain (C. cretensis), leaves and branches in a forest, like red deer (C. rethymnensis), leaf-like food and soft bushes in a forested terrain with many obstacles (Candiacervus sp. V and C. sp. VI).

A rather similar situation is met on Mallorca. The Pleistocene endemic goat Myotragus balearicus exhibits a great size variation, far exceeding that of mainland relatives. The largest metatarsal is three times the smallest from the same cave (e.g. Muleta) and the same level, and all intermediate sizes are also found. Grouping of bones into different size ranges or morphotypes is not possible (Sondaar, 1977).

Also the Pleistocene deer from the Ryukyu Islands (Japan) seem to follow the same pattern. On Kume Island four size-groups of metacarpal bones of Cervus astylodon can be distinguished (Matsumoto & Otsuka, 2000).

MECHANISM

At the end of the Mesozoic (66.5 Ma), there was again a mass-extinction, the nature of which is not yet fully understand, but which surely had a huge impact on life in the seas as well as on land (Smit, 1989). Most of the mesozoic reptiles disappeared, and the environment was "empty" again. All kinds of niches became available and the mammals could take their chance. The most primitive members of the derived mammalian groups initially show relatively few changes aside from dental specialisation. The success of the placental mammals is often attributed to the versatility of their teeth. It is true that mammals have been more adaptable and successful in their range of diet than any other amniote group: insectivores with small, sharp teeth, carnivores with enlarged canines, herbivores with blunt, grinding molars and huge tusks, piscivores with peg-like teeth, plankton-eaters with palate-born baleen sievs, and anteaters without teeth (Benton, 1991).

As in the cichlids, there is no evidence of significant new morphological changes that facilitated their differentiation into many trophic levels; rather they capitalised on structural and behavioural flexibility that was evident in their ancestors.

During the Miocene and Pliocene (24-2 Ma), the climate gradually became cooler and drier worldwide, due to continental movements and the development of an ice cap at the South Pole. The main global effect was the spread of grasslands, and the size reduction of tropical and subtropical forests. New kinds of landscapes evolved: prairies, steppes, savannas, and pampas. These large grasslands stimulated great adaptive radiation of various ungulate groups, as a new environment is by definition empty. The change in evolutionary direction from tridactyl padded feet to monodactyl ungulate feet coincides with the spread of grasses at that time (e.g. Simpson, 1951; Stebbins, 1981). The innovation consists, besides hypsodonty, of a true unguligrade foot with an elongated first phalanx (a pioneer structure according to Spaan et al., 1994, and Sondaar, 1994).

The late Middle - Late Pleistocene herbivorous macrofauna of Crete contains only cervids and elephants (De Vos, 2000, and Sondaar & Van der Geer, in press, for an overview), and its taxa are endemic to Crete. The island was submerged during the Pliocene, and got its present shape during the Pleistocene. When the ancestor of the cervids arrived on Crete, a variety on herbivore niches was available, except for those occupied by the elephants.

DISCUSSION

The major pattern of the evolution of biodiversity, termed adaptive radiation, is recognisable on all taxonomic levels. The processes behind all these adaptive radiations appears to be the same at all levels: innovation followed by a rapid, monophyletic radiation to occupy as much available ecological niches as possible. In a broader sense, this innovation can also consist of the mere fact that a taxa was a survivor of some kind of mass extinction caused by a natural disaster. A comparable situation is found when a taxon colonizes an empty region: an island for terrestrial taxa, an isolated lake for aquatic taxa. The end result is, however, in all cases the same, be it after an innovation, a mass extinction or an isolation: a radiation, in order to adapt to and to occupy new niches.

Although this sounds reasonably fair, and can be easily checked at the base of the fossil record as we saw above, it still doesn't explain what the biological mechanism is behind such a speciation event. Is it possible to explain the mechanism in terms of natural selection? Darwinian explanations surely fit for species that evolve from other species. But do they fit also for sisterspecies that are supposed to have evolved simultaneously from one and the same ancestor within one and the same habitat? In other words, though Darwinian natural selection is primarily meant to explain allopatric speciation events, can it be the drive behind sympatric speciation as well? If so, island chrono-morphotypes can be much easier explained by the same mechanism of natural selection as the one behind the evolution of mainland sistertaxa.

Allopatric species are by definition species whose dispersion areas are separated from each other, and who, as a result, have no possibilities to hybridize. If allopatric species happen to come into contact, there is a change of successful interbreeding, resulting in a species fusion, due to the fact that before this contact sexual barriers were superfluous as the species never met. Sympatric species on the other hand are species whose dispersion areas coincide or partly overlap, but who nevertheless do not interbreed. Sexual barriers to prevent interbreeding can be manyfold: differences in behaviour, breeding season, pheromones, anatomy and so on, plus the fact that bastards are at the best less fertile (Burton & Friedhoff, 1975).

In our view, the definition of sympatric and allopatric species indirectly means that if speciation on islands would have been allopatric (archipelagos, highly mountaineous islands) and they meet each other after migration, they would hybridize, resulting in one intermediate species, which is not what we observe. The trend on islands is on the contrary: the longer the isolation of the endemic insular genus from the mainland ancestral population, the more different morphotypes evolve, each occupying a different niche, and there is no such intermediate hybridizating clade to be found.

In fact, most examples to prove allopatric speciation can be as easily be explained in terms of sympatric speciation. For example, white tailed ground squirrel Citellus leucurus and the dark tailed ground squirrel Citellus harrisii live in the same habitat, but occupy different niches. At the moment they are therefore sympatric species. However, it is supposed that in the past they formed one population. When the Colorado river separated them, they gradually began to differ, and formed two species, the white tailed form at the north side of the river, the dark tailed form at the south side of the river. Long after this allopatric speciation, the white form somehow reached the other side, and nowadays they live in the same habitat, in different niches. This sounds not very convincing, as the canyon got deeper in the mean time, and crossing became even more difficult in the present than in the past. Why didn't the white tailed Citellus cross this boundary earlier? And how coincidental that at both sides of the river there is the same habitat, but different niches. Why didn't the ancestral form expand its trophic niche already in the past? Apparently it was empty, as the white-tailed squirrel could occupy it much later, after it had evolved into a new species at the other side of the canyon. Unfortunately such questions cannot be answered anymore, and the only thing we have are the bare facts. Fact is, that along the Colorado river two related ground squirrels are found, that do not interbreed. Fact is, that these sisterspecies each occupy a different niche. Fact is, that the Gran Canyon is a clear geological barrier between the north and the south side of the river. Fact is, that this barrier became steeper and broader in time, and thus less passable. Besides the classic explanation given above, another explanation is possible. Just because of the Gran Canyon the squirrel population could not expand its territorium more, and got limited on food. It was simply urged to explore other trophic niches. The radiation into two trophic types was facilitated/stimulated/initiated by the geological barrier. The squirrels lived so to say on a peninsula.

In which way the speciation took place during the Pleistocene in the Cretan deer genus Candiacervus is not clear. Gradual genetic changes may have played a role. Although we cannot exclude allopatric speciation, which means nothing else but repetitive invasions, it is more realistic to suppose that the speciation took place on Crete itself. Invasions of the same trophic type would earlier lead to extinction of the resident species or otherwise to an unsuccessful colonization. Invasions of a different trophic type would, according to the characterisation of allopatric species, gradually lead to hybridization and thus to a monomorphic taxon.

Notwithstanding the different opinions concerning the number of cervid species in Crete, it is clear that the morphotypes differ in dental, cranial as well as in postcranial morphology. This makes us assume that each type must have been adapted to a different niche, and maybe even to a different habitat. Just as recent sympatric species, they all originate from one and same ancestor, which implies that there was only one successful invasion (De Vos, 1996, 2000). If by any change the large and the small-sized lineages appear to be polyphyletic, two invasions might have occurred (Capasso Barbato & Petronio, 1986; Capasso Barbato, 1992b; Caloi & Palombo, 1996), though this model is not so realistic, regarding the antler morphology (De Vos, 1996). The speciation of the Cretan cervids is a clear case of adaptive radiation, similar to that observed in Darwin's finches of the Galápagos. Here, too, it is likely that they all descended from a single ancestral species: all Darwin's finch species are more similar to each other than anyone is to a continental species, including the suggested ancestors Melanospiza richardsonii, Volatinia jacarina and Coereba flaveola (Grant, 1986). Similarly, for the Cretan deer also possible continental candidates have been mentioned: a megacerine deer of the verticornis group, Cervus peloponnesiacus, Cervus sp., ?Pseudodama, Allocenelaphus - Nesoleipoceros, but none of these genera shares synapomorphologies with Candiacervus that are not shared with other genera.

Solving the problems of taxonomy for endemic, insular species is possible only when the mechanism behind their speciation is understood. This is not easy, as taxonomy and evolution are not independent (Lack, 1947). To explain the speciation of Darwin's finches, Grant (1986) discussed the different models of speciation: allopatric, parapatric and sympatric speciation, and speciation based on gradual genetic changes (genetic drift). The latter model can merely contribute to an already divergent evolution by means of new alleles, but is not considered a means of speciation on itself (Grant, 1986). Parapatric speciation cannot be demonstrated either in the case of the Galápagos, leaving open only the allopatric model, or, with reluctance, the sympatric model if a more stringent set of conditions were met (Grant, 1986). Amazing however, that examples of allopatric speciation given by Grant (1986) can be as easily be explained in terms of sympatric speciation.

A possible example of sympatric speciation is the medium sized ground finch Geospiza fortis of Santa Cruz, Galápagos Islands. In this area there is a diverse flora, providing a wide range of diets (Abbott et al., 1977), and as a consequence, differences are found within G. fortis. Ford et al. (1973) found differences in beak depth of the males, resulting in a clear bimodality, which they explained as disruptive selection. According to them, the population was halfway the process of splitting.

Another, related example is provided by Cactospiza on Isabela, Galápagos Islands. These finches are divided in two populations: one living in the mangrove forests along the coast, and the other living in dry areas and mesic woods, sporadically nestling in the mangroves (Gifford, 1919; Grant, 1986). The populations are separated, though not completely, and it can be easily imagined that they originated from a species that comprised contiguous populations over a habitat gradient. According to Grant (1986), the populations may have been already separated geographically by the time of the fission, and thus there is not enough evidence of parapatric speciation on the Galápagos. However, separation is not strictly necessary (Endler, 1977; Lande, 1982), and partly differentiated populations can be found in adjoining habitats on other places in the world. The fact that it's difficult to find two congeneric and sympatric species in the fossil record that would confirm the possibility of sympatric speciation might be explained by the possibly rapid speciation. It appears that micromammals at least need only a few thousand years (Clethrionomys, Apodemus), or even only seventy years (Mus) to change substantially in size (Heaney, 1978: 40).

The ability to explore slightly different ecological niches appears only possible for clades that are not too specialized. This is why deer can change their habitats, if necessary, whereas a panda cannot. The plasticity of deer is illustrated by the brow-antlered deer: Rucervus eldi from Manipur, India, lives in very wet swamps and on floating grass islands, whereas the same species from Thailand lives in higher, less wet areas (Van Bemmel, 1973). The same plasticity accounts for the easy fusion of deer species, for example, Schomburgks deer (R. duvauceli schomburgki) dissolved into the herds of R. eldi when their swampy habitats disappeared in the nineteen-twenties and they had to flee to the higher areas.

For lower, cold-blooded vertebrates the idea of sympatric speciation seems more acceptable. In fact, there is in some cases no alternative. For example, how else can the monophyly of cichlids in the African Lakes be explained, other than by sympatric speciation (Schliewen et al., 1994; Turner, 1994), since they share a close common ancestry, as suggested by mitochondrial DNA (Meyer et al., 1990) and the fact that interspecific hybrids can be fertile (Crapon de Caprona & Fritzsch, 1984)? But why would patterns of biodiversity be different for lower vertebrates than for higher, and why would these patterns be different for small-scale, short-term radiations than for large-scale, long-term radiations? Our classifications are artificial, and nature doesn't care about our taxonomy. Mammalian cells do not differ essentially from those of fishes, nor do our genes, except for their number and their amino-acid sequences.

Recent studies, however, give increasing evidence of the process of sympatric speciation, and it appears that reproductive isolation of species may occur without physical isolation. Higashi, Takimoto & Yamamura (1999) explain the process convincingly at the basis of variation in a male secondary sexual character with two conspicuous extremes and the corresponding variance in female mating preference. It appears that in this way the modal divergence of the male and female traits increase, resulting in a bimodal distribution, pulling the population apart into two prezygotically isolated populations. Interestingly enough, sympatric speciation may occur more readily if there is a decrease in the cost of male conspicuousness, which occurs if barrier-free or predator-free conditions arise (Higashi et al., 1999). In fact, this is the situation we meet on islands.

REFERENCES

ABBOTT, I., L.K. ABBOTT, P.R. GRANT,
1977. Comparative ecology of Galápagos Ground Finches (Geospiza Gould): evaluation of the importance of floristic diversity and interspecific competition. Ecological Monographs 47:151-184.

BAREL, C.D.N., M.J.P. VAN OIJEN, F. WITTE, E.L.M. WITTE-MAAS,
1977. An introduction to the taxonomy and morphology of the haplochromine Cichlidae from Lake Victoria. Netherlands Journal of Zoology 27: 333-389.

BENTON, M.J.,
1991. The Rise of the Mammals; The story of the mammal families, from their origins to the dawn of the age of man. Crescent Books, New York: 1-144.

BOLES, W.E.,
1995. The world's oldest songbird. Nature 374: 21-22. BOOKSTEIN, F.C., P.D. GINGERICH, & A.G. KLUGE,
1978. Hierarchical linear modeling of the tempo and mode of evolution. Paleobiology 4: 120-134.

BURTON, M., H. FRIEDHOFF (EDS),
1975. Encyclopedia of the Animal Kingdom. 6 vols. Lausanne: Elsevier Publishing Projects.

CALOI, L, M.R. PALOMBO, 1
996. Functional aspects and ecological implications in Hippopotami and cervids of Crete. In: Reese, D.S. (ed.). Pleistocene and Holocene Fauna of Crete and its First Settlers. Monographs in World Archaeology 28: 125-151.

CAPASSO BARBATO, L.,
1989. Cervidi endemici del pleistocene di Creta. Dottorato in Palaeontologia, Modena-Bologna, Vol. I and II.

CAPASSO BARBATO, L.,
1992a. Nuova specie di cervide del pleistocene di Creta. Atti della accademia Nazionale dei Lincei, Memorie Lincee, Scienze Fisische e Naturali Serie IX, 1, 7: 183-220.

CAPASSO BARBATO, L.,
1992b. Observations on the biostratigraphy of Cretan Pleistocene vertebrates. Il Quaternario 5, 1: 67-76.

CAPASSO BARBATO, L.C., C. PETRONIO,
1986. Cervus major n. sp. of Bate Cave (Rethymnon, Crete). Atti della Accademia Nazionale dei Lincei, Memorie, classe di Scienze fisischi, matematische e naturali, Serie VIII, vol. XVIII, Sez. IIa (Fisica, chimica, geologia, paleontologia e mineralogia), fascicolo. 2: 59-1-00.

CARROLL, R.L.,
1997. Patterns and Processes of Vertebrate Evolution, 448 p. Cambridge: University Press.

CHALINE, J., B. LAURIN,
1986. Phyletic gradualism in a European Plio-Pleistocene Mimomys lineage (Arvicolidae, Rodentia). Paleobiology 12, 2: 203-216.

CHIAPPE, L.M.,
1995. The first 85 millions years of avian evolution. Nature 378:

349-355.
CRAPON DE CAPRONA, M.D., B. FRITZSCH,
1984. Interspecific fertile hybrids of haplochromine Cichlidae (Teleostei) and their possible importance for speciation. Netherlands Journal of Zoology 34: 503-538.
DARWIN, C.,
1859. On the Origin of Species by Means of Natural Selection, or the Preservation of Favored Races in the Struggle for Life. London: John Murray.
DE GIULI, C., D. TORRE,
1984. Species interrelationships and evolution in the Pliocene endemic faunas of Apricena (Gargano Peninsula, Italy). Geobios, Mem. Spec. 8: 379-383.
DE GIULI, C., F. MASINI, D. TORRE,
1985. Effetto arcipelago: un esempio nelle faune fossili del Gargano. Bollettino della Società Paleontologica Italiana 24, 2-3: 191-93. Modena.
DE GIULI, C., F. MASINI, D. TORRE, V. BODDI,
1987. Endemism and bio-chronological reconstructions: the Gargano case history. Bull. della S.P.I. 25, 3:267-276. Modena
DE VOS, J.,
1979. The endemic Pleistocene deer of Crete. Proceedings of the Koninklijke Nederlandse Akademie van Wetenschappen, Series B 82, 1: 59-90.
DE VOS, J.,
1984. The endemic Pleistocene deer of Crete. Verhandeling der Koninklijke Nederlandse Akademie van Wetenschappen, afd. Natuur-kunde, eerste reeks 31. North-Holland Publishing Compa-ny, Amsterdam, Oxford, New York: 1-100.
DE VOS, J.,
1996. Taxonomy, Ancestry and Speciation of the Endemic Pleistocene Deer of Crete Compared with the Taxonomy, Ancestry and Speciation of Darwin's Finches. In Reese, D.S. (ed.). Pleistocene and Holocene Fauna of Crete and its First Settlers. Monographs in World Archeology 28: 111-124. Madison: Prehistoric Press.
DE VOS, J.,
2000. Pleistocene deer fauna in Crete: Its adaptive radiation and extinction. In: Otsuka, H. et al. International Symposium The Ryukyu Islands - The Arena of Adaptive Radiation and Extinction of Island Fauna. Tropics 10, 1: 125-134. Kagoshima.
DERMITZAKIS, M.D., J. DE VOS,
1987. Faunal Succession and the Evolution of Mammals in Crete during the Pleistocene. Neues Jahrbuch Geologische und Paläontologische Abhandlungen 173, 3: 377-408. Stuttgart.
ECHELLE, A.A., I. KORNFIELD (EDS),
1984. Evolution of Fish Species Flocks. Orono, Maine: University of Maine at Orono Press.
ELDREDGE, N.,
1974. Testing evolutionary hypotheses in paleontology: a comment on Makurath and Anderson (1973). Evolution 28: 479-481.
ELDREDGE, N., S.J. GOULD,
1972. Punctuated equilibria: an alternative to phyletic gradualism. In: Schopf (ed.). Models in Paleobiology: 82-115. San Fransisco: Freeman, Cooper, Co.
ELDREDGE, N., S.J. GOULD,
1988. Punctuated equilibrium prevails. Nature 332: 211-212.
ENDLER, J.A.,
1977. Geographic Variation, Speciation, and Clines. Princeton, N.J.: University Press.
FEDUCCIA, A.,
1994. Tertiary bird history: Notes and comments. In: Prothero, D.R., & R.M. Schoch (eds). Major Features of Vertebrate Evolution: 178-189. Knoxville: University of Tennessee Paleontological Society.
FEDUCCIA, A.,
1995. Explosive evolution in Tertiary birds and mammals. Science 267: 637-638.
FORD, H.A., D.T. PARKIN, A.W. EWING,
1973. Divergence and evolution in Darwin's Finches. Biological Journal of the Linnean Society 5: 289-295.
FREUDENTHAL, M.,
1976. Rodent stratigraphy of some Miocene fissure fillings in Gargano (prov. Foggia, Italy). Scripta Geologica 37. Leiden: Rijksmuseum van Geologie en Mineralogie.
FRYER, G., T.D. ILES,
1972. The Cichlid Fishes of the Great Lakes of Africa: Their Biology and Evolution. Edinburgh: Oliver, Boyd.
GIFFORD, E.W.,
1919. Field notes on the land birds of the Galápagos Islands and of Cocos Island, Costa Rica. Proceedings of the California Academy of Sciences Series 4, 2: 189-258.
GINGERICH, P.D.,
1974. Stratigraphic record of early Eocene Hyopsodus and the geometry of mammalian phylogeny. Nature 248: 107-109.
GINGERICH, P.D.,
1976. Paleontology and phylogeny: patterns of evolution at the species level in early Tertiary mammals. American Journal of Science 276: 1-26.
GOULD, S.J., N. ELDREDGE, 1977. Punctuated equilibria: the tempo and mode of evolution reconsidered. Paleobiology 3: 115-151.
GOULD, S.J., N. ELDREDGE,
1986. Punctuated equilibrium at the third stage. Systematic Zoology 35, 1: 143-148.
GRANT, B.R.,
1985. Selection on bill characters in a population of Darwin's Finches: Geospiza conirostris on Isla Genovesa, Galápagos. Evolution 39: 523-532.
GRANT, B.R., P.R. GRANT,
1983. Fission and fusion in a population of Darwin's Finches: an example of the value of studying individuals in ecology. Oikos 41: 530-547.
GRANT, P.R.
1986. Ecology and Evolution of Darwin's Finches Princeton University Press, Princeton, New Jersey: 1-458.
GREENWOOD, P.H.,
1974. The Cichlid Fishes of Lake Victoria, East Africa: the Biology and Evolution of a Species Flock. Bulletin British Museum for Natural History (Zoology) Supplement 6: 1-134.
GREENWOOD, P.H.,
1981. The Haplochromine Fishes of the East African Lakes. München: Kraus Int.
GRZIMEK, B.,
1968. Enzyklopädie des Tierreiches. Zürich: Kindler Verlag.
HEANEY, L.R.,
1978. Island area and body size of insular mammals: evidence from the tri-colored squirrel (Callosciurus prevosti) of Southeast Asia. Evolution 32, 1: 29-44.

HIGASHI, M., G. TAKIMOTO, N. YAMAMURA,
1999. Sympatric speciation by sexual selection. Nature 402: 523-526.
HOOGERHOUD, R.J.C.,
1984. A taxonomic reconsideration of the haplochromine genera Gaurochromis Greenwood, 1980, and Labrochromis Regan, 1920 (Pisces, Cichlidae). Netherlands Journal of Zoology 34: 539-565.
KEENLEYSIDE, M.H.A.,
1991. Cichlid Fishes, Behaviour, Ecology and Evolution. London: Chapman, Hall.
KUSS, S.E.,
1975. Die pleistozänen Hirsche der ostmediterranen Inseln Kreta, Kasos, Karpatos und Rhodos (Griechenland). Berichte der Naturforschenden Gesellschaft zu Freiburg im Breisgau 65, 1-2: 25-79.
LACK, D.,
1947. Darwin's Finches, Cambridge University Press, Cambridge: 1-208.
LANDE, R.,
1982. Rapid origin of sexual isolation and character divergence in a cline. Evolution 36: 213-223.
LIEM, K.F., J.W.M. OSSE,
1975. Biological versatility, evolution and food resource exploitation in African cichlid fishes. American Zoologist 15: 427-454.
MACFADDEN, B.J.,
1992. Fossil Horses: Systematics, Paleobiology, and evolution of the family Equidae. Cambridges University Press, 1-369.
MAKURATH, J.H., 1974. Evolution of Appalachian gypidulid brachiopods: a reply to Eldredge (1974). Evolution 28: 481-483.
MAKURATH, J.H., E.J. ANDERSON,
1973. Intra- and interspecies variation in gypidulid brachiopods. Evolution 27: 303-310.
MATSUMOTO, Y., H. OTSUKA,
2000. Morphotypes of fossil deer (Cervus astylodon) and its miniaturization process. Tropics 10, 1: 155-164.
MAZZA, P.,
1986a. Prolagus (Ochotonidae, Lagomorpha, Mammalia) from Neogene fissure fillings in Gargano (Southern Italy). Bolletino della Societa Paleontologica Italiana 25, 2: 159-185.
MAZZA, P.,
1986b. Further data on the Gargano (Southern Italy) Prolagus (Ochotonidae, Lagomorpha, Mammalia). Bolletino della Societa Paleontologica Italiana 25, 3: 203-211.
MEYER, A., T.D. KOCHER, P. BASASIBWAKI, A.C. WILSON,
1990. Monophyletic origin of Lake Victoria cichlid fishes suggested by mitochondrial DNA sequences. Nature 347: 550-553.
PATTERSON, C.,
1978. Evolution. London: British Museum (Natural History), 1-197.
PROTHERO, D.R., & R.M. SCHOCH (EDS),
1994. Major Features of Vertebrate Evolution. Knoxville: University of Tennessee Paleontological Society.
RAUP, D.M.,
1991. A kill curve for Phanerozoic marine species. Paleobiology 17: 37-48.
RENDERS, E.,
1984. The gait of Hipparion sp. from fossil footprints in Laetoli, Tanzania. Nature 308, 179-81.
RENDERS, E., P.Y. SONDAAR,
1987. Hipparion. In: Leakey, M.D., J.M. Harris (eds). Laetoli: A Pliocene Site in Northern Tanzania, 471-481. Oxford: Clarendon Press.
SCHLIEWEN, U.K., D. TAUTZ, & S. PAABO,
1994. Sympatric speciation suggested by monophyly of crater lake cichlids. Nature 368: 629-632.
SEPKOSKI, J.J.,
1989. Periodicity in extinction and the problem of catastrophism in the history of life. Journal of the Geological Society of London 146: 7-19.
SEPKOSKI, J.J.,
1990. The taxonomic structure of periodic extinction. In: Sharpton, V.L. & P.D. Ward (eds). Global Catastrophes in Earth History. Geological Society of America, Special Paper 247: 33-44.
SIMPSON, G.G.,
1951. Horses. The story of the horse family in the modern world and through sixty million years of history. New York: Oxford University Press, 1-247.
SMIT, J.,
1989. Een catastrofale milieucrisis 66.5 miljoen jaar geleden. In: Prins, R.A., H.M. Van Emden (eds). Het verdwijnen van soorten: 41-59. Amsterdam: KNAW.
SONDAAR, P.Y.,
1977. Insularity and its effect on mammal evolu-tion, In: M.N. Hecht, P.C. Goody, B.M. Hecht (eds). Major patterns in vertebrate evolution. New York: Plenum Publishing Corporation: 671-707.
SONDAAR, P.Y., 1994. Paleoecology and evolutionary patterns in horses and island mammals. Historical Biology 8, 1-13.
SONDAAR, P.Y., A.A.E. VAN DER GEER,
in press. Extinction and Evolution of Plio-Pleistocene Island Ungulates. In: Cregut, E. (ed.). The holarctic Ungulates of the Pliocene and Pleistocene, 19th to 22nd September, Avignon. Proceedings. Quarternair. Paris.
SPAAN, A., P.Y. SONDAAR, & W. HARTMAN,
1994. The structure of the evolutionary process. Geobios 27, 3: 385-390. Villeurbanne.
STANLEY, S.M.,
1975. A theory of evolution above the species level. Proceedings National Acadamy of Sciences of the USA 72, 2: 646-650.
STANLEY, S.M.,
1978. Chronospecies' longevities, the origin of genera, and the punctuated model of evolution. Paleobiology 4: 26-40.
STEBBINS, G.L.,
1981. Coevolution of grasses and herbivores. Annals of the Missouri Botanical Garden 68: 75-86.
SZALAY, R.W., M.J. NOVACEK, M.C. MCKENNA (EDS),
1993. Mammal Phylogeny. New York: Springer Verlag.
TURNER, G.F.,
1994. Speciation mechanism in Lake Malawi cichlids: A critical review. Arch. Hydrobiology 44: 139-160.
VAN BEMMEL, A.C.V.,
1973. The concept of superspecies applied to Eurasiatic Cervidae. Zeitschrift für Säugetierkunde 38, 5:295-302.
VAN OIJEN, M.J.P.
1991. A revision of the piscivorous haplochromine cichlids of Lake Victoria, Part 1. Zoologische Verhandelingen Leiden 261: 1-95.
WITTE, F.,
1981. Initial results of the ecological survey of the haplochromine

cichlid fishes from the Mwanza Gulf of Lake Victoria (Tanzania): breeding patterns, trophic and species distribution, with recommendations for commercial trawl-fisheries. Netherlands Journal of Zoology 31: 175-202.
WITTE, F., E.L.M. WITTE-MAAS,
1981. Haplochromine cleaner fishes: a taxonomic and eco-morphological description of two new species. Revision of the haplochromine species (Teleostei, Cichlidae) from Lake Victoria. Netherlands Journal of Zoology 31: 203-231.
WOODBURNE, M.O., & J.A. CASE,
1996. Dispersal, vicariance, and the Late Cretaceous to Early Tertiary land mammal biogeography from South America to Australia. Journal of Mammalian Evolution 3: 121-161.

35

THE ROLE OF ISLANDS IN PUSHING THE PLEISTOCENE EXTINCTION ENVELOPE: THE STRANGE CASE OF THE CYPRIOT PYGMY HIPPOS

ALAN H. SIMMONS

Department of Anthropology,
University of Nevada, Las Vegas
Las Vegas, Nevada, USA
EMAIL: simmonsa@univ.edu

INTRODUCTION

The considerable impacts that humans have on the environment are an issue of contemporary concern. There is clear evidence that this is not a new phenomenon, however; people have substantially and adversely affected their environments for thousands of years. This has been particularly severe in fragile ecosystems, such as deserts or islands. Substantial archaeological and historical evidence, especially from the Pacific (Anderson 1991; Steadman 1995), demonstrates that humans have had immediate and long-lasting affects on islands, including, in many cases, the extinction of endemics. This is usually associated with food-producing, or Neolithic, economies, however, and nearly always involves relatively small animals. When one attempts to extend human-induced extinction episodes back into time, invoking direct predation by hunters, the evidence becomes much more equivocal, whether looking at islands or continental land masses. Indeed, the massive and global extinction of huge mega-fauna that occurred at the end of the Pleistocene is one of the most controversial topics in contemporary archaeology and paleontology. There essentially are two schools of thought in the debateBone proposes that human hunters were the prime cause of the extinctions, while the other minimizes the human role and looks, instead, to ecological factors (e.g., Diamond 1989; Grayson 1991; Martin and Klein 1984). Most researchers tend to favor the latter. This essay examines this issue from an island perspective, presenting data that implicates humans in the extinction of an unique island megafauna at the beginning of the Holocene, some 10,000 years ago.

Both the Mediterranean islands and the Near East are regions where human induced impacts have been especially harsh from Neolithic times up to the present (Blondell and Aronson 1995; Blumler 1993; Kohler-Rollefson 1988; Patton 1996; Rollefson 1997). Until recently, however, there has been only limited evidence that pre-Neolithic peoples were even on most of the Mediterranean islands, let alone responsible for endemic extinctions (Cherry 1990). A few claims have been made for the association of humans and Pleistocene fauna on the islands, but these generally have not been well-substantiated (Cherry 1992).

Mallorca is one of the few examples where humans and such fauna have been associated in secure contexts (e.g., Burleigh and Clutton-Brock 1980; Waldren 1994) at Neolithic sites. Recent research, however, has drastically changed this perception, and the past 10 to 15 years have been exciting times in Mediterranean island prehistory. Many new discoveries have literally rewritten our conception of the dynamic cultural processes that we now know characterized the early settlement of at least one island, Cyprus. One of the principal, and quite controversial, sites involved in this recent research is Akrotiri Aetokremnos (FIGURE 1). At Aetokremnos, we have proposed that humans were a prime causal factor in the extinction of a unique island-adapted Amini mega fauna, the Cypriot pygmy hippopotamus (Phanourios minutus). Much of the following discussion is based on the detailed arguments made in the site=s final report (Simmons 1999) and in a recent summary of some of the controversial aspects of Aetokremnos (Simmons 2001).

RESEARCH BACKGROUND

Aetokremnos is unique from at least two perspectives. The first is paleontological and the second is archaeological. Numerous fossil sites dating to the Pleistocene in Cyprus and other Mediterranean islands contain endemic, island-adapted fauna, which often include both pygmy hippopotami and elephants (Held 1992; Reese 1989; Sondaar 1986; Swiny 1988). Never before, however, has an association of these animals with humans been clearly documented.

Neolithic has, until the important new discoveries from Shillourokambos (Guilaine et al. 1995; Vigne et al. 2000) and Mylouthkia (Peltenburg et al. 2000, 2001) been dated relatively late in the Near Eastern Neolithic sequence.

EXCAVATION SUMMARY

Akrotiri Aetokremnos is located on the southern coast of Cyprus. It is a collapsed rockshelter, dramatically perched over the Mediterranean Sea (FIGURE 2); much of it has

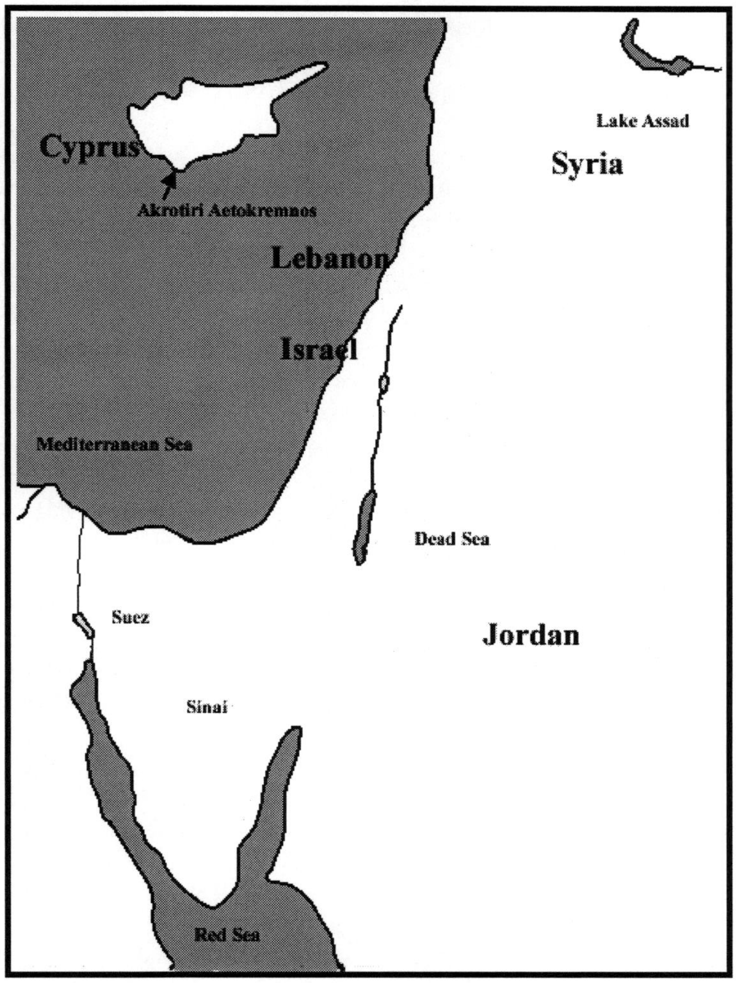

FIGURE 1. MAP OF THE MEDITERRANEAN ISLANDS, SHOWING LOCATION OF AKROTIRI AETOKREMNOS IN CYPRUS.

To fully understand the significance of Aetokremnos, one has to realize that for many years there have been claims for early (that is, pre-Neolithic) human occupation of many of the Mediterranean islands. Cherry (1990, 1992), however, has convincingly demonstrated that there are few, if any, compelling arguments for people being on most of the Mediterranean islands during the Late Pleistocene or Early Holocene. Cyprus has not been immune from claims for early antiquity (e.g., Adovasio et al. 1975; Stockton 1968; Vita-Finzi 1973). While these are intriguing declarations, they cannot be well-supported using the rigorous methods outlined by Cherry. Until the excavation of Aetokremnos, there simply have been no convincing sites documented that predate the Aceramic Neolithic. Indeed, the Aceramic

eroded to the waters below. Initially discovered by amateurs, its surface indications were far from impressive, and many questioned whether it was even an archaeological site or not. The locality contained a surface scatter of Phanourios bones, a layer of marine shell, and a few chipped stone artifacts. The likelihood of in situ deposits appeared slim, and there was no obvious connection between the bones and the artifacts. Numerous scatters of chipped stone artifacts occur on the area above the cliffs in which Aetokremnos is located, and could easily have washed down to the site, fortuitously mixing with a Pleistocene bone bed. Prior to our excavations, one radiocarbon determination on surface bone suggested a date of ca. 8-9,000 b.c., although two others indicated a more recent occupation.

Despite the apparent paucity of cultural materials at Aetokremnos, the early date was intriguing since it pre-dated the Neolithic for Cyprus. Even more tantalizing was the possible association of cultural remains with animals thought to have gone extinct far earlier. Thus, although both the chronology and the association of extinct animals with the cultural materials at Aetokremnos were questionable, the site piqued enough interest to warrant scientific excavations, which were conducted over three years. The results have convinced us that Aetokremnos is the oldest human occupation of Cyprus, and indeed, among the best-documented of any early site on the Mediterranean islands. Furthermore, we are equally convinced of the association of cultural materials at the site with extinct Pleistocene fauna. The following briefly summarizes these investigations.

Our excavations of around 40 square meters exposed nearly all that is left of Aetokremnos. When occupied it never was a large site. Nearly a meter of vertical deposition is present, with four major strata present. Both Strata 2 and 4 contain artifacts, bones, and features, while Stratum 1 is a mixed context and Stratum 3 is a sterile zone. The majority of the artifacts are located in Stratum 2, while the majority of the Phanourios remains occur in Stratum 4. Thus, even critics usually have not had a problem accepting that Stratum 2 was cultural; there was, however, more question about Stratum 4's anthropogenic origin.

We recovered over a thousand chipped stone artifacts, as well as numerous stone and shell beads. The chipped stone from Aetokremnos is so far unique in Cyprus, although it would not be out of place in Epipaleolithic or early Neolithic mainland assemblages. The debitage contains a large number of well-made blades as well as flakes. A variety of tools are present, with the most diagnostic being small thumbnail scrapers. In addition to chipped stone, numerous stone and shell beads were recovered, as were a small variety of ground implements. Eleven cultural features, most "casual hearths," also are documented.

Aetokremnos is one of the best absolute dated sites on any of the Mediterranean islands. Thirty-one radiocarbon determinations provide a coherent set of absolute dates, with a calibrated average of 10,529 b.p. (cal B.C. 9,825, single standard deviation) (this average omits five samples--Wigand and Simmons [1999] provide details). It is important to note that the time difference between Strata 2 and 4 is so small that it cannot be measured in radiocarbon years. We believe that the data indicate a relatively short-term occupation, perhaps of only a few hundred years, during the tenth millennium B.C. Certainly the radiocarbon determinations alone provide compelling evidence that people and Phanourios co-existed.

These dates make Aetokremnos the earliest site in Cyprus, and amongst the earliest on any Mediterranean island. To appreciate the context of Aetokremnos, conventional wisdom held that the first occupants of Cyprus dated to the Aceramic Neolithic, commencing around 7,000 BC (LeBrun et al. 1987). Thus Aetokremnos would appear to pre-date the (traditional) Aceramic Neolithic by some three thousand years. Recent work cited earlier in this paper, however, has now documented a ACypro-PPNB (Peltenburg et al. 2000. 2001) at ca. 8,000 BC, thus shortening the gap between what we have termed the AAkrotiri Phase and the Aceramic Neolithic.

Of paramount importance, of course, was the huge faunal assemblage, consisting of over 200,000 bones (Table 1). Most is Phanourios, with over 500 individuals of all ages represented. At least three dwarf elephants occur, as do a minimum of 70 birds, primarily great bustard. A few other terrestrial animals also are present, as is a huge assemblage of over 70,000 marine invertebrates. The majority of the Phanourios remains occurs in Stratum 4, while the bulk of the other fauna is in Stratum 2.

INTERPRETATION

The above discussion presented a concise outline of the empirical data from Aetokremnos. It is the interpretation of this information, however, that has caused the most controversy, and the remainder of this paper addresses these issues. Perhaps the major problem for many of the site=s skeptics relates to the association of the Phanourios remains with the cultural data. We believe that this is unequivocally the case, or at least as unequivocally as anything can be in archaeology. The following discussion outlines our rationale for these conclusions.

In assessing whether or not the Phanourios bones are actually directly associated with the cultural remains, several issues have had to be considered. These include: could the huge amount of bone be accounted for by natural means, such as a sink hole or fluvial deposition?; could not people have situated their site on top of a previously deposited fossil bed?; if these animals were eaten, why are there no cut marks on the generally well-preserved bone? These are good questions. Our answers to them are summarized below.

NATURAL VS. CULTURAL DEPOSITION

The issue of natural deposition is central to the argument that Aetokremnos is anthropogenic. Before proposing that the faunal remains from the site were the result of cultural activity, it was our task to first explain why this deposit was not natural, similar perhaps to many of the other paleontological sites known to exist in Cyprus. One scenario for explaining this deposit was that it reflected an accretional paleontological deposit. This supposes that individual animals or small groups of animals repeatedly died on the site. This could correspond to the natural death of, presumably, aging or sick animals or to the accidental occurrences of animals falling over the cliff or into a sinkhole. This interpretation is highly unlikely for the following reasons.

First, while accretional sites are well known they rarely if ever are found in locations as precipitous as Aetokremnos. The site also lacks both vegetation and water resources that would have been attractive to animals. It is doubtful that the shelter was ever deep enough to offer thorough protection from the elements and it is equally unlikely that older or sick animals could have easily climbed down the vertical upper portion of the cliffs to reach Aetokremnos. Furthermore, if Aetokremnos were a natural mass die site, one would expect to find an abundance of aged/sick individuals.

This is not the case; 27% of the Phanourios individuals analyzed are under one year old.

A second argument against a natural deposition is anatomical. In an undisturbed natural die site, one might expect some articulation, which is not the case at Aetokremnos--virtually none of the bones are articulated. There is no evidence for predators in Cyprus prior to the arrival of humans, so post-occupational disturbance by scavenging carnivores may be ruled out. Detailed geomorphic examination of the deposits also revealed no evidence of major disturbance, such as water transport (Mandel 1999; Mandel and Simmons 1997). Finally, there are many shelters all along the southern coast of the Akrotiri Peninsula that would have provided similar conditions and should therefore show evidence of Phanourios remains. As confirmed by later survey, this is not the case.

Third, while animals falling over the cliffs above Aetokremnos is possible, there is no feature of the plateau above the cliff that would account for a repeated jump at this very point and not elsewhere along the Akrotiri Peninsula. If accidental jumps or falls occurred, they should be distributed more or less randomly along the horizontal bottoms of the cliffs.

Finally, the sinkhole scenario can be discounted as the geomorphological nature of the Akrotiri Peninsula argues against their formation. Furthermore, in natural sinkholes, sedimentation occurs in between the various animals falls. At Aetokremnos, this is not the case (Mandel 1999).

If an accretional paleontological deposit can be discounted, what of the possibility of a natural single mass-death site? This scenario could suppose a mass-jump of one or several Phanourios herds, presuming that this type of behavior was possible for the species. There are a number of reasons that make this unlikely. First, it is likely that more than one major episode is responsible for the bone bed. Furthermore, the earlier observations that there are no features on the plateau that could account for several falls at just this point are equally valid. If this scenario were true, the animals would have rolled down the cliff and spread throughout the slope. There is no natural feature that could have stopped them from rolling down to the bottom of the cliffs and into the sea. This is particularly true for an animal as rotund as Phanourios. The distribution of bones at Aetokremnos was far too concentrated for such a phenomenon.

If such a situation occurred, one might expect that not all the bones would have been covered by sediment immediately, and at least some would show surface weathering. While many of the bones are differentially weathered, this appears to have been caused by post-depositional erosion of the deposit resulting in differential exposure to the elements.

FIGURE 2. PHOTOGRAPH OF AETOKREMNOS, SHOWING CLIFF-SIDE POSITION.

Finally, a mass death site does not explain the fact that the bones were located, primarily, inside the original shelter. If they had fallen or jumped from the cliffs en masse, it is physically impossible for them to have accumulated inside of the shelter. Of course, an alternate explanation involving a natural mass death is that the animals concentrated at the shelter under their own volition. This seems unlikely, since it is difficult to conceive that over 500 hippopotami, as well as a multitude of other animals, would have crammed into the relatively small confines of the Aetokremnos shelter. Furthermore, the radiocarbon determinations and stratigraphy do not indicate a long time span that could have allowed for smaller groups of Phanourios to have gradually accumulated inside of the shelter.

Very few natural agents can account for a concentration such as observed at Aetokremnos. A human origin for the bone deposit therefore becomes the most logical explanation, especially when considering other lines of evidence, such as the faunal remains themselves.

NATURE OF THE FAUNAL REMAINS

The faunal assemblage itself also provides evidence for our justification of the direct relationship between people and hippos at Aetokremnos. Lyman (1994:216-219) cites several criteria to distinguish cultural and natural bone: burning or charring; bone comminution; mineralization, weathering, and staining; butchering and technology marks; ethnological analogy; skeletal completeness; and context and associations. How well does Aetokremnos fit within these? We may brief examine each one of these topics in turn.

BURNING.

Approximately 30% of the bone at Aetokremnos is burned, much of it severely. In arguing against an anthropogenic origin, an obvious observation might be that the bone in Stratum 4 was burned by unrelated and later overlying fires in the clearly culturally derived Stratum 2. There are numerous arguments against this.

First, if this scenario were correct, burned bone would only, or primarily, occur under Stratum 2 hearths. This is not the case. Burned bone is distributed throughout the deposits, and highly concentrated areas of burned bone show no correlation with overlaying hearths. Furthermore, much of the Phanourios bone in Stratum 2 also is burned. If one argues the bone is burned because it is underneath hearths, then it cannot also be argued that the burning of Stratum 2 bone is due to the same process.

A second argument against the proximity scenario is purely stratigraphic. If the burned bone was the result of intense heat generated by overlaying hearths, then one would expect that the intervening Stratum 3 would exhibit signs of burning as well. It does not.

A third argument relates to the intensity of burning. It is unlikely that hearths in Stratum 2 could have generated enough heat to cause the degree of burning observed. In an experimental study examining the effects of fire on bone, David (1990) compared proportions of burning on bones subjected various types of fires. His results indicate that it is unlikely that bone beneath hearths, with only indirect exposure to heat, would be as severely burned as are those at Aetokremnos. Further evidence against the bone being burned by overlaying hearths is provided by an experiment by Stiner and colleagues. They point out that "...although bones were buried as deep as 15 cm below the coal bed, only those specimens in the first 5 cm were affected much by heat from the fire. Moreover, these shallowly buried bones were burned only to the point of carbonization..."(Stiner et al. 1995:230). At Aetokremnos, Stratum 4 often exceeds 50 cm. in thickness, and buried bone occurs throughout the stratum, not only on the top of it. Furthermore, Stratum 2 frequently is separated from Stratum 4 by several cm of Stratum 3. The separation between the bottom of fire hearths in Stratum 2 and bone in Stratum 4 almost always exceeds 15 cm, and yet much of this bone is burned. It is therefore exceedingly unlikely that the burning is the result of overlaying hearths.

Finally, the accidental burning scenario does not take into account that three of the 11 features at the site occur in Stratum 4. Thus these features are horizontally as well as vertically associated with burned Phanourios bone.

COMMUNUTION.

Comminution (small fragmentation) of bone, can result by boiling it to extract grease (Lyman 1994:217). This does not necessarily require that the bone fragments be burned. At Aetokremnos, much of the bone is both comminuted and burned. This is thus not direct evidence for grease extraction, but strongly suggestive of cultural intervention.

WEATHERING.

The degree of weathering, mineralization, or staining on bones can vary between naturally and culturally deposited bone. Lyman (1994:217) notes that cultural bone should be more heavily weathered than natural bone. At Aetokremnos, there is little variation in the degree of weathering on the bone.

ANALOGY.

Ethnographic analogy is useful in interpreting archaeofaunas, but has at least two problems. First, ethnozoological data may be ambiguous and second, cultures evolve and change, so what people were consuming when ethnographic data were compiled may not be what the ancestors of those people consumed (Lyman 1994:218). In interpreting the Aetokremnos fauna, we relied heavily on ethnographic and wildlife data, not so much for demonstrating whether or not people consume hippopotami (they obviously do), but rather to indicate the range of variation in deposition and butchering processes.

SKELETAL COMPOSITION.

The completeness of skeletal elements represented at a site may give a clue to human involvement. There is, however, a considerable behavioral range involved with how people process and consume animals, and they do not always accumulate and deposit only portions of carcasses (Lyman 1994:218). At Aetokremnos, complete, or nearly complete,

Phaniouros skeletal elements are present. While one could argue this indicates a natural deposition, one could just as well argue this reflects processing and consumption patterns.

CUT-MARKS

The presence of cut or butchery marks on faunal assemblages is a contentious issue, so important to Binford (2000), for example, that in his review of Aetokremnos, he essentially dismisses the cultural association of humans and Phanourios. I an assemblage of over 500 hippopotami, it seems reasonable that butchery or cut marks should be present if humans were consuming these animals. These, however, appear to be absent at Aetokremnos (Olsen 1999). There is an enormous, and contradictory, literature on the issue of cut-marks and other bone modifications in archaeological fauna (see Lyman [1994:294-353] for a thorough summary, and Morlan [1984] for rigid criteria for recognizing artificial bone modifications). After a review of the literature, it is apparent that there can be numerous arguments explaining either the lack or presence of cut-marks.

An important point is that cut-marks often are relatively rare, even in many well-documented butchery assemblages. Soft tissues frequently shield bone from being marked (Shipman and Rose 1983:86). Indeed, one might consider cut-marks on bone the signature of unskilled or inefficient butchers.

Unfortunately, there are few archaeological examples of hippo butchering in the literature, and virtually none of pygmy hippopotami. The exception to this is an intriguing article on the extinct Madagascar pygmy hippopotamus, where cut-marks on four bones are reported (MacPhee and Burney 1991). A convincing argument is made that these bones were cut with metal implements while green. The authors also conclude that it is unlikely that stone tools could have produced the types of marks observed (MacPhee and Burney 1991:701).

Lacking clear-cut archaeological data, there are ample ethnographic examples and experimental studies where it is apparent that cut-marks are not necessary for defining butchery. Crader (1983:135) has pointed out that the paucity of cut-marks is simply not a very reliable guide for understanding the degree of butchering of many carcasses. Frison (1989:778) and Haynes (1988) comment extensively on the lack of cut-marks in modern elephant butchery experiments. During the intense African herd culling that

Vertebrates	NISP	%	MNI
pygmy hippopotami-*Phanourios*	218,459	98.3	505+
pygmy elephants-*Elephas*	229	0.1	3
deer-*Dama*	4	--	1-4
pigs-*Sus*	13	--	c. 4
genets-*Genetta*	2	--	1
mice-*Mus*	5	--	1
birds	3,207	1.4	75+
snakes-*Vipera/Natrix*	c. 245	0.1	14-40
tortoises-*Testudo*	25+	--	9-14
toads-*Bufo*	1	--	1
fish	1	--	1
Total Vertebrates	222,190	99.9	--
Marine Invertebrates	73,365	--	21,576
Land Snails	90*	--	97
Total	295,645	--	--

TABLE 1.
SUMMARY OF FAUNAL REMAINS FROM AETOKREMNOS

Haynes summarizes, experienced culling crews did not, deliberately or accidentally, cut bones (Haynes 1991:185). If steel tools often do not produce cut-marks, how likely is it that stone ones will?

A related issue involves cooking processes (cf. Marshall 1989:17). Clearly, how an animal is cooked can relate to how it is butchered. If, for example, Phanourios were roasted whole at Aetokremnos, there may have been little need for butchering so invasive as to scar the bone, since the meat could have been pulled apart at articular joints with minimal cutting of tissue.

CONTEXT.

Finally, context and associations are critical is assessing the relationship of human and animal remains. At Aetokremnos, the constellation of associated data overwhelmingly point to a cultural origin for all of the fauna, including not only the hippopotami, but also the dwarf elephant, birds, and other faunal materials.

ARTIFACT ASSEMBLAGE

Some critics (e.g., Bunimovitz and Barkai 1996) are troubled that by the artifacts from Aetokremnos are too small to have been used in butchering animals such as Phanourios. We disagree, believing that it is dangerous to impute too much functional argument into chipped stone assemblages. Our argument that the Aetokremnos assemblage could easily have functioned as Phanourios processing implements is supported by Frison (1989), who demonstrates the facility of butchering modern elephants with the simplest of chipped stone implements. There is other archaeological evidence of the use of small, unspecialized tools associated with kill or butchery sites. This is particularly well documented North America Paleo-Indian sites (e.g., Frison 1991, especially pp. 289-325).

It is also important to recall that the chipped stone assemblage from Aetokremnos is unique in Cyprus. One cannot make the argument that the artifacts occurring with the Phanourios remains are simply fortuitous associations, representing intrusive items from a later occupation, since such an assemblage has not been documented on the island.

In summary, there are ample examples of the use of small, unsophisticated chipped stone implements that functioned quite efficiently as butchery tools. The assemblage from Aetokremnos easily fits within a multi-functional use pattern in which butchering was a major activity, both on Phanourios and the other fauna.

PALEONTOLOGICAL SITES

Another consideration relates to the numerous paleontological sites containing Phanourios (and elephants) on Cyprus (Reese 1995). If Aetokremnos were not cultural, it should resemble these. It does not. Very few of the paleontological sites contain the abundance of bones that Aetokremnos does, nor the other factors that make Aetokremnos stand out so dramatically (e.g., burning, artifacts, cultural features, stratigraphy, other fauna such as birds and shell). Clearly, however, this situation demands additional investigation of all sites containing these remains, which generally have not been scientifically studied.

CONCLUSIONS

In summary, we feel that the evidence overwhelmingly supports the direct association of Phanourios with cultural activities at Aetokremnos. This is demonstrated by:
(1) the dense accumulation of over 500 individuals with the small confines of a rockshelter;
(2) the near total disarticulation of all faunal remains;
(3) the burning of ca 30% of the hippo remains;
(4) the high amount of phosphorous in Stratum 4 (and Stratum 2); it is unlikely that a natural deposit of bone would be so high in phosphorous;
(5) the "clean" floor of shelter--no accumulation of sediments between the floor and Phanourios suggests a prepared floor;(6)the virtual contemporaneity of Stratum 2 and Stratum 4 based on abundant radiocarbon determinations;
(6) the presence of artifacts directly associated with Phanourios:
(7) the presence of an assemblage never before described in Cyprus
(8) the presence of cultural features;
(9) the stratigraphic integrity of Strata 2 and 4, making mixing unlikely ;
(10) and the presence of other associated fauna, including an inverse stratigraphic relationship in the number of birds and shells to Phanourios, suggesting, perhaps, that as the hippopotami became scarcer, the occupants of the site turned to a more diverse economic base.

So, when all aspects of the site are examined, the most parsimonious explanation is cultural in nature. This underlines the need for careful, interdisciplinary investigations of sites such as Aetokremnos. By combining all these data, we believe that the Akrotiri shelter was used for protection and storage, and functioned as a processing site and bone cache; much of the bone may have been used for fuel. We have modeled that a small group of humans could have, within a relatively short period, been the trigger to eradicate remnant Phanourios populations who may have already been suffering ecological stress due to climatic change, and were thus on the verge of extinction.

SIGNIFICANCE

The significance of Aetokremnos is belied by its small size. This site has changed several pre-existing perceptions of how humans colonize islands. In particular, Aetokremnos is significant from at least four perspectives. First, it is the oldest documented site in Cyprus, and amongst the oldest on any of the Mediterranean islands. Second, Aetokremnos has implications for how islands are colonized. Third, Aetokremnos is one of the few convincing demonstrations of a human role in the extinction of Pleistocene fauna. Finally, Aetokremnos has challenged research paradigms that exist on many of the Mediterranean islands on the actual nature of archaeological remains. That is, the concept that archaeology must equal architecture. The remains of hunter/gatherers are by definition ephemeral and have perhaps been ignored for far too long on the Mediterranean islands.

Aetokremnos represents an example of how even a small site can literally redefine prehistory. This is perhaps the ultimate significance of Aetokremnos, as it has caused a serious re-thinking of the nature of archaeological evidence, demonstrating that such data are variable and need not be found by any particular school of though as long as they can be defended with rigorous adherence to science method (Simmons 1991, 1998).

The site remains controversial, which pleases us immensely. We are, however, thoroughly convinced that Aetokremnos is, in fact, both cultural and dates to the 10th millennium b.c. Furthermore, we are equally convinced that humans, at least on Cyprus, played a major role in the extinction of an uniquely island-adapted Amini@ megafauna.

REFERENCES

ADOVASIO, J., G. FRY, J .GUNN, AND R. MASLOWSKI,
1975. Prehistoric and historic settlement patterns in western Cyprus (with a discussion of Cypriot Neolithic stone tool technology). World Archaeology, 6, 339-364.
ANDERSON, A.,
1991. Prodigious birds: moas and moa hunting in prehistoric New Zealand. Cambridge: Cambridge University Press.
BINFORD, L.,
2000. Review of Faunal extinctions in an island society: Pygmy hippopotamus hunters of the Akrotiri Peninsula, Cyprus by A. Simmons. American Antiquity 65, 771.
BLONDEL, J. AND J. ARONSON,
1995. Biodiversity and ecosystem function in the Mediterranean Basin: human and non-human determinants. In G. Davis and D. Richardson, eds. Mediterranean-type ecosystems: The function of biodiversity. Berlin: Springer-Verlag, 1995, 43-120.
BLUMLER, M..,
1993. Successional pattern and landscape sensitivity in the Mediterranean and Near East. In: D. Thomas and R. Allison, eds. Landscape Sensitivity. New York: John Wiley and Sons, 1993, 287-305.
BUNIMOVITZ, S. AND R. BARKAI,
1996. Ancient bones and modern myths: Ninth millennium BC hippopotamus hunters at Akrotiri Aetokremnos, Cyprus? Journal of Mediterranean Archaeology, 9, 85-96.
BURLEIGH, R., AND CLUTTON-BROCK, J.,
1980. The survival of Myotragus balearicus (Bate, 1909) into the Neolithic on Mallorca. Journal of Archaeological Science 7, 385-88.
CHERRY, J.,
1990. The first colonization of the Mediterranean islands: a review of recent research. Journal of Mediterranean Archaeology, 3, 145-221.
CHERRY, J.,
1992. Paleolithic Sardinians? Some questions of evidence and method. In: R. Tykot and T. Andrews, eds. Sardinia in the Mediterranean: A footprint in the sea. Monographs in Mediterranean Archaeology 3. Oxford: Sheffield Academic Press, 1992, 29-39.
CRADER, D.,
1983. Recent single-carcass bone scatters and the problem of "butchery" sites in the archaeological record. In J. Clutton-Brock and C. Grigson, eds. Animals and Archaeology: 1. Hunters and Their Prey. Oxford: British Archaeological Reports, International Series No. 163, 1983, 107-141
DAVID, B.,
1990. How was this bone burnt? In: S. Solomon, I. Davidson, and D. Watson, eds. Problem Solving in Taphonomy. Tempus--Archaeology and Material Culture Studies in Anthropology, vol. 2. St. Lucia: University of Queensland Anthropology Museum, 1990, 65-79.
DIAMOND, J.,
1989. Quaternary megafaunal extinctions: variations on a theme by Paganini. Journal of Archaeological Science, 16, 167-175.
FRISON, G.,
1989. Experimental use of Clovis weaponry and tools on African elephants. American Antiquity, 54, 766-784.
FRISON, G.,
1991. Prehistoric hunters of the High Plains. 2nd ed. San Diego: Academic Press.
GRAYSON, D.,
1991. Late Pleistocene mammalian extinctions in North America: taxonomy, chronology, and explanations. Journal of World Prehistory, 5, 193-231.
GUILAINE, J., F. BRIOIS, J. COULAROU, I. AND CARRÈRE,
1995. L'etablissement Neolithique de Shillourokambos (Parekklisha, Chypre). Premiers résultats. Report of the Department of Antiquities, Cyprus 1995, 11-32.
HAYNES, G.,
1988. Spiral fractures, cutmarks and other myths about early bone assemblages. In: J. Wilig; C. Aikens, and J. Fagan, eds. Early human occupation in far western North America: The Clovis-Archaic interface. Carson City: Nevada State Museum, Anthropological Paper 21, 1988, 145-151.
HAYNES, G.,
1991. Mammoths, mastodonts, and elephants: biology, behavior, and the fossil record. Cambridge: Cambridge University Press.
HELD, S., 1992. Pleistocene fauna and Holocene humans: A gazetteer of paleontological and early archaeological sites on Cyprus. Studies in Mediterranean Archaeology, Vol. 95. Jonsered, Sweden: Paul Åströms Förlag.
KÖHLER-ROLLEFSON, I.,
1988. The aftermath of the Levantine Neolithic revolution in light of ecologic and ethnographic evidence. Paleorient, 14, 87-93.
LEBRUN, A.; S. CLUZAN, S. DAVIS, J. HANSEN, AND J. RENAULT-MISKOVSKY,
1987. Neolithique preceramique de Chypre. L'Anthropologie, 91, 283-316.
LYMAN, R.,
1994. Vertebrate taphonomy. Cambridge Manuals in Archaeology. Cambridge: Cambridge University Press.
MACPHEE, R., AND D. BURNEY,
1991. Dating of modified femora of extinct dwarf Hippopotamus

from southern Madagascar: implications for constraining human colonization and vertebrate extinction events. Journal of Archaeological Science, 18, 695-706.

MANDEL, R.,
1999. Stratigraphy and sedimentology. In: A. Simmons. Faunal Extinctions in an Island Society: Pygmy Hippopotamus Hunters of the Akrotiri Peninsula, Cyprus. New York: Plenum/Klewer Academic Press, 1999, 49-69.

MANDEL, R. AND A. SIMMONS,
1997. Geoarchaeology of the Akrotiri Aetokremnos rockshelter, southern Cyprus. Geoarchaeology, 12, 567-605.

MARSHALL, L.,
1989. Bone modification and "the laws of burial." In: R. Bonnichsen and M. Sorg, eds. Bone Modification. Center for the Study of the First Americans. Orono: University of Maine, 1989, 7-24.

MARTIN, P. AND R. KLEIN, eds,
1984. Quaternary extinctions: A prehistoric revolution. Tucson: University of Arizona Press.

MORLAN, R.,
1984. Toward the definition of criteria for the recognition of artificial bone alternations. Quaternary Research, 22, 160-171.

OLSEN, S.,
1999. Investigation of the Phanourios bones for evidence of cultural modification. In: A. Simmons. Faunal Extinctions in an Island Society: Pygmy Hippopotamus Hunters of the Akrotiri Peninsula, Cyprus. New York: Plenum/Klewer Academic Press, 1999, 230-237.

PATTON, M.,
1996. Islands in time: island sociogeography and Mediterranean prehistory. London and New York: Routledge.

PELTENBURG, E., S. COLLEDGE, P. CROFT, A. JACKSON, C. MCCARTNEY, AND M. MURRAY,
2000. Agro-pastoralist colonization of Cyprus in the 10th millennium BP: initial assessments. Antiquity, 74, 844-53.

Peltenburg, E., S. Colledge, P. Croft, A. Jackson, C. McCartney, and M. Murray, 2001. Neolithic dispersals from the Levantine Corridor: a Mediterranean perspective. Levant, 33, 35-64.

REESE, D.,
1989. Tracking the extinct pygmy hippopotamus of Cyprus. Field Museum of Natural History Bulletin, 60, 22-29.

REESE, D.,
1995. The Pleistocene vertebrate sites and fauna of Cyprus. Geological Survey Department, Bulletin No. 9. Nicosia: Ministry of Agriculture, Natural Resources and Environment.

ROLLEFSON, G.,
1997. The Neolithic devolution: ecological impact and cultural compensation at 'Ain Ghazal, Jordan. In: Joseph Seger, ed. Retrieving the past. Mississippi State, Mississippi: Cobb Institute of Archaeology, 1997, 219-29.

SHIPMAN, P. AND J. ROSE,
1983. Early hominid hunting, butchering, and carcass-processing behaviors: approaches to the fossil record. Journal of Anthropological Archaeology, 2, 57-98.

SIMMONS, A.,
1991. One flew over the hippo's nest: Extinct Pleistocene fauna, early man, and conservative archaeology in Cyprus. In: G. Clark, ed. Perspectives on the Past. Philadelphia: University of Pennsylvania Press, 1991, 282-304.

SIMMONS, A.,
1998. Exposed fragments, buried hippos: Assessing surface archaeology. In: A. Sullivan, ed. Surface Archaeology. Albuquerque: University of New Mexico Press, 1998, 159-167.

SIMMONS, A.
1999. Faunal extinctions in an island society: Pygmy hippopotamus hunters of the Akrotiri Peninsula, Cyprus. New York: Plenum/Klewer Academic Press.

SIMMONS, A.,
2001. The first humans and last pygmy hippopotami of Cyprus. In: S. Swiny, ed. The earliest prehistory of Cyprus. From Colonization to Exploitation. Cyprus American Archaeological Research Institute Monograph Series, Vol. 12. Boston: American Schools of Oriental Research Reports, No. 5, 2001, 1-18.

SONDAAR, P.,
1986. The island sweepstakes. Natural History, 95, 50-57.

STEADMAN, D.,
1995. Prehistoric extinctions of Pacific island birds: biodiversity meets zooarchaeology. Science, 267, 1123-131.

STINER, M., S. KUHN, S. WEINER, AND O. BAR-YOSEF,
1995. Differential burning, recrystallization, and fragmentation of archaeological bone. Journal of Archaeological Science, 22, 223-237.

STOCKTON, E.,
1968. Pre-Neolithic remains at Kyrenia, Cyprus. Report of the Department of Antiquities, Cyprus, 16-19.

SWINY, S.,
1988. The Pleistocene fauna of Cyprus and recent discoveries on the Akrotiri Peninsula. Report of the Department of Antiquities, Cyprus 1988 1-14.

VIGNE, J-D, I. CARRÈRE, J-F SALIÉGE, A. PERSON, H. BOCHERENS, J. GUILAINE, AND J-F BRIOIS,
2000. Predomestic cattle, sheep, goat and pig during the late 9th and the 8th millennium Cal. BC on Cyprus: preliminary results of Shillourokambos (Parekklish, Limassol). In: M. Mashkour, A. Choyke, H. Buitenhuis, and F. Poplin, eds. Archaezoology of the Near East IVA, . Groningen: ARC, Publicatie 32, 2000, 83-105.

VITA-FINZI, C.,
1973. Paleolithic finds from Cyprus? Proceedings of the Prehistoric Society, 39, 453-454.

WALDREN, W.
1994. Survival and extinction: Myotragus Balearicus, an endemic Pleistocene antelope from the island of Mallorca. DAMARC 27. Oxford: Donald Baden-Powell Quaternary Research Center, Pitt Rivers Museum, University of Oxford and Baleares, Spain: Deìa Archaeological Museum and Research Centre, Deìa, Mallorca.

WIGAND, P. AND A. SIMMONS, 1999. The Dating of Akrotiri Aetokremnos. In: A. Simmons. Faunal Extinctions in an Island Society: Pygmy Hippopotamus Hunters of the Akrotiri Peninsula, Cyprus. New York: Plenum/Klewer Academic Press, 1999, 193-215.

36

THE WRANGEL DWARF MAMMOTHS WERE NO ISLAND ENDEMICS

JELLE W.F. REUMER (1), **DICK MOL** (1) & **JOHN DE VOS** (2)

(1) Natuurmuseum Rotterdam, P.O.Box 23452, NL-3001 KL Rotterdam, the Netherlands
(2) Naturalis, P.O. Box 9517, NL-2300 RA Leiden, the Netherlands
EMAIL: reumer@nmr.nl

SUMMARY

Island proboscideans (elephants, stegodonts, mammoths) are known from several places in the world where they evolved into endemic dwarfed forms. One supposed island form was described from the Arctic Wrangel Island in 1993, but doubts on its dwarfism soon emerged. Here we show that the mammoths from Wrangel Island were Holocene survivors in a refuge area, and that they were small, but no island endemics. The Wrangel Island population of mammoths is situated at the end of a general size-decreasing trend within the genus Mammuthus. The smallest specimens, on which the theory of island dwarfism was based, were female animals.

INTRODUCTION

Dwarfed or pygmy island Proboscidea (Elephantidae) are known from several areas covering the globe. Dwarfism is a typical island phenomenon among large mammals that became isolated; it is supposed to be an adaptation to either food shortage, rugged and inhospitable and often mountainous terrain, or to the general lack of large predators characterising unbalanced island faunas (Sondaar 1977), or to combinations of these parameters. Paleontological research during the last two centuries has provided us with a bestiary of dwarfed -or pygmy- larger mammals: hippo's, elephants, cattle and other bovids, and deer (see Sondaar 1977). Here we restrict ourselves to island proboscideans, the mammal group consisting of elephants, stegodonts, and mammoths. In 1993, small mammoths from the Siberian Arctic island of Wrangel were presented as dwarf forms by a group of Russian researchers (Garutt et al. 1993, Vartanyan et al. 1993). Many authors subsequently adopted this view, so that the Wrangel Island mammoths are now listed among the insular proboscideans (e.g., Lister & Bahn 1994, Agenbroad et al. 1999). As Mol (1995) has raised doubts against the insular character of the Wrangel mammoths, we decided to investigate the matter and to see whether or not these proboscideans were island endemics. Here we wish to revise the idea that the Wrangel Island population consisted of insular pygmy mammoths, which idea we consider not based on facts, and we will show that the mammoths from Wrangel Island -though indisputably small- were normal-sized animals from the mainland.

ISLAND PROBOSCIDEANS

The first known pygmy Elephantidae are found in islands in the Mediterranean Sea: e.g., Malta, Sicily, Crete, Tilos, Cyprus (Sondaar 1977). Their taxonomic attribution on the genus level is not always clear. The names Elephas (the genus to which the recent Indian elephant belongs), Palaeoloxodon, Loxodonta (the genus to which the recent African elephant belongs), Hesperoloxodon and Mammuthus (the mammoth) are used. Mol et al. (1996) concluded that the Cretan Early Pleistocene pygmy elephantids descended from the Southern Mammoth Mammuthus meridionalis and should hence be referred to as Mammuthus creticus. The same genus name (Mammuthus) could be used for the pygmy elephantids from Malta and Sicily (Agenbroad et al. 1999); at the same time it can not be ruled out that some of the Mediterranean island proboscideans are not mammoths.

In Southeastern Asia, pygmy Elephantidae of the genera Stegodon and/or Elephas are known from several islands: Java, Sulawesi (Celebes), and Flores in Indonesia (van den Bergh 1999), Luzon and Mindanao in the Philippines (von Koenigswald 1956). Other Elephantidae are known from Java that belong the mastodonts (Gomphotheriinae; van den Bergh 1999).

Pygmy mammoths are encountered again in N. America. Here a population of Pleistocene islands mammoths evolved on the former island of Santa Rosae, now an archipelago off the Californian coast (the so-called Channel Islands: Santa Rosa, San Miguel, Santa Cruz and Anacapa, separated from the Californian mainland by the Santa Barbara Channel). These pygmy mammoths Mammuthus exilis are considered derived from the Columbian Mammoth M. columbi (Agenbroad et al. 1999).

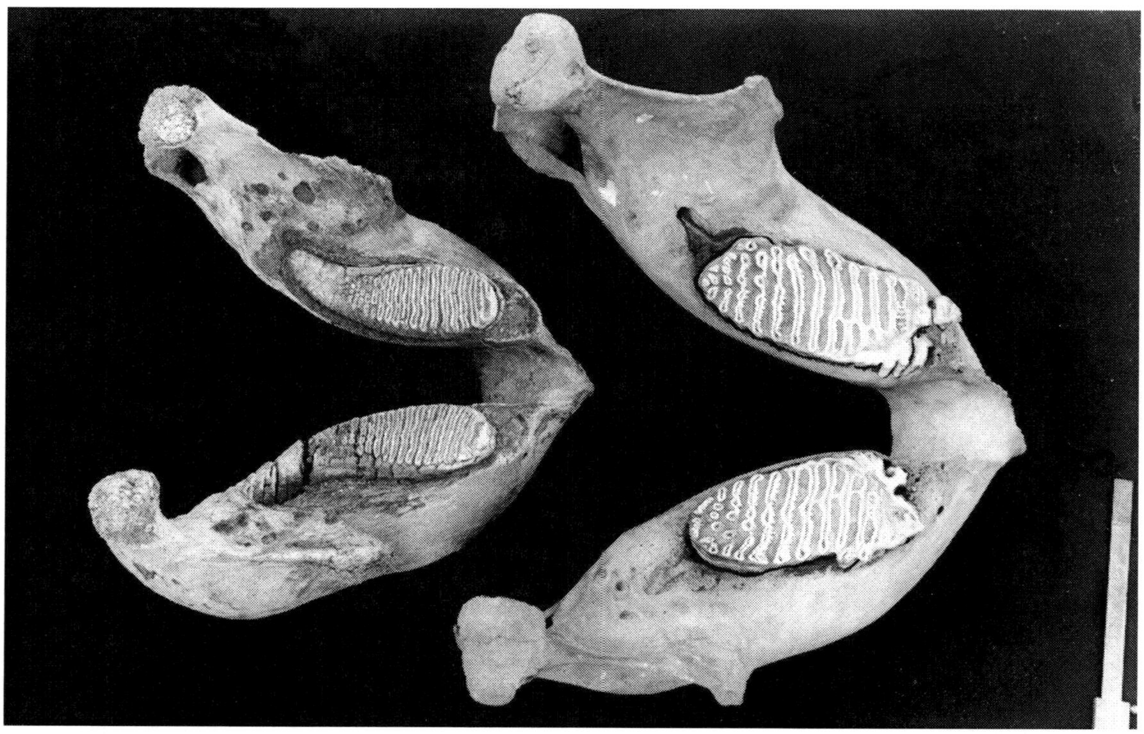

FIGURE 1. The male Jarkov mandible (right) and the small female Taimyr mandible (left). Photo copyright Cerpolex/Francis Latreille

Then, in 1993, the paleontological literature was enriched by a fourth mention of pygmy proboscideans, when Garutt et al. published the find of a Holocene small mammoth from the Arctic island of Wrangel. Apparently, the woolly mammoth Mammuthus primigenius survived until historic times on Wrangel Island, which is situated off the North Coast of Siberia in the Arctic Ocean or East Siberian Sea. 14C dates reported indicate extinction around 3,700 yrBP (Garutt et al. 1993; Vartanyan et al. 1993). These relatively recent remains, mostly third molars m3/M3, were ascribed to dwarf mammoths of a distinct subspecies, called M. primigenius vrangeliensis GARUTT, AVERIANOV ET VARTANYAN, 1993. Ever since, the Wrangel dwarf mammoths have appeared in the literature (e.g. Lister & Bahn 1994, Agenbroad et al. 1999) as another example of islands dwarfism among elephantids, and the island of Wrangel as the fourth place on earth (after the Mediterranean islands, islands in S.E. Asia, and islands off the Californian coast) where this phenomenon took place. Lister & Bahn (1994) mentioned Wrangel as the third place on earth where mammoths evolved into pygmy island forms: "Dwarf mammoths evolved independently on at least three islands across the globe. In each case, these mammoths were descended from normal-sized animals from the adjacent mainland" as they stated in a legend next to a map showing Wrangel Island, Malta/Sicily, and the Californian Channel Islands (Lister & Bahn 1994, p. 35).

THE WRANGEL MATERIAL

Vartanyan et al. (1993) first mentioned the presence of dwarf mammoths from Wrangel Island in a news-breaking article in Nature. They found 'numerous teeth of dwarf mammoth dated 7,000 - 4,000 Y BP'. At the same time, they described teeth from a 'large form' dated around 13,000 y BP, and stated that the island had become separated from the mainland around 12,000 y BP. The dwarfism is thus supposed to have arisen within a time-span of 5,000 years only.

Shortly after the Nature paper was published on 23 March

1993, Garutt et al. (1993, paper presented to the Russian Academy of Sciences on 22 April 1993, submitted on 7 May 1993, and published October 1993) formally named a new subspecies of small mammoth (Mammuthus primigenius vrangeliensis) from Wrangel Island, based on the material published in the Nature article. The rather mathematically formulated diagnosis (Garutt et al. 1993) of the new taxon was:

Small mammoth with small molars (width of corona dentis of M3: 63-72 , mean = 68.3 (mm) +/- 1.67, n=4; m3: 66-74 mm, mean = 70.3 (mm) +/- 1.43, n=4), with significant frequency of layering (M3: 9.5-13.0, mean = 11.0 +/- 0.64, n=4; m3: 8.5-9.0, mean = 8.9 +/- 0.11, n=4) and low enamel thickness (1.0-1.3 mm, mean = 1.1 +/- 004, n=7).

Leaving out all the numbers, this sentence reads: Small mammoth with small molars, with significant frequency of layering and low enamel thickness. This is a rather meagre diagnosis for a (sub)species. In a provided comparison, the new subspecies was distinguished from the nominal subspecies M. p. primigenius by the features of tooth structure mentioned in the diagnosis, and from M. p. compressus OSBORN, 1924 by small sizes. Comparison with other subspecies (e.g. M. p. minor GROMOV) was not given.

OTHER SMALL MAMMOTHS

Small mammoths have been reported before. A small molar (M3) was described from the North Sea by Van Essen (1986, in Dutch only) and again by Mol et al. (1999). In both publications this small M3 was depicted life size, indicating its relative smallness. Van Essen (1986) gave the following details: width of the crown (at lamella XVI): 75 mm, Lamellar Frequency Index: 12.2; he also mentioned the subspecific denomination M. primigenius minor GROMOV that was given to such small animals. Mol et al. (1999) gave details about other finds of small mammoth remains in Europe and attributed such small specimens to female animals. The molar width of this North Sea mammoth M3 of 75 mm is slightly above the range of (only) 4 molars of M. p. vrangeliensis given by Garutt et al. (1993): 63-72 mm, while the lamellar frequency of 12.2 falls within the range of M. p. vrangeliensis of 9.5-13.0.

In 1999, a lower jaw of a small mammoth (FIGURE 1) was discovered and rescued from the Taimyr Peninsula (Siberia). Mr. Y. Karbainov, the director of the Taimyr Nature Reserve, presented it to the Natuurmuseum Rotterdam as a gift. This jaw (now known as the Taimyr mandible) appears to be of an extremely small specimen, much like the material from the North Sea or from Wrangel Island. The jaw was

	female	male
M. meridionalis	3,50	4,20
M. trogontherii	3,50	4,00
M. primigenius	2,50	2,75
"vrangeliensis"	1,80	2,65

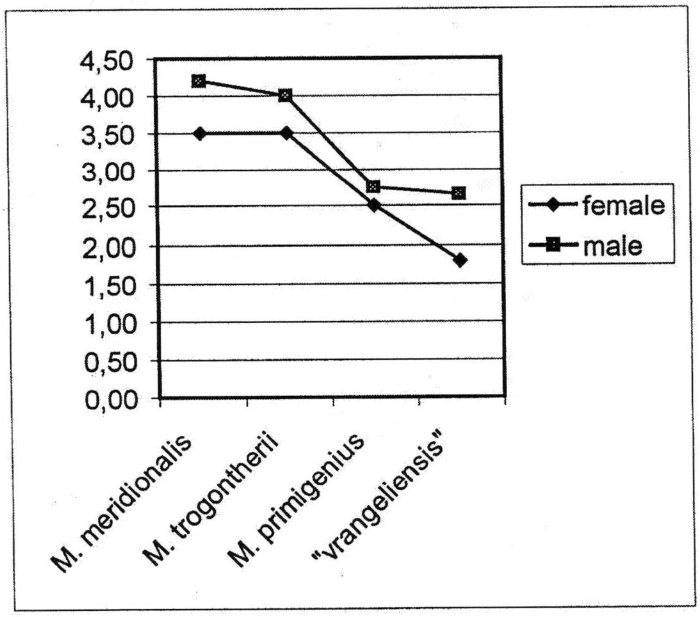

FIGURE 2. The average shoulder heights of male and female Mammuruthus species and the Wrangler Island population of M. primigenius.

dated by means of 14C at Groningen University and proved to be 9.920 +/- 60 y BP (Laboratory # GrA-17350/DM5, delta13C: -22.80 o/oo). This shows that small mammoths were geologically very young, and that mammoths were still present on the Eurasian continent at a time less then 10,000 years ago (i.e. 2,000 years after the supposed separation of Wrangel Island). A thorough description of this mandible is forthcoming, but it can here be stated that its morphology agrees with the diagnosis of the vrangeliensis subspecies.

DISCUSSION

The genus Mammuthus has undergone considerable size reduction since its first appearance in Eurasia in the Early Pleistocene (FIGURE 2). The oldest species, the Early Pleistocene Southern Mammoth Mammuthus meridionalis reached a shoulder height of 4.20 m in male and 3.50 m in female animals; the Middle Pleistocene Steppe Mammoth M. trogontherii is noted to have reached 4.00 m in males and 3.50 m in females; the Late Pleistocene Woolly Mammoth M. primigenius reached on average 2.75 m in males and 2.50 m in females. We observe both a trend of size decrease, and of sexual dimorphism. Could it therefore be possible that the small mammoths described above are in fact extremely young (i.e., Latest Pleistocene and Holocene) female specimens? It is in this respect interesting to observe the data of the well-studied Jarkov mammoth, an adult male mammoth found in the permafrost of the Taimyr Peninsula (Mol et al. 2001). This adult male animal is 14C dated at 20,380 yrBP, had an age of 47 +/- 2 AEY (African Elephant Years), possessed a large skull and enormous tusks of 294 and 298 cm, respectively. Yet it is calculated to have had a shoulder height of only 2.65 m! The same height of 2.65 m was estimated for the male Berezovska mammoth, while the female mammoths from Sevsk had heights of 2.30-2.40 m only.

Mol (1995) was the first to cast doubt on the dwarf aspect of the Wrangel mammoths, later followed by Tikhonov (1997); both publications however were not widely available: Mol wrote in Dutch in the journal Cranium, and Tikhonov wrote in a non-scientific context, viz. in an activities report published in the EuroMam Newsletter. Tikhonov (1997) wrote: “Alexei Tikhonov (…) organized an expedition to Wrangel Island in cooperation with Dr Ulrich Joger (…) and Sergei Vartanyian (Wrangel Island State Reserve) for investigating the Holocene mammoth. (…) The size of mammoth bones which have a Holocene age show that the Holocene mammoth of Wrangel Island was not a dwarf form as was firstly reported” (Tikhonov 1997).

Tikhonov et al. (1999) compared the Wrangel Island mammoths with the Californian Channel Island M. exilis. They found “that the Channel Island mammoths were dwarfed, whereas the Wrangel mammoths are no longer considered to be dwarfs”. The comparison of both islands / island groups showed that the time of isolation for the Channel Island population (leading to dwarfism) is supposed to exceed 40,000 years, but for the Wrangel population this was not more than 6,000-7,000 years. These authors also stated that “during the winters, Wrangel Island was connected to the mainland by ice, allowing migration in both directions”. Our conclusion can only be that Wrangel Island may be an island in a geographical sense (i.e. a surface of land surrounded by water), but that it was certainly not an island in a biogeographical sense for mammoths, as each winter migration to and fro was possible. There was, thus, no such thing as isolation and the mammoths from Wrangel Island can thus not have been just another case of insular dwarfism. The only scientifically sound conclusion about the Wrangel Island mammoths is that they are the youngest recorded specimens with an age of 3,700 yrBP, thus 6,200 years younger than the above-mentioned Karbainov jaw and some 6,000 years younger than other Late Pleistocene / Early Holocene mainland mammoths. Tikhonov et al.'s (1999) 6,000-7,000 years of isolation are in fact 6,000-7,000 years of survival in a remote refuge place along the edge of the species' former distribution area. The Wrangel mammoths were a relic population of typically Late Quaternary (Holocene) small mammoths. The smallest specimens recovered, i.e. the ones that were used to establish the subspecies M. p. vrangeliensis will most probably have been female specimens that were on the lower end of the sexually dimorphic size spectrum.

CONCLUSION

We now reach the following reconstruction and conclusions:

1 Mammoths lived on the Eurasian continent -at least in Northern parts such as the Taimyr peninsula- to a time somewhat less than 10,000 years ago, and subsequently disappeared. A remaining population of mammoths continued to survive on an even more remote and northern place, viz. on Wrangel Island.

2 The last Pleistocene and/or Holocene mammoths were the smallest known mammoths, thus continuing a size-decreasing trend that led from the large M. meridionalis to the slightly smaller M. trogontherii, and finally to the still smaller M. primigenius. This latter species, the woolly mammoth, continued its size decrease, which in combination with a marked sexual size dimorphism resulted in very small (female) individuals by the time of extinction.

3 These 'dwarf' mammoths found on Wrangel Island were relatively very small (a) due to their very late occurrence and (b) as a result of sexual dimorphism, and not as a result of endemic island evolution. The taxon M. primigenius vrangeliensis is not based on representative material, but on female teeth only, and must therefore be considered a nomen nudum.

ACKNOWLEDGEMENTS

We are grateful to Mr. Y. Karbainov (Taimyr Nature Reserve) for donating the small Taimyr mammoth mandible. Dr. J. van der Plicht (Centrum voor Isotopen Onderzoek, Groningen University) performed the 14C dating on this Taimyr jaw.

REFERENCES

AGENBROAD, L.D., D. MORRIS & L. ROTH,
1999 - Pygmy mammoths Mammuthus exilis from
Channel Islands National Park (USA). In: Haynes, G., J. Klimowicz & J.W.F. Reumer (eds.) - Mammoths and the Mammoth Fauna: Studies of an Extinct Ecosystem. Deinsea 6: 89-102

BERGH, G.D.VAN DEN,
1999 - The Late Neogene elephantoid-bearing faunas of Indonesia and their palaeozoogeographic implications. A study of the terrestrial faunal successions of Sulawesi, Flores and Java, including evidence for early hominid dispersal east of Wallace's Line. Scripta Geologica 117: 1-419

DOUKAS, C.S. & A. ATHANASSIOU, In Press - Fossil Proboscidea From Greece. In: J. de Vos & J.W.F. Reumer (eds.) - Advances in Mammoth Research. Deinsea 9

ESSEN, H. VAN,
1986 - Signalement van een diminutieve M3 sup. van een wolharige
mammoet (Identification of a diminutive M3 sup. of woolly mammoth). Cranium 3 (1): 6-7 (in Dutch)

GARUTT, V.E., A.O. AVERIANOV & S.L. VARTANYAN,
1993 - On the systematic position of
Holocene dwarf mammoths Mammuthus primigenius (Blumenbach, 1799) from Wrangel Island (North-Eastern Siberia). Doklady Akademii Nauk 332 (6): 799-801

LISTER, A. & P. BAHN,
1994 - Mammoths. Macmillan, New York, USA: 168 pp.

Mol, D., 1995 - Over dwergolifanten en dwergmammoeten (On dwarf elephants and dwarf
mammoths). Cranium 12 (1): 38-42 (in Dutch)

MOL, D., Y. COPPENS, A. N. TIKHONOV, L. D. AGENBROAD, R.D.E. MACPHEE, C. FLEMMING, A. GREENWOOD, B. BUIGUES, C. DE MARLIAVE, B. VAN GEEL, G.B.A. VAN REENEN, J.P. PALS, D.C. FISHER & D. FOX,
2001 - The Jarkov Mammoth: 20,000-Year-Old Carcass of a Siberian Woolly Mammoth Mammuthus primigenius (Blumenbach, 1799) - Proceedings of the 1st International Congress The World of Elephants, Rome, 16-20 October 2001: 305-309

MOL, D., J.W.F. REUMER, J. DE VOS & P. CLEVERINGA,
1999 - On remains of very small (female)
woolly mammoth, Mammuthus primigenius, from Western Europe. Abstract in: Reumer, J.W.F. & J. de Vos (eds.) - Official Conference Papers 2nd International Mammoth Conference, May 16-20, 1999, Rotterdam: 42-44

MOL, D., J. DE VOS, G.D. VAN DEN BERGH & P.Y. SONDAAR,
1996 - The Taxonomy and Ancestry
of the Fossil Elephants of Crete: Faunal Turnover and a Comparison with Proboscidean Faunas of Indonesian Islands. In: Reese, D.S. (ed.) - Pleistocene and Holocene Fauna of Crete and its First Settlers. Monographs in World Archaeology, Prehistory Press, Philadelphia: 81-98

SONDAAR, P.Y.,
1977 - Insularity and its effect on mammal evolution. In: Hecht, M.K., P.C.
Goody & B.M. Hecht (eds.) - Major patterns in Vertebrate Evolution. Plenum, New York: 671-707

TIKHONOV, A.,
1997 - untitled report about the Zoological Institute of the Russian Academy of
Sciences, St. Petersburg, Russia - EuroMam Newsletter 4: 14-15

TIKHONOV, A., L. AGENBROAD & S. VARTANYAN, 1999 - Comparative analysis of the mammoth
populations on the Wrangel Island and California Channel Islands. Abstract in: Reumer, J.W.F. & J. de Vos (eds.) - Official Conference Papers 2nd International Mammoth Conference, May 16-20, 1999, Rotterdam: 65-66

VARTANYAN, S.L., V.E. GARUTT & A.V. SHER,
1993 - Holocene dwarf mammoths from Wrangel
Island in the Siberian Arctic. Nature 362: 337-340

37

A LIFE HISTORY AND CLIMATE CHANGE SOLUTION TO THE EVOLUTION AND EXTINCTION OF INSULAR DWARFS: A CYPRIOT EXPERIENCE

TIMOTHY G. BROMAGE (1), **WENDY DIRKS** (2), **HEDIYE ERDJUMENT-BROMAGE** (1), **MATHIAS HUCK** (3), **OTTMAR KULMER** (4), **RASIME ÖNER** (5), **OLIVER SANDROCK** (3), **FRIEDEMANN SCHRENK** (6)

(1) Anthropology, Hunter College, 695 Park Avenue, New York, NY, USA, 10021, (2)Anthropology, University of Wisconsin Madison, 1180 Observatory Drive, Madison, WI, USA, 53706, (3)Palaeontology, Hessisches Landesmuseum, Friedensplatz 1, D-64283 Darmstadt, Germany, (4)Palaeoanthropology, Forschungsinstitut Senckenberg, Senckenberganlage 25 D-60325 Frankfurt am Main, Germany, (5)Antiquities, Turkish Republic of Northern Cyprus, Lefkosa, Mersin 10, Turkey, (6)Vertebrate Paleobiology, Johann Wolfgang Goethe-University, Siesmayerstrasse 70, D-60054 Frankfurt am Main, Germany

INTRODUCTION

One of the most intriguing and least understood marvels of Cypriot prehistory must be the existence of pygmy elephant and hippopotamus on the island. Even more curious, such animals (referred to as insular dwarfs) lived on a number of Mediterranean islands and, appear to have gone extinct on all of them near outset of, or during, the early Holocene. To date there have been various explanations offered for the origin and extinction of the pygmy fauna, but it is universally agreed that these phenomena remain one of the major puzzles in evolutionary biology and vertebrate paleontology.

Akrotiri Aetokremnos, situated on Cyprus' southern shore, is the most well known Cypriot fossil locality. A decade of systematic field and laboratory research on the geology, paleontology, and archeology of Aetokremnos has significantly increased our understanding of conditions pertinent to the existence and extinction of pygmy hippopotamus in the eastern Mediterranean (Simmons 1999). In the north of Cyprus, on the other hand, there have been no such formal studies of pygmy elephant and hippopotamus localities, previous surveys being mainly concerned with the salvage of fossils washing out of their containing sediments. In order to systematize this work and to establish research priorities, the authors formed the Northern Cyprus Pleistocene Research Project (NCPRP) in order to record the distribution of approximately a dozen northern fossil pygmy localities using Global Positioning System computer technology and to perform more formal studies of selected sites.

Here we draw inferences about the evolution and extinction of Cypriot pygmy faunas as permitted so far by laboratory and field research at localities in Northern Cyprus (cf. Boekschoten & Sondaar 1972; Reese 1995). These inferences, we emphasize, may be generally applicable to insular evolutionary phenomena and, potentially, extinction conditions on many islands around the world. Our specific laboratory research benefits from dental material recovered from a locality near the village of Tatlysu, meaning "Sweet Water" in Turkish, otherwise known by its Greek place name, Akanthou Arkhangelos Mikhail. The larger share of our field research to date has been performed at a locality located near to the village of Akdeniz. This name roughly translates as "White Sea" in Turkish, otherwise known by its Greek place name, Ayia Irini Dragontovounari.

LABORATORY RESEARCH AND INTERPRETATION

"If we cannot understand factors influencing the evolution of body size, we have little hope of interpreting the rest of the life history" (Boyce 1988, 20). Our contribution to the study of these factors derives from mammalian life history theory which establishes that the duration of dental developmental events will positively relate to the time required for growth to maturity. To investigate this we capitalize on the presence of histological growth markers, or incremental features, in enamel (Boyde 1963, 1964; cf. Dean 1987 for a recent review). These features, observable by conventional light microscopy, are the "striae of Retzius" which represent near-weekly periods of growth, and their "cross striations" representing daily periods. Such features reflect organismal life history, indicating the rate and amount of time required for tooth development and, hence, also the duration of other life history events such as the length of the growth period. If, for example, the duration of growth to adult body size was reduced, then body size itself should be predictable to be reduced.

We thus hypothesized: 1) that the rate of enamel growth, as revealed by measurements between striae of Retzius and cross striations, would be similar, but 2) that the pygmy elephant enamel would be reduced in thickness compared to their larger Plio-Pleistocene representatives. In other words, while the rate of tooth growth was expected by us to be the same amongst the large and dwarf elephants, the timing of growth was thought to be truncated such that tooth formation ceased earlier in the dwarf, resulting in a reduced tooth size.

To date our histological studies of one molar specimen of the Cypriot pygmy elephant (Elephas cypriotes) from the site of Tatlysu (UTM 4 945 14N, 39 117 13E), and one representative mega-scale elephant (Elephas recki) specimen from the African Late Pliocene (Bromage & Schrenk 1995), suggests that our life history predictions are very near to expectation (TABLE 1). The rate of molar enamel development, as measured by widths between successive striae of Retzius, is the same for both the large African and the Cypriot pygmy elephants. This is further supported by the similarity of widths between successive cross striations (this, incidentally, confirms that the number of daily increments between stria of Retzius must be the same in the two elephant taxa).

Enamel thickness, however, is much less in the pygmy elephant molar (Table 1). This, combined with our observed similarity in enamel developmental rate, demonstrates that tooth development was arrested at an earlier absolute age. This histological evidence demonstrates that the pygmy elephant growth period was abbreviated in comparison to the mega elephant and, this, according to mammalian life history theory, explains how body size was reduced. This is further indicated by a reduction in the number of characteristic lamellar "plates" making up the molar tooth, from approximately 15 in Elephas recki to 7 in Elephas cypriotes.

FIELD RESEARCH AND INTERPRETATION

Pygmy fossil localities abound on both the landward and seaward sides of the Girne (Kyrenia) Range. Sites adjacent to the Mediterranean Sea are variously associated with marine terraces uplifted during the Pleistocene and Holocene (Dreghorn 1978). The seaside site near Akdeniz (UTM 4 945 14N, 39 117 13E) measuring approximately 17x7x1 meters and situated above the Koupia Terrace was recently cleaned of brush and obscuring vegetation. The cleaned surface revealed an area of calcareous breccia (fossils embedded in a calcium carbonate cementing matrix) variously saturated with fossils embedded in the rock. The exposed surface was roughly mapped in its three-dimensions (3D) by numerous Theodolite surveying measurements and, in parts, by high resolution 3D imagery obtained by video imaging technology brought to the site (clearly revealing even the pointed cusps of pygmy hippopotamus molar teeth). We hope to follow the daily progress of our future excavations in 3D in order to provide the baseline for excavations that, in the future, will accurately portray all bone and tooth specimens in their original 3D relationships. This will contribute something to our understanding of site formation process and interpretations of the accumulating agents (e.g., by natural forces, human activity, or both).

The site physiognomy, as it exists on a terraced slope, is such that rainwater drains through the site from above. In the path of this drainage the calcareous matrix has demineralized to a significant extent and, for some time, has lost many of its containing fossils to a catchment area immediately below the site. These fossils, being out of context, were quarried from the soil over approximately 5x1x0.30 square meters and screened for all fossil fragments 1cm and larger. An estimated 10,000 fossil bones, teeth, and fragments thereof, were recovered from this operation. Several Neolithic stone tools were also recovered. All dental material acquired from the screening derived from pygmy hippopotamus, and so it is presumed, prior to any detailed analysis, that most or all postcranial material (i.e. non cranial) is also attributable to these animals. While their is a preponderance of pygmy hippopotamus within the intact breccia, Reese (1995) recovered a pygmy elephant tooth during his reconnaissance and we have, to date, left an exposed and intact elephant molar on the site surface.

Fossils were inspected as they were picked from the screening. In no case was a cutmark noticed which, if present, would demonstrably indicate animal processing behaviors by Neolithic hunters/scavengers. Their absence may constitute negative evidence for the lack of a human consumer relationship with these creatures and, thus, their relationship to the copious amounts of bone at the site remains moot.

It should be noted, however, that the absence of cutmarks might yet have another explanation. Animal tissues close to the bone may not have been sought. Skillful butchering of carcasses may leave no traces of cutmarks. Butchering marks may have been universally eroded from bone surfaces during transport. In respect of the latter, it is the case that all specimens scrutinized during the Akdeniz screening operation show evidence of being transported to the site by water and, thus, this site does represent a cave or fissure-fill deposit of natural origin. Until tangible evidence of animal processing can be established, we may surmise that all objects, be they the skeletons of animals or cultural artifacts, will have been expected to wash in from the surrounding watershed. This would be, therefore, of little help concerning the putative anthropogenic cause of the assemblage except to confirm what we already know: pygmy faunas and humans occupied the watershed vicinity in ancient times.

The situation at both Akdeniz and Tatlysu is similar to the material findings at Aetokremnos (Simmons 1999). However, while Aetokremnos has large amounts of skeletal material and many stone tools, additional observations are provided by Simmons and colleagues to suggest that a functional relationship exists between them (Simmons 1988, this volume; cf. Reese 1996).

EVOLUTION

We have the notion that elephants and hippopotami may have been swept from the northeast African coast during freshwater flooding of the eastern Mediterranean region. Good evidence exists for such flooding during the last glacial maximum-Holocene transition (Fontugne et al. 1994), and we suggest that such flooding may also have occurred during Pleistocene interglacial intervals prior to the 18,000-20,000 glacial maximum. Environmental conditions in North Africa would have been favorable for these taxa during such times and population pressures may well have led to behavioral inclinations to set out to sea. This explanation is offered only as a means of setting the scene, as it were, for small numbers of our Cyprus-bound elephants and hippopotami. We do not present it as a necessary condition for our discussion below, because colonization schemes may also include dispersals between islands after evolutionary dwarfing (cf. Gilbert et al. 1990; Wayne et al. 1991), but we like to think it establishes the likelihood that small numbers of large continental individuals would be involved.

TABLE 1

TAXON	ENAMEL THICKNESS	DISTANCE BETWEEN STRIAE OF RETZIUS	DISTANCE BETWEEN CROSS STRIATIONS
E. recki	Range: 3.00 - 3.34 mm	Range: .06 mm - .08 mm	Range: 2.9 µm - 4.6 µm
molar	Mean = 3.10 mm	Mean = .07 mm	Mean = 3.6 µm
E. cypriotes	Range: 1.67 - 2.13 mm	Range: .05 mm - .08 mm	Range: 2.8 µm - 4.6 µm
molar	Mean = 1.92 mm	Mean = .07 mm	Mean = 3.7 µm

We assume that, normally, individuals of a taxon immigrate to islands and that they represent a small founder sample, significantly less than the population from whence they came and certainly a minute proportion of the species. We further assume that the conditions ultimately responsible for dispersal do not contribute to the subsequent evolutionary dynamics of taxa once they settle on islands. While we begin our discussion with these two assumptions in mind, we agree that the former assumption is riddled with contingencies and that the latter assumption may require more thought.

Our argument begins by specifically acknowledging the importance of body size to general theories of mammalian life history variation. Our understanding of body size in relation to life history is greatly facilitated by a theoretical foundation that links body size with demographic consequences of life history, namely juvenile mortality (Charnov 1991). Harvey and Nee (1991) have provided a succinct narrative of this theory, which we quote here:

Adult mortality is determined by the environment. Mammals mature at the age which maximizes lifetime reproductive success. A growth law determines adult size as a function of age at maturity. At maturity, the energy which would have gone into growth is channeled into the production of offspring. In the long term, the birth rate must equal the death rate in natural populations. That equality is ensured by juvenile mortality being density dependent: at low population densities fecundity exceeds mortality, but at high density mortality exceeds fecundity (p. 23).

In this view a balance must be achieved through the evolutionary process between 1) elephant and hippopotamus phylogenetic history characterized by slow and prolonged growth to achieve larger body sizes (evolved on the mainland), and 2) juvenile mortality prior to reproductive age amongst a relatively small founder population sample. Slow growing mammals living at extremely low population founder densities on islands will experience directional selection for diminished life history schedules simply because slower maturing individuals have a higher probability of dying before reaching sexual maturity. Over some generations, average age at reproductive maturity will occur absolutely earlier in these populations, and due to the reduction in growth period, adults will achieve smaller body sizes. The resultant increase in fecundity will lead to sustainable population densities and a consequent increase in juvenile mortality, maintaining long-term population balance.

Consistent with the results of our enamel-based histological research, we surmise that foundling numbers of medium to large bodied immigrants are selected to reduce the length of the growth period according to rules governing mammalian life history (Pagel & Harvey 1990). Net reproductive losses arising from sources of extrinsic mortality (even accidental) may be devastating to such a small founder population. Thus elephant and hippopotamus immigrants from the mainland will have been under directional selection to precipitously reduce their respective growth periods and thus body sizes, two inextricably linked life history variables. Indeed, the fitness advantage of reproducing earlier may have been greatest while populations were increasing (Blanckenhorn 2000), the expected condition for fledgling populations on the island. As ages at sexual maturity and first breeding decline, however, fecundity will increase until population densities are at equilibrium with mortality. Population densities of dwarfed varieties would give rise to sustainable levels because of the lack of predators at even these smaller sizes, and because of the relatively open niche available to them (no other pre Neolithic-contact animals appear to have competed for the same resources as elephant and hippopotamus on the island).

Sondaar (1977, 1994) has taken the view that selection pressures maintaining large size in elephants and hippopotami, such as predation pressure, were absent on islands. The advantages of small size were thus explained to outweigh the disadvantages of remaining large. This is an insightful approach, and one not entirely incompatible with the life history model proposed here. However, we prefer to emphasize a positive selectionist argument for small size over one concerned mainly with the relaxation of selection. Widespread insular dwarfism, despite varied predator-prey conditions on islands around the world is, we believe, a further justification for our mortality-based hypothesis.

The argument made by Case (1978), that insular body size variation is regulated by food abundance and availability, is also intriguing. However, this hypothesis does not emphasize the costs of body size variation as directly as our predictions based on mortality-based selection. To be sure, these costs operate through the allocation of resources to growth and reproduction, but they do so in respect of selection favoring early reproduction and relatively higher probabilities of survival vs. future reproduction and relatively diminished probabilities of survival (Boyce 1988). From this perspective, and our example of insular megafaunal foundlings with long growth periods and late reproduction, the tradeoff between growth and reproduction is expected to result in an evolutionary trend toward smaller body size (Sadlier 1969).

Further, we are curious about how we might explain the relationships between food abundance and availability and body size for foundling individuals having, perhaps, the whole of an island's resources available to them. In respect of the temporal coexistence of elephants and hippopotamus on Cyprus, we have reason to expect that they would seasonally split their seasonal grazing and browsing activities on dune grasslands and riverine woodlands as they do today (O'connor & Campbell 1986). There is thus no reason to expect a significant degree of food resource competition between these taxa. Further, if considering the food resource argument alone, founder taxa of all body sizes would be universally expected to experience selection for increased body size during initial low population stages after colonization (which would be a relatively long time for very large mammals). However, this phenomenon occurs almost exclusively amongst small mammals (Foster 1964), and would vary from island to island depending upon local ecological conditions. Finally, food abundance and availability arguments, used in respect to evolutionary explanations for diminished body size, run counter to a substantive literature linking resource shortages to increased body size (cf. Zeveloff & Boyce 1988).

It is thus obvious (to us at any rate) that the evolution of insular dwarfs has not anything directly to do with causal selection for diminished body size per se. We believe that, for this reason, most attempts to explain island dwarfing have been looking at the problem askance and, most likely, inappropriately. It is a difficult search, of course, because what fundamental and large scale ecogeographic principle can be said to operate preferentially on islands? For instance, Zeveloff & Boyce (1988) provide evidence for significant correlation between body size and precipitation variables on the North American continent. However, the rather broad island dwarfing experience on many of the world's islands situated at various latitudes and experiencing varied seasonality conditions, argues against the primary role of such continental principles on island patterns.

Zeveloff & Boyce (1988) also note an inverse relationship between body size and species richness. Such a relationship is the antithesis of any attempt to explain the island dwarfing phenomenon. Islands are commonly observed to be relatively impoverished and thus this continental relationship cannot possibly be invoked to reconcile the evolution of pygmy elephants and hippopotami.

Natural selection must then be operating on other features of organismal life history that, in consequence, affect body size. Boyce (1988) has pointed out that many linked variables make up the life history of an organism (and of taxa), and it would be unwise to think that selection were acting on body size alone, if indeed at all. Looked at another way, the physiological consequences of significant changes in body size involve so many tissue and organ systems (Lindstedt & Swain 1988; Schmidt-Nielsen 1984), selection could be hypothesized to operate on any one or group of factors, independently of body size.

Indeed, this is our argument. Our evidence suggests that selection favored individuals whose growth potential was accelerated, meaning that the life history gamut was affected, including shorter gestation, smaller neonatal body and brain size, reduced age at weaning, sexual maturity, and first reproduction, earlier completion of the permanent dentition, shorter interbirth intervals, shorter lifespan, small adult brain and body size.

EXTINCTION

Our tentative explanation for the demise of the pygmy fauna, which may be our hypothesis for future testing, relies on the known behaviors of living hippopotamus and elephant. As conditions for these animals become dryer, available fresh water resources dwindle. Hippopotamus, being particularly sensitive to ultraviolet radiation, need to be submerged in this water during the day, and as waterholes become smaller and fewer, hippopotamus are forced to congregate into larger and larger groups. For instance, on one occasion was observed "...a single herd equal in number to the whole of the 1971 Lamai population (150 animals) concentrated in one of the group locations when it was the last remaining pool in the area during the 1970 dry season" (Olivier & Laurie 1974; 253; Note: Lamai is a location along the Mara River of the Serengeti National Park, Tanzania).

When environmental conditions cannot support such large groups dependent upon such small pools, these populations will be expected to, and do suffer catastrophic losses. We believe that this explanation can account for the accumulation we see at Akdeniz and, thus, their extinction on Cyprus (a view also held by Dreghorn 1978). We estimate, quite provisionally, that skeletal elements of well more than 100 animals are represented at the site. Evidence of transport suggests that animals died in significant, if not disastrous, numbers around parched streams and upstream lowland pools along the watershed. Skeletal elements were then concentrated at the site during subsequent rainy seasons. In the absence of evidence for animal butchering on such a scale, we thus feel that climate change over a relatively brief period in Cypriot prehistory, leading to increasingly arid conditions is a valid explanation for this catastrophic assemblage. This hypothesis may be generally supported by examination of climate records for the terminal Pleistocene and early Holocene. Though terminal dates assigned to Cypriot fossil localities vary, this period ca. 10 k.y. is reasonably well represented (Reese 1995; Simmons 1999; Simmons & Wigand 1994).

The Holocene is generally considered to have been reasonably warm and stable, thus lacking the significant climate variability documented for the Pleistocene ice age. But variability there was. The earth's orbit was such that summer insolation reached a maximum at 10 k.y. This strengthened the African monsoon over what is now the arid Sahara and subtropical North Africa, bearing wetter conditions and tropical taxa (e.g. elephant and hippopotamus), presently re-confined to sub-Saharan Africa, between 14.8 and 5.5 k.y. (deMenocal et al. 2000). Pygmy elephants and hippopotami inhabited Cyprus at this time.

Cooling and aridifying events nevertheless punctuated the African humid period. The most significant terminal Pleistocene event is referred to as the Younger Dryas cold oscillation occurring about 10.4 to 10.2 k.y., as determined from eastern Mediterranean pollen records (Baruch 1994) and reconstructed West African offshore sea-surface temperature (SST) variations (deMenocal, 2000) respectively. By this time, pollen evidence for the transition to a Mediterranean climate, characterized by winter rains and summer drought, had taken hold (Roberts & Wright, Jr. 1994). At 9 k.y., pollen data indicate that portions of the eastern Mediterranean were, indeed, drier than the present, presumably owing to increased temperature and evapotranspiration or meteorological circumstances (Roberts & Wright, Jr. 1994). Additional SST variations reveal other cooling and aridifying events at 8.0, 6.0, 4.6, 3.0, and 1.9 k.y., thus sufficient environmental hardships must have befallen the pygmy fauna restricted to the island.

It is our opinion that severe drought of only one or two years is sufficient to explain the utter demise of the pygmy fauna, let alone larger scale climate change during the Late Pleistocene and early Holocene. It is interesting to note that the cursorial adaptation observed for the pygmy hippopotamus (Houtekamer & Sondaar 1979) is a strong indication that these animals did, in fact, actively maintain their continental habitat preferences under the somewhat more seasonal and mountainous terrain of Cyprus. Hippopotamus living today have been observed to move significantly inland during the dry season in order to take advantage of riverine woodland food resources (O'connor & Cambell 1986). We suggest, however, that precipitous Late Pleistocene and early Holocene climate variability altered their habitat to such a degree that seasonally available woodland resources were no longer sufficient and extinction occurred.

This view is not completely contrary to speculation that pygmy faunas became extinct owing to limiting food availability (Sondaar 1977, 1994), but we suggest somewhat different circumstances leading to food resource limitations than previously proposed. The same lack of predators said to relieve constraints on large body size is described by Sondaar to unconstrain population growth potential, leading to their ultimate overpopulation and mass starvation. The life histories of elephants and hippopotami, being rather large mammals, are relatively prolonged and characterized by late age at first reproduction. Their population growth must therefore have been much slower at the outset of their arrival on Cyprus. Further to this, we must consider that many natural populations live at equilibrium on many islands of the world without evidence of such demise. Indeed, environmental modeling by Wassersug et al. (1979) shows us that "...dwarfs should be able to utilize the available resources more completely than giants and, at the same time, are less likely to exceed the environmental carrying capacity and crash" (292).

Something of a compromise is that food resource competition from domestic animals brought by humans to the island (e.g. deer, sheep, goat, pig, and fox; cf ªevketoðlu, this volume; cf. Frame, this volume) may have helped to bring about the extinction of the endemic fauna (Davis 1984). Grazing and browsing domestic animals, together with populations of feral derivatives, might themselves have caused significant environmental change and resultant stress on the pygmy fauna. Perceived competition for resources by Neolithic inhabitants may also have led to predation. A difficulty of assessing such a model is that ecological and social phenomena may be coincident in time. Climate change, while precipitating pygmy faunal extinction, might have been a factor related to Neolithic expansion to the island. The environmental consequences of domestic animals brought with them would have exacerbated the plight of the pygmy fauna. Sorting out this possible complex of events will be an objective of our future research.

At present our model emphasizes the difference between the effects of climate change on continents and islands. Climate change leads to habitat vicariance over significant latitude on continental domains. Under such continental circumstances animals have options. Studies of the distributions of African mammals have indicated that they may maintain their connections to preferred habitats by dispersing (i.e. by vicariously shifting with their habitat along latitude), they may remain in their geographic position and adapt to their new habitat conditions (i.e. evolve), or they may be unable to disperse or adapt and thus go extinct (Grubb 1999; Vrba 1999). The options animals take are the consequence of morphological, physiological, behavioral, ecological, and geographic factors. On islands, dispersal is tricky and the ability to adapt will be determined by generation time and the rapidity of climate change. During periods of precipitous climate change, we should expect that an inability to physiologically adapt to changing conditions, in addition to possible resource limitations, would be the leading cause of extinction. Indeed, instead of population demographic instability in the absence of climate change, we suggest that early Holocene climate variability is the most likely cause of pygmy elephant and hippopotamus extinctions, a view first expressed in simple terms nearly 100 years ago by Bate (1906).

CONCLUSION

Our preliminary research suggests to us that founder populations (i.e. small immigrant groups) of a megafauna were selected for an absolutely earlier reproductive age, hence abbreviated growth periods, in order to avoid extinction from relatively high levels of extrinsic juvenile mortality (in proportion to population size). The result was a transformation to smaller body sizes. We also suggest here that Cypriot insular dwarf extinction may be related to Late Pleistocene-to-Holocene climate variability.

These ideas on the evolution and extinction of the pygmy fauna of Cyprus are offered in the hopes that they may contribute to a more general discourse on body size variation in the island setting. For instance, our evolutionary model based on life history constraints, to be valid, must also explain microfaunal gigantism. Indeed, if island faunas are relatively impoverished, might this explain body size increase in small mammal immigrants? As natural predators are either missing or reduced in species numbers, juvenile mortality would be reduced. Our life history model predicts that this will lead to selection for increased duration of growth and size (cf Melton 1982) to achieve improved metabolic efficiency at larger sizes (McNab 1990; Schmidt-Nielsen 1984). Indeed, Foster (1964) has observed that island rodent species are more varied in color, thus implying reduced selection from predation.

Our original motivation to examine the history of insular dwarfs on Cyprus was driven by our search for a natural experimental "laboratory" within which we might find hints of explanations for evolutionary and extinction phenomena pertinent to the African Plio-Pleistocene (Bromage et al., 1995; Bromage and Schrenk, 1999; Schrenk et al., 1993). We approach our African studies from a perspective in which continental landscapes are observed to be a patchwork of habitat "islands" that influence the paleobiogeography of its inhabitants (Kingdon, 1989). Our Cypriot laboratory is an extreme example of an island in which the large scale ecogeographic relationships do not obviously exist. Thus we expected to find an exaggerated response in respect to "island" conditions existing on continents. What we tentatively understand from this natural experiment and our work on Cyprus is that vicariance, such as that exhibited by African faunas in response to climate change (Grubb, 1995), can lead to remarkably strong selection pressures resulting in life history adaptations to demographic and ecological conditions and, finally, allopatric speciation. The Cypriot experience appears to highlight the relevance of a severely reduced number of vicariant individuals to the evolutionary dynamics of insular dwarfism and, further, the relevance of an ecological barrier (the sea) to vicariance and the extinction of pygmy faunas during periods of precipitous climate change.

ACKNOWLEDGMENTS

This paper was originally presented to the Islands in Prehistory Conference, Deia, Mallorca, September 13-19, 2001. It was our privilege to discuss the ideas presented here with conference participants sharing our love of island biogeography and organismal life history. Special thanks go to Bill and Jackie Waldren for an outstanding conference and, most of all, for their warm hospitality. Out thanks also go to Mary Peterson, Janey Roberts, Aly Gardner-Shelby, and Sue Thompson for keeping the conference glued together and for including us in their Deia family. Chris Stray gets "warm" thanks for the loan of a jacket to TGB.

Our thanks go to David Reese for helping us to remain tethered to the work and literature concerned with the pygmy fauna of Cyprus, and to Bill Dreghorn (deceased 2001) and Ergün Olgun for their enthusiasm and help with our work in the field. We thank Fred Szalay for commenting on the manuscript. The indefatigable Erdjument Tahir and family kindly provided for our project needs. Our appreciation also goes to the Department of Antiquities, Turkish Republic of Northern Cyprus, for their support of and their dedication to our fieldwork program. Major funding for this work was provided by a PSC-CUNY Award of the City University of New York, and by funds provided by the Hessisches Landesmuseum Darmstadt.

REFERENCES

BATE, D.M.A.,

1906. The pygmy hippopotamus of Cyprus. The Geological Magazine, New Series, Decade V, 3: 241-245.

BARUCH, U.,

1994. The Late Quaternary pollen record of the Near East. In O. Bar-Yosef and R.S. Kra, eds. Late Quaternary Chronology and Paleoclimates of the Eastern Mediterranean. Arizona: Radiocarbon, 1994, 103-119.

BLANCKENHORN, W.U.,

2000. The evolution of body size: What keeps organisms small? The Quarterly Review of Biology, 75, 385-407.

BOEKSCHOTEN, G.J. & SONDAAR, P.Y., 1

972. On the fossil Mammalia of Cyprus. Koninkl. Nederl. Akademie van Wetensch. Proceed. Series B, 75, 306-338.

BOYCE, M.S.,

1988. Evolution of life histories: Theory and patterns from mammals. In M.S. Boyce, ed. Evolution of life histories of mammals. New Haven: Yale University Press, 1988, 3-30.

BOYDE, A.,

1963. Estimation of age at death of a young human skeletal remains from incremental lines in the dental enamel. Proc. 3rd. Int. Meet. Forensic Immunol. Med. Pathol. Toxicol. Lond. Excerpta

Medica Int. Cong. Ser., 80, 36-46.
BOYDE, A.,
1964. The structure and development of mammalian enamel. Ph.D. dissertation. London: University of London.
BROMAGE, T.G. AND SCHRENK, F. EDS.,
1995. Evolutionary History of the Malawi Rift. Journal of Human Evolution 28. London: Academic Press.
BROMAGE, T.G. AND SCHRENK, F. EDS., 1999. African Biogeography, Climate Change, and Early Hominid Evolution. Bromage, T.G. & Schrenk, F. Eds. New York: Oxford University Press.
BROMAGE, T.G., SCHRENK, F. AND JUWAYEYI, Y.M.,
1995. Palaeobiogeography of the Malawi Rift: Age and vertebrate palaeontology of the Chiwondo Beds, northern Malawi. Journal of Human Evolution, 28, 37-57.
CASE, T.J.,
1978. A general explanation for insular body size trends in terrestrial vertebrates. Ecology, 59, 1-18.
CHARNOV, E.L., 1991. Evolution of life history variation among female mammals. Proceedings of the National Academy of Sciences, 88, 1134-1137.
DAVIS, S.J.M.,
1984. Khirokitia and its mammal remains: A Neolithic Noah's Ark. In: A. LeBrun, ed. Fouilles Récentes à Khirokitia (Chypre), 1977-1981. Paris: Éditions Recherche sur les Civilisations, ADPF, 1984, 147-162.
DEAN, M.C.,
1987. Growth layers and incremental markings in hard tissues: A review of the literature and some preliminary observations about enamel structure in Paranthropus boisei. Journal of Human Evolution, 16, 157-172.
DEMENOCAL, P., ORTIZ, J., GUILDERSON, T. AND SARNTHEIN, M.,
2000. Coherent high- and low-latitude climate variability during the Holocene warm period. Science, 288, 2198-2202.
DREGHORN, W.D.,
1978. Landforms in the Girne Range, Northern Cyprus (No. 172). Ankara: The Mineral Research and Exploration Institute of Turkey.
FOSTER, J.B.,
1964. Evolution of mammals on islands. Nature, 202, 234-235.
FONTUGNE, M., ARNOLD, M. LABEYRIE, L., PATERNE, M., CALVERT, S.E. & DUPLESY, J-C.,
1994. Paleoenvironment, sapropel chronology and the Nile River discharge during the last 20,000 years as indicated by deep-sea sediment records in the eastern Mediterranean. In O. Bar-Yosef and R.S. Kra, eds. Late Quaternary Chronology and Paleoclimates of the Eastern Mediterranean. Arizona: Radiocarbon, 1994, 75-88.
FRAME, S.,
(this volume), Animal Exploitation in the Aceramic Neolithic of Cyprus.
GILBERT, D.A., LEHMAN, N., O'BRIEN, S.J. & WAYNE, R.K.,
1990. Genetic fingerprinting reflects population differentiation in the California Channel Island fox. Nature, 344,764-767.
GRUBB, P.,
1999. Evolutionary processes implicit in distribution patterns of modern African mammals. In T.G. Bromage and F. Schrenk eds. African Biogeography, Climate Change, and Human Evolution. New York: Oxford University Press, 1999, 150-164.
HARVEY, P.H. AND NEE, S.,
1991. How to live like a mammal. Nature, 350, 23-24.
HOUTEKAMER, J.L. AND SONDAAR, P.Y.,
1979. Osteology of the fore limb of the Pleistocene dwarf hippopotamus from Cyprus with special references to phylogeny and function. Proceedings of the Koninklijke Nederlandse Akademie van Wetenschappen, Series B, 82, 411-448.
KINGDON, J.,
1989. Island Africa: The Evolution of Africa's Rare Animals and Plants. Princeton: Princeton University Press.
Lindstedt, S.L. And Swain, S.D., 1988. Body size as a constraint of design and function. In M.S. Boyce, ed. Evolution of life histories of mammals. New Haven: Yale University Press, 1988, 93-105.
MCNAB, B.K.,
1990. The physiological significance of body size. In J. Damuth, J. and B.J. MacFadden, eds. Body Size in Mammalian Paleobiology: Estimation and Biological Implications. Cambridge: Cambridge University Press, 1990, 11-23.
MELTON, R.H.,
1982. Body size and island Peromyscus: A pattern and a hypothesis. Evol. Theory, 6, 113-126.
O'Connor, T.G. And Cambell, B.M., 1986. Hippopotamus habitat relationships on the Lundi River, Gonarezhou National Park, Zimbabwe. African Journal of Ecology, 24, 2-26.
OLIVIER, R.C.D. AND LAURIE, W.A.,
1974. Habitat utilization by hippopotamus in the Mara River. East African Wildlife Journal, 12, 249-271.
PAGEL, M.D. & HARVEY, P.H.,
1990. Diversity in the brain sizes of newborn mammals. BioScience, 40, 116-121.
REESE, D.S.,
1995. The Pleistocene vertebrate sites of fauna of Cyprus (Bulletin No. 9). Nicosia: Geological Survey Department, Ministry of Agriculture, Natural Resources and Environment, Republic of Cyprus.
REESE, D.S., 1996. Cypriot hippo hunters no myth. J. Mediterranean Arch., 9, 107-112.
ROBERTS, N. AND WRIGHT, JR., H.E., 1994. Vegetational, lake-level, and climatic history of the Near East and Southwest Asia. In H.E. Wright, Jr., J.E. Kutzbach, T. Webb III, W.F. Ruddiman, F.A. Street-Perrott, and P.J. Bartlein, eds. Global Climates Since the Last Glacial Maximum. Minneapolis: University of Minnesota Press, 1994, 194-230.
SADLIER, R.M.F.S.,
1969. The Ecology of Reproduction In Wild And Domestic Mammals. London: Methuen.
Schmidt-Nielsen K., 1984. Scaling: Why is Animal Size So Important? Cambridge: Cambridge University Press.
SCHRENK, F., BROMAGE, T.G., BETZLER, C.G., RING, U. AND JUWAYEYI, Y.,
1993 . Oldest Homo and Pliocene biogeography of the Malawi Rift. Nature, 365, 833-836.
SEVKETOGLU, M.,
(this volume), Akanthou-Arkosyko (Tatlysu-Ciftlikduzu) and the Anatolian Connections in the 9th Millenium BCE.
SIMMONS, A.H.,

1988. Extinct pygmy hippopotamus and early man in Cyprus. Nature, 333, 554-557.
SIMMONS, A.H.,
1999. Faunal Extinction In An Island Society: Pygmy Hippopotamus Hunters of Cyprus. New York: Kluwer Academic/Plenum Publishers.
SIMMONS, A.,
(this volume), The Role of Islands in Pushing the Pleistocene Extinction Envelope: the Strange Case of the Cypriot Pygmy Hippos.
SIMMONS, A.H. AND WIGAND, P.E.,
1994. Assessing the radiocarbon determinations from Akrotiri Aetokremnos, Cyprus. In O. Bar-Yosef and R.S. Kra, eds. Late Quaternary Chronology and Paleoclimates of the Eastern Mediterranean. Arizona: Radiocarbon, 1994, 247-254.
SONDAAR, P.Y.,
1977. Insularity and its effect on mammal evolution. In M. Hecht, P. Goody, and B. Hecht, eds. Major Patterns in Vertebrate Evolution. New York: Plenum Publishing, 1977, 671-707.
SONDAAR, P.Y., 1994. Paleoecology and evolutionary patterns in horses and island mammals. Historical Biology, 8, 1-13.
VRBA, E.S.,
1999. Habitat theory in relation to the evolution in African Neogene biota and hominids. In T.G. Bromage and F. Schrenk eds. African Biogeography, Climate Change, and Human Evolution. New York: Oxford University Press, 1999, 19-34.
WASSERSUG, R.J., YANG, H., SEPKOSKI, JR., J.J. &RAUP, D.M.,
1979. The evolution of body size on islands: A computer simulation. The American Naturalist, 114, 87-295.
WAYNE, R.K., GEORGE, S.B., GILBERT, D., COLLINS, P.W., KOVACH, S.D., GIRMAN, D. & LEHMAN, N.,
1991. A morphologic and gnetic study of the island fox, Urocyon littoralis. Evolution, 45, 1849-1868.
ZEVELOFF S.I. AND BOYCE, M.S.,
1988. Body size patterns in North American mammal faunas. In M.S. Boyce, ed. Evolution of life histories of mammals. New Haven: Yale University Press, 1988, 123-146.

38

PLEISTOCENE FAUNAS OF SICILY: A REVIEW

BONFIGLIO L.(1), **MARRA A. C.**(1), **MASINI F.**(2), **PAVIA M.** (3) **PETRUSO D.**(2)

(1)Dipartimento di Scienze della Terra, Università degli Studi di Messina, Via Sperone 31- casella postale 54 - 98166 Messina, Italy
(2)Dipartimento di Geologia e Geodesia , Università degli Studi di Palermo, Corso Tükory 131, I-90134 Palermo, Italy
(3) Dipartimento di Scienze della Terra, Università degli Studi di Torino, Via Accademia delle Scienze 5, 10123 Torino, Italy
E-MAIL: laura@labcart.unime.it; fmasini@unipa.it

INTRODUCTION

Sicily has a rich palaeontological heritage of Pleistocene continental vertebrates, that indicates different faunal complexes inhabited the island through time.
Most of the older studies on Quaternary vertebrates in Sicily have focused on evolutionary and taxonomic aspects (e.g. Pohlig 1893; Pohlig 1909; Vaufrey 1929; Accordi & Colacicchi 1962; Ambrosetti 1968; Aguirre 1968-69; Gliozzi & Malatesta 1984; Brugal 1987) with the aim to identify the continental species from which the Sicilian ones derived. Most of the known vertebrate remains came from cave deposits and little was known about the palaeoenvironmental conditions of the vertebrate-bearing deposits.

Chronological arrangements of the various Pleistocene mammal assemblages of Sicily were based on the assumption of the phyletic derivation of the dwarf elephant Elephas falconeri from the middle sized Elephas mnaidriensis, which was in turn considered a direct descendent of Elephas antiquus (Accordi 1963; Accordi 1965; Accordi & Colacicchi 1962; Ambrosetti et al. 1980). Vaufrey (1929) was inclined to assume a post-Tyrrhenian age for all the vertebrate faunas. Since the studies of Accordi (1957; 1963; 1965) it was thought that most of the size reduction of elephants in Sicily took place during the period preceding the Tyrrhenian. The smallest species (Elephas falconeri) was considered to be limited to the early Würm period and to have evolved as a consequence of environmental stress linked to the Würmian climatic cooling (Ambrosetti 1968; Kotsakis 1979). Stratigraphic and taphonomic features have been generally disregarded although the first valuable scientific paper dealing with the excavations of the Hippopotamus pentlandi bearing deposits of the S. Ciro cave (Palermo), carried out by the abbot Scinà (Scinà 1831), contains many interesting remarks concerning the relationship between the vertebrate-bearing deposits and the underlying marine sands which extend inside and outside the cave.

Vaufrey (1929), Accordi & Colacicchi (1962) and Accordi (1963; 1965) provide also interesting observations concerning the relationship between vertebrate-bearing deposits and littoral marine deposits.

Since 1985, a new synthesis has incorporated new stratigraphic and aminostratigraphic data (Belluomini & Bada 1985; Bada et al.1991; Burgio & Cani 1988; Bonfiglio 1987; Bonfiglio 1991; Bonfiglio 1992 a; Bonfiglio 1992 b; Bonfiglio & Burgio, 1992; Bonfiglio & Insacco 1992).
These data concern essentially the following aspects:

(1) Number and composition of the Pleistocene faunal complexes
(2) Stratigraphic, environmental and taphonomic data
(3) Chronological data
(4) Geographic distribution of vertebrate bearing deposits and palaeogeography.

NUMBER AND COMPOSITION OF PLEISTOCENE FAUNAL COMPLEXES (Table. 1)

On the basis of extensive new evidence, the Pleistocene vertebrate assemblages of Sicily can be arranged into 5 phases or Faunal Complexes (F.C.), spanning from the Early Pleistocene to the Late Glacial (Bonfiglio et al. 2000;

Bonfiglio et al. 2001; Bonfiglio et al. in press).

MONTE PELLEGRINO FAUNAL COMPLEX.

This is the oldest Quaternary fossil record so far known (late Villafranchian, Burgio & Fiore 1997) and it is documented only in the very restricted geographic area of Monte Pellegrino, close to the town of Palermo. The poorly diversified fauna includes species of the genera Hypolagus, Pannonictis, Leithia, Asoriculus, Pellegrinia, Apodemus.

Hypolagus and Pannonictis are moderately endemic taxa and the potential ancestors appear to be typical European forms, not found so far in northern Africa (Burgio & Fiore 1997); the endemic shrew Asoriculus burgioi is of uncertain provenance (Masini & Sarà 1998); Asoriculus is well documented in Europe from the Early Pliocene, and it has also been found recently in northern Africa, in Late Pliocene - Early Pleistocene deposits (Geraads 1995; Rzebik-Kowalska 1988).

The endemic dormouse Leithia (Thaler 1972), is regarded as one of the direct ancestor of the better known and better described Leithia and Maltamys from the Middle and Late Pleistocene of Sicily and Malta (Zammit Maempel & De Bruijn 1982). Leithia and Maltamys are considered by Daams & De Bruijn (1995) as relics of a scarcely known older (Messinian) population phase and these elements have remained isolated in the Sicilian Maltese archipelago for the entire Pliocene and most of the Quaternary.

According to Thaler (1972) Pellegrinia has derived characteristics of the African stock of Ctenodactilids, thus indicating a dispersal from that region.

The composition of the Monte Pellegrino fauna - unique for the Mediterranean islands - suggests it may have been derived in part from an older, not locally known, population phase (Messinian age ? Azzaroli 1974; Azzaroli & Guazzone 1979) and partially from younger dispersals from Europe. The different degree of endemism and the different geographical affinity of the taxa, indicate a polyphasic origin (Masini et al. in press). After the first report by Thaler (1972) some of the taxa have been revised more recently but our knowledge of this very peculiar and intriguing faunal assemblage is still not exhaustive.

"Monte Pellegrino" F.C.	**"*Elephas falconeri*" F.C.**	**"*Elephas mnaidriensis*" F.C.**	**"Grotta S. Teodoro – Pianetti" F.C.**	**"Castello" F.C.**	**Holocene fauna**
Testudo graeca *Asoriculus* sp. *Pannonictis arzilla* *Apodemus maximus* *Leithia* sp. *Maltamys* cf. *gollcheri* *Pellegrinia panormensis* *Hypolagus* sp.	*Discoglossus* cf. *pictus* *Bufo* cf. *viridis* *Hyla* sp. *Emys orbicularis* *Testudo hermanni* *Geochelone sp.* *Lacerta siculomelitensis* *Lacerta viridis* *Lacerta sp.* *Coluber* cf. *viridiflavus* *Natrix* sp. Several species of bats *Crocidura esuae* *Vulpes* sp. *Nesolutra trinacriae* *Elephas falconeri* *Leithia cartei* *Leithia melitensis* *Maltamys gollcheri*	*Discoglossus cf. pictus* *Emys orbicularis* *Testudo hermanni* *Lacerta siculomelitensis* Birds *Erinaceus europaeus* *Crocidura* aff. *esuae* *Crocuta crocuta* cf. *spelaea* *Panthera leo* *Canis lupus* *Nesolutra trinacriae* *Ursus* cf. *arctos* *Elephas mnaidriensis* *Sus scrofa* *Hippopotamus pentlandi* *Cervus elaphus siciliae* *Dama carburangelensis* *Bos primigenius siciliae* *Bison priscus siciliae* *Leithia* cf. *melitensis* *Maltamys* cf. *wiendincitensis*	Reptiles *Testudo* sp. Birds Chiroptera *Crocuta crocuta* cf. *spelaea* *Ursus* cf. *arctos* *Canis* cf. *lupus* *Vulpes vulpes* *Elephas mnaidriensis* *Sus scrofa* *Equus hydruntinus* *Cervus elaphus siciliae* *Bos primigenius siciliae* *Apodemus* cf. *silvaticus* *Microtus (Terricola)* ex gr. *savii* *Crocidura* cf. *sicula*	Birds *Erinaceus europaeus* *Crocidura* cf. *sicula* *Canis lupus* *Vulpes vulpes* *Equus caballus* *Equus hydruntinus* *Sus scrofa* *Cervus elaphus* *Bos primigenius* *Microtus (Terricola)* ex gr.*savii* *Apodemus* sp. *Lepus europaeus*	*Bufo bufo* *Bufo* cf. *viridis* Turtle indet. *Emys orbicularis* *Lacerta viridis* Birds Chiroptera indet *Erinaceus europaeus* *Crocidura* sp. *Canis* cf. *lupus* *Vulpes vulpes* *Felis silvestris* *Martes* sp. *Mustela* cf. *nivalis* *Ursus* sp. *Sus scrofa* *Cervus* sp. *Bos primigenius,* *Microtus (Terricola)* cf. *savii* *Apodemus* sp. *Arvicola terrestris* *Mioxus glis* *Monachus monachus*

TABLE 1. FAUNAL COMPLEXES OF SICILY (AFTER BONFIGLIO et al., 2001)

ELEPHAS FALCONERI FAUNAL COMPLEX. This Faunal Complex is even poorer than the preceding one in mammalian biodiversity. The poorly diversified fauna includes, besides the pigmy elephant, members of the genera Crocidura, Lutra, Leithia, Maltamys, a giant tortoise, and a rich typical endemic avifauna (Pavia 1999; Pavia 2000).

The shrew Crocidura esuae is an endemic species, of uncertain bio-geographic affinity (Kotsakis 1986). The genus Crocidura occurs widely in the Quaternary of Europe (Reumer 1984; Rzebik-Kowalska 1995) and in northern Africa (Butler 1980). Large mammals include the pigmy elephant Elephas falconeri, strongly reduced in size, and a member of the Lutrinae (genus Lutra).

Neither the ancestor nor the geographic provenance of E. falconeri has been determined unequivocally. Even though its derivation from a Paleoloxodontine elephant is accepted by several authors, its possible origin from European Elephas antiquus stock, or from a north African species is still a matter for discussion.

Moreover, also the composition of this faunal complex reveals a polyphasic origin; some taxa are relics from the preceding phase and others are 'newcomers' that probably entered the island through a strongly filtering barrier (Masini et al. in press).

ELEPHAS MNAIDRIENSIS FAUNAL COMPLEX: This Faunal Complex is almost completely renewed with respect to the preceding F. C. The pigmy E. falconeri is extinct, while the faunal composition is more balanced and includes top predators such as the lion and the spotted 'cave' hyena. The faunal composition is quite similar to that found in the southern Italian peninsula, and the continental forerunner of most of these taxa has been fairly accurately identified. The only significant absence is that of perissodactyls (rhinos and horses). The herbivorous taxa (bison, auroch, fallow deer, red deer, hippo) are moderately modified with respect to the cogeneric or cospecific taxa from the Italian mainland and the endemic nature of the fauna is apparent mainly from the modest reduction in size. E. mnaidriensis is a smaller, but not extremely modified, descendant of Elephas antiquus. The red deer Cervus elaphus siciliae is slightly reduced in size (about 20%). Dama carburangelensis (previously identified as Megaceroides carburangelensis) is probably derived from fallow deer populations (Dama dama tiberina) widespread in central and southern Italy during the late Middle Pleistocene (Abbazzi et al. 2001).

A peculiarity of this fauna is the very low diversity of small mammals, which are represented only by survivors from the E. falconeri F. C. (Leithia, Maltamys and Crocidura esuae; Petruso 2001; Petruso in progress; Masini et al. in press). The knowledge of the composition of the avifauna is still incomplete (Pavia 2000; Pavia 2001). One endemic species of birds, Cygnus falconeri occurs, while the endemic species of the previous F. C. became extinct, except Grus melitensis whose persistence is, however, questionable. Typical continental taxa (i. e. Galliforms) are still lacking. On the whole the avifauna assemblages are partially renewed alike the mammal fauna.

THE PIANETTI - SAN TEODORO CAVE FAUNAL COMPLEX

The Pianetti - San Teodoro Cave Faunal Complex dates to the last glacial cycle. The faunal history of this period is characterised by extinction events (hippopotamus, endemic dormice and Crocidura esuae), and by the dispersal of equids (Equus hydruntinus) and of mainland small mammals, which are represented by taxa still occurring in Sicily (Microtus (Terricola), Crocidura cf sicula, Apodemus cf. sylvaticus, Erinaceus europaeus) (Bonfiglio et al. 1997; Bonfiglio et al. 2001). Almost all of the large mammals belonging to this F. C. seem to be inherited from Elephas mnaidriensis F. C. endemic species of avifauna are lacking (Pavia 2000).

The faunal composition of this F. C. is probably not completely known, since only a few sites have been discovered so far and one of them (S. Teodoro cave) has not yet been completely investigated.

Castello Faunal Complex. Finally, the Late Glacial associations show a dramatic decrease in diversity, missing all endemic large mammals still occurring in the Pianetti-S. Teodoro F. C. Late glacial faunas, which are similar to continental ones, are associated to lithic artifacts and cultural evidences of late Upper Palaeolithic and, following some authors (Kotsakis 1979), include taxa present in one (Equus hydruntinus) or both (Sus scrofa) previous faunal complexes.

STRATIGRAPHIC, ENVIRONMENTAL AND TAPHONOMIC DATA

Taphonomic data show that Pleistocene vertebrates were distributed in both cave environments and broad, open environment (Bonfiglio 1987; Bonfiglio 1992 b; Bonfiglio 1995; Bonfiglio et al. 1993; Bonfiglio et al. 1996; Bonfiglio et al. 1997; Chilardi & Gilotti 1996) and numerous relationships have been found between the vertebrate bearing deposits and terraced marine deposits, which can be correlated with the d18O isotopic record and the main palaeogeographic events in Sicily (Agnesi et al. 1997; Bonfiglio 1991; Bonfiglio et al. 2000; Bonfiglio et al. in press; Di Maggio et al. 1999).

Actually most of the mammalian fossils of Sicily are found in caves and fissures - rather common in the carbonate mountain ranges of Northern Sicily and in the Hyblean plateau. Another important category of deposits is represented by ancient shorelines, beach deposits and marine terraces, frequently associated with a lagoon or swamp. Limnic deposits related to small freshwater basins also occur, often in relation to coastal and fully marine deposits.

The assemblages of the Monte Pellegrino F.C. and those of S. Teodoro Cave - Pianetti and Castello F.C.'xes are contained in caves and fissure-filling deposits and they have not relationships with marine deposits. The assemblages of the Elephas falconeri and of the Elephas mnaidriensis F.C. 'xes are contained in cave deposits as well as in coastal plain and/or in marine littoral deposits.

The biodiversity, the preservation conditions and the concentration of the skeletal remains are very diverse in the different environments (see FIGURE 1).

Littoral marine sands - In the late Middle Pleistocene sandy gravel outside the S. Ciro cave, the rare hippopotamus and elephant remains (molars, limbs) are disarticulated, encrusted by serpulid polychaetes worms and associated with a rich marine fauna containing gastropods, bivalves, ostracods, foraminifera, of littoral a euryhaline environment (Galletti & Scaletta 1991).

An isolated scapula of Elephas mnaidriensis comes from Thyrrhenian marine deposits at Maddalena peninsula (Siracusa) (Accordi 1963; Di Grande & Scamarda 1973).

Deltaic marine clay sands - At Contrada Cacaladritta (south eastern Sicily) rare skeletal remains of Hippopotamus pentlandi, Elephas cf. mnaidriensis and Bos primigenius siciliae are concentrated in deltaic Middle Pleistocene clay sands with Ostrea edulis (Roda 1965; Ruggieri 1973; Lentini 1984; Bonfiglio et al. 1997).

Deltaic marine gravels and sands - Disarticulated, fragmented, worn and mechanically selected remains of elephant (Elephas mnaidriensis), hippopotamus (Hippopotamus pentlandi), red deer (Cervus elaphus siciliae), bear (Ursus cf. arctos), tortoise (Testudo cf. hermanni) are contained in the deltaic marine gravel and sandy deposits underlying Thyrrhenian sands with Strombus bubonius at Cape Peloro. Serpulids encrust some skeletal elements (Bonfiglio & Berdar 1979; Bonfiglio & Violanti 1986; Marra 2001).

FLUVIAL AND LACUSTRINE COASTAL PLAIN DEPOSITS

At Comiso a regressive marine sequence begins with Early Pleistocene marine deposits and ends at the top with sands of brackish environment which, in turn, pass to deposits of limnic environment underlying Middle Pleistocene marine sands (Conti et al. 1980; Carbone et al. 1982). In the different levels of the limnic deposits skeletal elements of Elephas falconeri, Leithia melitensis, bats, fishs, birds, reptiles (Lacerta sp., Testudo sp., Emys orbicularis, Geochelone sp.) are contained (Bonfiglio & Insacco 1992).

At Contrada Fusco, near Siracusa, gravel, sandy and silty deposits of coastal plain environment contain a rich vertebrate fauna (an elephant of little reduced size, Elephas mnaidriensis, Hippopotamus pentlandi, Ursus cf. arctos, Crocuta crocuta spelaea, Lutra trinacriae, Leithia melitensis, Maltamys wiedincitensis, Crocidura esuae, Emys orbicularis, Testudo cfr. hermanni, Lacerta siculomelitensis, Natrix sp.) (Chilardi 1996; Kotsakis 1996 a; Kotsakis 1996 b; Cassoli & Tagliacozzo 1996 a; Cassoli & Tagliacozzo 1996 b). The coastal plain deposits overlie Early Pleistocene bathyal clays and underlie Tyrrhenian calcarenites (Chilardi & Gilotti 1996).

To the coastal plain environment belong the silty, laminated lacustrine deposits at Acquedolci (north eastern Sicily) containing very abundant skeletal elements of the endemic hippo Hippopotamus pentlandi associated with scarce remains of Cervus elaphus siciliae, Ursus cfr. arctos, Canis lupus, Elephas cfr. mnaidriensis, Testudo cfr. hermanni, birds (Bonfiglio 1992a; Bonfiglio 1995).

CAVE DEPOSITS CORRELATED WITH MARINE DEPOSITS

The vertebrate assemblage containing abundant remains of Hippopotamus pentlandi associated with rare remains of Elephas mnaidriensis, Cervus siciliae, Dama carburangelensis, Bos primigenius siciliae, Canis lupus, Ursus cfr. arctos, Crocuta spelaea of the S. Ciro cave overlie middle Pleistocene marine sands (Scinà 1831; Fabiani 1928a; Fabiani 1928b)

At Spinagallo cave the lower vertebrate deposits containing abundant remains of Elephas falconeri associated with Leithia melitensis, Maltamys gr. gollcheri-wiedincitensis, Crocidura esuae, Testudo hermanni overlie Middle Pleistocene littoral calcarenites (Accordi & Colacicchi 1962; Petronio 1970; Kotsakis 1977; 1986; Kotsakis & Petronio 1981; Di Grande & Raimondo 1984; Bonfiglio 1992 b; Petruso in progress).

Vertebrate bearing deposits overlying marine terraces - At San Vito Lo Capo peninsula mammal remains of the Elephas falconeri F.C. and of the Elephas mnaidriensis F.C. overlie marine terraces correlated with the IOS 11-15 and with the IOS 5a or 5c respectively. (Di Maggio et al. 1999). In eastern Sicily vertebrate bearing deposits with prevalent hippo overlie late Middle Pleistocene (Coste di Gigia; Bonfiglio 1992 b) or Tyrrhenian (Scodonì, Bonfiglio 1987) abrasion platforms.

KARST FISSURE AND CAVE DEPOSITS

Fossil vertebrates contained in fissure-filling deposits were accumulated by different ways. At Contrada Pianetti remains of Bos primigenius siciliae, Cervus elaphus siciliae, Canis cf. lupus, Ursus cf arctos, have been recovered in a small karst cavity (Bonfiglio et al, 1997; Marra & Bonfiglio, 2000). The taphomic features (fragmentary and disarticulated skeletal elements belonging to different species) suggest that detritus and bones scattered on surface filled the cavity by the action of gravity and/or washing. At Contrada Cimillà, abundant remains of Dama carburangelensis were associated to scarce remains of Cervus elaphus siciliae, Elephas mnaidriensis, Ursus cf. arctos (Bonfiglio et al., 1993; Abbazzi et al., 2001). The skeletal element representation shows that surface opening of the fissure may acted as a trap for living fallow deers, which were abundant at the site by ethological and/or ecological reasons.

In the bone assemblage of the San Teodoro cave large mammals skeletal elements were accumulated by Crocuta crocuta spelaea which damaged the main part of bones. A so impressing accumulation by hyena in a large cave (floor area: more than 1000 square meters) is an absolutely novelty for insular environments (Bonfiglio et al. 2001; Marra & Bonfiglio 2002).

CHRONOLOGICAL DATA

The record of fossil vertebrate faunas from Sicilian Pleistocene deposits allowed the reconstruction of a detailed biochronological scale, based on geochemical and radiometric dating also (Bada et al. 1991; Rhodes 1996).

Where correlated with marine sediments, the assemblages of the Elephas falconeri Faunal Complexe are associated with deposits dated early Middle Pleistocene; amino-acid racemization dating yielded an age of 455 ± 90 Ky for Elephas falconeri from the Spinagallo and Luparello cave deposits (Bada et al. 1991).

Where correlated with sediment of littoral environments, the assemblages of the E. mnaidriensis Faunal Complex are associated with deposits dated as late Middle Pleistocene and / or Late Pleistocene (Bonfiglio & Burgio 1992; Bonfiglio et al. 1996).

Bada et al. (1991) assigned an age of 200 ± 40Ky to this faunal complex. ESR dating for teeth enamel of Elephas mnaidriensis and Hippopotamus pentlandi from Contrada Fusco (Rhodes 1996) provided an age ranging between 146.8 ± 28.7 and 88.2 ± 19.5 Ky.

The assemblages of the Pianetti- S. Teodoro F. C., that are not correlated to marine deposits, have not still been dated radiometrically.

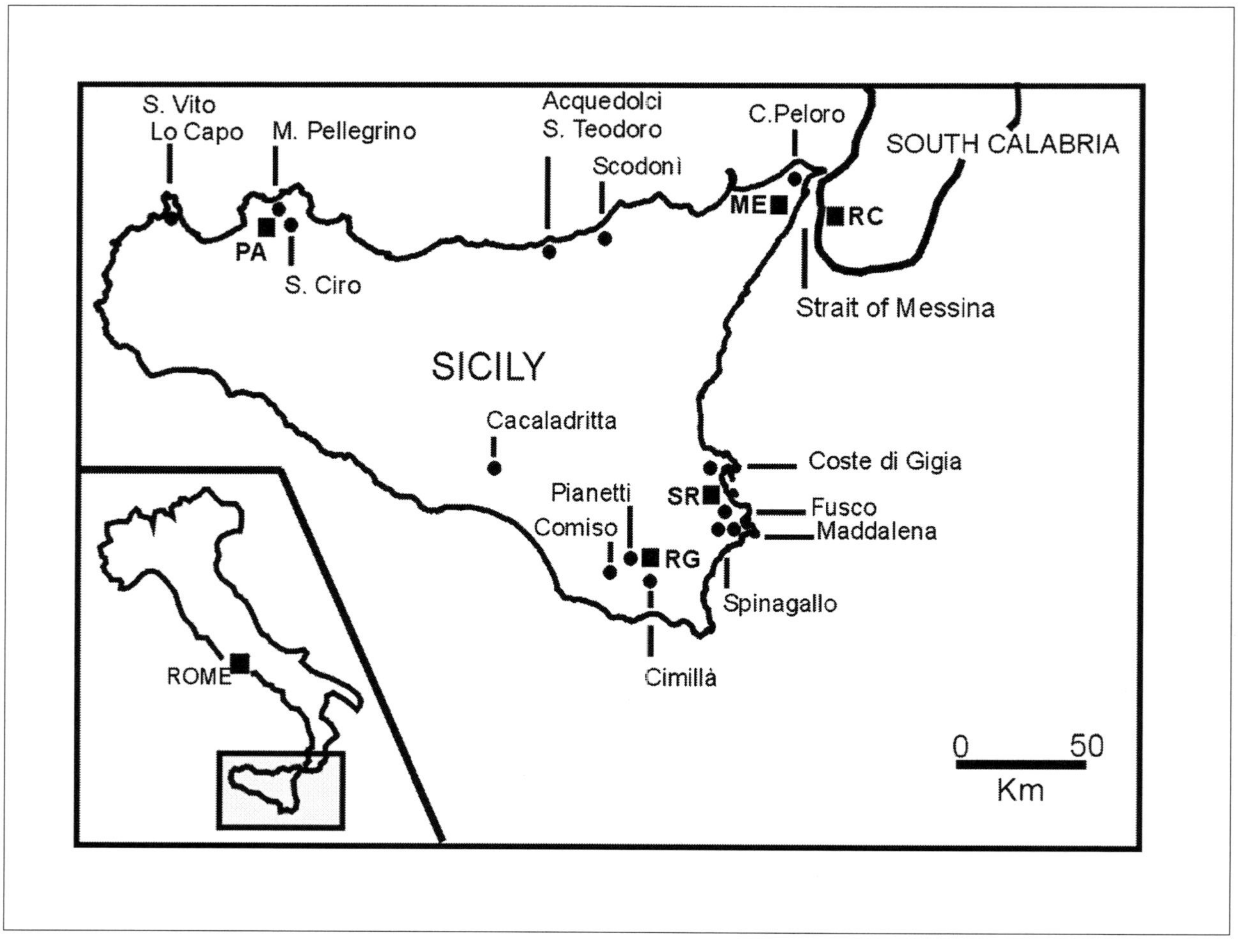

FIGURE 1. COMPOSITE MAP OF ITALY, SICILY AND RESEARCH SITES

GEOGRAPHIC DISTRIBUTION OF VERTEBRATE BEARING DEPOSITS AND PALAEOGEOGRAPHY.

Fossil documentation of Quaternary mammalian populations in Sicily is not evenly distributed in time and space. A trend is evident towards a richer fossil record from the Early Pleistocene to the Late Glacial, the latter fossils being most widespread. The geographic distribution could be related to taphonomical factors and to effective differences in paleogeography in the insular system (Bonfiglio & Burgio 1992), while the reason why the number of sites and stratigraphic levels varies so greatly with respect to time (younger sites are much more numerous) has never been investigated. This pattern, widespread in regions in which tectonic activity is very intense, is probably due to the destruction of older deposits by intense erosion (Masini et al. in press).

In Sicily the establishment of an extensional tectonic regime from the Early Pleistocene onward resulted in the collapse of peripheral zones of the island and led to the creation of a series of deep marine basins which occupied large areas until the end of the Early Pleistocene (Bonfiglio & Piperno 1996; Bonfiglio et al. in press; Di Geronimo 1979; Di Geronimo & Costa 1978; Di Geronimo et al. 1980; Grasso & Lentini 1982; Ruggieri et al. 1976; Ruggieri & Unti 1974; Ruggieri & Unti 1977; Sprovieri 1982). The Hyblean Plateau was separated from the northern sector of the island. The continental limnic succession at Comiso made up by paleosols, lacustrine and aeolian deposits, constitutes the evidence of the first connection of the Hyblean Plateau with the northern sector of Sicily in the Pleistocene (Conti et al. 1980).

South Calabria, which represents the continental connection of Sicily with southern Italian peninsula, in the Plio-Pleistocene was constituted also by three main blocks bounded by deep marine basins (Barrier et al. 1990; Barrier et al. 1993; Fabbri et al. 1980; Bonfiglio et al. in press).
Actually the E. falconeri F. C. denotes the occurrence of an insular system made up of strongly geographically isolated small islands, with very difficult and sporadic connections with the mainland.

According to Bonfiglio & Piperno (1996) the distribution of the E. falconeri assemblages in south-eastern and western Sicily and the scattered traces of Lower Palaeolithic artifacts in the Hyblean Plateau and along the area linking it to western Sicily, might suggest that the dispersal events of humans and of some faunal elements have followed the same route. According to the latter authors the hypothesis of an African provenance for both E. falconeri and human populations through the Sicilian-Tunisian isthmus, as a result of tectonic controlled changes of sea bottom depth during the Early Pleistocene (see Alimen 1975; Kelling & Stanley 1972; Vaufrey 1929), should at least merit a fresh and more detailed re-consideration.

From the beginning of the Middle Pleistocene onward, the evolution of Southern Calabria and Sicily was characterised by an uplifting tendency which led, in the late Middle Pleistocene, to the establishment of a palaeogeography very similar to the present one (Barrier et al. 1990; Barrier et al. 1993; Fabbri et al. 1980; Bonfiglio et al. in press). In the late Pleistocene the interaction of neotectonic with eustatic marine cycles probably led to the establishment of an intermittent filter barrier in the area of Messina Straits, which controlled the processes and timing of the Late Pleistocene-Late Glacial vertebrate faunal dispersion in Sicily. In fact the composition of the two younger F. C.'xes denotes that temporary connections with southern Italy occurred.

DISCUSSION AND CONCLUSION

Changes in faunal associations of islands are controlled by climatic and environmental factors, but mainly by geographical barriers (Azzaroli 1982; Azzaroli 1996; Dermitzakis & Sondaar 1978).

Biochronologic, taphonomic and palaeogeographic evidence regarding Pleistocene vertebrate faunas of Sicily leads to the following considerations:

The two oldest Faunal complexes (Monte Pellegrino and E. falconeri F.C.) characterised by strongly endemic faunal assemblages, denote the occurrence of an insular system made up of geographically isolated small islands, with very difficult and sporadic connections with the mainland. Palaeogeographic data do not provide more precise information on the route followed by dispersals for these complexes.

The three dispersals documented in Sicily from the late Middle Pleistocene to the Late Pleistocene prove the possibilities of connection between the island and the continent through the Strait of Messina area. However, these conditions do not fully explain the modifications recorded in the transition between faunal complexes.

As a matter of fact, variations in the three faunal complexes depend on three components:

(1) Dispersal events. Dispersals (or new occurrences) change obviously the faunal composition and are easier during lowstand episodes.

(2) Extinctions. It seems that most of the extinction events recorded in insular complexes corresponds to those occurring in Italian peninsula and in Western Europe (Elephas antiquus become globally extinct; lion and hyena disappear from Western Europe; the European bison is now present only in few residual populations living in Middle-Eastern Europe; Equus hydruntinus disappear from Italian peninsula at the beginning of the Holocene and became totally extinct; etc.) It is still unclear, however, if extinctions took place at the same time in the Southern Italian peninsula as in the island. Observations from S. Teodoro cave suggest that at least elephant survived in Sicily longer than elsewhere in Europe. A remarkable exception is, however, represented by the extinction of the endemic dormice (Leithia and Maltamys) and of the shrew Crocidura esuae. These endemics became extinct during the last glacial cycle, prior then the extinction of Elephas mnaidriensis, possibly owing to the interaction of several factors, such as dispersal of new small mammals and predators from the mainland, concomitant to a severe climatic deterioration (see Masini et al. this volume, for a discussion).

(3) Environmental changes related to climatic and physic variations. These changes (connection between the island and the mainland during the Glacial maximum, disappearance of coastal plains subsequent to deglaciation; rearrangement of vegetation in response to climatic changes; etc.) represent very probably direct and indirect causes of dispersals and extinctions.

Components 1 and 2 are basic parameters to define trends in faunal changes; component 3 refers to explanatory models of the variations themselves. The entity of impact in extinction directly linked to these changes is still unclear.

Many taxa identified in different sites and attributed to the considered faunal complexes are still little known. Exhaustive morphological and biometrical data are lacking and relationships with continental ancestors have not still been investigated in the details. In particular, data about the degree of endemism of some taxa (bear, spotted cave hyena, wolf, wild boar) have not still been collected and elaborated. Direct comparisons between the best known taxa of Elephas mnaidriensis F.C. (ox, deer) and those of Pianetti-S. Teodoro F.C. have not been done; Equus hydruntinus from the latter F. C. is still to be described. Detailed descriptions of micromammals from the three faunal complexes, in particular voles and dormice, are also partially lacking. The relationship between micromammals from Sicily and Malta has never been studied in the details (Petruso in progress), and can bring a new sight on dispersal dynamics internal to the Sicily-Malta archipelago, given that the latter island likely was a promontory of Sicily during low stand phases.

Careful taxonomic investigations on the recorded taxa, whose spreading on the island has been controlled by geographical barriers of different entity, and a correct evaluation of changes in faunal composition can also emerge from the recognition of differences between mainland and island taxa and a comparison with data from the continent (Italy).

These studies, still in progress, could enhance our knowledge on how and to what extent "discontinuous" insularity conditions interact with climatic and environmental changes in determining the faunal changes of Sicily during last phases of Pleistocene.

REFERENCES

ABBAZZI, L., ET AL.,
2001. A revision of medium and small sized deer from the Middle and Late Pleistocene of Calabria and Sicily. Bollettino della Società Paleontologica Italiana, 40(2), 115-126.

ACCORDI, B.,
1957. Nuovi resti di ippopotamo nano nel Pleistocene dei dintorni di Siracusa. Atti Accademia Gioenia di Scienze Naturali, 11, 99-109.

ACCORDI, B.,
1963. Rapporti tra il "Milazziano" della costa iblea (Sicilia sud-

orientale) e la comparsa di Elephas mnaidriensis. Geologica Romana, 2, 295-304.
ACCORDI, B.,
1965. Some data on the Pleistocene stratigraphy and related pigmy mammalian faunas of eastern Sicily. Quaternaria, 6, 415-430.
ACCORDI, B. AND R. COLACICCHI,
1962 - Excavations in the pygmy elephants cave of Spinagallo (Siracusa). Geologica Romana, 1, 217-230.
Agnesi, V., T. Macaluso T. and F. Masini,
1997 . L'ambiente e il clima della Sicilia nell'ultimo milione di anni. In S. Tusa, ed. Prima Sicilia. Alle origini della società siciliana, vol. I. Siracusa: Ediprint, 1997, 31-56
AGUIRRE, E.
1968-69. Revision sistematica del los Elephantidae por su morfologia y morfometria dentaria. Estudios Geologicos, 24 (1968), 109-168; 25 (1969), 123-177, 317-367.
ALIMEN, M. H.,
1975. Les isthmes hispano-marocain et siculo-tunisien aux temps acheuléens. L'Anthropologie, 79, 399-436.
AMBROSETTI, P.,
1968. The Pleistocene dwarf elephants of Spinagallo (Siracusa, south eastern Sicily). Geologica Romana, 7, 277-398.
AMBROSETTI, P., A. AZZAROLI & T. KOTSAKIS,
1980. Mammiferi del Plio-Pleistocene delle isole italiane. In "I Vertebrati fossili italiani", Catalogo della Mostra, Verona, 243-248.
AZZAROLI, A.,
1974. Late Miocene interchange of terrestrial fauna across the Mediterranean. Memorie della Soirtà Geologica Italiana, 13(suppl. 2), 261-265.
AZZAROLI, A.,
1982. Insularity and its effects on the terrestrial Vertebrates : evolutionary and biogeographic aspects. In "Palaeontology, essential of Historical Geology", First International Meeting, Venice, June 1981, 193-213.
AZZAROLI, A.,
1996. Processes of island colonization by terrestrial organisms. Biogeographia, 18, 33-50.
AZZAROLI, A. & G. GUAZZONE,
1979. Terrestrial mammals and land connections in the Mediterranean before and during the Messinian. Palaeogeography Palaeoclimatology Palaeoecology, 29, 155-167.
BADA, J. L. ET AL.,
1991. Isoleucine epimerization ages of Quaternary Mammals of Sicily. Il Quaternario, 4(1a), 5-11.
BARRIER, P. ET AL.,
1990. Le Détroit de Messine (Italie). Evolution tectono-sédimentaire récente (Pliocene et Quaternarie) et envirnment actuel. Documents et Travaux, IGAL, 11, 1-272.
BARRIER, P. ET AL.,
1993. La lagune Pléistocene à Gobius sp. du Monte Torre (Calabre Meridionale): signification paléogéographique. Rivista Italiana di Paleontologia e Stratigrafia, 99(1), 127-140.
BELLUOMINI, G. AND J. L. BADA,
1985. Isoleucine epimerization ages of the dwarf elephants of Sicily. Geology, 13, 451-452.
BONFIGLIO, L.,
1987. Nuovi elementi faunistici e stratigrafici del Pleistocene superiore dei Nebrodi (Sicilia nord-orientale). Rivista Italiana di Paleontologia e Stratigrafia, 93(1), 145-164.
BONFIGLIO, L.,
1991. Correlazioni tra depositi a Mammiferi, depositi marini, linee di costa e terrazzi medio e tardopleistocenici nella Sicilia orientale. Il Quaternario, 4(1b), 205-214.
BONFIGLIO, L.,
1992a. Campagna di scavo 1987 nel deposito pleistocenico a Hippopotamus pentlandi di Acquedolci (Sicilia nord-orientale). Bollettino della Società Paleontologica Italiana, 30(3), 157-173.
BONFIGLIO, L.,
1992b. Middle and Upper Pleistocene Mammal-bearing deposits in south-eastern Sicily: new stratigraphical records from Coste di Gigia (Syracuse). Geobios, 14, 189-199.
BONFIGLIO, L.,
1995. Taphonomy and depositional setting of Pleistocene mammal-bearing deposits from Acquedolci (North-Eastern Sicily). Geobios, 18, 57-68.
BONFIGLIO, L. AND A. BERDAR,
1979. Gli elefanti delle ghiaie pleistoceniche di Messina. Quaternaria, 21, 139-177.
BONFIGLIO, L. AND E. BURGIO,
1992. Significato paleoambientale e cronologico delle mammalofaune pleistoceniche della Sicilia in relazione all'evoluzione paleogeografica. Il Quaternario, 5(2), 223-234.
BONFIGLIO, L. AND G. INSACCO,
1992. Palaeoenvironmental, paleontologic and stratigraphic significance of Vertebrate remains in Pleistocene limnic and alluvial deposits from South Eastern Sicily. Palaeogeography, Palaeoclimatology, Palaeoecology, 95, 195-208.
BONFIGLIO, L., A. C. MARRA AND F. MASINI,
2000. The contribution of Quaternary vertebrates to the paleoenvironmental and paleoclimatic reconstructions in Sicily. In: Hart M. B. (ed.) Climates: Past and Present. Geological Society, London, Special Publications, 181, 169-182.
BONFIGLIO, L. AND M. PIPERNO,
1996. Early Faunal and Human populations. In: Leighton R. (ed.), Early Societies in Sicily, University of London, Accordia, 21- 29.
BONFIGLIO L. & D. VIOLANTI,
1986 – Prima segnalazione di Tirreniano ed evoluzione pleistocenica del Capo Peloro (Sicilia Nord-Orientale). Geogr. Fis. Dinam. Quat., 6, 3-15.
BONFIGLIO, L., ET AL.,
1996. Large mammal remains from late Middle Pleistocene deposits of Sicily: new stratigraphic evidence from the western edge of the Hyblean Plateau (Southeastern Sicily). Rivista Italiana di Paleontologia e Stratigrafia, 102(3), 375-384.
BONFIGLIO, L., ET AL.,
1993 - New Pleistocene fissure-filling deposits from the Hyblean Plateau (South Eastern Sicily). Rivista Italiana di Paleontologia e Stratigrafia, 98(4), 523-540.
BONFIGLIO, L. ET AL.,
1997. Large and small mammals, amphibians, reptiles from a new fissure filling deposit of the Hyblean Plateau (South Eastern Sicily). Bollettino della Società Paleontologica Italiana, 35(1), 97-122.
BONFIGLIO, L. ET AL.,
2001. A new Late Pleistocene vertebrate faunal complex from Sicily (S. Teodoro Cave, North-Eastern Sicily, Italy). Bollettino della Società Paleontologica Italiana, 40(2), 149-158.

BONFIGLIO, L. ET AL,
in press: Pleistocene calabrian and sicilian paleobioprovinces. Geobios Special Memoires
BRUGAL, J. P.
1987. Cas de "nanisme" insulaire chez l'aurochs. 112th Congrès National des Sociétés Savantes, Lyon, 2: 53-66.
BUGIO, E. & M. CANI,
1988. Sul ritrovamento di elefanti fossili ad Alcamo (Trapani, Sicilia). Il Naturalista Siciliano, 12(3-4), 87-97.
BUGIO, E. & M. FIORE,
1997. Pannonictis arzilla (De Gregorio 1886) a "Villafranchian" element in the fauna from Monte Pellegrino (Palermo, Sicily). Il Quaternario, 10, 65-74.
BUTLER, P.M.,
1980. The giant erinaceid insectivore Deinogalerix Freudenthal from the Upper Miocene of the Gargano, Italy. Scripta Geologica, 57, 1-72.
CARBONE, S., ET AL.,
1982. I terrazzi marini quaternari dell'area iblea (Sicilia sud-orientale). Contributi conclusivi per la realizzazione della Carta Neotettonica d'Italia, Progetto Finalizzato Geodinamica, Napoli, 506, 1-35.
CASSOLI, P.F. & A. TAGLIACOZZO,
1996a. L'avifauna. In: Basile, B. & S. Chilardi, eds. Le ossa dei Giganti. Lo scavo paleontologico di Contrada Fusco. Siracusa: Arnaldo Lombardi, 1996, 61-67.
CASSOLI, P.F. & A. TAGLIACOZZO,
1996. L'ittiofauna. In: Basile, B. & S. Chilardi, eds. Le ossa dei Giganti. Lo scavo paleontologico di Contrada Fusco. Siracusa: Arnaldo Lombardi, 1996, 55-54.
CHILARDI, S.,
1996. I macromammiferi. In: Basile, B. & S. Chilardi, eds. Le ossa dei Giganti. Lo scavo paleontologico di Contrada Fusco. Siracusa: Arnaldo Lombardi, 1996, 73-80.
CHILARDI, S. AND A. GILOTTI,
1996. A new important palaeontological site of Mediterranean basin: preliminary results from Contrada Fusco (Siracusa, Southeastern Sicily). Second International Symposium on the Conservation of our Geological Heritage, Rome, abstract book, 24.
CONTI, M. A., ET AL.,
1980. Il Pleistocene in facies limnica di Vittoria (Sicilia meridionale). Geologica Romana, 18, 93-104.
DAAMS, R. AND H. DE BRUIJN,
1995. A classification of the Gliridae (Rodentia) on the basis of dental morphology. Hystrix, 6(1-2), 3-50.
DERMITZAKIS, M. D. AND P. Y. SONDAAR,
1978. The importance of fossil mammals in reconstructing paleobiogegraphy with special reference to the Pleistocene Aegean archipelago. Annales Gèologiques des Pays Helléniques, 29, 808-840.
DI GERONIMO, I.,
1979. Il Pleistocene in facies batiale di Valle Palione (Grammichele, Catania). Bollettino di Malacologia, 15, 85- 156.
DI GERONIMO, I. AND B. COSTA,
1978. Il Pleistocene di Monte dell'Apa (Gela). Rivista Italiana di Paleontologia e Stratigrafia, 84(4), 1121-1158.
DI GERONIMO, I. ET AL.,
1980. Dati preliminari sulla tettonica della Sicilia sud-orientale. Fogli 273 (Caltagirone), 274 (Siracusa), 275 (Scoglitti), 276 (Ragusa) e 277 (Noto). Contributi preliminari alla realizzazione della Carta Neotettonica d'Italia, Progetto Finalizzato Geodinamica, Napoli, 356, 747-773.
DI GRANDE, A. AND W. RAIMONDO,
1984. Linee di costa plio-pleistoceniche e schema litostratigrafico del Quaternario siracusano. Geologica Romana, 21, 279-309.
DI GRANDE, A. AND G. SCAMARDA,
1973. Segnalazione di livelli a Strombus bubonius Lamarck nei dintorni di Augusta (Siracusa). Bollettino Accademia Gioenia Scienze Naturali in Catania, 11, s .4 (9-10), 157-172.
DI MAGGIO, C., ET AL.,
1999. Oscillazioni eustatiche, biocronologia dei depositi continentali quaternari e neotettonica nella Sicilia Nord - Occidentale (Penisola di San Vito Lo Capo - Trapani). Il Quaternario, 12(1), 25-50.
FABBRI, A., F. GHISETTI AND L. VEZZANI
1980. The Peloritani Calabria range and the Gioia basin in the Calabrian arc (Southern Italy): relationships between land and marine data. Geologica Romana, 19, 131-150.
FABIANI, R.,
1928a. Cenni sulle raccolte di Mammiferi quaternari del Museo Geologico dell'Università di Palermo e sui risultati di nuovi saggi esplorativi. Boll. Ass. Miner. Sicil., 4, 25-34.
FABIANI, R.,
1928 b. Aggiunte alla conoscenza dei mammiferi fossili del Ragusano in Sicilia. Istituto di Geoologia della Regia Università di Palermo.
GALLETTI, L. AND C. SCALETTA,
1991. Descrizione di una sequenza del Pleistocene superiore con fauna continentale a San Ciro-Maredolce (Palermo). Il Naturalista siciliano, 15(1-2), 3-10.
GERAADS, D.,
1995. Roungeurs et insectivores (Mammalia) du Pliocene final de Ahl Oughlam (Casablanca, Maroc). Geobios, 28(1), 99-115.
GLIOZZI, E. AND A. MALATESTA,
1984. A megacerine in the Pleistocene of Sicily. Geologica Romana, 21, 311-389.
GRASSO, M. AND F. LENTINI,
1982. Sedimentary and tectonic evolution of the eastern Hyblean Plateau (Southeastern Sicily) during late Cretaceous to Quaternary time. Palaeogeography, Palaeoclimatology, Palaeoecology, 39, 261-280.
KELLING, G. AND D. J. STANLEY,
1972. Sedimentation in the vicinity of the Strait of Gibraltar. In D. J. Stanley, ed., The Mediterranean Sea. A Natural Sedimentation Laboratory. Dowden, Hutchinson & Ross, Stroudsburg, 489-519.
KOTSAKIS, T.,
1977. I resti di Anfibi e Rettili pleistocenici della grotta di Spinagallo (Siracusa, Italia). Geologica Romana, 16, 211-229.
KOTSAKIS, T.,
1979. Sulle mammalofaune quaternarie siciliane. Bollettino del Servizio Geologico Italiano, 99, 263-276.
KOTSAKIS, T.,
1986. Crocidura esui n. sp. (Soricidae, Insectivora) du Pléistocène supérieur de Spinagallo (Sicilia orientale, Italie). Geologica Romana, 23, 51-64.
KOTSAKIS, T.,
1996a. Anfibi e rettili, In: Basile, B. & S. Chilardi, eds. Le ossa dei Giganti. Lo scavo paleontologico di Contrada Fusco. Siracusa: Arnaldo Lombardi, 56-60.

Kotsakis, T., 1996b. I micromammiferi. In: Basile, B. & S. Chilardi, eds. Le ossa dei Giganti. Lo scavo paleontologico di Contrada Fusco. Siracusa: Arnaldo Lombardi, 68-72

KOTSAKIS T. AND C. PETRONIO,
1981. I Chirotterri del Pleistocene superiore della grotta di Spinagallo (Siracusa, Sicilia). Bollettino del Servizio Geologico d'Italia, 101, 49-76.

LENTINI, F. (Ed.),
1984. Carta geologica della Sicilia sudorientale. Scala 1:100.000. Ente minerario siciliano. Comunità economica europea. Istituto di Scienze della Terra, Univ. degli Studi di Catania. S.E.L.C.A., Firenze.

MARRA, A. C.,
2001. Resti di vertebrati delle ghiaie pleistoceniche della Formazione di Messina (Sicilia Nord-Orientale). Il Naturalista Siciliano, s. 4, 25(1-2), 3-13.

MARRA, A. C. AND L. BONFIGLIO,
2000. Tafonomia dei depositi di fessura: due esempi dal Plateau Ibleo – Atti del I Seminario Internazionale di Studi su "il carsismo negli Iblei e nell'area sud-mediterranea", Speleologia Iblea, 8, 115-119.

MARRA, A. C. AND L. BONFIGLIO,
2002. Taphonomy of hyena dens: an "insular" perspective. In De Renzi et al., eds. Current Topics on Taphonomy and fossilization, Valencia: Ajuntament de Valencia, Collecciò Encontres 5, 2002

MASINI, F. AND M. SARÀ. M.,
1998. Asoriculus burgioi sp. nov. (Soricidae, Mammalia) from the Monte Pellegrino faunal complex (Sicily). Acta Zoologica Cracoviensia, 41(1), 111-124.

MASINI F., ET AL.,
in press. The role of coastal areas in the Neogene - Quaternary mammal island populations of the central Mediterranean . Biogeographia.

MASINI F., ET AL,,
this volume - Vertebrate assemblages of central-western Mediterranean islands during the Pliocene and Quaternary: reflecting on extinction events.

PAVIA, M.,
1999. The Middle-Pleistocene Avifauna of Spinagallo Cave (Sicily, Italy): Preliminary Report. Smithsonian Contributions to Paleobiology, 89, 125-127

PAVIA, M.,
2000. Le avifaune pleistoceniche dell'Italia Meridionale. Unpublished PhD Dissertation, University of Torino, XIII cycle, 150 pags.

Pavia, M.,
2001. Middle Pleistocene fossil avifauna from the Elephas mnaidriensis Faunal Complex of Sicily (Italy): preliminary results. The World of Elephants, 1st International Congress (Rome, October 16th-20th, 2001), Abstract book.

PETRONIO, C.,
1970. I roditori pleistocenici della Grotta di Spinagallo (Siracusa). Geologica Romana, 9, 149-194.

PETRUSO, D.,
2001. I ghiri della Sicilia antica. In Sarà, M. eds. Ghiri in Sicilia: ecologia e conservazione. WWF, 27-41.

PETRUSO, D.,
in progress. Stratigrafia e Paleogeografia del Quaternario continentale siciliano: il contributo dei vertebrati. PhD Dissertation in Geological Science, Universities of Naples and Palermo.

POHLIG, H.,
1893. Eine Elephantenhohle Siciliens und der erste Nachweis des Cranialdomes von Elephas antiquus. Aus den Abhandlungen der k. bayer. Akademie deer Wiss, 18, 73-100.

POHLIG, H.,
1909. Ueber zwei neue altpleistocäne formen von Cervus. Zeitschr. deutsch. geol. Ges., 61, 250-253.

Reumer, W. F. J., 1984. Plio-Pleistocene Soricidae from Tegelen and Hungary. Scripta Geologica, 73, 1-173.

RHODES, E. J.,
1996. ESR dating of tooth enamel. In: Basile, B. & S. Chilardi, eds. Le ossa dei Giganti. Lo scavo paleontologico di Contrada Fusco. Siracusa: Arnaldo Lombardi, 39-44.

RODA, C.,
1965. La sezione stratigrafica pleistocenica di Niscemi (Caltanissetta). Atti dell'Accademia Gioenia Scienze Naturali, s. 6, 17, 37-62.

RUGGIERI, G.,
1973. Gli Ostracodi e la stratigrafia del Pleistocene marino mediterraneo. Bollettino della Società Geologica Italiana, 92, 213-232.

RUGGIERI, G. AND M. UNTI,
1974. Pliocene e Pleistocene nell'entroterra di Marsala. Bollettino della Società Geologica Italiana, 93, 723-733.

RUGGIERI, G. AND M. UNTI,
1977. Il Quaternario del Pianoro di S. Margherita di Belice (Sicilia). Bollettino della Società Geologica Italiana, 96, 803-812.

RUGGIERI, G., ET AL.,
1976. La calcarenite di Marsala (Pleistocene inferiore) e i terreni contermini. Bollettino della Società Geologica Italiana, 94, 1623-1627.

RZEBIK-KOWALSKA, B.,
1988. Soricidae (Mammalia, Insectivora) from the Plio-Pleistocene and Middle Quaternary of Morocco and Algeria. Folia Quaternaria, 57, 51-90.

RZEBIK-KOWALSKA, B.,
1995. Climate and history of European shrews (family Soricidae). Acta Zoologica Cracoviensia, 38, 95-107.

SCINÀ, D.,
1831. Rapporto sulle ossa fossili di Mardolce e degli altri contorni di Palermo. Dalla Reale Tipografia di Guerra, Palermo.

SPROVIERI, R.,
1982. Considerazioni sul Plio-Pleistocene della Sicilia. In "Guida Geologica della Sicilia Occidentale", Società Geologica Italiana, 115-118.

THALER, L.,
1972. Les rongeurs (Rodentia et Lagomorpha) du Monte Pellegrino et la question des anciens isthmes de la Sicilie. Comptes rendus de l'Académie des Sciences, 274, 188-190.

VAUFREY, R.,
1929. Les éléphants nains des iles mediterranèennes et la question des isthmes pléistocènes. Archives de l'Institut de Paléontologie Humaine, 6, 1-220.

ZAMMIT MAEMPEL, G. AND H. DE BRUIJN,
1982. The Plio-Pleistocene Gliridae from the Mediterranean Islands reconsidered. Proceedings, Koninklijke Nedrlandse Akademie van Wetenschappen, 85, 113-128.

39

VERTEBRATE ASSEMBLAGES OF THE CENTRAL-WESTERN MEDITERRANEAN ISLANDS DURING THE PLIOCENE AND QUATERNARY: REFLECTING ON EXTINCTION EVENTS

FEDERICO MASINI (*1), **LAURA BONFIGLIO, LAURA ABBAZZI, MASSIMO DELFINO, FLAVIANO FANFANI, MARCO FERRETTI , TASSOS KOTSAKIS , DARIA PETRUSO , CINZIA MARRA , DANILO TORRE**

(*1)(*) Work supported by grants CoFin MURST 1999 " Central Mediterranean palaeogeography from the Miocene to the Quaternary" and by grants University of Palermo "Ateneo 1998" to Federico Masini,
Dipartimento di Geologia e Geodesia, C.so Tukory 131, Palermo, I-90134, Italy,
E-MAIL: fmasini@unipa.it

PREMISE

Theoretical issues drawn from models on faunal diversity are critically discussed in order to obtain a deeper insight on extinction events and processes in insular systems. These are then compared with actual case histories of vertebrates from some Mediterranean islands. For each case we propose and discuss extinction events and processes.

K.W. Islands, Models, Extinction.

INTRODUCTION

The fossil record, even with its severe limitations and uncertainties, often provides the only direct access to the chronological dimension of biological events and processes. It therefore supplies unique and valuable information on biodiversity changes in islands, which can be used to reconstruct palaeoecological patterns and to develop or test ecological models. Factors controlling faunal diversity and its changes are of great general interest for ecology and biogeography and many studies are devoted to this topic. As on the mainland, the faunal diversity of an island can be viewed as a balance between origination and extinction processes through time. In this contribution we combine conceptual elaboration of biodiversity models with observation and data drawn from the fossil record, allowing us to classify and interpret extinction patterns and events using a theoretical basis.

We address extinction processes for several reasons: a, present day extinctions are a serious problem in biosphere; b, in island biogeography much work has been done on the processes and mechanisms responsible for the presence of organisms on islands (see Azzaroli 1996, and literature therein) while lesser attention has been devoted on their extinction, and c, last but not least, extinction processes of endemic taxa dominated the history of faunal changes in islands after the arrival of man and particularly of Neolithic-Eneolithic colonisation. With regards to the last point, - a key point for students interested in human history - we feel we can learn more about island extinction if we consider also processes which occurred in "completely natural" conditions, i.e. without the disturbance of human impact, and compare them with extinctions caused, directly or indirectly, by humans. In practice this means also considering faunal populations of older geological periods (e.g. Pliocene and early Pleistocene).

Because they involve species that are no longer living and environments that have changed (without exact modern analogues), extinction processes, also are difficult study and, like many other natural processes, are not suitable for a strictly "experimental" methodological approach. Indeed, extinctions are the subject of many theoretical and speculative explanations, often contradictory. If the study of "historical" extinctions (extinction which occurred during historical times or occurring at present), can draw on many diverse potential sources of information, the extinctions of "fossil" taxa are much more problematic. Processes occurring in islands should follow the same 'general rules' and patterns as those occurring on the mainland, nevertheless islands have their own peculiarities worth investigating in their own right.

When facing such a complex problem, a model is a suitable starting point, since it has the advantage of simplicity. Models are by definition a simplification of reality, a simplification that allows to make some predictions and to compare these with observation; however one has to keep in mind that "although models are always true, since they are a construction of logical processes, they are not real" (Janes 1976). They are useful inasmuch as they reflect what we feel is a significant part or aspect of reality. Nevertheless, in criticising a model conceptually and in 'comparing' it with observations one can learn much about the real behaviour of the phenomenon that is the subject of interest.

Among a variety of models we have chosen the 'classical' dynamic model of Mac Arthur & Wilson (1963), i.e. the simplest model that assumes that faunal diversity tends to reach a "natural equilibrium" in insular ecosystems (Fig. 1). Indeed, the Mac Arthur & Wilson model is not restricted to "true geographic islands" but deals with all geographic areas that are isolated relative to the surrounding environment (e.g. lakes, humid areas in deserts, tops of mountain ranges...).

We summarise here some basic assumptions of the model developed for "oceanic-type" islands.

(1) Originations derive from immigration events evenly distributed in time.

(2) Immigration rate (i.e. the rate of successful dispersals) is related principally to the distance of the island from the continent and to the number of species already present in the island, and is therefore controlled by external and internal factors with respect to the island system.

(3) In turn, extinction rate is controlled by internal components (to a first approximation, the area of the island) and by interactions with immigrant taxa. The interactions among immigrants and residents are assumed on average to behave randomly over time.

(4) The resulting biodiversity is constrained by the availability and variety of resources present in each island. These factors are, in turn, linked primarily to the area of the island and to several other parameters such as: the variety of environments supported by the island (physiography), the climate, and variation of climate with respect to physiography, the geographic location, etc.

Bearing in mind the above issues we shall attempt some criticism and reflections in order to get further our argument.

With regard to the 'terrestrial' vertebrates of Mediterranean islands, and also other insular systems that can be classified as oceanic - like or continental, several lines of evidence indicate that the first assumption of the model (even distribution of dispersal over time) is usually not fully satisfied. Originations are seldom evenly distributed over time, but rather they are concentrated in specific time zones separated by longer periods of isolation. Indeed, an extremely dynamic paleogeography and the interrelation between paleogeographic changes and glacial eustatic fluctuation of sea level are the major controls on dispersal within insular systems in the Mediterranean (cf. Azzaroli 1996). Another important related aspect is that dispersals to islands usually occur by crossing sea barriers of various effectiveness that behave as filters for some taxa. Their effectiveness is not fixed but changes over geological time: ranging from dispersals when there are temporarily complete connections (land bridges) through sweepstake dispersals, to complete isolation (Azzaroli 1996, Masini et al. in press). Last but not least, in long lasting isolated systems the origination rate could be influenced by speciation events occurring within the island (evolutionary radiation, see De Giuli et al. 1986).

With regard to extinctions, firstly the Mc Arthur & Wilson model predicts that the number of successful dispersals to an island is less than the number of effective immigrant taxa, i.e. taxa which are able to survive in the island for a consistent time span. So, in principle, we would expect that the fossil record should include information on taxa that survived for a very short time on insular system, before becoming extinct.

However, another aspect of extinction processes is not explicitly considered by the model: i.e. the interactions between immigrant predators and resident herbivores (see above point 3). We feel that ignoring this fact is an oversimplification as if a predator disperses successfully on an island, it could produce dramatic disturbance of the community of resident herbivores.

Furthermore, the model assumes the conditions of an island to be constant: however, as observed above, we know that in some cases the geography of insular systems can change dramatically through time, and that global climatic fluctuations have a deep effect on biological communities.

To a first approximation, the expectation is that extinctions in insular systems may be related to:

(1) competitive interaction between new comers and residents;

(2) predatory interaction between new comers and residents;

(3) global climatic changes and their regional effects on the insular environment;

(4) changes in island geography (e.g. variation in area etc.) in response to geodynamic events (crustal movements, eustatic fluctuations or a combination of these);

Each factor or process may interact to a different extent in determining the 'real' behaviour of the biological system in each island and in affecting extinction events.

It is worth recalling here that competition and predatory interactions fall within the domain of the Lotka - Volterra model and related models. Perhaps a combined Mac Arthur & Wilson and Lotka - Volterra approach could produce a more 'realistic' and comprehensive model for biodiversity changes in insular ecosystems.

EXTINCTION EVENTS IN THE MEDITERRANEAN ISLANDS

Giving these premises we shall attempt a closer look at the characteristics of the extinction of insular taxa and discuss possible explanations. That is to say: we move from a theoretical approach towards historical 'real' cases as documented by the fossils from several Mediterranean islands

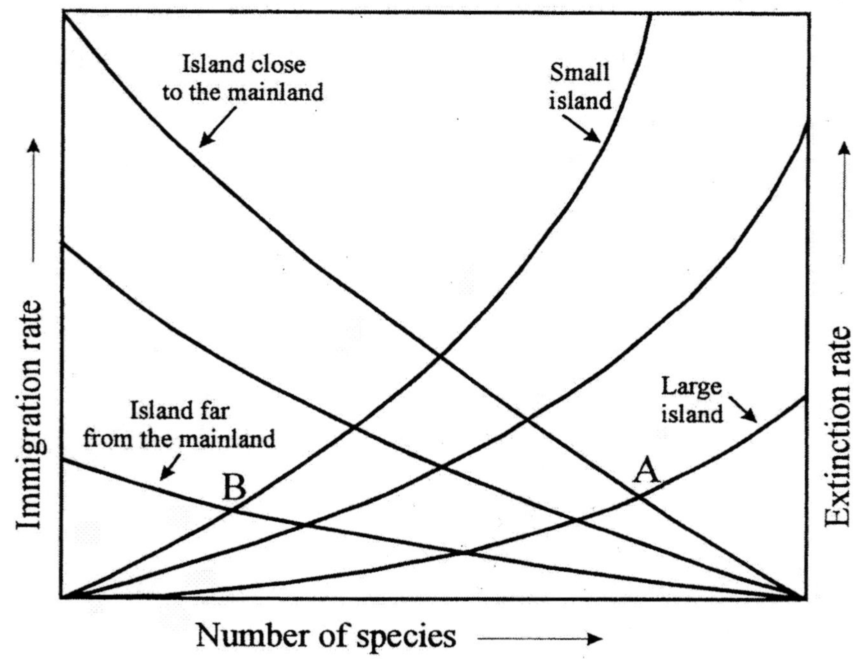

FIGURE. 1. MAC ARTHUR & WILSON (1963) MODEL FOR THE DYNAMIC EQUILIBRIUM OF THE BIOGEOGRAPHY OF AN ISLAND.

(FIGURE 2). Examples from Sicily, Gargano, Sardinia have been considered. Populations of these islands can be defined as 'polyphasic'. By this term we mean a faunal population in which, in a given time zone, faunal elements derived from more ancient population phases coexist more recent elements (Masini et al. in press).

At a first sight extinctions fall into three or four broad categories based on the timing of extinction of insular taxa with respect to their mainland relatives and on the time range that these taxa were present on the island.

(1a) Extinction events roughly coeval with events recorded on the mainland. (e.g. Equus hydruntinus)

(1b) Timing of extinction of taxa occurring both on the island and the mainland but generally occuring later on the island than on the mainland.

These categories of extinctions are probably related to global or regional environmental/climatic changes.

(2) 'Snapshot' events, where extinction strikes 'new comer' taxa that are apparently only present on the island for a geological brief time.

(3) Extinction events of taxa that have survived for a very long time on the island. In several cases these taxa, or their endemic descendants, survive much later than on the mainland: e.g. Prolagus, Nesiotites and Megaceroides in Sardinia, Myotragus, in the Balearics, and Leithia and Maltamys in Sicily.

EFFECT OF MAJOR CLIMATIC CHANGES ON EXTINCTION EVENTS

One might expect that changes in global or regional climate would have an amplified effect on islands since geographic isolation prevents faunal and floral reorganisation - the 'normal' response to climatic changes (stochastic individualistic model of Graham & Grimm 1990). In the continental environment, if a given area becomes ecologically unsuitable for a species, that species may become locally extinct but survive in a refuge area, or migrate to regions in which the favourable ecological condition persist. In both cases, the species can reinvade the former area when the appropriate ecological conditions are restored (e.g. by cyclic climatic change). In contrast, for an insular species such dynamic responses are not possible, thereby increasing the likelihood of extinctions. However, insularity can also act in the opposite manner, limiting the local effect of global changes. For instance, the island could be more 'conservative' than the mainland. Moreover the presence of a marine barrier could prevent colonisation by competitors and/or predators to the island (migrating in response to climatic changes), thus 'protecting' the island endemics. As a matter of fact the long lasting endemics become evidently so well adapted to the island biotome that are able to overcome global climatic changes that produce severe disturbances on mainland faunal assemblage (see section three). Therefore the following discussion will deal with taxa which are poorly modified by endemic evolution.

The "wild ass" Equus hydruntinus might represent a case of synchronic extinction. This equid was widespread in central – south Italian Peninsula during the Lateglacial, it entered also Sicily during the last glacial cycle then became extinct on the island in post Mesolithic times, as it did in the whole peninsula (Torre et al. 1996; Bonfiglio et al. 2001). The extinction of E. hydruntinus is possibly linked to the spreading of dense tree cover during the Holocene Climatic Optimum although the contribution of hunting by human's cannot be excluded (see section 2 and 3 for a discussion on human impact on insular faunas).

Strictly synchronic extinctions in islands and the mainland are, however, difficult to demonstrate since this would require a very high degree of temporal resolution. However a cies' Elephas mnaidriensis, a moderately modified endemic offshoot of Elephas antiquus, a widespread taxon in the Middle –Late Pleistocene of Europe, apparently survived in Sicily significantly later than it's continental relative (Bonfiglio et al. 2001). Conversely 'boreal -cool' species would be expected to have survived longer in northern islands. A well-known example of such northern survival is that of woolly mammoth Mammuthus primigenius on Wrangel Island, in the Arctic Ocean, more than six millennia after its last occurrence on the Siberian mainland (Vartanyan et al. 1993).

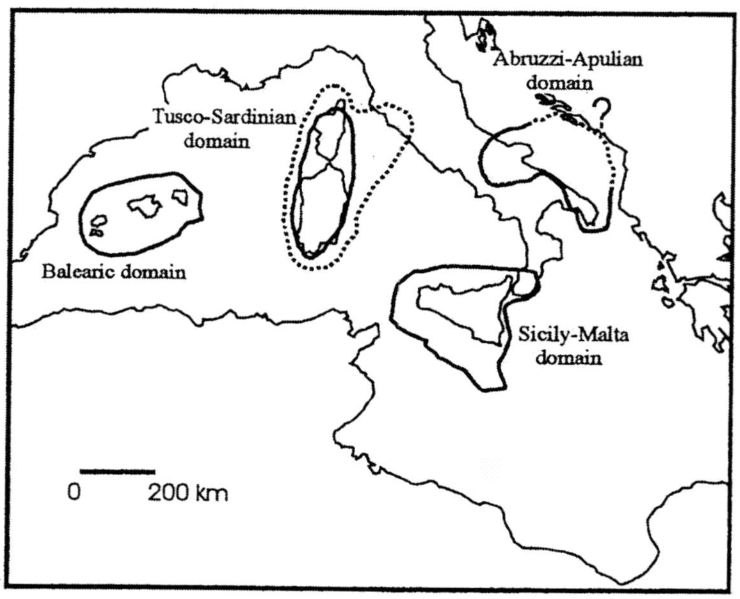

FIGURE 2. INSULAR DOMAINS OF WESTERN MEDITERRANEAN AREA DURING NEOGENE AND QUATERNARY. FOR THE TUSCO SARDINIAN DOMAIN THE DOTTED LINE INDICATES THE POSSIBLY EXTENSION OF INSULAR AREA DURING LATE MIOCENE (TORTONIAN).

number of clues suggest that, in some cases, extinctions were diachronic, i.e. either earlier or later on islands than on the mainland. Such diachronicity can related to the particular ecological requirements of the species in question (life history; Foufopoulos & Ives 1999) and to the peculiar characteristic of each insular system. Thus each extinction event must be carefully analysed in it's own right.

We feel that the geographic position of the island could be a key factor in explaining diachronic extinction.

Species adapted to 'warm – temperate' climatic landscapes, in which global extinction took place during a 'cold' climatic phase, would be expected to have survived longer in an island located in the south. In this case the island could operate temporarily as a refugium. For example, the extinction pattern in Amphisbaenians ('worm-like' burrowing reptiles) in Italy follows a clear N-S route, with the youngest remains from the Middle Pleistocene of Sicily (Delfino 1997, 2002). Similarly, in the south of Europe the 'temperate spe-

An example of earlier ('anticipated') extinction on islands is probably provided by the extinct bovid Bison priscus siciliae, a close relative of the steppe bison showing moderate size reduction. B. priscus siciliae is always very rare in the Sicilian fossil record. Bison priscus, widespread throughout northern Eurasia in the middle - late Pleistocene, is apparently the only true 'boreal' taxon to have colonised Sicily. It presumably entered during a phase of climatic deterioration when Sicily was temporarily connected to the mainland at a time of low sea level and was then 'trapped' as the climate ameliorated (Bonfiglio et al. 2000). A moderate degree of endemism could also be responsible for some delay in the timing of extinction.

The case of the local extinction of the water vole (genus Arvicola) in Sicily is a somewhat intermediate example falling within the second category, since it involves a taxon that survived for a geologically short time on the island (about 4000-5000 years, i.e. snapshot extinction). Its extinction was probably the result of a climatic deterioration. The remains of this vole have been found in archaeological deposits (Mesolithic and Neolithic). A rare fossil on the island, Arvicola apparently entered Sicily at a time close to the Pleistocene / Holocene boundary. Its extinction during the climatic deterioration following the Holocene Climatic Optimum is suggested by radiometric dating of its last occurrence (Upper Belice's basin, 5.700 - 6.600 yrs B.P., Tusa 1992; Burgio & Di Patti 1994). As this animal depends on the presence of water resources (rivers, pools, etc), the increasing aridity that of the second part of the Holocene in the Mediterranean regions (Watts et al, 1996) could have been a primary cause of extinction (Bonfiglio et al. 2000). In contrast Microtus (Terricola) ex gr. savii another arvicolid, which is a fossorial dweller nowadays well adapted to semi arid condition in the Southern Italian Peninsula and in Sicily, is still the most widespread rodent in the island.

SNAPSHOT EXTINCTIONS

This category involves taxa that survived for a very short time in the insular environment. Such cases are very difficult to detect in the fossil record. From a biological and, in palaeontological (morphological) stand point the expectation is that these taxa would be only slightly affected, if at all, by endemic evolutionary changes.

For some groups of organisms, (e.g. certain invertebrates, plants etc.) in which temporally random dispersal to islands is not strictly controlled by the effectiveness of sea barriers, extinction of a colonising species can occur soon after it's arrival (Zunino & Zullini 1995). One might expect some cases of 'newcomer' extinction in mammals also, although this is rarely observed because mammal, in which dispersal is much influenced by barriers.

The principal factor in these extinctions might be the problems experienced by a 'newcomer' in adapting to an integrated and co-evolved ecological system on an island which had existed for a long time. Competition with well-established resident species could also be an important factor. Such extinction might occur when the 'newcomers' arrived by random dispersal, e.g. on floating vegetation, and were subject to competition by well-established endemics (Masini et al. in press).

As stated above the palaeontological record offers very few cases of this category of extinction: examples could be the dormouse Eliomys and the shrew Crocidosorex from the endemic faunas of the Gargano Peninsula (De Giuli & Torre 1984; De Giuli et al. 1987, 1988, 1990; Fanfani 1999). These two taxa attract the attention of palaeontologists because they are the only 'normal' fossils (in size and morphology) within a highly endemic assemblage. Unlike other forms they apparently did not undergo morphological changes and their presence is limited to the oldest stratigraphic phase, thus suggesting an early extinction, while all the other components of the assemblage survived to undergo evolutionary changes.

In addition we feel that the extinction of large mammalian 'top' predators on islands (see below) could fall within this category.

EXTINCTION OF HIGHLY ENDEMIC TAXA THAT HAVE SURVIVED FOR A VERY LONG TIME ON THE ISLAND

This category includes taxa, which were well adapted to the ecological conditions on islands. Apparently many of them had successfully overcome the numerous climatic- environmental crises of the Middle and Late Pleistocene. In such cases a climatic explanation for extinction seems very unlikely, unless very severe climatic/environmental deterioration is invoked. However, the arrival on the island of predators or competitors could be an important factor in causing extinctions. In particular cases, a reduction of the area of the island could have reduced biodiversity and caused extinctions. For instance, De Giuli et al. (1987, 1988, 1990) suggested a reduction of the area of the Early Pliocene Gargano paleo island, caused by tectonic events, to explain the drastic reduction in faunal diversity seen in the last phase of the endemic population.

With regard to Holocene extinctions, several lines of evidence demonstrate that the increasing ecological impact of Neolithic and post Neolithic populations is the main factor in causing extinction of endemic taxa.

A detailed analysis of each case would be needed in order to understand each single event. In particular the cases of large and small mammals must be analysed separately since these two categories are related to very different ensembles of ecological niches and it is apparent from the pattern of extinction events in some islands (Sardinia, Corsica and Sicily) that the timings of extinctions of large and small mammals do not coincide (Vigne 1988; Bonfiglio et al. 2000, in press; Sondaar 2000). The arrival of predators can impact both groups, but owls, other predatory birds – raptors and snakes, which feed prevalently on small vertebrates, are common in island assemblages, reminding us that small mammals are continually subject to predatory pressure.

The predator – competitor explanation for the extinction of mammalian endemics may appear to partially contradict the discussion of category two. However, the possibility that some immigrant taxon might succeed in 'inserting' itself into the island ecological system cannot be ruled out a priori. One might expect that the resulting destabilisation of ecological equilibrium could result in the extinction of some endemics.

The "predator" hypothesis opens the question of the presence of large mammalian predators in islands and merits more detailed discussion since large mammalian predators, wild cat and most mustelids (with the exception of the aquatic or semi-aquatic otters) are usually considered absent in islands. Is this absence real, or is it a bias due to the incompleteness of the fossil record? The absence of mammalian large predators in the fossil record has been commonly considered as typical of the unbalanced faunal composition of islands. It has usually been interpreted as the result of filtering mechanisms that would not allow large predators to reach islands. Some clues – e.g. the presence of the Hyenid Chasmaportetes in the Early Pleistocene fauna of Sardinia (Ginesu & Cordy 1997) - might suggest a different line of reasoning. One could argue that, particularly large mammalian predators, may well have dispersed to the islands but then become extinct in a short time owing to their 'top' position in the trophic net. Their apparent absence in the fossil record would therefore not be real, but due to the incompleteness of the fossil record.

In any case, the arrival of a predator may well cause the extinction of endemics. Perhaps this is the explanation for the extinction of the Elephas falconeri in Sicily, the endemic dwarf elephant occurring in the island during the early Middle Pleistocene. This taxon probably become extinct during the upper part of the Middle Pleistocene (IOS 6?) at a time when an important faunal turnover occurred, due to the immigration of two 'top predators' (lion and spotted cave hyena), and several large mammalian taxa. It is worth stressing that this later Sicilian vertebrate faunal phase, known as the "Elephas mnaidriensis faunal complex", represents an interesting exception because of the well documented presence of large mammalian predator in an insular environment (e.g. spotted hyaena, Marra et al. 2001). This suggests that at that time the Sicilian archipelago became a continental-like island, able to support a well diversified trophic chain of large mammals.

In addition, according to some authors Palaeolithic hunter-gatherer populations in islands behaved as non - selective predators resulting in the extinction of some large mammalian taxa (see Sondaar & van der Geers 1999 and Sondaar 2000, on this topic).

Pre Neolithic human hunting or trapping is not a realistic explanation for the extinction of insular small mammals since these animals reproduce rapidly, and offer poor energy returns for the effort expended in procuring them, although there is evidence to suggest that some small mammals were a food resource exploited by man on islands (Vigne 1998). Case histories, such as the extinction of endemic glirids and Crocidura in Sicily, Prolagus, Nesiotites, Tyrrenicola and Ragamys in Sardinia, suggest that several – perhaps interacting - factors, might have been responsible.

The endemic dormice of Sicily were present in the island since at least the Early Pleistocene (Monte Pellegrino). They became extinct during the last glacial cycle (possibly IOS 4, or 2, Bonfiglio et al. 2000), probably coinciding with the reappearance of small-mammal assemblages, signalling a renewed full connection with the mainland. Even though we cannot give a definite explanation for their extinctions, it is likely that these dormice, which were unaffected by the arrival of several large herbivore and predator mammals on the island during IOS 6, succumbed to severe climatic deterioration during the last glacial cycle combined with the arrival of terrestrial predators (for instance the wild cat) via a land bridge connection.

The Holocene extinction of most small mammals in Sardinia and Corsica apparently has a different pattern and cause. These taxa, or their direct ancestors, were present on the island from the Pliocene and survived through the Pleistocene / Holocene boundary. Unlike Sicily, there is no evidence of a land bridge connection with the mainland during the Late Pleistocene, and these small mammals are found in the Holocene, within Meso-Neolithic anthropogenic deposits (Mezzabotta & Masini 1996, Sondaar & van der Geers 1999). The increasing impact of Neolithic and post Neolithic human colonisation of the island appears to be the primary cause of extinction (see also Vigne 1988, on this point). From the Neolithic colonisation onwards, humans directly or indirectly introduced several mainland taxa (including predators and potential competitors) and, even more importantly, drastically altered the vegetation cover. In such condition even small vertebrates had little chance of survival. There is only one exception: the case of Prolagus, a lagomorph, which, according to some sources became extinct very late within historical times in Sardinia, its last representatives being observed on a very small island (Tavolara) close to the eastern coast of Sardinia.

CONCLUSIONS

We wish to stress that this contribution is intended to highlight preliminary ideas and suggestions, rather than to present definite conclusions. Our ideas have developed from a comparison of models and consideration of historical cases, and include a large degree of speculation. Much work has still to be done to improve our knowledge of the fossil record in islands, and to refine models of extinctions and other aspects of case histories. Here we summarise some of the main issues.

Global climatic changes are probably the cause of extinction of moderately endemic taxa on islands. These extinctions may occur later or earlier with respect to the extinction of the 'continental' relatives, depending on the environmental - climatic affinities of the taxa and on the geographic location of the island.

More data are needed to document 'snapshot extinction' which, by their nature, are very difficult to detect in the fossil record. These extinctions, predicted by the Mac Arthur & Wilson model, are probably due to difficulties experienced by new comers in attempting to 'insert' within highly endemic and co-evolved communities. In some cases, however, e.g. Arvicola in Sicily, environmental changes (in this case towards a more arid climate), may also be the primary cause of extinction.

Climatic changes can hardly be invoked to explain the extinction of long lasting endemic taxa which were well adapted to the insular environment and to its changes thorough time. In some particular cases a cause of extinction could be the reduction of the area of the insular system caused by geodynamic evolution (ecologically this would result in an increased competition). More commonly, disruption due to immigration of predators and or competitors could be the general explanation of extinction events. The cases of large and small mammals must be considered separately.

Large mammals apparently are more prone to extinctions caused by predators. The presence of large predators in islands is however a matter of further investigation and could be very difficult to detect in the fossil record. Some clues indicate that large mammalian predators probably entered the islands more frequently than one might expect and that they had difficulties surviving in islands ("snapshot" extinctions). Owing to their 'high' position in the trophic net they are more prone to extinction than the primary consumer. Meanwhile, before becoming extinct themselves they may cause extinction in some herbivore. Paleolithics and Mesolitics hunter - gatherer, according to the literature, may be regarded as non-selective predators, the most dangerous category.

Endemic small mammals on the contrary are more robust and appear to be not much disturbed by the factors that result in the extinction of large mammalian species. Apparently their extinction is driven by simultaneously interacting factors (competition, predation, alteration of environment caused by anthropogenic activity and by climate changes...). In the case of Holocene the definitive cause of extinction is the increasing impact of anthropogenic activities including introduction of predators, competitors and domestic species, stock – rearing and alteration of the vegetation cover.

ACKNOWLEDGEMENTS.

Prof. Anthony Stuart (University College of London) critically read the manuscript and extensively improved the English language: we wish to thank him sincerely. We are grateful to the friend and colleague prof. Raffaele Guido Delle Valle (University of Bologna) for the interesting, critical, and frank discussion of the modellistic aspect of this work. We are deeply obliged to Prof Augusto Azzaroli (University of Florence) who kindly revised and criticised the draft. Last but not least, we wish to thank the friend and colleague Prof. Mario Zunino (University of Urbino), whose brilliant and sharp discussion lead us to reconsider ideas and interpretation on faunal diversity in islands under a fresh light.

ABOUT THE AUTHORS

Federico Masini (1), Laura Bonfiglio (2), Laura Abbazzi (3), Massimo Delfino (3), Flaviano Fanfani (3), Marco Ferretti (3), Tassos Kotsakis (4), Daria Petruso (1), Cinzia Marra (2), Danilo Torre (3)

(1) Dipartimento di Geologia e Geodesia, C.so Tukory 131, Palermo, I-90134, Italy,
E-MAIL: fmasini@unipa.it

(2) Dipartimento di Scienze della Terra, Via Sperone, 31, S. Agata di Messina, I-98166, Italy, e-mail: laura@labcart.unime.it

(3) Dipartimento di Scienze della Terra e Museo di Storia Naturale sezione Geo - Paleontologica, Via G. La Pira 4, Firenze, I-50121, Italy, e-mail: paleovert@cesit1.unifi.it

(4) Dipartimento di Scienze Geologiche, Largo San Leonardo Murialdo, Roma, I-00146,Italy, e-mail: kotsakis@uniroma3.it

REFERENCES

AZZAROLI, A.,
1996. Processes of island colonization by terrestrial organisms. Biogeographia, 18(1995), 33-50.

BONFIGLIO, L., MARRA, A.C., MASINI, F.,
2000. The contribution of Quaternary vertebrates to the paleoenvironmental and paleoclimatic reconstructions in Sicily. In: M. B. Hart, ed. Climates: Past and Present. London: Geological Society, Special Publications, 181, 169-182.

BONFIGLIO, L., MANGANO, G., MARRA, A.C., F. MASINI,
2001. A new late Pleistocene vertebrate faunal complex from Sicily (S. Teodoro Cave, North Eastern Sicily, Italy). Boll. Società Paleontologica Italiana, 40 (2), 149-158.

BONFIGLIO, L., MANGANO, G., MARRA, A.C., MASINI, F., PAVIA, M., D. PETRUSO, i
n press. Pleistocene Calabrian and Sicilian paleobioprovinces. Geobios.

BURGIO, E. AND C. DI PATTI,
1994. La fauna del fossato/trincea di contrada Stretto (Partanna). In: S. Tusa, ed. La preistoria del basso Belice e della Sicilia meridionale nel quadro della preistoria siciliana e mediterranea. Palermo, 201-109.

DE GIULI, C. AND D. TORRE,
1984. Species inter-relationships and evolution in the Pliocene endemic faunas of Apricena (Gargano Peninsula, Italy). Geobios, Mem. Spec. 8, 379-383.

DE GIULI, C., MASINI, F., D. TORRE,
1986. Effetto arcipelago, un esempio nelle faune fossili del Gargano. Boll. Società Paleontologica Italiana, 24 (2-3), 191-193.

DE GIULI, C., MASINI, F., G. VALLERI,
1987. Paleogeographic evolution of the Adriatic area since Oligocene to Pleistocene. Riv. It. Paleont. Strat., 93, 109-126.

DE GIULI, C., MASINI, F., TORRE, D., V. BODDI,
1988. Endemism and bio-chronological reconstructions: the Gargano case history. Boll. Soc. Paleontologica Italiana, 25(3), 267-276.

DE GIULI, C., MASINI, F., D. TORRE,
1990. Island endemism in the Eastern Mediterranean mammalian paleofaunas: radiation patterns in the Gargano Paleo-Archipelago. Atti Conv. Lincei, 85,247-162.

DELFINO, M.,
1997. Blanus from the Early Pleistocene of Southern Italy: another small tessera from a big mosaic. In: Böhme W., Bischoff W. and T. Ziegler, eds. Herpetologia Bonnensis, Bonn (SEH), 89-97.

DELFINO, M.,
2002. Erpetofauna italiane del Neogene e del Quaternario. Ph.D. dissertation, Modena and Reggio Emilia University, 382 pp.

FANFANI, F.
1999. Revisione degli Insettivori tardo Neogenici e Quaternari della penisola italiana. Ph.D. dissertation, of Modena and Reggio Emilia University, 282 pp.

Graham, W.R. and E.C. Grimm, 1990. Effect of Global Climate Change on the Patterns of Terrestrial Biological Communities. Trends in Ecology and Evolution, 5(9), 289-292.

FOUFOPOULOS J., A.R. IVES,
1999. Reptile extinctions on Land-Bridge Islands: Life History Attributes and Vulnerability to Extinction. The American Naturalist,

153(1),1-25.
JANES, J.,
1976. The origin of consciousness in the Breakdown of Bicameral Mind. [Il crollo della mente bicamerale e l'origine della coscienza]. Milano: Adelphi.
GINESU, S., AND J.M. CORDY,
1997. Il monte Tuttavista, Orosei-Galtelli. I riempimenti carsici e la scoperta delle nuove specie del Pleistocene. Sassari.
MAC ARTHUR, R.H., E.O. WILSON,
1963. An equilibrium theory of insular zoogeography. Evolution, 17, 373-387.
MARRA, A.C., VILLA, P., BEAUVAL, C., BONFIGLIO, L., P. GOLDBERG,
2001. Same predator, variable prey: taphonomy of two upper Pleistocene hyena dens in Sicily and SW France. XIVe Congres de l'Union Internationale des Sciences Prehistoriques et Protohistoriques. Liège, 2-8 septembre 2001, Abstracts Volume, pp. 69.
MASINI, F., BONFIGLIO, L., PETRUSO, D., MARRA, A.C., ABBAZZI, L., DELFINO, M., FANFANI, F., D. TORRE,
in press. The role of coastal areas in the Neogene - Quaternary mammal island populations of the central Mediterranean. Biogeographia.
MEZZABOTTA, C. AND F. MASINI,
1996. Influences of climatic environmental variations at the late Pleistocene - Holocene transition on the vole Microtus (Tyrrhenicola) henseli. Il Quaternario. 9(2), 721-724.
SONDAAR, P.Y., A.A.E VAN DER GEER.,
1999. Plio-Pleistocene terrestrial vertebrate faunal evolution on Mediterranean islands, compared to that of the Palearctic mainland. 1st International Workshop on Late Plio-Pleistocene Extinction and Evolution in the Palearctic: The Vatera site. 29th September - 2nd October 1999, Polychnitos, Lesbos.
SONDAAR, P.Y.,
2000. Early Human Exploration and Exploitation of Islands. Tropics, 10(1), 203-230.
TORRE, D., ABBAZZI, L., FICCARELLI, G., MASINI, F., MEZZABOTTA, C., L. ROOK,
1996. Changes in mammal assemblages during the Late-Glacial earliest Holocene. Il Quaternario, 9(2), 551-560.
TUSA, S.,
1992. La Sicilia nella Preistoria. Palermo: Sellerio.
Vartanyan, S.L., Garutt, V.E., A.V. Sher, 1993. Holocene dwarf mammoths from Wrangel Island in the Siberian Arctic. Nature, 362, 337-40.
VIGNE, J.D., 1988. Les mammmifères post-glaciaires de Corse. Etude Achéozoologique. Paris: Ed. Centre National de La Recerche Scientifique.
VIGNE, J.D., 1998. Preliminary results on the exploitation of animal resources in Corsica during the preneolithic. In: M.S. Balmuth and R.H. Tykot, eds. Sardinian and Aegean chronology. Studies in Sardinian archaeology, Sheffield, 5, 57-62.
WATTS, W.A., ALLEN, J.R.M., HUNTLEY, B., AND S.C. FRITZ,
1996. Vegetation history and climate of the last 15,000 years at Laghi di Monticchio, Southern Italy. Quaternary Science Review, 15, 113-132.
ZUNINO, M., A. ZULLINI,
1995.Biogeografia. La dimensione spaziale dell' evoluzione. Milano: Ambrosiana.

40

THE STRATIGRAPHIC SEQUENCE OF GONNESA (SW SARDINIA): PALAEOENVIRONMENTAL, PALAEONTOLOGICAL AND ARCHAEOLOGICAL EVIDENCE

RITA T. MELIS (1), **MARIA RITA PALOMBO** (2), **MARGHERITA MUSSI** (3)

(1) Dipartimento di Scienze della Terra, Università degli Studi di Cagliari, Via Trentino 51, ... Cagliari, Italia
(2) Dipartimento di Scienze della Terra, Università degli Studi di Roma "La Sapienza" CNR Centro Studi per il Quaternario e l'Evoluzione Ambientale, Piazzale Aldo Moro, 5, 00185 Roma, Italia
(3) Dipartimento di Scienze dell'Antichità, Università degli Studi di Roma "La Sapienza", Via Palestro, 63, 00185 Roma, Italia

SUMMARY

The paleopedological and palaeontological study of a stratigraphic section of aeolian and alluvial deposits outcropping in the Gonnesa Basin (south-western Sardinia) allow to better understand palaeoenvironmental changes in southern Sardinia during the Upper Pleistocene. At Funtana Morimenta, very red paleosols, that we refer to OIS5e, were affected by intense fluvial erosion (OIS4). The incision was subsequently filled by fluvial deposits dated at about 40,000 bp, after a C14-AMS date on charcoal fragments collected within the alluvium (43,000 ±1,400 bp, GX-25850) (OIS 3). In the coarsest alluvial levels, cervid remains were found, including four incomplete skulls with basal portions of antlers and a few fragments of long bones. The dimensions are comparable to those of the largest specimens from Grotta Dragonara (NW Sardinia) and Grotte Coscia (Corsica), dated to the Late Pleistocene.

The presence of large-sized Megaceroides cazioti (Depéret, 1897) in deposits correlated with OIS3, together with the presence of a smaller-sized population in the Last Glacial deposits of Grotta Cobeddu (Sardinia), indicates a tendency towards a more marked decrease in the size of the cervid populations of Sardinia during the advanced stages of the Last Glacial.

INTRODUCTION

In Sardinia, studies on the Quaternary have mostly focused on coastal areas and, in particular, on Tyrrhenian fossiliferous deposits (OIS5e) (Ulzega and Hearty 1986, Kindler et al. 1997). Little research has been done on continental deposits and on climate variations affecting the island during the Quaternary.

During our current research, aiming at the paleoenvironmental and paleoclimatic reconstruction of the Quaternary of southern Sardinia, we found the Gonnesa Basin (south-western Sardinia) particularly interesting, due to the presence of extensive aeolian deposits, attributable to various climatic stages of the Pleistocene, which include vertebrate remains. This basin has been well-known since the end of the 19th century, after the discovery, during railway construction, of the remains of an endemic elephant, perhaps a nearly-complete skeleton . Some tarsal, carpal and long bones, probably from Last Glacial aeolian deposits outcropping at Fontana Morimenta, were recorded and first reported by Acconci (1881). Later, Major (1883) described the specimens as the new species "Elephas lamarmorae", without illustrating it. After Comaschi Caria (1965), some other remains, that she referred to Elephas melitensis, as well as the Fontana Morimenta ones, all come from the close Cuccu de Cori locality. Unfortunately, the actual stratigraphic position of the remains is unknown, and it is not clear in which aeolian deposits outcropping in the Fontana Morimenta area the elephant was discovered. Following different lines of reasoning, the deposits have been dated either to the Upper Pleistocene (Comaschi Caria 1955), or to the Middle Pleistocene (Orrù and Ulzega 1986).

We will report preliminary results of stratigraphic, pedostratigraphic, paleontological and archaeological studies on a stratigraphic section outcropping on the front of a sand quarry near the spring at Funtana Morimenta. This allows to better understand past climatic conditions and the paleoenvironmental evolution of the area during the Late Pleistocene.

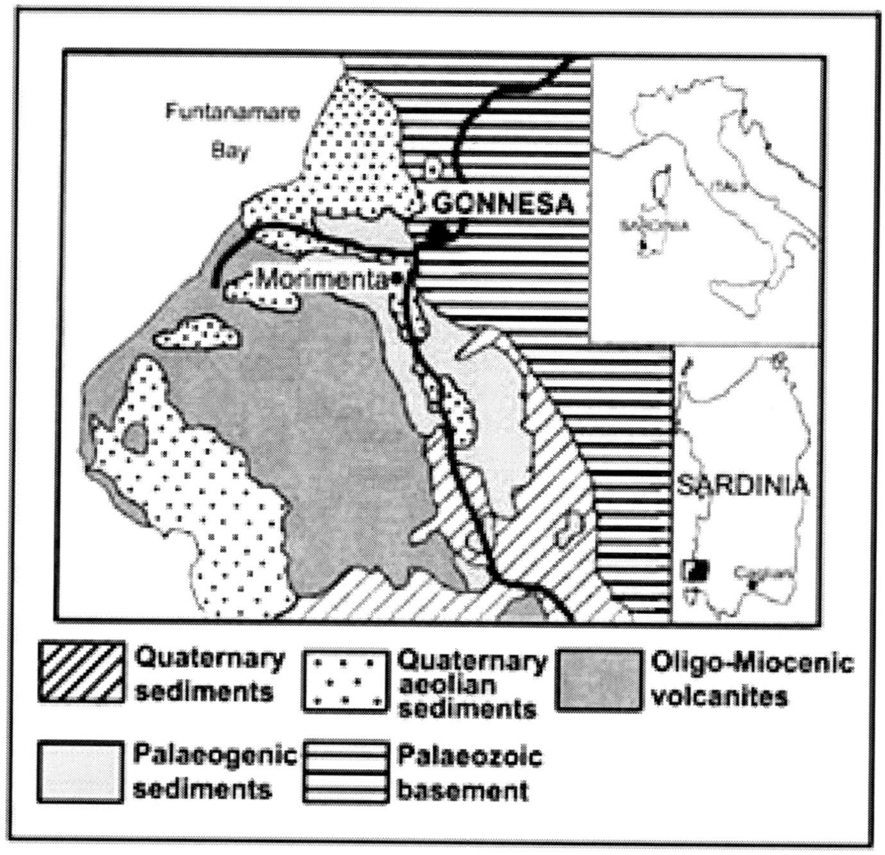

FIGURE 1 GEOLOGICAL SKETCH MAP OF THE GONNESA BASIN

..

GEOLOGICAL AND MORPHOLOGICAL SETTING

The Gonnesa Basin, whose genesis is linked to Tertiary distensive tectonics, is bordered on the South by the Giba graben and on the north and west by the Mare di Sardegna, while in the Eastern sector we find the Paleozoic rises of the Iglesiente and Northern Sulcis (Pillola et al. 1995) (FIGURE1).

During the Oligo-Miocene, the Gonnesa Basin started to fill up, with the depositon of calcaline vulcanites of prevalently tufaceous and ignimbritic facies. Such pyroclastic products partially cover the arenaceous-conglomeratic sediments of the Cixerri Formation of the Upper Eocene-Oligocene (Barca et al. 1973), outcropping prevalently in the southern sector. Volcanic activity and subsequent exogenous processes have originated in the western part of the basin characteristic cuestas rises , cut by deep valleys with broad, shelfed walls.

During the Quaternary, the area was affected by aeolian processes which produced areas of dunes extending several kilometres inland from the coast. Orrù and Ulzega (1986) have recognized several generations of dunes along the coast, in which we also collected cervid remains. The oldest aeolian deposits, called the "Funtana Morimenta Formation", have been attributed by the authors to the Middle Pleistocene on a stratigraphic basis, as the eolianites were cut by an erosional surface overlaid by a Tyrrhenian conglomerate (OIS5e), along the nearby Plage e Mesu beach. The most recent dunes, here and there weakly cemented and the intercalated dark paleosols, have been attributed to a generic Upper Pleistocene (Orrù and Ulzega, 1986). At the top of the aeolian sequence loose sands occur, which date back to the Holocene.

Thick outcroppings of eolianites, attributed to the post-Tyrrhenian (Orru and Ulzega, 1986), well-cemented and with cross bedding, are also found in the valley of Riu Crabiola, between Gonnesa and Bacu Abis (FIGURE 2), and about 5 km from the coast. In this area, aeolian deposits lie both on the Cixerri Formation (Paleocene) and on the Paleozoic base. Contact with the latter often consists of a sandy, very red paleosol.

The section described below is located about 200 metres south of the that locality, on the left of the Rio Crabiola, where a sand quarry has brought to light a long stratigraphic sequence that includes aeolian and alluvial deposits. In the latter, cervid remains and two possible lithic implements have been found in distinct stratigraphic levels.

FUNTANA MORIMENTA SECTION

The stratigraphic section (FIGURE 3), about 20 m thick, is characterised by a succession of aeolian deposits with intercalated paleosols. A deep incision, subsequently filled by alluvial deposits, affected the aeolian deposits at the base of the section.

Three units have been distinguished, from bottom to top : Unit A (5 m) (FIGURE4): at the bottom aeolian cross bedding deposits (A1), about 2 m thick, are present. Next is a paleosol (A2), 160 cm thick, with the following horizon sequence, starting from the top: yellow-red, sandy-clayey Btb (5YR5/8); reddish yellow Btgb, (5YR6/8); yellow BC (10YR7/6). Such horizons are what remains of a much thicker soil (horizon A is absent) which was subsequently cut by an erosional surface. It was later buried by a layer of cross bedding aeolian sand (A3), about 160 cm thick, on which a soil (A4) developed, characterised by an upper red (2,5YR/8) Btb horizon (60 cm) and a lower, dark brown (7,5YR5/8) BC horizon (100 cm).

The weakly cemented aeolian deposits are made up, for the most part, by rounded and sub-rounded quartz and feldspar granules. To a lesser degree, we find lithorelicts and foraminifer remains, as well as well-rounded fragments of sponges, at times greatly altered. Under the microscope, the granules appear to be bound by sparitic and microsparitic cement. Their mineralogical characteristics and particle size are similar to those found in outcropping eolianites, both along the coast (Funtana Morimenta Formation, sensu Orrù and Ulzega,1986) and near the spring at Morimenta.

The A2 and A4 soils are both deeply rubified. In thin section (described according to Bullock et al. 1985), microstuctures (bridget and pellicular grain), groundmass (undifferentiated) and c/f ratio (chitonic) are similar. The coarse fraction is prevalently composed of well-rounded quartz and plagioclase grains. The rock fragments, greatly altered, consist of volcanic and metamorphic rocks. Bt horizons present thin coverings of clay coating (ferriargillans) (FIGURE5) around dark reddish granules which developed in a moderately wet environment with alternating warm-dry and wet seasons. The coatings are weakly birefringent, with lamination features in the process of being dissolved. These characteristics can be interpreted as depending mainly on a long period of pedogenesis: in fact, dissolution of coating is indicative of old soil age. Furthermore, weak birefringence is described as a property of kaolinite (Cailler et al. 1982). Small iron nodules have been observed in all horizons.

In the Btb horizon of the A1 soil, there are areas in which calcite crystals occur within the clay coating. This is evidence of climatic evolution towards more arid conditions.

In addition, the presence in the Btgb horizon of extended ar-

FIGURE 2 FUNTANA MORIMENTA EOLIANITES.

eas with manganiferous mottles covering the clay coating, as well as carbonates, indicates hydromorphic conditions, probably linked to fluctuations in the water table.

Unit A was affected by a deep incision, subsequently filled by fine-gravelly alluvial deposits (Unit B).

Unit B: it is represented by about 6 m of gravelly, sandy and clayey alluvium deposited by a braided stream. The lower part of the deposit, about 4 m thick (B1), with large-scale cross bedding (FIGURE6), is characterised by alternating facies of abandoned channels made up of grey clayey-sandy deposits, with frequent charcoal remains, and by coarser bar deposits (FIGURE 6) . In the latter deposits, made up of gravelly and sandy sediments, radiocarbon dated to 43,000 ±1,400 bp (AMS: GX-25850), four incomplete cervid skull, described below, were found.

The upper part of Unit B is made up by 1.50 m (B2) of fine sediments in horizontal lamination, evidenced by sandy levels alternating with dark clayey silt with abundant organic compounds.

Unit C: it is made up of about 7-8 m of aeolian cross bedding deposits above Unit B. From bottom to top we recognized:

- 1 m (C1) of non-cemented aeolian deposits with weak cross bedding and Prolagus remains;
- 1.50 m (C2) of strongly-cemented, light coloured calcrete, made of well-rounded, prevalently quartz granules in a micritic, microsparitic matrix;
- 4 m of aeolian deposits in cross-bedding on which a yellowish-red (5YR5/8) soil about 2 m thick developed. Both field and micromorphological observations (i.e. rubification, undifferentiated b-fabric and red clay coating) show that soil formation occurred under a more or less mediterranean, seasonally contrasted climate (Duchaufour, 1983). Also, weak birefringence of clay may indicate the occurrence of non-swelling clay minerals (Embrechts and Sys 1988).

At the top of the section, we have about 1 m of well-cemented cross bedding eolianites on top of which the of the old railway is still visible.

As discussed above, the base of Unit A is composed of a

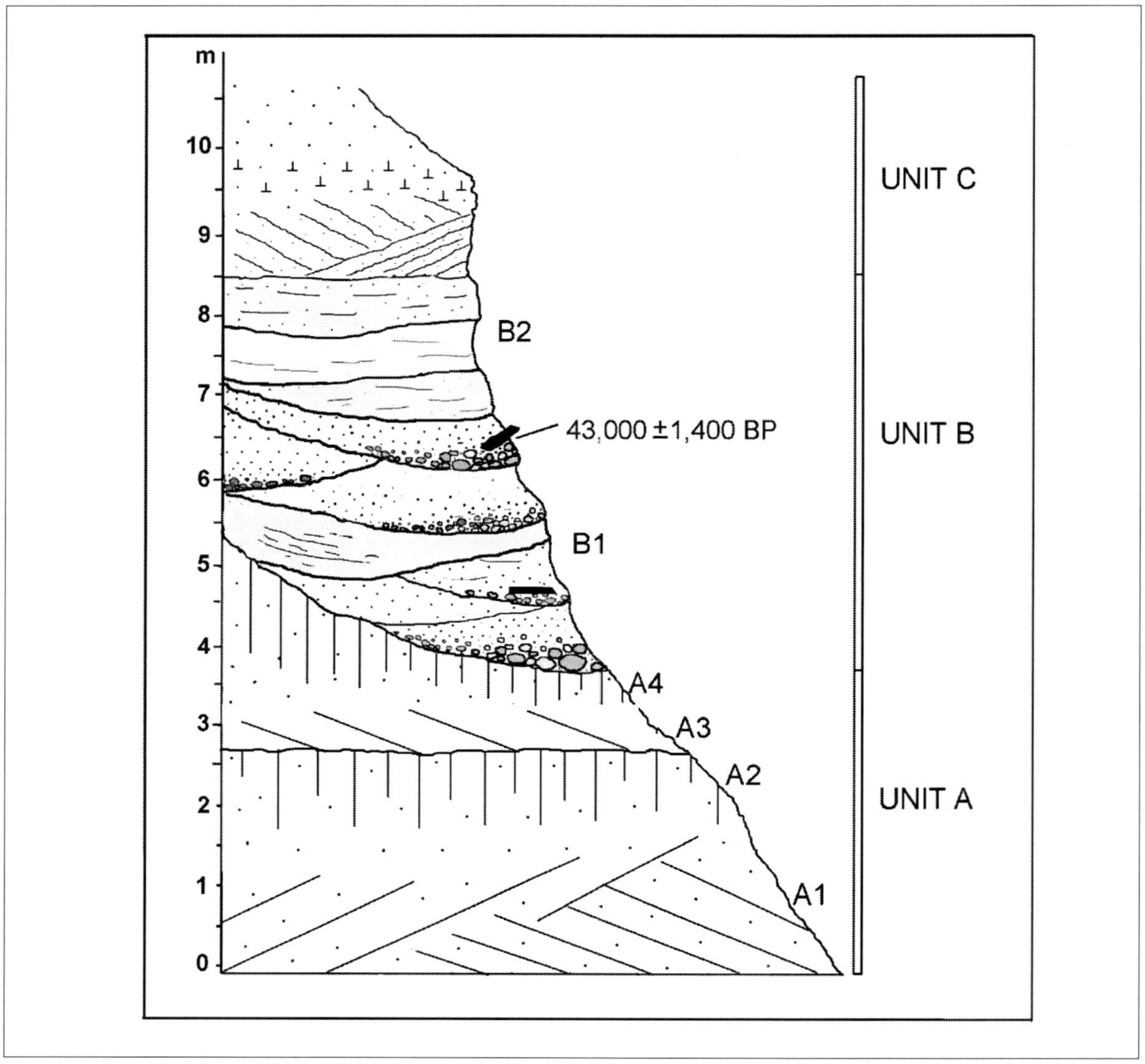

FIGURE 3 FUNTANA MORIMENTA SECTION.

thick layer of cross-bedded aeolian deposits, similar to the eolianites outcropping both along the coast and in the Riu Crabiola valley.

Between the aeolian deposits described above, and the Middle Pleistocene ones, outcropping along the coast and belonging to the "Funtana Morimenta Formation" (Orru and Ulzega 1986), no direct stratigraphic correlation exists. However, sedimentological and mineralogical studies show close affinity, and the hypothesis of their correlation to OIS6 is reasonable.

The formation of the soil occurring at the top of this unit, through rubification, accumulation of iron oxides and ferriargillans, required hot climatic conditions with seasonal contrast. This paleosol has not been dated directly, but we suggest that it formed throughout OIS5e. The aeolian sediments above it and the rubified soil which evolved on the latter deposit can be linked to arid and wet variations of middle and late OIS5.

Afterwards, intense erosional processes, probably starting during OIS4, partially eroded Unit A. Thus a deep incision was formed, subsequently filled by pebbly, fine alluvium during wet climatic conditions. The dating of charcoal remains found within the alluvium allows to relate them to OIS3. Accordingly, during that period, in the Funtana Morimenta area there was a braided-type watercourse, evidenced by characteristic sedimentary formations..

We refer respectively to OIS2 and to the Holocene the aeolian deposits, calcrete and paleosols on top of the alluvium. Obsidian lithic implements are also found, in a disturbed position, on the very top of the sequence.

MEGACEROIDES CAZIOTI (DEPÉRET, 1897)

In the Pleistocene of Sardinia, cervid remains, generally attributed to "Megaceroides" cazioti (Depéret, 1897), are relatively common, especially in eolianites and in cave deposits. They are for the most part attributed to the Late Pleistocene. The first reports date back to the second half of the 19th century: in 1856, Studiati illustrated and described the "Sardinian deer" for the first time, after teeth and vertebrae collected by Lamamora at Monreale (Cagliari). More remains were discovered later by Major in a karst cavity of the Nurra (Lovisato, 1884) and at Capo Figari (= Cervus elaphus corsicanus in Dehaut, 1911). In 1935 Tobien attributed remains of cervids found in the outskirts of Alghero to

FIGURE 4 EOLIANITES AND PALEOSOLS OF UNIT A.

FIGURE 5 THIN SECTION IMAGES 2,5X. FERRIARGILLANS IN BT HORIZON OF A2 PALEOSOLS UNIT A.

"Cervus" cazioti, a species established by Depéret in 1897 for the remains found in Corsica, in a cave near Nonza. In Sardinia, the two largest samples are those of the post-Tyrrhenian levels of Grotta Dragonara (Alghero) (Caloi and Malatesta, 1974), and of the Late Glacial levels of Grotta Corbeddu (Hofmejer,1997, and reference there in). Other reports refer for the most part to sporadic remains. Among the best-known are those of Capo Figari (Dehaut, 1911), Porto Vesme (Comaschi Caria, 1955), Alghero (Comaschi Caria, 1956), Grotta Inferno (Azzaroli, 1962), Maritza (Cordy and Ozer, 1973), San Giovanni in Sinis (Caloi et alii., 1980). Recently, more remains have been located in a Last Glacial breccia of Grotta Sommersa (Punta Giglio, Alghero) (Antonioli et al., 1998), and in Grotta Nurighe (Middle-Upper Pleistocene, Ginesu et al., 1998)). Specimens in partial anatomical connection were discovered in the eolianites outcropping at Porto Paglia (G. Pillola and M.R. Palombo, unpublished data) and in the Holocene deposits of Grotta Juntu (Siniscola). Cervids that have not yet been studied analytically (Megaceroides sp.), are also documented in the Rhagamys minor fauna of early Middle Pleistocene age from the karstic fissures of Monte Tuttavista (Ginesu and Cordy, 1997). So far, they are the oldest reported cervids on the island, since cervids are absent from the Rhagamys minor breccia at Capo Figari 1, whose age has been hypothesised at about 1,8 ± 20% Ma (ESR on a Nesogoral molar, Made, 1999),.

In Corsica, isolated remains have been reported at Maccinaggio (Sigogneau, 1960), Capo Sagro and Grotta al Margine (Azzaroli, 1962). Abundant remains were found in the post-Tyrrhenian deposits of Grotte Coscia (about 60,000-80,000 years, after U/Th dating) (Bonifay et al., 1998; Pereira and Bonifay, 1998).

Four incomplete skulls and a few fragments of post-cranial bones were also discovered within coarse, cross-bedded alluvium deposits (Unit B) at Funtana Morimenta. The cranial fragments belong to adult male individuals, of whom are variously preserved part of frontal bones, with basal segments of the antlers, occipital and parietal bones. Among the remains coming from basal alluvium levels, specimen n°2 is distorted by dorsoventral compression, more accentuated on the left side, with consequent dislocation of the orbital and occipital region; specimens n° 3 and n°4 are also deformed, even if less strongly. The three incomplete skulls above mentioned are covered by a millimetric encrustation of calcium carbonate, denoting that the remains lay in sediments affected by ground water. Skull n° 1 (FIGURE 7), the most complete and well preserved one, without any such encrustation, was found in a coarse level in the uppermost part of Unit B1. It is characterised by a broad forehead, longitudinally concave, with a protruding crista frontalis externa, an interpeduncular region slightly swollen and a frontal crest that rims the base of the peduncles. In the fragment of well preserved left antler, between the burr and the first tine, the beam has a subcircular section (DT=34.5; DAP=approx.32); the section becomes ovoid, with a reduced transversal diameter, at the base of the first tine (which is inserted at 67 mm from the burr), and tends to flatten out in the preserved portion above it (DT=approx. 33.5 ; DAP=22.7). Above the insertion of the tine, the beam bends down and slightly forward, forming an angle of 140°. This value, slightly less than the average one calculated for the Dragonara sample (post-Tyrrhenian, Caloi and Malatesta, 1974), is comparable to that of specimens from Maritza (Last Glacial, Cordy and Ozer, 1973), Il Cantaro (Last Glacial, Comaschi Caria, 1956) and San Giovanni in Sinis (?Last Interglacial, Caloi et al, 1980); it is inferior to the minimum values found in the large Corsican sample from Grotte Coscia (Pereira and Bonifay, 1998), whose specimens have the two initial segments of the beam reciprocally less inclined. Ongoing comparative studies of Sardinian and Corsican material will clarify the meaning

FIGURE 6 LARGE-SCALE CROSS BEDDING IN UNIT B1.

and importance of this difference. The dimensions of the antlers of specimen n° 1 from Gonnesa are comparable to those of the larger-sized specimens from Dragonara and Grotte Coscia, while the antler fragment from San Giovanni in Sinis is only slightly larger.

The relatively large size of the Gonnesa cervids is confirmed by values for the maximum transversal (49 mm), articular (43 mm) and antero-posterior (41.4 mm) diameters of the distal epiphysis of the humerus, found in the same level as skull n°1. These values are respectively close to, or above, the maximum values both of Grotta Dragonara (Caloi and Malatesta, 1974) and Grotte Coscia (Pereira and Bonifay, 1998) samples. The dimensions are also greater than those reported for the Tardiglacial specimens of Grotta Corbeddu (Hofmejier, 1997).

The endemic cervid of Sardinia and Corsica is well-known, but in spite of the many studies carried out, authors still do not agree on their systematic and phylogenic relationship. The hypothesis of a derivation from M. verticornis group representatives has been widely accepted; however, according to Sondaar (personal communication to M.R. Palombo), Sardinian cervids would have been phyletically related to the Dama genus, while Van der Made (1999) believes that they may have originated from Eucladoceros. The morphological characters that can be singled out in the skulls from Gonnesa seem to indicate a basic structural with the megacerines belonging to the M. verticornis group, especially in relation to the forehead conformation, the position and direction of the peduncles and the proportion of the neurocranium. The morphology of the Gonnesa specimens thus confirms the data already collected on the basis of the Dragonara, Maritza and Porto Vesme specimens.

CONCLUSIONS

The environmental changes documented in the Funtana Morimenta section can be reconstructed and dated by paleopedological, paleontological, and radiometric evidence.

The stratigraphic record is the result of aeolian, fluvial and pedogenic processes which took place during the Middle-Upper Pleistocene. The eolian deposit at the base of the section is most probably related to OIS6, on the basis of the stratigraphic context. The development of extremely rubified soils in aeolian sediments indicates that that deposit underwent intense pedogenetic processes in hot climatic conditions with an arid season, probably attributable to OIS5e. Alluvial deposits dated to 43,000 ± 1,400 bp are the consequence of a sharp increase in precipitation during OIS3. Analysis of cervid remains in these alluviums leads to hypothesising, in Sardinia, a substantial uniformity in Megaceroides cazioti size during most of the Late Pleistocene, while a consistent decrease in size seems to have characterised the latest Glacial cervids, as highlighted by the Grotta Corbeddu specimens. Good overlapping of dimensional and morphological variability fields has been evidenced in the Late Pleistocene Corsican and Sardinian samples, confirming that these populations belong to the same species, even if a subspecific (or ecophenotipical) separation cannot be ruled out.

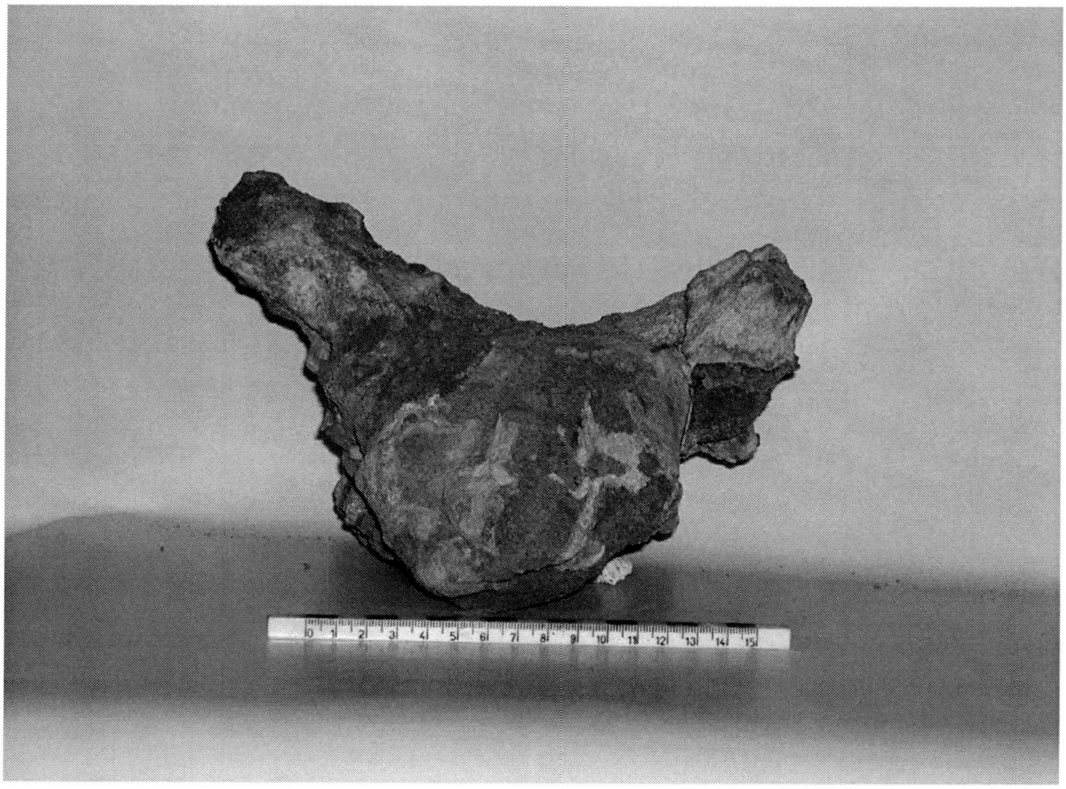

FIGURE 7 "MEGACEROIDES" CAZIOTI (DEPÉRET, 1897): FRONTAL VIEW OF SPECIMEN N.1.

REFERENCES

ACCONCI, L.,
1881. Sopra alcune ossa fossili di Elefante rinvenute nel Quaternario della zona di Movimenta in Sardegna. Proc.Verb. Atti Soc. Tosc. Sc. Nat., 2, 266-267.
ANTONIOLI, F., O. GIROTTI, P. ORRU AND M. VOLTAGGIO,
1998. Il deposito fossile a cervi nani nella
Grotta Sommersa di Punta Giglio (Alghero): considerazioni paleoambientali dallo stadio 5 al presente,78- 81
AZZAROLI, A.,
1962. Il nanismo dei cervi insulari. Paleontogr. Ital., 56, 1-32.
BARCA, S., C. MAXIA AND V. PALMERINI,.
1973. Sintesi sulle attuali conoscenze relative alla Formazione del Cixerri (Sardegna sud-occidentale). Boll. Serv. Geol. It., 94, 307-338.
BONIFAY, E., Y. BASSIAKOS, M.F. BONIFAY, A. LOUCHART, C. MOURER-CHAUVIRÈ, E. PEREIRA, Y. QUINIF AND M. SALOTTI,
1998. La grotte de la Coscia (Rogliano - Maccinaggio). Etude preliminaire d'un nouveau site du Pléistocène supérieur de Corse. Paleo, 10, 17-41.
BULLOCK, P. ET AL.,
1985. Handbook for Soil Thin Section Description. Albrighton: Waine Research Publication.
CAILLER, S., S. HÉNIN AND M. RAUTUREAU,
1982. Minéralogie des Argiles, II. Classification and Nomenclature. Paris: Masson.
CALOI, L., T. KOTSAKIS, M.R. PALOMBO AND C. PETRONIO,
1980. Il giacimento a vertebrati del Pleistocene superiore di San Giovanni in Sinis (Sardegna occidentale). Rend. Acc: Naz. Lincei, 8, 69, 185-197.
CALOI, L. AND A. MALATESTA,
1974. Il cervo pleistocenico di Sardegna. Mem. Ist. It. Pal. Um., 2, 163-247.
COMASCHI CARIA, I.,
1956. Specie nuova nel Quaternario di Alghero (Sardegna). Rendiconti Seminario della Facoltà di Scienze, Univ. Cagliari, 25(3-4),1-6.
COMASCHI CARIA, L.,
1955. Resti di cervidi nel Quaternario di Porto Vesme. Riv. It. Paleont. Strat., 61(1),17-27.
CORDY, J.M. AND A. OZER,
1973. Découverte d'un crane de cervidé mégacerin (Nesoleipoceros catzioti) dans le Quaternaire de la Sardaigne septentrionale. Ann. Soc. Géol. Fr., 95,425-449.
DEHAUT, E.G.,
1911. Matériaux pour servir à l'histoire zoologique et paléontologique des îles de Corse et de Sardaigne. III, Animaux fossiles de Capo Figari, 53-59.
DÉPÉRET, C.,
1897. Etude de quelques gisement nouveaux de Vertébrés pléistocènes de l'île de Corse. Ann. Soc. Linnéenne Lyon, 44,111-128.
DUCHAUFOUR, PH.,
1983. Pédologie: 1. Pédogenèse et Classification. Paris: Masson.
EMBRECHTS,J. AND C. SYS,
1988. Genesis of subsurface horizons of a soil catena in a humid

tropical climate (Yaounde, Cameroon). Catena, 15, 53-63.
GINESU, S. AND J.M. CORDY,
1997. Il Monte Tuttavista (Orosei-Galtellì). Sassari: Edizioni Poddighe.
GINESU, S., S. SIAS, J-M. CORDY, L. TREBINI, A. MELE, A. SERRA, P. VIRGILIO AND A. ZARA,
1998. Grotta di Nuraghe (Sassari), esplorando il passato. Speleologia, 38, 52-60.
HOFMEJER, G. K.,
1997. Late Pleistocene Deer Fossils from Corbeddu Cave. BAR Intern. Series, 432-663.
KINDLER, P., E. DUVAUD, AND A. STRASSER,
1997. Tyrrhenian coastal deposits from Sardinia (Italy): a petrographic record of high sea levels and shifting climate belts during the last interglacial (isotopicsub stage 5e). Palaeogeography, Palaeoclimatology, Palaeoecology, 133, 1-25.
LOVISATO, D.,
1884. Nota sopra il Permiano e il Triassico della Nurra in Sardegna. Boll. Com. Geol. It., 15, 87-122.
MADE VAN DER, J.,
1999. Biogeography and stratigraphy of the Mio-Pleistocene mammals of Sardinia and the description of some fossils. In J.W.F. Reumer and J. de Vos, ed. "Elephants have a Snorkel", Deinsea, 337-360.
MELIS, R. AND M.R. PALOMBO,
2000. Megaceroides cazioti (Dépéret) nel Pleistocene superiore di Gonnesa. In: "Tutela e Valorizzazione dei Siti paleontologici a Vertebrati", 1° Workshop nazionale di Paleontologia dei Vertebrati, Reggio Calabria, Messina -Acque Dolci, S. Vito Lo Capo, 20-21.
ORRÙ, P. AND A. ULZEGA,
1986. Geomorfologia Costiera e sottomarina della Baia di Funtanamare (Sardegna sud-occidentale). Geogr. Fis. Dinam. Quat., 9, 56-67.
PEREIRA, E. AND M.F. BONIFAY,
1998. Etude préliminaire des restes de Cervidés de la grotte de la Coscia (Roglian o - Haute-Corse). Paleo, 10, 43-60.
PILLOLA, G.L., F. LEONE AND A.LOI,
1995. General consideration on the Lower Cambrian of SW Sardinia. In: A. Cherchi ed., 6th Paleobenthos International Symposium: Vol. LXV, Rendiconti del Seminario della Facoltà di Scienze Univ. di Cagliari, 27-60.
SIGOGNEAU, D.,
1960. Etude d'un CervidéPléistocène de Corse. Ann. Paléont., 46, 49-77.
ULZEGA,A. AND P.J. HEARTY,
1986. Geomorphology, stratigraphy and geochhronology of Late Quaternary marine deposits in Sardinia. Z. Geomorphol. N.F., Suppl. Bd., 62, 119-129.

41

S. MARIA IS ACQUAS, A NEW PRE-NEOLITHIC SITE: SOUTH-WESTERN SARDINIA

RITA MELIS (1), **MARGHERITA MUSSI** (2)

(1) Dipartimento di Scienze della Terra, Università degli Studi di Cagliari, Via Trentino 51, 09100 Cagliari, Italia
(2) Dipartimento di Scienze dell'Antichità, Università degli Studi di Roma "La Sapienza", Via Palestro 63, 00185 Roma, Italia

SUMMARY

After recent archaeological research, island peopling was a typically "modern" behaviour. This allows to put into a proper perspective the recent discovery, in the Campidano of south-western Sardinia, of lithic industry in controlled stratigraphical position. Large blades, and bladelets, as well as flakes, were produced, and the typology fits well into the Upper Palaeolithic. The lithic implements were found blanketed by thick deposits of aeolian sands, and coated by illuvial clay developed during pedological processes. Both locally available chert and flint pebbles, and raw material outcropping at some distance, were used for knapping. Micromorphological and pedological analysis suggest that the implements were made during a phase of climatic instability, followed by increasing aridity, that we correlate with final OIS 3, and/or early OIS 2.

INTRODUCTION

While there is ample evidence of prehistoric Homo sapiens settling into remote islands (Gamble 1993; Keegan and Diamond 1987), there was so far no sound record of any Upper Palaeolithic industry from Sardinia. Upper Pleistocene evidence was limited to hypothetical limestone implements from Grotta Corbeddu (Klein Hofmeijer et al. 1987-1988); to an isolated human phalanx from the same site (Sondaar et al. 1995); and to a female figurine from Macomer, found well before the age of scientific archaeology (Mussi 2001).

We will present below the recent discovery, in the Campidano of south-western Sardinia, of lithic industry in controlled stratigraphical position. The Campidano is a graben which originated during the Pliocene (FIGURE 1). The vast depression currently develops over some 90 km, between the gulf of Oristano and the gulf of Cagliari, but was larger during the Last Glacial Maximum, when the whole continental shelf emerged. The axis of Campidano corresponds to the direction of the prevailing NW and SE winds (Maestrale and Scirocco), and allowed for the deposition, during the late Pleistocene (Maxia and Pecorini 1968) of windborn sands extending over some 25 km to Sardara spa. S. Maria Is Acquas is located in its central part, where the aeolian blanket is well developed, but close to the hills which border the Campidano to the west. The chert, which outcrops nearby, and is also found redeposited as cobbles and pebbles (see below), apparently attracted human groups, which left abundant evidence of knapping activity.

We will outline below the main typological and technological characteristics of the lithic industry, and we will discuss the geomorphological and pedological features which, together with the stratigraphical position of the implements, allow us to date them to late OIS 3-early OIS 2.

GEOMORPHOLOGICAL PERSPECTIVE

Distensive tectonic movements, starting in the Middle/Upper Pleistocene, played an important role in the geomorphological evolution of the Campidano graben. In fact, the western slope, characterized by Paleozoic metamorphic and intrusive rocks, appears relatively simple, with a piedmont strip that gradually degrades from the rise to the centre of the graben. The eastern slope, however, occupied for the most part by Miocene (Tertiary) marine sediments, has a complex structure with steps, strongly influenced by both endogenous and exogenous factors. The area under study, located in the east-central slope of the graben, is characterized by Plio-Pleistocecene lava flows located along a paleovalley set among Miocene rises, now partially eroded, as evidenced by the calcareous mound of Puisteris to the north, and the marly one of Sedda su Cardu to the southwest (FIGURE 2).

In the latter locality, intercalated in the marl, chalcalkaline products of underwater Oligomiocene volcanic activity have been found. During such activity, late hydrothermal events created flint lenses inserted into Miocene marine sediments, presently outcropping along the slope of the Sedda su Cardu hill. In the Lower and Middle Pleistocene, the progressive subsidence of the Campidano graben determined the erosion of surrounding rises and the formation of extensive coalescing alluvial cones (Seuffert 1970). In particular, in the area being investigated, on the low slope of the basalt plateau of Perdiana, the Riu Serri has created an extensive alluvial cone, subsequently affected by intense pedogenetic processes. During the regressive stages of the Upper Pleistocene, the alluvial cone was partially covered by a thick layer of aeolian deposits.

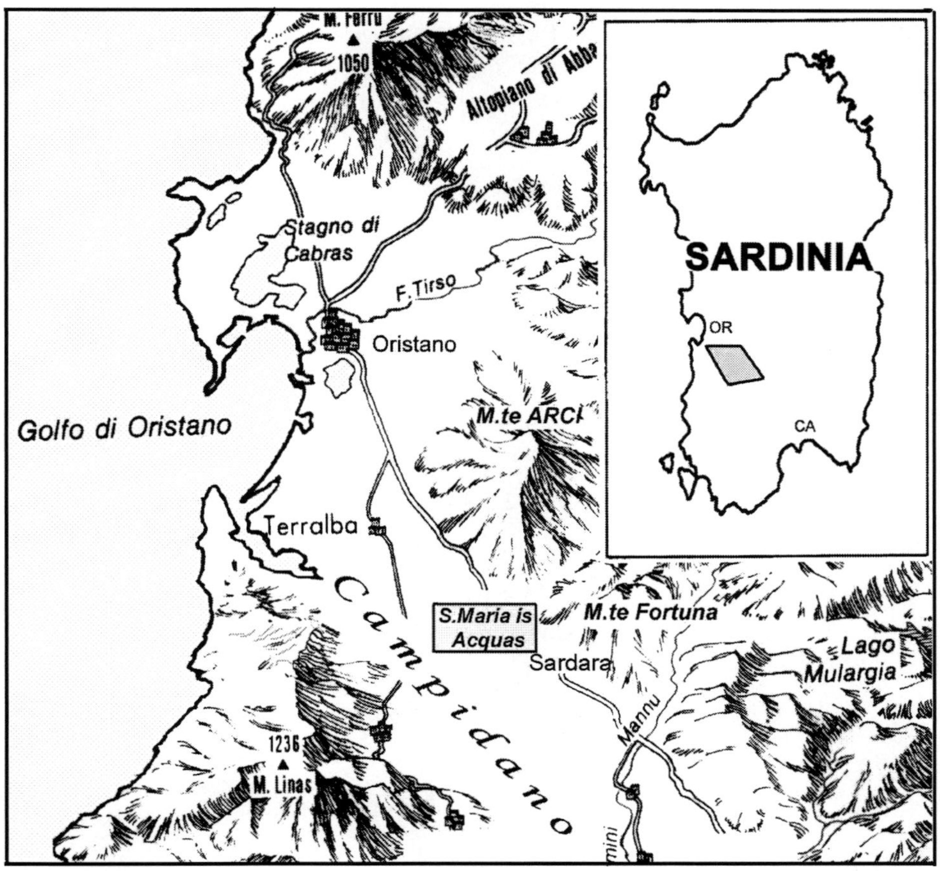

FIGURE 1 THE CAMPIDANO OF SW SARDINIA, WITH THE RESEARCH AREA

In recent times, the aeolian sediments have been affected by quarrying activity, and the underlying pebbly alluvium deposits have been exposed. Many chert and flint implements have been found both on the alluvium and in the sand. In the latter, we also observe countless obsidian implements in the uppermost 20-30 cm, close to the surface.

Detailed analyses have been carried out, as the quarry allowed to observe the relation of archaeological remains to deposits. We here report initial results of the pedostratigraphic study of two sections close to S. Maria Is Acquas, belonging to different morphological contexts. The first (Section A) is located at the top of the terraced old alluvium of Riu Serri; the second (Section B) is, on the contrary, in a small paleovalley, filled with aeolian sediments, on the same terrace (FIGURE 2).

STRATIGRAPHIC SECTION A

Section A is composed of two lithostratigraphic units: (FIGURE 3)

Lower Unit A1: old alluvium characterized by heterometric, well-rounded marl, flint, basalt and, rarely, obsidian pebbles. On this alluvium, a soil developed, of which now only the Bt illuvial horizon remains: very hard, decarbonated, with a clayey texture, light yellow-brown in colour (10YR6/4), with numerous red mottles (2,5YR5/8). Micromorphological analysis has indicated the presence of many areas with parallel striated b-fabric, typical of soils rich in 2:1 clay. Mottles and manganese nodules testify to conditions of temporary hydromorphosis. The genesis of this horizon is strictly connected with the presence of a sandy upper eluvial horizon (E), no longer present, impoverished in its clayey portion following leaching processes.

Upper Unit A2: aeolian sand with intercalated thin pebble levels, related to brief pluriometric events and intermittent running waters. The lower part of the aeolian deposit also contains flint and chert implements, testifying to a human presence in the area.

On the aeolian deposits a soil developed, characterized by a sandy to sandy loam texture, with a topsoil (Ap) of 30/35 cm of a yellow-brown colour (10YR5/6), changing to a Bt red clay horizon (7.5YR7/4), 30-40 cm thick.

Micromorphological study has shown that the sandy portion is characterized by well-rounded quartz and, secondarily, feldspar granules. The surface of sand granules in the Bt horizon is covered by a thin coating of illuvial clay, whose formation is compatible with mediterranean climatic conditions. Chert and flint implements enclosed in the aeolian deposit display the same coating.

In the horizon closer to the surface, we also find, for about 30 cm, a lot of obsidian implements.

The evolutionary level of the soil in the alluvium (Unit A1) indicates a long period of pedogenesis on a morphologically stable surface in mild, wet climatic conditions seasonally alternating with dry periods.

Environmental conditions determining some morphogenetic stability were followed by climate changes that ended in intense erosive processes, leading to the complete erosion of the eluvial horizon. Consequently, the most superficial part of the Bt horizon was subjected to processes of wind erosion, testified to by small streaks on the surface of pebbles, due to the impact of quartz granules, and identified microscopically. Among the pebbles on this level, the presence of lithic implements is noteworthy.

This leads one to conclude that dry environmental conditions became prevalent, characterized by active processes of deflation and aeolian sedimentation, that we date to the Late Glacial.

Aeolian activity was initially characterized by deflation processes in some areas, while in others sediments were deposited. The persistence of arid conditions subsequently favoured the prevalence of aeolian sedimentation processes, determining the progressive covering of Stratigraphic Unit A by a thick layer of sand.

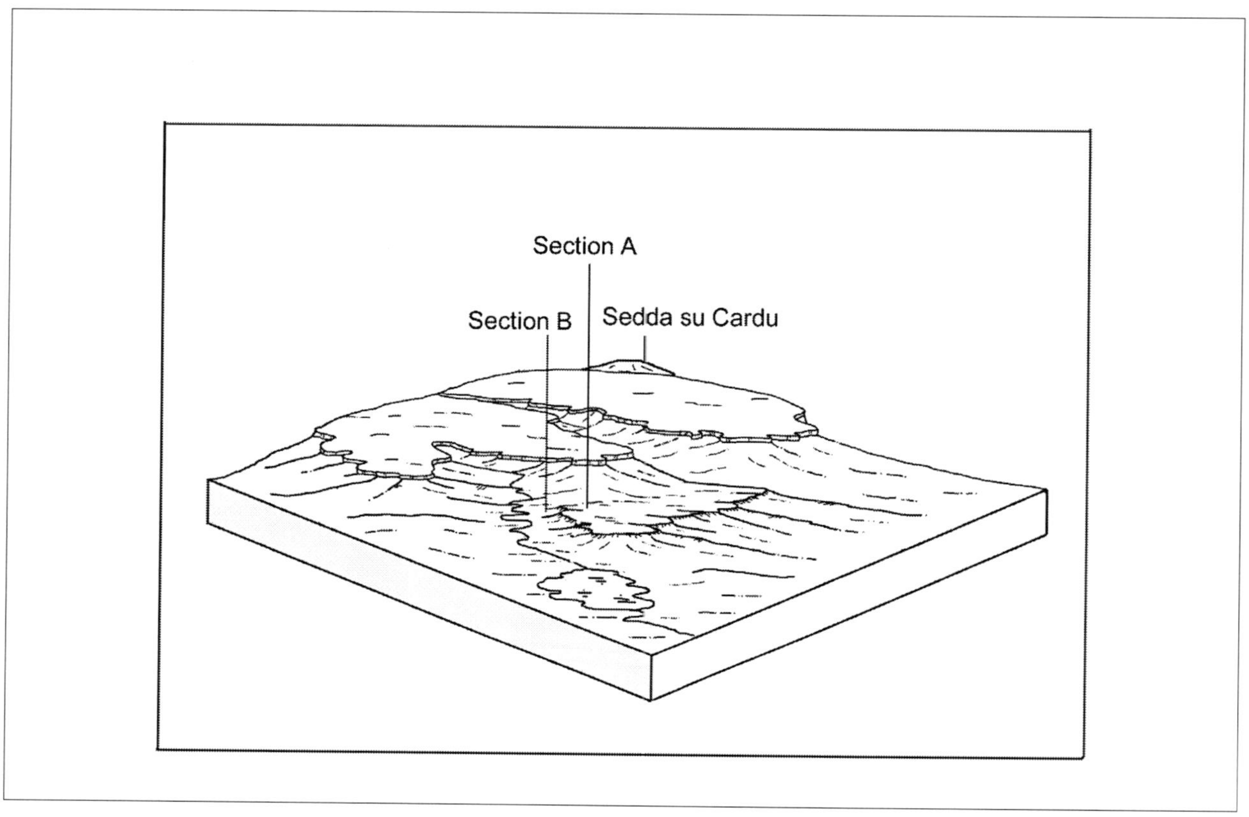

FIGURE 2 DIAGRAM BLOCK OF THE AREA NEXT TO SANTA MARIA IS ACQUAS

The aeolian deposition was followed by a period with prevailingly milder, wetter climatic conditions of Late Pleistocene and/or Early Holocene age, leading to intense pedogenetic processes.

STRATIGRAPHIC SECTION Bx

Section B is composed of a single lithostratigraphic unit, made up of a layer of aeolian sediments about 3 m thick (FIGURE 3).

The sedimentation of aeolian deposits was also favoured by the morphological context, characterized by a small valley oriented in the direction of north-west winds.

On these deposits developed a soil with a superficial Ap horizon about 40 cm thick, containing obsidian implements. Then we find an eluvial E horizon 80 cm thick, lying on a clayey Bt horizon of about 80 cm, which in turn covers a Btg horizon.

The E horizon is characterized by a sandy texture and a light grey colour (10YR7/2). An abrupt, slightly undulating boundary marks the transition to the underlying Bt horizon, composed of two sub-horizons, Bt1 e Bt2, with a sand loam texture, very dense and reddish yellow in colour (7,5YR6/8). The reddish yellow (7,5YR6/8) Btg horizon has the same characteristics as the Bt horizon, from which it differs in the presence of mottles and manganese nodules, indicating hydromorphic conditions linked to the formation of a temporary aquifer at the contact with the underlying old alluvium.

In the Bt horizon, the main pedogenetic characteristics are the thin, reddish-yellow, alluvial clay coating on granules. The coarse portion of the groundmass is well-classified and principally composed of well-rounded quartz and feldspar granules and fragments of lithorelics with much altered femic minerals.

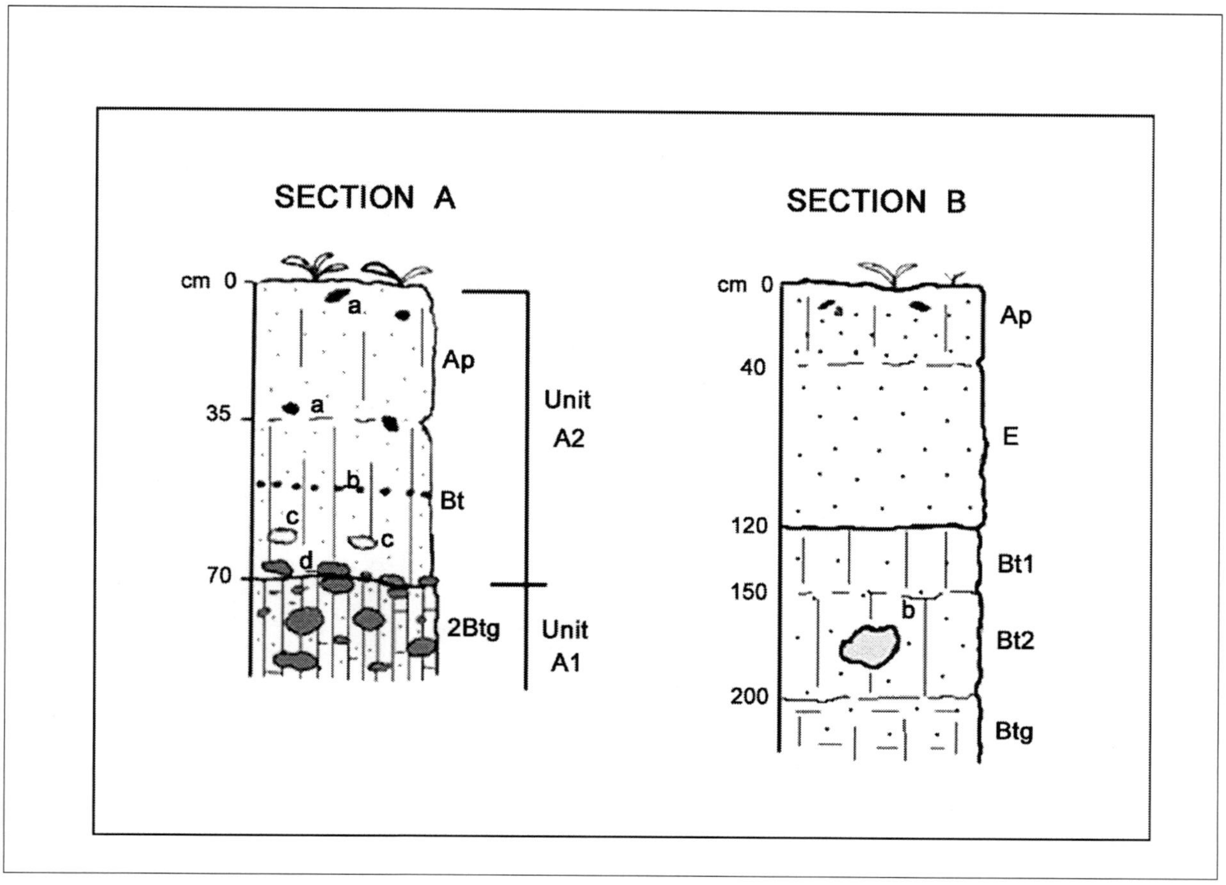

FIGURE 3 S. MARIA IS ACQUAS. PEDOSTRATIGRAPHY OF SECTIONS A AND B: (A) OBSIDIAN IMPLEMENTS; (B) PEBBLE LEVEL; (C) FLINT AND CHERT IMPLEMENTS; (D) DESERT PAVEMENT.

Chemical, and particularly micromorphological, analyses indicate that the soil was affected by decarbonation processes, illuviation and rubification. These processes denote a long pedogenesis during a period of warm climate with seasonal contrast, when the plant cover was well developed. The presence of ferriargillans around sand granules confirms warm conditions.

Encased in the Bt horizon, at a depth of about 150 cm, a flint blade core was found (FIGURE 4), covered with a thin film of reddish-yellow illuvial clay. This leads one to infer that the object was already encased in the sand before the beginning of pedogenetic and illuviation processes. Accordingly, the deposition of the lithic implement happened when the aeolian sand was being deposited.

Overall, there is evidence that, after a period of climatic aridity attributable to the Late Glacial, in which aeolian processes predominated, there was a period of pedogenesis in temperate, wet climatic conditions with seasonal contrast,

FIGURE 4 S. MARIA IS ACQUAS. BLADE CORE FROM SECTION B.
THE CORE, WHICH IS HEAVILY PATINATED, IS ON A LARGE FLAKE. TWO BLADES WERE PRODUCED BEFORE RESHARPENING WAS UNSUCCESSFULLY ATTEMPTED, AND THE CORE WAS DISCARDED.

which ended in the Holocene.

THE ARCHAEOLOGICAL EVIDENCE

As said above, obsidian implements of Neolithic age are frequently found in the fields, all over the study area. At Terralba, 10 km north of S. Maria Is Acquas, a detailed study led to their attribution to the early Neolithic (Lugliè 1999), which, after the available C14 dates, developed after 6700 bp (uncalibrated) (Depalmas et al. 1998). We will focus on the flint and chert implements, which come from well within the aeolian sand cover, with illuvial clay, produced by pedogenetic processes described above, often found adhering to their surface. Some were discovered in situ along artificial cuts; more were collected in field where the aeolian sand had been exploited and quarried away, so that the older alluvium outcropped. Our aim was to provide a first characterisation of the industry, and to frame it firmly into the local Quaternary depositional sequence. Accordingly, we have only collected a limited number of implements (n=70), either when found in situ, or when they provided important typological and/or technological information. Excavations are planned for the next future, in collaboration with the Soprintendenza archeologica per le province di Cagliari e Oristano. This explicitly unsystematic collection only allows for a preliminary description of the archaeological implements.

TECHNO-TYPOLOGICAL DESCRIPTION

Chert and flint were by far the preferred raw materials, even if there is limited evidence of the use of calcedony. The complex taphonomy of the deposit is reflected in implement preservation, as they range from fresh and slightly patinated, to rolled and heavily patinated.

The collected cores are 16, on pebbles or large flakes. Pebbles, up to 10cm in length, were knapped both by unipolar and by bipolar percussion. The large flakes were first produced by bipolar percussion, and then reduced by direct percussion. This is exemplified by the largest core of our sample (174x140x70mm), which was discarded after the production of a single blade of 110x30mm (FIGURE 5), because of fracture planes. Overall, only a limited number of flakes/blades were produced by each core, either single- or bi-platformed. The only exhausted core is a chalcedony one, used to produce bladelets. The reduction sequence is often interrupted because of hidden fracture planes, even if resharpening is attempted.

Our sample also includes 20 flakes and 10 blades, as well as 5 resharpening flakes, some of which are corticated, while some belong to a full débitage sequence. There are few complete blades, which are up to 96mm in length, even if scars on cores, as said, refer to longer blades; width ranges between 9 and 37mm; thickness between 4 and16mm. Bladelets were not collected, but bladelet production is suggested by the chalcedony core mentioned above, as well as by the dorsal scars of some resharpening flakes.

The following retouched tools are included in the sample: 2 endscrapers on a flake; 1 truncated blade; 1 perforator; 2 retouched blades; 2 denticulate flakes; 1 notched flake; 4 retouched flakes.

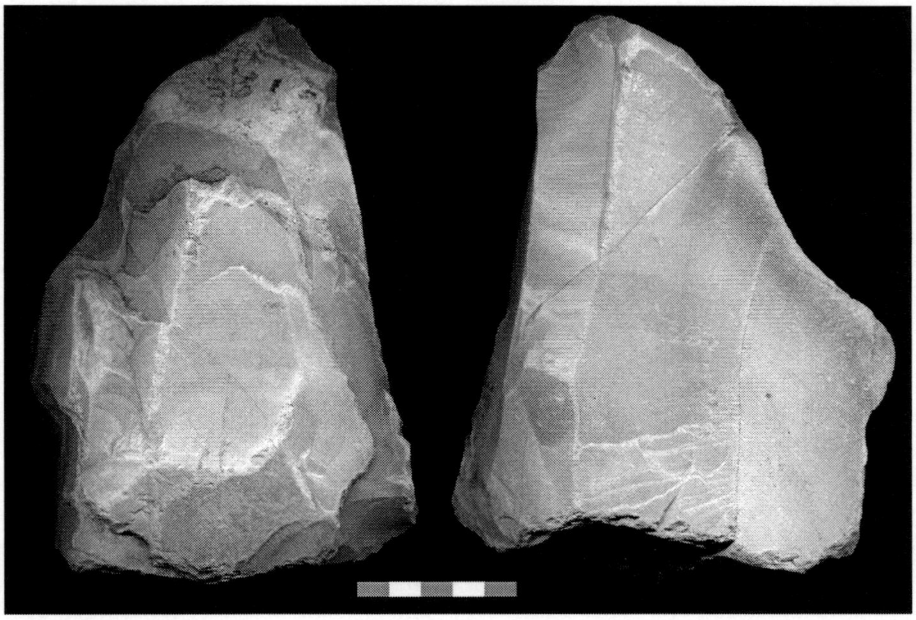

FIGURE 5 S. MARIA IS ACQUAS. BLADE CORE ON A LARGE FLAKE, ROLLED AND PATINATED. ONLY ONE BLADE WAS PRODUCED, PROBABLY BECAUSE OF A HIDDEN FRACTURE PLANE.

CONCLUSIONS

Implement preservation, stratigraphy and area of collection, all point to more than just a "site" as commonly understood. In the area of S. Maria Is Acquas, instead, remains of a palaeo-landscape are preserved below aeolian sands.

Lithic implements were knapped in part using chert and flint pebbles and cobbles collected locally in the alluvial deposits; in part preparing cores on large flakes extracted at outcrops at a distance of more than 4 km. More implements have been actually collected at Sedda su Cardu, a nearby small hill, where the outcropping chert is very similar to the chert discovered at the findspots of the graben. The implements of Sedda su Cardu are scarcely rolled, but are found on the surface, as the aeolian sands did not deposited on the hills and mountains bordering the Campidano.

The available information on the local stratigraphical sequences, as well as the technological and typological analysis, suggest that the chert industry of the area around S. Maria is Acquas is homogeneous and deposited during a relatively short period. The production of blades, of 100x30mm and more, was attempted again and again, even if the available pebbles were rather small. Alternatively, large flakes were produced on the hills where the raw material outcrops, and transported to lower-lying areas. Such flakes were expeditely reduced in blade cores, but the hidden fracture planes only allowed for a limited production. We suggest that the blade technology was originally developed in a different area, where better-quality raw materials were available. The lithic industry is fully Upper Palaeolithic, both in technology (blade and bladelet production), and in typology (endscrapers, truncated blades, retouched blades, borers etc...) (FIGURE 6).

Pedostratigraphical and micromorphological analysis consistently point to a period of morphodinamic stability, during which a soil developed on the old alluvium, which is related to a warm and humid climate. Increasing aridity followed, with rare rains, often stormy, and prevailingly aeolian processes, which eventually eroded the upper horizon of the soil. As elsewhere in Sardinia (Melis 1996), this led both to deflation and desert pavement formation, and to the deposition of aeolian sands. Later, pedogenesis developed again, during one or more warm and humid phases, affecting the sand cover. Then, Neolithic industries were produced, using obsidian, as well documented at Terralba. Sometimes, the Holocene settlements were dug into the pedogenised horizon, as at Terralba itself (Bertorino et alii 1999; Bertorino et alii 2000).

The Upper Palaeolithic chert implements were produced and discarded in the central Campidano during the climatic crisis, when aeolian sands had starter to deposit. This happened before a later pedogenetic phase, and well before the early Neolithic. The climatic sequence of the last interglacial/glacial cycle is known into some detail in Sardinia after the study of deep-sea cores from the Tyrrhenian Sea (Paterne et alii 1986; 1988). As far as the development of the Upper Palaeolithic is concerned, during final OIS3, light and heavy isotopic peaks quickly alternate. Then, at c. 30,000, there is a marked increase in d O18 , indicative of fully glacial conditions, and of the Last Glacial Maximum. A milder period develops c. 12,000 years ago, before the final crisis of Dryas 3 around 11,000 bp, followed by the Holocene climatic events.

This sequence can also be outlined at S. Maria Is Acquas. We assume that the lithic industry was produced during the final part of the climatic crisis of late OIS3, and/or during early OIS2. This happened before the pedogenesis which is indicative of milder climatic phase(s) during the late OIS2 and the early Holocene. This is also evidenced by the remains of illuvial clay, found adhering on the knapped implements: the clay deposited during the pedogenetic phase which affected the eolian sands, possibly during the mild Tardiglacial oscillation c.12,000 years ago, and/or during the Holocene if it was a polyciclic process. By then, the lithic tools had already laid buried in the sands during several millennia.

FIGURE 6 S. MARIA IS ACQUAS. SELECTED RETOUCHED TOOLS.

ACKNOWLEDGEMENTS

We thank Dr. Vincenzo Santoni (Soprintendenza archeologica per le provincie di Cagliari e Oristano) for constant backing, and Prof. Maria Rita Palombo (Università di Roma "La Sapienza") for countless discussions on the endemic faunas of Sardinia.

M.M. was funded by Ministero della Pubblica Istruzione (Ricerca scientifica 60% - Ricerche di Ateneo dell' Università degli Studi di Roma "La Sapienza", 1998 e 1999; and Co-finanziamento 1999-2001, programme Paleobiogeografia del Mediterraneo centrale dal Miocene al Quaternario. R.T.M. was funded by Ministero della Pubblica Istruzione (Ricerca scientifica 60% - Università degli Studi di Cagliari).

FIGURE 4-5 were graphically elaborated by F. Scarpelli (Laboratorio di Paletnologia del Dipartimento di Scienze dell'Antichità, Università di Roma "La Sapienza").

REFERENCES

BECCALUVA, L., G. MACCIOTTA AND G. VENTURELLI, 1976. Dati geochimici e petrografici dulle vulcaniti "pha Quaternaire" della Sardegna centro occidentale. Bollettino Soc. Geol. It., 94, 1437-1457.

BERTORINO G., C. LUGLIÈ, M. MARCHI AND R.T. MELIS, 1999. Insediamenti preistorici nell'Oristanese (Sardegna centro-occidentale), primo approccio archeometrico. In: C. D'Amico and C. Tampellini, eds. 6° Giornata "Le Scienze della Terra e l'Archeometria". Museo Nazionale Atestino, Este,185-192.

BERTORINO G., C. LUGLIÈ, M. MARCHI AND R.T. MELIS, 2000. Insediamenti all'aperto della Sardegna centro occidentale nel Neolitico Antico: Ambiente e produzione ceramica. Bollettino

Acc. Gioenia Sc. Nat., 33, 121-134.
DEPALMAS A., M.G. MELIS AND G. TANDA,
1998. La Sardaigne. In: J.Guilaine, ed. Atlas du Néolithique Européen, Vol. 2A. L'Europe Occidentale, ERAUL, 46, 343-394.
GAMBLE C.,
1993. Timewalkers. The prehistory of global colonisation. Allan Sutton Publisher.
KEEGAN W.F., AND J.M. DIAMOND,
1987. Colonization of Islands by Humans: A biogeographical perspective. Advances in Archaeological Methods and Theory, 10, 47-91.
KLEIN HOFMEIJER G., F. MARTINI, M. SANGES, P.Y. SONDAAR AND A.ULZEGA,
1987-1988. La fine del Pleistocene nella Grotta Corbeddu in Sardegna. Rivista di Scienze Preistoriche, 41, 29-64.
LUGLIÈ C.,
1999. L'industria litica a ritocco erto di Coddu Is Abionis, Terralba (OR). Annali della Facoltà di Lettere e Filosofia dell'Università di Cagliari, LIV (N.S. XVII), 5-74.
MACCIONI L.,
1969. Ialoclastiti e pillow-lave nel Miocene della Marmilla (Sardegna centro occidentale). Rendiconti Seminario della Facoltà di Scienze Univ. Cagliari, 39, 207-220.
MACCIONI L.,
1974. Nuovi rinvenimenti di lave a cuscino nel Miocene della Marmilla (Sardegna Sud occidentale). Rendiconti Seminario della Facoltà di Scienze Univ. Cagliari, 43, 277-282.
MAXIA C., AND G. PECORINI,
1968. In: Il Quaternario della Sardegna, Atti X° Congresso Internazionale Studi Sardi, 56-69.
MELIS R.,
1996. Suoli sepolti del Pleistocene superiore nella Sardegna sud-occidentale. Il Quaternario, 9, 509-512.
MUSSI M.,
2001. Earliest Italy. An overview of the Italian Palaeolithic and Mesolithic. New York: Kluwer Academic/Plenum Publishers.
PATERNE M., F.GUICHARD, J. LABEYRIE, P.Y. GILLOT, AND J.C. DUPLESSY,
1986. Tyrrhenian Sea tephrochronology of the Oxygen Isotope Record for the last 60,000 years. Marine Geology, 72, 259-285.
PATERNE M., F.GUICHARD, AND J. LABEYRIE,
1988. Explosive activity of the South Italian volcanoes during the past 80,000 years as determined by marine tephrochronology. J. of Volcanology and Geothermal Research, 34, 153-172.
SEUFFERT O.,
1970. L'evoluzione del rilievo nel Campidano e nel bacino del Cixerri. Wurzburg: Tesi di dottorato.
SONDAAR P, R. ELBURG, G. KLEIN HOFMEIJER, F. MARTINI, M. SANGES, A. SPAAN, AND H. DE VISSER, 1995. The human colonization of Sardinia: a Late-Pleistocene human fossil from Corbeddu cave. Comptes Rendus Acad. Sciences Paris, ser. IIa, 320, 145-150.

42

LOWER PALAEOLITHIC IN SARDINIA

DAVID TRUMP

Mac Donald Institute,
University of Cambridge, Cambridge, England
EMAIL: trump@defrave.freeserve.co.uk

There is a natural and understandable tendency with researchers to claim that their find is the ëearliestí of its type, whether a human species, a tool type, a technique or whatever. There is correspondingly a tendency among other scholars to view such claims with deep suspicion, to combat the effects of over enthusiasm. Both should, of course, be commended, provided that a balanced assessment is made of the evidence.

While an Upper or Final Palaeolithic in Sardinia is now generally accepted from the evidence of the Grotta Corbeddu, Oliena (Nuoro), claims for a Lower Palaeolithic in the Anglona region, the Altana valley at Perfugas (Sassasri), from discoveries first made in 1974, remain more controversial. Since John Cherry (1992) urged caution, a little more evidence has come to light, and Fabio Martini (1999) has presented his case in greater detail. The major contribution of the latter, together with colleagues like Dr Giuseppe Pitzalis of Perfugas, is hereby warmly acknowledged and deserves to be much more widely known and appreciated.

Though firm confirmation has yet to be found, the present paper will look at the new evidence, and spell out some of the quite drastic implications for Sardinia, the Mediterranean, and world prehistory.

Since Sardinia, though for long periods of the Pleistocene linked with Corsica, was never joined on to the Italian peninsula, and so to the European mainland (fig.1), such an early colonisation would indeed be surprising. Apart from understandable scepticism, what possible flaws to the claims might be expected?

Objections to similar claims elsewhere have come under three principle headings.

(1) By sorting through a vast quantity of natural flints, it is always possible to pick out a handful that suggest human workmanship - the procedure which produced the now discredited ëeolithsí of the 1920s and 30s.

(2) Primitive, archaic-looking, techniques of flint working can survive, or be reintroduced, into much later contexts - the Italian Gargano ëCampignianí core axes, later shown to be Copper Age rather than Acheulian, is but one example of many.

(3) Even geological contexts have on occasion been shown to be due to wishful thinking.

While working in the area in 1998-9, at a site to be mentioned shortly, the present writer was able to contribute in a small way to the discussion, although no expert in that early period. For most of the evidence, he willingly defers to those primarily responsible for that research, namely Prof. Martini and Dr Pitzalis.

The starting point was a visit to the site of Pantallinu-Sa Pedrosa, a hill in the angle between the Rio Altana and the Rio Anzos, tributaries of the Coghinas. It is 2km south of Perfugas, 35km east of Sassari, and stands at an altitude of 80m above sea level, 30m above the beds of the rivers.

A saddle between the two highest points is now occupied by cultivated fields, whose surface is littered with flint flakes, derived from immediately local deposits. The high proportion of these showing indubitable human workmanship immediately rules out the first objection above. This is even more apparent from the excavated material from this site and from Sa Coa Sa Multa, 2km to the south-west, both interpreted as workshops or chipping floors. The Perfugas Museum displays a truly remarkable quantity of struck flakes from these and lesser sites, and from surface finds in the area, particularly along the banks of the Rio Altana until these were concreted as a flood control measure.

As for the second possible objection, while the absence of any later archaeological material in the surface collections is hardly conclusive, it carries more weight with reference to the excavated sites, though still negative evidence.

Further, a sample of a dozen surface finds offered to Prof. Paul Mellars for comment were accepted as ëindubitably late Clactonian or proto-Levalloisíon the strength of their general form and striking platforms, some showing incipient multi-facetting. A selection of these is illustrated in fig.2.

On the geomorphology, while Dr A. Ulzegaís arguments in the Martini volume look convincing, the writer is not sufficiently qualified to judge. Here, further research would clearly be desirable. At least the contexts have not been shown to be demonstrably later.

An additional small observation is highly relevant here. In our excavations in the Grotta di Su Coloru, Laerru, 7km from Pantallinu and 4 from Sa Coa, many worked and unworked pieces of a great variety of flints were recovered in 1999, particularly from Impressed Ware levels of the Early Neolithic. Among them were three typical in both material and form of the Pantallinu type. These, presumably collected with other raw material, are therefore provided with a terminus ante quem by the context in which they were found. Furthermore, all showed the typical heavy grey-white patination, suggesting that they were already ancient at the time of burial, c.5000BC.

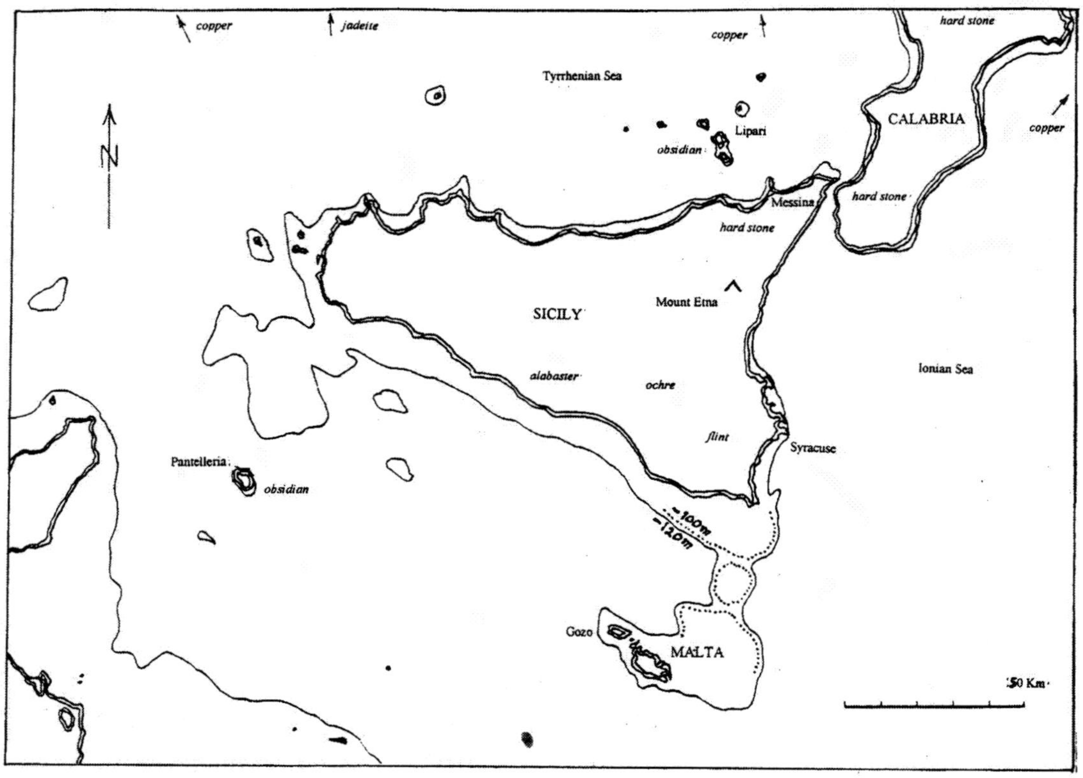

FIGURE 1. THEPERFUGAS SITES, SARDINIA, CORSICA AND MAINLAND ITALY, NOW AND AT MINIMUM SEA LEVEL.

Subsequent excavation has descended into an underlying Mesolithic level, though it is not known whether any more evidence has emerged pertinent to the present discussion. The cave remained in occupation down to the Nuragic Bronze Age.

If a Lower Palaeolithic in Sardinia is accepted, the implications are profound.

At no time in the Pleistocene was there a land bridge to the continent. The minimum sea crossing at lowest sea level, thought to be about -100m, is estimated at 15km, between Cap Corse in northern Corsica and the present island of Capraia, then joined to both Elba and the Tuscan coast. This, incidentally, is almost exactly the present separation of Sardinia and Corsica across the Straits of Bonifacio. A modest lowering of sea level would have eliminated these second straits, joining Sardina and Corsica, with the intermediate Maddalena Islands, into a single land mass.

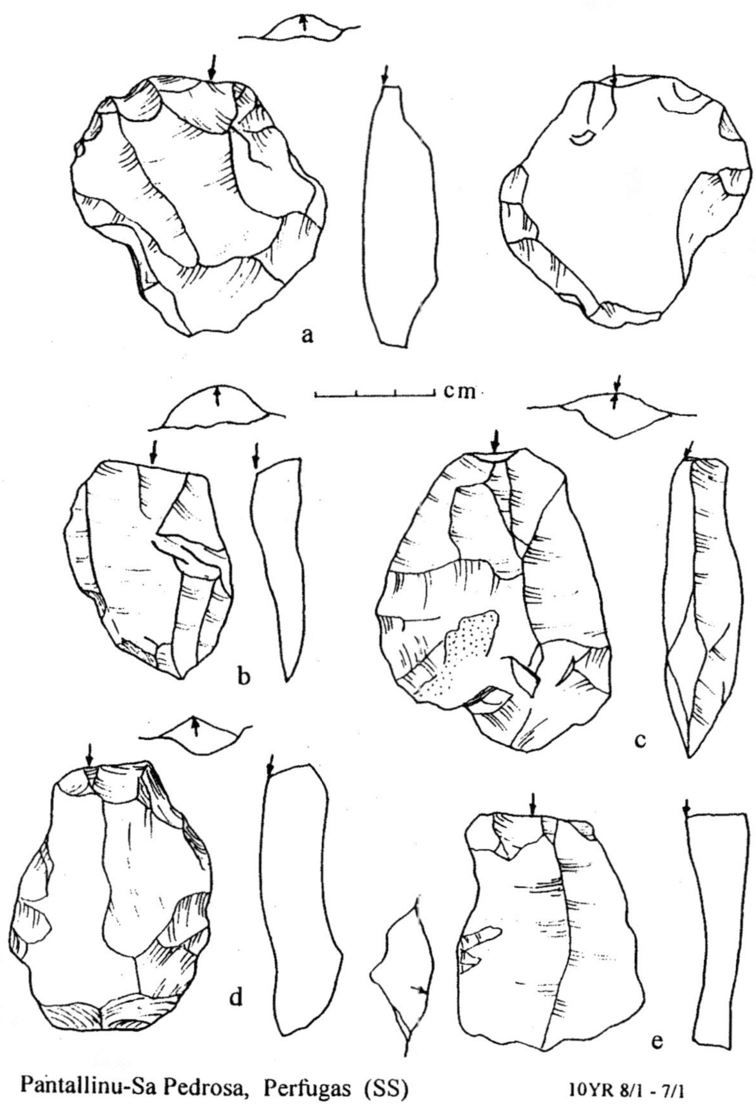

FIGURE 2. CLACTONIAN FLAKES FROM PANTALLINU-SA PEDROSA, PERFUGAS.

To colonise Sardinia, humans must have employed some navigational aid - ëboatí is perhaps putting it rather too strongly, a simple raft being much more likely. The casual arrival of an individual on, for example, flood debris would not have been sufficient to establish a viable population, nor would swimming over that distance have been practicable.

To date, the earliest generally accepted evidence for colonisation involving sea crossings is that of Australia, of the order of 40-50 KYA. The Sardinian material here reviewed cannot be less than four times as old, contained within a terrace attributed to the Mindel glaciation. In this it compares with the substantially earlier date claimed for Flores in Indonesia, referred to elsewhere in this conference.

At that time mankind was represented only by Homo erectus. Was he, not H. sapiens, the first navigator? If so, we must surely reconsider the recent tendency to play down the ëhumannessí of our forebears, erectus and neanderthalensis, claiming our own subspecies as the only true ëmení. It also reopens the controversy over the first colonization of Europe. It has for long been considered impossible that early man could have crossed from his African ancestral home by way of the Straits of Gibraltar. The new evidence suggests that that, albeit difficult, would not have posed an insuperable barrier.

REFERENCES

CHERRY J.F.,

1992) Palaeolithic Sardinians: some questions of evidence and method. In Tykot R. and Andrews T. Sardinia in the Mediterranean: Footprint in the Sea. Sheffield, Sheffield Academic Press.

MARTINI F.,

1999) Sardegna Paleolitica. Firenze, Museo Fiorentino di Preistoria ìPaolo Graziosiî.

43

THE ANIMALS OF PREHISTORIC CYPRUS

LILIAN KARALI

Assoc. Prof. Department of Archaeology, University of Athens, University Campus, Athens 157 84 – Greece.
EMAIL: ikarali@arch.uoa.gr

INTRODUCTION

Although thousands of pages have been dedicated to the history, the monuments and the artifacts of Cyprus, only a few pages focus on ecofacts and especially on skeletal remains. From the Paleolithic period to the end of the Bronze Age, a number of sites have been discovered, with burials of different types and quite abundant archaeozoological material .

Excavations in Prehistoric Cyprus do not always provide detailed information about skeletal remains. In the past, faunal and anthropological finds were not often considered interesting and worthy of a thorough examination. In addition the discovered ecofacts are not well preserved because of the climatic and the environmental conditions of the island which do not help the preservation of skeletal material.

As far as it concerns the bones in most of the cases they are not well preserved due to environmental and human factors. The soil and the subsoil are not stable and have reactive forces. As they are heterogeneous, they change and develop according to the climate, the vegetation and the microorganisms. Subsequently, the osteological remains suffer from the consequences of the natural and chemical environmental activities.

The geographical isolation of Cyprus influenced the natural environment, the population and therefore the civilisation. In Late Neolithic II an effort to escape from the isolation of the Ancient Neolithic period can be observed and this will be completed during the Bronze Age. In the meantime during the Chalcolithic period, important changes in the social organization, the economy and the technology, with main characteristic the demographic development occur. The Cypriot sites are characterized by a certain cultural uniformity, although local variations occur according to the special environment of each one. The choice of the area of the first installation is closely linked to the vicinity of the sea, to areas of strategic importance or to crossroads. The sites in the interior of the island are located near water sources, fertile soils and forests, to ensure hunting activities in case of bad harvest. They are quite often naturally protected, as it happens in the case of the site of Kantou – Koufovounos.

PRESENTATION OF THE ARCHAEOZOOLOGICAL MATERIAL:

PALEOLITHIC AND MESOLITHIC PERIOD

Depositions of fossil animals of the Pleistocene are discovered in 46 locations in Cyprus lengthwise the north and south coast and in the south slope of the mountain range of Pentadaktylos. This distribution is not accidental and it reffers to the semi-aquatic way of living of the proboscides (even the sites of the south slope of the mountain range of Pentadaktylos were coastal, because the plain of Messarias was still covered with water). Until now only the site of Acrotiri – Aetokremnos has been excavated.

During the Palaeolithic Age it seems that only a small number of animals were able to reach the island crossing the sea separating Cyprus from the continental coasts. With reference to the specialists of the Pleistocene fauna of Cyprus (Boekschoten and Sondaar 1972) among the animals living on the island were the hippopotamus, the elephants and field-mice. The hippopotamus of Cyprus (Phanourious minutus), was of the size of a pork and the Elephas Cypriotes of the height of a pony. Geological observations of the terraces yielding the bones suggest that they first appeared during the later Pleistocene, probably between 250 000 and 100 000 years ago. The Cyprus pygmy mammals were indeed very small. An average adult was 3,5 – 4,5 m long, over 1m tall, and weighted up to 4 and a half tones. The Cypriot pygmy hippo, currently known from over 30 sites, would have been less than 1,5 m long and 0,75 m tall. The Cypriot pygmy hippo was better adapted to walking than to swimming. Man has burned many of the bones collected from the site of Aetokremnos. This site is dated by C14 at about 10.000 years ago, that means 2000 years earlier than man was thought to appear in Cyprus. This suggests that these animals did not die long before man arrived, but were probably driven to extinction by the first human occupants of the island.

During the Mesolithic period (about 8500 BC), human presence is confirmed at the above site by tools and animal bones of Pygmy Elephant and Pygmy Hippopotamus. They were certainly hunted and consumed by nomads hunters (Hunt 1990, Simmons and Reese 1993).

PRE-NEOLITHIC AND NEOLITHIC PERIOD

In the same area of Aetokremnos in the peninsula of Akrotiri in Cyprus at 8500 BC thousands of bones from the same species were discovered, which do not exist nowadays, especially Pygmy Elephant and Pygmy Hippopotamus. Individuals of different ages were preserved and one quarter of them had combustion traces. Furthermore birds and snails were discovered. It seems that for a short period in the middle of the 9th millennium BC human groups of nomads hunters-gatherers lived there and hunted (Hunt 1990, Simmons and Reese 1993).

In neolithic Kissonerga – Mylouthkia and Parekklisia – Sylourokambos dear, pork and goat-sheep have been identified. Dear prevails and is followed by pig. It is worth mentionning the presence of cattle in the site Parekklisia – Sylourokampos. Cattle were thought to be introduced on the island of Cyprus at the beginning of Bronze Age. Bones of fishes like blackfish, two-banded bream, dentex, grouper and trident (Le Brun 1989), sea turtle from the site of Agios Epiktitos – Vrysi, reptiles, rodents of the species of Mus and birds, like Columba palumbus L, Corvus corone L, Ciconia ciconia L etc and shellfish have been identified. Numerous bones of Mus muscles come from the Aceramic Neolithic Khirokitia.

KALAVASSOS – TENTA

During the excavation of Kalavassos-Tenta (1976), 75 bones have been discovered: 52 from the species Dama mesopotamica, 3 from Cervus elaphus, 10 from Sus scrofa, 10 from Ovis or Capra (Todd 1979, 24).

Apostolos Andreas – Kastros

Mammals of the species Dama mesopotamica, Capra aegagrus, Ovis aries or Capra hircus, Sus scrofa, etc. have been discovered as well as a carpal bone of pygmy hippopotamus and a phalanx of Canis sp. (Ducos 1974, 89, Davis 1981, 189-210). Rodents are represented with the species of Mus musculus Domesticus (Helmer 1981, 91-92) and fishes with Epinephelus sp., Epinephelus quaza, Diplodus aff. Sargus (L), Epinephelus aff quaza, Odontasis sp., Sparus aurata L., Dplodus aff. Annularis (L), Labrus sp. (Garnier 1981, 93-94). Shellfish as dentalia were used like ornaments, .

KHIROKITIA

Many organic remains and animal bones, probably remains of lunch were discovered (Dikaios, 1953). The mammals' bones belong to the species Capra hircus aegagrus, Dama mesopotamica, Ovis aries, Capra hircus, Ovis orientalis orientalis, Sus scrofa (King 1953, Stanley-Price 1973, 30). Anomalies on the sheep's teeth are observed. Davis (1984, 147-162) mentions also remains of fox and the absence of cattle, horse and especially of dog, species common in the Near East. A cat's jaw (Davis 1989) belongs to the wild species of Felis silvestris lybica. Many teeth and bones of mice have been identified (Mus). There are three different species of birds: Columba palumbus L, Corvus corone L. and Ciconia ciconia L (Pichon, 1984, 163-165). Among the shellfish the species Cassis sulcosa Brug., Charonia lampas L., Collumella sp., Conus coronatus Gmelin, Cymatium olearium Desh., Cypraea lurida L., Monodonta turbinatus Born., Murex trunculus L., Pectunculus pilosus Lamk., Pectunculus sp., Spondylus gaederopus L. and Thais haemastoma Lamarck are identified (Demetropoulos 1977-1981, 169-182). The shell of the species Charonia lampas Linne could have been used as a vessel after the extraction of its internal spirals. Fossil shellfishes have also been found (Strombus coronatus Defrance, Crassostrea gryphoide Schloteim, Isocardia cor. L. var. maveriana Cocconi) and a fossil coral (Wilkins 1953). Various fishes have been recognized: Epinenephelus sp., Dicentrarchus labrax, Sparus aurata, Sparides, Mugiles sp., etc (Desse 1984, 167-168).

SOTIRA – TEPPES

The mammals represented are: dog, small canidae (probably jackal), pork, fallow deer, small deer, goat, sheep, donkey or mule, bird, fish and reptile (Zeuner 1961, 235-236). Dog and goat were certainly domesticated. The horn-cores of the goat refer to the pre-pottery Neolithic goat of Jericho. Fallow Deer belongs to the species Dama mesopotamica and not to Dama dama which occurs in Asia Minor opposite Cyprus and was probably herded. It is not clear if the pig is wild or domesticated and whether the donkey or the mule are represented. There is not definite evidence of sheep bones. This fauna is compared with the one from Neolithic Jericho (Zeuner and Grosvenor Ellis 1961), containing two definitely domesticated species (dog and goat), characteristic of the early Neolithic in the Middle East. The following species of mollusks are also recognised: Cassis sulcosa Brug., Purpura haemastoma Lin., Patella sp., Fusus sp., Cardium edule Lin, Conus mediterranens Brug, Murex trunculus Lin, Murex sp., Ostrea sp., Dentalium sp., Corithium sp., Cardium sp., Trochus turbinatus Born., Spondylus sp., Helix uncta Mull., Rumina decollata Lin., etc. (Dance

1961, 237-238).

AYIOS EPIKTITOS – VRYSI

The common big mammals are recognized: fallow deer, sheep, goat and pork. Dog, cat, fox and possibly roe deer are also present. More than 95% of the animal bones belongs to the Middle Period of occupation of the site (Davis 1989, 189-210). A small number of bones belongs to reptiles, rodents, birds, fishes. Late Neolithic Ayios Epiktitos – Vrysi produced also 14 remains of marine turtle (Legge 1982). It is important to refer to the horns of the species Dama mesopotamica (Legge 1982). Fallow deer and roe deer were hunted in the nearest forest, fish did not constitute an important part of the human diet and so fishing was not important. The goat offered various products: milk, wool and meat.

KANTOU – KOUFOVOUNOS

The intense corrosion of the soil and the chalky subsoil of the hill are responsible for the bad preservation of the ecofacts. Only a few bones are preserved mainly corroded fragments. There is a big percentage of splinters and teeth (Teeth were better persevered because of their histological composition). Many fragments of horn cores were found and the largest part of a goat's skull. The bones (species of animal and part of the body) are 268 (Karali 1986, 25-28).

The main species of animals recognized are goat, sheep, pig and deer (Capra hircus, Ovis ariae, Sus scrofa, Dama mesopotamica). All these are living until nowadays in the same area of the island. It is worth mentioning that in the area of Akrotiri and generally this of Lemesos two species of poisonous snakes, Vipera lebetina and Vipera ammodytis known also from ancient times as "koufos" and "koufi" are not encountered. Consequently one would expect to find some remains of these species. Animals known from most of the other sites of Cyprus, like, Oryctolucus cunniculus, Lepus capensis, smaller mammals and micro fauna which could be characteristic of the natural environment, the hygienic conditions and the way of living of the inhabitants of the settlement (ex. Lizards, mice, e.t.c.) have not been found1. Flotation provided modern insects and other small animals (ex. Ants) indicating the disturbance of the archaeological strata.

There have been identified 27 marine mollusks and 3 land snails, characteristic of deltaic depositions, rock coasts and deep waters (∫·Ú·Ï‹ 1995)2. The edible species are the same as in the other Cypriot sites. They are food remains and most of them were found together with animal bones. Some of them and specially the shells belonging to the species Murex trunculus Linne bear an irregular hole. Moreover many of the shells belonging to the species Patella sp. have the main cone cut off and they have a ring-shaped form and their periphery has many cuttings3.

CHALCOLITHIC PERIOD: ERIMI – PAMPOULA

Animal bones from all levels show that deer (dama mesopotamica), sheep/goat and pig were the main constituents of the meat diet. Judging from patterns of consumption in other sites, deer was probably preferred than pig or sheep/goat, consumed in equal proportions (Dikaios 1962, Croft 1981, 24-42).

Concerning the marine mollusks several edible varieties exist. Cockles (cardium echinatum), limpets (patellidae) and tritons (charonia tritonis) predominate.

LEMBA – LAKKOUS

Dear (Dama mesopotamica),cattle, pig, fox, cat, dog and horse represent the main species of the faunal remains. Some have traces of fire (Croft 1985, 98-100). Concerning the marine mollusks several edible species exist like: Cerithium vulgaturn, Charonia nodifera, Conus mediterraneus, Cypraca lacrimalis, Monodonta turbinate, Monodonta sp. Also land snails exist like Melanopsis praemorsa, Caccilianella aegyptiaca, Calaxis unidentata, Caracellina lenticula, Cecilioides sp., Helix aperta (Ridout-Sharpe 1985, 103-106).

KISSONERGA – MOSPHILIA

The faunal remains from this site represent the larger sample of prehistoric Cyprus. There have been counted and identified 11.302 fragments of mammals: deer, pig and the goat represent the 97,2% of the total (Croft 1998, 207-214). Fox, cattle, dog, cat, horse and pygmy hippopotamus are represented in small quantities. Birds (like Columba palumbur, Columba livia, Streptopelia turtur, Anas platyrynchos), fishes (like Clupidae, Sardina pichardus, Liza ramada, Scarus sp., Scomber scombus L., Sparidae, Triglidae, Serranidae, Epinephalus sp.) (Irving 1998, 230-232), and crabs and rodents (Mus musculus, Crocidura suavrolens) are represented in smaller quantities. Concerning the sea shells the following species exist: Patella coerulea, Patella rustica, Patella lusitanica, Monodonta turbinate, Glycymeris glycymeris, Glycymeris violascens, Callista chione, Arca barbata, Dentalium vulgare, D. Inacquicostatum, D. Rectum. Land snails also exist like: Monacha suriaca, Caracollina lenticula, Cochlicella acuta and Calaxis unidentata (Ridout-Sharpe 1998, 224-229).

KITION

Kition inhabited from the Chalcolithic period provided hippopotamus and elephant remains (Reese 1985, 391-408), astragaloi, a tortoise carapace (Reese 1985, 409-415), interesting shell remains (Reese 1985, 340-371) and ostrich eggshells (Reese 1985, 371-382).

BRONZE AGE: LAPITHOS VRYSI TOU BARBA

From the Bronze Age cemetery of the site there have been discovered animal remains accompanying the human skeletons. Dog, birds, horse, camel (?) and cattle have been identified. The animal bones when associated with ceramic vessels suggest funerary meals (Herscher 1978, 782-800).

MARKI ALONIA

The site was occupied during the Early and the Middle Bronze Age. The recognized animal remains belong mainly to caprines, cattle, fallow deer (Dama mesopotamica), pork, fox, horse, dog, cat and birds (Circus sp.) (Frankel D. Webb J. 1996, 218).

KALAVASSOS – PANAYIA CHURCH CEMETERY

From the excavation of the Middle Bronze Age settlement faunal remains of sheep/goat, dog, horse and pig were discovered. The animal bones are better preserved than the human bones. There is no evidence of butchering or cooking and a low degree of fragmentation, suggesting that these animals were not eaten. According to the stratigraphy these findings, except maybe of the equid bones, did not have any funerary function. They were deposited later on, around the beginnings of the christian era. Perhaps there was a need to dispose the animals after a catastrophic event, or they were killed and buried after a ritual process (Todd 1987).

ALAMBRA

There have been found 8.809 animal bones from this Middle Bronze Age settlement, one of the largest samples of the Cypriot Bronze Age. Sheep/goat, deer, cattle, pig, donkey or horse have been identified. They are badly preserved and fragmented and very few have cut marks. The animal bones found in 3 of the 6 tombs must have probably been funerary offerings (Reese 1996, 475-482).

AYIOS IAKOVOS AND ENKOMI

From Enkomi a significant find is a fragmented Testudo carapace from the Late Chalcolithic III (ca 1220-1100 B.C.) and the attached plastron cut in the interior (Courtois, Lagarce 1986).

KALAVASSOS – AYIOS DHIMITRIOS II

The excavations of the Late Bronze Age settlement provided animal remains like sheep/goat, pig, deer, dog, cattle, rock dove (Columba livia) and horse (South et al. 1989).

HALA SULTAN TEKKE

There have been identified about 2000 animal bones mainly food remains from the settlement and the wells. Those from the wells are better preserved. Horse, donkey, cattle, sheep/goat, pig and boar, dog, persian fallow deer (Dama mesopotamica), pigeon and rock dove have been identified. Less frequently appear the following: hedgehog, european white-toothed shrew, mice, fox and some species of birds (mallard, teal, goose, sparrow), fishes and mollusks. The pigeon frequently discovered, had a particular role in Bronze Age society being the deliverer of the elixir of life, it was later sacrificed to Aphrodite perhaps as a symbol of resurrection. Also among the marine and fresh water mollusks identified predominate the usual species of Patella sp., Monodonta ep., Cerihtium sp., Murex sp., etc. (Demetropoulos 1979, 134-144)

CONCLUSIONS

(a) The hippopotamus and elephant arrived on the island during one of the glacial phases of the Middle or the Final Pleistocene (400.000 – 100.000 BP). The geological profile of the island, the sites where these fossilized animals were discovered and the statigraphy of the coastline prove the above. They came by the sea, while some small ones were transported on floating tree trunks. Consequently the big mammals constituting small sized isolated populations became smaller. Meanwhile rodents and some kinds of birds augmented in size. The small size of the mammals gave them more agility in the mountainous soils. Therefore they were adapted perfectly in their new environment, especially the hippopotamus. The various stages of this evolution are not easily detected from the archaeozoological material.

These species have disappeared not before after the first human installation on the island of Cyprus. During the Aceramic Neolithic some fossil bones of hippopotamus are still found (cap Apostolos Andreas – Kastros, Akanthou – Arkosyko).

(b) From the Neolithic period are present the species known from Balkans and the Near East. While Dama mesopotamica (Persian or Mesopotamian fallow deer) is the most common deer on the island, other deer species have been reported. Red deer (Cervus elaphus) has been identified in several Cypriot Neolithic, Chalcolithic and Bronze Age sites. The (European) Fallow deer (Dama dama) has been reported by Croft from Palaepaphos – Teratsoudhia. Finally the roe deer (Capreolus capreolus) has been reported by Schwartz from Prehistoric to Iron Age Cyprus.

High percentages of deer, accompanied by caprines, pigs and occasionally by dog, cat and fox are characteristic of the Chalcolithic faunal assemblages from Cyprus although cattle is not known to have been present on the island before the Bronze Age (Croft 1985). After the Chalcolithic period fallow deer seems to have declined in importance as a subsistence resource, and by the Middle Ages diminishes.

(c) Horse is of particular interest in ancient Cyprus, as in ancient Greece (Kosmetatou 1993, Reese 1995) where cases of horse sacrifices or animal burials are encountered. From the area of Argolid (continental Greece), famous for its horses, there is evidence of horse burials. According to Homer, horses accompanied the burials of noblemen and warriors. Horses lead the chariot of the dead and after the ceremony they were sacrificed usually outside the tomb and buried together with the chariot. This is the case for the burial in Dendra (Medea). In Late Minoan Crete horses are also sacrificed accompanying rich women burials. The animal was cut in pieces during the funerary ceremony and put into the tomb. In addition, a similar burial deposit was found in the Vaulted Tomb A at Fourni - Archanes in Crete. Considering Cyprus it is not yet clear if this tradition came with the Greeks or it has already existed on the island. From the Bronze Age archaeological examples have been discovered in Lapithos, in Episkopi, in Hala Sultan Tekke (Reese 1995), Audimos (Vavouranakis et ales in press) etc. During Geometric and Archaic period this custom became common.

(d) Closely related to the natural environment of the island are the faunal remains of the sea world from the Neolithic period and onward. For example in Agios Epiktitos – Vrysi there have been identified 55 species of mollusks, followed by 52 in Apostolos Andreas, 22 in Khirokoitia, 29 in Kantou etc. The prevailing species during the Neolithic are Patella, Monodonta turbinata and articulata, Murex trunculus, Arcularia gibosula, Conus sp., Charonia nodifera etc.

The consumption of sea food seems to decrease somehow at the Bronze Age in the whole Aegean due to the development of new techniques in agriculture and breeding. The same phenomenon is not easily observed in associated material like the cephalopods, the remains of which are not easily detected. At most of these sites the sea is at a certain distance, but not so far away. The land snails present everywhere were also eaten quite often (for example Agios Epiktitos – Vrysi and Kantou).

Shells have been used in many ways (Karali 1997, 1999, 2000; Reese 1978, 1984, 1985, 1987, 1988, 1990, 1991, 1992, 1993, 1996). In comparison with the Aegean there are some similarities with the vessels and the ornaments. The absence of ornaments from worked spondylus is to be proved. However the following species Glycimeris, Monodoonta, Dentalium, Charonia, Pinna, Arcularia, Collumbella and Cerithium have been used. In the Aegean islands there is a diversification of species due to a microclimatic geographical variety. There is no evidence for the mollusk trade on the island of Cyprus. There have been observed similarities with the adjacent coast.

Unfortunately the bad preservation of bone remains caused their fragmentation, dispersion and distraction. This has been accentuated by catastrophic events (earthquake, war, fire, flood, public works). In addition different systems of recording have been used. The reader must take in consideration how difficult it is to make a perfect scientific comparison among different sites and periods of Cyprus. The extent to which animal economies were oriented towards hunting and animal breeding in prehistoric Cyprus has been demonstrated. It is apparent that according to the cultural phase there are some prevailing species and some others stay stable. It is important to have studies of pure zoologists as well as of archaeologists and environmentalists in order to reach a complete spectrum of animal life in the past.

NOTES

1 Considering the data disposed until now the absence of this kind of animals could be explained in many ways, for example:

(a) Environmental factors. The acid soil destroyed the organic remains and the settlement in the top of the hill, was influenced by corrosion factors.

(b) The "economy of the site": the killing of young animals. Their bones are very fragile and they are not easily preserved.

(c) In the cultural level and generally in the way of living of the inhabitants (for instance cleaning of the houses from the remains of food).

(d) The study has not yet been finished, especially for the material provided by flotation.

2 All marine and continental shellfish were collected. Entire shells and fragments were found. An effort was made of connecting the valves of the bivalves and of estimating the ¡.ª.π. (Minimal Number of Individuals) of gastropods. An exceptional importance was given to marine shells because they resist more in corrosion.

3 In the case of purple one can clearly distinguish the effort for the extraction of the animal and in the case of the limpet the fractures call to the use of a pointed tool for the cutting off from the rock. Consequently one can say that they surely consist food remains. Of course some very small shells belonging in many different species were transported accidentally, probably after sticking to others or because of their beautiful color or shape.

The molluskan material provided by flotation has not yet been studied. During the flotation we localized many microscopical fragments of shells, some of which belong to land snails (Helix sp.). Some others are modern and it seams that they interfere in the archaeological stratums by the plant's roots.

BIBLIOGRAPHY

BOLGER D.,

1988. Erimi Pamboula, a Chalcolithic Settlement in Cyprus, BAR International Series ......

BOEKSCHOTEN G. J. AND SONDAAR P. Y.,

1972. On the Fossil Mammalia of Cyprus, I & II, Proc. Koninklijke Nederlandese Akad. Wetenschap., Amsterdam (B), 75/4, 306-338.

CARTER P. L.,

1989. Fauna from the 1976 season. Pp. 244-58 in American Expedition to Idalion Cyprus 1973-1980, eds. L. P. Stager and A. Walker. Oriental Institute Communications 24. Chicago : Unoversity of Chicago.

COLEMAN J. E. ET AL..,

1996. Alambra. A Middle Bronze Age Site in Cyprus. SIMA CXVIII. Jomsered: P. Astromes Forlag.

CROFT P.,

1985, The Mammalian Fauna in Peltenburg E. J. Lembra archaeological project. Excavations at Lembra Lakkous 1976-1983, Studies in Mediterranean Archaeology Vol. LXX:1, Goteborg, 98-100, 202-208, 295-296.

CROFT P.,

1986. Fauna in Todd I. A. (ed) Vasilikos Valley Project 1: The Bronze Age Cemetery in Kalavasos Village, Studies in Mediterranaen Archaeology Vol. LXXI: 1, Goteborg, 179-182.

CROFT P.,

1991. The Animal Remains. Chapter 9 in E. J. Peltenburg, Lemba Archaeological Project II.2. SIMA LXX:3. Goteborg: P. Astroms Forlag.

CROFT P.,

1996. 9. Subsistence Economy A. Animal Remains in Frankel D. and Webb J. M. Makri Alonia, An Early and Middle Bronge Age Town in Cyprus, Excavations 1990-1994, Studies in Mediterranean Archaeology Vol. CXXIII: 1, Jonsered, 217-223.

CROFT P.,

1998. Animal Bones in South A., Russell P., and Schuster Keswani P. Vasilikos Valley Project 3: Kalavasos-Ayios Dhimitrios II, Ceramics, Objects, Tombs, Specialist Studies in Mediterranean Archaeology Vol. LXX: 3, Goteborg, 70-72.

CROFT P.,

1998. Animal remains: Synopsis in Peltenburg E., Bolger D., Croft P. et al. Lembra Archaeological Project Vol. II.1A, Excavations at Kissonerga-Mosphilia 1979-1992, Studies in Mediterranean Archaeology Vol. LXX: 2, Jonsered, 207-214.

DANCE S. P.,

1961. The Shells in Dikaios P. Sotira, Philadelphia, 237-238

DAVIS S. J. M.,

1984. Khirokitia and its mammal remains, A Neolithic Noah's Ark in Le Brun A. Fouilles recentes a Khirokitia (Cypre) 1977-1981, Tome I: Texte, Paris, 147-162

DAVIS S. J.,

1987. La faune, in. A. Le Brun, et al. Le Néolithique Préceramique de Chypre. L' Anthropologie 91/1 : 305-309

DAVIS S. J. M.,

1989. Some more animal remains from the Aceramic Neolithic of

Cyprus in Le Brun A. Fouilles recentes a Khirokitia (Cypre) 1977-1981, Tome I: Texte, Paris, 189-210
DEMETROPOULOS A.,
1984. Marine mollusks, land snails, etc. in Le Brun A. Fouilles recentes a Khirokitia (Cypre) 1977-1981, Tome I: Texte, Paris, 169-182
DESSE J.,
1984. L' ichthyofaune du site neolithique de Khirokitia (Cypre) in Le Brun A. Fouilles recentes a Khirokitia (Cypre) 1977-1981, Tome I: Texte, Paris, 167-168
DESSE J. AND DESSE-BERSET,
1989. Les poissons de Khirokitia (campagnes 1983, 1984 et 198...) in Le Brun A. Fouilles recentes a Khirokitia (Cypre) 1977-1981, Tome I: Texte, Paris, 223-233
DESSE J.,
1994. Strategie de pêche au 8e millenaire : Les poissons de Cap Andreas Kastros (Chypre). Chapitre 17 in A. Le Brun, . Fouilles recentes à Khirokitia (Cypre) 1977-1981, Tome I: Texte, Paris, 335-360
DIKAIOS P. AND STEWART J. R.,
1962. The Sewdish Cyprus Expedition: finds and results of excavations in Cyprus 1927-1931. Vol. IV Part 1A, The Stone Age and the Early Bronze Age in ... Lund
DUCOS P.,
1965. Le daim à Chypre aux époques préhistoriques. Report of the Department of Antiquities, Cyprus : 1-8.
DUCOS P.,
1981. Notes sur les grands Mammiferes du Cap Andreas-Kastros in Le Burn A. Un site neolithique preceramique en Cypre : Cap Andreas-Kastros, Paris, 89
FRANKEL D., WEBB J.,
1996. Marki-Alonia, An Early and Middle Bronze Age Town in Cyprus. Excavations 1990-1994, SIMA CXXIII: 1, Jonsered.
GARNIER J.,
1981. Les poissons du Cap Andreas-Kastros in Le Brun A. Un site neolithique preceramique en Cypre : Cap Andreas-Kastros, Paris, 93-94
HELMER D.,
1981. Les rongeurs du Cap Andreas-Kastros in Le Burn A. Un site neolithique preceramique en Cypre : Cap Andreas-Kastros, Paris, 91
HERSCHER E. C.,
1978. The Bronze Age Cemtery at Lapithos Vrysi tou Barba, Cyprus: Results, Part 2, University Microfilms Int.
KARALI L.,
1996. Paleoperivalontiko iliko apo ti Neolithiki Thesi Kantou-Koufovounos. Epimetro stin "Ekthesi apotelesmaton tis anaskafis stin Thesi Kantou-Koufovounos gia tis periodous 1994-1995, E. Mantzoyrani", Epistimoniki Epetiris tou Tmimatos Arxaiologias, Lefkosia, 25-28
KARALI L.,
2000. Kantou-Koufovounos palaioperivalolontiko iliko, Praktika Tritou Diethnous Kiprologikou Sinedriou (16-20 Apriliou 1996) Lefkosia, 237-244
KING J. E.,
1953. Mammal bones from Khirokitia and Erimi in Dikaios P. Khirokitia: Final report on the excavation of a Neolithic settlement in Cyprus on behalf of the Department of Antiquities, 1936-1946, Oxford, 431-437
LEGGE A. J.,
1982. Ayios Epiktitos : the recent farming economy / the vertebrate fauna / .... worked antler, and bone tools / Flotation samples in Peltenburg E. J. Vrysi: a subterranean settlement in Cyprus; excavation at Prehistoric Ayios Epiktitos Vrysi, 1969-1973, Warminster, 14-20, 76-90, 4.....
PELTENBURG E. J.,
1982. Vrysi: a subterranean settlement in Cyprus; excavation at Prehistoric Ayios Epiktitos Vrysi, 1969-1973, Warminster
PELTENBURG E. J.,
1985. Lembra archaeological project. Vol 1, Excavations at Lembra Lakkous 1976-1983, Studies in Mediterranean Archaeology Vol. LXX:1, Goteborg
PELTENBURG E. J.,
1990. Palaeolithic to Late Bronze Ages in Hunt S. D. (ed) Footprints in Cyprus: an illustrated history (rev. ed), London
PELTENBURG E. J. AND BAXEVANI E.,
1998. Lembra archaeological project. Vol 2. 1A, Excavations at Kissonerga-Mosphilia 1979-1992, Studies in Mediterranean Archaeology Vol. 70:2, Jonsered
PICHON J.,
1984. Les oiseaux du site de Khirokitia in Le Brun A. Fouilles recentes a Khirokitia (Cypre) 1977-1981, Tome I: Texte, Paris, 163-165
REESE D. S.,
1985. Appentixes in V. Karageorgis Excavations at Kition V/II, Nicosia, Department of Antiquities
REESE D. S.,
1978. Molluscs from Archaeological Sites in Cyprus: "Kastros" Cape Andreas, Cyprus a .... Pre Bronze Age Mediterranean sites. Fisheries Bulletin 5, Nicosia, Dept. of Fisheries
REESE D. S.,
1995. The Pleistocene Vertebrate Sites and Fauna of Cyprus. Bulletin 9. Nicosia: Geological Survey Department
REESE D. S.,
1996. Animal Bones and Shells in Coleman J. E., Barlow J. A. et al. Alambra, A Middle Bronze Age Settlement in Cyprus, Studies in Mediterranean Archaeology Vol. CXVIII, Jonsered, 475-482
REESE D. S.,
1998. Bibliography of Cyprus Zooarchaeological Reports.
RIDOUT J. S.,
1982. The Mollusca in Peltenburg E. J. Vrysi: a subterranean settlement in Cyprus; excavation at Prehistoric Ayios Epiktitos Vrysi, 1969-1973, Warminster, 93-95, 437-452
RIDOUT SHARPE J. S.,
1985. The Mollusca in Peltenburg E. J. Lembra archaeological project. Vol 1, Excavations at Lembra Lakkous 1976-1983, Studies in Mediterranean Archaeology Vol. LXX:1, Goteborg, 103-106, 212-216, 298-305
SIMMONS A. H. AND REESE D. S.,
1993. Hippo Hunters of Akrotiri, Archaeology Vol. 46 (5), 40-43
TODD I. A.,
1987. Vassilikos Valley project 6: Excavations at Kalavassos-Tenta I, Studies in Mediterranean Archaeology Vol LXXI:6, Goteborg
TSALIKI A.,
2000. Environmental remains and burials from Neolithic and Chalcolithic Cyprus. Ancient Cyprus Web Project.
VAVOURANAKIS G., MAGGINIS G., KARALI L., TSALIKI A.
(to be published), The Audimus faunal and skeletal remains.
WILKINS G. L.,
1953. Shells from Khirokitia and Erimi in Dikaios P. Khirokitia:

Final report on the excavation of a Neolithic settlement in Cyprus on behalf of the Department of Antiquities, 1936-1946, Oxford, 438-440

XENOPOULOS K.,
1985. Environment and Resources in Peltenburg E. J. Lembra archaeological project. Vol 1, Excavations at Lembra Lakkous 1976-1983, Studies in Mediterranean Archaeology Vol. LXX:1, Goteborg, 6-8

ZEUNER F. E. AND GROSVENOR ELLIS A.,
1961. Animal Bones in Dikaios P. Sotira, Philadelphia, 235-236

44

LOS GRUPOS DOMESTICOS EN LA PREHISTORIA DE MALLORCA. EL EDIFICIO ALFA DEL PUIG MORTER (SINEU, MALLORCA) Y LAS PRACTICAS SOCIALES DEL HORIZONTE DE SON FERRAGUT (c. 750/700 -525/475 cal ANE).

PEDRO V. CASTRO MARTÍNEZ (1), **TRINIDAD ESCORIZA MATEU** (2), **Mª ENCARNA SANAHUJA YLL** (3)

(1) Universitat Autònoma de Barcelona, Barcelona, Spain
(2) Universidad de Almería, Almiria, Spain
(3) Universitat Autònoma de Barcelona, Barcelona, Spain
EMAIL: PedroCcastro@uab.es

INTRODUCCION

LA HETEROGENEIDAD DE LAS COMUNIDADES DE MALLORCA.

Un territorio insular como el de las islas Baleares, cercanas entre sí y ubicadas en el centro del Mediterráneo Occidental, podría ser concebido como un espacio de confluencia de redes de relaciones intercomunitarias, dado el contexto colonial y las conocidas rutas marítimas fenicias, etruscas o focenses que atravesaban la región. Sin embargo, sólo la isla de Ibiza, con su asentamiento fenicio, resulta representativa del Horizonte Orientalizante, caracterizado por sociedades aristocráticas en las que la clase dominante utilizó, entre los siglos IX y V ANE, productos de lujo de estilo homogéneo en casi todas las áreas del Mediterráneo. Por el contrario, las grandes islas de Mallorca y de Menorca parecen apartadas de las prácticas sociales que favorecieron la circulación de esos productos orientalizantes y de los bienes cuya distribución se atribuye a comerciantes fenicios, griegos o etruscos. Mallorca y Menorca presentan algunas características comunes durante esa etapa, lo que ha permitido definir una cultura talayótica, cuyo fósil director lo constituyen los edificios turriformes denominados Talaiots. No obstante, las diferencias existentes entre los propios edificios y otros productos asociados a los mismos han posibilitado también un estudio separado de las comunidades de Mallorca y de Menorca.

Si nos centramos en una de las islas, concretamente en Mallorca, la heterogeneidad parece la mejor manera de describir la situación. A pesar de que el aislamiento insular podría sugerir lo contrario, las comunidades malloquinas no aparecen como un bloque único ni como un grupo social sujeto a normas que impongan ciertos modelos de asentamiento o de arquitectura. La situación es más compleja de lo que pudiera parecer. Resulta, pues, necesario abordar las condiciones de la vida social teniendo en cuenta que debió existir una cierta fragmentación político-social, es decir, una sociedad segmentaria con comunidades cuyas prácticas sociales, tradiciones, decisiones y redes de relaciones no dependían de pautas homogeneizadoras surgidas de centros supracomunitarios. Por lo tanto, debemos preguntarnos cuáles fueron las prácticas sociales compartidas por las distintas comunidades y las que las diferenciaron.

Dada tal situación, creemos que hay que contemplar el contexto insular en términos distintos a una mera definición de rasgos culturales, y que la ubicación cronológica no ha de suponer una periodización cultural, puesto que intentar una u otra cosa supone construir una norma que inevitablemente conduce a uniformar una realidad diversa. En consecuencia, pensamos que resulta más adecuado definir horizontes temporales, ajustándolos, en la medida de lo posible, a inflexiones que afecten de forma más o menos general a las diversas comunidades, con la finalidad de observar posteriormente qué ámbitos de prácticas sociales desarrollaron las mismas y qué relaciones sociales existieron entre ellas, si es que las hubo.

En los últimos años hemos llevado a cabo un estudio del registro de las excavaciones dirigidas por Jordi Estévez, Pepa Gasull, Vicente Lull, Mª Encarna Sanahuja Yll y Assumció Vila, durante los años 80, en el Edificio Alfa del

Puig Morter de Son Ferragut (Sineu, Mallorca). Dicho estudio nos permite plantear que el Edificio Alfa constituye un modelo de vivienda que hasta ahora no se había tomado en consideración como tal en los yacimientos de la Prehistoria Reciente de Mallorca y que nos permite definir el Horizonte ma). Todas estas edificaciones parecen constituir el conjunto urbanístico de la comunidad que construyó y utilizó el Edificio Alfa. Además existen varios tramos de un recinto amurallado con un torreón emplazado en la cota más elevada (Phi) y otros restos de estructuras, cuya sincronía respec-

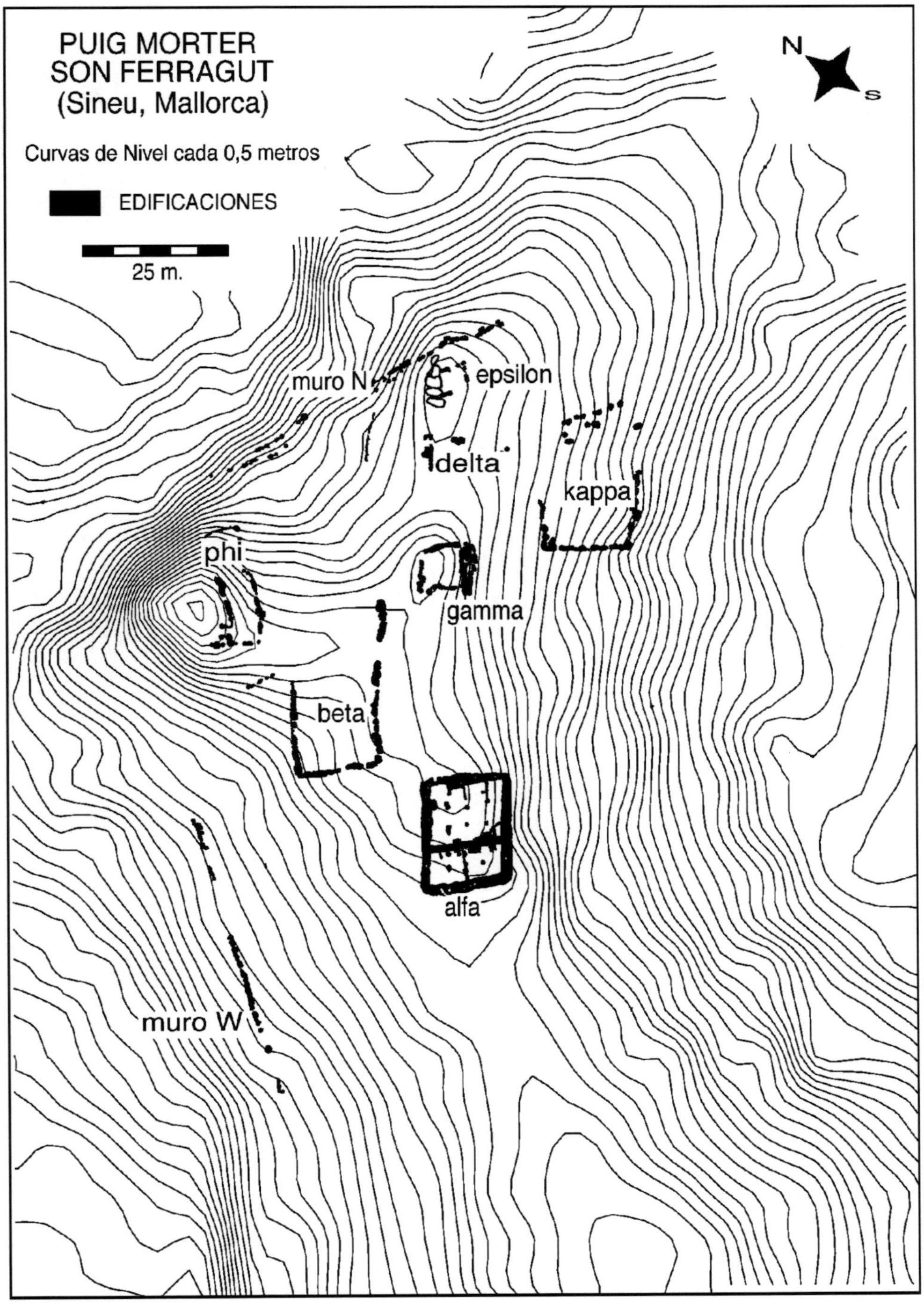

Figura 1
Plano topográfico del Puig Morter de Son Ferragut (Sineu, Mallorca).

de Son Ferragut (Castro Martínez, Escoriza Mateu y Sanahuja Yll, 2001, ep).

En el asentamiento del Puig Morter (FIGURA1), junto con el Edificio Alfa, se han identificado otras dos construcciones de similares características y tamaño (Edificios Beta y Kappa), así como un Talaiot de planta cuadrada (Gamto al Edificio Alfa no puede ser confirmada por el momento.

El Edifio Alfa está construido con muros de doble paramento de bloques de piedra y relleno interior, cuyo aparejo exterior presenta ortostatos dispuestos en hiladas horizontales sobre un zócalo (FIGURA 2). Presenta una planta con forma de paralelogramo que ocupa un área de 296 metros cuadrados, con unas dimensiones máximas de 19.50 m x 15.80 m. Constituye una unidad arquitectónica formada por diversos espacios estructurados, y en la que se distinguen dos agrupaciones separadas por un muro medianero. La áreas son las siguientes: El Area del Hogar, el Pasillo, el Area entre Tirantes, el Area del Angulo Este, el Area entre Columnas y el Area del Receptáculo. En cuanto a las dos habitaciones interiores, divididas por un murete de piedra, contaban probablemente con una azotea superior sustentada

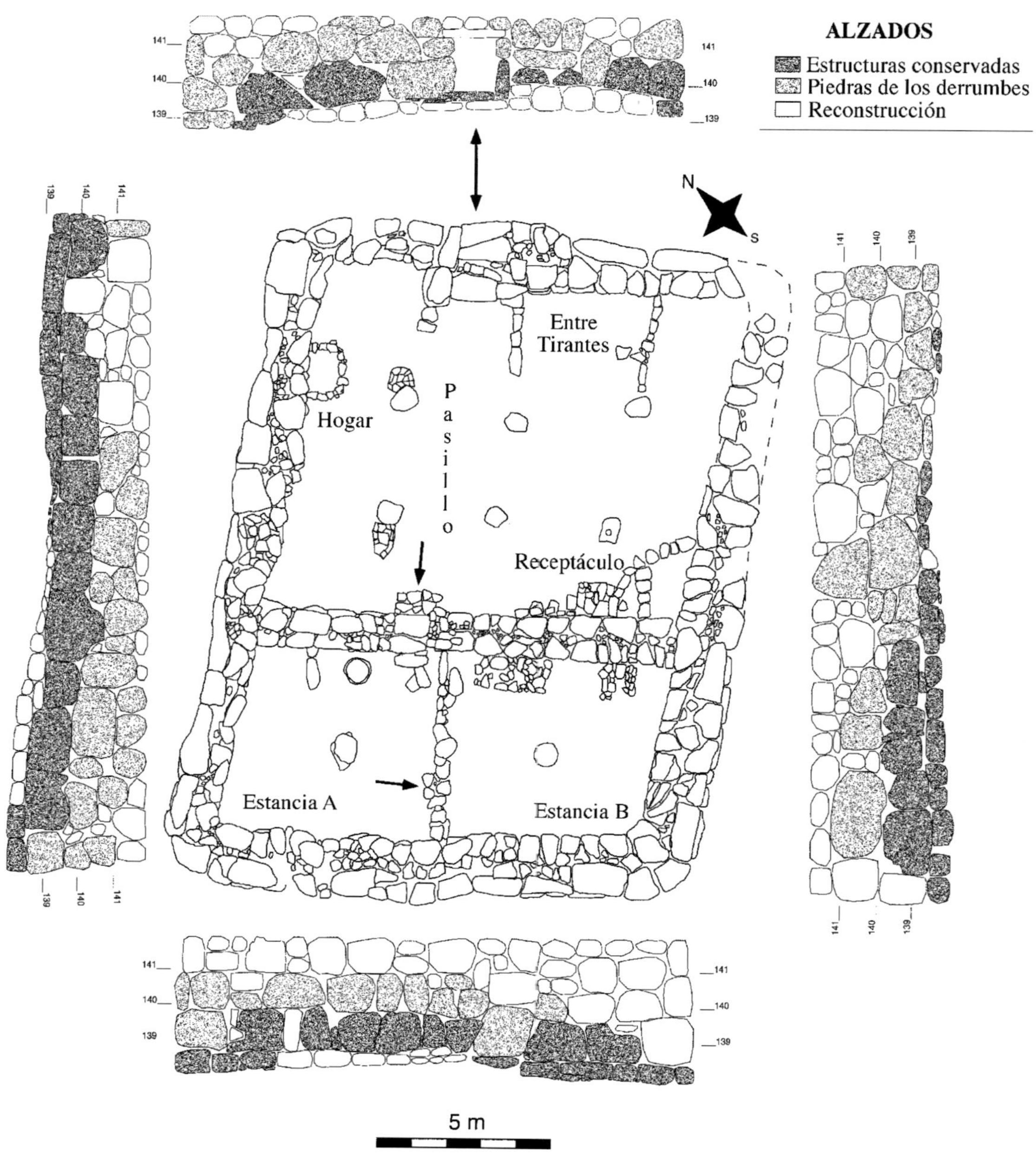

Figura 2
Edificio Alfa del Puig Morter de Son Ferragut.
Planta y alzados del paramento exterior del muro perimetral.

primera corresponde al área de un patio porticado y la segunda a dos habitaciones interiores, las Estancias A y B. El patio estaba delimitado por 6 columnas de postes de madera sobre basas de piedra y presentaba a su alrededor una serie de espacios cubiertos, separados, bien por las columnas, bien por tirantes que arrancaban del muro perimetral. Estas por tres columnas polilíticas. El acceso al Edificio desde el exterior se realizaba a través de un portal con umbral monolítico y tres escalones que descendían hasta el patio. Desde el patio se pasaba a la Estancia A a través de otro portal con escalera descendente. Por su parte, las dos habitaciones estaban comunicadas por un vano abierto en el muro de separación entre ambas.

Un edificio con las características del Alfa responde a las descripciones que se han venido utilizando para definir los llamados "santuarios" de los momentos postalayóticos de la isla de Mallorca. Por esta razón, el conjunto de Son Ferragut ha sido tratado como un centro de carácter ritual por algunos autores (Aramburu-Zabala, Garrido y Sastre 1994; Guerrero 1996: 47; Aramburu-Zabala 1998: 125-127). Sin embargo, como ya hemos señalado, el estudio del registro arqueológico de las excavaciones permite afirmar que se trata de una vivienda. Además, la serie de seis dataciones de carbono 14 obtenidas incide en una cronología inesperada para un edificio de este tipo, situándolo en una etapa en la que hubo de coexistir con la mayor parte de los asentamientos talayóticos. Dos de las fechas realizadas sobre vigas de Olea se remontan a c. 790/780 cal ANE. Las fechas más recientes corresponden a una muestra de vida corta de fauna hallada en el piso, c. 500 cal ANE, y a un tronco de Pinus que pudo emplearse como estante, banqueta, escalera de acceso a la azotea o para la reparación de alguna estructura constructiva, c. 480 cal ANE. La fecha de un peine de madera de Erica ofrece como resultado una cronología de c. 680 cal ANE. Finalmente, un hueso de fauna procedente de un contexto formado con posterioridad al abandono es fechado c. 620 cal ANE. En conclusión, podemos datar la construcción del Edificio Alfa entre c. 750/700 cal ANE, dada la vida larga de las muestras de cronología más alta, y el abandono de la ocupación c. 525/475 cal ANE.

La etapa entre c. 750/700 y c. 525/475, a la que vamos a denominar Horizonte de Son Ferragut, nos ofrece la coexistencia en la isla de Mallorca de diversas comunidades con unidades habitacionales de características arquitectónicas muy diferentes. Los edificios exentos de tipo Alfa del Puig Morter convivieron con viviendas de reducidas dimensiones, adosadas unas a otras o a Talaiots, tal como las conocemos en poblados como el cercano de Son Fornés (Gasull, Lull y Sanahuja Yll 1984a, 1984b, 1984c). Este tipo de unidades domésticas son las más conocidas en los asentamientos de Mallorca y configuran agregados que incorporan edificios turriformes de tipo Talaiot. Sin embargo, el modelo del Edificio Alfa, exento y con patio, tampoco resulta desconocido, ya que la Habitación 10 de Ses Païsses (Artà) responde a una unidad doméstica de similares características (Lilliu 1962-63).

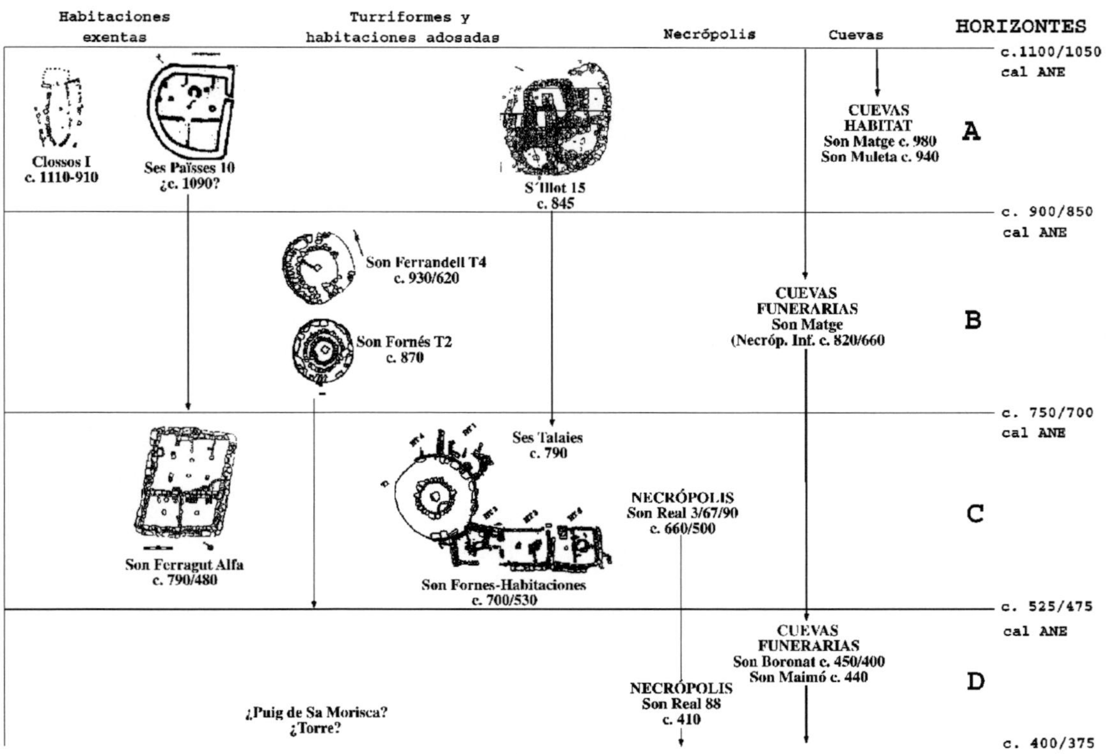

Figura 3
Situación histórica del Edificio Alfa del Puig Morter.
Horizontes temporales de la isla de Mallorca.

El Horizonte de Son Ferragut corresponde al intervalo correspondiente al tiempo de construcción y utilización del Edificio Alfa del Puig Morter. Pero, para enmarcar ese horizonte en la dinámica histórica de las comunidades mallorquinas, hemos delimitado también las etapas anterior y posterior al mismo. Para ello nos hemos basado en las muestras de Carbono 14 procedentes de contextos de uso social, calibradas dendrocronológicamente mediante el programa Calib (v. 4.3) de la Universidad de Washington (Stuiver y Reimer 1993; Stuiver et alii 1998). Se ha utilizado la fecha mediana del intervalo de máxima probabilidad de la calibración para 1 sigma, el método B del programa, que permite un mayor ajuste de las fechas en etapas de fluctuaciones de la curva de calibración (Castro Martínez, Lull y Micó 1996:27-28), tal como ocurre en los años que abarca el horizonte cronológico de Son Ferragut. Con ello proponemos la distinción de cuatro horizontes, que vamos a definir y a poner en relación con las más recientes propuestas de periodización que afectan a la isla de Mallorca (Lull et alii 1999; Guerrero 1999). Los horizontes definidos son los siguientes (FIGURA 3):

HORIZONTE A (C. 1100/1050 - 900/850 CAL ANE).

Esta etapa corresponde a la denominada Fase Prototalayótica (Lull et alii 1999) o a los primeros momentos del Bronce Talayótico (Guerrero 1999). La continuidad de comunidades con habitaciones naviformes, bien documentada en Clossos de Can Gaià (Calvo y Salvá 1999), coexiste con núcleos constituidos por turriformes con habitaciones radiales adosadas, documentadas en S'Illot. El turriforme de Trebalúger, en Menorca, también está fechado en este momento. Una confusa datación de la Habitación 10 de Ses

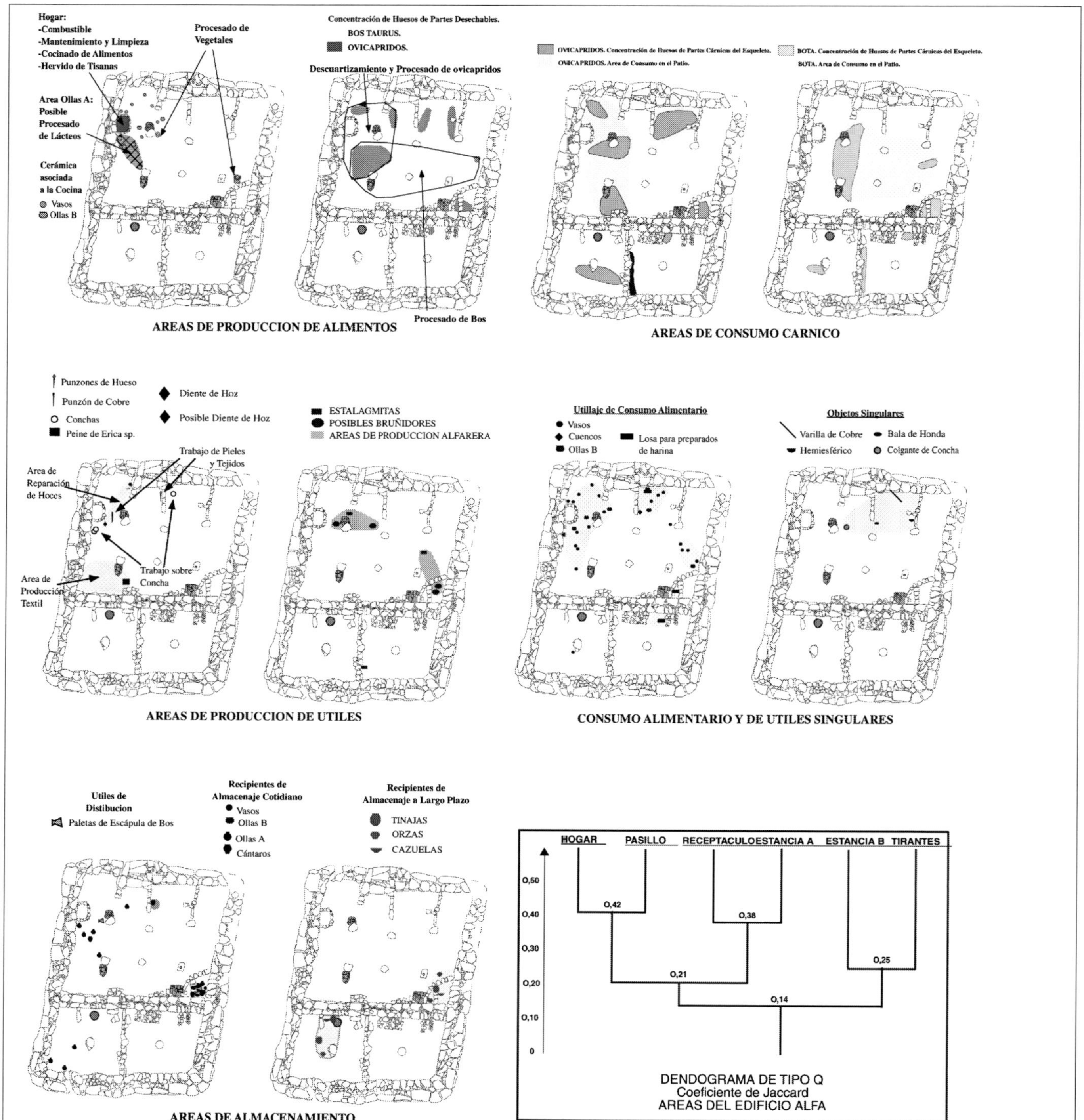

Figura 4
Areas de Actividad en el Edificio Alfa del Puig Morter de Son Ferragut.

Païsses podría remontar a esta etapa la construcción de unidades habitacionales similares al Edificio Alfa. En este horizonte se pone de manifiesto la coexistencia de distintas comunidades, que organizan su vida social en asentamientos con diferentes tipos de unidades domésticas. Las prácticas funerarias en Mallorca no son bien conocidas, aunque existen dataciones en los niveles funerarios de Son Matge, y en la vecina isla de Menorca este horizonte coincide con la construcción y uso de las navetas funerarias y con las últimas etapas de cuevas también funerarias, como Es Carritx, Es Mussol o Mongofre Nou (Lull et alii 1999).

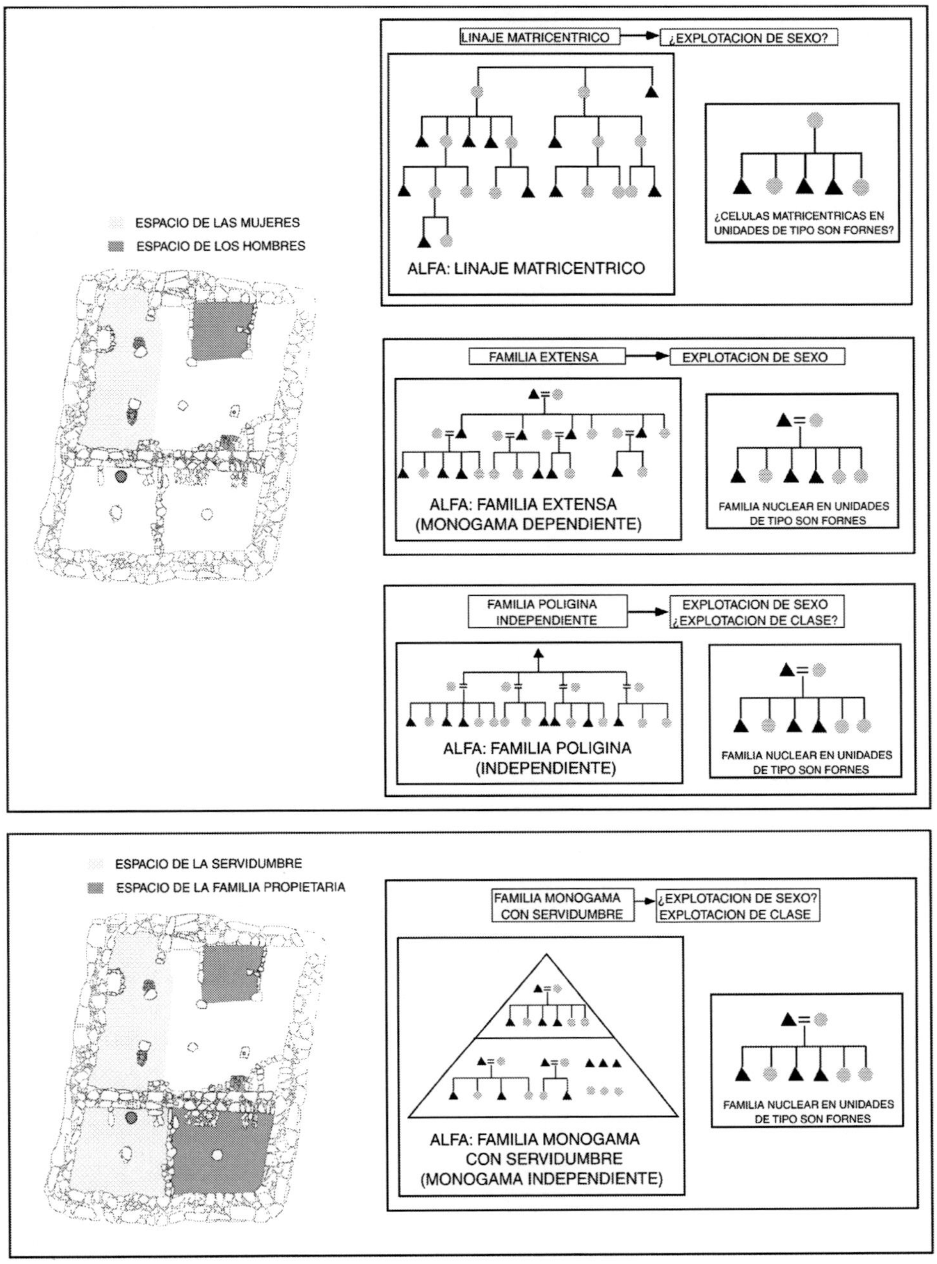

Figura 5

Hipótesis sociales para la Unidad Doméstica del Edificio Alfa del Puig Morter de Son Ferragut.

HORIZONTE B (C. 900/850 - 750/700 CAL ANE).

Se trata de una etapa inscrita en los primeros momentos de la Fase Talayótica de Lull et alii (1999) o en los momentos finales del Bronce Talayótico de Guerrero (1999). Es una etapa conocida muy deficientemente, pero que muestra algunos cambios importantes. Ya no existen asentamientos con habitaciones naviformes y la ideología que había sustentado las prácticas funerarias de las comunidades del II° milenio sufrió una quiebra que conocemos bien en Menorca, donde navetas y cuevas de enterramiento dejan de ser usadas como lugares de prácticas funerarias (Lull et alii 1999). En cuanto a los asentamientos mallorquines, se construyeron los Talaiots que ofrecen un modelo arquitectónico más homogéneo, según señalan las dataciones de Son Fornés o de Son Ferrandell-Oleza. Se trata de edificaciones con planta circular, cámara y columna polilítica central. Debemos esperar los datos que confirmen que también ahora se levantaron los Talaiots de planta cuadrada. La existencia de núcleos habitacionales en torno a edificios turriformes centrales está documentada por la continuidad del conjunto central de S'Illot. La posible presencia en estos momentos de unidades exentas del tipo Edificio Alfa podría, a su vez, contar con la citada habitación de Ses Païsses, lo que supondría la continuidad de la coexistencia de diversas formas de organización social.

HORIZONTE C (C. 750/700 - 525-475 CAL ANE).
Se trata del Horizonte de Son Ferragut. Corresponde a los momentos recientes de la Fase Talayótica (Lull et alii 1999) o a las fechas propuestas para la Edad del Hierro I (Guerrero 1999). El establecimiento del Puig Morter ilustra la creación de nuevos asentamientos, entorno a Talaiots de planta cuadrada, como el caso que nos ocupa. Los Talaiots siguen utilizándose como lugares de prácticas sociales, puesto que contamos con fechas que alcanzan c. 490-480 cal ANE en el Talaiot 2 de Son Ferrandell-Oleza (Waldren 1992), aunque desconocemos los objetos usados en ese nivel de ocupación. Además asistimos a la remodelación de poblados con Talaiots, donde éstos últimos siguen empleándose, y se construyen viviendas adosadas a ellos y entre sí, según nos indica la serie de dataciones de Son Fornés (Gasull, Lull y Sanahuja 1984a, 1984b, 1984c). Las habitaciones adosadas a Talaiots también se edificaron en el poblado de Can Jordi de Santanyi (Pons 1999:61). Aunque algunas fechas del interior del llamado Santuario de Son Mas corresponden a este horizonte o son incluso más antiguas (Van Strydonck, Waldren y Hendrix 1998; Waldren 1998), parecen corresponder a rellenos, y estamos a la espera de la publicación del registro del nivel de ocupación del edificio para evaluar el contexto de uso social del edificio. Las prácticas funerarias de las comunidades costeras de Mallorca muestran durante este horizonte la creación de nuevos lugares de enterramiento, como indican las dataciones de la necrópolis de Son Real (Santa Margalida). Los mismo parece ocurrir con el inicio de la necrópolis menorquina de Calescoves, según una fecha de huesos humanos del hipogeo XXI (Plantalamor y Van Strydonck 1997: 39), situada c. 620 cal ANE. Paralelamente, en Mallorca también se inicia el uso funerario de nuevos lugares, como documentan los registros de los niveles inferiores, anteriores a los niveles con sarcófagos, de la cueva funeraria de Son Maimó (Petra). La heterogeneidad de las comunidades mallorquinas del Horizonte de Son Ferragut se expresa en los asentamientos tanto como en los lugares funerarios.

HORIZONTE D (C. 525/475 - 400/375 CAL ANE).
Corresponde a la primera etapa del llamado periodo Postalayótico, en un momento previo al establecimiento en Mallorca de un sistema comercial basado en asentamientos coloniales costeros. Coincide con los últimos momentos de lo que se está denominando comercio precolonial (Guerrero y Calvo 2001). Es en esta etapa cuando se inició la contratación de tropas mercenarias de honderos por parte del ejército de Cartago. En la región central de Mallorca, se abandonaron o remodelaron la mayor parte de los asentamientos de etapas anteriores (Puig Morter, Son Fornés), aunque otras comunidades, sobre todo en las regiones litorales de la isla continuaron desarrollando su vida social, a juzgar por la continuidad de la necrópolis de Son Real. Esa perduración de los lugares funerarios también se aprecia en diversas cuevas funerarias, aunque ahora se constata la utilización de sarcófagos de madera (Son Maimó, Son Boronat). Alguna de las comunidades litorales mantuvieron estrechos contactos con las redes de intercambios transmediterráneas, según indica la cerámica griega aparecida en los rellenos previos a la construcción de una torre en Puig de Sa Morisca (Guerrero y Calvo 2001).

LAS ACTIVIDADES SOCIALES DESARROLLADAS EN EL EDIFICIO ALFA.
El análisis de la materialidad social del Edificio Alfa del Puig Morter nos ha permitido detectar las diferentes actividades que tuvieron lugar en esta unidad habitacional, emplazarlas en las distintas áreas del edificio y establecer las redes de las prácticas sociales en el marco del grupo doméstico que ocupó este lugar social (FIGURA 4).

ARQUITECTURA.
La propia construcción del Edificio Alfa supuso una actividad productiva de gran envergadura. La estimación del volumen de material constructivo necesario para elevar el edificio se ha observado que resulta similar al empleado en la construcción del turriforme de mayores dimensiones que conocemos en Mallorca, el Talaiot 1 de Son Fornés (Gasull, Lull y Sanahuja Yll 1984a). Se utilizaron mayoritariamente conglomerados procedentes del propio cerro del Puig Morter (estudios de D. Gómez-Gras y R Risch).

PRODUCCIÓN DE ALIMENTOS.
La producción ganadera del grupo doméstico se circunscribe a la producción de ovicápridos, destinados a la obtención de carne y productos lácteos (análisis de J. Estévez y M. Montero). Los ovicápridos eran sacrificados, descuartizados y cocinados en el Area del Hogar, con extensiones de procesado en el Pasillo. Los estudios faunísticos han permitido concluir que la carne de vacuno y la escasamente representada de cerdo llegaba preparada para ser cocinada y consumida. Su procesado final tenia lugar en el Area del Hogar y Pasillo. Al parecer, la ganadería de bóvidos y de suidos era gestionada en un marco social exterior al grupo doméstico, bien fuera en el seno de la propia comunidad de Son Ferragut, bien en otra comunidad. Por otra parte, el elevado grado de fragmentación de los restos óseos indica un tipo de cocinado que implicaba el troceado de las porciones cárnicas, a juzgar por la densa concentración de esquirlas en el Area del Hogar. La estructura de edades de la fauna representada en el edificio Alfa apunta a un aprovechamiento de productos lácteos en el caso de los ovicapridos y quizás también de las vacas. El procesado de lácteos se llevaría a cabo en el área del Hogar, donde también se localiza una

concentración importante de Ollas tipo A.

Por primera vez en el estudio de un grupo doméstico del Horizonte de Son Ferragut contamos con datos directos de la producción de alimentos vegetales. La existencia de almidón de Triticum aestivum (análisis de D. Zurro) sobre una losa del Area entre Tirantes confirma la disponibilidad de cereal. Se empleó en la preparación de tortas, en un cocinado a baja temperatura sobre losas, que se llevaban hasta el lugar de consumo, donde han sido encontradas. Además, la presencia en el Area del Hogar de un diente de hoz de sílex con huellas de uso (análisis de I. Briz y J. Pijoan) puede relacionarse con el mantenimiento de hoces que pudieron destinarse a la cosecha del cereal. El estudio del utillaje macrolítico (análisis de R. Risch) sugiere que un molino del Area del Hogar y un mortero del Area del Receptáculo pudieron relacionarse con la fabricación de harina de cereal.

El imprescindible transporte de agua, que debió obtenerse fuera del Puig Morter, tuvo que contar con recipientes apropiados, entre los que destacan Cántaros y Ollas A, cuya presencia en el Area del Hogar puede relacionarse con la disponibilidad de agua para las tareas de cocinado. Estas tareas se llevaron a cabo posiblemente mediante el empleo de Vasos y de Ollas tipo B, los recipientes más idóneos para tal actividad.

Finalmente, hay que sumar a las actividades mencionadas, la busqueda de combustible, el mantenimiento de la propia infraestructura del Hogar, el cocinado de los alimentos y el hervido de tisanas. El único resto carpológico identificado corresponde al género Mercurialis (Stika, com. pers). Precisamente la especie Mercurialis annua es una planta usada en la medicina tradicional por sus cualidades laxo-purgantes. Otra especie, Mercurialis perennis, se emplea actualmente para preparados homeopáticos. Los principios activos de las plantas del género Mercurialis se encuentran en la parte aérea, de manera que la presencia de una semilla sería coherente con la recolección de esta planta para un uso terapeútico.

PRODUCCIÓN DE ARTEFACTOS.

La producción doméstica de cerámica en el Edificio Alfa está atestiguada por evidencias indirectas de la presencia de materia prima: la llegada a la vivienda de arcillas de cueva es deducible de la presencia de estalagmitas. Además existen instrumentos de trabajo alfarero tales como como bruñidores confeccionados en cerámica (estudio de R. Risch), que aparecieron en el Area del Hogar y Pasillo, o en el Area del Receptáculo. Se han detectado dos tendencias de producción cerámica en el Edificio Alfa, que implican la existencia de dos técnicas diferenciadas en la elaboración, acabados y cocción de los recipientes (análisis de R. Risch y D. Gómez Gras). Ese mismo análisis ha permitido definir producciones alfareras llegadas de otros centros de producción, aunque en número reducido. Los productos cerámicos fabricados en el Edificio Alfa corresponden a los mismos tipos que conocemos en otros asentamientos del mismo momento.

La utilización de lana, procedente del rebaño doméstico, permitiría el empleo de la misma en la fabricación de tejidos. La aparición en un sector del Area del Hogar de un peine de madera de Erica (análisis de R. Piqué) podría asociarse a su empleo en la fabricación de tejidos de lana (probablemente mediante telares de cintura), uso que también pudo tener un punzón de hueso. Otro punzón, en este caso de cobre, pudo servir para trabajar materiales más consistentes, como el cuero.

La presencia de diversas conchas de moluscos marinos en el Area del Hogar (análisis de A. Bonet), en diversos estados de procesado para la fabricación de colgantes, indica trabajos esporádicos de manufacturación de ornamentos

Por último, la identificación de dos balas de honda de piedra, una de ellas contextualizada en el Area entre Tirantes, ha sido confirmada en el análisis del utillaje macrolítico (R. Risch), y pone en evidencia la existencia de armas de guerra especializadas. Las balas con forma fusiforme tienen un objetivo bélico que conlleva la existencia de algún tipo de situación de conflicto intercomunitario, materializado finalmente en la destrucción de los asentamientos de la región de Es Pla de Mallorca, hecho que coincide con el final del Horizonte de Son Ferragut.

EL ALMACENAJE.

Hemos podido distinguir dos tipos de almacenaje. El primero, un almacenaje a largo plazo, localizado en el Area del Receptáculo y asociado a Orzas, Tinajas y Cazuelas Carenadas, algunas de las cuales contenían porciones de carne, tal como lo indica la elevada concentración de huesos de ovicápridos y sobre todo de bóvidos hallados junto a ellas. Los mismos tipos cerámicos aparecieron en la Estancia A, donde se ubicaba otro espacio destinado al mismo fin.

El segundo tipo de almacenaje consiste en una reserva a corto plazo, cotidiana. Se vincula a Vasos, Ollas A, Ollas B y Cántaros. Serviría para el acopio de agua, líquidos y otros productos alimentarios. Se localiza en el Area del Hogar, entre Tirantes, en el Receptáculo y en la Estancia A.

Por último, cabe mencionar la presencia de escápulas de Buey trabajadas (J. Estévez y M. Montero), cuya función como paletas o cucharones para trasladar productos alimentarios de un recipiente a otro resulta factible.

EL CONSUMO DE ALIMENTOS Y EL USO SOCIAL DE OBJETOS SINGULARES.

A partir del estudio de los residuos faunísticos, han podido ser circunscritas tres áreas principales de consumo alimentario, dos de ellas de ovicápridos y la tercera de bóvidos. El Area del Hogar y la de entre Tirantes constituyen áreas de consumo cotidiano de carne de ovicápridos. Esta situación implica la existencia de dos grupos en el seno de la unidad doméstica, que se reunían habitualmente en las comidas. El primero lo hacía en el Area del Hogar, el espacio donde se concentra la mayor parte de actividades productivas, de manera que este grupo incluía a quienes destinaban su trabajo a dichas actividades. Por el contrario, el segundo grupo se reunía en el Area de Tirantes, en la que no existe ningun indicio de prácticas productivas y donde los únicos elementos ajenos al propio consumo alimentario son objetos singulares (no asociados a tareas de producción en la vivienda), de manera que debió tratarse de un grupo apartado de la producción registrada en el patio porticado del Edificio Alfa.

El consumo comunitario de carne de vacuno tenía lugar en el espacio central del patio, en torno al Area entre Columnas. Aquí se consumían las porciones (tercio superior) de las canales de vacuno que llegaban al edificio. Probablemente todo el grupo doméstico se reunía para comer carne de vacuno.

La Estancia B, el espacio más interior del Edificio Alfa, únicamente ofrece restos de consumo alimentario esporádico. Puede plantearse, por lo tanto, que se trata de un espacio especializado en el descanso. Lo mismo puede decirse de la Estancia A, donde además había recipientes de almacenaje. La dualidad de espacios de consumo se refuerza con esta la dualidad de espacios de descanso, apuntando hacia la existencia en el interior de la unidad doméstica de dos grupos sociales diferenciados.

A partir de la distinción entre el utillaje de consumo alimentario y los que denominamos objetos singulares (ajenos a las actividades de producción económica), puede observarse claramente que estos últimos se localizan casi exclusivamente en el Area entre Tirantes. Se trata de un fragmento de varilla de cobre, de la bala de honda antes mencionada, de un Cuenco Hemiesférico con perforaciones a lo largo del borde y de un colgante de concha. Es importante insistir en que dichos objetos se localizan en un área donde no se lleva a cabo ningún proceso de trabajo y en el que, al parecer, únicamente se consume.

Por su parte, entre el utillaje de consumo alimentario, contamos con la evidencia de tres losas de piedra detectadas, respectivamente, en el Area entre Tirantes, el Receptáculo y la Estancia B. El estudio de los fitolitos realizado sobre aquéllas testimonia, como ya se ha señalado, la presencia de almidón procedente de Triticum aestivum como resultado del procesado de harina. También debemos relacionar con las prácticas de consumo la presencia de contenedores cerámicos utilizables como vajilla -Vasos, Cuencos y Ollas- en el Area del Hogar, entre Tirantes y junto al Receptáculo.

LAS DIFERENCIAS EN EL USO SOCIAL DE LOS ESPACIOS DEL EDIFICIO ALFA.

El dendograma efectuado a partir del coeficiente de Jaccard nos ilustra sobre la similitud existente entre los distintos espacios del Edificio Alfa, teniendo encuenta el tipo de actividades que se realizaba en cada uno de ellos (FIGURA 4). Como puede observarse, se hace patente una clara dicotomía entre los espacios en función a los trabajos que se realizan y el lugar en el que se consume.

Así, resulta claro que el Area del Hogar y del Pasillo se muestran muy cercanas, ya que son lugares donde se ejecutan la mayor parte de las actividades productivas. El Area del Receptáculo y la Estancia A también aparecen relacionadas por la actividad de almacenamiento que las caracteriza.

En el otro extremo del uso social del espacio, se detecta una estrecha relación entre la Estancia B y el Area entre Tirantes debido a su excepcionalidad como lugares alejados de cualquier tipo de producción de objetos. En ellas tiene lugar el consumo de alimentos y la concentración de objetos singulares asi como el descanso de los individuos.

GRUPO DOMESTICO Y DISTANCIA SOCIAL

Teniendo en cuenta el registro arqueológico del Horizonte de Son Ferragut, con la coexistencia de diversas formas de organización de grupos domésticos de diferentes tamaños, así como las actividades documentadas en el marco de prácticas sociales del Edficio Alfa y en los demás lugares sociales conocidos, no estamos en condiciones de cerrar el debate sobre cuáles eran las relaciones sociales en y entre las comunidades mallorquinas de esa etapa. No obstante, consideramos que deben contemplarse varias hipótesis que se ajustan al apoyo empírico disponible, dan cuenta de las relaciones sociales internas de los grupos domésticos y de las comunidades y permiten abordar las condiciones de reproducción de la vida social.

Sin embargo, antes de entrar en el terreno de las lecturas sociológicas, creemos necesario insistir en que no hay que confundir una unidad doméstica con una falimia, ni tampoco una familia con una familia nuclear. El actualismo de la familia nuclear ha invadido, en muchas ocasiones, la bibliografía arqueológica, y debemos movernos con mucha precaución ante generalizaciones basadas en la universalidad de lo propio. Definimos unidad doméstica como un grupo con relaciones de consaguineidad, afinidad y/o imposición, de tipo familiar o no, cuyos vínculos se establecen a partir de la convivencia cotidiana y proporcionan el contexto idóneo para los cuidados implicados en la producción de cuerpos y su mantenimiento. En cambio, el concepto de familia se relaciona con un grupo de personas vinculadas a partir del matrimonio, adscrito a la procreación y cuidado de los hijos e hijas (San Román y González 1994: 27), y con una referencia de residencia común.

En definitiva, la familia no es la forma universal de unidad doméstica, y hace ya mucho tiempo que se ha cuestionado tal presunción generalizadora (Levi-Straus, Spiro y Gough 1987). Es por ello que, para la arqueología prehistórica, preferimos la denominación de unidad doméstica, intentando matizarla en lo posible, a fin de no homogeneizar ni naturalizarlo todo, homogeneización y naturalización que tanto atrae al poder dominante y coercitivo.

Los diversos criterios empleados para establecer el tamaño del grupo que vivía en el Edificio Alfa del Puig Morter, aplicados anteriormente a Son Fornés, en especial a la HT5, la úniva vivienda completa, convergen en un numero que oscila entre 20 y 26 individuos (Castro Martínez, Escoriza Mateu y Sanahuja Yll 2001 e.p.) frente a los 6-9 individuos de las unidades domésticas de Son Fornés. Ello permite plantear sólo cuatro posibilidades, según los criterios establecidos por las Ciencias Sociales, en especial la Antropología, de cara a explicar el sistema de parentesco, y por extensión las relaciones sociales y políticas, al menos en la región de Es Pla del centro de Mallorca (FIGURA 5):

LINAJES MATRICÉNTRICOS.

Esta hipótesis descarta la existencia de familia y la exogamia. Así, un linaje matricéntrico en viviendas como el Edificio Alfa, puede incluir a hermanos y hermanas, junto con las hijas e hijos de éstas. Paralelamente, podríamos encontrar células matricéntricas en pequeñas unidades domésticas de tipo Son Fornés, de manera que en el Horizonte de Son Ferragut no tendrían lugar estructuras patriarcales ni familiares. Las distintas comunidades, al menos en Es Pla de Mallorca, compartirían unos mismos criterios de socialización y similares referentes ideológicos. Si, por contra, se defendiera que las unidades domésticas de Son Fornés estuvieran compuestas por familias nucleares, la distancia social entre comunidades acarrearía profundas brechas entre ellas. La divergencia en la socialización de los individuos, en las prácticas políticas y en las concepciones ideológicas, dificultaría el establecimiento de sistemas de circulación de individuos, lo cual, sin duda, podría implicar unas difíciles relaciones de coexistencia en los reducidos espacios territoriales intensamente ocupados por las comunidades mallorquinas. Estas condiciones no parecen las idóneas para comunidades matricéntricas. Estas están basadas en el control por las mujeres de la producción de cuerpos y probablemente en una división del trabajo escasamente marcada o fundamentalmente orientada a asegurar la redistribución equitativa entre toda la comunidad y a evitar relaciones de explotación en las tres producciones de la vida social (producción de cuerpos, producción de objetos y producción de mantenimiento de cuerpos y objetos), ya que la reproducción social de unidades matricéntricas exige mantener una estabilidad y continuidad de los grupos domésticos.

COMUNIDADES DE FAMILIAS MONÓGAMAS.
Esta hipótesis presupone una familia de tipo extenso, monógama y dependiente, en viviendas de tipo Son Ferragut, con la convivencia de tres generaciones y con una media de cuatro hijos/hijas por matrimonio, de acuerdo con una ratio que se ajusta a las estimaciones de una unidad familiar nuclear en viviendas de tipo Son Fornés. La sexuación de la fuerza de trabajo en una familia extensa basada en matrimonios monógamos ofrece una composición por sexos equiparable a la de una familia monógama independiente, de manera que la presencia de un hombre y de una mujer en cada unión matrimonial acarrea una situación similar a la de una familia nuclear con división sexuada de tareas. Ambas comparten una misma concepción de la socialización y del trabajo que facilita la exogamia entre grupos. Pero las diferencias entre comunidades de familias extensas y de familias nucleares repercuten en el potencial para organizar la fuerza de trabajo, que es mayor en las familias extensas. En esta situación, la división del grupo doméstico detectada en el Edificio Alfa puede explicarse en términos de división por sexos, con las mujeres, y las criaturas, vinculadas al trabajo y al consumo en el Area del Hogar, y con los hombres asociados al Area entre Tirantes, a la guerra y a actividades ejecutadas fuera de las viviendas. Si estas últimas no compensaban el trabajo realizado en el propio espacio doméstico nos encontraríamos con una situación de explotación interna en los grupos domésticos, que beneficiaría al colectivo masculino.

GRUPOS DOMÉSTICOS POLÍGINOS.
Según esta hipótesis, la unidad doméstica del Edificio Alfa debería estar constituida por una familia formada por un hombre y, al menos, tres o cuatro mujeres con sus hijos/as. La existencia de familias políginas independientes establece unas condiciones en las que difícilmente puede evitarse la explotación del trabajo de las mujeres en beneficio de los hombres. Además, la propia exigencia de la poligamia supone que no todos los hombres pueden disponer de varias mujeres, de forma que, si no existe una elevada mortalidad masculina por alguna razón, se establecen unas condiciones para que algunos hombres, a través del matrimonio, se beneficien del trabajo de las mujeres en mayor medida que otros, abocados a la monogamia. Ese podría ser el caso de las unidades domésticas de tipo Son Fornés, donde el registro no permite avalar una situación de poligamia, dado el pequeño tamaño de las células domésticas. Unicamente si el cabeza de familia residiera fuera de la unidad doméstica podría explicarse una situación equiparable a la que puede sugerirse para Son Ferragut. En el ámbito interno del Edificio Alfa, los dos grupos identificados serían fruto de una división sexual del trabajo, del consumo y del descanso, con un espacio específico para los hombres (padre e hijos), desvinculados de los trabajos productivos dentro de la unidad habitacional.

SERVIDUMBRE FAMILIAR.
El Edificio Alfa podría corresponder también a una unidad doméstica formada por una familia monógama independiente y por individuos a su servicio. Según esta hipótesis, la división del trabajo, del consumo y del descanso en espacios diferenciados se explicaría a partir de la propia configuración de clases sociales. Los componentes de la familia propietaria ocuparían el Area de Tirantes y la Estancia B, mientras que hombres y mujeres dependientes estarían en el Area del Hogar y en la Estancia A. El panorama social resultante implicaría que en el Horizonte de Son Ferragut existía una clase social dominante, que residía en comunidades como el Puig Morter, y que pudo haberse apropiado en beneficio propio de la fuerza de trabajo de sectores dependientes, asentados en otras comunidades, como Son Fornés. Este modelo social no excluye, sino que refuerza, la probable explotación de las mujeres.

Como ya hemos ido comentando al mencionar cada una de las cuatro posibilidades en relación con la composición del grupo doméstico que habitaba el Edificio Alfa, la explicación al hecho de que el consumo cotidiano, derivado de la ganadería de ovicápridos gestionada por el propio grupo doméstico, esté fragmentado, implica que el propio grupo también lo estaba. Es decir, que existía una situación de diferencia social dentro del propio grupo, diferencia que podría suponer una desigualdad social. Esta segmentación del grupo doméstico podría responder a dos situaciones:

(a) Hombres y mujeres del Edificio Alfa se reunían para comer y trabajar en espacios distintos. Si el grupo doméstico estaba formado por unos 23 individuos, el sector masculino contaría con 4-6 hombres, que tendrían cabida en el Area entre Tirantes, y el resto estaría formado por mujeres y niños/as de diferentes edades, que trabajarían y comerían en al Area del Hogar.

(b) La familia propietaria del Edificio Alfa e individuos dependientes serían los que ocuparían espacios diferenciados a la hora de las comidas cotidianas. El primer grupo se ubicaría en el Area de Tirantes, mientras el segundo permanecería en los espacios de trabajo habitual. Este segundo grupo de individuos dependientes podría estar constituido por parientes sin derechos de propiedad o por criados/as, siervos/as o esclavos/as.

En ambas hipótesis, el consumo de carne de bóvidos respondería a un reagrupamiento de la totalidad de la unidad doméstica, seguramente en el marco de ciertas celebraciones o ceremonias.

En el caso de aceptar la primera hipótesis, es decir, que los hombres comieran en el Area entre Tirantes y las mujeres y los niños/as en el Area del Hogar, podríamos plantear que existía una clara distancia social sobre el colectivo femenino y los jóvenes si los hombres que consumían el alimento en el Area entre Tirantes se beneficiaban de todo el trabajo productivo realizado por las mujeres y los niños sin ningún tipo de contrapartida material. En el Area entre Tirantes no aparece ningún elemento que nos permita hablar de prácticas económicas. Sin embargo, si suponemos que los hombres realizaban tareas fuera del Edificio Alfa (pastoreo) y quizás estaban también asociados la actividad bélica (presencia de la bala de honda) y a la gestión política, la explotación absoluta deja de tener sentido y habría que matizar la existencia de una explotación parcial (más horas de trabajo por parte de un sector frente al otro) por parte del colectivo masculino, lo que resulta difícil de precisar. Sólo en el caso de la familia polígina independiente cabría asegurar la explotación de sexo por la propia composición política de la poliginia y quizás también la de clase, ya que el matrimonio con varias mujeres responde a un ideal social que sólo pueden realizan algunos hombres del grupo, no todos.

Si, por el contrario, aceptamos la segunda posibilidad, quedaría claramente evidenciada una explotación absoluta de clase de una familia monógama propietaria del Edificio Alfa sobre la servidumbre que trabaja para ella. La explotación de sexo, en este caso, no podría ser precisada.

NOTA

Las investigaciones realizadas en el Edificio Alfa han sido posibles gracias a la financiación del proyecto PB98-0889 de la DGESIC y de las subvenciones de la Comissió de Cultura i Patrimoni del Consell Insular de Mallorca. El autor y las autoras pertenecen al "Grup de Recerca d´Arqueoecologia Social del Mediterrani" de la Generalitat de Catalunya (SGR2000-00051). Agradecemos a nuestros compañeros V. Lull, R. Micó y R. Risch y a nuestras compañeras S. Gili y C. Rihuete Herrada el interés mostrado por nuestro trabajo y el tiempo invertido en las discusiones a lo largo del camino que nos ha permitido llegar a los resultados expuestos, y en particular a V. Lull, codirector de las excavaciones, que nos ha proporcionado toda la información que hemos necesitado.

BIBLIOGRAFIA.

ARAMBURU-ZABALA, J. ,
1998. El patrón de asentamiento de la cultura talayótica. Palma de Mallorca, El Tall.

ARAMBURU-ZABALA, J., GARRIDO, C. Y SASTRE, V.
1994. Guía arqueológica de Mallorca, Palma de Mallorca, La Foradada.

CALVO TRIAS, M. y SALVA, B. ,
1999), "Aproximació a la seqüencia cronocultural de la naveta i del jaciment de Closos de Can Gaià", Mayurqa, 25: 59-82.

CASTRO MARTÍNEZ, P.V., ESCORIZA MATEU, T. y SANAHUJA YLL, M.E. ,
200., e.p.), El Edificio Alfa del Puig Morter de Son Ferragut (Sineu, Mallorca), en prensa.

CASTRO MARTINEZ, P.V., LULL, V. y MICO, R.,
1996), Cronología de la Prehistoria Reciente de la Península Ibérica y Baleares (c. 2800-900 cal ANE), Tempus Reparatum, Oxford, "British Archaeological Reports, International Series", 652.

GASULL, P., LULL, V. y SANAHUJA Yll, M.E.,
1984a. Son Fornés I: La Fase Talayótica. Ensayo de recxonstrucción socio-económica de una comunidad prehistórica de la isla de Mallorca, B.A.R., Oxford, "British Arcaheological Reports, International Series", 209.

GASULL, P., LULL, V. y SANAHUJA Yll, Mª E.,
1984b. "Estudio comparativo de los talaiots nº 1 y 2 de Son Fornés (Montuiri, Mallorca)", en WALDREN W. H., CHAPMAN, R. W., LEWTHWAITE, J. G. y KENNARD, R.-C. eds, The Deià Conference of Prehistory. Early Settlement in the Western Mediterranean Islands and Peripherial Areas, Oxford, "British Arcaheological Reports, International Series", 229, iv, pp. 1239-1258.

GASULL, P., LULL, V. and SANAHUJA Yll, M.E.,
1984. "La habitación no.5 de Son Fornés (Montuiri-Mallorca). Modelo de una vivienda talayótica", in WALDREN W. H., CHAPMAN, R. W., LEWTHWAITE, J. G. and KENNARD, R.-C. (eds), The Deià Conference of Prehistory. Early Settlement in the Western Mediterranean Islands and Peripheral Areas, Oxford, "British Arcaheological Reports, International Series", 229, iv, pp. 1259-1297.

GUERRERO, V.M. ,
1996. "Son Corró. Restauración (destrucción) del santuario prehistórico mallorquín más emblemático". Revista de Arqueología, nº 178, febrero: 42-51.

GUERRERO, V.M. ,
1999. Arquitectura y Poder en la Prehistoria de Mallora, El Tall, Palma de Mallorca, "El Tall del Temps", 31.

GUERRERO, V.M. y CALVO, M. ,
2001. URL: http://www. talaiots.com (febrero 2001).

LEVI-STRAUSS, C., SPIRO, M.E. y GOUGH, K.,
1987. Polémica sobre el origen y la universalidad de la familia. Barcelona, Anagrama.

LILLIU, G. ,
1962-63. "Cenno sui piu recenti scavi del villaggio talaiotico di Ses Païsses ad Artà-Maiorca (Baleari)", Studi Sardi, XVIII: 22-52.

LULL, V., MICO, R., RIHUETE HERRADA, C. y RISCH, R.,
1999. Ideología y Sociedad en la Prehistoria de Menorca. La Cova des Carritx y la Cova des Mussol , Consell Insular de Menorca, Barcelona.

PONS, G. ,
1999. Anàlisi espacial del poblament al Pretalaiòtic final i al Ta-

laiòtic I de Mallorca, Consell de Mallorca-Cultura i Patrimoni, Palma, "Col.lecció La Deixa", 2.

SAN ROMAN, T. y GONZALEZ ECHEVARRIA, A.,
1994), Las Relaciones de Parentesco, Universitat Autònoma de Barcelona, Bellaterra.

STUIVER, M. y REIMER, P.J.,
1993), "Extended 14C Data Base and Revised CALIB 3.0 14C Age Calibration Program", Radiocarbon, 35: 215-230.

STUIVER, M., REIMER, P.J., BARD, E., BECK, J.W., BURR, G.S., HUGHEN, K.A, DROMER, B., MCCORMAC, F.G., van der PLICHT, J. and SPURK, M. (1998), "INTCAL98 Radiocarbon Age Calibration, 24,000-0 cal BP", Radiocarbon, 40:1041-1083.

VAN STRYDONCK, M., WALDREN, W.H. y HENDRIX, V.,
1998. "The 14C Chronology of the Son Mas Sanctuary Site (Valldemosa, Mallorca, Spain)", Radiocarbon, 40 (2): 735-748.

WALDREN, W. H.,
1992. Radiocarbon and Other Isotopic Age Determinations Fron the Balearic Islands. A Comprehensive Inventory, Deià Archaeological Museu, Deià, DAMARC, 26.

WALDREN, W. H. ,
1998. Beaker Culture of the Balearic Islands, Oxford, BAR International Series, 709, Western Mediterranean Series, 1.

45

BALANCE Y NUEVAS PERSPECTIVAS EN LA INVESTIGACIÓN PREHISTÓRICA DE LAS ISLAS PITIUSAS

BENJAMÍ COSTA RIBAS (1)
VÍCTOR M. GUERRERO AYUSO (2)

(1) Conservador del Museo Arqueologico de Ibiza y Formentera
EMAIL: maef@jet.es
(2) University of the Balearic Islands, Palma de Mallorca, Baleares, Spain
EMAIL: vmguerrero@uib.es

SUMMARY

Archaeological research on the Prehistory of Ibiza and Formentera (Pityusic Islands) suffered a considerable delay during the 20th century in relation to that carried out in Majorca and Minorca. On one hand, the lack of monumental remains, together with the richness and spectacularity of Punic finds in Ibiza, have traditionally darkened the study of other periods of the Pityusic History. On the other hand, despite the promising beginnings of the seventies, with impressive discoveries such as the megalithic chamber-tomb of Ca na Costa in Formentera, continuity in the research and long-term programmes was slack after that. Thus, radiocarbon dating test (14C) to build up our own sequence for the Prehistory of Pityusic Islands are currently scarce. Apart from that, there are no analysis to know the natural environment that met the first humans who arrived, and to evaluate its transformation due to the impact of settlement by permanent inhabitants. These facts make it difficult to draw the lines of the social, economic and cultural evolution of the prehistoric communities dwelling in these islands, of which we offer here the present state. However, a refocusing of the guidelines and research strategies makes it possible for us to expect successful results in future investigation.

INTRODUCCIÓN

Los estudios de la prehistoria de las Pitiusas han constituido, tradicionalmente, la hermana pobre de la investigación arqueológica balear. Ello ha sido así por varias razones que tienen su explicación lógica, aunque no justificarían que esta situación continuara así por más tiempo. Por un lado, la importancia del poblamiento fenicio-púnico de estas islas oscureció durante muchas décadas una historia anterior ("prehistoria") tan interesante y rica como pudiera serlo la de cualquiera de las otras islas. Por otro, la falta de una arquitectura ciclópea de carácter tan monumental como la de Mallorca y Menorca no atrajo el interés por la prehistoria pitiusa de los investigadores que trabajaban en aquellas islas, quienes a la sazón estaban embarcados en sesudos debates sobre cuestiones tipológicas, funcionales y cronológicas de esta arquitectura monumental (Rosselló 1979, Plantalamor 1991). Tampoco los que investigaban en las Pitiusas, fascinados por los espectaculares hallazgos púnicos que se estaban realizando en Ibiza desde los inicios de la actividad de la Sociedad Arqueológica Ebusitana en 1903, prestaron alguna atención al pasado prehistórico de estas islas.

De este modo, durante muchas décadas todos los esfuerzos investigadores en las Pitiusas fueron consumidos en reconstruir su floreciente pasado como centro clave de la cultura púnica. Todos los hallazgos prehistóricos realizados desde los inicios del siglo XX pasaron casi desapercibidos, de manera que gran parte de los investigadores nacionales y extranjeros concluyeron que las Pitiusas debieron permanecer deshabitadas, o muy poco pobladas, durante la prehistoria. Fue en los años setenta cuando se produjo un punto de inflexión en esta situación. Los primeros trabajos de Jordi H. Fernández intentando presentar pruebas materiales de poblamiento prehistórico, a partir del estudio de elementos metálicos de la Edad del Bronce desgraciadamente siempre descontextualizados (Fernández, 1973 y 1974), junto a las sagaces consideraciones de Miquel Tarradell y Matilde Font, acerca de la improbabilidad de que unas islas como las Pitiusas se hubiesen mantenido despobladas durante la prehistoria (Taradell y Font, 1975), culminaron con el descubrimiento y excavación del sepulcro megalítico de Ca na Costa en Formentera dirigidas por el equipo formado por Jordi H. Fernández, Lluís Plantalamor y Celia Topp (Fernández, Plantalamor y Topp, 1976; Ibidem, 1988).

A las excavaciones de Ca na Costa siguieron otros hallazgos de yacimientos importantes en Formentera, como los del Cap de Barbaria y la muralla de sa Cala en la Mola, así como el de Can Sargent en Ibiza (Fernández, 1977; Topp, Fernández y Plantalamor, 1979; Fernández y Topp, 1984), cuyas excavaciones fueron dirigidas por el mismo equipo. Todo ello por fin hizo despertar a muchos del sueño irreal de unas islas mediterráneas próximas a las costas continentales pero supuestamente deshabitadas hasta los inicios de la Historia, que había durado demasiado tiempo. Pero aún en el año 1978, en una revista de contenido divulgativo, Historia 16, J.H. Fernández publicaba un artículo que con el significativo título de "Formentera salta a la Prehistoria" (Fernández, 1978: 59-66), reflejaba aún la cruda realidad de un desconocimiento generalizado en los ámbitos científicos de la naciente prehistoria pitiusa.

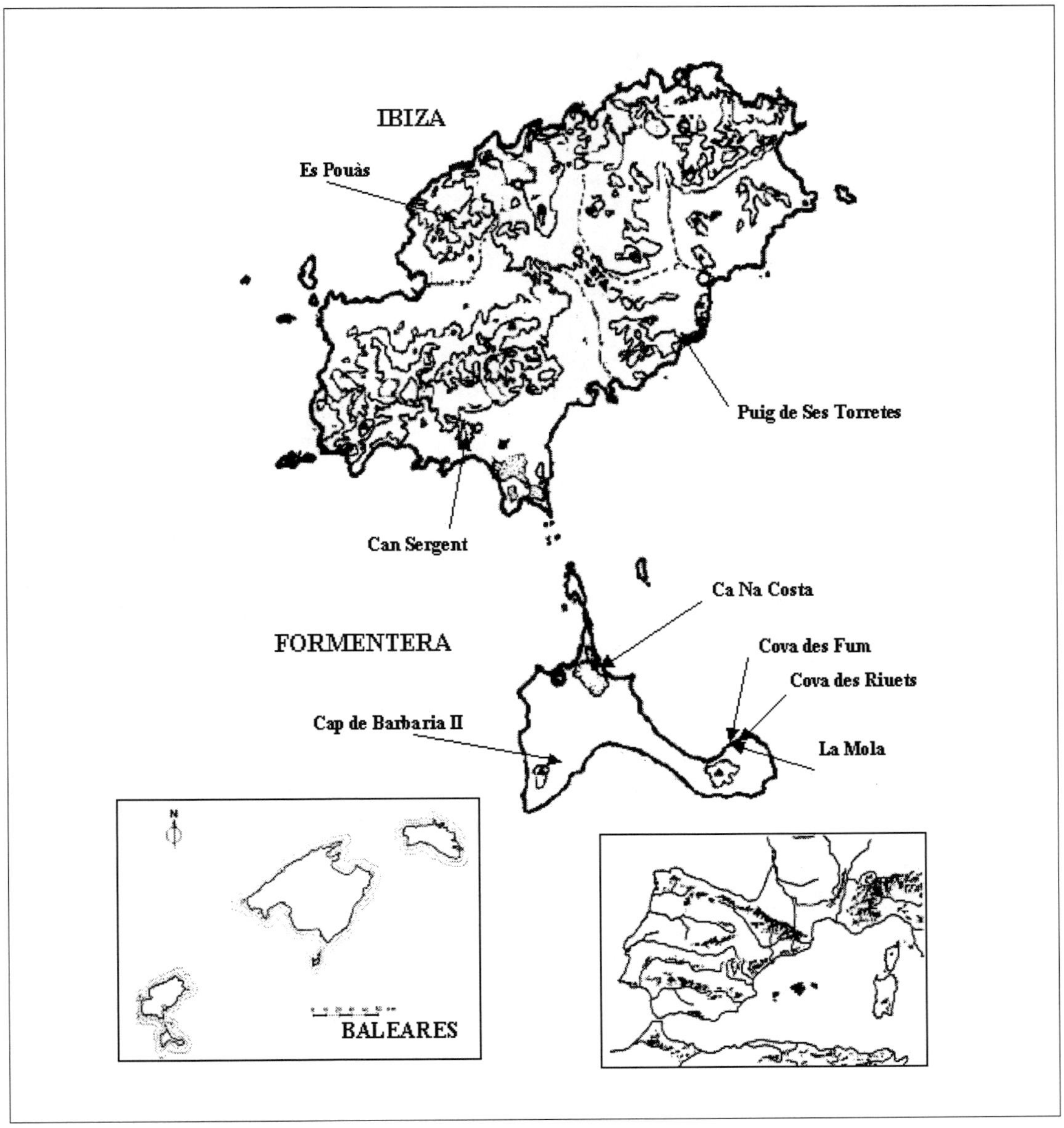

FIGURE 1. MAPA DE IBIZA Y FORMENTERA CON LOS YACIMIENTOS CITADOS EN EL TEXTO.

A mediados de la década siguiente hubo significativos esfuerzos debidos a J. Ramón (1985), a C. Gómez y Mª. P. San Nicolás (1988) de elaborar estados de la cuestión que fueron en su momento de una utilidad notable. A todo ello podemos sumar también las contribuciones de Celia Topp, centradas en el problema que planteaba la presencia de cerámicas incisas en las Pitiusas, discutiendo su filiación campaniforme (Topp, 1987a; Eadem 1987b y Eadem, 1988). Una síntesis muy completa del pasado prehistórico y protohistórico de las Pitiusas la aportaron B. Costa y J.H. Fernández a las X Jornades d'Estudis Històrics Locals celebradas en Palma en el otoño de 1991 (Costa y Fernández, 1992), que hasta la reciente publicación de nuevos trabajos poniendo al día el estado de la investigación (Costa, 2000;

Costa y Benito, 2000), ha sido una útil herramienta de trabajo, sobre todo para la docencia universitaria ante el clamoroso déficit de textos asequibles a los estudiantes que se enfrentan por primera vez a estos estudios.

A pesar de todo, el conocimiento de la prehistoria pitiusa continuaba lastrada por dos carencias de la investigación muy significativas que la hacían distanciarse de los avances que ya se estaban produciendo en las otras islas mayores. Por un lado, no se disponía de una secuencia de dataciones absolutas que permitiese ordenar el desarrollo cultural de estas islas de forma autónoma (Costa y Fernández, 1992: 280); es decir, sin tener que depender de las dataciones obtenidas en Mallorca, o en otros lugares, como marco de referencia cronológico para datar, a partir de supuestos paralelismos culturales, los materiales y contextos pitiusos (Costa y Benito 2000: 228-229). Y, por otro lado, no había estudios paleoambientales para conocer los rasgos principales del ecosistema de las Pitiusas durante la prehistoria, ni tampoco se habían planificado jamás estudios destinados a esclarecer las bases subsistenciales de las comunidades que vivieron en las islas antes de su colonización por nuevos grupos de gentes semitas (Costa y Benito 2000: 227).

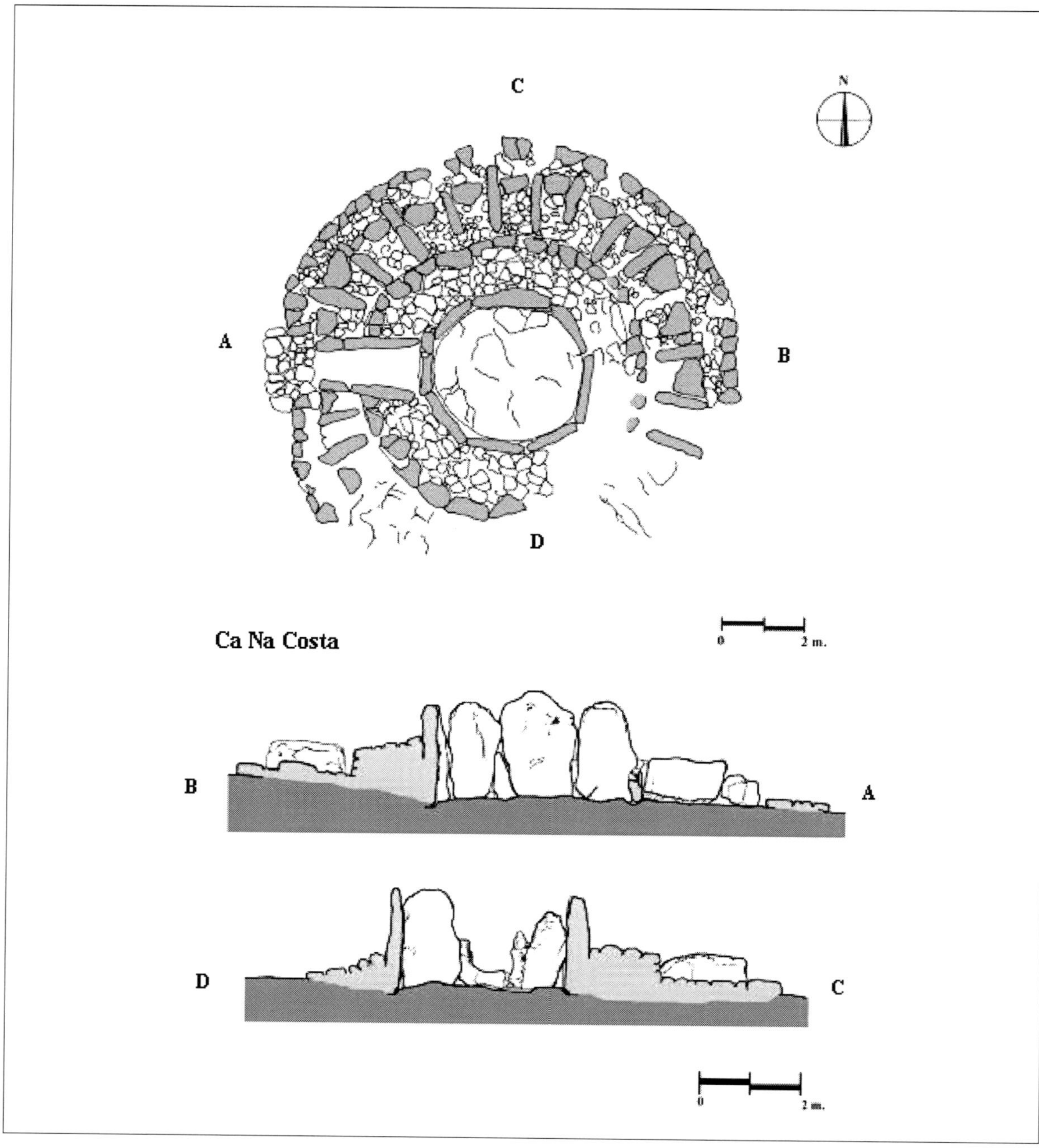

FIGURE 2. DOLMEN DE CA NA COSTA (SEGÚN FERNÁNDEZ, PLANTALAMOR Y TOPP 1987 RETOCADO).

Estas carencias están lejos de haber sido subsanadas aún. Sin embargo, recientes trabajos han abierto nuevas perspectivas que enfocan la interpretación del pasado prehistórico de las islas desde planteamientos que toman en consideración variables biogeográficas que hasta ahora pocas veces habían sido tenidas en cuenta. Al mismo tiempo, también se están iniciando análisis paleoambientales y nuevas dataciones radiocarbónicas que a buen seguro compensarán más pronto que tarde el retraso crónico de estas islas en materia de investigación prehistórica.

Otro problema que cabe resaltar es la falta de conexión entre las investigaciones que se desarrollaban en cada isla, cosa que no favorecía la comunicación y el conocimiento entre los diferentes grupos investigadores, dando lugar a un autismo que hasta hace muy poco no había permitido tener una visión coherente de conjunto de todo el archipiélago balear (Costa y Guerrero, 2001: 29).

En este contexto, la presente comunicación no pretende ser otra cosa que una breve reflexión sobre la situación en la que nos encontramos y las perspectivas que pueden abrirse, en el caso de que la tradicional trayectoria ya comentada se rectifique y se preste a la prehistoria de las islas menores, tanto por parte de los investigadores como de la Administración competente, la atención que merece. Precisamente, su escasa dimensión territorial convierte Ibiza y Formentera en casos de estudio de extraordinario interés para la investigación de los procesos de adaptación de pequeñas comunidades humanas en un medio ambiente caracterizado por su reducida biodiversidad y, en general, su escasa productividad (Costa y Benito, 2000: 238-239).

ALGUNAS CUESTIONES ACERCA DEL PROCESO DE POBLAMIENTO EN MEDIOS INSULARES

Es bien conocida la dificultad que entraña hablar con seguridad del origen de algo en prehistoria. Como toda ciencia, su investigación requiere de documentación empírica que no siempre se muestra disponible ni suficientemente sólida. Planteamientos teóricos e hipótesis de trabajo han de suplir entonces la falta de suficientes evidencias empíricas. Sin embargo, por muy razonables y hasta probables que aquellos puedan parecer, no pueden ser utilizados como demostración de hechos no contrastados positivamente en el registro arqueológico. Ello no debe impedir que se trabaje siempre a partir de modelos teóricos de los que emanan las hipótesis que manejamos, las cuales deben ser continuamente confrontadas con la documentación disponible en cada momento. Pero es justamente en los casos que existen pocos datos, cuando las hipótesis deben estar dotadas de mayor flexibilidad. Las afirmaciones rotundas y taxativas no sólo son arriesgadas, sino nocivas incluso para el avance de la propia ciencia. Y aún más en arqueología prehistórica, donde la pérdida de documentación con el paso del tiempo es notable.

Por lo que a los procesos de poblamiento se refiere, en la actualidad está bien contrastado que la llegada del hombre a ecosistemas vírgenes no ha sido nunca neutra. Incluso en el caso de grandes masas continentales como América o Australia su primera presencia se ha saldado con las extinciones de muchas especies de animales (con carácter general: Martín y Klein 1984; MacPhee y Sues 1999; Roberts et alii 2001; Alroy 2001), que pueden ser especialmente rápidas entre los grandes vertebrados. Actuaciones como son, por ejemplo, los incendios forestales para favorecer la caza, más que la propia caza, han sido uno de los factores más decisivos en los cambios medioambientales que siempre siguen a la presencia del hombre en un territorio (Smith 1970; Mellars 1976; Edwards 1982; Edwards y Ralston 1984; Kershaw 1986; Riera 1996). Las islas, precisamente por su limitado territorio, acusan con mayor intensidad este fenómeno, de forma que la rapidez e intensidad del cambio ecológico va en relación directa con su extensión. Hoy podemos afirmar que la presencia humana, antes que el cambio climático del Holoceno, ha sido la principal causa , cuando no la única (Schüle 1993, Vigne 1999), de la extinción de las especies faunísticas autóctonas y de profundos cambios en la cubierta vegetal, particularmente cuando la primera colonización es protagonizada por grupos humanos con economías productoras que, además, continúan explotando la caza como importante estrategia de subsistencia complementaria (Stuart 1991).

Tanto la arqueología como la etnología comparada permiten saber que existen distintas formas de explotación de territorios marginales, y las islas siempre lo son con respecto a las tierras continentales. Una de estas formas de utilización de un territorio es la frecuentación más o menos intensa del mismo para el aprovechamiento de algunos recursos, sin que ello implique asentamientos permanentes de población. Este sistema de interacción entre hombre y ecosistema aporta algo tan valioso o más que los propios recursos materiales: información contrastada del territorio insular, sin la cual no es viable ninguna colonización no azarosa del territorio. De esta forma, la implantación de población permanente va siempre precedida de un tiempo, más o menos largo según los casos (Schwartz 1970; Masseti and Darlas 2000; Schüle 2001), que podríamos denominar genéricamente "precolonización". El análisis de estos procesos ha dado lugar a la definición de modelos teóricos, como el de Graves y Addison (1995), inspirado a su vez en el que previamente diseñó Irving (1992), y recogido y complementado por otros investigadores (Patton, 1996; Costa, 2000). Este modelo y sus variantes se basan en establecer una secuencia continua del proceso de poblamiento de un medio insular, definiendo una serie de fases sucesivas – descubrimiento, frecuentación y/o colonización temporal y, finalmente, establecimiento permanente de población- que en función de diversas variantes (Keegan y Diamond, 1987) pueden tener diferente duración en cada caso, pero siempre se suceden en el mismo orden.

El sistema de vida cazador recolector está caracterizado, entre otras cosas, por una bajísima densidad de población, de forma que bandas de entre 25 y 40 individuos explotan territorios muy extensos (Binford 1978; Silberbauer 1981; Lee y Daly 1999). Por esta razón los cazadores recolectores no han colonizado las decenas de pequeñas islas que jalonan el Mediterráneo. Solo algunas de gran extensión como Sicilia (Tusa 1999), Córcega (Camps 1988), Cerdeña (Klein Hohmeijer 1997) o Chipre (Simmons 1999) tuvieron poblaciones cazadoras recolectoras estables antes de la implantación del Neolítico. No obstante, existen evidencias contrastadas de la frecuentación sin asentamientos permanentes por cazadores recolectores de pequeñas islas mediterráneas como es el caso muy claro de Melos (Perlès 1979; Idem 1995; Cherry y Torrence 1982), y probablemente de Skyros (Efstratiou 1985), entre otras (Patton 1996:48-52).

La adopción de economías productoras cambió esta relación entre los grupos humanos y el territorio que explotaban. De esta forma comunidades neolíticas terminaron asentándose desde principios del V milenio a.C. de forma permanente y colonizando hasta sus últimas consecuencias islas con una extensión superficial menor que Ibiza y con ecosistemas muy poco productivos debido a su aridez y dificultad para abastecerse de agua potable como Malta (Bonanno, 2000). Y más aún, pequeñas islas, mucho más alejadas del continente y con menor extensión que Formentera, como Pantelaria (Courtin 1983), Lípari (Tusa 1999: 145-49) o Lampedusa (Radi 1972) fueron colonizadas por pastores que fundaron aldeas permanentes entre el VI y el V milenio BC. Estas dos últimas no superan un tercio y un cuarto respectivamente de la extensión de la propia Formentera. Algo similar ocurre también en muchas islas egeas (Broodbank 2000), e incluso pequeños islotes, como Panarea, al Norte de Sicilia, o Tremiti en el Adriático, fueron ocupados a principios del Neolítico (Cherry, 1984: 15).

Es necesario recordar todo esto por que lo verdaderamente raro no es que durante el V o IV milenio BC hubiese intentos de colonizar ("precolonizar") las islas, sino precisamente lo contrario, a tenor de lo que ocurre en el hinterland más inmediato de las Baleares. Pese a todo, ¿con qué documentación arqueológica contamos verdaderamente en la actualidad para presentar un panorama coherente y fundamentado de la primera llegada de humanos a las islas Pitiusas?. Los pocos datos que poseemos para plantear una aproximación a la primera ocupación humana de estas islas son extremadamente escasos y dispersos. No obstante, debe advertirse que jamás se ha llevado a cabo un plan con estrategias de estudio metódicamente proyectadas para llenar este vacío de la investigación. Por tanto, la primera afirmación que puede establecerse es que carecemos de fundamentos sólidos para plantear otra cosa que no sean hipótesis de trabajo formuladas a partir de modelos teóricos (Costa 2000; Benito y Costa, 2000: 263-270), en espera de confirmación suficiente mediante datos empíricos y contrastados en el registro arqueológico.

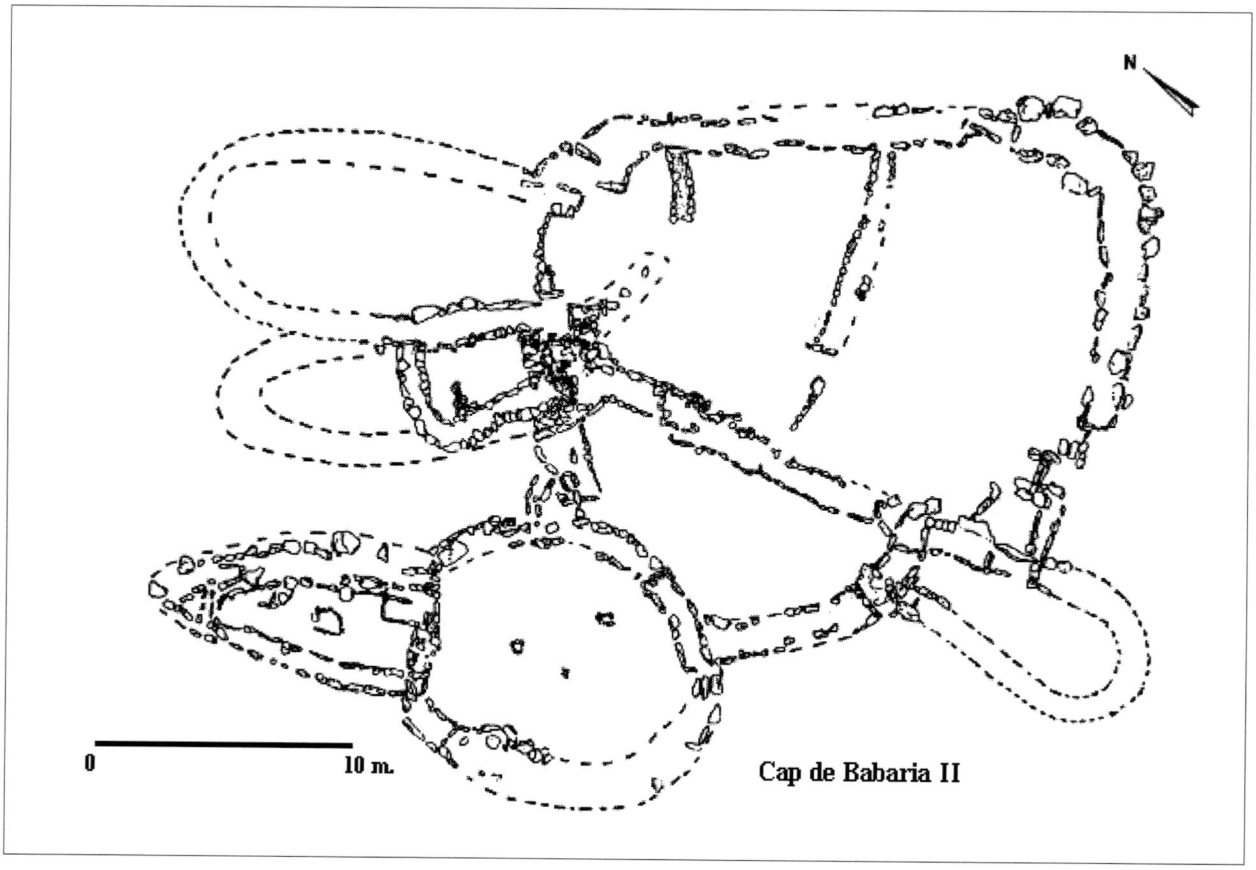

FIGURE 3. ASENTAMIENTO DEL CAP DE BARBARIA II (SEGÚN COSTA Y BENITO 2000).

Hace unos años se realizaron excavaciones en la gruta del Pouàs de Sant Antoni, aunque los trabajos sólo tenían por objetivo estudios de carácter paleontológico, en los niveles superiores a dicho depósito fue descubierta una secuencia estratigráfica en la que había trazas de ocupación humana consistentes en carbones, restos de fauna doméstica y algunos fragmentos cerámicos. La caracterización cronocultural de esta ocupación antrópica de la gruta es, por el momento, desconocida. En los estratos subyacentes fueron encontrados numerosos restos de ornitofauna holocénica endémica, entre ellos dos huesos que aparentemente presentaban signos de haber sido quemados . Las dataciones radiocarbónicas de este horizonte holocénico de la gruta del Pouàs (Alcover et alii 1994), a partir tanto de los huesos aparentemente quemados (UtC-6222 y UtC-6516), así como de otros también de ornitofauna extinta (CSIC-870), se jalonan en un intervalo cronológico que va desde c. 5100 a 4450 cal. BC (ver tabla) Nada de particular tendría que en este intervalo de tiempo las Pitiusas hubiesen sido frecuentadas para la explotación, más o menos ocasional, de determinados recursos –sal, pesca, colonias de aves- por poblaciones de la costa Levantina peninsular (Guerrero, 1999: 566; Eadem, 2001; Costa, 2000: 41; Costa y Benito, 2000: 266), que a la sazón constituían poblaciones con una economía neolítica plenamente desarrollada (Bernabeu y Martí 1992) y extendida por vía marítima a toda la cuenca del Mediterráneo central y occidental. Recordemos que Ibiza es visible en los días sin calimas desde la costa de Denia, donde, precisamente, se localiza la cueva de las Cendres (Llobregat et alii 1981) que tiene una larga ocupación durante el neolítico valenciano.

Sin embargo, los datos del Pouàs necesitan ser verificados y contrastados. Debe asegurarse que los huesos de aves extintas con señales de cremación no son otra cosa que termoalteraciones postdeposicionales. Es decir, que no se quemaron accidentalmente, ni fueron utilizados como combustible por la frecuentación humana de la cueva documentada en los niveles inmediatamente superiores. Fenómeno relativamente frecuente y bien contrastado tanto en Mallorca como en Menorca sobre restos de fauna extinta (Guerrero 2000). Por otro lado, como ya se ha apuntado, queda por ver el estudio, aún inédito, de los mencionados niveles arqueológicos (Alcover et alii 1994) que, al parecer, cubrían los paleontológicos. En cualquier caso, una crítica más detenida no puede hacerse hasta que la secuencia sedimentaria y las relaciones entre las unidades estratigráficas sean dadas a conocer a la comunidad científica, mientras tanto sólo estamos ante hipótesis más o menos razonables, e incluso probables, pero que aún no tienen, por mucho que se quiera (Alcover et alii 1999: 178), confirmación empírica contrastada, ni un correlato claro en otros yacimientos de las islas.

EL ASENTAMIENTO DEFINITIVO DE POBLACIÓN: LOS DATOS ARQUEOLÓGICOS

Ya se ha argumentado que en el proceso lógico de colonización de un territorio siempre hay unas fases que preceden al establecimiento de poblamiento permanente. El ritmo y el momento en que cada fase del proceso tendrá lugar depende de una combinación de factores tanto biogeográficos como de los derivados del desarrollo demográfico, económico y social de las comunidades humanas involucradas. Por ello, dichas fases pueden ser, según las circunstancias, más o menos dilatadas y las visitas y estancias más o menos frecuentes e intensas en función de la accesibilidad y del uso que se haga de los recursos del territorio en cuestión.

Las evidencias que muestran una ocupación permanente y, por lo tanto, un verdadero asentamiento de población estable, son la presencia de necrópolis y poblados duraderos o de asentamientos multifuncionales. La introducción de fauna doméstica en las islas es también una buena señal de que los pobladores que la provocan no están de paso, sino que, por el contrario, han establecido ya estrategias de uso y aprovechamiento integral del ecosistema.

Llegados a este punto podemos adelantar que los datos, escasos aunque bien contrastados, de presencia humana estable en las Pitiusas se remontan a la segunda mitad del III milenio cal BC. Sin que ello, naturalmente, descarte la posible existencia de episodios precoloniales y coloniales más tempranos, como ya hemos apuntado. Las evidencias incuestionables de un poblamiento estable a fines del tercer milenio cal. BC, tanto en Ibiza como en Formentera, se han obtenido recientemente a partir de dataciones radiocarbónicas (ver tabla) en Puig de ses Torretes (Ibiza) y en Ca na Costa (Formentera).

La datación radiocarbónica del Puig de ses Torretes, hace poco dada a conocer (Costa y Benito, 2000), documenta actividad humana en este yacimiento ibicenco hacia 2100/1900 cal BC. La datación tiene un alto margen de fiabilidad técnica por varias razones. Por un lado, ha sido obtenida a partir del colágeno de un hueso de bóvido doméstico, se trata de una muestra de vida "corta" y como todo el mundo sabe, a diferencia de lo que pasa con los carbones, estamos datando exactamente el momento en que murió el animal. Por otro, tiene una desviación típica de +42 años que es realmente poco.

En cuanto a las nuevas dataciones de Ca na Costa (ver tabla), cuya noticia preliminar está a punto de salir a la luz en el momento de redactar estas líneas (Costa y Guerrero, 2001), han sido obtenidas de dos huesos astrágalos del interior de la cámara del sepulcro , procedentes de las antiguas excavaciones (Fernández, Plantalamor y Topp, 1976; Ibidem, 1988). Las desviaciones típicas de ambas, de + 35 y + 40 años respectivamente, aseguran también un elevado margen de precisión en ambas dataciones. En cualquier caso, estas nuevas fechas confirman que este sepulcro megalítico formenterés estaba en uso a fines del tercer milenio, o a lo sumo en el primer siglo del segundo milenio BC y que, en consecuencia, es el dolmen más antiguo del archipiélago balear (Costa y Guerrero, 2001; Guerrero e.p.).

Así pues, las nuevas dataciones permiten plantear, ahora ya con un alto índice de probabilidad, que a fines del tercer milenio había población estable en las Pitiusas. En el caso ibicenco se trataría de una comunidad de pastores instalada en el Puig de Ses Torretes, que ya habían introducido los bóvidos como elementos significativos de su cabaña ganadera, y no sólo las cabras y ovejas como ocurre en comunidades aún más inestables y en proceso de implantación. La cría de bóvidos es, pues, un signo de estabilidad y desarrollo que de ninguna manera se conjuga con un establecimiento efímero o temporal, pues el mantenimiento de estos animales es difícil y gravoso para una comunidad primitiva en un medioambiente no demasiado productivo. En este caso, el pastoreo de bóvidos probablemente se vio favorecido por la existencia de algunas extensiones de prado y marjal en las cercanías de Cala Llonga, donde diversos torrentes que bajan de las montañas del entorno, atraviesan el valle y vienen a desembocar en la cala. Es difícil saber aún si a estas fechas tan tempranas pueden asociarse ya los primeros fondos de cabaña que se observan en la ladera de este puig, pero es bastante probable.

Por lo que respecta al conocido sepulcro megalítico de Ca na Costa (fig. 2), yacimiento paradigmático de la prehistoria de Formentera, las excavaciones realizadas a principios de los años setenta pusieron al descubierto unos ajuares funerarios que no se diferencian en absoluto de lo que encontramos en los registros arqueológicos de los otros dólmenes excavados hasta ahora en Mallorca y Menorca (Costa y Benito, 2000: 278, taula 3). La primera datación radiocarbónica obtenida de una muestra de restos humanos, corresponde a la serie realizada por el British Museum que tuvo que revisarse posteriormente por problemas técnicos (Bowman et alii 1990), aunque la de Ca na Costa no se incluyó jamás en esta revisión. Pese a todo, la datación aún sin corregir nos daría una fecha calibrada a un sigma que se situaría en el intervalo de 1652 y 1465 BC. En cualquier caso, la misma podría suponer una corrección al alza de unos doscientos años. De ser así, esta primera datación de Ca na Costa tampoco desdice en absoluto de las series de dataciones conocidas para s'Aigua Dolça (Calvo et alii 1997; Guerrero, et alii 1997) y Son Bauló (Pons 1999) en Mallorca y Montplè (Hedges et alii 1996; Plantalamor y Strydonck 1997; Mestres y Nicolás 1999) en Menorca.

puede establecerse sin dificultad entre c. 2400/2300 y 2000 cal BC a partir de más de una treintena de dataciones absolutas (Castro et alii 1996; van Stridonck et alii 1998; Waldren 1998; Calvo y Guerrero e.p.) obtenidas en estos contextos campaniformes baleáricos. Espectro cronológico que coincide plenamente también con el vigor de las cerámicas campaniformes continentales y de difusión marítima (Harrison 1977, Delibes 1977, Garrido-Pena 2000, Salanova 2000, entre otros). No obstante, la datación absoluta inicialmente obtenida en el megalito formenterés parecía poco coherente con la presencia de estos fragmentos incisos claramente campaniformes, y ello generó no pocas dudas y

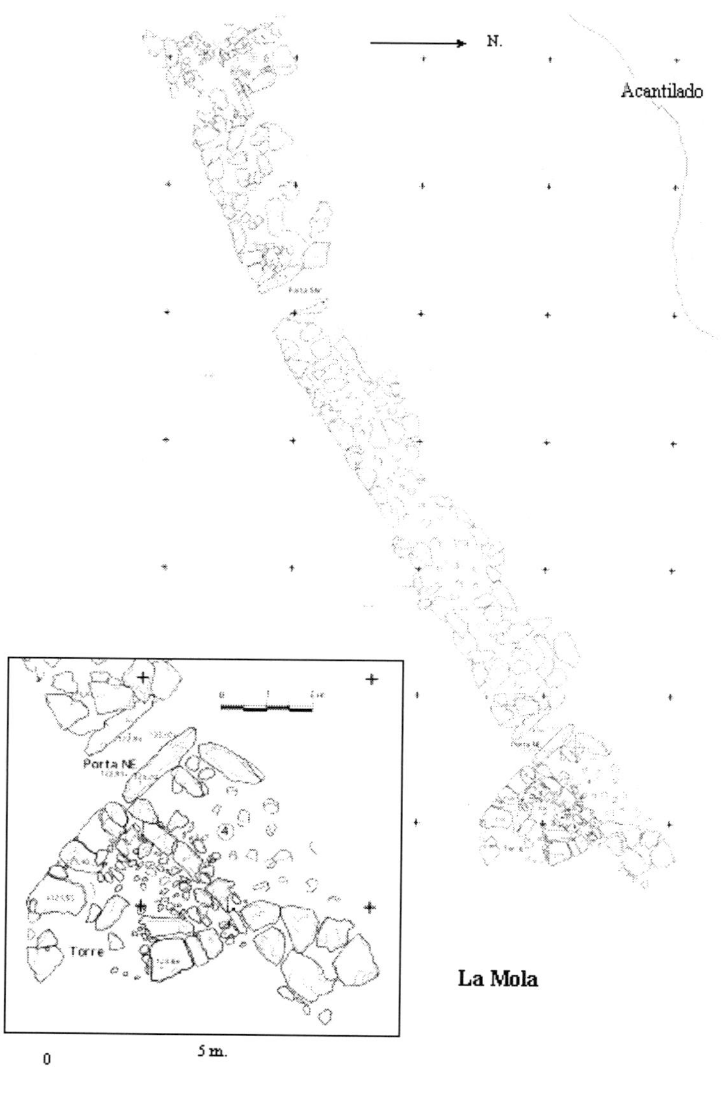

FIGURE 4. MURALLA DE LA MOLA (SEGÚN RAMÓN Y COLOMAR 1999).

Sin embargo, el sector III de la cámara de Ca na Costa proporcionó dos fragmentos de cerámica incisa de estilo campaniforme, que tienen paralelos perfectos en las cerámicas del mismo estilo de los yacimientos mallorquines de Ca na Cotxera (Cantarellas 1972; 1972a), Son Matge, Son Ferrandell-Oleza o Son Mas (Waldren 1998). El horizonte cronológico que cubre este fósil director en Mallorca

posturas encontradas sobre este aspecto.

El alto índice de fragmentación y la ausencia de otros elementos muebles claramente contemporáneos de estas cerámicas incisas, nos inducen a pensar que se trata de un testimonio del momento de construcción y/o de una ocupación más antigua que ahora, con las nuevas dataciones recientemente obtenidas, sabemos que podemos situar con seguridad al menos a fines del tercer milenio BC. En el continente son bien conocidas las reutilizaciones que sufren algunos dólmenes tras varios siglos de abandono o inactividad (p.e. Pérez Arrondo 1987; López de Calle y Ilarraza 1997), de forma que las últimas ocupaciones corresponden a secuencias correspondientes con el Bronce Antiguo, aunque hay evidencias arqueológicas de ocupaciones en época calcolítica bien patentes precisamente por la presencia de cerámica campaniforme. En la necrópolis menorquina de Biniai Nou se ha podido documentar igualmente un proceso similar. Las siete dataciones radiocarbónicas obtenidas (Rubinos 2000, Strydonck y Maes 2001) nos indican que la sepultura nº 2 tuvo dos claros periodos de actividad: uno desde el último cuarto del tercer milenio hasta c. 1900 BC. y otro desde c. 1650 hasta 1500 cal. BC, coincidiendo, el más moderno con las últimas fases de ocupación de los otros dólmenes baleáricos. Por lo tanto, no puede descartarse que Ca na Costa haya sufrido un proceso similar: construcción y primera fase de uso durante la segunda mitad del III milenio, de la que serían testigo material las cerámicas incisas, y una reutilización durante el Bronce Antiguo al que correspondería el grueso de los materiales conocidos.

Otro posible testimonio de poblamiento en este período en Formentera son las cerámicas incisas aparecidas en la Cova del Fum, ya estudiadas hace años por Celia Topp (1988), las cuales se corresponden estilísticamente con sus homólogas "campaniformes" de los paradigmáticos yacimientos mallorquines ya citados. Por lo tanto nada impide pensar que esta cueva tuvo una ocupación que podría situarse entre 2500/2400 y 2000/1900 BC. La circunstancia en que fueron obtenidos los materiales de la Cova del Fum no permite mucha precisión para otros hallazgos descontextualizados que tienen una perduración mayor que las cerámicas campaniformes. Sin embargo, el magnífico ejemplar de placa-afilador perforada que fue encontrada junto con las cerámicas incisas no tiene, en principio, por qué desentonar del contexto de las cerámicas campaniformnes de esta cueva. Como se sabe, este tipo de instrumentos acompañan en el continente a las cerámicas campaniformes (p.e. Leisner et alii 1961: pl. V; Cornaggia 1962-63; Harrison 1977), aunque no es menos cierto que también son frecuentes en horizontes argáricos (Siret 1890; Schubart y Ulreich 1991). Fenómeno que de igual forma se repite en Mallorca (Guerrero e.p.a).

La existencia de comunidades humanas plenamente asentadas en las islas de Ibiza y Formentera en la segunda mitad del tercer milenio es un fenómeno perfectamente coherente con lo que pasa en el resto de las islas del archipiélago balear, donde igualmente una colonización intensa y extensa de todos los ecosistemas isleños no parece producirse antes del 2500 BC (Guerrero 2000; Eadem 2001; Calvo y Guerrero e.p.). En Menorca la necrópolis de Biniai Nou ha podido demostrar muy recientemente (Gómez 2000, Plantalamor y Marqués 2001) que en la segunda mitad del III milenio BC en la isla estaba ya plenamente colonizada.

En Mallorca, como ya se ha indicado, el complejo cultural asociado a la presencia de cerámicas incisas de estilo campaniforme se extiende, grosso modo, entre 2600/2400 y 2000 BC con la presencia bien contrastada de asentamientos al aire libre tan importantes como Son Ferrandell-Oleza o Son Mas (Waldren 1998, Cantarellas 1972), en grutas y abrigos rocosos como Son Gallard, Son Matge (Waldren 1982) o Coval Simó (Coll 2000). En algunos de estos yacimientos, como en Son Ferrandell-Oleza, se documenta también la existencia de fondos de cabaña delimitados por zócalos de piedra (Waldren 1982: fig. 88) que son muy similares a los del Puig de ses Torretes, los cuales, con toda probabilidad, precedieron al conocido asentamiento con naviformes protegidos por una cerca rectangular (Waldren 1998: 96).

Tal vez Mallorca conoció episodios de colonización más tempranos (Costa 2000, Guerrero 2000; Eadem 2001), como así parecen indicarlo algunas dataciones absolutas de carbones procedentes de los abrigos rocosos de Son Gallard (c. 3970 BC) y Son Matge (c. 3400 BC), al parecer, asociados a la presencia de fauna doméstica y grandes contenedores toneliformes de cerámica muy parecidos a los que también encontramos en el yacimiento menorquín de Biniai Nou. En Mallorca todo parece indicar que estas formas cerámicas preceden a la introducción de los estilos incisos campaniformes de difusión marítima que, curiosamente, no tienen presencia en Ibiza, seguramente por una insuficiencia de la investigación, ni en Menorca.

En este último caso, hasta el conocimiento de las recientes dataciones de Biniai Nou la ausencia de cerámica campaniforme era generalmente atribuida a la inexistencia de una presencia humana estable anterior a c. 1700 BC. Sin embargo, hoy puede ya comenzar a plantearse que Menorca está caracterizada por un calcolítico con personalidad propia que la diferencia sensiblemente de Mallorca y, seguramente, de Ibiza y Formentera. Un fenómeno similar ya se había detectado en la isla de Córcega, cuyos horizontes calcolíticos están muy bien documentados, con desarrollos regionales tan relevantes como el Terriniense (Camps 1988a), no obstante, tampoco se conocen los tipos cerámicos campaniformes de gran difusión, aunque algunas formas decoradas, como el cuenco aparecido en el taffonu n. 6 de Calanchi (LGQ 279: 3910 + 150 bp), podían remitirnos a corrientes estilísticas de influencia campaniforme (Camps y Cesari 1989; Cesari 1992; 1994: 40).

LAS PITIUSAS A LO LARGO DEL SEGUNDO MILENIO BC: NUEVOS ELEMENTOS DE JUICIO

En la isla de Ibiza los todavía escasos testimonios de poblamiento durante el Bronce Antiguo-Medio se circunscriben al pleno desarrollo del establecimiento del Puig de ses Torretes (lám. 1a-b), ahora con numerosas construcciones con gruesos zócalos de doble hilada de piedras, así como en la ocupación de algunas cuevas, como la cueva Xives y la cueva des Culleram, aunque la tipología de algunos elementos cerámicos no permiten descartar la posibilidad de alguna ocupación más antigua (Costa y Fernández, 1992; Costa y Benito, 2000).

En Formentera, el poblamiento de este período está mucho mejor documentado. Además de la última fase de ocupación del dolmen Ca na Costa, son bien conocidos los yacimientos del Cap de Barbaria, además de otras agrupaciones, como Can Marroig, Punta Prima, es Ram y sa Cala, repartidas por el resto de la isla (Costa y Fernández 1992, Costa y Benito 2000). Del impresionante conjunto de veintiún yacimientos que se agrupan en la zona del Cap de Barbaria, resulta particularmente interesante el denominado Cap de Barbaria II (fig.3), pues gracias a las diversas campañas de excavación realizadas en el mismo (Topp, Fernández y Plantalamor, 1979; Fernández y Topp, 1984; Costa y Fernández, 1992; Costa y Benito, 2000) sabemos tensas remodelaciones. También la erosión natural, la acción de la vegetación y, sobretodo, la acción antrópica, utilizando el yacimiento como cantera para la obtención de piedra, incidieron negativamente sobre su conservación. De esta forma, hoy sólo un naviforme de los que han sido completamente excavados conserva el recorrido perimetral de sus muros completo. Su mejor estado de conservación permitió equipararlo claramente con los edificios naviformes mallorquines y menorquines, pues incluso, al igual que

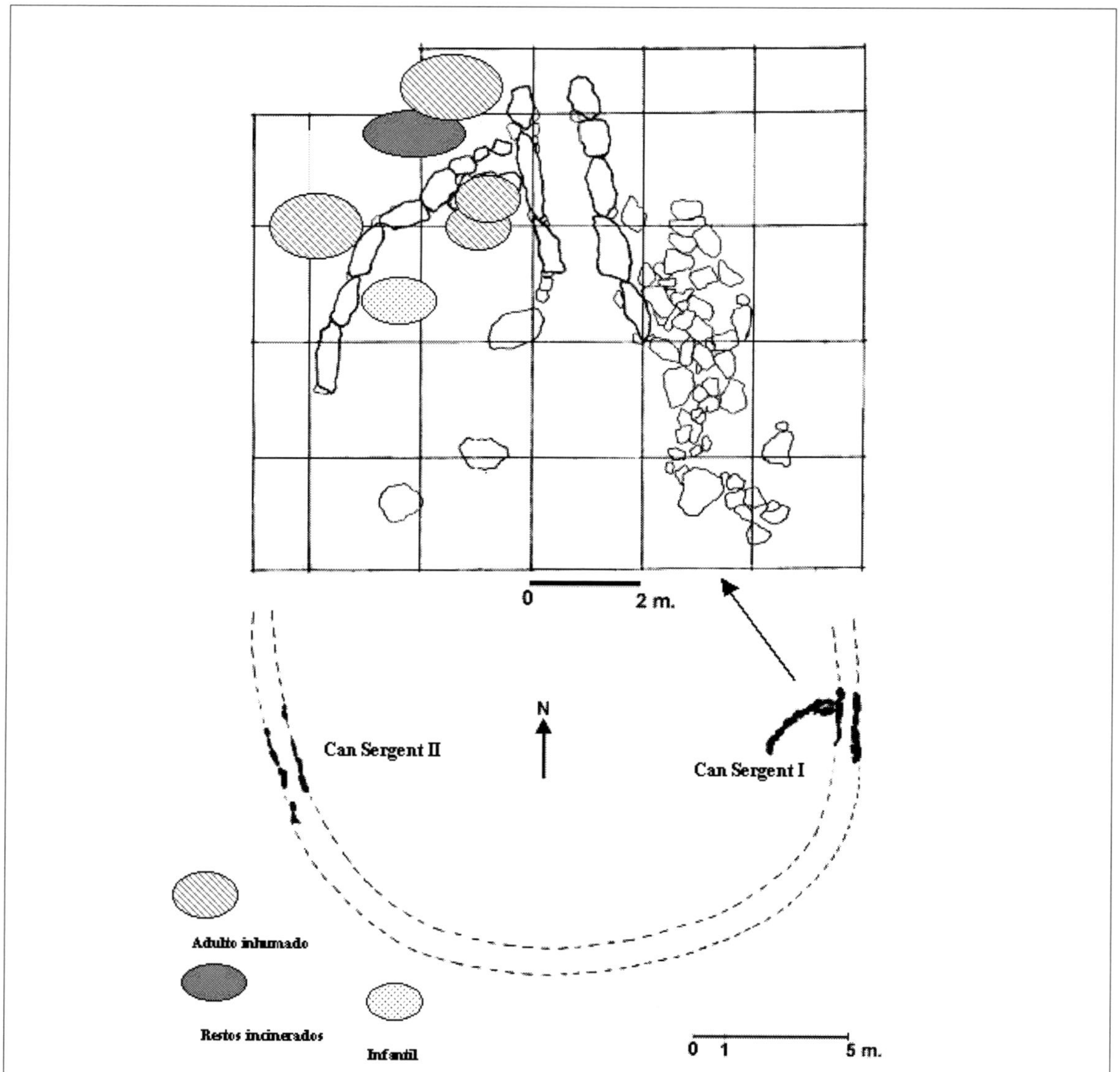

FIGURE 5. YACIMIENTO DE CAN SERGENT (SEGÚN COSTA Y BENITO 2000, GONZÁLEZ Y LALUEZA 2000).

que constituye un asentamiento con varias unidades naviformes, como las que se conocen en Mallorca y Menorca. Además de otras habitaciones con plantas de morfologías diversas, formando un complejo constructivo constituido por al menos nueve ámbitos o estancias. Por tanto, en el caso pitiuso, los naviformes no aparecen como construcciones aisladas, sino integradas en conjuntos más amplios, como ocurre también en Mallorca y Menorca (Calvo et al. 2001).

El asentamiento, cuya excavación no está culminada por completo, sufrió a lo largo de su existencia numerosas e in-

aquellos, aparece provisto de una hogar bien delimitado por losetas, aunque de elaboración más simple que algunos de los documentados en naviformes mallorquines. Pese a todo, pueden identificarse el arranque de dos naviformes más y tal vez un cuarto reconvertido en una pequeña estancia pseudorectangular en un momento indeterminado de la dilatada existencia de este yacimiento.

La organización espacial comunitaria de este asentamiento parece estructurarse a partir de un espacio central, que probablemente puede ser interpretado como un corral y que posteriormente fue compartimentado en distintos recintos o cercas, entorno a los que se distribuyen los naviformes con sus portales afrontados y con salida directa a los distintos corrales. Sin embargo, no se trata de un caso único, pues dentro del amplio conjunto de yacimientos prehistóricos del Cap de Barbaria, existen otros cuatro más que por su complejidad y superficie son equiparables al Cap de Barbaria II, mientras que el resto ofrecen plantas mucho más sencillas (Costa y Fernández, 1992; Costa y Benito, 2000).

Al igual que en los asentamientos de las otras islas, la actividad metalúrgica en el Cap de Barbaria II parece que tuvo una presencia notable, al menos en la última fase de su existencia, detectándose restos de fundición que contienen un 85,02% de cobre con un 14,97% de estaño (Costa y Benito, 2000), lo que, en un contexto de ausencia de materias primas, indica unos contactos muy fluidos con el comercio ultramarino de metales.

Documentación reciente de distintos yacimientos mallorquines y menorquines (Calvo et alii 2001), sobre todo la proporcionada por el poblado de naviformes de Closos de Can Gaià en Mallorca (Calvo y Salvà 1999), actualmente en curso de excavación, obliga a revisar el encuadre cronológico que tradicionalmente se ha venido proponiendo para Cap de Barbaria II. La datación de este asentamiento, como se había hecho (Ramón 1985; Gómez Bellard 1995), entre el 2000 y 1400 bc (lo que en dataciones calibradas podría equivaler a 2500-1700 BC) es hoy inaceptable y, con toda seguridad, es una visión errónea que se arrastra de una deficiente interpretación de los contextos antiguos del yacimiento mallorquín, ya citado, de Son Ferrandell-Oleza. Ningún contexto claro de hábitat naviforme remonta con seguridad el 1650 BC y, si aceptamos que su expansión definitiva a todos los ecosistemas isleños no se produce hasta c.1400 BC, sería más razonable pensar que los momentos iniciales de Cap de Barbaria II se sitúen, como mucho, entre 1400 y 1300 BC (Costa y Guerrero, 2001).

Más difícil resulta, hoy por hoy, evaluar la perduración temporal de este tipo de establecimientos. Admitir hoy que estos complejos hábitats formenterenses con estructuras navetiformes estaban habitados hacia el 900/850 BC no constituye nada asombroso cuando sabemos (Calvo et alii 2001) que en este tipo de construcciones en Mallorca, y seguramente en Menorca, esto es un hecho relativamente frecuente. A juzgar por las reformas que a lo largo de su existencia sufrió, su pervivencia pudo ser muy larga (Costa y Fernández, 1992). Las últimas remodelaciones y ocupaciones del lugar, como nos indica una datación absoluta recientemente obtenida de un hueso de ovicáprido asociado a la cimentación de uno de los muros que dividían el posible corral comunitario, puede fechar esta reforma hacia 445 BC (Costa y Benito, 2000). Esta cuestión no está exenta de interés pues supondría un interesante caso de posible coexistencia de comunidades indígenas con el desarrollo de la sociedad urbana ebusitana. Hecho, por otro lado, nada sorprendente si nos atenemos al registro etnográfico que nos proporcionan múltiples casos de colonizaciones modernas.

Además de algunas cuevas en los acantilados de la Mola con amplias evidencias de ocupación, como la cova des Riuets, Formentera registra también la presencia de un interesante yacimiento cercado o fortificado, conocido de antiguo (Fernández, 1977 y 1978; Fernández y Topp, 1984; Ramón 1985). Sin embargo, las recientes excavaciones (Ramón y Colomar 1999) han permitido conocer con detalle la planta del lienzo de muro conservado (fig. 4), dos puertas y el refuerzo a modo de pequeño torreón macizo que aún conserva una de ellas. En la situación actual de las investigaciones es difícil concretar una atribución cronocultural clara de este yacimiento. Construcciones similares son conocidas en Mallorca ligadas al cierre del espacio comunal ocupado por naviformes (Calvo et alii 2001). Sin embargo, la singularidad de Sa Mola radica es que delimita un espacio adosado a un acantilado en el que, por otro lado, se localizan cuevas naturales, como las del Fum y Riuets, entre otras. Por lo tanto no es descartable que este conjunto arquitectónico se articule con algunos de los momentos de ocupación de estas cuevas. En opinión de Ramón y Colomar (1999: 36) es difícil una explicación funcional de este recinto si no se relaciona con la actividad desarrollada en las cuevas del acantilado.

En el momento actual se carece aún de dataciones absolutas y las cerámicas halladas en las excavaciones (Ramón y Colomar 1999) no permiten mucha precisión, aunque pueden señalarse dos aspectos relevantes: por un lado, la ausencia de cerámicas campaniformes y otros materiales asociados, como las placas-afiladores perforadas, que aparecen en la vecina cueva del Fum (Topp 1988) y, por otro, sí están presentes grandes vasijas toneliformes que en Mallorca y Menorca (Juan y Plantalamor 1997, Calvo y Salvà 1997, Calvo et alii 2001) se asocian a la fase de uso de los naviformes que va aproximadamente del 1400/1300 al 1200/1100 BC. Todo ello podría constituir un indicio sólido de que, al menos, una fase de ocupación del lugar pudo situarse en el Bronce Antiguo de forma sincrónica a la extensión del hábitat naviforme, aunque dicho sea esto con todas las reservas que el estado actual de los trabajos aconseja.

EL BRONCE FINAL Y LA COLONIZACIÓN FENICIA.

Es evidente que durante la primera mitad del primer milenio BC se produjo una profunda diferenciación entre los procesos históricos de los dos grupos de islas que conforman el archipiélago balear, la cual ha dado lugar a la tradicional distinción entre Pitiusas por una parte y Gimnésias o Baleares por otra, que encontramos en la mayor parte de fuentes antiguas griegas y romanas. Por un lado, en las Pitiusas no se desarrolló la denominada "cultura Talayótica" y, por otro lado, Ibiza fue objeto de colonización por parte de grupos semitas. Mallorca y Menorca no se vieron afectadas por esta corriente colonizadora que tanta incidencia tuvo en el Mediterráneo Occidental (Aubet, 1994), pero sí que la evolución económica, social y cultural de las comunidades talayóticas se vio fuertemente influida, des de fines del siglo VI a.n.e. en adelante, a partir de las relaciones cada vez más estrechas con Ibiza (Guerrero 1997).

En este contexto, el principal interrogante que se plantea a la investigación actual es por qué en Ibiza y Formentera no se desarrolló la cultura Talayótica y qué pasó con su poblamiento cuando en Mallorca y Menorca se construían los talayots y poblados fortificados, así como el resto de manifestaciones arquitectónicas (Guerrero 1999a) tan espectaculares que caracterizan el final de su prehistoria?. Para responder a estas cuestiones, hoy por hoy contamos solamente con dos grupos de documentos: los enterramientos de Can Sergent (Sant Josep, Eivissa) y los diversos hallazgos de objetos metálicos del Bronce Final, que desde las primeras décadas del siglo XX se han venido sucediendo en Formentera e Ibiza, en algunos casos agrupados en depósitos o escondrijos intencionales.

LÁMINA 1. PUIG DE SES TORRETES.

El yacimiento de Can Sergent, situado al pie de la ladera del Puig Palleu, que a partir de su excavación (Topp, Fernández y Plantalamor, 1979; Fernández y Topp 1984) había sido identificado como un sepulcro megalítico, parece poderse interpretar más adecuadamente (Costa y Benito 2000) como un recinto de planta curvada (fig. 5), en muy mal estado de conservación, con un muro perimetral hecho de doble hilada de ortostatos y un relleno interior de tierra y piedras, de trazado casi semicircular. A este muro, por su parte interna, se le adosó al menos otra estructura menor (cabaña?), de planta también curvada, delimitada con una única hilera de grandes losas de piedra. No conocemos la función original de este recinto, tal vez similar a la de alguno de los observados en el Cap de Barbaria. No obstante, en su interior y ya en un momento claramente posterior al de su construcción (tal vez incluso cuando ya había sido abandonado), una pequeña comunidad lo utilizó como área sepulcral de sus difuntos. El espacio funerario ocupado incluía (fig. 5) tanto el ámbito interno del pequeño recinto interior, como el exterior de dicha construcción. Los individuos inhumados (seguramente en inhumación secundaria), fuero dos hombres adultos y un niño dentro del pequeño espacio interno, y un adulto masculino y otro femenino, además de un cráneo adulto quemado que no correspondía a ninguno de ellos, en el exterior (González y Lalueza 2000). Una muestra de uno de los enterramientos exteriores dio una datación de 720 BC, mientras que otra del interior dio 550 BC (ver tabla).

Vemos, por tanto, que estos datos apoyan la idea de una continuidad de poblamiento que se mantendría hasta bien entrado el primer milenio BC, incluso en fechas cuando la presencia fenicio-púnica en la isla estaba ya bien asentada. Así parecería indicarlo, también, la reciente datación obtenida en el Cap de Barbaria II, que es sólo un siglo más moderna que la más reciente de las de Can Sergent. Ello, no obstante, de ninguna manera significa que no se produjeran evoluciones en los aspectos económicos y sociales respecto a la situación en el segundo milenio. La presencia de un niño compartiendo el espacio funerario con adultos podría suponer la adquisición de una mayor importancia de los linajes, en los que sus miembros son reconocidos como tales e integrados desde el momento de su nacimiento (Costa y Benito 2000). También, el estudio de las piezas dentales de Can Sergent indica un predominio de harinas y sémolas en la dieta (González y Lalueza, 2000), a diferencia de los individuos de Ca na Costa, cuya alimentación se basaba en las proteínas de origen animal y las verduras frescas. Ello podría ser indicio de un incremento de la importancia relativa de la agricultura dentro del modelo de producción, frente al claro predominio de la ganadería en el milenio anterior. La aparición de recintos fortificados, como el antes mencionado de sa Cala (Formentera) y otro todavía prácticamente inédito en el Puig Rodó (Ibiza), podría indicar también problemas territoriales o de competición por determinados recursos entre los diferentes grupos o linajes que habitaban las islas. Por otra parte, la presencia de objetos metálicos demuestra la conexión de las Pitiusas con las redes de intercambio de los productos metalúrgicos, que en el primer milenio BC conocieron una impresionante expansión gracias a las mejoras de las técnicas y medios de navegación (Guerrero, 1993) y de las comunicaciones en general, lo que permitió conectar el Atlántico y la Europa continental con toda la cuenca mediterránea.

En cualquier caso, lo que sí puede afirmarse con seguridad, es que el proceso de transformaciones que, mediante una progresiva intensificación de la producción y profundos cambios sociales, dio lugar al desarrollo de las comunidades talayóticas en Mallorca y Menorca, en Ibiza y Formentera fue muchísimo más lento y mucho menos intenso. Parece, pues, como si en las Pitiusas dicho proceso tan solo hubiese dados sus primero pasos y nunca hubiese llegado a culminarse.

Seguramente, el día que se obtengan nuevas dataciones absolutas, las piezas de este rompecabezas irán encajando mejor y se irán desvaneciendo los supuestos hiatus de poblamiento que hace años se habían planteado (Gómez y San Nicolás, 1988; Gómez, 1995). En cualquier caso, con las todavía escasas dataciones actuales, junto con una revisión profunda de los registros arqueológicos ibicencos y formenterenses como la que estamos proponiendo, adquieren también más sentido y encajan mucho mejor los hallazgos de piezas y depósitos de bronces conocidos de antiguo.

LÁMINA 2. PUIG DE SES TORRETES.

En relación con dichos hallazgos, conviene recordar que las hachas de talón del depósito de Can Gallet en Formentera (Fernández, 1973), pueden ser fechadas hacia el año 1000 BC, y algo después los lingotes con apéndices laterales del mismo depósito (González Prats, 1985), así como las piezas de Can Pere Joan (Costa y Fernández, 1992) y las del depósito de la Sabina (Almagro, 1962; Fernández, 1974), donde aparece un hacha de cubo con asa lateral, que en las Baleares podría datarse en cronología calibrada entre 860 y 790 BC –en buena sintonía, por tanto, con los enterramientos más antiguos de Can Sergent-. Tanto en Mallorca como en Menorca, estos depósitos, que en algunos casos pueden ser amortizaciones de carácter votivo de piezas suntuarias de bronce (Lull et alii 1999: 119-126), son bien conocidos y pueden datarse, justamente, entre el 1000 y el 850 BC.

El origen de estas piezas metálicas halladas en las Pitiusas, sin descartar totalmente que alguna pudiera ser de producción local, se sitúa, sobre todo en la región costera del SE peninsular y su hinterland. La producción de hachas de apéndices laterales como las de la Sabina y Can Pere Joan, y sobretodo los lingotes de apéndices laterales –que recuerdan por su forma las hachas, pero cuyas láminas son tan delgadas que no pueden ser consideradas ya como verdaderos instrumentos-, está bien documentada en la región alicantina, sobretodo en Penya Negra de Crevillent (González Prats 1983: 177). De más interés, si cabe, es la presencia de moldes de fundición de hachas de apéndices laterales en la fase III del asentamiento fenicio de La Fonteta de Alicante (González Prats 1998, González Prats y Ruiz 1999) que se fecha entre 670 y 635 aC. Sin embargo, este tipo de hachas de bronce planas con apéndices laterales fueron también fabricadas por las comunidades indígenas de las otras islas, como lo indican los moldes de fundición hallados en los poblados de Torelló (Plantalamor 1991:192) y Biniparratxet (Guerrero et alii en prensa), lo que parece reforzar la idea de una fuerte influencia de la metalurgia de la costa del SE peninsular sobre la insular.

Dataciones Radiocarbónicas referidas a las Pitiusas citadas en el texto

Yacimiento	Lab. Ref. y nº de muestra	Edad convencional del C14 sin calibrar BP	Edad calibrada a 1 sigma BC (68,3%)	Muestra	Contexto, comentario y referencias bibliográficas.
Es Pouàs	UtC-6222	6130 ±80	5210-5205 5141-4938	Hueso de ornitofauna	"Es Pouàs" (Ibiza). Contexto no publicado. Aparentemente quemado (com. personal).
	CSIC-870	5770 ±100	4780-4510	Hueso ornitofauna	*"The lower levels in what remains of the archaeological deposit"* (Alcover et al. 1994)
	UtC-6516	5650 ±60	4536-4453 4417-4401	Hueso de ornitofauna	"Es Pouàs" (Ibiza). Contexto no publicado. Aparentemente quemado (com. personal).
Puig de Ses Torretes	UtC-8319	3645 ±42	2110-2088 2037-1936	Hueso de bóvido	Puig de Ses Torretes (Ibiza), Estructura 01, Nivel-3 (Costa y Benito 2000).
Ca Na Costa	KIA-14329	3595 ±35	2010 (0,08) 2000 1980 (0,92) 1880	Hueso humano	Dolmen de Ca Na Costa. Calcáneo izquierdo de un individuo(Costa y Guerrero 2001).
	KIA-14330	3535 ±40	1920 (1,00) 1770	Hueso humano	Dolmen de Ca Na Costa. Calcáneo izquierdo de un individuo (Costa y Guerrero 2001).
	BM - 1677	3270 ±75	1610-1440 1710-1360	Hueso humano	Dolmen de Ca Na Costa. Datación antigua del BM no rectificada (Bowman *et al.* 1990).
Can Sergent	BM - 1511	2670 ±60	860-770 910-740	Hueso humano	Exterior de la estructura(Fernández y Topp 1984).
	BM - 1510	2500 ±100	770-440 850-370	Hueso humano	Interior de la estructura (Fernández y Topp 1984).
Cap de Barbaria II	UtC -8320	2393 ±43	495-437 425-396	Hueso de cabra	Estancia 5 A, nivel 2 b (Costa y Benito 2000)

La relación de estas piezas metálicas y la presencia de comerciantes fenicios en las Pitiusas ya hace tiempo que ha sido planteada. No obstante, creemos que el planteamiento que hoy por hoy resulta más verosímil es que, justamente, la existencia de unas comunidades isleñas que participaban de las redes regionales de intercambios metalúrgicos, sobretodo entre las islas y las costas peninsulares hispanas, habría sido uno de los estímulos que atrajeron la presencia fenicia.

Las dataciones radiométricas de las primeras factorías fenicias en el Sur peninsular (Torres 1998), como por ejemplo Morro de Mezquitilla, sitúan los primeros establecimientos coloniales fenicios de Occidente entre mediados y el último cuarto del siglo IX BC, mientras que las primeras importaciones fenicias que aparecen en contextos indígenas han sido documentadas en Acinipo (Ronda) hacia el 910 BC (Carrilero 1992, Torres 1998). Estas fechas coinciden también, grosso modo, con la primera constatación de hierro y de cuentas de collar de pasta vítrea en Son Matge (Waldren 1982: 129). Mientras que en Menorca la presencia de brazaletes de hierro en la cueva des Càrritx (Lull et alii 1999) está documentada con una datación ante quem de c. 800 BC, y algo antes (c. 1000 BC) en la necrópolis del Forat de ses Aritges.

En este marco revisionista, conviene recordar también que algunos materiales anfóricos fenicios (Ramón 1996), la datación de los cuales podría remontarse al 850-825 BC (equivalente en cronología sin calibrar a la segunda mitad del siglo VII bc), han aparecido en sa Caleta (Ibiza), en un momento en el que la presencia fenicia en la desembocadura del Segura está plenamente demostrada. No obstante, la definitiva ocupación de la isla de Ibiza por colonos fenicios ha de situarse, en dataciones radiométricas calibradas, aproximadamente hacia el 800-750 BC, rango de edad radiocarbónica en la que podría encajar la fecha fundacional histórica de 654-53 a.C. transmitida por Diodoro Sículo (V 16, 2-3).

Vemos, pues, como a medida que avanzan las investigaciones es posible comenzar a plantear, por una parte, que en las Baleares el pleno desarrollo de la cultura Talayótica se aproxima más a una sociedad del Hierro Antiguo que del Bronce Final. Y, por otra parte, que en las Pitiusas la menor extensión territorial, la menor biodiversidad y la consiguiente menor productividad del ecosistema, hicieron que, tal vez por ello, el proceso de evolución económica y social fuese mucho más lento, y que aquí no llegaran a darse las condiciones que dieron lugar a una gran arquitectura turriforme, ni el contexto socioeconómico que la hizo posible. Pero en ese caso, si el proceso evolutivo estaba en marcha, aunque fuese más lentamente, ¿podría también plantearse que la colonización fenicia colapsó el desarrollo de las comunidades ibicencas y formenterenses hacia formas sociales más complejas y desarrolladas?. Ésta, hoy por hoy, es una hipótesis de trabajo que debe permanecer abierta, pero que no deberemos dejar de tener en cuenta, pues sobre ella, sin ninguna duda, la investigación futura tendrá mucho que decir.

BIBLIOGRAFÍA

ALCOVER, J.A.; MCMINN, M. y ALTABA, C.R.,
1994. Eivissa: A Pleistocene Oceanic-like Island in the Mediterrane-an, National Geographic Research & Exploration, 10(2), p.236-248.
ALCOVER, J.A., SEGUÍ, B, y BOVER, P.,
1999. Extinctions and local disappearances of vertebrates in the western Mediterranean islands, en MAcPHEE, R.D.E. y SUES, H.D. (eds.), cit. infra.
ALMAGRO, M., 1962. Depósito de La Sabina. Isla de Formentera. Inventaria Archaeologica fasc. 6, E. 12 1-(1). Madrid.
ALROY, J.
2001, A multispecies overkill simulation of the End-Pleistocene megafaunal mass extinction, Science 292, p.1893-1896.
AUBET, M.E.,
1994. Tiro y las colonias fenicias de Occidente, Ed. Crítica. Barcelona.
BERNABEU, J. y MARTÍ, B.,
1992. El País Valenciano de la aparición del Neolítico al horizonte campaniforme, en UTRILLA, P.
(coord..) Aragón/litoral mediterráneo: Intercambios culturales durante la Pehistoria, "Homenaje a J. Maluquer de Motes", Institución Fernando el Católico, Zaragoza, p.213-234.
BINFORD, L.R.,
1978. Dimensional analysis of behaviour and site structure: learning from an Eskimo hunting stand. American Antiquity 43: 330-61.
BONIFAY, E.,
1998. La grotte de la Coscia (Macinaggio/Rogliano, Cap Corse) et le problème du peuplement des îles de Méditerranée Occidentale par l'homme de néandertal, en L'Homme préhistorique et la mer, 120 congrès CTHS, Aix-en-Provence (1995), p.133-140.
BONIFAY, E./ BASSIAKOS, Y./ BONIFAY, M.-F./ PEREIRA, E./ QUINIF, Y./ SALOTTI, M.,
1998. La Grotte de la Coscia (Rogliano, Macinaggio): Étude préliminaire d'un nouveau site du Pléistocène Supérieur du Corse, Paleo, 10, p.17-41.
BOWMAN, S.,
1990. Radiocarbon dating, British Museum Pub. 48, London.
BOWMAN, S. G. E.; AMBERS, J. C. y LEESE, M. N.,
1990. Re-Evaluation of British Museum Radiocarbon Dates Issued between 1980 and 1984. Radiocarbon, 32, 1, pp. 59-79.
BROODBANK,M C., 2000. An island archaeology of Early Cyclades, Cambridge University Press, Cambridge.
CALVO, M.; COLL, J. y GUERRERO, V. M.,
1997. El dolmen de s'Aigua Dolça, sepulcro colectivo del Pretalayótico. Revista de Arqueología, nº 191, pp. 18-29. Zugarto Ediciones, S. A. Madrid.
CALVO, M. y GUERRERO, V.M.
e.p. Los inicios de la metalurgia en el archipiélago balear. El Calcolítico 2500-1700 cal. BC, Palma.
CALVO, M.; GUERRERO, V.M. y SALVÀ, B.,
2001. Arquitectura ciclópea del Bronce Balear. Análisis morfofuncional y desarrollo secuencial, Ed. EL Talll, col. "El Tall del Temps", Palma.
CALVO, M. y SALVÀ, B.,
1997. El bronze final a les Balears. La transició cap a la cultura Talaiòtica. Quaderns Arca núm. 14. ARCA. Palma de Mallorca.
CALVO, M. y SALVÁ, B.,
1999. Aproximaciò a la seqüència cronocultural de la naveta I del jaciment de Closos de Can Gaià, Mayurqa, 25, p.61-82.
CAMPS, G.,
1988. Préhistoire d'une Île. Les origines de la Corse, París.
CAMPS, G.
1988a, Terrina et le terrinien. Recherches sur le Chalcolithique de la Corse, Collection de l'École Française de Rome 109, Rome.
CAMPS, G. y CESARI, J.
1989, Découverte d'un tesón campaniforme en Corse, Travaux du Laboratoire d'Anthropologie et de Prehistorire des pays de la Máditerranée occidentale, Aix.en-Provence, p.213-216.
CANTARELLAS, C.
1972, Excavaciones en Ca Na Cotxera (Muro, Mallor-ca), Not. Arq. Hisp. 1, p.179-226.
CANTARELLAS, C.
1972a, Cerámica incisa en Mallorca, Mallorca.
CASTRO, P.V., LULL, V. y MICÓ, R.
1996, Cronología de la Prehistoria reciente de la Península Ibérica y Baleares (c. 2800-900 cal ANE), British Archaeological Reports, Intenational, International Series 652, Oxford.
CESARI, J.
1992, Nouveaux acquis de la prehistoire Corse, en La Prehistoria de les Illes de la Mediterrània Occidental, X Jornades d'Estudis Històrics Locals, Palma, 55-81.
COLL, J.,
2000, Excavación arqueológica del Coval Simó. Estado actual y perspectivas de futuro, en GUERRERO y GORNÉS (coords.), cit. infra, p.371-400.
CORNAGGIA, O.
1962-63, Ricerche sulla problemática degli "pseudo-brassards" preistorici, Bulletino di Paletnologia Italiana, XIV, p. 7-71.
COSTA, B.,
2000. Plantejaments per l'anàlisi del procés d'establiment humà en petits medis insulars. El cas de l'arxipèlag balear, en GUERRERO y GORNÉS (coords.), cit. infra, p.11-72.
COSTA, B. y BENITO, N.,
2000. El poblament de les illes Pitiüses durant la Prehistoria. Estat actual de la investigació, en GUERRERO y GORNÉS (coords.), pp.215-322.
COSTA, B. y FERNÁNDEZ, J.H.,
1992. Les Illes Pitiüses: de la Prehistòria a la fi del'època Púnica, en ROSSELLÓ, G. (ed.): La Prehistòria de les Illes de la Mediterrània Occidental. X Jornades d'Estudis Històrics Locals, pp. 277-355.
COSTA, B. y GUERRERO, V.M.,
2001. La Prehistòria Pitiüsa: Avenços, rectificacions i perspectives de futur, Fites 2, Associació d'Amics del Museu Arqueològic d'Eivissa i Formentera, pp. 27-40
COURTIN, J.,
1983. Le trafic de l'obsidienne en Méditerranée Occidentale aux 5e et 4e millénaires, en Séminaire sur les structures d'habitat I: Circulation et échanges, 44-49. Paris: Collège de France.
CHERRY, J. F.,

1984. The initial Colonisation of the West Mediterranean Islands in the light of island Biogeography and Palaeogeography, en WALDREN, W. H. ET ALII (eds.): The Deya Conference of Prehistory: Early Settlement in the Western Mediterranean and their Peripheral Areas. British Archaeological Reports, Intenational International Series 229 (i), pp. 7-23. Oxford.
CHERRY, J.F. y TORRENCE, R.,
1982. The earliest Prehistory of Melos, en RENFREW, C. y WAGSTAFF, M. (eds.), An island polity. The archaeology of exploitation in Melos, Cambridge University Press, Cambridge, pp. 24-34.
DELIBES, G.
1977, El vaso campaniforme en la Meseta Norte española, Valladolid.
DELIBES, G. y FERNÁNDEZ- MIRANDA, M.,
1988. Armas y utensilios de bronce en la Prehistoria de las Islas Baleares. Studia Arqueológica, 78. Universidad de Valladolid.
EDWARDS, K.J., 1982. Man, space and woodland edge: Speculations on the detection and interpretation of human impact in pollen profiles, en BELL, A.M. Y LIMBREY, S. (eds.), Archaeological Aspects of Woodland Ecology. British Archaeological Reports, Intenational, International Series 146: 5-22. Oxford.
EDWARDS, K.J. y RALSTON, I.,
1984. Postglacial hunter-gatherers and vegetational history in Scotland. Proceeding Society of Antiquaries of Scotland 114: 15-34.
EFSTRATIOU, N.,
1985. Agios Petros: A Neolithic site in the Northern Sporades, British Archaeological Reports, Intenational, int. Series 241, Oxford.
FERNÁNDEZ, J. H.,
1973. Nuevo depósito de hachas de bronce descubierto en la isla de Formentera. Pyrenae, 9, pp. 177-183. Universidad de Barcelona.
FERNÁNDEZ, J.H.,
1974. Hachas de bronce de Ibiza y Formentera. VI Symposium de Prehistoria Peninsular (Palma 1972), pp. 63-71. Universidad de Barcelona.
FERNÁDEZ, J.H.,
1977. Últimos descubrimientos prehistóricos de la isla de Formentera, Actas del XIV Congreso Nacional de Arqueología (Vitoria, 1975), p.471-477.
FERNÁNDEZ, J.H., 1978. Formentera salta a la Prehistoria. Historia 16, núm. 25 (Mayo), pp.59-66.
FERNÁNDEZ, J.H.; PLANTALAMOR, L.; TOPP, C.; GÓMEZ BELLARD, F. y REVERTE, J.M.,
1988. El sepulcro megalítico de Ca Na Costa. Trabajos del Museo Arqueológico de Ibiza, 19.
FERNÁNDEZ, J. H. y TOPP, C.,
1984. Prehistoric activities in the Pitiussae Islands, en WALDREN, W. H. et alii (eds.): The Deya Conference of Prehistory: Early Settlement in the Western Mediterranean Island and the Peripheral Areas. British Archaeological Reports, Intenational International Series 229, vol. III, pp. 76-784. Oxford.
GARRIDO-PENA, R.
2000, El Campaniforme en la Meseta Central de la Península Ibérica (c. 2500-2000 AC.), British Archaeological Reports, Intenational Series 892, Oxford.
GÓMEZ, J.L.
2000, Memoria de excavación del monumento nº 2 de Biniai Nou (campañas 1997-1998), en GUERRERO, V.M. y GORNÉS, S. (coords.), cit. infra, p. 417-445.
GÓMEZ BELLARD, C.,
1995, The first colonization of Ibiza and Formentera (Balearic Islands, Spain): some more islands out of the stream?, en CHERRY, J.F. Colonization of islands, World Archaeology, vol. 26, n.3, p.442-455.
GÓMEZ BELLARD, C. y SAN NICOLÁS, P.,
1988. La Prehistoria de Ibiza y Formentera: Estado actual de la cuestión. Trabajos de Prehistoria, 45, pp. 201-228.
GONZÁLEZ, A. y LALUEZA, C.,
2000. Informe antropológico de los restos humanos de Can Sergent (Ibiza), en GUERRERO y GORNÉS (coords.), cit. infra, p.318-321.
GONZÁLEZ PRATS, A.
1983, Estudio arqueológico del poblamiento antiguo de la sierra de Crevillente (Alicante), Univ. de Alicante, Anejo I de la revista Lucentum,
GONZÁLEZ PRATS, A.,
1985. Sobre unos elementos materiales del comercio fenicio en tierras del Sudeste peninsular. Lucentum, IV, p. 97-106.
GONZÁLEZ PRATS, A.
1998, La Fonteta. El asentamiento fenicio de la desembocadura del río Segura (Guardamar, Alicante, España). Resultados de las excavaciones de 1996-97, Studi Fenici vol. XXVI, 2, p. 191-228.
GONZÁLEZ PRATS, A. y RUIZ, E.,
1999. Una zona metalúrgica de la primera mitad del siglo VII en la ciudad fenicia de la Fonteta (Guardamar, Alicante), en Actas del XXIV Congreso Nacional de Arqueología, (Cartagena 1997), vol. 3, p. 355-357.
GORNÉS, S.; GUAL, J.M.; LÓPEZ, A y DE NICOLÁS, J.V.,
1992. Nous monuments funeraris del món pretalaiòtic de Menorca, en ROSSELLÓ, G. (ed.): La Prehistòria de les Illes de la Mediterrània Occidental. X Jornades d'Estudis Històrics Locals, pp. 419-452.
GRAVES, M. W. y ADDISON, D. J.,
1995. The Polynesian settlement of the Hawaiian Archipelago: integrating models and methods in archaeological interpretation, en CHERRY, J. F. (ed.): Colonization of Islands. World Archaeology vol. 26 nº 3, pp. 380-399. Rouletge.
GUERRERO, V.M.,
1993. Navíos y navegantes en las rutas de Baleares durante la Prehistoria, El Tall editorial nº 17, Palma.
GUERRERO, V.M.,
1997. Colonización púnica de Mallorca. La documentación arqueológica y el contexto histórico. Ed. El Tall-U.I.B., Palma.
GUERRERO, V. M.,
1999. Rectificaciones y nuevos enfoques al tránsito de la caza-recolección a una economía productora en Baleares. II Congrés del Neolític a la Península Ibèrica. Saguntum-PLAV, Extra-2, pp. 565-570.
GUERRERO, V.M.,
1999 a. Arquitectura y poder en la prehistoria de Mallorca, "El Tall editorial" nº 31, Palma.
GUERRERO, V.M.,
2000. La colonización humana de Mallorca en el contexto de las islas occidentales del Mediterráneo: Una revisión crítica, en GUERRERO y GORNÉS (eds.), cit. infra, p. 99-190.

GUERRERO, V.M.,
2001. The Balearic Islands: Prehistoric colonization of the furthest Mediterranean islands from the mainland, Journal of Mediterraneam Archaeology 14(2) en prensa.
GUERRERO, V.M.
(e.p.). El megalistismo mallorquín en el contexto del Mediterráneo Central, en GUERRERO, V.M.; CALVO, M. y COLL, J.
(coords.) El Dolmen De S'aigua Dolça (Colonia De Sant Pere, Mallorca), Palma.
GUERRERO, V.M.
(e.p. a). Los afiladores o placas perforadas ("brazaletes de arquero") en GUERRERO, V.M.; CALVO, M. y COLL, J.
(coords.) El Dolmen De S'aigua Dolça (Colonia De Sant Pere, Mallorca), Palma.
GUERRERO, V.; SANMARTÍ, J.; HERNÁNDEZ, J.; GORNÉS, S.; GUAL, J.; LÓPEZ PONS, A. y DE NICOLÁS, J.,
(e.p.). Biniparratx Petit (Sant Lluís): Un proyecto de investigación y revalorización del patrimonio arqueológico en el Sureste de la isla de Menorca.
GUERRERO, V.M., COLL, J. y CALVO, M.,
1997. Estado actual del megalistismo en Mallorca. El yacimiento arqueológico de S'Aigua Dolça, actas del II° Congreso de Arqueología peninsular, (Zamora, 26-20 set., 1996), Fundación Rei Afonso Henriques, vol. II, Zamora, p.359-370.
GUERRERO, V. M. y GORNÉS, S.
(coords), 2000. Colonización humana en medios insulares. Interacción con el medio y adaptación cultural, UIB, Palma. HARRISON, R. J.
1977, The Bell Beaker cultures of Spain and Portugal, American School of Prehistoric Research. Peabody Museum, Harvard University, Bulletin 35, Cambridge & Massachusetts.
HEDGES, R.E.M., PETTITT, P.B., BRONK RAMSEY, C. y VAN KLINKEN, J.
1996, Radiocarbon dates from the Oxford AMS System: Archaeometry datelist 22, Archaeometry, 31 (2), p.207-234.
IRWIN, G, 1992. The Prehistoric Exploration and Colonization of the Pacific. Cambridge: Cambridge University Press.
JUAN i BENEJAM, G. y PLANTALAMOR, L.,
1997. Memòria de les excavacions a la naveta de Cala Blanca.1986-1993, Treballs del Museu de Menorca, 21, Maó.
KEEGAN, W. F.,
1995. Modeling dispersal in the prehistoric West Indies, en CHERRY, J. F. (ed.): Colonization of Islands. World Archaeology vol. 26 n° 3, pp. 400-420. Rouletge.
KEEGAN, W. F. y DIAMOND, J. M.,
1987. Colonisation of Islands by Humans: A Biogeographical Perspective, a SHIFFER, M. B. (ed.): Advances in Archaeological Method and Theory, 10, pp. 49-92.
KERSHAW, A.P.,
1986. Climatic change and aboriginal burning in northeast Australia during the last two glacial/interglacial cycles. Nature 322: 479.
KIRCH, P. V. y ELLISON, J.,
1994. Palaeoenvironmental evidence for human colonization of remote Oceanic islands. Antiquity, 68, pp. 310-321.
KLEIN HOFMEIJER, G.,
1997. Late Pleistocene Deer Fossils from Corbeddu Cave. Implications for human colonization of the island of Sardinia, British Archaeological Reports, International Series, 663, Oxford.
KRA, R.,
1986. Standardizing procedures for collecting, submitting, recording, and reporting radiocarbon samples, Radiocarbon, 28, n. 2A, p.765-775.
LEE, R.B., y R. DALY (eds.),
1999. The Cambridge Encyclopaedia of Hunters and Gatherers. Cambridge: Cambridge University Press.
LEISNER, V.; ZBYSZEWSKI, G, y VEIGA FERREIRA, O.
1961, Les Grottes artificielles de Casal do Pardo (Palmela) et la Culture du Vase Campaniforme, Serviços Geológicos de Portugal, Lisboa, Memória n° 8 (Nova Serie).
LÓPEZ DE CALLE, C. y ILARRAZA, J.A.
1997, Condenaciones y remodelaciones. Una respuesta a las estratigrafías de los sepulcros megalíticos de Cameros, en BALBÍN, R. y BUENO, P. (ed.) II° Congreso de Arqueología Peninsular, Tomo II, Neolítico, Calcolítico y Bronce, Fundación Rei Afonso Henriques, Zamora, p. 309-21.
LULL, V.; MICÓ, R.; RIHUETE, C. y RISCH, R.,
1999. Ideología y sociedad en la Prehistoria de Menorca. La Cova des Càrritx y la Cova des Mussol. Barcelona.
LLOBREGAT, E.; MARTÍ, B.; BERNABEU, J. et alii, 1981. Cova de les Cendres (Teulada, Alicante). Informe prelminar, Revista del Instituto de Estudios Alicantinos 34, p. 87 y ss.
MASSETI, M. y DARLAS, A.,
2000. Pre-Neolithic man and other mammals on the Eastern Mediterranean islands. Mediterranean Prehistory Online (30/06/2000). <<www.med.abaco-mac.it/issue001/articles/arkeos/masseti.htm>>
MAcPHEE, R.D.E. y SUES, H.D.
(eds.), Extinctions in Near Time: Causes, Contexts, and Consequences, 165-88. New York: Plenum.
MARTIN, P.S. y KLEIN, R. G. (eds.),
1984, Quaternary extinctions a prehistoric revolution, The University of Arizona Press, Tucson & London.
MELLARS, P.A.,
1976. Fire ecology, animal populations and man: study of some ecological relationships in prehistory. Proceedings of the Prehistoric Society 42: 15-45.
MESTRES, J.S.,
1995. La datació per radiocarboni i el calibratge de les dates radiocarbòniques. Objectiu, problemes i aplicacions, Revista d'Arqueología de Ponent 5, p.260-275.
MESTRES, J.S. y NICOLÁS, J.C. de,
1999, Contribución de la datación por radiocarbono al establecimiento de la cronología absoluta de la prehistoria de Menorca, Caesaraugusta, 73, Zaragoza, p.327-341.
PATTON, M.,
1996. Islands in Time. Island sociogeography and Mediterranean prehistory. Routledge. London-New York.
PÉREZ ARRONDO
1987, El fenómeno megalítico en la margen derecha del Ebro: La Rioja. Estado de la cuestión y principales problemas, en El megalitismo en la Península Iberica, Madrid, p. 159-80.
PERLÈS, C.,
1979. Circulation de l'obsidienne en Méditerranée Orientale: peut-on appliquer les modèles? In Séminaire sur les structures d'habitat I: Circulation et échanges, 128-39. Paris: Collège de France.
PERLÈS, C.,
1995. La transition Pléistocène/Holocène et le problème du Mésolithique en Grèce, en VILLAVERDE, V. (ed.), Los últimos cazadores. Transformaciones culturales y económicas durante el Tardi-

glaciar y el inicio del Holoceno en el ámbito mediterráneo, 179-209. Alicante: Instituto de Cultura JuanGil-Albert, Diputación de Alicante.
PLANTALAMOR, L.,
1991. L'Arquitectura prehistòrica i pro-tohistòrica de Menorca i el seu marc cultural, Maò.
PLANTALAMOR, L. y MARQUÉS, J. (coords.)
2001, Biniai Nou: el megalitisme mediterrani a Menorca, Treballs del Museu de Menorca, 24, Maó.
PLANTALAMOR, L. y STRYDONCK, L. Van,
1997, La cronología de la prehistòria de Menorca (Noves dtacions de C14). Treballs del Museu de Monorca, 20. Mahón, p.29-31.
PONS, B.
1999, Anàlisi espacial del poblament al Pretalaiòtic Final y Talaiòtic I de Mallorca, col. La Deixa, 2, Palma.
RADI, G.,
1972. Tracce di un insediamiento neolítico nell'isola di Lampedusa. Tai della Società Toscana di Scienze Naturale 79: 197-205.
RAMÓN, J.,
1985. Els monuments antics de les Illes Pitiüses, Eivissa.
RAMÓN, J.,
1996. Las relaciones de Eivissa en época fenicia con las comunidades del Bronce Final y Hierro Antiguo de Catalunya, en ROVIRA, J. (ed.): Models d'ocupació, transformació y explotació del territori entre el 1600 y el 500 A.N.E. a la Catalunya meridional y zones limítrofes de la depressió de l'Ebre. Taules Rodones d'Arqueologia de Sant Feliu de Codines. Gala 3-5 (1994-1996), pp. 399-422.
RAMÓN, J. y COLOMAR, M.,
2000. Investigacions arqueòlogiques a la fortificació prehistòrica de Sa Cala (La Mola, Formentera), Mayurqa, 25, p.31-41.
RIERA, S.,
1996. Incendis i pertorbacions forestalls d'origen antròpic Durant el Neolític Antic al pla de Barcelona (Sector central de la costa catalana). In Actas del I Congrés del Neolític a la Península Ibèrica, Gavà-Bellaterra, 1995, Vol. 1. Rubicatum 1: 35-42. Barcelona: Revista del Museu de Gavà.
ROBERTS, R.G.; FLANNERY, T.F.; AYLIFFE, L.K.; YOSHIDA, H.
et alii New Ages for the last Australian megafauna: Continent-wide extinction about 46000 years ago, Science 292, p.1888-1892.
ROSSELLÓ, G.,
1979. La cultura Talayótica en Mallorca, (2ª ed.) Palma.
RUBINOS, A.
2000, Las fechas de carbono-14 del monumento nº 2 de Biniai Nou (Menorca), Apéndice III a GÓMEZ, J. L. Memoria de excavación del monumento nº 2 de Biniai Nou (campañas 1997-1998), en GUERRERO, V.M. y GORNÉS, S. (coord.), cit. supra, p. 437-440.
SALANOVA, L.
2000, La question du Campaniforme en France et dans les îles anglo-normandes. Productions, chrinologie et roles d'un standard céramique. París, Société Préhistorique Française.
SCHUBART, H.S. y ULREICH, H.
1991, Die funde der Südostspanischen Bronzezeit aus der Sammlung Siret, Madrider Beiträge, Band 17, Mainz am Rhein.
SCHÜLE, W.,
1993. Mammals, vegetation and the initial human settlement of the Mediterranean Islands: a palaeoecological approach. Journal of Biogeography 20: 399-412.
SCHÜLE, W.
2001, Preneolithic navigation in the Mediterranean: a palaeoecological approach. Mediterranean Prehistory Online (21/01/2001). <www.med.abaco-mac.it/issue002/articles/doc/003.htm>>
SCHULTZ, H.,
1997. Estratigrafía y líneas costeras durante el Holoceno en la isla de Ibiza, a Prospecciones geo-arqueològicas en las costas de Ibiza. Treballs del Museu Arqueològic d'Eivissa y Formentera, pp. 11-31.
SCHWARTZ, D.W.,
1970. The postmigration culture: A base for archaeological inference, en LONGACRE, W. (ed.), Reconstructing Pueblo societies Alburquerque (Nuevo México), University of New México Press.
SIMMONS, A.H. (ed.),
1999. Faunal Extinction in an Island Society: Pygmy Hippopotamus Hunters of Cyprus, New York: Kluwer Academic/Plenum Publishers.
SILBERBAUER, G.,
1981. Hunter and Habitat in the Central Kalahari Desert. Cambridge: Cambridge University Press.
SMITH, A.G.,
1970. The influence of Mesolithic and Neolithic man on British vegetation: a discussion, en D. Walker and R.G. West (eds.), Studies in the Vegetational History of British Isles, 81-96. Cambridge: Cambridge University Press.
SIRET, L.
1890, Las primeras edades del metal en el S.E. de España (vol. I y II.) Barcelona.
STRYDONCK, M van y MAES, A.
2001, Les datacions i les análisis isotòpiques de Biniai Nou, en PLANTALAMOR, L. y MARQUÉS, J. (coords.), Biniai Nou: el megalitisme mediterrani a Menorca, Treballs del Museu de Menorca, 24, Maó. p.163-174.
STRYDONCK, M.J., WALDREN, W. y HENDRIX, V.,
1998. The 14C chronology of the Son Mas Sanctuary site (Valldemosa, Mallorca, Spain), en MOOK, W.G./ PLICHT, J. Var der, (ed.), Proceedings of the 16th International 14C Conference, Radiocarbon, vol.40 nº2 p.735-748.
STUART, A.,
1991. Mammalian extinctions in the late Pleistocene of Northern Eurasia and North America. Biological Review 66, p. 453-562.
TARRADELL, M. y FONT, M.,
1975. Eivissa cartaginesa, Barcelona.
TOPP, C.,
1987a. Further Considerations on the Beaker Phenomenon/ Problem especially in relation to Wares of the Balearic Archipielago, en WALDREN, W. H. y KENNARD, R.C. (eds.): Bell Beakers of the Western Mediterranean. Definition, Interpretation, Theory and New Site Data. The Oxford International Conference 1986, part i, pp. 19-22. British Archaeological Reports, International Series 331 (i). Oxford.
TOPP, C., 1987 b. Incised vs Beaker Wares in the Balearic Islands, en WALDREN, W. H. y KENNARD, R.C. (eds.): Bell Beakers of the Western Mediterranean. Definition, Interpretation, Theory and New Site Data. The Oxford International Conference 1986, part i, pp. 267-282. British Archaeological Reports, International Series 331 (i). Oxford.
TOPP, C.,

1988. "Incised" or "Beaker" Wares in the Balearic Islands?, Bulletin of the Institute of Archaeology, 25, p.67-85.
TOPP, C.; FERNÁNDEZ, J.H. y PLANTALAMOR, L.,
1979. Recent archaeological activities in Ibiza and Formentera, Bulletin XVI. Inst. of Arch. of the Univ. of London, p.215-231.TUSA, V., 1999. La Sicilia nella prehistoria. Palermo: Sellerio Editore.
TORRES, M.
1998 , La cronología absoluta europea y el inicio de la colonización fenicia en Occidente. Implicaciones cronológicas en Chipre y el Próximo Oriente, Complutum, 9: 49-60.
VIGNE, J.D.,
1987. L'extinction holocène du fond de peuplement mammalien indigène de îles de Méditerranée occidentale, Mém. Soc. Géol. France, n° 150, p.167-177.
VIGNE, J.D.,
1995. Faunes sauvages et sociétés humaines sur les îles Méditerranéennes: exemples corses, en CHAIX, L./ OLIVE, C./ ROGUIN, L./ MAAMAR, H.S./ STUDER, J., L'animal dans l'espace humain, l'home dans l'espace animal, Anthropozoologica, 21, París, p.41-54.
VIGNE, J.D., 1999. The large "true" Mediterranean islands as a model for the Holocene human impact on the European vertebrate fauna? Recent data and new reflections, en BENECHE, N.
(ed.), The Holocene history of the European vertebrate fauna. Modern aspects of research (Berlin 1998), Rhaden/Westf., Leidorf, pp. 295-322.
WALDREN, W.
1982, Balearic Prehistoric Ecology and Culture, British Archaeological Reports, Intenational Series, 149, Oxford.
WALDREN, W.
1998, The Beaker Culture of the Balearic Islands, British Archaeological Reports, Intenational Series 709, "Western Mediterranean Series 1, Oxford.

46

BINIPARRATX PETIT (SANT LLUÍS): A RESEARCH AND RE-EVALUATION PROJECT IN THE SOUTHEAST OF THE ISLAND OF MINORCA

VÍCTOR GUERRERO AYUSO (1), **JOAN SANMARTÍ GREGO** (2), **JORDI HERNÁNDEZ GASCH** (2) **J. SIMÓN GORNÉS HACHERO** (1), **JOANA Mª. GUAL CERDÓ** (3), **ANTONI LÓPEZ PONS** (3) **JOAN DE NICOLÁS MASCARÓ** (3)

(1) University of the Balearic Islands, Palma de Mallorca, Baleares, Spain
(2) University of Barcelona, Barcelona, Spain
(3) Institut Meoriqui d'Estudis (IME)

RESUMEN.

El proyecto de investigación y revalorización emprendido por las Universidades de las Islas Baleares y Barcelona, con la colaboración del Institut Menorquí d'Estudis, conforma un proyecto que pretende ser una experiencia multidisciplinar de investigación centrada en el estudio de una comunidad prehistórica que se asentó en el lugar denominado Biniparratxt Petit, yacimiento arqueológico situado, en su mayor parte, en el interior de las actuales instalaciones del aeropuerto de Menorca.
La iniciativa nació del acuerdo marco subscrito el 26 de mayo de 1999 entre AENA, el Consell Insular de Menorca-IME, el Ayuntamiento de Sant Lluís y las Universidades de las Illes Balears y de Barcelona.
Es objetivo básico de este proyecto es la reconstrucción del paleoambiente en el que se ubicó esta pequeña comunidad prehistórica menorquina y, por extensión, conocer a fondo sus bases de subsistencia y organización social.

Entre los objetivos terminales se contemplan también tareas de señalización y preparación del sitio para su uso didáctico y como oferta turístico-cultural de la zona.

Además, de acuerdo con las instituciones competentes, se potenciará la restauración y exposición de los bienes muebles. Igualmente se pretende la publicación periódica de los resultados a distintos niveles, sin olvidar el divulgativo.

ABSTRACT
This research and re-evaluation project undertaken by the University of the Balearic Islands and the University of Barcelona, with the collaboration of the Institut Meorquí d'Estudis (IME), is a multidisciplinary investigation focusing on the prehistoric community that was established in a place known as Biniparratx Petit, an archaeological site which is largely within the perimeter of the Minorca airport.

The initiative emerged from the framework agreement signed on 26th May 1999 between the Spanish airport authorities (AENA), the Island Council for Minorca-IME, the Municipality of Sant Lluís and the Universities of the Balearic Islands and Minorca.

The main objective of the project is the reconstruction of the palaeoenvironment of this small Minorcan prehistoric community, and, by extension, to achieve an in depth understanding of its basic structures of subsistence and social organisation.

Among the final objectives is the provision of signs and preparation of the site for the purposes of education and cultural tourism.

In addition, and in agreement with the competent institutions, found objects will be restored and exhibited. The results of the project will be published in various contexts, including the provision of information for the general public.

I-. Introduction and justification of motives

From the early days of scientific archaeology in Minorca almost to the present day, the spectacular monumental construction of certain talayotic monuments has stimulated a focus on aspects of the "prestige architecture" of talayotic society. These include large "talaiots", "taulas" (native sanctuary), etc., or settlements and complexes including major monuments, such as Trepucó, Torre d'en Gaumés,

Torralba, and others which are only partially known many years after their investigation was commenced. As a result the scientific literature relating to talayotic culture has, with very few exceptions, discussed the origins, parallels and functions of these architectural elements. The greater part of this historiographic production, moreover, is restricted to simple description, and hardly any documentation is provided which could be of relevance to anthropological interpretation, which without question constitutes the goal of the study of prehistory.

Despite this body of work, and a large number of archaeological excavations undertaken, in particular since the last quarter of the 20th century, we are still ignorant of the general aspects of the cultural origins and development of the prehistoric populations of Minorca. To give only one example, it should be recognised that we do not know the function and exact chronology of the talayots of Minorca. These are perhaps the most emblematic monuments of the prehistory of the island, but we understand very little about them. The same is true, to a greater or lesser extent, of the "taulas" sanctuaries, the hypostyles, hypogaea of the post-talayotic period, etc.

R. Chapman, an author who is eminently qualified to analyse the state of research into the prehistory of the Balearic Islands, gave this assessment in 1991:

"Los análisis de las culturas de la Prehistoria reciente en estas islas se han centrado en sus particulares construcciones arquitectónicas... Los estudios sobre el cambio medioambiental, intensificación medioambiental y complejidad social han recibido una menor atención... Pese a los intensos debates sobre la tipología arquitectónica, distribución, función y cronología relativa y absoluta de estos monumentos [talayóticos] son pocos los análisis que se centran en los procesos socieconómicos experimentados por las culturas que los edificaron. ¿Cómo se transformó el patrón de asentamiento y la densidad de población a lo largo del tiempo? ¿Qué papel pudieron haber desempeñado en el ámbito de la demarcación territorial? ¿Es posible discernir el grado de expansión demográfica y de aumento de la complejidad patente en los poblados fortificados del Ier milenio". (Chapman, 1991: 343).

The situation has, with some exceptions, changed little since then. In the words of Lull et alii (1999: 11):

"The public archaeological heritage of the island of Minorca, and the Balearic Islands in general, is one of the richest and best in Europe. As a result the opportunities for studying the social, political and economic organisation of prehistoric societies are excellent... The very "monumentality" of a significant part of the archaeological record has no doubt been a factor in limiting most research to an architectonic discourse, whose principal objective has been the definition of a chronological sequence on the basis of typological and functional analysis of variation in the constructions. Less attention has been given to the study of items found in association with the monumental constructions, and above all, investigation of the social and economic structures of the past has been systematically ignored. The greater significance ascribed to the container over and above the contents has meant that even today a large part of the functional and chronological interpretation of the settlements is limited to more or less reasonable and probable hypotheses, lacking sufficiently solid empirical referents."

The programme of work proposed here sets out to follow more deeply a line of research with the potential to reinvigorate research into the prehistory of the Balearic Islands, based on precisely those aspects which, as described above, have been accorded less significance in previous research. In other words, equal significance will be given to container and to content, and issues related to the talayotic settlement as a whole will be addressed, and in particular, its habitat, communication network, production and storage areas, without of course ignoring its prestige constructions.

The objective, therefore, is an extended excavation of the settlement, with a corresponding recovery and analysis of the archaeological evidence, which should provide evidence to support a reconstruction of the palaeoenvironment and the social and economic history of the community which inhabited the settlement.

In the opinion of the present authors, it is only possible for valid outcomes to result from such a project if a multidisciplinary approach is taken, opening the project the participation of all those university departments and research institutes whose methods of study have the potential to throw light on the questions set out in the objectives of the project.

In parallel with the specific archaeological research goals, it is intended to provide practical training in field archaeology for students of the University of Balearic Islands (UIB) and University of Barcelona (UB), and in particular those resident in the Balearic Islands wish to undertake research in their home area. In addition students from other institutions will have the opportunity to be involved, through European programs (Erasmus) or Hispano-American programs (Intercampus), or short term collaboration agreements, such as that established in 2001 with the University of Durham (England).

The goals of the project which we describe here are not limited to research into archaeology and prehistory, but rather encompass the recovery and re-evaluation of a landscape which has deteriorated severely as a result of human intervention, but which can yet be partially restored so that it may be enjoyed by society. In this work the project will collaborate with the competent institutions: the Island Council of Minorca, the Municipal authorities of Sant Lluís, and the Spanish aviation authority AENA.

An important social gaol, established at an early stage of the project, is the creation of a cultural attraction which can contribute to the tourist activities offered by Minorca, in precisely the place where most visitors enter and leave the island. When it has been restored it is also intended that the site be educationally stimulating, so that it can function as a teaching resource contributing to the understanding of our prehistoric communities.

DELIMITATION OF THE PROJECT SITE

The site where the intervention will take place is situated at the southern end of the runway of Minorca Airport, in land owned by AENA, although the settlement also extends into some neighbouring private properties. The archaeological excavations undertaken in the project will, in the short term, only involve those parts of the settlement which fall within the area owned by AENA, but it would nevertheless be desirable in the medium term to unify all the architectonic units and integrate them within a single archaeological park.

The area to be studied and investigated is at present completely surrounded by a high security double fence, and falls within the present installations of the airport. At the southern edge it is bounded by the rural track known as Es Milord, and to the north by the airport runway. As we have mentioned above, the settlement extends to the east and west, taking in the various private properties in the area, and lying in two municipalities, Sant Lluís and Mahón.

THE CURRENT STATE OF KNOWLEDGE

PRECEDENTS.

The Biniparratx Petit site contains the remains of a talayotic settlement which has been partially destroyed. It is located at beside the southern end of the runway of Minorca Airport, near the house of the same name (coordinates UTM: 604100-4412100). It lies within the municipality of Sant Lluís of the Island of Minorca. It is included in the Catàleg de "Bens d'Interés Cultural" (catalogue of items of cultural

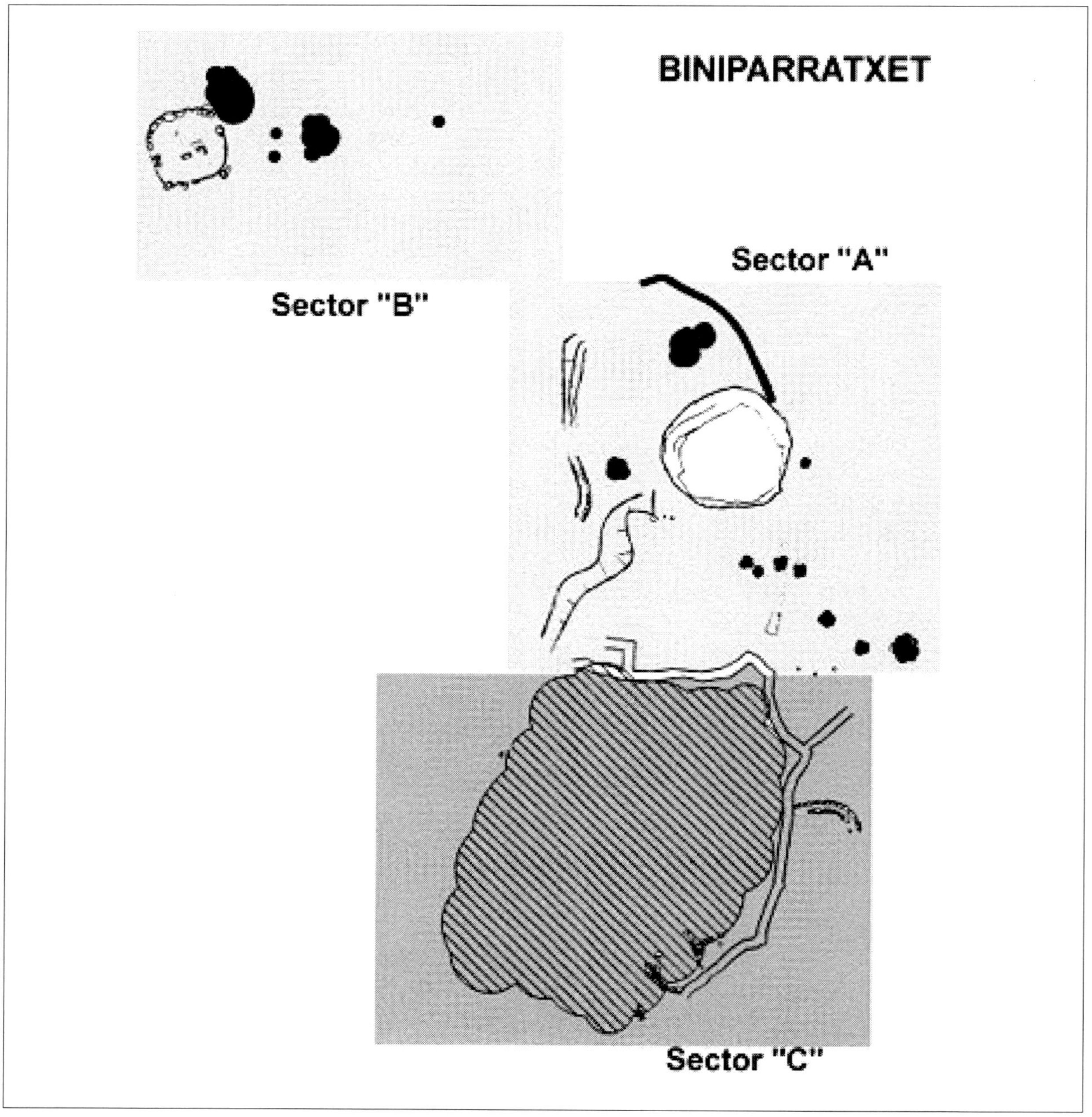

FIGURE 1. EXCAVATION AREAS

significance) as entry 07052-2-2-51-002124-0, in the inventory of J. Mascaró Pasarius as number 2088. The same author gave a different number to the remains in the neighbouring farm of Binparratxet, which probably form part of the same group.

The settlement in question may be related to other sites belonging to the same cultural complex, and it is some 625 m. to the west of the settlement of Biniparratx, 1,250 to the north-east of Bincalaf, 1,250 m. to the south-east of the settlement of Binimainut, and a similar distance from the settlement of Torelló.

There are few references to this complex in the archaeological literature. J. Mascaró Pasarius (1982) cited it in his inventories, and from him we know that in 1964 most of the structures adjoining the talayot were destroyed, and that in

FIGURE 2.

1968 Mª Luisa Serra Belabre carried out excavations at some location in the settlement, probably in a circular habitation at its southern end which is at present being re-excavated by the team from the University of Barcelona. We have no knowledge of any reports resulting from this activity, but some of the material found may be that which is catalogued in the collections of the Museum of Minorca as being from this site.

According to the information provided by the survey of the archaeological areas of the airport carried out by the technical service of AENA, it seems that in 1980 workers from the farm demolished a small apsidal building situated to the south of the talayot, which was still roofed, although it is not clear if the roof was original or the result of later adaptation of the building as a shelter for livestock.

The last documented intervention in the site was in 1995, when work was carried out to level the terrain. A complex made up of a circular house and hypostyle was excavated and moved by the company Balear d'Arqueologia (under the direction of J.C. de Nicolás Mascaró) to an area adjoining the airport car park. This is at present the only element of the settlement which is indicated and prepared for visitors, and it is an important educational resource within the network of visitable settlements in Minorca.

FIGURE 3. ARCHITECTURE

The remains located in the neighbouring "lloc" of Biniparratxet have generated a larger number of bibliographical references, since at some time in the late 19th century two bronze statuettes were found on the site. These are the Odysseus and Isis-Fortuna which were formerly part of the Pons and Soler collection, now property of the De Olives family. A number of authors have discussed these figures (Albertini: 1905; Garcia y Bellido, 1935, 1936, 1948), but the most complete and specific study is that of M. Orfila Pons (1983), whose undergraduate thesis classifies the Odysseus as a Hellenistic work, without going into greater detail, and the Isis-Fortuna as a Roman work of the first century AD.

On the same site, apparently in the late 19th century, an impressive collection of Hellenistic terracottas was also found (Mascaró, 1983), suggesting that there must have been a major sanctuary in Biniparratx Petit which has not yet been identified. We do not know exactly where the find was made, nor the circumstances, but in the 2000 campaign we accidentally found a fragment of a terracotta figure in a toga, on the surface by the side of the security fence which encircles the site.

SEMIMICROSPATIAL DESCRIPTION, AND HYPOTHESIS FOR THE FUNCTIONAL INTERPRETATION OF THE SITE.

The site is constituted by the remains of a talayotic settlement which extends over a surface of approximately 10,500 m.2. At present it is possible to distinguish:

(a) A conical talayot constructed of irregular stones of varying sizes, which in some sections make up regular courses. The upper platform is relatively well preserved, with a south facing door, and it has significant archaeological potential.

(b) Remains of the walls of constructions adjoining the talayot, no longer standing as a result of uncontrolled activities in 1964, together with some water cisterns, which in all probability form part of the complex of talayotic dwellings which still survives, particularly to the north-east. In this area trenches were dug in the year 2000 by the team from the University of the Balearic Islands, directed by Victor Guererro and Simón Gornés.

(c) A group of circular or semicircular talayotic dwellings to the east of the talayot. Walls built with two faces with courses of regular blocks and an interior of gravel and rubble. Pilasters and other supporting elements are also visible. The area suffered substantial damage in the construction of the security fence of the airport.

(d) Set of large cisterns situated to the west of the talayot. Some of these were covered in the levelling work carried out in 1995. This was the site of the dwelling moved during the levelling work.

(e) Remains of a circular house situated to the south of the talayot, and excavated in 1968 by Mª Luisa Serra Belabre and re-excavated in 2000 by the team from the University of Barcelona directed by Joan Sanmartí and Jordi Hernández.

(f) Remains of a small apsidal construction situated to the south of the talayot. J. Mascaró Pasarius suggests that this might be the precinct of a taula sanctuary. Today no parts of this construction remain standing.

(g) Remains of constructions dispersed around the whole area, together with a large quantity of Punic and Roman ceramics on the surface, and other items such as hand mills and hammers.

(h) Largely destroyed and thus difficult to interpret remains, situated in the neighbouring form of Biniparratxet, where the above mentioned bronze statues were found.

From a functional point of view these remains should be classified as those of a small talayotic settlement, very similar in structure and size to that of Binissafullet. The settlement is composed of the architectonic units which characterise these complexes of habitation. It remains to be established whether or not this settlement had a taula sanctuary precinct.

THE 2000 CAMPAIGN OF EXCAVATION.

The first campaign of excavation of the prehistoric settlement of Biniparratx Petit, corresponding to the first phase of the project, took place between 15th August and 25th September 2000. In all 20 people were involved, including senior technicians, students fulfilling the practical requirements of their courses, collaborators and volunteers.

Objectives

1. Document the present state of the settlement following the interventions which have been made there for various reasons.
2. Survey as precisely as possible the limits and state of the architectural complexes.
3. Identify the phases of occupation of the site, and the degree of intensity of each.
4. Present an evaluation of the archaeological and monumental state of the settlement.

Zones of the site

The settlement covers an extensive area adjoining the runway of Minorca Airport. In their present state, however, the architectonic structures are dispersed, and, in many areas, without apparent physical continuity.

Because of this, and only for the purposes of planning the present and future excavation, the settlement has been divided into the following zones (FIGURE 1):

Sector A: The structures adjoining and near to the talayot. Prior to excavation it was possible to distinguish with clarity the existence of the remains of a second talayotic house, partially destroyed some years ago by work with a mechanical excavator which took place on the site.

This area was excavated in 2000 by the team of the University of the Balearic Islands and the Institut Menorquí d'Estudis.

Sector B: A rectangular area, adjoining the road, which includes a circular talayotic house. It has been excavated by the team from the University of Barcelona.

Sector C: An area of the settlement demarcated by a "tanca", or dry stone wall, built on an irregular plan, situated to the south of sector A. This zone retains a high density of visible constructions, among them a number of talayotic houses and others which are difficult to delimit because they are overgrown and half covered by accumulated stones.

THE EXCAVATION OF SECTOR A: METHODOLOGY.

The area to be excavated in the campaign of 2000 did not have any architectonic structures clearly delimited on the surface. Only three aligned blocks suggested the possible existence of a wall.

The terrain is noticeably inclined, with the highest point

at the base of the talayot, and the lowest at the eastern limit of the sector selected for excavation. The surface consisted of cultivated land, with an abundant presence of pottery fragments. Because of the characteristics of the archaeological deposits the most appropriate excavation methods were those developed by Harris and Laplace-Meroc.

Mechanical procedures to remove the superficial levels were ruled out, since while these levels might lack prehistoric archaeological material it was clear that they were extraordinarily rich in medieval and modern archaeological remains and ethnographic material.

FIGURE 4. PLAN AND ARCHITECTURE

Taking reference points from the general survey of the site, sector "A" was divided into 4x4 metre squares, as shown in Figure 2.

The area of excavation was delimited by the squares C-4, C-5, D-4 and D-5, which had not been disturbed by mechanical excavators. In the area which had been partially disturbed, on the other hand, it was planned to open a stratigraphic trench which would be partly situated in square B-4.

The above mentioned squares were divided by an aerial grid into 1 metre squares, which were identified respectively as 12 to 21 on the south-north axis, and 11 to 22 on the west-east axis.

Before commencing the excavation elevations were taken to prepare for the topographic reconstruction of the terrain. Similarly maps of elevations were made at the end of each day's excavation, independently of the triangulation notes corresponding to each find of an artefact, osseous remains and carbon particles.

All the information emerging from the archaeological record was collected simultaneously on paper and in a computer data base developed by the team from the University of the Balearic Islands, using Microsoft Office-2000, and File Maker Pro.3.

All the documentation will eventually be processed using GIS (ESRI) and Autocad, to provide a geospatial analysis of all the finds and to provide greater rigor in the analysis of contextual relationships.

In parallel the excavation was exhaustively documented in graphic form, using video, traditional slide photography, and digital photography processed with Adobe Photoshop 5.0 and stored as jpeg files for PC-IBM and TIFF format for IBM and Macintosh.

The plans and sections were prepared in the traditional format of tracing paper and archived digitally.

Progress of work and preliminary results from the 2000 campaign in Sector A.

The material from the excavations is still being studied and analysed, so it is not possible to provide more than a provisional approximation of the historical context. The picture will be completed when the range of analyses (remains of fauna, charcoal, and sedimentological) are completed, and the excavation of these sectors is carried out in the 2001 campaign.

The present description should therefore be understood as referring to provisional results, but, with due prudence and from an initial examination of the complex of artefacts we may advance the following observations:

The historic contexts:

In the light of the excavation of sector "A" the following periods of occupation of the site have been identified:

THE CONTEMPORARY PERIOD:

The plot occupied by the settlement has been used for cultivation and grazing until relatively recently.

On the surface itself there are abundant remains of the traditional pottery of Minorca.

We believe that this phase of occupation of the site has an evident ethnographic interest which should not be undervalued. Because of this all the pottery fragments have been collected using exactly the same methodology as that employed in the documentation of other historic and prehistoric periods, and mechanical procedures for clearing the upper strata have been specifically rejected.

LATE MEDIEVAL AND MODERN PHASE:

These historic periods are richly documented in the excavated sector. The talayot itself was no doubt used as a watch tower because of its commanding view of the sea. At its foot, in the eastern sector a level of occupation has been documented by the presence of abundant pottery (white tin glaze with blue decoration) –which can probably be dated from the 16th century, together with monochrome pottery with honey coloured and green glazes.

In this sector there was clearly a small house of which only a wall remains (UE-2), built repurposing blocks from the talayot. It has not been possible to establish the shape of the house or hut. A burnt area no doubt corresponds to this phase of the occupation of the site.

This occupation had hardly any effect on the lower levels.

ISLAMIC AND MEDIEVAL: The presence of some fragments attributable to these periods suggests that there was some degree of frequentation of the site. It has not been possible to determine its intensity, and it certainly did not constitute permanent occupation.

REPUBLICAN AND IMPERIAL ROME: The excavated sector does not have any sign of activity from any part of the Roman administration of the island. This statement is only valid for the immediate environs of the talayot, as we know that in other areas of the talayotic settlement finds of Roman pottery are common.

In the excavated area only one fragment of sigillata pottery has been identified, which corroborates the low or null use of this space in the Roman period.

LATE PROTOHISTORIC OCCUPATION: One of the best documented chronocultural phases thus far has been that of the occupation of the site which appears to have taken place between the mid and late 3rd century BC.

It is to this phase that the stratigraphic units UE-13; UE-3/17 and UE-20 belong, corresponding to elements of construction which may have been in use during this phase. The walls are insubstantial, built on earlier strata, and demarcate a long room which adjoined the talayot. It is difficult to define the type of architectonic structure until excavation proceeds in adjacent sectors. At present we clearly have only a very partial view of the spatial organisation.

Functionally we may suppose that the structure is a dwelling, perhaps in only occasional use, and not inhabited for a long period.

The sratigraphic unit UE-4, which is sedimentary, contains the archaeological record which corresponds to this context of habitation. The most abundant is pottery is late talayotic jars and turned pots. The latter are mainly Punic amphorae from Ibiza (PE-16 y 22), some fragments of cooking pots and table ware from the same source, and some fragments of amphorae from metropolitan Carthage and Iberia. No Greco-Italian amphorae have been found, nor black varnished pottery.

Although it is clearly impossible to establish what occurred in such an extensive settlement from the documentation provided by a few square metres of excavation, we may nevertheless propose a working hypothesis. In our opinion, the period of occupation corresponds to a period of crisis in the indigenous settlement related to the Second Punic War. That moment had a special historic significance for Minorca, as it coincided with Magon spending the winter in the island (Tito Livio, XXVIII, 37,3), the recruitment of the last indigenous mercenary reinforcements from the Balearic Islands, and the eventually in the foundation of the city of Mahon itself.

A B C

FIGURE 5. AREAS AND ARCHITECTURE

This occupation of sector "A" of the settlement of Biniparratx Petit in the late 3rd century bc, appears, in any event, to be very precarious. The walls are not mortared, and they rest on older sediments which are hardly effected.

The reuse of older semi-ruined walls (UE-17) can also be detected, and these were extended using a cruder construction technique.

UNDETERMINED PROTO-HISTORIC CONTEXTS

In various sectors of the excavation stratigraphic units have been revealed which predate the chronocultural context described above, but for the moment we lack clear diagnostic elements which would enable us to determine their precise chronology. UE-8 has been excavated, and no turned pottery has been found in this sedimentary unit. This suggests that the context is prior to the first massive importations of amphorae and other pottery, which in general terms occurred in the late 5th and 4th centuries bc.

This sedimentary unit clearly rests on the radial wall UE-9, (FIGURE 2) which will provide us with a precise chronology relative to the previous phase.

TALAYOTIC

Naturally the architectonic structure which may with confidence ascribed to this phase is the talayot itself. Two stratigraphic units (FIGURE 3) of an architectonic nature (UE-9 and UE-17) which were revealed during the excavation may also belong to this phase of Minorcan prehistory.

In the first place, wall UE-9 (FIGURE 3) springs from the wall of the talayot itself, and its blocks are very well engaged with the external face of the talayot. Everything suggests that this is a radial room, adjoining the main monument, of uncertain chronology but certainly old (circa 800-700 BC).

The excavation has reached the upper limits of a stratigraphic unit (UE-10) which could correspond to the period of activity in this room, which at present is known only as the above mentioned wall UE-9. No turned pottery to be found in association with this sedimentary unit. The base of a hand mill (nº105) has been found in situ resting on this wall (FIGURE 3).

Secondly, wall UE-17 is a very similar architectonic structure to that just described, and is perhaps the opposite wall of the same room. It is also a radial wall springing from the talayot. In any event, the way in which its stones engage with the external face of the talayot supports the hypothesis that it was built simultaneously with the talayot itself.

If these suppositions are confirmed when excavation proceeds on this area, then this would be an adjoining chamber of the talayot, which was in use at the same time as the main construction. Pending the availability of radiocarbon dates, the period of use may be provisionally placed at between 850/800 and 650 BC.

THE EXCAVATION OF SECTOR B: OBJECTIVES AND METHODOLOGY.

The excavation carried out in the 2000 campaign by the team from the University of Barcelona, directed by Joan Sanmartí and Jordi Hernández-Gasch, concentrated on the demarcation and excavation of talayotic house 1 (FIGURE4) of sector 1 of the settlement. This structure, clearly visible after the clearing of vegetation, had been partially excavated by Ma Lluïsa Serra-Belabre in 1968. Despite this, it was evident that intact sediment remained, and the excavation itself has confirmed the great stratigraphic and structural interest of the site.

Structural elements

The excavation left the perimeter walls of the house exposed, since they had been exposed during the 1968 excavation and covered since then. The plan of the building is roughly circular, and it covers a surface of approximately 114.5 m2. The exterior wall occupies an area of 35.5 m2, which reduces the interior space to 79 m2. Even this is not all usable space, as there are numerous internal subdivisions, which have not yet all been explored.

These areas are similar to those which Joan de Nicolàs (1997: 48) calculated for the house which he excavated and moved from Biniparratxet, situated a few dozen metres from C-1: 128 m2 in total surface area, and 75 m2 of usable internal space.

The wall is seated on the bedrock, which very probably served as the quarry from which blocks were hewn for the construction of the house itself. This may be seen in the incisions in the rock which are to be seen in the south-west of the sector, between the house and the edge of the settlement. In the western part of the wall there is an enormous block (155 x 115 cm) which takes up the entire width of the wall. It is probable that it was extracted from the rock platform, and was raised in the same place, without the need for traction.

For the most part the wall is faced on both sides. In the first course the face is formed of large, regular, worked blocks tending to a parallelepiped form. The interior of the wall consists of more irregular blocks of stone, of medium size and sometimes small. No infill has been found. Inserted in the internal wall are three surviving pilasters, at intervals of 3.35 and 5.35 m, which protrude slightly from the line of the wall. The central pilaster has been displaced by a wild olive tree, but its base can be identified with precision.

This type of design is well documented among examples of prestige domestic architecture known from Minorcan protohistory.

The door of the house has been identified on the southern side of the structure, where the walls are rather ruinous. The bedrock seems to have been cut and lowered at this point. There is also a area with flagstones in the entrance. Both these features require confirmation in the next campaign.

INTERNAL SUBDIVISIONS

So far two areas have been clearly identified, the northern room, and the kitchen, but there are signs of others.

THE NORTHERN ROOM

The northern room has a surface area of 24.5 m2, and is roughly semicircular, with the head opposite the entrance door. It is closed to the south by a straight wall 9m. long (UE 30). The wall is in ruins, and in some sections only the cuts in the bedrock enable us to establish its continuity (FIGURE 5B). There is a large fallen slab (no. 79) which no doubt came from the median wall, separating this room

from the western rooms. The wall (UE 26) remains in contact with the perimeter wall.
In the interior the floor is grooved and smooth in many places (UE 25). In some parts paved areas have survived (UEE 28 and 29), filling some depressions in the bedrock, and scored by the previous excavation work.

To 40 cm from the central pilaster of the room, and only 10 cm from the internal face of the perimeter wall, there is a hemispherical cavity excavated in the bedrock (UE 6). We can only speculate as to its function, perhaps as a food or liquid container, or possibly with sanitary functions, as suggested for the small carved cavity in the N. room of the house moved in 1995.

1.9 m from the base of the pilaster there is a more or less circular perforation in the bedrock of 35 x 30 cm. It narrows towards the interior and opens, it appears, into a cavity of unknown dimensions, which may possibly be related to the mouth of a cistern (FIGURE 5C), situated in the central patio of the house, near the wall of the northern room. Since the perforation is not wide enough to permit extraction of the contents of a cistern it may be a ventilation hole or light source.

To the west the floor of the room is elevated. A groove in the rock delimits the upper and lower floors. It may be that there was a partition between them, (possibly constructed of perishable materials), and what seems to be the seating of a door has been identified, a groove in the bedrock, about 80 cm long and of variable width (from 10 to 23 cm).

In the upper floor area, near the groove which separates it from the lower, there is a small hemispherical excavation in the rock (18 x 20cm), which may be related to a wooden structure.

THE KITCHEN

This room has an apsidal plan, which is clearly recognisable as its base was excavated from the bedrock. Its location in the house and its morphology denote its purpose, as similar structures opening onto the patios of the talayotic houses excavated in Minorca have repeatedly been identified as the site of hearths.
Although excavation is still pending, we can state that there were two periods of use (perhaps with different functions) in this space. The horizontal pilasters which close the kitchen to the west seem to be a later adaptation, as these structures were not originally closed. Moreover this seems to be repurposed material, no doubt from the monoliths which surrounded the patio.

DYNAMIC STRATIGRAPHY

Excavation has enabled us to identify a series of stratigraphic units (sedimentary, structural, negative), and establish a number of stratigraphic facts. The documented strata may be grouped in three categories: superficial, ruins, and abandonment/final use.

The superficial strata (UE 1) have a certain thickness, due to two factors: previous excavations and possible agricultural work in clearing stones from neighbouring fields (perhaps UE 2). During the excavation it was possible to distinguish the limits of the sounding taken in the 1968 excavations, together with the spoil heaps which were created (UU EE 3 and 9), some of which covered the walls and areas previously exposed.

The ruins are to be found in UUEE 14 and 21, in the area of the patio and the western rooms; UU EE 16, 17 and 18, in the interior o the kitchen; and UU EE 4,7, 10 and 11 in the western part of the northern room. Structural elements have been identified in these areas (partition walls, fallen pilasters), but it was not possible to complete excavation in the present campaign.

The strata of abandonment/final use are few in number and low in stratigraphic potential. This is because they have only been excavated in the parts of the site effected by previous excavations, and also because the floor of the house (except in the paved areas) is composed of the bedrock, and this must have been kept relatively clean of debris until the abandonment of the dwelling.
The strata which have been clearly identified as abandonment/final use are UU EE 12, 5, 8 and 20. The last three are situated in the interior of the hemispheric cavity. UE 20 contains a significant quantity of wood, covered by a layer of grey / light yellow clay. This could be the remains of a cover for the cavity, but the poor conservation of the wood makes this impossible to establish.

THE ARCHAEOLOGICAL MATERIAL

The greater part of the archaeological material found is pottery. There are some lithic specimens, fragments of mortars and hand mills. There are very few osseous remains.
It is evident that there are relatively large quantities of imported pottery, and much smaller amounts of indigenous pottery. Most of the imported pottery derives from Punic amphorae from Ibiza, with the oldest examples dating from the 4th century bc. (PE 14).

The scarcity of Iberian and campanian black varnish pottery is noteworthy, (even though Italic amphorae are plentiful). There is also little thin walled table ware, and few sigillatas, although the presence of Italic sigillatas together with Roman amphorae Dressel 2/4 provides chronological indicators from the first century AD.

BIBLIOGRAPHY

ALBERTINI, E.,,
1905: Statuettes de bronze trouvés à Minorque in Melanges d'Archeologie de l'Ecole Française de Rome XXV, p. 346. Roma.
CHAPMAN, R.,
1991. La formación de las sociedades complejas. Ed. Crítica. Barcelona.
GARCÍA Y BELLIDO, A., 1935: Figuras griegas de bronce y barro halladas en las Islas Baleares in Anales de la Universidad de Madrid. Letras, p. 148. Madrid.
GARCÍA Y BELLIDO, A.,
1936: Hallazgos griegos en España. p. 16. Madrid.
GARCÍA Y BELLIDO, A., 1948: Hispania Graeca. Tomo II, pp. 107-108. Barcelona.
LULL, V., MICÓ, R., RIHUETE, C., RISCH, R.,
1999. La Cova d'Es Càrritx y la Cova d'Es Mussol. Ideologia y sociedad en la prehistoria de Menorca. Publishers: Consell Insular

de Menorca, Ajuntament de Ciutadella, Fundació Rubió i Tudurí Andrómaco, Govern Balear y Univeristat Autònoma de Barcelona.
MASCARÓ PASARIUS, J..
1982. Carta Arqueológica de Sant Lluís, in Ib. Geografía e Historia de Menorca, Tomo III, pp. 148, 154 y 160. Ciutadella.
NICOLÁS MASCARÓ, J.C. DE,
1983: Romanización de Menorca in MASCARÓ, J. (coord.) Geografia e Historia de Menorca, Tomo IV. Pp. 201-288. Ciutadella.
ORFILA PONS, M., 1983: Estatuillas de bronce antiguas, in MASCARÓ, J. (coord.) Geografía e Historia de Menorca, Gráfica Al.lés. Tomo IV, pp. 85-158. Ciutadella.

LA CAMPAÑA DE EXCAVACIÓN DE 2001 EN BINIPARRATX PETIT (MENORCA): AVANCE DE LOS PRIMEROS ANÁLISIS RADIOCARBÓNICOS

RESUMEN
En este escrito se presentan ocho dataciones C14 procedentes del poblado talaiòtico de Biniparratx Petit (Menorca). Varias de las cuales son especialmente interesantes para fechar los orígenes del asentamiento y la construcción del talaiot.

INTRODUCCIÓN

En las siguientes líneas damos a conocer la primera serie de análisis radiocarbónicos procedentes de muestras obtenidas en distintos contextos arqueológicos excavados durante la campaña de 2001 en el yacimiento arqueológico de Biniparratx Petit (Sant Lluís, Menorca). Se trata de una primera serie de análisis que se obtienen cuando aún se están dando los primeros pasos para el estudio de los materiales arqueológicos que acompañaban a las muestras datadas y todavía están en proceso otra serie de dataciones. Por lo tanto, no permiten aún conclusiones definitivas y deben entenderse como un simple avance.

Creemos oportuno publicarlos con prontitud para que puedan ser utilizados y contrastados por otros colegas, ya que en dos casos estimamos que sirven de referencia a procesos que aún están en discusión en los estudios de prehistoria menorquina. En un caso, se trata de un contexto de hábitat o de frecuentación del lugar en los momentos anteriores a la construcción del talaiot, por lo tanto los resultados pueden ser considerados como un elemento inmediatamente ante quem de la erección del mismo. En el otro, representa la primera datación absoluta que se tienen como referencia del posible momento de construcción de una casa talayótica subcircular con patio central.

CONTEXTOS LIGADOS AL TALAIO, EL SECTOR D

Es un área de excavación adosada al exterior Oeste del talaiot. Los contextos datados, correspondientes a las UE 204, 205 y 206 se refieren a ámbitos diferenciados que pasamos a describir someramente:

La UE-206, con tres muestras datadas, está directamente relacionada con las piedras basamentales del talaiot ya que los sedimentos de este contexto, en contacto con la roca madre, fueron recuperados al excavar algunos huecos angostos de la hilada inferior, formando parte de su relleno. Constituye por tanto una datación ante quem.

La UE-205 (una muestra datada) estaba formada por unos sedimentos arcillosos de escasa potencia, de color marrón rojizo, con fragmentos óseos de fauna doméstica i algunos fragmentos presumiblemente de cerámica pretalaiótica, que ocupaban las oquedades irregulares de la roca madre frente a la hilada basamental del talaiot. Se trata también en este caso de una datación ante quem correspondiente a restos alimentarios procedentes de una zona ocupada antes de la construcción del talaiot y que seguramente fue amortizada en el momento de su erección.

Finalmente la UE-204 (con una sola muestra de fauna doméstica datada) estaba constituida por un nivel de cierta potencia de cenizas y tierra quemada con abundantes restos de fauna doméstica, carbones y cerámica exclusivamente fabricada a mano. Este nivel, que ocupaba una zona perfectamente delimitada por el color negruzco del sedimento que ha afectado también a la roca base. Todo parece indicar que esta zona aneja al talaiot siguió en uso tras la erección del mismo, aunque no sabemos qué estructuras arquitectónicas pudieron estar relacionadas con el área de actividad pues hace décadas fueron eliminadas con máquina excavadora, prácticamente al nivel de la roca madre.

Las cuatro primeras dataciones consideramos que son de gran interés por contribuir a fechar la construcción del talaiot, mientras que la quinta, también del sector D, anejo como se ha dicho al lateral Oeste del talaiot, nos fecha, con menor precisión por desgracia, una gran fogata que afectó incluso a varias piedras de la base del talaiot, quizá realizada allí con fines rituales como lo sugiere la presencia de los restos de un ovicáprido parte de cuyos huesos presentaban conexión anatómica en el momento de la excavación, junto a abundantes fragmentos de cerámica talaiótica. La continuidad de uso de esta área está por otro lado ampliamente documentada con materiales de importación aunque sus sedimentos han sido muy afectados por los destrozos causados por el rebaje del terreno efectuado con máquinas hace ya varias décadas, cuando se construyó la pista de aterrizaje del aeropuerto de Menorca.

DATACIÓN 1:
KIA-15219: 2685 ±30BP (Colágeno de fauna doméstica)
d13C = -19.97‰, δ15N = +6.65‰ en C/N = 2.72

68.2% confidence
895 BC (0.22) 880 BC
835 BC (0.78) 800 BC 95.4% confidence
900 BC (1.00) 800 BC

Comentario: Sector D. UE 205. Conjunto 17 (x: (1) 4,45; y (3) 6,60. Tierra compacta, de color marrón rojizo, en contacto con la roca madre, justo delante de la grieta formada por las piedras 6-7 de la base del talaiot. Aparecen numerosos huesos de fauna doméstica, básicamente ovicápridos y cerdo (19 fragmentos) y un fragmento de piedra arenisca ("pedra de cot") quemada.

DATACIÓN 2:
KIA-15220: 2735 ±25BP (Colágeno de fauna domésti-

ca)
δ13C = -20.23‰, δ15N = +5.66‰ en C/N = 5.32

68.2% confidence
900 BC (1.00) 835 BC 95.4% confidence
930 BC (1.00) 820 BC

Comentario: Sector D. UE 206. Conjunto 14. (x: (1) 5,00; y (3) 5,25). Grieta entre las piedras basamentales 9-10 del talaiot, extraída a 60-90 cm del interior de la misma. Tierra muy fina, casi polvo, de color grisáceo. Aparece acompañada de fragmentos de huesos de fauna doméstica, básicamente ovicápridos i cerdo (31 fragmentos) y un solo fragmento de cerámica indígena indeterminada.

DATACIÓN 3
KIA-15222: 2775 ±40BP (Colágeno de fauna doméstica)
δ13C = -21.56‰, δ15N = +5.55‰ en C/N = 2.58

68.2% confidence
980 BC (1.00) 830 BC 95.4% confidence
1010 BC (1.00) 830 BC

Comentarios: Sector D. UE 206. Conjunto 12 (x (1): 5,40; y (3): 4,90). Grieta entre las piedras 10-11 de la base del talaiot, que profundiza unos 40 cm hacia el interior, formado por tierra muy fina, casi polvo, de color amarronado, ligeramente cenicienta. Aparecen abundantes huesos de fauna doméstica, básicamente ovicápridos (unos 20 fragmentos) y escasos restos de carbón y muy pocos restos de cultura material (2 fragmentos cerámicos, uno de apariencia pretalaiótica y otro de talaiótica).

DATACIÓN 4
KIA-15221: 2825 ±25BP (Colágeno de fauna doméstica)
δ13C = -20.14‰, δ15N = +4.52‰ en C/N = 2.77

68.2% confidence
1005 BC (0.50) 970 BC
960 BC (0.50) 925 BC 95.4% confidence
1050 BC (1.00) 900 BC

Comentarios: Sector D. UE 206. Conjunto 12 (x (1): 5,40; y (3): 4,90). Grieta entre las piedras 10-11 de la base del talaiot, que profundiza unos 40 cm hacia el interior, formado por tierra muy fina, casi polvo, de color amarronado, ligeramente cenicienta. Aparecen abundantes huesos de fauna doméstica, básicamente ovicápridos (unos 20 fragmentos) y escasos restos de carbón y muy pocos restos de cultura material (2 fragmentos cerámicos, uno de apariencia pretalaiótica y otro de talaiótica).

DATACIÓN 5
KIA15245: 2475 ±30BP (Colágeno de fauna doméstica)
δ13C = -20.22‰, δ15N = +6.34‰ en C/N = 2.80

68.2% confidence
760 BC (1.00) 520 BC 95.4% confidence
770 BC (1.00) 410 BC

Comentario: Sector D. UE 204. Conjunto 8 (x (2): 6,70; y (5): 6,80; z: -103). Nivel de tierra color ceniza, grisácea, que se caracteriza por la presencia abundante de restos óseos de fauna doméstica, sobre todo ovicápridos, muchos fragmentos de pequeños carbones y también de cerámica talaiótica, junto a algunos de pretalaiótica. La muestra analizada pertenece a un conjunto de huesos de ovicáprido algunos de los cuales presentaban conexión anatómica.

EL SECTOR A

Se trata de un área de excavación localizada en la zona opuesta del "sector D" y adosada al propio talaiot. Se ha documentado una secuencia ocupacional compleja que va desde la fundación hasta el abandono de este sector localizado a partir de unas estructuras domésticas con materiales muy característicos del final de la segunda Guerra Púnica, es decir hacia 220-200 aC. Se ha estimado poco necesario proceder a datar por radiocabono este horizonte tardío pues el complejo artefactual cerámico permite una aproximación calendárica muy ajustada.

Por el contrario, el secuencia de ocupación arcaica no registraba la presencia de materiales a torno y se hacía necesario una determinación cronométrica por C14. Se han datado dos muestras, ambas a partir colágeno de fauna doméstica:

DATACIÓN 6
KIA15700: 2735±30BP (colágeno de fauna doméstica)
δ13C=-21.77‰, C/N=2.66

68.2% confidence
900 BC (1.00) 835 BC 95.4% confidence
970 BC (0.03) 960 BC
930 BC (0.97) 810 BC

Procedente de la UE-8 o unidad estratigráfica sedimentaria dura y compacta con abundante fauna doméstica en forma de detritus alimenticios y escasa cerámica, toda ella de origen talaiòtico

DATACIÓN 7
KIA-15728: 2765±30BP (colágeno de fauna doméstica)
δ13C y C/N (en proceso de cálculo)

68.2% confidence
970 BC (0.09) 960 BC
930 BC (0.91) 830 BC 95.4% confidence
1000 BC (1.00) 830 BC

La presente muestra fue recogida de la UE-30, la cual es una unidad sedimentaria que cubre y rellena las irregularidades de la roca base. Sobre ella se asientan los restos de un enlosado muy mal conservado (UE-24) de alguna estructura adosada al talaiot en gran parte destruida por las ocupaciones posteriores.

Cronoestratigráficamente corresponde a un momento de frecuentación del lugar en sus momentos iniciales. Es posible que pueda ligarse con las dataciones del sector D ante quem de la construcción del talaiot. El lugar exacto de recogida de la muestra no permite esgurar este extremo de manera tan segura.

CÁMARA SUPERIOR DEL TALAIOT
Las dataciones 8 y 9 de Biniparratx Petit corresponden a fragmentos de madera carbonizada de la base de dos pilastras (la número 3 y la número 2) que estaban integradas en el paramento interno de la cámara superior del talaiot. Pudieron localizarse hasta cuatro huecos de pilares, todos integrados en el muro de piedra de la cámara, de planta rectangular, que en un momento todavía no concretado –en cualquier caso anterior a la introducción de cerámicas a torno en la isla- sufrió un incendio que destruyó la construcción erigida en la parte superior del talaiot. Estas dataciones presentan el previsto problema de que están realizadas sobre muestras de vida larga, puesto que datan un momento de la edad del árbol y no exactamente la fecha en que fue construida la cámara superior del turriforme, por lo que deben considerarse como fechas post-quem. Desgraciadamente los contextos arqueológicos de la cámara superior (UE-106 y UE-107) fueron muy pobres en hallazgos arqueológicos in situ y no pudieron documentarse niveles de frecuentación, por lo que poco más podemos decir hasta que no estén concluidos los estudios de estos contextos. Hacia el cambio de era la cima del talaiot, y concretamente el corredor de acceso a la cima, fue ocupado de forma persistente.

DATACIÓN 8
KIA15698: 3065±35BP

68.2% confidence
1390BC (1.00) 1260BC 95.4% confidence
1420BC (1.00) 1190BC

Muestra de carbón extraído de los restos de la pilastra nº 3 de la cámara superior del talaiot..

DATACIÓN 9
KIA15699: 2965±30BP

68.2% confidence
1260BC (1.00) 1120BC 95.4% confidence
1310BC (1.00) 1050BC

Muestra de carbón extraído de los restos de la pilastra nº 2 de la cámara superior del talaiot.

CASA TALAYÓTICA
Conocida como "Casa Sera-Belabre" por haber sido esta investigadora menorquina la que realizó un pequeño sondeo en la zona central de la misma. Es un tipo de vivienda característica de la cultura talayótica menorquina, de planta aproximadamente circular, de muros ciclópeos con grandes losas ortostáticas incorporadas al paramento externo. Dispone de patio central con grandes columnas monolíticas y dependencias perimetrales. Se sabía que mucho de estos edificios domésticos habían llegado en uso, aunque con reformas, hasta bien entrada la dominación romana, sin embargo, el momento de su aparición en la cultura talaiótica menorquina resta por conocer. Por esta razón, entre los objetivos básicos de esta campaña, estaba el intentar obtener información sobre este aspecto.

DATACIÓN 10
KIA15218: 2370 ±30BP (Colágeno de fauna doméstica)
d13C = -22.06‰, d15N = +4.55‰ en C/N = 2.99

68.2% confidence
520BC (1.00) 390BC 95.4% confidence
760BC (0.07) 700BC
540BC (0.93) 380BC

Comentario: Casa Serra-Belabre. UE 36. X=206 (sec). Z=0,9. Trinchera muro norte. Sedimento que rellena la trinchera excavada en la roca que sirvió para asentar los cimientos de la casa monumental de planta circular. El sedimento aparece muy compacto, mezclado con fragmentos machacados de cerámica talaiótica y pretalaiótica.

DISCUSIÓN
Para la obtención de esta serie de dataciones radiocarbónicas se han desechado, pese a la abundancia, muestras de vida larga, como el carbón, por la incertidumbre que se genera a la hora de asegurar la conexión del resultado cronométrico con el hecho arqueológico que se quiere fechar. Por esta razón todas las determinaciones radiométricas han sido obtenidas a partir de colágeno de fauna doméstica.

Todas las dataciones presentan desviaciones típicas que oscilan entre ±25 y ±40 lo que ha proporcionado intervalos de calibración dendocronológica relativamente ajustados ganando gracias a ello precisión y exactitud.

Por lo que respecta a los aspectos estrictamente arqueológicos dos cuestiones deben resaltarse: En primer lugar, la única referencia de cronología absoluta, hasta el momento disponible, relacionada con la posible fecha fundacional de la casas talayótica, nos sitúa el evento hacia el 450/460 cal. m. BC., sobre esta cuestión sin duda hay que insistir pues una sola determinación cronológica es insuficiente para hacer inferencias de carácter más general.

En segundo término, es importante resaltar que las dataciones referidas a los contextos demarcadores de una referencia ante quem para la construcción del talaiot de Biniparraxt Petit vienen a confirmar una tendencia que se consolida poco a poco. Las investigaciones de los últimos años (Castro et al. 1996; Strydonck et al. 1998; Lull et al. 1999, Guerrero 1999; Mestres y Nicolás 1999) vienen señalando que los contextos bien documentados (como Son Fornés y Son Ferrandell en Mallorca o Sant Agustí en Menorca), ligados a los momentos constructivos y de uso prístino de los talaiots, se sitúan en un intervalo de tiempo que oscila entre 950 y 850 cal. BC, por lo que, en la medida que se profundiza en el tema, el desarrollo pleno de la cultura talaiótica se inserta mejor en la Edad del Hierro que en el Bronce.

BIBLIOGRAFÍA

CASTRO, P.V., LULL, V. y MICÓ, R.
1996. Cronología de la Prehistoria reciente de la Península Ibérica y Baleares (c. 2800-900 cal ANE), British Archaeological Reports, Intenational, International Series 652, Oxford.
GUERRERO, V.M.
1999. Arquitectura y poder en la prehistoria de Mallorca, "El Tall editorial" Palma.
LULL, V.; MICÓ, R.; RIHUETE, C. y RISCH, R.
1999. La Cova des Càrritx y la Cova des Mussol. Ideología y sociedad en la prehistoria de Menorca, Barcelona.
MESTRES, J.S. y NICOLÁS, J.C. de,
1999. Contribución de la datación por radiocarbono al establecimiento de la cronología absoluta de la prehistoria de Menorca, Caesaraugusta, 73, Zaragoza, p.327-341.
STRYDONCK, M.J., WALDREN, W. y HENDRIX, V.,
1998. The 14C chronology of the Son Mas Sanctuary site (Valldemosa, Mallorca, Spain), en MOOK, W.G./ PLICHT, J. Var der, (ed.), Proceedings of the 16th International 14C Conference, Radiocarbon, vol.40 nº2 p.735-748.

47

ANÁLISIS COMPARATIVO DEL PATRÓN DE ASENTAMIENTO DE LA CULTURA BALEÁRICA.

JAVIER ARAMBURU-ZABALA

Instituto Balear de Arqueologia
Palma de Mallorca, Baleares, Spain
EMAIL: jaramburu@talayots.com

INTRODUCCIÓN

El patrón de asentamiento de la cultura baleárica (c.1.100-123 a.C.) en las dos islas conocidas en la antigüedad como "baleares" (Mallorca y Menorca) ha sido objeto de sendos recientes estudios de detalle (Aramburu-Zabala, 1998; Aramburu-Zabala, e.p.) lo que permite abordar un análisis comparativo entre ambos, que, en este caso, vamos a centrar en su fase inicial (1000 - 600 a.C.), fácilmente legible en el territorio por la abundancia de sus construcciones.

El interés de este análisis radica en el hecho de que la hipótesis que se plantea para su explicación cuenta con dos campos de comprobación en unas condiciones excepcionales: en primer lugar, ambas islas se encuentran relativamente aisladas de su entorno geográfico (la costa continental más cercana se encuentra a 175 km (FIGURA 1), lo que permite el desarrollo de procesos internos con el menor grado posible de influencias externas, que, por otra parte, serían recibidas teóricamente por ambas islas de manera casi simultánea. En segundo lugar, ambas islas se encuentran separadas por un peligroso canal de 31 km de ancho, de forma que, aunque los contactos existían y eran lógicamente mayores que con el exterior, garantiza para cada isla también un cierto aislamiento. En tercer lugar, partimos del supuesto, que comentaremos, de que el punto de arranque en ambas islas era similar, la cultura pretalayótica, lo que facilita realizar la comparación posterior entre ambas. Finalmente, disponemos de otras islas (Ibiza-Formentera y Cabrera) donde no se reproducen los mismos fenómenos que en las baleares.

De esta manera, disponemos de dos sociedades sometidas a procesos similares en un espacio cuasi cerrado, lo que permite aislar las variables que intervienen en mayor medida que en los vastos espacios continentales (Evans, 1973)

El modelo que planteamos en los estudios parciales que hemos mencionado (Aramburu-Zabala, 1998; Aramburu-Zabala, e.p.) es que ambas islas estaban sometidas a un intenso crecimiento de la población. En ambos casos esta presión demográfica no está correlacionada, inicialmente, con un cambio de las bases subsistenciales sino con una concentración de la población en asentamientos más grandes, sin jerarquía entre ellos. En una segunda fase la continuación de la presión se correlaciona con la necesidad de marcar los diferentes territorios y el desarrollo de una sociedad en proceso de jerarquización.

Lo que tratamos de probar ahora es que esa correlación tiene como causa la presión demográfica y para ello compararemos lo que sucede en ambas islas, partiendo de la premisa de que, con un mismo origen y en condiciones similares, si el territorio es más pequeño, se producirá una intensificación de los procesos sociales. Esta interpretación rechaza la posibilidad de una influencia exterior decisiva, del tipo de la llegada de nuevos grupos humanos por invasión.

MÉTODO

Para la realización del presente análisis nos hemos basado en los datos obtenidos en la prospección llevada a cabo en las dos islas. En el caso de Mallorca la prospección no sólo ha abarcado los yacimientos anteriormente conocidos, sino que se ha comprobado la existencia de yacimientos en áreas que, de acuerdo al patrón general, debían contenerlos. En todo caso, algunas zonas se han prospectado con carácter más intensivo como el pie de monte de la Serra de Tramuntana y los valles interiores de ésta. En el caso de Menorca, la prospección se ha centrado en los yacimientos conocidos ante la imposibilidad de explorar la isla en detalle, aunque se ha de tener en cuenta que la superior visibilidad de los yacimientos menorquines y la gran tradición en prospecciones entre los investigadores insulares permiten corregir la diferente intensidad de la prospección entre las dos islas.

cuadrangulares, talayots subcirculares y taulas. Además hay otros tipos menos frecuentes.

El resto de los datos, no espaciales, provienen de las excavaciones e investigaciones de diferentes autores que, en su caso, son citados.

El análisis consiste en identificar las variables relevantes en función de la hipótesis y de las características del registro arqueológico y, en la medida de lo posible, cuantificarlas. A continuación se compara lo que sucede con dichas variables en ambas islas.

Fig. 1.
ISLAS BALEARES

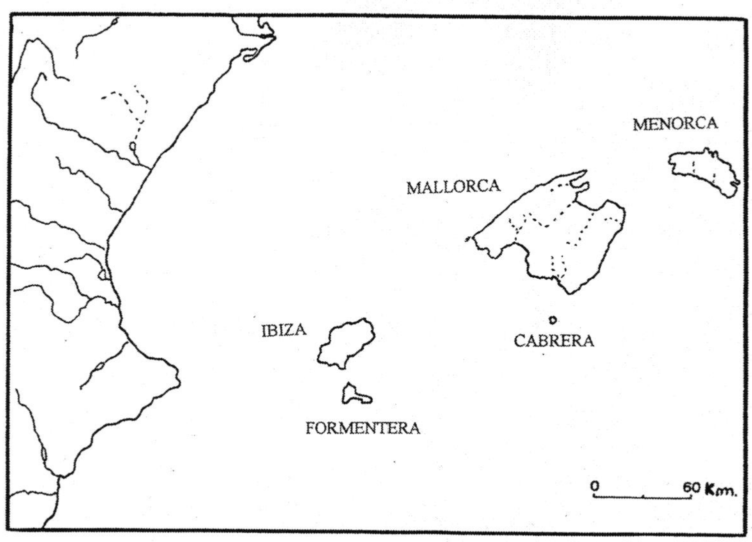

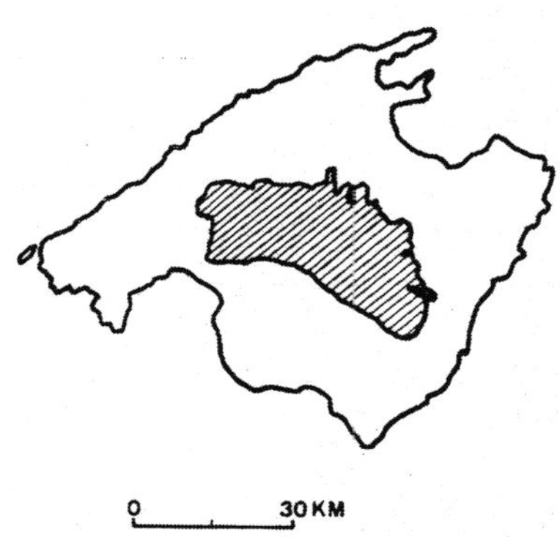

Fig.2. Tamaño comparativo de Mallorca y Menorca.

También se ha prospectado el archipiélago de Cabrera (Aramburu-Zabala, 2001)

La tipología de yacimientos aplicada es la contenida en las obras ya mencionadas: en ambas islas hay, básicamente, poblados, centros ceremoniales y turriformes aislados Las dos islas tienen talayots escalonados, túmulos escalonados y plataformas escalonadas. En Mallorca hay, además, talayots circulares, talayots cuadrados y santuarios, mientras que en Menorca hay talayots-torre, talayots absidales, talayots sub-

ANÁLISIS

El punto de partida: la cultura pretalayótica o naviforme.

En ambas islas el patrón de asentamiento del sistema sociocultural anterior, conocido como "pretalayótico" o "naviforme", presenta unas características similares: poblados de navetiformes no amurallados extendidos por todo el territorio con el único límite de las alturas superiores a 900 m en Mallorca y con una dedicación al parecer básicamente ganadera y sin jerarquización social no derivada de la edad y el sexo.

A falta de un análisis en profundidad, parece que la densidad de poblados de naviformes podía ser similar en ambas islas: en Menorca se han contado 79 yacimientos con naviformes, de los cuales 64 son seguros (López Pons, 1979: 106) lo que ofrece una media de 8,8 km2 por grupo de naviformes que, lógicamente, hay que reducir por los núcleos destruidos y no localizados y que también es un número similar al de poblados baleáricos, algunos de los cuales vado, parece que la distancia entre núcleos de naviformes en llano también es de unos 2,5 km, que es una distancia similar a la que mantienen los poblados baleáricos, lo que supondría un número también similar o ligeramente superior de asentamientos, que podríamos suponer, a efectos de cálculo, en 325.

En cuanto al tamaño de los poblados de naviformes, algunos de Mallorca eran más grandes que los de Menorca, pues los hay que llegan a tener hasta 50 naviformes, pero la may-

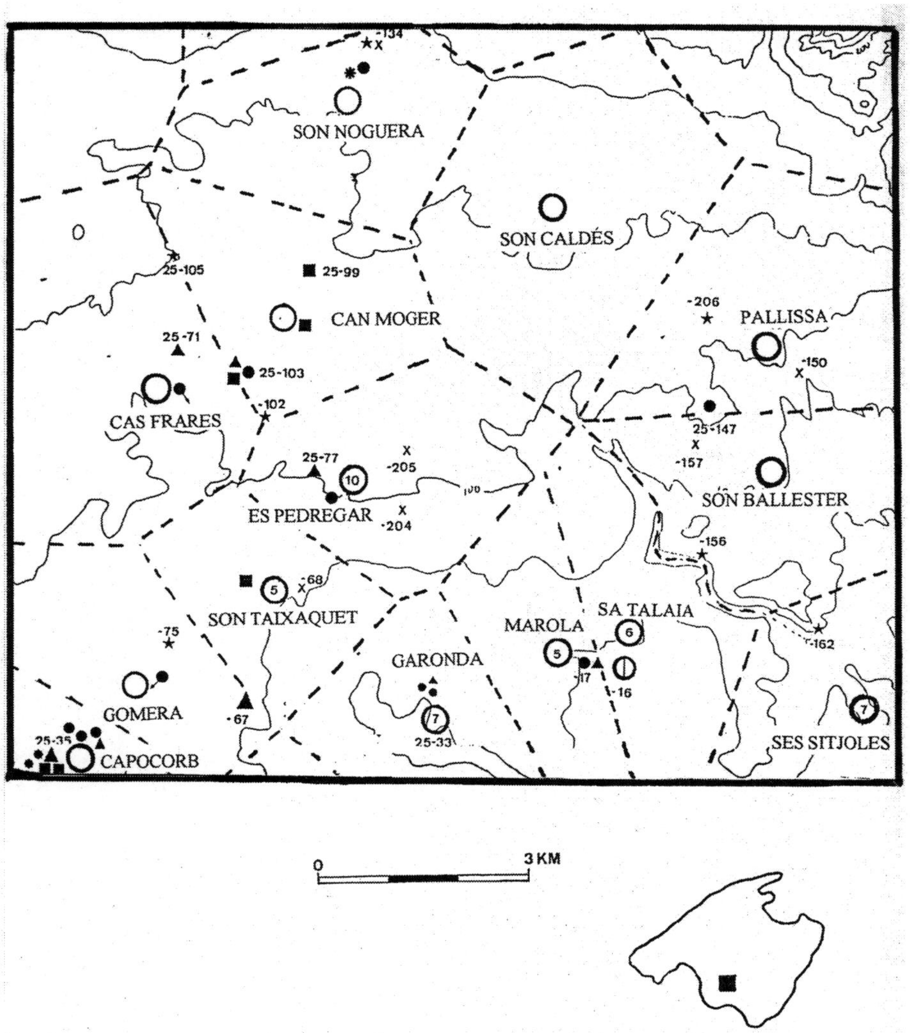

Fig.3. Ejemplo de distribución de asentamientos talayóticos en Mallorca.

pueden haberse sobrepuesto a otros pretalayóticos. De hecho la distancia media entre asentamientos pretalayóticos más próximos es de unos 2 km, mientras que entre poblados baleáricos es de 1,84 km. Dado que con este tipo de distribución en la isla hay unos 75 poblados baleáricos, podemos suponer que el número de poblados pretalayóticos sería ligeramente superior, con un número que a efectos de cálculo fijamos en 115.

En Mallorca no tiene interés el número absoluto, dada la destrucción generalizada de los yacimientos del centro de la isla , pero observando las áreas en que mejor se han conser-

oría se agrupan en pequeños poblados de menos de una decena de naviformes, aunque en algunos casos también los encontremos aislados.

El otro testimonio arquitectónico de esta cultura son los hipogeos funerarios: en Mallorca están muy extendidas las cuevas artificiales de planta alargada, mientras que en Menorca, suelen ser de planta simple, muy difíciles de diferenciar de las posteriores baleáricas. Los ajuares, tanto cerámicos (lisos, con formas globulares) como metálicos (cuchillos y punzones) son similares.

VARIABLES NEUTRAS

Denominamos variables neutras a las que son similares en las dos islas y se pueden desechar como determinantes de las diferencias entre ambas. Dada su importancia en otros contextos, las analizaremos brevemente.

EL AGUA

En Mallorca, las precipitaciones se hacen más escasas según se pasa del noroeste al sudeste, al igual que los manantiales, que son más abundantes en la Sierra de Tramuntana que en el resto de la isla, lo que sin duda influye en el patrón de asentamiento. Sin embargo, problemas de acceso a agua sólo se encuentran en la marina sudeste de la isla. Se ha documentado que los poblados baleáricos no se disponen cerca de manantiales importantes (Aramburu-Zabala, 1998: 177)

En Menorca las fuentes son escasas y se concentran en la periferia sur de la isla, pero tampoco parece que los poblados se dispusieran en sus inmediaciones. Se puede pensar, en las dos islas, que había agua suficiente para abrevar ganados, que necesitan mucha menos agua que los cultivos.

RECURSOS MINERALES.

Las dos islas tienen escasos yacimientos minerales: algo de cobre y plomo y, al tratarse de zonas costeras, sal. En ambos casos, no se ha podido documentar un interés por situarse cerca de este tipo de recursos, por lo que esta variable puede anularse.

IMPORTANCIA EN LAS RUTAS COMERCIALES

Las dos islas tuvieron una importancia similar en cuanto a lugar de paso del comercio marítimo mediterráneo este – oeste o norte - sur. Ambas disponen de buenos puertos (más abundantes en Mallorca) Ninguna de las dos islas disponía de bienes que las hiciera particularmente interesantes como objetivo comercial.

VARIABLE INDEPENDIENTE: LA DENSIDAD RELATIVA DE POBLACIÓN

Mientras que la densidad absoluta de población consta de dos factores: el número de personas y la superficie, la densidad relativa de población tiene en cuenta, en este caso, la diferente productividad de los suelos, que debe, por tanto, uniformarse.

LA SUPERFICIE RELATIVA

Mallorca tiene una superficie de 3651 km2 y dispone de cuatro unidades fisiográficas diferentes: la Serra de Tramuntana, la Serra de Llevant, el Pla y las marinas. En la primera, formada por rocas calizas, se encuentran las mayores alturas de la isla (Puig Major, 1443 m) y pequeños valles con suelos fértiles y bien regados. La segunda está formada por montes de poca altura (500 m), con suelos no tan fértiles y sequedad más acusada. El Pla ocupa todo el centro de la isla y está formado por terrenos llanos, muy fértiles o fértiles, de clases agrológicas III (laboreo permanente) y IV (laboreo ocasional, pastoreo) . Las marinas, que se extienden por la costa sur y este son terrenos llanos, pobres y secos. Si aplicamos una proporción a las clases agrológicas donde las clases II-III tengan un valor 1, las clases IV y VI un valor de 0,5 y las clases VII y VIII (pastoreo de cabras y estéril) un valor de 0,25, obtenemos una superficie ponderada de Mallorca de 2.546 km2 lo que supone una reducción al 0,7.

TABLA 1

	MALLORCA		MENORCA	
A os A.C.	Poblaci n	Densidad	Poblaci n	Densidad
600	50.830	20,0	12903	39,1
900	29.900	11,7	7.590	23,0
1.300	13.000	5,1	3.300	10,0
1.700	5652	2,2	1.434	4,3
2.500	2093	0,8	531	1,6
3.400	805	0,3	204	0,6

Menorca, situada a 31 km de Mallorca, es una isla de 701 km2 (5 veces más pequeña que Mallorca, FIGURA.2) que presenta dos zonas geológicas, morfológicas y litológicas claramente diferenciadas: Es Mitjorn, que ocupa el sur y sudoeste, es en general una llanura de roca calcarenita, cortada en su parte sur por torrentes paralelos que forman profundos barrancos. La zona de Tramuntana, al norte, formada por materiales paleozoicos y mesozoicos (pizarras, arcillas y calizas) con un relieve de valles separados por colinas o montes de escasa altura (El Toro, con 358 m es la altura dominante en la isla).

6-8 naviformes (Son Mut tiene 16, Na Mera 15, Son Rossinyol 11, Can Quiam 9, Es Clossos y Es Velar de Sencelles 8, Son Bugadelles, al menos 7, al igual que Cals Reis y Sa Rota Paparrí 7, Na Mora y Son Ferrer 6, Cas Toro 5, Mossa y Es Burotell 3). Otros sólo tienen 1 ó 2, pero únicamente de los situados en zonas de montaña marginales se puede asegurar que no tenían otros naviformes cerca, ahora destruidos. No tenemos en cuenta tampoco los grandes poblados, de hasta 50 naviformes porque son escasos y concentrados en la zona de Pollença.

-Naviformes en Menorca: Número de poblados: 115.

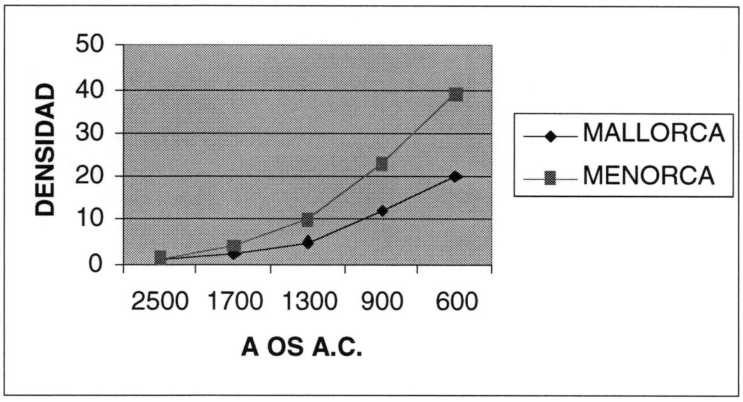

Esta división ha condicionado el poblamiento de Menorca, que se ha asentado en todas las épocas en la parte sur, debido a la mayor fertilidad de las tierras, la facilidad para obtener agua y, a su vez, a la ausencia de tierras encharcadas, abundantes en el norte.

Los suelos de Menorca son de menor calidad que los de Mallorca (la clase VI es la más abundante). Si aplicamos el mismo cálculo que a Mallorca, los kilómetros cuadrados ponderados son ahora 330, lo que supone una reducción al 0,47. De esta manera, es fácil retener un primer e importante dato: Menorca es, agrológicamente, más pobre que Mallorca, lo que hace que su territorio se haga, para el uso humano, más pequeño aún.

EL NÚMERO DE HABITANTES

Calcular el número de habitantes en ambas islas es un campo en el que los datos son escasos y escurridizos. Sin embargo, podemos hacer aproximaciones cruzando varios tipos de datos.

El método que nos parece más completo es contar el número de cabañas en tres momentos diferentes: cuando se produce la mayor expansión de los naviformes, hacia 1300 A.C., cuando se construyen los talayots hacia 900 A.C. y cuando los poblados talayóticos alcanzan la mayor expansión intramuros, hacia 600 A.C.

Comenzamos contando el número de poblados y el número de cabañas por poblado:

-Naviformes en Mallorca: Número estimado de poblados: 325. Parece que los poblados de naviformes que se han conservado en buen estado están formados habitualmente por

Cala'n Morell tiene 9 naviformes, Son Mercer, 5, y Clariana al menos 3, por lo que parece que los poblados eran ligeramente más pequeños que en Mallorca.

Parece que no todos los naviformes eran lugares de habitación, por lo que hay que reducir su número a una media aproximada de 4 lugares de habitación en Mallorca y 3 en Menorca.

-Poblados baleáricos en Mallorca: Tras las últimas prospecciones, hemos calculado 275 poblados. El número es igual tanto en el periodo talayótico (1000-600 A.C.) como en los inicios del postalayótico (600-123 A.C.). Sin embargo, lo que varía es el número de cabañas: en el periodo talayótico, como vemos en S' Illot, cada poblado tendría unas 12 cabañas, con 10 usadas simultáneamente, cifra que aumentaría a 30 en el postalayótico, de las cuales 22 podrían estar en uso simultáneamente (Aramburu-Zabala, 1998: 111)

-Poblados baleáricos en Menorca: Son en torno a 75 (Aramburu-Zabala, e.p.) Desconocemos el número de cabañas que tenían en el periodo talayótico. En cambio sí se pueden contar los "círculos" o viviendas que fechamos tentativamente hacia 600 A.C. Son una media de 13 por poblado, que, dada su monumentalidad estarían en uso sin excepción. Es de suponer que, además de los círculos, habría otras viviendas de peor calidad. No podemos, sin embargo, cuantificarlas.

Tanto los naviformes como las cabañas talayóticas tienen el mismo espacio habitable (alrededor de 35 m2) Este espacio se reduce en casi 10 m2 a partir del postalayótico.

Cada cabaña debía estar habitada por una familia. Tanto en el periodo naviforme como en el talayótico adoptamos para el cálculo 10 personas por familia (3,5 m2 por persona), mientras que en el postalayótico lo reducimos a 8, de acuerdo también con la tendencia constatada hacia la familia nuclear (Hernández, 1998: 210). Los círculos menorquines disponen de 75 m2 de espacio techado, sin embargo, alguna de las estancias parece claramente dedicada a almacén o taller, lo que puede dejar el espacio reducido a 50 m2, con lo que, manteniendo la misma proporción, permitiría vivir a 14 personas.

De esta manera, hacia 1300 A.C. serían 40 los habitantes en cada poblado naviforme mallorquín y 30 en los menorquines. El número total en ambas islas sería (redondeando a la centena) de 13.000 y 3.300. Hacia 900 A.C. los habitantes de los poblados talayóticos mallorquines serían 100 y el total en la isla 27.500, mientras que en Menorca carecemos de datos directos. Hacia 600 A.C. el número de habitantes en cada poblado mallorquín sería de 154 y el número total en la isla de 48.400. En Menorca, el número de habitantes en los "círculos" de cada poblado sería de 182 y el total para la isla, al menos, de 13.600.

Observemos ahora los datos que nos han trasmitido los autores clásicos:

Según Diodoro Sículo, autor del siglo I A.C. que cita a Timeo, del siglo V-IV A.C, Mallorca y Menorca contarían con 30.000 habitantes, muy alejado de los 62.000 que hemos calculado. Desde luego no es improbable una reducción de la población causada por la fuerte presión demográfica anterior, pero no vemos posible que esa cifra de 30.000 habitantes corresponda al momento en que están en uso los poblados y talayots, alrededor del 700 A.C. porque supondría que en cada uno de los alrededor de 350 poblados que había en las islas vivirían 86 personas, cifra que ni como media puede ser aceptable.

Hay que tener en cuenta, además, que si el dato se refiere al siglo IV a.C. es totalmente especulativo, porque parece dificil que en esa época un foráneo pudiera obtener esas cifras, mientras que si se refiere al siglo I a.C. nos encontramos en las décadas posteriores a la conquista romana, con las transformaciones que dicho cambio debió producir.

Por otra parte desde mediados del primer milenio A.C. habitantes de ambas islas actúan como mercenarios, en una señal de que se producía un exceso de población. Por ejemplo, el 206 a.C. Magón recluta en Menorca 2.000 mercenarios, según las fuentes clásicas. Si esta cifra fuera cierta, si se hubiera reclutado al 100% de la población masculina entre, aproximadamente, 18 y 30 años, estaríamos hablando de una población mínima de 10.000 personas. Como no parece lógico que la recluta afecte a la totalidad de varones, debemos pensar que el número de habitantes se acercaría, como mínimo, al que hemos calculado más arriba.

A continuación vamos a comparar estos datos con los índices de crecimiento usuales en sociedades prehistóricas (Hassan,1978: 68-69). El punto de partida es la población calculada en 1300 A.C. Aplicamos un índice de crecimiento del 0,2 para el Bronce y Hierro (desde 1.700 A.C.), de 0,15 para el calcolítico (desde 2.500 A.C.) y de 0,1 para el Neolítico (3.400 A.C.)

Hay que precisar que la densidad es ponderada, calculada sobre las superficies también ponderadas. Las densidades absolutas resultan de dividir las relativas por 1,41 en Mallorca y por 2,12 en Menorca (densidades en 600 A.C. de 14 y 18 hab/km2 respectivamente)

Estas cifras se adaptan bien a los datos anteriormente expuestos. Hemos incluido también las cifras anteriores al pretalayótico, que se adaptan también a lo que se conoce:
En ambas islas no está clara la fecha de llegada del hombre: en Mallorca las primeras evidencias de ocupación humana, (Son Matge, estrato 28) encuadrables culturalmente en el Neolítico, no se fechan antes de 3400 A.C. (Waldren, 1998: 75), aunque son más claras a partir de 2700 A.C. En Menorca los restos de los primeros pobladores localizados hasta el momento, en Biniai Nou, no son anteriores a fines del tercer milenio A.C. (Rubinos, 2000: 437-440), aunque el tema no puede darse por cerrado. De todas maneras, dado el número bajo absoluto de habitantes antes del 2.000 A.C. en Menorca, se puede considerar cualquier fecha de arribada anterior.

Los restos conocidos calcolíticos son escasos, mientras que los del pretalayótico o naviforme son muy abundantes, lo que coincide con la explosión demográfica que situamos entre 2.000 y 1.300 A.C. También los cambios en la vegetación indica la acción del hombre desde inicios del cuarto milenio A.C. y la presencia de cereales y ganadería desde el tercer milenio A.C. (Yll y otros, 1999:49-50)

En este modelo rechazamos la posibilidad de un crecimiento discontinuo, a escala amplia, antes de 1.300 A.C. En ninguna de las dos islas (y probablemente tampoco entonces) se alcanzó un nivel de presión sobre el territorio que pudiera hacer estancarse o retroceder la población, aunque sí imponer cambios en el sistema de subsistencia. Los cálculos efectuados (Aramburu-Zabala 1998: 104-108) parecen indicar que la ganadería sólo podría aportar la mitad de las calorías necesarias, pero la agricultura tenía aún terreno suficiente para expandirse. De hecho, en época romana encontramos que la mayoría de los poblados continuaban habitados y, sin embargo, se añaden quizás entre 20 y 30 asentamientos nuevos entre ellos, además de dos ciudades.

Se observará en la tabla 1 que el análisis lo detenemos a partir de mediados del primer milenio A.C.. El motivo es que pensamos que la jerarquización de la sociedad y los contactos exteriores pudieron introducir nuevos factores que distorsionan el crecimiento que venimos indicando.

Como elemento de comparación, podemos señalar que la isla de Gran Canaria tenía en el siglo XIV, con una sociedad a caballo entre la jefatura y el estado, una densidad de 27-30 hab/km2 (Jimenez, 1999: 194-195) mientras que la isla de Pascua, con un organización de estado incipiente, alcanzaría los 40 hab/km2 en los momentos anteriores a la conquista europea (Metraux, 1995: 65)

VARIABLES DEPENDIENTES

CAMBIOS EN LA VEGETACIÓN

Consideramos a la vegetación variable dependiente porque un aumento de la población debe provocar una transformación del medio. Sin embargo, hay que tener en cuenta que los cambios climáticos también parecen ir en dirección de una sustitución de los elementos más propios de climas húmedos y fríos.

Los análisis polínicos realizados en las dos islas muestran resultados similares: Desde el 2000 A.C. y de forma más acentuada en Menorca, se advierte una apertura notable de la vegetación, con una disminución progresiva de la frecuencia absoluta del polen arbóreo y un predominio de los pastizales (88% en Turó de ses Beies, situado en la costa de Mallorca a fines del primer milenio; aumento de Plantago, típico de pastos, a partir de 2.000 a.C. en Menorca) y presencia de bosques de encina, pino y, sobre todo, acebuche. Por otra parte, los cereales no vuelven a aparecer en el registro polínico antes de mediados del primer milenio.

EL FACTOR TIEMPO

Partimos de la hipótesis de que si se aumenta la presión demográfica en una isla, el proceso de evolución cultural debe acelerarse, aunque las bajas densidades existentes en el momento de transición no harían posible una clara diferenciación inicial.

En Menorca, las últimas dataciones procedentes de naviformes, incluido el intervalo a 1 sigma, no bajan de 1150 cal A.C, con fechas más probables entre 1350 y 1200 según los datos de Clariana, Cala Blanca (Plantalamor/ Strydonck, 1997) y Torre del Ram (Mestres, 1996).

En Mallorca, hay fechas de abandono de naviformes, con el mismo intervalo, que llegan a 1029 (Son Oms) o hasta 843 cal A.C. (Clossos de Can Gaiá, Calvo/Salvá, 1999: 68), con fechas medias de 1137 y 907 cal. A.C., respectivamente.

"La Fase de Transición", continuación de la anterior, se documenta en Menorca en las cabañas de Torralba, con una ocupación datada hacia 1300-1200 A.C. (Fernández-Miranda, 1991: 41) y Trebeluger, con fechas de abandono comprendidas entre 1380 y 1000 (Plantalamor/ Strydonck, 1997)

En Mallorca para esta fase de transición tenemos el Figueral de son Real, abandonado entre 1332-1054 A.C. y, quizás, S'Illot, con una estructura absidal también de contexto dudoso, fechada en el abanico entre 1445 y 1042 cal A.C.

En esta "Fase de Transición" los asentamientos están formados por cabañas con cierto parecido con las naviformes y con cerámica transicional. En cuanto al registro funerario de esta fase, en Menorca es posible que se utilicen las navetas, edificios monumentales inspirados en los naviformes de habitación, pero que, en algunos casos como Biniac, Llumena d'En Montanyés, Torralbet o Torre Llisá Vell, también adoptan formas circulares más típicas de los talayots, aunque su uso sólo está demostrado para la fase siguiente. En Mallorca cabe pensar que continúan utilizando los hipogeos y cuevas naturales de la fase anterior.

Hacia el cambio de milenio nos encontramos con que en ambas islas comienzan a levantarse turriformes: es la fase que denominamos "talayótico". En Menorca las primeras fechas corresponen al talayot de san Agustí Vell (870 cal A.C, Lull y otros, 1999:62) y a las navetas de enterramiento, utilizadas hasta 900-800 cal A.C. (Plantalamor/ Strydonck, 1997) . En Son Ferrandell-Oleza (Mallorca) una revisión de los datos llevada a cabo por sus excavadores recientemente (Strydonck/ Waldren/ Hendrix, 1998) les permite concluir que el primer uso de los talayots de este centro ceremonial debe irse ca. 800 cal. A.C. y no 1100 cal. A.C. como inicialmente se había comunicado, lo cual coincide con la fecha del talayot 2 de Son Fornés: 935 ± 45 cal A.C. y con la torre de observación de Son Ferrandell-Oleza: 834 ± 29 cal A.C.

En resumen, parece que los procesos de transición en ambas islas se inician simultáneamente, pero tardan más en completarse en Mallorca, especialmente por una mayor perduración en el uso de los naviformes, aunque la poca seguridad de los datos no permite afirmarlo de manera clara.

LA DISTRIBUCIÓN DE LOS ASENTAMIENTOS

La mayor presión demográfica debe provocar una distribución diferenciada: o bien poblados más cercanos unos a otros o bien de diferentes tamaños.

En Mallorca tenemos unos 275 poblados distribuidos por toda la isla (identificados 261) salvo las partes más secas de las marinas. La distribución es muy regular, con distancias medias entre poblados de 2,47 km, encontrándose el 40% de los mismos entre 2 y 3 km y el 80% entre 1 y 3 km. De esta manera el territorio medio por poblado sería de 12 km2, con la gran mayoría comprendidos entre 5,1 y 15 km2 (FIGURA.3) Naturalmente, los espacios entre poblados son mayores en la Serra de Tramuntana debido a la existencia de grandes macizos rocosos. Por ello, si sólo comparamos a los situados en llano, la media baja a 9,5 km2

En Menorca, el número de poblados localizados es de 71, con una distancia media de 1,84 km, encontrándose el 60% entre 1 y 2 km (FIGURA.4). Sin embargo en Menorca hay que señalar la existencia de una importante diferencia regional entre Tramuntana y Mitjorn, por lo que la distribución de los territorios no es uniforme: 6,6 km2 de territorio para los poblados de Mitjorn y el doble para los poblados no costeros de Tramuntana.

EL TAMAÑO DE LOS ASENTAMIENTOS

En Mallorca, la media de superficie de los poblados (excepto los situados en la Serra de Tramuntana, que son más pequeños) es de 6.123 m2, encontrándose el 60% entre 4.000-8.000 m2, alcanzando el mayor de todos, Vista Alegre Vell, 12.000 m2, mientras que en Menorca la media de los poblados de Es Mitjorn es de 13.000 m2, algo más del doble, alcanzando el mayor de todos (Son Catlar) los 36.000 m2, 2,5 veces mayor que el primero de Mallorca (FIGURA.5). De esta manera, a cada metro cuadrado de poblado le corresponde en Mallorca, de media, 1332 m2 de territorio, mientras que en Menorca le corresponden tan solo 508 m2.

LA JERARQUIZACIÓN DE LOS ASENTAMIENTOS

Consideramos que se produce jerarquización entre los asentamientos cuando un gran poblado está rodeado de otros más pequeños y su territorio de explotación es claramente insuficiente para su autoabastecimiento. No podemos considerar testimonio de jerarquización el simple hecho de que los poblados, independientemente de su distancia, tengan superficies diferentes porque la literatura etnográfica presenta testimonios de poblados de diferentes tamaños que no están jerarquizados, dentro de estructuras tribales (Tait, 1979:185-198). Simplemente son agregados más grandes, donde ni tan siquiera existe una unidad política de poblado, permaneciendo los vínculos a través de los linajes, que pueden distribuirse por varios poblados.

En Mallorca es difícil hallar asentamientos próximos con diferencias importantes de superficie (sólo hay 15 casos de vecinos que tengan una diferencia de superficie igual o superior al doble) y de ellos solo Es Pedregar tiene su corona de poblados muy cerca, tanto como para pensar que el gran poblado era incapaz de autoabastecerse, mientras que en Menorca tampoco son frecuentes, pero hay un caso muy claro: son Catlar y sus vecinos Egipte, Son Marc y Son Olivaret. Además, tenemos los casos de Binicalaf y su vecino Torralbenc y, finalmente, Torre d'en Lozano y su vecino Son Bernardí.

Por tanto, los indicios de jerarquización apenas se manifiestan en la distribución territorial, porque el origen de ésta corresponde a un estadio más próximo a la organización tipo tribu que tipo jefatura. Sin embargo es de nuevo Menorca la que presenta los casos más abundantes de organización jerárquica.

(FIGURA.6)

Conviene en este momento hacer notar que la tipología de los turriformes es diferente en ambas islas. Nos encontramos con tipos similares como los túmulos escalonados y las plataformas escalonadas y, con diferencias, los talayots escalonados, aunque estos son mucho más frecuentes en Mallorca. El resto de los tipos son muy diferentes: el talayot circular, que es muy abundante en Mallorca, es casi completamente desconocido en Menorca, al igual que el talayot cuadrado. Sin embargo, en Menorca es frecuente el talayot de tipo A o talayot-torre, con cámara elevada y el talayot subcuadrangular con esquinas redondeadas, que denomimanos de tipo C. Menos frecuente en Menorca y totalmente ausente en Mallorca es el talayot de tipo B o absidal.

Además, mientras que en los talayots circulares mallorquines la presencia de una amplia cámara inferior parece poder indicar una actividad de reunión o consejo, los abundantes talayots-torre menorquines, con cámara elevada parecen haber tenido una función más ceremonial con separación más clara oficiante-pueblo.

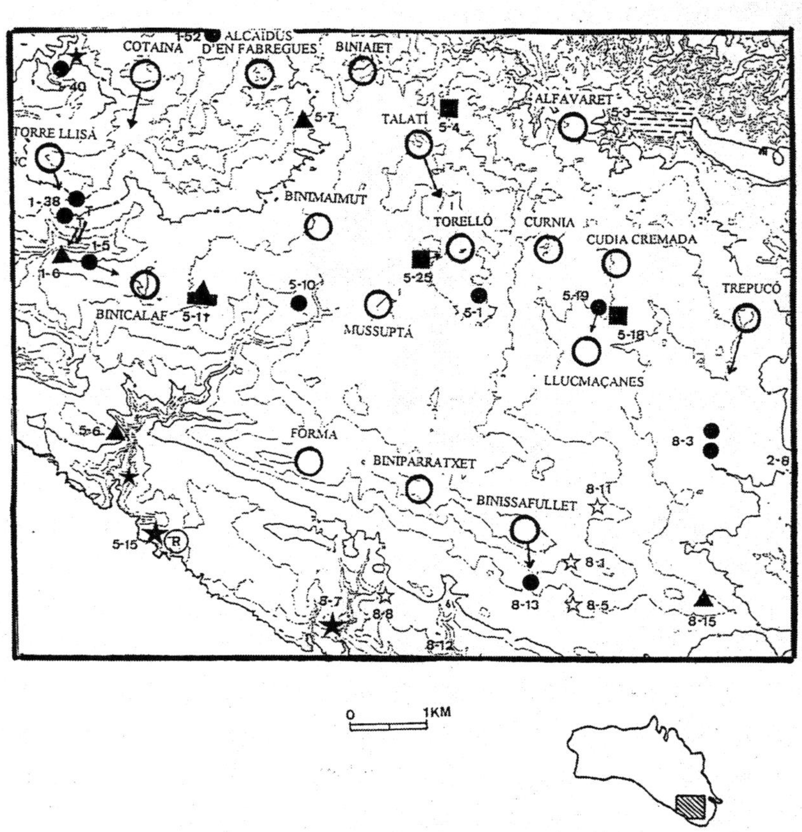

Fig.4. Ejemplo de distribución de asentamientos talayóticos en Menorca.

EL TAMAÑO DE LOS MONUMENTOS.

En Mallorca los turriformes suelen tener entre 10 y 12 metros, alcanzando el mayor de todos los 18 m, mientras que la altura conservada no suele exceder de los tres metros, aunque algún túmulo escalonado alcanza la veintena de metros de diámetro y los 10 m de altura. En Menorca, por el contrario, la mayoría tiene de 12 a 18 m, pero algunos tienen entre 24 y 30 metros, alcanzando los 7 m de altura

También encontramos diferencias de tamaño en los santuarios o taulas, tanto en el recinto, mucho más monumental en Menorca, como en la propia presencia de la imponente taula menorquina, en nada comparable a las columnas-betilos de los santuarios mallorquines.

Pero no sólo hay diferencias de tamaño en los monumentos talayóticos: también Menorca tiene los naviformes más monumentales, tipo son Mercer de Baix, con grandes columnas interiores de piedra, que carecen de paralelos similares en Mallorca.

DISTRIBUCIÓN DE LOS MONUMENTOS

Podemos pensar que a mayor presión territorial, aparecerán monumentos marcando los límites, pero también que si ese proceso se intensifica, desaparecerá el carácter modular del territorio, donde cada poblado tiene su territorio y podría darse una agrupación de poblados que separan su territorio de otra agrupación.

El carácter territorial del proceso que sucede en ambas islas parece claro si observamos la distribución de turriformes: en Mallorca muchos señalan el límite del territorio de los poblados (el 34% de los talayots circulares, el 40% de los talayots cuadrados, el 54% de los túmulos escalonados y el 71 % de las plataformas escalonadas fuera de poblados, lo que suma el 36% de los turriformes, se sitúan en la distancia al límite del polígono de Thiessen 75-100%), mientras que en Menorca eso sucede en el 53% de los casos. En Mallorca, el porcentaje es más bajo por la gran cantidad de centros ceremoniales próximos a los poblados. Si se consideran únicamente los turriformes aislados, el porcentaje se acerca al 50%.

La orientación de las puertas de los talayots circulares de Mallorca y de los talayots de tipo A y, sobre todo, C de Menorca es un indicador también de esa territorialidad, al relacionar diversos monumentos entre sí y con los poblados.

En Mallorca, la mayoría de los poblados se encuentran separados por turriformes, especialmente túmulos escalonados y talayots cuadrados, mientras que en Menorca no. La alta densidad de poblados en esta última isla hace que en la mayoría de los casos los talayots no puedan separar territorios, debiendo pensarse, como antes explicamos, en formas de cooperación más estrecha entre los poblados.

EL NÚMERO DE TURRIFORMES

Nos interesa saber si la densidad de turriformes es una variable relacionada con el tamaño de los poblados y su densidad.

En términos absolutos, en Mallorca se cuentan 458 turriformes (uno cada 7,8 km2), mientras que en Menorca son unos 270 (uno cada 2,5 km2)

El análisis realizado para Mallorca .(Aramburu, 1998: 205-207) muestra que los poblados más grandes tienen, en su interior, el mayor número de turriformes (básicamente talayots circulares), con una media de 3.472 m2 de poblado por turriforme y alcanzando hasta los 6 ejemplares por poblado y que esta relación se mantiene de forma aceptablemente proporcional, también si contamos el número de turriformes fuera del poblado pero dentro de su polígono de Thissen.

En Menorca vuelve a haber diferencias zonales: al este, los grandes poblados tienen pocos turriformes en su interior, aunque de gran tamaño, mientras que en la zona oeste tienen hasta seis turriformes. En el exterior hay pocos turriformes, salvo en la zona centro-sur de la isla, donde aparecen en ocasiones en gran número.

LUGARES DE ENTERRAMIENTO

Puede suponerse, en principio, que a un aumento de población corresponde un aumento de lugares de enterramiento o número de inhumados.

No hay datos del número de necrópolis menorquinas atribuibles a después del 800 cal. A.C., aunque sólo entre las grandes necrópolis de Cales Coves, Cala Morell y Biniparratxet suman más de un centenar. De todas formas, resulta fácil comprobar que si a estas necrópolis se le suman las cuevas que aparecen en los poblados, son, proporcionalmente, mucho más numerosas que en Mallorca, donde suele haber una por poblado y donde los conjuntos son extremadamente raros.

LOS ÍTEMS DE RANGO

Podemos suponer que el aumento de la población va unido, en algún momento, a la jerarquización social. Uno de sus testimonios es la aparición de objetos o estructuras asociados a personajes de diferentes rangos.

Las primeras pruebas de jerarquización social no aparecen claramente, en ambas islas, hasta después del 800 A.C., especialmente en la segregación de los lugares de enterramiento. Sin embargo, en la fase que nos ocupa hay algunas evidencias de diferencias de rango:

En Menorca la utilización de las navetas como recintos funerarios finaliza hacia el 800 A.C. A la misma época corresponde el fin de la utilización de la cueva des Càrritx. Estos yacimientos son lugares de inhumación colectiva secundaria, por lo que cabe suponer que no todas las familias podían permitirse el construir una naveta, debiendo recurrir a otros lugares de enterramiento. En Mallorca los testimonios son muy escasos, pero muestran una menor inversión de trabajo para las familias de mayor rango: en son Matge se practica también el rito de inhumación colectiva secundaria, con una zona separada por un muro que puede interpretarse como lugar de enterramiento de mayor rango. También los ajuares testimonian un rango más elevado para algunas personas, a juzgar por la aparición de espadas, lanzas, hachas y adornos como pectorales.

ANÁLISIS Y CONCLUSIONES:

Pocas dudas caben de que durante el segundo milenio A.C. se produjo en las dos islas un fuerte incremento de población. El dato del número y superficie de los asentamientos así lo avala. También parece factible que el crecimiento continuara durante la primera mitad del primer milenio A.C., tal como atestigua el desarrollo de algunos poblados.

Ahora bien, una cosa es afirmar ese crecimiento demográfico y otra demostrar que ese fenómeno es responsable del cambio cultural, especialmente cuando, con carácter general, no es una causa que goce de aceptación entre los investigadores (Hassan, 1978: 83-87) Sin embargo, al tratarse de islas y por tanto sin posibilidad de salida fácil, creemos que es un factor que puede llegar a ser determinante.

La única forma de comprobar si en efecto el crecimiento demográfico es la variable independiente del proceso es estudiar los efectos que se producen si se cambia su intensidad, cosa que sucede en las islas: la densidad de población era mayor en Menorca que en Mallorca.

Si analizamos las variables dependientes vemos que parecen responder a la relación de partida de que en Menorca los fenómenos tendrían que ser de más intensidad: la vegetación parece más abierta en Menorca, indicando una mayor presión sobre el medio, el cambio pretalayótico-talayótico parece más rápido; sólo en Menorca hay naviformes monumentales tipo Son Mercer; los poblados talayóticos son más grandes y están más cercanos unos a otros que en Menorca; también parecen estar más jerarquizados; los turriformes y otros monumentos como las taulas son más grandes y alcanzan mayor densidad que en Mallorca; también las viviendas son más grandes y complicadas y, finalmente, las cuevas de enterramiento son más abundantes en Menorca.

También nos podemos plantear porqué la vecina isla de Ibiza no vivió un fenómeno similar, siendo así que tiene un tamaño y unas condiciones agrológicas similares a Menorca. Incluso, históricamente, la población ha sido igual de numerosa que en Menorca y, por tanto, al ser Ibiza más pequeña, con mayor densidad . Quizás la respuesta haya de tener en cuenta que si bien es verdad que las pitiusas no tuvieron un fenómeno constructivo similar al talayótico de las baleares, tampoco tuvieron unas edificaciones similares a las pretalayóticas. Por tanto, podemos decir, simplificando, que si no hubo una etapa talayótica, es porque no hubo cultura pretalayótica y, por tanto, no se dio el crecimiento demográfico que condujo a aquella. La causa de la ausencia de la cultura pretalayótica escapa a este trabajo, aunque pudiera tener que ver con un poblamiento inicial no tan intenso debido a la inexistencia de mamíferos de interés (aunque disponían de grandes cantidades de aves marinas), o bien al hecho de que, al ser la isla más cercana al continente, podía

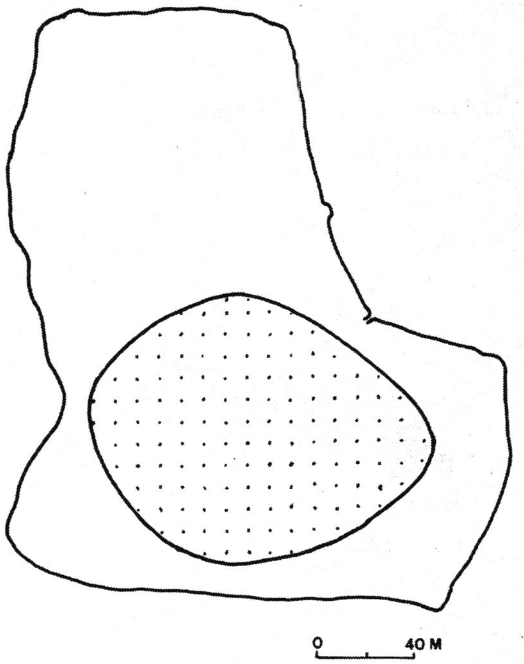

Fig.5.
Comparación del tamaño de los poblados de Son Catlar (Menorca) y Ses Paīsses (Mallorca) (punteado)

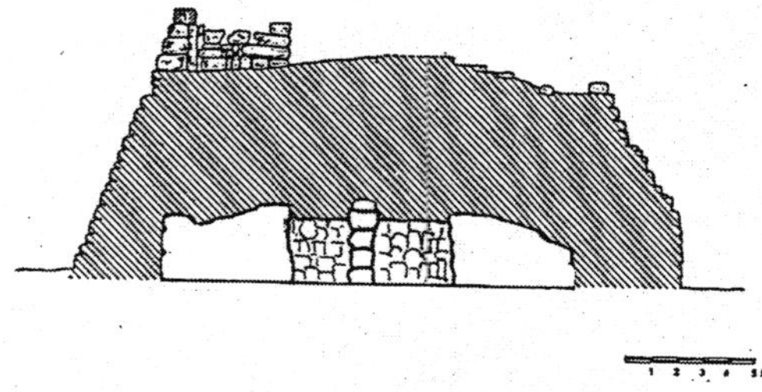

Fig. 6.
Comparación del tamaño de los talayots de Torelló (Menorca) (rayado.Según Mascaró,1986) y Son Fornés (Mallorca) (a partir de Gasull y otros, 1984)

haber una comunicación que sirviera de válvula de escape al crecimiento demográfico, cosa que en Mallorca y Menorca no era posible. En cualquier caso, el crecimiento en condiciones de aislamiento fue abortado por el dominio púnico de la isla, en la segunda mitad del siglo VII A.C.

Ibiza comparte características, en este sentido, con el archipiélago de Cabrera, que parece sólo fue habitado en muy escasa medida y con hábitat fundamentalmente en cuevas, hasta la época romana. Este hecho parece indicar que cada isla era un mundo casi cerrado, con escasos contactos entre ellas.

En conclusión, hemos observado una correlación entre la densidad de población y la intensidad del desarrollo de la cultura baleárica en las dos islas, Mallorca y Menorca. La explicación de esta correlación parece deberse al aumento de las tensiones intergrupales en un marco de competición por el territorio, provocando esto a su vez el agrupamiento de la población y el inicio de la jerarquización social. La continuación de este proceso hizo necesario el cambio en la base productiva, con un aumento de la importancia de la agricultura. El retraso en adoptar este cambio, especialmente en Mallorca, pudo deberse a las características edafológicas de ambas islas, con suelos pesados, arcillosos y muy pedregosos, que no gararantizaban la producción cerealística imprescindible, sometida además al vaivén de la sequía estival. En cualquier caso, los procesos en Menorca parecen más intensos que en Mallorca, lo que parece avalar la hipótesis inicial de la relación con la presión demográfica.

BIBLIOGRAFÍA:

ARAMBURU-ZABALA, J.,
1998. El patrón de asentamiento de la cultura talayótica en Mallorca. Palma: El Tall.
ARAMBURU-ZABALA, J,
2001. Prospección arqueológica en Cabrera. [en linea] http://www.talayots.com
ARAMBURU-ZABALA, J.
(e.p.) El poblamiento baleárico de Menorca. Un análisis espacial.
BURJACHS, F. ET AL.,
1994. Dinámica de la vegetación durante el Holoceno en la isla de Mallorca. En: A.A.V.V. Trabajos de palinología básica y aplicada. X Simposio de Palinología. Valencia, 199-210.
CALVO,M., SALVÁ,B.,
1999. Aproximació a la seqüència cronocultural de la naveta i del jaciment de clossos de Can Gaiá. Mayurqa, 25, 59-82.
CAMPS,J., VALLESPIR,A.,
1999. El turó de Les Abelles. Palma: C.I.M.
CASTRO,P., LULL,V., MICÓ,R.,
1996. Cronología de la Prehistoria reciente de la Península Ibérica (c.2.800-900 cal ANE).Oxford: B.A.R. Int. Series 652.
EVANS,J.D.
,1973. Islands as Laboratories for the Study of Cultural Change. En Renfrew,C. (ed.) The explanation of cultural change: Models in Prehistory, Londres, 1973, 517-520.
FERNÁNDEZ-MIRANDA,M.,
1991. La transición hacia la cultura talayótica en Menorca. Trabajos de Prehistoria, 48: 37-50.
HASSAN,F.A.,
1978. Demographic archaeology. En Schiffer,M. (Ed.) Advances in archaeological Method and Theory. New York. Vol. 1, 49-103.
HERNÁNDEZ GASCH,J.,
1998. Son Real. Necrópolis talayótica de la Edad del Hierro. Estudio arqueológico y análisis social. Arqueomediterránea 3 (II) Barcelona: U.A.B y Ayuntamiento de Santa Margalida
JIMENEZ, J.J.,
1999. Gran Canaria prehistórica. Un modelo desde la Arqueología antropológica. Las Palmas: Centro de Cultura Popular Canaria.
LOPEZ SÁEZ,J.A. ; LOPEZ GARCÍA,P.,
2000. Análisis palinológicos del yacimiento de Biniai Nou (Mahón, Menorca) En Guerrero,V.M.; Gornés,S. (coord.) Colonització humana en ambients insulars. Interacció amb el medi i adapatació cultural. Palma: U.I.B., 2000, 433-436.
LULL,V. ET AL.,
1999. Ideología y sociedad en la prehistoria de Menorca. La cova des Cárritx y la Cova des Mussol. Barcelona.
MENÉENDEZ AMOR, J./ FLORSCHUTZ, F.,
1961. La concordancia entre la composición de la vegetación durante la segunda mitad del Holoceno en la costa de Levante (Castellón de la Plana) y en la costa W. De Mallorca. Bol. De la R. Soc. Esp. Hist. Nat.,59, 97-100.
MESTRES, J., NICOLÁS, J.C.,
1999. Contribución de la datación por radiocarbono al establecimiento de la cronología absoluta de la prehistoria de Menorca. Caesaraugusta, 73, 327-341.
METRAUX,A.,
1995. La isla de Pascua. Barcelona.
Parra, I et al.,1992. A Holocene pollen analyses from Mallorca (Core SP1) En A.A.V.V. Abstracts of the 8 International Palynological Congress. Aix- en- Provence.
PÉRZ OBIOL ,R. ET AL.,
2000. Evaluación de los impactos antrópicos y los cambios climáticos en el paisaje vegetal de las Islas Baleares durante los últimos 8000 años. En Guerrero,V.M., Gornés,S. (coord.) Colonització humana en ambients insulars. Interacció amb el medi i adapatació cultural. Palma: U.I.B,73-98.
PLANTALAMOR,LL./STRYDONCK,M.VAN,
1997. La cronología de la Prehistòria de Menorca. Maó: Museo de Menorca.
PONS I HOMAR,G.,1997. Anàlisi espacial del poblament al pretalaiòtic final i al talaiòtic I de Mallorca. Palma: C.I.M.
RUBINOS PÉREZ, A,
2000. Las fechas Carbono 14 del monumento nº2 de Biniai Nou (Menorca) En Guerrero,V.M., Gornés, S. (coord.) Colonització humana en ambients insulars. Interacció amb el medi i adapatació cultural. Palma: U.I.B, 2000, 437-444.
STRYDONCK, M.J. VAN, WALDREN,W. Y HENDRIX, V.,
1998.The 14C Chronology of the Son Mas Sanctuary Site (Valldemossa, Mallorca, Spain) Radiocarbon, 40 nº 2, 735-748
TAIT ,D.,
1979. El sistema político de los konkomba. En: Llobera,J.R. (Ed.) Antropología Política. Barcelona, 185-198.
WALDREN,W.,1998. The Beaker Culture of the Balearic Islands. Oxford: BAR Int. Series 709, 75.
WALDREN,W.,1992. Radiocarbon and other Isotopic Age Determination from Balearic Islands: A Comprehensive Inventory.. Deiá: DAMARC 26.
YLL, E.I.,

1984. análisis polínico del poblado talayótico de Son Fornés. EN P. Gasull, V. Lull, MªE. Sanahuja. Son Fornés I: La fase talayótica. Oxford: B.A.R. Int. Series 209, 133-135.

YLL, E.I ET AL.,

1999. Cambio climático y transformación del medio durante el Holoceno en las Islas Baleares. II Congreso del Neolítico en la Península Ibérica. Saguntum Extra-2, 45-51.

48

AEGEAN ISLAND CAVES: DIACHRONIC CULT PLACES

ANNA PAPAMANOLU-QUEST

Archéologue Ethnologue, Membre de la Société Archéologique d' Athénes,
58 Rue de Babylone, 75007 Paris, France
10 Rue Amynta 116 35 Athénes,Greece

INTRODUCTION

At the East end of the Mediterranean, the islands of the Aegean Sea form a highly diversified and deeply uniform geographic entity.

The overwhelming tourist tidal wave of the last 40 years made of clear water, whitewashed houses, friendly hosts, discos, bars and carefree life, overshadows the real character of the life and history of the Aegean islands: a continuous struggle for survival in extremely hard conditions.

They have been named "Stepping stones" by archeologists. Indeed. Nevertheless, the Aegean is not a peaceful stream but a droughty corridor, an unpredictable, violent, dangerous sea, forming the natural frontier between Europe and Asia.

Though each island is a microcosm with its own flora, fauna and character, often strikingly different from its closest neighbor – as is the case of Lesbos/Chios or Thera/Ios-common natural features, as well as history, have formed an Aegean islander identity.

Theodore Bent's "Life Among Insular Greeks", written soon after 1885, is in may respects a present day reality. The same applies to accounts written when the Aegean was under Ottoman rule, by Western European travelers such as Randolph, Tournefort or Sonnini, specially as far as religious customs are concerned and, of course, natural curiosities of which the most exciting seem to be the caves.

Due to a 65% calcareous sub-stratum, Greece has over 7,000 caves, most of which are karstic formations – Crete alone has about 3,000, a number of which are now explored and studied.

DISCOVERIES AND RESEARCH

Speleological and archeological research in Aegean caves has developed over the last 40 years. In the past, caves were known as local hide-places, shepherd's refuge, and/or sacred spots linked to ancestral beliefs in an Orthodox cult. Important caves have often been discovered by chance, either because an animal or a human disappeared into the ground, or because an out-law was spotted there.

Cretan caves have particularly attracted Western European travelers because of the mentions in antique texts from Hesiod to Diodore of Sicily, of Zeus' birthplace in the Idaion and/or the Diktaeon Andron. The search for the Labyrinth was an even stronger motivation for visiting Cretan caves in 17th and 18th centuries. The travelers were guided by local people to whom interest in such places were either the sheltering of flock or the presence of a chapel dedicated to an Orthodox Saint. The quest of the "mylords" for antiquities was simply incomprehensible, that is, intriguing.

STALAGMITIC CULT STATUES

Western Europeans' search for antiquities has been Adriadne's thread for the discovery of the cavern of Antiparos, French Ambassador to the Ottomans Marquis de Nointel followed by a very large escort of painters, builders, Jesuits, corsairs, local guides and all sorts of other people, were visiting Naxos in order to collect antiquities.

He then was informed that in the neighbouring island of Antiparos, lying in the countryside, there was a very large statue. When the French reached the spot, they found a human shaped stalagmite next to the entrance of a cave, into which they ventured to descend. They found themselves in an underground cathedral. It was on 24 December 1673. The Ambassador and his numerous suite, in religious awe, celebrated Christmas mass around the stalagmitic altar, with torches, candles and choirs. To discover such a cave on the birthday of the Christian God who has also been born in a cave, is rather uncommon. That day, European visitors and Greek folks shared the same religious feeling.

FIGURE 1 THE 'SACRED WATER-POINT'
(Lera Cave, Western Crete)

FIGURE 2 THE 'VIRGIN-BEAR' or 'BEAR THE VIRGIN' ?
(Arcondia Cave, Western Crete)

The event is commemorated by a Latin inscription.

Later, were found ancient Greek inscriptions, which inform us that the poet Archilochos had been there in the beginning of the 7th century B.C., that a group of Alexander's contemporaries such as Antipatros, Aristeas, Kilokratos etc., had left their signatures and that Artemis was worshipped there. Today, the place belongs to St. John, whose chapel is at the entrance. Lots of people from the neighboring islands make the trip for the festival of his nameday. (1)

Another stalagmitic statue, surrounded by offerings, bones and ashes was found in the Kefala cave in the island of Erifos. The place has been plundered, filled-in with the neighboring quarry material and forgotten. This is a very exceptional case in the Aegean.

A large stalagmite in the shape of a pregnant woman as well as hieratic female figures are the center of myth, religious beliefs and cult in the famous cave of Eileuthyia, in Crete, located above the "dangerous port of Amnissos"-according to Ulysses' narration in Odyssey XIX. This cave has been excavated by Chadjidakis in 1886, by Marinatos in 1929-30 and studied by a great number of archeologists. (2) Ceramic finds range from Middle Neolithic to Sub-Neolithic, Early Minoan, Late Minoan, Geometric, Hellenistic and Roman Byzantine. More intensive use was in Neolithic, Geometric and Roman times. Four Lin B tablets from Knossos of the 14th century B.C. mentions the Goddess "E-re-u-ti-ja", in archaic Cretan Greek, "........" Erutija, that is "the one who sets free". Today, pregnant women make the pilgrimage to the cave hoping a "......." Which means "good freedom" in other words: to give birth easily. In current Greek the word for "freedom", "......." is almost the same with the name of the Proto-historic Goddess presiding an easy birth.

Richard Pococke, visiting Crete in the summer of 1739, describes "a large round grot…and a figure which resembles a sitting bear, from which it is called the Cave of the Bear; at the entrance of it, there is a chapel to the Virgin Mary". To the Cretan people, then and now, this cave is known as "Panagia Arcoudia" that is the Madona-Bear. By various myths they justify their behavior of veneration towards the huge stalgamitic bear on which have been carved small basins, while a larger, stone-built one, at the foot of the Bear, collects the pure water dropping from the roof, which is known to be holy (3).

FIGURE 3 ORTHODOX MASS IN ST JOHN'S CAVE (Western Carpathos)

HOLY WATER OR "AGIASMA"

A brief non exhaustive excavation in 1966 revealed elements of cult since EM1, but better documented for LM III B. Two clay votive offerings representing Apollo playing his lyre and Artemis the Huntress – who links with bears are well known-belong to the Archaic period.

In the same area, near the top of a soft hill overlooking a small natural harbor and a previously fertile plain, is the cavern of Lera. It takes its name from Lera, a Greek resistant, who hid there during the War of Independence of 1821, to be caught by the Turks in 1833 and hanged. The entrance of the cave is visible from the whereabouts. In the past, it must have been much higher. It is well known that the whole area is slowly and surely sinking into the sea. Byzantine and Roman buildings are now found at a depth of c. 3m under the sand. This cave has no particular impressive stalagmites – except a wing-shaped one which sometimes beats – and some human-shaped ones, in its third room where there is also the holy water-point.

An urgency-excavation in 1966 revealed an exceptionally long sequence from Early Neolithic to Byzantine. The first inhabitants of Crete left in the cave of Lera their Early Neolithic ceramics identical to those of the John Evans' dig in the central court of Knossos. Still, Neolithic people of Western Crete did not import their ceramics to the cave of Lear from Knossos. Microscopical examinations of the clays and comparison with the soils of the nearby plain indicate local production (4).

There are no clear data about cultural practice in the cave during the three millenniums from early Neolithic (c. 6,000 B.C.) to Early Minoan (c. 3,000 B.C.). As these ceramics have been found in a low level of the first part of the cave, it appears that the place was a shelter. Traces of religious beliefs and cult during this long period are not evident. This is also the case in other Neolithic cave dwellings of Crete, but it must be due either to our lack of elements or a better insight of that period.

From early Minoan to the beginning of Late Minoan, have been found shards of all kind of Minoan ceramics, from Pyrgos to Vassilike and to Kamares, included parts of vessels imported from the mainland and supposedly used in cult-like sauceboats. In Late Minoan I, Cretans seem to have forgotten the Lera cave, only to come back in Lake Minoan III B, when the major city of the area, Kydonia, was flourishing. The numerous votive offerings of that period have been found either inside or around the "sacred lake", which is c. 1,50m large, very shallow and muddy but still a proof of divine presence.

At the same holy water-point, but also in the third room at the foot of human-shaped figures, were found votive offerings, dating from Archaic to Roman times, which indicate a cult linked to fertility and clearly addressed to Pan and the Nymphs (5). The vitality of this cult in Late Antiquity is such that most caves of the Aegean islands, but also of the mainland, were dedicated to it.

Only a few shards let us know that the cave was visited in Byzantine times. Today, people climb the slope to take "agiasma" from the holy water point.

The special interest of the Lera cave lies in three points: (1) It proves that human presence in Western Crete existed since early Neolithic, a fact insufficiently documented until the dig of the cave. (2) It provides data for continuous cult linked to the water-point from Mycenean to early Christian times, and (3) it illustrates, through today's beliefs concerning its modest water-point, the length and the strength of cult linked to the holiness of the "agiasma", or water present in the womb of the earth.

FIGURE 3.1 ORTHODOX MASS IN ST JOHN'S CAVE
(Western Carpathos 1995 1995)

HEALING WATER

Holy water is healing water. This is the case of a cave named "Agio Gala" (Holy Milk) in Northern Chios. There, a Byzantine Emperor sent off his leprous daughter. The girl guided by an unknown lady, followed a dry river bed, reached a cave and with the "agiasma" got cured. After three years, remorse made the Emperor (whose name was never known) sent a boat and brought back his daughter. He found her miraculously cured. Then, to thank Agios Galas, the local saint who makes the water curative, he dedicated to him a chapel, in the entrance of the cave. The chapel is of the 12th –13th century and has beautiful Byzantine frescoes.

The well-known specialist of "agiasma" which cures eye diseases is Saint Paraskevi. In Lesbos, where every chasm between rocks is a sacred place, her church is clinging on the rocks of a grotto from which springs "agiasma". The neighboring town is named after the church, where is performed a regular Orthodox cult.

It often happens that the church is built separately from the "agiasma" found in the cave. Beliefs and cult addressed to the holy water are more important than the Christian mass. A good example is the cave Agiasma of St. Myron, in Crete.

HEALING BONES

Not only water but also bones have healing powers. An interesting example is the cave of Agioi Saranta in Cyprus. The name Saranta refers to forty anonymous Christian martyrs who all died together standing up for their faith. In popular mind the number 40 has become a name: Saranta. Heaps of calcified bones in that cave, but also in other Cypriot caves have received for centuries and up to now, an intense popular cult, with oil lamps, candles, icons and prayers. A piece of bone reduced to powder and mixed with water will cure malaria fever but also any other disease.

These bones are identified as those of pygmy hippopotamus, extinct on the island just before, or just after, the arrival of the first humans (6). Strangely enough, the people who absorb the bone powder are cured. Should one look for an explanation into the components of the bones, into the component of belief, or into those of both?

Animal bones found together with archeological items suggesting sacrifice are a common feature in most explored and /or excavated Cretan caverns. These date range from Early Minoan to Late Roman times. Of course, local shepherds have always slaughtered animals for food, either for small groups or for numerous people attending the religious festival which, once a year, takes place in the chapel sheltered in the cave.

Human burials have been found in some caverns in Crete, but they are rather rare.

HEAPS OF BONES AND HISTORICAL EXPERIENCE

Heaps of bones are, in the collective memory of Greek people, synonymous of war and massive death. In the cavern of Melidoni in Western Crete, in 1823, during the War of Independence, 370 persons, the population of the nearby village, resisting the Turkish attack, had taken refuge in the caverns. The enemy set fire at the exit and they all died asphyxiated. At the entrance of that cave is a chapel dedicated to the Virgin.

PILGRIMAGES ARE ADDRESSED TO THE BONES OF THE HEROES.

Archeological research in that cave indicates cultural practices dating from middle Minoan, Late Minoan I, Late Mi-

FIGURE 3.2 ORTHODOX CHURCH WITH ANTIQUE COLUMNS IN ST JOHN'S CAVE (Western Carpathos)

noan III, Protogeometric, Archaic and Roman times. A stalagmitic formation looking like the Madonna holding the Child, is next to one whose form is that of horns of consecration.

The cave of Melidoni has been visited by Cyriaco di Ancona in the 15th century as well as Venetian 16th and 17th century travelers. They first mention the inscription on its walls relative to the goddess Maia, her son Hermes of Talos, but also Zeus of Talos-the-Giant.

CAVES: BIRTHPLACE OF GODS

As everyone knows, at least since Hesiod, large Gaia received the secretly born Zeus on the fertile land of Crete where Rhea hid the child "in the secret depths of divine earth" (7) Two major Cretan caverns have been competing in the past, today and most certainly in the future, for the title of Zeus' cradle. The Diktaeon Antron on Mount Dikte and the Idaion Antron on mount Ida. Since the 1880's both caves have been excavated and studied.

The finds from the Diktaeon Antron – among which libation tables with three L in A inscriptions now in the Ashmolean Museum – indicate that at least since Middle Minoan was taking place, a vigorous cult, which continued during Late Minoan I, II, and III and which was probably addressed to a sun-god and to a Nymph (8) . Numerous later ceramics prove that the cave was a place of cult to the end of Antiquity. If there is not proof of Rhea hiding there baby-Zeus, near the "agiasma", a stalagmitic form representing Virgin Mary holding child-Jesus. This statue has been mentioned by Bochini in 1653 and still revered today by numerous Greek visitors, the same ones who nostalgically admire the room into which Rhea gave birth to Zeus.

If the Diktean cave is an object of dreams, the Idaion Antron is an object of passion. Since the first excavations in 1885, archeological research has been on and off (9). From 1993 on, the almost yearly campaigns in the Idaion Antron are on the focus of Greek mass media. (10). The famous shields, tripods and other votive objects of the 8th century B.C. are still unique, but huge amounts of offerings form an uninterrupted sequence from Early Minoan to late Roman times. Late Minoan I, II, and III, Geometric, Archaic, Hellenistic are periods of intense cult in the cave. A huge alter and many other features illustrate some aspects of those religious practices. P. Faure, base on Porphyriou's "Life of Pythagoras" and some Roman statuettes with Ploutonic characteristics, would like future archeologists to look, in a dark unexplored hole of a cavern, for the ritual "tomb of Zeus". Who knows if Gods die in the same place they were born... Still, an Underworld ritual is very probable in the depths of a cavern, as is also a cult linked to the death and rebirth of a young God.

Orthodox Cretans have of course recognized, at a hardly accessible part of the cave, a stalagmitic Virgin with the Child and left a bone cross her feet. (11)

In the 1986 excavations has amongst others, been discovered a Late Neolithic level, which brings the first use of the cavern around 3,000 B.C. There, traces of religion are absent. Like in the Lera cave, whose early Neolithic finds date from c. 6,000 B.C., it seems that the place was a shelter.

CAVES: BIRTHPLACE OF MYTH

Shepherds of this rough mountainous area understand very clearly the history of the Idaion Antron. So much more that they have their own versions of stories: King Minos was out hunting and thought he had shot a deer to find out he had shot his dear Shepherdess. Desperate, he buried her near the spring which is on the platform in front of the cave. The story which is sung by shepherds of Ida, in long narrative songs, is so common throughout the world that would not be worth mentioning if the tomb of the young woman were not next to a cave and to a fresh water spring – places traditionally linked to Nymphs.

Whether Zeus, or any other God, was born in the above Cretan caves, is a matter beyond the aim and the possibilities of this paper. What is obvious is that throughout the ages, myth is born and reborn in Aegean caves.

Zeus is certainly at home with the shepherds of Naxos, we are informed by the inscription "........(Greek lettering)" (Zeus sheep-protector), on the mountain named: Zas. In the homonymous cavern – from where antique remains are not reported – the story is entirely Christian. During the centuries of foreign occupation, local people were taking refuge deep in the cavern, near a spring, where they had built a chapel dedicated to the Holy Spirit – in Orthodox terms "Spring of Life". Behind that area, two stalagmites are the ".....(Greek Letters)" (the priest and his wife) whom God has petrified in order to save them from the Ottomans. (12)

It is to be saved from pirates that seven lovely young girls were hidden in a cave in Kalymnos. But they never found their way out. This is why the place is named: "Cave of the Seven Virgins". There, archeological finds prove the existence of an antique cult addressed to the Nymphs.

CAVES AND FEMININE NATURE

Most caves are believed to be the home of "Neraides", the Greek fairies who also haunt springs, tress and desert seashores. Their supernatural beauty is impossible to resist. If a lonely man lets himself to be lured by a Neraida, he loses his mind.

Neraides want mortals to bear them children. Soon after they disappear, leaving the man mute for the rest of his life. They are often found in deep caverns nursing their babies – white as the foam of waves. This is what happens in Neraido-Spilios in South-West Crete. But if they do not have their own babies, they steal those of mortal women, an evil act which can happen in the famous cave of birth-presiding goddess Eileithya, named also "cave of the Neraides".

No matter what, Neraides are always dancing and most of the time they are singing heavenly songs. This they do in the cave named the Fountain of the Virgin in Meskla in Western Crete. There, the chapel of the Virgin is built on the ruins of a 6th century Byzantine basilica, itself founded on antique remains suggesting a cult addressed to Artemis and Nymphs. (13) Cave chapels are mainly dedicated to women Saints, such as St. Sophia, St. Foteini, St. Paraskevi, St. Marina, or to the Virgin herself. The presence of the Christian feminine divine power linked to healing, child bearing and nourishing keeps away evil Neraidas.

It is obvious that behind the personality of the moderns

Neraides profile the antique Nymphs, but also the sea-born Nereids. The name "Neraides" is the same to that of the wave-daughters of Nireus. Their nature and deeds are those of tree-nymphs, spring-nymphs, cave-nymphs-like Kalypso. Both antique and modern are linked to love, child bearing and nourishment. But, if antique Nymphs and Nereids were desirable and good, modern Neraides are irresistible and evil.

Their difference reflects the love-hate relation of Christian Greeks with their antique past. Not only desire is contrary to ethics, but female supernatural beings with such a strong pagan nature are the Devil in person. But they are everywhere. Inside caves they are even more dangerous. The only possible protection is prayer.

HERMITAGES AND HIDDEN CHAPELS

Most caves, whether important cult places in Antiquity or not, are now dedicated to an Orthodox Saint. The historical life of the Saint to whom the chapel is dedicated has no connection with the beliefs concerning the supernatural nature. The Antique counterpart of the beliefs of the Greek people is often striking. There is always a story justifying cult in caves. The simplest explanation given for the holiness of the place is that a Christian ascet was once upon a time retired there.

The world renowned Apokalypse grotto in Patmos is the best example. Still, which is the part of history and that of myth in St. John's story? Grottos and rock shelters converted to chapels following the presence there of an Orthodox ascet are very common in Northern Aegean islands.

CHAPELS HIDDEN INSIDE CAVES

Even though there is no record of an ascets presence, permanent insecurity, desperate need for protection, are good enough reasons for the building of chapels in inaccessible caves, on steep cliffs.

On the island of Kythera, 60m. above the sea, among barren rocks, is hiding a cave. Inside, c20m. from the entrance, is a chapel dedicated to Saint Sophia. It is said that her body has been found there. Next to the scared water-point, well hidden chapel, dated 1785, is covered with high quality frescoes . No antique remains are reported. On religious occasions, local people climb the cliff animated by long habit and strong faith.

The chapel "Zoodochos Pigi" (Spring of life) is hidden in the caves of the Zas mountain in Naxos. Deep inside the cavern is also hidden the chapel of St. Foteini in central Crete. In the cavern of Skoteino, were Late Minoan cult is documented since A. Evans' time, the ruins of a 17th Century Orthodox chapel are lying behind large protective rocks. In the cavern of Topolia, in Western Crete, where shards of Neolithic, Classical and Roman periods have been found, a chapel dedicated to St. Sophia and St. John the Baptist is hidden deep down the earth (14). Orthodox chapels arranged in the depth of a cave are not rare.

RURAL GROTTO SANCTUARIES

Chapels are not always necessary to cultural practice. In Lesbos and many other islands, a local Saint appears in a rock shelter or just amongst some barren rocks. There, icons, candles, incense are addressed to the holy entity and, on a certain day, people walk for hours to attend the festivities of the rural sanctuary.

THE "PANYGIRI", A TOKEN OF IDENTITY

All protector Saints have, at their nameday, their religious festival, the "panygiri". Usually, the festivities last three days and nights. An Orthodox mass at sun down, another at sun rise are the basic time marks regulating ritual festivities. Very important is the time before, in between and after. A religious "panygiri" is the mirror of the communities' structure. It, of course, varies according to the island, the locality, the Saint and a multitude of local beliefs but is, in any case, a token of identity. "Panygiris" take place in all chapels. When the saint's chapel belongs to a cave, the links are stronger than in any other case. In Antiquity, each town had its sacred cave. The holy place is the cave.

The historical life of the Saint to whom the chapel is dedicated has no connection with the beliefs concerning his supernatural nature. This is, of course, a world wide phenomenon.

The Aegean island cave sanctuaries, the antique counterpart of the Orthodox cult is often strikingly vivid.

SAINT JOHN THE BAPTIST OF THE GROTTOS

Of the men saints, the unquestionable master of cave chapels is St. John the Baptist – to Orthodox Greeks: St. John the Prodrome.

In the small island of Irakleia, the Saint's icon printed on the back of a shepherd who lay under a bush, next to a hole, which proved to be one of the most impressive Aegean island caverns. The icon of the Prodrome has been put inside. At his nameday takes place a very important "panygiri". The Antiparos cave, mentioned at the beginning of this paper, is also dedicated to St. John, whose chapel is at the entrance. The same St. John shares the cave of Topolia, in Crete, where two other caves have chapels dedicated to him. (15)

THE "PANYGIRI" AT THE CAVE OF SAINT JOHN

In the island of Karpathos, a two hours walk through pricky rocks from the village of Olymbos, which is at two hours boat of the main port, on a steep cliff next to the rough Aegean, beaten by wild North winds, lies the small unimpressive grotto of St. John the Prodrome.

Here the story of that of a baby having fallen into the watery dark depth of the cave and St. John bringing it back alive and safe, to its mother. St. John, The Baptist has lived here. This is why, his head, cut on the evil woman's demand, rises under the form of blood- shot moon from the end of the sea, when, one after another, fall on the horizon the colors of Salome's veil.

It happens sometimes on the 29th of August, at a time of the year when light declines and when the malaria fevers, very common in the past, were in paroxysm. That is when he is supposed to have been decapitated. His cult is linked to the fear of death and to the hope of healing – expressed through a number of rituals and of epithets adjacent to his name. At his nameday, in this forlorn area, takes place the symbolically riches and socially most important "panygiri" of the Aegean island cave chapels.

The iconostase of the chapel, inside the cave, is supported by Doric columns. Antique marbles are everywhere. Here the whole area is that of the ancient city of Vrykous, whose inhabitants, in the 7th century A.D., when pirate attacks were a common event, fled away from the sea and built the mountain village Olymbos. They kept their customs, their archaic dialect and their sacred cave.

Once a year the entire local population, carrying rich wardrobe and household on donkey-back, is transported at St. John the Prodrome's cave of Vrykous, for three days. Socially regulated traditional dancing will go on without pause from the last ray of sunset to the first ray of sunrise. Immigrants to Baltimore – that is half the male population of Olymbos-come back for that day, bringing their babies to be baptized with the "agiasma" of the cave.

In this desert, where danger comes from the sea, where winds are merciless and stone the only refuge, it is understandable that local people cling to their identity, whose highest statement is the Orthodox religious festival at St. John's the Prodrome's cave. They say that their cult in the cave has existed since "always"-the antique marbles being the irrefutable proof.

If antique marbles built in walls of Christian chapels are just as common as chapels inside caverns, the combination of both is a rarity. But the "panygiri" at this cave-church is only exceptional because of the social and natural environment. The same intensity and the same rituals exist in all island cave religious festivals.

CONCLUSION

As shown above, evidence of cult in Aegean island caves is discontinuous. It varies according to the historical period and the locality. In the best studied Cretan caves, the bulk of the material related to cult is dated to Late Minoan III, a time of Mycenean presence on the island, though in major caverns, it is documented since middle Minoan, the time of the First Palaces, from Geometric, to Archaic, Classical, Hellenistic, and Late Roman cultural practices, evident in most caves, are addressed to well-known deities: Zeus, Hermes, Apollo, Artemis, Pan and most of all the Nymphs. Early Christian times are considered as a period of abandonment. P. Faure, who is the undisputable authority on Cretan caves, is right in pointing out that our knowledge of that period is insufficient and influenced by official data such as the 4th century edits of Theodose prohibiting antique cult (15). It certainly must have been much easier to control the application of the emperor's orders on mainland plains than in the Aegean islands.

The example of the Karpathos cave suggests that conversion to Chritianism is not a good enough reason for giving up a local sacred cave sheltering cult, beliefs and myth. The same suggested by the presence of a 6th century basilica's ruin at the Meskla cave in Crete. On the other hand, when today, local people take "agiasma" from the hardly accessible sacred lake of the Lera cave, there is no trace left. Who can assert that such religious acts did not exist in early Christian centuries?

Speaking of beliefs concerning holy water in a chapel in Amorgos, the very Catholic botanist, Tournefort, who has spent many months in the Aegean islands in 1700, is shocked and writes: "le Christianisme n'a pas change' ;'esprit fabuleux des Grecs".

That statement could as well have been written by early Byzantine times or today. Since antique gods had to be turned to demons to leave the place to Christianity, Christian Greeks understand religion through their own mind. Since antique sanctuaries have become residence of demons, Christian Saints had to replace them – in caves as well.

Much more destructive than any new religion is the modern materialist behavior. In our days, island caves, which are not dedicated to a Saint, are turned into rubbish deposits – unless tourism comes to their rescue.

Clinging – up to ancestral faith as a token if identity, is a healthy reaction.

Throughout tormented history, the Aegean islanders stuck to the protected space of caves, where idols are made by an unknown divine sculptor, where God sends miraculous holy healing-water, where invisible but omnipresent feminine beings spring constantly from fresh myth bearing minds. A diachronic, deeply religious relation to environment, which has helped survival in the past and allows some hope for the future.

REFERENCES

TOURNEFORT DE J.P.
1985, Voyage du Levant, 1727, Paris
DAVARS C.,
1983, Fonctions des Caverns Cretoises, Paris 1964, Cretica Amnissos, B.C.H.
POCOCKE A.
1745, A Description Of The East And Some Other Countries. London
PASHLEY R., 1837 Travels in Crete, London
FAURE P.' 1966, 1967
GUEST-PAPAMANOLI A.,
1976, Les grottes de Lera el de l'Arcoudi a en Crete Occidentale aux.
LAMBRAKI A., Epoques Prehistoriques et Historiques
31, pp. 178-243. (Abrev: G.L.A.C.O.)
GUEST-PAPAMANOLI A. -G.L.A.C.O. p. 200.
LAMBRAKI A., -G.L.A.C.O. p. 236
BATE D.,
1904, On The Ossiferous Cave Deposits Of Cyprus,
Geological Magazine, V.
(8) BROMAGE T. & AL.
(this volume), A Life History and Climate Change Solution to the Evolution and Extinction of Insular Dwarfs : The Cyprus Experience.
REESE D.,
1975, Men, Saints or Dragons?
Expedition 17, 4.
SIMMONS A.,
(this volume) The Role of Islands in Pushing the Pleistocene Extinction Envelope: The Strange Case of Cypriotic Pygmy Hippos.
HARBHERR F., ORSI P.,
1888 Scoperte nell' antro di Psychro, Museo Ital. di ant. class. II,
EVANS A.,
1897, Inscribed libation table from the Diktean Cave
J.H.S. 17, Palace of Minos I, 1921.

49

INSULARITY AND IDENTITY IN PREHISTORIC CYPRUS

J. CLARKE

School of World Art Studies and Museology, University of East Anglia, UK
EMAIL:

INTRODUCTION

Throughout the early prehistoric period Cyprus was characterised by cultural insularity, observable as idiosyncrasies in the material culture, which suggest there was little contact with the mainland. This insularity is so conspicuous that Cyprus has been termed a "particular regional entity from the outset" (McCartney and Peltenburg 2000, 9). Not long after the first evidence of the arrival of colonising populations from the mainland in the 10th millennium b.p. cultural signifiers representative of mainland connections were diluted by adaptations to the local island environment, specifically the rapid decline in the numbers of arrowheads and the initial disappearance of cattle (McCartney and Peltenburg 2000, 9). Nevertheless, many believe that contact with the mainland continued sporadically but continuously throughout the prehistoric period (Broodbank 2000; Broodbank and Strasser, 1990; Clarke 2001; McCartney and Peltenburg 2000; Peltenburg et al. 2000; Peltenburg et al. 2001). Although primary evidence for continued contact with the mainland is slim there is now sufficient circumstantial evidence to support this premise, significantly, the substantial increase in the number of obsidian artefacts in the Cypro-MPPNB, the shift to covering pressure retouch at Kissonerga-Mylouthkia and Khirokitia-Vounoi documented in Late and Final PPNB assemblages on the mainland (McCartney and Peltenburg 2000, 9) and later, the reappearance of cattle at the beginning of the Khirokitia phase at Kritou Marottou-Ais Yiorkis (Simmons 1998). In light of the growing shift in thinking about island insularity as a cultural construct (Broodbank 2000, 19-21) and following latest theories on human agency, the object of this paper is to examine whether insularity on Cyprus, and the Ceramic Neolithic period in particular, was the result of "a divergent insular evolution caused by external isolation and internal limitation combined to inhibit the growth of a maritime interaction sphere" as argued by Held (1992), or whether it could be argued that the Neolithic population acted in ways that maintained cultural and economic isolation from the mainland.

Such a theoretical position brings up issues about identity in prehistoric groups – a difficult topic to tackle methodologically – and one that fails to be adequately addressed through debates on style. Recently, however, methodological approaches to the analysis of material culture have focused on the examination of formal variation in technological attributes rather than on formal variation in stylistic attributes. Schiffer's Behavioural Chains (LaMotta and Schiffer 2001) and Bourdieu's "Chaîne Opératoires" (Bourdieu 1977) are methodologies designed around the study of ancient technology in order to illuminate processes and activities of past societies. Studies of this kind, that focus less on formal variation and more on variability in technological processes, and therefore in overall human behaviour, can help to locate identity within normal patterns of behaviour. Ethno-archaeological studies undertaken using these methodologies (see papers in Chilton 1999; Dobres and Hoffman 1999; LaMotta and Schiffer 2001; Stark 1998) have suggested that social identity is more likely to be observable in every day behavioural patterns than in ornamental styles such as decoration on pottery. Specifically, Dietler and Herbich (1998, 237) advise that "it is now clear from ethnographic studies that decoration is of highly variable significance in relation to other physical attributes and its meaning cannot be comprehended in isolation". However, the study of technological processes has yet to be more generally applied to archaeological data. It is only recently that archaeologists have begun to appreciate the importance of using behavioural and technological systems in the search for more enduring facets of social identity (Gosselain 1998, 82). But, as far as I am aware, in prehistoric Cypriot studies only Frankel and Webb (1998,) and Frankel individually (2000) have considered the problem of identifying ethnicity in any formal sense in prehistoric contexts.

Although Frankel and Webb (1998, 2) would argue that habitus – defined as all aspects of society, as lived through everyday experience (Bourdieu 1977; Jones 1997) – can be regarded as closely related to ethnicity as "it is through shared experience of the world that members of a group identify themselves as having a common identity" (Frankel and Webb 1998, 2), I prefer to see the concept of habitus as being more complex than this. Bentley's definition of a shared habitus that engenders feelings of identification among people similarly endowed (Bentley 1987, 173) defines a looser, less socially distinct concept of group identity than the purely ethnic type looked for by Frankel and Webb. This is an important difference and is highly relevant to the present argument because ethnicity assumes a degree of cultural identity that is not only self-defining but also defined from the outside. In the Cypriot Ceramic Neolithic there is no evidence for such a prescribed sense of identity but rather the concept of habitus as I argue it, was represented by broad cultural homogeneity coupled with ceramic diversity suggesting a subconscious tendency toward a shared identity – 'The Same, But Different' scenario observed in ethnographic studies of modern island societies (Cohen 1978, 1987).

Therefore, identifying identity in prehistoric material culture requires a holistic approach that focuses as much on the "sameness" of the Ceramic Neolithic cultural assemblage, as it does on the specific differences – i.e. the variation in the painting schemes on pottery. Although arguably it would be more rigorous to examine in detail more than one cultural assemblage from the Ceramic Neolithic period, or the evolution of one cultural assemblage over time, the pottery assemblage lends itself well to this type of study because first, I have previously examined the ceramics from a traditional perspective of stylistic variability and found that the theoretical models used to define style do not adequately explain observed material culture patterning, and secondly, as mentioned above, it is not variability in the ceramic assemblage that is significant but rather the homogeneity in material culture overall. It is this that traditional studies fail to adequately explain and why other methodologies are now needed to examine the unusual material culture patterning.

Regional Variation in Ceramic Neolithic Cyprus: an Overview

The Ceramic Neolithic phase was characterised by a considerable degree of uniformity observable in most facets of the material culture. Leaving aside chronological considerations, already discussed in detail elsewhere (Clarke 2001), it is really only in the ceramic repertoire that variation can be detected and this seems to be primarily in the way pottery was decorated. Although previous work (Clarke 2001) has shown that a significant amount of this variation can be explained using stylistic models created by Wiessner (1983, 1985) and Sackett (1985) that define variation in terms of symbolic behaviour, these go only part way towards understanding the full gambit of social behaviour which is observable as elusive patterns in the material culture. In these earlier studies it was suggested that formal variation in painting schemes on pottery could be divided into two broad forms of style: 1) assertive style, or style which is personally based and which carries information about individual identity (Wiessner 1983, 257) – visible in the types and variety of decorative schemes used within and between closely located sites and 2) emblemic style, or style that transmits a clear message to a defined target population about conscious affiliation or identity (Wiessner 1983, 258) – represented by distinct variation between the primarily red-on-white painted tradition in the north, east and west of the island and the primarily combed and monochrome tradition found in the south-central region.

Using a range of independent ceramic attributes such as manufacturing technique, primary vessel form, secondary vessel form (rim shape, spout shape etc.), surface treatment and decorative scheme, I undertook a number of rudimentary statistical tests to establish whether non-random patterning was attributable to factors other than normative variation (such as variation in types of decorative motifs used to decorate different parts of a vessel). The results indicated that some attributes, such as overall vessel morphology displayed little formal variation, whereas attributes such as decorative schemes and some component designs (bases, spouts etc) displayed quite formal patterns of variation. On the basis of these results it was argued that stylistic variation existed in the morphology and painting schemes on pottery and that it probably reflected the interactive processes that took place between individuals and groups involved in the day-to-day activities that ensured continued survival. Exchange systems, based upon kinship and intermarriage, would have led to the creation of complex social relations that would have been negotiated on both an individual and group level. Competition for resources within site catchment areas probably led to the creation and maintenance of social and economic boundaries. Stylistically, these boundaries may have been represented in the ceramic assemblage as differences in the type and distribution of motifs between sites. This sort of variation can be equated with assertive style, or style which is personally based and which is therefore more common between closely located sites where strong familial relationships existed and regular contact occurred. Between regions, social interaction networks may have been more structured, involving only sporadic but important exchanges of information and goods. As distances were greater, communities would have been less likely to meet on a regular basis and therefore meetings between groups and individuals may have been more formalised. In these cases variation would have been more likely to attributable to emblemic style – that is style that contains information about group affiliation – because contact would have been rare, groups would have been less familiar with each other, or may have been in competition for the same hunting areas.

The outcome of this study illustrated that there existed quite developed social dynamics between villages and between regions, apparently based on scales of personal and group identification and interaction. However, in terms of the Ceramic Neolithic as a whole formal patterning in material culture was considerably more uniform than would be expected in a diverse environmental and ecological setting such as Cyprus, suggesting an overlying sense of group cohesion or a wider sense of group identity, that may have been subconsciously projected outward to possible incoming groups.

LOCATING IDENTITY USING MATERIAL CULTURE PATTERNING

The first step in tackling the issue of identity in prehistoric societies – assuming other explanations for material culture patterning have been ruled out – is to ascertain what sort of identity might be present in a given prehistoric context. Considering that anthropologists studying modern societies speak of symbols as being "imprecise, ambiguous, and vague, not because people are sloppy in their use but because (a) there are some things that can only be expressed in such ways; and (b) it is socially expedient to have imprecise means of expression available" (Cohen 1987, 12) the task of recognising identity becomes all the more difficult when another level of cultural data is removed. In archaeology we are left with the examination of formal material culture patterning and the behavioural systems that account for this patterning. Pre-historians have an advantage over other archaeologists studying later periods in that they can exclude a number of types of identity almost immediately because there is just not enough evidence to indicate either presence or absence in the archaeological record. In the Ceramic Neolithic there are no markers in the cultural record that would help us to recognise self-identity (examples of which would be potter's marks, or recognising makers through manufacturing techniques or decoration). There are also no cultural markers to indicate the presence of ethnic or political identity (for example see Frankel and Webb 1998), both of which are usually dependent upon a degree of social complexity. On the other hand, there does appear to be clearly visible cultural markers to indicate group identity.

The second step is to define exactly what group identity is and in what context the term is being used. Woodward (1997, 29) argues, "Identity is forged through the marking of difference. This marking of difference takes place both through symbolic systems of representation and through forms of social exclusion. Identity then is not the opposite of but depends on difference". In the case of group identity this suggests that members of a group have something in common with each other, which distinguishes them from members of other groups. In prehistoric societies identity may manifest itself as differences in material culture patterning between small familial or clan groups such as would exist in prehistoric villages, or between larger groups such as different communities on Cyprus or between the Cyprus and the wider Mediterranean. In each case the symbolic boundaries that characterise these differences are self-defining and constantly changing within the consciousness of the group or community that created them. However, they also have to be recognised by a receiver group and therefore must be robust enough to be understood outside of a local context.

Finally, there needs to be a formal mechanism or methodology for identifying group identity in the material culture, therefore, the third step is to locate cultural indicators in the archaeological record that reflect behavioural systems (or ways of doing things) and which could be indicative of shared experience or common identity. This effectively means that instead of considering material culture insofar as it displays formal or ornamental variation such as decoration, archaeologists should be looking for variation in behavioural systems.

Returning then to Neolithic pottery, archaeologists can study the life histories of artefacts, or the sequence of behaviours that lead from the procurement of raw material and manufacture of an object, through various stages of use re-use and recycling to eventual discard (LaMotta and Schiffer, 2001, 21). Recording the life history of an object is in effect, the same as studying the chaîne opératoire or technological process behind the creation use, re-use and discard of an object. There are nine elementary operations in the technological process used in the making pottery. Each of these technological operations can be viewed as part of a larger behavioural system that reflects "similar ways of doing things" or habitus. The elementary operations are: 1. Procurement of raw materials; 2. Cleaning; 3. Kneading.; 4. Modelling; 5. Finishing; 6. Firing ; 7. Use; 8. Re-use and repair; 9. Discard.

PROCUREMENT

At most Neolithic sites clay is collected from locations around the village, most often clay beds associated with local rivers or springs (Shiels 1993, f.c.). Although Arnold has noted that potters may travel distances of 1-50km from camp (Arnold 1980, 149; 1985, 39-49) there is no evidence of this in Neolithic Cyprus. Philia-Drakos A is the only site where large numbers of clay types have been recognised and this may be simply the result of deep downgrading of rivers during the wet season exposing different types of clay. At other sites such as Ayios Epiktitos-Vrysi the excavators were able to identify local clay and temper sources (Peltenburg 1982, 62). According to Arnold (1980, 149) again, temper procurement usually happens between distances of 1 to 24km form camp whereas distances travelled for slips can be much larger. In Neolithic Cyprus evidence indicates that clays, temper and paints were all locally procured. Temper ranged from vegetal matter to mineral temper to grog but the underlying technological decision on the part of the potter was to temper pottery with whatever was at hand and whatever best suited the vessel being constructed. Red ochre, the primary mineral used in paint would have been acquired from copper bearing deposits. It would have always been mixed with a binder (Shiels, f.c.), often animal fat or oil. Shiels has shown that the ratio of binding material to paint is particularly important when achieving the high burnished surfaces characteristic of the Neolithic fabrics. Therefore, Neolithic potters hoping to achieve the high gloss characteristic of all sites would have procured the same elements, mixed them in the same quantities and presumably in the same fashion.

CLEANING

Cleaning is the method by which clay is processed for use – either through levigation which is a method of "washing" the clay in order to separate the finer clay suspension from the courser clay suspension, or by drying the clay and then sieving it to remove larger inclusions. In Neolithic Cyprus there is no evidence to suggest the method used but it is known that temper was intentionally added to the paste (Peltenburg 1982, 62) and that there was a common tendency at most sites for potters to use whatever material was closest to hand as temper. There is also evidence that good quality clay that needed no levigation was sometimes available. Potters familiar with local clay beds would have exploited these clays when available. Therefore some shared rudimentary patterns of clay procurement and preparation existed.

KNEADING

Kneading is a process by which clay is homogenised and aerated to give it greater workability for shaping and moulding. It also helps to reduce the likelihood of breakages during firing. I know of no examples of wasters found at Neolithic sites but then I also know of no examples of firing sites in the Neolithic or even the Chalcolithic. Studies by Shiels (1993, f.c.) have shown that even in open pits the amount of wastage is negligible. There is, however an overall high degree of quality in the Neolithic ceramics, which is observable in the homogeneity of the clay paste, the uniform distribution of inclusions and the hardness of the fabric and which suggests wider similarity in technical knowledge.

MODELLING

Modelling includes scraping, moulding and shaping the vessel into its final form. Decisions made at this stage are, in the majority, functional decisions but there is also a greater opportunity for ornamental variation to be included in the making process at this stage than at earlier stages. As a consequence it is at the modelling stage that we begin to see variation in the shape of vessels. Even so, there is very little variation in the technological operations used for modelling vessels. Where methods of moulding can be observed pottery is always coil or slab built depending on the malleability of the clay. Scraping or wet smoothing may be interchanged depending on the vessel form but both techniques are used at all sites. Bowls appear to have been dried upside down because rims often show flattening that cannot be intentional as it often occurs in conjunction with a rounded rim. Vessel form is homogeneous across the island. The decorated ceramic repertoire is limited to three primary shapes - bowls, bottles and holemouth jars (Figure 1). Bowls may be large or small, and frequently large bowls have spouts. All sites have relatively similar frequency distributions of these shapes. There can sometimes be variation in the percentage of one shape over another at different sites, but the differences are small and presumably reflect slight variations in subsistence tasks (Clarke 2001, 65-67). Effectively, there is no differentiation between ceramic form and function across the island.

FINISHING

The Ceramic Neolithic period was characterised by a predominance of pottery decorated in variations of red on white painted design. In the south-central region pottery was decorated in negative painted design whereby the red painted surface was combed away to produce a pattern in reserve known as Combed Ware (Figures 1.6, 1.7, 2.1, 2.5). In the north, west and east of the island the predominant style was positive painted design or Red on White Ware (Figures 3.1, 3.4, 3.5, 3.6, 3.7, 3.8), where red decoration was painted onto a white or self slip. There was also a Red Monochrome Painted version, which is commonly found at sites in the south central region but also occurs in some quantities at all sites. Although decorative schemes display marked regional diversity there is a sense that the duel techniques of positive and negative design can be seen as the makers trying to achieve the overall effect of a contrast of red-on-white. This similarity was noted by Peltenburg (1978, 66) and again suggests a wider sense of uniformity overlying local regional differences (compare Figures 1, 2 and 3 for overall similarity, particularly Figures 1.4, 1.5, 1.8 where the two techniques appear together on the same vessel). In other respects too the decorative process displays uniformity across the island. In almost all instances a slip was added to the clay surface before the vessel was painted. Sometimes the slip is lighter in colour than the paste and sometimes it is the same colour, depending on whether it was a self-slip or a slip produced with a different type of clay. It has been suggested that the use of a lighter clay slip produces a more desirable contrast with the red paint in regions where clays fired a darker red. Another reason might be that a slip was added to provide a smoother surface for painting. Prior to firing and while the clay was leather hard it was burnished to a very high sheen with a small pebble or some other type of burnishing tool. Again, Shiels (f.c.) has shown that the very high burnishing characteristic of the Neolithic can only be achieved when a combination of factors are at their optimum. Interestingly, the tradition of burnishing to a high sheen of the sort seen in Neolithic pottery, and also characteristic of the early Monochrome Burnished Wares from Philia-Drakos A and Dhali-Agridhi, was a tradition that is known to have occurred widely on the mainland. As the technology needed to produce high burnishes required specific knowledge about the proportion of paint to bonding agent might indicate continued contact with the mainland, at least in the formative stages of the Ceramic Neolithic phase.

In addition to the painted wares, all sites produce similar quantities of a Course Ware – the function of which is unknown but has been suggested to be either for cooking or for the sorting of grain (Peltenburg 1978, 61). There is a considerable degree of homogeneity in Course Wares across the island and examples from the north would generally be indistinguishable from examples in the south of the island.

FIRING

There are no confirmed kilns or fire pits from Neolithic sites. However, Shiels (1993, f.c.) has shown that vessels are fired at similar temperatures, using similar oxidising agents and that this is primarily due to the technological requirements for achieving certain finishes.

USE

Pottery is found exclusively in settlement contexts, although mortuary contexts are rare in the Neolithic period and there are no deposits that could be interpreted as definitively ritual. Therefore, in the absence of other evidence pottery appears to have had a utilitarian function, primarily for use in the preparation and serving of food and other domestic products. As there is little variation in the type of contexts in which pottery is found, and also in the quantity and variety of pottery found, it is possible to assume that each village community produced, used and discarded similar quantities of pottery each season.

RE-USE

There is some evidence of repair to broken vessels and at most sites pottery is occasionally re-used as temper in pottery (grog). Some decorated vessels have been discovered with drill holes positioned close to a break indicating intentional repair. Whether this method was successful is not known but in most instances the drill holes show little sign of wear suggesting that the mend did not last long.

DISCARD

Finally, pottery was discarded on the floors of habitation units, in courtyard spaces and sometimes in middens. At most sites down-washing of material from the top of the site has meant that pottery was often found deposited en masse in gullies and ditches. There is no obvious difference in the distribution of pottery within settlements, although very little research has been conducted on special distributions of material culture. Where work has been done there is no appreciable difference between sites (but for evidence of incipient variation in distribution see Peltenburg 1985).

The analysis confirms that there was a considerable degree of uniformity in the production of pottery. It is not just manufacturing techniques that were similar but also the behavioural processes of the potters themselves. Variation is observable at both the modelling and finishing stages but it is "ornamental variation" of the type described by Gosselain (1998, 82) not variation in behaviour. Therefore, the technological choices that comprise the behavioural systems for pottery manufacture are broadly similar across the island. Technological choices – such as collection of clay and temper from locales close to the site, decisions about morphology, detailed knowledge of how to achieve high burnishes, similarity in vessel use, re-use and discard, all indicate a single behavioural system. When this is combined with the similarity observable in other facets of material culture – such as the similarity in subsistence strategies (Peltenburg 1978, 1979; Croft 1991), settlement arrangements and built structures (Peltenburg 1978, 1985; Stanley Price 1979), similarities in other material items and discard patterns of artefacts and ecofacts, the evidence is overwhelmingly in support of a single behavioural system – or habitus for the Ceramic Neolithic population.

WAS CYPRUS A SINGLE HABITUS IN THE CERAMIC NEOLITHIC?

Returning to the question of group identity and habitus. If Cyprus was, at the very least, visited by mainland peoples on a sporadic basis throughout the Neolithic, why do we not see more examples of cultural influence from external sources in the archaeological record? My own feeling is that the Ceramic Neolithic population, while displaying internal differentiation in the form of stylistic variation in the painting schemes on pottery, maintained a sense of group identity, which manifested itself in the material culture as homogeneity, and symbolically as the construction of boundaries in socio-cultural space. This concept can be best described by a term coined by Anthony Cohen in regard to his work on the Scottish island of Whalsay - "the same but different" (Cohen 1978). Cohen describes this phenomenon as being characteristic of remote communities, whether their remoteness is geographical, cultural, or both, he says, "they have a keen sense of their own anomalousness, yet beneath the surface there is a powerful sense of self" (Cohen 1987, 16). This is not the forum to address whether or not prehistoric peoples actually had a sense of self, and some archaeologists would argue that they may not (Chapman 1996, 216), however, amongst social psychologists and anthropologists there is consensus that selfhood is not licensed by society or defined by others but is inherent in individuals (Tajfel, 1978, 1982; Wiessner 1989, 57). Therefore, to have a sense of group identity one must first have a sense of self. Ethnographic work by Cohen and others has looked at the way in which island populations negotiate their social and economic space with more imposing surrounding populations. The results suggest that island populations, whether real or perceived, maintain a sense of isolation from their more imposing neighbours in order to preserve their distinctive ways of life, in essence – their insularity.

This idea is not unknown in prehistoric island research. Recently Broodbank (2000, 19-21) has argued that "thinking of insularity as a cultural construct sheds intriguing light on some of the processes that commonly occur during the formation of and island society" (2000, 20). In particular he highlights the example of Malta noting that throughout most phases of the island's early settlement it maintained contacts with neighbouring areas. However, for unexplained reasons contact with the outside was arrested at the beginning of the Temple period. The most plausible explanation, Broodbank suggests, is closure, agreed upon or imposed, as presumably one element in a set of social strategies than enabled colossal monument building. Obviously each island scenario is different but Broodbank also highlights the insularity of the Aceramic phase in Cypriot prehistory commenting that the new data from the 10th and 9th millennium imply that the idiosyncrasies of the Khirokitia phase resulted from decisions made well after the inception of farming, and that the distinctive island identity that these idiosyncrasies reveal was forged at least in part via a rejection not only of mainland contacts but also of mainland traits, including some already introduced (Broodbank 2000, 21).

Whether the Neolithic Cypriots were cognisant of living on an island is uncertain but again, as Broodbank has noted, islands can be perceived in many different ways and do not necessarily denote landmasses surrounded by water. In fact, almost any circumscribed area is effectively a candidate (Broodbank 2000, 16). Certainly, in the example of Neolithic Cyprus there must have been a sense of being part of a geographic and cultural entity that was not the same as the geographical and cultural entities of others that presumably came via boat from long distances across the sea. Therefore, there must have been a sense of the "Other" and that this collective "Other" did not do things the same as the people who populated the geographical space of the Neolithic Cypriots.

ETHNOGRAPHIC PARALLELS

There are two ethnographic parallels I would like to mention here to illustrate my point. The first is the work of Anthony Cohen on group identity on the small Scottish Island of Whalsay off the coast of Shetland. The other is the work of Peter Mewett (1982) who undertook a similar study of group identity on the Hebridian Island of Lewis. Although both are now effectively islands of Scotland each has a totally separate historical and cultural background - the Whalsay peoples originating from Scandinavian colonisation whereas Celtic peoples populate Lewis. Leaving aside the problems associated with the use of ethnographic analogy in archaeological interpretation, I think the results of both studies are relevant in a general sense to the argument being made in this paper.

Cohen's research on Whalsay showed that the island was internally very prescriptive in the types of behaviour that could be expressed. Individuality was suppressed in favour of group identity and therefore individuality was manifested more specifically in symbolic ways. This maintenance of the integrity of a collective identity Cohen says is important for two reasons - the mere continuity of the society and its structures, and also to provide a front mask, which serves as a collective identity when the community engages with the outside world. Within the community itself he showed that the island was not a crucial referent of identity but rather a person is known by their lineage or clan. Further, Cohen noted that in Whalsay, the exigencies of communal life provided sound reasons for retraining the exhibition of one's individuality. This is where the concept of "the same but different too" was formed.

Peter Mewett's study looked at the village of Clachan on Lewis. His work showed that even though Clachan was only a short distance from other villages on the island, social constructions existed that set the village apart as different from neighbouring villages. He concluded that this was an example of socio-cultural processes being reified as territorial boundaries, symbolic of Clachan's perceived distinctiveness. However, the paradox here is that substantial links in the form of kin ties that connect Clachan with neighbouring villages ensure that the Clachanites are not distinct from these neighbours but rather it is that the Clachanites who wish to differentiate themselves from others while being essentially the same. In this case it is "different but the same too".

IS MODERN SCOTLAND RELEVANT TO ANCIENT CYPRUS?

Two points seem relevant to Neolithic Cyprus: 1) that geographically bounded societies will often portray an external image of homogeneity and cohesiveness to the outside world. Within an internal social system, this homogeneity exists to protect the social structures of the community while still allowing for individual expression through the use of highly stylised socio-cultural symbolism. As Wiessner (1989, 57) points out "in situations of extreme conformity, individuals strive to differentiate themselves from similar others". 2) It is the very boundedness of island societies that produces a unique set of social systems that work to preserve the society's culture from being eroded by outside influences.

I maintain that something similar to these phenomena was happening on Cyprus during the Ceramic Neolithic phase. The evidence suggests that there existed some peculiar sort of structured social interaction that produced very specific types of material culture patterning. The absence of any evidence for external influences on the other hand appears to be more socially prescribed rather than any real absence of contact. The limitations of ethnographic analogy are well known and although Cohen's work provides us with information about visual symbols it tells us very little about how these symbols might be manifested in material culture. I would advocate similar studies of modern island societies but from an ethno-archaeological viewpoint that will provide us with information about whether identity is actually observable in material culture remains and whether we can assign it meaning.

BIBLIOGRAPHY

ARNOLD, D.E.
1980. Localized exchange: an ethnoarchaeological perspective. Pp. 147-50 in R.E. Fry (ed.), Models and Methods in Regional Exchange. SAA Papers 1, Washington D.C., Society for American Archaeology.
ARNOLD, D.E.
1985. Ceramic Theory and Cultural Process. Cambridge, Cambridge University Press.
BARTH, F.
2000. Boundaries and connections. Pp. 15-36 in A. P. Cohen, Signifying Identities: Anthropological Perspectives on Boundaries and Contested Values, New York, Routledge.
BENTLEY, G.C.
1987. Ethnicity and practice. Comparative Studies in Society and History 29, 24-55.

BOURDIEU, P.
1977. Outline of a Theory of Practice. Cambridge. Cambridge University Press.
BROODBANK, C.
1999. The insularity of island archaeologists: comments on Rainbird's islands out of time. Journal of Mediterranean Archaeology 12.2, 235-239.
BROODBANK, C.
2000. An Island Archaeology of the Early Cyclades. Cambridge, Cambridge University Press.
BROODBANK, C., AND STRASSER, T.F.
1990. Migrant farmers and the colonization of Crete. Antiquity 65, 233-245.
CHAPMAN, J.
1996. Enchainment, commodification, and gender in the Balkan Copper Age. Journal of European Archaeology 4, 203-242.
CHILTON, ELIZABETH
1999. Material meanings and meaningful materials: an introduction. Pp. 1-6 in Elizabeth S. Chilton (ed.), Material Meanings: Critical Approaches to the Interpretation of Material Culture, Salt Lake City, University of Utah Press.
CLARKE, J.T.
2001. Style and society in Ceramic Neolithic Cyprus. Levant 33, 65-80
COHEN, A.P.
1994. Self Consciousness: an Alternative Anthropology of Identity. New York, Routledge.
COHEN, A.P.
1978. 'The same – but different': the allocation of identity in Whalsay, Shetland. Sociological Review 26, 449-469.
COHEN, A.P.
1987. Whalsay, Symbol, Segment and Boundary in a Shetland Island Community. Manchester, Manchester University Press.
CROFT, P.
1991. Man and Beast in Chalcolithic Cyprus. Bulletin of the American Schools of Oriental Research 282/283, 63-79.
DIETLER, M., AND HERBICH, INGRID.
1998. Habitus, techniques, style: an integrated approach to the social understanding of material culture and boundaries, in Miriam T. Stark (ed.) The Archaeology of Social Boundaries, Washington, Smithsonian Institution Press.
DOBRES, M-A., AND HOFFMAN, C.R.
1999. The Social Dynamics of Technology: Practice, Politics, and World Views. Washington and London, Smithsonian Institution Press.
ELLEMERS, N., SPEARS, R., AND DOOSJE, B. 1999. Social Identity: Context, Commitment, Content. Oxford, Blackwell Publishers.
FRANKEL, D. AND WEBB, J.M.
1998. Three faces of identity: ethnicity, community and status in the Cypriot Bronze Age. Mediterranean Archaeology 11, 1-12.
FRANKEL, D.
2000. Migration and ethnicity in prehistoric Cyprus: technology and habitus. European Journal of Archaeology 3/2, 167-187.
GOSSELAIN, O.P.
1998. Social and technical identity in a clay crystal ball. Pp. 78-106 in Miriam T. Stark (ed.), The Archaeology of Social Boundaries, Washington, Smithsonian Institution Press.
HEGMON, MICHELLE.
1998. Technology, style and social practices: archaeological approaches. Pp. 264-280 in Miriam T. Stark (ed.), The Archaeology of Social Boundaries, Washington, Smithsonian Institution Press.
JENKINS, R.
1992. Social Identity. London, Routledge.
JONES, S.
1997. The Archaeology of Ethnicity. London, Routledge.
LAMOTTA, V.M. AND SCHIFFER, M.B.
2001. Behavioral archaeology: toward a new synthesis. Pp. 14-64 in I. Hodder (ed.), Archaeological Theory Today. Cambridge, Polity Press.
MCCARTNEY, C., AND PELTENBURG, E.
2000. The colonization of Cyprus: questions of origins and isolation. Neolithics 1/00, 8-11.
MEWETT, P.
1982. Exiles, nicknames, social identities and the production of local consciousness in a Lewis crofting community. Pp. 222-46 in A.P. Cohen (ed.), Belonging: Identity and Social Organisation in British Rural Cultures. Manchester, Manchester University Press.
PELTENBURG, E.J.
1978 . The Sotira Culture: Regional Diversity and Cultural Unity in Late Neolithic Cyprus. Levant 10, 55-74.
PELTENBURG E.J.
1979. Troulli Reconsidered. Pp. 21-45 in V. Karageorghis (ed.) Studies Presented in Memory of Porphyrios Dikaios. Lions Club: Nicosia.
PELTENBURG, E.
1982. Recent developments in the later prehistory of Cyprus', Paul ≈str^m (ed), Studies in Mediterranean Archaeology, Pocketbook 16. Paul ≈str^m's F^rlag, G^teborg.
PELTENBURG, E.J.
1982. Vrysi: a Subterranean Settlement in Cyprus: Excavations at Prehistoric Ayios-Epikititos Vrysi, 1969-1973, Warminster.
PELTENBURG, E.J.
1985. Pattern and Purpose in the Prehistoric Cypriot Village of Ayios Epiktitos Vrysi. Pp. 46-64 in Chypre, La Vie Quotidienne de l'Antiquité à nos jours, Actes du Colloque. Musée de l'Homme: Paris.
PELTENBURG, E., COLLEDGE, S., CROFT, P., JACKSON, A., MCCARTNEY, C., AND MURRAY, M-A.
2000. Agro-pastoralist colonization of Cyprus in the 10th millennium BP: initial assessments. Antiquity 74, 844-53.
PELTENBURG, E., COLLEDGE, S., CROFT, P., JACKSON, A., MCCARTNEY, C., AND MURRAY, M-A.
2001. Neolithic dispersals from the Levantine Corridor: a Mediterranean perspective. Levant 33, 35-64.
SACKETT, J.R.
1985. Style and ethnicity in the Kalahari: a reply to Wiessner, American Antiquity 50, 154-159.
SHIELS, J.N.
1993. A Fabric Analysis of Prehistoric Cypriot Pottery Wares. Unpublished undergraduate dissertation: University of Edinburgh.
SHIELS, J.N.
(F.C.). The experimental replication of Chalcolithic pottery. Forthcoming in E. Peltenburg (et al.) Excavations at Kissonerga-Mylouthkia.
SIMMONS, A.H.
1998. Of Tiny Hippos, Large Cows and Early Colonists in Cyprus. Journal of Mediterranean Archaeology 11.2, 232-241.

STANLEY PRICE, N.P.
1977. Colonization and Continuity in the Early Prehistory of Cyprus. World Archaeology 9/1, 27-4.
STARK, MIRIAM, T.
1998. Technical choices and social boundaries in material culture patterning: an introduction. Pp. 1-11 in Miriam T. Stark (ed.), The Archaeology of Social Boundaries, Washington, Smithsonian Institution Press.
STARK, MIRIAM, T.
1999. Social dimensions of technical choice in Kalinga ceramic traditions. Pp. 24-43 in E.S. Chilton (ed.), Material Meanings: Critical Approaches to the Interpretation of Material Culture, Salt Lake City, University of Utah Press.
TAJFEL, H.
1978. Differentiation Between Social Groups. New York, Academic Press.
TAJFEL, H.
1982. Introduction. Pp. 1-11 in H. Tajfel (ed.), Social Identity and Intergroup Relations. New York, Academic Press.
TURNER, J.C.
1999. Some current issues in research on social identity and self-categorization theories. Pp 6-34 in N. Ellemers, R. Spears and B. Doosje, Social Identity: Context, Commitment, Content, Oxford, Blackwell Publishers.
VIGNE, J-D., CARRÈRE, I., SALIÉGE, J-F., PERSON, A., BOCHERENS, H., GUILAINE, J., AND BRIOIS, J.F.
2000. Predomestic cattle, sheep, goat and pig during the late 9th and 8th millennium cal. BC on Cyprus: preliminary results of Shillourokambos (Parekklisha, Limassol). Pp. 83-106 in M. Mashkour, A.M. Choyke, H. Buitehuis and F. Poplin (eds) Archaeolozoology of the Near East IVA. Proceedings of the Fourth International Symposium on the Archaeozoology of Southwestern Asia and Adjacent Areas. ARC Publicatie 32 Groningen, The Netherlands.
WIESSNER, P.
1983. Style and social information in Kalahari San Projectile Points, American Antiquity 48, 253-276.
WIESSNER, P.
1985. Style or isochrestic variation: A reply to Sackett, American Antiquity 50/1, 160-166.
WIESSNER, P.
1989. Style and changing relations between the individual and society. Pp. 56-63 in I. Hodder (ed.), The Meaning of Things: Material Culture and Symbolic Expression, London, Unwin Hyman.
WOBST, M.H.
1999. Style in archaeology or archaeologists in style, in Elizabeth S. Chilton (ed.), Material Meanings: Critical Approaches to the Interpretation of Material Culture, Salt Lake City, University of Utah Press.
WOODWARD, K.
1997. Concepts of identity and difference. Pp. 7-120, in K. Woodward (ed.), Identity and Difference, Open University Milton Keynes, Sage Publications.

50

INDICATIONS OF A POSSIBLE ASTRONOMICAL ORIENTATION OF AN ARCHITECTURAL FEATURE IN THE PREHISTORIC SANCTUARY OF SON MAS, MALLORCA.

M. VAN STRYDONCK (1), **W. WALDREN** (2) **M. HOSKIN** (3) **J. ENSENYAT ALCOVER, J. ORVAY** (4)

(1) Royal Nuseum of Cultural Heritage, Jubelpark 1, B-1000, Brussels, Belgium
(2) Baden Powell Quaternary Research Centre, Pitt Rivers Museum, Unioversity of Oxford, Oxford, UK
(3) Churchill College, University of Cambridge, Cambridge, UK
(4) Deia Archaeological Museu m Research Centre, Deia, Mallorca, Baleares, Spain
EMAIL william. waldren@prm.ox.ac.uk

INTRODUCTION

Radiocarbon dating combined with astronomical considerations and archaeological excavations have shown that in the Prehistoric Sanctuary of Son Mas on the island of Mallorca a possible sighting stone is orientated towards a V-shaped saddle between two mountain peaks where the Southern Cross was visible until ca. 1700 BC. This is the oldest indication of the possible astronomical orientation of a Chalcolithic-Initial Bronze Age (Pretalayotic) in this end of the Western Mediterranean. The age, importance and function of the site are supported by the large quantities of Bell Beaker pottery and skeletal remains of young domestic animals slaughtered in the last half of the first and second year, strongly suggesting extensive ritual or religious activities. The overwhelming majority of the intricate, geometrically, decorated Beaker pottery are highly indicative of solar and lunar motifs, further supporting the idea of astral religious activity at the site (Waldren 1995).

The Son Mas sanctuary site is located near the present-day village of Valldemosa (Figure 1) in a mountain basin known as the Pla del Rei (Plain of the king) (39°45'N, 6°66'E). The valley is surrounded by a mountain range on three sides, except to the north, where there is a 400m-high cliff that overlooks and drops steeply to the sea. In and around this fertile flat-land, more than 25 archaeological sites are located. Besides Son Mas, there are two of these numerous sites that are especially pertinent to this report: the Prehistoric Settlement Complex of Son Ferrandell-Oleza and the Rock Shelter of Son Matge.

Discovered in 1972, the Son Ferrandell-Oleza complex consists of a Chalcolithic-Initial Bronze Age (Pretalayotic), open-air settlement, known as the Chalcolithic Old Settlement (SFO-OS) (Waldren et al. 1994), and a Bronze and Iron Age (Talayotic-Post Talayotic) linear settlement, consisting of five talayots (tower-like structures from which the local culture gets its name) and other associated structures, known as the Son FerrandellYounger Settlement (SFO-YS) (Chapman et al.1993; Stuiver and Waldren 1974; Waldren and Van Strydonck 1993). The Chalcolithic Old Settlement has an exceptional chronological continuum from c. 2500 calBC until c. 1300 calBC, after which the site is abandoned and activity begins at the younger settlement from c 1300 calBC to 300 AD, as shown by an continuous series of 55 radiocarbon dates (TABLES 1-3) .

Discovered in 1968, the Rock Shelter of Son Matge is located at the mouth of a narrow mountain pass connecting the Pla del Rei mountain basin with the Central Plain of the island. In Matge another large series of more than 50 radiocarbon dates have shown that the shelter was used throughout the entire prehistory of the island from c. 5800 BC onwards, first as a corral, later as an habitation site and finally

FIGURE 1. LOCATION OF THE RESEARCHSITES IN THE PLA DEL REI, MALLORCA.

as a cemetery (Waldren 1982; Van Strydonck and Waldren 1990, 1995).

The Prehistoric Sanctuary of Son Mas was discovered in 1987 and has been under investigation since then. The main features at this site are a horseshoe or apsidal-shaped sanctuary with a concave frontal aspect (ca. 13 x 13 m), an apsidal naviform dwelling, and a large free-standing boulder (175 cm high by a 100 cm in diameter). This stone (Figure 2) has a worn circular groove that may be artificial. The groove looks upwards at an angle of about 13° to the horizontal, and points in a southern direction towards a V-shaped saddle in the mountains. It is the only stone of that size found on the site. In the area to the east of this stone and in front of the (concave) sanctuary precinct and stands isolated in the field , large quantities of beaker pottery have been found (Figure 3). The stone is in its original location and state, as it was prior to the excavation.

The basic feature of the site consists of large areas of bedrock that surface in the interior of the existing sanctuary. Although the bedrock is composed of a loose conglomerated limestone that is easy to work, no attempt was made by the builders of the existing sanctuary to level the inside of the precinct, and the structure we see has been built around the protruding bedrock. This has been interpreted (Waldren, 1996) as indicating that this was a significant feature of an earlier sanctuary that was left unmodified by the builders of the Late Bronze-Iron Age sanctuary that we see today.

THE CHRONOLOGY OF THE PREHISTORIC SANCTUARY OF SON MAS (SMSS)

On the radiocarbon inventory of the site.

Here, as in the other sites, the site has undergone extensive radiocarbon dating and more than 53 radiocarbon dates have been made on bone (mainly animal) and charcoal. Figure 3 shows the Son Mas site with indications of the different architectural features and archaeological areas and the locations of the radiocarbon samples as well as the Bell Beaker pottery finds. Incorporated in this figure are also the median dates from the radiocarbon analyses. The detailed list of these analyses has been published elsewhere (Van Strydonck et al. 1998 and Van Strydonck, Waldren and Hoskin, in press). Figure 4 represents the summed probabilities of all the radiocarbon dates of the site. The graph shows that the site was first used in the latter part of the 3rd millennium BC, a date that is somewhat later than the corresponding dates of the Son Ferrandell-Oleza Old Settlement and the Rock Shelter of Son Matge.

The chronology of the site has been discussed extensively elsewhere (Waldren and Van Strydonck 1992; Waldren 1996a nd 1997; Van Strydonck et al. 1998). However, some further discussion is needed to provide a context for the topic that presently concerns us, that of the site's possible astromonical orientation and evidence supporting it.

At Son Mas, excavations have taken place on a terrain that has been heavily eroded by annual ploughing and otherwise since the Middle Ages and that has undergone natural weathering during that time. This has led in most cases to the re-depositioning of soil and artefacts (and especially the earliest artefacts) that mainly have found their way into deep bedrock crevices all over the site. While this has been beneficial for the preservation of such artefacts and the like from severe surface deterioration by differential preservation, it has led to other distributional difficulties (Waldren 1997). This is illustrated by a set of six paired charcoal and bone samples. The differences obtained from the radiocarbon dating of the paired samples was 80±100 14C years. Furthermore, site-activity zones both inside and outside the

FIGURE 2: THE SIGHTING STONE AT THE SON MAS SANCTUARY SITE.

sanctuary precinct and other buildings were periodically cleaned out in prehistoric times and the materials redistributed by later occupants. At different stages in the site's history certain architectural elements were altered and reused. This indicates that some of the samples have varying histories of redistribution. On the other hand, the redistributed charcoal and bone artefacts are anthropogenic and not the result of natural fires or modern field clearance; therefore, although they cannot be attributed in every case to a specific cultural phase, other than when they are directly associated with artefracts of known age and typology, they still can be used and shown in a dispersion diagram that represents the total chronology of the site (eg. as with the Beaker pottery distribution).

THE QUESTION OF THE CHRONOMETRIC HIATUS.

A hiatus (Figure 5) of circa half a millennium (400-600 years at 68.2% confidence) exists in the radiocarbon inventory. This gap in the otherwise unbroken chronology of the site is puzzling, but belief explicable when applicable evidence is considered. At first it was assumed that the hiatus appearing in the middle of the second millennium BC was the result of a lack of radiocarbon results from the appropriate archaeological horizons. But the gap was very consistent, and today, after more than 10 years of extensive fieldwork, the hiatus is still present, so affirming itself as something other than the result of insufficient sample collection.

CONCERNING THE AGE OF THE ARCHITECTURAL FEATURES.

The possible renovation and reconstruction of a section of wall of the existing, horseshoe-shaped sanctuary, during the local Late Bronze Age, was radiocarbon dated by means of both charcoal and bone from inside and under the west wall of the monument at c 800 calBC (Van Strydonck et al. 1998). The alteration date of this section of the sanctauery wall is marked by the very sharp peak in Figure 4. However, several dates demonstrate the extensive use of the site in a period from 1300 BC onwards to c 2200 cal BC, which strongly suggests that the present sanctuary was probably not the first building or area of activities on the site.

To the southeast of the sanctuary precinct proper, there is evidence of a large stone building located in the so-called Agora area, which is younger and its construction and which can be dated to circa the 5th century BC. In and around this building, extensive evidence of a local production of lead-cast ornaments was encountered and although no actual moulds were found, a plethora of casting sprews, ingots and miscast pieces, as well as completed items show that considerable lead casting was carried out on site (Waldren and Grimalt Cubì 1995). It is interesting to note that a number of the completed lead pectorals exactly match those encountered in at least three known cemeteries, scattered over the island (Cova de'n Alova, Muertos Gallard, Son Matge) where they served as personal belongings or offerings as funerary paraphenalia.

obvious position and seeming importance and early artefatcs surrounding it may well have been set up much earlier, in Chalcolithic (Pretalayotic) times. It is actually firmly set on bedrock and the red earth deposits diagnostic of that earlier period and has a shape and character totally unlike the highly worked (shaped and seated) stones of the exist-

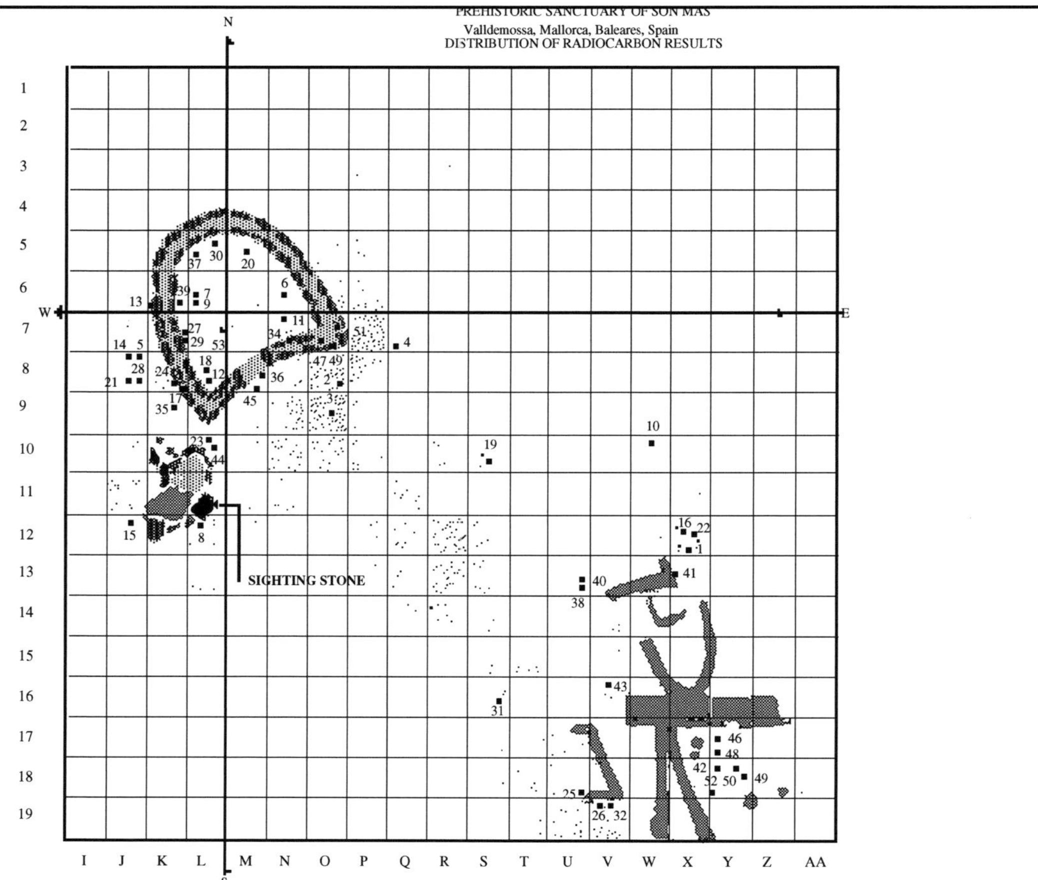

CALIBRATED RADIOCARBON DATES AS OF 1992-2001

1.	2170 BC	10.	1167 BC	19.	835 BC	28.	786 BC	37.	616 BC	46.	278 BC
2.	1926 BC	11.	1157 BC	20.	832 BC	29.	786 BC	38.	613 BC	47.	274 BC
3.	1923 BC	12.	1086 BC	21.	814 BC	30.	716 BC	39.	477 BC	48.	271 BC
4.	1816 BC	13.	1040 BC	22.	799 BC	31.	691 BC	40.	467 BC	49.	259 BC
5.	1325 BC	14.	1026 BC	23.	794 BC	32.	666 BC	41.	465 BC	50.	126 BC
6.	1256 BC	15.	979 BC	24.	788 BC	33.	645 BC	42.	464 BC	51.	62 BC
7.	1195 BC	16.	915 BC	25.	788 BC	34.	629 BC	43.	458 BC	52.	46 AD
8.	1172 BC	17.	872 BC	26.	787 BC	35.	629 BC	44.	454 BC	53.	1420 AD
9.	1169 BC	18.	852 BC	27.	786 BC	36.	617 BC	45.	282 BC	54.	

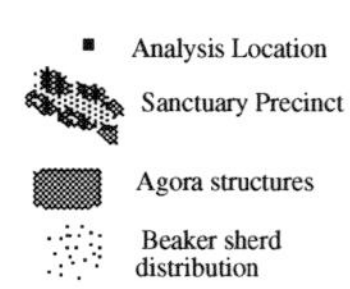

FIGURE 3: LOCATION OF THE SIGHTING STONE IN RELATION TO THE HE SANCTUARY PRECINCT, THE APSIDAL NAVIFORM DWELLING, THE LOCATION OF THE RADIOCARBON SAMPLES AND THE DISTRIBUTION OF THE BELL BEAKER POTTERY FINDS.

The erection of the large 'sighting stone' approximately 20 meters to the south of the concave entrance of the sanctuary edcited curiousity because of its obvious isolation was eventually radiocarbon dated to between 1310 and 1050 cal BC (at 1F range) from animal bone found near the base of this stone. This erection date however is by no means certain other than by association and the stone because of its

ing later Bronze Age sanctuary. It has the unfinished and natural appearance of the interior protruding bedrock and other extemely large stones forming the apsidal rear of the sanctuary. Since the first evaluations and despite of the late dates of the samples selected at the base of the sighting stone, it is now evident that its age and function has a much earlier origin.

THE ARCHAEOASTRONOMICAL INVESTIGATION

During a field trip in 1997, one of the authors (Hoskin) (who was quite unaware of the hiatus in the radiocarbon record at the Son Mas Sanctuary) was invited to investigate whether or not in prehistoric times there might have been an astronomical purpose served by the site similar to that of another Minorcan Taula Sanctuary. He had investigated a similar situation in collaboration with another of the present authors (Waldren) to see as was demonstrated by Hoskin in the Taula Sanctuary of Torralba on Menorca whether or not there may have been some dominant celestial feature in Son Mas in astromomical alignment trally. In the present situation in the direction indicated by the sighting stone of Son Mas. It transpired that the sighting stone faced towards a valley to the south, and that celestial bodies at a suitable elevation would have emerged from behind the saddle between two mountains and framed in the saddle, before eventually passing from sight behind the ridge of the saddle. It was found that at the suggested dates a particularly brilliant spectacle would have been afforded by the Southern Cross, which would have been visible throughout the whole of the third millennium BC. However, eventually, because of the long-term wobble of the earth's axis ("precession") caused by the gravitational pull of the sun and the moon on the earth, the Southern Cross would have been seen lower and lower in the sky as the centuries passed, until around 1700 BC the lowest star of the Cross would have disappeared from view, to be followed in due course by the rest of the Cross. It has been shown with high probability that the hiatus demonstrated by the radiocarbon dating sequence coincides with this theory.

DISCUSSION

In the light of the evidence equally archaeological as well as astronomical and chronometric that a number of interesting factors have emerged that affect the interpretation of the site and support current theory as to the function of the Balearic sanctuaries as astromomical, ritual and probably religious in purpose.

1 -On the basis of 14C-dating of the various Chalcolithic- Initial Bronge Age (Pretalayotic) levels and the abnormally large quantities of Beaker pottery and other contextual evidence, directly associated with them, supports the beginning of the hiatus at 3510±60 BP (1920-1740 calBC) (one sigma confidence range). This date predates, only slightly, the time that the lowest star of the Southern Cross would have been visible above the skyline of the mountains, before it disappeared behind the saddle in the hills, circa 1700 BC.

2-A second charcoal sample dated 3060±35 [1390-1260 calBC (one sigma confidence range)] marks the end of the hiatus and the ev entual reoccupation of the site by a later culture several hundred yeras later . This date coincides with what is traditionally accepted as the transition from the local Pretalayotic to the Talayotic period.

3-The artefacts and 14C-inventory of 44 dates from the nearby Son Ferrandell-Oleza Old Settlement show an unbroken chronological continuum from c 2500 BC until c 1300 BC, after which time the chronometric sequence is taken over by another unbroken continuum spanning the Bronze and Iron Age Younger Settlement and is represented by artefacts and another dating serious of over 30 dates between 1300 BC and the 3rd century AD. This transition from the old to the younger settlement, occuring around 1300 BC, is in total agreement with the traditional chronology of the island. This in turn implies as well as demonstrates that the site was never abandoned, except perhaps during a very brief period or transitional stage untraceable by radiocarbon dating.

4-In the Rock Shelter of Son Matge, which has a representative extensive intersite radiocarbon inventory and vertical stratigraphical sequence of some 34 levels, there is also no hiatus in the inventory of approximately 64 dates. Although at Matge, there is at least two significant cultural and burial changes in the stratigraphy, according to the 14C dates. Again the first of the cultural changes appear, as they do in the other sites reviewed here, at the transitional stage between the local Pretalayotic and the Talayotic period. The second of these cultural changes appear in the transitional stage between the Bronze and Iron Age (between the Talayotic and Post Talayotic Periods) . This latter transition is marked by the introduction and increasing use of iron and the employment of markedly different burial techniques (Stuiver and Waldren 1974, Waldren 1982 and Van Strydonck and Waldren, 1990).

5- The high percentage (6%) of symbolically decorated Bell Beaker pottery, present on the Son Mas sanctuary site and accreditable to c 2200 BC to c 1700 BC (Waldren, 1997) and additional (11%) fine wares in the artefact repertoire along with the high percentage (over 80%) of bones from young animals, strongly suggests that the site had a ritual and religious function as well as astronomical one in Chalcolithic (Pretalayotic) times. Although usually dated to a younger Bronze- Iron Age (local Talayotic and Post-Talayotic) period, the construction of the horseshoe-shaped sanctuary and other factors, such as the presence of the lead ornament production on the site during the later Iron

Age is still further indication that the site's religious and ritual character persisted a remarkably long time throughout prehistory, through to the second or third centuries AD. A period lasting well over 2500 years, spanning the largest part of the local prehistoric cultural sequence

From the combined evidence, we see that these chronological data clearly indicate that the abandonment of Son Mas in the 18th century BC is not related nor coincides to an abandonment of any other site in the plain, and that the reuse of the site coincides, despite the hiatus at Son Mas, with a cultural change at both Son Matge and Son Ferrandell-Oleza that is also reflected in the Son Mas sequence at a different level. Furthermore, there is a striking synchronism between the abandonment of the Son Mas site and the disappearance of the Southern Cross beneath the skyline of the mountains. One which may even have caused a cultural as well as religious crisis. Similarily, both factors as well as evidense strongly suggest that the use of the site and the presence of the sighting stone is somehow related, even depended on the appearance of the Southern Cross. Also, strongly implying that the site already had a religious function and astronomical importance in earlier Chalcolithic- Intitial Bronze Age (Pretalayotic) times. Most of the artefacts, including the rich earlier Beaker pottery as well as later wealth of the artefact, pottery and metal during later Brfonze and Iron Age (certainly equally religiously related objects) give direct archaeological proof for this claim of the special importance and function of the site. Furthermore this has been , equally, demonstrated on the culturally related island of Menorca that the Southern Cross played an important role in the orientation of the local Talayotic Taula sanctuaries (Hoskin 1991, Hoskin 2001), although, so far, the Taula sanctuaries are dated well after 1700 BC, but similarly, strongly and clearly indicates the important role of this constellation in the prehistoric religion and thinking of the Balearic islands.

As far as we can see, there are only two possible objections to the proposition of the function of these sanctuary sites. The first derives from the discrepancy between the time span of the hiatus and the radiocarbon date of the bone found at the base of the sighting stone. In our opinion, while there is little chronological relation between the single date given by the bone and the possible original erection of the sighting stone, and the presence of the dated artefact is no more than a coincidence and the result of latter redistribution. A possibility not to be taken lightly, considering the agricultural activity in and around the sighting stone over the interim millennia. A second much more serious problem as well as criticism lies in the prehistorical chronology proposed in recent years by some archaeologists. Guerrero (Guerrero, 1998) for Mallorca and Lull (Lull et al.

RADIOCARBON DATES: SON OLEZA CHALCOLITHIC OLD SETTLEMENT and DOLMEN AREA 2002

nos	site designation	years BP	95.4 probability range	date cal BC	lab.inv.no
1.	**SFO-OS-CH1**	2990 ± 160 yrs	1610- 820 (3559-2769)	***1261 cal BC***	QL-4044
2.	**SFO-OS-CH1**	2990 ± 160 yrs	1610- 820 (3559-2769)	***1261 cal BC***	IRPA-1053
3.	**SFO-OS-EXW**	3070 ± 50 yrs	1440-1250 (3389-3199)	***1357 cal BC***	HAR-3490
4.	**SFO-OS-C2**	3095 ± 70 yrs	1520-1250 (3469-3199)	***1403 cal BC***	BM-1698
5.	**SFO-OS-WC**	3150 ± 300 yrs	2192-2161 (4141-4110)	***1428 cal BC***	BM-1988
6.	**SFO-OS-SWF-A**	3180 ± 80 yrs	1670-1668 (3619-3617)	***1445 cal BC***	QL-4191
7.	**SFO-WOS-DOL1: 17-2M8 EX**	3130 ± 35 yrs	1520-1300 (3420-3250)	***1450 cal BC***	KIA-12274
8.	**SFO-OS-WC**	3230 ± 30 yrs	1670-1410 (3619-3359)	***1516 cal BC***	BM-1567
9.	**SFO-OS-CSS**	3240 ± 30 yrs	1611-1441 (3560-3390)	***1519 cal BC***	QL-1224
10.	**SFO-OS-C2**	3240 ± 30 yrs	1611-1441 (3560-3390)	***1519 cal BC***	IRPA-1054
11.	**SFO-OS-EXW-UW**	3260 ± 100 yrs	1854-1849 (3803-3798)	***1523 cal BC***	QL-4040
12.	**SFO-OS-WC-IN**	3280 ± 120 yrs	1880-1310 (3829-3259)	***1549 cal BC***	QL-1896
13.	**SFO-OS-H4**	3300 ± 60 yrs	1738-1708 (3687-3657)	***1565 cal BC***	UtC-3192
14.	**SFO-OS-WC**	3350 ± 330 yrs	2569-2538 (4518-4487)	***1673 cal BC***	BM-1988R
15.	**SFO-OS-WC**	3390 ± 80 yrs	1884-1682 (3833-3631)	***1712 cal BC***	BM-2312R
16.	**SFO-WOS-DOL1: 16-2M4**	3445 ± 40 yrs	1880-1610 (3830-3660)	***1745 cal BC***	KIA-12275
17.	**SFO-OS- EW-INT-B1**	3460 ± 45 yrs	2115-2085 (4064-4034)	***1745 cal BC***	UtC-9021
18.	**SFO-WOS-DOL3: 15-EEP**	3450 ± 30 yrs	1880-1680 (3880-3630)	***1760 cal BC***	KIA-12273
19.	**SFO-OS-CH1-INT4**	3450 ± 110 yrs	1878-1839 (3827-3788)	***1749 cal BC***	QL-4043
20.	**SFO-WOS-DOL1: 12- 4I3**	3440 ± 30 yrs	1880-1640 (3830-3590)	***1770 cal BC***	KIA-12267
21.	**SFO-WOS-DOL1: 11- 5L7**	3440 ± 35 yrs	1877-1840 (3826-3789)	***1770 cal BC***	KIA-10559
22.	**SFO-WOS-DOL1: 9- 5L8**	3445 ± 30 yrs	1930-1680 (3879-3629)	***1770 cal BC***	KIA-10558
23.	**SFO-OS-WCS**	3430 ± 30 yrs	1880-1630 (3830-3586)	***1755 cal BC***	KIA-16267
24.	**SFO-OS-SW-AG**	3405 ± 35yrs	1880-1600 (3830-3556)	***1740 cal BC***	KIA-16268
25.	**SFO-WOS-DOL3: 14-EWEX**	3445 ± 30 yrs	1880-1680 (3850-3590)	***1760 cal BC***	KIA-12270
26.	**SFO-WOS-DOL3: 13- NEX**	3460 ± 30 yrs	1880-1680 (3850-3590)	***1780 cal BC***	KIA-12268
27.	**SFO-OS-EW-UF-S**	3470 ± 50 yrs	1901-1742 (3850-3691)	***1819 cal BC***	QL-4100
28.	**SFO-OS-AT-QB**	3490 ± 30 yrs	1901-1742 (3850-3691)	***1827 cal BC***	QL-1859
29.	**SFO-OS-CH1-INF3**	3490 ± 30 yrs	1946-1768 (3895-3717)	***1829 cal BC***	QL-4042
30.	**SFO-WOS-DOL1: 2- 5M3**	3540 ± 30 yrs	2198-2159 (4147-4108)	***1860 cal BC***	KIA-10557
31.	**SFO-OS-EWF**	3550 ± 110 yrs	2007-2004 (3956-3953)	***1884 cal BC***	UtC-2742
32.	**SFO-WOS-DOL1: 12-5L7**	3550 ± 30 yrs	2030-1988 (3979-3937)	***1884 cal BC***	KIA-10560
33.	**SFO-WOS-DOL1: 1**	3565 ± 35 yrs	2030-1770 (3980-3720)	***1925 cal BC***	KIA-1228
34.	**SFO-OS-EXW**	3585 ± 35 yrs	2040-1880 (3990-3930)	***1929 cal BC***	UtC-8952
35.	**SFO-WOS-DOL1: 10-6M4**	3605 ± 35 yrs	2137-2076 (4086-4025)	***1970 cal BC***	UtC-10079
36.	**SFO-OS-EW-INT-B2**	3620 ± 50 yrs	2123-2097 (4072-4046)	***1985 cal BC***	UtC-9022
37.	**SFO-WOS-DOL1: 4-5M9**	3625 ± 30 yrs	2123-2097 (4072-4046)	***1995 cal BC***	KIA-10583
38.	**SFO-WOS-DOL1: 7-5M6**	3625 ± 30 yrs	2279-2231 (4229-4180)	***1995 cal BC***	KIA-10582
39.	**SFO-WOS-DOL1 J3-G13**	3640 ± 50 yrs	2199-2158 (4148-4107)	***2030 cal BC***	BM-1981R
40.	**SFO-OS-EW- INT-CC**	3680 ± 45 yrs	2199-2158 (4148-4107)	***2074 cal BC***	UtC-9023
41.	**SFO-WOS-DOL1: J9 W17**	3700 ± 30 yrs	2290-1930 (4240-3880)	***2110 cal BC***	QL-1592
42.	**SFO-WOS-DOL1: G13 FLT**	3720 ± 35 yrs	2275-2141 (4224-4198)	***2138 cal BC***	BM-1981
43.	**SFO-OS-NW-BB**	3790 ± 90 yrs	2480-1970 (4429-3919)	***2239 cal BC***	QL-1636
44.	**SFO-OS-EW-BB**	4030 ± 60 yrs	2585-2568 (4537-4512)	***2533 cal BC***	BM-1843R

TABLE 2

RADIOCARBON DATES: ROCK SHELTER OF SON MATGE
Valldemossa, Mallorca, Baleares, Spain
2002

nos.	site designation	date cal BC	date BP	lab. inv. no.
1.	ABSMPT-L	244 cal AD	1706 BP± 30 yrs	IRPA-710L
2.	ABSM-PT-L	25 cal BC	1925 BP± 120 yrs	IRPA-710
3	ABSM-PT-L	101 cal BC	2050 BP± 65 yrs	QL-7a
4	ABSM-PT-L15	196 cal BC	2145 BP± 65 yrs	IRPA-695L
5	ABSM-PT-L	266 cal BC	2215 BP± 65 yrs	IRPA-778L
6	ABSM-LIA-T	278 cal BC	2215 BP± 100 yrs	QL-9
7	ABSM-PT-L	379 cal BC	2244 BP± 60 yrs	QL-22
8	ABSM-PT-GT	379 cal BC	2328 BP± 190 yrs	BM-1409
9	ABSM-PT-L	385 cal BC	2328 BP± 65 yrs	IRPA-789L
10	ABSM-PT-L	391 cal BC	2336 BP± 100 yrs	QL-5c
11.	ABSM-PT-L30	400 cal BC	2340 BP± 55yrs	IRPA-808L
12.	ABSM-PT-L	400 cal BC	2349 BP± 55 yrs	IRPA-710
13.	ABSM-PT-L	400 cal BC	2349 BP± 55 yrs	IRPA-789
14.	ABSM-PT-L	402 cal BC	2349 BP± 140 yrs	UtC-1003
15.	ABSM-PT-L	402 cal BC	2351 BP± 60 yrs	IRPA-803L
16	ABSM-PT-L45	408 cal BC	2352 BP± 55 yrs	IRPA-776
17.	ABSM-PT-L65	408 cal BC	2358 BP± 80 yrs	Y-2669
18.	ABSM-PT-L45	641 cal BC	2590 BP± 40 yrs	IRPA-790L
19.	ABSM-PT-L15	651 cal BC	2595 BP± 60 yrs	QL-10a
20.	ABSM-PT-L1	651 cal BC	2600 BP± 70 yrs	QL-10b
21.	ABSM-PT-MIA	655 cal BC	2604 BP± 70 yrs	QL-10
22.	ABSM-PT-L	687 cal BC	2635 BP± 60 yrs	IRPA-751
23.	ABSM-PT-L	770 cal BC	2719 BP± 80 yrs	QL-6
24.	ABSM-PT-MIA	779 cal BC	2728 BP± 80 yrs	QL-4
25.	ABSM-PT-EIA	779 cal BC	2728 BP± 80 yrs	QL-24
26	ABSM-PT-MIA	779 cal BC	2728 BP± 60 yrs	IRPA-752
27.	ABSM-PT-EIA	786 cal BC	2735 BP± 60 yrs	IRPA-676
28.	ABSM-PT-EIA	786 cal BC	2735 BP± 60 yrs	IRPA-751
29.	ABSM-PT-EIA	789 cal BC	2738 BP± 60 yrs	IRPA-803
30.	ABSM-PT-EIA	786 cal BC	2736 BP± 85 yrs	IRPA-811L
31.	ABSM-PT-L90	786 cal BC	2735 BP± 60 yrs	IRPA-752L
32	ABSM-PT-L	791 cal BC	2741 BP± 55 yrs	IRPA-676L
33.	ABSM-PT-L	793 cal BC	2743 BP± 60 yrs	IRPA-1012L
34.	ABSM-PT-L	795 cal BC	2744 BP± 100 yrs	IRPA-790L
35	ABSM-PT-L	795 cal BC	2744 BP± 100 yrs	QL-20
36.	ABSM-PT-L	801 cal BC	2750 BP± 160 yrs	IRPA-695
37.	ABSM-PT-L	805 cal BC	2754 BP± 60 yrs	IRPA-695B
38.	ABSM-PT	809 cal BC	2759 BP± 100 yrs	QL-27
39.	ABSM-PT	813 cal BC	2763 BP± 60 yrs	IRPA-811
40.	ABSM-PT-L10	826 cal BC	2775 BP± 60 yrs	IRPA-806L
41.	ABSM-PT-T	838 cal BC	2787 BP± 80 yrs	QL-11
42	ABSM-T	897 cal BC	2847 BP± 100 yrs	QL-7
43.	ABSM-T	993 cal BC	2943 BP± 50 yrs .	QL-986
44.	ABSM-T	993 cal BC	2942 BP± 50 yrs	BM-2140N
45.	ABSM-T	1144 cal BC	3098 BP± 60 yrs	IRPA-750L
46.	ABSM-T	1314 cal BC	3263 BP± 110 yrs	BM-2140
47.	ABSM-PRT	1485 cal BC	3435 BP± 100 yrs	QL-4191
48.	ABSM-PRT	1673 cal BC	3622 BP± 60 yrs	QL-5
49.	ABSM-PRT	1685 cal BC	3634 BP± 50 yrs	BM-1995
50.	ABSM-PRT	1710 cal BC	3659 BP± 100 yrs	BM-2312R
51,	ABSM-PRT	1740 cal BC	3689 BP± 80 yrs	QL-5A
52.	ABSM-PRT	1833 cal BC	3774 BP± 80 yrs	CSIC-180
53.	ABSM-PRT	1923 cal BC	3872 BP± 80 yrs	CSIC-179
54.	ABSM-PRT	1999 cal BC	3948 BP± 80 yrs	UtC-902
55.	ABSM-PRT	2075 cal BC	4026 BP± 45 yrs	QL-24
56.	ABSM-PRT-CE	2090 cal BC	4049 BP± 60 yrs	IRPA-835
56.	ABSM-PRT	2288 cal BC	4237 BP± 120 yrs	Y-2359
58.	ABSM-PRT	2470 cal BC	4419 BP± 100 yrs	QL-23
59.	ABSM-PRT	2483 cal BC	4432 BP± 100 yrs	QL-5B
60.	ABSM-PRT	2523 cal BC	4472 BP± 120 yrs	CSIC-178
61.	ABSM-PRT-EE	2675 cal BC	4625 BP± 40 yrs	UtC-9269
62.	ABSM-PRT	2735 cal BC	4689 BP± 392 yrs	BM-1408
63.	ABSM-PRT-EE	3375 cal BC	5324 BP± 120 yrs	QL-988
64.	ABSM-PRT-EE	3656 cal BC	5520 BP± 50 yrs	BM-1994
65.	ABSM-PRT-EE	4635 cal BC	6592 BP± 115 yrs	I-5516
66.	ABSM-PRT-EE	4722 cal BC	6671 BP± 360 yrs	CSIC-176
67.	ABSM-PRT-EE	5591 cal BC	7538 BP± 120 yrs	QL-29

TABLE 3

1999) for Mallorca and Menorca recognise in assessing the date of 1750 BC as the end of the Bell Beaker phase of the Pretalayoticum, which from the present day intersite inventory of dates as early as c 2500 BC to 2200 BC (Waldren 1997 and Waldren 2001. If this more recent chronology is correct, the beginning of the hiatus could be a cultural marker, and independent of any stellar phenomenon. We believe however that the hiatus is not caused by any cultural change because there is continuous human activity throughout the Pretalayotic period (2500 – 1300 BC) at the other related, nearby research sites of Son Matge as well as at Son Ferrandell-Oleza.

CONCLUSION

On the basis of the evidence we have reported above, we believe that the erection of the sighting stone, because of its isolated position on bedrock and red soils, as well as physical character, took place in the Pretalayotic Period and

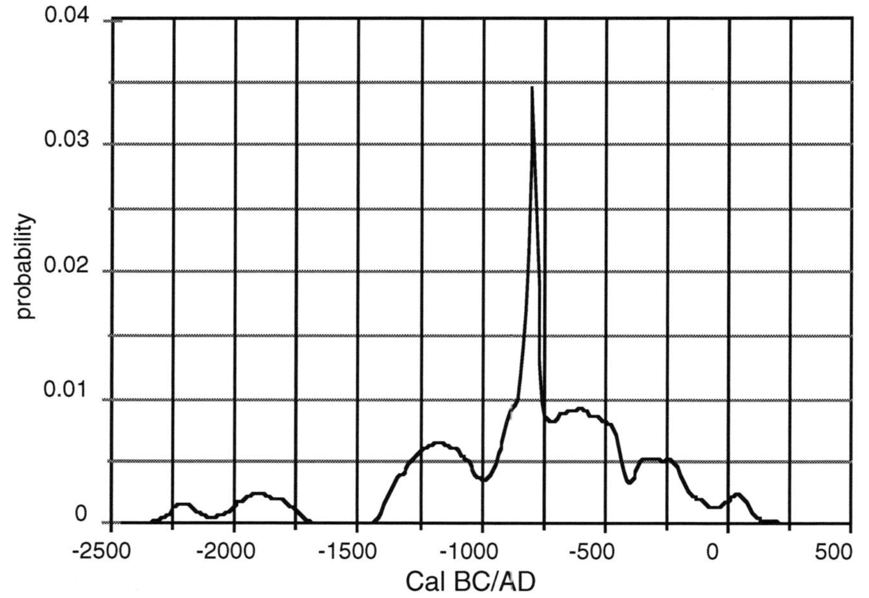

FIGURE 4. SON MAS: SUMMED PROBABILITIES (100-YR. MOVING AVERAGE) OF 53 RADIOCARBON DATES. THE DATING OF THE CONSTRUCTION OF THE SANCTUARY BY SEVERAL INDEPENDENT SAMPLES CAUSES THE SHARP PEAK AT 800 BC.

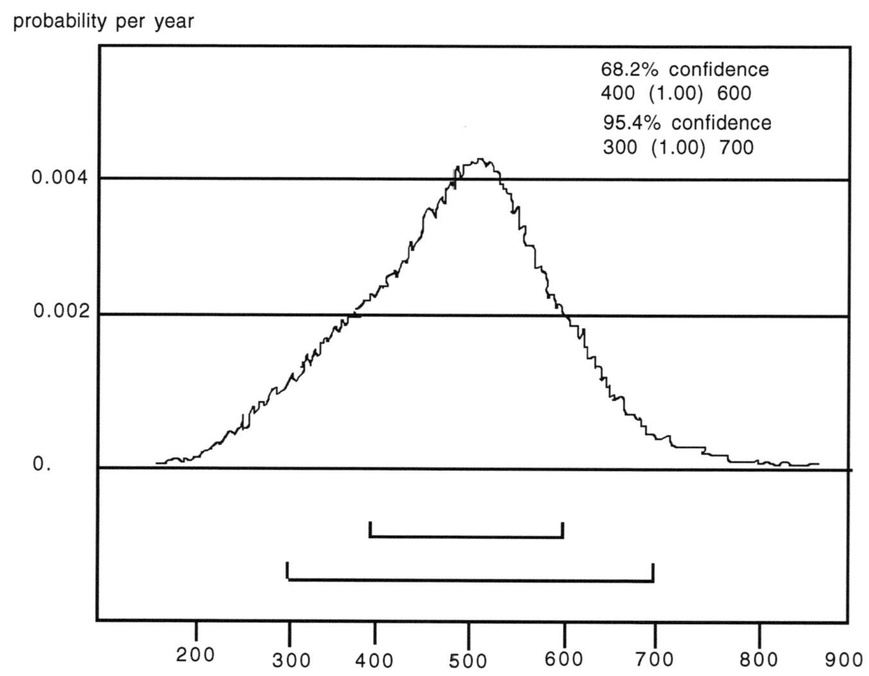

FIGURE 5. PREHISTORIC SANCTUARY OF SON MAS: PROBABILITY DISTRIBUTION OF THE DURATION OF THE HIATUS

bears no relationship to the existing later Bronze and Iron Age Talayotic structure. The high percentage of early Bell Beaker pottery and other fine ware of the period, along with the high percentage of bones from young animals suggests that the site had a ritual as well as religious function. Although no direct evidence has been found on site of an astral religion, there is a strong indication that the erection of a sighting stone in Pretalayotic times was related to the striking appearance of the Southern Cross in a V-shaped saddle between two mountain peaks overlooking the Pla del Rei. After the disappearance from view of this constellation in the southern sky, the site was abandoned for several hundred and by recent findings moved to an area further up on a ridge about 500 meters to the north of the Son Mas sanctuary. The later reuse of the site for ritual activities, perhaps for similar, but not exact, purposes in the Bronze and Iron Ages, circa 1400- 1300 BC also as a sanctuary coincides with what is traditionally accepted as the transition from the Pretalayotic to the Talayotic period and arrival of new and different cultural influences than those of the former period. Recent findings in the Ferrandell-Oleza Prehistoric Settlement Complex further substantiate this hypothesis of new arrivals and different cultural presence.

If our hypothses are correct the Son Mas sanctuary site represents the oldest indication of a star-orientated monument in the Balearic Islands.

REFERENCES

CHAPMAN, R., VAN STRYDONCK, M. & WALDREN, W.H.
1993. Radiocarbon dating and Talayots: the example of Son Ferrandell Oleza, Antiquity 67 (254): 108-116.

HOSKIN, M.
1991. The Taulas of Menorca, in Waldren et al. (ed.) IInd Deya International Conference of Prehistory, recent developments in western Mediterranean prehistory: Archaeological Techniques, Technology and Theory, BAR International series 573 (2): 221-236.

LULL, V., MICÓ, R., RIHUETE-HERRADA, C. & RISCH, R.
1999. Ideología y Sociedad en la Prehistoria de Menorca: La Cova des Càrritx y la Cova des Mussols. Ed. Consell Insular de Menorca, Ciutadella: 699.

GUERRORO AYUSO, V.M.
1998. La Mallorca Prehistòrica, des dels inicis al Bronze Final, Conèixer Mallorca 3:106.

STUIVER, M. & WALDREN, W.H.
1974. 14C Carbonate dating and the age of Post Talayotic lime burials in Mallorca, Nature, Vol. 255: 475-476.

VAN STRYDONCK, M. & WALDREN, W.H.
1990. Radiocarbon dating of Lime Burials, PACT 29, v. 5: 403-414.

VAN STRYDONCK, M. & WALDREN, W.H.
1995. Radiocarbon dating of the Son Matge rock shelter, in Waldren et al. (ed.) IIIrd Deya International Conference of Prehistory, Ritual, Rites and Religion in Prehistory, Bar International series 611 (1): 164-182.

VAN STRYDONCK, M., WALDREN, W.H. & HENDRIX, V.
1998. The 14C chronology of the Son Mas Sanctuary site (Valldemosa, Mallorca, Spain) after 10 yr. of excavating, Radiocarbon 40(2): 735-748.

WALDREN, W.H.
1982. Balearic Prehistoric Ecology and culture: the excavation and study of certain caves, rock shelters and settlements, Bar International series 149(i & ii): 1-773.

WALDREN, W.H. & VAN STRYDONCK, M.
1992. Prehistoric sanctuary of Son Mas: a radiocarbon analysis survey dating the activity sequences of the sanctuary, Damarc, 24: 1-66.

WALDREN, W.H. & VAN STRYDONCK, M.
1993. Talayot T1, Dating the activity sequence of the structure / radiocarbon analyses survey, Damarc, 22: 1-21.

WALDREN, W.H., ENSENYAT ALCOVER, J. AND CUBÌ GRIMALT, C. 1994. Prehistoric architectural elements, Ferrandell-Oleza Chalcolithic old settlement, Damarc, 20: 1-45.

WALDREN, W.H. AND GRIMALT CUBÌ, C.
1995. A case history: Function and origin of Mallorcan cast lead, votive funerary jewellery , in Waldren et al. (ed.) IIIrd Deya International Conference of Prehistory, Ritual, Rites and Religion in Prehistory, Bar International series 611 (1): 264-280.

WALDREN, W.H.
1996. The prehistoric sancuary of Son Mas 1995: a radiocarbon analysis survey, Complutum Extra, 6(1): 191-215.

WALDREN, W.H.
1997. The definition and duration of the Beaker culture in the Spanish Balearic Islands: a radiocarbon survey, Oxford journal of archaeology, 6(1): 25-48.